Nanoelectronic Mixed-Signal System Design

Nanoelectronic Mixed-Signal System Design

Saraju P. Mohanty, Ph.D.

New York Chicago San Francisco
Athens London Madrid
Mexico City Milan New Delhi
Singapore Sydney Toronto

McGraw-Hill Education books are available at special quantity discounts to use as premiums and sales promotions or for use in corporate training programs. To contact a representative, please visit the Contact Us page at www.mhprofessional.com.

Nanoelectronic Mixed-Signal System Design

Copyright © 2015 by McGraw-Hill Education. All rights reserved. Printed in the United States of America. Except as permitted under the United States Copyright Act of 1976, no part of this publication may be reproduced or distributed in any form or by any means, or stored in a data base or retrieval system, without the prior written permission of the publisher.

1 2 3 4 5 6 7 8 9 0 QVS/QVS 1 2 1 0 9 8 7 6 5

ISBN 978-0-07-182571-9
MHID 0-07-182571-1

The pages within this book were printed on acid-free paper.

Sponsoring Editor	**Copy Editor**
Michael McCabe	Garima Sharma, MPS Limited
Editorial Supervisor	**Proofreader**
Stephen M. Smith	Margaret S. Haywood
Production Supervisor	**Indexer**
Lynn M. Messina	Edwin Durbin
Acquisitions Coordinator	**Art Director, Cover**
Amy Stonebraker	Jeff Weeks
Project Manager	**Composition**
Ruchika Abrol, MPS Limited	MPS Limited

Information contained in this work has been obtained by McGraw-Hill Education from sources believed to be reliable. However, neither McGraw-Hill Education nor its authors guarantee the accuracy or completeness of any information published herein, and neither McGraw-Hill Education nor its authors shall be responsible for any errors, omissions, or damages arising out of use of this information. This work is published with the understanding that McGraw-Hill Education and its authors are supplying information but are not attempting to render engineering or other professional services. If such services are required, the assistance of an appropriate professional should be sought.

To Uma Choppali, my parents, and my sisters.

About the Author

Saraju P. Mohanty, Ph.D., is a faculty member in the Department of Computer Science and Engineering at the University of North Texas, where he directs the NanoSystem Design Laboratory (NSDL). He obtained a Ph.D. in computer science and engineering from the University of South Florida in 2003, a master's degree in systems science and automation from the Indian Institute of Science, Bangalore, India, in 1999, and a bachelor's degree (honors) in electrical engineering from Orissa University of Agriculture and Technology, Bhubaneswar, India, in 1995. Dr. Mohanty's research is in low-power, high-performance nanoelectronics. He is an author of hundreds of peer-reviewed journal and conference publications. Dr. Mohanty holds many U.S. patents in the areas of VLSI, nanoelectronics, and security hardware. He has advised/co-advised many Ph.D. dissertations and numerous master's theses. Dr. Mohanty currently serves as the chair of the Technical Committee on Very Large Scale Integration (TCVLSI), IEEE Computer Society (IEEE-CS). He serves on the editorial board of many peer-reviewed international journals, including *IET-CDS Journal, Elsevier Integration Journal,* and *Journal of Low Power Electronics*. Dr. Mohanty has served as a guest editor for many journals, including *ACM Journal on Emerging Technologies in Computing Systems (JETC)* for an issue titled "New Circuit and Architecture-Level Solutions for Multidiscipline Systems," August 2012, and *IET Circuits, Devices & Systems (CDS)* for an issue titled "Design Methodologies for Nanoelectronic Digital and Analog Circuits," September 2013. He serves on the organizing and program committees of several international conferences. Dr. Mohanty is a senior member of the IEEE and ACM.

Contents

Preface xxiii

Acknowledgments xxv

Acronyms xxvii

Notation xxxv

1 Opportunities and Challenges of Nanoscale Technology and Systems 1
   1.1 Introduction 1
   1.2 Mixed-Signal Circuits and Systems 3
       1.2.1 Different Processors: Electrical to Mechanical 3
       1.2.2 Analog versus Digital Processors 3
       1.2.3 Analog, Digital, Mixed-Signal Circuits and Systems 3
       1.2.4 Two Types of Mixed-Signal Systems 4
   1.3 Nanoscale CMOS Circuit Technology 5
       1.3.1 Developmental Trend 5
       1.3.2 Nanoscale CMOS Alternative Device Options 5
       1.3.3 Advantages and Disadvantages of Technology Scaling 8
       1.3.4 Challenges in Nanoscale Design 8
   1.4 Power Consumption and Leakage Dissipation Issues in AMS-SoCs 9
       1.4.1 Power Consumption in Various Components in AMS-SoCs 9
       1.4.2 Power and Leakage Trend in Nanoscale Technology 9
       1.4.3 The Impact of Power Consumption and Leakage Dissipation 10
   1.5 Parasitics Issue 11
       1.5.1 Types of Parasitics 11
       1.5.2 The Impact of Parasitics 11
       1.5.3 Challenges to Account Parasitics during Design 12
   1.6 Nanoscale Circuit Process Variation Issues 12
       1.6.1 Types of Process Variation 12
       1.6.2 The Impact of Process Variation 12
   1.7 The Temperature Variation Issue 13
       1.7.1 The Issue of Temperature 13
       1.7.2 The Impact of Temperature 14
       1.7.3 Challenges to Account through PVT-Aware Design 14
   1.8 Challenges in Nanoscale CMOS AMS-SoC Design 15
       1.8.1 AMS-SoC Design Flow 15
       1.8.2 AMS-SoC Unified Optimization 15
   1.9 Tools for Mixed-Signal Circuit Design 16
       1.9.1 The AMS-SoC Design Issue 16
       1.9.2 Languages for AMS-SoC Design 16
       1.9.3 Tools for AMS-SoC Design and Simulation 17
       1.9.4 Transistor Models 18
   1.10 Questions 18
   1.11 References 19

2 Emerging Systems Designed as Analog/Mixed-Signal System-on-Chips 25
- 2.1 Introduction ... 25
- 2.2 Atomic Force Microscope ... 25
  - 2.2.1 What Is It? ... 25
  - 2.2.2 Background ... 25
  - 2.2.3 What Is Inside? ... 26
- 2.3 Biosensor Systems ... 27
  - 2.3.1 What Is It? ... 27
  - 2.3.2 Background ... 29
  - 2.3.3 What Is Inside? ... 30
- 2.4 Blu-Ray Player ... 31
  - 2.4.1 What Is It? ... 31
  - 2.4.2 Home Video Systems Background: From Video Cassette Player to Blu-Ray Player ... 31
  - 2.4.3 What Is Inside? ... 31
- 2.5 Drug-Delivery Nano-Electro-Mechanical Systems ... 32
  - 2.5.1 What Is It? ... 32
  - 2.5.2 Background ... 32
  - 2.5.3 What Is Inside? ... 33
- 2.6 Digital Video Recorder ... 34
  - 2.6.1 What Is It? ... 34
  - 2.6.2 Background ... 34
  - 2.6.3 What Is Inside? ... 35
- 2.7 Electroencephalogram System ... 35
  - 2.7.1 What Is It? ... 35
  - 2.7.2 Background ... 35
  - 2.7.3 What Is Inside? ... 35
- 2.8 GPS Navigation Device ... 36
  - 2.8.1 What Is It? ... 36
  - 2.8.2 Background ... 36
  - 2.8.3 What Is Inside? ... 37
- 2.9 GPU-CPU Hybrid System ... 37
  - 2.9.1 What Is It? ... 37
  - 2.9.2 Background ... 38
  - 2.9.3 What Is Inside? ... 38
- 2.10 Networked Media Tank ... 40
  - 2.10.1 What Is It? ... 40
  - 2.10.2 Background ... 40
  - 2.10.3 What Is Inside? ... 41
- 2.11 Net-Centric Multimedia Processor ... 41
  - 2.11.1 What Is It? ... 41
  - 2.11.2 Background ... 42
  - 2.11.3 What Is Inside? ... 43
- 2.12 Radiation Detection System ... 44
  - 2.12.1 What Is It? ... 44
  - 2.12.2 Background ... 45
  - 2.12.3 What Is Inside? ... 45
- 2.13 Radio Frequency Identification Chip ... 46
  - 2.13.1 What Is It? ... 46
  - 2.13.2 Background ... 47
  - 2.13.3 What Is Inside? ... 47

Contents

- 2.14 Secure Digital Camera 49
  - 2.14.1 What Is It? 49
  - 2.14.2 Background 49
  - 2.14.3 What Is Inside? 50
- 2.15 Set-Top Box 50
  - 2.15.1 What Is It? 50
  - 2.15.2 Background 50
  - 2.15.3 What Is Inside? 51
- 2.16 Slate Personal Computer 52
  - 2.16.1 What Is It? 52
  - 2.16.2 Background: The Developmental Trend of General-Purpose Computer Reaches Slate PC. 52
  - 2.16.3 What Is Inside? 53
- 2.17 Smart Mobile Phone 54
  - 2.17.1 What Is It? 54
  - 2.17.2 Background 55
  - 2.17.3 What Is Inside? 55
- 2.18 Software-Defined Radio 56
  - 2.18.1 What Is It? 56
  - 2.18.2 Background 56
  - 2.18.3 What Is Inside? 56
- 2.19 TV Tuner Card for PCs 57
  - 2.19.1 What Is It? 57
  - 2.19.2 Background 58
  - 2.19.3 What Is Inside? 58
- 2.20 Universal Remote Control 59
  - 2.20.1 What Is It? 59
  - 2.20.2 Background 59
  - 2.20.3 What Is Inside? 59
- 2.21 Questions 60
- 2.22 References 60

3 Nanoelectronics Issues in Design for Excellence 65

- 3.1 Introduction 65
- 3.2 Design for eXcellence 65
- 3.3 Different Types of Nanoelectronic Devices 67
  - 3.3.1 Nanoscale Classical SiO_2/Polysilicon FET 68
  - 3.3.2 High-κ and Metal-Gate Nonclassical FET 70
  - 3.3.3 Multiple Independent Gate FET 72
  - 3.3.4 Carbon Nanotube FET 79
  - 3.3.5 Graphene FET 81
  - 3.3.6 Single-Electron Transistor 84
  - 3.3.7 Thin-Film Transistor 85
  - 3.3.8 Tunnel FET 86
  - 3.3.9 Vibrating Body FET 88
  - 3.3.10 Memdevices: Memristor, Memcapacitor, and Meminductor 88
- 3.4 Nanomanufacturing: The Origin and Source of Process Variations ... 91
  - 3.4.1 Classical CMOS Fabrication Process 93
  - 3.4.2 Carbon Nanotube FET Fabrication Process 95
  - 3.4.3 FinFET Fabrication Process 95
  - 3.4.4 Graphene FET Fabrication Process 96

		3.4.5 Tunnel FET Fabrication Process	97
		3.4.6 Memristor Fabrication Process	98
3.5	The Issue of Process Variation		98
	3.5.1	Types of Process Variation	99
	3.5.2	Impact on Device Parameters	100
	3.5.3	Design Phase Incorporation of Process Variation	103
3.6	The Yield Issue		107
3.7	The Power Issue in Nanoelectronic Circuits		109
	3.7.1	Power Dissipation in Nanoscale Classical CMOS Circuits	111
	3.7.2	Power Dissipation in Nanoscale High-κ and Metal-Gate FET	120
	3.7.3	Power Dissipation in Double-Gate FinFET	121
3.8	The Issue of Parasitics in Nanoelectronic Circuits		122
	3.8.1	Different Types of Parasitics	122
	3.8.2	Device Parasitics	122
	3.8.3	Interconnect Parasitics	125
3.9	The Thermal Issue		133
3.10	The Reliability Issue		135
	3.10.1	Hot Carrier Injection	135
	3.10.2	Negative Bias Temperature Instability	137
	3.10.3	Latchup Effect	137
	3.10.4	Time-Dependent Dielectric Breakdown	138
	3.10.5	Electromigration	139
	3.10.6	Thermal Stress	141
3.11	The Trust Issue		141
	3.11.1	Information Protection Issue	142
	3.11.2	Information Leakage Issue	143
	3.11.3	Chip Intellectual Property Protection Issue	143
	3.11.4	Malicious Design Modifications Issue	144
3.12	Questions		145
3.13	References		145

4 Phase-Locked Loop Component Circuits ... 157

4.1	Introduction		157
4.2	Phase-Locked Loop System Types		158
4.3	Phase-Locked Loop System: A Broad Overview		160
	4.3.1	Definition	160
	4.3.2	Block-Level Representation	160
	4.3.3	Characteristics, or Performance Metrics	161
	4.3.4	Theory in Brief	163
4.4	Oscillator Circuits		164
	4.4.1	Oscillator Types	164
	4.4.2	Oscillator Characteristics, or Performance Metrics	165
	4.4.3	Comparison of Oscillators	170
4.5	Ring Oscillators		171
	4.5.1	Basics	171
	4.5.2	45-nm CMOS	171
	4.5.3	Multigate FET	173
	4.5.4	Carbon Nanotube	175
4.6	Current-Starved Voltage Controlled Oscillators		177
	4.6.1	Basics	177
	4.6.2	Circuit Design	177
	4.6.3	90-nm CMOS	179

			4.6.4	50-nm CMOS	179
			4.6.5	45-nm CMOS	179
			4.6.6	45-nm Double-Gate FinFET	183

4.7	LC-Tank Voltage-Controlled Oscillator	184
	4.7.1 Basics	184
	4.7.2 180-nm CMOS	186
	4.7.3 CNTFET	187
	4.7.4 Memristor	188
4.8	Relaxation Oscillators	190
	4.8.1 Low-Power Relaxation Oscillator	191
	4.8.2 Memristor Relaxation Oscillator	191
	4.8.3 Memristor-Based Schmitt Trigger Oscillator	192
4.9	Phase-Frequency Detectors	193
	4.9.1 D Flip-Flop-Based PFD	194
	4.9.2 XOR Gate-Based PFD	194
4.10	Charge Pumps	195
	4.10.1 Basics	195
	4.10.2 180-nm CMOS	196
4.11	Loop Filters	198
4.12	Frequency Dividers	200
	4.12.1 Basics	200
	4.12.2 DFF-Based 180-nm CMOS	200
	4.12.3 JK Flip-Flop-Based 45-nm CMOS	200
4.13	Design and Characterization of a 180-nm CMOS PLL	203
4.14	All Digital Phase-Locked Loop	204
	4.14.1 Basics	204
	4.14.2 A Simple ADPLL Using an NCO	204
	4.14.3 A High-Resolution ADPLL Using Double DCO	205
4.15	Delay-Locked Loop	206
	4.15.1 Basics	206
	4.15.2 An Analog DLL for Variable Frequency Generation	207
	4.15.3 A Digital DLL	208
4.16	Questions	209
4.17	References	210

5 Electronic Signal Converter Circuits 215

5.1	Introduction	215
5.2	Types of Electronic Signal Converters	216
	5.2.1 Concrete Applications	216
	5.2.2 Signal Converter Types	217
5.3	Selected ADC Architectures: Brief Overview	218
	5.3.1 Overview	218
	5.3.2 Ramp-Compare ADC or Ramp Run-Up ADC	219
	5.3.3 Flash ADC or Direct Conversion ADC	219
	5.3.4 Successive-Approximation ADC	220
	5.3.5 Integrating ADC	220
	5.3.6 Pipeline ADC or Subranging ADC	221
	5.3.7 Sigma-Delta ADC or Oversampling ADC	221
	5.3.8 Time-Interleaved ADC	222
	5.3.9 Folding ADC	222
	5.3.10 Tracking ADC or Counter-Ramp ADC or Delta-Encoded ADC	223
	5.3.11 Architecture Selection	224

5.4	Selected DAC Architectures: Brief Overview	225
	5.4.1 Binary-Weighted DAC	225
	5.4.2 Thermometer-Coded DAC	227
	5.4.3 Pulse-Width Modulator DAC	227
	5.4.4 R-2R Ladder DAC	227
	5.4.5 Segmented DAC	228
	5.4.6 Oversampling or Interpolating DAC	229
	5.4.7 Sigma-Delta DAC	229
	5.4.8 Successive-Approximation or Cyclic or Algorithmic DAC	229
	5.4.9 Multiplying DAC	230
	5.4.10 Pipeline DAC	230
5.5	Characteristics for Data Converters	231
	5.5.1 Characteristics for ADC	231
	5.5.2 Characteristics for DAC	236
5.6	A 90-nm CMOS-Based Flash ADC	238
	5.6.1 Comparator Bank	238
	5.6.2 1 of N Code Generator	240
	5.6.3 NOR ROM	240
	5.6.4 Physical Design and Characterization of 90-nm ADC	241
	5.6.5 Post-Layout Simulation and Characterization	241
5.7	A 45-nm CMOS-Based Flash ADC	245
	5.7.1 Comparator Bank	245
	5.7.2 1 of N Code Generator	245
	5.7.3 NOR ROM	246
	5.7.4 Functional Simulation and Characterization	246
5.8	Single-Electron-Based ADC	249
	5.8.1 Single-Electron Circuitry-Based ADC	249
	5.8.2 Single-Electron Transistor-Based ADC	249
5.9	Organic Thin-Film Transistor-Based ADCs	250
	5.9.1 Organic Thin-Film Transistor VCO-Based ADC	250
	5.9.2 Complementary Organic Thin-Film Transistor-Based Successive-Approximation ADC	250
5.10	Sigma-Delta Modulator-Based ADC	252
	5.10.1 Broad Prospective	252
	5.10.2 Architecture Overview	252
	5.10.3 Architecture Components	254
5.11	Sigma-Delta Modulator-Based Digital-to-Analog Converter	256
5.12	Single Electron Transistor-Based Digital-to-Analog Converter	257
5.13	Questions	258
5.14	References	259

6 Sensor Circuits and Systems 263

6.1	Introduction	263
6.2	Nanoelectronics-Based Biosensors	264
	6.2.1 Spintronic-Memristor-Based Biosensors	264
	6.2.2 Tunnel-FET-Based Biosensors	265
	6.2.3 Graphene-FET-Based Biosensors	265
6.3	Thermal Sensors for Mixed-Signal Circuits and Systems	266
	6.3.1 Performance Metrics for Thermal Sensors	267
	6.3.2 A Concrete Example: A 45-nm CMOS Ring Oscillator-Based Thermal Sensor	268
	6.3.3 A Concrete Example: Spintronic-Memristor Temperature Sensor	271

	6.4	Solar Cells		272
		6.4.1	Operation and Performance of Cells	273
		6.4.2	Selected Solar Cell Designs	275
		6.4.3	Solar Cell Models for Circuit Simulations	276
	6.5	Piezoelectric Sensors		278
	6.6	Image Sensors		280
		6.6.1	Types of Image Sensors	281
		6.6.2	Characteristics of the Image Sensors	285
		6.6.3	A Concrete Example: 32-nm CMOS APS Design	289
		6.6.4	Smart Image Sensors	291
		6.6.5	Secure Image Sensors	293
	6.7	Nanoelectronics-Based Gas Sensors		294
		6.7.1	CNTFET-Based Gas Sensor	294
		6.7.2	CNTFET-Based Chemical Sensor	294
	6.8	Body Sensors		295
	6.9	Epileptic Seizure Sensors		297
	6.10	Humidity Sensors		297
		6.10.1	A Diode-Based Humidity Sensor	298
		6.10.2	A CMOS Device–Based Humidity Sensor	298
	6.11	Motion Sensors		299
	6.12	Sense Amplifiers		300
		6.12.1	Types of Sense Amplifiers	301
		6.12.2	Performance Metrics for the Sense Amplifiers	302
		6.12.3	A Concrete Example: 45-nm CMOS Clamped Bitline Sense Amplifier	304
	6.13	Questions		307
	6.14	References		308
7	**Memory in the AMS-SoCs**			**315**
	7.1	Introduction		315
	7.2	Static Random-Access Memory		317
		7.2.1	SRAM Array	317
		7.2.2	Different Types of SRAM	318
		7.2.3	Traditional Six-Transistor SRAM	318
		7.2.4	Four-Transistor SRAM	321
		7.2.5	Five-Transistor SRAM	321
		7.2.6	Seven-Transistor SRAM	322
		7.2.7	Eight-Transistor SRAM	324
		7.2.8	Nine-Transistor SRAM	326
		7.2.9	Ten-Transistor SRAM	326
		7.2.10	Performance Metrics of SRAM	327
		7.2.11	Characterization of Specific SRAMs	331
	7.3	Dynamic Random-Access Memory		338
		7.3.1	DRAM Array	339
		7.3.2	Different Types of DRAM	339
		7.3.3	Selected DRAM Designs Based on Topology	340
		7.3.4	DRAMs Based on Modes of Operation	343
		7.3.5	Synchronous DRAMs	344
		7.3.6	Video or Graphics DRAM	344
		7.3.7	Ferroelectric DRAM	345
		7.3.8	Characteristics of DRAM	346
	7.4	Twin-Transistor Random-Access Memory		347
	7.5	Thyristor Random-Access Memory		348

7.6	Read-Only Memory	349
	7.6.1 Programmable Read-Only Memory	349
	7.6.2 Erasable Programmable Read-Only Memory	349
	7.6.3 Electrically Erasable Programmable Read-Only Memory	350
7.7	Flash Memory	351
7.8	Resistive Random-Access Memory	352
	7.8.1 Nonvolatile Resistive RAM for Storage	352
	7.8.2 Conductive Metal-Oxide Memory	353
	7.8.3 Memristor-Based Nonvolatile SRAM	354
7.9	Magnetic or Magnetoresistive Random-Access Memory	355
7.10	Phase-Change RAM	356
7.11	Questions	358
7.12	References	359

8 Mixed-Signal Circuit and System Design Flow ... 365

8.1	Introduction	365
8.2	AMS-SoC: A Complete Design Perspective	365
8.3	Integrated Circuit Design Flow: Top-Down versus Bottom-Up	369
8.4	Analog Circuit Design Flow	371
	8.4.1 Behavioral Simulation	372
	8.4.2 Transistor-Level Design or Schematic Capture	374
	8.4.3 Transistor-Level Simulation and Characterization	374
	8.4.4 Physical Design or Layout Design	374
	8.4.5 Design Rule Check	375
	8.4.6 Parasitic (RCLK) Extraction	376
	8.4.7 Layout versus Schematic Verification	377
	8.4.8 Electrical Rule Check	377
	8.4.9 Physical Design Characterization	378
	8.4.10 Variability Analysis	379
	8.4.11 Performance Optimization	380
8.5	Digital Circuit Design Flow	382
	8.5.1 System-Level Design	384
	8.5.2 Architecture-Level Design	384
	8.5.3 Logic-Level Design	385
	8.5.4 Transistor-Level Design	386
	8.5.5 Physical Design	386
	8.5.6 Physical Verification	387
	8.5.7 Design Signoff	387
	8.5.8 Engineering Change Order	388
	8.5.9 Circuit Fabrication, Packaging, and Testing	388
8.6	Analog and Mixed-Signal Circuit Design Flow	389
	8.6.1 Mixed-Signal Design Flow	389
	8.6.2 Analog and/or Mixed-Signal Circuit Synthesis Techniques	391
8.7	Design Flow Using Commercial Electronic Design Automation Tools	393
	8.7.1 Selected Commercial EDA Tools	393
	8.7.2 For Analog Design	394
	8.7.3 For Digital Design	395
	8.7.4 For Mixed-Signal System Design	398
8.8	Design Flow Using Free or Open-Source EDA Tools	399
	8.8.1 Selected Free or Open-Source EDA Tools	399
	8.8.2 For Analog Design	401

		8.8.3	For Digital Design	402
		8.8.4	For Mixed-Signal Design	402
	8.9	Comprehensive Design Flows		403
		8.9.1	For Analog/Mixed-Signal Circuits and Systems	403
		8.9.2	For Digital Circuits and Systems	406
	8.10	Process Design Kit and Libraries		407
	8.11	EDA Tool Installation		409
		8.11.1	Client-Server Platform	409
		8.11.2	Workstation-Based Platform	410
		8.11.3	Mixed-Configuration Platform	410
	8.12	Questions		411
	8.13	References		412

9 Mixed-Signal Circuit and System Simulation 421

	9.1	Introduction		421
	9.2	Simulation Types and Languages for Circuits and Systems		422
		9.2.1	Simulations Based on Abstraction Levels	422
		9.2.2	Simulations Based on Signal Types	423
		9.2.3	Simulations Based on System Models	423
		9.2.4	Simulations Based on Design Tasks	423
		9.2.5	Simulation Languages	424
	9.3	Behavioral Simulation using MATLAB®		425
		9.3.1	System- or Architecture-Level Simulations	426
		9.3.2	Circuit-Level Simulations	430
		9.3.3	Device-Level Simulations	432
	9.4	Simulink®- or Simscape®-Based Simulations		433
		9.4.1	System- or Architecture-Level Simulations	434
		9.4.2	Circuit-Level Simulations	438
		9.4.3	Device-Level Simulations	440
	9.5	Circuit-Level and/or Device-Level Analog Simulations		445
		9.5.1	SPICE Analog Simulation Background	446
		9.5.2	Commercial Accurate Analog Circuit Simulators	447
		9.5.3	Free and/or Open-Source Accurate SPICE	448
		9.5.4	Fast SPICE	449
		9.5.5	Analog-Fast SPICE	450
		9.5.6	High-Speed SPICE	450
		9.5.7	Different Types of Analysis using SPICE	450
		9.5.8	SPICE-Based Simulation Examples	452
		9.5.9	Inside of SPICE	455
		9.5.10	SPICE Simulation Flow	460
	9.6	Verilog-A-Based Analog Simulation		464
		9.6.1	Verilog-A-Based Circuit-Level Simulations	465
		9.6.2	Verilog-A-Based Device-Level Simulations	468
	9.7	Simulations of Digital Circuits or Systems		474
		9.7.1	SystemVerilog-Based Simulation	474
		9.7.2	VHDL-Based Simulation	475
		9.7.3	MyHDL-Based Simulation	477
		9.7.4	SystemC-Based Simulation	478
	9.8	Mixed-Signal HDL-Based Simulation		480
		9.8.1	Verilog-AMS-Based Simulation	481
		9.8.2	VHDL-AMS-Based Simulation	489

		9.8.3	OpenMAST™-Based Simulation.	491

- 9.8.3 OpenMAST™-Based Simulation ... 491
- 9.8.4 SystemC-AMS-Based Simulation ... 494
- 9.9 Mixed-Mode Circuit-Level Simulations ... 495
  - 9.9.1 Nanoelectronics Analog versus Mixed-Signal Simulation: A Comparative Perspective ... 496
  - 9.9.2 Mixed-Mode with Individual Analog and Digital Engine ... 497
  - 9.9.3 Mixed-Mode with Unified Analog and Digital Engine ... 498
- 9.10 Models for Circuit Simulations ... 498
  - 9.10.1 Compact Model Generation Flow ... 498
  - 9.10.2 Types of Compact Models ... 500
  - 9.10.3 Automatic Device Model Synthesizer (ADMS) ... 501
- 9.11 Questions ... 501
- 9.12 References ... 502

10 Power-, Parasitic-, and Thermal-Aware AMS-SoC Design Methodologies ... 513

- 10.1 Introduction ... 513
- 10.2 Power Dissipation: Key Design Constraint ... 513
  - 10.2.1 The Effects of High-Power Dissipation ... 514
  - 10.2.2 Power Dissipation Sources ... 515
  - 10.2.3 Power or Energy Dissipation Metrics ... 516
  - 10.2.4 Energy/Power Dissipation: Application Perspectives ... 518
  - 10.2.5 Limits to Low-Power Design ... 520
- 10.3 Different Energy or Power Reduction Techniques for AMS-SoC ... 521
  - 10.3.1 AMS-SoC Energy or Power Reduction Techniques: An Overview ... 521
  - 10.3.2 Analog Circuit Power Optimization: An Overview ... 523
  - 10.3.3 Digital SoC Power or Energy Optimization Procedures: An Overview ... 525
- 10.4 Presilicon Power Reduction Techniques ... 527
  - 10.4.1 Brief Discussion ... 527
  - 10.4.2 Dual-Threshold-Based Circuit-Level Optimization of a Universal Level Converter ... 528
  - 10.4.3 Dual-Oxide-Based Logic-Level Optimization of Digital Circuits ... 531
  - 10.4.4 Dual-Oxide-Based RTL Optimization of Digital Circuits ... 534
- 10.5 Hardware-Based Postsilicon Power Reduction Techniques ... 536
  - 10.5.1 Brief Discussion ... 536
  - 10.5.2 Dynamic or Variable Frequency Clocking for Power Reduction ... 538
  - 10.5.3 Adaptive Voltage Scaling for Power and Energy Reduction ... 540
- 10.6 Dynamic Power Reduction Techniques ... 541
  - 10.6.1 Brief Discussion ... 541
  - 10.6.2 Dual-Voltage and Dual-Frequency-Based Circuit-Level Technique ... 542
  - 10.6.3 Multiple Supply Voltage-Based RTL Technique ... 544
- 10.7 Subthreshold Leakage Reduction Techniques ... 548
  - 10.7.1 Brief Discussion ... 548
  - 10.7.2 Dual-Threshold-Based Circuit-Level Optimization of Nano-CMOS SRAM ... 550
- 10.8 Gate-Oxide Leakage Reduction Techniques ... 553
  - 10.8.1 Brief Discussion ... 553
  - 10.8.2 Dual-Oxide-Based Circuit-Level Optimization of a Current-Starved VCO ... 554
  - 10.8.3 Dual-Oxide-Based RTL Optimization of Digital ICs ... 558
- 10.9 Parasitics: Brief Overview ... 560
- 10.10 The Effects of Parasitics on Integrated Circuits ... 562
  - 10.10.1 Parasitics in Real-Life Example Circuits ... 562
  - 10.10.2 Effects of the Parasitics ... 562

	10.11	Modeling and Extraction of Parasitics	565
		10.11.1 Signal Propagation: In a Real Wire	565
		10.11.2 Parasitics Modeling and Simulation: The Key Aspects	566
		10.11.3 Circuit (Device+Parasitic) Extraction Process	566
		10.11.4 Parasitics Extraction Techniques	568
		10.11.5 Parasitics Modeling	569
		10.11.6 Parasitics Model Order Reduction	570
	10.12	Design Flows for Parasitic-Aware Circuit Optimization	574
		10.12.1 Parasitic-Aware Analog Design Flow with Multilevel Optimizations	574
		10.12.2 A Rapid Parasitic-Aware Design Flow for Analog Circuits	574
		10.12.3 Single-Manual Iteration Fast Design Flow for Parasitic-Optimal VCO	576
		10.12.4 Parasitic-Aware Low-Power Design of the ULC	578
	10.13	Temperature or Thermal Issue: An Overview	582
	10.14	Thermal Modeling	584
		10.14.1 Heat Dissipation: Structure View	584
		10.14.2 Compact Thermal Modeling	586
	10.15	Thermal Analysis or Simulation Techniques	588
		10.15.1 Heat Transfer Basics	588
		10.15.2 Thermal Analysis Basics	589
		10.15.3 Thermal Analysis Types	589
		10.15.4 A Runge-Kutta-Based Method	590
		10.15.5 An Integrated Space-and-Time-Adaptive Chip Thermal Analysis Framework	590
		10.15.6 A Fast Asynchronous Time Marching Technique	592
		10.15.7 Green Function-Based Method	592
		10.15.8 Thermal Moment Matching Method	594
	10.16	Temperature Monitoring or Sensing	594
		10.16.1 Hardware-Based Thermal Monitoring	594
		10.16.2 Software-Based Temperature Monitoring	594
		10.16.3 Hybrid Hardware- and Software-Based Thermal Monitoring	594
	10.17	Temperature Control or Management	595
		10.17.1 Basic Principle	595
		10.17.2 Types	596
	10.18	Thermal-Aware Circuit Optimization	596
		10.18.1 A Thermal-Aware SRAM Optimization	596
		10.18.2 A Thermal-Aware VCO Optimization	599
	10.19	Thermal-Aware Digital Design Flows	602
		10.19.1 Thermal-Aware Digital Synthesis	602
		10.19.2 Thermal-Aware Physical Design	603
	10.20	Thermal-Aware Register-Transfer-Level Optimization	604
	10.21	Thermal-Aware System-Level Design	605
	10.22	Questions	606
	10.23	References	607
11	**Variability-Aware AMS-SoC Design Methodologies**		**619**
	11.1	Introduction	619
	11.2	Methods for Variability Analysis	621
		11.2.1 Monte Carlo Method	622
		11.2.2 Design of Experiments Method	628
		11.2.3 Corner-Based Method	633
		11.2.4 Fast Monte Carlo Methods	638

	11.3	Tool Setup for Statistical Analysis	643
	11.4	Methods for Variability-Aware Design Optimization	644
		11.4.1 Brief Concept	644
		11.4.2 Variability-Aware Schematic Design Optimization Flow	645
		11.4.3 Single Manual Layout Iteration Automatic Flow for Variability-Aware Optimization	646
	11.5	Variability-Aware Design of Active Pixel Sensor	648
		11.5.1 Impact of Variability on APS Performance Metrics	648
		11.5.2 Variability-Aware APS Optimization	649
	11.6	Variability-Aware Design of Nanoscale VCO Circuits	654
		11.6.1 A Conjugate-Gradient-Based Optimization of a 90-nm CMOS Current-Starved VCO	654
		11.6.2 A Particle Swarm Optimization Approach for a 90-nm Current-Starved VCO	657
		11.6.3 Process Variation Tolerant LC-VCO Design	661
	11.7	Variability-Aware Design of the SRAM	662
	11.8	Register-Transfer-Level Methods for Variability-Aware Digital Circuits	667
		11.8.1 Brief Overview	667
		11.8.2 A Simulated-Annealing-Based Statistical Approach for RTL Optimization	668
		11.8.3 A Taylor-Series Expansions Diagram-Based Approach for RTL Optimization	669
		11.8.4 Variability-Aware RTL Timing Optimization	673
		11.8.5 RTL Postsilicon Techniques for Variability Tolerance	675
	11.9	System-Level Methods for Variability-Aware Digital Design	677
	11.10	An Adaptive Body Bias Method for Dynamic Process Variation Compensation	678
	11.11	Parametric Variation Effect Mitigation in Clock Networks	679
	11.12	Statistical Methods for Yield Analysis	681
	11.13	Questions	683
	11.14	References	685
12	**Metamodel-Based Fast AMS-SoC Design Methodologies**		**689**
	12.1	Introduction	689
	12.2	Metamodel: An Overview	689
		12.2.1 Concept	689
		12.2.2 Types	691
		12.2.3 Generation Flow	693
		12.2.4 Metamodel versus Macromodel	696
	12.3	Metamodel-Based Ultrafast Design Flow	697
	12.4	Polynomial-Based Metamodeling	698
		12.4.1 Theory	698
		12.4.2 Generation	699
		12.4.3 Ring Oscillator	701
		12.4.4 LC-VCO	702
		12.4.5 Verilog-AMS Integrated with Polynomial Metamodel for an OP-AMP	702
		12.4.6 Verilog-AMS Integrated with Polynomial Metamodel for a Memristor Oscillator	707
		12.4.7 Verilog-AMS Integrated with Parasitic-Aware Metamodel	711
	12.5	Kriging-Based Metamodeling	714
		12.5.1 Theory	715
		12.5.2 Generation	717

		12.5.3 Simple Kriging Metamodeling of a Clamped Bitline Sense Amplifier	718

- 12.5.3 Simple Kriging Metamodeling of a Clamped Bitline Sense Amplifier 718
- 12.5.4 Ordinary Kriging Metamodeling of a Sense Amplifier 720
- 12.5.5 Universal Kriging Metamodeling of a Phase-Locked Loop 720

12.6 Neural Network–Based Metamodeling 722
- 12.6.1 Theory 722
- 12.6.2 Generation 725
- 12.6.3 Neural Network Metamodel of PLL Components 726
- 12.6.4 Intelligent Verilog-AMS 727
- 12.6.5 Kriging Bootstrapped Training for Neural Network Metamodeling 731

12.7 Ultrafast Process Variations Analysis Using Metamodels 733
- 12.7.1 Kriging-Metamodel-Based Process Variation Analysis of a PLL 733
- 12.7.2 Neural Network Metamodel-Based Process Variation Analysis of a PLL 733
- 12.7.3 Kriging-Trained Neural Network–Based Process Variation Analysis of a PLL 736

12.8 Polynomial-Metamodel-Based Ultrafast Design Optimization 737
- 12.8.1 Polynomial-Metamodel-Based Optimization of a Ring Oscillator 737
- 12.8.2 Polynomial-Metamodel-Based Optimization of a PLL 739
- 12.8.3 Polynomial-Metamodel-Based Optimization of an OP-AMP 744

12.9 Neural Network Metamodel-Based Ultrafast Design Optimization 747
- 12.9.1 Neural Network Metamodel-Based Optimization of an OP-AMP 747
- 12.9.2 Neural Network Metamodel-Based Variability-Aware Optimization of a PLL 750

12.10 Kriging Metamodel-Based Ultrafast Design Optimization 752
- 12.10.1 Simple Kriging Metamodel-Based Optimization of a Thermal Sensor ... 752
- 12.10.2 Ordinary Kriging Metamodel-Based Optimization of a Sense Amplifier 755

12.11 Questions 758

12.12 References 760

Index **765**

Preface

Consumer electronics such as mobile phones, smart phones, and tablets have a profound impact on society. They have revolutionized the way people deal with information in various forms including pictures, text, video, and audio. Modern consumer electronics systems are essentially built as heterogeneous multicores consisting of diverse analog and digital cores and their interfaces on a single board or die. The hardware components of these systems are built using nanoelectronic technology such as nano-complementary metal-oxide semiconductor (nano-CMOS) and multigate transistors. At the same time, the industry is seriously exploring the use of other nanoelectronic technologies such as graphene transistors, tunnel transistors, and memristors. This book constitutes a unified text that covers mixed-signal design based on existing and futuristic nanoelectronic technologies. The book covers mixed-signal systems using nanoelectronic technologies instead of either digital-only or analog-only applications. The author's objective is to provide nanoelectronic very large-scale integration (VLSI) design training requiring the shortest possible learning curve.

From a system perspective, mobile phones and other similar portable systems are embedded systems designed using analog/mixed-signal system-on-a-chip (AMS-SoC) technology. These systems consist of several heterogeneous components including (1) digital circuits, (2) memory circuits, (3) analog circuits, (4) mixed-signal circuits such as analog-to-digital converters (ADCs), (5) radio frequency (RF) components, (6) power management systems including the battery, and (7) system and application software. This book deals with these components of system design when the building blocks are tiny nanoscale transistors. However, software design aspects of the AMS-SoC are beyond the scope of this book. This self-contained text is directed to nanoelectronic VLSI design engineers, mixed-signal computer-aided design (CAD) engineers, students, and researchers. It provides a sufficient amount of fundamentals for the reader to become familiar with the terminology of the field. The main objective of this book is to provide in-depth knowledge of some topics and a bigger picture of the contexts of others.

For nanoelectronic technology-based design of mixed-signal systems, there are several issues and challenges that academia and industry are working very hard to solve. The main challenges include (1) power consumption for battery-powered devices, (2) leakage dissipation, (3) process variation tolerance, (4) high-performance requirement, (5) small form-factor requirement, (6) low design-cycle requirement, (7) low-cost requirement, and (8) system security. The book covers these issues so that design and CAD engineers are aware of them and researchers can try to solve them more efficiently. The techniques proposed in this book can be applied by the industry to enhance the yield of electronic systems and ultimately reduce their cost.

Special features of this book include the following:

- Coverage of mixed-signal circuit and system design instead of digital-only or analog-only design.
- Coverage of current and futuristic nanoelectronic technologies instead of a nano-CMOS-only focus.
- Coverage of nanoelectronic technology-based mixed-signal circuit and system design.
- Coverage of key issues and of solutions for nanoelectronic challenges that the industry is striving to address.
- Treatment of the science, engineering, and technology of nanoelectronic system design with special emphasis on emerging applications.
- A unified text that provides top-down analysis of all stages from design to manufacturing.

This book is written and organized in such a way that it can be used as a primary reference for a first course in areas such as nanoelectronic system design and mixed-signal system design. It can also serve as a primary reference for a second course in VLSI design, computer engineering, and electrical engineering. The book targets senior undergraduate, master's, and Ph.D. students in addition to practicing engineers in the industry. The author's decade-long experience in research and teaching and training graduate and undergraduate students in the United States has been utilized to build a strong text specifically addressing all aspects of modern consumer electronics systems. The author has advised and co-advised several dissertations and theses primarily focused on research in nanoelectronic system design.

A unified scientific, engineering, and technology perspective of nanoelectronic mixed-signal system design is presented. A balanced treatment of what, why, and how is presented by judiciously combining theory and practical perspectives. An attempt has also been made to include device and system-level views. A significant number of practical examples are presented. The discussions include a large number of figures to provide a pictorial representation of concepts motivated by the fact that a picture is worth a thousand words.

This book is organized in the following manner for gradual reading of the material presented:

1. Chapter 1 discusses the state-of-the-art nanoelectronic technology including nanoscale CMOS technology with an introduction to mixed-signal circuits.
2. Chapter 2 discusses some examples of systems that have been used in day-to-day life or are being conceptualized for future development.
3. Chapter 3 discusses the most important issues of nanoscale CMOS or nanoelectronic-based circuit and system design including process variation, power and leakage dissipation, and parasitics.
4. Chapter 4 presents the design and simulation of phase-locked loops (PLLs) and various PLL components.
5. Chapter 5 discusses the design of sigma-delta modulators, ADCs, and digital-to-analog converters (DACs).
6. Chapter 6 discusses various types of sensors including imaging sensors that are used in mobile phones or digital cameras and thermal sensors that are used for temperature sensing.
7. Chapter 7 presents the design of various types of memories for computing systems.
8. Chapter 8 presents an easy-to-follow tutorial perspective for mixed-signal circuit design that lucidly discusses all the steps.
9. Chapter 9 discusses the different simulation methodologies possible for mixed-signal circuits.
10. Chapter 10 presents power-, parasitic-, and thermal-aware design methods for nanoelectronic technology-based mixed-signal systems.
11. Chapter 11 discusses the variability-aware or -tolerant design approaches.
12. Chapter 12 focuses on the metamodel-based fast AMS-SoC design methodologies.

Some of the key benefits of this text for different audiences follow:

- Undergraduate students will learn the basics of nanoelectronic technology and the structure of popular electronic systems.
- Graduate students will learn key techniques for nanoelectronic technology-based system design.
- Computer engineering students will learn various techniques for nanoelectronic-based system design methodology.
- Electrical engineering students will obtain a large picture of the design and manufacturing aspects of nanoelectronic-based systems.
- Practicing engineers will learn various techniques needed for design for excellence and manufacturability.

Saraju P. Mohanty, Ph.D.

Acknowledgments

The author acknowledges the full support and encouragement of the department chair, faculty, and staff of the Department of Computer Science and Engineering at the University of North Texas. He also acknowledges the past and present student members (e.g., Dr. Dhruva Ghai, Dr. Garima Thakral, Dr. Oleg Garitselov, Dr. Geng Zheng, and Dr. Oghenekarho Okobiah) of the NanoSystem Design Laboratory of the Department of Computer Science and Engineering at the University of North Texas.

The author is immensely grateful to Dr. Uma Choppali, who helped immensely in the preparation of figures in the book, and to Dr. Elias Kougianos, for providing valuable feedback that was helpful in enriching the content.

The author acknowledges Cadence for the tool that they provide for academic purposes and the National Science Foundation (NSF) and Semiconductor Research Corporation (SRC) for funding his research and education activities.

The publication of this book would not have been possible without the support of the publishing and editorial staff. The author would like to thank them for their promptness and patience in answering all his questions.

Acronyms

1T-1C-DRAM	One-Transistor-One-Capacitor DRAM
1T-DRAM	One-Transistor DRAM
3T-DRAM	Three-Transistor DRAM
4T-DRAM	Four-Transistor DRAM
4T-SRAM	Four-Transistor SRAM
5T-SRAM	Five-Transistor SRAM
6T-SRAM	Six-Transistor SRAM
7T-SRAM	Seven-Transistor SRAM
8T-SRAM	Eight-Transistor SRAM
9T-SRAM	Nine-Transistor SRAM
10T-SRAM	Ten-Transistor SRAM
AAF	Anti-Aliasing Filter
ABB	Adaptive Body Bias
ABC	Artificial Bee Colony
ACF	Auto-Correlation Function
ADC	Analog-to-Digital Converter
ADLL	Analog Delay-Locked Loop
ADPLL	All-Digital PLL
AFE	Analog Front End
AFM	Atomic Force Microscope
AHDL	Analog Hardware Description Language
ALD	Atomic Layer Deposition
AMS	Analog/Mixed-Signal
AMSHDL	Analog/Mixed-Signal Hardware Description Language
AMS-SoC	Analog/Mixed-Signal System-on-Chip
APC	Advanced Power Controller
APLL	Analog PPL
APS	Active Pixel Sensor
ARAM	Audio DRAM
ARF	Autoregressive Filter
ASIC	Application-Specific Integrated Circuit
ASIP	Application-Specific Instruction Processor
ATE	Automated Test Equipment
AVS	Adaptive Voltage Scaling
BCE	Branch Constitutive Equation
BCO	Bee Colony Optimization
BD	Blu-Ray Disc
BEDO	Burst Extended Data Out
BEM	Boundary Element Method
BEOL	Back-End of Line
BJT	Bipolar Junction Transistor
BPF	Band Pass Filter
BPR	Bayesian Process Regression
BTBT	Band-to-Band Tunneling
C4	Controlled Collapse Chip Connection
CA	Conditional Access
CAD	Computer-Aided Design

CAM	Content Addressable Memory
CANCER	Computer Analysis of Nonlinear Circuits Excluding Radiation
CBGA	Ceramic Ball Grid Array
CCD	Charge Coupled Device
CDF	Cumulative Distribution Function
CDFG	Control Data Flow Graph
CDMA	Code Division Multiple Access
CDRAM	Cache DRAM
CHE	Channel Hot Electron
CIFF	Cascade of Integrators with Feed Forward
CLM	Channel Length Modulation
CMOS	Complementary Metal-Oxide Semiconductor
CMOX	Conductive Metal Oxide
CMP	Chemical Mechanical Polishing
CNT	Carbon Nanotube
CNTFET	Carbon Nanotube FET
CPM	Circuit Parameter Metamodel
CPPLL	Charge-Pump-Based PLL
CPU	Central Processing Unit
CS-VCO	Current-Starved Voltage-Controlled Oscillator
CSO	Cuckoo Search Optimization
CVD	Chemical Vapor Deposition
DAC	Digital-to-Analog Converter
DAE	Differential Algebraic Equation
DAG	Directed Acyclic Graph
DAHC	Drain Avalanche Hot Carrier
DCO	Digitally Controlled Oscillator
DCT	Discrete Cosine Transformation
DCU	Dynamic Clocking Unit
DCVS	Differential Cascode Voltage Switch
DDLL	Digital Delay-Locked Loop
DDNEMS	Drug-Delivery Nano-Electro-Mechanical System
DDR	Double Data Rate
DDR2	Double Data Rate Type-2
DDR3	Double Data Rate Type-3
DDR4	Double Data Rate Type-4
DfC	Design for Cost
DFG	Data Flow Graph
DFM	Design for Manufacturing
DfV	Design for Variability
DFX	Design for eXcellence, Design for X
DGFET	Double-Gate Field-Effect Transistor
DIBL	Drain-Induced Barrier Lowering
DINOR	Divided Bitline NOR
DIP	Dual In-Line Package
DLCMOS	Dual-Channel-Length CMOS
DLL	Delay-Locked Loop
DMP	Digital Media Player
DNA	Deoxyriboucleic Acid
DNL	Differential Nonlinearity
DNM	Dynamic Noise Margin
DOCSIS	Cable Service Interface Specifications
DOE	Design of Experiments
DOE-ILP	DOE-Assisted Integer Linear Programming

DOE-MC	DOE-Assisted Monte Carlo
DOE-TSO	DOE-Assisted Tabu Search Optimization
DOXCMOS	Dual-Oxide CMOS
DPLL	Digital PLL
DPS	Digital-Pixel Sensor
DR	Dynamic Range
DRAM	Dynamic Random-Access Memory
DRC	Design Rule Check
DRDRAM	Direct Rambus DRAM
DRM	Digital Rights Management
DSLAM	Digital Subscriber Line Access Multiplexer
DSLR	Digital Single-Lens Reflex
DSNU	Dark Signal Nonuniformity
DSP	Digital Signal Processor
DTCMOS	Dual-Threshold CMOS
DTM	Dynamic Thermal Management
DVB	Digital Video Broadcasting
DVD	Digital Video Disc
DVR	Digital Video Recorder
DVS	Dynamic Voltage Scaling
ECAD	Electronic CAD
ECB	Electron Tunneling from Conduction Band
ECG	Electrocardiogram
ECO	Engineering Change Order
EDA	Electronic Design Automation
EDO	Extended Data Out
EDP	Energy Delay Product
EDRAM	Enhanced DRAM
EEG	Electroencephalogram
EEPROM	Electrically Erasable Programmable Read-Only Memory
EFIE	Electric Field Integral Equation
EKV	Enz-Krummenacher-Vittoz Model
EM	Electromigration
EMG	Electromyogram
EMU	Energy Management Unit
ENIAC	Electronic Numerical Integrator and Computer
EOG	Electrooculogram
EPC	Electronic Product Code
ERG	Electroretinogram
ESD	Electrostatic Discharge
ESDRAM	Enhanced Synchronous DRAM
ESL	Electronic System Level
EVB	Electron Tunneling from Valence Band
EVT	Extreme Value Theory
EWF	Elliptic Wave Filter
FBB	Forward Body Bias
FDA	Fully Differential Amplifier
FDM	Finite Difference Method
FEM	Finite Element Method
FEOL	Front-End of Line
FeRAM	Ferroelectric DRAM
FET	Field-Effect Transistor
FG-FET	Floating-Gate FET

FIR	Finite Impulse-Response Filter
FM	Frequency Modulation
FN	Fowler-Nordheim
FoM	Figure-of-Merit
FPGA	Field-Programmable Gate Array
FPM	Fast Page Mode
FPN	Fixed Pattern Noise
FRAM	Ferroelectric DRAM
FS	Full-Scale
FSR	Full-Scale Range
FU	Functional Unit
GBP	Gain-Bandwidth-Product
GCH	GPU-CPU Hybrid
GDS	Graphic Data System
GFET	Graphene (Nanoribbon) FET
GIDL	Gate-Induced Drain Leakage
GNR-FET	Graphene Nanoribbon FET
GNR-TFET	Graphene Nanoribbon Tunneling FET
GPD	Generalized Pareto Distribution
GPDK	General Process Design Kit
GPR	Gaussian Process Regression
GPS	Global Positioning System
GPU	Graphics Processing Unit
GSM	Global System for Mobile Communications
GSS	General-Purpose Semiconducor Simulator
GTFET	Green Tunneling FET
HAL	HAL Differential Equation Solver
HCE	Hot Carrier Effects
HCI	Hot Carrier Injection
HD-DVD	High-Definition DVD
HDL	Hardware Description Language
HDMI	High-Definition Multimedia Interface
HDVL	Hardware Description and Verification Language
HF	High Frequency
HIL	Hardware in the Loop
HKMGFET	High-κ/Metal-Gate FET
HLS	High-Level Synthesis
HNM	Hold Noise Margin
HPM	Hardware Power Monitor
HVB	Hole Tunneling from Valance Band
HVL	Hardware Verification Language
IC	Integrated Circuit
IDCT	Inverse DCT
IF	Intermediate Frequency
IGFET	Insulated Gate FET
IIR	Infinite Impulse-Response Filter
ILP	Integer Linear Programming
IMG	Independent Multigate
INL	Integral Nonlinearity
IP	Internet Protocol
IP-TV	IP Television
IR	Infrared
ISFET	Ion-Sensitive FET
iVAMS	Intelligent Verilog-AMS

JVD	Jet Vapor Deposition
KCL	Kirchhoff Current Law
KFL	Kirchhoff Flow Law
kHz	Kilohertz
KPL	Kirchhoff Potential Law
KVL	Kirchhoff Voltage Law
JPEG	Joint Photographic Experts Group Format
LCD	Liquid Crystal Display
LD	LaserDisc
LDP	Leakage Delay Product
LED	Light-Emitting Diode
LER	Line Edge Roughness
LF	Low Frequency
LHS	Latin Hypercube Sampling
LHS-MC	LHS-Based Monte Carlo
LMS	Linear Mean Square
LPF	Loop Filter, Low-Pass Filter
LPLL	Linear PLL
LRM	Language Reference Manual
LSCVD	Liquid Source Misted CVD
LUE	Latchup Effect
LVLP	Low Voltage Low Power
LVS	Layout versus Schematic
MARS	Multivariate Adaptive Regression Spline
MC	Monte Carlo
MDRAM	Multibank DRAM
MEMS	Micro-Electro-Mechanical System
MIGFET	Multiple Independent Gate FET
MISFET	Metal Insulator Semiconductor FET
MLCMOS	Multiple-Channel-Length CMOS
MLE	Maximum Likelihood Estimation
MLHS	Middle LHS
MLHS-MC	Middle LHS–Based Monte Carlo
MMV	MPEG Motion Vector
MOR	Model Order Reduction
MOSFET	Metal-Oxide Semiconductor FET
MOXCMOS	Multiple Oxide Thickness CMOS
MPEG	Motion Picture Export Group
MPSoC	Multiprocessor System-on-a-Chip
MNA	Modified Nodal Analysis
MRAM	Magnetic RAM
MRRAM	Magnetoresistive RAM
MSE	Mean Square Error
MTCMOS	Multiple Threshold CMOS
MTTF	Mean Time to Failure
MuGFET	Multiple Independent-Gate FET
MWCNT	Multiwall Carbon Nanotubes
Nano-CMOS	Nanoscale Complementary Metal-Oxide Semiconductor
NAS	Network Attached Storage
NBTI	Negative Bias Temperature Instability
NCO	Numerically Controlled Oscillator
NDR	Negative Differential Resistance
nDRAM	Next-Generation DRAM
NEMS	Nano-Electro-Mechanical System

NIM	Network Interface Module
NMP	Net-Centric Multimedia Processor
NMT	Networked Media Tank
NN	Neural Network
NoC	Network-on-a-Chip
NP	Nondeterministic Polynomial Time
NTF	Noise Transfer Function
NVRAM	Nonvolatile Random Access Memory
OASIS	Open Artwork System Interchange Standard
OBG	Out-of-Band Gain
ODE	Ordinary Differential Equation
OS	Operating System
OTP	One-Time Programmable
PC	Personal Computer
PCB	Printed Circuit Board
PCR	Polymerase Chain Reaction
PCRAM	Phase-Change RAM
PDA	Personal Digital Assistant
PDE	Partial Differential Equation
PDK	Process Design Kit
PE	Processing Element
PEEC	Partial Element Equivalent Circuit
PFD	Phase-Frequency Detector
PID	Proportional-Integral-Differential
PLL	Phase-Locked Loop
PMM	Performance Metric Metamodel
PMP	Personal Media Player
PMR	Personal Media Recorder
PRAM	Phase-Change RAM
PRNU	Photo Response Nonuniformity
PROM	Programmable ROM
PSD	Power Spectral Density
PSDRAM	Pseudostatic DRAM
PSNR	Peak-Signal-to-Noise Ratio
PSO	Particle Swarm Optimization
PSR	Power-to-SNM Ratio
PSRAM	Pseudostatic SRAM
PTM	Predictive Technology Model
PVD	Physical Vapor Deposition
PVR	Personal Video Recorder
PVT	Process, Voltage, and Temperature Variations
PWI	Powerwise Interface
QCA	Quantum Dot Cellular Automata
QFG-FET	Quasi Floating-Gate FET
QoS	Quality of Service
Qucs	Quite Universal Circuit Simulator
R^2	R-Square
RAAE	Relative Average Absolute Error
RAM	Random-Access Memory
RBB	Reverse Body Bias
RBF	Radial Basis Function
RCO	RC Oscillator
RDRAM	Rambus DRAM
RDS	Radiation Detection System

ReRAM	Resistive RAM
RF	Radiofrequency
RFID	Radiofrequency Identification
RG-FET	Resonant-Gate FET
RMAE	Relative Maximum Absolute Error
RMSE	Root Mean Square Error
RNM	Read Noise Margin
RO	Ring Oscillator
ROM	Read-Only Memory
RPECVD	Rapid Plasma-Enhanced CVD
RRAM	Resistive RAM
RTCVD	Rapid Thermal CVD
(S/(N + D))	Signal-to-Noise and Distortion Ratio
SA-ADC	Successive-Approximation ADC
SAD	Sum of the Absolute Difference
SAM	Sequential-Access Memory, Serial-Access Memory
SAR	Successive-Approximation Register
SCBE	Substrate Current-Induced Body Effect
SCE	Short-Channel Effect
SCM	Statistical Compact Model
SD	Secure Digital
SDC	Secure Digital Camera
SDR	Software-Defined Radio
SDRAM	Synchronous DRAM
SEM	Scanning Electron Microscope
SE-SRAM	Single-Ended Static Random-Access Memory
SET	Single-Electron Transistor
SFDR	Spurious-Free Dynamic Range
SFM	Scanning Force Microscope
SGRAM	Synchronous Graphics DRAM
SHC	Substrate Hot Carrier
SHE	Substrate Hot Electron
SHH	Substrate Hot Hole
SIMD	Single Instruction, Multiple Data
SINAD	Signal-to-Noise and Distortion
SINM	Static Current Noise Margin
SLDRAM	SyncLink DRAM
SLR	Single Lens Reflex
SMP	Secure Media Processor
SNDR	Signal-to-Noise and Distortion Ratio
SNM	Static Noise Margin
SNR	Signal-to-Noise Ratio
SoC	System-on-a-Chip
SOI	Silicon-on-Insulator
SPC	Slate Personal Computer
SPI	Serial Peripheral Interface
SPICE	Simulation Program with Integrated Circuit Emphasis
SPLL	Software PLL
SPM	Scanning Probe Microscope
SRAM	Static Random-Access Memory
STA	Statistical Timing Analysis
STB	Set-Top Box
STF	Signal Transfer Function
STT-MRAM	Spin-Transfer Torque MRAM

STU	Set-Top Unit
SVM	Support Vector Machine
SWCNT	Single-Walled Carbon Nanotube
SysML	Systems Modeling Language
TCAD	Technology CAD
TDDB	Time-Dependent Dielectric Breakdown
TDFDM	Time-Domain Finite-Difference Method
TDP	Thermal Design Power
TED	Taylor-Series Expansions Diagram
TEM	Transmission Electron Microscopy
TFET	Tunneling FET
TFT	Thin-Film Transistor
TGFET	Triple-Gate FET, Trigate FET
THD	Total Harmonic Distortion
TIM	Thermal Interface Material
TMM	Thermal Moment Matching Method
TMR	Tunnel Magnetoresistance
TNL	Time Nonlinearity
TRAM	Thyristor RAM
TS	Thermal Stress
TTRAM	Twin-Transistor Random-Access Memory
UART	Universal Asynchronous Receiver/Transmitter
UHF	Ultra-High Frequency
ULC	Universal-Level Converter, Universal Voltage-Level Converter
ULS	Universal-Level Shifter, Universal Voltage-Level Shifter
UPC	Universal Product Code
URC	Universal Remote Control
USB	Universal Serial Bus
VBFET	Vibrating Body FET
VCD	Video Compact Disc
VCDL	Voltage-Controlled Delay Line
VCO	Voltage-Controlled Oscillator
Verilog-AMS-PAM	Verilog-AMS Integrated with Parasitic-Aware Metamodel
Verilog-AMS-PoM	Verilog-AMS Integrated with Polynomial Metamodel
VHDL	Very High-Speed Integrated-Circuit Hardware Description Language
VHF	Very High Frequency
VHIC	Very High-Speed Integrated-Circuit
VHS	Home Video System
VLC	DC-DC Voltage-Level Converter
VLSI	Very Large-Scale Integration
WDAG	Weighted Directed Acyclic Graph
WINM	Write Current Noise Margin
WLAN	Wireless Local Area Network
WNM	Write Noise Margin
WORM	Write-Once, Read-Many
WSN	Wireless Sensor Network
WTV	Write Trip Voltage
ZRAM	Zero-Capacitor RAM

Notation

A_{DC}	Open-loop DC gain of OP-AMP
A_{tunnel}	Active source tunneling cross-sectional area in TFET
A_{chip}	Overall area of the chip
BW_{OP-AMP}	Bandwidth of OP-AMP
c	Velocity of light in $\left(\frac{m}{s}\right)$
C_A	Capacitance per unit area in $\left(\frac{F}{m^2}\right)$
$C_{CNT,ES}$	Electrostatic capacitance of CNT bundle interconnect
$C_{CNT,tot}$	Total capacitance of CNT bundle interconnect
$C_{CNT,QM}$	Quantum capacitance of CNT bundle interconnect
C_{db}	Drain diffusion capacitance in farads (F)
C_{fringe}	Fringe capacitance of interconnect
C_{gb}	Gate capacitance to body in farads (F)
C_{gd}	Gate capacitance to drain in farads (F)
C_{ground}	Capacitance to the ground
C_{gs}	Gate capacitance to source in farads (F)
$C_{interwire}$	Interwire or intermetal capacitance of interconnect
$C_{interlayer}$	Interlayer or interlevel capacitance of interconnect
C_L	Output load capacitance in farads (F)
C_M	Mutual capacitance
C_{sb}	Source diffusion capacitance in farads (F)
C_{side}	Sidewall fringing capacitance in farads (F)
C_{top}	Top fringing capacitance in farads (F)
$C_{tun,eff}$	Effective capacitive load due to gate-oxide tunneling
C_{ox}	Gate capacitance in $\left(\frac{F}{m^2}\right)$
cfi_c	Cycle frequency index
d_h	Horizontal distance between two wires
d_{CNT}	Diameter of a single CNT
D_{prop}	Propagation delay in seconds
$D_{prop,c}(\mu_{D,c}, \sigma_{D,c})$	Statistical distribution of a clock cycle delay
$D_{prop,FU_{k,i,v}}(\mu_{D,k,i}, \sigma_{D,k,i})$	Statistical distribution of propagation delay of a $FU_{k,i,v}$
d_v	Vertical distance between two wires of two metal layers
E_A	Activation energy
$e(i,j)$	Edge or arc task v_i and task v_j
E_G	Bandgap energy in electronvolts (eV)
f_{base}	Base clock frequency
f_c	Center frequency
f_{co}	Cutoff frequency or -3 dB frequency or bandwidth
f_o	Operating frequency
f_{osc}	Oscillating frequency
f_s	Sampling frequency
$\mathbf{F}_{wind}$	Electron wind force
$\mathbf{F}_{direct}$	Direct force
$FU_{k,i}$	Functional unit of type k and design corner i
$FU_{k,i,v}$	Functional unit of type k and design corner i used by vertex v
G_{DC}	DC gain
g_m	Transconductance of a MOSFET in ohm^{-1}
g_{mi}	Transconductance of input stage of OP-AMP

$G(V,E)$	Data flow graph (DFG) or task graph of vertex set V and edge set E
h	Planck's constant
H_{fin}	Height of the Fin in a FinFET
$H(s)$	Closed loop transfer function of PLL
I_{cp}	Channel punch through current
I_{dark}	Dark current of a photodiode
I_{DSSat}	Saturation drain current in A
I_{hc}	Hot carrier gate current
$I_{jun,BTBT}$	Junction band-to-band-tunneling (BTBT) current
I_{max+}	Maximum available positive current of OP-AMP
I_{max-}	Maximum available negative current of OP-AMP
I_{ox}	Gate-oxide leakage current due to direct tunneling in A
I_{photo}	Photocurrent of a photodiode
$I_{pump}(s)$	Charge-pump (CP) current
I_{sb}	Substrate current
I_{subth}	Subthreshold current in a MOSFET (A)
I_{subth}	Subthreshold leakage current
J_{DT}	Direct tunneling current density in $\left[\frac{A}{m^2}\right]$
J_{GIDL}	Current density of the GIDL
J_{hc}	Hot carrier gate current density
J_{ox}	Direct tunneling current density
J_{TAT}	Current density due to trap assisted tunneling in $\left[\frac{A}{m^2}\right]$
k or k_B	Boltzmann's constant in $\left[\frac{J}{K}\right]$
K_{PFD}	Gain of the phase frequency detector (PFD) of PLL
K_{VCO}	Gain of the VCO
$L_{CNT,K}$	Kinetic inductance of a SWCNT
$L_{CNT,M}$	Magnetic inductance of a SWCNT
L_{eff}	Effective channel length of MOSFET in nanometers (nm)
L_{ext}	Extension length of the Fin in a FinFET
L_{gate}	Physical length of the gate in nm
$L_{int,rec}$	Inductance of a rectangular cross-section interconnect
L_M	Mutual inductance
$L_{overlap}$	Gate overlap length in nm
$LPF(s)$	Loop filter impedance
L_{phy}	Geometrical or physical channel length of MOSFET in nm
$LG(s)$	Open loop gain
m^*	Effective mass of the carrier
N_{FB}	Dividing factor of the feedback divider of the PLL
N_{cc}	Number of clock cycles
N_{chan}	Channel doping concentrations in cubic centimeters (cc)
N_{DFG}	Total number of clock cycles
N_g	Number of logic gates in the logic block
N_p	Number of pins or the number of external connections
$N_{particle}$	Number of particles or population size
N_{poly}	Polysilicon gate doping concentrations in cc
N_{stage}	Number of inverter stages in a VCO
N_{sub}	Substrate doping concentrations in cc
N_{int}	Intrinsic doping concentrations in cc
n_{fin}	Total number of Fins in a FinFET
n_s	Electron density at the surface
$P_{dyn,c}(\mu_{dyn,c}, \sigma_{dyn,c})$	Distribution of dynamic power of each clock cycle c
$P_{dyn,FU_{k,i,v}}(\mu_{dyn,k,i}, \sigma_{dyn,k,i})$	Distribution of dynamic power of a $FU_{k,i,v}$
PD_{OP-AMP}	Power dissipation of OP-AMP
PM_{OP-AMP}	Phase margin of OP-AMP

$P_{\text{OP-AMP}}$	Function for poles of OP-AMP
$P_{\text{ox},c}(\mu_{\text{ox},c}, \sigma_{\text{ox},c})$	Distribution of gate-oxide leakage power of each clock cycle c
$P_{\text{ox,FU}_{k,i,v}}(\mu_{\text{ox},k,i}, \sigma_{\text{ox},k,i})$	Distribution of gate-oxide leakage power of a $FU_{k,i,v}$
$P_{\text{sub},c}(\mu_{\text{sub},c}, \sigma_{\text{sub},c})$	Distribution of subthreshold leakage power of each clock cycle c
$P_{\text{sub,FU}_{k,i,v}}(\mu_{\text{sub},k,i}, \sigma_{\text{sub},k,i})$	Distribution of subthreshold leakage power of a $FU_{k,i,v}$
$P_{\text{total},c}(\mu_{\text{total},c}, \sigma_{\text{total},c})$	Statistical distribution of total power of each clock cycle c
$P_{\text{total,DFG}}(\mu_{P,\text{DFG}}, \sigma_{P,\text{DFG}})$	Statistical distribution of of the overall DFG
P_{switch}	Power consumed due to capacitive switching
P_{sc}	Short circuit power dissipation
$P_{\text{peak,sust}}$	Maximum sustained power
q	Electronic charge in coulombs
q_{id}	Normalized inversion charge density calculated at the drain
q_{is}	Normalized inversion charge density calculated at the source
R_{channel}	MOSFET channel resistance in Ω
$R_{\text{contact,bundle}}$	Contact resistance of a CNT bundle
R_{doped}	Resistance of the doped layer of a length L_{physical}
R_{drain}	MOSFET drain resistance in Ω
$R_{\text{scattering,bundle}}$	Intrinsic scattering resistance of a CNT bundle
R_{source}	MOSFET source resistance in Ω
$R_{\text{therm,sink}}$	Thermal resistance of the heat sink
R_{undoped}	Resistance of the undoped layer of length L_{physical}
$SR_{\text{OP-AMP}}$	Slew rate of OP-AMP
tCL	Strobe or column address select latency
t_{fall}	Fall time in ns
t_{int}	Integration time in sec
tRAS	Row address strobe active time
tRCD	Row address strobe to column address strobe delay
t_{rise}	Rise time in ns
tRP	Row precharge time
T	Temperature in kelvins (K)
T_{epi}	Epitaxial layer thickness TFET
T_{fin}	Width of the Fin in a FinFET
T_{ox}	Electrical equivalent oxide thickness in nm
$T_{\text{ox,b}}$	Thickness of the bottom/back oxide in nm
$T_{\text{ox,f}}$	Thickness of the front oxide in nm
T_{oxn}	Electrical equivalent oxide thickness of N-type NMOS in nm
T_{oxp}	Electrical equivalent oxide thickness of P-type NMOS in nm
$T_{\text{ox,t}}$	Thickness of the top oxide in nm
T_{CPU}	CPU execution time in seconds
T_{GPU}	GPU execution time in seconds
T_{Si}	Silicon thickness in nm
V_{body}	Body voltage in V
V_{bs}	Body-to-source voltage in V
V_{DD}	Supply voltage in volts (V)
V_{ds}	Drain-to-source voltage in V
$V_{\text{dsSat}}, V_{\text{ds,SAT}}$	Drain-to-source voltage value at which saturation is attained
v_F	Fermi velocity for CNTs
V_{fb}	Flat-band voltage in V
V_{gb}	Voltage between back-gate and source of DGFET
V_{gf}	Voltage between front-gate and source of DGFET
V_{gs}	Gate-to-source voltage in V
V_{Th}	Threshold voltage in V
V_{Th0}	Threshold voltage in V at zero body bias
V_{ox}	Voltage across the gate dielectric in V
V_{poly}	Voltage across the polysilicon in V

v_{therm}	Thermal voltage calculated as $\left(\frac{kT}{q}\right)$
V_{PT}	Punchthrough voltage
V_{pz}	Piezopotential or piezoelectric voltage
v_{sat}	Electron saturation velocity in $\left(\frac{cm}{s}\right)$
W	Width of MOSFET in nm
W_n	Width of N-type MOSFET in nm
W_p	Width of P-type MOSFET in nm
W_{fin}	Width of the Fin in a FinFET
W_{phy}	Geometrical or physical channel width of MOSFET in nm
Y_{int}	Young's modulus of the interconnect material
Z_{OP-AMP}	Function for poles of OP-AMP
α	Physical constant modeling carrier saturation velocity
α_{R-C}	Rent coefficient
α_{int}	Coefficients of linear thermal expansion of the interconnect material
α_{sw}	Average activity factor
α_{sub}	Coefficients of linear thermal expansion of the silicon substrate
β_{R-E}	Rent exponent
δ	Skin depth during skin effect in an interconnect
ε_0	Permittivity of the free space
ε_{IL}	Permittivity of the interlayer dielectric
ε_{IM}	Permittivity of the intermetal dielectric
ε_{ox}	Permittivity of gate dielectric in $\left(\frac{F}{m}\right)$
ε_{Si}	Permittivity of Si in $\left(\frac{F}{m}\right)$
η_{DIBL}	Drain-induced barrier lowering (DIBL) coefficient
η_{QE}	Quantum efficiency
γ_{body}	Body effect coefficient
$\hbar$	Reduced Planck's constant
κ	Dielectric constant of insulators
λ	Electron mean free path of the CNT
λ_{tight}	Wavelength of light in μm
μ_{eff}	Effective carrier mobility in $\left(\frac{cm^2}{V-s}\right)$
μ_o	Permeability of the free space, which is $4\pi \times 10^{-7}\frac{H}{m}$
μ_r	Relative permeability
μ_{sub}, μ_0	Bulk mobility in $\left(\frac{cm^2}{V-s}\right)$
ν_{int}	Poisson ratio
ω_{-3dB}	Closed-loop bandwidth of the PLL
ω_L	Locking range of the PLL
ω_N	Natural angular frequency of the PLL
ω_{osc}	Angular oscillation frequency of the VCO/PLL
ω_{ug}	Unity-gain frequency
ϕ_B	Barrier height for the gate dielectric in eV
ϕ_F	Fermi level in V
ψ_S	Surface potential in V
ρ	Resistivity of the interconnect material in $\left(\frac{\Omega}{m^2}\right)$
τ	Subthreshold slope/swing factor
Θ	Temperature
$\Theta_{ambient}$	Ambient temperature
Θ_{chip}	Average operating temperature of the chip
Θ_{peak}	Peak/maximum temperature
Θ_{therm}	Thermal stress in $\left(\frac{N}{m^2}\right)$
ξ	Electric field in $\left(\frac{V}{m}\right)$

Nanoelectronic Mixed-Signal System Design

CHAPTER 1

Opportunities and Challenges of Nanoscale Technology and Systems

1.1 Introduction

Consumer electronics systems such as mobile phones, digital cameras, digital television, high-definition content players, health-monitoring systems, and DVD/MP3 players have profound impact on society. The main component of these appliances is a tiny integrated circuit (IC) [37, 43, 141]. ICs are used everywhere, from kitchen appliances, to automobiles, to aircrafts. A system in modern consumer electronics is built as an analog/mixed-signal system-on-chip (AMS-SoC) [17, 43, 61, 72, 85]. A representative AMS-SoC is illustrated in Fig. 1.1. It has image sensors for the camera. A general-purpose digital processor is programmable and executes the system and application software. A digital signal processor (DSP) performs the signal processing in the system. Analog circuits are necessary in AMS-SoCs at least as interface elements even when the functions are being performed by digital processors. In a smartphone, the baseband telecommunication chip performs communication operations using GSM or CDMA protocols. It is the main chipset in smartphones and is directly interfaced to other hardware such as speakers. The wireless and bluetooth connections take place through the wireless component. Analog-to-digital converters (ADCs) and digital-to-analog converters (DACs) are two examples of data converters that are intrinsic mixed-signal circuits [18, 71, 127]. For a portable electronic system, preferably a rechargeable battery is included.

The hardware in the present-day AMS-SoCs is of gigascale complexity and consists of transistors of nanoscale process technology (Fig. 1.2). Nanoscale complementary metal-oxide semiconductor (CMOS) technology, such as classical silicon dioxide/polysilicon bulk MOSFET and high-κ/metal-gate MOSFET, is used to build such hardware. With the growth of nanotechnology, a number of nanodevices have emerged to replace the classical MOSFET. Representative nanodevices include tri-gate field-effect transistors (TGFETs) and graphene FET (GFET) [35, 40, 147]. The TGFET is being used for ultra-low-power designs. The GFET can operate at high frequencies (e.g., 100 GHz) and has potential for high-speed nanoelectronics. Nanoscale size has reduced power dissipation of each transistor, cost per transistor, and cost per function and has allowed high packing density. These nanoscale technology-based AMS-SoCs (a.k.a. nanoelectronic systems) must be able to perform under severe process variations, in the presence of highly dense interconnect with large parasitic elements while satisfying strict power and performance requirements [67, 68, 92]. The nanoscale AMS-SoCs must be tolerant of voltage fluctuations and operate under a wide range of environmental and on-chip temperatures.

A nanoelectronic system designer should consider nanoscale issues while making the design decision in order to ensure that high yield is achieved. Nanotechnology (which includes nanoelectronics) has revolutionized numerous industry sectors, including information technology, energy, medicine, homeland security, and transportation [6, 8, 118]. An estimation in a recent study shows that the nanotechnology industry will require 2 million workers by 2015, with 800,000 in the United States [119]. At the same time, it is estimated that the total market for nanotechnology-related products will be $3 trillion and the industry will need 6 million personnel by 2020.

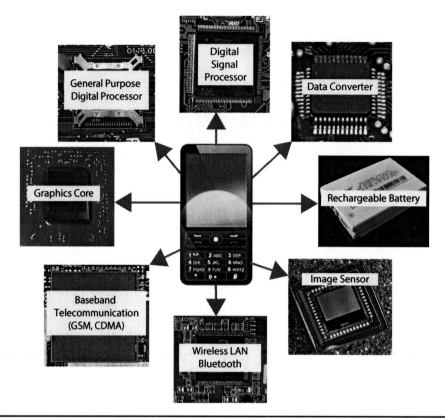

Figure 1.1 Components of a typical electronic system.

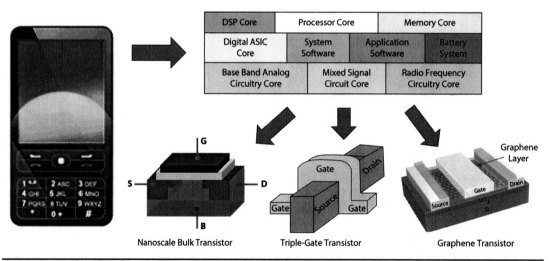

Figure 1.2 A nanoelectronic system with components made from nanodevices.

When AMS-SoCs are fabricated using nanoscale technology, their circuits are strongly affected by the imperfections in manufacturing processes, such as subwavelength lithography, lens aberration, and chemical-mechanical polishing [29, 134, 150, 151]. The design cycle of circuits is complicated by the impact of process variation on the performance of the circuits due to the use of state-of-the-art nanoscale CMOS (nano-CMOS) technologies for fabrication [80]. Process variations originating from fabrications based on nanoscale technology have a profound impact on power, leakage, delay, and performance [36]. For example, in a voltage-controlled oscillator (VCO), they affect the oscillating frequency and phase noise [33, 71, 83, 127].

For sub-65-nm CMOS AMS-SoCs, the power dissipation components consisting of dynamic power, subthreshold leakage, gate-oxide leakage, and gate-induced drain leakage (GIDL) pose challenges to battery life, heat dissipation, packaging costs, and reliability [30, 34, 102, 105, 106, 121, 130]. In addition, power dissipation due to band-to-band tunneling (BTBT) is important for sub-45-nm circuits. Higher power dissipation means less battery life and higher energy cost. Also, as power dissipation increases, heat dissipation increases, which, in turn, affects reliability.

Temperature is an important parameter because it has direct impact on reliability, packaging, and delay, and can lead to thermal runaway [38, 106]. The increasing temperature has three negative effects on nanoscale devices [23, 69, 73]: (1) increases device leakage, (2) reduces active current strength, and (3) increases propagation delay, thus making the device slow. If a positive feedback loop is established between leakage and temperature, a complete thermal breakdown can occur.

1.2 Mixed-Signal Circuits and Systems

1.2.1 Different Processors: Electrical to Mechanical

A processor that is realized as either a circuit or system is an entity that receives an input signal, processes the signal using the built-in instructions, may store some of the signals, and sends the output signal. Figure 1.3 shows various types of processors. A processor can be a fixed circuit that is designed right from the beginning for an application, known as application-specific ICs (ASIC). The fully flexible general-purpose central processing unit (CPU), which is fabricated on a single die called a microprocessor, is programmed by the users to execute diverse applications. The field-programmable gate array (FPGA) are ICs designed to be configured by the customer after manufacturing. The application-specific instruction processors (ASIP) have instruction sets customized for a specific application which can be programmed either in the field (such as FPGA) or during the chip synthesis [123]. In addition, there are micro-electro-mechanical systems (MEMS) or nano-electro-mechanical systems (NEMS) designed for different purposes [98, 120, 136].

1.2.2 Analog versus Digital Processors

Analog circuits and systems are those that process a continuously variable signal. On the other hand, in the digital circuits and systems, the signals usually take only two different levels (binary). The term *analog*, in essence, is a proportional relationship between an input signal (voltage or current) and an output signal (voltage or current). The information is encoded differently in analog and digital circuit and systems, and hence the way they process a signal is consequently different. All operations that are performed on an analog signal, including amplification and filtering, can also be duplicated in the digital signal. A single transistor can be either analog or digital depending on its biasing condition. A transistor when biased to operate in cutoff or saturation mode has two states: OFF (open circuit) or ON (short circuit), thus realizing the binary logic. On the other hand, when the transistor is biased to operate in linear or transconductance mode, it is an analog device.

1.2.3 Analog, Digital, Mixed-Signal Circuits and Systems

The transistors are put together in a chip to construct analog circuits, digital circuits, mixed-signal circuits, or mixed-signal systems (see Fig. 1.4 and 1.5). An analog circuit considers an analog signal as an

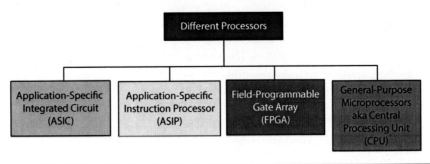

FIGURE 1.3 Different types of digital processors.

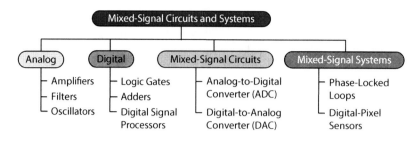

Figure 1.4 Different types of mixed-signal circuits and systems.

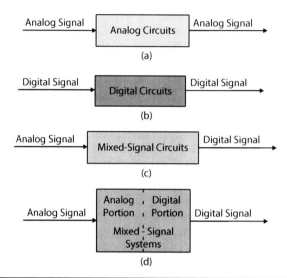

Figure 1.5 Analog, digital, mixed-signal circuits, and mixed-signal systems.

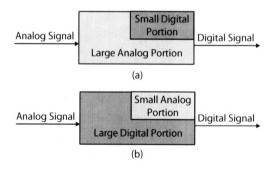

Figure 1.6 Mixed-signal systems: big "A" small "d" versus small "a" big "D".

input and generates it as an output (see Fig. 1.5(a)). A digital circuit considers a digital signal as an input and generates it as an output (see Fig. 1.5(b)). A mixed-signal circuit or system takes an analog (or digital) signal as an input and produces a digital (or analog) signal as an output (see Fig. 1.5(c)). A mixed-signal circuit is a single monolithic entity, while a mixed-signal system is an analog and a digital circuit clubbed together. The analog and digital circuits are two domains of a mixed-signal system with the analog circuit in one domain and the digital circuit in the other. Both are connected through interfacing circuit elements (see Fig. 1.5(d)).

1.2.4 Two Types of Mixed-Signal Systems

The integration of analog and digital portions on the same die provides cost-performance tradeoffs [43, 61, 72, 85]. The mixed-signal system can be either A/d (i.e., big "A" small "d") or a/D (i.e., small "a" big "D"; see Fig. 1.6). In A/d-based mixed-signal systems, the transistor count for the analog portion is much higher than that of the digital portion. On the other hand, in a/D-based mixed-signal systems, the transistor count for the analog portion is much lower than that of the digital portion. However, the size of the analog-portion transistors is larger than the size of digital-portion transistors.

1.3 Nanoscale CMOS Circuit Technology

1.3.1 Developmental Trend

The first IC was built by Jack Kilby at Texas Instruments in 1958 [126]. The first commercial planar IC was built by Fairchild Semiconductor in 1960 [4, 112]. Since their first demonstration, the planar silicon transistors (metal-oxide semiconductor FET, MOSFET) have had a steady exponential decline in their critical dimensions. Over a period of 45 years, the printed gate lengths of the MOSFETs have been scaled down from 100 nm to 32 nm. Because of a steady improvement in their performances through scaling, MOSFETs have become the leading IC technology for high-performance and low-power logic applications. Over this long period of development, the technology has faced numerous challenges, which were always solved by vigorous research, ingenious design, and brilliant engineering. The exponential scaling down of the feature sizes, and hence the exponential increase of the transistor count in an IC, was first observed by Gordon Moore in 1965 [108]. His observation, which was later known as Moore's law, states that the number of transistors per IC doubles every 24 months. Moore's law has been serving as the guiding principle for the semiconductor industry for over 30 years. During this period, the very large-scale integration (VLSI) technology has had compound growth of 53% [148]. It is estimated that a total of 340 billion transistors were manufactured in 2006 when the world population was only 6.5 billion [12]. An estimated 1 billion transistors per person were manufactured in 2010. In order to meet the growth of ICs, the semiconductor industry has resorted to aggressive technology scaling with higher packaging density. Various semiconductor houses are working on different nanoscale technology nodes including 45, 32, 28, and 16 nm [7]. Table 1.1 presents a change in device and circuit performance with technology scaling [49]. A dimensionless scaling factor κ is used to present the changes in device and circuit performance parameters.

1.3.2 Nanoscale CMOS Alternative Device Options

The feature sizes of transistors have shrunk dramatically with technology scaling, and the value of gate-oxide thickness (T_{ox}) has reached the range of 12 to 16Å, which is just a few monolayers of SiO_2 in nano-CMOS devices (see Fig. 1.7). The use of alternatives for SiO_2 as a gate dielectric has led to the

Device and Circuit Parameters	Change Trend
Device Geometry (L, W, T_{ox})	$\frac{1}{\kappa}$
Doping Concentration	κ
Gate Capacitance	$\frac{1}{\kappa}$
Supply Voltage	$\frac{1}{\kappa}$
Drain Current	$\frac{1}{\kappa}$
Power Dissipation	$\frac{1}{\kappa^2}$
Power Density	1
Propagation Delay	$\frac{1}{\kappa}$

TABLE 1.1 Device and Circuit Parameter Change with CMOS Technology Scaling [89]

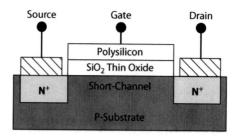

FIGURE 1.7 Nano-CMOS classical transistor with short-channel and thin-gate oxide.

construction of nonclassical devices. The use of high-κ serves the dual purpose of scaling the device and reducing gate-oxide leakage, which was a major concern for nano-CMOS technology [19, 31, 100]. Another nano-CMOS device, double-gate FET (DGFET) has shown strong potential [107]. The industry has already adopted classical nanoscale CMOS, is in the process of adopting high-κ nano-CMOS for the sub-45-nm technology node, and is anticipated to adopt DGFET for the sub-8-nm technology node. Other nanodevices, such as carbon nanotube FET (CNTFET) [3, 20, 132], quantum dots [9, 95], deoxyribonucleic acid (DNA) transistors [46], tunneling field effect transistors (TFET) [111], and single-electron transistors (SET) [152], have been introduced to replace the traditional silicon transistor. The next stage in this progression is the assembly of nanoscale devices to build integrated multidiscipline systems; however, these emerging technology devices are at their initial stages of development.

CMOS technology using P-type MOS (PMOS) and N-type MOS (NMOS) FETs became popular due to low-power dissipation, smaller feature size, and high packing density. The transistor-feature sizes have shrunk dramatically with the technology scaling (see Fig. 1.7). The channel length of a device is dictated by the technology node that a particular process follows. The length of a device is lower bound by the technology node. At these nano-CMOS technology regimes, the channel is much shorter than the conventional channel length, and the device characteristics are very much affected by these shorter channel lengths. A number of adverse effects peculiar to scaling that result in a significantly reduced channel length are identified as short-channel effects (SCEs). The most important parameter for scaling is the drastic change in the gate-oxide thickness. Aggressive scaling of SiO_2 has been going on for the past 15 years for low-power, high-performance CMOS transistor applications [21]. SiO_2 with a physical thickness of 1.2 nm was implemented in the 90-nm logic technology node. The typical dimension of 12 to 16Å is a few monolayers of SiO_2 in the nano-CMOS devices. Research transistors with 0.8 nm (physical thickness) SiO_2 have been demonstrated in the laboratory.

Systems that used to operate at 3.3 or 2.5 V now need to operate at 1.8 V or lower without causing any performance degradation. The smaller devices provide higher packing density and lower overall power consumption, due to lower parasitics and lower supply voltages. This shortening of the minimum channel length has resulted in the reduction of power supply voltage to the 1 to 0.7 V range. The AMS-SoC trend also forces analog circuits to be integrated with digital circuits. To keep up with the scaling of the minimum channel length and AMS-SoC trend, analog circuits need to be operated at low voltages. However, the minimum supply voltage for analog circuits predicted in the semiconductor road map [3] does not follow the reduction of digital supply voltage. Analog supply voltages between 1.8 and 2.5 V are still being used with channel lengths of 0.18 and 0.13 μm [17]. Hence, it is a great challenge to design low supply voltage-operating mixed-signal circuits while considering the relatively high-threshold voltage of short-channel length transistors. Another important consideration for an AMS-SoC is that the mixed-signal circuits should be designed by using a nano-CMOS digital process without having process options such as deep n-well or on-chip inductors or varactors.

Aggressive scaling of nano-CMOS devices has led to a drastic change in the leakage components of the device, both in its active and inactive states. Consequently, gate-oxide leakage has emerged as the most prominent form of leakage in a nano-CMOS device, particularly in the 65 nm and below regime. It has consequently become desirable to find suitable alternatives for SiO_2 as the gate

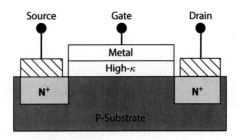

FIGURE 1.8 Structure of high-κ/metal gate transistor.

dielectric [57, 79, 145]. Sustaining Moore's law, however, depends crucially on the gate insulator scaling; consequently, replacing SiO_2 with high-κ dielectric-metal gate stack is within the realm of possibility [22]. This has led to the construction of nonclassical transistors (Fig. 1.8) [84, 101, 140]. The use of high-κ serves the dual purpose of scaling the device and reducing gate leakage.

The CMOS fabrication process technology using high-κ dielectrics and its compact modeling have been developed. Several materials have been investigated for use in nano-CMOS technology, such as ZrO_2, TiO_2, BST, HfO_2, Al_2O_3, SiON, and Si_3N_4 [19, 20, 100]. It is a challenging task in itself to integrate these materials into the conventional CMOS process [149]. Progress has been made in the development of various technologies for high-κ gate dielectric deposition [56]. This includes the extension of chemical vapor deposition (CVD), rapid thermal CVD (RTCVD), rapid plasma-enhanced CVD (RPECVD), liquid source misted CVD (LSCVD), physical vapor deposition (PVD) [116], jet vapor deposition (JVD) [90], oxidation of metallic films [87], and molecular beam epitaxy [81]. In this scenario, it is necessary to study the effect of high-κ dielectrics on other device parameters, including the effective gate dielectric thickness, the threshold voltage, the gate capacitance, the carrier saturation velocity and mobility, the gate polysilicon depletion depth, gate leakage, and delay to provide guidelines for design engineers.

Dual-gate devices show promise for scaling beyond the planar bulk MOSFET limit [132]. The electrostatic integrity of the nanoscale CMOS devices improves considerably when additional gates are included, such as for dual-gate or FinFET devices. Because these nonplanar devices are inherently resistant to SCEs, it is widely believed that one of them will form the basic device architecture for future generations of CMOS devices. General MOSFET at a submicron level suffers from several submicron issues such as SCEs and threshold voltage variation. FinFET is proposed to overcome the SCEs. The silicon-on-insulator (SOI) process is used to fabricate the FinFET. This process ensures the ultrathin specifications of the device regions. In FinFET, electrical potential throughout the channel is controlled by the gate voltage. This control is possible because of the proximity of the gate control electrode to the current conduction path between the source and the drain. These characteristics of the FinFET minimize the SCE [125]. Advantages of the FinFET over its bulk-Si counterpart include the following: Conventional MOSFET-manufacturing processes can also be used to fabricate the FinFET. FinFET provides better area efficiency than MOSFET. Mobility of the carriers can be improved by using the FinFET process in conjunction with the strained Si process. The structure of FinFET is shown in Fig. 1.9.

The SOI technology is used for the fabrication of FinFET. In SOI technology, an insulator, SiO_2, isolates the bulk from the substrate. An extremely shallow junction is formed because of the depth limitation put by the insulator. The dielectric isolation and elimination of the latch-up problem are the advantages of the SOI process. In the FinFET fabrication process, silicon nitride (Si_3N_4) and SiO_2 are deposited on a thin SOI layer. Electron beam lithography is used to form a Si-Fin. Channel length and width are determined by the accuracy of the Fin. Poly-Si with pentavalent impurities and the oxide layer are deposited over the Si-Fin. The source and drain regions are separated and insulator spacers are formed. Then the etching process is carried out on a spacer until a Si-Fin is reached. A gate is formed by depositing the gate layer. Silicidation is carried out to decrease the high-source drain resistance, which is formed because of very thin layers of source and drain. A single polysilicon layer is deposited over a Fin. Thus, polysilicon straddles the Fin structure to form perfectly aligned gates.

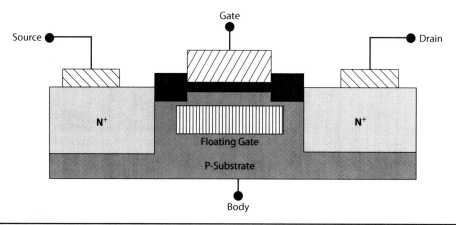

FIGURE 1.9 Structure of the double-gate FinFET transistor.

Here, Fin itself acts as a channel and it terminates on both sides of source and drain. In general, like any MOSFET device, poly-Si gate is formed over the Si substrate, which controls the channel. Straddling of a poly-Si gate over an Si-Fin gives efficient gate-controlled characteristics compared to MOSFET. Since the gate straddles the Fin, the length of the channel is the same as that of the width of the Fin. As there are two gates effectively around the Fin, we propose that the width of the channel is equivalent to twice the height of the Fin, that is, $W = 2 \times h$. A term called "Fin pitch" is used to define the space between two Fins. The height of the FinFET is equivalent to the width of the MOSFET. In order to attain the same area efficiency, if W is the Fin pitch, then the height of the Fin should be $W/2$. But practical experiments have shown that Fin height can be greater than $W/2$ for a Fin pitch of W; thus, FinFET achieves more area efficiency than MOSFET.

1.3.3 Advantages and Disadvantages of Technology Scaling

The multifold benefits of device scaling are as follows [32, 70]. First, the decreasing device size allows increasing chip and component density, thus more logic and functionality in the same die size. Second, the device miniaturization has helped in increasing its performance and speed. Third, due to high packing density (i.e., more number of transistors per area), the cost per transistor has been reduced. Finally, the technology scaling allows operating the chips or transistors at decreasing supply voltage, thereby reducing the power dissipation per transistor and power consumption per operation.

The technology scaling has several disadvantages as well. First, the manufacturing cost has increased significantly. Second, the cost of mask has increased for the development of the nano-manufacturing process. This has forced many companies to go on fabless mode. Third, in general, the tremendous increase in complexity and decrease in design flexibility have led to an increase in design cost. Fourth, uncertainty introduced during the nanomanufacturing process has caused a gap between the performance and post-fabrication performance of the chip. This leads to loss of yield and cost of chip [68, 70]. These uncertainties have resulted in device mismatch and process variations [69, 70]. Fifth, the power dissipation of the whole chip has increased as a result of an increase in the number of transistors per chip. Sixth, the worst scenario is the power density (i.e., power dissipated for unit area) that has increased much faster than the power dissipation [96, 106]. Finally, issues of reliability and robustness and an increased likelihood of soft errors are of concern for nano-CMOS circuits and systems [70].

1.3.4 Challenges in Nanoscale Design

Three different categories of designers—analog (including mixed-signal), digital, and MEMS or NEMS—are at different points of nanoscale technology trend. The digital technology follows the technology trend most aggressively. This goes with the digital portion of mixed-signal and electrical portions of NEMS. The analog circuits too follow technology aggressively. MEMS technology follows the trend less aggressively. However, eventually, for monolithic fabrication of mixed-signal circuits and systems and NEMS, all the portions need to be designed in the same nanoscale technology.

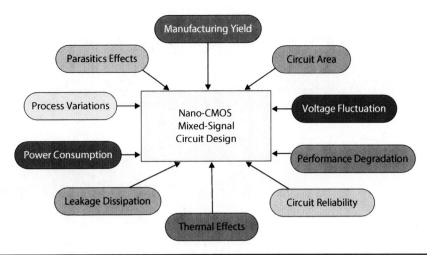

FIGURE 1.10 Different challenges for nanoscale mixed-signal design.

Hence, NEMS designers need to be aware of nanoscale issues [120]. However, there are several challenges that nanoscale circuit and system designers must address (Fig. 1.10) [68, 97]. These challenges need not be independent of each other. They are interdependent and hence quite difficult to address during the design phase. For example, process variation affects power, leakage, and parasitics; parasitics affect power and delay.

1.4 Power Consumption and Leakage Dissipation Issues in AMS-SoCs

1.4.1 Power Consumption in Various Components in AMS-SoCs

The distinct portions of a portable system designed as an AMS-SoC are the hardware, software (may include application and system), and firmware. A hardware component directly consumes power while performing its operation. On the other hand, when software (or firmware) performs its operation, it does not consume power directly. The software is executed by the hardware which, in turn, consumes power on behalf of the software. Therefore, power dissipation in hardware needs discussion. A user is primarily worried about the applications and maybe the software because the user does not directly use the system hardware.

1.4.2 Power and Leakage Trend in Nanoscale Technology

As the dimensions of the CMOS device reach nanometer ranges, dynamic power consumption remains almost unchanged, while leakage power dissipation increases significantly and thus with a change in technology becomes a large portion of the total power dissipation [97, 100, 106]. With the accompanying shrinkage of feature size, a paradigm shift occurs in the power profiles of the devices. The leakage components of the fundamental CMOS device change dramatically and each component of the total leakage gains relative importance. Thus, a drastic change occurs in the leakage components of the device, both in the active and the standby modes of operation. Figure 1.11 shows directions of different current flow in a nanoscale CMOS device [36, 100, 106, 121]. The current components are as follows: I_1, drain-to-source active current; I_2, subthreshold leakage current; I_3, short-circuit current; I_4, gate-oxide leakage; I_5, gate current due to hot-carrier injection; I_6, channel punch through current; I_7, GIDL; I_8, BTBT current; and I_9, reverse-biased PN junction leakage.

The relative magnitude of these current components depends on various aspects of the nano-CMOS technology, such as technology node, nano-CMOS process, and states of device operation [69, 97]. For example, for above 90-nm technology, dynamic power and subthreshold leakage are prominent and the gate-oxide leakage comes into play from sub-90 nm. For sub-65 nm, the gate-oxide leakage is significant, and for sub-45 nm, BTBT current becomes important. Figure 1.12 shows major components of SiO_2/polysilicon classical nano-CMOS and high-κ/metal-gate nonclassical nano-CMOS.

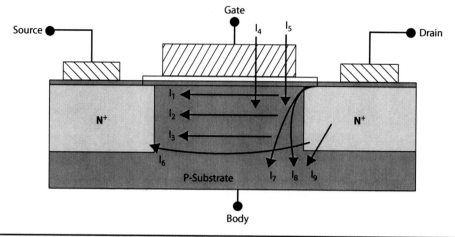

FIGURE 1.11 Flow of different current components in a nano-CMOS transistor.

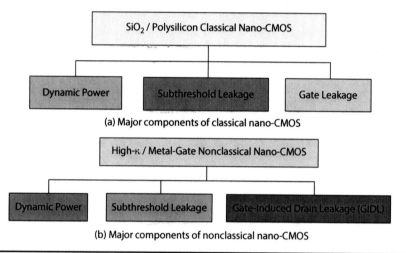

FIGURE 1.12 Major components of power dissipation in nano-CMOS technology.

1.4.3 The Impact of Power Consumption and Leakage Dissipation

The significant demand for portable systems combined with other issues, such as reliability, thermal considerations, and environmental concerns, has driven the need for low-power designs [96, 102]. The power dissipation of a mixed-signal circuit or system affects various aspects of the system, including battery life, reliability, system noise, system performance, cooling costs, and packaging costs [70, 96, 99, 106]. It has been shown that gate-oxide leakage can affect analog circuit performance. For example, in a VCO, the frequency is degraded by 10% [83]. Power consumption with a significant component of leakage dissipation is one of the major issues in nanoscale circuits and system design [99]. Digital designers dealing with millions of devices in a single chip have been working on the issue for the last decade [104, 106, 115]. Analog designers are addressing the issue at present while considering their traditional specification-driven priorities [65, 66]. The power issue has, however, become more and more difficult as devices dissipate power during both active and standby modes.

Power conservation affects every budget, whether technological or financial. Product acceptability, reliability, and profitability depend as much on power efficiency as they do on performance. Designers resort to low-power design and/or power-aware design to address the power issue [70, 106, 115]. There is a difference between low-power design and power-aware design. While a low-power design refers to minimizing power with or without a performance constraint, a power-aware design refers to maximizing some other performance metric, subject to a power budget. Two of the biggest challenges to maintaining performance are the increasing power specifications owing to power leakage and potential yield loss caused by increasing process variations.

1.5 Parasitics Issue

1.5.1 Types of Parasitics

An IC is made up of active and passive elements. The active elements include diodes, bipolar junction transistors (BJT), and MOSFET. The passive elements include interconnect metal wires, which have several layers. With present technology, up to nine layers of copper are used in the nanoscale CMOS chips. The numerous parasitic effects induced by the physical design of the circuit (see Fig. 1.13), especially for high-performance and high-speed circuits, the pose a problem for designers. The transistors have resistance (R) and capacitance (C) as parasitics. The interconnect parasitic elements are resistance (R), capacitance (C), and inductance (L). In addition, mutual inductance (L_m) and mutual capacitance (C_m) are other parasitics. Together they are grouped as *RLCK* [142].

1.5.2 The Impact of Parasitics

The parasitic elements have profound impact on the circuit performance. They also affect circuit power consumption and reliability. Let us consider Fig. 1.14 as a particular case study that shows the frequency and voltage characteristics of a 90-nm CMOS VCO [66]. First, the logical design of a 90-nm VCO has been carried out and its frequency-voltage characteristics are recorded. Then, the physical design of this VCO is set up and its frequency-voltage characteristics are recorded. Due to parasitics, the frequency-voltage characteristics of the physical design show a large discrepancy up to 25% compared to the logical design of a 90-nm CMOS-based VCO [66].

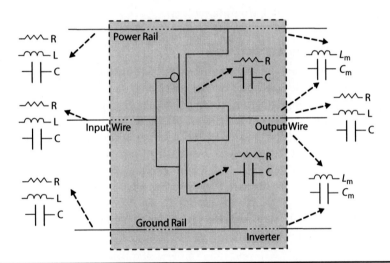

FIGURE 1.13 Different types of parasitics. An inverter circuit is shown with wires.

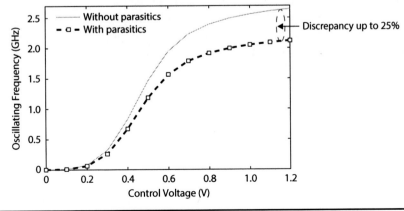

FIGURE 1.14 Comparison of frequency-voltage characteristics of the logical design and parasitic extracted physical design of the VCO.

1.5.3 Challenges to Account Parasitics during Design

In IC design, lack of exact layout information during circuit sizing leads to long design iterations involving time-consuming runs of complex tools. The traditional IC design flow involves repetitive iterations of circuit sizing, layout generation, parasitic values extraction, and performance evaluation. Considerable experience is needed from the designer to manually produce good layout for a given sized circuit. Redesign is needed whenever the final performance does not conform to the specification. In order to improve design efficiency and reduce the time-to-market, it is crucial to be able to predict the parasitic effects for accurate performance.

Parasitic-aware optimization methodologies require that the parasitics be considered at the beginning of the design [113]. Where IC components are designed assuming ideal components, it is observed that parasitics have serious degrading effects at high frequencies. The only way to overcome these effects is by considering parasitics as an integral part of the circuit (IC). Thus, parasitic-aware design and optimization are essential. If parasitics have an acute effect on the design, an early layout needs to be created so that the parasitics can be extracted and their effects estimated. Without the early layout-parasitic information, designers rely mostly on experience. If a design is understood well enough to know the sensitive nodes, dummy elements can be placed on those nodes to mimic the effects of real parasitics. As this process is tedious and error-prone, a methodology is required that can achieve the required performance while accounting for the parasitics. Hence, there is a need for a design methodology accounting for parasitics and process variation of general IC components.

1.6 Nanoscale Circuit Process Variation Issues

1.6.1 Types of Process Variation

To facilitate fabrication of circuits using nanoscale emerging technologies, more and more sophisticated lithographic, chemical, and mechanical processing steps are adopted. The uncertainty in the process steps, such as ion implantation, chemical mechanical polishing (CMP), and CVD involved in the fabrication, causes parameter variations [29, 53, 97, 106, 110, 131, 134] (see Fig. 1.15). The fluctuation of device characteristics caused by process variation has considerably increased in nanometer technologies. Process variations can be classified into inter- and intra-die. Inter-die variation, which comes from lot-to-lot, wafer-to-wafer, and within wafer, affects every device on a single chip equally. On the other hand, intra-die variation (also known as mismatch) refers to device characteristics, such as device geometry change, dopant density change, threshold voltage, gate-oxide thickness, and circuit timing change, which vary from device to device within the same die. Some of the variations are random and some are systematic.

1.6.2 The Impact of Process Variation

The process variations explained previously have profound effects on electrical parameters and overall performance in a VLSI circuit and are manifested in the variations in power, delay, and other attributes of CMOS circuits. The magnitude of each leakage component of the device is mostly dependent on the device geometry, doping profiles, and temperature. They affect design margins and yield and may lead to loss of money in the ever-reducing time-to-market scenario. Thus, it is absolutely critical that these variations are accounted for in any design decisions to make the circuits robust and improve yield targeted for design for manufacturing (DFM). Capturing and modeling the intra-die process variation become essential to device and interconnect extraction tools for accurate timing and power analysis.

As a concrete example of effects of process variation, let us examine a 90-nm VCO. Monte Carlo simulations have been carried out on the parasitic extracted netlist of the VCO to determine the effect of process variation on the oscillation frequency of the VCO. The lower curve in Fig. 1.16 shows that the discrepancy between the logical and physical design increases even more [66]. In this cycle, the parasitic parameterized netlist is subjected to worst-case process variations. In the parasitic and process variation–aware design cycle, a performance degradation of 43.5% is observed when the parasitic extracted netlist is subjected to worst-case process variation in a 90-nm CMOS VCO.

Opportunities and Challenges of Nanoscale Technology and Systems 13

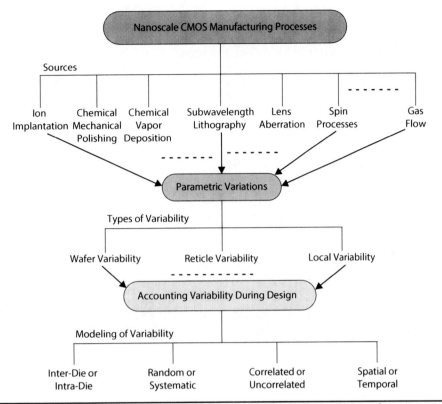

FIGURE 1.15 Source, origin, and types of process variation due to nanoscale CMOS manufacturing processes.

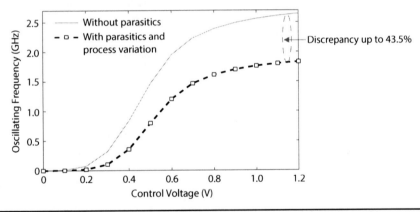

FIGURE 1.16 Comparison of frequency-voltage characteristics of the logical design, parasitic extracted physical design, and parasitic extracted physical design subjected to worst-case process variation.

1.7 The Temperature Variation Issue

1.7.1 The Issue of Temperature

Another emerging critical issue due to technology scaling is the effect of on-die temperature variation [41]. What was previously a second-order effect that could be adequately addressed with a few corner cases and guard bands has now become a first-order effect. Temperature interacts with a number of these other issues in ways that make analysis difficult. In general, with an increase in temperature, the effect on the nanoscale device has several negative aspects [36, 71, 73] as follows:

1. It increases the device leakage.
2. It reduces the active current strength.

3. It increases the propagation delay, thus making the device slow.
4. In case a positive feedback loop is established between the leakage and the temperature, a complete thermal breakdown can occur.

An increase in temperature will lead to an increase in leakage, which, in turn, can contribute to a higher temperature. Thus, there is a need for both thermal-aware and low-power design. The design, in turn, needs models at various levels of circuit abstraction (device level to system level) and different design decisions at those levels.

1.7.2 The Impact of Temperature

To study the effects of temperature, a 45-nm CMOS-based static random access memory (SRAM) circuit is examined (see Fig. 1.17) [109, 139]. Different portions of an SRAM circuit may experience different temperature profiles depending on their proximity to other logic units [93]. When the maximum temperature that a chip can attain during its operation increases, it affects reliability and cooling costs. In ambient temperature analysis, it is observed how the SRAM behaves in operating or environmental-temperature conditions. The impact of ambient temperature variation (measured at 25°C, 50°C, 75°C, 100°C, and 125°C) is observed in the 7-transistor SRAM circuit for two figures of merit, i.e., average power and static noise margin (SNM). The increase in leakage in the SRAM circuit increases the temperature (ambient temperature and on-chip temperature) because of the strong temperature dependence of subthreshold leakage. As leakages increase there is an increase in the total power dissipation of SRAM circuit. SNM is degraded as the temperature increases. SNM is the maximum amount of voltage a circuit can handle but with an increase in temperature it becomes unstable.

1.7.3 Challenges to Account through PVT-Aware Design

There is a need for new temperature-aware design methodologies in order to produce proper-functioning and reliable first silicon. Process, voltage, and temperature (PVT) variations are very critical in practical circuit designs intended for high-volume manufacturing [42]. Both power dissipation and operating frequency worsen at high temperatures because of the increase of leakage currents and the reduction of carrier mobility. The challenge for radiofrequency (RF) design is the centering of a design while accounting for PVT variations. By integrating temperature-aware capabilities into today's design flows, there is no need to rewrite the golden analysis standards that have been established in the past decade. Instead, through the use of tools that incrementally retrofit today's flows with temperature-aware data, the temperature effects can be fully accounted for. The application of thermal analysis reduces pessimism or risk associated with the assumption of a uniform on-chip temperature. A temperature-aware design flow is useful for existing technologies down to 90 nm and is required for technologies below 90 nm.

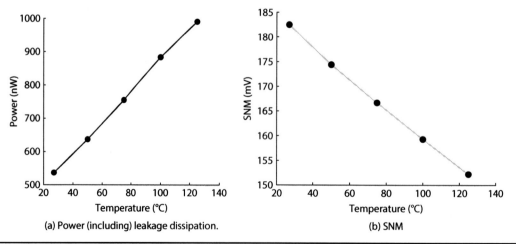

Figure 1.17 Effects of temperature on performance of a nano-CMOS-based SRAM circuit.

1.8 Challenges in Nanoscale CMOS AMS-SoC Design

1.8.1 AMS-SoC Design Flow

A typical AMS-SoC is a combination of heterogenous components, such as software, firmware, analog hardware, and digital hardware (see Fig. 1.18) [17, 70, 128]. Design of these components needs diverse skills and can be designed in parallel. Development of system software, application software, and firmware involves skills of computer science and information technology. Design of digital circuit, analog circuit, and RF circuit involves skills from computer engineering and electrical engineering. Once the design of the hardware components is completed, they are manufactured in the scopes of materials science and chemistry. Then, all the subsystems are assembled and tested by engineering technology personnel to manufacture the final product that is used by consumers.

Design of analog and digital circuits involves two different approaches (see Fig. 1.19). The digital design is performance-driven, whereas the analog design is specification-driven. The digital blocks of an SoC are generally designed "top-down," concentrating on abstract descriptions (e.g., architecture level and logic level). The analog blocks are usually built "bottom-up" and hence consume significant amounts of design cycle time. The computer-aided design (CAD) tools for digital circuits are quite advanced and can be used to automatically generate digital circuits, whereas, CAD tools for analog circuit design are not that advanced. For the automatic design of mixed-signal circuits through synthesis, these two styles need to interact while processing an analog/mixed-signal hardware description language (AMSHDL, VHDL-AMS or Verilog-AMS) of a circuit. Conversely, existing mixed-signal CAD tools treat and synthesize the analog and digital portions of the circuit separately [28, 138]. As the analog and digital synthesis engines of the mixed synthesis do not interact with each other during the optimization process, the solution may not be globally optimal from the overall AMS-SoC point of view. For accurate simulation and verification of these system designs, the digital part of a circuit is being treated as a special case of the analog part, which forces the simulation of the entire design in the analog domain, which is an intractable situation for today's multimillion transistor AMS-SoCs, demanding more computational resources. Moreover, there is a need for true mixed-signal synthesis tools that deal with AMS circuits through a single synthesis or optimization process equipped with facilities to address the challenges of the nanoscale transistor era.

1.8.2 AMS-SoC Unified Optimization

Unified optimization of a mixed-signal circuit is shown in Fig. 1.20; the vertically integrated synthesis flow takes various inputs, such as AMS, battery, technology libraries, power, performance, process,

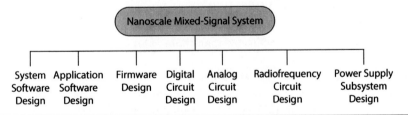

Figure 1.18 Different subsystem design in a AMS-SoC.

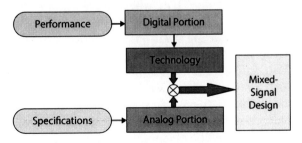

Figure 1.19 Design flow of analog and mixed-signal ICs.

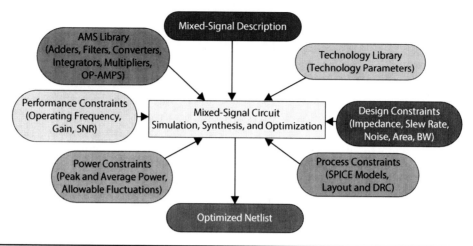

Figure 1.20 Optimization of analog and mixed-signal ICs considering different design constraints.

and design constraints and produces manufacturable optimized target netlists from AMSHDL specifications. A technology library contains information—such as oxide and metal layer thicknesses, sheet resistances, threshold voltages, and their statistical distributions—about the technology used to implement a specific process. The process constraints contain information on the manufacturability (i.e., design rule check or DRC) and performance of the process (SPICE models). Process variation severely affects analog circuits more than digital circuits. However, the digital circuits adopt technology scaling trend more aggressively than the analog circuits.

1.9 Tools for Mixed-Signal Circuit Design

1.9.1 The AMS-SoC Design Issue

Automatic design of ICs involves behavioral, logic, and physical synthesis that together translate a given algorithmic description of a system to physical level design [45, 64, 117, 133]. Digital behavioral synthesis is defined as the transformation of behavioral descriptions into register transfer level (RTL) structural descriptions that implement the given behavior for various constraints [96, 103]. In logic synthesis, a logic level netlist is generated from RTL description under various constraints. In addition to the constraints of logic synthesis, the corresponding process for analog circuits, called analog synthesis, deals with constraints, such as jitter, voltage offset, slew rate, noise restrictions, etc. This makes the primitive steps of analog synthesis, such as transistor sizing, component value optimization, and bias setting, more complicated than those of logic synthesis.

1.9.2 Languages for AMS-SoC Design

At present, there are two major AMSHDLs in common use, VHDL-AMS and Verilog-AMS, for textual-based design of AMS-SoCs [16, 39, 44, 63, 86, 94, 114] (see Fig. 1.21).

VHDL-AMS is an AMSHDL for analog, mixed-signal, and digital applications and is composed of two IEEE standards: VHDL 1076-1993 and VHDL 1076.1-1999 [5, 14, 44, 114]. As an AMSHDL, VHDL-AMS is a superset of VHDL, which models purely digital domains and an earlier analog-only HDL named HDLA. Figure 1.21(a) shows the relationship of these standards. VHDL-AMS supports hierarchical modeling at various abstraction levels of electrical and nonelectrical (such as mechanical, electromechanical, chemical, thermodynamical, biological, and optical) disciplines. The system descriptions can be continuous or discrete in time and can be conservative or nonconservative. Similar to digital VHDL, VHDL-AMS models comprise entities that specify the interface of the module to the overall system being simulated via the mechanism of ports. On the other hand, the architecture specifies the implementation semantics of the model. The architecture can be in one or a combination of more than one modeling styles, such as behavioral, structural, or behavioral/structural.

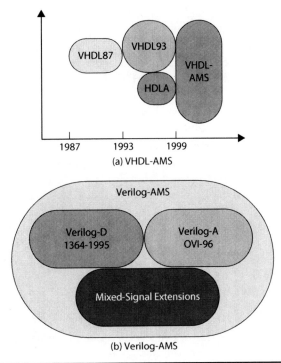

FIGURE 1.21 Languages for textual description of mixed-signal ICs.

Verilog-AMS is very similar to VHDL-AMS in concept and goals but uses different semantics, coding techniques, and modeling style [16, 63, 86, 94, 114] (see Fig. 1.21(b)). Verilog-AMS allows designers to create and use modules that encapsulate high-level behavioral descriptions as well as structural descriptions of systems and modules. Their behavior is described mathematically in terms of their interface ports and constituent relations. These descriptions are also multidisciplinary and encompass electrical, mechanical, fluid dynamics, and thermodynamics, among others. It supports both conservative and signal flow descriptions. The relationships of its various subset languages (digital Verilog and purely analog Verilog-A) are shown in Fig. 1.21(b). The current version of the Verilog-AMS LRM is stable, but the language is evolving [16]. Another effort is also underway to integrate Verilog-AMS into SystemVerilog, a system-level extension of Verilog aimed at SOC designs [13]. These extensions will provide nanosystem designers with a framework capable of modeling and simulating systems of great complexity.

1.9.3 Tools for AMS-SoC Design and Simulation

Tools such as CADICS, HiFaDiCC, AZTECA, CATALYST, OASYS, MIDAS, IDAC, OPASYN, ARIADNE, ARCHGEN, KANSYS, GPCAD, and VASS have been proposed for text-based design of analog circuits [19, 24, 47, 48, 50, 52, 74, 75, 76, 77, 78, 82, 124, 135, 144]. Mainly the tools are simulation-based but few are knowledge-based. OASYS and IDAC are two examples of knowledge-based tool. Each of these tools has its own advantages and disadvantages [21, 122]. In some cases, the computational formulation of the problem is solved using iterative numerical procedures, which does not always lead to physically implementable results. Moreover, these systems are also very specific in the sense that some handle only filters and some deal with only ADCs or DACs, and so on. More importantly, they are targeted neither for low-power analog circuits nor for nanoscale circuits. Hence, they are not suitable to produce circuits with nanoscale transistors for modern systems in which battery life is critical. In summary, the automatic synthesis of analog circuits remains a challenging task, particularly for nanoscale analog circuits.

The research in automatic synthesis of mixed-signal circuits still needs to evolve. The MxSICO compiler targets high-performance mixed-signal circuits [25, 26, 27, 28]. One of the earliest efforts for automatic synthesis of mixed-signal circuits from VHDL-AMS is presented in [49, 51, 143]. Techniques have been explored for behavioral modeling of mixed-signal circuits [129]. Techniques have been investigated for modeling and optimization of mixed-signal circuits operating in mixed-frequency ranges [137]. Methodologies have been investigated for the automation of interconnect design of mixed-signal systems [91]. The tool DAISY produces optimal $\Sigma-\Delta$ modulator topologies with building blocks for specified accuracy and bandwidth [62]. The tools such as MCAST and ADMS have been developed for compiling VHDL-AMS or Verilog-AMS to generate device code in C for SPICE [88, 146]. As evident from this discussion, there is no methodology available for nanoscale mixed-signal synthesis. No research addresses process or battery-aware synthesis of such circuits. Thus, existing tools and methodologies are simply not adequate to take on the design challenges of low-power nanoscale size and gigascale integration mixed-signal era.

Mixed-signal commercial solutions include tools offered by Mentor Graphics® and Cadence Design Systems®, Inc. These two major electronic design automation (EDA) companies have been active in circuit and digital simulation for decades and have significant resources and large customer bases in both industry and academia. Mentor's solution, called Advance MS, provides support for mixed-signal and mixed-language (VHDL, VHDL-AMS, Verilog, Verilog-AMS, and SPICE) simulation [1]. Cadence's solution, called AMS Designer, supports the same languages [15]. Another tool in this arena is Dolphin Integration with its SMASH simulator, which offers Verilog, Verilog-A, VHDL-AMS, and SPICE support [11].

1.9.4 Transistor Models

At the transistor and physical-design level of the circuit design, it is important to use accurate transistor and interconnect models. For commercial quailing design, these models can be used from different semiconductor warehouses whose process technology will be used for the manufacturing of the circuits. For academic purposes and wider spread of the technology, different institutes provide general process design kits (GPDK) which are used for fabless design. Berkeley short-channel insulated gate FET model (BSIM), which are physics-based, accurate, and scalable, are available for transistor-level simulation of the circuits [2]. The charge-based Enz-Krummenacher-Vittoz (EKV) MOS transistor model is developed to provide a clear understanding of the device properties that can be used for high-performance analog circuit design to describe static and dynamic behaviors, noise and matching limitations, and temperature variations [54].

1.10 Questions

1.1 List five leading conferences in the area of VLSI and VLSI design automation. Provide Web addresses and venues (previous and current) for the last two years.

1.2 List five leading journals in the area of VLSI and VLSI design automation. Provide the publishers' names and their Web addresses.

1.3 List five free and/or commercial EDA/CAD tools. Explain the capabilities of each.

1.4 List three EDA/CAD tools used for designing the following: (1) digital circuits, (2) analog circuits, (3) digital systems, (4) analog/mixed-signal systems.

1.5 List the hardware description languages (HDLs) needed for the modeling and simulation of the following: (1) digital circuits, (2) analog circuits, (3) digital systems, (4) analog/mixed-signal systems.

1.6 Discuss the different levels of digital circuit abstraction and tools used in each of the abstraction levels.

1.7 Discuss the advantages and disadvantages of SystemC-AMS versus Verilog-AMS as a mixed-signal HDL.

1.8 What is the latest technology used for fabricating analog and digital circuits?

1.9 Name three chips that have or will have more than a billion transistors. Discuss their architecture and process technology.

1.10 Comment on similarities and differences among a central processing unit (CPU), graphics processing unit (GPU), application-specific integrated circuit (ASIC), and field-programmable gate array (FPGA) chip.

1.11 References

1. Advance MS Simulator. http://www.mentor.com/products/ic_nanometer_design/simulation/advance_ms/index.cfm.
2. Berkeley Short-channel IGFET Model (BSIM). Accessed on 14 November 2012.
3. Carbon Nanotubes. http://www.research.ibm.com/nanoscience/nanotubes.html.
4. Computer history museum. http://www.computerhistory.org/.
5. Definition of Analog and Mixed Signal Extensions to IEEE Standard VHDL. IEEE Standard 1076.1-1999.
6. Filling Nanotech Jobs. http://cen.acs.org/articles/88/i29/Filling-Nanotech-Jobs.html.
7. Intel Developer Forum. http://www.intel.com/idf/.
8. National Nanotechnology Initiative (NNI). http://www.nano.gov/. Accessed on 15 May 2013.
9. QCADesigner. http://www.qcadesigner.ca/.
10. Semiconductor Industry Association, International Technology Roadmap for Semiconductors. http://public.itrs.net.
11. Smash simulator. http://www.dolphin.fr/medal/smash/smash_overview.html.
12. Taiwan semiconductor manufacturing company limited. http://www.tsmc.com.
13. The SystemVerilog Language. http://www.systemverilog.org/overview/overview.html.
14. VHDL Language Reference Manual. IEEE Standard 1076-1993.
15. Virtuoso AMS Designer Simulator. http://www.cadence.com/products/custom_ic/ams_designer/index.aspx.
16. Accelera International, Inc. *Verilog-AMS Language Reference Manual*, November 2004.
17. O. Adamo, S. P. Mohanty, E. Kougianos, M. Varanasi, and W. Cai. VLSI Architecture and FPGA Prototyping of a Digital Camera for Image Security and Authentication. In *Proceedings of the IEEE Region 5 Technology and Science Conference*, pages 154–158, 2006.
18. A. K. Ale. *Comparison and Evaluation of Existing Analog Circuit Simulators Using Sigma-Delta Modulator*. Master's thesis, University of North Texas, 2006.
19. B. Antao and A. Brodersen. ARCHGEN: Automated Synthesis of Analog Systems. *IEEE Transactions on Very Large Scale Integration Systems*, 3(2):231–244, June 1995.
20. R. Ashraf, M. Chrzanowska-Jeske, and S. G. Narendra. Carbon nanotube circuit design choices in the presence of metallic tubes. In *Proceedings of the IEEE International Symposium on Circuits and Systems*, pages 177–180, 2008.
21. S. Balkir, G. Dundar, and A. S. Ogrenci. *Analog VLSI Design Automation*. CRC Press, 2003.
22. S. Basu, B. Kommineni, and R. Vemuri. Variation Aware Spline Center and Range Modeling for Analog Circuit Performance. In *Proceedings of the International Symposium on Quality Electronic Design*, pages 162–167, 2008.
23. S. Basu, P. Thakore, and R. Vemuri. Process Variation Tolerant Standard Cell Library Development Using Reduced Dimension Statistical Modeling and Optimization Techniques. In *Proceedings of the 8th International Symposium on Quality Electronic Design (ISQED)*, pages 814–820, 2007.
24. G. Beenker, J. Conway, G. Schrooten, and A. Slenter. *Analog CAD for Consumer ICs*, pages 347–367. Kluwer Academic Publishers, 1993.
25. E. Berkcan. MxSICO: A Mixed Analog Digital Compiler: Application to Oversampled A/D Converters. In *Proceedings of the IEEE Custom Integrated Circuits Conference*, page 14.9.1, 1990.
26. E. Berkcan. MxSICO: A Silicon Compiler for Mixed Analog Digital Circuits. In *Proceedings of the IEEE International Conference on Computer Design*, pages 33–36, 1990.
27. E. Berkcan and B. Currin. Module Compilation for Analog and Mixed Analog Digital Circuits. In *Proceedings of the IEEE International Symposium on Circuits and Systems*, pages 831–834, 1990.
28. E. Berkcan and F. Yassa. Towards Mixed Analog / Digital Design Automation: A Review. In *Proceedings of the IEEE International Symposium on Circuits and Systems*, pages 809–815, 1990.
29. K. Bernstein, D. J. Frank, A. E. Gattiker, W. Haensch, B. L. Ji, S. R. Nassif, E. J. Nowak, D. J. Pearson, and N. J. Rohrer. High-Performance CMOS Variability in the 65-*nm* Regime and Beyond. *IBM Journal of Research and Development*, 50(4/5):433–449, July-September 2006.
30. A. J. Bhavnagarwala, B. L. Austin, K. A. Bowman, and J. D. Meindl. A Minimum Total Power Methodology for Projecting Limits of CMOS GSI. *IEEE Transactions on VLSI Systems*, 8(3):235–251, June 2000.
31. Mark T. Bohr, Robert S. Chau, Tahir Ghani, and Kaizad Mistry. The High-κ Solution. *IEEE Spectrum*, October 2007.
32. S. Borkar. Design Challenges of Technology Scaling. *IEEE Micro*, 19(4):23–29, July 1999.
33. S. Borkar, T. Karnik, and V. De. Design and Reliability Challenges in Nanometer Technologies. In *Proceedings of the Design Automation Conference*, pages 75–75, 2004.
34. K. A. Bowman, L. Wang, X. Tang, and J. D. Meindl. A Circuit-Level Perspective of the Optimum Gate Oxide Thickness. *IEEE Transactions on Electron Devices*, 48(8):1800–1810, August 2001.
35. A. A. Breed and K. P. Roenker. A Small-Signal, RF Simulation Study of Multiple-Gate and Silicon-on-Insulator MOSFET Devices. In *Topical Meeting on Silicon Monolithic Integrated Circuits in RF Systems*, pages 294–297, 2004.
36. J. A. Butts and G. S. Sohi. A Static Power Model for Architects. In *Proceedings of the 33rd Annual IEEE/ACM International Symposium on Microarchitecture (MICRO-33)*, pages 191–201, 2000.

37. O. Campana, R. Contiero, and G. A. Mian. An H.264/AVC video coder based on a multiple description scalar quantizer. *IEEE Transactions on Circuits and Systems for Video Technology (TCSVT)*, 18(2):268–272, February 2008.
38. T. Chantem, R. P. Dick, and X. S. Hu. Temperature-Aware Scheduling and Assignment for Hard Real-Time Applications on MPSoCs. In *Proceedings of the Design, Automation, and Test in Europe Conference (DATE)*, pages 288–293, 2008.
39. J. J. Charlot and A. Napieralski. Multi-Technology with VHDL-AMS. In *Proceedings of the International Conference on Modern Problems of Radio Engineering, Telecommunications and Computer Science*, pages 331–337, 2002.
40. Robert S. Chau. Integrated CMOS Tri-Gate Transistors. http://www.intel.com/technology/silicon/integrated_cmos.htm. Accessed on 15 May 2013.
41. K. Choi and D. J. Allstot. Post-optimization design centering for RF integrated circuits. In *Proceedings of the International Symposium on Circuits and Systems*, pages 956–959, 2004.
42. K. Choi, J. Park, and D. J. Allstot. *Parasitic-aware Optimization of CMOS RF Circuits*. Kluwer Academic Publishers, 2003.
43. E. Y. Chou and B. Sheu. System-on-a-chip design for modern communications. *IEEE Circuits and Devices Magazine*, 17(6):12–17, November 2001.
44. E. Christen and K. Bakalar. VHDL-AMS – A Hardware Description Language for Analog and Mixed-Signal Applications. *IEEE Transactions on Circuits and Systems – II: Analog and Digital Signal Processing*, 46(10):1263–1272, October 1999.
45. R. Composano and W. Wolf. *High Level Synthesis*. Kluwer Academic Publishers, 1991.
46. S. D'Amico, G. Maruccio, P. Visconti, E. D'Amone, R. Cingolani, R. Rinaldi, S. Masiero, G. P. Spada, and G. Gottarelli. Transistors Based on the Guanosine Molecule (a DNA base). *Microelectronics Journal*, 34(10):961–963, 2003.
47. M. G. R. Degrauwe, O. Nys, E. Dijkstra, J. Rijmenants, S. Bitz, B. L. A. G. Goffart, E. A. Vittoz, S. Cserveny, C. Meixenberger, G. van der Stappen, and H. J. Oguey. IDAC: An Interactive Design Tool for Analog CMOS Circuits. *IEEE Journal of Solid-State Circuits*, 22(6):1106–1116, 1987.
48. M. del Mar Hershenson, S. P. Boyd, and T. H. Lee. GPCAD: A Tool for CMOS OP-AMP Synthesis. In *Proceedings of the International Conference on Computer Aided Design*, pages 296–303, 1998.
49. N. R. Dhanwada and R. Vemuri. Constraint Allocation in Analog System Synthesis. In *Proceedings of the International Conference on VLSI Design*, pages 253–258, 1998.
50. A. Doboli and R. Vemuri. A VHDL-AMS Compiler and Architecture Generator for Behavioral Synthesis of Analog Systems. In *Proceedings of the Design Automationa and Test in Europe (DATE) Conference*, pages 338–345, 1999.
51. A. Doboli and R. Vemuri. Behavioral Modeling for High-Level Synthesis of Analog and Mixed-Signal Systems from VHDL-AMS. *IEEE Transactions on CAD of Integrated Circuits*, 22(11):1504–1520, 2003.
52. A. Doboli and R. Vemuri. Exploration-Based High-Level Synthesis of Linear Analog Systems Operating at Low/Medium Frequencies. *IEEE Transactions on CAD of Integrated Circuits*, 22(11):1556–1568, 2003.
53. P. G. Drennan and C. C. McAndrew. Understanding MOSFET Mismatch for Analog Design. *IEEE Journal of Solid-State Circuits*, 38(3):450–456, March 2003.
54. Christian C. Enz and Eric A. Vittoz. *Charge-Based MOS Transistor Modeling: The EKV Model for Low-Power and RF IC Design*. John Wiley & Sons., 2006.
55. C. Sandner et al. A 6-bit 1.2Gs/s Low-Power Flash ADC in $0.13\mu m$ Digital CMOS. *IEEE Journal of Solid State Circuits*, 40(7):1499–1505, July 2005.
56. H. R. Huff et al. Integration of High-k Gate Stack Systems into Planar CMOS Process Flows. In *International Workshop on Gate Insulator*, pages 2–11, 2001.
57. L. Manchanda et al. High-K Gate Dielectrics for the Silicon Industry. In *Proceedings of International Workshop on Gate Insulator*, page 56–60, 2001.
58. M. Yang et al. Performance Dependence of CMOS on Silicon Substrate Orientation for Ultrathin and HfO_2 Gate Dielectrics. *IEEE Electron Device Letters*, 24(5):339–341, May 2003.
59. R. Chau et al. 30 nm physical gate length CMOS transistors with 1.0ps n-MOS and 1.7ps p-MOS gate delays. *IEDM Technical Digest*, pages 45–48, 2000.
60. R. Choi et al. Fabrication of high quality ultra-thin HfO_2 gate dielectric MOSFETs using deuterium anneal. *IEDM Technical Digest*, pages 613–616, 2002.
61. S. Okada et al. System on a Chip for Digital Still Camera. *IEEE Transactions on Consumer Electronics*, 45(3):1689–1698, August 1999.
62. K. Francken, P. Vancorenland, and G. Gielen. DAISY: A Simulation-Based High-Level Synthesis Tool for $\Delta\Sigma$ Modulators. In *Proceedings of International Conference on Computer Aided Design*, pages 188–192, 2000.
63. P. Frey and D. O'Riordan. Verilog-AMS: Mixed-Signal Simulation and Cross Domain Connect Modules. In *Proceedings of the IEEE/ACM International Workshop on Behavioral Modeling and Simulation*, pages 103–108, 2000.
64. D. Gajski and N. Dutt. *High Level Synthesis: Introduction to Chip and System Design*. Kluwer Academic Publishers, 1992.

65. D. Ghai, S. P. Mohanty, and E. Kougianos. A Dual Oxide CMOS Universal Voltage Converter for Power Management in Multi-V_{DD} SoCs. In *Proceedings of the International Sympoisum on Quality Electronic Design*, pages 257–260, 2008.
66. D. Ghai, S. P. Mohanty, and E. Kougianos. Parasitic Aware Process Variation Tolerant VCO Design. In *Proceedings of the International Sympoisum on Quality Electronic Design*, pages 330–333, 2008.
67. D. Ghai, S. P. Mohanty, and E. Kougianos. Design of Parasitic and Process Variation Aware RF Circuits: A Nano-CMOS VCO Case Study. *IEEE Transactions on Very Large Scale Integration Systems*, 2009.
68. D. Ghai, S. P. Mohanty, and E. Kougianos. Unified P4 (Power-Performance-Process-Parasitic) Fast Optimization of a Nano-CMOS VCO. In *Proceedings of the Great Lakes Symposium on VLSI*, pages 303–308, 2009.
69. D. Ghai, S. P. Mohanty, E. Kougianos, and P. Patra. A PVT Aware Accurate Statistical Logic Library for High-κ Metal-Gate Nano-CMOS. In *Proceedings of 10th International Symposium on Quality of Electronic Design (ISQED)*, pages 47–54, 2009.
70. D. Ghai. *Variability Aware Low-Power Techniques for Nanoscale Mixed-Signal Circuits*. PhD thesis, University of North Texas, 2009.
71. D. Ghai, S. P. Mohanty, E. Kougianos, and P. Patra. A PVT aware accurate statistical logic library for high-κ metal-gate nano-cmos. In *Proceedings of the 10th International Symposium on Quality of Electronic Design*, pages 47–54, 2009.
72. G. G. E. Gielen and R. A. Rutenbar. Computer-aided Design of Analog and Mixed-Signal Integrated Circuits. *Proceedings of the IEEE*, 88(12):1825–1854, 2000.
73. A. Golda and A. Kos. Temperature Influence on Power Consumption and Time Delay. In *Proceedings of the Euromicro Symposium on Digital Systems Design*, pages 378–378, 2003.
74. S. K. Gupta and M. M. Hasan. KANSYS: a CAD Tool for Analog Circuit Synthesis. In *Proceedings of the International Conference on VLSI Design*, pages 333–334, 1996.
75. R. Harjani, R. Rutenbar, and L. Carley. OASYS: A Framework for Analog Circuit Synthesis. *IEEE Transactions on Computer-Aided Design of Integrated Circuits and Systems*, 8(12):1247–1266, 1993.
76. N. Horta, J. Franca, and C. Leme. *Automated High Level Synthesis of Data Conversion Systems*. Peregrinus, London, 1991.
77. G. Jusuf, P. R. Gray, and A. L. Sangiovanni-Vincentelli. CADICS—Cyclic Analog-to-Digital Converter Synthesis. In *Proceedings of the International Conference on Computer Aided Design*, pages 286–289, 1990.
78. G. Jusuf, P. R. Gray, and A. L. Sangiovanni-Vincentelli. A Performance-Driven Analog-to-Digital Converter Module Generator. In *Proceedings of the IEEE International Symposium on Circuits and Systems*, pages 2160–2163, 1992.
79. A. Karamcheti, V. H. C. Watt, H. N. Al-Shareef, T. Y. Luo, G. A. Brown, M. D. Jackson, and H. R. Huff. Silicon Oxynitride Films as Segue to the High-K Era. *Semiconductor Fabtech*, 12:207–214, 2000.
80. D. Kim et al. CMOS Mixed-Signal Circuit Process Variation Sensitivity Characterization for Yield Improvement. In *Proceedings of the IEEE Custom Integrated Circuits Conference*, pages 365–368, 2006.
81. A. I. Kingon, J. P. Maria, and S. K. Streifferr. Alternative Dielectrics to Silicon Dioxide for Memory and Logic Devices. *Nature*, 406:1021–1038, 2000.
82. H. Y. Koh, C. H. Sequin, and P. R. Gray. OPASYN: A Compiler for CMOS Operational Amplifiers. *IEEE Transactions on Computer-Aided Design of Integrated Circuits and Systems*, 9(2):113–125, 1990.
83. E. Kougianos and S. P. Mohanty. Impact of Gate-Oxide Tunneling on Mixed-Signal Design and Simulation of a Nano-CMOS VCO. *Elsevier Microelectronics Journal (MEJ)*, 40(1):95–103, January 2009.
84. E. Kougianos and S. P. Mohanty. A Comparative Study on Gate Leakage and Performance of High-κ Nano-CMOS Logic Gates. *Taylor & Francis International Journal of Electronics (IJE)*, 97(9):985–1005, September 2010.
85. K. Kundert, H. Chang, D. Jefferies, G. Lamant, E. Malavasi, and F. Sendig. Design of Mixed-Signal Systems-on-a-Chip. *IEEE Transactions on Computer-Aided Design of Integrated Circuits and Systems*, 19(12):1561–1571, 2000.
86. K. Kundert and O. Zinke. *The Designer's Guide to Verilog-AMS*. Kluwer Academic Publishers, 2004.
87. B. H. Lee et al. Thermal Stability and Electrical Characteristics of Ultrathin Hafnium Oxide Gate Dielectric Reoxidized with Rapid Thermal Annealing. *Applied Physics Letters*, 77:1926–1928, 2000.
88. L. Lemaitre, C. C. McAndrew, and S. Hamm. ADMS—Automatic Device Model Synthesizer. In *Proceedings of the IEEE Custom Integrated Circuits Conference*, pages 27–30, 2002.
89. L. L. Lewyn, T. Ytterdal, C. Wulff, and K. Martin. Analog Circuit Design in Nanoscale CMOS Technologies. *Proceedings of the IEEE*, 97(10):1687–1714, October 2009.
90. T. P. Ma. Making Silicon Nitride a Viable Gate Dielecrtric. *IEEE Transaction on Electron Devices*, 45:680–690, 1999.
91. Y. Massoud. CAREER: Integrated Automation Strategy for Interconnect Design: A New Paradigm for Mixed-Signal Nanoscale Integrated Circuits, 2005. NSF-CSE Directorate, CCF Division, Award Abstract #0448558.
92. V. Menon, S. Das, B. Jayaraman, and V. Govindaraju. Transitioning from Microelectronics to Nanoelectronics. *Computer*, 44(2):18–19, January 2011.
93. M. Meterelliyoz, J. P. Kulkarni, and K. Roy. Thermal Analysis of 8-T SRAM for Nano-Scaled Technologies. In *Proceeding of the 13th international symposium on Low power electronics and design*, pages 123–128, 2008.
94. I. Miller. Verilog-A and Verilog-AMS Provides a New Dimension in Modeling and Simulation. In *Proceedings of the Third IEEE International Caracas Conference on Devices, Circuits and Systems*, pages C49/1 – C49/6, 2000.

95. V. Mlinar and F. M. Peeters. Theoretical study of inas/gaas quantum dots grown on [11k] substrates in the presence of a magnetic field. *Microelectronics Journal*, 37(12):1427–1429, 2006.
96. S. P. Mohanty. *Energy and Transient Power Minimization during Behavioral Synthesis*. PhD thesis, Department of Computer Science and Engineering, University of South Florida, USA, 2003.
97. S. P. Mohanty. Unified Challenges in Nano-CMOS High-Level Synthesis. In *Proceedings of the 22nd International Conference on VLSI Design*, pages 531–531, 2009.
98. S. P. Mohanty, D. Ghai, E. Kougianos, and B. Joshi. A Universal Level Converter Towards the Realization of Energy Efficient Implantable Drug Delivery Nano-Electro-Mechanical-Systems. In *Proceedings of the International Sympoisum on Quality Electronic Design*, 2009.
99. S. P. Mohanty and E. Kougianos. Simultaneous Power Fluctuation and Average Power Minimization during Nano-CMOS Behavioural Synthesis. In *Proceedings of the 20th IEEE International Conference on VLSI Design*, pages 577–582, 2007.
100. S. P. Mohanty, E. Kougianos, D. Ghai, and P. Patra. Interdependency Study of Process and Design Parameter Scaling for Power Optimization of Nano-CMOS Circuits under Process Variation. In *Proceedings of the 16th ACM/IEEE International Workshop on Logic and Synthesis*, pages 207–213, 2007.
101. S. P. Mohanty, E. Kougianos, and R. N. Mahapatra. A comparative analysis of gate leakage and performance of high-κ nanoscale cmos logic gates. In *Proceedings of the 16th ACM/IEEE International Workshop on Logic and Synthesis (IWLS)*, pages 31–38, 2007.
102. S. P. Mohanty and N. Ranganathan. A Framework for Energy and Transient Power Reduction during Behavioral Synthesis. *IEEE Transactions on VLSI Systems*, 12(6):562–572, June 2004.
103. S. P. Mohanty and N. Ranganathan. Energy efficient datapath scheduling using multiple voltages and dynamic scheduling. *ACM Transactions on Design Automation of Electronic Systems*, 10(2):330–353, April 2005.
104. S. P. Mohanty, N. Ranganathan, and K. Balakrishnan. A Dual Voltage-Frequency VLSI Chip for Image Watermarking in DCT Domain. *IEEE Transactions on Circuits and Systems II*, 53(5):394–398, May 2006.
105. S. P. Mohanty, N. Ranganathan, and S. K. Chappidi. ILP Models for Simultaneous Energy and Transient Power Minimization during Behavioral Synthesis. *ACM Transactions on Design Automation of Electronic Systems (TODAES)*, 11(1):186–212, January 2006.
106. S. P. Mohanty, N. Ranganathan, E. Kougianos, and P. Patra. *Low-Power High-Level Synthesis for Nanoscale CMOS Circuits*. Springer, 2008. 0387764739 and 978-0387764733.
107. A. F. Mondragon-Torres, M. C. Schneider, and E. Sanchez-Sinencio. Well-Driven Floating Gate Transistors. *IEE Electronics Letters*, 38(11):530–532, May 2002.
108. G. E. Moore. Cramming More Components onto Integrated Circuits. In *Proceedings of the IEEE*, pages 82–85, 1998.
109. V. Mukherjee, S. P. Mohanty, E. Kougianos, R. Allawadhi, and R. Velagapudi. Gate leakage current analysis in read/write/idle states of a sram cell. In *Proceedings of the IEEE Region 5 Technology and Science Conference*, pages 196–200, 2006.
110. S. R. Nassif. Within-chip variability analysis. In *Proceedings of the International Electron Devices Meeting*, pages 283–286, 1998.
111. T. Nirschl, P. F. Wang, W. Hansch, and D. Schmitt-Landsiedel. The tunnelling field effect transistors (TFET): the temperature dependence, the simulation model, and its application. In *Proceedings of the IEEE International Symposium on Circuits and Systems*, pages 713–716, 2004.
112. Last R. Norman and I. J. Haas. Solid-State Micrologic Elements. In *Digest of Technical Papers IEEE International Solid-State Circuits Conference*, pages 82–83, 1960.
113. J. Park, K. Choi, and D. J. Allstot. Parasitic-aware design and optimization of a fully integrated CMOS wideband amplifier. In *Proceedings of the Asia South Pacific Design Automation Conference*, pages 904–907, 2003.
114. F. Pêcheux, C. Lallement, and A. Vachoux. VHDL-AMS and Verilog-AMS as Alternative Hardware Description Languages for Efficient Modeling of Multidiscipline Systems. *IEEE Transactions on Computer-Aided Design of Circuits and Systems*, 24(2):204–225, February 2005.
115. M. Pedram and J. M. Rabaey. *Power Aware Design Methodologies*. Springer, 2002.
116. W. J. Qi et al. Ultrathin Zirconium Silicate Film with Good Thermal Stability for Alternative Gate Dielectric Application. *Applied Physics Letters*, 77:1704–1706, 2000.
117. J. Rabaey. *Digital Integrated Circuits: A Design Perspective*. Prentice Hall, Inc., Upper Saddle River, NJ, 1996.
118. M. C. Roco, C. A. Mirkin, and M. C. Hersam. *Nanotechnology Research Directions for Societal Needs in 2020: Retrospective and Outlook*. Springer, 2011.
119. M. C. Roco. The Long View of Nanotechnology Development: The National Nanotechnology Initiative at 10 Years. In *Nanotechnology Research Directions for Societal Needs in 2020*, volume 1 of Science Policy Reports, pages 1–28. Springer Netherlands, 2011.
120. M. Roukes. Nanoelectromechanical systems: A new opportunity for microelectronics. In *Proceedings of the European Solid State Device Research Conference*, pages 20–20, 2009.
121. K. Roy, S. Mukhopadhyay, and H. M. Meimand. Leakage Current Mechanisms and Leakage Reduction Techniques in Deep-Submicrometer CMOS Circuits. *Proceedings of the IEEE*, 91(2):305–327, Feb. 2003.

122. R. A. Rutenbar, G. G. E. Gielen, and B. A. Antao. *Computer-Aided Design of Analog Integrated Circuits and Systems*. Wiley-IEEE Press, 2002.
123. M. B. Rutzig, A. C. S. Beck, and L. Carro. Dynamically adapted low power asips. In *Proceedings of the 5th International Workshop Reconfigurable Computing: Architectures, Tools and Applications (ARC)*, pages 110–122, 2009.
124. S. G. Sabiro, P. Sen, and M. S. Tawfik. HiFADiCC: A Prototype Framework of a Highly Flexible Analog to Digital Converters Silicon Compiler. In *Proceedings of the IEEE International Symposium on Circuits and Systems*, pages 1114–1117, 1990.
125. T. Sairam, W. Zhao, and Y. Cao. Optimizing FinFET technology for high-speed and low-power design. In *Proceedings of the ACM Great Lakes Symposium on VLSI*, pages 73–77, 2007.
126. Brian R. Santo. 25 Microchips That Shook the World. *IEEE Spectrum*, May 2009.
127. G. Sarivisetti. *Design and Optimization of Components in a 45 nm CMOS Phase Locked Loop*. Master's thesis, University of North Texas, 2006.
128. G. Sarivisetti, E. Kougianos, S. P. Mohanty, A. Palakodety, and A. K. Ale. Optimization of a 45 nm cmos voltage controlled oscillator using design of experiments. In *Proceedings of IEEE Region 5 Technology and Science Conference*, pages 87–90, 2006.
129. C. Shi. CAREER: Behavioral Modeling and Simulation of Mixed-Signal/Mixed-Technology VLSI Systems: An Integrated Research and Education Program, 2000. NSF - CSE Directorate, CCF Division, Award Abstract #9985507.
130. D. Singh, J. M. Rabaey, M. Pedram, F. Catthoor, S. Rajgopal, N. Sehgal, and T. J. Mozdzen. Power Conscious CAD Tools and Methodologies: A Perspective. *Proceedings of the IEEE*, 83(4):570–594, 1995.
131. K. Singhal and V. Visvanathan. Statistical device models from worst case files and electrical test data. *IEEE Transanction on Semiconductor Manufacturing*, 12(4):470–484, November 1999.
132. S. Sinha, A. Balijepalli, and Y. Cao. A Simplified Model of Carbon NanoTube Transistor with Applications to Analog and Digital Design. In *Proceedings of the International Symposium on Quality Electronic Design*, pages 502–507, 2008.
133. M. J. S. Smith. *Application-Specific Integrated Circuits*. Addison Wesley Professional, 1997.
134. S. Sundareswaran, J. A. Abraham, A. Ardelea, and R. Panda. Characterization of Standard Cells for Intra-Cell Mismatch Variations. In *Proceedings of the International Symposium on Quality Electronic Design*, pages 213–219, 2008.
135. K. Swings and W. Sansen. ARIADNE: A Constraint-Based Approach to Computer-Aided Synthesis and Modeling of Analog Integrated Circuits. *Analog Integrated Circuits and Signal Processing, Kluwer Publications*, 3:197–215, 1993.
136. Sassan Tabatabaei and Aaron Partridge. Silicon mems oscillators for high-speed digital systems. *IEEE Micro*, 30(2):80–89, 2010.
137. S. X. D. Tan. CAREER Development Plan: Behavioral Modeling, Simulation and Optimization for Mixed-Signal System on a Chip, 2005. NSF - CSE Directorate, CCF Division, Award Abstract #0448534.
138. H. Tang, H. Zhang, and A. Doboli. *Towards High-Level Analog and Mixed-Signal Synthesis from VHDL-AMS Specifications*, pages 201–216. Languages for System Specification. Kluwer Academic Publishers, 2004.
139. G. Thakral, S. P. Mohanty, D. Ghai, and D. K. Pradhan. A combined doe-ilp based power and read stability optimization in nano-cmos sram. In *Proceedings of the 23rd International Conference on VLSI Design*, pages 45–50, 2010.
140. G. Thakral, S. P. Mohanty, D. Ghai, and D. K. Pradhan. A doe-ilp assisted conjugate-gradient based power and stability optimization in high-κ nano-cmos sram. In *Proceedings of the 20th ACM Great Lakes Symposium on VLSI*, pages 323–328, 2010.
141. F. Tobajas, G. M. Callic, P. A. Perez, V. D. Armas, and R. Sarmiento. An efficient double-filter hardware architecture for h.264/avc deblocking filtering. *IEEE Transactions on Consumer Electronics*, 54(7):131–139, 2008.
142. K.-Y. Tsai, W.-J. Hsieh, Yuan-Ching Lu, Bo-Sen Chang, Sheng-Wei Chien, and Yi-Chang Lu. A new method to improve accuracy of parasitics extraction considering sub-wavelength lithography effects. In *Proceedings of the 15th Asia South Pacific Design Automation Conference*, pages 651–656, 2010.
143. R. Vemuri, N. Dhanwada, A. Nunez, and P. Campisi. VASE: VHDL-AMS Synthesis Environment—Tools for Synthesis of Mixed-Signal Systems from VHDL-AMS. In *Proceedings of the Analog and Mixed-Signal Applications Conference*, pages 1C:77–1C:84, 1997.
144. J. C. Vital and J. E. Franca. Synthesis of High-Speed A/D Converter Architectures with Flexible Functional Simulation Capabilities. In *Proceedings of the IEEE International Symposium on Circuits and Systems (ISCAS)*, pages 2156–2159, 1992.
145. E. M. Vogel, K. Z. Ahmed, B. Hornung, P. K. McLarty, G. Lucovsky, J. R. Hauser, and J. J. Wortman. Modeled Tunnel Currents for High Dielectric Constant Dielectrics. *IEEE Transactions on Electron Devices*, 45(6):1350–1355, June 1998.
146. B. Wan, B. P. Hu, L. Zhou, and C. J. R. Shi. MCAST: An Abstract-Syntax-Tree Based Model Compiler for Circuit Simulation. In *Proceedings of the IEEE Custom Integrated Circuits Conference*, pages 249–252, 2003.
147. Z. Wang, Z. Zhang, H. Xu, L. Ding, S. Wang, and L.-M. Peng. A High-Performance Top-Gate Graphene Field-Effect Transistor Based Frequency Doubler. *Applied Physics Letters*, 96(17), 2010.

148. N. H. E. Weste and D. Harris. *CMOS VLSI Design: A Circuit and Systems Perspective*. Addison Wesley, 2005.
149. X. Guo and T. P. Ma. Tunneling Leakage Current in Oxynitride: Dependence on Oxygen/Nitrogen Content. *IEEE Electron Device Letters*, 19(6):207–209, June 1998.
150. L. Xie and A. Davoodi. Robust estimation of timing yield with partial statistical information on process variations. In *Proceedings of the International Symposium on Quality Electronic Design*, pages 156–161, 2008.
151. L. Xie, A. Davoodi, J. Zhang, and T. H. Wu. Adjustment-based modeling for statistical static timing analysis with high dimension of variability. In *Proceedings of the IEEE International Conference on Computer-Aided Design (ICCAD)*, pages 181–184, 2008.
152. C. Zhu, Zhenyu (Peter) Gu, L. Shang, R. P. Dick, and R. G. Knobel. Towards an Ultra-Low-Power Architecture Using Single-Electron Tunneling Transistors. In *Proceedings of the Design Automation Conference (DAC)*, pages 312–317, 2007.

CHAPTER 2
Emerging Systems Designed as Analog/Mixed-Signal System-on-Chips

2.1 Introduction

Due to the ever-decreasing cost of electronics hardware and software in last several years, more and more people are able to afford consumer electronics. Consumer electronics have a profound impact on society. The transfer of information around the globe is possible in no time and without any cost. People around the globe are staying connected every moment through social networks. This chapter discusses some examples of systems that have been used in day-to-day life or are being conceptualized for future development. It is difficult to discuss all of these systems in the limited space, though an attempt has been made to discuss the selected systems. Examples of systems discussed are biosensor systems, tablet PC, smart mobile phone, Blue-ray player, multimedia tank, TV tuner card, secure digital camera, net-centric multimedia processor, drug-delivery nano-electromechanical systems, radio frequency (RF) universal remote control, RF identification tag, and global positioning systems. These are typically designed as analog/mixed-signal system-on-chips (AMS-SoCs) containing analog, digital, and RF circuits, field-programmable gate array (FPGA), firmware, and software components.

2.2 Atomic Force Microscope

2.2.1 What Is It?

Atomic force microscope (AFM) is a characterization instrument of nanoscience that is used to determine topography and other properties of surfaces [15, 16, 57, 73, 86]. The AFM can analyze the thick and thin films, metals, semiconductors, polymers, and composites. The AFM is also known as scanning force microscope (SFM) or scanning probe microscope (SPM). AFM technique requires minimal specimen preparation and can be operated without vacuum. The high signal-to-noise ratio provided by this instrument allows submolecular features to be discerned. AFM can be used to image and manipulate single molecules.

2.2.2 Background

Study of surfaces has been very important right from microscale film and remains the same for the nanoscale films. Of the available techniques, SPM is most popular. It forms images of surfaces using a physical probe that scans the specimen under observation. In 1980, when the scanning tunneling microscope was invented by Gerd Binnig and Heinrich Rohrer, which earned them the Nobel Prize for Physics, it marked the beginning of the scanning microscopes. AFM is a very high-resolution type of SPM that can have a resolution on the order of fractions of a nanometer [76]. Several other microscopes frequently used in nanoscale characterization include scanning electron microscope (SEM) and transmission electron microscopy (TEM). While SEM scans the samples using a high-energy beam of electrons in a raster-scan fashion [64], TEM sends a beam of electrons through an

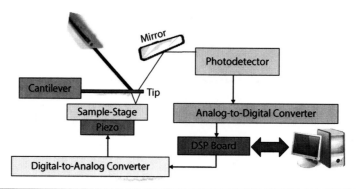

FIGURE 2.1 The AFM system with a desktop computer and control software.

ultra-thin specimen [152]. The interactions of the electrons are then used to create images. However, AFM is still widely used because of simplicity and low cost. From the system design point of view, AFM is presented here; however, system design for TEM or SEM can be discussed in a similar way.

2.2.3 What Is Inside?

The schematic representation of the operational parts of an AFM system is shown in Fig. 2.1 [57, 86, 104, 135]. In this AFM system, the samples are scanned with a tip attached to a reflective cantilever. A diode laser is focused onto the back of the reflective cantilever. The laser beam is deflected off the back of the cantilever into a segmented photodetector, as the tip follows the contours of the sample. The position-sensitive photodetector is thus used to measure the probe motion. The photodetector converts light-intensity differences into voltage. The samples are placed on a piezoelectric transducer, which senses the force between the tip and the sample. The digital signal processor (DSP) board provides interfacing among the scanning system, piezoelectric transducer, and the computer as shown [47]. It supplies the voltages to control the piezoelectric transducer through a feedback control system. The feedback loop provides a correction signal to the piezoelectric transducer through the software control from the computer to keep constant force or constant height between the sample and tip. The rate of data acquisition depends on the speed of the feedback loop corrections. AFM feedback loops tend to have a bandwidth of about 10 kHz, resulting in image acquisition times of about 1 minute. The computer stores the local height position at each point and assembles the image. Three-dimensional topographical images of the surfaces are constructed by plotting the local sample height versus the horizontal probe tip position. There are three common modes of operations depending upon the separation of the tip from the sample. The above desktop computer-based AFM system has the following drawbacks [104]:

1. A separate computer with several monitors is necessary to operate the system, store images, and do postacquisition analysis.

2. The interface between the DSP and the host desktop computer limits the speed of operation to a great extent due to slower data transfer [57].

3. The use of additional desktop computer and corresponding monitors makes system portability a difficult task.

4. The AFM system including the feedback process is heavily dependent on the software installed in the host desktop computer.

5. The software used for characterization purposes needs high-skilled human intervention, and has relatively low reliability compared to the hardware components of the system.

6. The total power consumption of a typical AFM system is very large and is in the range of 600 to 1800 W [74].

A portable very large-scale integration (VLSI) system that can replace the host desktop computer and provide a self-sufficient smart controller to operate and characterize and make low-power portable AFM systems is shown in Fig. 2.2 [74, 104]. The smart controller–based portable AFM will

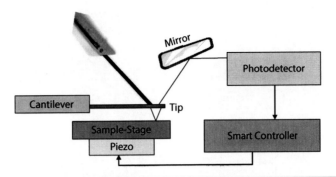

Figure 2.2 The portable all hardware AFM which is a mixed-signal system.

not only perform the conventional control operations, but will also have data acquisition, data storage, user interface, and output display. In other words, it will be a self-sufficient unit that will not need the use of separate computer and monitor systems for sample characterization. The AFM characterization system that includes such a smart controller as shown in Fig. 2.2 can be designed as an AMS-SoC. Such a system will reduce the overall power consumption of the system to a great extent. The smart controller of the portable AFM system will include at least the following components: a liquid-crystal display (LCD) for user interface, on-chip memory to store images in JPEG form for a user session, a universal serial bus (USB) port for data transfer for external storage or printing, a compact flash memory for additional storage, low-power image processor to perform various image processing functions for analysis and characterization, and a JPEG codec for compression and decompression of images.

2.3 Biosensor Systems

2.3.1 What Is It?

The formal definition of biosensor is quite diverse. For simplicity, the biosensor can be defined as follows. A biosensor is an analytical device incorporating a deliberate and intimate combination of a specific biological element that creates a recognition event and a physical element that transduces the recognition event [72, 81, 108]. A specific "bio" element, which is called enzyme, recognizes a specific analyte and the "sensor" element transduces the change in the biomolecule into an electrical signal. The bio-element is very specific to the analyte to which it is sensitive. It does not recognize other analytes.

The name "biosensor" signifies that the device is a combination of two parts: bio-element and sensor element (see Fig. 2.3). The bio-element may be an enzyme, antibody, antigen, living cells, tissues, etc. The large variety of sensor elements includes electric current, electric potential, intensity, and phase of electromagnetic radiations, mass, conductance, impedance, temperature, viscosity, and so on. A detailed list of these elements is presented in Fig. 2.4. However, modern-day biosensors are typically built as biosensor systems in which one or more biosensors are used together with lots of additional components for complex sensing and multiuser facility. A single-sensor biosensor system is shown in Fig. 2.5 [37]. The biocatalyst converts the substrate to product. The transducer determines this reaction and converts it to an electrical signal. The amplifier amplifies the signal for better sensing. The analog-to-digital converter (ADC) converts the analog signal to digital signal. The DSP processes this signal. The display unit displays the signal obtained from the DSP.

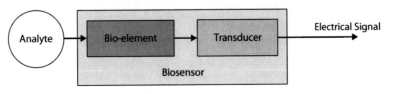

Figure 2.3 Basic concepts of biosensor [108].

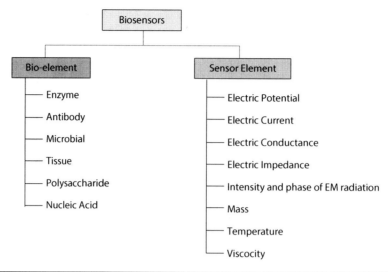

Figure 2.4 Elements of a biosensor [108].

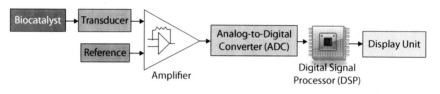

Figure 2.5 Basic concepts of biosensor systems [37].

Depending on the transducing mechanism used, the biosensors can be of many types, as follows [93, 108]:

1. Resonant Biosensors. In this type of biosensors, an acoustic wave transducer is coupled with an antibody (bio-element). The resonant frequency of the transducer is measured.
2. Optical-Detection Biosensors. The output-transduced signal that is measured is a light signal for this type of biosensors.
3. Thermal-Detection Biosensors. This type of biosensors is constructed combining enzymes with temperature sensors.
4. Ion-Sensitive FETs (ISFETs) Biosensors. These are basically semiconductor field-effect transistors (FETs) having an ion-sensitive surface whose electrical potential changes when the ions and the semiconductor interact. The change is then measured.
5. Electrochemical Biosensors. The underlying principle for this class of biosensors is that many chemical reactions produce or consume ions or electrons, which, in turn, cause some change in the electrical properties of the solution that can be sensed out and used as a measuring parameter.

The electrochemical biosensors based on the parameter measured can be further classified in the following types [138]:

1. Conductometric. The measured parameter is the electrical conductance/resistance of the solution.
2. Amperometric. In this case, the measured parameter is current.
3. Potentiometric. In this type of sensors, the measured parameter is oxidation/reduction potential of the electrochemical reaction.

A comparative discussion of these three types is given in Fig. 2.6.

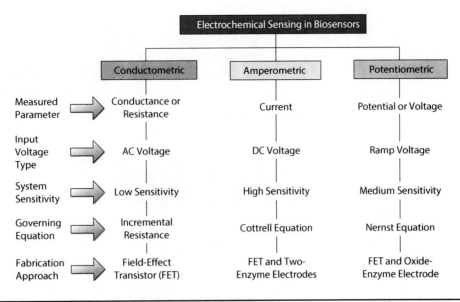

FIGURE 2.6 Different types of electrochemical sensing [108].

2.3.2 Background

There has always been tremendous interest in improving the quality of human life and life expectancy. The history of biosensors starts in 1950 with the first experiment by L.C. Clark (who is addressed as the father of biosensors) to measure the dissolved oxygen in blood [60, 108, 126]. This experiment became a commercial reality in 1973 with the development of Glucose Analyser in Ohio, USA. This is considered as the first of many biosensor-based laboratory analyzers. The first bioaffinity biosensor was developed in 1980 in which radiolabeled receptors were immobilized on a transducer surface. The blood-glucose biosensors were launched in 1987. In 1992, the handheld blood biosensors were developed. Since then the biosensors have strived for the simple solutions to measurement in complex matrices. The individual biosensors have been doing an excellent job; however, pragmatic solutions to many problems involve the construction of biosensor systems that include associated electronics, fluidics, and separation technology.

In the last decade, tons of biosensors have been developed for diverse applications. The biosensors can have a variety of biomedical and industry applications [91, 132, 134]. Some of the possible applications are shown in Fig. 2.7. Though the major application so far is in blood glucose sensing, because of abundant market potential [81] biosensors have tremendous opportunities for commercialization in other fields of application as well. Depending on the field of applications,

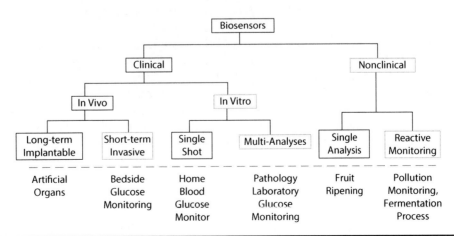

FIGURE 2.7 Potential applications of biosensors [81, 108].

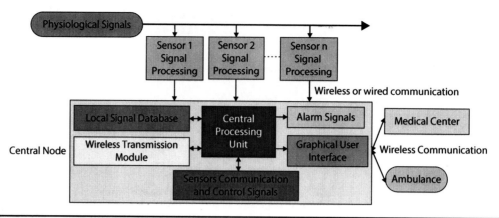

Figure 2.8 Components of a typical biosensor system [127].

biosensors are known by many other names, including immunosensors, optrodes, chemical canaries, resonant mirrors, glucometers, biochips, and biocomputers [56].

2.3.3 What Is Inside?

The type of biosensors and their systems are quite diverse in nature. So, it is difficult to present a common architecture for biosensor system. However, several representative biosensor systems are presented for better understanding.

2.3.3.1 A Wearable Biosensor System with Multiple Biosensors

The architecture of a biosensor system is shown in Fig. 2.8. The wearable biosensor system may include a variety of components such as sensors, wearable materials, smart textiles, actuators, power supplies, wireless communication modules, control units, processing units, user interface, and firmware. These biosensor systems can measure several physiological parameters such as heart rate, blood pressure, body temperature, oxygen saturation, respiration rate, and electrocardiogram. The measured parameters are communicated using a wireless or wired link. The parameters are displayed in a personal digital assistant (PDA) or a microcontroller board-based display.

2.3.3.2 DNA Detection Biosensor System with a Single Biosensor

The category of biosensors used for DNA detection is also known as biodetectors. The biodetectors are used to identify a small concentration of DNA of microorganism such as virus or bacteria. This relies on comparing sample DNA with DNA of known microorganisms, which is called probe DNA. Because the sample solution may contain only a small number of bio-organic molecules, multiple copies of the sample DNA need to be created for proper analysis. This is achieved with the help of polymerase chain reaction (PCR). PCR starts by splitting samples of double-helix DNA into two parts by heating it at about 95°C. If the reagents contain proper growth enzymes, then each of these strands will grow the complementary missing part and form a double-helix structure again. This happens when temperature is lowered. In one heating/cooling cycle, the amount of sample DNA is doubled (one cycle time is about 1 minute). Thus, for n cycles, 2^n copies are made. Twenty-five to 40 cycles are needed to produce approximately 1 billion copies. This amount is sufficient enough to be detected optically. With the use of fluorescent DNA probes, the identification of DNA is also possible while copying DNA in PCR. PCR is very power-consuming because of successive heating/cooling cycles that take about 30 minutes. Hence, it was previously not possible to fabricate portable battery-operated biodetectors that can do PCR. But, by using micro-electro-mechanical system (MEMS), such kinds of biodetectors, which are basically lab-on-a-chip systems, have been developed. In these MEMS-based devices, the amount of reagent used is scaled down. Figure 2.9 shows a microfluidic device [45, 130]. This lab-on-a-chip system contains channels, valves, and chambers [56].

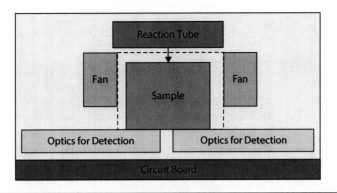

FIGURE 2.9 Biodetector designed as a microfluidic device [45, 130].

2.4 Blu-Ray Player

2.4.1 What Is It?

Blu-ray players are the players that can play the Blu-ray disc (BD; and other old formats such as DVD) along with online content streaming. The name Blu-ray is derived from the underlying technology, which utilizes a (blue-)violet laser to read and/or write data. The name comes from "Blue" (blue-violet laser) and "Ray" (optical ray) [2, 17]. The BD format was developed by the Blu-ray Disc Association (BDA), a large group of leading consumer electronics, personal computer, and media manufacturers [17].

2.4.2 Home Video Systems Background: From Video Cassette Player to Blu-Ray Player

The first popular home video system, which is abbreviated as VHS, was introduced in 1976. The first commercial optical home video disc storage medium known as the LaserDisc (LD) was introduced in 1978. In 1978, the video compact disc (VCD), which was digital video recorded on a compact disc (CD), came to market. This could support only low-resolution Motion Picture Export Group-1 (MPEG-1). In 1995, the digital video disc (DVD) started the era of high-quality video. DVD discs could store much more information than CDs while having the same dimensions. The high-quality video further improved to high-definition DVD (called HD-DVD) and the corresponding HD-DVD player was invented in 2006. The Blu-ray player was officially released in 2006 and emerged as a successor of the HD-DVD [17, 41]. This is known to be supporting the full-HD resolution. The various home video media and systems are listed in Fig. 2.10. In addition, recently, ultra-HD resolution TVs, media, and media players, which support much higher resolution than full HD, have started entering the market.

2.4.3 What Is Inside?

The various components of a typical Blu-ray player are shown in Fig. 2.11. The most important and unique component of a Blu-ray player is the "Secure Media Processor (SMP)." It is a high-performance and low-power chip designed using system-on-a-chip (SoC) technology. In addition, a wide variety of

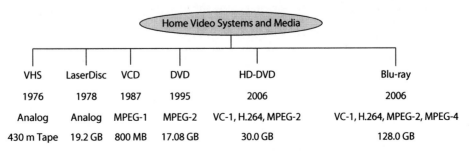

FIGURE 2.10 Different types of home videos discs with the supported format and maximum size.

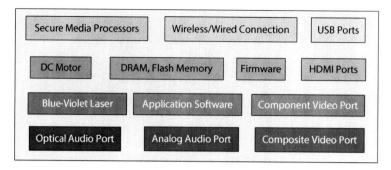

FIGURE 2.11 Different components of a Blu-ray player.

digital rights management (DRM) and Conditional Access (CA) solutions are supported in it [34]. The processor supports a large number of video and audio codec including H.264, MPEG-4, DTS, Dolby Digital. The Blu-ray player supports different ports including a high-definition multimedia interface (HDMI) port for full HD, i.e., 1080p display. The "Wireless/Wired Connections" is used by the player for streaming of the online contents in the player using the "Application Software."

2.5 Drug-Delivery Nano-Electro-Mechanical Systems

2.5.1 What Is It?

Drug-delivery nano-electro-mechanical systems (DDNEMS) are implantable drug-delivery systems designed using NEMS. The function of DDNEMS is to administer drugs in predetermined targets and doses using implantable chips that are controlled or programmed externally through an RF interface. These can be visualized as "bioactuators" that are complementary to "biosensors" in terms of functionality.

2.5.2 Background

There has always been a strong interest in improving the quality of human life and increasing life expectancy. This has led to multifold research and development in the area of self-health management and care giving. Efficient and reliable drug delivery is an effort in this direction. Drug delivery is the process of administering a pharmaceutical compound to achieve a therapeutic effect in living beings [95]. The type and variety of drug delivery systems are as many in number as the number of diseases. Thus, a comprehensive classification is difficult and beyond the scope of mixed-signal system design. The drug delivery systems can either be passive or active in nature.

In 1970, the first drug delivery system was introduced that was based on lactic acid polymers [78, 100, 149]. Since then, various drug delivery systems, such as insulin pump, self-adhesive skin patch, and sublingual drop, have been developed. The field of controlled drug delivery uses mechanisms such as transdermal patches, polymer implants, bioadhesive systems, and microencapsulation [100]. Conventional controlled-release formulations are designed to deliver drugs at a predetermined, preferably constant rate. Most of the conventional drug delivery schemes suffer from drawbacks that can seriously limit their effectiveness in the area of self-health management. The drawbacks of the conventional drug delivery system include the following [21, 140, 150, 154]:

1. Drug release rate typically decreases exponentially with time.
2. Effective long-term treatment is difficult and costly.
3. Over-dosage and under-dosage.
4. It may take a long time to inject the drug, hence diminishing its effect.
5. A narrow therapeutic window.
6. A complex dosage schedule.
7. Difficult to establish individualized or emergency-based dosing regimens.

There is an urgent need to develop novel drug delivery systems that will have profound impacts on the health care industry. Future generations of the drug delivery systems are inherently expected to have a high degree of multifunctional activities. This implies that these systems, consisting of electrical, mechanical, and chemical subsystems, will be highly complex and will require a shift in design paradigm. Recent advances in nanotechnology and implantable devices suggest that it is an opportune time to framework a toolset that will enable design of future generations of drug delivery systems. They can be implemented using miniaturization technology of MEMS or more recent NEMS that are essentially microfluidic devices [56, 158]. They are called DDNEMS [115, 116]. One of the earliest attempted MEMS-based drug delivery systems is presented in [147]. NEMS are a technological solution for building miniature systems that can be deployed in small spaces [69, 70, 90, 140, 143, 154]. Such devices can be beneficial in terms of safety, efficacy, or convenience.

2.5.3 What Is Inside?

One of the earliest MEMS-based drug delivery systems is shown in Fig. 2.12 [29, 100, 147]. This microchip drug reservoir contains multiple sealed compartments, which are opened on demand to deliver a dose of a drug [147]. The drug reservoirs are pyramid-shaped square holes in shape. The hole size is from 20 to 50 μm and the depth is from 200 to 500 μm. On one side, the reservoir is covered with an approximately 100-nm thick gold membrane. In total, an array of approximately 500 holes is available. The backside of the reservoirs is sealed with either biocompatible solder materials or polymers.

An SoC-based architecture that consists of several vital components is shown in Fig. 2.13 [115, 116]. The typical components of a DDNEMS architecture are presented later. The most important and active component of the DDNEMS is the drug delivery subsystem, which is typically nonelectrical

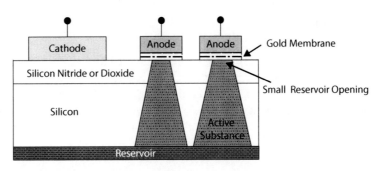

Figure 2.12 Microchip drug reservoir [147].

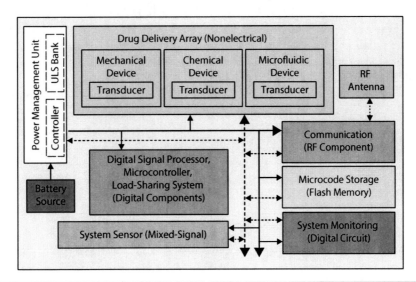

Figure 2.13 The system architecture of a DDNEMS [115, 116].

in nature (e.g., mechanical, chemical, and microfluidic). The array could be homogeneous containing all elements of the same kind or heterogeneous. The array elements may include micropumps, microfluidic devices, stents, and microneedles. The array elements also include appropriate transducers to allow their control and interfacing to the electronic portion of the DDNEMS. The data processing, controlling, and interfacing aspects of the DDNEMS are handled by electrical integrated circuit (IC) subsystems. These circuits are analog, digital, or mixed-signal in nature. The monitoring and control of the drug array elements are performed by the system sensor subsystem that receives information from and sends control signals to the transducers. Its front-end (transducer side) is analog but the back-end, interfacing to the DSP, is digital. The DSP subsystem analyzes and processes the on-line data generated by the sensors and, under the control of the program stored in the flash memory subsystem, generates control signals to affect the drug delivery, load sharing, and drug mixing. The power management unit (PMU) is one of the most important components of the entire DDNEMS. Under the control of the DSP and the stored microcode, it manages the power distribution to the various subsystems to optimize energy consumption. It has built-in timers that put the system to "sleep" or "wake-up" mode and can be induced to activate the system via external signals received by the RF subsystem (e.g., to force an emergency drug delivery). The RF subsystem, comprising an antenna and transmitter/receiver, will be built using well-established RFID principles for the shape and placement of the antenna and communication protocol. Its function is to allow noninvasive maintenance of the system, e.g., modification of the microcode stored in the flash memory. It allows remote collection of data, such as the amount of drug remaining in the reservoir, drug array element failures, and battery status. It also allows emergency drug delivery or system deactivation.

2.6 Digital Video Recorder

2.6.1 What Is It?

The digital video recorder (DVR) is a system that records digital video to enable permanent storage. The permanent (nonvolatile) memory may include hard disk, USB flash drive, secure digital (SD) memory card, or networked mass storage [141]. The DVR is also known as personal video recorder (PVR). The different types of DVR include the following (see Fig. 2.14): stand-alone DVR with camera (e.g., surveillance system), set-top boxes (STB) with recording facility, personal media players (PMP) with recording facility, personal media recorder (PMR; e.g., digital camcorder), and desktop/laptop computer containing software for video capturing and playback.

2.6.2 Background

In the beginning of 1999, consumer digital video recorders, ReplayTV [12] and TiVo [39], were launched at the Consumer Electronics Show [26]. Microsoft demonstrated a DVR unit and made it available by the end of 1999. A TV set with built-in DVR facilities was introduced by LG in 2007, and other manufacturers followed. An alternative to a stand-alone DVR is a PC-based DVR. There are several free and open-source DVR applications available for Microsoft Windows, Linux, and Mac OS-based personal computers. In Microsoft Windows, the applications include GB-PVR [3] and MediaPortal [4]. Freevo is an open-source DVR for Linux-based computers [11]. The open-source DVR called MythTV is useful for both Linux- and Mac-OS-based computers [30].

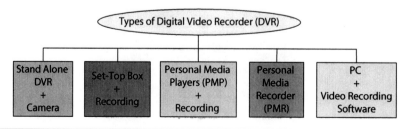

FIGURE 2.14 Different types of DVR or PVR.

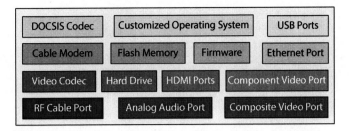

FIGURE 2.15 Different components of a stand-alone DVR.

2.6.3 What Is Inside?

As evident from the discussions of the previous subsection, the DVR system is quite diverse in nature. The DVR system can be software based in a computer. The DVR can be a stand-alone system with major hardware components. The schematic representation of a stand-alone DVR is shown in Fig. 2.15. One of the main components of the DVR, the cable modem streams the data and video on demand. Data over cable service interface specifications (DOCSIS) codec is the main receiver chip, which decodes the DOCSIS and is the main network interface. The DVR has other components and ports as shown in the figure.

2.7 Electroencephalogram System

2.7.1 What Is It?

The electroencephalogram (EEG) is a noninvasive method for measuring brain waves of a person. EEG measures and records the electrical activity of the brain over a desired period of time [99, 133]. In an EEG, special electrodes are attached to the head and hooked by wires to a computer. The computer records electrical activity of the brain. An EEG can detect certain conditions, such as seizures in the brain, which are characterized by the changes in the normal pattern of the electrical activity. An EEG is used in a large number of fields such as epilepsy, sleep disorder diagnosis, and brain-computer interfaces [55].

2.7.2 Background

The electrogram is derived from Electro (in Greek), i.e., electricity and gram (in Greek), i.e., measurement. The electrograms, including electrocardiogram (ECG), EEG, electrooculogram (EOG), electromyogram (EMG), and electroretinogram (ERG), are used to measure physical and cognitive function for diagnosis and monitoring of health [133]. Electrograms are thus very crucial for effective health care.

2.7.3 What Is Inside?

In an EEG, the neurophysiological electrical activity is measured in the brain using the electrodes placed on the scalp, subdurally, or in the cerebral cortex. The resulting voltage signal, which is the EEG, is the summation of postsynaptic potentials from a large number of neurons. The basic idea is that when the wave of ions reaches the electrodes on the scalp, they can move electrons on the metal electrodes that are measured as a voltage. These voltages when recorded over time represent the EEG. A block diagrammatic representation of an EEG system is shown in Fig. 2.16 [151].

In the EEG system presented above, 32 channels with single- or dual-pole measurements are included [151]. The driven right leg (DRL) circuit is connected to a reference (REF) circuit to provide the desired common-mode gain. DRL circuit is a biological signal amplifier that reduces common-mode interference by actively canceling the interference. As the common-mode noise voltage of the human body is much higher than the EEG signals, the preamplifier is used to measure the EEG signals by offsetting the common-mode signals. The EEG signals are of different frequencies, thus needing an adjustable amplifier and filter. An ADC provides DC voltages for further processing. The signal is then processed through DSP to reduce the occurrence of error and to achieve better measurement results. The universal asynchronous receiver/transmitter (UART) is used to interface with a PC for visual display and recording.

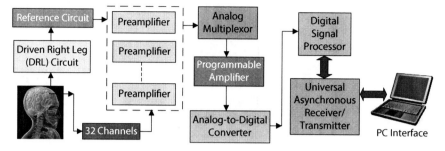

Figure 2.16 Block diagram of a EEG [151].

2.8 GPS Navigation Device

2.8.1 What Is It?

Overall, the GPS consists of three basic components [5]:

1. Space component such as satellites.
2. Control component such as control stations.
3. User-end component such as GPS receivers.

The space component of the GPS consists of at least 24 satellites. The control components are monitoring stations at different parts of the globe that are set up for monitoring the satellites. A GPS navigation device, or GPS device, is any device that receives GPS signals to determine its current position on the earth [159]. The modern GPS devices are able to find much more information, e.g., paths or roads for travel, safer or faster routes for travel, and nearby gas stations and hotels.

2.8.2 Background

Autonomous geo-spatial positioning has been of tremendous importance for a quite long time. During World War II, ground-based radio navigation systems, such as LORAN and the Decca Navigator, were developed. In 1960, the first satellite navigation system, Transit, was used by the US Navy. In 1967, the US Navy developed the Timation satellite that proved the ability to place accurate atomic clocks in space. Different types of Global Navigation Satellite System (GNSS) are presented in Fig. 2.17. In the 1970s, the first worldwide radio navigation system, the Omega Navigation System, was developed. In 1994, the US GPS became the only fully operational GNSS. Though GPS is a commonly used term, it is accurately described as Navigation System for Timing and Ranging (NAVSTAR) [5, 66]. In 2007, GLObal NAvigation Satellite System (GLONASS) of Russia became available for civilian use. The BeiDou (or Compass) Navigation System of China is in its initial phase of covering

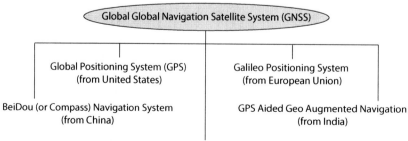

Figure 2.17 Different GNSS.

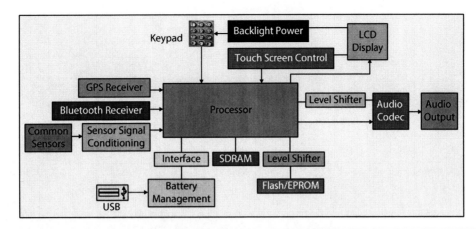

FIGURE 2.18 Different components of GPS portable navigation devices (GPND) [84, 159].

the entire globe. The Galileo positioning system from the European Union is in its initial deployment phase too. GAGAN, which stands for GPS-Aided Geo-Augmented Navigation or GPS and Geo-Augmented Navigation, is the system under development by India. *Gagan* is the Hindi word of Sankrit origin for *sky*. GPS receivers or GPS navigation devices that capture GPS signals are integral parts of mobile phones, notebooks, automobiles, ships, aircrafts, and so on.

2.8.3 What Is Inside?

The schematic representation of a GPS receiver device is shown in Fig. 2.18 [84, 159]. The overall AMS-SoC of the GPS device is quite complex as it consists of analog, digital, RF, firmware, and software components. Several communication components including GPS receiver, Bluetooth receiver, frequency-modulation (FM) transmitter/receiver, and wireless local area network (WLAN) technology (WiFi) receiver are integral parts of GPS device. The digital processor is typically an enhanced advanced RISC machines (ARM)-based architecture mobile entertainment application-specific processor. The GPS AMS-SoC has many sensors including altimeter, humidity, temperature, Gyro, and accelerometers to calculate position. Audio features such as music players are integrated entertainment applications. Special DC-to-DC converters with advanced dynamic voltage scaling meet the needs of the newest processors. Touchscreen LCD or light-emitting diode (LED) display with electrostatic discharge (ESD) protection is the key user interface. Phase-locked loop (PLL)-based programmable clock synthesizers generate multiple clocks from a single input frequency. Firmware and software ensure the control of the components and user interfacing. The GPS receiver has a built-in map stored in a flash or magnetic drive.

2.9 GPU-CPU Hybrid System

2.9.1 What Is It?

The GPU-CPU hybrid (GCH) system is still at its early stage of full-fledged realization for common use. The GCH system is a system that uses GPU and CPU together in a heterogeneous coprocessing computing model (see Fig. 2.19) [1]. In this system, the sequential parts of the applications run on the CPU and the computationally intensive parts are accelerated by the GPU. This model is followed by applications such as medical imaging and electromagnetics due to excellent floating point performance in GPUs. This model is also used in mini home theater PCs that uses a high-end GPU for the high-definition video processing (its target application) and a low-end CPU for holding the operating system. However, with the emergence of nanoscale complementary metal oxide semiconductor (CMOS) technology and billion transistor packing era, it has been possible to make chips using the wafer-scale integration that has both CPU and GPU. For example, Intel i7 Sandy Bridge chip contains both CPU and GPU (see Fig. 2.20) [25].

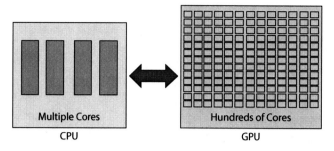

FIGURE 2.19 GCH multicore system for high-performance computing [1].

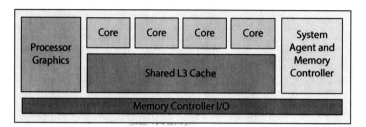

FIGURE 2.20 Schematic representation of a chip that has both GPU and CPU in one die [25].

2.9.2 Background

High-quality display has been one of the desired and routine operations of a typical computing device. However, the graphic processing through general-purpose microprocessor has been quite slow, which was affecting the overall system performance. In 1991, ATI introduced its product, the Mach8, which performed the process of graphics without the CPU. In 1999, NVIDIA marketed GeForce 256 as "the world's first Graphics Processing Unit (GPU)." In 2002, ATI used the term "Visual Processing Unit (VPU)" while releasing Radeon 9700. However, GPU is a more widely used term [1, 71]. A GPU (or VPU) is a special-purpose processor that performs the graphics rendering without the use of CPU or central microprocessor. The GPUs are designed to perform a specific number of operations on large amounts of data [71, 146]. The GPUs are an integral part of embedded systems, mobile phones, personal computers, workstations, and game consoles. The GPU is either part of the graphics card or part of the motherboard or chip set.

The complexity of GPUs is increasing at an extraordinary rate. The GPU IC consists of a couple of billion nanoscale transistors. The performance of the GPUs is increasing at a higher rate. The computing power of the GPUs is increasing much faster than Moore's law. The performance gap of GPU and CPU is widening. The GeForceFX 5900 GPU operating at 20 GigaFlopsis is equivalent to a 10 GHz Pentium 4 processor [53]. As the GPUs are becoming faster and evolving to incorporate additional programmability, the challenge is to provide new functionality without sacrificing the performance advantage over conventional GPUs [106, 110]. GPUs use a different computational model than the classical von Neumann architecture used by the CPUs [125]. In this context, the question that arises is whether the GPU and CPU of a desktop computer can be used together for high-performance and low-cost computing [1, 106, 110].

2.9.3 What Is Inside?

2.9.3.1 A GPU-CPU Architecture

A GCH architecture in the context of digital video broadcasting (DVB), which is a far more computationally intensive application, is shown in Fig. 2.21. In this application, a server containing GPU and CPU performs tasks, such as video compression/decompression (e.g., MPEG-4 and H.264) and DRM. In one working model of the GPU-CPU–based video processing, the video data that arrives at shared memory is directed by the CPU and sent to the GPU, which is then mapped into GPU memory. The GPU processes the video data. After the GPU completes the processing, the control from the CPU initiates copying the data back to the CPU, and the video is then stored in a database.

Emerging Systems Designed as Analog/Mixed-Signal System-on-Chips

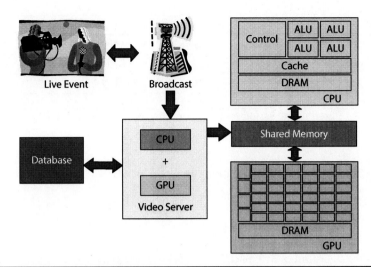

FIGURE 2.21 The IP-TV broadcasting scenario showing the shared architecture for GPU-CPU multicore processing in a video server [1, 106, 110].

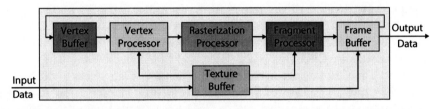

FIGURE 2.22 The programmable graphics pipeline [106, 110].

This GPU-CPU coprocessing approach ensures faster video processing for vast amounts of data and will not add extra hardware cost to either video providers or receivers. The video data is sent to the GPU as textures; in other words, the video data is treated as a group of pictures.

In order to get a detailed perspective of high-performance capability of the GPU, a block diagram of a modern programmable GPU is shown in Fig. 2.22 [71, 106, 110, 125]. The GPU architecture offers a large degree of parallelism at a relatively low cost through the well-known vector-processing model known as single instruction, multiple data (SIMD). A GPU includes two types of processing units: vertex and pixel (or fragment) processors. The programmable vertex and fragment processors execute a user-defined assembly-level program having four-way SIMD instructions. The vertex processor performs mathematical operations that transform a vertex into a screen position. This result is then pipelined to the pixel or fragment processor that performs the texturing operations.

2.9.3.2 Theoretical Speed-Up Using GPU-CPU Hybrid

The overall best performance can be achieved when both GPU and CPU work together and complete the execution at the same time so that both are available at the same time for other computations. For GPU-CPU as multicore architecture, the following relationship can be derived [106, 110]. Let us assume the following: T_{CPU}, CPU execution time; T_{GPU}, GPU execution time; α, ratio of execution time of GPU and CPU; P, total number of primitive operation in an algorithm; β, a fraction; and Q, the penalty due to Amadahl's law in terms of number of primitive operations due to parallel execution of GPU and CPU. The throughput of a GPU is higher than that of the CPU due to long pipelining of the GPU. However, the execution time of GPU is lower than CPU, thus, $\alpha < 1$. So, for best hybrid system performance with GPU-CPU multicore, the following relation must be satisfied [106, 110]:

$$\alpha = \left(\frac{T_{GPU}}{T_{CPU}}\right) \qquad (2.1)$$

$$\beta \times (P+Q) \times T_{GPU} = (1-\beta) \times (P+Q) \times T_{CPU} \qquad (2.2)$$

40 Chapter Two

Test Cases	CPU Time	GPU Time	CPU Release Time
CPU only	4.376 ms	Free	Fully Occupied
GPU only	Free	44.789 ms	Fully Free
GPU and CPU together	2.15 ms	21.16 ms	50.87% Free

TABLE 2.1 Experimental Results for Execution Time for GCH System [106, 110]

Algebraically, the following relation can be deduced:

$$\beta = \left(\frac{1}{1+\alpha}\right) \tag{2.3}$$

Thus, the algorithm of an application needs to be partitioned to $\left(\frac{1}{1+\alpha}\right)$ and $\left(\frac{\alpha}{1+\alpha}\right)$ primitive operations to run in GPU and CPU, respectively, for optimal hybrid system performance.

For experimental analysis, a case study is performed for the following: (1) using CPU only, (2) using GPU only, and (3) using both GPU and CPU [106, 110]. The simulation environment computes 100 iterations of discrete cosine transformation (DCT). In the experimental setup, the CPU is a Pentium 4, 3.20 GHz with random access memory (RAM) of 1 GB and the GPU is NVIDIA GeForce FX 5200. The simulation results are reported in Table 2.1. The three parameters used to express results are CPU Time, GPU Time, and CPU Release Time. The CPU Time refers to the total "Elapsed Time," in the system for case (1). The GPU Time refers to the total "Elapsed Time" in the system for case (2). In case (3), these refer to the total "Elapsed Time" in the system when one half is executed by CPU and the other half by GPU. It is observed from Table 2.1 that CPU is free for more than 50% of the time when at least half of the computation is performed by the GPU. Although the computation power of GPU is much higher than CPU in terms of throughput, the fastest DCT technique on GPU is still slower than the implementation on CPU. The reason is that the speed of GPU is limited by memory access. However, the scenario may change with the use of more advanced GPU and interface.

2.10 Networked Media Tank

2.10.1 What Is It?

The networked media tank (NMT) is an all-in-one network-connected media system [31]. It is a state-of-the-art integrated digital multimedia entertainment system that allows a user to play, access, store, and share digital multimedia contents in a computer network. As far as the operations are concerned, NMT is a combination of a digital media player (DMP), a network-attached storage (NAS) device, and a media server (see Fig. 2.23). The NMT allows for seamless integration among digital media, entertainment systems, and the Internet.

2.10.2 Background

The consumer electronics devices are no longer stand-alone, single-tasking systems like old DVD players. Selected Blu-ray players can now be connected to the Internet for streaming video. One key development in the consumer electronics multimedia content player is the "Digital Media Receiver,"

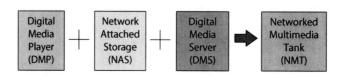

FIGURE 2.23 Conceptual representation of a NMT.

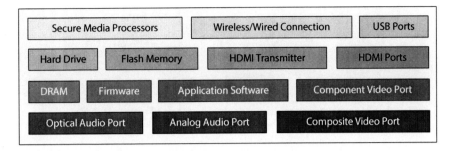

FIGURE 2.24 The schematic representation of a NMT.

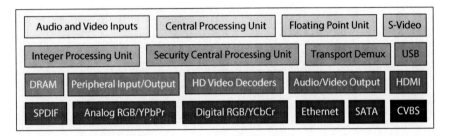

FIGURE 2.25 Architecture of SMP [67].

"Media Player," "Multimedia Jukebox," "Networked Media Jukebox," or "Networked Media Tank." This device can do all the operations of DVD players and Blu-ray players in addition to their new capabilities. In some cases additional external peripherals may need to be connected. The battle among different companies for this very important consumer electronics product is in full swing. It is difficult to accurately point out who developed the first NMT. The major products include Apple TV®, Popcorn Hour® players, and WD TV® HD Media Player [54]. However, Asus® O!Play HD2 is the first to support USB 3.0. Xbox® from Microsoft® and PlayStation® from Sony®, which supports a variety of different media formats, including high-definition video, can be considered ancestors of the NMT [54].

2.10.3 What Is Inside?

The typical components of NMT are shown in Fig. 2.24. The SMP is the most important workhorse of the NMT. This supports decoding of a wide variety of video and audio. More powerful the SMP, more the multimedia formats that the player supports. The dynamic RAM (DRAM) is quite small compared to a PC and may be spread around different parts of the motherboard supporting the chips. The HDMI transmitter enables the delivery of rich digital video and audio content.

The architecture of a typical SMP is shown in Fig. 2.25. The high-definition video decoders handle the decoding needed to play the compressed video in real time. At the same time, audio decoder handles the audio portion of the multimedia. The security processor handles several decryption algorithms.

2.11 Net-Centric Multimedia Processor

2.11.1 What Is It?

The NMP is a SoC that performs Internet Protocol (IP) packet processing, video processing, and DRM without using the main CPU [107, 116, 144]. This is essentially possible due to hardware-amenable DRM algorithms, efficient architectures for net-centric operations, and their efficient VLSI implementation. To understand the application perspective of NMP see Fig. 2.26. The major motivation is its use in DVB over an IP network, the IP-TV service. Figure 2.26(a) shows the existing and widely used scenario in which three different bills are paid for three different appliances, and overall a high-energy bill. Figure 2.26(b) shows the future broadband Internet scenario using the NMP in which one

42 Chapter Two

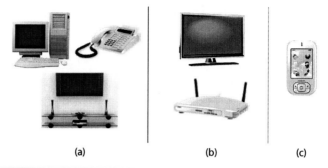

FIGURE 2.26 Motivation for DVB over an IP network using NMP.

bill and one appliance is used. Figure 2.26(c) shows the portable mobile TV application, which can also have similar applications as Fig. 2.26(b) in which battery life is a crucial constraint of the system, demanding energy-efficient solutions [6]. The major advantage of merging digital TV transmission and IP network technologies in a common framework is that accessibility will be provided in a single box. Transmission of digital TV signals through the IP network will potentially provide several advantages, such as better quality of service (QoS), low cost, and single (low) energy bill. The NMP is still in the conceptual stage [113]. The NMP can completely offload computationally intensive multimedia processing and network packet processing from the main CPU of a computer, thus making the CPU available for other applications. The NMP can be a part of line cards of common routers, network cards, STBs, smart-home gateways, or other high-speed communications devices in the IP network cloud.

2.11.2 Background

IP-TV is a system in which digital TV services (digital video streams) are delivered to subscribing consumers using the IP over a broadband connection [43, 58, 85]. In essence, it is a service, not a protocol, which offers the traditional aggregate of a retail video product that is available on digital broadcast TV using an IP network as the transport medium as well as the platform. This, in turn, creates the potential for digital cable and satellite replacement. IP-TV requires two-way communication and therefore typically uses broadband technology over the local loop. The typical network topology is of "star" form, which is different from the traditional cable TV systems that use a ring network topology. The hub of the star, called a head end, uses a "digital subscriber line access multiplexer (DSLAM)." IP-TV needs a new STB with storage mechanisms for providing the full scope of services. Even though IP-TV is available from different vendors, several issues and challenges still need the attention of researchers. The NMP, proposed to solve the security and copyright issues, takes video streams of digital data standardized by a compressed video, processes the stream, inserts DRM attributes, and then broadcasts them. Many entertaining and educating applications, such as commercial TV, video on demand, time-shifted TV, video phones, game portals, personal digital libraries, etc., will be supported by IP-TV.

A possible deployment scenario for an NMP in an IP network cloud is shown in Fig. 2.27. This figure shows a scenario by which digital video can be transported from the source to the receiver's STB while using real-time DRM through a judicious combination of encryption, scrambling, invisible-robust watermarking, and visible watermarking for video. The selective and judicious use of such DRM is needed for security, power dissipation, and real-time performance trade-offs. The NMPs at different nodes of the IP network cloud would perform different operations and may be needed to have slightly different hardware capabilities. In particular, different on-chip memory and throughput requirements may be needed at different nodes and different DRM operations. For example, NMPs in the core/edge network are different from NMPs close to access, curb, and user's home in terms of bandwidth and amount of data-processing capabilities. However, the proposed research activities intend to build a single NMP with the DRM features, on-chip memory, and throughput capability, which can be programmed or configured through the use of a firmware. This solution is expected to be cost effective for mass production. In Fig. 2.27, the enterprise server, or a particular channel, or a vendor for a certain number of channels injects the compressed video in MPEG format into the IP network via the NMP, which is maintained by the core/edge network owner. The bandwidth requirements of this NMP are in gigabits/s because the NMP processes data from several channels for

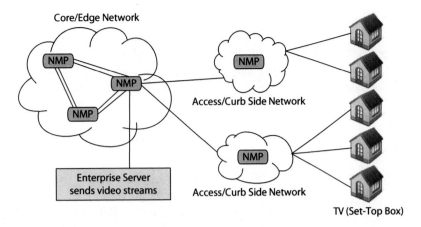

FIGURE 2.27 Deployment of the NMP for secure and copyrighted broadcasting for IP-TV.

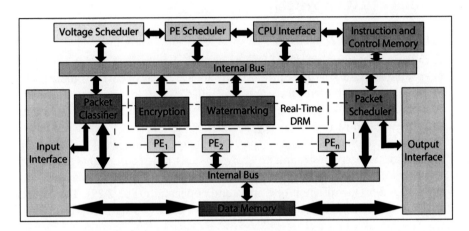

FIGURE 2.28 Architecture for the NMP to be realized as a low-power SoC [107, 113, 116, 144].

each enterprise server and may process several enterprise servers at the same time. Therefore, the buffer and the queue size requirements are very high. The packets received are then processed and broadcast to the users through the IP network. The STB contains a user-end NMP. Even when a single user requests video (video on demand), it is broadcast through the network by the NMP. Once the packets reach the user, the STB splits Internet data and digital TV into two different streams so that the user can view them separately.

2.11.3 What Is Inside?

The architecture proposed for the NMP to be designed as a SoC is shown in Fig. 2.28. The SoC consists of several processing elements (PEs), each with dedicated operational capabilities and all of them connected through an internal bus. The internal bus forms the physical communication channel between the PEs and other components of the NMP. Packet classification is an intensive task in an NMP and is carried out by the packet classifier. It reads the header of an incoming packet, determines the stream to which the packet belongs, decides the outgoing interface using routing lookup, and passes the packet to the appropriate PE for further processing. The outgoing packet is dynamically buffered by the packet scheduler until it is sent to the outgoing link. The instruction and control memory is used to store the instructions corresponding to the program that will be executed using the NMP. The data memory is used to store or buffer the data, and an appropriate mechanism is needed to avoid data conflict among the PEs. Input and output interface are two ports through which the proposed NMP will communicate with the external environment. A brief description of some of the important units of the NMP is as follows:

1. *Packet classifier:* The design of a low-power-consuming packet classifier that performs classification in real time is needed. Such a design needs to exploit structure and characteristics of packet classification rules [92, 124].

2. *Packet scheduler:* The packet scheduler is needed to control different traffic streams and to determine the stream's quality [157]. Wide ranges of scheduling algorithms whose hardware implementation is needed for the NMP are described in the literature.

3. *PE scheduler:* The PE scheduler will activate and deactivate each PE, depending on the application requirement to be executed. The inactive PEs will be shut off with a switching mechanism to reduce standby power consumption [83].

4. *Voltage (frequency) scheduler:* It will dynamically assign the operating voltage of each PE depending on the traffic load and application requirements so that power and performance specifications are met [111].

5. *Encryption and watermarking units for real-time DRM:* These units together form the set of units to provide real-time DRM facility in the NMP [109, 112]. The sequence in which they will be used depends on the application and location of the NMP in the IP network cloud.

2.12 Radiation Detection System

2.12.1 What Is It?

A radiation detection system (RDS) or particle detector is a device used to detect and track high-energy particles, such as those produced by nuclear decay or cosmic radiation. If their primary purpose is radiation measurement, then they are called radiation detectors. However, for photons that are massless particles, the term *particle detector* is still correct [128]. Modern detectors are also used as calorimeters to measure the energy of the detected radiation. They may also be used to measure other attributes including momentum, spin, and charge of the particles. Detectors designed for modern accelerators are large in size and expensive. When the detector counts the particles but does not resolve its energy or ionization, the term *counter* is often used instead of *detector*. The detectors are of many types based on the different operation and application (Fig. 2.29). The ionization detectors such as gaseous ionization detectors and semiconductor detectors are more common.

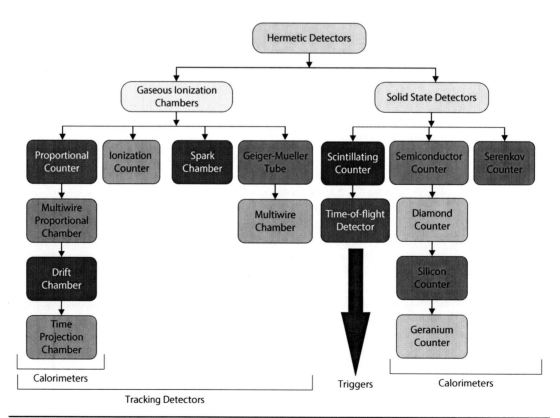

FIGURE 2.29 Various types of particle detectors.

2.12.2 Background

The history of nuclear particle detectors starts way back in 1908, when Hans Geiger and Ernest Rutherford developed the Geiger counter [137]. The Geiger counter was only detecting the alpha particles; however, after undergoing several modifications, its particle detection efficiency and other features were enhanced. In 1947, the updated version of the Geiger counter that was known as the *halogen counter* was developed by Sidney H. Liebson [98]. This halogen-based Geiger counter has a much longer life and lower operating voltage than its predecessors. Currently, a widely used device called "Radiation Dosimeter" can measure exposure to ionizing radiation, such as X-rays, alpha rays, beta rays, and gamma rays [44, 49, 77].

2.12.3 What Is Inside?

A specific and simple example of an RDS is a Geiger-Mueller counter (Fig. 2.30) [119]. In a Geiger-Mueller counter, an inert gas-filled tube briefly conducts electricity when a particle or photon of radiation makes the gas conductive. The tube amplifies this conduction by a cascade effect and generates a current pulse, which is then often displayed by a needle or lamp and/or audible sound. The inert gas used is typically helium, neon, or argon with halogens added.

The schematic representation of a portable dosimeter reader is provided in Fig. 2.31 [94]. The components of a typical radiation dosimeter reader include the following: radiochromic film readout head, LED-based dosimeter readout head, SRAM socket, and electronic systems. The main component of the electronic system is the AVR microcontroller that is responsible for the user interface handling, USB interface, and control of the readout system. For an accurate measurement, the radiation dose is calculated on the basis of the ratio between light intensity of the LED before irradiation and after irradiation. In this process, the new LEDs are labeled, their light intensity is measured with the readout system, and the data is stored in the system memory. The irradiated LEDs are also read out using the same readout system. The driving circuit of the LED consists of ADC, DAC, and operational amplifier (OP-AMP). The current of LED is monitored using a current-sensing circuit and a shunt resistor. The initial value of the DAC is set by the microcontroller. The LED current measurement and the DAC

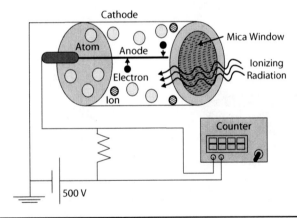

FIGURE 2.30 Schematic representation of a basic Geiger-Mueller counter.

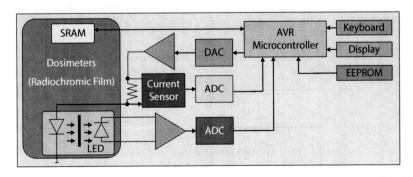

FIGURE 2.31 Schematic representation of dosimeter reader system [94].

setting correction are performed by the microcontroller. The voltage output of the readout circuit is proportional to the light intensity of the LEDs.

2.13 Radio Frequency Identification Chip

2.13.1 What Is It?

The RFID chip is a small microchip that performs wireless data transmission for automated identification of objects and people [33, 82, 88]. It is also known as RFID device, RFID tag, RFID transponder, smart tag, smart label, or radio barcode. The RFID tag may be present in any single product, case, pallet, or container. The cost of an RFID tag is decreasing significantly, and hence it is expected that several trillion RFID tags will be part of every consumer product. Several leading retailers around the world are in the process of evaluating the use of RFID.

The broad concept of RFID is shown in Fig. 2.32 [82]. The crucial component of an RFID tag embedded in an item contains unique information about the item and is always ready to communicate through its antenna. The RFID reader or scanner is used to send and receive RFID data to the RFID tag as well from the RFID tag using its built-in antennas. The host computer gets the data from the RFID reader, which uses specialist RFID software to filter the data and route it to the correct applications.

The RFID tags can be classified into different types based on various criteria (Fig. 2.33). Based on the source of power supply, they can be passive, semi-passive, or active [88]. The semi-passive RFID tags (a.k.a. battery-assisted passive) have a battery which is used to initiate communication when they are interrogated. The active RFID tags use batteries to initiate communication and transmissions. Depending on the operating frequency, RFID tags are low frequency (LF), high frequency (HF), or ultra-high frequency (UHF). The LF tags operate in the range of 124 to 135 kHz. The HF tags operate at 13.56 MHz. The UHF RFID tags operate in the range of 850 to 950 MHz. The microwave RFID tags operate in the range of 2.4 to 5.8 GHz. RFID tags can either be read-only or read-write type based on their data storage mode [33, 82, 88]. The read-only RFID tags are programmed with unique information

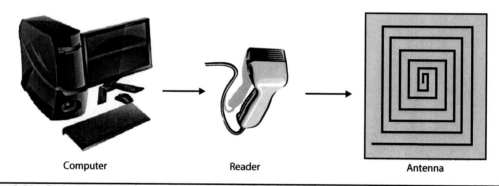

Figure 2.32 Basic operations of RFID.

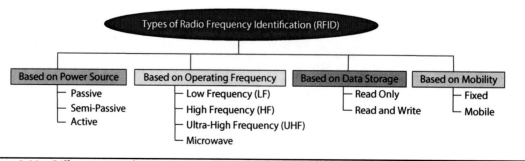

Figure 2.33 Different types of RFID tags [82, 88].

stored on them during the manufacturing process that cannot be changed. The read-only RFID tag is also known as "write-once, read-many" (WORM) RFID tags [33]. The read-write RFID tags allow writing over existing information when the tag is within the range of the RFID reader. Depending on the mobility, RFID readers are classified into two different types: fixed RFID and mobile RFID. In fixed RFID, the reader reads tags in a stationary position. In mobile RFID, the reader is mobile when the reader reads tags.

2.13.2 Background

A specific type of barcode called universal product code (UPC) is widely used to identify objects. The UPC barcode was invented in 1973 [96, 97]. In 1974, the first UPC-marked object was scanned at a retail checkout (Marsh's supermarket in Troy, Ohio) [7, 122]. These are omnipresent and are used almost everywhere on the planet. However, they have several disadvantages that are as follows:

1. UPC only identifies the type of object; no other information is provided.
2. UPC is optically scanned, hence line-of-sight contact with the reader is required.
3. UPC needs careful physical positioning for reading.
4. UPC cannot provide information on the location of the item.
5. UPC reading is a slow process as only one item is scanned at a time.
6. UPC requires human intervention.

In order to overcome the above-mentioned limitations and to deploy it in diverse applications, RFID technology is being further explored. During World War II, an RFID tag was first used by Britain to identify friend and foe aircraft [82, 136]. The commercial application of RFID started during mid-1980. The RFID technology has been more widely accepted now. Applications including automated toll payment and the ignition keys in automobiles are common [88]. However, several issues involving privacy, authentication, and volume of data handling need to have advanced solutions for universal acceptability of RFID [82, 88]. An RFID tag contains a unique identification number called an electronic product code (EPC). The standardization of RFID tags is handled by EPCglobal Inc. [22, 88]. In addition to EPC, the RFID tag potentially can contain other information that is of interest to manufacturers, health care organizations, military organizations, logistics providers, retailers, and other users who need to track the physical locations of goods or equipment.

2.13.3 What Is Inside?

The simplified view of the block diagram of an RFID is shown in Fig. 2.34 [155]. The RFID chip needs to be a single integrated monolithic structure; however, for clarity each component is described in this subsection [28, 50]. The RFID chip consists of an RF antenna, RF analog front end (AFE), the base band circuit, and a clock control. The antenna receives information from the RFID and also translates information to the RFID. One antenna is used for transmission and reception at the same frequency and at the same time. The AFE circuit of the RFID is responsible for transmission and reception of the RF signals. The baseband circuit is responsible for performing coding/decoding, modulation/demodulation, communication protocol, and the anticollision mechanism. The clock control circuit provides the system clock and also resets the signal. The PC is not a part of the RFID chip; it is used for communication of a user with the RFID chip.

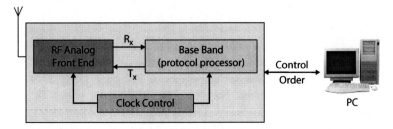

FIGURE 2.34 Block level representation of a RFID chip.

48 Chapter Two

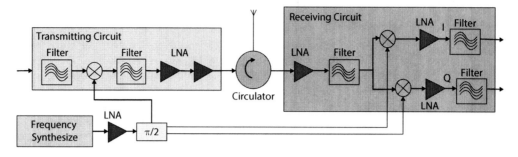

FIGURE 2.35 A RF AFE circuit of the RFID.

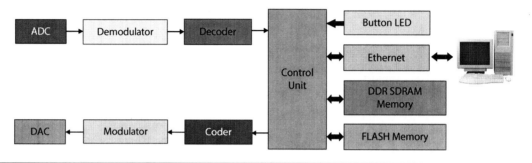

FIGURE 2.36 The block diagram of the baseband circuit of the RFID.

An RFID AFE circuit is shown in Fig. 2.35 [155]. The primary components of the AFE circuit include the following:

1. Transmission circuit.
2. Reception circuit.
3. Frequency synthesizer circuit.
4. The circulator.

The transmission circuit is used for RF signal transmission and the reception circuit is used for receiving RF signals. The frequency synthesizer circuit is responsible for generating appropriate frequency. The circulator determines the performance of the RFID reader system. A good-quality circulator is required to isolate transceivers.

The block diagram of the baseband circuit of the RFID is shown in Fig. 2.36 [155]. The baseband circuit contains various units including the coding/decoding unit, modulator/demodulator unit, DRAM, flash memory, ADC, and DAC, which are needed for managing the communication protocols.

An industrial standard RFID chip is now discussed (Fig. 2.37) [28, 50]. This RFID chip can be used for various applications as follows:

1. Tracking of temperature-sensitive products.
2. Temperature monitoring of medical products.
3. Monitoring of fragile goods transportation.

The temperature sensor that is fully integrated has a maximum accuracy of 0.5°C with an operating temperate range from −20°C to 60°C. The optional external sensor can be connected through the analog sensor interface. The serial peripheral interface (SPI) is used for parameter setting and connection of other external circuits. The RFID has an on-chip 8k bit electrically erasable programmable read-only memory (EEPROM). The real-time clock is used for data logging. Data logging control coordinates the RFID chip components. The RFID chip is powered using a thin and flexible battery or electromagnetic waves from an RFID reader.

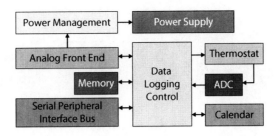

Figure 2.37 Schematic representation of structure of a RFID chip.

2.14 Secure Digital Camera

2.14.1 What Is It?

SDC is an appliance with the standard features of a digital camera and a built-in facility for real-time, low-cost, and low-power DRM and is typically designed as an SoC. SDC performs many DRM-related tasks including copyright information, extent of tampering, source of multimedia, and owner's, creator's, or camera operator's information.

2.14.2 Background

The history of photography can be traced back to the 12th century with the invention of simple glass lenses [123, 129, 142]. In 1787, the first photograph was taken by Nicephore Niepce using in-house cameras [129, 142]. In 1885, George Eastman introduced the use of photographic film. In 1888, he built the first camera called the "Kodak" [129, 142]. In 1975, a scientist from Kodak presented the world's first digital camera [20]. In 1988, Fuji DS-1P became the first true digital camera that recorded images as a computerized file in 16 MB internal memory. In 1990, Logitech Fotoman became the first commercially available digital camera that used a CCD image sensor, stored pictures digitally, and connected directly to a computer for download. In 1991, Kodak DCS-100 marked the beginning of a long line of professional Kodak DCS SLR cameras [19].

The existing digital cameras are quite diverse in terms of size, price, and capability. They can be classified into different types as shown in Fig. 2.38. The simplest type of digital cameras are the webcams that are typically compact CMOS sensors–based cameras and are used along with the computing appliances, such as PCs, PDAs, or slate-PCs, etc. Compact cameras have portable designs that are particularly suitable for casual use. Bridge digital cameras are mid-end digital cameras with a more classical single-lens reflex (SLR) look and with advanced features. Digital SLR cameras (DSLRs) are digital cameras that are based on classical SLR cameras. Mirrorless interchangeable-lens cameras are the ones that combine the larger sensors and interchangeable lenses of DSLRs with the live-preview system of compact cameras.

The explosive growth of the Internet has made it possible to transfer multimedia information worldwide without any effort and time. Flexibility in using digital multimedia has many issues including copyright protection and enforcement of intellectual property rights. In order to provide real-time DRM, SDC is introduced [52, 117].

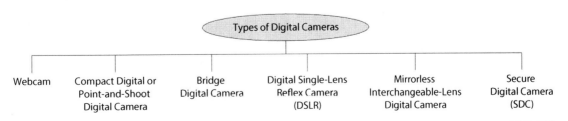

Figure 2.38 Different types of digital cameras.

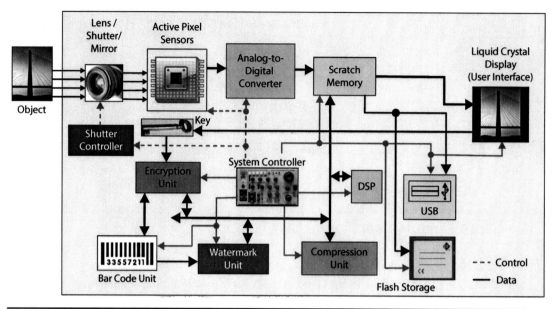

Figure 2.39 Block-level representation of an SDC [114, 116].

2.14.3 What Is Inside?

The block-level representation of the SDC is shown in Fig. 2.39 [105, 114, 116, 118]. The system presented for SDC is in a generic context and the exact architecture depends on specific applications. The main components of the SDC include active pixel sensor (APS), LCD, memory unit (including volatile and nonvolatile), encryption unit, compression unit, bar code unit, and watermarking unit. In the SDC, the multimedia is captured by a sensor and converted to a digital signal by the ADC. While there are other alternatives, a nanoscale CMOS pixel sensor with an embedded ADC is preferred. The captured multimedia is stored temporarily in the scratch memory, after which it is displayed on the LCD panel. The multimedia is then further transmitted over the network or transferred to flash memory, a computer hard drive, or optical discs. The flash memory is a nonvolatile memory used for permanent storage of multimedia in SDC. The controller unit is responsible for controlling the entire sequence of events. The invisible-robust, visible watermarking algorithms are used along with encryption and data compression for different purposes. The choice of operations performed on the multimedia depends on the user of the SDC.

2.15 Set-Top Box

2.15.1 What Is It?

An STB is an appliance that is connected with the TV (or other similar display devices) to receive the external signal and to display the signal content [59, 79]. STBs receive and select broadcast signals, decode and decompress the signals, and convert them into a format that can be displayed by the end-user devices. It is also known as a set-top unit (STU).

2.15.2 Background

In the 1980s, the use of the cable converter box marked the beginning of the STB for TV and other display devices [9]. The cable converter box received the additional analog cable TV channels and converted them to frequencies that could be seen on a regular TV, thus bringing cable TV to the masses. These boxes could be used to shift one selected channel to a low-range very high frequency (VHF). In the mid-1990s, the digital STBs were introduced. These STBs could be used to descramble paid cable channels along with using interactive services such as video on demand, pay per view, and home shopping through TV.

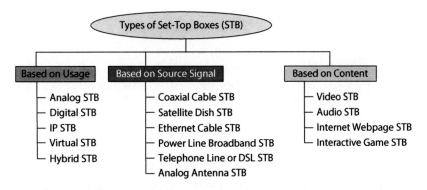

FIGURE 2.40 Different types of STBs.

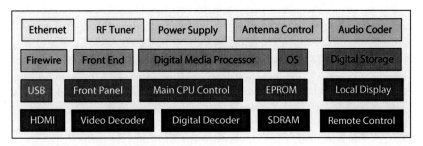

FIGURE 2.41 Main components of a typical STB.

The STBs are of diverse types as shown in Fig. 2.40. Based on the usage, the types of STBs include analog STB, digital STB, IP STB, virtual STB, and hybrid STB. The analog STB is essentially a converter box; however, as broadcasting is mostly digital, this has now limited usage. The digital STBs are the current standard boxes used in digital TV receiving. An IP STB converts IP-TV data into a format that can be used by user-end display [27]. The virtual STB (V-STB) [75] is an application that enables specific subscription channels and packages to be viewed online that can be used with regular STB service. A hybrid STB handles DVB and IP-based video to allow users to view digital cable channels and Internet videos from an IP network.

2.15.3 What Is Inside?

The schematic representation of a typical STB is shown in Fig. 2.41 [42, 59, 79]. The main components are now briefly discussed. The "power supply" provides the voltage to run various peripherals inside an STB. The "front panel" is a small microcontroller that takes the user input and interacts with the main processor. The front panel also contains an infrared (IR) receiver to listen to remote control input. The "front end" that is a "network interface module" (NIM) has two important modules, the "tuner" and the "demodulator." The NIM is a peripheral device that interacts with the input signal source to receive the signal and provides the signal to the "digital decoder" in the form of a transport stream. The "digital decoder" receives the transport stream from the front end and then demultiplexes and decompresses. The many types of data include graphics, audio, video, and software programs. The "CPU (or processor)" takes care of interactions of all the hardware peripherals and software modules inside an STB. The operating system (OS) of the STB is different from a PC's OS in that it operates with limited memory and a low-end processor. It responds to services in real time, and is highly reliable. An STB contains various types of "memory" for different roles, including dynamic RAM (DRAM), nonvolatile RAM (NVRAM; including EEPROM and Flash), and read-only memory/one-time programmable (ROM/OTP). The most important module of STB, the "conditional access module" (CAM) is used for descrambling an encrypted signal and provides a smart card interface for various security features. The "return path" such as cable/DSL modem and Ethernet jack is used to communicate with the head end and send data packets. An STB can contain various peripherals including RS-232 port, USB port, WiFi enablers, and Bluetooth devices for enhanced interactivity and user experience.

2.16 Slate Personal Computer

2.16.1 What Is It?

The slate personal computer (Slate PC), also called Slate Tablet PC, is a small form factor portable PC equipped with touchscreen as a primary input device. The Slate PC is a lighter and slimmed down PC without a dedicated physical keyboard and other extra components [48, 80]. This is different from a convertible tablet with rotating screen and a dedicated key board, the concept presented by Microsoft in 2001 [63]. The advantages of Slate PC include instant-on, ease of use, speedy operation, and wireless Ethernet.

At present, each and every semiconductor and/or consumer electronics industry has or is in the process of having a Slate or Tablet PC. They are of different form factors, processors, and mobile OS (refer to Fig. 2.42). The majority of them are based on the Android OS and ARM processor. The vast use of ARM processors can be attributed to the fact that, unlike other semiconductor companies (e.g., Intel and AMD), ARM licenses its technology as intellectual property rather than manufacturing its own. Thus, there are several companies making processors based on ARM's designs. Apple iPad has its own Apple A4/A5 custom-designed, high-performance, low-power SoC and iOS and hence has longer battery life. A few Slate PCs are in the market with some Intel processors (e.g., Intel Atom, Intel Dual-Core, Intel i3, and Intel i5) and Window 7 OS [46]. Research in Motion (RIM) and BlackBerry Tablet OS are making a Slate PC called PlayBook. Several other variants of Slate PCs are going to come in the next few years.

2.16.2 Background: The Developmental Trend of General-Purpose Computer Reaches Slate PC

The computer that is highly essential for automation and accurate repetitive computation is a machine that takes inputs and processes them using the predefined instructions and generates outputs. The history of automated computing starts from 3000 BC with the invention of the abacus in Babylonia [103]. With the growth of civilization and consequent increase in the demand for faster, high-performance computations, various computing machines have been tried to meet the requirements. The different forms of computing machines attempted include slide rules, mechanical calculators, analytical machines. The first digital computer, complex numerical calculator was demonstrated in Bell Laboratory in 1940. The first general purpose computer, "electronic numerical integrator and computer" (ENIAC) was made in 1946 using vacuum tubes during the World War II. The modern computer age started with invention of transistors in the Bell Laboratory in 1959. After that the computers have undergone rapid changes in speed, size, and other features. The computer's size has changed from the dimensions of square feet as big as a bedroom to the size of a palm. Various types of computers are presented in Fig. 2.43. The current personal computing devices are slate computers and tablet computers.

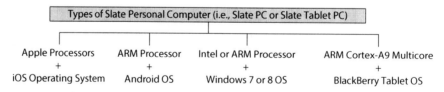

Figure 2.42 The slate PCs.

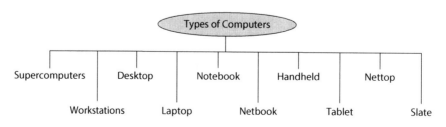

Figure 2.43 Different types of computers.

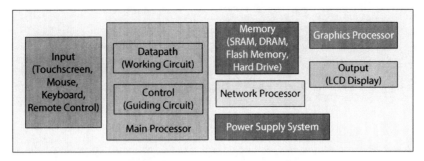

FIGURE 2.44   Typical components of a computer.

The typical components of a computer include the following (see Fig. 2.44):

1. *Input:* Touchscreen, stylus, mouse, keyboard, track ball, remote control.
2. *Memory:* Static RAM (SRAM), DRAM, flash memory, magnetic drive (a.k.a. hard disk), and optical disc.
3. *Datapath:* The main workhorse that does the computation in the computer.
4. *Control:* The intelligent portion that makes the datapath work.
5. *Output:* LCD monitor, CRT monitor, and Digital TV display.

Both datapath and control together are called "central processing unit" (CPU), or processor. A "microprocessor" is a CPU on a single chip [62]. In 1971, the first microprocessor Intel 4004 was built with only 2300 transistors. In 1993, the Pentium processor with 3.1 million transistors was a major milestone, and now microprocessors are available with billions of transistors.

2.16.3   What Is Inside?

There are several variants of the slate PC tablet that have been designed keeping in mind the target cost and market share. Besides some common features such as touchscreen and small form factor, the components of slate PCs differ from one type to the other [46]. The typical components of a slate PC are shown in Fig. 2.45. The touchscreen and virtual keyboard are input devices [139]. The multimedia devices include camera, microphone, and speaker. The processors include general-purpose and GPU. The "accelerometers" are used to align the screen depending on the direction in which the device is held, i.e., switching between portrait and landscape modes. GPS is built-in to have a navigation system. DRAM, solid state flash drive, SD expandable drive, and magnetic hard drive are various types of memory devices used in the slate PC. The network connectivity includes wireless 802.11 b/g/n, cellular 3G/4G, and Bluetooth. Ambient light sensors are used in the slate PCs for adjusting the brightness of the screen based on the environment. USB and HDMI ports are used for data transfer in the slate PC. An LCD (which is also the input for touchscreen) is also the main output of the slate PC. One rechargeable battery is a lithium polymer battery that provides very good operation life to the slate PC [61].

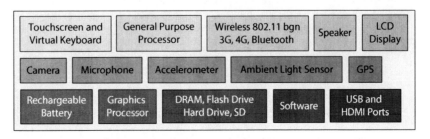

FIGURE 2.45   Typical components of a slate PC.

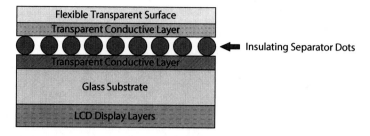

FIGURE 2.46 Schematic representation of construction of a touchscreen of a slate PC.

The unique aspect of the slate PC is the touchscreen input [139, 153]. There are two key aspects of a touchscreen:

1. Sensing methodology to sense the touching of the LCD.
2. Special construction of the LCD screen compared to a regular LCD screen.

The above two are related on the basis of the type of sensing method on which the LCD has to be constructed. Typical sensors and circuitry are used to monitor changes in a particular state, e.g., to monitor changes in electrical current. This can be performed by capacitive and resistive approaches. In addition, the touchscreen following the capacitive or resistive approaches have the following layers (see Fig. 2.46):

1. Layer 1: Flexible transparent surface layer.
2. Layer 2: Transparent metallic conductive coating on the bottom of layer 1.
3. Layer 3: Adhesive spacer consisting of nonconducting separator dots.
4. Layer 4: A transparent metallic conductive coating on the top of layer 5 below.
5. Layer 5: Glass substrate.
6. Layer 6: Adhesive layer on the backside of the glass for mounting.
7. Layer 7: LCD layers.

The capacitive touchscreens use a capacitive material layer to hold an electrical charge that is initiated by touching the screen at a specific point of contact. In resistive touchscreens, the pressure from touching causes conductive and resistive layers of circuitry to touch that changes the circuit resistance.

2.17 Smart Mobile Phone

2.17.1 What Is It?

The smartphone can be conceptualized as a handheld computer integrated with a mobile phone as presented in Fig. 2.47. The smartphones allow the users to install new applications as new applications are made available in the market. Smartphones may contain complete operating systems that provide platforms for application developers and high-end users.

As the market for smartphones is huge, various kinds of these devices are available. The semiconductor houses, IC design industries, and software companies are coming together to get

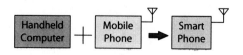

FIGURE 2.47 The concept of smart phones.

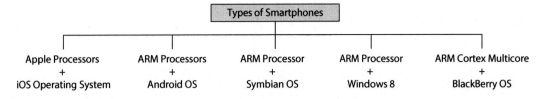

FIGURE 2.48 Different types of smart phones.

a share of this ever-growing market. At high-level, the smartphones can be grouped into various types as presented in Fig. 2.48 [35]. Many hardware and software combinations build different smartphones with diverse capabilities [120]. The primary processors in the smartphone include different types of ARM processors, Apple custom SoC processors, and Qualcomm processors. The main operating systems include Apple iOS, Android OS, Symbian OS, and Windows Phone.

2.17.2 Background

Invention of the telephone by Alexander Graham Bell in 1876 started the age of long-distance speech communication. In 1973, Martin Cooper of Motorola presented the first mobile phone. A mobile phone, which is also called mobile, cellular telephone, or cell phone, is used to make mobile telephone calls across a wide geographic area unlike regular corded or cordless phones. Since then the development of mobile phones is in progress on a competitive basis. The current generation of mobile phones are smartphones that have more advanced computing ability, applications, and connectivity than a regular mobile phone.

2.17.3 What Is Inside?

The different components of a typical smartphone are shown in Fig. 2.49. The touchscreen and the LCD display are essentially the typical user input/output interface of the smart phone. The touchscreen uses the similar technology as in the case of the slate PC. The processors are variants of ARM core. Other processors that are either custom built or built on the basis of an ARM core can be present. The WiMAX chip that supports mobile WiMAX baseband and three frequency bands—2.3–2.4 GHz, 2.5–2.7 GHz, and 3.3–3.8 GHz—is an important component of the smartphone. The radio chip has functionality like 802.11n WiFi, Bluetooth, and FM. An RF transceiver with GPS is an integral part of smartphones. The Li-ion rechargeable battery with a power management unit provides power to the components. The smartphone system has NAND Flash for permanent storage of data and volatile RAM to be used by the processors. The ambient light sensor adjusts the brightness of the display based on the environment and disables the touchscreen when it is held close to the ear. The accelerometer aligns the smartphone in a portrait or a landscape orientation based on the direction of the device. A speaker and microphone are built-in audio devices in a smartphone. There are two cameras, comprising low-power-consuming CMOS sensors, in a smartphone, one in front and one in back, to facilitate video conferencing. The mobile operating systems coordinate the processing of the different components. Application software is available in large numbers depending on the OS of the smartphone.

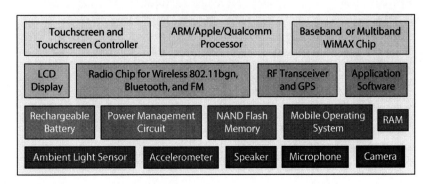

FIGURE 2.49 The typical components of a smart phone.

2.18 Software-Defined Radio

2.18.1 What Is It?

A software-defined radio (SDR) system, or simply SDR, is a set of hardware and software components that constitute a reconfigurable system for wireless communication [36, 38, 65, 68, 89, 156]. In an ideal scenario, the SDR consists of a wideband antenna, wideband ADC, wideband DAC, and a programmable processor (Fig. 2.50) [148]. SDR is an efficient multimode, multiband, multifunctional communications system that can be enhanced using software upgrades. The SDR essentially leads to universal radio terminals (URTs) while allocating as many hardware functions as possible to software.

2.18.2 Background

In 1984, the term "software radio" was coined by the Garland Texas Division of E-Systems Inc. that is now known as Raytheon Company [8, 87]. This "software radio" consisted of a digital baseband receiver of thousands of adaptive filter taps using multiple array processors accessing shared memory that provided programmable interference cancellation and demodulation for broadband signals. In 1991, the term "Software Defined Radio" was coined [101, 102]. By performing as many large portions as possible of radio functions in a software, the SDR presents many advantages over specialized hardware approaches:

1. SDR has the ability to receive and transmit different modulation schemes using a common set of hardware.
2. SDR has the possibility of adaptively choosing an operating frequency and a mode best suited for prevailing conditions.
3. SDR is fully field upgradeable and extensible as better controlling software becomes available.
4. SDR provides the opportunity to recognize and avoid interference with other communications channels.
5. SDR becomes the basis of the next generation of mobile communications, replacing a multitude of incompatible standards, thus providing true universality.
6. SDR users realize significant time and cost savings as the same device replaces a number of competing technologies and protocols.
7. SDR makes the world roaming of wireless devices become a reality.
8. SDR eliminates analog hardware and its cost that results in simplification of radio architectures and improved performance.

2.18.3 What Is Inside?

In order to give a detailed and comparative perspective of the SDR, this section starts discussions from the traditional radio. The schematic representation of the structure of a traditional and hardware-based radio system is shown in Fig. 2.51 [68]. In the traditional radio schematic, the AFE consists of a significant portion of the system and it performs the reception using an antenna and down-conversion of RF signals to intermediate frequency (IF) and then to baseband. Once the signal has been down-converted, the special-purpose hardware processes the signal. The software involved in this case is minimal. In an SDR system, a significant portion of the processing is transferred from specialized hardware to general hardware under software control. The schematic representation of the SDR is provided in Fig. 2.52. In the SDR, the entire baseband processing and even the conversion

Figure 2.50 The system architecture of an ideal SDR.

Emerging Systems Designed as Analog/Mixed-Signal System-on-Chips

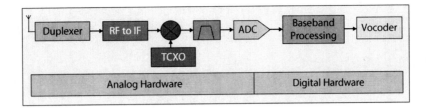

FIGURE 2.51 The structure of a conventional hardware-based radio system.

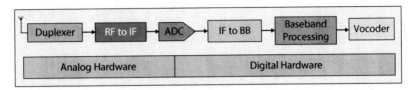

FIGURE 2.52 The structure of an SDR system showing a significant portion of processing is performed in general hardware under software control.

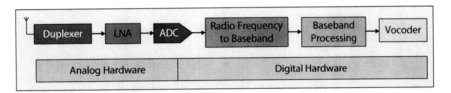

FIGURE 2.53 The pure software radio in which the majority of functions are performed by software.

from IF to baseband is performed in the software. In a pure software radio, the majority of functions are performed by a software that has moved as close to the antenna as possible. This SDR is shown in Fig. 2.53. In this system, direct conversion from RF to baseband is performed by using software.

2.19 TV Tuner Card for PCs

2.19.1 What Is It?

A TV tuner card is a component that allows the personal computers to receive, capture, process, and record TV signals [51, 121, 131]. It is similar to the TV tuner of a digital TV. Most TV tuners also function as video capture cards, allowing them to record TV programs onto a hard disk. The schematic representation of a PC with TV tuner card is shown in Fig. 2.54. In essence, the monitor becomes a

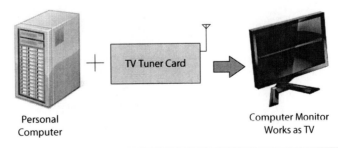

FIGURE 2.54 TV tuner card once installed in PC allows viewing TV in a monitor.

regular TV that displays a TV signal [18]. The advantage of using a TV tuner card to watch TV in a PC is the removal of the actual geographical hurdle that would generally obstruct a typical TV.

2.19.2 Background

In the early 1990s, the NEC laptop TV with a built-in television tuner and an antenna marked the beginning of watching TV signals in a digital computer [10]. The NEC laptop TV did not record and was only for watching TV. In 1992, the Mac TV was introduced primarily for watching TV and not for video recording. In 1992, the TV tuner card was introduced to watch TV signals in a PC by Animation Technologies Inc. [13]. In the same year, Hauppauge introduced the TV tuner cards for the PC [24]. In 1996, ATI Technologies Inc. introduced its TV tuner cards [14].

The existing varieties of TV tuner cards can be classified into the following types (see Fig. 2.55) [18]:

1. *Analog TV Tuner*: Analog TV tuners generate a raw video stream on receiving a signal from the antenna. They typically need a data integrated compression strategy for further recording.
2. *Digital TV Tuner*: Digital TV tuners generate compressed video (e.g. MPEG2) that displays high-quality videos and audio. These cards do not need additional encoding chips.
3. *Hybrid TV Tuner*: Hybrid TV tuners are the cards that can be reconfigured to operate as either an analog TV tuner or a digital TV tuner.
4. *Combination TV Tuner*: The combination TV tuner cards are the cards that simultaneously operate as an analog TV tuner and a digital TV tuner.

2.19.3 What Is Inside?

The schematic representation of a typical TV tuner card is shown in Fig. 2.56. The antenna that is external to (not a part of) the TV tuner card receives terrestrial signals from the air. The signal may also come from a cable TV or other video-generating devices when the TV tuner is used for capturing and digitization purposes. The tuner is essentially PLLs. The RF signal is demodulated to an IF signal using the demodulators. The ADCs then convert the signals to digital forms for further processing using digital processors. The video is then processed and MPEG2 compressed video is sent to the PC through an interface. A ghost reducer is responsible for reducing ghost pictures. EEPROM nonvolatile memory is used to store data that must be saved when power is OFF. DDR volatile memory is used for intermediate storage.

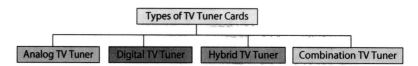

FIGURE 2.55 Different types of TV tuner cards.

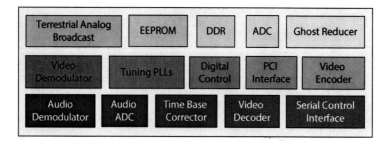

FIGURE 2.56 Different components of a TV tuner card.

2.20 Universal Remote Control

2.20.1 What Is It?
A universal remote control is a remote control that can be programmed to communicate with various brands of one consumer electronics device or various types of consumer electronics devices. The low-end universal remote controls can only communicate with a few consumer electronics devices. The high-end universal remote controls can communicate with large number of consumer electronics devices and can learn from new remote controls as they become available.

2.20.2 Background
In 1898, the earliest remote control was conceptualized by Nikola Tesla and described in his patent [145]. In 1939, the battery-operated LF radio transmitter called "Philco Mystery Control" became the first wireless remote control for a consumer electronics device [40]. In 1995, "Flashmatic" became the first wireless TV remote [23]. A large variety of remote controls have been designed since then. The different types of remote controls are presented in Fig. 2.57. In the early 1980s, the IR technology-based remote controls became popular. By the early 2000s, the proliferation of remote controls happened with an increasing number and variety of remote controls to communicate with the ever-increasing number of consumer electronics appliances. The RF remote controls provide longer and wider range as compared to the IR remote. The RF remote was first used in World War II. However, IR remote became a household name because of the cheaper price. The universal remote controls can communicate with various brands of one or more types of consumer electronic appliances. The programmable remote controls can be programmed to accommodate any system configuration to provide the control as if the systems are fully integrated. The universal and programmable remote controls are becoming immensely popular these days. The touchscreen remote controls have an LCD screen in which the buttons are actually images on the screen that send signals to the electronic appliance when touched.

2.20.3 What Is Inside?
The different components of a universal programmable touchscreen remote control is shown in Fig. 2.58 [32]. The main workhorse of the remote control is a low-power-embedded processor. This processor is designed using SoC technology. The processor is powered by a battery and battery management unit. Of note, the battery for this kind of remote is rechargeable. The signal communication component is RF or IR. The RF portions contain RF circuits and RF antenna. The signal from the IR is sent through the LED and its drivers. The remote control's display is an LCD touchscreen. The touchscreen has different control buttons as pictures. In addition, different physical keyboard buttons may be provided. The remote may contain a limited amount of permanent storage. The high-end versions of the remote may include a voice-control mechanism.

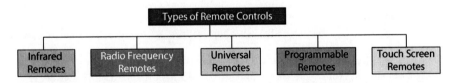

FIGURE 2.57 Different types of remote control.

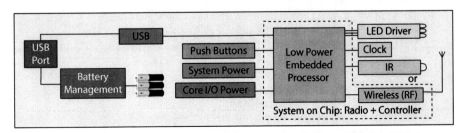

FIGURE 2.58 Different components of a universal programmable touchscreen remote control.

2.21 Questions

2.1 Pick a consumer electronics media player system of your choice. Identify its hardware and software components. Briefly describe their functionalities.

2.2 Pick a consumer electronics system of your choice. Identify the different types of chips presented in the system and discuss their functionalities.

2.3 Pick a human health-monitoring system of your choice. Identify its hardware and software components. Briefly describe their functionalities.

2.4 Pick a security system of your choice. Identify its hardware and software components. Briefly describe their functionalities.

2.5 Describe the working principle of a touchscreen in an electronic system.

2.6 Identify different types of analog chips present in a smartphone.

2.7 Identify different types of digital chips present in a smartphone.

2.8 Briefly discuss different pixel sensors used in present-day cameras.

2.9 Identify a system that has multiple cores. Discuss what software and hardware components constitute such a multicore system.

2.10 Discuss the differences between a sensor and actuator. Give three examples of each.

2.22 References

1. http://www.nvidia.com/object/GPU_Computing.html. Accessed on 08 March 2011.
2. http://www.blu-ray.com. Accessed on 22 February 2011.
3. http://www.gbpvr.com/. Accessed on 07 April 2011.
4. http://www.team-mediaportal.com/. Accessed on 07 April 2011.
5. http://www.kowoma.de/en/gps/. Accessed on 06 March 2011.
6. http://mobilementalism.com/. Accessed on 20 March 2011.
7. http://www.wired.com/science/discoveries/news/2008/06/dayintech_0626?currentPage=2 Accessed on 23 March 2011.
8. http://www.raytheon.com. Accessed on 18 May 2011.
9. A Brief History of Set-Top Box Innovation. http://ubuntunation.org/?tag=set-top-box. Accessed on 18 May 2011.
10. A Nutshell Early History of PC-based TV Video Recording. http://ruel.net/pc/tv.tuner.video.recording.history.htm. Accessed on 18 May 2011.
11. About freevo. http://freevo.sourceforge.net/. Accessed on 07 April 2011.
12. About replaytv. http://www.digitalnetworksna.com/about/replaytv/. Accessed on 07 April 2011.
13. Animation Technologies Inc. http://www.lifeview.com.tw. Accessed on 18 May 2011.
14. ATI Technologies Inc. http://www.ati.amd.com/. Accessed on 18 May 2011.
15. Atomic Force Microscopy. http://www.chembio.uoguelph.ca/educmat/chm729/afm/firstpag.htm. Accessed on 07 April 2011.
16. Atomic Force Microscopy. http://www.weizmann.ac.il/Chemical_Research_Support/surflab/peter/afmworks/. Accessed on 07 April 2011.
17. Blu-ray disc association. http://www.bluraydisc.com/. Accessed on 22 February 2011.
18. Computer TV Tuner. http://www.computertvtuner.net/. Accessed on 18 May 2011.
19. DigiCam History Dot Com. http://www.digicamhistory.com/. Accessed on 18 May 2011.
20. Digital Photography Milestones from Kodak. http://www.womeninphotography.org/Events-Exhibits/Kodak/EasyShare_3.html. Accessed on 18 May 2011.
21. Drug Delivery Systems - Markets and Applications for Nanotechnology Derived Drug Delivery Systems.
22. EPCglobal Inc. http://www.EPCglobalinc.org. Accessed on 07 April 2011.
23. Flashmatic: The First Wireless TV Remote. http://web.archive.org/web/20080116212531/http://www.zenith.com/sub_about/about_remote.html. Accessed on 18 May 2011.
24. Hauppauge Computer Works, Inc. http://www.hauppauge.com/. Accessed on 18 May 2011.
25. Intel Microarchitecture Codename Sandy Bridge. http://www.intel.com/technology/architecture-silicon/2ndgen/index.htm. Accessed on 10 March 2011.
26. The international conumer electronic show (ces). http://www.cesweb.org/. Accessed on 07 April 2011.
27. IP Set-Top Box (IP STB). http://www.iptvmagazine.com/iptvmagazine_directory_ip_stb.html. Accessed on 18 May 2011.

28. ISO 15693 sensory tag chip identifies, monitors and logs. http://www.ids-microchip.com/prod3_IDS-SL13A.htm. Accessed on 07 April 2011.
29. MicroCHIPS, Inc. http://www.mchips.com/. Accessed on 26 February 2011.
30. Mythtv. http://www.mythtv.org/detail/mythtv. Accessed on 07 April 2011.
31. Networked multimedia tank. http://rpddesigns.com/Documents/NetworkedMediaTankBrochure.pdf. Accessed on 11 March 2011.
32. RF4CE Remote Control. http://focus.ti.com/docs/solution/folders/print/518.html. Accessed on 18 May 2011.
33. RFID Tags. http://www.rfidjournal.com/faq/18. Accessed on 23 March 2011.
34. Secure Media Processor Overview. http://www.sigmadesigns.com/media_processor_overview.php. Accessed on 23 February 2011.
35. Smart Phone Reviews. http://www.smartphonereviews.info/. Accessed on 18 May 2011.
36. Software Defined Radio for All. http://www.sdr4all.org/. Accessed on 18 May 2011.
37. What are biosensors? http://www.lsbu.ac.uk/biology/enztech/biosensors.html. Accessed on 20 February 2011.
38. What are Software Defined Radios? http://www.flex-radio.com/. Accessed on 18 May 2011.
39. What is tivo? http://www.tivo.com/what-is-tivo/tivo-is/index.html. Accessed on 07 April 2011.
40. Philco Mystery Control. *Collier's Magazine*, October/November 1938.
41. Blu-ray disc association. http://www.blu-raydisc.com/Assets/Downloadablefile/BD-ROMwhitepaper 20070308-15270.pdf, 2010. Accessed on 23 February 2011.
42. Inside Set Top Box. *Electronics For You*, (1), Jan 2010. Accessed on 01 January 2010.
43. B. Alfonsi. I Want My IPTV: Internet Protocol Television Predicted a Winner. IEEE Distributed Systems Online, February 2005.
44. National Areonautics and Space Admnistration (NASA). Radiation Equipment. http://spaceflight.nasa.gov/shuttle/reference/shutref/crew/radiation.html. Accessed on 21 April 2011.
45. C. Aston. Biological Warfare Canaries. October 2001.
46. ASUS. ASUS CES Show Room – Video: Roundup of ASUS products at CES 2011. Video, January 2011.
47. D. R. Baselt, S. M. Clark, M. G. Youngquist, C. F. Spence, and J. D. Baldescahwieler. Digital Signal Processor Control of Scanned Probe Microscopes. *AIP Review of Scientific Instruments*, 64(7):1874–1882, 1993.
48. H. Beck, A. Mylonas, J. Harvey, R. L. Rasmussen, and A. Mylonas. *Business Communication and Technologies in a Changing World*. Macmillan Education Australia, 2009.
49. N. A. Bertoldo, S. L. Hunter, R. A. Fertig, G. W. Laguna, and D. H. MacQueen. Development of a Real-time Radiological Area Monitoring Network for Emergency Response at Lawrence Livermore National Laboratory. *IEEE Sensors Journal*, 5(4):565–573, 2005.
50. A. Bindra. RFID labeling and tracking gets smarter. http://mobiledevdesign.com/hardware_news/rfid-labelling-tracking-smarter-0501/. Accessed on 07 April 2011.
51. S. Birleson, J. Esquivel, P. Nelsen, J. Norsworthy, and K. Richter. Silicon Single-Chip Television Tuner Technology. In *Proceedings of the International Conference on Consumer Electronics*, pages 38–39, 2000.
52. P. Blythe and J. Fridrich. Secure Digital Camera. In *Proceedings of Digital Forensic Research Workshop (DFRWS)*, 2004.
53. I. Bucks. Data Parallel Computing on Graphics Hardware, July 27 2003.
54. M. Butrovich. Western digital's wd tv hd media player: Break out the popcorn. http://techreport.com/articles.x/16565. Accsssed on 20 March 2011.
55. A. Casson, D. Yates, S. Smith, J. Duncan, and E. Rodriguez-Villegas. Wearable Electroencephalography. *IEEE Engineering in Medicine and Biology Magazine*, 29(3):44–56, 2010.
56. K. Chakrabarty and T. Xu. *Digital Microfluidic Biochips: Design Automation and Optimization*. CRC Press, Boca Raton, FL, 2010. ISBN: 9781439819159.
57. P. K. Cheng, K. Yackoboski, G. C. McGonigal, and D. J. Thomson. A Digital Singal Processor Based Atomic Force Microscope Controller. In *Proceedings of IEEE Communications, Power, and Computing Conference*, pages 456–461, 1995.
58. S. Cherry. The battle for broadband [Internet protocol television]. *IEEE Spectrum*, February 2005.
59. W. S. Ciciora. Inside the Set-Top Box. *IEEE Spectrum*, 32(4):70–75, 1995.
60. L. C. Clark. *Trans. Am. Soc. Artif. Intern. Organs*, pages 41–48, 1956.
61. Apple Corporation. iPad Specification. Technical report, 2009. Accessed on 13 February 2011.
62. Intel Corporation. The Journey Inside^{SM}. http://educate.intel.com/en/TheJourneyInside.
63. Microsoft Corporation. Tablet PC: An Overview. Technical report, June 2002.
64. A. V. Crewe, M. Isaacson, and D. Johnson. A Simple Scanning Electron Microscope. *Review of Scientific Instruments*, 40(2):241–246, 1969.
65. M. Cummings and T. Cooklev. Tutorial: Software-Defined Radio Technology. In *Proceedings of the 25th International Conference on Computer Design*, pages 103–104, 2007.
66. P. Daly. Navstar GPS and GLONASS: Global Satellite Navigation Systems. *Electronics & Communication Engineering Journal*, 5(6):349–357, 1993.
67. Sigma Designs. Secure Media Processors. http://www.sigmadesigns.com/uploads/documents/SMP8640_br.pdf. Accessed on 11 March 2011.

68. M. Dillinger, K. Madani, and N. Alonistioti. *Software Defined Radio: Architectures, Systems, and Functions*. Wiley, 2003.
69. K. L. Ekinci and M. L. Roukes. Nanoelectromechanical Systems. *Review of Scientific Instrumentation*, 76, 2005.
70. M. Gad el Hak, editor. *MEMS: Introduction and Fundamentals*. Taylor & Francis, 2006.
71. K. Fatahalian and M. Houston. GPUs: A Closer Look. *ACM Queue*, 6(2):18–28, 2008.
72. D. M. Fraser. Biosensors: Making Sense of Them. *Medical Device Technology*, 5(8):38–41.
73. P. L. T. M. Frederix, B. W. Hoogenboom, D. Fotiadis, D. J. Muller, and A. Engel. Atomic Force Microscopy of Biological Samples. *Materials Research Society (MRS) Bulletin*, 29(7):449–455, 2004.
74. B. J. Furman, J. Christman, M. Kearny, and F. Wojcik. Battery Operated Atomic Force Microscope. *AIP Review of Scientific Instruments*, 69(1):215–220, 1998.
75. J. L. Gilmour, C. G. Hooks, G. Jenkin, M. C. Liassides, and D. J. Evans. Virtual set-top box. http://www.freepatentsonline.com/y2010/0064335.html. Accessed on 18 May 2011.
76. F. J. Giessibl. Advances in Atomic Force Microscopy. *Reviews of Modern Physics (RMP)*, 75(3):949–983, 2003.
77. M. E. Goulder. *A Geiger Muller Counter Circuit for x-ray Intensity Measurement*. Bachler's thesis, The Massachutes Insitute of Technology, 1942.
78. W. Greatbatch and C. F. Holmes. History of Implantable Devices. *IEEE Engineering in Medicine and Biology Magazine*, 10(3):38–41, 1991.
79. L. Harte. *Introduction to TV STB*. Althos Publishing, Fuquay Varina, NC, 2011.
80. K. Haven. *100 Greatest Science Inventions of All Time*. Libraries Unlimited, 2006.
81. S. P. J. Higson, S. M. Reddy, and P. M. Vadgama. Enzyme and Other Biosensors: Evolution of a Technology. *Engineering Science and Education Journal*, 41–48, 1994.
82. S. Holloway. RFID: An Introduction. http://msdn.microsoft.com/en-us/library/aa479355.aspx, June 2006. Accessed on 21 March 2011.
83. Z. Hu, A. Buyuktosunoglu, and V. Srinivasan. Microarchitectural Techniques for Power Gating of Execution Units. In *Proceedings of the International Symposium Low Power Electronics and Design*, 2004.
84. Texas Instruments. GPS: Personal Navigation Device. http://focus.ti.com/docs/solution/folders/print/413.html. Accessed on 06 March 2011.
85. R. Jain. I Want My IPTV. *IEEE Multimedia*, July 2005.
86. N. Jalili and K. Laxinarayana. A Review of Atomic Force Microscopy Imaging Systems: Application to Molecular Metrology and Biological Sciences. *Elsevier Mechatronics*, 14(8):907–945, 2004.
87. P. Johnson. New Research Lab Leads to Unique Radio Receiver. *E-Systems Team*, 5(4):6–7, 1985.
88. A. Juels. RFID Security and Privacy: A Research Survey. *IEEE Journal on Selected Areas in Communications*, 24(2):381–394, 2006.
89. P. B. Kenington. *RF and Baseband Techniques for Software Defined Radio*. Artech House, 2005.
90. P. Khanna, J. A. Storm, J. I. Malone, and S. Bhansali. Microneedle-Based Automated Theraphy for Diabetes Mellitus. *Journal of Diabetes Science and Technology*, 2:1122–1129, 2008.
91. P. T. Kissinger. Biosensors–A Perspective. *Biosensors and Bioelectronics*, 20(12):2512–2516, 2005.
92. M. E. Kounavis et al. Directions in Packet Classification for Network Processors. In *Proceedings of the Second Workshop on Network Processors*, 2003.
93. G. Kovacs. *Micromachined Transducers: Sourcebook*. McGraw Hill, Inc., 1998. ISBN: 978-0072907223.
94. P. Krasinski, D. Makowski, and B. Mukherjee. Portable gamma and neutron radiation dosimeter reader. In *IEEE Nuclear Science Symposium Conference Record*, pages 2048–2051, 2008.
95. M. N. V. Ravi Kumar. *Handbook of Particulate Drug Delivery*. American Scientific Publishers. ISBN 1-58883-123-X.
96. G. J. Laurer. http://bellsouthpwp.net/l/a/laurergj/. Accessed on 23 March 2011.
97. G. J. Laurer. *Engineering Was Fun*. Lulu.com, 2008.
98. S. H. Liebson. The Discharge Mechanism of Self-quenching Geiger-Mueller Counters. *Physical Review*, 72(7): 602–608, 1947.
99. R. Martins, S. Selberherr, and F. A. Vaz. A CMOS IC for Portable EEG Acquisition Systems. *IEEE Transactions on Instrumentation and Measurement*, 47(5):1191–1196, 1998.
100. A. A. Mhatre. *Implantable Drug System with an in-Plane Micropump*. Master's thesis, The University of Texas at Arlington, May 2006.
101. J. Mitola. Software Radios - Survey, Critical Evaluation and Future Directions. In *Proceedings of the IEEE National Telesystems Conference*, pages 13/15–13/23, 1992.
102. J. Mitola. The Software Radio Architecture. *IEEE Communications Magazine*, 33(5):26–38, 1995.
103. S. P. Mohanty. Intel Pentium Processors. Technical report, Dept. of Computer Science and Engineering, University of South Florida, 2000.
104. S. P. Mohanty. A Low Power Smart VLSI Controller for Nano-Characterization in Atomic Force Microscope (AFM). Junior Faculty Summer Research Fellowship, University of North Texas, 2005.
105. S. P. Mohanty. Methods and Devices for Enrollment and Verification of Biometric Information in Identification Documents. US Patent filed on 24th April 2008, U.S. Serial No. 12/150,009, 2008.
106. S. P. Mohanty. GPU-CPU Multi-Core for Real-Time Signal Processing. In *Proceedings of the 27th IEEE International Conference on Consumer Electronics*, pages 55–56, 2009.

107. S. P. Mohanty, D. Ghai, E. Kougianos, and P. Patra. A Combined Packet Classifier and Scheduler Towards Net-Centric Multimedia Processor Design. In *Proceedings of the 27th IEEE International Conference on Consumer Electronics (ICCE)*, pages 11–12, 2009.
108. S. P. Mohanty and E. Kougianos. Biosensors: A Tutorial Review. *IEEE Potentials*, 25(2):35–40, March/April 2006.
109. S. P. Mohanty and E. Kougianos. Real-Time Perceptual Watermarking Architectures for Video Broadcasting. *Elsevier Journal of Systems and Software (JSS)*, 84(5):724–738, 2011.
110. S. P. Mohanty, N. Pati, and E. Kougianos. A Watermarking Co-Processor for New Generation Graphics Processing Units. In *Proceedings of 25th IEEE International Conference on Consumer Electronics*, pages 303–304, 2007.
111. S. P. Mohanty, N. Ranganathan, and K. Balakrishnan. A Dual Voltage-Frequency VLSI Chip for Image Watermarking in DCT Domain. *IEEE Transactions on Circuits and Systems II (TCAS-II)*, 53(5):394–398, 2006.
112. S. P. Mohanty, R. Sheth, A. Pinto, and M. Chandy. CryptMark: A Novel Secure Invisible Watermarking Technique for Color Images. In *Proceedings of the 11th IEEE International Symposium on Consumer Electronics (ISCE)*, pages 1–6, 2007.
113. S. P. Mohanty. Apparatus and Method for Transmitting Secure and/or Copyrighted Digital Video Broadcasting Data Over Internet Protocol Network, 2008.
114. S. P. Mohanty. A Secure Digital Camera Architecture for Integrated Real-time Digital Rights Management. *Journal of Systems Architecture - Embedded Systems Design*, 55(10-12):468–480, 2009.
115. S. P. Mohanty, D. Ghai, E. Kougianos, and B. Joshi. A Universal Level Converter Towards the Realization of Energy Efficient Implantable Drug Delivery Nano-electro-mechanical-systems. In *Proceedings of the 10th International Symposium on Quality of Electronic Design*, pages 673–679, 2009.
116. S. P. Mohanty and D. K. Pradhan. ULS: A Dual-V_{th}/high-κ Nano-CMOS Universal Level Shifter for System-level Power Management. *ACM Journal on Emerging Technologies in Computing Systems (JETC)*, 6(2):8:1–8:26, 2010.
117. S. P. Mohanty, N. Ranganathan, and R. Namballa. VLSI Implementation of Visible Watermarking for a Secure Digital Still Camera Design. In *Proceedings of the International Conference on VLSI Design*, pages 1063–1068, 2004.
118. S. P. Mohanty, N. Ranganathan, and R. Namballa. A VLSI Architecture for Visible Watermarking in a Secure Still Digital Camera (S^2DC) Design. *IEEE Transactions on VLSI Systems*, 13(8):1002–1012, 2005.
119. G. A. Morton. Nuclear Radiation Detectors. *Proceedings of the IRE*, 50(5):1266–1275, 1962.
120. G. Murphy and S. Clow. Smartphones in the Enterprise. http://www.contextis.co.uk/resources/white-papers/smartphones/Context-Smartphone-White_Paper.pdf. Accessed on 18 May 2011.
121. L. Nederlof. One-Chip TV. In *Proceedings of the 42nd International Solid-State Circuits Conference*, pages 26–29, 1996.
122. B. Nelson. *Punched Cards to Bar Codes*. Helmers Publishing, 1997. 0-911261-12-51997.
123. B. Newhall. *The History of Photography*. The Museum of Modern Art, New York, 1982.
124. M. Nourani and M. Faezipour. A Single-Cycle Multi-Match Packet Classification Engine Using TCAMs. In *Proceedings of the IEEE Symposium on High Performance Interconnects*, pages 73–78, 2006.
125. J. D. Owens, D. Luebke, N. Govindaraju, M. Harris, J. Kruger, A. Lefohn, and T. J. Purcell. A Survey of General-Purpose Computation on Graphics Hardware. In *Prodeings of the Eurographics*, pages 21–51, 2005.
126. I. Palchetti and M. Mascini. Biosensor Technology: A Brief History. *Lecture Notes in Electrical Engineering*, 54(1):15–23, 2010.
127. A. Pantelopoulos and N. G. Bourbakis. A Survey on Wearable Sensor-Based Systems for Health Monitoring and Prognosis. *IEEE Transactions on Systems, Man, and Cybernetics, Part C: Applications and Reviews*, 40(1):1–12, 2010.
128. D. Passeri. Characterization of CMOS Active Pixel Sensors for Particle Detection: Beam Test of the Four-Sensors RAPS03 Stacked System. *Nuclear Instruments and Methods in Physical Research*, A 617(1-3):573–575, 2010.
129. M. R. Peres. *The Focal Encyclopedia of Photography*. Focal Press, 4th edition, 2007.
130. K. Peterson. Biomedical Applications of MEMS. In *IEEE Electron Devices Meeting*, pages 239–242, 1996.
131. R. Powell. Getting Started with a TV Tuner Card. http://www.linuxjournal.com/article/8116. Accessed on 18 May 2011.
132. A. R. A. Rahman, C.-M. Lo, and S. Bhansali. A Micro-electrode Array Biosensor for Impedance Spectroscopy of Human Umbilical Vein Endothelial Cells. *Sensors and Actuators B: Chemical*, 118(1-2):115–120, 2006.
133. R. B. Reilly and T. C. Lee. Electrograms (ECG, EEG, EMG, EOG). *Technology and Health Care*, 18(6):443–458, 2010.
134. R. L. Rich and D. G. Myszka. Survey of the year 2007 commercial optical biosensor literature. *Wiley J. Mol. Recognit.* 21(6):355–400, 2008.
135. O. M. El Rifai and K. Youcef-Toumi. Design and Control of Atomic Force Microscopes. In *Proceedings of the IEEE American Control Conference*, pages 3714–3719, 2003.
136. M. Roberti. The History of RFID Technology. http://www.rfidjournal.com/article/view/1338. Accessed on 23 March 2011.
137. E. Rutherford and H. Geiger. An Electrical Method of Counting the Number of γ Particles from Radioactive Substances. *Proceedings of the Royal Society (London)*, 81(546):141–161, 1908.
138. R. S. Sethi and C. R. Lowe. Electrochemical Microbiosensors. In *IEE Colloqium on Microsensors*, pages 911–915, 1990.
139. B. Shneiderman. Touch Screens Now Offer Compelling Uses. *IEEE Software*, 8(2):93–94, 1991.

140. M. Staples, K. Daniel, M. Cima, and R. Langer. Application of Micro- and Nano-electromechanical Devices to Drug Delivery. *Pharmaceutical Research*, 23(5):847–863, 2006.
141. J. Strickland and J. Bickers. How DVR Works. http://electronics.howstuffworks.com/dvr.htm. Accessed on 12 March 2011.
142. L. Stroebel and R. D. Zakia. *The Focal Encyclopedia of Photography*. Focal Press, 3rd edition, 1993.
143. F. Su, K. Chakrabarty, and R. B. Fair. Microfluidics-Based Biochips: Technology Issues, Implementation Platforms, and Design-Automation Challenges. *IEEE Transactions Computer-Aided Design of Integrated Circuits and Systems*, 25(2):211–223, 2006.
144. S. Tarigopula. *A CAM Based High-Performance Clssifier-Scheduler for a Video Network Processor*. Master's thesis, University of North Texas, 2007.
145. N. Tesla. Method of an Apparatus for Controlling Mechanism of Moving Vehicle or Vehicles. http://www.google.com/patents?vid=613809. Accessed on 18 May 2011.
146. C. J. Thompson, S. Hahn, and M. Oskin. Using Modern Graphics Architectures for General-Purpose Computing: A Framework and Analysis. In *Proceedings of the 35th International Symposium on Microarchitecture*, pages 306–317, 2002.
147. J. T. Santini, A. C. Richards, R. A. Scheidt, M. J. Cima, and R. S. Langer. Microchip Technology in Drug Delivery. *Annals of Medicine*, 32:377–379, 2000.
148. W. H. W. Tuttlebee. Software-Defined Radio: Facets of a Developing Technology. *IEEE Personal Communications*, 6(2):38–44, 1999.
149. C. T. Vogelson. Advances in drug delivery systems. http://www.drugdel.com/ddsci.htm. Accessed on 25 February 2011.
150. C. T. Vogelson. Advances in Drug Delivery Systems. *ACS Modern Drug Discovery*, 4(4):49–50, 2001.
151. C.-S. Wang. Design of a 32-Channel EEG System for Brain Control Interface Applications. *Journal of Biomedicine and Biotechnology*, 2012(Article ID 274939):10 pages, 2012.
152. D. B. Williams and C. B. Carter. *Transmission Electron Microscopy*. Springer, 1st edition, 2004.
153. T. V. Wilson. How the iPhone Works. http://electronics.howstuffworks.com/. Accessed on 13 February 2011.
154. G. Wolbring. Nanoscale drug delivery systems. http://www.innovationwatch.com/choiceisyours/choiceisyours-2007-12-15.htm. Accessed on 14 November 2012.
155. C. Ying. A Verification Development Platform for uhf rfid Reader. In *International Conference on Communications and Mobile Computing*, pages 358–361, 2009.
156. G. Youngblood. A Software Defined Radio for the Masses, Part 1. http://www.flex-radio.com/Data/Doc/qex1.pdf. Accessed on 18 May 2011.
157. L. L. Zhang et al. A Scheduler ASIC for a Programmable Packet Switch. *IEEE Micro*, 20(1):42–48, 2000.
158. T. Zhang, K. Chakrabarty, and R. B. Fair. *Microelectrofluidic Systems: Modeling and Simulation*. CRC Press, Boca Raton, FL, 2002.
159. W. Zhuang and J. Tranquilla. Digital Baseband Processor for the GPS Receiver Modeling and Simulations. *IEEE Transactions on Aerospace and Electronic Systems*, 29(4):1343–1349, 1993.

CHAPTER 3
Nanoelectronics Issues in Design for Excellence

3.1 Introduction

One nanometer is equal to one billionth of a meter or, simply put, 10^{-9}. To comprehend how small this dimension is, even the width of a human hair is approximately 80,000 nm (i.e., 80 µm) [10]. Generally, nanotechnology deals with structure sizes less than 100 nm. The study of nanotechnology encompasses nanoscale science, engineering, and technology. It involves design, imaging, measuring, modeling, manipulating, simulation, and characterization of nanoscale matters and devices as well as nanotechnology-based circuits and system. Nanotechnology can be electrical or nonelectrical in nature (e.g., nano-electro-mechanical systems, NEMS). The electronics at nanoscale dimensions (called nanoelectronics) are the focus of this book. In nanoelectronics, devices, circuits, and systems can be designed and fabricated. The nanoelectronic devices may include nanoscale CMOS or nano-CMOS field-effect transistor (FET; i.e., classical bulk CMOS FET), tri-gate FET (TGFET), graphene FET (GFET), carbon nanotube FET (CNTFET), etc. Selected key issues faced by nanoelectronics design engineers will be discussed in detail in this chapter. To give a strong understanding of design issues and challenges in nanoelectronics-based circuits and systems, in particular process variation, this chapter includes discussions on various devices. In addition, fabrication processes are also presented.

3.2 Design for eXcellence

While technology scaling facilitates the use of smaller devices, it also allows the packing of more of those small devices in the same die area, thus effectively reducing the cost of computation. The following are the advantages of the technology scaling [5, 207]:

1. Technology scaling increases packing density of the devices in a die (or chip). As the metal-oxide semiconductor FET (MOSFET) size decreases, more of such devices can be packed in the same die area.

2. The current drive that is manifested by the transconductance (g_m) enhances due to technology scaling. This will be evident from the following discussion. In general, transconductance is defined as follows for the saturation region:

$$g_m = \left(\frac{\partial i_d}{\partial v_{gs}}\right) = \left(\frac{W}{L}\right)\mu\left(\frac{\varepsilon_{ox}}{T_{ox}}\right)(V_{gs} - V_{Th}) \qquad (3.1)$$

From the above expression, it is evident that the current drive of the transistor can increase in the following ways: (1) reducing the gate length, (2) reducing the oxide thickness, (3) increasing dielectric constant (by using high-κ dielectrics), and (4) decreasing the threshold voltage.

3. The technology scaling has reduced power dissipation per computation. This simply can be explained by the fact that smaller devices are being switched to perform operations and operating voltage has been reduced with scaling.

4. The technology scaling results in smaller capacitances.
5. The technology scaling improves the frequency response.
6. The technology scaling improves performance due to combined effect of high current drive, smaller capacitance, and reduced gate delay.

On the other hand, the disadvantages of technology scaling are as follows [5]:

1. *Short-Channel Effect:* Several types of short-channel effects (SCEs) or second-order effects may arise due to technology scaling [103]. In a MOSFET, the reverse-biased junction between the drain and body forms a depletion region. The length of this depletion region increases with the drain to body voltage and shortens the effective channel length, which is called "channel length modulation" (CLM). For short-channel MOSFET, the effect of "drain-induced barrier lowering" (DIBL) takes place that results in reduction of threshold voltage with increase in drain voltage. For short-channel MOSFET, the threshold voltage decreases with the decrease in channel length, whereas for longer devices, the threshold voltage is not affected by channel length.

2. *Gate Leakage Current:* Ultra-thin gate oxide has increased the electric field across the gate oxide [166]. This has resulted in the quantum mechanical tunneling of the electrons. The tunneling between substrate and gate can be either direct tunneling or Fowler-Nordheim tunneling. For short-channel and ultra-thin oxide transistors, Fowler-Nordheim tunneling is negligible and direct tunneling is dominating. This has significant effects on the current distribution of the device.

3. *Threshold Voltage Fluctuation:* Short-channel MOSFET suffers from threshold voltage fluctuation, which, in turn, has several effects on the performance of transistor. The dopant density fluctuation and distribution, gate length fluctuation, and oxide thickness variations lead to threshold voltage fluctuations. The threshold voltage fluctuation leads to cell-to-cell variations, reduction on the voltage margin, and subthreshold leakage.

4. *Polysilicon Gate Depletion:* This happens due to increases in the electric field across the gate and due to boron penetration from P^+ polysilicon-gate into silicon substrate [134, 247]. This may cause deactivation of dopants near polysilicon-SiO_2 interface and lessen charge inversion. These have negative effects on the I-V characteristics due to the shifting threshold voltage and the reduction in the current driving capability of the device.

5. *Junction Capacitance:* The junction capacitance increases due to higher doping density and abrupt junction. This may result in an increase in gate delay.

6. *Mobility Degradation:* The mobility degradation can happen due to high electric field across the channel, increase in channel doping, and boron penetration from polysilicon gate. The carrier mobility degradation reduces the current drive of the MOSFET.

7. *Junction Leakage:* The junction leakage increases in nanoscale MOSFET due to the use of shallow junction and silicide metallization. The junction leakage results in increase in standby power dissipation. It may be noted that the use of shallow junction is a trade-off between junction leakage and gate sheet resistance.

8. *Source and Drain Resistance:* Use of shallow junction increases the source and drain resistance of nanoscale MOSFETs. This reduces the current drive of the MOSFET.

9. *Gate Sheet Resistance:* Narrow gate length in nanoscale MOSFET results in an increase in the gate sheet resistance. This results in the gate voltage drop and may affect the yield and current drive of the MOSFET.

With the technology scaling, the complexity of nanoscale circuits and systems has increased exponentially. The tasks of design engineers have intensified due to the emergence of additional challenges in the nanoelectronics. There is an ever-increasing gap between the chip technology and engineers' skills. To make the situation worse for such highly complex systems, the time-to-market has been reduced significantly. It is essential to accommodate the nanoelectronics challenges during the design flow to reduce design errors so as to improve yield and make electronics affordable.

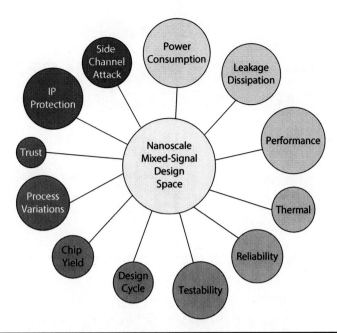

Figure 3.1 Selected important challenges in mixed-signal circuit and system design.

A selected set of nanoelectronics challenges is presented in Fig. 3.1 [154]. These challenges will be discussed in the rest of this chapter.

Design for eXcellence (DFX) is the new design paradigm that collectively accommodates the nanoelectronics challenges in the design flow [8, 9, 49, 94, 179]. DFX ensures timely production of nanoelectronics chips with highest possible yield. There are two views on the meaning of "X" in DFX. In "Design for X," the "X" stands for many terms including assembly, cost, environment, fabrication, maintainability, manufacturability, obsolescence, procurement, reliability, safety, serviceability, test, usability, and yield. For example, with manufacturability, it is design for manufacturability (DFM). However, the "Design for eXcellence" suggests the use of possible "Design for" approaches to achieve excellence—for example, an organizationwide culture, a concurrent engineering-based methodology, and a continuous improvement tool [9].

3.3 Different Types of Nanoelectronic Devices

A variety of nanoelectronics devices is being explored as an alternative for classical bulk CMOS technology (i.e., with Si/SiO$_2$/polysilicon structure). Selected different types of nanoelectronics devices are presented in Fig. 3.2. A select few of them have been discussed in the rest of this section. They have been grouped into three categories: A nanoscale CMOS, post-nanoscale CMOS, and other nanoelectronic devices. Nanoscale MOSFET are classical Si-body/SiO$_2$/polysilicon structures of dimensions less than 100 nm. From sub-45-nm CMOS onward, many semiconductor houses have migrated to Si-body/high-κ/metal structures. Post-nano-CMOS are coexisting technologically with nano-CMOS. Several alternative structures, including multiple independent-gate FET (MIGFET), CNTFET, and graphene nanoribbon FET (GNR-FET), integrate well with nano-CMOS technology. Other nanoelectronic devices, including single-electron transistor (SET) and vibrating body transistor (VBT), have been explored to meet the technology scaling, ultra-lower-power operation, and high-density integration.

At the same time, nanostructure-based devices such as quantum cellular automata (QCA) have been explored for future computing [2, 123, 186, 257]. In QCA, the information is processed by coulombic interactions rather than current flow or voltage difference as is the case of CMOS FET. The quantum dots in a QCA respond to the charge state of their neighbours [123]. The quantum-dot nanostructures can be created from semiconductor materials such as GaAs and InAs. Metal tunnel junctions QCA with single electron properties and molecular QCA with a single molecule as the computing element are other forms of QCA fabrication.

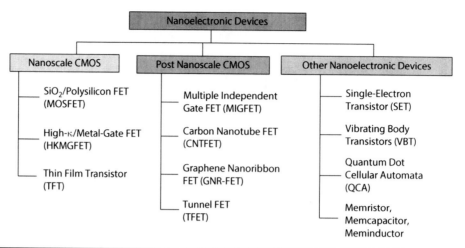

FIGURE 3.2 Selected nanoelectronic and nanoscale devices.

An entirely different application scope that uses MOSFET (Power MOSFET) is not included in this book. A Power MOSFET is a specific type of MOSFET built to handle significant power levels [33, 185]. It has the same operational principles as a standard MOSFET. The main advantages of power MOSFET include high computation speed and good efficiency at low voltages as compared to the other power semiconductor devices such as Thyristor and bipolar junction transistor (BJT). The power MOSFET is the most widely used as a switch for low voltage such as less than 200 V.

3.3.1 Nanoscale Classical SiO₂/Polysilicon FET

Classical MOSFET are present in all modern-day microprocessors and memories. These are also known as insulated-gate FET (IGFET) or metal insulator semiconductor FET (MISFET). IGFET is due to the fact that there is no gate current. They already have been scaled up to a feature size of 32 nm and are still going strong to meet the high-density integration and low-power computation. It is estimated that 340 billion transistors were manufactured in 2006 and 1 billion transistors per person in 2010. The MOSFET is a four-terminal device, which is classified as N-type MOSFET (NMOS) or P-type MOSFET (PMOS) based on the doping type of substrate (or well). The basic structure of an NMOS is shown in Fig. 3.3(a). This classical SiO₂/polysilicon FET (a.k.a. MOSFET or MOS) is shown with typical dimensions for a 45-nm technology node, such as, $L_{eff} \approx 45$ nm, $T_{ox} \approx 1$ nm, $V_{Th} \approx 0.4$ V, $V_{DD} \approx 1.0$ V. The NMOS consists of a P-type semiconductor substrate with two highly doped N-type regions, called source and drain. The third terminal is a polysilicon plate called gate separated from the substrate through an oxide layer. The fourth terminal forms an ohmic contact with the substrate and is kept at the reference voltage. Generally, the materials used for gate are heavily doped polysilicon, the oxide layer is thermally grown silicon dioxide (SiO_2), and the substrate is of P-type silicon [252]. Silicon is doped with arsenic and phosphorous for negative charge carriers; when doped with boron, silicon has excess positive charge carriers. Hence, the NMOS structure has two PN junctions connected back to back. A PMOS has dual structure as a NMOS with N-type materials swapped with P-type material and vice versa as shown in Fig. 3.3(b).

The MOSFETs can be classified as enhancement type or depletion type depending on the type of the inversion layer [283]. The enhancement-mode MOSFETs are used as switches and the depletion-mode MOSFETs are used as resistors [283]. An enhancement-type NMOS is always in "OFF" state and is turned "ON" when a positive voltage is applied between the gate and the source, V_{gs}. On applying V_{gs}, an electric field is created across the oxide layer. This electric field causes the positive charges in the P-region to repel and attracts the electrons from the base inducing an n-channel in its place. In this scenario, the MOS substrate gets inverted, forming a channel between the two N⁺ regions. The channel connects the source and the drain through which majority carriers or current can flow. The conduction of current then takes place among the N⁺ (drain), enhanced n-channel, and the N⁺ (source). The conduction can be controlled by changing the size of the induced n-channel by increasing or decreasing the V_{gs}. In NMOS, the current flows only from drain to source, whereas it flows from source to drain in PMOS transistors. The depletion-type MOSFET is identical to the enhancement-type MOSFET with an

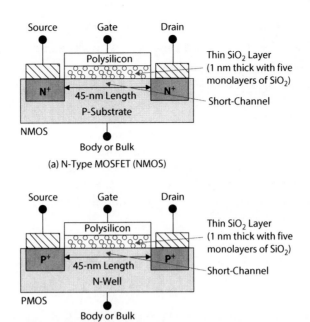

FIGURE 3.3 Classical SiO$_2$/polysilicon FET (a.k.a. MOSFET) with typical dimensions for a 45-nm technology node.

exception that the channel is physically implanted. Therefore, the depletion MOSFET is always ON even when V_{gs} is zero. When V_{gs} is positive, the channel is further enhanced and is depleted when V_{gs} is negative. The negative value of V_{gs} at which the channel is completely depleted is the threshold voltage, V_{Th}, for a depletion NMOS device. The polarity of V_{Th} is positive for enhancement-type NMOS transistors, but negative for depletion-type NMOS transistors and vice versa for the PMOS transistors.

The current between source and drain is represented by the current and voltage relation (or I-V characteristics) of a MOSFET. During design and simulation of the nano-CMOS circuits, the I-V characteristics are modeled in the compact models and other physics-based models. The compact models or SPICE models that may be empirical in nature are used to simulate large circuits during analog simulations through numerical routines [170]. Ideally, the cutoff state has zero current and saturation state has maximum I_{DS} current. However, in a typical scenario, weak, moderate, and strong inversion happens. In weak inversion, which is the OFF state, the current flow may happen due to diffusion. In the strong inversion case, the device has fully ON saturation state, and the current is due to the drifting. In the moderate inversion case, the device has ON linear state, and both drifting and diffusion mechanisms may contribute to the current flow. The current and voltage relation (or I-V characteristics) during the ON-state linear region of a traditional MOSFET is expressed as follows [205]:

$$I_{ds} = \mu_{eff}\left(\frac{\varepsilon_{ox}}{T_{ox}}\right)\left(\frac{W}{L_{eff}}\right)\left((V_{gs} - V_{Th})V_{ds} - \left(\frac{V_{ds}^2}{2}\right)\right) \tag{3.2}$$

where μ_{eff} is effective carrier mobility, ε_{ox} is dielectric constant of a gate dielectric, T_{ox} is gate oxide thickness, W is channel width and L_{eff} is effective channel length. The threshold voltage (V_{Th}) that is a process parameter is expressed as follows:

$$V_{Th} = V_{fb} + 2\phi_F + \left(\frac{T_{ox}}{\varepsilon_{ox}}\right)\sqrt{2q\varepsilon_{Si}N_{sub}(2\phi_F + V_{bs})} \tag{3.3}$$

$$= V_{fb} + 2\phi_F + \left(\frac{T_{ox}}{\varepsilon_{ox}}\right)\sqrt{2q\varepsilon_{Si}N_{sub}}\sqrt{(2\phi_F + V_{bs})} \tag{3.4}$$

$$= V_{fb} + 2\phi_F + \gamma_{body}\sqrt{(2\phi_F + V_{bs})} \tag{3.5}$$

where V_{fb} is the flat-band voltage, V_{bs} is the body bias, γ_{body} is the body effect coefficient that is calculated as $\left(\frac{\sqrt{2q\varepsilon_{Si}N_{sub}}}{\varepsilon_{ox}/T_{ox}}\right)$, ϕ_F is the Fermi-level, q is the electronic charge, ε_{Si} is permittivity of silicon, and N_{sub} is substrate doping concentration.

CMOS technology using PMOS and NMOS FETs has become popular due to low-power dissipation, smaller feature size, and high packing density. The transistor-feature sizes have shrunk dramatically with the technology scaling. The channel length of a device is dictated by the technology node that a particular process follows. The length of a device is lower bound by the technology node. A number of adverse effects peculiar to scaling that result in a significantly reduced channel length are identified as SCEs. The nonideal effects such as velocity saturation, mobility degradation in short channel transistors affect the current flow in it. For example, alpha power law [41, 205] models the current more accurately. For the linear region with the velocity saturation index α, the alpha-power law is expressed as follows [205]:

$$I_{ds} = \left(\frac{I_{Dsat}}{V_{Dsat}}\right)\left(\frac{V_{gs} - V_{Th}}{V_{DD} - V_{Th}}\right)^{\left(\frac{\alpha}{2}\right)} V_{ds} \tag{3.6}$$

For short channels, the drain current varies linearly with gate voltage V_{gs}. On the other hand, for long-channel MOSFET, the drain current varies quadratically with gate voltage V_{gs}.

At the nano-CMOS technology regime, the channel is much shorter than the conventional channel length, and the device characteristics are very much affected by these shorter channel lengths. The most important parameter in scaling is the drastic change in the gate-oxide thickness. Aggressive scaling of gate-oxide thickness has been going on for the past decade for the applications needing low-power, high-performance CMOS FETs [21]. For example, gate-oxide with a physical thickness of 1.2 nm was implemented in the 90-nm CMOS technology node. The typical dimension of 12 to 16Å is a few monolayers of SiO_2 in the nano-CMOS devices. The shortening of the minimum channel length has resulted in the reduction of power supply voltage to the 1 to 0.7 V range. The mixed-signal design trend also pushes analog circuits to be integrated with digital circuits [71]. To keep up with the scaling of the minimum channel length and mixed-signal trend, analog circuits need to be operated at low voltages. However, the minimum supply voltage for analog circuits predicted in the semiconductor road map [3] does not follow the reduction of digital supply voltage. Analog supply voltages between 1.8 and 2.5 V are still being used with channel lengths of 0.18 and 0.13 µm [17].

3.3.2 High-κ and Metal-Gate Nonclassical FET

The scaling of the dimensions has reduced the thickness of the SiO_2 layer to 1.2 nm or its physical limit. This gate-oxide thickness is just five atomic layers of silicon compound crystal [39]. The thin oxide layer induces gate-oxide leakage currents due to quantum effects. For SiO_2 of 1.2 nm thick, the gate-oxide leakage current density reaches 100 A/cm² at 1 V. For a 1 V V_{gs} and 1 nm gate-oxide thickness, the electric field across the oxide E_{gs} is $\left(\frac{1V}{1 nm}\right)$, i.e., 1 GV/m. The alternate gate oxides have been investigated to provide more insulation for the same thickness. Aluminum oxide (Al_2O_3), tantalum pentoxide (Ta_2O_5), titanium dioxide (TiO_2), hafnium dioxide (HfO_2), and zirconium dioxide (ZrO_2) were some of the promising candidates for the replacement. The high-κ/metal-gate FET (HKMGFET) with typical dimensions for a 32-nm technology node, e.g., $L_{eff} \approx 32$ nm, $T_{ox} \approx 5$ nm, $V_{Th} \approx 0.4$ V, $V_{DD} \approx 0.7$ V, is shown in Fig. 3.4 [100, 116, 161].

Aggressive scaling of nano-CMOS devices has led to a drastic change in the leakage components of the device, both in its active and passive states. Consequently, gate-oxide leakage has emerged as the most prominent form of leakage in a nano-CMOS device, particularly in the 65 nm and below regime. It has consequently become a necessity to obtain suitable alternatives for SiO_2 as the gate dielectric [19, 100, 116, 256]. Sustaining Moore's law depends crucially on the gate insulator scaling; consequently, replacing SiO_2 with high-κ dielectric metal-gate stack is inevitable [22]. This has led to the construction of nonclassical transistors demonstrated in Fig. 3.4. The use of high-κ serves the dual purpose of scaling of the CMOS device and reducing of gate-oxide leakage [39].

The CMOS fabrication process technology using high-κ dielectrics as well as their compact modeling is not widely available. In the course of research, several materials have been investigated for

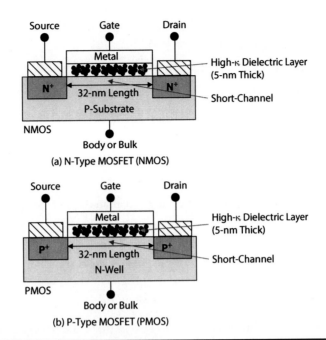

FIGURE 3.4 HKMGFET with typical dimensions for a 32-nm technology node.

use in nano-CMOS technology, such as Al_2O_3, BST, HfO_2, SiON, Si_3N_4, TiO_2, and ZrO_2 [19, 20, 100, 200]. Selected dielectrics are listed in Table 3.1. It has been a difficult task to integrate these materials into the conventional CMOS process [269]. Progress has been made in the development of various technologies for high-κ gate dielectric deposition [18]. This includes the extension of chemical vapor deposition (CVD), jet vapor deposition (JVD), liquid source misted CVD (LSCVD), physical vapor deposition (PVD), rapid thermal CVD (RTCVD), rapid plasma-enhanced CVD (RPECVD), oxidation of metallic films, and molecular beam epitaxy [112, 121, 140, 189]. Only selected semiconductor houses have access to such process technology, and still it is not a mainstream process technology [39].

In the context of high-κ dielectrics, finding the structure of gate has been a challenging task [39]. Polysilicon has been a legacy of the last several decades; however, it is degenerately doped

Dielectric Names	Dielectric Constant or Relative Permittivity
Aluminium Oxide (Al_2O_3)	9.3–11.5
Barium Strontium Titanate (BST or $Ba_{1-x}Sr_xTiO_3$)	100–185
Hafnium Oxide (HfO_2)	25
Hafnium Silicate ($HfSiO_4$)	15–18
Lanthanum Oxide (La_2O_3)	30
Silicon Dioxide (SiO_2)	3.9
Silicon Nitride (Si_3N_4)	10
Silicon Oxynitride (SiON)	5–34
Strontium Titanate ($SrTiO_3$)	2000
Tantalum Pentoxide (Ta_2O_5)	22
Titanium Dioxide (TiO_2)	86–173
Yttrium Oxide (Y_2O_3)	15
Zirconium Dioxide (ZrO_2)	25
Zirconium Silicate ($ZrSiO_4$)	10–15

TABLE 3.1 Selected High-κ Dielectrics with Their Permittivity [113, 116, 146, 200]

polycrystalline silicon and is not the best metal [200, 252]. In the case of polysilicon and high-κ structure, following issues may arise [39]:

1. Charge trapping due to uneven dielectric surface.
2. Scattering of electrons in channel by the phonons.
3. Poor bonding between dielectric and polysilicon makes turning ON of a MOSFET difficult.

The use of thicker high-κ gate oxide along with a metal gate resolved these problems in MOSFET, and hence the structure of HKMGFET shown in Fig. 3.4. The advantages of metal gate include the following [5]:

1. The gate resistance reduces even better than the silicide gate.
2. The metal gate results in suppressing the gate depletion, and hence removes the electrical scaling limit of polysilicon-gated dielectric.
3. The metal gate is more suitable for high-κ dielectrics. Poly gate is not suitable for most high-κ dielectrics.
4. The metal gate stops boron penetration.

However, metal gate may lead to an increase in threshold voltage and a reliability issue as metal ion may also degrade dielectric.

For this new structure, it is necessary to study the effect of high-κ dielectrics on other device parameters, including the threshold voltage, the gate capacitance, the carrier saturation velocity and mobility, the gate polysilicon depletion depth, and current. For proportionate dimension scaling and to ensure that the electrical properties are not affected, an equivalent oxide thickness (EOT) can be calculated as follows [200]:

$$T_{ox} = \left(\frac{3.9}{\kappa}\right) T_\kappa \quad (3.7)$$

where 3.9 is the static dielectric constant of SiO_2 and T_κ is the thickness of high-κ gate dielectric in a MOSFET. This EOT can then be used in the expressions such as Eqn. 3.2 to calculate I-V characteristic of HKMGFET [171]. However, a much more accurate way is to get the HKMGFET SPICE models from a semiconductor house.

3.3.3 Multiple Independent Gate FET

Both the classical MOSFET or HKMGFET discussed in the previous sections are planar devices with a single gate. The leakage components of these single-gated transistors are still higher in OFF state even though high-κ reduces the gate leakage. In a multiple-gate FET (MuGFET), the channel is surrounded by multiple gates that are used to reduce OFF-state leakage of the device [199, 232]. The double-gate transistor is made either in a planar fashion or nonplanar fashion (which is called FinFET). On the other hand, a TGFET is a nonplanar 3D device. The multiple gates in the transistor can be controlled either by a single-gate electrode or by gate electrodes. In the first case, the multiple gate surfaces act electrically as a single gate. In the second case, the device is called a multiple independent gate FET (MIGFET). The following are the advantages of the MIGFET/MuGFET devices [221, 232, 267]:

1. Effective reduction of OFF-state leakage as the channel is surrounded by multiple gate surfaces.
2. Increase of ON-state drive current.
3. Lower power dissipation.
4. Increased device performance.
5. Higher transistor density in a chip as a nonplanar MIGFET/MuGFET is more compact than conventional planar transistors.

MIGFET can provide specific advantages in analog and digital circuits due to the above explained points.

Nanoelectronics Issues in Design for Excellence 73

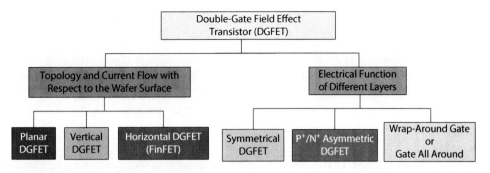

Figure 3.5 Three types of DGFETs [266, 267].

3.3.3.1 Double-Gate FET

The double-gate FET (DGFET) can be of various types as presented in Fig. 3.5 [70, 232, 266, 267]. Depending on the structure or topology of the DGFET and correspondingly the direction of current flow, the DGFET is of three types. Moreover, DGFET can be of three types, based on the electrical function of the different layers present, their control of the channel dimensionality, and electric field. The second set of three types can be used as any of the first set of three types and hence in effect nine types of DGFETs can be conceptualized [267]. The DGFET can be with either symmetric gates or asymmetric gates. For two gates, asymmetric DGFET has the same oxide thickness, the same voltage application, and the same work function.

The DGFET is of the following types based on their structure [70, 87, 118, 232]:

1. *Planar Double-Gate FET*: The planar DGFET has one gate at the top and another gate at the bottom. The gate and channel are horizontal in this DGFET. The planar DGFET is similar to classical planar MOSFET, however, the bottom gate layer is difficult to fabricate [118].

2. *Vertical Double-Gate FET*: The two gates are placed vertical in this DGFET. The current flows perpendicular to the silicon wafer surface in this DGFET. Vertical DGFET needs difficult lithography for sub-50-nm nodes as the channel region is very narrow. In addition, the vertical DGFET has drain contact problems [118].

3. *Horizontal Double-Gate FET (FinFET)*: The channel and gate are perpendicular to the surface in this DGFET. The current flows parallel to wafer surface in this DGFET. This type of DGFET is called FinFET due to its look. The horizontal DGFET or FinFET is relatively easy to fabricate and standard processes can be used [118].

The structures are shown in Fig. 3.6 for a visual perspective. An easier way to perceive the structure is as follows. Let us start from a planar DGFET shown in Fig. 3.6(a). Let us assume that the two gates are a set of structure and source/drain are another structure. If both gate and source/drain structures are rotated 90° counter clockwise, then the vertical DGFET structure is obtained (Fig. 3.6(b)). Starting from a vertical DGFET, if the source/drain structure is rotated only 90° inward, then the horizontal DGFET or FinFET is obtained (Fig. 3.5(c)).

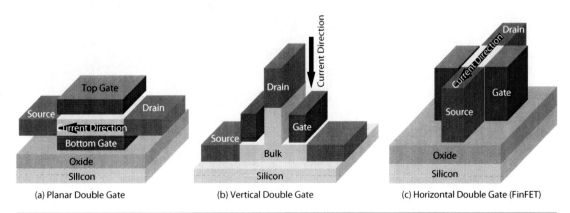

(a) Planar Double Gate   (b) Vertical Double Gate   (c) Horizontal Double Gate (FinFET)

Figure 3.6 Types of DGFETs [70, 118, 267].

In the second type of classification, on the basis of electrical function of the different layers, the following are the type of DGFETs [267]:

1. *Symmetrical DGFET*: The symmetrical DGFET is the ideal DG structure. It uses separate control for the individual gates. It has high channel mobility and it does not exhibit dopant fluctuation. The disadvantages of this DGFET include larger area due to additional gate contact, channel thickness limitation due to quantum effects, and inflexibility in threshold voltage control.

2. *P+/N+ Asymmetric DGFET*: In this case, top and bottom gates are made of P+ and N+ materials, respectively. The P+ and N+ gates are similar to that of the standard bulk CMOS transistors. The channel in this DGFET is better confined than for an equivalent symmetrical DGFET. In this DGFET, silicon thickness requirement is not a problem due to better electron confinement. The field distribution across the channel is, however, asymmetrical that is similar to a bulk CMOS FET. The mobility in asymmetric DGFET is lower than that of symmetrical DGFET.

3. *Wrap-Around Gate or Gate All Around DGFET*: In a wrap-around gate structure in a DGFET, the gate is wrapped around a silicon beam. It is also called gate all around structure. This provides better control than the ideal DGFET. This kind of DGFET has variable device width and can have very compact structure. The disadvantages are that they have small current-carrying capability and are difficult to fabricate.

The planar DGFET is discussed in the rest of this subsection. The FinFET will be discussed in the next subsection.

The cross-section of DGFET is shown in Fig. 3.7 [266, 267]. The concept of an ideal DGFET is shown in Fig. 3.7(a). The ideal DGFET has a uniform thin silicon channel as compared to the gate length. Dimensionally, $L_{gate} \approx 4 \times T_{si}$ with channel thickness T_{si}. It has a thick source/drain fan-out

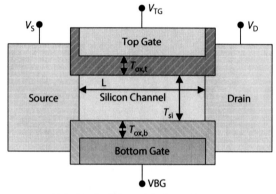

(a) An ideal DGFET with independent gates.

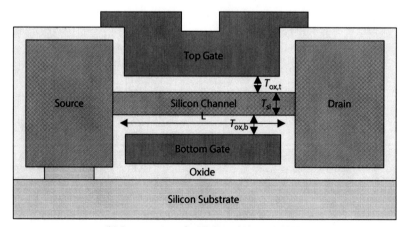

(b) Cross-section of a SOI-based Planar DGFET.

Figure 3.7 Cross-section of an SOI DGFET with typical geometry marked [266, 267].

structure to provide reduced series resistance. The top and bottom gates are perfectly aligned to each other as well as to source/drain fan-out to provide high speed and driving current. Additionally, a misalignment may result in additional source/drain overlap capacitances with the gate and reduced driving current. A fabricated planar DGFET with self-aligned top and bottom gates that is fabricated from an ultra-thin silicon-on-insulator (SOI) process is presented in Fig. 3.7(b).

In the DGFET, the upper and lower gates control the channel region and the ultra-thin body acts as a rectangular quantum well. The channel that is lightly doped in a DGFET results in a negligible depletion charge. In a most common mode of operation of a DGFET, both the gates switch simultaneously. In another mode, only front gate is switched and back gate is set to a bias voltage. Separate control of each gate allows dynamic control of threshold voltage in the DGFET.

For a symmetric DGFET in which the same voltage is applied to the two gates with the same work function, the drain source current is modeled as follows [80]:

$$I_{ds} = \left(\frac{2\mu W_{gate} C_{ox}}{L}\right)\left(\frac{kT}{q}\right)^2\left[\left(\frac{q_{is}^2 - q_{id}^2}{2}\right) + (q_{is} - q_{id})\right] \tag{3.8}$$

where W_{gate} is the gate width. The q_{is} and q_{id} are inversion charge densities calculated at the source and drain, respectively. A factor 2 is included in the equation as essentially due to double-gate control. By relaxing the assumption of constant mobility in the above model, a more accurate model for I_{ds} is presented as follows [13]:

$$I_{ds} = \frac{\left[2\mu_o W_{gate} C_{ox}\left(\frac{kT}{q}\right)^2\left\{\left(\frac{q_{is}^2 - q_{id,eff}^2}{2}\right) + (q_{is} - q_{id,eff})\right\}\right]}{\left[L + \left(\frac{\mu_o}{2v_{sat}}\right)\left(\frac{kT}{q}\right)(|q_{is} - q_{id,eff}|)\right]} \tag{3.9}$$

where $q_{id,eff}$ is the effective drain inversion charge density. μ_0 is the low field mobility.

3.3.3.2 Double-Gate FinFET

The term "FinFET" has been used to describe a nonplanar, double-gate transistor built on an SOI substrate [66, 81, 86, 132, 136, 204, 275]. A distinct characteristic of the FinFET is that the channel is wrapped by a thin silicon "Fin," which forms the body of the device. The thickness of the Fin is measured in the direction from source to drain that determines the effective channel length of the device. The structure of a double-gate FinFET is shown in Fig. 3.8 [240]. There are two different forms of FinFET: one where two gates are independent (Fig. 3.8(a)) and another where two gates are tied or connected (Fig. 3.8(b)). In the latter case, the two gates are essentially controlled by one voltage. The cross-section of a FinFET with a dimension for 32-nm node is shown in 3.8(c) [240].

Typically, an SOI process is used to fabricate FinFET. The SOI process ensures that the ultra-thin specifications are achieved for the device regions. In SOI-based FinFET, the thin silicon body and an insulating buried oxide (BOX) help to reduce the leakage. The scaling of BOX thickness can help to reduce SCEs of the device. The heavily doped back gate in the DGFET serves as a ground plane and effectively reduces the electrostatic coupling present in the drain to the channel through the BOX. The planar DGFET employs conventional planar manufacturing processes to create the two gates. The drain-source channel is sandwiched between two independently fabricated gate and gate-oxide stacks. The key challenge in fabricating such FinFET is achieving proper self-alignment between the upper and lower gates. In the FinFET, the electrical potential throughout the channel is controlled by the gate voltage. The advantages of the FinFET over its bulk-Si counterpart are the following [66, 204]:

1. The FinFET due to its structure provides better area efficiency than MOSFET.
2. The FinFET can have enhanced mobility of the carriers when used in conjunction with the strained Si process.
3. The FinFET intrinsic structure and characteristics minimize the SCEs.

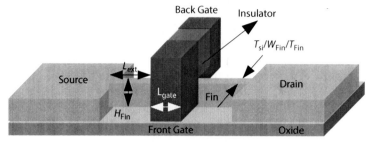
(a) DG FinFET with independent gates.

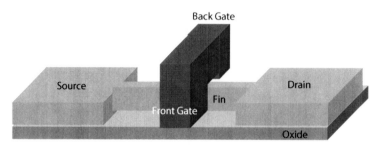
(b) DG FinFET with unified gates.

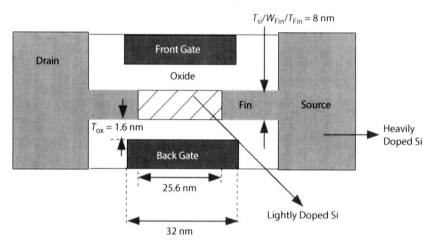
(c) Cross-section of a DG FinFET with example dimensions for a 32-nm node.

Figure 3.8 Structure of a DG FinFET [240].

In a FinFET, the Fin itself acts as a channel and it terminates on both sides of source and drain. In a FinFET, the unique position of gates over the Si-Fin gives efficient gate-controlled characteristics compared to traditional MOSFET [66, 204]. As the gates extend over the Fin, the effective length of the channel is same as that of the width of the Fin. The width of the channel is equivalent to twice the height of the Fin as there are two gates around the Fin. The height of the FinFET is equivalent to the width of the MOSFET. The geometrical or physical dimensions of the FinFET are expressed as follows [76, 181]:

$$\text{Geometrical Channel Length } L_{\text{phy}} = L_{\text{gate}} + 2 L_{\text{ext}} \tag{3.10}$$

$$\text{Geometrical Channel Width } W_{\text{phy}} = T_{\text{Fin}} + 2 H_{\text{Fin}} \tag{3.11}$$

In the above expression L_{gate} is physical gate length. H_{Fin} is the height of the Fin. T_{Fin} or W_{Fin} is the width of the Fin. L_{ext} is the extension length. Typically, $H_{\text{Fin}} \gg T_{\text{Fin}}$ and the top gate-oxide thickness is way larger than the sidewall oxide thicknesses. The space between two Fins is called "Fin pitch" in a FinFET circuit. If W is the Fin pitch, then Fin height is maintained at $\left\lfloor \frac{W}{2} \right\rfloor$ for same area efficiency. However, typically the Fin height can be greater than $\left\lfloor \frac{W}{2} \right\rfloor$ for a Fin pitch of W; thus, the FinFET provides better area efficiency than a MOSFET.

In a FinFET, the current flow can happen due to the thermionic emission across the potential barrier of the channel or directly via tunneling through the barrier from source to drain [199]. It is observed that the gate control is effective for FinFET for a range of channel length and gate voltage. The drain to source is a function of gate voltage and band-to-band tunneling (BTBT). For longer gate lengths, the tunneling contribution is of the order of the current over the potential barrier. For shorter gate length, the current is increased by two orders of magnitude by tunneling. For very short gate lengths, there are still OFF currents of reasonable values.

For a DG FinFET with two gates connected together and switched simultaneously, the drain current can be described by the following expression [224, 239]:

$$I_{ds} \approx \mu \left(\frac{W}{L}\right)\left(\frac{4\varepsilon_{Si}}{T_{Si}}\right)\left(\frac{2kT}{q}\right)^2 \left[\left(\frac{q_{is} - q_{id}}{2}\right) + r\left(q_{is}^2 - q_{id}^2\right)\right] \qquad (3.12)$$

where r is the structural parameter of DGFET and is defined as follows [239]:

$$r = \left(\frac{\varepsilon_{Si} T_{ox}}{\varepsilon_{ox} T_{Si}}\right) \qquad (3.13)$$

For inversion charge density Q_i, the normalized charge sheet density is defined as follows [239]:

$$q_i = \frac{Q_i}{\left(8\varepsilon_{Si} kT/q T_{Si}\right)} \qquad (3.14)$$

This is in range of the normalized charge density at the drain (q_{id}) and source (q_{is}). In multiple-gate MOSFET compact model presented by BSIM-CMG 106.1.0, the drain to source current is calculated as follows [254]:

$$I_{ds} = I_{ds0} I_{DS0MULT} \mu_0 C_{ox} NFIN_{total} \left(\frac{W_{eff}}{L_{eff}}\right)\left(\frac{M_{oc} M_{ob} M_{nud}}{D_{mob} D_r D_{vsat}}\right) \qquad (3.15)$$

where $I_{ds0} I_{DS0MULT}$ is the multiplier to source-drain channel current. $NFIN_{total}$ is the total number of Fin per finger. D_{mob} is mobility. D_r is a degradation factor. D_{vsat} is the current degradation factor due to velocity saturation. M_{oc}, M_{ob}, and M_{nud} different modeling factors used in the model.

3.3.3.3 Triple-Gate FinFET or Tri-Gate FinFET

In general, FinFETs can be fabricated using three different technologies as shown in Fig. 3.9 [53, 65, 83, 274]. FinFET based on bulk CMOS is shown in Fig. 3.9(a). FinFET based on SOI is shown in Fig. 3.9(b). Nanowire FinFET is shown in Fig. 3.9(c). The bulk CMOS-based FinFET has excellent process compatibility, low wafer cost, and high heat transfer rate to substrate. In the SOI technology,

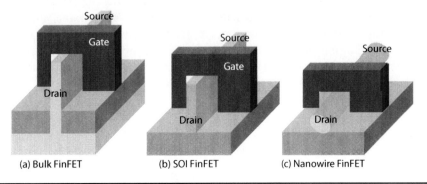

FIGURE 3.9 Three types of FinFET structures [53, 65, 83, 274].

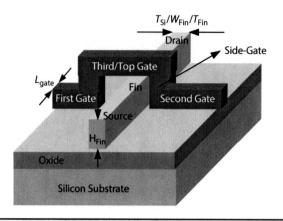

Figure 3.10 A TGFET [122].

an insulator isolates the bulk from the substrate [81]. SOI-based FinFET has less complex doping implementation, easier variability control, small footprint, and better technology-scaling. Therefore, SOI-based FinFET is getting more acceptability to sustain technology-scaling trends compared to the bulk FinFET. Nanowire FinFET provides the better gate control compared to the bulk and SOI FinFET as gate surrounds the silicon nanowire.

TGFET is another follower of a classical MOSFET. TGFET that is also called 3D transistor is being used as an effective nonplanar transistor architecture in the next-generation commercial chips. These transistors employ a single gate stacked on top of two vertical gates allowing for essentially three times the surface area for electrons to travel. It has been reported that TGFETs reduce leakage and consume far less power than predecessor transistors. This allows up to 37% higher speed, or a power consumption under 50% of the previous type of transistors [28, 45]. The structure of a TGFET is presented in Fig. 3.10 [122]. The geometric channel length and width can be calculated as it is performed in DG FinFET in the previous subsection.

The TGFET has very low DIBL. It has steep subthreshold slopes that show that TGFET has good gate control. Thus, TGFET can be deployed to reduce leakage, improve performance, enable lower operating voltage, and reduce active power [28, 199]. The lower threshold voltage (of the order 100 mV) when combined with the strain enhancement leads to a reasonable increase of drain saturation current. The lower threshold voltage significantly increases the drive current in both NMOS and PMOS.

The I-V characteristics of a triple-gate FinFET is expressed as the follows [63]:

$$I_{ds} = 2\mu_{eff} W \left(\frac{\varepsilon_{ox}}{T_{ox}}\right) \left(\frac{2kT}{q}\right)^2 \left[\frac{q_{is} - q_{id}}{L} + \frac{1}{2}\left(\frac{q_{is}^2 - q_{id}^2}{L_{eff}}\right)\right] \quad (3.16)$$

where the electrical gate length L_{eff} is the difference of the channel length L and depletion depth ΔL. This accounts the effects of CLM in the drain to source current. q_{is} and q_{id} are the normalized inversion charge densities calculated at the source and drain, respectively. q_{is} dominates the subthreshold region and q_{id} dominates the above threshold region of the transistor operation. The quantum-mechanical effects become significant for the silicon Fin width W_{Fin} smaller than 10 nm. The quantum-mechanical effects change the gate-oxide thickness and the new value is expressed as follows:

$$T_{ox,eff} = T_{ox} + \Delta z \left(\frac{\varepsilon_{ox}}{\varepsilon_{Si}}\right) \quad (3.17)$$

$$= T_{ox} + 1.2 \left(\frac{\varepsilon_{ox}}{\varepsilon_{Si}}\right) \quad (3.18)$$

In the above expression, Δz is the average distance inside the silicon from the interface. The quantum-mechanical effects also change the threshold voltage and the change in T_{ox} also affects the threshold voltage. The modified threshold voltage can thus be expressed as follows:

$$V_{Th,eff} = V_{Th} + \Delta V_{Th,QM} + \Delta V_{Th,T_{ox,eff}} \tag{3.19}$$

Accounting for the quantum-mechanical effects, the drain to source current can be expressed as follows [63]:

$$I_{ds} = 2\mu_{eff,QM} W \left(\frac{\varepsilon_{ox}}{T_{ox}}\right)\left(\frac{2kT}{q}\right)^2 \left[\frac{q_{is,QM} - q_{id,QM}}{L} + \frac{1}{2}\left(\frac{q_{is,QM}^2 - q_{id,QM}^2}{L_{eff}}\right)\right] \tag{3.20}$$

This is a more accurate model of the drain to source current in a tri-gate/triple-gate FinFET accounting for the SCEs, CLM, quantum-mechanical effects, mobility degradation, and series resistance.

3.3.4 Carbon Nanotube FET

Due to promising properties, the carbon nanotubes (CNTs) are important candidates for future electronic devices. The structure of a CNT is conceptually presented in Fig. 3.11 [164]. A single-wall CNT (SWCNT) is essentially a cylinder formed by the rolling of a single layer of graphite called a graphene layer (Fig. 3.11(a)). A multiwall CNT (MWCNT), as shown in Fig. 3.11(b), is a coaxial structure of cylinders of SWCNTs with the separation between the tubes equal to that of the separation between the graphite layers. CNTs have a well-defined direction along the axis of the nanotubes. Different types of CNTs may be produced by rolling the graphene sheet in various ways. The direction in which the graphene sheet is rolled is called chirality or chiral angle. CNTs may behave as metallic or semiconductor depending on the chiral angle of the nanotube. The chiral angle is the angle between the axis of its hexagonal pattern and the axis of the tube. The way the graphene sheet is wrapped to form a CNT is represented by a pair (n, m) with n and m being integers. The n and m denote the number of unit vectors along two directions in the honeycomb crystal lattice of graphene. If $n = m$, then the CNTs are called armchair CNTs (Fig. 3.12(a)), if $m = 0$, the CNTs are called zigzag CNTs (Fig. 3.12(b)), otherwise, the CNTs are called chiral CNTs (Fig. 3.12(c)). The unit vectors are shown as a_1 and a_2 in the figure. The diameter of an ideal CNT is calculated from its (n, m) indices as follows [59, 190]:

$$d = \left(\frac{a}{\pi}\right)\sqrt{(n^2 + nm + m^2)} \tag{3.21}$$

$$= \left(\frac{0.246}{\pi}\right)\sqrt{(n^2 + nm + m^2)} \; nm \tag{3.22}$$

$$= 0.0783\sqrt{(n^2 + nm + m^2)} \; nm \tag{3.23}$$

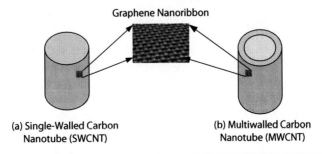

(a) Single-Walled Carbon Nanotube (SWCNT)
(b) Multiwalled Carbon Nanotube (MWCNT)

FIGURE 3.11 Structure of CNTs: single versus multiwall.

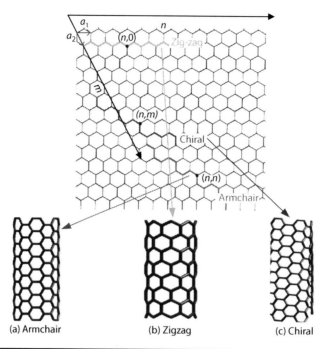

Figure 3.12 Structure of CNTs: armchair, zigzag, or chiral.

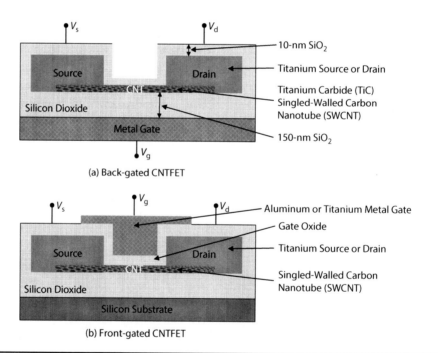

Figure 3.13 Cross-section of CNTFET [149, 263].

The SWCNT FET (CNTFET) is considered as one of the most promising candidates for enhancing functionality of silicon-based CMOS circuits and extending Moore's law. The cross-section of a CNTFET is shown in Fig. 3.13 [149, 263]. Depending on the location of the gate in its structure, CNTFET is of two types. The back-gated CNTFET is shown in Fig. 3.13(a) and the top-gate CNTFET is shown in Fig. 3.13(b). Back-gated CNTFET is the earlier one to be fabricated and later top-gated CNTFET is fabricated. Top-gated CNTFETs are generally preferred over back-gated CNTFETs, despite their more complex fabrication process. Top-gated CNTFET provides a larger electric field for a lower gate voltage due to the thin-gate dielectric.

Both diffusion and drift carrier transport mechanisms contribute to the current in a CNTFET, particularly, if CNT is longer than the magnitude of the optical phonon mean free path (i.e., 100 nm) [142]. A simplified drain to source current calculation formula for CNTFET is expressed as follows [141]:

$$I_{ds} = \beta(V_{gs} - V_{Th}) \left(\frac{V_{ds}}{R_c + \alpha + \left(qR_c \left(\frac{1}{\hbar\Omega_1} + \frac{1}{\hbar\Omega_2} \right) \right) V_{ds}} \right) \qquad (3.24)$$

In the above expression, β and α are curve fitting model parameter, R_c is the contact resistance, and Ω is the angular frequency due to phonon scattering. There are two types of scattering processes such as zone boundary and optical scattering in CNTFET. A more detailed expression for I-V characteristics follows [142, 226]:

$$I_{ds} = \gamma \left(\frac{\mu_{eff} C_{ox,front}}{L^2} \right) \left[f(\psi_{CNT,s}(L), V_{gs}) - f(\psi_{CNT,s}(0), V_{gs}) \right] (1 + \lambda V_{ds}) \qquad (3.25)$$

In the above expression, function f is of the following form:

$$f(\psi_{CNT,s}(x), V_{gs}) = \left(V_{gs} + V_{sb} - V_{fb} + \frac{kT}{q} \right) \psi_{CNT,s}(x) - \frac{1}{2} \psi_{CNT,s}^2(x) \qquad (3.26)$$

In the above expression, γ is the conversion factor (varying from 0 to 1) for CNT from graphite. μ_{eff} is the mobility that is calculated as $\mu_{eff} = \gamma \mu_{graphite}$, as each CNT (n, m) will have different values for the mobility. $C_{ox,\,front}$ is the C_{ox} for the front gate of the CNTFET. $\psi_{CNT,s}$ is the potential across the CNT. λ is the CLM parameter. Equation 3.26 is evaluated with $x = L$ and $x = 0$ and used in Eqn. 3.25 to determine the I_{ds} for CNTFET. Compact models of CNTFET are available for use in SPICE and perform both analog and digitial circuit designs [222].

3.3.5 Graphene FET

GFET is a device that uses a GNR for its channel. Graphene is the primary unit of the graphitic structure. Multilayered graphite is used as pencil lead and essentially graphene is a single layer of it. The individual carbon atoms in graphene are linked into a hexagonal lattice that is laid out in a flat sheet [184, 250]. The synthesis of the single-atom-thick crystallites of graphene from bulk graphite rewarded the inventor Andre Geim and Kostya Novoselov the Nobel Prize [69, 137, 184]. In terms of the structure, a graphene sheet is a two-dimensional array of carbon atoms. While three valence atoms of each carbon atom form the strong in-plane sigma bonds, the fourth valence atom lies above and below the plane of graphene, which overlaps with neighboring atoms to form pi bonds. Thus, the intrinsic graphene is a semi-metal or zero band gap semiconductor. The graphene nanoribbon exhibits either metallic or semiconducting properties depending on its structure. The electronic properties of the GNRs depend on the shape of the edge of nanoribbons, namely armchair and zigzag orientations, in the transport direction. The armchair GNR is similar to a zigzag CNT but has different geometry and quantum confinement boundary conditions in the transverse direction [177].

Graphene has extremely high electron mobility compared to silicon, which is typically used in semiconductors. The GNRs have higher mobilities than the CNT. This makes GFET more suitable than CNTFET and other CMOS transistors for high-speed applications. GFET is very suitable for fast- and low-voltage nanoelectronics. GFET has greatly reduced SCEs such as DIBL and subthreshold swing. The GFET can operate at high frequencies in the range of 300 GHz [128]. A few distinct properties of graphene are summarized as follows [69, 211]:

1. *Ambipolar Electric Field Effect:* Graphene has a significant ambipolar electric field effect so that charge carriers are tuned continuously between electrons and holes with concentration as high as $10^{13}/cm^2$. The graphene is a zero band gap material. Graphene exhibits a *linear energy*

spectrum of the following form: $\hbar v_F m_q$, where v_F is Fermi velocity and m_q quasi particle momentum.

2. *Very High Carrier Mobility:* The carrier mobility (μ) in graphene can exceed 15,000 cm² V⁻¹ s⁻¹ even under ambient conditions. The carrier mobility is weakly dependent on temperature in a graphene. Therefore, the mobility can be increased significantly up to 200,000 cm² V⁻¹ s⁻¹ [128, 216]. For a comparative perspective, in a typical silicon at room temperature, the electron mobility is 1400 cm² V⁻¹ s⁻¹ and the hole mobility is 450 cm² V⁻¹ s⁻¹. In silicon, the carrier mobility decreases as the temperature increases due to scattering (i.e., collision) of the carriers, thus reducing the drift velocity. This feature makes graphene particularly attractive for ultra-high-speed circuits.

3. *Room Temperature Quantum Hall Effect:* Graphene exhibits quantum Hall effect even at room temperature. This is also an indicator for very good electronic quality of the graphene even at room temperature. In general, the Hall effect refers to the voltage (called the Hall voltage) produced across an electrical conductor that is transverse (i.e., perpendicular) to an electric current in the conductor when a magnetic field is present in perpendicular to the current. In such a system, two resistances can be measured: the standard longitudinal resistance (in the direction of current) and the transverse resistance (i.e., Hall resistance) that is perpendicular to the direction of the current. The Hall resistance (R_{Hall}) is expressed as follows [75]:

$$R_{Hall} = \left(\frac{h}{ne^2}\right) \quad (3.27)$$

where n is an integer for integer quantum Hall effect and a fraction for fractional quantum Hall effect. Graphene can be a very good Hall effect sensor even at room temperature.

4. *Low $\left(\frac{I_{ON}}{I_{OFF}}\right)$ Ratio:* For digital circuits that essentially use transistors as switches, $\left(\frac{I_{ON}}{I_{OFF}}\right)$ ratio is a very important property. A large $\left(\frac{I_{ON}}{I_{OFF}}\right)$ ratio manifested by order of magnitude larger I_{ON} compared to I_{OFF} and sizable band gap is key to ON-OFF switching behavior needed for digital electronics [216]. However, graphene is a zero band gap material with typically low $\left(\frac{I_{ON}}{I_{OFF}}\right)$ ratio and is not well suited for digital circuits in such a situation.

The structure of the graphene layer can lead to different behaviors in the GFET. Different types of GFETs follow [214, 215, 216, 259, 260]:

1. *Monolayer GFET:* By basic definition, the graphene is a flat monolayer of carbon atoms tightly packed into a two-dimensional (2D) honeycomb lattice. The monolayer GFET has low $\left(\frac{I_{ON}}{I_{OFF}}\right)$ ratio. The monolayer GFET cannot be switched OFF and hence will have very high leakage compared to a silicon-CMOS transistor. Hence, gapless large-area monolayer GFET cannot be used in digital circuits [214]. The situation is different in a radio frequency (RF) circuit in which switching off the transistor is an absolute requirement for circuit operation.

2. *Bilayer GFET:* The bilayer GFET uses two layers of graphene [210]. Though the $\left(\frac{I_{ON}}{I_{OFF}}\right)$ ratio for a bilayer GFET is higher than that for the monolayer GFET, it is not as high as of the silicon FET to be used in digital electronics. The voltage gain is better than the monolayer GFET and hence is a better transistor than monolayer GFET for analog and RF circuits.

3. *GNR-FET:* Several approaches are explored to create a band gap in graphene as the large-area graphene has zero band gap [214]. The structural approach is to constrain one dimension of the large-area graphene, thus forming narrow GNRs. The graphene FET using GNR is called GNR-FET. The GNR-FET has the highest $\left(\frac{I_{ON}}{I_{OFF}}\right)$ ratio compared to the monolayer GFET and bilayer GFET. GNR-FET can therefore be used for digital logic gates. The channel geometry of a GNR-FET is defined by lithography that has better control over a CNTFET.

In addition, depending on the structure of the gate, single-gate and double-gate GFETs as well as GNR-FETs are being built [215].

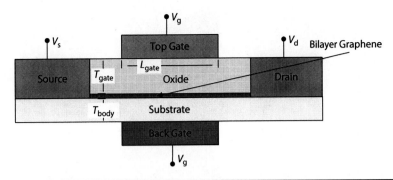

FIGURE 3.14 Cross-section of a double-gate bilayer GFET [251].

The cross-section of a double-gate bilayer GFET is shown in Fig. 3.14 [251]. The gate length L_{gate} is the active region of the channel. The top-gate dielectric is of thickness T_{gate}. The oxide material used is HfO_2 to prevent phonon and Coulomb scattering to preserve the carrier mobility. The I-V characteristics of a double-gate bilayer GFET can be expressed as follows [251]:

$$I_{ds} = \left(\frac{1}{4R_s}\right)\left[V_{ds} - E_c L_{gate} + I_0 R_s + \sqrt{(V_{ds} - E_c L_{gate} + I_0 R_s)^2 - 4I_0 R_s V_{ds}}\right] \quad (3.28)$$

In the above expression, E_c is the critical electric field. R_s is the series resistance. I_0 is calculated using the following expression:

$$I_0 = 2\left(\frac{W}{L_{gate}}\right)\mu E_c L_{gate} \left(\frac{C_e C_q}{C_e + C_q}\right)\left(V_{gs} - V_{Th} - \frac{V_{ds}}{2}\right) \quad (3.29)$$

where C_e is capacitance due to top-gate dielectric and C_q is the quantum capacitance. For a large-area double-gate GFET, the I-V characteristics can be expressed as follows [276]:

$$I_{ds} = \left(\frac{1}{2}\right)\left(\frac{W}{L}\right)q\mu_o n_{so} V_{dsat}\left(1 - exp\left(-\frac{2V_{ds}}{V_{dsat}}\right)\right) \quad (3.30)$$

In the above expression, n_{so} is the carrier density near the source and μ_o is the low-field carrier mobility.

The cross-section of the double-gate GNR-FET with the edge-terminated graphene sheet GNR as the channel structure is shown in Fig. 3.15 [177]. The double-gate GNR-FET has a GNR between the two gates [177]. As an example, the dimension of a GNR-FET is as follows: $L = 20$ nm, $T_{ox} = 2$ nm, and $\varepsilon_{ox} = 4$. The source and drain are directly connected to the GNR channel. The Schottky barrier (SB) height between the source/drain and the channel is half of the GNR bandgap. Wider ribbons will give lower band gap values, which will cause the device to suffer from BTBT and degrade the $\left(\frac{I_{ON}}{I_{OFF}}\right)$ ratio that is the case of monolayer and bilayer GFET.

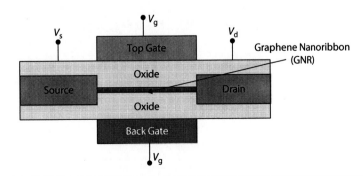

FIGURE 3.15 Cross-section of a double-gate GNR-FET [177].

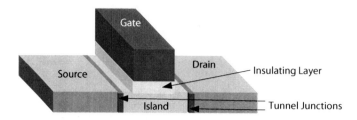

Figure 3.16 Simplified representation of an SET [58].

3.3.6 Single-Electron Transistor

The SETs have been considered for ultra-low power and high-density integrated circuits (ICs). SETs have potential for ultra-low power operation involving only a few electrons. Using lithography, it is possible to confine electrons to the smallest possible dimensions to observe their charge and energy quantizations. In essence, an SET is formed if the confined electrons are allowed to tunnel to metallic leads. The SET turns ON and OFF every time one electron is added to the isolated region. In an SET, the operations are quantum mechanical in nature as compared to a silicon MOSFET that uses classical concepts [104]. The SETs must be operable at room temperature to be useful in typical circuit design. They will be more suitable for ultra-low-noise analog circuits than the conventional digital electronics [58]. Graphene-based SETs have lots of potential as the carbon atoms form a stable and clean 2D crystal structure [88].

The 3D structure of an SET is shown in Fig. 3.16 [58]. For a comparative perspective with the traditional silicon MOSFET, one can analyze the SET structure as follows. Let the channel of the silicon MOSFET be replaced by a sandwich consisting of a nanoscale metal electrode (which is shown as island) connected to the drain and the source by two tunnel junctions. The island size has to be smaller than 10 nm as for larger sizes the capacitance and thermal fluctuation are manifested. Similar to silicon MOSFET, the gate electrode influences the island electrostatically. The tunnel junctions are characterized by their capacitance and tunnel resistance.

The current of the tunneling junction is expressed as follows [52]:

$$I_{\text{tun},i} = \left(\text{SIGN}(V_d - V_s)\right)\left(\frac{I_i(0)I_{i+1}(1)}{I_i(0)+I_{i+1}(1)}\right) \tag{3.31}$$

The direction of current is determined by the difference $V_d - V_s$. I_i is the current through the ith junction and I_{i+1} is the current through the $(i+1)$th junction. The current through any junction i is expressed as:

$$I_i(x) = \frac{V_{\text{eff},i}(x)}{R_i\left[1-exp\left(-\frac{eV_{\text{eff},i}(x)}{k_BT}\right)\right]} \tag{3.32}$$

where x is either 0 or 1. The effective voltage across the ith channel is calculated using the following expression:

$$V_{\text{eff},i}(x) = \left(V_{i-1} - V_i - \frac{e(2x-1)}{2C_i}\right) \tag{3.33}$$

In general, for a multi-island SET with N islands and $N+1$ tunneling junctions, the overall current of the SET is the harmonic mean of the currents of all tunneling junctions. Thus, the drain to source current of the SET is calculated as [52]:

$$I_{ds} = \frac{N}{\left(\frac{1}{I_{\text{tun},1}} + \frac{1}{I_{\text{tun},2}} + \cdots + \frac{1}{I_{\text{tun},N+1}}\right)} \tag{3.34}$$

3.3.7 Thin-Film Transistor

Liquid crystal, which was discovered in 1888, is neither solid nor liquid and can be compared to soapy water [11]. Around 1965, it was observed that liquid crystals could change the properties of light passing through the crystals when stimulated by an external electrical charge by applying supply voltage. In 1972, the first active-matrix liquid crystal display (LCD) was invented [43]. At present, LCD-based computer monitors and televisions are widely used as a replacement for old cathode ray tube (CRT)–based bulky predecessors [111, 178]. Essentially, an LCD uses picture elements (pixels) made of liquid crystal cells that change the polarization direction of the light passing through them in response to an external voltage [11, 43]. The power dissipation of an LCD is way less than the CRT displays.

Many of these color LCD monitors and TVs use thin-film transistor (TFT) technology. Additionally, TFT panels are used in digital radiography as a base for the image receptor. A TFT-based display is composed of a matrix of pixels as shown in Fig. 3.17 [133]. Millions of these pixels together create an image on the display. The TFTs are the active elements on the LCD that are arranged in a matrix, and hence often called "active-matrix TFT." TFTs act as switches to individually turn ON or OFF each pixel. The ON produces a "light" and OFF produces a "dark." The most beneficial aspect of TFT technology is a separate transistor for each pixel on the display. As each transistor is small, the amount of charge needed to control it is also small, and hence fast rendering of the display.

A TFT is a special kind of field-effect transistor. For illustration, the cross-section diagram is shown in Fig. 3.18 [91, 187]. It shows two types of TFTs, one with top gate and another with bottom gate. A TFT is typically made by depositing thin films of a semiconductor active layer and the dielectric layer. The metallic contacts are deposited over a supporting substrate. Typically, the common substrate is glass as the primary application of TFTs is in LCD design. This differs from the conventional MOSFET in which the semiconductor material in the substrate is silicon. The I-V characteristics of a TFT has well-defined linear and saturation regions and during ON state can be described by the standard MOSFET equations [91, 187]. The drain current increases exponentially with gate voltage for small V_{gs} and then increases linearly for larger V_{gs}.

Flexible TFT can be a fabricated SWCNTs [92]. In such an application, randomly oriented SWCNTs are synthesized. The same fabrication method can be used for wide range of SWCNT-based flexible electronics including sensors, organic light-emitting diode (OLED), and solar cells.

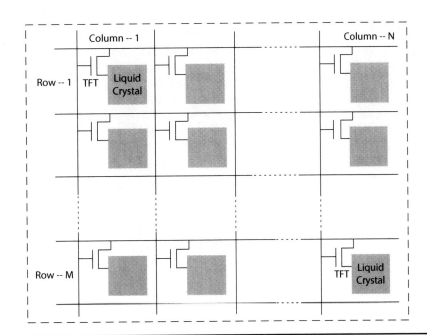

Figure 3.17 A display array that uses TFTs and liquid crystals.

86 Chapter Three

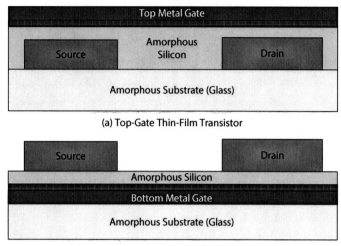

Figure 3.18 Cross-section of TFTs [91, 187].

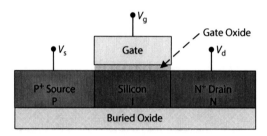

Figure 3.19 Cross-section of an N-type P-I-N structure TFET [93].

3.3.8 Tunnel FET

The tunnel FET (TFET) is based on ultra-thin semiconducting films or nanowires to achieve a 100X power reduction over traditional MOSFET. In nanoelectronic ICs, power dissipation has been a major issue. At a room temperature, the classical MOSFET needs minimum gate voltage of 60 mV for reasonable current flow. The TFET overcomes this limitation by using quantum-mechanical BTBT rather than using the thermal injection process to inject charge carriers into the device channel [89, 93]. The use of TFET technology will significantly improve power dissipation nanoelectronic ICs.

The cross-section of a simple N-type TFET that is essentially a P-I-N diode structure is shown in Fig. 3.19 [89, 93]. In the TFET, the source and drain are highly doped. For example, TFET has $L = 50$ nm, $T_{ox} = 3$ nm, source P$^+$ doping concentration is of the order 10^{20} atoms per cm^3, and drain N$^+$ doping concentration is of the order 10^{18} atoms per cm^3. In this device, the tunneling takes place between the intrinsic (I-channel) and P$^+$ regions. The gate controls the BTBT between the P$^+$/N$^+$ region and I-channel region. In effect, an N$^+$ channel is formed at the surface of the I-channel region and creates a P$^+$/N$^+$ junction at the source to channel interface [93].

The current flow for a specific TFET called sandwich tunnel barrier FET is calculated as follows [27, 279]:

$$I_d = \left(\frac{A_{tunnel} q^3 \sqrt{\frac{2m^*}{E_G}}}{4\pi^2 \hbar^2} \right) E_{vert} V_{eff} \exp\left(-\frac{\left(4\sqrt{m^*}\sqrt{E_G^3}\right)/3q\hbar}{E_{vert}} \right) \quad (3.35)$$

$$= \left(\frac{W_{gate} L_{gate} q^3 \sqrt{\frac{2m^*}{E_G}}}{4\pi^2 \hbar^2} \right) E_{vert} V_{eff} \exp\left(-\frac{\left(4\sqrt{m^*}\sqrt{E_G^3}\right)/3q\hbar}{E_{vert}} \right) \quad (3.36)$$

In the above expression, A_{tunnel} active source tunneling cross-sectional area is quantified as the area ($W_{gate} \times L_{gate}$) of the gate. m^* is the effective mass of the carrier. E_G is the energy band gap. V_{eff} is the tunnel-junction bias voltage. E_{vert} is the vertical electric field that can be obtained from the device simulator or can be calculated as follows [27]:

$$E_{vert} = \left(\frac{V_{gs} - V_{Th} + V_{tunnel}}{T_{epi} + 3T_{ox}} \right) \quad (3.37)$$

where V_{tunnel} is the voltage across the tunnel and T_{epi} is the epitaxial layer thickness.

The cross-section of a double-gate (DG TFET) is shown in Fig. 3.20 [40]. The additional gate potentially doubles the current flow in the DG TFET as compared to its single-gate counterpart. For ultra-thin SOI structures, the current boosting can be even higher when volume inversion takes place. Due to the DG structure, both the ON-current and OFF-current doubles thus, maintaining same $\left(\frac{I_{ON}}{I_{OFF}} \right)$ ratio. However, the OFF-current that is in the range of femtoamperes or picoamperes is extremely low.

The vertical pocket tunnel FET, or green TFET (GTFET), is a specific version of TFET. The cross-section of a GTFET is shown in Fig. 3.21. GTFET, an ultra-thin, fully depleted highly doped pocket, is used in addition to the intentional gate-to-source overlap. This structural change helps to achieve superior turn-ON characteristics compared to a general TFET. The pocket ensures a smooth BTBT vertically from the source to the pocket. The turn-ON voltage of GTFET can be adjusted with the dose of charge in the pocket similar to the threshold voltage adjustment using dopants. In the GTFET, the tunneling current is proportional to the pocket of overlap area.

The GNR can be used for TFETs due to its symmetric band structure, low effective mass, and monolayer-thin body. Such a FET is called GNR-TFET whose cross-section is shown in Fig. 3.22 [62, 278]. For example, dimensions of the GNR-TFET are the following: $L_{gate} = 10$ nm, $T_{GNR} = 5$ nm,

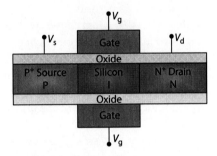

Figure 3.20 Cross-section of a double-gate N-type P-I-N structure TFET [40].

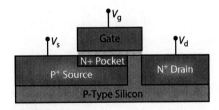

Figure 3.21 Cross-section of an N-type GTFET [93].

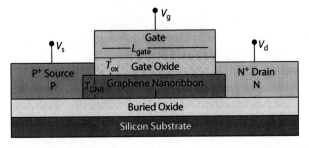

Figure 3.22 Cross-section of an N-type GNR-TFET [62, 278].

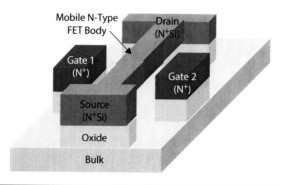

Figure 3.23 An independent double-gate VBFET [77].

and $T_{ox} = 1$ nm. The GNR-TFET has low OFF-state leakage current and low turn ON voltage, and hence will be useful for low-power nanoelectronics.

3.3.9 Vibrating Body FET

A resonant-gate FET (RG-FET) was invented in 1965 in which a movable cantilever rod, which was essentially a mechanical resonator, was used as the gate electrode [77, 248]. The electrostatic force acts on the gate rod upon the application of the gate voltage. In effect, the gap between the gate rod and RG-FET channel decreases, which results in an increase in drain current. This idea was extended in which a mechanical component moves in parallel to the substrate instead of perpendicular to it, as is the case in the RG-FET [77, 248]. The movable component in this case is a body. The new FET is thus called vibrating body FET (VBFET). In the VBFET, both piezoresistive stress and field effect and capacitive amplitude are used simultaneously.

An independent double-gate VBFET is shown in Fig. 3.23 [77]. The structure has a suspended double-clamped beam with an integrated transistor channel in the center of the beam (i.e., mobile FET body). The independent gates are located on either side of the channel beam. The channel length L_{chat} is shorter than the beam length L_{beam}. As the channel is placed on the lateral sides, its width is limited by the substrate thickness T_{body}. The junction depth of the source and drain implants is given by the width of the beam W_{beam} and not by the depth of the implantation.

The current flow in a VBFET is quite complex. The current is an effect of mechanisms (such as a typical silicon FET) as well as mechanisms. The overall current in the VBFET can be summarized as follows:

$$I_{ds} = I_{\text{Electrostatic Capacitive}} + I_{\text{Field-Effect Modulation}} + I_{\text{Piezoresistive Modulation}} \qquad (3.38)$$

The I-V characteristic of a VBFET has a linear behavior for low-drain voltages. At the higher drain voltages, the saturation of the carrier velocity starts to limit the drain current. The drain current to some extent is affected by the gate voltage.

3.3.10 Memdevices: Memristor, Memcapacitor, and Meminductor

The fourth fundamental element, called memristor, was conceptually invented in 1971 [56]. However, this fourth fundamental element memristor is only recently fabricated [198, 231, 262]. This new two-terminal passive element is named memristor as it combines the behavior of a memory and a resistor (i.e., **memory+resistor**) [198]. Its resistance depends on the magnitude, direction, and duration of the applied voltage. The memristor remembers its most recent resistance when voltage was turned off and until the next time the voltage is turned on. The memristor is characterized by its memristance. This element has the promising properties that can revolutionize nanoelectronics.

For a fundamental perspective, the new element can be linked with basic electrical or electronics variables. In the fundamentals of electronics, there are four basic variables: voltage (v), current (i), charge (q), and flux (ϕ). The three fundamental passive elements that are used to describe typical electronics circuits are resistors, capacitors, and inductors. The inter-relations of fundamental

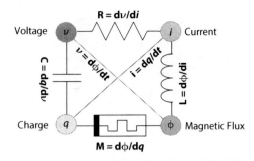

FIGURE 3.24 The four fundamental variables and passive elements of electronics [105, 157].

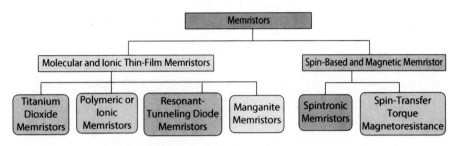

FIGURE 3.25 Different types of memristors [157].

elements and variables are presented in Fig. 3.24 for a clear understanding [105, 157]. The memristor is characterized by its memristance (*M*). The memristance is described by the charge-dependent rate of change of flux with charge as follows [105]:

$$\text{Memristance } M = \left(\frac{d\Phi}{dq}\right) \tag{3.39}$$

This is like the other fundamental element resistor, which is characterized by resistance (*R*). Similarly, the fundamental element inductor has inductance (*L*) and capacitor has capacitance (*C*) as their basic characteristics.

Memristors are of various types depending on how they are built as shown in Fig. 3.25 [157]. Three primary different memristors are [148, 157, 188]:

1. TiO_2 thin-film memristors,
2. Spintronic memristors, and
3. Ion-conductor chalcogenide-based memristors.

TiO_2 thin-film memristors are the earlier ones to be fabricated and widely considered for modeling, design, and deployment. The structure of a TiO_2 thin-film memristor is schematically presented in Fig. 3.26(a) [157, 183, 188]. The titanium dioxide thin-film memristor consists of the following five distinct layers of material structures:

1. Layer – 1: This is the bottom titanium/platinum (Ti/Pt) bilayer electrode.
2. Layer – 2: This is the active TiO_2 layer.
3. Layer – 3: This is the active TiO_2 with excess oxygen (TiO_{2+x}) layer.
4. Layer – 4: This is the top titanium/platinum (Ti/Pt) bilayer electrode.
5. Layer – 5: This is the silicon substrate.

The TiO_{2+x} with excess oxygen provides charge carriers when voltage is applied across the top and bottom electrodes. The charge carriers then flow toward the active TiO_2 layer, thus changing the resistance of the TiO_2 layer and that of the memristor. On the other hand, if the current direction is

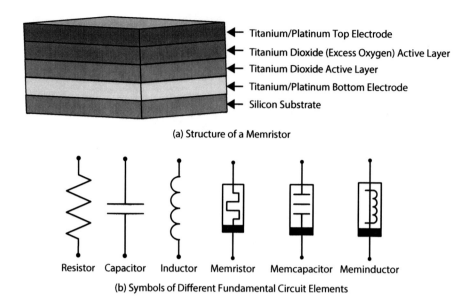

Figure 3.26 Structure and symbol of memristor [157, 183, 188].

reversed through the memristor electrodes then the excess charge carriers from the TiO_2 layer moves toward the TiO_{2+x} layer. The top and bottom Ti/Pt electrodes are metal connections.

The idea of memristor can be generalized to other two-terminal devices. The most logical extension of this approach is the memory-integrated capacitor and inductor. These are called the memcapacitor and meminductor. The three new elements, memristor, memcapacitor, and meminductor, are together known as "memdevices" [182]. All the passive elements are shown in Fig. 3.26(b) for a visual perspective [182, 183]. It may be noted that memristance is like a variable resistance. A battery can be considered to have memristance. However, the battery is an energy source and an active element; on the other hand, memristor is a passive element. One of the important characteristics of the memristor is the exhibition of *nonlinear negative resistance*. This has created the opportunity of using the memristor as an oscillator to create high-speed sustained oscillations [64]. The concept of "negative resistance" is not new. Negative resistance is a property of some electric circuits where an increase in the current results in a decreased voltage. Negative resistance diodes have been designed which have an "S"-shaped transfer curve [228].

The operation principle of the memristor is presented in Fig. 3.27 [125, 157, 194]. The bias voltage causes a drift of the dopants and electrically divides the memristor to doped, transitions, and undoped regions as shown in Fig. 3.27(a). The doped region has a large number of carriers and is the conduction region of the device. The undoped region is called the insulating region. The I-V characteristic of the memristor is shown in Fig. 3.27(b). It essentially demonstrates the pinched hysteresis effect of the memristor. The change in slope of the I-V characteristic demonstrates a switching between the different states of the memristor. The resistance is positive when the applied voltage increases and it negative when it decreases. The symmetrical voltage bias results in I-V double-loop characteristic that may collapse to a straight line for high frequencies. The I-V characteristic of the memristor can be expressed as follows:

$$i(t) = \frac{v(t)}{\left(R_{\text{doped}} - R_{\text{undoped}}\right)\left(\dfrac{L_{\text{doped}}(t)}{L_{\text{physical}}}\right) + R_{\text{undoped}}}$$ (3.40)

In the above expression, L_{physical} is the total physical length of the memristor. This remains fixed once manufactured. $L_{\text{doped}}(t)$ is the doped active length of the memristor. This changes with the voltage applied across the two terminals. R_{doped} is resistance of the doped layer of length L_{physical} that is equivalent

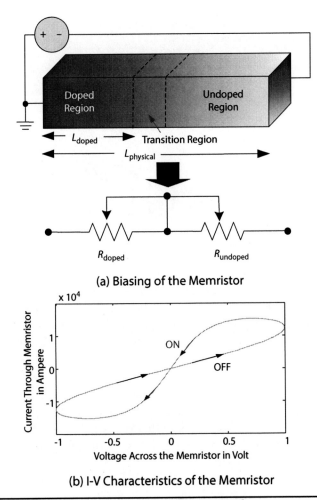

FIGURE 3.27 Biasing and characteristics of the memristor [125, 157, 194].

to ON-state resistance of the memristor. $R_{undoped}$ is resistance of the undoped layer of length $L_{physical}$ that is equivalent to OFF-state resistance of the memristor.

3.4 Nanomanufacturing: The Origin and Source of Process Variations

To closely understand the process variations in nanoscale circuits, knowledge of manufacturing is essential. A broad view of lithography process is presented in Fig. 3.28 [51, 242]. To facilitate fabrication of circuits using nanoscale technology, more and more sophisticated lithographic, chemical, and mechanical processing steps are adopted. It is anticipated that a subset or a superset of these processes will be used in manufacturing nanoscale chips. These processes introduce different types of process variations, hence need to be appropriately modeled and accounted for during the design phase.

The different phases of manufacturing involve various processes such as the following [7, 106, 154]:

1. *Chemical Mechanical Polishing (CMP)*: The CMP uses a combination of chemical and mechanical forces to smooth surfaces. The CMP process in a CMOS process polishes away excess material from the metal layer by mechanical polishing to reveal a specific pattern of metal such as copper.

2. *CVD*: In the CVD process, compounds contained in a gas are deposited onto a heated substrate. The materials in the vapor (or gas) react with the high-temperature substrate and produce chemical reaction to form new atoms or molecules that are deposited on the substrate. Many materials including single crystal silicon, SiO_2, silicon epitaxial layer, and silicon nitride can be deposited using CVD.

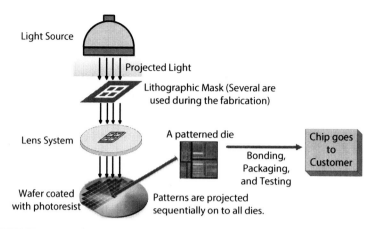

Figure 3.28 The overall perspective of photolithography for circuit manufacturing.

3. *Diffusion Process:* In the diffusion process, the silicon wafer is inserted in a high temperature (900°C–1100°C) furnace containing appropriate dopant impurities. In the process, the impurity atoms are introduced onto the silicon wafer. The diffusion takes place into the lattice because of the tendency of the impurity atoms to move from regions of high to low concentration. Typically, the source doping, drain doping, and N-well doping are performed using diffusion process. Diffusion process is typically slow and less controllable.

4. *Etching Process:* The etching process involves removal of materials from the wafer's surface during the manufacturing. The materials to be etched are exposed to chemical or physical forces for removal. Etching can either be dry etch or wet etch. However, modern semiconductor processes deploy dry etching with plasma. A typical semiconductor process involves several etching steps.

5. *Ion Implantation:* In an ion implantation process, a high-energy beam of ions is bombarded on a wafer. The range of implantation in this process depends on the type of impurity and beam energy. This is an alternative to the diffusion process in making source and drain of MOSFETs and N-well of dies as it provides better control of dopant density. Even with ion implantation it is difficult to uniformly implant a wafer.

6. *Metallization Processes:* The metallization process involves fabrication of metal portions of the chip. There are a few different forms of metallization process based on its application, such as interconnects, contacts, and gates. For example, in the modern manufacturing process, electroplating is used to deposit copper interconnects. In this electroplating, the wafer is placed in a copper sulfate solution. The wafer is made of cathode (negative) terminal, and copper ions travel from the anode (positive) terminal to the wafer surface for a thin-film metal layer deposition. Metallization is also performed using CVD and PVD. A specific PVD process called sputtering is used for aluminium metallization. Metallization is also used to produce pads around the periphery of the die to create metalized areas for wire bonding.

7. *Oxidation Process:* The oxidation process involves inserting the silicon wafer in an oxidation furnace at a temperature of 900°C to 1200°C. The oxidation furnace contains water or oxygen. As a result, a layer of SiO_2 (a.k.a. silica) is formed on the surface of the silicon wafer. The thickness of SiO_2 depends on the heating temperature and the time of oxidation. The oxidation process is very important in semiconductor manufacturing as SiO_2 has multiple roles as follows:

   a. It is the active dielectric material for the gate.

   b. It acts as an isolator among the different devices in the die.

   c. It can work as an isolator among different levels of metallization.

   d. It is used as a passive layer during manufacturing to protect the areas not supposed to be processed at any step.

   e. It protects the junction from atmospheric contaminants such as moisture.

8. *Photolithography*: Photolithography is the key process to produce electronic circuits consisting of millions/billions of transistors on the silicon wafer. Photolithography helps to selectively remove oxide from desired area for further processing in manufacturing (Fig. 3.28). In the photolithography, lithographic mask and light-sensitive photoresist work together to make the windows for further processing. Photoresist that is essentially a polymer is selectively melted by passing a light beam through the mask. The light source and lens system produce the right intensity light needed for the melting of photoresist. Several types of photolithography are ultraviolet lithography, electron-beam lithography, X-ray lithography, ion-beam lithography, extreme ultraviolet lithography, and ion projection lithography. Ion beam lithography uses heavier particles (ions) than electron-beam lithography and can create very small nanostructures and nanoelectronic circuits.

9. *PVD*: PVD involves condensation of a vaporized form of a material onto a wafer. It uses physical processes such as high-temperature vacuum or plasma sputter bombardment rather than a chemical reaction as is the case in the CVD. The PVD can be electron-beam PVD, evaporation deposition, or sputtering.

10. *Spin Process*: Spin process or spin coating involves dropping a liquid on the wafer surface. Then, the wafer is rotated on a turntable at a high speed to produce a thin coating of the liquid. Typically, photoresist coating is done using spinning. The wafer is rotated at a velocity in the range of 3000 to 7000 rpm for 30 to 60 seconds. As a result, the photoresist solution spreads out to form a thin and uniform coat. The thickness of the photoresist coating is inversely proportional to the square root of the rotational velocity.

3.4.1 Classical CMOS Fabrication Process

The classic CMOS-manufacturing process is well known. The steps of the process are presented here with the example of a CMOS inverter. The schematic representation of the CMOS inverter is shown in Fig. 3.29(a) and the cross-section view of the CMOS on a semiconductor wafer is shown in Fig. 3.29(b). Discussion of this CMOS process serves as basic understanding for nanoscale transistors being used or considered as an alternative for the classic MOSFETs that are discussed in the rest of this section.

The process of fabrication of a CMOS inverter is depicted in Fig. 3.30 [189]. The P-type silicon substrate obtained from dicing the electronic grade silicon ingot is the beginning of circuit fabrication on the die. The N-well is created for the PMOS devices using photolithography, diffusion, or

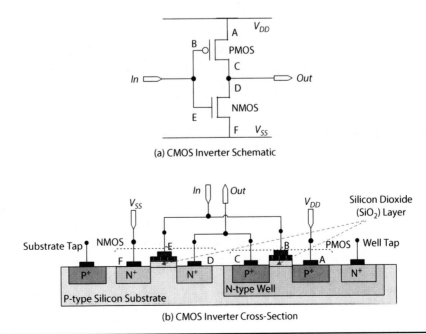

FIGURE 3.29 CMOS inverter schematic and cross-section [193].

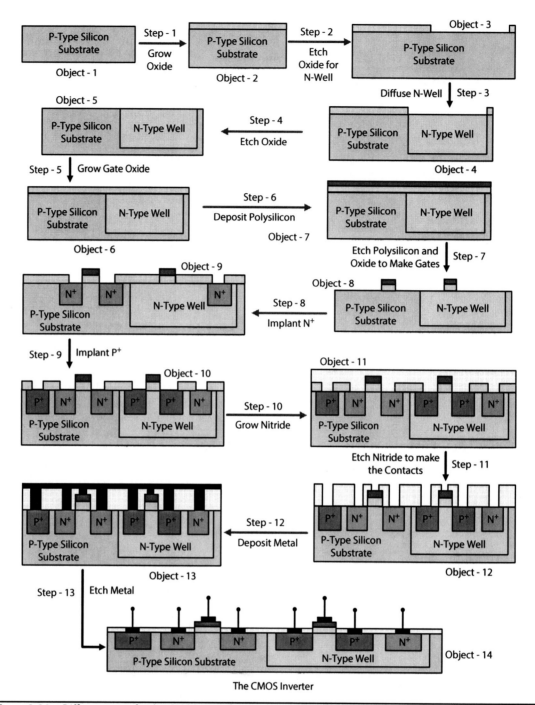

Figure 3.30 Different steps for the CMOS manufacturing with inverter example.

ion-implantation of well region with oxide used as a passive later. For gate creation, thin gate oxide is grown using the oxidation process and then polysilicon is deposited using a CVD process. Gates are then patterned using lithography. At this stage, N^+ regions and P^+ regions are created using ion implantation. The polysilicon gate that can withstand high temperature effectively self-aligns the source and drain doping. In other words, polysilicon gate is used as a mask for the source and drain doping process. The metallization process then deposits the metals and the circuit is fabricated.

The process is different when HKMG CMOS technology is used [29, 90, 106, 150]. In HKMG fabrication, "replacement metal gate" or "gate last" approach is followed as the metal gate cannot withstand high-temperature processing. A sacrificial (a.k.a. dummy or temporary) polysilicon gate is first made. This helps in self-alignment and creation of source and drain. The high-κ-like HfO_2 is deposited using atomic layer deposition (ALD). ALD is similar to CVD. The metal gate electrodes are formed using a lithography step. The metal gates for NMOS and PMOS are different and are created sequentially. For example, titanium nitride (TiN) for PMOS and then titanium aluminium nitride (TiAlN) for NMOS can be deposited using PVD. In the end, electroplating forms the metallization of interconnects (e.g., copper).

3.4.2 Carbon Nanotube FET Fabrication Process

CNTFET structure can be back-gated, front-gated, or even have multiple gates. However, top-gated has been preferred over the back-gated CNTFET. A top-gated CNTFET fabrication is schematically presented in Fig. 3.31 [126, 263].

In CNTFET fabrication, the reparation of silicon substrate and a solution with CNTs can happen in parallel. A layer of SiO_2 is deposited on silicon using an oxidation process as is the case in CMOS. At this point, the CNTs are grown on the surface of SiO_2 using a CVD process. The individual CNTs can be located using scanning electron microscope (SEM). Then the source and drain for the CNTFET are patterned using high-resolution electron beam lithography. In CNTFET, a high-temperature annealing process may be used to improve adhesion between the contacts and CNT, which, in turn, helps in reducing the contact resistance. For the gate insulator, a thin dielectric is deposited on the top of the CNT using ALD. Finally, the top gate contact is deposited on the gate dielectric.

3.4.3 FinFET Fabrication Process

FinFET can be made using bulk CMOS, SOI technology, or nanowires, as discussed in Sec. 3.3.3.3. In this section, the fabrication of the FinFET using an SOI technology is discussed [81]. The specific steps of the process flow for FinFET are presented in Fig. 3.32 [275].

At the first step of the FinFET process flow, the BOX is grown on the silicon substrate. The BOX can be formed using a combination of oxygen ion beam implantation process and high-temperature annealing. For the patterning of the Fin, a SiO_2 cap, SiN layer, and a photoresist layer are deposited. The SiN layer works a hard mark. The SiO_2 cap is used to relieve the stress. The silicon Fin is patterned using an electron beam lithography process. The accuracy of the Fin dictates the length and width of the channel. At this point, a thin sacrificial SiO_2 is grown. The sacrificial SiO_2 is stripped completely to remove etch damage until the silicon Fin is reached. Gate oxide is formed by depositing a thin layer of

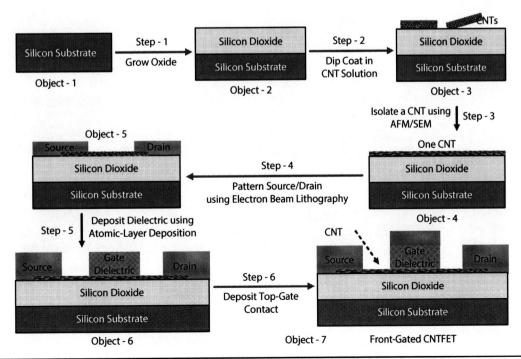

FIGURE 3.31 Different steps for the front-gated CNTFET manufacturing [126, 263].

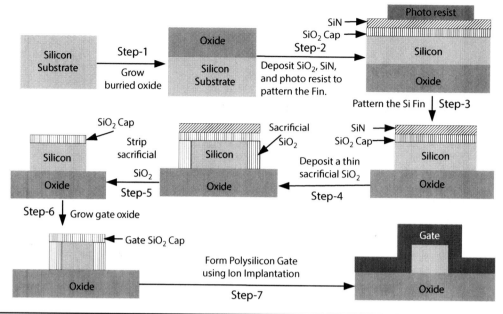

Figure 3.32 Different steps for the DG FinFET manufacturing [275].

it over the silicon Fin. This is then followed with polysilicon deposition using CVD process and lithography. Using ion implantation, thin layers of source and drain are formed in a self-aligned fashion. However, silicidation is necessary to reduce the source and drain resistance. It may be noted that in the case of HKMG, the process will be somewhat different. A "gate-last" like processing flow in that case can be deployed as discussed in Sec. 3.4.1 [106].

3.4.4 Graphene FET Fabrication Process

Since the invention of graphene isolation, many processes have been explored for GFET fabrication. Different steps of a fabrication process for graphene FET are shown in Fig. 3.33 [184].

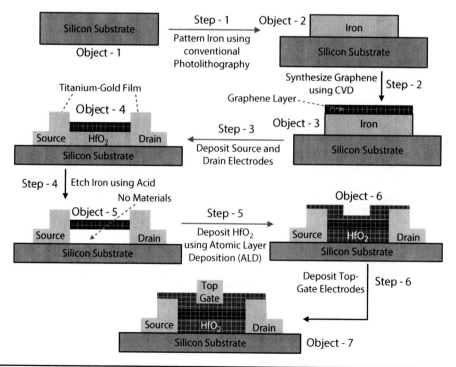

Figure 3.33 Different steps for GFET manufacturing [184].

Starting with a silicon substrate, an iron catalyst is formed into the target channel shape. A conventional photolithographic process can be used for iron deposition. The graphene layer is synthesized over the iron layer using a CVD process. At this point, the source and drain electrodes are formed at both ends of the graphene. The source and drain electrodes are made from titanium-gold film. The iron catalyst at this point is removed using an acid. This results in a graphene bridge between the source and drain electrodes as the graphene layer essentially hangs between the two electrodes. A layer of HfO_2 is now grown using ALD to stabilize graphene as well as for gate dielectric. A gate electrode is formed on top of the graphene, thus completing the graphene transistor.

3.4.5 Tunnel FET Fabrication Process

The simplest TFET is a P-I-N structure. The steps of a TFET fabrication process is presented in Fig. 3.34 [93]. Many other TFET processes are also available.

The P-I-N-based TFET has an asymmetric structure. Thus, it is *not possible to fabricate the TFET in a self-aligned process*, as is the case in a classical MOSFET fabrication [93]. In P-I-N-based TFET, the source and drain are formed separately. Starting from a silicon substrate, a BOX is formed using an oxygen ion beam implantation process and high-temperature annealing as typical for the SOI process. Compared to other isolation techniques of SOI (such as LOCOS, MESA, and trench), mesa isolation was chosen as the method for isolation as well as active region formation due to its simplicity and lesser number of steps. The gate structure consists of the high-κ gate stack along with a mid-gap work

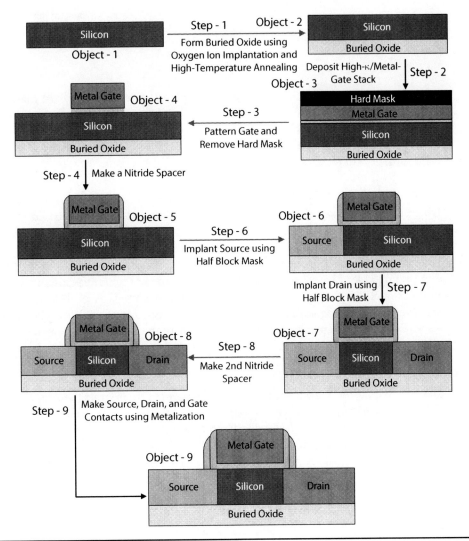

FIGURE 3.34 Different steps for TFET manufacturing [93].

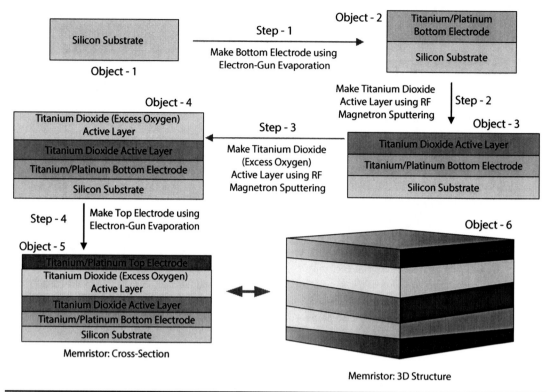

Figure 3.35 Different steps for thin-film memristor manufacturing.

function metal and polysilicon. The gate is then patterned and the hard mask is removed. Now, a nitride spacer is formed to protect the sidewalls of the gate. The source is then formed using ion implantation with half block mask and high-temperature annealing. The same step is then repeated for drain implantation. At this point, a second spacer is formed. Finally, a self-aligned silicide process is performed with nickel followed by metallization for contacts.

3.4.6 Memristor Fabrication Process

Different types of memristors such as TiO_2 thin-film and spintronic-based memristors may follow different processes for fabrication. The fabrication flow of TiO_2 thin-film memristor is presented in Fig. 3.35 [148, 188].

Starting from a silicon substrate, the first step is making the bottom electrode using electron-gun evaporation [148, 188]. The bottom electrode is made of titanium/platinum. The TiO_2 active layer is made using RF magnetron sputtering. At this point, an active layer of TiO_2 with excess oxygen (TiO_{2+x}) is made using RF magnetron sputtering. The top electrode of titanium/platinum constructed using electron-gun evaporation completes the structure of the TiO_2 thin-film memristor.

3.5 The Issue of Process Variation

Designer engineers are facing numerous challenges as they migrate their existing designs or start new designs in a 65 nm and finer technology node. While 22-nm technology node using TGFET is available, the research is in full swing to go to as small as 8 nm. Process variations thus remain and will remain the single biggest challenge for DFX of nanoelectronics. Many other challenges for DFX, such as increasing leakage power, thermal and yield loss, are further aggregated by increasing process variations. The process variations have profound effects on electrical parameters and overall performance of a chip and are manifested in the variation in power and delay and other attributes of the chip. The magnitude of each leakage component of the device is mostly dependent on the device geometry, doping profiles, and temperature. They affect design margins and yield and may lead to loss of money in the ever-reducing time-to-market scenario.

3.5.1 Types of Process Variation

In general, the process variations can be classified into various different types (Fig. 3.36) [154]. The nanoscale process variations are a combination of wafer, reticle, and local variations. The wafer variations are combinations of global, linear, and radial variations. The reticle (or photomask) variations are caused by photo processes. The local variations are due to random microscopic processes. Global variations originate from wafer, lot (i.e., a set of wafers processes in bulk), or fabrication-plant processes. The wafer process variation is the change in device geometry and structure from one wafer to another in a lot. The lot process variation is the change from one lot to another in a fabrication plant. The fab process variation is the change from one plant to another plant. It can be said that every post-manufactured transistor has different geometry and characteristics even though it may be of the same size and characteristic during the design phase.

A visual perspective of the process variation effects on the die and wafer is presented in Fig. 3.37 [38, 220]. The global uniform variation, as shown in Fig. 3.37(a), causes the same change in all the transistors of a wafer. The linear variation, as shown in Fig. 3.37(b), causes different changes in the transistors of a wafer and the change varies in a linear fashion across the wafer. The radial variation,

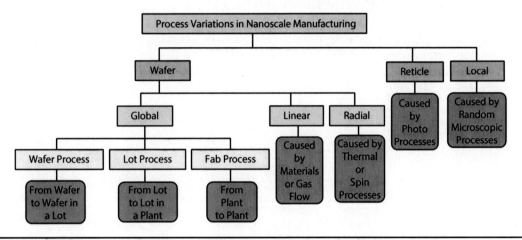

FIGURE 3.36 Different types of process variations in nanoelectronics technology.

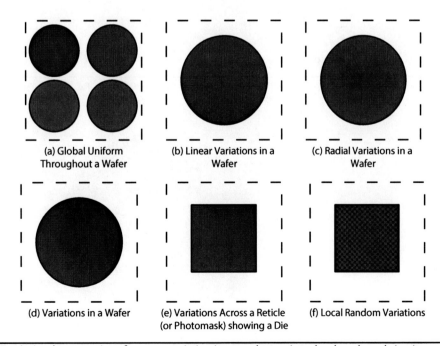

FIGURE 3.37 A visual perspective of process variation in nanoelectronic technology-based circuits and systems [38, 220].

as shown in Fig. 3.37(c), causes different changes in the transistors of a wafer where the change varies across the wafer diameter. The overall wafer-level process variation, as shown in Fig. 3.37(d), is the combination of global, linear, and radial variations from Fig. 3.37(a), Fig. 3.37(b), and Fig. 3.37(c), respectively. The variation caused by reticle or photomask is shown in Fig. 3.37(e). The local variations that are due to random microscopic processes are shown in Fig. 3.37(f). The transistor geometry, structure (impurity level), and characteristics are affected by all of the above process variations. For accurate statistical design exploration in DFX, the design engineers need to consider these variations during the design phase (when even the chip is not built).

For the purpose of incorporation during the design phase in the DFX, the process variations are modeled as presented in Fig. 3.38. These process variations are categorized as inter-die or intra-die. The intra-die process variations are the variations in the transistors in a die. On the other hand, inter-die process variations are variations in different dies, wafers, lots, and fabrication plants. As the device technology scaling continues to the nanometer region, intra-die process variations such as gate oxide thickness (T_{ox}) and threshold voltage (V_{Th}) variations have as much impact as inter-die variation on the performance and the yield of a chip [35, 36, 42, 175].

3.5.2 Impact on Device Parameters

The effects of process variations in a die and wafer are depicted in Fig. 3.39 [38, 154]. In essence, if a die is picked at random from a chip then its resistance, capacitance, inductance, and mutual inductances will be different from another chip. Thus, no two chips are alike in a die. The device geometry and doping profiles are affected in different ways in different process technologies, e.g., classical CMOS, CNTFET, GFET, and memristor.

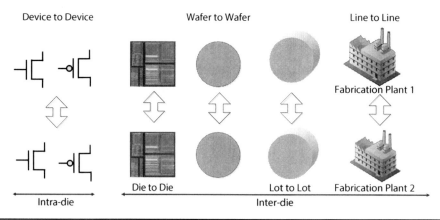

Figure 3.38 Intra-die versus inter-die process variation in nanoelectronic technology-based circuits.

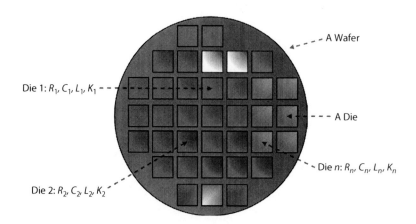

Figure 3.39 Nanoscale process variations cause different characteristics for different dies in a wafer and affects yield.

3.5.2.1 For Classical Nanoscale CMOS Transistor

For classical CMOS transistors, the uncertainty in the processes has caused variations in process and design parameters [33, 38] such as the following: (1) NMOS threshold voltage (V_{ThN}), (2) PMOS threshold voltage (V_{ThN}), (3) NMOS gate dielectric thickness (T_{oxN}), (4) PMOS gate dielectric thickness (T_{oxP}), (5) NMOS channel length (L_{effN}), (6) PMOS channel length (L_{effP}), (7) NMOS channel width (W_{effN}), (8) PMOS channel width (W_{effP}), (9) NMOS gate doping concentration (N_{gateN}), (10) PMOS gate doping concentration (N_{gateP}), (11) NMOS channel doping concentration (N_{chN}), (12) PMOS channel doping concentration (N_{chP}), (13) NMOS source or drain doping concentration (N_{sdN}), (14) PMOS source or drain doping concentration (N_{sdP}), (15) metal wire thickness, and (16) via resistance. In addition, supply voltage variation (V_{dd}) is also a concern. The standard deviation of the threshold voltage variation of a MOSFET can be analytically expressed as follows [101, 230]:

$$\sigma_{V_{Th}} = \left(\frac{\sqrt[4]{4q^3 \varepsilon_{Si} \phi_B}}{2} \right) \left(\frac{T_{ox}}{\varepsilon_{ox}} \right) \left(\frac{\sqrt[4]{N_{ch}}}{\sqrt{W_{gate} L_{gate}}} \right) \quad (3.41)$$

$$= \left(\frac{\sqrt[4]{4q^3 \varepsilon_{Si} \left(2\kappa_B T \ln\left(\frac{N_{ch}}{n_i} \right) \right)}}{2} \right) \left(\frac{T_{ox}}{\varepsilon_{ox}} \right) \left(\frac{\sqrt[4]{N_{ch}}}{\sqrt{W_{gate} L_{gate}}} \right) \quad (3.42)$$

In the above expression, ϕ_B is barrier height and N_{ch} is the channel dopant concentration, n_i the intrinsic carrier concentration, T the absolute temperature, q the elementary charge, and ε_{ox} and ε_{Si} are the permittivity of oxide and silicon, respectively. All these parameters affect threshold voltage variations in different ways. It may be noted that the $\sigma_{V_{Th}}$ is inversely proportional to the square root of the device area. A summary of chip features that are affected by process variations is presented in Table 3.2 [147]. It may be noted that the gate line edge roughness (LER) is a random change of the gate line edges from the target value [272]. The LER may be caused by various sources including photoresist LER and gate polysilicon etching condition. High-frequency LER may cause a decrease in channel length due to source/drain diffusion extensions. The LER severely affects OFF-state leakage current of the MOSFET.

Feature Affected	Variation Description	Variation Sources	Variation Scope
Device	Lot-to-Lot Wafer-to-Wafer	Focus, dose and etch changes	Global
Device	Wafer-Level	Nonuniform gas flow resist coat, oxide growth	Global
Parasitics	Interconnect geometry	Metal layer width, thickness and inter-layer dielectric	Global
Device	Across Chip Line-width Variation of gates	Nonuniform processes on die	On-Chip Intra-Die
Length	Critical Dimension Variation	Lithographic interferences Transistor proximity	On-Chip Systematic
Mobility	Layout-dependent	Stress Effect	On-Chip Systematic
Interconnect	Litho-induced variability CMP-induced thickness variations	Dependence on pitch, density dishing, erosion	On-Chip Systematic
Threshold Voltage	Random dopant fluctuation	LER	On-Chip Random
Interconnect	Metal layer	LER	On-Chip Random

TABLE 3.2 Summary of Process Variations in Nanoelectronic Circuits and Systems [147]

3.5.2.2 For Carbon Nanotube FET

In the CNTFET, the variations in the diameter of CNTs and misalignment of CNTs are additional issues that can affect manufacturing yield [54, 180]. The threshold voltage and diameter of a CNTFET are determined by the chirality of the CNTs used in it. In a CNTFET, the diameter variation is more severe compared to the length and width variations. In a CNTFET, the gate width and oxide thickness variations are less severe compared to a classical MOSFET. In CNTFET, the dopant fluctuation does not affect device parameters, unlike the case of MOSFET [180]. Overall, CNTFETs have four times less performance variation than classical MOSFETs.

3.5.2.3 For FinFET

In a FinFET, variations can happen in gate length (L_{gate}), gate width (W_{gate}), gate oxide thickness (T_{ox}), and Fin thickness (T_{Si}). In addition, random dopant fluctuation and work function variation also happens. All of these can cause variations in threshold voltage (V_{Th}) of the FinFET [180]. The work function variation has the strongest impact on threshold voltage (V_{Th}) variation of the FinFET [60, 180]. The V_{Th} variation in turn causes I_{ON} current variation. The standard deviation of the threshold voltage variation of a FinFET can be expressed as follows [60]:

$$\sigma_{V_{Th}} \propto \left(\frac{1}{\sqrt{W_{gate} L_{gate}}} \right) \tag{3.43}$$

This is similar to the case of classical MOSFET as discussed above.

3.5.2.4 For Tunnel FET

For the TFET, variations in device parameters such as the channel length, the thickness of the silicon thin film, and gate oxide thickness impact its performances [209]. In addition, source, drain, and channel doping fluctuations can cause further variation issues in the TFET. For TFET, the relation among the tunneling current and various parameters can be expressed as follows:

$$I_{ON} \propto \Delta\phi \, exp\left(-\frac{4\sqrt{2m^*}\sqrt{E_G^3}}{3q\hbar(\Delta\phi + E_G)} \sqrt{T_{ox} T_{Si} \left(\frac{\varepsilon_{Si}}{\varepsilon_{ox}}\right)}\right) \tag{3.44}$$

where E_G is the band gap energy and $\Delta\phi$ is the energy range over which the tunneling can take place in a TFET. The variations in the silicon-body and gate-oxide thicknesses cause variations in the spatial extent of the tunneling region. The impact of the device parameter variations on the ON current variation is more severe in a TFET than in a classical MOSFET.

A summary of the impact of process variation of characteristic of different devices is presented in Table 3.3 [180, 209].

Device Parameters	Characteristics of Different Devices							
	Classical MOSFET		CNTFET		FinFET		TFET	
	I_{ON}	C_{gate}	I_{ON}	C_{gate}	I_{ON}	C_{gate}	I_{ON}	V_{th}
Channel Length	Strong	Strong	None	Weak	Weak	Strong	Weak	Weak
Channel Width	Strong	Strong	None	Strong	Strong	Strong	NA	NA
Oxide Thickness	Strong	Strong	Weak	Weak	Strong	Strong	Strong	Strong
Dopant Fluctuation	Strong	Weak	None	None	Strong	Weak	NA	NA
Tube Diameter	NA	NA	Strong	Weak	NA	NA	NA	NA
Fin Width	NA	NA	NA	NA	Strong	Strong	NA	NA

TABLE 3.3 Impact of Process Variation on Characteristic of Different Devices [180, 209]

3.5.2.5 For Graphene FET

In graphene FET (GFET), the variability of the ON current can be due to variations in gate length, gate width, and threshold voltage as evident from its I-V characteristic discussed in Sec. 3.3. The variations in GNR dimensions can also cause severe variability in performance of GFET. In GNR-TFET, the variations can be introduced due to imperfect control of the width of nanoribbon and LER [138].

3.5.2.6 For Memristor

The parameters that are affected by process variations are different for three memristors, TiO$_2$ thin-film memristors, spintronic memristors, and ion-conductor chalcogenide-based memristors [148, 188]. For TiO$_2$ thin-film memristors, the process parameters include length (L), width (W), and thickness (T). For spintronic memristors, relevant parameters are hard anisotropy (H_p), easy anisotropy (H_k), and domain wall velocity coefficient Γ_v [125]. For memristors, the possible sources of geometry variations are the following [84]: (1) LER, (2) thickness fluctuation, and (3) random doping. For thin-film memristors, thickness fluctuation is the major source of process variation. On the other hand, for spintronic memristors, LER is the major source of process variation. The LER is most difficult to model for accurate characterization of memristor performance. A model for variation due to LER is the following [84]:

$$\Delta L = L_{\text{low-freq}} \sin\left(f_{\max} \overline{X}\right) + L_{\text{high-freq}} \overline{P} \tag{3.45}$$

where ΔL is the LER noise. The model has two components: low frequency and high frequency. The low-frequency range regular disturbance is modeled as a sinusoidal function. $L_{\text{low-freq}}$ is the amplitude of the sinusoidal noise. $f_{\max}$ is the mean of the low-frequency range. $\overline{X}$ is the random variable that represents the equal distribution of all frequency components of the low-frequency range. $\overline{X}$ follows uniform distribution $\mathcal{U}(-1,1)$. $L_{\text{high-freq}}$ is the amplitude of Gaussian white noise. $\overline{P}$ has Gaussian distribution $\mathcal{N}(0,1)$. The values of $L_{\text{low-freq}}$, $L_{\text{high-freq}}$, and $f_{\max}$ are calculated from autocorrelation function (ACF) and power spectral density (PSD).

3.5.3 Design Phase Incorporation of Process Variation

It is evident from the discussions in the previous sections that there is a discrepancy between the design time data and post-manufacturing data due to the process variation. Simplistically speaking, the geometry and process parameters are not precisely controllable to get the desired device parameters. This leads to differences in the estimated performances of the chip during the design phase and measured performances of the chip after it is manufactured. The first question that arises for DFX, in particular for DFM, is how to accommodate the process and geometry variations during the design phase so that the estimations are accurate. The second question that arises is how to compensate for the effects of the process variations to obtain target performance to enhance yield. This section discusses the high-level idea. Detailed discussions of the techniques integrating process awareness in design flows will be presented in different chapters.

3.5.3.1 For Analog Circuits

Analog circuits and system are typically designed to meet multiple stringent performance metrics (much more than the digital circuits). The process variations can thus lead to suboptimal design solutions and loss of yield. Proactive approaches are needed to handle the performance and yield during the design phase not after the manufacturing of the first silicon. The simulation, estimation, and optimization tools based on nominal data do not optimize parametric yield of the analog circuits and systems. Optimization of analog design yield along with the sensitivity analysis and knowledge of the process variation distributions are critical to achieve the maximum yield.

The process-variation aware design problem is quite complicated in analog circuits in contrast to the digital circuits due to the lack of abstraction levels. A broad perspective of process-variation aware analog design is depicted in Fig. 3.40. It represents three levels such as system-level design (including behavioral simulation), transistor-Level (schematic or logical design), and physical-level (or layout-level). The key points of the process-variation aware design are the following:

1. Identification of target performance characteristics (i.e., figure of merit or FoM) to be optimized for a specific circuit; for example, oscillating frequency or phase noise for oscillator circuits.

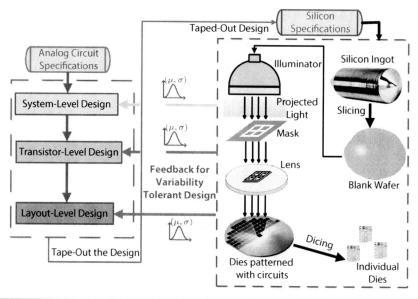

Figure 3.40 The key idea of process-variation aware design of analog and mixed-signal circuits.

2. Identification of the tuning or design parameters of the specific circuit and target FoM; for example, transistor width for power and frequency.

3. Identification of device geometry and process parameters that affect the target FoM. For example, length, width, oxide thickness, and threshold voltage can affect the FoMs. The tuning parameter is a subset of device geometry and process parameters.

4. Identification of manufacturing processes that impact the device geometry and process parameters under consideration; for example, diffusion or implantation for drain and source dimensions.

5. Identification of appropriate statistical distribution functions for device geometry and process parameters based on their manufacturing process sources. For simplicity, Gaussian (or normal distribution) can be considered; however, non-Gaussian distributions can be used for more accurate modeling. At the same time, correlations and locality need to be considered to ensure accuracy.

6. Statistical modeling of target FoMs based on statistical distribution of device geometry and process parameters. This involves determination of statical distribution of FoMs based on the statistical distribution of device parameters.

7. Optimization of target FoMs that are in the form of probability density functions (PDFs) for specific constraints. At this point $(\mu + \sigma)$ or $(\mu + 3\sigma)$ of the PDF is considered for optimization.

The above steps can be directly used at the schematic-level and layout-level of analog circuits. At both of these levels, a SPICE netlist is created over which the statistical simulation, modeling, and optimization can be performed. The SPICE netlist at the layout-level contains parasitics of the circuits that are originated from transistors and interconnects. Thus, the design flow steps are significantly slow as compared to the schematic level. However, for analog circuits, the system-level or behavioral-level process-variation aware optimization is not straightforward. This is due to the fact that behavioral-level or system-level design does not contain any circuit-level information for process variation awareness. However, the concepts such as Verilog-AMS-POM and Verilog-AMS-PAM in which circuit-level metamodels are integrated in Verilog-AMS can make such design possible [280, 288]. These techniques will be discussed in later chapters.

For mixed-signal circuits (such as ADC), the above analog flow can be used. On the other hand, for mixed-signal systems with comparable analog and digital portions, the process variation optimization flow needs to follow separate routes. The above optimization can be used for analog portions, whereas the digital portion will follow the following flow discussed for digital circuits.

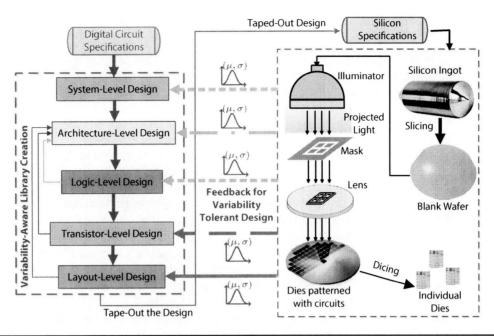

FIGURE 3.41 The key idea of process variation aware design of digital circuits.

3.5.3.2 For Digital Circuits

The process-variation aware digital circuit and system design flow are presented in Fig. 3.41. As evident from the flow, the process-variation awareness can be introduced at one or more levels of design abstractions (such as layout, schematic, or architecture level). At the layout and schematic level, the circuit is essentially a netlist as is the case of analog circuits and systems discussed in the previous subsection. However, the number of elements in the netlist such as transistors is much larger than its analog counterpart. Thus, flat netlist optimization for digital circuits is quite slow or even infeasible for larger circuits and systems. The presence of design abstractions such as logic, architecture, and system level facilitates the optimization of large digital circuits and systems. At the higher levels of abstractions, the granularity of circuit and system components are larger, but the number of components to deal with is small.

The higher the levels of design abstractions are, the lesser the degree of freedom is, and hence the faster the optimization process is. There are different sets of challenges in digital design compared to the analog design. At the same time, the higher the abstraction level is, the lesser the accuracy of estimation and optimization. For nanoelectronics-based digital circuit and system DFX many challenges arise such as:

1. How can one identify the sources of process variations for different nanoelectronic technology?
2. How can statistical models for process variations of different nanoelectronic technologies be determined?
3. How can statistical modeling of power, leakage, delay, and area for different logic, architecture, or system level components be performed?
4. How can one model process variations at lower levels of circuit abstraction and propagate them to higher levels?
5. How can power, leakage, delay, area, and yield be estimated during logic-, architecture-, or system-level design exploration in the presence of process variations?
6. How can one consider both device and interconnect process variations at higher levels of digital design abstraction?
7. How can one consider correlations in the process variations at different levels of digital abstraction?
8. How can leakage, power, delay, area, and yield trade-offs be performed for design space exploration in the presence of process variations?

9. How can one fix the clock cycle width for multicycle or pipelined data path in the presence of process variations?
10. How can one quantify digital circuit and system characteristics for a specific cycle and minimize costs, such as current-delay-product (CDP), energy-delay-product (EDP), or leakage-delay-product (LDP), which are functions of PDFs?

Many different techniques for digital circuits are available for process-variation aware logic synthesis, high-level (register-transfer level or architecture-level) synthesis, and system-level design to maximize the yield [149, 154, 162]. The idea of process-variation aware hierarchical statistical modeling and estimation for digital circuits and systems is depicted in Fig. 3.42. The flow essentially presents two alternative ways for process-variation aware statistical hierarchical modeling and estimation in digital circuits. In the first-principle physics approach, the input-output descriptions of device primitives are obtained by self-consistently solving the governing partial differential equations (PDEs). The main challenges in this approach are the generation of the input-output physics-based model and the capability to simulate the hardware description language (HDL) netlist. This approach is very accurate; however, it can only be applied to a very small number of elements concurrently. A circuit with many devices and interconnects, which is the flattened representation of an architecture component, is very complex and its modeling and simulation at this level is infeasible. In the process-logic-RTL hierarchical simulation approach, the logic gates are modeled for different input-output states from first-principle physics. Because a logic gate contains a very small number of devices, irrespective of technology or structure of the logic-gate itself, this modeling is time consuming but feasible. The generated models are much more accurate than compact models and can be used with traditional HDL simulation media. In the process-transistor-logic-RTL (partial-physics approach), two methods can be used such as device-circuit cosimulation method and circuit simulation method. In the device-circuit cosimulation method, new compact models are obtained by using technology computer-aided design (TCAD) simulations, in which nano-devices are built from processes recipe. The new compact models are then used to simulate a circuit level netlist of register transfer level components. Integration of the new compact models to SPICE is required. The circuit simulation method is applied for technologies whose compact models are available (e.g., BSIM4), which can be used in a circuit-level netlist of architecture components. This solution is fast but semi-empirical and less accurate.

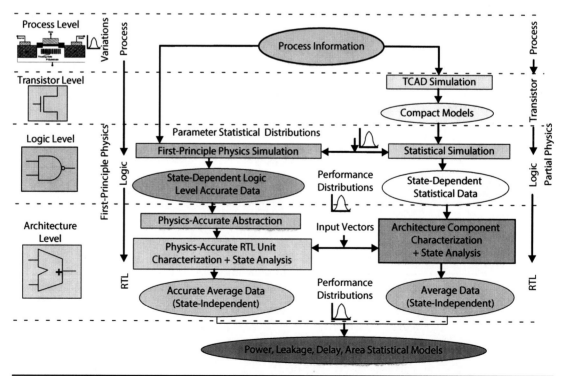

Figure 3.42 Process-variation aware statistical hierarchical modeling and estimation in digital circuits.

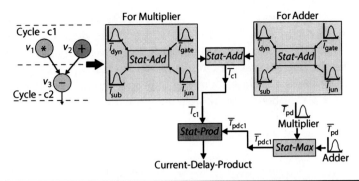

FIGURE 3.43 Statistical computations over PDFs of performances for modeling and estimation in digital circuits.

For the computation of performances (e.g., power dissipation or energy dissipation) at the system-level, architecture-level, and logic-level, statistical approaches are needed. The number and type of components involved at each of the levels are of course different. At the architecture-level, the components used to design circuits include adder, subtractor, multiplier, and divider. A mechanism to perform statistical computation of CDP at the architecture-level is presented in Fig. 3.43. The left subfigure shows a data flow graph (DFG) representation of a hypothetical algorithm whose chip is to be designed. The total current dissipation of a resource (e.g., adder) that performs computations (e.g., addition) is the statistical summation of all the components of the current dissipation; for example, $Stat - Add$ (I_{dyn}, I_{sub}, I_{gate}, I_{jun}). For a clock cycle with two operations, the overall current dissipation is the statistical addition of the current dissipation of the individual resources active in a specific clock cycle. Similarly, the current dissipation estimation of the complete DFG spanning over all the clock cycle can be performed. On the other hand, the delay calculation may involve statistical maximum over all the resources active in a clock cycle. For statistical computation, one way is to quantify a PDF as a weighted sum of mean and variance; that is ($\alpha \mu + \beta \sigma$), where α and β are scalars. The mean is representative of average cost, and the variance is representative of yield.

3.6 The Yield Issue

The bottom line of DFX is to get maximum possible yield of the chip. The best scenario is for the chip to have no defects while meeting all the target specifications. It may be noted that the missing materials and extra patterns, which may be either intralayer or interlayer, are called defects. Some of the defects that disconnect a continuity or connect two disjoints become a fault and all the defects may not result in a fault. The faults lead to yield losses. The importance of the yield is evident from the fact that yield is the trade secret of a semiconductor house. The yield of a chip and a process is typically not made public.

A variety of manufacturing control factors and production indices such as work-in-process level, line balancing programs, cycle time, rework rates, and statistical process controls are used to ensure the success of chip fabrication [57]. Yield models that essentially present the probability distributions of the faults are used to estimate the yield. Assuming constant work-in-process over a period of time in a manufacturing line, the overall yield is expressed as follows [57]:

$$Y_{overall} = Y_{line} Y_{die} Y_{assembly} Y_{finaltest} Y_{quality} \qquad (3.46)$$

The two yields Y_{line} and Y_{die} have maximum impact on the production cost of the chip. For constant work-in-process, the line yield is defined as follows:

$$\text{Line Yield}(Y_{line}) = \left(\frac{\text{Number of Patterned Wafers Out over a Given Time}}{\text{Number of Blank Wafers In over a Given Time}} \right) \qquad (3.47)$$

Specifically, two different forms of yield can be discussed for ICs as follows: (1) process (Die) yield and (2) circuit yield [277].

Process (die) yield is the commonly referred to yield that is associated with the fabrication of the chips. The net-die-per-wafer yield, or simply die yield (Y_{die}), can be the primary factor in the success or failure of a particular manufacturing process and plant. Physical defects on the wafer that cause the devices and the circuits to fail affect the process yield. Processing phase problems such as cleanliness of a clean room and improper diffusion of impurities can cause defects such as dislocations and lithographic defects and pinholes in insulator layers [227]. Electrically active defects including short circuits and open circuits also cause failures of the chips. Process (die) yield is thus the ratio of number of good chips to total number of dies in a given wafer:

$$\text{Process Yield}(Y_{die}) = \left(\frac{\text{Number of Good Dies in a Wafer}}{\text{Total Number of Dies in a Wafer}}\right) \quad (3.48)$$

Die yield models are important for the following reasons [57]:

1. To predict the yield so that measures can be taken to improve it.
2. To determine the maximum level of integration for a given design implementation.
3. To predict the manufacturing cost of the chip.

A die yield model is typically a function of the average number of defects per unit area (D_{avg}), the die area (A_{die}), and empirical correction factors (γ_{corr}). Thus, the die yield model has the following form:

$$Y_{die} = \text{function}(D_{avg}, A_{die}, \gamma_{corr}) \quad (3.49)$$

Many different types of models are available in the literature. Out of these, the most accurate models are the Poisson and the negative binomial. The negative binomial is more accurate for larger die sizes. The Poisson model for die yield analysis is expressed as follows:

$$Y_{die,\text{Poisson}} = exp(-D_{avg}, A_{die}) \quad (3.50)$$

The negative binomial model for die yield analysis is expressed as follows:

$$Y_{die,\text{binomial}} = \left(1 + \frac{D_{avg} A_{die}}{\alpha}\right)^{-\alpha} \quad (3.51)$$

where α is called the cluster parameter and it is calculated as $\left|\frac{\lambda}{\lambda_{var}}\right|$. λ is the defects per die, λ_μ is the mean of λ, and λ_{var} is the variance of λ.

Circuit yield (a.k.a. design yield or parametric yield) is the sensitivity of a design to nanoelectronics process variations in the device parameters. In general, the operating margin of a design is the range of device parameters for which the design is functional [277]. For example, for a single design parameter, the range of the device parameter for which the design is functional is (Para_{min}, Para_{max}). Assuming that this device parameter follows a Gaussian distribution, the optimal average value of the parameter Para_{avg} is halfway between Para_{min} and Para_{max} as presented in Fig. 3.44 [277]. The probability that the device parameter (Para) of N devices lies within the range (Para_{min}, Para_{max}) is expressed as follows if there is no statistical correlation among (Para) [277]:

$$\text{Prob}_N = 1 - N\sqrt{\frac{2}{\pi}}\left(\frac{exp\left(-\frac{\alpha^2}{2}\right)}{\alpha}\right) \quad (3.52)$$

where the parameter α in the expression is calculated as $\left|\frac{\text{Para}_{max} - \text{Para}_{min}}{2\sigma}\right|$. In a typical scenario, the nominal design point is not necessarily the best operating point for a process. Due to the process variations of the device parameters and the nonlinear nature of analog, mixed-signal, and RF circuits,

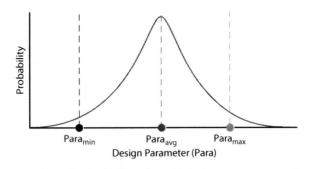

FIGURE 3.44 Concept of circuit yield in the context of nanoelectronics process variations [277].

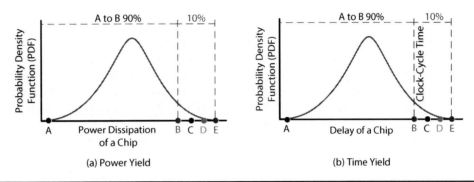

FIGURE 3.45 Power and timing yield affected by process variations.

the design must operate across the entire range of design parameter. Designing for yield is important for digital, analog, mixed-signal, or RF circuits.

In a typical scenario, the design yield problem is addressed after a design is manufactured. The design engineers track a low-yielding wafer to see what process variations possibly resulted in the yield loss [65]. Then, simulations are performed to identify where the design should be tweaked to improve yield. This redesign approach to address yield issues is very costly. On the contrary, handling design yield at the front end of the design flow using a specific DFX called *design for yield* (DFY) technique can be cost effective. The use of DFY techniques in the front-end design flow can accelerate the design cycle and get the chip to market faster and with higher yield. In such a situation, an automated flow is needed to maximize design yield such that the design operates as specified across the entire process and operating environment. When an analog, mixed-signal, or RF design is optimized for multiple performance metrics or FoM (such as gain, bandwidth, and jitter), the worst-case condition may be different for each FoM [50]. Trying to simultaneously optimize a design for each such performance parameter is very difficult. However, the net yield of the design is determined by simultaneously meeting all those requirements. For the DFX, the concept of trade-off for optimization using a statistical approach is presented in Fig. 3.45 for propagation delay versus timing yield trade-offs [31]. The propagation delay for a specific timing yield is calculated by finding the area under the PDF curve. For a 100% timing yield, the delay of the data path component corresponds to point E. However, for a 90% timing yield (10% yield is sacrificed), the propagation delay corresponds to point B. Similarly, point C represents propagation delay for 94% timing yield and D represents the propagation delay value corresponding to 97% of timing yield. In a similar fashion, power yield trade-offs can be performed.

3.7 The Power Issue in Nanoelectronic Circuits

A study of the energy usage in U.S. homes suggests that energy consumption for appliances and electronics continues to rise. The energy used for appliances, electronics, and lighting was 24% of the total energy consumption in 1993. In 2009, this energy consumption was increased to 34.5%. The scenario is depicted in Fig. 3.46 in which a quadrillion British Thermal Units (BTUs), or quad, is

10^{15} BTU = 1.055 Exa Joule (EJ) [15]. The reasons why power dissipation has become an issue in electronic system design can be summarized as follows [63, 65, 162]:

1. Electronic systems consume a significant amount of energy at home. Wasting precious energy resources affects the environment.
2. Like many other technologies, the battery technology is not expected to keep pace with the growth of electronics.
3. The packaging and cooling costs increase due to higher energy consumption that may lead to noisy, bulky, and costly systems.
4. With scaling, new and ever-more obtrusive forms of power leakage become prevalent and continuously consume battery life even when no useful work is performed by an electronic system.
5. Energy consumption has made reliability and robustness of the electronic system major concerns.

The proliferation of portable systems such as tablets and smart mobile phones in which battery life is a major selling point, strongly suggests the importance of the power issue. The hardware portions of the electronic systems are at present made up of a nanoelectronic technology such as nano-CMOS or TGFET. Other nanoelectronic technologies are aggressively explored for future use. The power dissipation remains the major issue in nanoelectronic-based circuits and systems [63, 65, 162]. Component-wise power or leakage dissipation breakdown of a chip is presented in Fig. 3.47 [48]. The devices and interconnects consume energy. In the nanoelectronic circuits, the interconnects have significant parasitics and more than 50% of chip power dissipation is due to the interconnects.

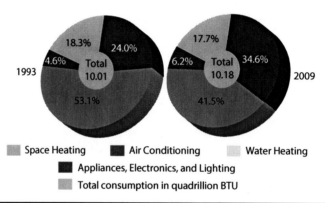

FIGURE 3.46 Energy consumption breakdown by end uses in the United States [15].

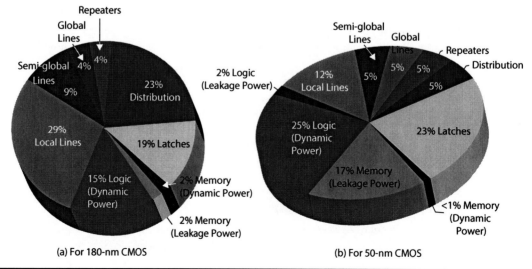

FIGURE 3.47 Component-wise power or leakage dissipation in a chip [48].

In nanoelectronic systems, both the active power dissipation and standby power (leakage) dissipation are important issues. The aggressive scaling of CMOS technology has enabled higher performance and integration levels and has originated new problems in IC design. To match with the technology trend, the supply voltage has continually scaled down to reduce the dynamic power consumption. The threshold voltage and gate-oxide thickness have been reduced along with the supply voltage. As a result, an alarming increase in subthreshold source-drain leakage and gate leakage power dissipation in both active and standby modes of operation has happened. The relative prominence of the power components depends on a specific nanoelectronic technology. In the remaining section, selected nanoelectronic technologies will be discussed. The DFX for nanoelectronics need to handle these to meet the tight power budget of the designs.

3.7.1 Power Dissipation in Nanoscale Classical CMOS Circuits

In nanoscale CMOS transistors, SCEs including the DIBL, large V_{Th} roll-off, diminishing on-to-off current ratio, and BTBT have manifested. Consequently, drastic change has occurred in the leakage components of the device, both in the inactive and the active states. The different current components are shown in Fig. 3.48(a) [154, 158, 162, 201]. The current components are as follows: I_{ds}, drain to source active current (ON state); I_{SC}, drain to source short-circuit current (ON state); I_{sub}, subthreshold leakage (OFF state); I_{ox}, gate leakage (ON and OFF states); I_{hc}, gate current due to hot carrier injection (ON and OFF states); I_{cp}, channel punchthrough current (ON state); I_{GIDL}, gate induced drain leakage (OFF state); and I_{jun}, reverse-bias PN junction leakage (ON and OFF states). The various forms of currents in a nanoscale CMOS device have several forms and origins. The current components flow between different terminals and in different operating conditions of a transistor. For example, the reverse-biased diode leakage and SiO_2 tunnel currents flow during both ON and OFF states, the other currents flow during the OFF state only. The current components are grouped into two categories, static and dynamic, as shown in Fig. 3.48(b). Gate-oxide leakage is a major component of power dissipation in nanoscale CMOS transistor using SiO_2 as the gate dielectric and is present during OFF, ON, and transient states [159]. Thus, it is included in both static and dynamic categories.

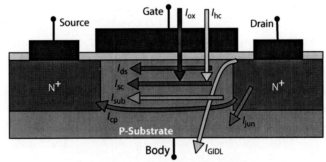

(a) Various forms of current in a nano-CMOS device.

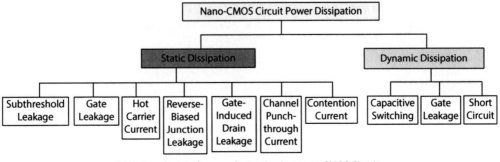

(b) Various forms of power dissipation in a nano-CMOS Circuit.

FIGURE 3.48 The various forms of current in a nano-CMOS device and corresponding power dissipation mechanisms in a nano-CMOS circuit.

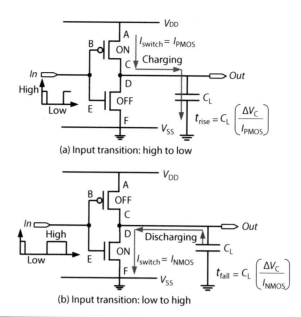

FIGURE 3.49 Concept of capacitive switching power dissipation.

3.7.1.1 Capacitive Switching Power Dissipation

Capacitive switching power dissipation is caused by charging and discharging of parasitic capacitances in the circuit as depicted in Fig. 3.49. The time to charge and discharge is affected by load capacitor along with the charging and discharging currents, respectively [191, 192]. The capacitance may include device capacitances and interconnect capacitances. The capacitance may include the internal capacitances as well as a load capacitance. The average power consumed due to capacitive switching of a CMOS circuit is calculated as follows [141, 171]:

$$P_{\text{switch}} = \frac{1}{2}\alpha_{\text{sw}} C_L V_{\text{DD}}^2 f \qquad (3.53)$$

where the α_{sw} term is an average activity factor that captures how many devices are active on any particular clock cycle, C_L is the total switched capacitive load (including internal and load capacitances), V_{DD} is the supply voltage, and f is the frequency of the clock. P_{switch} term is derived from the equations for energy consumed in charging and discharging a capacitor. For a binary input sequence S of length L_S with L_1 number of "1"s, the average switching activity can be estimated as follows [141, 217]:

$$\alpha_{\text{sw}} = 2\left(\frac{L_1}{L_S}\right)\left(\frac{L_S - L_1}{L_S}\right) \qquad (3.54)$$

For example, for the input binary sequence $S = \{0110110000\}$, L_S is 1 and L_1 is 4. At the logic-level, the average capacitive switching power is calculated using the following expression [171]:

$$P_{\text{switch}} = \frac{1}{2} V_{\text{DD}}^2 f \sum_g \alpha_{\text{sw}}(g) C_L(g) \qquad (3.55)$$

where $\alpha_{\text{sw}}(g)$ is the average switching activity at the output of gate g and $C_L(g)$ is the loading capacitance of gate g. The capacitive switching power dissipation thus depends on the loading capacitance and input vectors and not on the device features.

3.7.1.2 Subthreshold Leakage

Subthreshold leakage current is the OFF-state current of the device. During OFF state of the MOSFET, weak inversion happens and the channel surface potential is constant across the channel. In this

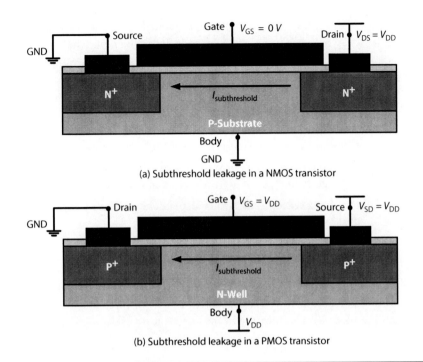

FIGURE 3.50 Subthreshold leakage in NMOS and PMOS Transistor.

situation, the current flow is the result of the diffusion of minority carriers due to a lateral concentration gradient. The condition is depicted for both NMOS and PMOS transistors in Fig. 3.50. The subthreshold leakage current through a device is modeled as follows [1, 119, 220]:

$$I_{\text{subth}} = I_{\text{DS},0} \exp\left(\frac{V_{gs} - V_{\text{Th}}}{\tau v_{\text{therm}}}\right)\left(1 - \exp\left(\frac{-V_{ds}}{v_{\text{therm}}}\right)\right) \tag{3.56}$$

where V_{gs} is the gate-to-source voltage, V_{ds} is the drain-to-source voltage, and v_{therm} is the thermal voltage. The threshold voltage V_{Th} is expressed as follows that has a strong dependence on temperature T [158, 201]:

$$V_{\text{Th}} = V_{\text{fb}} + \left(\frac{2kT}{q}\right)\ln\left(\frac{N_{\text{sub}}}{N_{\text{int}}}\right) + \left(\frac{T_{\text{ox}}}{\varepsilon_{\text{ox}}}\right)\sqrt{2q\varepsilon_{\text{Si}}N_{\text{sub}}\left(\frac{2kT}{q}\ln\left(\frac{N_{\text{sub}}}{N_{\text{int}}}\right) + V_{bs}\right)} \tag{3.57}$$

where N_{int} is intrinsic doping concentration and the Fermi-level ϕ_F is calculated as $\left[\frac{2kT}{q}\ln\left(\frac{N_{\text{chan}}}{N_{\text{int}}}\right)\right]$ [158]. $I_{\text{DS},0}$ is drain current when $V_{gs} = V_{\text{Th}}$ that can be calculated using the following expression:

$$I_{\text{DS},0} = e^{1.8}\mu_0\left(\frac{\varepsilon_{\text{ox}}}{T_{\text{ox}}}\right)\left(\frac{W}{L}\right)v_{\text{therm}}^2 \tag{3.58}$$

where μ_0 is the zero-bias mobility and ε_{ox} is the oxide dielectric constant or relative permittivity. The thermal voltage v_{therm} is calculated as $\left(\frac{kT}{q}\right)$ with k as Boltzmann constant, T temperature, and q electronic charge. τ is the subthreshold slope/swing factor which is calculated as $\left(1 + \frac{C_{\text{body}}}{C_{\text{gate}}}\right)$. For V_{ds} much larger compared to thermal voltage, the subthreshold current is expressed as follows [91]:

$$I_{\text{subth}} = I_{\text{DS},0} 10^{\left(\frac{V_{gs} - V_{\text{Th}}}{S}\right)} \tag{3.59}$$

where $I_{DS,0}$ is drain current when $V_{gs} = V_{Th}$ and S is the subthreshold swing that is calculated as $[\tau v_{therm} \ln(10)]$ which has a unit of mV/decade. It may be noted that $\left(\frac{1}{S}\right)$ is the slope of I-V characteristic in the subthreshold region and is called subthreshold slope. Thus, subthreshold swing is the inverse of subthreshold slope.

The following can be observed from Eqns. 3.56–3.58 about the subthreshold leakage current:

1. Subthreshold current depends on V_{gs}, V_{ds}, and V_{bs}.
2. Subthreshold current depends on threshold voltage V_{Th} and various doping concentrations.
3. Subthreshold current depends on T_{ox} and ε_{ox} as V_{Th} and $I_{DS,0}$ are affected by them. Dependence on ε_{ox} suggests that use of high-κ dielectric will affect the subthreshold current.
4. Subthreshold current has strong dependency on the temperature as $I_{DS,0}$, v_{therm}, and V_{Th} are affected by it.
5. Subthreshold current depends on device geometry W and L.

As barrier lowering (BL), for short-channel MOSFET, V_{Th} decreases as L decreases and consequently the subthreshold current increases [91]. As DIBL, for short-channel MOSFET, an increase in V_{ds} results in a decrease in V_{Th} and consequently the subthreshold current increases. The subthreshold current increases with an increase in the temperature as v_{therm} has a strong effect on it.

3.7.1.3 Gate-Oxide Leakage

Gate oxide leakage arises due to tunneling current through the gate dielectric. The quantum mechanical tunneling between substrate and gate can be either direct tunneling or Fowler-Nordheim tunneling. These two differ in the form of potential barrier. The direct tunneling has trapezoidal potential barrier as shown in Fig. 3.51(b). For a comparative perspective, the flat-band condition is shown in Fig. 3.51(a). For short channel and ultra-thin oxide transistors, Fowler-Nordheim tunneling is negligible [16, 158]. The tunneling probability of an electron is affected by the barrier height, structure, and thickness of the oxide. Three mechanisms are involved in direct tunneling as follows (see Fig. 3.51 [16]): (1) electron tunneling from conduction band (ECB), (2) electron tunneling from valence band (EVB), and (3) hole tunneling from valance band (HVB). In the case of an NMOS transistor, ECB controls gate-to-channel tunneling in inversion, EVB controls gate-to-body tunneling in depletion-inversion, and ECB controls gate-to-body tunneling in accumulation. On the other hand, for a PMOS transistor, HVB controls the gate-to-channel tunneling in inversion, EVB controls gate-to-body tunneling in depletion-inversion, and ECB controls gate-to-body tunneling in accumulation. Φ_{ox} for HVB is higher than Φ_{ox} for ECB, thus the tunneling current associated with HVB is less than that with ECB. Subsequently, PMOS transistors have less gate-oxide leakage than the NMOS transistor.

The gate-oxide current due to direct tunneling is modeled using five components as shown in Fig. 3.52 [1, 16, 171]. The I_{gs} and I_{gd} components are due to the overlap of gate and diffusions. The I_{gcs}, I_{gcd} components are due to tunneling from the gate to the diffusions via the channel. The I_{gb} component is due to tunneling from the gate to the bulk via the channel. The relative significance of

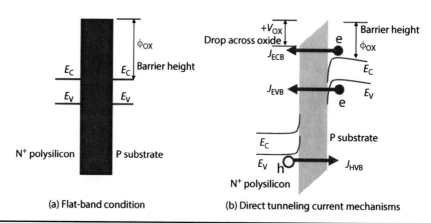

FIGURE 3.51 The three mechanisms of quantum mechanical tunneling in a Nano-CMOS transistor.

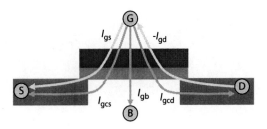

FIGURE 3.52 The components of direct tunneling current in a nano-CMOS transistor.

the components depends on the biasing condition, state of the device, and type of the device. For example, I_{gcs} and I_{gcd} components are present during the ON state only when the channel is formed. The bulk component of the direct tunneling current is typically the smallest component in magnitude. For complete and accurate characterization of gate-oxide tunneling current, three metrics are needed to quantify it during ON, OFF, and transition states of the device [115, 220]. The set of three metrics is ($I_{ox,ON}$, $I_{ox,OFF}$, $C_{tun,eff}$). $I_{ox,ON}$ is the gate-oxide leakage during ON state that is calculated from the above five components. $I_{ox,OFF}$ is the gate-oxide leakage during OFF state which is also calculated from the above five components. During the low-to-high and high-to-low transitions due to the introduction of time-varying components, the effect of direct tunneling is quantified as an effective capacitive load that is defined as follows:

$$C_{tun,eff} = \left| \frac{I_{ox,ON} - I_{ox,OFF}}{\left(\frac{dV_{gs}(t)}{dt}\right)} \right| \tag{3.60}$$

For simplicity, it can be assumed that the rise and fall time of the signal is equal, i.e., $t_{rise} = t_{fall}$. Under this assumption, the above expression can be rewritten as follows:

$$C_{tun,eff} = \left(\frac{|I_{ox,ON} - I_{ox,OFF}|}{V_{DD}} \right) t_{rise} \tag{3.61}$$

$I_{ox,ON}$ affect the ON-state current, $I_{ox,OFF}$ contributes to OFF-state leakage, and $C_{tun,eff}$ increases the intrinsic gate capacitance of the MOSFET.

The current density due to the direct tunneling through gate oxide is calculated as follows [16, 158, 201]:

$$J_{ox} = \frac{q^3}{16\pi^2 \hbar \phi_{ox}} \left(\frac{V_{ox}}{T_{ox}}\right)^2 \exp\left[-\frac{4\sqrt{2m_{eff}\phi_{ox}^3}}{3\hbar q}\left(\frac{T_{ox}}{V_{ox}}\right)\left[1-\left(1-\frac{V_{ox}}{\phi_{ox}}\right)^{\frac{3}{2}}\right]\right] \tag{3.62}$$

where I_{ox} is the direct-tunneling current, W is the width of the transistor, L is the channel length, V_{ox} is the potential drop across the thin oxide, T_{ox} is the oxide thickness, ϕ_{ox} is the barrier height for the tunneling particle (hole or electron), and m_{eff} is the effective mass of the tunneling particle. In a CMOS circuit with a supply voltage V_{DD} and effective gate oxide thickness T_{ox}, the gate-oxide direct tunneling current dissipation is given by [108, 167]:

$$I_{ox} \propto \left(\frac{V_{DD}}{T_{ox}}\right)^2 \exp\left(-\gamma \frac{T_{ox}}{V_{DD}}\right) \tag{3.63}$$

where, γ is an experimentally derived factor. It is evident from the above expressions that the gate oxide leakage is strongly affected by the oxide thickness and supply voltage.

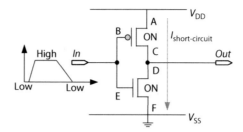

FIGURE 3.53 The components of direct tunneling current in a nano-CMOS transistor.

3.7.1.4 Short Circuit Power

During transitions in the input signal, there may be a small time interval for which both NMOS and PMOS are in ON state in a static CMOS circuit. The condition is depicted in Fig. 3.53. This is not desired in static CMOS technology in which either NMOS (pull-down) network or PMOS (pull-up) network is expected to be ON. However, the situation arises if the transistors are not properly sized. This results in a direct path from power supply to ground for the small time interval that may lead to high current and power dissipation. The short-circuit power dissipation of nanoscale MOSFETs is a nonlinear function of signal transition time, capacitance, and device properties [249].

Under the assumption of equal rise and fall time of the signal, i.e., $t_{rise} = t_{fall}$, in one model, the short circuit power dissipation is expressed as follows [141, 220, 253]:

$$P_{sc} = \frac{\beta}{12}(V_{DD} - V_{Th})^3 t_{rise} f \tag{3.64}$$

where β is the transistor gain factor in $\left[\frac{\mu A}{V^2}\right]$ and f is the clock frequency. This is one of the earliest models and was developed when CMOS was in micron technology node [253]. It is evident from the above equation that the supply voltage, threshold voltage, rise/fall time, and clock frequency affect the short-circuit power dissipation. Another model for short-circuit power dissipation is to express it using the equivalent capacitance concept as follows [249]:

$$P_{sc} = C_{sc} V_{DD}^2 f \tag{3.65}$$

where C_{sc} is the equivalent capacitance that is estimated from the mean accumulated charge that existed during the short-circuit period.

3.7.1.5 Reversed Biased Junction Leakage

In order to get a clear understanding of the parasitic diodes in a MOSFET, let us revisit the cross-section of a nano-CMOS inverter presented in Fig. 3.29. Due to the structure of the MOSFET, several junction diodes are formed as depicted in Fig. 3.54 [162, 169, 261]. In essence, wherever there is a junction between N-type and P-type materials a diode is formed. The junctions appear among the source or drain diffusion, substrate, and well of the structure. For each of the junctions, one can visualize a side junction and a bottom junction. For example, consider one diffusion structure of NMOS marked "D" that is the drain. Then one junction is at the bottom of the drain with substrate (as shown in figure), and there can be a side junction (again with substrate) but toward the left side

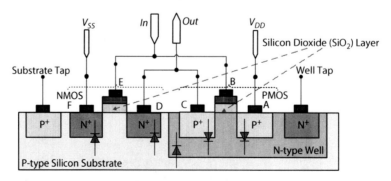

FIGURE 3.54 Junction diodes in a nano-CMOS inverter [162, 169, 261].

of the drain. However, from the behavior point of view it is one diode. The current in the junction diodes, of course, depends on the biasing conditions. In a typical situation, the junction diodes are reversed biased as the substrate is connected to ground and the well is connected to V_{DD}.

The junction leakage current has the following components [139]: (1) diffusion current, (2) generation current, (3) avalanche current, and (4) Zener tunnel current. The diffusion current, generation current, and avalanche current are standard currents that are well known in literature and typically quantified by classic exponential equations. However, for sub-65-nm technology node, the reverse-biased drain-to-substrate and source-to-substrate junction BTBT current are becoming prominent as channel doping densities increase [162, 169, 261]. The BTBT current due to the tunneling in the drain-substrate junction during the OFF state of the MOSFET is expressed as follows [168, 169]:

$$I_{jun,BTBT} = \alpha_{phy} W L_{side} V_{DD} \left(\frac{\xi_{side}}{E_G^{1/2}} \right) \exp\left(-\frac{\beta_{phy} E_G^{3/2}}{\xi_{side}} \right) + \alpha_{phy} W L_{bot} V_{DD} \left(\frac{\xi_{bot}}{E_G^{1/2}} \right) \exp\left(-\frac{\beta_{phy} E_G^{3/2}}{\xi_{bot}} \right) \quad (3.66)$$

In the above expression, $\alpha_{phy} = \left| \frac{2m^{*1/2} q^3}{4\pi^3 \hbar^2} \right|$ and $\beta_{phy} = \left| \frac{4(2m)^{*1/2}}{3q\hbar} \right|$ are physical constants. ξ_{side} and ξ_{bot} are electric fields at the side and bottom junctions, respectively. L_{side} and L_{bot} are lengths of the side and bottom junctions, respectively. E_G is the band gap. A similar expression can be used for drain-substrate junction BTBT current calculation. The overall BTBT current in a MOSFET is the summation at the drain and source junctions. It is observed from the above expression that the junction leakage is affected by the device geometry and supply voltage.

3.7.1.6 Gate-Induced Drain Leakage

A depletion region is formed in the gate and drain overlap region due to the high electric field across it [49, 74, 130]. The situation is depicted in Fig. 3.55 [46, 139]. For an NMOS transistor, GIDL takes place when the gate is at a lower potential than the drain (i.e., $V_{gs} < V_{DD}$) [74, 139]. This results in a significant change in the energy band that allows electron and hole pair generation through avalanche multiplication and BTBT.

In one model, the current density of the GIDL is represented using the following expression [46, 55]:

$$J_{GIDL} = \alpha_{exp} \left(\frac{V_{DD} - V_{gs} - V_{fb} - 1.2}{3T_{ox}} \right) \exp\left(-\frac{3\beta_{phy} T_{ox}}{V_{DD} - V_{gs} - V_{fb} - 1.2} \right) \quad (3.67)$$

where α_{exp} is a preexponential parameter and β_{phy} is a physically based exponential parameter. As evident from the above expression, the device parameters such as V_{DD}, V_{gs}, V_{fb}, and T_{ox} affect GIDL.

3.7.1.7 Hot Carrier Current

Carriers (electrons and holes) are referred to as 'hot carriers" when they have very high kinetic energy due to acceleration by a high electric field [61, 191, 192, 233, 258]. The hot carriers have

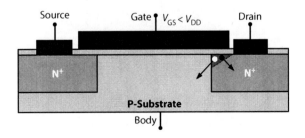

FIGURE 3.55 Concept of GIDL.

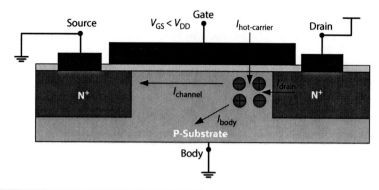

Figure 3.56 Concept of GIDL [61, 233].

enough energy needed to generate electron-hole pairs by impact ionization. The generated bulk minority carriers can either be collected by the drain or injected into the gate oxide while the generated majority carriers create a bulk current that can be used as a measurable quantity to determine the impact ionization. More discussions on the hot carriers will be presented in the section on reliability in the later part of this chapter.

The drain avalanche hot-carrier (DAHC) is significant at stress conditions when V_{gs} is low and V_{ds} is high. The situation is depicted in Fig. 3.56 [61, 233]. The carriers gain their energy due to a high electric field in the drain region. In this situation, the hot carriers lead to impact ionization that generates electron-hole pairs. In the DAHC injection regime, hot electrons and hot holes are injected into the gate oxide. It is caused by the injection of holes and electrons generated by avalanche multiplication. Accurate quantification of DAHC is difficult as both carrier types (electrons and holes) are injected simultaneously. Some of the generated carriers lead to a bulk current. The hot carrier gate current density is expressed as follows [172]:

$$J_{hc} = q n_s \left(\frac{kT_e}{2\pi m^*} \right)^{1/2} \exp\left(-\frac{q\phi_B}{kT_e} \right) \tag{3.68}$$

where n_s is electron density at the surface, T_e is the electron temperature, and ϕ_B is the barrier height between the silicon substrate and gate. Empirically, the hot carrier gate current is calculated as follows [1, 162]:

$$I_{hc} = \frac{ALPHA0 + ALPHA1 L_{eff}}{L_{eff}} (V_{DS} - V_{DSeff}) \exp\left(\frac{BETA0}{V_{DS} - V_{DSeff}} \right) I_{dsNoSCBE} \tag{3.69}$$

where $ALPHA0$, $ALPHA1$, and $BETA0$ are model parameters, V_{DSeff} is an internally computed effective drain-source voltage, and $I_{dsNoSCBE}$ is the drain-source current ignoring the substrate current induced body effect (SCBE).

3.7.1.8 Channel Punchthrough Current

For a better understanding of channel punchthrough, the discussion of CLM in the saturation region of the MOSFET I-V characteristic is needed. In the saturation region, the linear region I-V characteristic of Eqn. 3.2, the nano-CMOS classical MOSFET, changes to the following [205]:

$$I_{ds,SAT} = \left(\frac{1}{2} \right) \mu_{eff} \left(\frac{\varepsilon_{ox}}{T_{ox}} \right) \left(\frac{W}{L_{eff}} \right) (V_{gs} - V_{Th})^2 \tag{3.70}$$

In other words, the drain current in saturation region of the MOSFET is independent of the drain voltage V_{ds}.

In a typical scenario, the following types of depletion region (also known as depletion layer, depletion zone, junction region, or the space charge region) can be visualized: (1) reverse-biased source diode, (2) reverse-biased drain diode, and (3) depletion layer just below the channel. This condition is depicted in Fig. 3.57. The depletion regions at the source and drain ends are symmetrical. For a

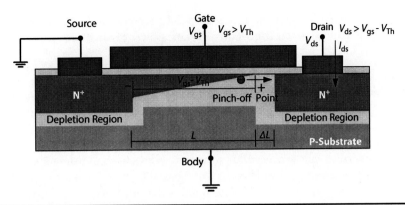

FIGURE 3.57 Concept of CLM and punchthrough.

constant positive gate voltage more than threshold voltage, as the drain voltage (V_{ds}) increases beyond the drain voltage, at which saturation ($V_{ds,SAT}$) is attained (i.e., ($V_{ds} < V_{ds,SAT}$)), pinch-off (i.e., to constrict or squeeze forcefully) of the channel happens. In other words, depletion regions at the source and drain ends start becoming unsymmetrical with the increase of the drain depletion region. At this stage of saturation, the channel is pinched off by a distance ΔL at the drain end of the MOSFET as shown in Fig. 3.57. In this condition, the electrons in the inversion layer move through this depletion region as they have enough energy and cause additional drain current. As a result, I_{ds} increases slightly with V_{ds} instead of being constant at the saturation region. The additional saturation current is modeled by the expression $(1+\lambda V_{ds})$ with λ being the CLM coefficient or CLM parameter. The drain saturation current with CLM effect is expressed as follows:

$$I_{ds,SAT} = \left(\frac{1}{2}\right)\mu_{eff}\left(\frac{\varepsilon_{ox}}{T_{ox}}\right)\left(\frac{W}{L_{eff}}\right)(V_{gs}-V_{Th})^2(1+\lambda V_{ds}) \qquad (3.71)$$

In other words, the drain current of the MOSFET is again dependent on the drain voltage V_{ds}.

If drain voltage (V_{ds}) is further increased, the depletion regions at the source and drain ends become much more unsymmetrical. Then at some point, the drain depletion region physically touches the source depletion region [26]. This phenomenon is called punchthrough. The punchthrough leads to lowering of the energy barrier of the source, and as a result, a large amount of current flow can happen even with a gate bias to turn off the MOSFET [26]. Punchthrough decreases the output resistance and limits the maximum operating voltage of the MOSFET in order to prevent its damage due to the high electron acceleration. Punchthrough in a MOSFET can be considered as an extreme case of CLM. The drain voltage at which punchthrough happens is called punchthrough voltage (V_{PT}) which is modeled as follows [26, 34]:

$$V_{PT} = \left(\frac{\pi\varepsilon_{ox}L_{eff}}{\varepsilon_{Si}T_{ox}}\right)(V_{gs}-V_{Th}) \qquad (3.72)$$

The channel punchthrough current is modeled as follows [82, 223]:

$$I_{cp} = I_0\left(\frac{W}{L_{eff}}\right)\exp\left(\frac{\phi_{min}+V_{bs}}{V_{Th}}\right)\left(1-\exp\left(-\frac{V_{ds}}{V_{Th}}\right)\right) \qquad (3.73)$$

In the above expression, I_0 is an experimental factor and is approximated as the diode reverse saturation current. ϕ_{min} is the potential minimum along the minimum-barrier path. Thus, punchthrough current is dependent on device geometry, process parameters, and applied voltage at the drain and body.

3.7.1.9 Contention Current

Contention current can be present in a CMOS circuit in specific conditions. The fully or static complementary CMOS circuit is ratioless as the output signal does not depend on the size of the transistors. However, this is not the case in ratioed circuits such as dynamic or domino logic circuits.

The ratioed circuits dissipate power in a fight among the ON transistors [120, 241]. This is called contention current and it contributes to static power dissipation.

3.7.2 Power Dissipation in Nanoscale High-κ and Metal-Gate FET

Different types of power dissipation in a HKMG-based CMOS circuit have several forms as presented in Fig. 3.58. Due to easier growth of SiO_2 from silicon substrate and other important properties, SiO_2 has been the best choice for several decades. SiO_2 worked well with MOSFET with SiO_2 layers as thin as 1.5 nm, which is equivalent to seven layers of it. However, further scaling of SiO_2 layer thickness leads to several gate leakages due to quantum mechanical tunneling. One of the solutions to this problem is to use a gate insulator with a higher dielectric constant κ than that of SiO_2 and metal gate that is referred to as HKMGFET. This reduces the gate leakage and improves the reliability of the gate. However, the use of HKMG causes a significant GIDL current (I_{GIDL}) in addition to subthreshold leakage (I_{sub}) [74, 116, 167]. I_{GIDL} is high mainly because [49, 74, 131]:

- The metal gate introduces a high gate effective work function that leads to a high electric field and a high GIDL current.
- The high-κ gate dielectric and SiO_2 spacers meet at the surface of the drain region, causing a high electric field leading to a high GIDL current.

The GIDL has a strong impact in HKMG transistors [74]. The BSIM4 expression used to model GIDL is [1, 74]:

$$I_{GIDL} = AW_{effCJ} N_f \left(\frac{V_{ds} - V_{gs} - E}{3T_{gate}} \right) \exp\left(\frac{-3T_{gate} B}{-V_{ds} - V_{gs} - E} \right) \left(\frac{V_{db}^3}{C + V_{db}^3} \right) \tag{3.74}$$

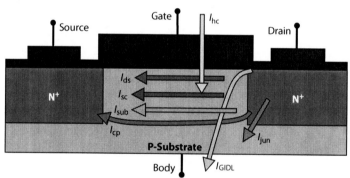

(a) Various forms of current in a HKMG nano-CMOS device.

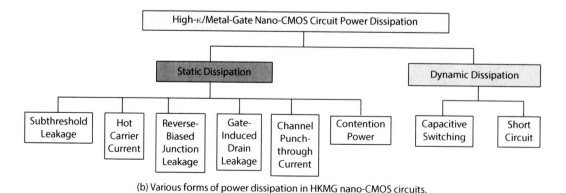

(b) Various forms of power dissipation in HKMG nano-CMOS circuits.

Figure 3.58 The various forms of current in a Nanoscale HKMG device and corresponding power dissipation mechanisms in a nano-CMOS circuit.

In the above expression, V_{db} is the drain to body voltage. W_{effCJ} is the effective width of the drain diffusions. N_f is the number of fingers of the transistor and A, B, C, and E are all BSIM4 GIDL fitting parameters that are obtained from existing data [49, 74, 130].

In general, the conduction mechanism in dielectrics include [265]: (1) direct tunneling, (2) Fowler-Nordheim tunneling, (3) field ionization of trapped electrons, (4) hopping of thermally excited electrons, (5) Poole-Frenkel effect, (6) shallow trap-assisted tunneling, (7) space charge limited current, and (8) trap-assisted tunneling. These conduction mechanisms have similar field dependencies and I-V characteristics cannot differentiate them [265]. In a good thin-field dielectric, the current conduction is generally governed by Poole-Frenkel effect and Fowler-Nordheim tunneling. In a good ultra-thin dielectric (i.e., thickness less than 3 nm), the direct tunneling is more prominent. For thin dielectrics with high trap density, trap-assisted conduction, hopping, and space charge limited conduction may be involved in the conduction mechanisms. However, it seems that only considering those mechanisms is inadequate for the ionic or heteropolar high HKMG.

Different forms of the current and power dissipation, which are shown in Fig. 3.58(b), for HKMG can be calculated similar to the way the classical nano-CMOS circuits were in the above section. For example, the capacitive switching power and short-circuit power dissipation are calculated using corresponding capacitances similar to the classical nano-CMOS circuits as in the previous section. The value of capacitances may be different in the two technologies. Similarly, the parameters used in subthreshold leakage, BTBT current, etc. may be different from classical nano-CMOS.

3.7.3 Power Dissipation in Double-Gate FinFET

Various forms of current flow and the corresponding power dissipation for DG FinFET are shown in Fig. 3.59. The capacitive switching and short circuit power dissipation of FinFET-based circuits can be calculated in a similar manner as in classical nano-CMOS circuits using switching activity parameters. The origin and nature of contention power dissipation is the same as other nano-CMOS circuits in the previous sections.

In the FinFET, the diffusion mechanism is mainly responsible for current transport in the subthreshold region, as opposed to drift in the strong inversion region. The subthreshold current in a DG FinFET is expressed as follows [238]:

$$I_{sub} = \alpha \left(\frac{H_{Fin}}{L_{eff}}\right) \exp\left(\frac{V_{gs} - V_{Thnf}}{\beta}\right)\left(1 - \exp\left(\frac{-qV_{ds}}{kT}\right)\right) \quad (3.75)$$

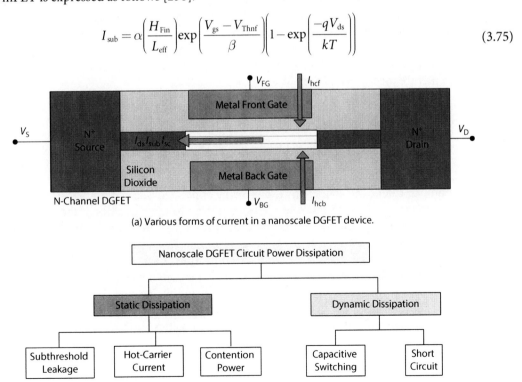

(a) Various forms of current in a nanoscale DGFET device.

(b) Various forms of power dissipation in nanoscale DGFET circuits.

FIGURE 3.59 The various forms of current in a DG FinFET and the corresponding power dissipation mechanisms in nanoelectronic circuits.

where k is Boltzmann constant, T is the temperature in kelvins, α and β are fitting parameters, q is the electronic charge, H_{Fin} is the Fin height, and V_{Thnf} is the front-gate threshold voltage.

In the DG FinFET, two prominent components of tunneling are the current from the channel region at surface inversion condition and the sum of currents from the overlap regions [68]. The latter component is predominant in the subthreshold mode. The first component is due to the direct tunneling of carriers and the second component is due to the trap states through the dielectric material and its interface with silicon. In the DG FinFET, the gate current can be calculated by the following expression [68]:

$$I_{\text{gate}} = \left[\frac{n_{\text{Fin}}(2H_{\text{Fin}} + W_{\text{Fin}})}{2}\right](L_{\text{eff}} J_{\text{DT}} + L_{\text{overlap}} J_{\text{TAT}}) \quad (3.76)$$

where n_{Fin} is the total number of Fins. H_{Fin} is the height of the Fin in a FinFET. W_{Fin} is the width of the Fin in a FinFET. L_{overlap} is the gate overlap length. J_{DT} is the current density due to the direct tunneling effect. J_{TAT} is the current density due to trap-assisted tunneling mechanism at the overlap regions of FinFET structure.

Hot carriers can be a problem in DG FinFET as well. In an n-channel DG FinFET, the energetic electrons generated by impact ionization near the drain can be trapped at the gate-oxide interface (or in the gate oxide), while the generated holes flow to the region of lowest potential. The hot-carrier immunity of DG FinFET improves as the Fin width or body thickness decreases, which facilitates gate-length scaling. On the other hand, the hot-carrier immunity of DG FinFET degrades at the elevated temperatures due to the self-heating effects.

3.8 The Issue of Parasitics in Nanoelectronic Circuits

Parasitics are the single most important challenge that consumes a large portion of the design cycle time. All the real-circuit elements present in nanoelectronic-based chips such as diodes, transistors, and even wires have significant parasitics that create discrepancies between the ideal and actual performances of the chips. In all the circuit types, whether digital, analog, or mixed-signal, the parasitics have a strong effect. However, in the high-frequency circuits, the effects of the parasitics are felt most. For example, parasitic capacitance in an operational amplifier (OP-AMP) can reduce its bandwidth. The parasitics in a voltage-controlled oscillator (VCO) affect its characteristics such as center frequency and phase noise.

3.8.1 Different Types of Parasitics

For nanoscale circuits, the parasitics can arise from various elements [195]. For a broad overview, the parasitics types are summarized in Fig. 3.60. The nanoelectronic devices, nanoscale interconnects (like Cu), nanoelectronic interconnects (such as CNTs), on-chip inductor coils, on-chip capacitor, and pad can be the source of various forms of parasitics. In nanoscale circuits, the parasitics due to interconnects is much more severe as compared to the parasitics of the devices. The following can be a rule of thumb for prominence of the interconnect versus transistor parasitics. For short interconnect length, the transistor parasitics dominate, hence the transistor speed determines the circuit speed. For medium interconnect length, the transistor resistance and interconnect capacitance govern the circuit speed. However, for the long interconnect length, the interconnect resistance and interconnect capacitance govern the circuit speed [98, 99, 208]. The parasitics can be passive such as resistance, capacitance, inductance, mutual capacitance, and mutual inductance. The active parasitics include PN junctions and BJTs. The parasitics are dependent on the layout implementation of the elements.

3.8.2 Device Parasitics

The devices in a typical mode of operations exhibit parasitic capacitances and parasitic resistances. At a high-frequency mode of operations, the devices can also exhibit parasitic inductances.

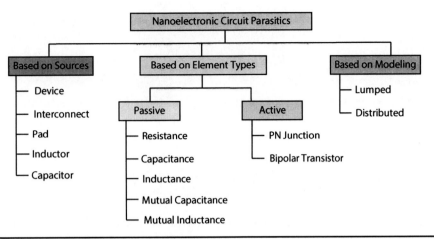

FIGURE 3.60 Different types of parasitics in nanoscale circuits.

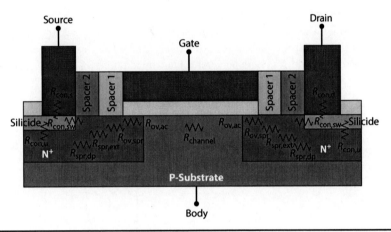

FIGURE 3.61 Resistance components in a MOSFET structure.

3.8.2.1 Device Parasitic Resistances

Another important parasitic that is present in the MOSFET is resistance. In general, the overall resistance offered by the MOSFET can be calculated as follows [237]:

$$R_{MOS} = R_{source} + R_{drain} + R_{channel} \qquad (3.77)$$

where R_{source} is parasitic resistance of the source, R_{drain} is parasitic resistance of the drain, and $R_{channel}$ is the resistance offered by the channel. The schematic representation of different resistances is shown in Fig. 3.61 [109, 110, 237]. The extrinsic resistance of either source or drain has the following components [109, 110]:

1. Contact resistance due to metal and silicide interface at source ($R_{con,s}$) as well as at the drain ($R_{con,d}$).

2. The resistance of the metal ($R_{met,s}$) as well as silicide ($R_{sil,s}$) at the source can present resistances. These are typically smaller than the contact resistances. The metal and silicide resistances are also present at the drain.

3. The contact resistance at the silicide-to-silicon interface R_{con}. It has two components, ($R_{con,u}$) and ($R_{con,sw}$), which are interfaces at upper and sidewall, respectively. R_{con} is the parallel connected value of ($R_{con,u}$) and ($R_{con,sw}$).

4. The spreading resistance under the sidewall spacers (R_{spr}). It is the summation of ($R_{spr,dp}$) and ($R_{spr,ext}$), which are due to deep junction and extension portions.

5. The resistance in the extension-to-gate overlap (R_{ov}). It has two components in series $R_{ov,spr}$ and $R_{ov,ac}$ due to the spreading of overlap and accumulation part.

The source resistance R_{source} is calculated in the following manner:

$$R_{source} = R_{con,s} + R_{met,s} + R_{sil,s} + \frac{R_{con,u} R_{con,sw}}{R_{con,u} + R_{con,sw}} + R_{spr,dp} + R_{spr,ext} + R_{ov,spr} + R_{ov,ac} \qquad (3.78)$$

The same expression applies for the drain resistance R_{drain}. The calculation of channel resistance is more involved as its state changes significantly for various biasing conditions and operating regions. The channel resistance for the linear regions is calculated as follows [237]:

$$R_{channel} = \frac{L_{eff}}{\mu_{eff} C_{ox} W (V_{gs} - V_{Th})} \qquad (3.79)$$

In the saturation region, several physical mechanisms affect the channel resistance of the MOSFET [165]. These mechanisms include the following [165]: CLM, DIBL, and the SCBE. One of these mechanisms dominates in the order CLM, DIBL, and SCBE with the increase of V_{ds} in the I-V characteristic.

3.8.2.2 Device Parasitic Capacitances

For a MOSFET, the parasitic capacitances are shown in Fig. 3.63. In the figure, T_{spacer} and T_{ge} are spacer width and gate electrode thickness, respectively [163]. The capacitances, C_{gs}, C_{gd}, C_{gb} are gate capacitance to source, drain, and body, respectively. The capacitances, C_{sb} and C_{db} are source and drain diffusion

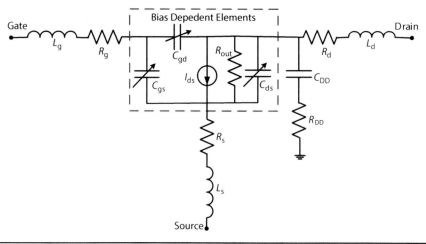

FIGURE 3.62 A MOSFET model showing parasitics in RF domain including inductors [47].

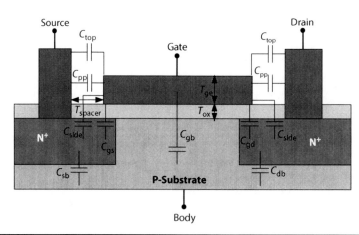

FIGURE 3.63 Capacitance components in a MOSFET structure.

capacitances, respectively. C_{side} is the sidewall fringing capacitance associated with electric field lines emerging from the sides of gate electrode, going through the sidewall spacer and ending at source or drain regions of the MOSFET [163]. C_{pp} is the parallel plate capacitance associated with electric field lines arising from the sides of gate electrode, going through the side wall spacer and ending at source or drain electrodes (a.k.a contact plugs) [163]. C_{top} is the top fringing capacitance associated with field lines emerging from the top surface of the gate electrode and going through the first layer of passivation or planarization dielectrics and ending at source or drain electrodes (a.k.a. contact plugs) [163].

3.8.2.3 Device Parasitic Inductances

In addition to the above discussed parasitic capacitances and resistances, the MOSFET can have parasitic inductances that are manifested in RF domain [47]. A transistor model that includes parasitics for its RF domain is schematically presented in Fig. 3.62 [47]. It shows the capacitance, the bias depended capacitances, the resistances, and the bias-dependent resistances. Most importantly, the parasitics inductors that cannot be ignored in high frequency have been presented. Three inductances shown as lumped values are L_g, L_s, and L_d for gate, source, and drain terminals, respectively.

3.8.3 Interconnect Parasitics

For the purpose of illustration, the layers of interconnects are presented in Fig. 3.64. For example, it shows nine layers of interconnects for a 45-nm HKMG technology [150]. It illustrates the concepts of vias and contacts needed by the interconnects. The cross-sectional view of the above interconnect layers is depicted in Fig. 3.65. It may be noted that the dielectric for the gates of the transistors may be high-κ. However, the dielectric among the interconnects that provides physical support to them is preferred to have low-κ in order to have small interwire parasitic capacitances. The interconnects can be global, semiglobal, or local based on their length, diameter, and current-carrying capability [208]. The longer interconnects are of higher conductivity material such as copper. On the other hand, the local, shorter interconnects can be of lower conductivity material such as silicides or tungsten.

The design constraints such as current density and propagation delay force a larger number of interconnect layers. The interconnect layers are dictated by the upper limits on the length of local, of semiglobal interconnects. These lengths are decided by the above design constraints. Typically, more interconnects are used at the lower layers. The Rent Rule is used for a priori wire length estimation during the design process [198, 255]. The wire length distribution for a whole chip can be determined by the recursive use of the Rent Rule. The Rent Rule, in essence, helps designers to reasonably predict the various characteristics such as number of routing layers needed, area of chip, power dissipation, and clock frequencies of chips for nanoelectronic technologies. The Rent Rule is an empirical power law relationship of the following form:

$$N_p = \alpha_{R-C} N_g^{\beta_{R-E}} \tag{3.80}$$

where N_p is the number of pins or the number of external connections to a logic block. N_g is the number of logic gates in the logic block. α_{R-C} is called "rent coefficient," that is, a proportionality

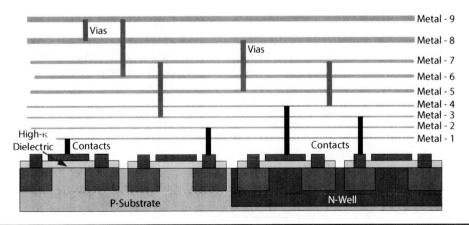

FIGURE 3.64 Illustration of interconnect metal layers in a nanoscale IC.

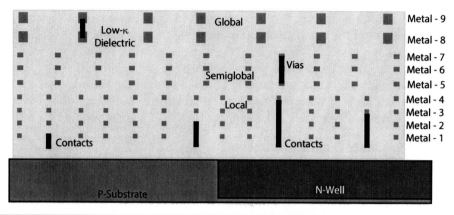

Figure 3.65 Cross-section of the interconnect metal layers in a nanoscale IC.

constant or multiplier that corresponds to the average number of ports. β_{R-E}, called "rent exponent," is another constant. A high rent exponent indicates that there are many global interconnections, and hence a high interconnection complexity. For a static random access memory (SRAM), which is a device size limited with regular compact structure and needs fewer interconnects, $\beta_{R-E} = 0.12$ with $\alpha_{R-C} = 6$. For a microprocessor, which is interconnect pitch limited with irregular structures and needs more levels of interconnects, $\beta_{R-E} = 0.45$ with $\alpha_{R-C} = 0.82$. However, it may be noted that the Rent Rule was originally proposed for logic layout and routing in mainframe computers.

There are a few important aspects of the interconnects in the nanoscale ICs:

1. The interconnect is more of a major contributor of parasitics of an IC than the active transistors for nanoscale technologies.

2. For the interconnects resistance, capacitance, inductance, and mutual capacitance (RCLK) are present in typical frequency-operating frequency range. In such a frequency range, transistor has RC as main parasitics.

3. For wide global wires, the inductance is prominent. For RF operations, the wire can have mutual inductance and the transistors can have inductance parasitics.

4. For the interconnects, the RCLK are all distributed throughout the length of the wire. However, this is not so in the case of transistors. The distributed effect is illustrated in Fig. 3.66 in which r_i or r_j is per unit resistance, c_i or c_j is per unit capacitance to ground, c_{ij} is per unit interwire capacitance, and l_i or l_j is per unit inductance. The distributed nature of interconnect parasitics makes it more difficult to model than the transistor parasitics.

5. The dielectric preferred for the interconnects is a low-κ so the interwire capacitance is minimum and in turn will have minimal crosstalk. However, the dielectric preferred for the transistors is high-κ in order to reduce the gate-oxide leakage that has been a problem for sub-90-nm CMOS technology nodes.

The major effects of interconnect parasitics on performances of the integrated circuit include the following: power loss, signal delay, signal noise, and data transmission reliability. The I^2R loss due to resistances and CV^2f loss due to capacitances add to the overall power dissipation of the circuits. The parasitic resistances and capacitances can introduce RC delays in the circuit. The inductance also affects delay when the interconnect length is under critical length [98]. Noises in the circuit due to interconnects can arise from many sources such as power supply noise, crosstalk, shot noise, thermal noise, and flicker noise. The noises can then lead to issue of data transmission reliability. The parasitics inductance can introduce a noise voltage of $L\left(\frac{di}{dt}\right)$. The parasitics inductance can introduce a noise current signal of $C\left(\frac{dv}{dt}\right)$. Similarly, IR drop can have impact on voltage signal and is a noise. The crosstalk is noise due to mutually coupled parasitics in the wires. The interwire capacitance leads to crosstalk in which a switching in one interconnect affects signal in the other interconnect. The mutual parasitic inductance also affects the crosstalk.

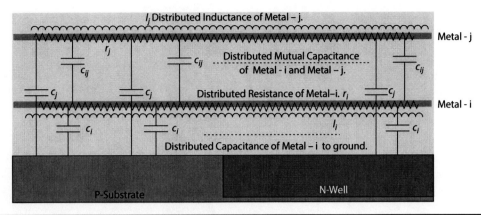

FIGURE 3.66 Parasitics components of interconnects in a nanoscale IC.

Material Properties	Different Materials			
	Copper (Cu)	Silver (Ag)	Gold (Au)	Aluminum (Al)
Resistivity (Ωm)	1.67×10^{-8}	1.59×10^{-8}	2.35×10^{-8}	2.66×10^{-8}
Thermal Conductivity (W/m)	398	425	315	238
Melting Point (°C)	1085	962	1064	660
Resistance to Air Corrosion	Poor	Poor	Excellent	Good
Adhesion to SiO_2	Poor	Poor	Poor	Good

TABLE 3.4 Selected Interconnect Materials [107]

The interconnects thus have many requirements for reduced effects at the nanoscale dimensions from the electrical, reliability, and material processing point of view [208]. From electrical properties point of view, interconnect materials must have low capacitance, low mutual capacitance, low inductance, and low resistance. From reliability point of view, resilience to electromigration (EM), stability of electrical contacts to silicon and other layers, and good adhesion to other layers are important requirements. From a materials processing point of view, the important requirements include easy deposition of thin films of the material and ability to withstand the chemicals and high temperatures. To fulfill the above requirements, copper has been a preferred interconnect material. In the past, aluminum has been extensively used due to low cost and inertness to silicon when copper was contaminating silicon due to its high diffusivity in silicon. However, due to advanced processing technology, copper became popular as it has very low resistivity compared to aluminum. Selected materials that can be used as interconnects are presented in Table 3.4 [107].

The dual-damascene process played an important role in making copper the preferred interconnect material. The simplified steps of dual-damascene process for copper interconnect are shown in Fig. 3.67 [107]. In a dielectric layer, interconnect trench is first made using etching for the inlaying of copper for the main metal interconnects. At this point, etching is performed again for the copper inlaying for vias. A diffusion barrier layer is now deposited in the etched area of trench and via. The barrier layer is very important to stop the diffusion of copper into surrounding materials that otherwise may degrade their properties. At this point, one metal deposition is used to form both the main interconnect copper line as well as the copper vias. Both trenches and vias are formed in a single dielectric layer, hence the name dual damascene [264]. At the end step, a chemical mechanical polishing (CMP) can be performed to obtain a polished surface. The above process can be similarly repeated to build multilevel interconnections.

In the nanoelectronic circuits, CNT interconnects are being considered [143, 235, 273]. The current density limits the current-carrying capabilities of the copper (Cu) conductors that may restrict its scaling. CNTs have much higher current densities, in the order of 10^{10}A/cm^2, which is double that of the Cu. For narrow Cu interconnects, resistivity also increases electron surface scattering and grain

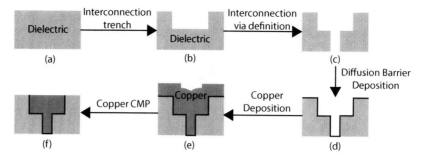

FIGURE 3.67 The steps of the dual-damascene process for copper interconnects [107].

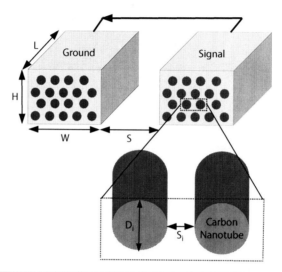

FIGURE 3.68 Parasitics components of CNT interconnects in a nanoscale circuits [143].

boundary scattering. These effects can also lead to EM problems in Cu interconnects. CNT interconnects do not have such issues. In addition, CNT interconnects can provide lower capacitances and inductances, and higher performances. Both SWCNTs and MWCNTs are being considered as possible replacements for on-chip Cu interconnects. It is observed that MWCNT is a better interconnect than SWCNT and Cu. The idea of CNT-based interconnects is depicted in Fig. 3.68 [143]. In this, a CNT interconnect or bundles of CNTs run in parallel to have lower resistance and higher current-carrying capabilities. CNT can be a preferred structure for local as well as global interconnects. Detailed modeling of the interconnects with parasitics will be presented in a later chapter.

3.8.3.1 Interconnect Parasitic Resistances

For an interconnect, the resistance depends on its cross-section and length and on the materials used. The structure of an interconnect with rectangular cross section is depicted in Fig. 3.69(a). The dimensions are as follows: L is the length of the interconnect, W is the width, and T is the thickness, thus making the cross-sectional area WT for a rectangular cross-section of the interconnect. Therefore, for an interconnect, the resistance is calculated as follows [4]:

$$R = \rho \left(\frac{L}{WT} \right) \tag{3.81}$$

where ρ resistivity of the interconnect material is expressed in (Ω/m^2) and R is expressed in Ω. The resistance of the interconnect increases as its length increases. On the other hand, the resistance of the interconnect decreases as its area of cross-section increases. With low resistivity, copper has been the choice of interconnects when aluminum was the low cost in the past. Aluminum was not

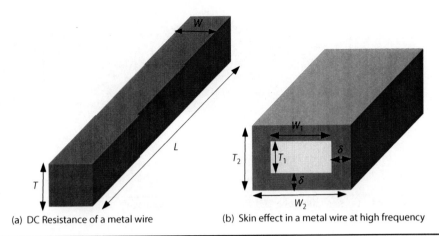

FIGURE 3.69 Parasitics components of interconnects in a nanoscale IC.

only low cost but was also easily processed during fabrication flow. Cu is the interconnect of choice after the dual-damascene process was invented. Along with the low-κ dielectric, the Cu interconnect essentially provided low resistance, low capacitance, and low delay as compared to the aluminum interconnect.

For the metal interconnects operating in high-frequency, such as in RF circuits, the skin effect is a cause of concern. At high-frequency operations, the electron flow is not uniform throughout the cross-section of the interconnect. The electron concentration is dense at the periphery (or outer surface) and sparse at the center, thus the term *Skin Effect*. Due to the skin effect, the resistance of the wire increases as the effective cross sectional area in which electrons flow decreases. The idea of skin effect is demonstrated in Fig. 3.69(b) [4]. The electrons in essence flow in the δ region at the periphery. In this figure, the cross-sectional area effectively decreases from $W_2 T_2$ to $W_1 T_1$. Thus, the effective skin depth area is $(W_2 T_2 - W_1 T_1)$. The skin depth δ is the depth below the surface of the conductor at which the current density is $\left|\frac{1}{e}\right|$ times the surface current density. The skin depth can be calculated using the following expression:

$$\delta = \sqrt{\frac{\rho}{\pi f \mu}} \quad (3.82)$$

where ρ is resistivity of the interconnect material, f is the frequency of the current flow, and μ is the absolute magnetic permeability of the interconnect material. The skin depth is larger in aluminum, followed by gold and then copper. In other words, skin depth is minimal for the copper among the three materials.

The cross-section of a CNT-based interconnect is shown in Fig. 3.70(a) [235]. It represents the bundles of CNTs that form the interconnect. The CNT bundle is then surrounded by the dielectric material and the grounded shield. The diameter of the inner CNT bundle is d and the inner diameter of surrounding grounded shield is D. The dielectric material has thickness of $\left|\frac{D-d}{2}\right|$. The diameter of a single CNT is d_{CNT}. There are air spaces among the individual strands of the CNT bundle as the individual CNT strands do not fill it tightly. This is illustrated in the cross-section of Fig. 3.70(b) [235]. The number of tubes in a CNT bundle interconnect is approximated by the following expression:

$$n = n_F n_G \left(\frac{d}{d_{\text{CNT}}}\right)^2 \quad (3.83)$$

where n_F is the fill factor and n_G is the growth density factor of the CNT bundle.

For a long CNT interconnect consisting of a SWCNT bundle, the total resistance has two components, contact resistance and the intrinsic scattering resistance. However, for a short CNT interconnect that has length shorter than the ballistic limit, there may be additional contact resistance

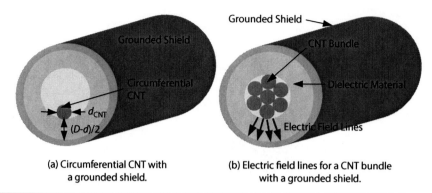

(a) Circumferential CNT with a grounded shield.

(b) Electric field lines for a CNT bundle with a grounded shield.

Figure 3.70 Cross-section of a CNT interconnect in a nanoscale IC [235].

due to imperfect electrode contacts. For a CNT bundle with n SWCNTs, the contact resistance at each contact end is expressed by the following:

$$R_{\text{contact,bundle}} = \left(\frac{h}{8e^2 n}\right) \tag{3.84}$$

Based on the Eqn. 3.83, the above expression can be rewritten as follows:

$$R_{\text{contact,bundle}} = \left(\frac{h}{8e^2}\right)\left(\frac{1}{n_F n_G}\right)\left(\frac{d_{\text{CNT}}}{d}\right)^2 \tag{3.85}$$

The above expression suggests that the CNT bundle contact resistance increases with an increase in CNT tube diameter d_{CNT} and decrease in CNT bundle inner diameter. For the same CNT bundle with n tubes, the scattering resistance is calculated using the following expression:

$$R_{\text{scattering,bundle}} = \left(\frac{h}{4e^2 \lambda n}\right) \tag{3.86}$$

where λ is the electron mean free path of the CNT. Based on the Eqn. 3.83, the above expression can be rewritten as follows:

$$R_{\text{scattering,bundle}} = \left(\frac{h}{4e^2 \lambda}\right)\left(\frac{1}{n_F n_G}\right)\left(\frac{d_{\text{CNT}}}{d}\right)^2 \tag{3.87}$$

Similar to the $R_{\text{contact,bundle}}$, the CNT bundle scattering resistance $R_{\text{scattering,bundle}}$ also increases with an increase in CNT tube diameter d_{CNT} and a decrease in CNT bundle inner diameter.

3.8.3.2 Interconnect Parasitic Capacitances

Parasitic capacitances manifested by the complex interconnects of the ICs are of various forms [208, 218, 225]. An illustration of three-layer metal interconnects on the silicon substrate is presented in Fig. 3.71(a). The figure depicts silicon substrate and SiO_2/high-κ as gate dielectric with either polysilicon or metal gate. The dielectric among interconnect layers is SiO_2/low-κ. A selected number of low-κ dielectrics are presented in Table 3.5. A simplified view of the cross-sections of three-layers of metal for the calculation of different types of interconnect parasitics capacitances is presented in Fig. 3.71(b). The interconnect capacitance has the following components [32]:

1. The lateral or horizontal capacitance between two intralayer interconnects. This will be called the interwire or intermetal capacitance $C_{\text{intermetal}}$.

2. The vertical capacitance due to the overlap between two interconnects in two different metal layers. This will be called the interlayer or interlevel capacitance $C_{\text{interlayer}}$.

3. The fringe capacitance is between the sidewall of one interconnect and the top or bottom of another interconnect of a different interconnect layer. The fringe capacitance will be denoted as C_{fringe}.

Nanoelectronics Issues in Design for Excellence 131

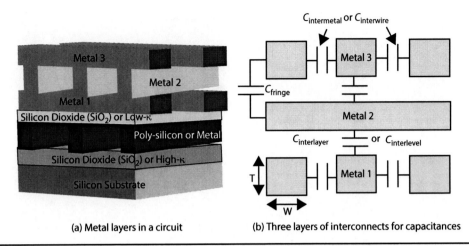

(a) Metal layers in a circuit (b) Three layers of interconnects for capacitances

FIGURE 3.71 Illustration of the trends of interconnect parasitic capacitances in an IC.

Dielectric Names	Dielectric Constant or Relative Permittivity
Air	1.0
Carbon Dioxide	1.0
Carbon-Doped Silicon Dioxide	3.0
Silicon Dioxide (SiO_2)	3.9
Fluoropolymer	2.24
Fluorine-Doped Silicon Dioxide	3.5
Nanoglass	≈2.0
Polyethylene	2.4
Polypropylene	2.3
Porous Silicon Dioxide	<2.0
Organic Polymer – Benzo-cyclo-butane	2.7
Organic Polymer – Polyimide	3.5
Organic Polymer – Polytetrafluoroethylene	2.1
Xerogels	1.2

TABLE 3.5 Selected Low-κ Dielectrics with Their Permittivity [107, 219]

The interwire or intermetal capacitance is the capacitance between any two wires in the same layer or level. Using parallel plate approach, this capacitance is calculated as follows:

$$C_{\text{intermetal}} = \varepsilon_{\text{IM}} \left(\frac{LT}{d_h} \right) \quad (3.88)$$

where T is the thickness of a metal layer, ε_{IM} is the dielectric constant of the intermetal dielectric and d_h is the horizontal distance between two wires on a same layer. This can be calculated from the horizontal pitch of the wire. $C_{\text{intermetal}}$ increases with increase in length and thickness of the interconnects. The interlayer or interlevel capacitance is the capacitance between any two wires in two different layers or levels. Using the parallel plate approach, this capacitance is calculated as follows:

$$C_{\text{interlayer}} = \varepsilon_{\text{IL}} \left(\frac{LW}{d_v} \right) \quad (3.89)$$

where W is the width of a metal layer. ε_{IL} is the dielectric constant of the interlayer dielectric. d_v is the vertical distance between two wires of two different metal layers. This can be calculated from

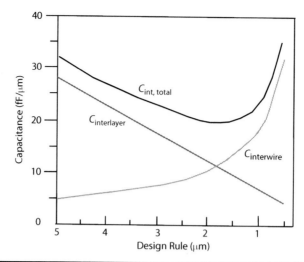

Figure 3.72 Illustration of the trends of interconnect parasitic capacitances in an IC.

the vertical pitch of the wire. $C_{interlayer}$ increases with an increase in the length and width of the interconnects. As a specific case of the interlayer capacitance, the capacitance to the ground C_{ground} (a.k.a. self-capacitance) can be calculated, which is capacitance of a layer to the grounded substrate.

With the technology scaling, the interconnect width W has decreased, and at the same time, the interconnect width T has increased. Thus, the technology scaling has resulted in a significant increase in the lateral and fringe capacitances [32]. For the accurate calculation of the interconnect capacitances, the fringe capacitance needs to be accounted for. However, calculation of the fringe capacitances of complex interconnects is a very difficult problem [32]. The trend of different capacitances with the technology change is presented in Fig. 3.72 [135, 208]. With the technology scaling, the interwire capacitance has increased significantly, whereas the interlayer capacitance has reduced. However, the overall capacitance of the interconnects $C_{int,total}$ is increasing significantly with technology scaling.

In the case of CNT-based interconnects, shown in Fig. 3.70, two forms of capacitances are manifested [44, 235]. These are quantum and electrostatic capacitances that will be denoted as $C_{CNT,QM}$ and $C_{CNT,ES}$, respectively. The total capacitance per unit length of the CNT bundle-based interconnect with n SWCNTs is then expressed by the following [235]:

$$C_{CNT,tot} = \left(\frac{C_{CNT,QM} C_{CNT,ES}}{C_{CNT,QM} + C_{CNT,ES}} \right) \quad (3.90)$$

The (quantum) capacitance per unit length is calculated as follows [44]:

$$C_{CNT,QM} = \left(\frac{2e^2}{hv_F} \right) \quad (3.91)$$

where v_F is the Fermi velocity for CNTs. The electrostatic capacitance of n SWCNTs in a bundle can be expressed by the following:

$$C_{CNT,ES} = \left(\frac{\pi(d - d_{CNT})}{d_{CNT}} \right) \frac{2\pi\varepsilon_{ox}}{\ln\left(\frac{D-d}{2d_{CNT}}\right)} \quad (3.92)$$

From the above expression, it is evident that the CNT bundle geometry has complex dependency with the CNT capacitances.

3.8.3.3 Interconnect Parasitic Inductances

For the calculation of inductances of an interconnect, see Fig. 3.69(a) [4]. The inductance of an interconnect with a rectangular cross-section can be calculated using the following expression [4]:

$$L_{\text{int,rec}} = 2.0 \times 10^4 L \left(\ln\left(\frac{2L}{W+T}\right) + 0.5 - \alpha_L \right) \quad (3.93)$$

where $L_{\text{int,rec}}$ has the unit μH. L, W, and T are length, width, and thickness, respectively, of the interconnects expressed in nm. α_L is an empirical parameter that is affected by W and T of the interconnects, where $0 \leq \alpha_L \leq 0.0025$. From the above expression, it is evident that the interconnect length has maximum effect on the inductances of the interconnects. Thus, interconnect inductance increases with an increase in its length. On the other hand, the interconnect inductance slightly decreases with an increase in the width and thickness of the interconnect.

For the CNT bundle interconnect presented in Fig. 3.70, inductance can be calculated. There are two contributing components such as kinetic inductance ($L_{\text{CNT,K}}$) and magnetic inductance ($L_{\text{CNT,M}}$) that contribute to the total inductance of the CNT bundle interconnect. The total inductance of the CNT bundle interconnect with n SWCNTs is calculated as follows [235]:

$$L_{\text{CNT,tot}} = \frac{L_{\text{CNT,K}} + L_{\text{CNT,M}}}{n} \quad (3.94)$$

The kinetic inductance of one SWCNT is calculated as follows [44]:

$$L_{\text{CNT,K}} = \frac{h}{2e^2 v_F} \quad (3.95)$$

The magnetic inductance of one circumferential SWCNT is calculated using the following expression [235]:

$$L_{\text{CNT,M}} = \left(\frac{\mu_o \mu_r}{2\pi}\right) \ln\left(\frac{D-d}{2d_{\text{CNT}}}\right) \quad (3.96)$$

where μ_o permeability of the free space is $4\pi \times 10^{-7}$ H/m and μ_r is the relative permeability. Thus, the magnetic inductance of the CNT bundle decreases slightly with an increase of a single CNT diameter.

3.9 The Thermal Issue

Nanoscale size and gigascale complexity go hand in hand in a nanoelectronic technology-based chip. The power dissipation is taking place in an ever-smaller chip area. This leads to a significant increase in the power density of the chip. In other words, the high integration density chips have high power density that is manifested as heat, which increases on-chip temperature [30, 50, 206]. The on-chip power density has reached an uncontrollable level, for example, 100 W/cm² for 50-nm technology [50]. The temperature across the die can be as high as 50°C or even more. A nonuniform power profile across the chip can result in hotspots where the temperature is more than 100°C. To provide a big picture of the thermal profile, a chip floor plan, its power density profile, and the corresponding temperatures are presented in Fig. 3.73 [85]. DFX needs to deal with these thermal issues in order to achieve reliable chip design.

The heat increase is a result of activities going on in the large number of devices as well as the multiple layer interconnects [30, 50]. For nanoelectronic technology, the switching capacitance is high. This results in high switching activity power dissipation. The switching capacitance can be due to the transistor parasitics as well as to the interconnect parasitics. The various forms of leakage components such as subthreshold leakage and junction leakage also contribute to the total power dissipation of the chip and thermal issues. The self-heating or Joule heating due to the result of I^2R loss caused by the flow of current in the interconnect is also an important contributor to the thermal profile. Temperature rise in the interconnects due to this self-heating is significant. The location of the

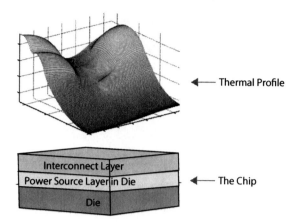

Figure 3.73 Thermal profile of a chip [85].

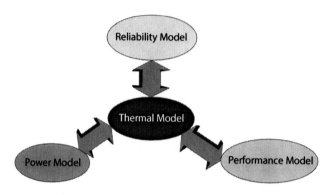

Figure 3.74 Interactions among power, temperature, performance and reliability for thermal-aware design [50].

interconnects causes such temperature rise. The interconnects are located away from the silicon substrate as well as the heat sink by several layers of insulating materials. The insulating materials that have lower thermal conductivities do not transfer the heat. The temperature peaks in a chip occur at the top of the chip where the global interconnects are located [30].

The average temperature of a chip can be estimated using the following expression [30]:

$$T_{\text{chip,avg}} = T_{\text{ambient}} + R_{\text{therm}}\left(\frac{P_{\text{avg,chip}}}{A_{\text{chip}}}\right) \tag{3.97}$$

where T_{ambient} is the ambient temperature, R_{therm} is the thermal resistance of the silicon substrate layer and package, $P_{\text{avg,chip}}$ is the average power dissipation of the chip, and A_{chip} is the chip area. R_{therm} is 4.75 cm^2°C/W for 180-nm CMOS technology node. However, it can be used to estimate on-chip temperatures for other technology nodes [30]. From the above expression, it is evident that the temperature increases with an increase in the average power dissipation and thermal resistance, and decreases in the chip area. There is a strong connection among the power dissipation, temperature, reliability, and performance of a chip. A thermal model can act as a bridge for accurate estimation of power, performance, and reliability of a chip as depicted in Fig. 3.74 [50]. The thermal modeling along with the design for temperature, in other words, temperature-aware design methodology, will be discussed in a later chapter.

The rising temperature in the chip has many negative effects. The current drive capability of a MOS transistor decreases by 4% for every 100°C rise in the temperature. Temperature puts a limit on the maximum allowable average as well as root mean square (RMS) current density in the interconnects. The subthreshold leakage has an exponential dependency with the chip temperature that may lead to thermal runaway. The increase in leakage current may increase the IR drop in the power rails that may lead to longer switching delays. For example, an IR drop in the power rail can reduce V_{DD} by 10% that can lead to degrading the switching delay by 30% to 40% [50]. The EM effect of interconnects that affects interconnect lifetime and chip reliability has an exponential dependency on temperature [30]. The increased temperature difference between victim and aggressor wires can cross talk noise. A temperature difference of 100°C between victim and aggressor can increase the cross talk noise by 25% [6, 50]. In general, the temperature has a severe adverse effect on the expected lifetime of a microprocessor. The mean time to failure (MTTF) of an integrated circuit can be empirically estimated from the Arrhenius Equation that is expressed as [206]:

$$\text{MTTF} = \alpha \exp\left(-\frac{E_A}{k_B T}\right) \quad (3.98)$$

where α the preexponential factor is an empirical constant, E_A is the activation energy of the failure mechanism, K_B is the Boltzmann constant, and T is the temperature.

3.10 The Reliability Issue

In general, "reliability of an IC" is the probability of a circuit being available for use in a specified availability schedule window. In other words, reliability is the lifetime of an IC [37]. The related term "aging of an IC" is the performance degradation of the circuit over time [107, 282]. The span of time can be a few months to a few years. Aging may lead to reliability problems over time. On the other hand, "soft errors" in circuits are random failures not related to physical defects but rather introduced by external sources such as α-particle radiations [151]. Soft errors are important for sequential circuits such as registers, and memory where information loss is caused by it. The different aspects of IC reliability are presented in Fig. 3.75 [37, 107, 282]. Device and interconnect reliability issues have been discussed. In addition, there may be package-level reliability issues that are not discussed in this book. The DFX needs to handle the reliability challenges that are manifested from various sources.

3.10.1 Hot Carrier Injection

The HCI is the process of carrier injection into the channel or gate dielectric that are produced by impact ionization near the drain end of the channel [124]. The following four distinguished mechanisms are possible for the injection of hot carriers into the MOSFET device [61, 233].

1. *Channel Hot-Electron (CHE) injection*: CHE injection can occur when both gate and drain voltage are very high compared to the source voltage, in which case some electrons are driven toward gate oxide.
2. *Drain Avalanche Hot-Carrier (DAHC) injection*: DAHC injection can occur when the drain voltage is much higher than the gate voltage. In such a condition, the acceleration of channel carrier can cause impact ionization. The avalanche multiplication then causes the injection of holes and electrons. In DAHC injection, the generated electron-hole pairs gain enough energy to break the barrier in Si-SiO$_2$ interface.

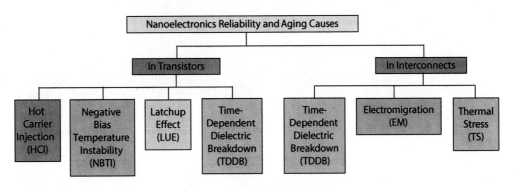

FIGURE 3.75 Different phenomena that can cause reliability issues and aging in nanoelectronic circuits.

3. *Substrate Hot-Carrier (SHC) injection*: The carrier may refer to either electron or hole, i.e., substrate hot-electron (SHE) injection or substrate hot-hole (SHH) injection. SHE or SHH injection can happen when the substrate back bias is very positive or very negative. In such a condition, the carriers of one type in the substrate are driven by the substrate electric-field toward the Si-SiO$_2$ interface.

4. *Secondary Generated Hot-Electron (SGHE) injection*: SGHE injection can happen due to photon emission. In a photo-induced process, the photons are generated in the high electric field near the drain, which, in turn, generated electron-hole pairs. At the same time, avalanche multiplication near the drain region may inject electrons and holes into the device oxide. SGHE injection is supported by the substrate bias that drive carriers to the interface.

The above different types of HCI are illustrated in Fig. 3.76 [61, 234]. At the saturation region of the MOSFET, the hot carriers traveling with saturation velocity may cause parasitic effects at the drain side of the channel that are called "hot carrier effects."

The HCI into the gate oxide may lead to hot carrier effects, such as mobility degradation, threshold voltage changes, subthreshold swing deterioration, transconductance degradation, drain current reduction, speed reduction, and stress-induced drain. The HCI is manifested as a change in the form of ΔI_{DS}, Δg_m, or ΔV_{Th} [271]. In general, the degradation is expressed as follows [271]:

$$\Delta D = \text{AGE}^n \tag{3.99}$$

where D is a device parameter like I_{DS}, g_m, or V_{Th}. The parameter AGE quantifies the amount of hot carrier stress. For an NMOS, the parameter AGE is calculated using the following quasistatic approximation [271]:

$$\text{AGE} = N \int_0^T \frac{1}{\alpha_N} \left(\frac{I_{ds}}{W}\right) \left(\frac{I_{sb}}{I_{ds}}\right)^{\beta_N} dt \tag{3.100}$$

where N is the number of periods and T is a specified period. I_{ds} is the drain current. I_{sb} is the substrate current. W is the device channel width. α_N is a parameter for NMOS. β_N is an acceleration factor. Both α_N and β_N are determined experimentally for a specific technology. AGE is calculated similarly for a PMOS; the difference is that I_{gate} is used instead of substrate current.

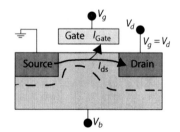

(a) Channel hot-electron (CHE) injection

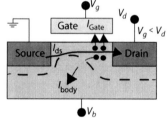

(b) Drain avalanche hot-carrier (DAHC) injection

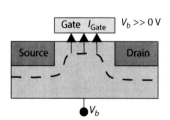

(c) Substrate hot-electron (SHE) injection

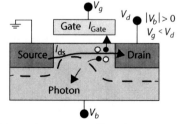

(d) Secondary generated hot-electron (SGHE) injection

Figure 3.76 Different types of HCI.

The hot carrier effects affect the lifetime of a transistor and need to be reduced. The bulk currents due to hot carrier injection are not comparatively significant as long as the parasitic series resistance of the bulk does not establish a drastically increased bulk potential that can lead to threshold voltage reduction or even more severe effects such as avalanche breakdown, snapback breakdown, and latch-up [61]. The avalanche breakdown happens when the drain voltage is too high such that the PN junction at the drain-to-substrate interfaces breaks down due to avalanche. The situation is similar to reverse-biased breakdown of a typical PN junction. On the other hand, the snapback breakdown or near-avalanche breakdown happens due to the excess electric field across gate dielectric. The latch-up effect will be discussed in a different subsection. The hot carrier effects result in performance degradations in RF circuits such as low-noise amplifier (LNA) and VCO [270]. For example, gain degradation of an LNA can happen due to the decrease in the transconductance of the transistors. In the case of a VCO, the tuning range reduces, gain reduces, and phase noise degrades due to the HCEs. Several device-level modifications can be made to reduce the hot carrier effects, for example, double diffusion of source and drain, longer channel lengths, and graded drain junctions. Similarly, circuit level modifications can be made to make the RF circuits less sensitive to HCEs.

3.10.2 Negative Bias Temperature Instability

Negative bias temperature instability (NBTI) originates due to the creation of interface traps and oxide charge by a negative gate voltage at high temperature [23, 212, 213]. An interface trap is an Si/SiO$_2$-interface trapped charge that is a trivalent silicon atom with an unpaired valence electron, in other words, with one empty bond. NBTI is more of a concern for a PMOS device as it operates at negative gate-to-source voltage. NBTI can also happen in NMOS when biased in the accumulation regime with a negative gate voltage. However, NBTI is much lower in NMOS compared to the PMOS for same gate voltage. The physical mechanism of NBTI is illustrated in Fig. 3.77. In a typical CMOS manufacturing, the hydrogen is introduced in the device and its interfacial areas to grow the SiO$_2$. The chemical reaction of the hydrogen at the interface and a subsequent diffusion of it through the gate-oxide leads to NBTI [212, 213]. NBTI can lead to an increase in the threshold voltage, mobility degradation, drain current reduction, transconductance degradation, and increase in delay. For example, the threshold voltage change due to NBTI can be presented by the following expression [212]:

$$\Delta V_{Th} = \alpha \exp(\varepsilon_{ox}\gamma) \exp\left(-\frac{E_A}{k_B T}\right) t^n \quad (3.101)$$

where α is a constant, γ is the electric field factor, E_A is the activation energy, k_B is the Boltzmann constant, T is temperature, t is time, and n is the time exponent.

3.10.3 Latchup Effect

In a CMOS circuit, the latchup effect (LUE), or just latchup, is the inadvertent creation of a low impedance path between power supply rails due to the triggering of parasitic devices [197]. The latchup can cause some sort of circuit malfunction, and in the worst case, destruction of the circuits.

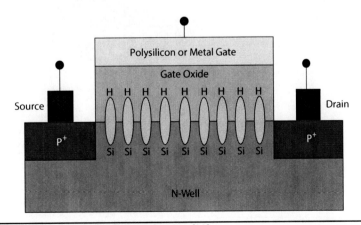

FIGURE 3.77 Demonstration of NBTI physical mechanism [23].

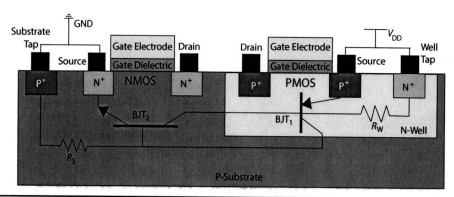

Figure 3.78 Parasitic BJT in a MOSFET structure that can cause latchup.

A bulk CMOS inverter structure with the parasitics that can create LUE is depicted in Fig. 3.78. The CMOS structure has two parasitic BJTs as follows: BJT_1 is the vertical PNP and BJT_2 is the lateral NPN. The collector terminal of one BJT is connected to the base of the other BJT in a positive feedback fashion. The BJTs together are analogous to the parasitic structure and are usually equivalent to a thyristor or silicon-controlled rectifier (SCR) that is a PNPN structure. The latchup occurs when both the BJTs conduct and create a low impedance path between the power supply rails and the product of gains of the BJTs is greater than one. This kind of gain makes a positive feedback among the parasitics which increase the current until the circuit fails. The triggering of the BJT feedback loop can happen for several reasons: (1) power supply voltage exceeding the absolute maximum ratings; (2) input or output pin voltage exceeding the power supply rail by more than a diode drop; (3) poorly managed multiple power supplies; and (4) ionizing radiation, which is a significant issue for circuits designed for very high-altitude or space applications. Latchup can be prevented in many ways. The straightforward approach is to stick to the absolute maximum ratings. Use of an oxide insulator layer surrounding NMOS and PMOS can break the parasitic thyristor structure. A separate tap connection for each transistor can prevent latchup. Use of a latchup protection circuit—for example, use of a Schottky diode between the power supply rails—can prevent latchup. The SOI devices are inherently resistant to latchup so their use instead of bulk technology can prevent latchup.

3.10.4 Time-Dependent Dielectric Breakdown

Time-dependent dielectric breakdown (TDDB) is the degradation of the dielectric over time [236]. TDDB of a gate dielectric insulator may result due to the application of a moderately high electric field, for example, approximately 5 MV/cm, over a period of time [212]. This may not be an immediate breakdown of oxide, which is caused by very high electric field. Simplistically speaking, TDDB is due to the charges that remain in the gate oxide. Stronger electric fields across the gate oxide can cause several types of quantum tunneling such as Fowler-Nordheim tunneling, direct tunneling, and trap-assisted tunneling as illustrated in Fig. 3.79 [236]. In Fowler-Nordheim

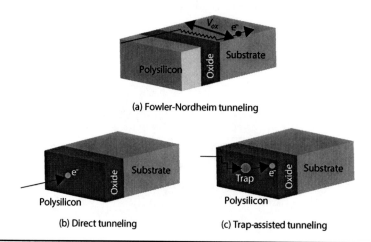

Figure 3.79 Various types of quantum tunneling through gate oxide [236].

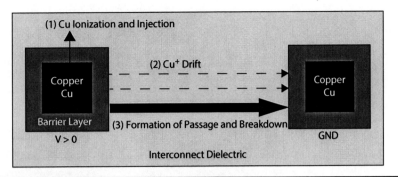

FIGURE 3.80 TDDB in interconnects [268].

tunneling, electrons penetrate through the oxide barrier to the conduction band of the oxide. In direct tunneling, the electrons penetrate through the gate oxide directly to the silicon channel. In trap-assisted tunneling, the electrons pass through the gate oxide into interface traps and then move into the silicon substrate. TDDB in transistors depends on gate dielectric material, gate-oxide field, gate area, supply voltage, operating frequency, temperature, and time.

TDDB is also exhibited in copper interconnects in the nanoelectronic circuits [174, 268]. The diffusion of copper in the dielectric among the interconnects can lead to degradation of the dielectric. The physical process of the TDDB in interconnects is depicted in Fig. 3.80. The TDDB caused by copper starts with the copper ionization and injection of copper-ion (Cu^+) from the higher voltage interconnect. Copper ions (Cu^+) then drift through the interconnect dielectric; eventually, leakage passages are formed and copper ions (Cu^+) are accumulated at the lower voltage interconnect. This process leads to higher leakage current and subsequently dielectric breakdown. The existence of traps in the interconnect dielectric facilitates copper ionization. TDDB of the interconnects is affected by the material characteristics, the electric field, and the temperature.

Several models exist to predict the time to failure due to TDDB [236]. The thermochemical model (also known as E model) for mean MTTF prediction is represented by the following expression [174]:

$$\text{MTTF} = \alpha \exp(-\gamma E_{\text{ox}}) \exp\left(-\frac{E_A}{k_B T}\right) \qquad (3.102)$$

where α is a constant, γ is field acceleration factor, E_{ox} gate-oxide electric field, E_A is activation energy, k_B is Boltzmann constant, and T is temperature. Thus, MTTF of TDDB strongly depends on the gate-oxide field and temperature. This lifetime reduces with lower dielectrics and smaller dimensions. Use of thicker oxide, use of high-κ dielectrics, and lowering of supply voltages are some options to deal with the TDDB problem [236].

3.10.5 Electromigration

EM is a major reliability concern in the metal interconnects but not in the transistors. In general, EM is migration of materials due to an electrical field [49, 129, 176]. The copper or aluminum interconnects are polycrystalline in structure. They consist of grains containing crystal lattices of identical structure but different orientations. As the current flows through the interconnects, there is an interaction between the moving electrons (referred to as "electron wind") and the metal ions in these lattice structures. The condition is depicted in Fig. 3.81(a). The flow of current introduces two different forces on the individual metal ions of the interconnects. The first force is a result of momentum transfer between the conducting electrons and the metal ions. This force is called "electron wind force" and is denoted by $\mathbf{F}_{\text{wind}}$. The electron wind force ($\mathbf{F}_{\text{wind}}$) works in the direction of the current flow. Another force is an electrostatic force that is the result of the electric field strength in the metallic interconnect. This force is called "direct force" and is denoted by $\mathbf{F}_{\text{direct}}$. It may be noted that

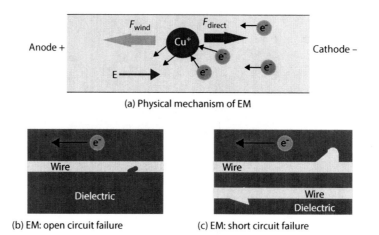

FIGURE 3.81 Illustration of EM in interconnects.

$\mathbf{F}_{wind} > \mathbf{F}_{direct}$ and $\mathbf{F}_{wind}$ is the primary force for the metal EM. The total driving force acting on a metal ion is vector sum of the wind force and direct force that is expressed as follows [176]:

$$\mathbf{F}_{EM} = \mathbf{F}_{wind} + \mathbf{F}_{direct} \tag{3.103}$$

So to produce EM, a lot of electrons and electron scattering are needed. Thus, EM happens in metal interconnects and not in semiconductors. However, it may occur in some heavily doped semiconductor materials.

EM can cause two types of failures: open circuit and short circuit. If the outgoing flux on metal ions exceeds the incoming flux, then the metal line can open to form "voids." This void in the worst case causes open circuit failures. Once a void is initiated, the current density in its vicinity increases due to the reduction of the cross-sectional area of the interconnect. The local current density leads to a local temperature rise around the void due to the joule heating that further accelerates the void growth. The whole process continues in a positive feedback fashion till the void is large enough to completely open the line. The acceleration process of EM due to the increase in local temperature is presented in Fig. 3.82 [229]. The open circuit situation is illustrated in Fig. 3.81(b).

The second type of failure due to EM is short circuit failure. If the incoming flux on metal ions exceeds the outgoing flux, then the metal ions accumulate at individual grain boundaries, forming "hillocks" in the direction of the current. The hillocks when large can cause short circuit failures. The short circuit situation is illustrated in Fig. 3.81(c).

The empirical model to estimate MTTF of circuits accounting for EM is [124, 129]:

$$\mathrm{MTTF} = \left(\frac{A}{J^n}\right) \exp\left(\frac{E_A}{k_B T}\right) \tag{3.104}$$

where A is a constant based on the cross-sectional area of the interconnect, J is the current density, n is a scaling factor, E_A is activation energy for electromigration, k_B is the Boltzmann constant, and T is

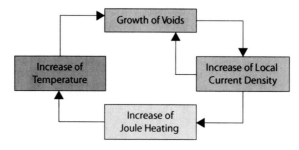

FIGURE 3.82 Thermal acceleration during EM in interconnects.

the temperature. The above empirical formula is called Black equation. Another model for the estimation of MTTF accounting for EM is based on Eyring equation [127]. The current density and temperature are the major parameters for electromigration. If the length of the interconnects is smaller than a "Blech Length," then it will not fail by EM [229]. In the interconnect material point of view, copper is intrinsically less susceptible to electromigration.

3.10.6 Thermal Stress

Due to a difference in the coefficient of thermal expansion between the interconnect metal and the silicon substrate, the thin interconnect lines are subjected to thermal stress [24]. Thermal stress can have negative impact on reliability. The thermal stress can be expressed as follows [24]:

$$\sigma_{therm} = \left(\frac{Y_{int}}{1-\nu_{int}}\right)(\alpha_{sub} - \alpha_{int})\Delta T \quad (3.105)$$

where Y_{int} is Young modulus of the interconnect material, ν_{int} is the Poisson ratio, α_{sub} coefficients of linear thermal expansion of the silicon substrate, α_{int} coefficients of linear thermal expansion of the interconnect material, and ΔT is the change in the temperature. Thus, the thermal stress is dependent on the materials as well as the temperature gradient. It may be noted that copper has a smaller coefficient of thermal expansion than aluminum, so it is supposed to have advantages in terms of thermal stress. However, this advantage is not fully manifested due to the higher stiffness of copper material [24]. In the worst-case scenario, the thermal stress can create voids and lead to open circuit failures as is the case of EM voids discussed in the previous subsection.

3.11 The Trust Issue

A typical embedded system that is the actual realization of any consumer electronic systems is subjected to various attacks, as depicted in Fig. 3.83 [153, 173]. For these systems, in a generic sense, the question arises as to whether a chip, a system-on-chip (SoC), or an analog/mixed-signal SoC (AMS-SoC) can be trusted. In other words, is a specific chip executing the exact operations that it is intended to perform? The nonexecution of operations should not necessarily be treated as a fault. Trust has various diverse aspects [14, 101, 153, 155]. The hardware must have the following attributes to be trustworthy:

1. It must maintain the integrity of information it is processing. For example, in the case of encryption and watermarking, it must perform all the steps to protect the information.

2. It must conceal any information about the computation performed through any side channels such as power analysis or timing analysis.

3. It must perform only the functionality for which it was designed, nothing more and nothing less.

4. It must not malfunction during critical operations such as battlefields.

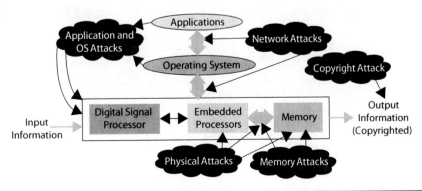

FIGURE 3.83 Different forms of attacks on a typical embedded system [153, 173].

5. It must be transparent in terms of design details and states only to its own designer while remaining opaque to others.

6. It must protect itself from reverse engineering and counterfeit designs of the same hardware.

DFX with "X" denotes "trust," i.e., design for trust needs to address one or more of the above as a target design objective. Various aspects of design for trust will be discussed in the following subsections.

3.11.1 Information Protection Issue

Numerous hardware has been deployed in security and copyright protection applications to perform various operations of cryptography, watermarking, or digital rights management (DRM) [117, 153]. Watermarking chips are being used in embedded systems for copyright protection of the information or media they process [80]. Various types of watermarking are shown in Fig. 3.84 [155]. The choice of a specific watermarking depends on target applications and media. Cryptographic chips are ubiquitous as encryption is needed almost everywhere [114]. For example, password authentication, secure communications, pay-TV, set-top box, and bank transactions are all heavily dependent on cryptography [25, 79]. Encryption chips are used now for data protection in USB flash drives and hard drives. Cryptography processors are being developed for trusted calls in smart phones. Various types of cryptography are presented in Fig. 3.85. DFX needs to pick right algorithms and architectures for appropriate cryptography and watermarking chip design. Preferably, energy-efficient watermarking and cryptography chips are needed for embedded portable systems.

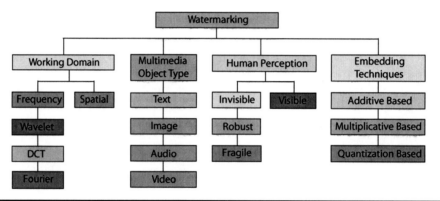

FIGURE 3.84 Different types of watermarking [155].

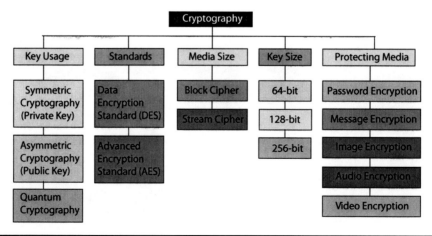

FIGURE 3.85 Different types of cryptography.

3.11.2 Information Leakage Issue

Watermarking and cryptography hardware provides low-power consumption, real-time performance, higher reliability, higher availability, and low cost, along with easy integration in multimedia hardware, as compared to the watermarking and cryptography software programs [117, 153]. However, the watermarking and encryption hardware that is implemented using static CMOS technology is susceptible to side-channel attacks [101, 144, 145, 196]. The physical or side-channel attacks have various forms as presented in Fig. 3.86 [196]. The side-channel attacks gather information from the physical implementation of a cryptosystem or watermarking system rather than from the theoretical weaknesses in the algorithms. Various techniques including differential power analysis (DPA), timing analysis, and electromagnetic analysis are used for information gain. These attacks take advantage of the switching activities of the circuits and systems that take place during the execution of the static CMOS circuits in which different capacitances are switched. These can be determined through power and timing analysis from which the encryption and watermarking keys can be determined, thus breaking the security of the electronic system. DFX for information leakage proof is very critical for electronic systems that have reached a stage where smart phones process bank and credit card information.

3.11.3 Chip Intellectual Property Protection Issue

Intellectual property blocks or reusable virtual components are used as a cost-effective solution for realization of the complex SoCs in the current highly competitive and short time to market [12, 95, 96, 203]. Sharing of intellectual property blocks for reusability in SoC designs poses severe security and ownership issues, as depicted in Fig. 3.87. Typical intellectual property blocks need significant time and effort for design and verification; however, they can be easily modified or even copied, in order to hide the ownership proof and use without paying proper royalty. The owners of intellectual property blocks desire that their designs are not illegally redistributed by consumers. At the same-time, the genuine consumers (who may be the SoC designers) want assurance that the intellectual property core designs they buy are legitimate [12].

DFX needs to consider this important issue for the protection of circuit or system design intellectual property such that revenue loss can be reduced. The watermarking of intellectual property designs is considered as solution [12, 95, 96, 203]. Watermarking has been a known technique in a

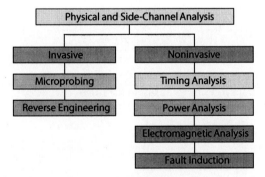

FIGURE 3.86 Different types of physical attacks on security chips [196].

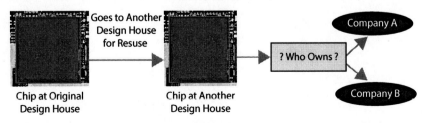

FIGURE 3.87 Chip Intellectual property protection issue.

different context as an approach for multimedia content protection [117, 153, 155]. The watermarking for intellectual property block protection is an identification code, imperceptible to human or machine analysis, that is permanently embedded as an integral part within a design. The desired characteristics of the intellectual property watermarking are [96, 203]: (1) maintenance of functional correctness, (2) transparency to existing design flows, (3) proportional component protection, (4) enforceability to ensure it can be used effectively, (5) flexibility in providing a spectrum of protection levels, (6) persistence throughout the life of the intellectual property block, (7) invisibility to unauthorized users, and (8) minimal overhead cost to keep the overall intellectual property cost more or less the same.

3.11.4 Malicious Design Modifications Issue

Fabrication of ICs happens around the globe and may include genuine as well as unauthentic fabrication plants for many reasons, including economic [14, 102, 243]. The malicious modifications of the designs become an issue when the chips manufactured in unauthentic fabrication plants are used in critical applications such as military and power grid. Such unauthentic fabrication plants might have deliberately introduced additional components in the chip so that the chip fails or works improperly during critical needs. The term "Hardware Trojans" explains such additional components or modifications. The hardware Trojans are defined as modifications to an original IC that are inserted by adversaries to exploit the hardware, to use the hardware mechanisms to gain access to data or software running on the hardware [243].

Illustration of a simple hardware Trojan is shown in Fig. 3.88 [102]. In normal operations, the watermarking or cryptography chip generates watermarked or encrypted information at the output. When a hardware Trojan is present in the watermarking or cryptography chip, some or all of the processing steps in the watermarking and encryption may be bypassed. Thus, the output information may not be watermarked or encrypted as desired by the user of the watermarking or cryptography chip. Thus, the security is compromised. In the worst-case situation, the functions of the critical chips can be completely stopped by the hardware Trojan. A comprehensive view of the different types of hardware Trojans is presented in Fig. 3.89 [102, 243]. As is evident, the hardware Trojan can be in any portion of a SoC. It can also be inserted in any step of the design and manufacturing process. Thus, it is a very difficult problem to handle under DFX.

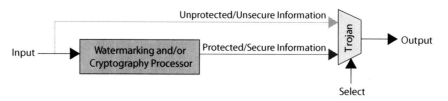

Figure 3.88 Illustration of simple Trojans in hardware [102].

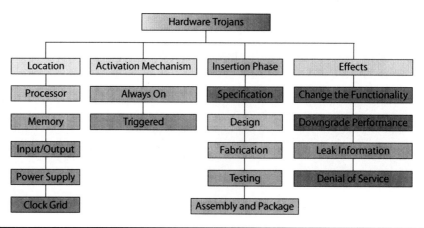

Figure 3.89 Illustration of simple Trojans in hardware [102, 243].

3.12 Questions

3.1 Identify five important challenges in mixed-signal design and discuss them in your own words.

3.2 Identify five alternative transistors and discuss them in your own words.

3.3 Plot I-V characteristics of the five devices using any EDA tool and discuss them.

3.4 Discuss why multiple gate devices and high-κ devices are used for chip design.

3.5 Discuss graphene as a material for high-speed nanoelectronics. In other words, what are its interesting characteristics for use in analog, digital, or RF?

3.6 Discuss memristor as an electronic element: what is it, why is it?

3.7 Simulate a memristor using an EDA tool and comment on its characteristics.

3.8 Briefly discuss the overall manufacturing process of a CMOS inverter.

3.9 Discuss the role of SiO_2 in the design and manufacturing of CMOS circuits.

3.10 Discuss the issue of process variation and how it affects nanoelectronic circuits.

3.11 Discuss different components of power dissipation in a nanoscale CMOS circuit.

3.12 Discuss gate-oxide leakage: what it is, why it has gotten so much attention, and how it is handled.

3.13 Discuss subthreshold leakage: what it is, why it is important, and how it is kept under control.

3.14 Discuss different parasitics of a nanoelectronic device.

3.15 Discuss different parasitics of a nanoelectronic interconnect.

3.16 Discuss the thermal issue of nanoelectronic circuits and systems.

3.17 Discuss the reliability issues of nanoelectronic circuits and systems.

3.18 Discuss the trust issues of nanoelectronic circuits and systems.

3.19 Discuss the commonalities and differences between multimedia watermarking and intellectual property watermarking.

3.20 Discuss hardware Trojans: what they are and why they are important.

3.13 References

1. BSIM4 MOS Models, Release 4.5.0. http://www-device.eecs.berkeley.edu/bsim/?page=BSIM4_Arc. Accessed on 16th January 2013.
2. Designing Ultra-Dense Computers with QCAs. http://www.cse.nd.edu/~cse_proj/qca_design/. Accessed on 05th October 2012.
3. Semiconductor Industry Association, International Technology Roadmap for Semiconductors. http://public.itrs.net.
4. Tutorial C2: A General and Comparative Study of RC(0), RC, RCL and RCLK Modeling of Interconnects and Their Impact on the Design of Multi-Giga Hertz Processors. In: *Proceedings of the 3rd International Symposium on Quality Electronic Design, ISQED '02*, pp. 10–10. IEEE Computer Society, Washington, DC, USA (2002). URL http://www.oea.com/assets/files/isqed2002.pdf. http://dl.acm.org/citation.cfm?id=846240.850304.
5. CMOS Transistor Process Technology. http://pasargad.cse.shirazu.ac.ir/hard/CMOS-technology.pdf (2003). Accessed on 25th November 2012.
6. Thermally Aware Design Methodology. Tech. rep., Gradient Design Automation Inc. (2005). URL http://www.cdnusers.org/community/encounter/resources/resources_imp/route/dtp_cdnlive2005_1349_moynihan.pdf. Accessed on 14th August 2012.
7. IC Fabrication Techniques. http://www.circuitstoday.com/ic-fabrication-techniques (2010). Accessed on 25th November 2012.
8. Design for Excellence. http://www.besttest.com/Courses/00001-DFX.cfm. Accessed on June 2012.
9. The DfX Concept. http://www.ami.ac.uk/courses/topics/0248_dfx/. Accessed on June 2012.
10. National Nanotechnology Initiative. http://www.nano.gov/. Accessed on October 2012.
11. What is TFT LCD TV and LCD Monitor Panel. http://www.plasma.com/classroom/what_is_tft_lcd.htm. Accessed on October 2012.
12. Abdel-Hamid, A.T., Tahar, S., Aboulhamid, E.M.: IP Watermarking Techniques: Survey and Comparison. In: *Proceedings of the 3rd IEEE International Workshop on System-on-Chip for Real-Time Applications*, pp. 60–65 (2003). DOI 10.1109/IWSOC.2003.1213006.

13. Abo-Elhadeed, A.F., Fikry, W.: Compact Model for Short and Ultra Thin Symmetric Double Gate. In: *Microelectronics (ICM), 2010 International Conference on*, pp. 24–27 (2010). DOI 10.1109/ICM.2010.5696130.
14. Adee, S.: The Hunt For The Kill Switch. *IEEE Spectrum* **45**(5), 34–39 (2008). DOI 10.1109/MSPEC.2008.4505310.
15. Administration, U.E.I.: Heating and Cooling No Longer Majority of US Home Energy Use. http://www.eia.gov/consumption/residential/index.cfm (2013). Accessed on 13th May 2013.
16. Agarwal, A., Mukhopadhyay, S., Kim, C., Raychowdhury, A., Roy, K.: Leakage Power Analysis and Reduction: Models, Estimation and Tools. *IEE Proceedings – Computers and Digital Techniques* **152**(3), 353–368 (2005). DOI 10.1049/ip-cdt:20045084.
17. Sandner, C., Clara, M., Santner, A., Hartig, T., Kuttner, F.: A 6-bit 1.2Gs/s Low-Power Flash ADC in $0.13\mu m$ Digital CMOS. *IEEE Journal of Solid State Circuits* **40**(7), 1499–1505 (2005).
18. Huff, H. R., Agarwal, A., Kim, Y., Perrymore, L., Riley, D., Barnett, J., Sparks, C., Freiler, M., Gebara, G., Bowers, B., Chen, P. J., Lysaght, P., Nguyen, B., Lim, J.E., Lim, S., Bersuker, G., Zeitzoff, P., Brown, G. A., Young, C., Foran, B., Shaapur, F., Hou, A., Lim, C., AlShareef, H., Borthakur, S., Derro, D. J., Bergmann, R., Larson, L. A., Gardner, M. I., Gutt, J., Murto, R. W., Torres, K., Jackson, M. D.: Integration of high-k Gate Stack Systems into Planar CMOS Process Flows. In: *International Workshop on Gate Insulator*, pp. 2–11 (2001).
19. Manchanda, L., Busch, B., Green, M. L., Moms, M., van Dover, R. B., Kwo, R., Aravamudhan, S.: High-K gate Dielectrics for the Silicon Industry. In: *Proceedings of International Workshop on Gate Insulator*, p. 5660 (2001).
20. Yang, M., Gusev, E. P., Ieong, M., Gluschenkov, O., Boyd, D. C., Chan, K. K., Kozlowski, P. M., D'Emic, C. P., Sicina, R. M., Jamison, P. C., Chou, A. I.: Performance Dependence of CMOS on Silicon Substrate Orientation for Ultrathin and HfO2 Gate Dielectrics. *IEEE Electron Device Letters* **24**(5), 339341 (2003).
21. Chau, R., Kavalieros, J., Roberds, B., Schenker, R., Lionberger, D., Barlage, D., Doyle, B., Arghavani, R., Murthy, A., Dewey, G.: 30nm physical Gate Length CMOS Transistors with 1.0ps n-MOS and 1.7ps p-MOS Gate Delays. *IEDM Technical Digest* 45–48 (2000).
22. Choi, R., Onishi, K., Kang, C. S., Gopalan, S., Nieh, R., Kim, Y. H., Han, J. H., Krishnan, S., Cho, H.-J., Shahriar, A., Lee, J. C.: Fabrication of High Quality Ultra-thin HfO_2 Gate Dielectric MOSFETs Using Deuterium Anneal. *IEDM Technical Digest* 613–616 (2002).
23. Alam, M.A., Mahapatra, S.: A Comprehensive Model of PMOS NBTI Degradation. *Microelectronics Reliability* **45**(1), 71–81 (2005). DOI 10.1016/j.microrel.2004.03.019. URL http://www.sciencedirect.com/science/article/pii/S0026271404001751.
24. An, J. H.: Thermal Stress Induced Voids in Nanoscale Cu Interconnects By In-Situ Tem Heating. Ph.D. thesis, The University of Texas at Austin, TX, USA (2007). Accessed on 06th June 2013.
25. Anderson, R., Bond, M., Clulow, J., Skorobogatov, S.: Cryptographic Processors – A Survey. Tech. Rep. UCAM-CL-TR-641, University of Cambridge, Computer Laboratory (2005). URL http://www.cl.cam.ac.uk/techreports/UCAM-CL-TR-641.pdf.
26. Arora, N.: *MOSFET Modeling for VLSI Simulation: Theory And Practice*. World Scientific Publishing Company, 5 Toh Tuck Link, Singapore 596224 (2007).
27. Asra, R., Shrivastava, M., Murali, K.V.R.M., Pandey, R.K., Gossner, H., Rao, V.R.: A Tunnel FET for V_{DD} Scaling Below 0.6 V With a CMOS-Comparable Performance. *IEEE Transactions on Electron Devices* **58**(7), 1855–1863 (2011). DOI 10.1109/TED.2011.2140322.
28. Auth, C., Allen, C., Blattner, A., Bergstrom, D., Brazier, M., Bost, M., Buehler, M., Chikarmane, V., Ghani, T., Glassman, T., Grover, R., Han, W., Hanken, D., Hattendorf, M., Hentges, P., Heussner, R., Hicks, J., Ingerly, D., Jain, P., Jaloviar, S., James, R., Jones, D., Jopling, J., Joshi, S., Kenyon, C., Liu, H., McFadden, R., McIntyre, B., Neirynck, J., Parker, C., Pipes, L., Post, I., Pradhan, S., Prince, M., Ramey, S., Reynolds, T., Roesler, J., Sandford, J., Seiple, J., Smith, P., Thomas, C., Towner, D., Troeger, T., Weber, C., Yashar, P., Zawadzki, K., Mistry, K.: A 22nm High Performance and Low-Power CMOS Technology Featuring Fully-Depleted Tri-Gate Transistors, Self-Aligned Contacts and High Density MIM Capacitors. In: *Proceedings of the Symposium on VLSI Technology (VLSIT)*, pp. 131–132 (2012). DOI 10.1109/VLSIT.2012.6242496.
29. Auth, C., Buehler, M., Cappellani, A., hing Choi, C., Ding, G., Han, W., Joshi, S., McIntyre, B., Ranade, P., Sandford, J., Thomas, C.: 45nm High-k+Metal Gate Strain-Enhanced Transistors. *Intel Technology Journal* **12**(2), 77–86 (2008). DOI 10.1535/itj.1202.01.
30. Banerjee, K., Pedram, M., Ajami, A.H.: Analysis and Optimization of Thermal Issues in High-Performance VLSI. In: *Proceedings of the International Symposium on Physical Design*, pp. 230–237. ACM, New York, NY, USA (2001). DOI 10.1145/369691.369779. URL http://doi.acm.org/10.1145/369691.369779.
31. Banerjee, S., Mathew, J., Mohanty, S.P., Pradhan, D.K., Ciesielski, M.J.: A Variation-Aware Taylor Expansion Diagram-Based Approach for Nano-CMOS Register-Transfer Level Leakage Optimization. *Journal of Low Power Electronics* **7**(4), 471–481 (2011).
32. Bansal, A., Paul, B.C., Roy, K.: An Analytical Fringe Capacitance Model for Interconnects Using Conformal Mapping. *Computer-Aided Design of Integrated Circuits and Systems, IEEE Transactions on* **25**(12), 2765–2774 (2006). DOI 10.1109/TCAD.2006.882489.
33. Barkhordarian, V.: Power MOSFET Basics. http://www.irf.com/technical-info/appnotes/mosfet.pdf. Accessed on October 2012.
34. Barnes, J.J., Shimohigashi, K., Dutton, R.W.: Short-Channel MOSFET's in the Punchthrough Current Mode. *IEEE Transactions on Electron Devices* **26**(4), 446–453 (1979). DOI 10.1109/T-ED.1979.19447.

35. Basu, S., Kommineni, B., Vemuri, R.: Mismatch Aware Analog Performance Macromodeling Using Spline Center and Range Regression on Adaptive Samples. In: *Proceedings of the International Conference on VLSI Design*, pp. 287–293 (2008).
36. Basu, S., Kommineni, B., Vemuri, R.: Variation Aware Spline Center and Range Modeling for Analog Circuit Performance. In: *Proceedings of the International Symposium on Quality Electronic Design*, pp. 162–167 (2008).
37. Bernstein, J.B., Gurfinkel, M., Li, X., Walters, J., Shapira, Y., Talmor, M.: Electronic Circuit Reliability Modeling. *Microelectronics Reliability* **46**(12), 1957–1979 (2006). DOI 10.1016/j.microrel.2005.12.004. URL http://www.sciencedirect.com/science/article/pii/S0026271406000023.
38. Bernstein, K., Frank, D.J., Gattiker, A.E., Haensch, W., Ji, B.L., Nassif, S.R., Nowak, E.J., Pearson, D.J., Rohrer, N.J.: High-Performance CMOS Variability in the 65-*nm* Regime and Beyond. *IBM Journal of Research and Development* **50**(4/5), 433–449 (2006).
39. Bohr, M.T., Chau, R.S., Ghani, T., Mistry, K.: The High-*k* Solution. *IEEE Spectrum* **44**(10), 29–35 (2007). DOI 10.1109/MSPEC.2007.4337663.
40. Boucart, K., Ionescu, A.M.: Double-Gate Tunnel FET With High-*k* Gate Dielectric. *IEEE Transactions on Electron Devices* **54**(7), 1725–1733 (2007). DOI 10.1109/TED.2007.899389.
41. Bowman, K.A., Austin, B.L., Eble, J.C., Tang, X., Meindl, J.D.: A Physical Alpha-Power Law MOSFET Model. *IEEE Journal of Solid-State Circuits* **34**(10), 1410–1414 (1999). DOI 10.1109/4.792617.
42. Bowman, K.A., Meindl, J.D.: Impact of within-die parameter fluctuations on future maximum clock frequency distributions. In: *Proceedings of the IEEE Custom Integrated Circuits Conference*, pp. 229–232 (2001).
43. Brody, T.P.: The Birth and Early Childhood of Active Matrix – A Personal Memoir. *Journal of the Society for Information Display* **4**(3), 113–127 (1996). DOI 10.1889/1.1985000. URL http://dx.doi.org/10.1889/1.1985000.
44. Burke, P.J.: An RF Circuit Model for Carbon Nanotubes. *IEEE Transactions on Nanotechnology* **2**(1), 55–58 (2003). DOI 10.1109/TNANO.2003.808503.
45. Cartwright, J.: Intel Enters The Third Dimension (2011). DOI 10.1038/news.2011.274. URL http://www.plasma.com/classroom/what_is_tft_lcd.htm. Accessed on October 2012.
46. Chan, T.Y., Chen, J., Ko, P.K., Hu, C.: The Impact of Gate-induced Drain Leakage Current on Mosfet Scaling. In: *Electron Devices Meeting, 1987 International, vol. 33*, pp. 718–721 (1987). DOI 10.1109/IEDM.1987.191531.
47. Chan, Y.J., Huang, C.H., Weng, C.C., Liew, B.K.: Characteristics of Deep-Submicrometer MOSFET and its Empirical Nonlinear RF Model. *IEEE Transactions on Microwave Theory and Techniques* **46**(5), 611–615 (1998). DOI 10.1109/22.668671.
48. Chandra, G., Kapur, P., Saraswat, K.C.: Scaling Trends for the on Chip Power Dissipation. In: *Proceedings of the IEEE International Interconnect Technology Conference*, pp. 170–172 (2002). DOI 10.1109/IITC.2002.1014923.
49. Chang, S., Shin, H.: Off-state Leakage Currents of MOSFETs with High-*k* Dielectrics. *Journal of the Korean Physical Society* **41**(6), 932–936 (2002).
50. Chaudhury, S.: A Tutorial and Survey on Thermal-Aware VLSI Design: Tools and Techniques. *International Journal of Recent Trends in Engineering* **2**(8), 18–21 (2009).
51. yu Chen, H., wen Chang, Y.: Routing for Manufacturability and Reliability. *IEEE Circuits and Systems Magazine* **9**(3), 20–31 (2009). DOI 10.1109/MCAS.2009.933855.
52. Chi, Y., Sui, B., Fang, L., Zhou, H., Zhong, H., Sun, H.: A Compact Analytical Model For Multi-Island Single Electron Transistors. In: *Proceedings of the IEEE 8th International Conference on ASIC*, pp. 662–665 (2009). DOI 10.1109/ASICON.2009.5351333.
53. Chiarella, T., Witters, L., Mercha, A., Kerner, C., Rakowski, M., Ortolland, C., Ragnarsson, L., Parvais, B., Keersgieter, A.D., Kubicek, S., Redolfi, A., Vrancken, C., Brus, S., Lauwers, A., Absil, P., Biesemans, S., Hoffmann, T.: Benchmarking SOI and Bulk FinFET Alternatives for Planar CMOS Scaling Succession. *Solid-State Electronics* **54**(9), 855–860 (2010). DOI 10.1016/j.sse.2010.04.010. URL http://www.sciencedirect.com/science/article/pii/S0038110110001279.
54. Cho, G., Kim, Y.B., Lombardi, F.: Assessment of CNTFET Based Circuit Performance and Robustness to PVT Variations. In: *Proceedings of the 52nd IEEE International Midwest Symposium on Circuits and Systems*, pp. 1106–1109 (2009). DOI 10.1109/MWSCAS.2009.5235961.
55. Choi, Y.K., Ha, D., King, T.J., Bokor, J.: Investigation of Gate-Induced Drain Leakage (GIDL) Current in Thin Body Devices: Single-Gate Ultra-Thin Body, Symmetrical Double-Gate, and Asymmetrical Double-Gate MOSFETs *Japan Journal of Applied Physics* **42** (Part 1, No. 4B), 2073–2076 (2003).
56. Chua, L.O.: Memristor – The Missing Circuit Element. *IEEE Transactions on Circuit Theory* **18**(5), 507–519 (1971).
57. Cunningham, J.A.: The Use and Evaluation of Yield Models in Integrated Circuit Manufacturing. *IEEE Transactions on Semiconductor Manufacturing* **3**(2), 60–71 (1990). DOI 10.1109/66.53188.
58. Devoret, M.H., Schoelkopf, R.J.: Amplifying Quantum Signals With The Single-Electron Transistor. *Nature* **406**, 1039–1046 (2000). DOI http://dx.doi.org/10.1038/35023253.
59. Dresselhaus, M.S., Dresselhaus, G., Avouris, P. (eds.): *Carbon Nanotubes: Synthesis, Structure, Properties and Applications*, 1st ed. Springer (2001).
60. Endo, K., O'uchi, S.I., Ishikawa, Y., Liu, Y., Matsukawa, T., Sakamoto, K., Tsukada, J., Yamauchi, H., Masahara, M.: Variability Analysis of TiN Metal-Gate FinFETs. *IEEE Electron Device Letters* **31**(6), 546–548 (2010). DOI 10.1109/LED.2010.2047091.

61. Entner, R.: *Modeling and Simulation of Negative Bias Temperature Instability*. Ph.D. thesis (2007). http://www.iue.tuwien.ac.at/phd/entner/node21.html. Accessed on 18th Jan 2013.
62. Fahad, M. S., Srivastava, A., Sharma, A.K., Mayberry, C.: Current Transport in Graphene Tunnel Field Effect Transistor for RF Integrated Circuits. In: *Proceeding of the IEEE International Wireless Symposium*, pp. 1-4 (2012).
63. Fasarakis, N., Tsormpatzoglou, A., Tassis, D., Pappas, I., Papathanasiou, K., Dimitriadis, C.: Analytical Compact Modeling of Nanoscale Triple-Gate FinFETs. In: *Proceedings of the 16th IEEE Mediterranean Electrotechnical Conference (MELECON)*, pp. 72–75 (2012). DOI 10.1109/MELCON.2012.6196383.
64. Fei, W., Yu, H., Zhang, W., Yeo, K.S.: Design Exploration of Hybrid CMOS and Memristor Circuit by New Modified Nodal Analysis. *IEEE Transactions on Very Large Scale Integration (VLSI) Systems* **20**(6), 1012–1025 (2012). DOI 10.1109/TVLSI.2011.2136443.
65. Fried, D., Hoffmann, T., Nguyen, B.Y., Samavedam, S.: Comparison study of FinFETs: SOI vs. Bulk. URL http://www.soiconsortium.org/pdf/Comparison study of FinFETs – SOI versus Bulk.pdf. Accessed on 07th November 2012.
66. Fried, D.M.: *The Design, Fabrication and Characterization of Independent-Gate FINFETs*. Ph.D. thesis, Cornell University (2004).
67. Gao, W.: *Low Power Design Methodologies In Analog Blocks Of CMOS Image Sensors*. Ph.D. thesis, Computer Science And Engineering Department, York University, Toronto, Ontario, Canada (2011).
68. Garduno, S.I., Cerdeira, A., Estrada, M., Kylchitska, V., Flandre, D.: Analytic Modeling of Gate Tunneling Currents for nano-scale Double-Gate MOSFETs. In: *Devices, Circuits and Systems (ICCDCS), 2012 8th International Caribbean Conference on*, pp. 1–5 (2012). DOI 10.1109/ICCDCS.2012.6188938.
69. Geim, A.K., Novoselov, K.S.: The Rise of Graphene. *Nature Materials* **6**(3), 183–191 (2007). DOI http://dx.doi.org/10.1038/nmat1849.
70. Geppert, L.: The Amazing Vanishing Transistor Act. *IEEE Spectrum* **39**(10), 28–33 (2002). DOI 10.1109/MSPEC.2002.1038566.
71. Ghai, D.: *Variability Aware Low-Power Techniques for Nanoscale Mixed-Signal Circuits*. Ph.D. thesis, University of North Texas (2009).
72. Ghai, D.: *Variability Aware Low-Power Techniques for Nanoscale Mixed-Signal Circuits*. Ph.D. thesis, Department of Computer Science and Engineering, University of North Texas, Denton (2009).
73. Ghai, D., Mohanty, S.P., Kougianos, E.: Variability-Aware Optimization of Nano-CMOS Active Pixel Sensors using Design and Analysis of Monte Carlo Experiments. In: *Proceedings of the International Sympoisum on Quality Electronic Design* (2009).
74. Ghai, D., Mohanty, S.P., Kougianos, E., Patra, P.: A PVT Aware Accurate Statistical Logic Library for High-k Metal-Gate Nano-CMOS. In: *Proceedings of 10th International Symposium on Quality of Electronic Design (ISQED)*, pp. 47–54 (2009).
75. Goerbig, M.O.: Quantum Hall Effects. *ArXiv e-prints* (2009).
76. Gossmann, H.J.L., Agarwal, A., Parrill, T., Rubin, L.M., Poate, J.M.: On the FinFET Extension Implant Energy. *IEEE Transactions on Nanotechnology* **2**(4), 285–290 (2003). DOI 10.1109/TNANO.2003.820783.
77. Grogg, D., Ionescu, A.: The Vibrating Body Transistor. *IEEE Transactions on Electron Devices* **58**(7), 2113–2121 (2011). DOI 10.1109/TED.2011.2147786.
78. Hajimiri, A., Limotyrakis, S., , Lee, T.H.: Jitter and Phase Noise in Ring Oscillators. *IEEE Journal of Solid State Circuits* **34**(6), 790–804 (1999).
79. Hamalainen, P., Hannikainen, M., Hamalainen, T.D.: Review of Hardware Architectures for Advanced Encryption Standard Implementations Considering Wireless Sensor Networks. In: *Lecture Notes in Computer Science*, vol. 4599, pp. 443–453 (2007).
80. He, J., Xuemei, X., Chan, M., Lin, C.H., Niknejad, A., Hu, C.: A Non-Charge-Sheet Based Analytical Model of Undoped Symmetric Double-Gate MOSFETs using SPP Approach. In: *Proceedings of the 5th International Symposium on Quality Electronic Design*, pp. 45–50 (2004). DOI 10.1109/ISQED.2004.1283648.
81. Hisamoto, D., Lee, W.C., Kedzierski, J., Takeuchi, H., Asano, K., Kuo, C., Anderson, E., King, T.J., Bokor, J., Hu, C.: FinFET – a Self-Aligned Double-Gate MOSFET Scalable to 20 nm. *IEEE Transactions on Electron Devices* **47**(12), 2320–2325 (2000). DOI 10.1109/16.887014.
82. Hsu, F.C., Muller, R., Hu, C., Ko, P.K.: A Simple Punchthrough Model for Short-Channel MOSFET's. *IEEE Transactions on Electron Devices* **30**(10), 1354–1359 (1983). DOI 10.1109/T-ED.1983.21298.
83. Hu, C.: FinFET and other New Transistor Technologies (2011). URL http://www.eecs.berkeley.edu/~hu/FinFET-and-other-New-Transistor-Tech-Hu.pdf. Accessed on 01th November 2012.
84. Hu, M., Li, H., Chen, Y., Wang, X., Pino, R.: Geometry Variations Analysis of TiO_2 Thin-Film and Spintronic Memristors. In: *Design Automation Conference (ASP-DAC), 2011 16th Asia and South Pacific*, pp. 25–30 (2011). DOI 10.1109/ASPDAC.2011.5722193.
85. Huang, P.Y., Lee, Y.M.: Full-chip Thermal Analysis for the Early Design Stage via Generalized Integral Transforms. *Very Large Scale Integration (VLSI) Systems, IEEE Transactions on* **17**(5), 613–626 (2009). DOI 10.1109/TVLSI.2008.2006043
86. Huang, X., Lee, W.C., Kuo, C., Hisamoto, D., Chang, L., Kedzierski, J., Anderson, E., Takeuchi, H., Choi, Y.K., Asano, K., Subramanian, V., King, T.J., Bokor, J., Hu, C.: Sub 50-nm FinFET: PMOS. In: *Technical Digest International Electron Devices Meeting*, pp. 67–70 (1999). DOI 10.1109/IEDM.1999.823848.

87. Ieong, M., Wong, H.S.P., Nowak, E., Kedzierski, J., Jones, E.C.: High Performance Double-Gate Device Technology Challenges and Opportunities. In: *Proceedings of the International Symposium on Quality Electronic Design*, pp. 492–495 (2002). DOI 10.1109/ISQED.2002.996793.
88. Ihn, T., Güttinger, J., Molitor, F., Schnez, S., Schurtenberger, E., Jacobsen, A., Hellmüller, S., Frey, T., Dröscher, S., Stampfer, C., Ensslin, K.: Graphene Single-Electron Transistors. *Materials Today* **13**(3), 44–50 (2010). DOI 10.1016/S1369-7021(10)70033-X. URL http://www.sciencedirect.com/science/article/pii/S13697021 1070033X.
89. Ionescu, A.M., Riel, H.: Tunnel Field-Effect Transistors as Energy-Efficient Electronic Switches. *Nature* **479**, 329–337 (2011). DOI doi:10.1038/nature10679.
90. James, D.: High-k/Metal Gates in Leading Edge Silicon Devices. In: *Proceedings of the 23rd Annual SEMI Advanced Semiconductor Manufacturing Conference (ASMC)*, pp. 346–353 (2012). DOI 10.1109/ASMC.2012.6212925.
91. Jang, S.J., Ahn, J.H.: Material Properties Controlling The Performance of Amorphous Silicon Thin Film Transistors. In: *Materials Research Society Symposium Proceedings*, vol. 33, pp. 259–273 (1984).
92. Jang, S.J., Ahn, J.H.: Flexible Thin Flim Transistor using Printed Single-Walled Carbon Nanotubes. In: *Proceedings of 3rd International Nanoelectronics Conference (INEC)*, pp. 720–721 (2010). DOI 10.1109/INEC.2010.5424575.
93. Jeon, K.: Band-to-Band Tunnel Transistor Design and Modeling for Low Power Applications. Tech. Rep. UCB/EECS-2012-86, Electrical Engineering and Computer Sciences, University of California at Berkeley (2012).
94. Kahng, A.B.: DfX and Signoff: The Coming Challenges and Opportunities. In: *Proceedings of the IEEE Computer Society Symposium on VLSI*, pp. xix–xix (2012).
95. Kahng, A.B., Lach, J., Mangione-Smith, W.H., Mantik, S., Markov, I.L., Potkonjak, M., Tucker, P., Wang, H., Wolfe, G.: Watermarking Techniques for Intellectual Property Protection. In: *Proceedings of the Design Automation Conference*, pp. 776–781 (1998).
96. Kahng, A.B., Lach, J., Mangione-Smith, W.H., Mantik, S., Markov, I.L., Potkonjak, M., Tucker, P., Wang, H., Wolfe, G.: Constraint-Based Watermarking Techniques for Design IP Protection. *IEEE Transactions on Computer-Aided Design of Integrated Circuits and Systems* **20**(10), 1236–1252 (2001). DOI 10.1109/43.952740.
97. Kao, J., Narendra, S., Chandrakasan, A.: Subthreshold Leakage Modeling and Reduction Techniques. In: *Proceedings of IEEE/ACM International Conference on Computer Aided Design*, pp. 141–148 (2002). DOI 10.1109/ICCAD.2002.1167526.
98. Kapur, P., Chandra, G., McVittie, J.P., Saraswat, K.C.: Technology and Reliability Constrained Future Copper Interconnects – Part II: Performance Implications. *IEEE Transactions on Electron Devices* **49**(4), 598–604 (2002). DOI 10.1109/16.992868.
99. Kapur, P., McVittie, J.P., Saraswat, K.C.: Technology and Reliability Constrained Future Copper Interconnects – Part I: Resistance Modeling. *IEEE Transactions on Electron Devices* **49**(4), 590–597 (2002). DOI 10.1109/16.992867.
100. Karamcheti, A., Watt, V.H.C., Al-Shareef, H.N., Luo, T.Y., Brown, G.A., Jackson, M.D., Huff, H.R.: Silicon Oxynitride Films as Segue to the High-K Era. *Semiconductor Fabtech* **12**, 207–214 (2000).
101. Karri, R., Makris, Y., Sinanoglu, O.: ICCD Tutorial: Hardware Security and Trust. http://www.iccd-conf.com/2012/File_index/ICCD-trusttutorial.pdf (2012). Accessed on 06th June 2013.
102. Karri, R., Rajendran, J., Rosenfeld, K., Tehranipoor, M.: Trustworthy Hardware: Identifying and Classifying Hardware Trojans. *Computer* **43**(10), 39–46 (2010). DOI 10.1109/MC.2010.299.
103. Kasemsuwan, V.: An Analytical Transit Time Model For Short Channel MOSFET's. In: *Proceedings of the IEEE International Conference on Semiconductor Electronics*, pp. 72–75 (2000). DOI 10.1109/SMELEC.2000.932436.
104. Kastner, M.A.: The Single Electron Transistor and Artificial Atoms. *Annalen der Physik* **512**, 885–894 (2000). DOI 10.1002/1521-3889(200011)9:11/12<885:AID-ANDP885>3.0.CO;2-8.
105. Kavehei, O., Kim, Y.S., Iqbal, A., Eshraghian, K., Al-Sarawi, S., Abbott, D.: The Fourth Element: Insights Into The Memristor. In: *Proceedings of the International Conference on Communications, Circuits and Systems*, pp. 921–927 (2009). DOI 10.1109/ICCCAS.2009.5250370.
106. Kelton, R.: From Sand to Silicon: The Making of a Chip. http://blogs.intel.com/jobs/2012/02/28/from-sand-to-silicon-the-making-of-a-chip/ (2012). Accessed on 25th November 2012.
107. Kim, D.Y.: *Study on Reliability of VLSI Interconnection Structures*. Ph.D. thesis, Department of Materials Science and Engineering, Stanford University, CA (2003). Accessed on 16th May 2013.
108. Kim, N.S., Austin, T., Blaauw, D., Mudge, T., Flautner, K., Hu, J.S., Irwin, M.J., Kandemir, M., Vijaykrishnan, N.: Leakage Current - Moore's Law Meets Static Power. *IEEE Computer* **36**(12), 68–75 (2003).
109. Kim, S.D., Narasimha, S., Rim, K.: An Integrated Methodology for Accurate Extraction of S/D Series Resistance Components in Nanoscale MOSFETs. In: *Technical Digest IEEE International Electron Devices Meeting*, pp. 149–152 (2005). DOI 10.1109/IEDM.2005.1609291.
110. Kim, S.D., Woo, J.C.S.: Detailed Modeling of Source/Drain Parasitics and Their Impact on MOSFETs Scaling. In: *Proceedings of the Third International Workshop on Junction Technology*, pp. 1–4 (2002). DOI 10.1109/IWJT.2002.1225186.
111. Kim, S.S.: LCD: Future Prospects and Impact on Human Lifestyle. In: *Proceedings of Pacific Rim Conference on Lasers and Electro-Optics*, p. 1 (2007). DOI 10.1109/CLEOPR.2007.4391207.
112. Kingon, A.I., Maria, J.P., Streifferr, S.K.: Alternative Dielectrics to Silicon Dioxide for memory and Logic Devices. *Nature* **406**, 1021–1038 (2000).

113. Konofaos, N., Evangelou, E.K., Aslanoglou, X., Kokkoris, M., Vlastou, R.: Dielectric Properties of CVD Grown SiON Thin Films on Si for MOS Microelectronic Devices. *Semiconductor Science and Technology* **19**(1), 50 (2004).
114. Kosaraju, N.M., Varanasi, M., Mohanty, S.P.: A High-Performance VLSI Architecture for Advanced Encryption Standard (AES) Algorithm. In: *Proceedings of the 19th International Conference on VLSI Design*, pp. 481–484 (2006). DOI 10.1109/VLSID.2006.9.
115. Kougianos, E., Mohanty, S.P.: Metrics to Quantify Steady and Transient Gate Leakage in Nanoscale Transistors: NMOS vs. PMOS Perspective. In: *Proceedings of the 20th IEEE International Conference on VLSI Design*, pp. 195–200 (2007). DOI 10.1109/VLSID.2007.107.
116. Kougianos, E., Mohanty, S.P.: A Comparative Study on Gate Leakage and Performance of High-κ Nano-CMOS Logic Gates. *Taylor & Francis International Journal of Electronics* **97**(9), 985–1005 (2010).
117. Kougianos, E., Mohanty, S.P., Mahapatra, R.N.: Hardware Assisted Watermarking for Multimedia. *Computers & Electrical Engineering* **35**(2), 339–358 (2009). DOI 10.1016/j.compeleceng.2008.06.002 . URL http://www.sciencedirect.com/science/article/pii/S004579060800061X.
118. Kretz, J., Dreeskornfeld, L., Hartwich, J., Rosner, W.: 20 nm Electron Beam Lithography and Reactive Ion Etching for the Fabrication of Double Gate FinFET Devices. *Microelectronic Engineering* **67–68**(0), 763–768 (2003). DOI 10.1016/S0167-9317(03)00136-9. URL http://www.sciencedirect.com/science/article/pii/S0167931703001369. *Proceedings of the 28th International Conference on Micro- and Nano-Engineering*
119. Kumar, A., Anis, M.: Dual-Threshold CAD Framework for Subthreshold Leakage Power Aware FPGAs. *IEEE Transactions on Computer-Aided Design of Integrated Circuits and Systems* **26**(1), 53–66 (2007).
120. Kursun, V., Friedman, E.G.: Variable Threshold Voltage Keeper for Contention Reduction in Dynamic Circuits. In: *ASIC/SOC Conference, 2002. 15th Annual IEEE International*, pp. 314–318 (2002). DOI 10.1109/ASIC.2002.1158077.
121. Lee, B.H. et al.: Thermal Stability and Electrical Characteristics of Ultrathin Hafnium Oxide Gate Dielectric Reoxidized with Rapid Thermal Annealing. *Applied Physics Letters* **77**, 1926–1928 (2000).
122. Lemme, M., Mollenhauer, T., Henschel, W., Wahlbrink, T., Gottlob, H., Efavi, J., Baus, M., Winkler, O., Spangenberg, B., Kurz, H.: Subthreshold Characteristics of P-Type Triple-Gate MOSFETs. In: *Proceedings of the European Solid-State Device Research*, pp. 123–126 (2003). DOI 10.1109/ESSDERC.2003.1256826.
123. Lent, C.S., Tougaw, P.D., Porod, W., Bernstein, G.H.: Quantum Cellular Automata. *Nanotechnology* **4**(1), 49–57 (1993). URL http://stacks.iop.org/0957-4484/4/i=1/a=004.
124. Lewyn, L.L., Ytterdal, T., Wulff, C., Martin, K.: Analog Circuit Design in Nanoscale CMOS Technologies. *Proceedings of the IEEE* **97**(10), 1687–1714 (2009). DOI 10.1109/JPROC.2009.2024663.
125. Li, H., Hu, M.: Compact Model of Memristors and its Application in Computing Systems. In: *Proceedings of the Design, Automation Test in Europe Conference Exhibition*, pp. 673–678 (2010).
126. Li, S.: *Carbon Nanotube High Frequency Devices*. Master's thesis, Department of Electrical and Computer Engineering, University of California, Irvine (2004).
127. Li, W., Tan, C.M.: Black's Equation For Today's ULSI Interconnect Electromigration Reliability – A Revisit. In: *Proceedings of the International Conference of Electron Devices and Solid-State Circuits (EDSSC)*, pp. 1–2 (2011). DOI 10.1109/EDSSC.2011.6117717.
128. Liao, L., Lin, Y.C., Bao, M., Cheng, R., Bai, J., Liu, Y., Qu, Y., Wang, K.L., Huang, Y., Duan, X.: High-Speed Graphene Transistors With A Self-Aligned Nanowire Gate. *Nature* **467**, 305–308 (2010). DOI doi:10.1038/nature09405.
129. Lienig, J.: Invited Talk: Introduction to Electromigration-Aware Physical Design. In: *Proceedings of the ACM International Symposium on Physical Design*, pp. 39–46 (2006). DOI 10.1145/1123008.1123017. URL http://doi.acm.org/10.1145/1123008.1123017.
130. Lim, T., Kim, Y.: Effect of Band-To-Band Tunnelling Leakage on 28 nm MOSFET Design. *IEE Electronic Letters* **44**(2), 157–158 (2008).
131. Lim, T., Jang J., Kim Y.: Source/Drain Design for 16 nm Surrounding Gate MOSFETs, In: *Proceedings of the International Semiconductor Device Research Symposium (ISDRS)*, pp. 1–2 (2009).
132. Lin, C.H., Dunga, M., Balasubramanian, S., Niknejad, A.M., Hu, C., Xi, X., He, J., Chang, L., Williams, R.Q., Ketchen, M.B., Haensch, W.E., Chan, M.: Compact Modeling of FinFETs Featuring Independent-Gate Operation Mode. In: *Proceedings of the IEEE VLSI-TSA International Symposium on VLSI Technology*, pp. 120–121 (2005).
133. Lin, C.W., Huang, J.L.: A built-in tft array charge-sensing technique for system-on-panel displays. In: *Proceedings of 26th IEEE VLSI Test Symposium*, pp. 169–174 (2008). DOI 10.1109/VTS.2008.22.
134. Lin, W.W., Liang, C.: Polysilicon gate depletion effect on deep-submicron circuit performance. In: *Numerical Modeling of Processes and Devices for Integrated Circuits, 1994. NUPAD V., International Workshop on*, pp. 185–188 (1994). DOI 10.1109/NUPAD.1994.343461.
135. Lin-Hendel, C.G.: Accurate interconnect modeling for high frequency lsi/vlsi circuits and systems. In: *Computer Design: VLSI in Computers and Processors, 1990. ICCD '90. Proceedings, 1990 IEEE International Conference on*, pp. 434–442 (1990). DOI 10.1109/ICCD.1990.130273.
136. Lu, D.D., Dunga, M.V., Lin, C.H., Niknejad, A.M., Hu, C.: A Computationally Efficient Compact Model For Fully-Depleted SOI MOSFETS With Independently-Controlled Front- and Back-Gates. *Solid-State Electronics* **62**(1), 31–39 (2011). DOI 10.1016/j.sse.2010.12.015. URL http://www.sciencedirect.com/science/article/pii/S0038110110004351.

137. Lu, W., Soukiassian, P., Boeckl, J.: Graphene: Fundamentals and Functionalities. *Material Research Society Bulletin* **37**(12), 1119–1124 (2012).
138. Luisier, M., Klimeck, G.: Performance Analysis of Statistical Samples of Graphene Nanoribbon Tunneling Transistors With Line Edge Roughness. *Applied Physics Letters* **94**(22), 223505 (2009). DOI 10.1063/1.3140505. URL http://link.aip.org/link/?APL/94/223505/1
139. Lundstrom, M.: MOSFET Leakage. http://nanohub.org/resources/5690/download/2008.10.28-ece612-l16.pdf (2008). Accessed on 15th January 2013.
140. Ma, T.P.: Making Silicon Nitride a Viable Gate Dielecrtric. *IEEE Transaction on Electron Devices* **45**, 680–690 (1999).
141. Marulanda, J.M., Srivastava, A.: I-V Characteristics Modeling and Parameter Extraction for CNT-FETs. In: *Proceedings of the International Semiconductor Device Research Symposium*, pp. 38–39 (2005). DOI 10.1109/ISDRS.2005.1595965.
142. Marulanda, J.M., Srivastava, A., Yellampalli, S.: Numerical Modeling of the I-V Characteristic of Carbon Nanotube Field Effect Transistors (CNT-FETs). In: *Proceedings of the 40th Southeastern Symposium on System Theory*, pp. 235–238 (2008). DOI 10.1109/SSST.2008.4480228.
143. Massoud, Y., Nieuwoudt, A.: Modeling and Design Challenges and Solutions For Carbon Nanotube-Based Interconnect in Future High Performance Integrated Circuits. *Journal of Emerging Technology in Computing Systems* **2**(3), 155–196 (2006). DOI 10.1145/1167943.1167944. URL http://doi.acm.org/10.1145/1167943.1167944.
144. Mathew, J., Banerjee, S., Rahaman, H., Pradhan, D.K., Mohanty, S.P., Jabir, A.M.: On the Synthesis of Attack Tolerant Cryptographic Hardware. In: *Proceedings of the 18th IEEE/IFIP International Conference on Very Large Scale Integration of System-on-Chip*, pp. 286–291 (2010).
145. Mathew, J., Mohanty, S.P., Banerjee, S., Pradhan, D.K., Jabir, A.: Attack Tolerant Cryptographic Hardware Design by Combining Galois Field Error Correction and Uniform Switching Activity. *Computers & Electrical Engineering* (2013). DOI 10.1016/j.compeleceng.2013.01.001. URL http://www.sciencedirect.com/science/article/pii/S0045790613000062.
146. McCormick, M.A., Roeder, R.K., Slamovich, E.B.: Processing Effects on the Composition and Dielectric Properties of Hydrothermally Derived $Ba_xSr_{(1-x)}TiO_3$ Thin Films. *Journal of Materials Research* **44**(10), 1200–1209 (2001). DOI 10.1557/JMR.2001.0166.
147. McGuinness, P.: Variations, Margins, and Statistics. In: *Proceedings of th International Symposium on Physical Design, ISPD '08*, pp. 60–67. ACM, New York, NY, USA (2008). DOI 10.1145/1353629.1353643. URL http://doi.acm.org/10.1145/1353629.1353643.
148. Michelakis, K., Prodromakis, T., Toumazou, C.: Cost-Effective Fabrication of Nanoscale Electrode Memristors With Reproducible Electrical Response. *IET Micro Nano Letters* **5**(2), 91–94 (2010). DOI 10.1049/mnl.2009.0106.
149. Misewich, J.A., Martel, R., Avouris, P., Tsang, J.C., Heinze, S., Tersoff, J.: Electrically Induced Optical Emission from a Carbon Nanotube FET. *Science* **300**(5620), 783–786 (2003). DOI 10.1126/science.1081294. URL http://www.sciencemag.org/content/300/5620/783.abstract.
150. Mistry, K., Allen, C., Auth, C., Beattie, B., Bergstrom, D., Bost, M., Brazier, M., Buehler, M., Cappellani, A., Chau, R., Choi, C.H., Ding, G., Fischer, K., Ghani, T., Grover, R., Han, W., Hanken, D., Hattendorf, M., He, J., Hicks, J., Huessner, R., Ingerly, D., Jain, P., James, R., Jong, L., Joshi, S., Kenyon, C., Kuhn, K., Lee, K., Liu, H., Maiz, J., McIntyre, B., Moon, P., Neirynck, J., Pae, S., Parker, C., Parsons, D., Prasad, C., Pipes, L., Prince, M., Ranade, P., Reynolds, T., Sandford, J., Shifren, L., Sebastian, J., Seiple, J., Simon, D., Sivakumar, S., Smith, P., Thomas, C., Troeger, T., Vandervoorn, P., Williams, S., Zawadzki, K.: A 45nm Logic Technology with High-k+Metal Gate Transistors, Strained Silicon, 9 Cu Interconnect Layers, 193nm Dry Patterning, and 100% Pb-free Packaging. In: *Proceedings of the IEEE International Electron Devices Meeting*, pp. 247–250 (2007). DOI 10.1109/IEDM.2007.4418914.
151. Mitra, S., Zhang, M., Seifert, N., Mak, T.M., Kim, K.S.: Soft Error Resilient System Design through Error Correction. In: *Proceedings of the IFIP International Conference on Very Large Scale Integration*, pp. 332–337 (2006). DOI 10.1109/VLSISOC.2006.313256.
152. Mohanty, S.P.: *Energy and Transient Power Minimization during Behavioral Synthesis*. Ph.D. thesis, Department of Computer Science and Engineering, University of South Florida, USA (2003).
153. Mohanty, S.P.: A Secure Digital Camera Architecture for Integrated Real-Time Digital Rights Management. *Journal of Systems Architecture* **55**(10-12), 468–480 (2009). DOI 10.1016/j.sysarc.2009.09.005. URL http://www.sciencedirect.com/science/article/pii/S1383762109000617.
154. Mohanty, S.P.: Unified Challenges in Nano-CMOS High-Level Synthesis. In: *Proceedings of the 22nd International Conference on VLSI Design*, pp. 531–531 (2009).
155. Mohanty, S.P.: ISWAR: An Imaging System with Watermarking and Attack Resilience. The Computing Research Repository (CoRR) **abs/1205.4489** (2012).
156. Mohanty, S.P.: DfX for Nanoelectronic Embedded Systems. http://www.cse.unt.edu/~smohanty/Presentations/2013/Mohanty_CARE2013_Keynote.pdf (2013). Accessed on 25th May 2014.
157. Mohanty, S.P.: Memristor: From Basics to Deployment. *IEEE Potentials* **32**(3), 34–39 (2013).
158. Mohanty, S.P., Kougianos, E.: Modeling and Reduction of Gate Leakage during Behavioral Synthesis of NanoCMOS Circuits. In: *Proceedings of the 19th International Conference on VLSI Design*, pp. 83–88 (2006).

159. Mohanty, S.P., Kougianos, E.: Steady and Transient State Analysis of Gate Leakage Current in Nanoscale CMOS Logic Gates. In: *Proceedings of the 24th IEEE International Conference on Computer Design (ICCD)*, pp. 210–215 (2006).
160. Mohanty, S.P., Kougianos, E.: Simultaneous Power Fluctuation and Average Power Minimization during Nano-CMOS Behavioural Synthesis. In: *Proceedings of the 20th IEEE International Conference on VLSI Design*, pp. 577–582 (2007).
161. Mohanty, S.P., Kougianos, E., Mahapatra, R.N.: A Comparative Analysis of Gate Leakage and Performance of High-κ Nanoscale CMOS Logic Gates. In: *Proceedings of the 16th ACM/IEEE International Workshop on Logic and Synthesis*, pp. 31–38 (2007).
162. Mohanty, S.P., Ranganathan, N., Kougianos, E., Patra, P.: *Low-Power High-Level Synthesis for Nanoscale CMOS Circuits*. Springer (2008). 0387764739 and 978-0387764733.
163. Mohapatra, N.R., Desai, M.P., Narendra, S.G., Ramgopal Rao, V.: Modeling of Parasitic Capacitances in Deep Submicrometer Conventional and High-κ Dielectric MOS Transistors. *Electron Devices, IEEE Transactions on* **50**(4), 959–966 (2003). DOI 10.1109/TED.2003.811387.
164. Monthioux, M., Kuznetsov, V.L.: Who Should Be Given the Credit for the Discovery of Carbon Nanotubes? *Carbon* **44**(9), 1621–1623 (2006). DOI 10.1016/j.carbon.2006.03.019. URL http://www.sciencedirect.com/science/article/pii/S000862230600162X.
165. Morshed, T.H., Lu, D.D., Wenwei Yang, M.V.D., Xi, X., He, J., Liu, W., Kanyu, Cao, M., Jin, X., Ou, J.J., Chan, M., Niknejad, A.M., Hu, C.: *BSIM4v4.7 MOSFET Model – User's Manual*. Department of Electrical Engineering and Computer Sciences University of California, Berkeley, CA 94720. (2011). 22nd March 2013.
166. Mukherjee, V.: *A Dual Dielectric Approach for Performance Aware Reduction of Gate Leakage in Combinational Circuits*. Master's thesis, Department of Computer Science and Engineering, University of North Texas, Denton (2006).
167. Mukherjee, V., Mohanty, S.P., Kougianos, E.: A Dual Dielectric Approach for Performance Aware Gate Tunneling Reduction in Combinational Circuits. In: *Proceedings of the 23rd IEEE International Conference of Computer Design (ICCD)*, pp. 431–436 (2005).
168. Mukhopadhyay, S., Raychowdhury, A., Roy, K.: Accurate Estimation of Total Leakage in Nanometer-Scale Bulk CMOS Circuits Based on Device Geometry and Doping Profile. *IEEE Transactions on Computer-Aided Design of Integrated Circuits and Systems* **24**(3), 363–381 (2005).
169. Mukhopadhyay, S., Roy, K.: Modeling and Estimation of Total Leakage Current in Nano-scaled-CMOS Devices Considering the Effect of Parameter Variation. In: *Proceedings of the International Symposium on Low Power Electronics and Design*, pp. 172–175 (2003).
170. Nagel, L., McAndrew, C.: Is SPICE Good Enough for Tomorrow's Analog? In: *Proceedings of the IEEE Bipolar/BiCMOS Circuits and Technology Meeting*, pp. 106–112 (2010). DOI 10.1109/BIPOL.2010.5668096.
171. Nayak, A., Haldar, M., Banerjee, P., Chen, C., Sarrafzadeh, M.: Power Optimization of Delay Constrained Circuits. In: *Proceedings of 13th Annual IEEE International ASIC/SOC Conference*, pp. 305–309 (2000). DOI 10.1109/ASIC.2000.880754.
172. Ng, K.K., Taylor, G.W.: Effects of hot-carrier trapping in n- and p-channel mosfet's. *Electron Devices, IEEE Transactions on* **30**(8), 871–876 (1983). DOI 10.1109/T-ED.1983.21229.
173. Nimgaonkar, S., Gomathisankaran, M., Mohanty, S.P.: TSV: A Novel Energy Efficient Memory Integrity Verification Scheme for Embedded Systems. *Journal of Systems Architecture* **59**(7), 400–411 (2013). DOI 10.1016/j.sysarc.2013.04.008. URL http://www.sciencedirect.com/science/article/pii/S138376211300057X.
174. Noguchi, J.: Dominant Factors in TDDB Degradation of Cu Interconnects. *IEEE Transactions on Electron Devices* **52**(8), 1743–1750 (2005). DOI 10.1109/TED.2005.851849.
175. Okada, K., Yamaoka, K., Onodera, H.: A statistical gate delay model for intra-chip and inter-chip variabilities. In: *Proceedings of the Asia and South Pacific Design Automation Conference*, pp. 31–36 (2003).
176. Orio, R.L.D.: *Electromigration Modeling and Simulation*. Ph.D. thesis, Faculty of Electrical Engineering and Information Technology, Vienna University of Technology, Vienna, Austria (2010). Accessed on 06th June 2013.
177. Ouyang, Y., Yoon, Y., Guo, J.: Scaling Behaviors of Graphene Nanoribbon FETs: A Three-Dimensional Quantum Simulation Study. *IEEE Transactions on Electron Devices* **54**(9), 2223–2231 (2007). DOI 10.1109/TED.2007.902692.
178. Pan, L.Y., Chang, S.C., Liao, M.Y., Lin, Y.T.: The Future Development of Global LCD TV Industry. In: *Portland International Center for Management of Engineering and Technology*, pp. 1818–1821 (2007). DOI 10.1109/PICMET.2007.4349508.
179. Parulkar, I., Anandakumar, S., Agarwal, G., Liu, G., Rajan, K., Chiu, F., Pendurkar, R.: DFX of a 3rd Generation, 16-core/32-thread UltraSPARC-CMT Microprocessor. In: *Proceedings of the IEEE International Test Conference*, pp. 1–10 (2008). DOI 10.1109/TEST.2008.4700552.
180. Paul, B.C., Fujita, S., Okajima, M., Lee, T.H., Wong, H.S.P., Nishi, Y.: Impact of a Process Variation on Nanowire and Nanotube Device Performance. *IEEE Transactions on Electron Devices* **54**(9), 2369–2376 (2007). DOI 10.1109/TED.2007.901882.
181. Pei, G., Kedzierski, J., Oldiges, P., Ieong, M., Kan, E.C.: FinFET Design Considerations Based on 3-D Simulation and Analytical Modeling. *IEEE Transactions on Electron Devices* **49**(8), 1411–1419 (2002). DOI 10.1109/TED.2002.801263.

182. Pershin, Y.V., Di Ventra, M.: Memristive Circuits Simulate Memcapacitors and Meminductors. *Electronics Letters* **46**(7), 517–518 (2010).
183. Pershin, Y.V., Di Ventra, M.: Teaching memory circuit elements via experiment-based learning. *IEEE Circuits and Systems Magazine* **12**(1), 64–74 (2012).
184. PhysOrg.com: Fujitsu Develops Technology for Low-Temperature Full-Service Direct Formation of Graphene Transistors on Large-Scale Sub (2009). http://phys.org/news178552799.html.
185. Podrzaj, J., Sesek, A., Trontelj, J.: Intelligent Power MOSFET Driver ASIC. In: *Proceedings of the 35th International Convention*, pp. 107–111 (2012).
186. Poolakkaparambil, M., Mathew, J., Jabir, A.M., Mohanty, S.P.: An Investigation of Concurrent Error Detection over Binary Galois Fields in CNTFET and QCA Technologies. In: *Proceedings of the IEEE Computer Society Annual Symposium on VLSI*, pp. 141–146 (2012).
187. Powell, M.J.: The physics of Amorphous-Silicon Thin-Film Transistors. *IEEE Transactions on Electron Devices* **36**(12), 2753–2763 (1989). DOI 10.1109/16.40933.
188. Prodromakis, T., Michelakis, K., Toumazou, C.: Fabrication and Electrical Characteristics of Memristors with TiO_2/TiO_{2-x} Active Layers. In: *Proceedings of the IEEE International Symposium on Circuits and Systems*, pp. 1520–1522 (2010). DOI 10.1109/ISCAS.2010.5537379.
189. Qi, W.J. et al.: Ultrathin Zirconium Silicate Film with Good Thermal Stability for Alternative Gate Dielectric Application. *Applied Physics Letters* **77**, 1704–1706 (2000).
190. Qin, L.C.: Determination of the chiral indices (n,m) of carbon nanotubes by electron diffraction. *Physical Chemistry Chemical Physics Journal* **9**, 31–48 (2007). DOI 10.1039/B614121H. URL http://dx.doi.org/10.1039/B614121H.
191. Quader, K.N., Fang, P., Yue, J.T., Ko, P.K., Hu, C.: Hot-Carrier-Reliability Design Rules for Translating Device Degradation to CMOS Digital Circuit Degradation. *IEEE Transactions on Electron Devices* **41**(5), 681–691 (1994). DOI 10.1109/16.285017.
192. Quader, K.N., Minami, E.R., Huang, W.J., Ko, P.K., Hu, C.: Hot-Carrier-Reliability Design Guidelines for CMOS Logic Circuits. *IEEE Journal of Solid-State Circuits* **29**(3), 253–262 (1994). DOI 10.1109/4.278346.
193. Rabaey, J.M., Chandrakasan, A., Nikolić, B.: *Digital Integrated Circuits*, second edn., Prentice-Hall Publishers (2003).
194. Radwan, A.G., Zidan, M.A., Salama, K.N.: On the Mathematical Modeling of Memristors. In: *Proceedings of the International Conference on Microelectronics*, pp. 284–287 (2010). DOI 10.1109/ICM.2010.5696139.
195. Ramos, J., Francken, K., Gielen, G.G.E., Steyaert, M.S.J.: An Efficient, Fully Parasitic-Aware Power Amplifier Design Optimization Tool. *IEEE Transactions on Circuits and Systems I: Regular Papers* **52**(8), 1526–1534 (2005). DOI 10.1109/TCSI.2005.851677.
196. Ravi, S., Raghunathan, A., Kocher, P., Hattangady, S.: Security in Embedded Systems: Design Challenges. *ACM Transactions on Embedded Computing Systems (TECS)* **3**(3), 461–491 (2004).
197. Redmond, C.: Winning the Battle Against Latch-up in CMOS Analog Switches. *Analog Dialogue* **35**(05), 2453–2458 (2001).
198. Rent, T.M.: Rent's Rule: A Family Memoir. *IEEE Solid-State Circuits Magazine* **2**(1), 14–20 (2010). DOI 10.1109/MSSC.2009.935294.
199. Risch, L.: Pushing CMOS Beyond the Roadmap. In: *Proceedings of the 31st European Solid-State Circuits Conference*, pp. 63–68 (2005). DOI 10.1109/ESSCIR.2005.1541558.
200. Robertson, J.: High Dielectric Constant Oxides. *The European Physical Journal Applied Physics* **28**(12), 265–291 (2004). DOI 10.1051/epjap:2004206.
201. Roy, K., Mukhopadhyay, S., Meimand, H.M.: Leakage Current Mechanisms and Leakage Reduction Techniques in Deep-Submicrometer CMOS Circuits. *Proceedings of the IEEE* **91**(2), 305–327 (2003).
202. Mohanty, S.P., Velagapudi, R., Kougianos, E.: Physical-Aware Simulated Annealing Optimization of Gate Leakage in Nanoscale Datapath Circuits. In: *Proceedings of the 9th IEEE International Conference on Design Automation and Test in Europe (DATE)*, pp. 1191–1196 (2006).
203. Saha, D., Sur-Kolay, S.: A Unified Approach for IP Protection across Design Phases in a Packaged Chip. In: *Proceedings of the 23rd International Conference on VLSI Design*, pp. 105–110 (2010). DOI 10.1109/VLSI.Design.2010.52.
204. Sairam, T., Zhao, W., Cao, Y.: Optimizing Finfet Technology for High-speed and Low-power Design. In: *Proceedings of the ACM Great Lakes symposium on VLSI*, pp. 73–77 (2007).
205. Sakurai, T., Newton, A.: Alpha-Power Law MOSFET Model and its Applications to CMOS Inverter Delay and Other Formulas. *IEEE Journal of Solid-State Circuits* **25**(2), 584–594 (1990). DOI 10.1109/4.52187.
206. Sankaranarayanan, K.: Thermal Modeling and Management of Microprocessors. Ph.D. thesis, Computer Science, School of Engineering and Applied Science, University of Virginia (2009). Accessed on 14th May 2013.
207. Saraswat, K.: MOS Device Scaling. http://www.stanford.edu/class/ee316/MOSFET_Handout5.pdf (2002). Accessed on 25th November 2012.
208. Saraswat, K.: Interconnect Scaling. http://www.stanford.edu/class/ee311/NOTES/InterconnectScalingSlides.pdf (2006). Accessed on 01st May 2013.
209. Saurabh, S., Kumar, M.J.: Estimation and Compensation of Process-Induced Variations in Nanoscale Tunnel Field-Effect Transistors for Improved Reliability. *IEEE Transactions on Device and Materials Reliability* **10**(3), 390–395 (2010). DOI 10.1109/TDMR.2010.2054095.

210. Savage, N.: Gain From Graphene. http://spectrum.ieee.org/semiconductors/nanotechnology/gain-from-graphene. Accessed on 19th November 2012.
211. Savage, N.: Graphene Makes Transistors Tunable. http://spectrum.ieee.org/semiconductors/materials/graphene-makes-transistors-tunable. Accessed on 19th November 2012.
212. Schroder, D.K.: Negative Bias Temperature Instability: What do we Understand? *Microelectronics Reliability* **47**(6), 841–852 (2007). DOI 10.1016/j.microrel.2006.10.006. URL http://www.sciencedirect.com/science/article/pii/S002627140600374X.
213. Schuster, C.M.: *Negative Bias Temperature Instability (NBTI) Experiment*. Master's thesis, Department of Electrical and Computer Engineering, Naval Postgraduate School, Monterey, California (2006). Accessed on 04th June 2013.
214. Schwierz, F.: Graphene for Electronic Applications – Transistors and More. In: *Proceedings of the IEEE Bipolar/BiCMOS Circuits and Technology Meeting (BCTM)*, pp. 173–179 (2010). DOI 10.1109/BIPOL.2010.5668069.
215. Schwierz, F.: Graphene Transistors. *Nature Nanotechnology* **5**, 487–496 (2010). DOI doi:10.1038/nnano.2010.89.
216. Schwierz, F.: Graphene Transistors – A New Contender for Future Electronics. In: *Proceedings of the IEEE International Conference on Solid-State and Integrated Circuit Technology*, pp. 1202–1205 (2010). DOI 10.1109/ICSICT.2010.5667602.
217. Kyoung Shin, M., Lin, C.H.: An Efficient Resource Allocation Algorithm with Minimal Power Consumption. In: *Proceedings of IEEE Region 10 International Conference on Electrical and Electronic Technology*, vol. 2, pp. 703–706 (2001). DOI 10.1109/TENCON.2001.949683.
218. Sim, S.P., Krishnan, S., Petranovic, D.M., Arora, N.D., Lee, K., Yang, C.Y.: A Unified RLC Model for High-Speed On-Chip Interconnects. *IEEE Transactions on Electron Devices* **50**(6), 1501–1510 (2003). DOI 10.1109/TED.2003.813345.
219. Singh, R., Ulrich, R.K.: High and Low Dielectric Constant Materials. *The Electrochemical Society Interface* pp. 26–30 (1999). Accessed on 04th June 2013.
220. Singhal, K.: Parametric Process Variations (June 2007). *Personal Communication and Synopsys Booth at Design Automation Conference (DAC)*.
221. Sinha, S., Balijepalli, A., Cao, Y.: A Simplified Model of Carbon NanoTube Transistor with Applications to Analog and Digital Design. In: *Proceedings of the International Symposium on Quality Electronic Design*, pp. 502–507 (2008).
222. Sinha, S., Balijepalli, A., Cao, Y.: Compact Model of Carbon Nanotube Transistor and Interconnect. *IEEE Transactions on Electron Devices* **56**(10), 2232–2242 (2009). DOI 10.1109/TED.2009.2028625.
223. Skotnicki, T., Merckel, G., Pedron, T.: A New Punchthrough Current Model Based on the Voltage-Doping Transformation. *IEEE Transactions on Electron Devices* **35**(7), 1076–1086 (1988). DOI 10.1109/16.3367.
224. Song, J., Yu, B., Yuan, Y., Taur, Y.: A Review on Compact Modeling of Multiple-Gate MOSFETs. *IEEE Transactions on Circuits and Systems I: Regular Papers* **56**(8), 1858–1869 (2009). DOI 10.1109/TCSI.2009.2028416.
225. Souri, S.J.: *3D ICs Interconnect Performance Modeling and Analysis*. Ph.D. thesis, Electrical Engineering, Stanford University (2003).
226. Srivastava, A.: Transistor and Interconnect Modeling For Design of Carbon Nanotube Integrated Circuits. In: *Proceedings of the 19th International Conference Mixed Design of Integrated Circuits and Systems*, pp. 30–36 (2012).
227. Stamenkovic, Z., Mitrovic, S.: Integrated Circuit Yield Prediction. In: *Proceedings of the 20th International Conference on Microelectronics*, vol. 2, pp. 479–483 (1995). DOI 10.1109/ICMEL.1995.500913.
228. Steiner, N.: Zinc Negative Resistance Oscillator. http://www.sparkbangbuzz.com/els/zincosc-el.htm (2001). Accessed on 24th November 2012.
229. Stiller, B., Bocek, T., Hecht, F., Machado, G., Racz, P., Waldburger, M.: Understand and Avoid Electromigration (EM) & IR-drop in Custom IP Blocks. Tech. rep., University of Zurich, Department of Informatics (2010). Accessed on 06th June 2013.
230. Stolk, P.A., Widdershoven, F.P., Klaassen, D.B.M.: Modeling Statistical Dopant Fluctuations in MOS Transistors. *IEEE Transactions on Electron Devices* **45**(9), 1960–1971 (1998). DOI 10.1109/16.711362.
231. Strukov, D.B., Sinder, G.S., Stewart, D.R., Williams, R.S.: The Missing Memristor Found. *Nature* **453**, 80–83 (2008).
232. Subramanian, V.: Multiple Gate Field-Effect Transistors for Future CMOS Technologies. *IETE Technical Review* **27**(6), 446–454 (2010). DOI 10.4103/0256-4602.72582.
233. Takeda, E., Suzuki, N., Hagiwara, T.: Device performance degradation to hot-carrier injection at energies below the si-sio2energy barrier. In: *Electron Devices Meeting, 1983 International*, vol. 29, pp. 396–399 (1983). DOI 10.1109/IEDM.1983.190525.
234. Takeda, E., Yang, C.Y., Miura-Hamada, A.: *Hot-Carrier Effects in MOS Devices*. Academic Press, Inc., San Diego, CA (1995).
235. Tan, C.W., Miao, J.: Transmission Line Characteristics of a CNT-based Vertical Interconnect Scheme. In: *Proceedings of the 57th Electronic Components and Technology Conference*, pp. 1936–1941 (2007). DOI 10.1109/ECTC.2007.374065.
236. Tarog, E.S.: *Design Techniques to Improve Time Dependent Dielectric Breakdown Based Failure for CMOS Circuits*. Master's thesis, Department of Electrical Engineering, California Polytechnic State University, San Luis Obispo, California (2010). http://digitalcommons.calpoly.edu/theses/229/. Accessed on 04th June 2013.

237. Taur, Y.: MOSFET Channel Length: Extraction and Interpretation. *IEEE Transactions on Electron Devices* **47**(1), 160–170 (2000). DOI 10.1109/16.817582.
238. Taur, Y., Liang, X., Wang, W., Lu, H.: A Continuous, Analytic Drain-Current Model for DG MOSFETs. *Electron Device Letters, IEEE* **25**(2), 107–109 (2004). DOI 10.1109/LED.2003.822661.
239. Taur, Y., Song, J., Yu, B.: Compact Modeling of Multiple-Gate MOSFETs. In: *Proceedings of the IEEE Custom Integrated Circuits Conference*, pp. 257–264 (2008). DOI 10.1109/CICC.2008.4672073.
240. Tawfik, S., Kursun, V.: Low-Power and Compact Sequential Circuits With Independent-Gate FinFETs. *IEEE Transactions on Electron Devices* **55**(1), 60–70 (2008). DOI 10.1109/TED.2007.911039.
241. Taylor, G., Wong, K.: Power-On Contention Elimination. In: *Digest of Technical Papers Symposium on VLSI Circuits*, pp. 22–23 (1996). DOI 10.1109/VLSIC.1996.507701.
242. Tüzel, V.H.: *A Level Set Method for an Inverse Problem Arising in Photolithography*. Ph.D. thesis, The University of Minnesota (2009).
243. Tehranipoor, M., Koushanfar, F.: A Survey of Hardware Trojan Taxonomy and Detection. *IEEE Design Test of Computers* **27**(1), 10–25 (2010). DOI 10.1109/MDT.2010.7.
244. Thakral, G.: *Process-Voltage-Temperature Aware Nanoscale Circuit Optimization*. Ph.D. thesis, Department of Computer Science and Engineering, University of North Texas, Denton (2010).
245. Thakral, G., Mohanty, S.P., Ghai, D., Pradhan, D.K.: A DOE-ILP assisted conjugate-gradient based power and stability optimization in High-κ Nano-CMOS SRAM. In: *Proceedings of the ACM Great Lakes Symposium on VLSI*, pp. 323–328 (2010).
246. Thakral, G., Mohanty, S.P., Ghai, D., Pradhan, D.K.: P3 (Power-Performance-Process) Optimization of nano-CMOS SRAM using Statistical DOE-ILP. In: *Proceedings of the 11th International Symposium on Quality of Electronic Design*, pp. 176–183 (2010).
247. Tuinhout, H.P., Montree, A.H., Schmitz, J., Stolk, P.A.: Effects of Gate Depletion and Boron Penetration on Matching of Deep Submicron CMOS Transistors. In: *Technical Digest of Papers International Electron Devices Meeting*, pp. 631–634 (1997). DOI 10.1109/IEDM.1997.650463.
248. Ueki, S., Nishimori, Y., Imamoto, H., Kubota, T., Kakushima, K., Ikehara, T., Sugiyama, M., Samukawa, S., Hashiguchi, G.: Modeling of Vibrating-Body Field-Effect Transistors Based on the Electromechanical Interactions Between the Gate and the Channel. *IEEE Transactions on Electron Devices* **59**(8), 2235–2242 (2012). DOI 10.1109/TED.2012.2199758.
249. Ulman, S.: Macromodel for Short Circuit Power Dissipation of Submicron CMOS Inverters and its Application to Design CMOS Buffers. In: *Proceedings of the International Symposium on Circuits and Systems*, vol. 5, pp. V-269–V-272 (2003). DOI 10.1109/ISCAS.2003.1206250.
250. Umadevi, D., Sastry, G.N.: Quantum mechanical study of physisorption of nucleobases on carbon materials: Graphene versus carbon nanotubes. *The Journal of Physical Chemistry Letters* **2**(13), 1572–1576 (2011). DOI 10.1021/jz200705w. URL http://pubs.acs.org/doi/abs/10.1021/jz200705w.
251. Umoh, I.J., Kazmierski, T.J.: VHDL-AMS Model of a Dual Gate Graphene FET. In: *Proceedings of the Forum on Specification and Design Languages (FDL)*, pp. 1–5 (2011).
252. Vadasz, L.L., Grove, A.S., Rowe, T.A., Moore, G.E.: Silicon-gate technology. *IEEE Spectrum* **6**(10), 28–35 (1969). DOI 10.1109/MSPEC.1969.5214116.
253. Veendrick, H.J.M.: Short-Circuit Dissipation of Static CMOS Circuitry and its Impact on the Design of Buffer Circuits. *IEEE Journal of Solid-State Circuits* **19**(4), 468–473 (1984). DOI 10.1109/JSSC.1984.1052168.
254. Venugopalan, S., Paydavosi, N., Lu, D., Lin, C.H., Dunga, M., Yao, S., Morshed, T., Niknejad, A., Hu, C.: *BSIM-CMG 106.1.0 Multi-Gate MOSFET Compact Model*. Department of Electrical Engineering and Computer Sciences, University of California, Berkeley, CA 94720 (2012). URL http://www-device.eecs.berkeley.edu/bsim/Files/BSIMCMG/BSIMCMG106.1.0/BSIMCMG106.1.0_TechnicalManual_20120911.pdf.
255. Verplaetse, P.: Refinements of Rent's Rule Allowing Accurate Interconnect Complexity Modeling. In: *Proceedings of the International Symposium on Quality Electronic Design*, pp. 251–252 (2001). DOI 10.1109/ISQED.2001.915235.
256. Vogel, E.M., Ahmed, K.Z., Hornung, B., McLarty, P.K., Lucovsky, G., Hauser, J.R., Wortman, J.J.: Modeled Tunnel Currents for High Dielectric Constant Dielectrics. *IEEE Transactions on Electron Devices* **45**(6), 1350–1355 (1998).
257. Walus, K., Dysart, T., Jullien, G., Budiman, R.: QCADesigner: A Rapid Design and Simulation Tool for Quantum-Dot Cellular Automata. *IEEE Transactions on Nanotechnology* **3**(1), 26–31 (2004). DOI 10.1109/TNANO.2003.820815.
258. Wang, T., Chiang, L.P., Zous, N.K., Hsu, C.F., Huang, L.Y., Chao, T.S.: A Comprehensive Study of Hot Carrier Stress-Induced Drain Leakage Current Degradation in Thin-Oxide n-MOSFETs. *IEEE Transactions on Electron Devices* **46**(9), 1877–1882 (1999). DOI 10.1109/16.784188.
259. Wang, Z., Zhang, Z., Xu, H., Ding, L., Wang, S., Peng, L.M.: A High-Performance Top-Gate Graphene Field-Effect Transistor Based Frequency Doubler. *Applied Physics Letters* **96**(17) (2010).
260. Wang, Z., Zheng, H., Shi, Q., Chen, J.: Emerging Nanocircuit Paradigm: Graphene-Based Electronics for Nanoscale Computing. In: *Proceedings of the IEEE International Symposium on Nanoscale Architectures*, pp. 93–100 (2007). DOI 10.1109/NANOARCH.2007.4400863.
261. Weste, N.H.E., Harris, D.: *CMOS VLSI Design: A Circuit and Systems Perspective*. Addison Wesley (2005).
262. Williams, R.: How We Found The Missing Memristor. *IEEE Spectrum* **45**(12), 28–35 (2008).

263. Wind, S.J., Appenzeller, J., Martel, R., Derycke, V., Avouris, P.: Vertical Scaling of Carbon Nanotube Field-Effect Transistors Using Top Gate Electrodes. *Applied Physics Letter* **80**(20), 3817–3819 (2002). DOI http://dx.doi.org/10.1063/1.1480877.
264. Wolf, S.: *Silicon Processing for the VLSI Era*. Lattice Press, Sunset Beach, CA (2004).
265. Wong, H.: The Current Conduction Issues in High-*k* Gate Dielectrics. In: *Proceedings of the IEEE Conference on Electron Devices and Solid-State Circuits*, pp. 31–36 (2007). DOI 10.1109/EDSSC.2007.4450055.
266. Wong, H.S.P., Chan, K.K., Taur, Y.: Self-Aligned (Top and Bottom) Double-Gate MOSFET with a 25 nm Thick Silicon Channel. In: *Technical Digest of International Electron Devices Meeting*, pp. 427–430 (1997). DOI 10.1109/IEDM.1997.650416.
267. Wong, H.S.P., Frank, D.J., Solomon, P.M., Wann, C.H.J., Welser, J.J.: Nanoscale CMOS. *Proceedings of the IEEE* **87**(4), 537–570 (1999). DOI 10.1109/5.752515.
268. Wu, W., Duan, X., Yuan, J.S.: A Physical Model of Time-Dependent Dielectric Breakdown in Copper Metallization. In: *Proceedings of the 41st Annual IEEE International Reliability Physics Symposium*, pp. 282–286 (2003). DOI 10.1109/RELPHY.2003.1197758.
269. Guo, X., Ma, T.P.: Tuenneling Leakage Current in Oxynitride: Dependence on Oxygen/Nitrogen Content. *IEEE Electron Device Letters* **19**(6), 207–209 (1998).
270. Xiao, E., Yuan, J.S.: RF Circuit Performance Degradation Due to Hot Carrier Effects and Soft Breakdown. In: *Proceedings of the IEEE 45th Midwest Symposium on Circuits and Systems*, pp. I–17–20 vol. 1 (2002). DOI 10.1109/MWSCAS.2002.1187142.
271. Xiao, E., Yuan, J.S., Yang, H.: Hot-Carrier and Soft-Breakdown Effects on VCO Performance. *IEEE Transactions on Microwave Theory and Techniques* **50**(11), 2453–2458 (2002). DOI 10.1109/TMTT.2002.804632.
272. Xiong, S., Bokor, J., Xiang, Q., Fisher, P., Dudley, I.M., Rao, P.: Gate Line Edge Roughness Effects in 50-nm Bulk MOSFET Devices. In: D.J. Herr (ed.) Society of Photo-Optical Instrumentation Engineers (SPIE) Conference Series, *Society of Photo-Optical Instrumentation Engineers (SPIE) Conference Series*, vol. 4689, pp. 733–741 (2002).
273. Xu, Y., Srivastava, A., Sharma, A.K.: A Model of Multi-Walled Carbon Nanotube Interconnects. In: *Proceedings of the 52nd IEEE International Midwest Symposium on Circuits and Systems*, pp. 987–990 (2009).
274. Yang, F.L., Lee, D.H., Chen, H.Y., Chang, C.Y., Liu, S.D., Huang, C.C., Chung, T.X., Chen, H.W., Huang, C.C., Liu, Y.H., Wu, C.C., Chen, C.C., Chen, S.C., Chen, Y.T., Chen, Y.H., Chen, C.J., Chan, B.W., Hsu, P.F., Shieh, J.H., Tao, H.J., Yeo, Y.C., Li, Y., Lee, J.W., Chen, P., Liang, M.S., Hu, C.: 5 nm-Gate Nanowire FinFET. In: *Digest of Technical Papers Symposium on VLSI Technology*, pp. 196–197 (2004). DOI 10.1109/VLSIT.2004.1345476.
275. Yu, B., Chang, L., Ahmed, S., Wang, H., Bell, S., Yang, C.Y., Tabery, C., Ho, C., Xiang, Q., King, T.J., Bokor, J., Hu, C., Lin, M.R., Kyser, D.: FinFET Scaling to 10 nm Gate Length. In: *Proceedings of the International Electron Devices Meeting*, pp. 251–254 (2002). DOI 10.1109/IEDM.2002.1175825.
276. Zebrev, G.I., Tselykovskiy, A.A., Turin, V.O.: Physics-Based Compact Modeling of Double-Gate Graphene Field-Effect Transistor Operation Including Description of Two Saturation Modes. ArXiv e-prints (2011).
277. Zeghbroeck, B.V.: *Principles of Semiconductor Devices*, 1st edn. Colorado Press (2011).
278. Zhang, Q., Fang, T., Xing, H., Seabaugh, A., Jena, D.: Graphene Nanoribbon Tunnel Transistors. *IEEE Electron Device Letters* **29**(12), 1344–1346 (2008). DOI 10.1109/LED.2008.2005650.
279. Zhang, Q., Zhao, W., Seabaugh, A.: Low-Subthreshold-Swing Tunnel Transistors. *IEEE Electron Device Letters* **27**(4), 297–300 (2006). DOI 10.1109/LED.2006.871855.
280. Zheng, G., Mohanty, S.P., Kougianos, E.: Metamodel-Assisted Fast and Accurate Optimization of an OP-AMP for Biomedical Applications. In: *Proceedings of the IEEE Computer Society Annual Symposium on VLSI*, pp. 273–278 (2012).
281. Zheng, G., Mohanty, S.P., Kougianos, E., Garitselov, O.: Verilog-AMS-PAM: Verilog-AMS Integrated with Parasitic-Aware Metamodels for Ultra-Fast and Layout-Accurate Mixed-Signal Design Exploration. In: *Proceedings of the ACM Great Lakes Symposium on VLSI*, pp. 351–356 (2012).
282. Zheng, R.: *Aging Predictive Models and Simulation Methods for Analog and Mixed-Signal Circuits*. Master's thesis, Department of Electrical Engineering, Arizona State University, AZ (2011). http://repository.asu.edu/attachments/56674/content/zheng_asu_0010n_10586.pdf. Accessed on 06th June 2013.
283. Zoolfakar, A., Hashim, H.: Comparison between Experiment and Process Simulation Results for Converting Enhancement to Depletion Mode NMOS Transistor. In: *Proceedings of the Second Asia International Conference on Modeling Simulation*, pp. 1061–1064 (2008). DOI 10.1109/AMS.2008.133.

CHAPTER 4
Phase-Locked Loop Component Circuits

4.1 Introduction

In this chapter, a widely used analog/mixed-signal system, the phase-locked loop (PLL), is presented. Different types of PLLs will be introduced and the various typical components of the PLL system will be discussed in detail. Earliest discussion of the PLL dates back to 1923 [55]. PLLs got major attention in 1970 and eventually integrated circuits for PLLs were realized. The PLLs are omnipresent in the day-to-day circuits and systems. PLLs have diverse application in various digital, analog, and mixed-signal circuits and systems. For example, frequency synthesis, frequency demodulation, synchronization, tracking, and television sweep circuits are a few of the diverse applications [34]. Different applications of PLL are categorized in Fig. 4.1. With the heavy usage of smart devices with high-speed wireless communication facilities, the performance of PLLs is becoming increasingly important for the overall system performance [33].

A simple usage of the PLL in the synchronous circuits and systems is depicted in Fig. 4.2. In these systems, there is a global clock signal that is understood and followed by all portions of the synchronous circuits and systems. Sequential elements such as flip-flops, latches, and registers need the clock signal. PLLs drive clock distribution of synchronous circuits and systems. The clock signals are typically efficiently generated by a PLL for any target frequency.

PLLs are heavily used as clock multipliers in microprocessors that essentially allow internal processor elements to run faster than external connections while at the same time maintaining precise timing relationships. PLLs are used to synthesize new frequencies that are a multiple of a reference frequency in the radio transmitters. PLLs are used to demodulate frequency-modulated signals. A specific application of a PLL in a graphic signal digitizer system, which is used in liquid

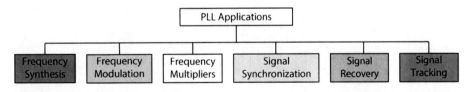

FIGURE 4.1 Diverse applications of PLLs.

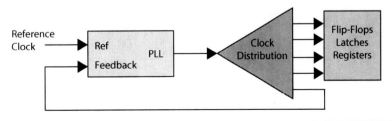

FIGURE 4.2 Application of a PLL in a graphic signal digitizer.

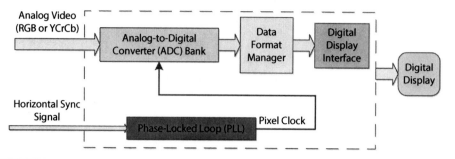

Figure 4.3 Application of a PLL in a graphic signal digitizer [70].

crystal display (LCD) monitors and digital television (DTV), is illustrated in Fig. 4.3 [70]. In this application, the PLL recovers the pixel clock based on input reference horizontal sync signal. PLLs are used to recover small signals in lock-in amplifiers; without PLLs such signals will be lost as noise. PLLs are used for recovery of clock timing information from a data stream such as from a disk drive.

4.2 Phase-Locked Loop System Types

PLL is broadly used for its analog version. However, in practice, there are several variants of PLLs depending on signal types being processed, application, and architecture. An overview of different variants of PLLs is presented in Fig. 4.4 [13, 87]. Besides the above, the oscillators of the PLL can be a simple LC-tank, simple ring oscillator (RO), or current-starved (CS) oscillators, which provide different architectures to the PLL along with the different types of loop filters (LPFs) shown. This section will briefly give an introduction about PLL components; details of selected PLL components will be presented in a later part of this chapter.

Based on the signal types, the PLL variants include the following [13, 32]: (1) analog PLL (APLL) also known as the linear PLL (LPLL), (2) digital PLL (DPLL), (3) all digital PLL (ADPLL), and (4) software PLL (SPLL).

1. *APLL or LPLL*: In this PLL, all the signals and circuits within the PLL are analog in nature. It uses a voltage-controlled oscillator (VCO) for signal generation. The phase detector is an analog multiplier and the LPF may be a passive or active circuit.

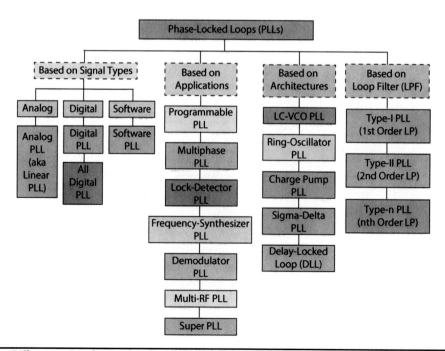

Figure 4.4 Different types of PLLs.

2. *DPLL:* The DPLL is essentially an analog PLL with a digital phase-detector such as XOR, phase-frequency detector (PFD), or edge-trigger JK flip-flop. DPLL may also have a digital divider in the loop.

3. *ADPLL:* In the ADPLL, all signals and circuits are digital in nature. All signals are clocked. The phase detector, filter, and oscillator are digital in ADPLL. ADPLL uses a numerically controlled oscillator (NCO).

4. *SPLL:* In a SPLL, the functional blocks are implemented by software rather than the specialized hardware. The input signals are analog-to-digital converted and the output signals are digital-to-analog converted. A software program performs the PLL treatment in between them.

A large variety of PLL architectures are available that can also be broadly categorized according to the application [87]. The PLLs are categorized according to their application as follows:

1. *Programmable PLL:* The programmable PLL can be programmed for a wide range of signals.
2. *Multiphase PLL:* The multiphase PLL can control many phases of signals. These are widely used in digital clock networks.
3. *Lock-Detector PLL:* The lock-detector PLL uses a lock on one of the pins and is used in frequency modulation.
4. *Frequency-Synthesizer PLL:* The frequency-synthesizer PLLs are used to synthesize the frequencies of various different ranges and bands.
5. *Demodulator PLL:* The demodulator PLL is used to modulate and demodulate FM/AM radio frequencies.
6. *Multi-RF PLL:* The multi-RF PLLs are used to control single or multiple radiofrequencies.
7. *Super PLL:* The super PLL is used for frequency synthesizing of radios, Global System for Mobile Applications (GSM) networks, and cordless phones.

The above PLLs that are classified as per applications have widely diverse architectures. The different architectures of PLLs include:

1. *LC-VCO PLL:* The LC-VCO PLL uses an LC-tank as its oscillator. LC-VCO PLLs typically have a limited frequency tuning range. However, techniques such as switched tuning have enhanced the range [79]. LC-VCO PLLs are used in wireless communications and high-speed serial data communications.
2. *RO PLL:* The RO PLL uses ROs for signal generation. The LC-VCO PLL has very good phase-noise performance, but it is bulky due to the large passive resonators needed to improve the tuning range. RO PLLs have small higher tuning ranges, small die areas, and easy integration with digital system, but they have low phase noise [75]. RO PLLs are used in wireless or optical transceivers.
3. *Charge-Pump PLL:* The Charge-pump (CP) PLL is a specific PLL in which the PFD controls a pulse current generator (i.e., charge pump) filtered by an RC filter to eventually control the VCO [32, 71]. The CP is essentially a current mirror that keeps the output current constant regardless of its load.
4. *Sigma-Delta PLL:* The Sigma-Delta PLL uses the principles of sigma-delta modulation that converts the analog voltage into a pulse frequency by pulse density modulation or pulse frequency modulation [58]. Sigma-Delta PLLs can provide high-frequency resolution and are useful for high data rate communications.
5. *Delay-Locked Loop (DLL):* The DLL is a type of PLL in which the VCO is replaced by a delay chain [32, 95]. PLL uses a variable frequency block; the DLL uses a variable phase or delay block. A DLL can be used to change the phase of a clock signal as well as for clock recovery.

PLLs are also classified according to the type of LPF used in their architectures [87]. The order of LPF dictates the type of the PLLs. For example, if a first order LPF is used in a PLL, then it is called a type-I PLL. If a second-order filter is used in a PLL, then it is called a type-II PLL.

4.3 Phase-Locked Loop System: A Broad Overview

4.3.1 Definition

Simplistically speaking, the PLL is a system that causes one signal to track another signal. In the process, the PLL keeps the output signal synchronized with a reference input signal in terms of frequency and phase. The definition representation is shown in Fig. 4.5 [30, 55]. PLL is a negative feedback servo system that controls the phase of its output signal in such a way that the phase error between output phase and reference phase reduces to a minimum [30, 55]. For example, let us assume the following input signal ($v_{in}(t)$) and output signal ($v_{out}(t)$) for the PLL:

$$v_{in}(t) = A_{in} \sin(2\pi f_{in} t + \theta_{in}) \tag{4.1}$$

$$v_{out}(t) = A_{out} \sin(2\pi f_{out} t + \theta_{out}) \tag{4.2}$$

where A_{in} and A_{out} are the amplitudes of the input and output signals, respectively. f_{in} and f_{out} are the frequencies of the input and output signals, respectively. θ_{in} and θ_{out} are the phases of the input and output signals, respectively. The PLL is locked when the following condition is satisfied:

$$f_{out} = N_{FB} f_{in} \tag{4.3}$$

where N is the dividing factor of the feedback divider (FBDIV) of the PLL.

4.3.2 Block-Level Representation

In the previous section, several types of PLLs with different architecture and signal handling capabilities were briefly discussed. A typical PLL system is presented in Fig. 4.6 [30, 32, 48, 56, 57, 93]. It provides a good example of a mixed-signal system. It is a closed-loop feedback control system that consists of the following components: (1) PFD, (2) CP, (3) LPF, (4) VCO, (5) level shifter (LS), and (6) FBDIV. The components in the PLL may vary depending on the PLL architecture and signal types. However, the above are generic components. Their functionalities are briefly discussed in this section. However, they will be discussed in more detail in different sections of this chapter.

1. *PFD:* The PFD detects the difference in phase and/or frequency between the reference signal and the feedback signal. PFD produces an output signal proportional to the difference in phase between the two signals. In other words, the PFD's output's signal width is proportional to the sampled phase error of reference and feedback signals.

Figure 4.5 Definition of the PLL.

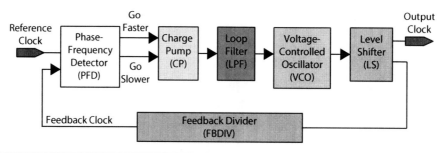

Figure 4.6 Block diagram of a typical PLL [30, 32, 56].

2. *CP*: The output signal of the PFD goes through the CP that supplies the current to the LPF. The CP converts the digital error pulse from PFD to analog error current. Simplistically, the CP is V to I converter. The CP works closely with the PFD and both circuits can be considered as one entity and can be simulated together during the design and simulation of the PLL.

3. *LPF*: The LPF is typically a passive continuous time low-pass filter. The LPF filters out the pulses generated by the CP to limit frequency jumps. In the process, the LPF integrates the error current from the CP to generate the control voltage for the VCO. LPF works with current CP for the I to V conversion. The LPF is used in PLL stability as it introduced poles and zeros in PLL transfer function. However, the LPF is the most silicon area-consuming component of the PLL.

4. *VCO*: The VCO is the core element of the PLL. It generates different output frequencies by changing its input frequency. It is essentially a low-swing oscillator with frequency proportional to control voltage. The three main types of VCO typically used are LC-tank oscillators, ring oscillators (ROs), and multi-vibrators.

5. *LS*: The LS amplifies VCO levels to full swing. The LS works closely with the VCO and can be seen as a unified entity along with it.

6. *FBDIV*: The FBDIV divides VCO clock to generate the feedback clock for phase comparison with the reference clock.

The reduction of phase difference between a locally generated signal and a reference signal is the basic working principle of a PLL [34, 48, 93]. The VCO oscillates according to its input control voltage at its input that is adjusted. Depending upon the frequency of the reference signal, the frequency of the output signal from the VCO is divided using a frequency divider. The output of the frequency divider is supplied to the PFD to get back to the reference frequency range. In essence, the PLL behaves like a frequency multiplier that stays in phase with the reference signal. The typical practice is to use the divider in powers of two for easy implementation. Once the PLL is locked, the frequency of the VCO is equal to the average frequency of the input signal. Therefore, the PLL responds both to frequency and phase of the input signal by automatically raising or lowering the frequency of the VCO until it is matched to the reference in both frequency and phase. The locking behavior of a PLL is shown in Fig. 4.7 [34]. The response of the PLL for an ideal reference VCO's input voltage is shown in Fig. 4.7(a). The response of the PLL for a real scenario is presented in Fig. 4.7(b).

4.3.3 Characteristics, or Performance Metrics

Typically, the target application domain with specifications drives the design metrics that are needed to begin the design for the components of a PLL. This is essential to ensure that the PLL design meets the target specification. In the rest of the subsection, selected performance metrics that are typically used to characterize a PLL will be discussed [30, 32, 34, 94].

1. *Tuning Range*: The tuning range of the PLL is essentially the tuning range of its VCO. It is the range of frequency that can be obtained from the VCO by varying its input voltage.

2. *Lock Range*: The lock range of the PLL is the frequency range for which the PLL is able to stay locked. The lock range is primarily defined by the tuning range of its VCO.

3. *Capture Range*: The capture range is the frequency range for which the PLL is able to lock-in when starting from its unlocked condition. This range is usually smaller than the lock range and will depend on the PFD.

4. *Loop Bandwidth*: Loop bandwidth is the modulation frequency at which the PLL begins to lose lock with the changing reference signal. It is the frequency where the closed loop gain decreases by 3dB and is denoted as ω_{-3dB}. The loop bandwidth defines the speed of the closed control loop.

5. *Transfer Function*: Transfer function of the PLL is essentially a relationship between output and input as any typical linear time invariant system. The transfer function of the PLL describes how PLL responds to excess reference phase. The poles and zeros of the transfer functions dictate the stability of the PLL.

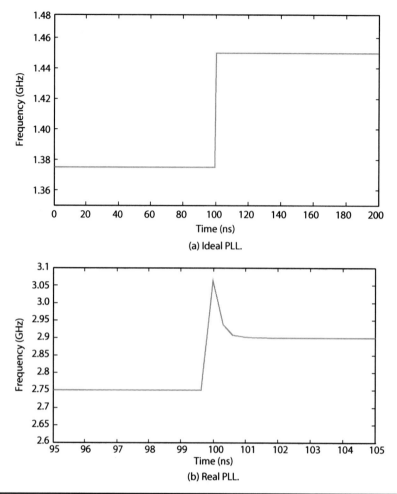

Figure 4.7 Response of the PLL to change in frequency [34].

6. *Loop Gain:* Gain in PLL has various forms depending on how it is measured, e.g., forward loop gain, reverse loop gain, open loop gain, and closed loop gain. The loop gain of the PLL is the product of the individual gains of the individual blocks.

7. *Operating Frequency or Center Frequency:* Center frequency is the frequency of operation when the input voltage to VCO is zero. This is also known as the operating frequency or fundamental frequency.

8. *Lock Time or Acquisition Time:* Lock time or acquisition time is the time needed by the PLL to generate the desired frequency within the required precision. The lock time is limited by the loop bandwidth of the PLL.

9. *Jitter:* Jitter of the PLL is the timing variation of the output clock edge versus an ideal clock edge. Jitter provides information about the frequency stability of the PLL.

10. *Phase Noise:* The phase noise of the PLL is defined by noise energy in a certain frequency band. The phase noise is highly dependent on bandwidth of the PLL and the phase-noise of its VCO.

11. *Input Frequency Range:* The input frequency range of the PLL is the input signal frequencies for which the PLL works correctly.

12. *Output Frequency Range:* The output frequency range of the PLL is the output signal frequencies that can be provided by the PLL.

13. *Multiplication Ratio:* The multiplication ratio of the PLL is the ratio between input and output frequencies of the PLL.

14. *Steady-State Errors:* The steady-state errors of the PLL are like typical phase or timing errors in a closed control system.
15. *Output Amplitude:* The output amplitude of the PLL is the change in the output voltage values during its operation.
16. *Supply Voltage Range:* The supply voltage range of the PLL is the range of supply voltage values within which the PLL can stay locked and operate.
17. *PLL Transient Response:* Transient response of the PLL is the overshoot and settling time to certain accuracy.
18. *Output spectrum purity:* The output spectrum purity of the PLL is essentially the strength of the main frequency of the PLL versus the sidebands generated from a certain VCO tuning voltage ripple.
19. *Power Consumption:* The power consumption of the PLL is the total power consumption of the PLL including leakage power over a period of time. This metric is important for low-power design of the PLL.
20. *Silicon Area:* The silicon area of the PLL is the silicon area consumed by the PLL when realized as physical design. This metric is important for PLL designs that are targeted for portable electronics.

4.3.4 Theory in Brief

The PLLs can be described by using either linear or nonlinear theory [70]. However, the nonlinear theory is quite complicated for the real-world designs of the PLL. So, the APLLs are modeled by the linear control theory. The linear control theory model is adequately accurate for most electronic applications. In order to provide mathematical models of the PLL, some important parameters are presented in Fig. 4.8 [30, 32, 34, 94].

The open loop gain of the above example system is calculated as follows [30, 94]:

$$\mathrm{LG}(s) = \left(\frac{I_{\mathrm{pump}}(s)}{2\pi}\right)\mathrm{LPF}(s)\left(\frac{K_{\mathrm{VCO}}}{s}\right)\left(\frac{1}{N_{\mathrm{FB}}}\right) \quad (4.4)$$

where $I_{\mathrm{pump}}(s)$ is the CP current, $\mathrm{LPF}(s)$ is the loop filter impedance, K_{VCO} is the gain of the VCO, and N_{FB} is the dividing factor of the FBDIV. The closed loop transfer function of the same PLL is calculated using the following [30]:

$$H(s) = \left(\frac{\phi_{\mathrm{FB}}}{\phi_{\mathrm{ref}}}\right) = \left(\frac{\mathrm{LG}(s)}{1+\mathrm{LG}(s)}\right) \quad (4.5)$$

For a second-order PLL with LPF resistance R and capacitance C_1, the closed loop transfer function is calculated as follows [34]:

$$H(s) = \left(\frac{K_{\mathrm{PFD}}K_{\mathrm{VCO}}(1+sRC_1)}{s^2 + \dfrac{K_{\mathrm{PFD}}K_{\mathrm{VCO}}R}{N_{\mathrm{FB}}} + \dfrac{K_{\mathrm{PFD}}K_{\mathrm{VCO}}}{N_{\mathrm{FB}}C_1}}\right) \quad (4.6)$$

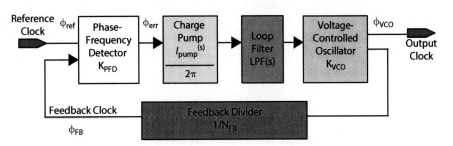

Figure 4.8 Linear model of the PLL [30, 32, 56, 94].

where K_{PFD} is the gain of the PFD. The natural frequency is derived from the transfer function presented as follows:

$$\omega_N = \left(\sqrt{\frac{K_{PFD} K_{VCO}}{N_{FB} C_1}} \right) \quad (4.7)$$

The damping factor of the PLL system is calculated using the following expression [27]:

$$\xi = \left(\frac{\omega_N}{2} RC_1 \right) \quad (4.8)$$

When the damping factor is set to $\xi = 1$, the pull in time of the PLL can be calculated as follows [34]:

$$T_{pull} = 2RC_1 \ln \left(\frac{\left(\frac{K_{VCO}}{N} \right) I_{pump}}{s^2 + \frac{K_{PFD} K_{VCO} R}{N} + \frac{K_{PFD} K_{VCO}}{N_{FB} C_1}} \right) \quad (4.9)$$

The above is based on the tuning range of the VCO in the PLL. By manipulating the expressions, the natural frequency of the PLL can be rewritten as follows [27]:

$$\omega_N = \left(\sqrt{\frac{\left(\frac{I_{pump}}{2\pi} \right) K_{VCO}}{2C_1}} \right) \quad (4.10)$$

The locking range of the PLL is calculated using the following expression:

$$\omega_L = 4\pi \xi \omega_N \quad (4.11)$$

For a third-order LC-VCO CP-based PLL, the closed-loop bandwidth is approximated by the following expression [94]:

$$\omega_{-3dB} \cong \left(\frac{I_{pump}}{2\pi} \right) R_2 \left(\frac{K_{VCO}}{N_{FB}} \right) \quad (4.12)$$

where R_2 is the LPF resistor.

4.4 Oscillator Circuits

The VCO is the most important block of the PLL system [10]. Its importance can also be evident from the fact that a simple PLL can be made from a VCO and a PFD put in a closed loop. The VCO generates the necessary frequency in the PLL used to synchronize with the reference signal and change its frequency of operation, depending on the control voltage supplied to it [11, 34]. In this section, different types of VCOs will be introduced along with their performance metrics and comparative perspectives. Detailed discussions of selected VCOs will be presented in some later sections of this chapter.

4.4.1 Oscillator Types

Oscillators and amplifiers are the two earliest circuits that have been explored since the invention of electronic devices. Around 1912, the first electronic oscillator was designed using a valve [74]. Since then, many varieties of oscillators have been designed, as they are either a main component or

Phase-Locked Loop Component Circuits

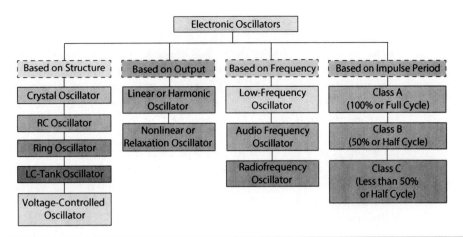

FIGURE 4.9 Different types of oscillators.

subcomponent of every modern electronic system. In general, the oscillators are circuits or devices that can create alternating current (AC) signals [27]. Specifically, an electronic oscillator is an electronic circuit that produces a repetitive electronic signal, such as a sine wave or a square wave. A broad view of the different types of the oscillators is shown in Fig. 4.9. They are classified in the following four ways: (1) based on the structures of the VCOs, (2) based on output of the VCOs, (3) based on the types of frequency generated by the VCOs, and (4) based on the impulse period of the VCOs.

Based on the structure, the oscillators can be as simple as a crystal oscillator to a complex VCO. Quartz is commonly used in oscillator crystals. The piezoelectric quartz crystal uses mechanical resonance to create an electrical signal [54]. The crystal oscillator can generate a frequency of kHz to MHz range and is used in clocks, wrist watches, radios, mobile phones, and computers. A crystal oscillator is also used as a reference clock or base clock in the PLL systems. The RC oscillator (RCO) is a network of capacitors and resistors that are used to generate low-frequency signals. However, the RCOs can be a cost-effective solution for on-chip oscillators with complete digital processes [47, 104]. The basic RO is a feedback connection of odd numbers of inverter stages. The RO can be either single-ended or differential in its architecture and is known as a single-ended RO or differential RO, respectively [27]. A basic LC-tank oscillator is a tank consisting of an inductor (L) and capacitor (C) connected together. The frequency of this LC-tank oscillator is adjusted by variable capacitor (varactor). LC oscillators are used when a tunable frequency source is necessary, such as in signal generators and tunable radio transmitters [51, 72]. The VCO is an electronic oscillator in which the oscillation frequency is adjusted by a voltage input. The VCO can generate a very wide range of frequencies and is used in most modern electronic systems. RO and LC-tank oscillators can be made voltage controlled with additional circuitry. The CS oscillator is an example of a VCO [68].

Based on the output signal, the oscillators can be either linear (harmonic) or nonlinear (relaxation) oscillators. The linear oscillators produce sinusoidal signals. The nonlinear oscillators produce nonsinusoidal signals, such as a square and sawtooth waves [16]. Depending on the frequency, the output signal of the oscillators can be low-frequency oscillators, audio-frequency oscillators, or radiofrequency oscillators. Depending on the impulse period that is applied to the oscillator to maintain the sustained oscillations, they are classified as class-A, B, or C oscillators [28].

4.4.2 Oscillator Characteristics, or Performance Metrics

The PLL systems mainly differ based on their oscillator blocks. The oscillator block of the PLL determines the frequency capture range. Thus, the oscillator is the main block that is responsible for the output signal of the PLL. The oscillators of the PLL need to be selected carefully for each design for its characteristics, target applications, and area overhead. In this subsection, several figures of merit (FoM) are discussed that can be used to characterize VCOs [7, 12, 27].

1. *Oscillation Frequency:* Various terms such as natural frequency, resonant frequency, center frequency, oscillation frequency, and operating frequency are used interchangeably in the literature. However, they have similar or different meaning depending on the operating

conditions and structure of the oscillators. While frequency is the number of occurrences per unit time, the natural, resonant, or resonance frequency is the frequency at which the response amplitude is maximum. For example, for a loss-less LC-tank, when the magnetic energy in L and electric energy in C feed each other, resonance happens when the reactants are equal in magnitude [12]. This frequency is the natural, resonant, or resonance frequency of the LC tank. In this specific case, this is the center frequency as well. In the generic scenario, voltage is applied to change the C of the LC-tank varactor and other parameters for ROs, and then center frequency is approximately the middle of the band covered by the oscillator. It may not be exactly middle frequency as system response may not always be linear, but it can be the frequency at mid-supply voltage. The oscillating or operating frequency of the oscillator is the frequency for its specific operating conditions. The oscillating frequency of a differential RO can be approximately calculated using the following expression [50, 56]:

$$f_{osc,ring} = \frac{1}{2N_{stage}T_{D,stage}}$$
$$= \frac{1}{\eta N_{stage}(T_{R,stage} + T_{F,stage})}$$
$$= \frac{1}{2\eta N_{stage}T_{R,stage}}$$
$$= \frac{I_{bias}}{2\eta N_{stage}q_{max}} \qquad (4.13)$$

where N_{stage} is number of stages of the oscillator, $T_{D,stage}$ delay in each stage, η proportionality constant, $T_{R,stage}$ rise time in each stage, $T_{F,stage}$ fall time in each stage, I_{bias} bias current flow through each stage, and q_{max} maximum charge at each stage. The bias current is also referred to as tail current I_{tail}. The oscillating frequency of a differential LC oscillator can be approximately calculated using the following expression [56, 92]:

$$f_{osc,LC} = \frac{1}{2\pi\sqrt{L_{tank}C_{total}}} \qquad (4.14)$$

where L_{tank} is the inductance of a spiral inductor in the LC tank, C_{total} is the capacitance of the varactors including all parasitic capacitance from transistors and the spiral inductor.

2. *Tuning Range:* The tuning range of a VCO is the range of frequencies within which the frequency of the VCO can be adjusted by applying a voltage. For an ideal VCO, the operating frequency, natural frequency, and applied voltage have the following linear relationship [7, 12]:

$$f_{VCO} = f_N + K_{VCO}V_{tune} \qquad (4.15)$$

where f_{VCO} is the oscillating frequency, f_N natural or free-running frequency, K_{VCO} gain or sensitivity of the VCO in Hz/V, and V_{tune} tuning voltage. The concept is presented in Fig. 4.10 [7, 12]. The tuning range is the possible output frequency ($f_{VCO,2} - f_{VCO,1}$). The nanoelectronic-based VCOs are designed with much larger tuning range as compared to the target application specification to compensate for PVT variations.

3. *Tuning Sensitivity:* The tuning sensitivity or gain of the VCO is the change in frequency per unit volt of tuning voltage. It is essentially the slope of the tuning curve. This is not constant along the entire tuning range as tuning range is not linear. It is denoted by K_{VCO} and measured as $\left(\frac{Hz}{Volt}\right)$, $\left(\frac{MHz}{Volt}\right)$, or $\left(\frac{GHz}{Volt}\right)$.

4. *Tuning Linearity:* In an ideal situation, the output frequency of the VCO has a linear relationship with the tuning voltage [7, 12]. In other words, the gain K_{VCO} that is the slope of the tuning curve is constant for an ideal VCO. Linear tuning is very much desired so that K_{VCO} is constant

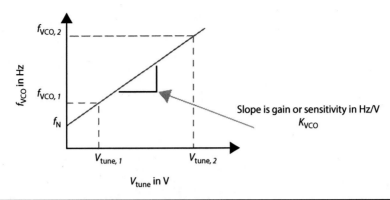

FIGURE 4.10 Illustration of tuning range of the oscillators [7, 12].

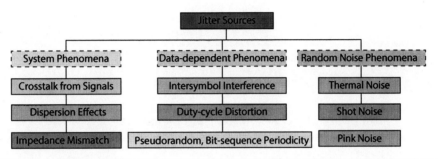

FIGURE 4.11 Various sources of Jitter [52].

over the tuning range and the PLL performs well throughout the tuning range. However, real VCO designs, the K_{VCO} is not constant and the tuning curve is nonlinear in nature.

5. *Frequency Pushing*: The frequency pushing of the VCO is the change of the output frequency of the VCO due to supply voltage changes [12]. It is thus the oscillating frequency sensitivity of the VCO with supply voltage variations, i.e., $\left(\frac{\delta f_{osc}}{\delta v_{dd}}\right)$, and is expressed as $\left(\frac{Hz}{Volt}\right)$.

6. *Frequency Pulling*: The frequency pulling of the VCO is the change of the output frequency of the VCO due to nonideal load changes [12]. It is thus the oscillating frequency sensitivity of the VCO with load impedance variations and is expressed as Hz peak to peak.

7. *Spectral Purity*: In a real-world VCO, sidebands (or harmonics) appear around the oscillation frequency of the VCO. These sidebands are measured as jitter in time domain and phase noise in frequency domain [7, 27]. The jitter is expressed in second and phase noise in $\left(\frac{dBc}{Hz}\right)$. The phase noise of a VCO represents its short-term stability. The estimation of phase noise of the VCO is a difficult task because of the complex mechanisms of the origin of phase noise. Various sources of jitter are presented in Fig. 4.11. However, several models and experimental approaches are available for the estimation of phase noise and jitter [24, 27, 50, 53]. For a simplistic concept of jitter see Fig. 4.12 [53]. If any nth period of VCO clock is T_n and the mean time period is T, then the difference $\Delta T_n = T_n - T$ is an indication of jitter in the signal.

One type of jitter, called *absolute jitter* or *long-term jitter*, is demonstrated in Fig. 4.12(a) [53]. The absolute jitter is calculated using the following expression [53]:

$$\Delta T_{abs}(N) = \sum_{n=1}^{N} \Delta T_n \quad (4.16)$$

where N is the number of clock cycles. The absolute jitter is used to quantify the jitter of PLLs. However, absolute jitter is not suitable for describing the performance of VCOs as its variance

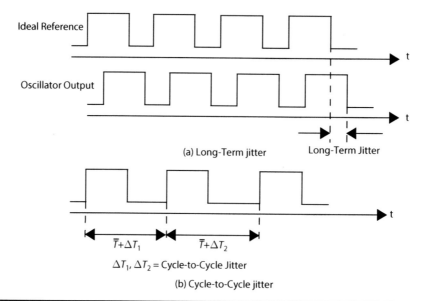

Figure 4.12 Illustration of jitter of the oscillators [53].

diverges with time. The second type of jitter is called *cycle jitter*. The cycle jitter is calculated as the root-mean-square (RMS) value of the errors in every cycle as follows [53]:

$$\Delta T_{cycle}(N) = \lim_{n \to \infty} \sqrt{\frac{1}{N}\sum_{n=1}^{N}\Delta T_n^2} \tag{4.17}$$

The cycle jitter describes the magnitude of the period errors; however, it provides no information about its dynamics. The third type of jitter is called *cycle-to-cycle jitter*. The cycle-to-cycle jitter is calculated as the RMS difference between two consecutive periods as follows [53]:

$$\Delta T_{cc}(N) = \lim_{n \to \infty} \sqrt{\frac{1}{N}\sum_{n=1}^{N}(T_{n+1} - T_n)^2} \tag{4.18}$$

An important time domain visual tool for jitter estimation is the eye diagram, which can be used to diagnose signal quality issues such as attenuation, noise, jitter, and dispersion [3, 52]. An illustrative view of the eye diagram is presented in Fig. 4.13 [3, 52]. The eye diagram is essentially a composite view of all the bit periods of a captured waveform superimposed upon each other. For example, the waveform trajectory from the start of period 2 to the start of period 3 is overlaid on the waveform trajectory from the start of period 1 to the start of period 2 [52]. This is continued for all the waveforms for all bit periods. The large wide-open eye at the center is the ideal location for sampling each bit as the bit error is least there because each value is either low or high at this point.

For the white noise present in an RO, the single-sideband phase noise with respect to the carrier is expressed as follows [53]:

$$\phi_{SSB,ring} = \frac{\left(\dfrac{\omega_{osc}^3}{4\pi}\right)\Delta T_{cc}^2}{(\omega-\omega_{osc})^2 + \left(\dfrac{\omega_{osc}^3}{8\pi}\right)^2 \Delta T_{cc}^4} \approx \frac{\left(\dfrac{\omega_{osc}^3}{4\pi}\right)\Delta T_{cc}^2}{(\omega-\omega_{osc})^2} \tag{4.19}$$

It may be noted that ω is angular frequency. The above expression can calculate phase noise from given cycle-to-cycle jitter. The reverse calculation can also be done, in other

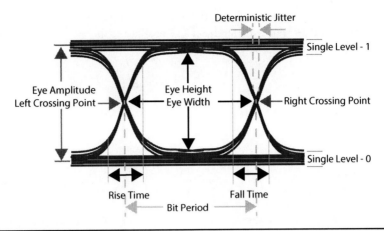

Figure 4.13 Illustration of Jitter of the oscillators [3, 52].

words, given the phase noise calculate the cycle-to-cycle jitter using the following expression [53]:

$$\Delta T_{cc} = \sqrt{\left(\frac{4\pi}{\omega_{osc}^3}\right)(\omega - \omega_{osc})^2 \phi_{SSB,ring}} \quad (4.20)$$

Similarly, different models can be used to calculate jitters due to other sources of noise such as the supply noise, the substrate noise, and the thermal noise [53].

For the LC-tank oscillator, the phase-noise is calculated based on the well-known Leeson's model. This is a heuristic model. The simplified single-band phase noise model is expressed as follows [27, 69, 92]:

$$\phi_{SSB,LC} = F\left(\frac{k_B T}{2 P_s}\right)\left(\frac{\omega_c^2}{Q^2 \Delta\omega^2}\right) \quad (4.21)$$

where F is an empirical device noise excess factor or, simply, noise factor, T is the temperature, P_s is signal level at the oscillator's active element input, ω_c is the center frequency of the LC-VCO, Q is the loaded quality factor of the LC-tank, and $\Delta\omega$ is the angular frequency offset. If V_{peak} is the peak amplitude voltage across the capacitor, then the above expression can be rewritten to the following [92]:

$$\phi_{SSB,LC} = F\left(\frac{k_B T}{V_{peak}^2}\right)\left(\frac{R_S^3}{L^2 \Delta\omega^2}\right) \quad (4.22)$$

where R_s is the serial AC resistance of the inductor in the LC-tank.

8. *Output Level:* The output level of an oscillator is typically specified as an RMS voltage that can be delivered into a particular load. The oscillator output power is specified as a power level delivered by the oscillator into a particular load that is typically measured in power ratio in decibels (dBm) (i.e., dB relative to 1 mW).

9. *Power Consumption:* The power consumed by the oscillators during their oscillations. The power dissipation for a differential RO is expressed using the following [50, 56]:

$$\begin{aligned} P_{ring} &= I_{tail} V_{DD} N_{stage} \\ &= \left(\frac{\eta N_{stage} q_{max}}{T_{D,stage}}\right) V_{DD} \\ &= 2\eta N_{stage} q_{max} V_{DD} f_{osc,ring} \end{aligned} \quad (4.23)$$

where the notations have the same meaning as presented for the oscillating frequency calculations. For a differential LC oscillator, the power dissipation is calculated using the following expression [51, 56, 92]:

$$\begin{aligned} P_{LC} &= 2R_S I_{peak}^2 \\ &= 2\left(\frac{CR_S}{L}\right) V_{peak}^2 \\ &= 2R_S C^2 \omega_{osc}^2 V_{peak}^2 \\ &= \left(\frac{2R_S}{L^2 \omega_{osc}^2}\right) V_{peak}^2 \end{aligned} \tag{4.24}$$

where I_{peak} is the peak current through the inductor. L is the inductance of a spiral inductor in LC-tank. C is the capacitance of the varactor and all parasitic capacitances from MOS devices as well as the spiral inductor. R_s is the serial AC resistance of the inductor in the LC-tank. For a load quality factor Q, this resistance can be calculated as: $\left(\frac{\omega_{osc} L}{Q}\right)$.

10. *Silicon Size:* Silicon area or chip layout area is an important design metric for the VCO, in particular for mobile applications. In a typical scenario, the LC-tank oscillators occupy more silicon area than the ring oscillators. Eventually, the oscillator area then dictates the silicon area of the PLL.

4.4.3 Comparison of Oscillators

The basic crystal oscillators made up of quartz are used for low-frequency applications. These are present in most of the real-life applications, such as PC, phones, and radios, as native signal generators or as reference/base clock for high-frequency clock multiplier electronic oscillators. For achieving higher stability, the quartz crystals are used to produce a general class of oscillator circuit known as the quartz crystal oscillator (XO). A comparative perspective of two integrated circuit oscillators, the RO and LC-tank oscillator, is presented in Table 4.1 [6, 27, 30, 56, 59, 76]. Between the RO and LC-tank oscillator, a broad tradeoff point is that, for a given power budget, ROs are compact but noisy; on the other hand, the LC oscillators are bulky but less noisy.

The ROs made up of MOSFETs are easier to design as integrated circuits. ROs monolithically integrate with larger circuits and occupy minimal silicon area. The ROs that are designed using MOSFETs are more suitable for nanoelectronics as they have low on-chip area. The ROs, while compact and easy to design, are noisy and unstable for very high-frequency oscillations. The RO's high-frequency oscillations with low-noise level are made possible by deploying different architectures. The LC-tank oscillators have better frequency performance and phase noise as these use passive resonating elements (L and C). The area of LC-tank is quite large and can have other complications such as eddy currents and magnetic coupling in substrates containing the LC oscillators. The inductance value

Performance Metrics	Ring Oscillator	LC-Tank Oscillator
Tuning Range	Wide	Narrow
VCO Gain	High	Low
Phase Noise	Good	Very Good
Silicon Area	Low	High
Power Dissipation	Low	High
System Integration	Easy Monolithic	Difficult Monolithic
Stability	Poor at High Frequency	Good Stability
Design Effort	Low	High

TABLE 4.1 Broad Comparison of Ring Oscillator and LC-tank Oscillator

cannot be easily controlled, making the LC-tank difficult to design. All these make it difficult for LC oscillators to be integrated in a circuit monolithically.

4.5 Ring Oscillators

The RO is a simple architecture made of MOS devices. Ring oscillators are useful in die and new technology testing. These are used to find the delay times of logic gates. These are the main element in on-chip temperature sensors.

4.5.1 Basics

The basic RO consists of an odd number of single-ended inverters with a feedback loop as shown in Fig. 4.14. The feedback loop creates oscillations that are derived from the propagation delay of each inverter. An odd number of stages (N) are needed to create sustained oscillations. The odd number of inverter stages oscillate as if input node is excited and then the signal propagates to reverse its polarity. For an even number of stages, the signal will propagate to the initial stage with the same polarity, and hence result in system stability and produce no oscillations. The idea of oscillation is depicted in Fig. 4.15 for three stages. An input of "1" produces an output of "0" that becomes input of reverse polarity due to feedback as shown Fig. 4.15(a). An input of "0" produces an output of "1" that becomes input of reverse polarity due to feedback as shown Fig. 4.15(b).

4.5.2 45-nm CMOS

The schematic diagram of a three-stage RO is shown in Fig. 4.16. As a specific case study, 45-nm RO design and characterization is now discussed in this section [34, 35]. In general, the number of stages can also be of different odd numbers to obtain different specifications including frequency. For a given technology node, the designers can adjust the widths of the NMOS and PMOS to obtain the desired specifications. For example, for 45-nm CMOS technology node, the baseline sizes are as follows [34, 35]: $L_n = 45$ nm, $L_p = 45$ nm, $W_n = 120$ nm, and $W_p = 240$ nm.

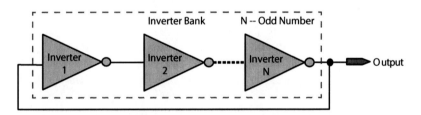

Figure 4.14 High-level schematic of the RO.

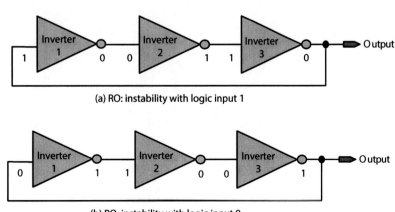

Figure 4.15 Oscillation principle in a RO.

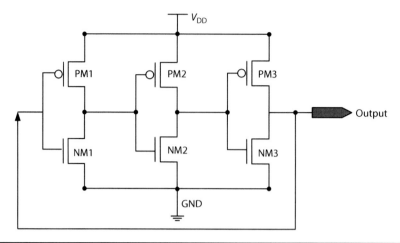

Figure 4.16 Transistor-level schematic of the RO.

If all the inverters have the same size transistors, then their fall and rise times may be the same. Under equal rise and fall time assumption, the frequency of oscillation of the RO is calculated using the following expression [34, 63]:

$$f = \left(\frac{1}{2N_{\text{stage}} T_{\text{D,stage}}}\right) \quad (4.25)$$

where N_{stage} is the number of inverters, which is odd, and $T_{\text{D,stage}}$ is the propagation delay of each inverter. As a case study, the design variables chosen for this RO circuit are the width of NMOS $W_n = 4L = 120$ nm and the width of PMOS $W_p = 8L = 240$ nm at a nominal operating voltage $V_{\text{DD}} = 1$V and minimal technology length of $L = 45$ nm. The ambient temperature of 27°C is assumed to be constant during the design and characterization as the temperature difference can affect the output dramatically.

For real-world design, the layout of the RO circuit needs to be created. The parasitics in analog nanoscale CMOS circuits have a very dramatic effect on their performance. The physical design makes it possible to estimate parasitic effects. Various commercial libraries can be used for physical designs that are manufacturable. However, for a rapid case study, a process design kit (PDK) provided by different industrial houses can be used. The 45-nm PDK provided by Cadence is used to produce the physical design presented in Fig. 4.17 [34, 35]. The physical design that is time consuming involves tedious labor-intensive tasks and hence consumes a large portion of the design cycle. For the parasitic-accurate characterization of the RO, the full SPICE netlist is generated from RCLK parasitic extraction of this physical design. A comparison of the number of components between the regular schematic and the parasitic netlist is shown in Table 4.2 [34, 35].

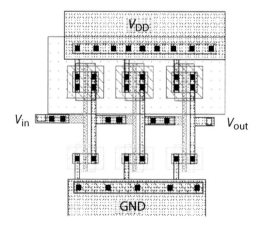

Figure 4.17 Baseline RO layout for a 45-nm CMOS technology [34, 35].

Ring Oscillator Circuit	Transistors	Capacitors	Resistors	Inductors	Total
Without Parasitics	6	0	0	0	6
With Parasitics	6	82	19	0	107

TABLE 4.2 Number of Elements in the 45-nm RO Circuits [34, 35]

Circuit Type	Power Consumption	Oscillation Frequency
Schematic	27.17 μW	16.21 GHz
Parasitic	26.96 μW	9.88 GHz

TABLE 4.3 Characterization of the 45-nm CMOS RO [34, 35]

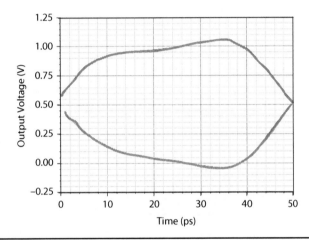

FIGURE 4.18 Eye diagram obtained from the physical design of the 45-nm RO [34, 35].

The characterization results of the RO with and without parasitics are shown in Table 4.3 [34, 35]. Even for this simple circuit, the run time of simulation has increased by a factor of 3 due to the presence of parasitics. Contemporary complex real-world analog circuits with thousands of transistors will definitely have the simulation time in days to weeks, depending on the complexity of the circuit. A dramatic decrease in frequency of the order of 40% is observed in simulations using the parasitics from the physical design versus the regular schematic simulations. The total power consumption that includes all the switching capacitance as well as leakage has not been altered and only changed by merely 1%. No drastic effects are observed in frequency and power from adjusting the widths of the MOSFET sizes to $W_n = 360$ nm and $W_p = 720$ nm in the physical design. From these results, it is obvious that the extraction of parasitics is necessary to calculate the desired characteristics such as the frequency for this circuit. The eye diagram for the jitter effect of physical design is shown in Fig. 4.18 [34, 35]. From the eye diagram, it can be seen that the jitter effect is negligible for a 100-ns period. The effect on jitter is minimal for this specific nanoscale circuit even when full-blown parasitics are taken into account during the characterization.

4.5.3 Multigate FET

Before implementation of double-gate (DG) FinFET-based circuits, a quick idea of its different configuration is essential. The DG FinFET can have three different configurations as shown in Fig. 4.19 [17, 18, 43, 67]. The shorted-gate (SG) DG FinFET is shown in Fig. 4.19(a), the independent-gate (IG) DG FinFET is shown in Fig. 4.19(b), and the low-power (LP) DG FinFET is shown in Fig. 4.19(c). The voltage between back-gate and source of DG FinFET is denoted as V_{gb} and the voltage between front-gate and source of DGFET is denoted as V_{gf}. In the SG DG FinFET the front and back gates are tied together. In the IG DG FinFET, the top gate is etched out making two independent gates [43, 67]. For IG DG FinFET, the V_{gb} is set to ground. The LP DG FinFET is a IG DG FinFET with V_{gb} reverse biased

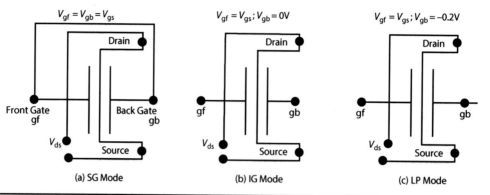

Figure 4.19 Different configurations for an n-type DG FinFET [17, 43, 67].

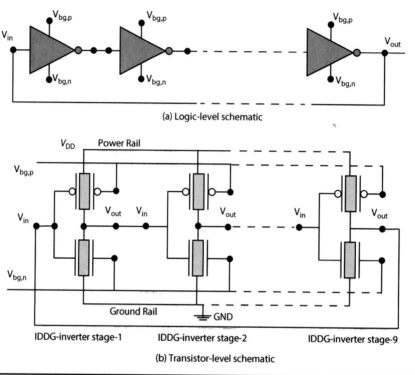

Figure 4.20 The nine-stage RO using IDDG CMOS inverters [64].

to reduce the subthreshold leakage. The SG-DG FinFET has the smallest delay followed by the IG-DG FinFET and LP-DG FinFET. The LP-DG FinFET has the lowest power consumption followed by IG-DG FinFET and SG-DG FinFET.

A nine-stage independently driven DG (IDDG) CMOS inverter-based ROs is shown in Fig. 4.20(a) [64]. The key element of this RO is the IDDG CMOS inverter. Using this IDDG CMOS inverter, the transistor-level schematic design of the RO is presented in Fig. 4.20(b) [64]. As a specific case, the transistor sizes are selected as follows for a target specifications: $L_n = 50$ nm, $L_p = 50$ nm, $W_n = 50$ nm, and $W_p = 100$ nm. The voltage control of this RO is performed in either of the following ways:

1. *Joint Biasing*: In the joint biasing, the back gates of all the MOSFETs are together. For if $V_{gb,n}$ is the back-gate voltage of NMOS DGFET and $V_{gb,p}$ is the back-gate voltage of PMOS DGFET, then $V_{gb,n} = V_{gb,p}$.

2. *Separate Biasing*: In the separate biasing, the NMOS DGFET and PMOS DGFET are in two separate groups, i.e., $V_{gb,n} = -V_{gb,p}$.

The joint biasing is simple and easy to implement. In this case, the transfer curve changes proportional to the applied voltage and as a result both the oscillation frequency and duty cycle of the oscillation change. The separate biasing option groups the back gates of N-type and P-type

	Jointly Biasing			Separately Biasing		
	f_{osc}	ΔT_{abs}	ΔT_{cc}	f_{osc}	ΔT_{abs}	ΔT_{cc}
	4.3 GHz	2.0%	3.1%	14.2 GHz	1.9%	3.3%

TABLE 4.4 Characterization of the Nine-Stage 50-nm DGFET CMOS RO [64]

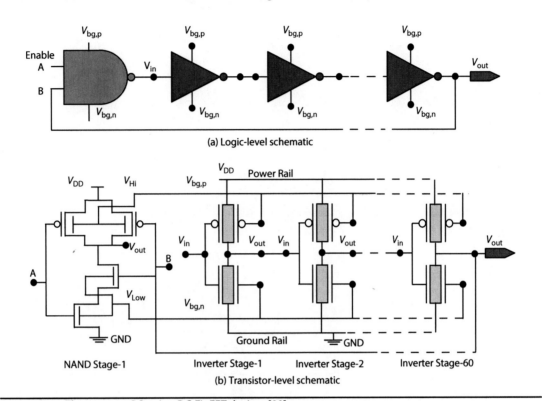

FIGURE 4.21 The 61-stage RO using DG FinFET devices [80].

FinFETs into two groups and is more complex to implement. The devices are then separately controlled, the inverters' gain can be changed to wider a range, and the higher range of frequencies is obtained. The characterization of the nine-stage 50-nm FinFET RO is presented in Table 4.4 for operating frequency, absolute jitter, and cycle-to-cycle jitter [64].

A 61-stage RO consisting of 60 inverters and a NAND stage was fabricated for 180-nm foundry [80]. The logic-level schematic and transistor-level schematic representations are shown in Figs. 4.21(a) and (b), respectively [80]. For a 140-nm gate length, 2.2-nm oxide thickness, and operating voltage of 1.5 V, the RO delay is 19 ps. The frequency versus voltage characteristics of the 61-stage RO using DG FinFET devices are shown in Fig. 4.22 [80].

A 41-stage RO using triple-gate FETs (TGFETs) is designed and fabricated [22]. The overall RO consists of 41 inverter stages, a divider, and an output buffer. In the RO, the gate length for the N-type is 25 nm and for P-type is 35 nm. For the triple-gate devices used in the circuit, the Fin widths are in the 10 to 65 nm range and Fin heights are in the 60 to 80 nm range. The relationships among the device geometry for the TGFET are as follows: $W_n = 60\,(W_{Fin} + 2H_{Fin})$ and $W_p = 92\,(W_{Fin} + 2H_{Fin})$. The TGFET-based 41-stage RO has a stage delay of 60 ps for 1.5 V supply voltage.

4.5.4 Carbon Nanotube

A single-wall carbon nanotube (SWCNT) RO consisting of five CMOS inverter stages has been designed and manufactured [9, 19, 20]. The additional inverter stage on the right is the stage included for measurement purposes only. The proposed SWCNT-based RO on one nanotube has much improved frequency performance as compared to the previous ROs fabricated on multiple nanotubes. The voltage-frequency characteristics of the five-stage RO using single-walled CNTFET devices fabricated on one nanotube is shown in Fig. 4.24 [19, 20]. Specifically, the RO at $V_{DD} = 0.92$ V has $f_{osc} = 52$ MHz with the stage delay of 1.9 ns. Similarly, at $V_{DD} = 1.04$ V, the RO has $f_{osc} = 52$ MHz with the stage delay of 1.4 ns.

176 Chapter Four

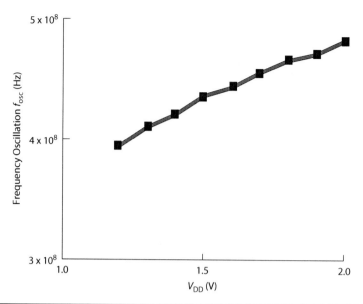

Figure 4.22 Characteristics of the 61-stage RO using DG FinFET devices [80].

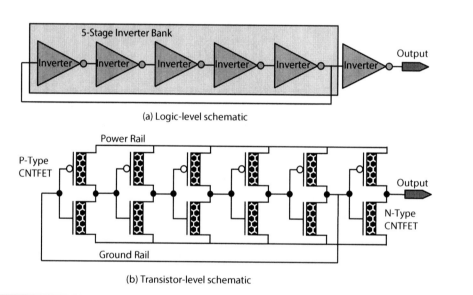

Figure 4.23 The five-stage RO using single-wall CNTFET devices [19].

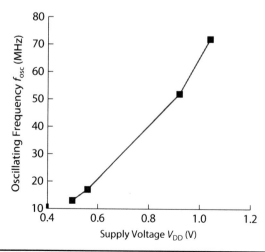

Figure 4.24 The voltage-frequency characteristics of the five-stage RO using single-wall CNTFET devices.

4.6 Current-Starved Voltage Controlled Oscillators

4.6.1 Basics

The basic idea of a CS-RO or CS-VCO is depicted in Fig. 4.25 [100]. The CS-VCO consists of an odd number of inverter stages as typically needed to provide unstable states for sustained oscillations. In CS-VCO, the low-to-high propagation-delay (T_{pLH}) and high-to-low propagation-delay (T_{pHL}) of each of the inverter stages are controlled by current source and current sink [39]. The current source is connected toward the power supply and can provide a current of $I_{CS,source}$. The current sink is connected toward the ground and can sink a current of $I_{CS,sink}$. Typically, the source current and the sink current are matched so that the following relation holds: $I_{CS,source} = I_{CS,sink} = I_{bias}$. The propagation delays can be calculated using the following expression:

$$T_{pLH} = \left(\frac{C_{eff,stage} V_{trip,stage}}{I_{CS,source}} \right) \tag{4.26}$$

$$T_{pHL} = \left(\frac{C_{eff,stage} (V_{DD} - V_{trip,stage})}{I_{CS,sink}} \right) \tag{4.27}$$

where $C_{eff,stage}$ is the effective load capacitance of each inverter stage. $V_{trip,stage}$ is the inverter trip voltage. The matching source and sink current cancels the $V_{trip,stage}$ term in the propogation delay calculation. The propagation delay of the overall CS-VCO is calculated as the summing of stage delays and expressed as follows:

$$T_{osc} = \left(\frac{N_{stage} C_{eff,stage} V_{DD}}{I_{bias}} \right) \tag{4.28}$$

This T_{OSC} then determines the oscillating frequency (f_{OSC}) of the VCO.

4.6.2 Circuit Design

The transistor-level design of CS-VCO is now presented. The transistor-level schematic of the CS-VCO is presented in Fig. 4.26. It consists of three high-level stages: input stage, CS stage, and output buffer stage. The input stage consists of two high-impedance transistors. The CS contains an odd-numbered chain of inverters. Each inverter has one current source transistor and one current sink transistor. The CS transistors (i.e., source or sink transistors) limit the current flow to the inverter, in other words, they are starved for current. For determination of the oscillation frequency of the VCO, the total effective capacitance ($C_{eff,stage}$) on the drain of each inverter stage is calculated.

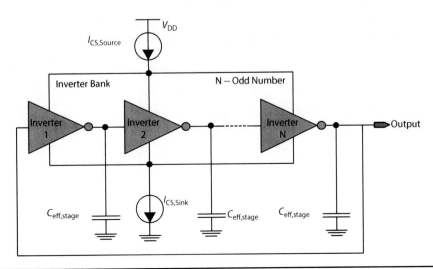

FIGURE 4.25 CS-RO basic concept [100].

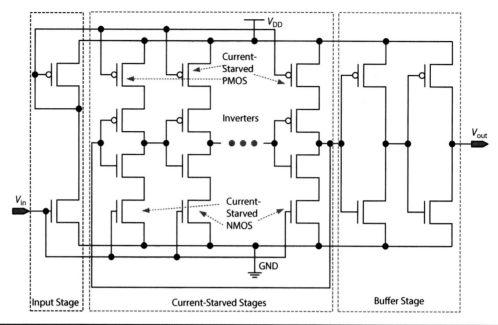

FIGURE 4.26 Circuit schematic of a CS-VCO.

The $C_{\text{eff,stage}}$ is estimated as the sum of the input capacitances (C_{in}) and output (C_{out}) capacitances of the inverter [39, 41, 42]:

$$C_{\text{eff,stage}} = C_{\text{out}} + C_{\text{in}}$$
$$= C_{\text{oxide}}(W_p L_p + W_n L_n) + \frac{3C_{\text{oxide}}}{2}(W_p L_p + W_n L_n)$$
$$= \frac{5C_{\text{oxide}}}{2}(W_p L_p + W_n L_n) \quad (4.29)$$

where C_{oxide} is the gate-oxide capacitance per unit area, W_n and W_p are the widths, and L_n and L_p are the lengths of the inverter NMOS and PMOS transistors, respectively. The gate-oxide capacitance per area C_{oxide} is calculated as follows:

$$C_{\text{oxide}} = \left(\frac{\varepsilon_{r_{\text{ox}}} \times \varepsilon_0}{T_{\text{ox}}}\right) \quad (4.30)$$

where $\varepsilon_{r_{\text{ox}}}$ is the relative dielectric constant of SiO_2, ε_0 is vacuum dielectric constant, and T_{ox} is the gate-oxide thickness of the NMOS and PMOS. The total time required to charge and discharge the capacitance of an inverter stage can be calculated by the following expression:

$$T_{\text{D,stage}} = \left(\frac{C_{\text{eff,stage}} V_{\text{DD}}}{I_{\text{bias}}}\right) \quad (4.31)$$

This is essentially $\left(\frac{1}{N_{\text{stage}}}\right)$ of the T_{OSC} discussed in the previous subsection. The operating frequency of the VCO can be determined by using this simple capacitance-charging estimate [39, 41, 42]:

$$f_{\text{osc}} = \left(\frac{1}{N_{\text{stage}} T_{\text{D,stage}}}\right)$$
$$= \left(\frac{1}{T_{\text{osc}}}\right)$$
$$= \left(\frac{I_{\text{bias}}}{N_{\text{stage}} C_{\text{eff,stage}} V_{\text{DD}}}\right) \quad (4.32)$$

where V_{DD} is the supply voltage, I_{bias} is the current flowing through the inverter, and N_{stage} is the odd number of inverters in the VCO circuit. The operating frequency can be mainly controlled by an applied DC input voltage, which adjusts the current I_{bias} through each inverter stage. In other words, the oscillation frequency is determined by the number of inverters, size of the transistors, and the current flowing through the inverter, which is determined by the input voltage to the VCO.

4.6.3 90-nm CMOS

As a specific example, a 90-nm CMOS CS-VCO design and characterization are now discussed [39, 40, 41]. The oscillation frequency is considered as the target specification for the VCO design. The target oscillation frequency is kept at a minimum of 2 GHz for this VCO design [39, 40, 41]. For the input stage, $L_n = L_p = 100$ nm, $W_n = 500$ nm, and $W_p = 10 \times W_n = 5000$ nm. The number of stages (N) is fixed to 13 to meet high-frequency requirements. Minimum-sized transistors have been used to design the inverter stages of the VCO. The length is kept constant for all devices. Hence, for the inverter stages, $L_n = L_p = 100$ nm, $W_n = 250$ nm, and $W_p = 2 \times W_n = 500$ nm. Choosing minimum-width transistors also ensures a reduced area design of the VCO. The $C_{eff,stage}$ is calculated using these values. The I_{bias} requirement is calculated for the desired f_{OSC}. The CS devices are sized to provide the required current I_{bias}. Thus, the following device geometry is obtained: $L_{ncs} = L_{pcs} = 100$ nm, $W_{ncs} = 500$ nm, $W_{pcs} = 10 \times W_{ncs} = 5$ μm. W_{ncs} and W_{pcs} are the widths and L_{ncs} and L_{pcs} are the lengths of the CS NMOS and PMOS transistors, respectively. For the output buffer stage of the VCO, $L_n = L_p = 100$ nm, $W_n = 500$ nm, and $W_p = 2 \times W_n = 1000$ nm. The schematic design of the 90-nm CMOS VCO is presented in Fig. 4.27(a) [39, 40, 41]. The minimum sizes of transistors needed for successful operation are calculated and used in this design. For the same sizes, the physical design is also performed for 90-nm CMOS PDK and the physical design is shown in Fig. 4.27(b) [39, 40, 41]. The oscillating frequency versus tuning voltage characteristics of the 90-nm 13-stage VCO is shown in Fig. 4.28. Some selected specifications of this VCO are presented in Table 4.5 [39, 40, 41]. It may be noted that it is difficult to find the optimum design by minimizing the phase noise and maximizing the tuning range simultaneously. To maximize the tuning range, it is necessary to minimize the parasitic capacitances, and hence reduce the size of transistors. If the size becomes too small, however, flicker noise can dominate, which, in turn, would increase phase noise.

4.6.4 50-nm CMOS

A 21-stage CS-VCO designed using 50-nm CMOS technology is now presented. The transistor-level schematic design is shown in Fig. 4.29 [45, 46]. In this VCO design, the currents at the input stage transistors are mirrored in each of the inverter stages including the current-source and current-sink transistors. The supply voltage for this VCO is kept at 1V. For a target oscillating frequency of 100 MHz, the bias current and effective capacitance are the following: $I_{Bias} = 10$ μA and $C_{eff,stage} = 4.7 fF$. The sizes of each transistor are then calculated using the basic equations presented in the previous subsection. For simplicity, the length of NMOS and PMOS are kept at nominal values and the width is calculated. The sizes of individual transistors are shown in Fig. 4.29 [45, 46]. The 21-stage VCO circuit is then characterized for selected specifications. The characterization results are presented in Table 4.6 [45, 46]. The frequency versus voltage tuning characteristics of the VCO are shown in Fig. 4.30 [45, 46].

4.6.5 45-nm CMOS

The transistor-level schematic of a five-stage CS-VCO is presented in Fig. 4.31 [66, 85]. The CS-VCO consists of two high-impedance input stage transistors and a chain of five inverters. Each inverter has two current sources or sinks transistors that limit the current flow to the inverter for current starving. The transistor sizes that provided proper sensitivity of frequency of the VCO to the DC voltage is shown in the schematic diagram. The transient analysis of the 45-nm CMOS VCO for a V_{DD} of 0.7 V is shown in Fig. 4.32 [66, 85]. The frequency-voltage characteristic of the VCO is shown in Fig. 4.33 [66, 85]. For this characteristic, the V_{DD} of 0.7 V and the input voltage is varied from 0.4 to 1.0 V. Selected characterization results are provided in Table 4.7 [66, 85].

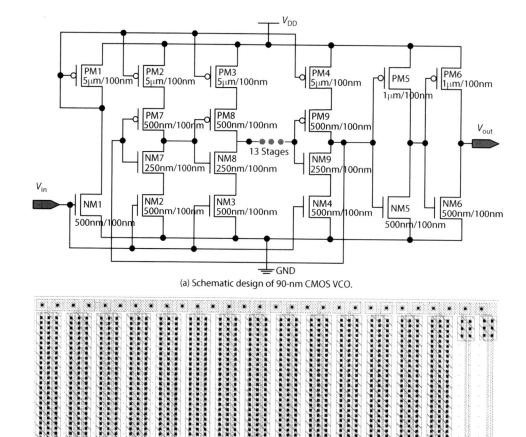

Figure 4.27 Design of the CS-VCO for 90-nm CMOS [39, 40, 41].

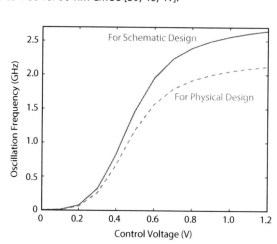

Figure 4.28 Frequency versus voltage characteristics of 90-nm CMOS CS-VCO [39, 40, 41].

Phase-Locked Loop Component Circuits

VCO Attributes	Actual Values
Technology	90-nm CMOS 1P 9M
Supply Voltage (V_{DD})	1.2 V
Oscillation Frequency (f_{osc}) – Schematic Design	1.95 GHz
Oscillation Frequency (f_{osc}) – Physical Design	1.56 GHz
Number of Design Variables	4 (W_n, W_p, W_{ncs}, W_{pcs})

TABLE 4.5 Characteristics of the 90-nm CMOS VCO [39, 40, 41]

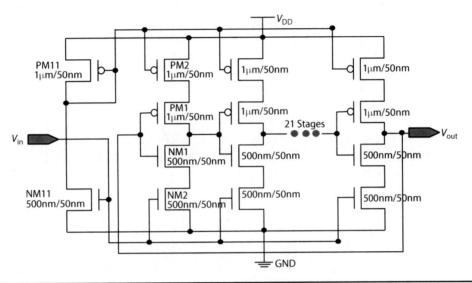

FIGURE 4.29 Schematic design of the 50-nm CMOS 21-stage CS-VCO [45, 46].

VCO Attributes	Actual Values
Minimum Oscillating Frequency ($f_{osc,\,min}$)	18.2 MHz
Maximum Oscillating Frequency ($f_{osc,\,max}$)	418.5 MHz
VCO Frequency Range	400.3 MHz
Average Power Dissipation	60 μW
VCO Tuning Voltage Range (V_{in})	0 to 1 V

TABLE 4.6 Characterization of 21-Stage 50-nm CMOS VCO [45, 46]

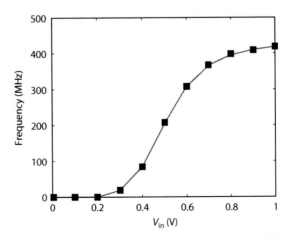

FIGURE 4.30 Frequency-voltage characteristics of the 50-nm CMOS CS-VCO [45, 46].

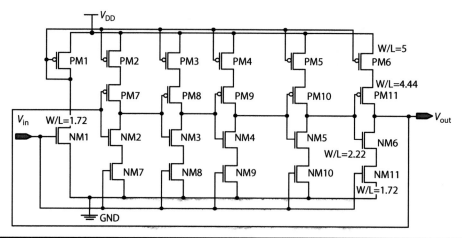

Figure 4.31 A 45-nm CMOS-based five-stage CS-VCO [66, 85].

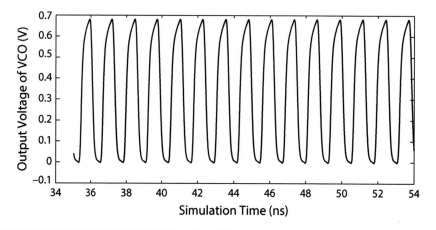

Figure 4.32 Simulation of the 45-nm CMOS five-stage CS-VCO [66, 85].

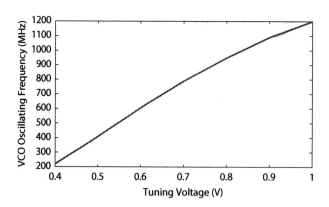

Figure 4.33 Frequency-voltage characteristic of the 45-nm CMOS five-stage CS-VCO [66, 85].

Technology Node	Oscillating Frequency	Power Dissipation
45-nm CMOS	718 MHz	28 µW

Table 4.7 Characterization Results Frequency-Voltage Characteristics of the 45-nm CMOS Five-Stage CS-VCO [66, 85]

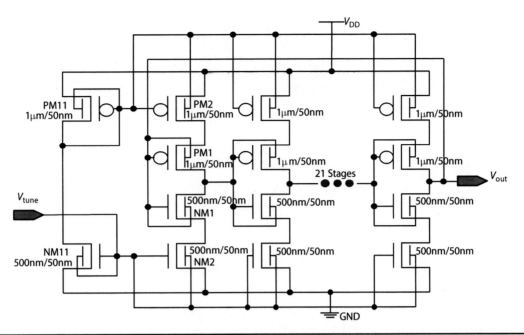

FIGURE 4.34 Schematic design of the 45-nm DG FinFET-based 21-stage CS-VCO [44].

4.6.6 45-nm Double-Gate FinFET

In this subsection, again the 21-stage VCO is discussed but the design is based on DG FinFET technology instead of bulk MOSFET. The FinFET device is used in SG configuration in which front and back gates are connected together. The FinFET used is a silicon-on-insulator (SOI) structure in which the bottom of the FinFET rests on top of a layer of SiO_2. In a typical FinFET process, SOI body thickness (T_{Si}) is so thin that the silicon body is fully depleted. For the FinFET devices in the VCO circuit, the size of the widths is quantized into units of the dimensions of the fins (H_{Fin}). Each Fin provides a device width of $2H_{Fin}$ and large width of FinFET device can be obtained by using multiple fins. The shorted DG FinFET with the following parameter has been used: oxide thickness, 1.4 nm; threshold voltage, $V_{Thn} = 0.31$ V and $V_{Thp} = -0.25$ V, channel doping concentration, 2×10^{16}; Fin height, 50 nm; and body thickness (T_{Si}), 8.4 nm.

The transistor-level schematic design is presented in Fig. 4.34 [44]. The parameter such as voltage are same as the VCO of previous subsection, for example, supply voltage of 1 V, bias current of 10 µA, and $C_{eff,stage}$ of 4.7 fF. The effective per stage capacitance is calculated as follows:

$$C_{eff,stage} = \left(\frac{1}{\dfrac{1}{C_{Si,ch}} + \dfrac{1}{C_{gate,dp}} + \dfrac{1}{C_{ox}}} \right) \tag{4.33}$$

where $C_{Si,ch}$ is the capacitance to the carriers in the channel, and $C_{gate,dp}$ is the depletion capacitance of the gate electrode. This leads to smaller intrinsic gate capacitance in the DG FinFET that results in higher oscillation frequencies. The nominal transistor sizes for the DG FinFET VCO are shown in Fig. 4.34 [44]. For these specific device sizes, the VCO characteristics are presented in Table 4.8 [44]. The tuning range for the DG FinFET 21-stage VCO is presented in Fig. 4.35 [44]. For a comparative perspective, tuning characteristics of the bulk CMOS 21-stage VCO and the DG FinFET 21-stage VCO are shown. It is evident from the plot that DG FinFET VCO has much larger oscillating frequency and linear tuning range for the same transistor sizes and inverter stages.

FinFET VCO Parameters	Values
Oscillating Frequency	775.6 MHz
Average Power Dissipation	65.4 μW
VCO Gain	$2.331 \left(\dfrac{GHz}{V}\right)$
Minimum Oscillating Frequency	363.2 MHz
Maximum Oscillating Frequency	1.165 GHz
Tuning Voltage Range	0 to 1 V

TABLE 4.8 Characterization Results of the 45-nm DG FinFET-Based 21-Stage CS-VCO [44]

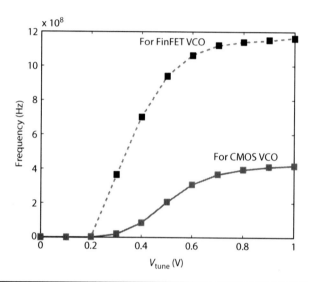

FIGURE 4.35 Voltage versus frequency characteristic of the 45-nm DG FinFET-based 21-stage CS-VCO [44].

4.7 LC-Tank Voltage-Controlled Oscillator

4.7.1 Basics

The core of the LC-tank VCO—a.k.a. the LC-tank oscillator or LC-VCO—is an LC-tank. The simplest form of LC-tank is depicted in Fig. 4.36 [2]. The basic principle of oscillation is the energy transfer between the inductor and the capacitor. When connected to a battery, the capacitor stores energy in the form of an electrostatic field and produces a static voltage across it (see Fig. 4.36(a)). The charged capacitor, when connected across the inductor, starts discharging itself (see Fig. 4.36(b)). The voltage across the capacitor starts reducing and current through the inductor increases. The increasing current induces an electromagnetic field around the inductor, which, in turn, opposes the current flow. When the capacitor fully discharges the energy from its electrostatic field, it is completely stored in the electromagnetic field of the inductor. At this point, when the capacitor-inductor loop does not have a voltage to maintain the current in the inductor, the current starts to decrease so does the electromagnetic field. The back electromotive force (EMF) in the inductor keeps the current flowing in the original direction. The current flow now charges the capacitor in the opposite polarity to its original charge. In the process, all the energy in the electromagnetic field of the inductor is again converted to the electrostatic field in the capacitor (see Fig. 4.36(c)). The capacitor now starts discharging and the process repeats between capacitor and inductor without the need of the capacitor charging from the battery. As the polarity of the voltage changes due to the back and forth energy transfers between the capacitor and inductor, AC voltage and current waveforms are produced.

Phase-Locked Loop Component Circuits 185

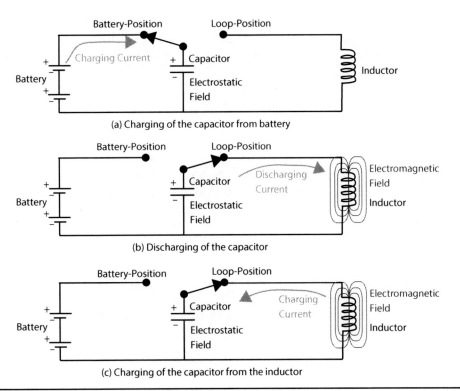

Figure 4.36 A basic inductor and capacitor tank [2].

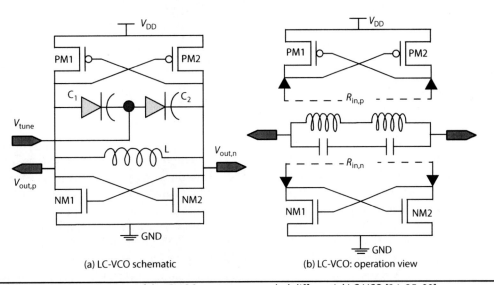

Figure 4.37 Schematic diagram of the CMOS core cross-coupled differential LC-VCO [34, 35, 99].

Theoretically, the energy-transfer process can continue indefinitely; however, it does not happen due to lossy inductors and capacitors that can have parasitic resistors in them. The resonance in the LC-tank happens when reactances due to inductor and capacitor are equal, which gives a resonance frequency of $\left(\frac{1}{2\pi\sqrt{LC}}\right)$ with C as capacitance of the capacitor and L as inductance of the inductor.

The basic idea of the above LC-tank is realized in the real-world as LC-tank oscillators. The transistor-level schematic design of a LC-VCO is presented in Fig. 4.37(a) [34, 35, 99]. Depending on the topology and the actual elements used for realization, the LC-VCO can have different variants or topologies [99]. However, CMOS core cross-coupled differential topology of the LC-VCO is shown in Fig. 4.37(a). The cross-coupled CMOS differential LC-VCO has better phase noise performance as compared to the cross-coupled NMOS only topology or PMOS only topology. In the LC-VCO, the capacitor is variable capacitor (varactor) that can be changed to get different C values. There are two NMOS devices and two PMOS devices in cross-coupled fashion. This CMOS pair provides gain that is more positive. The total negative resistance provided by a CMOS cross-coupled pair is calculated as follows [99]:

$$R_{\text{total}} = \left(\frac{R_{\text{in,n}} R_{\text{in,p}}}{R_{\text{in,n}} + R_{\text{in,p}}}\right) = -\left(\frac{2}{g_{\text{m,n}} + g_{\text{m,p}}}\right) \quad (4.34)$$

where $R_{\text{in,n}}$ is the negative resistance provided by the NMOS pair and $R_{\text{in,p}}$ is the negative resistance provided by the PMOS pair. $g_{\text{m,n}}$ and $g_{\text{m,p}}$ are transcoductances of NMOS and PMOS pair, respectively. For the LC-tank, the frequency of oscillations is calculated as follows [34, 35]:

$$f_{\text{osc}} = \left(\frac{1}{2\pi\sqrt{L_{\text{tank}} C_{\text{tank}}}}\right) \quad (4.35)$$

where L_{tank} is the inductance of the tank inductor. The tank capacitor is calculated as follows:

$$C_{\text{tank}} = C_1 + C_2 + C_{\text{other}} \quad (4.36)$$

where C_1 and C_2 are the capacitance of the varactors. C_{other} is the summation of NMOS and PMOS gate to drain or source and other parasitic capacitances of the circuit. The gain of LC-VCO is calculated as follows:

$$K_{\text{VCO}} = 2\pi \left(\frac{\text{Change in Frequency}}{\text{Change in } V_{\text{tune}}}\right) \quad (4.37)$$

$$= 2\pi \left(\frac{f_{\text{max}} - f_{\text{min}}}{V_{\text{max}} - V_{\text{min}}}\right) \quad (4.38)$$

where the gain is essentially the slope of the tuning range converted to radians as discussed in the subsection on oscillator characteristics.

4.7.2 180-nm CMOS

The design of a 180-nm CMOS-based CMOS differential LC-VCO is discussed in this subsection [34, 35, 36, 102]. The design of LC-VCO depends on target applications [37]. The size of the inductor was chosen to be large. The varactor can be implemented in many ways, such as PN junctions or MOS gate capacitors (e.g., when source and drain shorted become MOS capacitor) [15]. The PN-junction varactors have typical quality factor of 20 or more; however, they can be forward biased by large-amplitude voltage swings that are undesired in LC-VCO circuit. MOS capacitors or MOS varactors can be used to realize varactors with the highest possible quality factor in LC-tank. Therefore, in this design, MOS transistors are used as variable capacitors or varactors. Once preliminary decisions on the L and C values are made, the sizing of cross-coupled transistors is considered. All the components mirror each other to produce $V_{\text{out,p}}$ and $V_{\text{out,n}}$ with the same frequency to make the LC-VCO circuit symmetric in nature. The symmetry of LC–VCO circuit provides up-conversion and the even-order distortion in the differential output waveform [49]. Therefore, the sizes of the two NMOS and two PMOS transistors are made equal. The schematic design of the 180-nm CMOS differential LC-VCO is shown in Fig. 4.38(a) with baseline sizes.

The physical design or layout of the LC-VCO is performed once the schematic design is completed and thoroughly characterized; and once the schematic design is performed and characterized, it is designed. The physical design of the 180-nm CMOS LC-VCO is shown in Fig. 4.38(b) [34, 35]. In the physical design of the LC-VCO, the wire widths were maximized so as to reduce the interconnect resistances. The wide interconnects may also provide some flexibility to change the specification of the transistors in the future. The full-blown parasitic netlist is obtained from this physical design of the LC-VCO that can be used for silicon-accurate characterization. The parasitic effects of the circuit

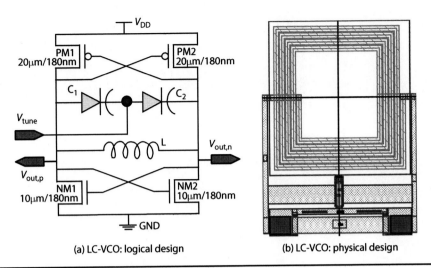

(a) LC-VCO: logical design   (b) LC-VCO: physical design

Figure 4.38 The 180-nm CMOS-based LC-VCO [34, 35].

LC-VCO Designs	Transistors	Capacitors	Resistors	Inductors	Total
Logical Design	4	2	0	1	7
Physical Design	4	108	600	14	726

Table 4.9 Number of Elements in the 180-nm CMOS LC-VCO Circuits [34, 35]

LC-VCO Designs	Frequency	Power	Phase Noise
Physical Design	2.15 GHz	0.153 mW (at 2.23 GHz)	−117 dBc/Hz (at 2.6 GHz)

Table 4.10 Characterization Results of the Differential 180-nm CMOS LC-VCO Circuits [34, 35]

were taken into consideration and the layout was made symmetrical. The number of elements active in the schematic and physical design of the LC-VCO are presented in Table 4.9 [34, 35]. As evident from this table, the physical design introduces additional parasitics to the circuit that will affect the LC-VCO performances.

The effect of the parasitics on the LC-VCO can be observed by simulating the parasitic-extracted netlist of the LC-VCO, with the same sizing of the devices as in the schematic design of the LC-VCO. The characterization results of the 180-nm CMOS LC-VCO physical design are presented in Table 4.10 [34, 35]. The tuning characteristics of the LC-VCO are shown in Fig. 4.39 [34, 35]. The target frequency or center frequency is the middle of the tuning characteristics corresponding to $\left(\frac{V_{tune}}{2}\right)$. The center frequency from this curve is 2.5 GHz. The phase noise for the center frequency of 2.5 GHz is shown in Fig. 4.40 [34, 35]. At 1-MHz offset from the carrier, the phase noise is −117 dBc/Hz.

4.7.3 CNTFET

A cross-coupled differential CMOS LC-VCO that uses CNT bundle wire and multiwall CNT (MWCNT) wire as inductor is shown in Fig. 4.41 [89]. The cross-coupled CMOS transistors provide better phase noise than a NMOS only or PMOS only topology. The overall CNT-inductor-based LC-VCO has symmetrical design for better phase noise performance and larger voltage swing. The varactor in CNT-inductor-based LC-VCO is implemented from an NMOS transistors with source and drain tied together for better Q factor. Most importantly, the inductor of the LC-tank circuit is designed using a CNT bundle wire and MWCNT wire. The Q factor of SWCNT bundle can be as high as 600% more than the copper inductor, whereas for MWCNT it is as high as 200% more than the

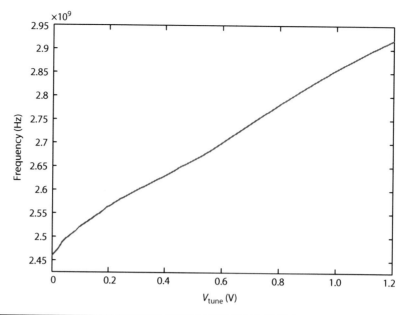

Figure 4.39 The frequency tuning characteristic of the LC-VCO [34, 35].

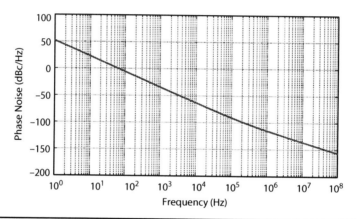

Figure 4.40 The phase noise of 180-nm CMOS LC-VCO for the center frequency of 2.5 GHz [34, 35].

copper inductor. The negligible skin effect in the CNT interconnects lowers the resistance that contributes to a high Q factor. The sizes of the devices for a target frequency of 2 GHz is shown in Fig. 4.41. For this specification, the inductance of the inductor is 6 nH and the capacitance of the capacitor is 0.5 pF. A brief characterization results for the cross-coupled differential 180-nm CMOS LC-VCO is presented in Table 4.11 [89]. The oscillation frequency versus control voltage characteristic of the LC-VCO is shown in Fig. 4.42 [89]. It is evident from the figure that the CNT-bundle inductor-based LC-VCO has a much higher oscillation frequency than that of a LC-VCO with copper inductor.

The MWCNT inductor-based differential LC-VCO is shown in Fig. 4.43 [88]. The inductors of the LC-tank are realized as MWCNT network-based inductors. The variable capacitors C_A, C_B, and C_C are microscale varactors. The varactors are realized with two fixed and two movable plates. Overall, the four-plate varactor structure is controlled by two tuning voltages. The tail current transistors are replaced with a MWCNT network-based inductor to operate at low-supply voltage and reduce noise. The characterization results of the MWCNT-inductor based LC-VCO is presented in Table 4.12 [88].

4.7.4 Memristor

As evident from the discussions in the previous subsections, crystal, LC-VCO, and ROs have their own advantages and disadvantages. Crystal oscillators are stable, produce clean signal, but produce low-frequency only. The LC-VCOs produce cleaner output; however, they occupy a significant chip

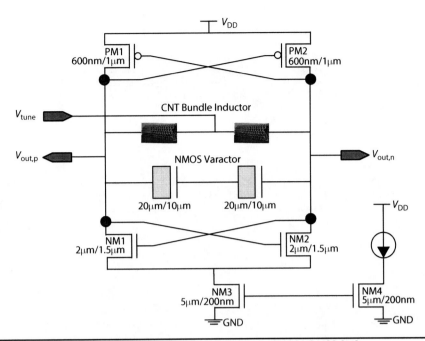

Figure 4.41 Carbon nanotube inductor-based differential 180-nm CMOS LC-VCO [89].

Oscillating Frequency	Frequency Range	Phase Noise
2.0 GHz	1.6–3.3 GHz	−123 dBc/Hz (at 2.0 GHz for 10 MHz offset)

Table 4.11 Characterization Results of the CNT Bundle Inductor-Based Differential 180-nm CMOS LC-VCO [89]

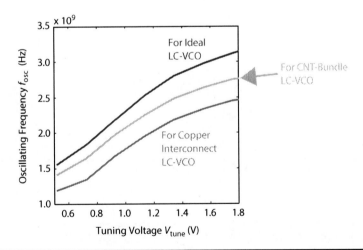

Figure 4.42 Frequency versus voltage characteristic of CNT bundle inductor-based differential 180-nm CMOS LC-VCO [89].

area and consume more power. The ROs have low-area and low-power consumption requirements; however, they have high phase noise and poor stability at high-oscillation frequencies. Therefore, there is always need for a better oscillator that has low-power dissipation, low-area requirement, low phase noise, high-frequency oscillations, and stable operation. As a mitigation of the above issues, memristor-VCO is being explored [23, 29, 77, 91]. Nevertheless, a concrete LC-tank oscillator with memristor still needs further research. The research on memristor-based circuits

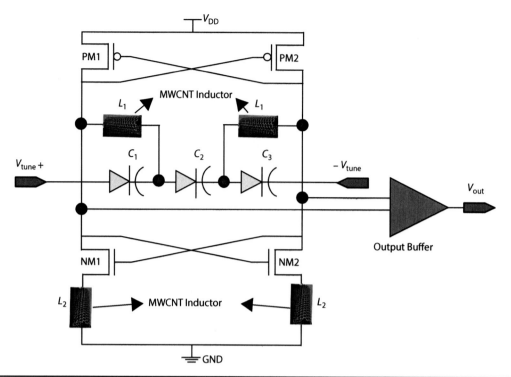

Figure 4.43 MWCNT inductor-based differential CMOS LC-VCO [88].

Frequency	Power Consumption	Phase Noise	Output Power
2.4 GHz	1.7 mW at 0.6 V	−133 dBc/Hz (at 2.4 GHz for 1MHz offset)	−10 dBm

Table 4.12 Characterization Results of the MWCNT-Inductor-Based Differential CMOS LC-VCO [88]

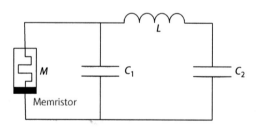

Figure 4.44 Memristor-based LC-tank oscillator [29].

and systems is in full swing [5]. The schematic representation of the memristor-VCO is shown in Fig. 4.44 [29]. The nonlinear and negative differential resistance of the memristor makes it well suited for oscillator designs. In the memristor LC-VCO, the functions of the memristor are that of an active device that can provide sustained oscillations without an external power supply.

4.8 Relaxation Oscillators

The relaxation oscillators are oscillators that alternate between two states. The relaxation oscillators store energy in a capacitor during one phase of oscillation-cycle and release energy in the next phase of the cycle. The second phase of oscillation cycle is called *relaxation phase* [38]. The periods of alternation between the store and release depend on the charging of a capacitor. An example of relaxation oscillator is a multivibrator. In this section, selected relaxation oscillators are discussed.

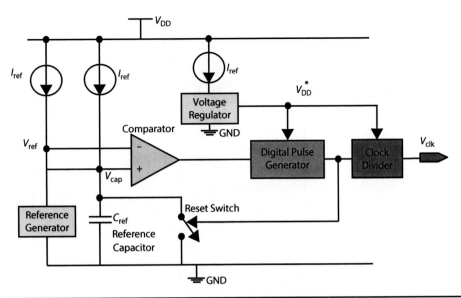

FIGURE 4.45 A low-power relaxation oscillator [25].

Relaxation oscillators can be easily implemented in CMOS technology and are attractive for low-power applications. However, jitter is an issue for the relaxation oscillators. Another issue of the relaxation oscillators is sensitivity to temperature variations [25].

4.8.1 Low-Power Relaxation Oscillator

A low-power relaxation oscillator realized with various system-level blocks is shown in Fig. 4.45 [25]. The majors blocks of the relaxation oscillator are the following: (1) reference voltage generator, (2) reference capacitor, (3) reset switch, (4) comparator, (5) voltage regulator, (6) digital pulse generator, and (7) clock divider. The reference voltage generator provides reference voltage V_{ref}. The reference capacitor with capacitance value C_{ref} provides a voltage of V_{cap}. Both reference voltage generator and the reference capacitor are charged by the same bias current I_{ref}. These two voltages V_{ref} and V_{cap} are compared using the comparator. When $V_{cap} > V_{ref}$ comparator outputs high to trigger the digital pulse generator. The digital pulse generator generates a reset pulse signal to close the reset switch that discharges the reference capacitor. The frequency of oscillation ($f_{osc,rxo}$) of the relaxation oscillator can be calculated using the following expression:

$$f_{osc,rxo} = \left(\frac{I_{ref}}{C_{ref} V_{ref}} \right) \quad (4.39)$$

The characterization results of low-power relaxation-oscillator are shown in Table 4.13 [25].

4.8.2 Memristor Relaxation Oscillator

A memristor-based relaxation oscillator is presented in Fig. 4.46 [31]. The memristor-based relaxation oscillator has two memristors, two comparators, one NAND gate, and one inverter. In this circuit, the two memristors emulate the behavior of the charging and the discharging of the capacitor. The two comparators ensure that the memristances (M_A or M_B) vary between two limits to achieve the oscillation function at the output. The two memristances change their values depending on the voltage polarities of the memristors. The oscillation frequency of this memristor oscillator is 745.4 Hz.

Technology	Supply Voltage	Frequency	Power Consumption	Area
350-nm CMOS	1 V	3.3 kHz	0.011 μW	0.1 mm²

TABLE 4.13 Characterization Results of Low-Power Relaxation-Oscillator [25]

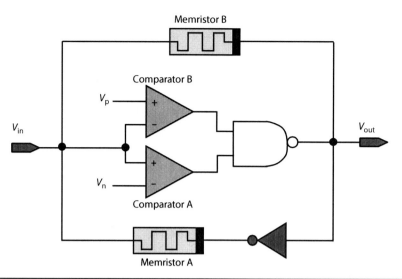

FIGURE 4.46 A memristor-based relaxation oscillator [31].

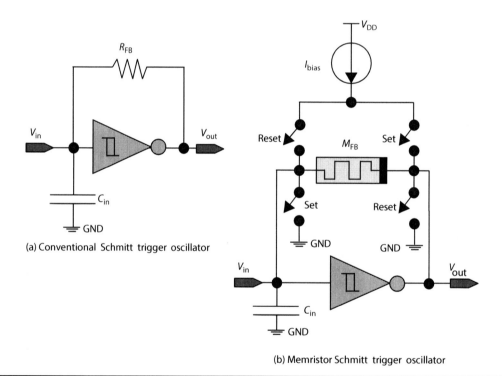

FIGURE 4.47 Memristor-based Schmitt Trigger oscillator – block-level diagram [103].

Memristor-based relaxation oscillators have been called reactance-less oscillators as an active capacitor has not been used in them [31, 105].

4.8.3 Memristor-Based Schmitt Trigger Oscillator

The relaxation oscillator that can be implemented by the Schmitt trigger is a bistable multivibrator. The simple idea is to connect a single RC integrating circuit between the output and the input of an inverting Schmitt trigger. The Schmitt trigger oscillators are used in many applications, including capacitive sensing, inductive sensing, pressure sensing, and implantable devices. The Schmitt trigger oscillator exhibits the least phase noise as compared to digitally controlled oscillators (DCO) and differential ROs [103]. The conventional Schmitt trigger oscillator is shown in Fig. 4.47(a) [103].

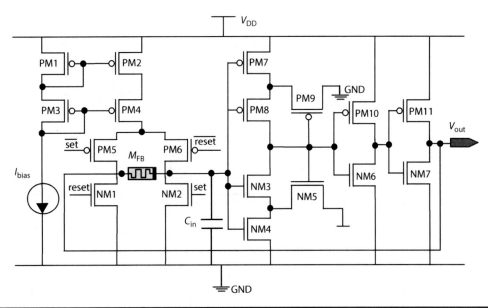

FIGURE 4.48 Memristor-based Schmitt Trigger oscillator – circuit-level schematic diagram [103].

Technology	Supply	Memristance	Oscillating Frequency	Jitter	Power Dissipation
90 nm	1 V	100 kΩ (0)	34.49 MHz	536 ps	130.4 μW
90 nm	1 V	10 kΩ (1)	251.46 MHz	69 ps	161.2 μW

TABLE 4.14 Characterization of Memristor-Based Programmable Schmitt Trigger Oscillator [103]

The oscillation frequency of the Schmitt trigger oscillator is determined by the feedback resistance (R_{FB}), input capacitance (C_{in}), and the Schmitt trigger's hysteresis.

A memristor-based programmable Schmitt trigger oscillator is shown in Fig. 4.47(b) [103]. It contains a feedback memristor with memristance value M_{FB}. There is an additional circuitry to facilitate the programmability of this memristor-based Schmitt trigger oscillator. The programming is performed by four switching, such as two set switches and two reset switches, and a programming current source (I_{bias}). All four switches are open during the normal mode of operation of the memristor-based Schmitt trigger oscillator. For the reset phase of the oscillator, the reset switches are closed, the set switches are open, and constant current I_{bias} flows through the feedback memristor to set a state. For the set phase of the oscillator, the set switches are closed, the reset switches are open, and constant current I_{bias} reversely flows through the feedback memristor to reach another state. The transistor-level schematic of the memristor-based Schmitt trigger oscillator is shown in Fig. 4.48 [103]. The characterization results for 90-nm technology at 1 V supply voltage is shown in Table 4.14 for two states "0" and "1" [103].

4.9 Phase-Frequency Detectors

The PFD compares the phase of two input signals to produce phase error as well as frequency error signals [34, 35, 36, 78]. On the other hand, a phase comparator only produces phase error signals and a comparator compares only two signal levels and produces a logical "high" or "low" result. The PFD is deployed for better accuracy at small phase differences and to lock phase signals at high frequencies. In a PLL, the PFD gets two inputs on from external source and another from a VCO. The PFD then produces two outputs that change the signal phase in PLLs. In the PLLs, the PDF directs the CP to supply charge amounts proportional to the phase error detected. In general, the PFD is an analog mixer or an asynchronous sequential logic circuit. The gain of the PFD is expressed as follows:

$$K_{PFD} = \left(\frac{V_{DD}}{4\pi}\right) \quad (4.40)$$

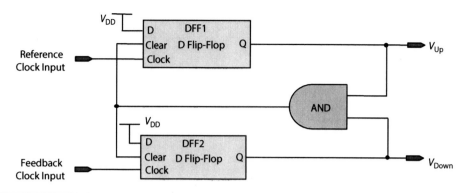

Figure 4.49 High-level representation of the phase detector [34, 36].

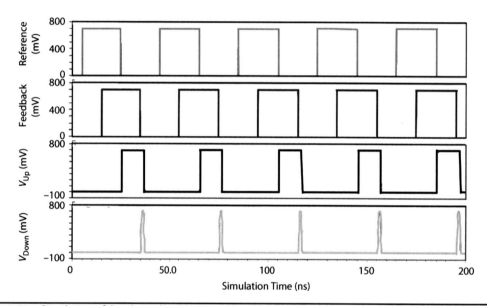

Figure 4.50 Simulation of the phase detector [34, 85].

The design of a PFD may be simple XOR gate-based to complex logic circuit consisting of flip-flops. Two different PFD designs are discussed in this section.

4.9.1 D Flip-Flop-Based PFD

The logic-level design of the phase-detector is shown in Fig. 4.49 [34, 36, 73]. It consists of two D flip-flops (DFFs) and one AND gate connected in a specific topology. The signals that go as input are the reference clock signal and the feedback clock signal. The PFD needs to detect the phase and frequency of the feedback signal to output V_{Up} and V_{Down} for adjusting the VCO signal accordingly. In the PFD, the DFFs are triggered by the inputs (reference clock signal and feedback clock signal) and both the outputs are low. When one of the inputs of the PFD rises, the corresponding output becomes high [73]. The functional simulation of the PFD is shown in Fig. 4.50 [34, 86]. The physical design of the PDF for 180-nm CMOS technology is shown in Fig. 4.51 [34,36]. The power consumption of the PFD is 9.3 nW.

4.9.2 XOR Gate-Based PFD

A PFD using two-XOR gates is shown in Fig. 4.52 [62]. It takes three inputs, $Clk_{Input,1}$, $Clk_{Input,2}$, and "Data," and generates two outputs—$Data_{Lead}$ and $Data_{Lag}$. In the context of the PLL, "Data" input is the reference clock and $Clk_{Input,1}$ and $Clk_{Input,2}$ are feedback clocks. In particular, $Clk_{Input,1}$ is the signal

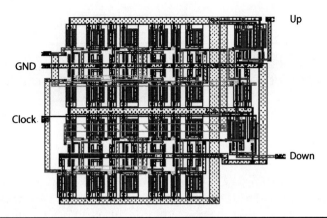

Figure 4.51 Physical design of phase detector for 180-nm CMOS [34, 36].

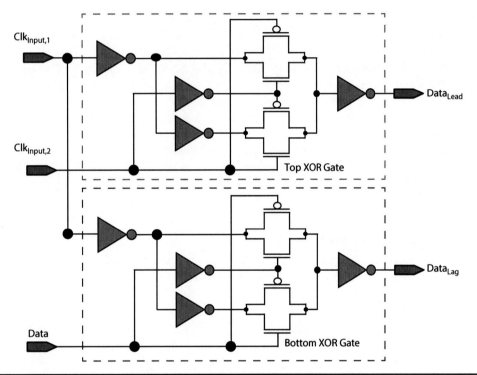

Figure 4.52 Phase-frequency detector using two XOR gates [62].

from the first inverter stage of the VCO and $Clk_{Input,2}$ is from the third inverter stage of the VCO. The PFD essentially nullifies the phase error and frequency error between the reference clock signal and input data clock signals. In the PFD, the top XOR gate uses $Clk_{Input,1}$ and $Clk_{Input,2}$ to determine the lead control signal. The resulted $Data_{Lead}$ signal then acts as a charge-down signal to drive the loop-filter output voltage low and increase the clock frequency. The bottom-XOR gate uses $Clk_{Input,1}$ signal and reference data signal to generate $Data_{Lag}$ signal which is the opposite of $Data_{Lead}$ signal.

4.10 Charge Pumps

4.10.1 Basics

In the PLL, the CP stabilizes the fluctuation of currents and switching time to minimize the spurs in the input of the VCO [34, 35, 36, 78, 85]. The CPs are used in many diverse electronic systems and their components such as RF antenna switch controllers, switched capacitor circuits,

196 Chapter Four

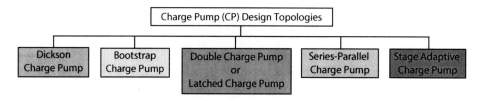

FIGURE 4.53 The CP types taxonomy.

piezoelectric actuators, LCD drivers, smart power, voltage regulators, and nonvolatile/volatile memories [81, 85]. The function of a CP is to convert a DC input to another DC output voltage several times higher than the power supply voltage (V_{DD}). Simplistically speaking, the CP is a DC to DC level-up converter generating output higher than V_{DD}. The gain for the CP (K_{CP}) is calculated using the following expression [34, 85]:

$$K_{CP} = \left(\frac{I_{CP}}{2\pi}\right) \tag{4.41}$$

where I_{CP} is the current through the CP. The CP circuits can have different topologies based on the way the switches are implemented [81]. Selected types of charge pumps are listed in Fig. 4.53. The Dickson CP was invented in 1976 and used diodes instead of switches. The bootstrap charge pump has a bootstrap circuit at each stage that is realized by a capacitor and MOS transistor and is widely used. The latched CP or double CP has a latch in each stage and is useful for very high-frequency applications. The series-parallel CP charges all the capacitors in parallel in the first half of a clock cycle and then uses series connection of capacitors in the second half of the cycle. The stage adaptive charge pump dynamically adapts its number of stages as needed for different applications.

4.10.2 180-nm CMOS

The transistor-level schematic of a CP design is presented in Fig. 4.54 [34, 36, 78, 85]. This charge pump is utilized to convert timed logic levels into analog quantities for controlling the locked VCO [34, 85]. The inputs to the CP are Up/Down or Lead/Lag. The signals are the two outputs of the PFD

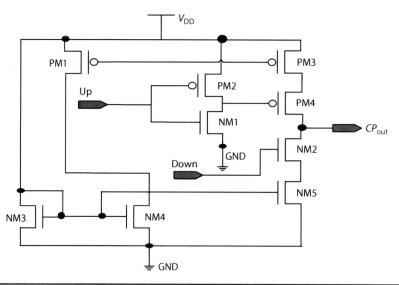

FIGURE 4.54 Schematic diagram of the CP [34, 36, 78, 85].

of the PLL. The CP generates a voltage (CP_{out}) that goes to the LPF of the PLL. The CP charges its capacitors when based on the signal at Up or Down input. When Up signal is high and Down signal is low, the charge pump deposits charge onto the capacitor, which raises the voltage. On the other hand, when Down signal is high and Up signal is low, the CP sinks current and lowers the output voltage. When both Up and Down signals are low, the net current is zero and the output voltage remains constant. Essentially, the CP manipulates the amount of charge on the filters of the capacitors depending upon the signals from the Up/Down or Lead/Lag of the PFD. The physical design of the CP for 180-nm CMOS is shown in Fig. 4.55 [34, 36, 78]. The simulation of the charge pump for 45-nm CMOS for 0.7 V supply is shown in Fig. 4.56 [34, 85]. The output current versus output voltage relationship of the charge pump is shown in Fig. 4.57 [34, 85].

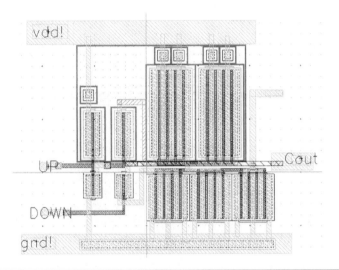

FIGURE 4.55 Physical design of the CP for 180-nm CMOS [34, 36, 78].

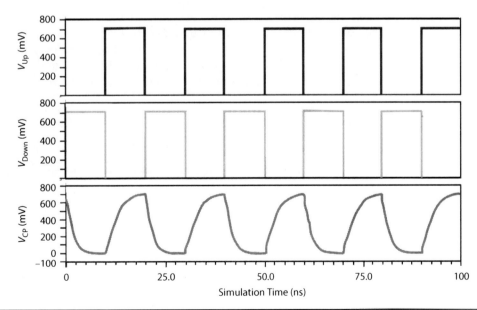

FIGURE 4.56 Simulation of the CP for a 0.7 V supply voltage [34, 85].

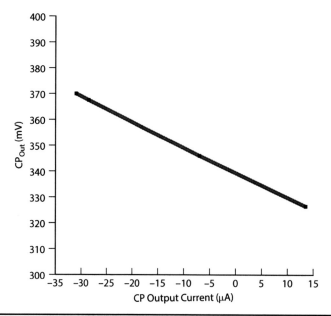

Figure 4.57 Output voltage versus output current characteristic of the 45-nm CMOS CP [34, 85].

4.11 Loop Filters

In general, the filter circuits of various types are used in almost every signal processing application. In the case of the PLL, the LPF is a low-pass filter, and a basic RC topology is shown in Fig. 4.58 [34, 36, 85]. The input of the LPF receives a signal from the PFD and the output of the LPF is connected to the input of the VCO. The LPF smooths the phase difference of the signal generated from the PFD before sending the signal to the VCO. Thus, the LPF determines the dynamic characteristics of the PLL. If the PLL is to be applied to acquire and track signals, then the bandwidth of the LPF has to be large for fixed-input frequency. Once the PLL is locked in a capture range and starts tracking a signal, the range of frequencies the PLL will follow is called the tracking range. The LPF determines how fast the signal frequency can change, that is, the maximum slew rate, and still maintain lock. The narrower the bandwidth of the LPF, the smaller the achievable phase error. This comes at the expense of slower response and reduced capture range [85].

In the PLL, the cutoff frequency of the LPF $f_{LPF,cutoff}$ is dictated by the maximum frequency of the VCO. This ensures that the LPF rejects signals at frequencies higher than the maximum frequency of the VCO. The RC-based LPF circuit shown in Fig. 4.58 is an AC voltage divider circuit that

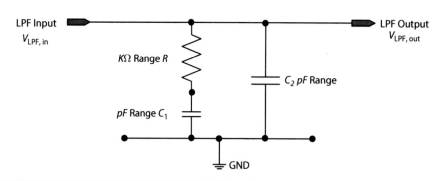

Figure 4.58 Circuit schematic of LPF with parameter values for a typical frequency of few GHz [34, 36, 85].

discriminates against high frequencies due to the decrease of the capacitive reactance with the increase in frequency. The cutoff frequency of the LPF can be calculated at the time domain using the following expression:

$$f_{\text{LPF,cutoff}} = \left(\frac{1}{2\pi RC_2}\right) \quad (4.42)$$

where C_2 is the capacitance at the output of the LPF. Thus, the capacitor C_2 prevents the CP voltage from causing voltage jumps on the input of the VCO and subsequently causing any frequency jumps in the PLL output and stability issues. If the phase of the clock signal changes slowly, then the current I_{pump} linearly charges C_1 and C_2. This gives an average effect and therefore small deviation in the input voltage of the VCO. If the phase of the clock signal changes fast, then the charge pump of the PLL drives the resistor R of the LPF that eliminates the averaging and allows the VCO to track quickly moving variations in the input clock signal. The transfer function of the LPF is expressed by the following:

$$\text{LPF}(s) = \left(\frac{1 + sRC_1}{sC_1}\right) \quad (4.43)$$

The design of the LPF is dictated by the characteristics required by the target application of the PLL. The resistance and capacitance of the LPF can also be determined by the following expression [34, 36]:

$$RC_1 = \left(\frac{2}{\omega_N}\right) \quad (4.44)$$

In general, C_2 is set to roughly one tenth of C_1. For a 2.5-GHz target frequency, the capacitors and resistors of the LPF are calculated to be $C_1 = 10$ pF, $C_2 = 1$ pF, and $R = 1.7$ KΩ [34, 36]. The simulation results for another LPF are shown in Fig. 4.59 [85], which has a cutoff frequency of 788 MHz.

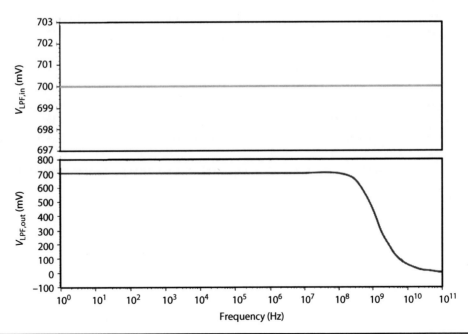

Figure 4.59 Simulation of the low-pass RC LPF [34, 36, 85].

4.12 Frequency Dividers

4.12.1 Basics

A frequency divider or a clock divider is a circuit or subsystem that can reduce the frequency of a signal passing through it. If $f_{\text{in,FB}}$ is the input frequency and N_{FB} is the dividing factor of the frequency divider, then the frequency of the output signal generated by it is expressed as follows [8, 21, 83, 84]:

$$f_{\text{out,FB}} = \left(\frac{f_{\text{in,FB}}}{N_{\text{FB}}}\right) = \left(\frac{f_{\text{osc}}}{N_{\text{FB}}}\right) \tag{4.45}$$

where N_{FB} is an integer. In a PLL, the frequency dividers or frequency FBDIVs are used to generate a signal frequency that is a multiple of the reference-signal frequency. In such a situation, the input frequency to the frequency FBDIV is the oscillating frequency f_{OSC} generated from the oscillator. The output of the frequency divider is fed as a signal to the PFD. The frequency dividers must have proper speed and locking range while maintaining input capacitance, input swings, and common-mode level [83].

The frequency dividers have many diverse topologies and implementations. They can be either analog or digital circuits. Selected types of frequency dividers are listed in Fig. 4.60 [8, 21, 83, 84]. The digital frequency dividers (DFD), made up of DFFs or JK flip-flops essentially functioning as counters, can perform simple power-of-2 integer divisions [34, 85]. The injection-locked frequency divider structurally is an oscillator with one or more terminals for signal injection [21]. The regenerative frequency divider, also known as the Miller frequency divider as introduced by Miller in 1939, is essentially a nonlinear feedback circuit made of a mixer and a loop-filter [84]. Fractional-N frequency divider can be implemented by using two integer dividers, a divide-by-N and a divide-by-(N + 1) frequency divider, to generate finer frequencies [14].

4.12.2 DFF-Based 180-nm CMOS

A simple frequency divider that is realized using true single-phase clock logic is shown in Fig. 4.61 [34, 36, 78]. The circuit is essentially comparable to a two-phase static DFF operation. It performs a simple division by two operations. When a continuous train of pulse waveforms at fixed frequency is given as input to this circuit, it generates an output signal of approximately half the frequency of the input signal. If the proposed circuit is used in a cascaded manner, then it can perform divisions that are power of 2, such as division-4 and division-8. The physical design of the divide by 2 circuit that was realized using 180-nm CMOS is shown in Fig. 4.62 [34, 36, 78].

4.12.3 JK Flip-Flop-Based 45-nm CMOS

The transistor-level schematic circuit of a simple frequency divider using JK flip-flops is presented in Fig. 4.63 [66, 85]. At the logic-level, it consists of two three-input NAND gates and two two-input NAND gates. This frequency divider works on the mechanism as follows: it counts two pulses and then resets and continues this process at each reset. The circuit is implemented for 45-nm CMOS technology. The transistor sizings are performed for a supply voltage of 0.7 V. The 45-nm CMOS-based design simulation result is shown in Fig. 4.64 [66, 85] for a supply voltage of 0.7 V. The top waveform

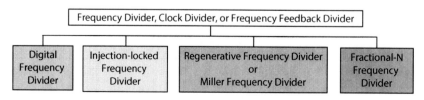

Figure 4.60 Different types of frequency dividers.

Phase-Locked Loop Component Circuits 201

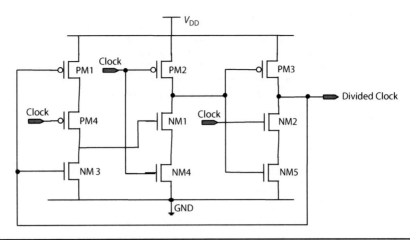

Figure 4.61 Schematic representation of a DFF-based frequency divider circuit [34, 35].

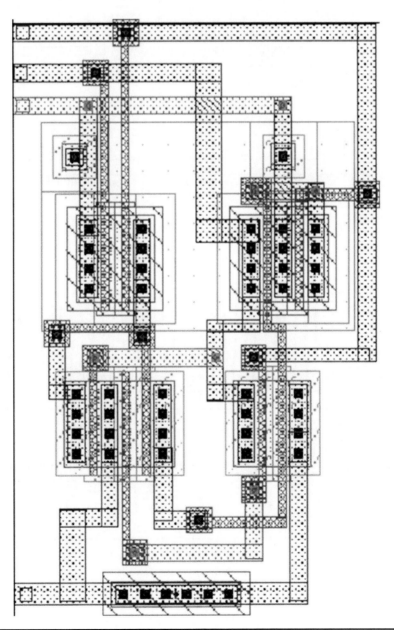

Figure 4.62 Physical design of a DFF-based frequency divider circuit [34, 35].

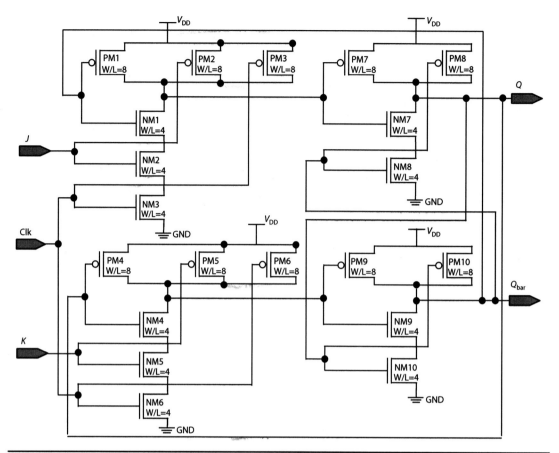

Figure 4.63 Schematic representation of a JK flip-flop-based frequency divider circuit [66, 85].

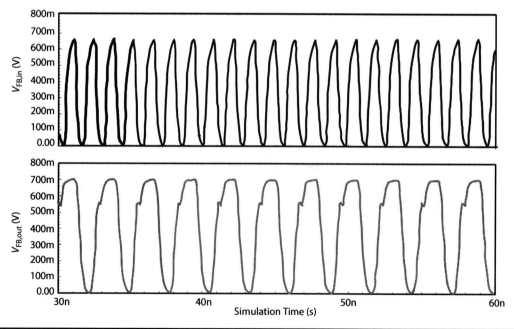

Figure 4.64 Simulation of the JK FF-based frequency divider circuit implemented using 45-nm CMOS [66, 85].

is the input voltage to the frequency divider and the bottom is the output voltage from it. As can be seen from the figure, the clock cycle width has doubled in the output waveform. In other words, the frequency has reduced.

4.13 Design and Characterization of a 180-nm CMOS PLL

The physical design of a 180-nm CMOS technology-based PLL is shown in Fig. 4.65 [34, 36, 78]. Typically, the final physical design like this is performed after the design subcomponents and the complete design are verified at different levels of design abstraction such as behavioral, transistor, and

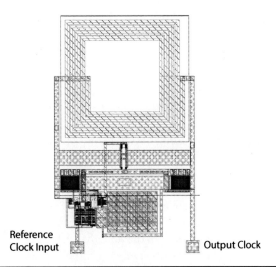

Figure 4.65 Baseline physical design of a complete PLL for 180-nm CMOS [34, 36].

Oscillating Frequency	Phase Noise	Locking Time	Horizontal Jitter	Vertical Jitter	Power Dissipation	Silicon Area
2.6 GHz	−162 dBc/Hz	1.9 μs	189 ps	168 μV	9.29 mW	525 × 326 μm²

Table 4.15 Characterization of the Physical Design of a Complete PLL for 180-nm CMOS [34, 36]

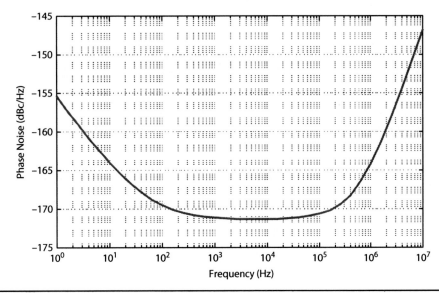

Figure 4.66 Phase noise of PLL [34, 36].

layout levels. The overall layout is then created by combining the individual layouts. The PLL uses the LC-VCO design from Fig. 4.38(b) as the oscillator that essentially occupies most of the silicon area. The DFF-based phase detector from Fig. 4.51 is used. The CP circuit from Fig. 4.55 is used. This PLL uses a simple low-pass LPF with RC topology from Fig. 4.58. The frequency FBDIV of the PLL is based on the divide by two circuit of Fig. 4.62. The PLL design is characterized using analog simulations with full-blown parasitics containing resistances, capacitances, inductances, and mutual capacitances originated from both transistors and interconnects. Selected characterization information is presented in Table 4.15 [34, 36]. The phase noise diagram of the PLL is shown in Fig. 4.66 [34, 36, 78]. A phase noise of -162 dBc/Hz is observed at 1 MHz offset for this PLL design.

4.14 All Digital Phase-Locked Loop

4.14.1 Basics

The ADPLL is a complete digital system that can generate or synthesize clock contrary to an APLL that has all analog components [101]. There is also an in-between system called DPLL that may have partial digital blocks or digital subsystems. A specific example of a DPLL is a frequency-locked loop (FLL) that uses a digital LPF [4].

For a simplistic understanding of APLL and ADPLL, block-level representations are provided in Fig. 4.67 [82, 90]. The APLL has all analog components such as PFD, CP, analog LPF, VCO, and analog frequency divider. On the contrary, the ADPLL has all digital components such as time-to-digital converter (TDC), digital loop filter (DLF), DCO, and DFD. The use of TDC instead of a CP suppresses any reference spurs and facilitates the DLF to be set at an optimal performance point between the phase noise of VCO and reference clock [90]. An APLL is typically susceptible to noise and nanoscale process variations. On the other hand, the digital PLLs can generate very high frequency while maintaining higher stability and low jitter. In addition, the ADPLL can provide design modularity, portability, and easy technology scaling as well as integration in nanoscale digital microprocessors [90]. The ADPLL has applications in Bluetooth and GSM.

4.14.2 A Simple ADPLL Using an NCO

The block diagram representation of a simple ADPLL is provided in Fig. 4.68 [1]. A digital multiplier has been used in the place of the phase detector. The digital multiplier generates a DC output that is proportional to the phase difference and series of higher-frequency components. The high-frequency components from the digital multiplier are filtered out of the DLF. The LPF is typically a first or second order low-pass infinite impulse response (IIR) filter. The output of the filter then goes to an

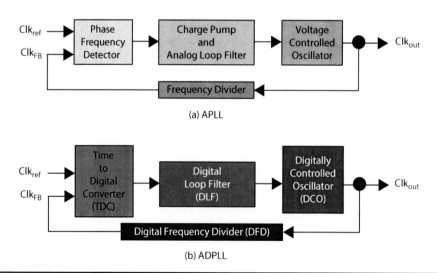

Figure 4.67 ADPLL versus classical APLL [82, 90].

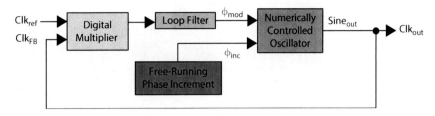

FIGURE 4.68 A simple ADPLL using a NCO [1].

NCO. The NCO adjusts frequency and phase in order to reduce phase error to zero, thus achieving the locking condition in the ADPLL.

The NCO is essentially a digital signal generator that generates a discrete-time, discrete-valued representation of a sinusoidal signal. The NCOs can be of several types including the following [26, 61]:

1. *Direct Form Digital Oscillator System*: This type of NCO is designed using a feedback circuit in which the poles are on the unit circle to produce a sinusoid signal. These are simple hardware but have stability issues.

2. *Look Up Table Based NCO (LUT-NCO)*: A typical approach to implement the NCO is to use a LUT for generating the waveforms. The LUT in a ROM essentially stores the samples of sinusoid that is read out at appropriate time intervals to produce a sinusoid signal. The size of the ROM is a design issue for this type of NCO.

3. *Coordinate Rotation Digital Computer Based NCO (CORDIC-NCO)*: The CORDIC is an iterative algorithm that performs transformations using only shifts and adds. For a comparative perspective, it can be stated that in a CORDIC-NCO, the sinusoids are generated by the iterative algorithm rather than stored in a ROM as in the case of LUT-NCO. Thus, much efficient hardware design is possible using CORDIC and there is no need to use silicon area-consuming ROM for LUT.

4.14.3 A High-Resolution ADPLL Using Double DCO

The architecture of a high-resolution ADPLL is presented in Fig. 4.69 [86]. It has one PFD, two DCOs, three dividers, and an overall controller (ADPLL controller). Two DCOs are track-DCO and average-DCO. The three dividers are predivider, DCO divider, and output divider. The track-DCO is used to track the reference clock (clk_{ref}). The average DCO is used to generate the output clock (clk_{out}). The PFD uses two input signals—ref_N and DCO_M. The ref_N is the clk_{ref} divided by N and DCO_M is the clk_{FB} divided by M. The PFD generates two output signals "Lead" and "Lag" depending on the frequency and phase difference between ref_N and DCO_M. Lead signal is high when DCO_M leads ref_N that then causes slowing-down of the track-DCO. Lag signal is high when DCO_M lags ref_N that

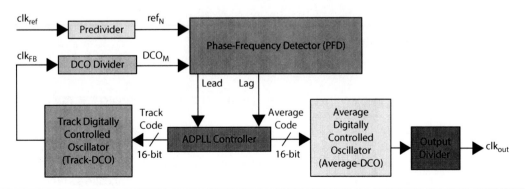

FIGURE 4.69 A high-resolution ADPLL using double DCO [86].

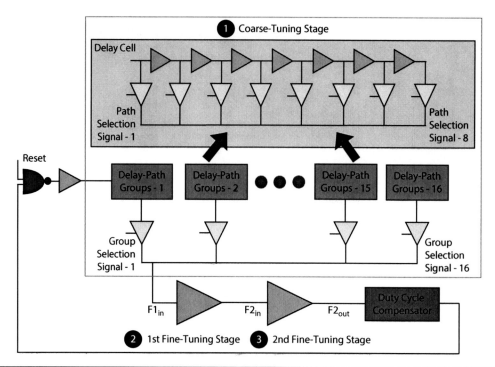

Figure 4.70 Block diagram of the DCO [86].

causes speeding-up of the track-DCO. The ADPLL controller helps in controlling the DCOs based on Lead and Lag signals. The ADPLL controller generates two 16-bit outputs "track code" and "average code." The track code controls the track-DCO to generate the clk_{FB} signal. The average code controls the average-DCO to generate the clk_{out} signal. The average-DCO helps to generate low-jitter output clock.

The block-diagram representation of the DCO used in the above ADPLL is shown in Fig. 4.70 [86]. The DCO has three stages as follows: (1) coarse-tuning stage, (2) first fine-tuning stage, and (3) second fine-tuning stage. The coarse-tuning stage has 128 (i.e., 16 × 8) different paths. There are 16 delay path groups and each group has eight delay paths. Each delay path or delay cell is designed using buffers and tristate buffers working as multiplexers. The path selection signal selects one path in a specific delay path group of the coarse-tuning stage. The group selection signal selects one group in the coarse-tuning stage. The first fine-tuning stage of the DCO is made up of 32 shunted tri-state buffers and inverters. The frequency of the output of the first fine-tuning stage depends on the number of ON-state tri-state buffers. The output frequency increases with the increase in the number of ON-state tri-state buffers. The second fine-tuning stage is made up of 32 three-input NOR gates. The second fine-tuning stage can improve the resolution of the DCO as the gate capacitance of NOR gate changing causing the changes in the delay.

4.15 Delay-Locked Loop

4.15.1 Basics

The DLL is a system similar to PLL in which the VCO component is replaced with a delay line [60, 96]. A delay line consists of a chain of variable delay elements that are controlled by either voltages or currents. The delay is adjusted by changing effective resistance and capacitance, which in turn, changes the RC time constant. The delay line can be either analog or digital in nature. A simple delay line can be designed from buffers or inverters [96]. The DLLs have better stability and low jitter compared to the PLLs. The DLLs can be used for various applications including clock recovery, enhance clock-timing characteristics, and change phase of clock signals.

Phase-Locked Loop Component Circuits 207

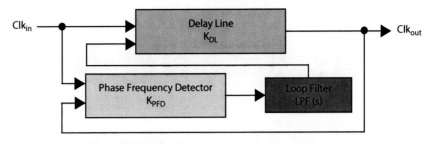

FIGURE 4.71 Block diagram of a basic DLL [96].

The block diagram of a simple DLL is shown in Fig. 4.71 [96]. In this DLL, the PFD compares the phase of the reference input signal and delay line output signal. A signal is produced by the PFD that is proportionate to the phase error. The LPF uses this error signal and produces voltage or current signal to adjust the delay of the delay line. Let us assume that the K_{PFD} is the gain of the PFD, LPF(s) is the transfer function of the LPF, and K_{DL} is the gain of the delay line. The open loop transfer function of the DLL is the following:

$$\text{DLL}_{OL}(s) = K_{PFD}K_{DL}\text{LPF}(s) \quad (4.46)$$

The first-order closed-loop transfer function is the following:

$$\text{DLL}_{CL}(s) = \left(\frac{1}{1+\left(\dfrac{s}{K_{PFD}K_{DL}\text{LPF}(s)}\right)} \right) \quad (4.47)$$

The DLLs can be either analog or digital in design [97]. The analog DLLs (ADLLs) have better jitter and noise performances; however, they have larger silicon area and longer locking time. The all-digital DLLs (ADDLLs) have shorter locking time, technology portability, and process insensitivity; however, they have poor noise performance.

4.15.2 An Analog DLL for Variable Frequency Generation

The block-level representation of an ADLL that can generate variable clock frequency is shown in Fig. 4.72 [65]. The ADLL has the following components: (1) phase detector, (2) CP with replica bias, (3) LPF, (4) voltage regulator, (5) voltage-controlled delay line (VCDL), (6) frequency multiplier (with transition detector and edge combiner), (7) multiplication factor controller (MFC), and (8) last-bit selector multiplexer. The phase detector compares reference clock signal and last-bit signal from the buffer at VCDL output. The replica bias controls the current in CP and that current along with gain of VCDL determines the bandwidth of the DLL. The LPF cleans the signal before the signal goes to the voltage regulator. The voltage regulator is made up of a two-stage current mirror. The voltage regulator adjusts the supply voltage of the VCDL, thus adjusting the delay of the VCDL. The output signal of the VCDL is provided as input to the frequency multiplier through a buffer. The programmable MFC generates signals for both the transition detector of the frequency multiplier as well as for the last-bit selector multiplexer. The transition detector of the frequency multiplier generates the pulses of short period on each rise of the input pulse. The edge combiner of the frequency multiplier combines these short pulses together while toggling the phase of output signal at every negative edge of the short pulses from the transition detector. Thus, the output clock signal of the frequency multiplier toggles at every rising edge of signal. The last-bit selector multiplexer selects the last bit from the buffered signal of the VCDL that connects to the phase detector according to the control signal. The frequency of the output clock signal generated from this ADLL is expressed as follows:

$$f_{clk,out} = \left(\frac{N_{MFC}}{2}\right) f_{clk,in} \quad (4.48)$$

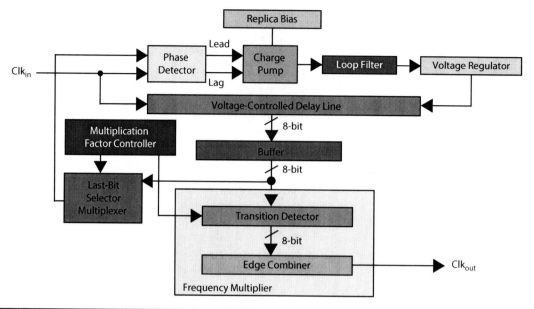

Figure 4.72 Block diagram of an DLL [65].

Technology	Voltage	Frequency	Cycle-to-Cycle Jitter	Power	Area
350-nm CMOS	3.3 V	0.12–1.8 GHz	1.8 ps at 1.3 GHz	86.6 mW at 1.6 GHz	0.07 mm²

Table 4.16 Characterization of the ADLL [65]

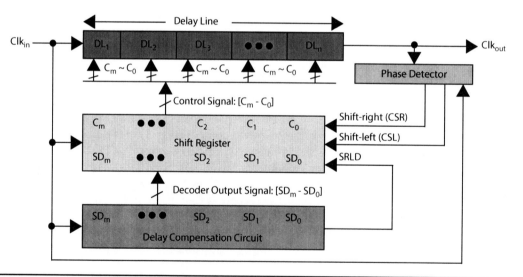

Figure 4.73 Block diagram of a DDLL [98].

where N_{MFC} is the multiplication factor that is selected dynamically by the MFC and $1 \leq N_{\text{MFC}} \leq N_{\text{stage}}$. N_{stage} is the number of delay cells in the VCDL. A selected characterization results for the ADLL is presented in Table 4.16 [65].

4.15.3 A Digital DLL

The architecture of a DDLL is shown in Fig. 4.73 [98]. This DDLL has the following components: (1) phase detector, (2) shift register, (3) delay compensation circuit (DCC), and (4) delay line. The phase detector is a simple DFF-based digital detector. The phase detector uses reference clock and

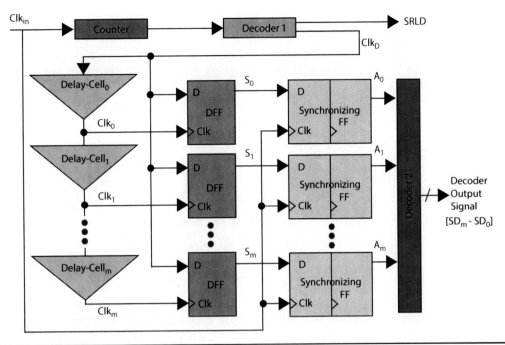

FIGURE 4.74 Block diagram of DCC the DDLL [98].

Technology	Supply Voltage	Frequency	Locking Time
130-nm CMOS	1.35 V	400 MHz	8 ns

TABLE 4.17 Characterization of the DDLL [98]

the delay line output clock to generate shift-right (CSR) and shift-left (CSL). The DLL is locked when both the shift-right (CSR) and shift-left (CSL) are at logic "0." The delay line consists of n-number of delay cell groups, $DL_1, DL_2, ..., DL_n$. Each delay cell group DL has a group of small delay cells, $D_1, D_2, ..., D_m$. The size of the small delay cells essentially determine the resolution of the DLL. The shift register takes the decoder output signal of the DCC $[SD_m-SD_0]$ as input and generates control signal $[c_m-c_0]$ for the delay line. The DCC uses the input clock signal and generates a control signal "SRLD" for the shift register. The structure of the DCC is shown in Fig. 4.74 [98]. It consists of counter, decoders, delay cells, DFF, and synchronizing flip-flops. The synchronizing flip-flops that are driven by input clock signal avoids the meta-stability condition in this digital DLL. A digital DLL, characterization data is presented in Table 4.17 [98].

4.16 Questions

4.1 Identify three applications of the PLL and discuss how the PLL operates in these applications.

4.2 Provide a block-level representation of an APLL and discuss the functionalities of each of the blocks.

4.3 What is clock jitter? What are the various sources of jitter? How is jitter calculated?

4.4 Provide a comparative discussion of RO and LC-tank oscillator.

4.5 Design a five-stage RO for a target application. Perform the simulation of a transistor-level schematic design for two different CMOS technology nodes. Characterize it for three metrics.

4.6 Perform the physical design of the above five-stage RO. Characterize and compare its characteristics with that of the transistor-level schematic design.

4.7 Perform the simulation and characterization of the above five-stage RO for the following configurations of the DG FinFET a specific technology node: shorted-gate, independent gates, and low-power.

4.8 Design an LC-tank oscillator for a target application. Perform the simulation of it transistor-level schematic design for two different CMOS technology nodes. Characterize it for three metrics.

4.9 Perform the physical design of the above LC-tank oscillator. Characterize and compare its characteristics with that of the transistor-level schematic design.

4.10 Perform the simulation and characterization of the above LC-tank oscillator for the following configurations of the DG FinFET a specific technology node: shorted-gate, independent gates, and low-power.

4.11 Design and simulate the memristor-based relaxation oscillator for a target application.

4.12 Design and simulate the memristor-based Schmitt trigger oscillator for a target application.

4.13 Design and simulate a DFF based PFD for the following technology: single-gate FET and DG FinFET for a specific technology node.

4.14 Design and simulate a CP for the following technology: single-gate FET and DG FinFET for a specific technology node.

4.15 Design and simulate a low-pass LPF for the following cutoff frequency: 800 MHz and 2.2 GHz.

4.16 Design and simulate a DFF-based frequency divider for the following technology: single-gate FET and DG FinFET for a specific technology node, for dividing factors of 2 and 4.

4.17 Perform design and simulation of a CORDIC-NCO using design technology of your choice.

4.18 Perform design and simulation of a DCO using design technology of your choice.

4.19 Perform design and simulation of a frequency multiplier using design technology of your choice.

4.20 Perform design and simulation of a DCC using design technology of your choice.

4.17 References

1. All Digital Phase-Locked Loop (ADPLL). http://www.altera.com/products/ip/altera/t-alt-adpll.html. Accessed on 25th June 2013.
2. LC Oscillator. http://www.electronics-tutorials.ws/oscillator/oscillators.html. Accessed on 18th June 2013.
3. Using Eye Diagrams. http://na.tm.agilent.com/plts/help/WebHelp/Analyzing/Analyzing_Data_using_Eye_Diagrams.html. Accessed on 13th June 2013.
4. Wolfson Frequency Locked Loop (FLL). http://www.wolfsonmicro.com/documents/uploads/misc/en/WAN0209.pdf (2009). Accessed on 25th June 2013.
5. Special Issue on Memristors: Theory and Applications (2013). DOI 10.1109/MCAS.2013.2256252.
6. Abidi, A.A.: Phase Noise and Jitter in CMOS Ring Oscillators. *IEEE Journal of Solid-State Circuits* **41**(8), 1803–1816 (2006). DOI 10.1109/JSSC.2006.876206.
7. Aktas, A., Ismail, M.: *CMOS PLLs and VCOs for 4G Wireless*. Kluwer Academic Publishers, Norwell, MA, USA (2004).
8. Apostolidou, M., Baltus, P.G.M., Vaucher, C.S.: Phase Noise In Frequency Divider Circuits. In: *IEEE International Symposium on Circuits and Systems, 2008 (ISCAS 2008)*, pp. 2538–2541 (2008). DOI 10.1109/ISCAS.2008.4541973.
9. Appenzeller, J.: Carbon Nanotubes for High-Performance Electronics: Progress and Prospect. *Proceedings of the IEEE* **96**(2), 201–211 (2008). DOI 10.1109/JPROC.2007.911051.
10. Azais, F., Ivanov, A., Renovell, M., Bertrand, Y.: A Methodology and Design for Effective Testing of Voltage Controlled Oscillators (VCOs). In: *Proceedings of the Seventh Asian Test Symposium*, pp. 383–387 (1998).
11. Baker, R.J.: CMOS Circuit Design, Layout and Simulation, 2nd edn. Wiley-IEEE (2008).
12. Banerjee, D.: PLL Fundamentals – Part 1: PLL Building Blocks. http://sva.ti.com/assets/en/other/National_PLL_Building_Blocks.pdf (2009). Accessed on 12th June 2013.
13. Best, R.E.: *Phase-Locked Loops: Design, Simulation and Applications*. McGraw-Hill (2007).
14. Brennan, P.V., Jiang, D., Wang, H.: Memory-Controlled Frequency Divider For Fractional-N Synthesisers. *Electronics Letters* **42**(21), 1202–1203 (2006).
15. Bunch, R.L., Raman, S.: Large-Signal Analysis of MOS Varactors in CMOS $-G_m$ LC VCOs. *IEEE Journal of Solid-State Circuits* **38**(8), 1325–1332 (2003). DOI 10.1109/JSSC.2003.814416.
16. Casaleiro, J., Lopes, H., Oliveira, L.B., Fernandes, J.R., Silva, M.M.: A 1 mW Low Phase-Noise Relaxation Oscillator. In: *Proceedings of the IEEE International Symposium on Circuits and Systems (ISCAS)*, pp. 1133–1136 (2011). DOI 10.1109/ISCAS.2011.5937770.

17. Chaudhuri, S., Jha, N.K.: 3D vs. 2D Analysis of FinFET Logic Gates Under Process Variations. In: *Proceedings of the 29th International Conference on Computer Design*, pp. 435–436 (2011).
18. Chaudhuri, S., Mishra, P., Jha, N.K.: Accurate Leakage Estimation for FinFET Standard Cells Using the Response Surface Methodology. In: *Proceedings of the 25th International Conference on VLSI Design*, pp. 238–244 (2012).
19. Chen, Z., Appenzeller, J., Lin, Y.M., Sippel-Oakley, J., Rinzler, A.G., Tang, J., Wind, S.J., Solomon, P.M., Avouris, P.: An Integrated Logic Circuit Assembled on a Single Carbon Nanotube. *Science* **311**(5768), 1735–1735 (2006).
20. Chen, Z., Appenzeller, J., Solomon, P.M., Lin, Y.M., Avouris, P.: High Performance Carbon Nanotube Ring Oscillator. In: *Proceedings of the 64th Device Research Conference*, pp. 171–172 (2006). DOI 10.1109/DRC.2006.305170.
21. Cheng, S., Tong, H., Silva-Martinez, J., Karsilayan, A.I.: A Fully Differential Low-Power Divide-by-8 Injection-Locked Frequency Divider Up to 18 GHz. *IEEE Journal of Solid-State Circuits* **42**(3), 583–591 (2007). DOI 10.1109/JSSC.2006.891448.
22. Collaert, N., Dixit, A., Goodwin, M., Anil, K.G., Rooyackers, R., Degroote, B., Leunissen, L.H.A., Veloso, A., Jonckheere, R., De Meyer, K., Jurczak, M., Biesemans, S.: A Functional 41-Stage Ring Oscillator Using Scaled FinFE Devices With 25-nm Gate Lengths and 10-nm Fin Widths Applicable for the 45-nm CMOS Node. *IEEE Electron Device Letters* **25**(8), 568–570 (2004). DOI 10.1109/LED.2004.831585.
23. Corinto, F., Ascoli, A., Gilli, M.: Nonlinear Dynamics of Memristor Oscillators. *IEEE Transactions on Circuits and Systems I: Regular Papers* **58**(6), 1323–1336 (2011). DOI 10.1109/TCSI.2010.2097731.
24. Demir, A., Mehrotra, A., Roychowdhury, J.: Phase Noise in Oscillators: A Unifying Theory and Numerical Methods for Characterization. *IEEE Transactions on Circuits and Systems I: Fundamental Theory and Applications*, **47**(5), 655–674 (2000). DOI 10.1109/81.847872.
25. Denier, U.: Analysis and Design of an Ultralow-Power CMOS Relaxation Oscillator. *IEEE Transactions on Circuits and Systems I: Regular Papers* **57**(8), 1973–1982 (2010). DOI 10.1109/TCSI.2010.2041504.
26. Deo, S., Menon, S., Nallathambhi, S., Soderstrand, M.A.: Improved Numerically-Controlled Digital Sinusoidal Oscillator. In: *Proceedings of the IEEE 45th Midwest Symposium on Circuits and Systems*, vol. 2, pp. II–211–II–214 (2002). DOI 10.1109/MWSCAS.2002.1186835.
27. Eken, Y.A.: *High Frequency Voltage Controlled Oscillators in Standard CMOS*. Ph.D. thesis, School of Electrical and Computer Engineering, Georgia Institute of Technology, GA, USA (2003). Accessed on 11th June 2013.
28. Elwakil, A.S., Murtada, M.A.: All Possible Canonical Second-Order Three-Impedance Class-A and Class-B Oscillators. *Electronics Letters* **46**(11), 748–749 (2010). DOI 10.1049/el.2010.0082.
29. Fei, W., Yu, H., Zhang, W., Yeo, K.S.: Design Exploration of Hybrid CMOS and Memristor Circuit by New Modified Nodal Analysis. *IEEE Transactions on Very Large Scale Integration (VLSI) Systems* **20**(6), 1012–1025 (2011). DOI 10.1109/TVLSI.2011.2136443.
30. Fischette, D.: First Time, Every Time Practical Tips for Phase-Locked Loop Design. http://www.delroy.com/PLL_dir/tutorial/PLL_tutorial_slides.pdf (2009). Accessed on 10th June 2013.
31. Fouda, M.E., Khatib, M.A., Mosad, A.G., Radwan, A.G.: Generalized Analysis of Symmetric and Asymmetric Memristive Two-gate Relaxation Oscillators. *IEEE Transactions on Circuits and Systems I: Regular Papers* **60**(10), 2701–2708 (2014) DOI 10.1109/TCSI.2013.2249172.
32. Frederic, C., Gilles, J.: Tutorial 1. Fundamentals About PLL. http://www.same-conference.org/same_2008/images/documents/SLIDE/TECHNICAL/Tutorial_1_TI_Polytech.pdf (2008). Accessed on 09th June 2013.
33. Galton, I., Razavi, B., Cowles, J., Kinget, P.: CMOS Phase-Locked Loops for Frequency Synthesis. In: *Proceedings of the IEEE International Solid-State Circuits Conference Digest of Technical Papers (ISSCC)*, pp. 521–521 (2010). DOI 10.1109/ISSCC.2010.5433853.
34. Garitselov, O.: *Metamodeling-Based Fast Optimization Of Nanoscale AMS-SoCs*. Ph.D. thesis, Computer Science and Engineering, University Of North Texas, Denton, TX, Denton, TX 76207 (2012).
35. Garitselov, O., Mohanty, S.P., Kougianos, E.: A Comparative Study of Metamodels for Fast and Accurate Simulation of Nano-CMOS Circuits. *IEEE Transactions on Semiconductor Manufacturing* **25**(1), 26–36 (2012).
36. Garitselov, O., Mohanty, S.P., Kougianos, E.: Accurate Polynomial Metamodeling-Based Ultra-Fast Bee Colony Optimization of a Nano-CMOS PLL. *Journal of Low Power Electronics* **8**(3), 317–328 (2012).
37. Garitselov, O., Mohanty, S.P., Kougianos, E., Okobiah, O.: Metamodel-Assisted Ultra-Fast Memetic Optimization of a PLL for WiMax and MMDS Applications. In: *Proceedings of the 13th International Symposium on Quality Electronic Design*, pp. 580–585 (2012).
38. Gates, E.: *Introduction to Electronics*. Delmar Cengage Learning, Clifton Park, NY 12065 (2011). URL http://books.google.com/books?id=JS_c1ttnwn8C.
39. Ghai, D.: *Variability Aware Low-Power Techniques for Nanoscale Mixed-Signal Circuits*. Ph.D. thesis, Department of Computer Science and Engineering, University of North Texas, Denton (2009).
40. Ghai, D., Mohanty, S.P., Kougianos, E.: Parasitic Aware Process Variation Tolerant Voltage Controlled Oscillator (VCO) Design. In: *Proceedings of the 9th International Symposium on Quality of Electronic Design*, pp. 330–333 (2008).
41. Ghai, D., Mohanty, S.P., Kougianos, E.: Design of Parasitic and Process-Variation Aware Nano-CMOS RF Circuits: A VCO Case Study. *IEEE Trans. VLSI Syst.* **17**(9), 1339–1342 (2009).

42. Ghai, D., Mohanty, S.P., Kougianos, E.: Unified P4 (power-performance-process-parasitic) Fast Optimization of a Nano-CMOS VCO. In: *Proceedings of the 19th ACM Great Lakes symposium on VLSI*, pp. 303–308. ACM (2009).
43. Ghai, D., Mohanty, S.P., Thakral, G.: Comparative Analysis of Double Gate FinFET Configurations for Analog Circuit Design. In: *Proceedings of the IEEE 56th International Midwest Symposium on Circuits and Systems (MWSCAS)*, pp. 809–812 (2013).
44. Ghai, D., Mohanty, S.P., Thakral, G.: Double Gate FinFET Based Mixed-Signal Design: A VCO Case Study. In: *Proceedings of the IEEE 56th International Midwest Symposium on Circuits and Systems (MWSCAS)*, pp. 177–180 (2013).
45. Ghai, D., Mohanty, S.P., Thakral, G.: Fast Analog Design Optimization Using Regression-Based Modeling and Genetic Algorithm: A Nano-CMOS VCO Case Study. In: *Proceedings of the International Symposium on Quality Electronic Design*, pp. 406–411 (2013)
46. Ghai, D., Mohanty, S.P., Thakral, G.: Fast Optimization Of Nano-CMOS Voltage-Controlled Oscillator Using Polynomial Regression and Genetic Algorithm. *Microelectronics Journal* **44**(8), 631–641 (2013). DOI http://dx.doi.org/10.1016/j.mejo.2013.04.010. URL http://www.sciencedirect.com/science/article/pii/S0026269213001067.
47. Ghidini, C., Aranda, J.G., Gerna, D., Kelliher, K., Baumhof, C.: A Digitally Programmable On-Chip RC-Oscillator In 0.25 μm CMOS Logic Process. In: *Proceedings of the IEEE International Symposium on Circuits and Systems*, Vol. 1, pp. 400–403 (2005). DOI 10.1109/ISCAS.2005.1464609.
48. Gupta, S.C.: Phase-Locked Loops. *Proceedings of the IEEE* **63**(2), 291–306 (1975).
49. Lee, H., Choi, T., Mohammadi, S., Katehi, L.P.B.: An Extremely Low Power 2 GHz CMOS LC VCO for Wireless Communication Applications. In: *Proceedings of the European Conference on Wireless Technology*, pp. 31–34 (2005). DOI 10.1109/ECWT.2005.1617647.
50. Hajimiri, A., Limotyrakis, S., Lee, T.H.: Jitter and Phase Noise in Ring Oscillators. *IEEE Journal of Solid-State Circuits* **34**(6), 790–804 (1999). DOI 10.1109/4.766813.
51. Ham, D., Hajimiri, A.: Concepts and Methods in Optimization of Integrated LC VCOs. *IEEE Journal of Solid-State Circuits* **36**(6), 896–909 (2001). DOI 10.1109/4.924852.
52. Hancock, J.: Jitter Understanding It, Measuring It, Eliminating It Part 1: Jitter Fundamentals. http://www.highfrequencyelectronics.com/Archives/Apr04/HFE0404_Hancock.pdf (2004). Accessed on 13th June 2013.
53. Herzel, F., Razavi, B.: A Study of Oscillator Jitter Due to Supply and Substrate Noise. *IEEE Transactions on Circuits and Systems II: Analog and Digital Signal Processing* **46**(1), 56–62 (1999). DOI 10.1109/82.749085.
54. Hosaka, K., Harase, S., Izumiya, S., Adachi, T.: A Cascode Crystal Oscillator Suitable for Integrated Circuits. In: *Proceedings of the IEEE International Frequency Control Symposium and PDA Exhibition*, pp. 610–614 (2002). DOI 10.1109/FREQ.2002.1075954.
55. Hsieh, G.C., Hung, J.C.: Phase-Locked Loop Techniques: A Survey. *IEEE Transactions on Industrial Electronics* **43**(6), 609–615 (1996). DOI 10.1109/41.544547.
56. Hsieh, M., Sobelman, G.E.: Comparison of LC and Ring VCOs for PLLs in a 90 nm Digital CMOS Process. http://mountains.ece.umn.edu/sobelman/papers/mthsieh_isocc06.pdf (2010). Accessed on 10th June 2013.
57. Jacquemod, G., Geynet, L., Nicolle, B., de Foucauld, E., Tatinian, W., Vincent, P.: Design and Modelling of A Multi-Standard Fractional PLL in CMOS/SOI Technology. *Microelectronics Journal* **39**(9), 1130–1139 (2008). DOI 10.1016/j.mejo.2008.01.069. URL http://www.sciencedirect.com/science/article/pii/S0026269208000992.
58. Jiang, S., You, F., He, S.: A Wideband Sigma-Delta PLL Based Phase Modulator with Pre-Distortion Filter. In: *Proceedings of the International Conference on Microwave and Millimeter Wave Technology (ICMMT)*, vol. 4, pp. 1–4 (2012). DOI 10.1109/ICMMT.2012.6230316.
59. Joeres, S., Kruth, A., Meike, O., Ordu, G., Sappok, S., Wunderlich, R., Heinen, S.: Design of a Ring-Oscillator with a Wide Tuning Range in 0.13 μm CMOS for the use in Global Navigation Satellite Systems. In: *Proceedings of the Annual Workshop on Circuits, Systems and Signal Processing (ProRISC)*, pp. 529–535 (2004).
60. Jung, Y.J., Lee, S.W., Shim, D., Kim, W., Kim, C., Cho, S.I.: A Dual-Loop Delay-Locked Loop Using Multiple Voltage-Controlled Delay Lines. *IEEE Journal of Solid-State Circuits* **36**(5), 784–791 (2001). DOI 10.1109/4.918916.
61. Kadam, S., Sasidaran, D., Awawdeh, A., Johnson, L., Soderstrand, M.: Comparison of Various Numerically Controlled Oscillators. In: *Proceedings of the IEEE 45th Midwest Symposium on Circuits and Systems*, vol. 3, pp. III–200–III–202 (2002). DOI 10.1109/MWSCAS.2002.1187005.
62. Kang, J.K., Kim, D.H.: A CMOS Clock and Data Recovery with Two-XOR Phase-Frequency Detector Circuit. In: *Proceedings of the IEEE International Symposium on Circuits and Systems*, vol. 4, pp. 266–269 (2001). DOI 10.1109/ISCAS.2001.922223.
63. Kang, S., Leblebici, Y.: *CMOS Digital Inegrated Circuits*, 3rd edn. McGraw-Hill, New York (2003).
64. Kaya, S., Kulkarni, A.: A Novel Voltage-Controlled Ring Oscillator Based On Nanoscale DG-MOSFETs. In: *Proceedings of International Conference on Microelectronics*, pp. 417–420 (2008). DOI 10.1109/ICM.2008.5393792.
65. Kim, J.H., Kwak, Y.H., Kim, M., Kim, S.W., Kim, C.: A 120-MHz-1.8-GHz CMOS DLL-Based Clock Generator for Dynamic Frequency Scaling. *IEEE Journal of Solid-State Circuits* **41**(9), 2077–2082 (2006). DOI 10.1109/JSSC.2006.880609.

66. Kougianos, E., Mohanty, S.P.: Impact of Gate-Oxide Tunneling on Mixed-Signal Design and Simulation of A Nano-CMOS VCO. *Microelectronics Journal* **40**(1), 95–103 (2009).
67. Kranti, A., Armstrong, G.A.: Design and Optimization of FinFETs for Ultra-Low-Voltage Analog Applications. *IEEE Transactions on Electron Devices* **54**(12), 3308–3316 (2007).
68. Lee, T.H., Abshire, P.A.: Design Methodology for a Low-Frequency Current-Starved Voltage-Controlled Oscillator with a Frequency Divider. In: *Proceedings of the IEEE 55th International Midwest Symposium on Circuits and Systems (MWSCAS)*, pp. 646–649 (2012). DOI 10.1109/MWSCAS.2012.6292103.
69. Leeson, D.B.: A Simple Model of Feedback Oscillator Noise Spectrum. *Proceedings of the IEEE* **54**(2), 329–330 (1966). DOI 10.1109/PROC.1966.4682.
70. Li, W., Meiners, J.: Introduction to Phase-Locked Loop System Modeling. *TI Analog Applications Journal* **SLYT015**, 198–212 (2000)
71. Loke, A.L.S., Barnes, R.K., Wee, T.T., Oshima, M.M., Moore, C.E., Kennedy, R.R., Gilsdorf, M.J.: A Versatile 90-nm CMOS Charge-Pump PLL for SerDes Transmitter Clocking. *IEEE Journal of Solid-State Circuits* **41**(8), 1894–1907 (2006). DOI 10.1109/JSSC.2006.875289.
72. Mansour, M.M., Mansour, M.M.: On the Design of Low Phase-Noise CMOS LC-tank Oscillators. In: *Proceedings of the International Conference on Microelectronics*, pp. 407–412 (2008). DOI 10.1109/ICM.2008.5393840.
73. Mansuri, M., Liu, D., Yang, C.K.K.: Fast Frequency Acquisition Phase-Frequency Detectors for Gsamples/s Phase-Locked Loops. *IEEE Journal of Solid-State Circuits* **37**(10), 1331–1334 (2002). DOI 10.1109/JSSC.2002.803048.
74. Mathis, W., Bremer, J.: Modelling and Design Concepts for Electronic Oscillators and Its Synchronization. *The Open Cybernetics and Systemics Journal* **3**, 47–60 (2009). Accessed on 11th June 2013.
75. Min, S., Copani, T., Kiaei, S., Bakkaloglu, B.: A 90-nm CMOS 5-GHz Ring-Oscillator PLL with Delay-Discriminator-Based Active Phase-Noise Cancellation. *IEEE Journal of Solid-State Circuits* **48**(5), 1151–1160 (2013). DOI 10.1109/JSSC.2013.2252515.
76. Miyazaki, T., Hashimoto, M., Onodera, H.: A Performance Comparison Of PLLs For Clock Generation Using Ring Oscillator VCO And LC Oscillator in A Digital CMOS Process. In: *Proceedings of the Asia and South Pacific Design Automation Conference*, pp. 545–546 (2004). DOI 10.1109/ASPDAC.2004.1337641.
77. Mohanty, S.P.: Memristor: From Basics to Deployment. *IEEE Potentials* **32**(3), 34–39 (2013). DOI 10.1109/MPOT.2012.2216298.
78. Mohanty, S.P., Kougianos, E., Garitselov, O., Molina, J.M.: Polynomial-Metamodel Assisted Fast Power Optimization of Nano-CMOS PLL Components. In: *Proceeding of the Forum on Specification and Design Languages*, pp. 233–238 (2012).
79. Nonis, R., Da Dalt, N., Palestri, P., Selmi, L.: Modeling, Design and Characterization of A New Low-Jitter Analog Dual Tuning LC-VCO PLL Architecture. *IEEE Journal of Solid-State Circuits* **40**(6), 1303–1309 (2005). DOI 10.1109/JSSC.2005.848037.
80. Nowak, E.J., Ludwig, T., Aller, I., Kedzierski, J., Leong, M., Rainey, B., Breitwisch, M., Gernhoefer, V., Keinert, J., Fried, D.M.: Scaling Beyond the 65 nm Node with FinFET-DGCMOS. In: *Proceedings of the IEEE Custom Integrated Circuits Conference*, pp. 339–342 (2003). DOI 10.1109/CICC.2003.1249415.
81. Palumbo, G., Pappalardo, D.: Charge Pump Circuits: An Overview on Design Strategies and Topologies. *IEEE Circuits and Systems Magazine* **10**(1), 31–45 (2010). DOI 10.1109/MCAS.2009.935695.
82. Perrott, M.H.: Tutorial on Digital Phase-Locked Loops. http://cppsim.org/PLL_Lectures/digital_pll_cicc_tutorial_perrott.pdf (2009). Accessed on 18th June 2013.
83. Razavi, B.: Design of Millimeter-Wave CMOS Radios: A Tutorial. *IEEE Transactions on Circuits and Systems I: Regular Papers* **56**(1), 4–16 (2009). DOI 10.1109/TCSI.2008.931648.
84. Safarian, A.Q., Heydari, P.: A Study of High-Frequency Regenerative Frequency Dividers. In: *Proceedings of the IEEE International Symposium on Circuits and Systems*, pp. 2695–2698 (2005). DOI 10.1109/ISCAS.2005.1465182.
85. Sarivisetti, G.: *Design and Optimization of Components in a 45 nm CMOS Phase Locked Loop*. Master's thesis, Computer Science and Engineering, University of North Texas, Denton (2006). URL http://books.google.com/books?id=U0m3uAAACAAJ.
86. Sheng, D., Chung, C.C., Lee, C.Y.: An All-Digital Phase-Locked Loop with High-Resolution for SoC Applications. In: *Proceedings of the International Symposium on VLSI Design, Automation and Test*, pp. 1–4 (2006). DOI 10.1109/VDAT.2006.258161.
87. Solanke, S.V.: *Design and Analysis of Novel Charge Pump Architecture for Phase Locked Loop*. Master of technology thesis, National Institute Of Technology, Rourkela, Rourkela, India (2009).
88. Sreeja, B.S., Radha, S.: Optimized 2.4 GHz Voltage Controlled Oscillator with a High-Q MWCNT Network-Based Pulse-Shaped Inductor. *Microelectronics Journal* **43**(1), 1–12 (2012). DOI 10.1016/j.mejo.2011.10.005. URL http://dx.doi.org/10.1016/j.mejo.2011.10.005.
89. Srivastava, A., Xu, Y., Liu, Y., Sharma, A.K., Mayberry, C.: CMOS LC Voltage Controlled Oscillator Design Using Carbon Nanotube Wire Inductors. *ACM Journal of Emerging Technologies in Computing Systems* **8**(3), 15:1–15:9 (2012).
90. Staszewski, R.B., Balsara, P.T.: All-Digital PLL With Ultra Fast Settling. *IEEE Transactions on Circuits and Systems II: Express Briefs* **54**(2), 181–185 (2007). DOI 10.1109/TCSII.2006.886896.

91. Sun, W., Li, C., Yu, J.: A Simple Memristor Based Chaotic Oscillator. In: *Proceeding of the International Conference on Communications, Circuits and Systems*, pp. 952–954 (2009). DOI 10.1109/ICCCAS.2009.5250350.
92. Tiebout, M.: Low-Power Low-Phase-Noise Differentially Tuned Quadrature VCO Design in Standard CMOS. *IEEE Journal of Solid-State Circuits* **36**(7), 1018–1024 (2001). DOI 10.1109/4.933456.
93. Wang, Z.: An Analysis of Charge-Pump Phase-Locked Loops. *IEEE Transactions on Circuit and Systems-I: Fundamental Theory and Applications* **52**(10), 2128–2138 (2005).
94. Wu, T., Hanumolu, P.K., Mayaram, K., Moon, U.K.: Method for a Constant Loop Bandwidth in LC-VCO PLL Frequency Synthesizers. *IEEE Journal of Solid-State Circuits* **44**(2), 427–435 (2009). DOI 10.1109/JSSC.2008.2010792.
95. Xia, L., Chen, H., Huang, Y., Hong, Z., Chiang, P.: 100-Phase, Dual-Loop Delay-Locked Loop for Impulse Radio Ultra-Wideband Coherent Receiver Synchronisation. *IET Circuits, Devices Systems* **5**(6), 484–493 (2011). DOI 10.1049/iet-cds.2011.0112.
96. Yang, C.K.K.: Delay-Locked Loops An Overview. Wiley-IEEE Press (2003). Accessed on 26th June 2013.
97. Yang, R.J., Liu, S.I.: A 2.5 GHz All-Digital Delay-Locked Loop in 0.13 μm CMOS Technology. *IEEE Journal of Solid-State Circuits* **42**(11), 2338–2347 (2007). DOI 10.1109/JSSC.2007.906183.
98. Ye, B., Li, T., Han, X., Luo, M.: A Fast-Lock Digital Delay-Locked Loop Controller. In: *Proceedings of the 8th IEEE International Conference on ASIC*, pp. 809–812 (2009). DOI 10.1109/ASICON.2009.5351573.
99. Yoon, S.: *LC-tank CMOS Voltage-Controlled Oscillators Using High Quality Inductors Embedded in Advanced Packaging Technologies*. Ph.D. thesis, Electrical and Computer Engineering, Georgia Institute of Technology, GA, USA (2004).
100. Zhang, X., Apsel, A.B.: A Low-Power, Process-and-Temperature–Compensated Ring Oscillator with Addition-Based Current Source. *IEEE Transactions on Circuits and Systems I: Regular Papers* **58**(5), 868–878 (2011). DOI 10.1109/TCSI.2010.2092110.
101. Zhao, J.: *A Low Power CMOS Design of an All Digital Phase Locked Loop*. Ph.D. thesis, Department of Electrical and Computer Engineering, Northeastern University, Boston (2011). Accessed on 28th June 2013.
102. Zheng, G., Mohanty, S.P., Kougianos, E., Garitselov, O.: Verilog-AMS-PAM: Verilog-AMS integrated with Parasitic-Aware Metamodels for Ultra-Fast and Layout-Accurate Mixed-Signal Design Exploration. In: *Proceedings of the ACM Great Lakes Symposium on VLSI*, pp. 351–356 (2012).
103. Zheng, G., Mohanty, S.P., Kougianos, E., Okobiah, O.: Polynomial Metamodel Integrated Verilog-AMS for Memristor-Based Mixed-Signal System Design. In: *Proceedings of the IEEE 56th International Midwest Symposium on Circuits and Systems (MWSCAS)*, pp. 916–919 (2013).
104. Zhou, Q., Li, L., Chen, G.: An RC Oscillator with Temperature Compensation for Accurate Delay Control in Electronic Detonators. In: *Proceedings of the IEEE International Conference of Electron Devices and Solid-State Circuits*, pp. 274–277 (2009). DOI 10.1109/EDSSC.2009.5394266.
105. Zidan, M.A., Omran, H., Radwan, A.G., Salama, K.N.: Memristor-Based Reactance-Less Oscillator. *Electronics Letters* **47**(22), 1220–1221 (2011). DOI 10.1049/el.2011.2700.

CHAPTER 5
Electronic Signal Converter Circuits

5.1 Introduction

In general, signals are symbols or values with some ordering [15]. A signal is a function that conveys information or transfers energy [63]. At almost every point of life, everyone handles one or the other form of electronic signal. The signals may either be available naturally or synthesized. The signal concept is depicted in Fig. 5.1. An electronic signal may be in video, image, or audio format used for entertainment as well as exchange of information. Signals may be the power supply signal in the form of alternating current or direct current. They may be communication signals in mobile communications, Wi-Fi, or Bluetooth. These signals are of different types such as visible, nonvisible, or sound. They are of different dimensions, such as 1D audio, 2D images, or 3D video. The signals may have different periods or clock cycle time. They can be of various frequencies such as radiofrequency and audio frequency. The signals may be of diverse shapes, e.g., square, sawtooth, or sine. The electronic signals may be either analog or digital in nature as depicted in Fig. 5.2. Analog signals are continuous in both value and time [15]. The discrete-time signals have values only at certain time stamps, i.e., continuous in value and discrete in time. The discrete-amplitude signals

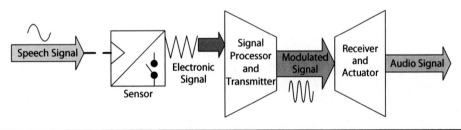

FIGURE 5.1 Concept of a typical signal.

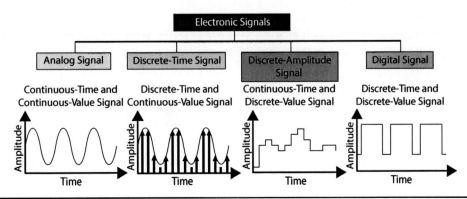

FIGURE 5.2 Different types of electronic signals.

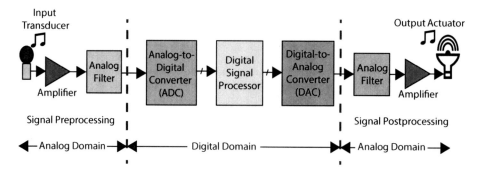

Figure 5.3 Digital signal processing using ADC and DAC.

have only discrete values, i.e., discrete in value and continuous in time. The digital signals have discrete values and discrete time. Both analog and digital electronic signals are encountered and processed all the time. Digital and discrete signals are obtained by sampling analog signals.

The analog and digital signals are processed by different electronic circuits and systems. The analog circuits, such as amplifiers and analog filters, process analog signals. The digital circuits and systems, such as digital signal processors, process digital signals. All modern microprocessors that perform most of the computations operate on digital signals. It is necessary to convert signals between analog and digital to perform the computations in digital while interface using analog signals. The analog-to-digital converter (ADC) and digital-to-analog converter (DAC) perform such signal conversions [41, 58]. A real-life example of the use of ADCs and DACs is presented in Fig. 5.3 [76]. Signal preprocessing and postprocessing are performed in analog domain using amplifiers and filters. The signal processing is performed at digital domain using a digital signal processor [44].

5.2 Types of Electronic Signal Converters

In a high-level perspective, different types of signal converters, which are used in an analog/mixed-signal system-on-a-chip (AMS-SoC) for various purposes, are presented in Fig. 5.4. The signal converters may convert data signal and power signal. The data converters (more often called signal converters) are either ADCs or DACs. The DC-to-DC voltage-level converters (VLCs) also knowns as voltage-level shifters (VLSs) convert one voltage-level to another and are used in AMS-SoCs.

5.2.1 Concrete Applications

As a concrete example of the applications of an ADC, the digital camera system is presented in Fig. 5.5 [8, 27, 29, 32, 59]. This system may be considered as a representation of the other similar portable electronic systems, e.g., mobile phones, satellite phones, and health-monitoring systems that are in essence AMS-SoCs, used in day-to-day life. In the digital camera, the ADC converts the analog picture sensed by the sensing element through the optical system to digital images at maximum possible speed. The situation is the same in digital video capture cards and TV tuner cards used in PCs [50]. The ADCs are deployed in the integration as low-data rate analog components for complete SoC technology-based applications. High-speed ADCs are used in digital storage oscilloscopes as well as in emerging systems such as software-defined radio. When analog recording is used for music, the ADCs convert music to the pulse-code modulation (PCM) data streams that are eventually saved as

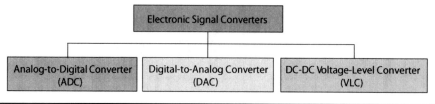

Figure 5.4 Different types of signal converters or data converters.

Electronic Signal Converter Circuits 217

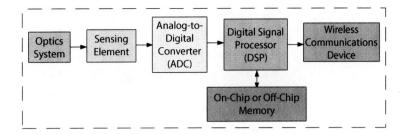

Figure 5.5 A concrete application of ADC: A CMOS sensor-based digital camera.

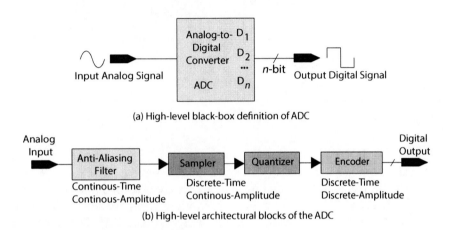

Figure 5.6 High-level representation of the ADC.

digital music files or burned onto compact discs. In summary, the data converters or signal converters are essential circuits in SoC designs that bridge the gap between the analog and digital world as an interfacing element.

5.2.2 Signal Converter Types

The high-level idea of the ADC is depicted in Fig. 5.6. An ADC considers an analog signal as input and generates a digital signal as output. The typical blocks of an ADC include the following: anti-alias filtering, sampler, quantizer, and encoder [32, 54, 62]. The anti-aliasing filter is used before the sampler to protect the information in the signals and reject interferences outside the target frequency band. The sampler cannot accurately sample the analog input signal if the input signal frequency is higher than the sample rate. In such a situation, aliasing occurs that makes the different signals indistinguishable, i.e., aliases of each other when sampled. The Nyquist-Shannon sampling theorem introduces a minimum bound on the sampling rate, which is called Nyquist rate or Nyquist frequency [16, 85]. The anti-aliasing filter essentially ensures that the input signal frequency at the input of the sampler is less than half of the sampling frequency. The sampler converts continuous-time analog signals to sampled discrete-time signals while preserving all the information contained in the analog input signals. The quantizer discretizes the values of the sampled signal from the sampler, thus converting to discrete-time and discrete-amplitude signal. The encoder at the final stage converts the digital signals to binary formats.

The high-level idea of the DAC is depicted in Fig. 5.7 [54, 62, 88]. The DAC considers a digital signal of certain bit size and converts it to an analog signal. The typical blocks of an ADC include the following: transcoder, sample and hold, and reconstruction filter. The transcoder converts digital input signal to an equivalent analog signal. The sample and hold component of the DAC captures (or samples) the signal at a constant value and locks (or holds) the value for a specified time period, thus eliminating variations in the input signal that can potentially corrupt the signal conversion process. The reconstruction filter of the DAC filters the out-of-band noise in the analog output signal.

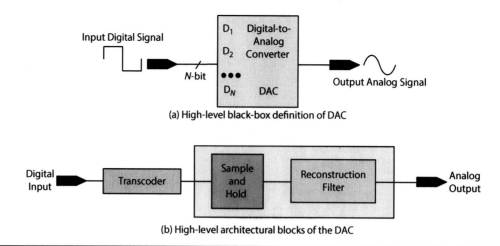

Figure 5.7 High-level representation of the DAC [54, 62].

5.3 Selected ADC Architectures: Brief Overview

5.3.1 Overview

An exhaustive list of different types of ADCs is presented in Fig. 5.8. This may not be a complete list of ADCs available in the literature. The ADCs are classified in terms of architectures, resolutions, and applications. The ADCs can be of three other types based on implementations such as hardware only, software only, and hardware-software hybrid type [23]. The ADCs can be of two categories depending on the sampling rate, undersampling, or oversampling converters [16, 74]. The undersampling ADCs sample the analog input signal below the Nyquist frequency. The oversampling ADCs sample the analog input signal at much higher frequencies than the Nyquist frequency. An example of oversampling ADC is a sigma-delta ($\Sigma - \Delta$) architecture. Selected ADCs will be discussed in detail in the rest of this chapter.

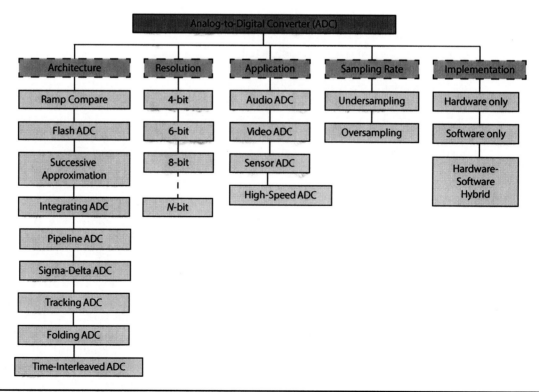

Figure 5.8 Different types of ADCs.

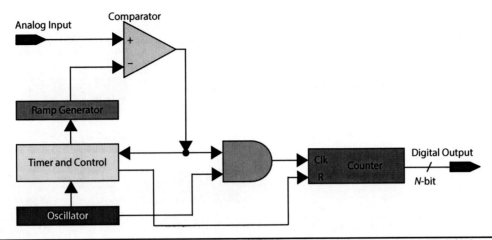

FIGURE 5.9 Architecture of the Ramp-Compare or Ramp Run-Up analog-to-digital converter (ADC) [41].

5.3.2 Ramp-Compare ADC or Ramp Run-Up ADC

The schematic representation of a ramp-compare or ramp run-up ADC is presented in Fig. 5.9 [25, 41]. In the ramp-compare ADC, a rising ramp waveform is generated by ramp generator and oscillator. The ramp waveform can also be generated using a DAC and counter. The input analog voltage is compared with the ramp waveform by the comparator. When the comparator outputs high, timer value is recorded, the counter value is generated, and then reset gets the signal. The counter output is essentially proportional to the input voltage. The accuracy of the ramp-compared ADC depends on the performance of the oscillator and the ramp generator. While the ramp-compare ADC is simpler in design, it is slow for higher input voltages as it takes longer to generate ramp in such situations.

5.3.3 Flash ADC or Direct Conversion ADC

The high-level architecture of the flash or direct conversion ADC is shown in Fig. 5.10 [21, 73, 84]. The bank of comparators in the flash ADC sample the input signals in a parallel fashion. There are $(2^n - 1)$ comparators in the comparator bank for an n-bit ADC. Each of the comparators fires at its specified switching voltages. The output of the comparator bank is read like that of a liquid thermometer and hence called thermometer code. The thermometer code or unary code or monotonic code is an entropy encoding in which there is one transition between 1 and 0. In thermometer code, a number is represented by n 1s followed by a "0"; for example, 5 is represented as 111110. Other

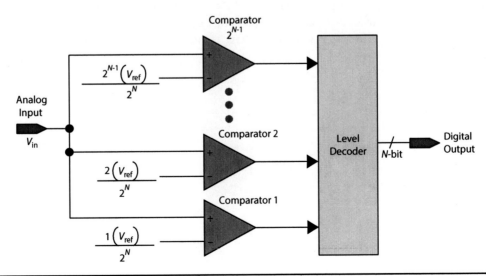

FIGURE 5.10 Architecture of the flash ADC [21].

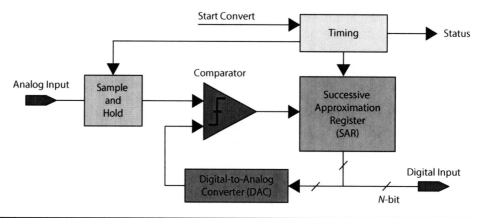

FIGURE 5.11 Architecture of the SA-ADC [21, 41].

forms of presentation, such as $n-1$ 1s followed by a "0" or n 0s followed by a "1," are possible. The level-decoder block of the flash ADC converts the thermometer code to binary form, the digital output. The flash ADC is a very fast circuit but is complex in design and manufacturing. The flash ADC will be discussed in a later part of this chapter in detail.

5.3.4 Successive-Approximation ADC

The schematic representation of successive-approximation ADC (SA-ADC) is shown in Fig. 5.11 [21, 41, 53, 93]. The SA-ADC can be thought of as an orthogonal of the flash ADC in terms of working principle. While the flash ADC uses many comparators in one cycle, the SA-ADC uses a single comparator for many cycles. In the SA-ADC in Fig. 5.11, the "Start Convert" signal is high and the sample-and-hold is kept at hold mode. In the same cycle, the most significant bit (MSB) of successive approximation register (SAR) is set to "1" with all the rest of the bits reset to "0." The DAC is driven by the output of the SAR. The comparator compares the sample-and-hold output and the DAC output. If the comparator output is high, when the sample-and-hold output (i.e., analog input) is higher than the DAC output, then the corresponding bit of SAR is set to "1." The process is repeated with each bit to complete the conversion. The SA-ADC is simple to implement. However, it needs an additional sample and hold amplifier compared to other ADCs. The SA-ADC is slow as it requires n comparison cycles for n-bit resolution as compared to one cycle of a flash ADC and few cycles of a pipelined ADC.

5.3.5 Integrating ADC

The architecture of a basic integrating ADC is presented in Fig. 5.12 [5, 28]. This basic integrating ADC consists of an integrator, a comparator, and a counter. The input analog voltage is integrated by the integrator. The integrated input voltage is then compared with a reference voltage (V_{ref}). Based on

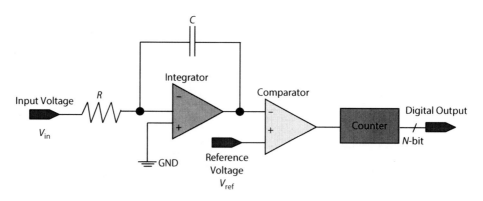

FIGURE 5.12 Architecture of the integrating ADC [5].

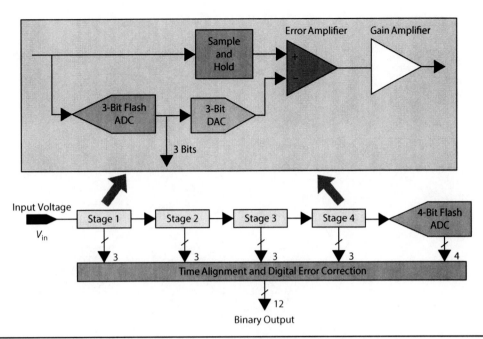

FIGURE 5.13 Architecture of the pipeline or subranging ADC [3, 21].

the output of the comparator, a counter is incremented. This is a single-slope ADC architecture whose accuracy depends on the tolerances of the R and C values of the integrator. In a real-life environment, it is difficult to maintain constant R and C values and a slight difference can have different conversion results. To reduce the sensitivity of the single-slope ADC, more complex integrating architecture, such as dual-slope or multi-slope ADC, is used [39].

5.3.6 Pipeline ADC or Subranging ADC

The architecture of a pipeline or subranging ADC is shown in Fig. 5.13 [3, 21, 92]. In a pipeline ADC architecture, the conversion task is divided into several consecutive stages. Each stage consists of a sample and hold, an m-bit ADC, and an m-bit DAC. The error amplifier finds the residue of the input signal (from the sample and hold) and DAC output signal. The residue signal is then amplified by the gain amplifier. The amplified residue signal then continues through the pipeline till it reaches the last ADC in the pipeline that is of the same bit size as the number of pipeline stages (p). The overall resolution of the pipelined ADC is ($m \times p$). In this particular pipelined architecture in Fig. 5.13, there are four stages, each stage has a 3-bit ADC, an additional 4-bit ADC, and overall resolution of 12. The time-alignment and digital-error correction time synchronize and correct bit streams that are from different pipeline stages and generated at different points of time. The pipeline ADC has higher resolution as compared to the flash ADC with the same number of comparators. However, the processing time of the pipeline ADC is of p cycles as compared to one cycle of the flash ADC [21].

5.3.7 Sigma-Delta ADC or Oversampling ADC

The high-level architecture of sigma-delta ($\Sigma - \Delta$), delta-sigma ($\Delta - \Sigma$), or oversampling ADC is shown in Fig. 5.14 [21, 89]. In general, the sigma-delta ADC oversamples the input analog signal by a large factor and filters the desired signal band. The sigma-delta architecture includes the following blocks: error amplifier, integrator, comparator, 1-bit DAC, digital filter, and decimator. The error amplifier calculates the difference between input signal and feedback signal from the 1-bit ADC. The difference signal from the error amplifier is then integrated by the integrator. The comparator is responsible for generating 1 or 0 from the integrator output voltage. The output bit from the comparator now goes as input to the DAC whose output becomes the new feedback signal for the error amplifier to repeat the above steps in integrator and comparator. The steps of the signal conversion process are carried out at high oversampled rate. At the final step, the bit stream from the comparator is then digitally filtered and decimated to obtain the digital output signal in binary format.

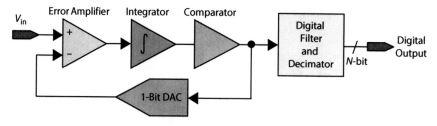

Figure 5.14 Architecture of the sigma-delta, delta-sigma, or oversampling ADC [21].

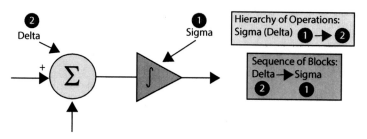

Figure 5.15 Sigma-delta ($\Sigma - \Delta$) versus delta-sigma ($\Delta - \Sigma$) nomenclature [41].

The terms sigma-delta ($\Sigma - \Delta$) and delta-sigma ($\Delta - \Sigma$) have been used interchangeably. The obvious question is which nomenclature is more accurate to describe the architecture. This can have a never-ending debate. However, an attempt has been made to clarify the same (see the structure presented in Fig. 5.15 [41]). In terms of structure of the architecture, first comes delta and then comes sigma. However, in terms of hierarchy of the operations, first comes sigma and then comes delta. Therefore, much existing literature uses sigma-delta ($\Sigma - \Delta$) [41]. This text will consistently use sigma-delta ($\Sigma - \Delta$). For clear understanding, let us discuss "sum-of-product" (SOP) in Boolean digital function. It is sum of products; in other words, first comes products and then sum in the hierarchy of the operations. Similarly, in the "root-mean-square" (RMS), first comes square, then comes mean, and then comes root in the hierarchy of the operations.

5.3.8 Time-Interleaved ADC

The architecture of the time-interleaved ADC is depicted in Fig. 5.16 [37, 82]. The overall time-interleaved ADC architecture contains N identical ADC channels. Each of these ADC channels is driven by a clock of frequency $\left(\frac{f_s}{N}\right)$. However, the phase of each of the ADC channel clocks is different, $\phi_1, \phi_2, \ldots \phi_N$. Thus, the individual ADC channels sample the input signal every Nth cycle of the sample clock. Thus, the overall ADC has N times more sampling rate than the individual ADC channels. Thus, the overall throughput of the time-interleaved ADC is quite high. However, the difference between the individual ADC channels may degrade the performance that needs to be addressed for quality designs.

5.3.9 Folding ADC

The high-level architecture of a folding or serial Gray ADC is presented in Fig. 5.17 [41, 80, 81]. The key block of the folding ADC is the folding processor. The folding processor is responsible for generating voltages in the forms that can be easily converted to Gray code. The folding block FB_1 converts the input voltage to a signal that changes its sign every $\left(\frac{V_{in}}{FF}\right)$, where FF is the folding factor. The FB_1 has two input voltages V_{in} and $V_{ref,1}$, where $V_{ref,1} = \left(\frac{0}{FF(n-1)}\right) = 0$, for n-bit ADC. The folding block FB_2 converts the input voltage into another signal that is similar to the output of the FB_1, but at a different phase. The FB_2 has two input voltages V_{in} and $V_{ref,2}$, where $V_{ref,2} = \left(\frac{1}{FF(n-1)}\right)$. For a specific example, FF = 8

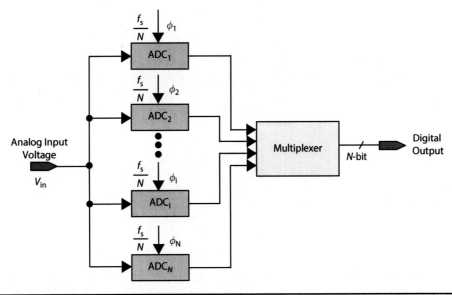

FIGURE 5.16 Architecture of the time-interleaved ADC [82].

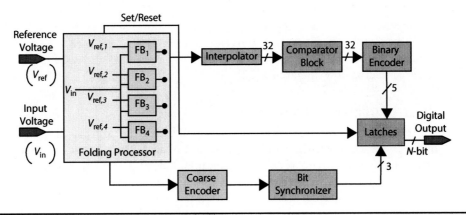

FIGURE 5.17 Architecture of the folding ADC [80].

and $n = 8$, $V_{ref,1} = 0$ and $V_{ref,2} = \left(\frac{1}{32}\right)$. The bus width is shown for an 8-bit resolution. The interpolator is used to increase the resolutions of the outputs from the folding processors, when the n is higher than the number of folding blocks in the folding processor. Two different sets of analog processing are performed to generate the digital signals. For the MSBs, an encoder called "Coarse Encoder" and a "Bit Synchronizer" is used to generate three bits. The Bit Synchronizer matches the delay in processing of the five bits versus three bits to reduce the code error of the ADC. For the least significant bits (LSBs), the comparator and the "Binary Encoder" are used to generate five bits. The output latches combine these bits from the coarse and binary encoders to generate the final digital binary output. The folding ADC requires many fewer comparators than the flash ADC. The folding ADC does not need sample and hold or DAC as the coarse and fine encoders work independently of each other.

5.3.10 Tracking ADC or Counter-Ramp ADC or Delta-Encoded ADC

The block-diagram of the tracking, counter-ramp, or delta-encoded ADC is shown in Fig. 5.18 [41, 72]. The tracking ADC consists of a comparator, an Up/Down counter, and a DAC. Essentially, the tracking ADC uses negative feedback from the comparator to adjust the Up/Down counter until

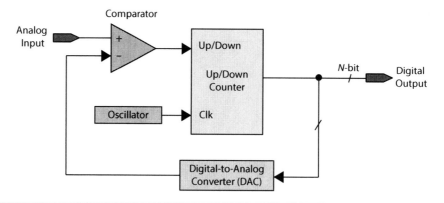

FIGURE 5.18 Architecture of the tracking or counter-Ramp or Delta-encoded ADC [41].

the output of the DAC is almost equal to the input signal. In the tracking ADC, the input analog voltage is compared with the feedback signal from the DAC. The Up/Down counters count up or down depending on the output of the comparator. If the input analog is higher than the feedback signal from the DAC, then the Up/Down counter counts up until the two signals are equal. If the input analog is lower than the feedback signal from the DAC, then the Up/Down counter counts down until the two signals are equal. The tracking ADC is simple and can be efficiently designed for minimal silicon area as it needs only one comparator. However, tracking ADC may be slow, as it needs more than one clock cycle to respond to a signal larger than the LSB voltage. For flat input voltages, the counter changes state at every clock cycle that may lead to higher switching activity power as well as slow operations [72].

5.3.11 Architecture Selection

The selection of a specific ADC architecture depends on the target applications. Each one of the ADC architectures has its own advantages and disadvantages. The sigma-delta ADC ($\Sigma - \Delta$ADC) due to oversampling can provide very-high resolution, but moderate bandwidth. The conversion speed of the sigma-delta ADC is a tradeoff between data conversion rate and noise. The SA-ADCs provide compact implementation with medium-high resolution. The SA-ADC has moderate speed that decreases with resolution. The pipeline ADC has good speed with medium-high resolution in which the conversion speed decreases with resolution. The flash ADC has the highest speed but has high power dissipation. It is suitable for low-medium resolution due to an increase of area with resolution. The conversion speed of the flash ADC does not change with resolution. A summary of resolution and bandwidth of selected ADCs is shown in Fig. 5.19 [6, 18]. A comparative perspective of selected ADCs is presented in Table 5.1.

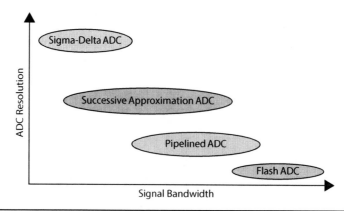

FIGURE 5.19 ADC architecture selection [18, 21].

Attributes	$\Sigma - \Delta$ ADC	SA-ADC	Pipeline ADC	Flash ADC
Conversion Method	Oversampling	Binary Search	Small Parallel Stages	Comparator Bank
Encoding Method	Oversampling Modulator	Successive Approximation	Digital Correction Logic	Thermometer Code
Speed	Low-Medium	Moderate	High	Ultra-High
Resolution	High	Medium-High	Medium-High	Low-Medium
Power	Medium	Low	Low	High
Size	Medium	Small	Medium	Large

TABLE 5.1 Comparative Perspective of Selected ADCs

5.4 Selected DAC Architectures: Brief Overview

An exhaustive list of different types of DACs is presented in Fig. 5.20 [41, 76]. This may not be a complete list of DACs available in the literature. Similar to the ADCs, the DACs are also classified in terms of architectures, outputs, and applications. Selected DAC architectures are briefly discussed in the rest of this section.

5.4.1 Binary-Weighted DAC

The schematic representation of the binary-weighted DAC is presented in Fig. 5.21 [41, 60]. In general, the binary-weighted DAC can be a voltage-mode binary-weighted resistor DAC, current-mode binary DAC, or capacitive binary-weighted DAC based on the electrical components that connect input digital signals to the amplifier [41]. As a specific example, a binary-weighted DAC using tunable resistors is shown in Fig. 5.21 [60]. This DACs input resistors, R_0, R_1, R_2, and R_3 switch between 0 V (ground) and reference voltage V_{ref}. A larger output voltage range is created by introducing the offset

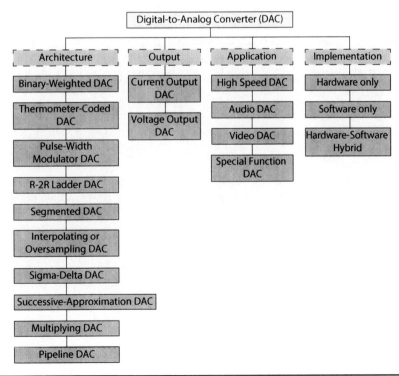

FIGURE 5.20 Different types of DACs.

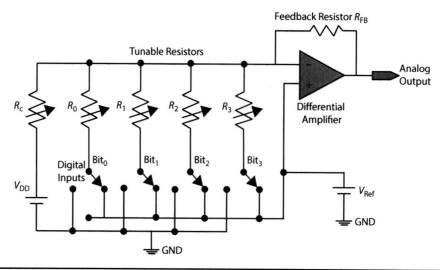

Figure 5.21 Architecture of the binary-weighted DAC [60].

current using V_{DD} and R_c. In general, if there are N resistors ($R_0, R_1, \ldots, R_{N-1}$), the voltage produced at the output is

$$V_{out} = -R_{FB}\left(\frac{V_0}{R_0} + \frac{V_1}{R_1} + \cdots + \frac{V_{N-1}}{R_{N-1}}\right) \quad (5.1)$$

where $V_0, V_1, \ldots,$ and V_{N-1} are voltages corresponding to the binary bits 0, 1, 2, and $(N-1)$, respectively. The value of the output voltage depends on these voltages (which are dictated by input digital bits) and the resistor values. The binary-weighted DACs, while simple and fast, can be poor in terms of accuracy. The binary-weighted DACs are nonmonotonic and their operation is insensitive to parasitics originated due to physical realization of the components.

As a specific example of a binary-weighted DAC, the schematic representation of the resistor-network binary-weighted summing amplifier DAC is provided in Fig. 5.22 [9]. In the simplest form of DAC, a summing amplifier can be easily used as a DAC. As a specific example, a 4-bit DAC is presented. It consists of an inverting operational amplifier (OP-AMP) and feedback resistor (R_{FB}). The 4-bit digital inputs are connected through resistors. The resistors essentially calculate the weighted sum of the digital signals for one input of the OP-AMP. The voltage produced at the output of the OP-AMP is

$$V_{out} = -R_{FB}\left(\frac{V_0}{8R} + \frac{V_1}{4R} + \frac{V_2}{2R} + \frac{V_3}{R}\right) \quad (5.2)$$

where $V_0, V_1, V_2,$ and V_3 are voltages corresponding to the binary bits 0, 1, 2, and 3, respectively. The exact value of the output voltage depends on the digital input voltages and resistor values.

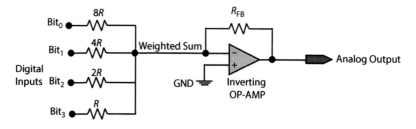

Figure 5.22 Architecture of the resistor-network binary-weighted summing amplifier-based DAC [9].

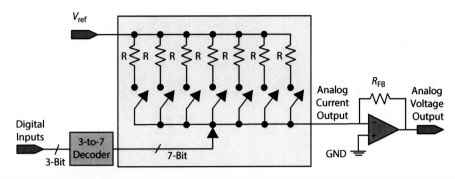

FIGURE 5.23 Architecture of the thermometer-coded DAC [38, 41].

5.4.2 Thermometer-Coded DAC

The schematic representation of a thermometer-coded DAC is shown in Fig. 5.23 [38, 41]. The thermometer-coded DAC consists of a decoder, resistor-network connected to the reference voltage and the output of the decoder, and an inverting amplifier performing current to voltage conversion at the output. There may be different versions of the thermometer-coded DAC depending on this input network [41]. It may contain resistors and a voltage reference as shown in the above figure or, maybe, active current sources to provide high-speed operations. The thermometer-coded DAC is quite fast and has the highest accuracy, but at the expense of high cost. For example, a 3-bit thermometer-coded DAC needs seven segments, an 8-bit needs 255 segments, and a 16-bit needs 65,535 segments.

5.4.3 Pulse-Width Modulator DAC

The high-level architecture of the pulse-width modulator (PWM) DAC is depicted in Fig. 5.24 [41, 68]. The PWM is the key component of this type of DAC. The PWM generates a signal whose width is proportional to its input binary signal. The input signal of the PWM is preprocessed by an oversampling filter, a cross-point detector, and a noise-shaping filter. The oversampling filter increases the sampling rate that reduces the signal distortion. The cross-point detector linearizes the signal that also helps to reduce signal distortion. The noise-shaping filter increases the signal-to-noise ratio (SNR) of the signal. The PWM DAC architecture provides both high resolution as well as high accuracy and they are monotonic.

5.4.4 R-2R Ladder DAC

The schematic architecture of an R-2R ladder DAC is presented in Fig. 5.25 [41, 55, 66]. The R-2R ladder DAC uses two different values with a ratio of 2:1. In general, an N-bit DAC needs 2N number of resistors. Depending on the R-2R ladder connections, there are two modes of R-2R ladder DAC: current mode and voltage mode. The two modes of R-2R ladder DAC are shown in Fig. 5.25(a) and

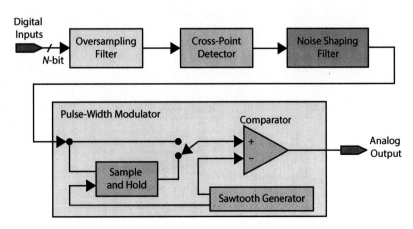

FIGURE 5.24 Architecture of the PWM DAC [68].

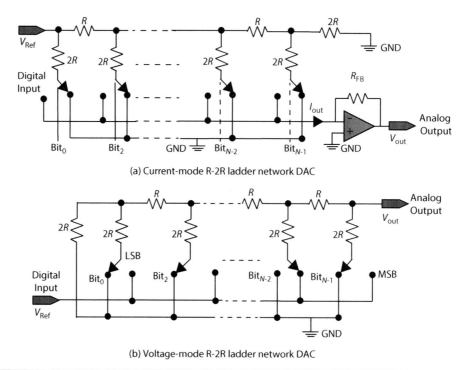

Figure 5.25 Architecture of the R-2R ladder DAC [41, 55].

Fig. 5.25(b), respectively. The current-mode R-2R ladder DAC is nonmonotonic and is sensitive to parasitics. The stabilization of the output current-voltage converter OP-AMP can be an issue due to impedance variation with digital code. The current-mode R-2R ladder DAC has a larger switching glitch than the voltage-mode R-2R ladder DAC as the switches are connected directly to the output. The design of the current-mode R-2R ladder DAC is simpler as the switches are always grounded and their voltage rating does not affect the reference voltage rating. The voltage-mode R-2R ladder DAC has constant output impedance, as it is independent of the input digital code. Thus, the stabilization of an output amplifier of a voltage-mode R-2R ladder DAC is not an issue. It may be noted that the resistors in the above resistor ladders can be realized using MOS transistors for a high-speed, high-resolution R-2R ladder DAC that has linear characteristics and consumes minimal power [26].

5.4.5 Segmented DAC

The schematic high-level representation of a segmented DAC is presented in Fig. 5.26 [41, 52, 69]. As depicted in the architecture, the segmented DAC combines the thermometer-coded principle and binary-weighted principle to achieve accuracy and speed tradeoffs. This process is called segmentation.

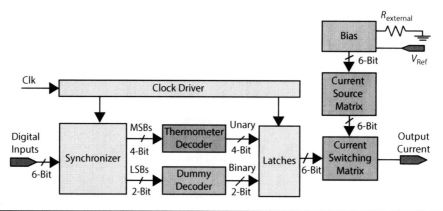

Figure 5.26 Architecture of the segmented DAC [52].

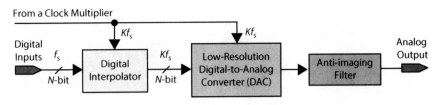

FIGURE 5.27 Architecture of the interpolating DAC [41].

The thermometer-coded principle is used for the MSBs while the binary-weighted principle is used for LSBs. In the segmented architecture shown, the synchronizer block aligns the input digital bits. The dummy decoders avoid time mismatch between the two paths of the thermal decoder (for MSBs) and the binary-weighted (for LSBs). The bias circuits generate the currents and are mirrored to the current source matrix. The current switching matrix consists of an array of current cells.

5.4.6 Oversampling or Interpolating DAC

The schematic high-level architecture of the oversampling or interpolating DAC is depicted in Fig. 5.27 [41, 48]. In this architecture, the N-bit digital data is received at f_s. The digital interpolator is clocked at an oversampling frequency of Kf_s. The internal DAC used in the oversampling DAC is of low resolution. The anti-imaging or reconstruction filter is used to smooth the output analog signal. The anti-imaging filter design is made simpler due to oversampling operation. The interpolating DAC has high speed and high SNR.

5.4.7 Sigma-Delta DAC

The architecture of a sigma-delta ($\Sigma - \Delta$) DAC is depicted in Fig. 5.28 [10, 35, 41]. The sigma-delta ($\Delta - \Sigma$) DAC is an oversampling DAC. The sigma-delta DAC is similar to a sigma-delta ADC. However, the sigma-delta DAC has more digital blocks as compared to the sigma-delta ADC. For example, in sigma-delta DAC, the noise-shaping function is followed by a digital modulator not an analog modulator. The basic block of an N-bit sigma-delta DAC consists of a digital interpolation filter (or digital interpolator), digital multibit sigma-delta modulator, an M-bit DAC, and analog output filter [41]. The internal DAC is of much lower resolution than the sigma-delta DAC, thus $M < N$. The digital interpolation filter increases the input digital signal rate from a lower rate by padding zeros and then applying a digital filter algorithm. The sigma-delta modulator converts the input digital data to a high-speed data stream. The sigma-delta modulator operates like a low-pass filter to the input digital bits and, at the same time, performs high-pass filtering operation for the quantization noise. The M-bit DAC converts the digital bit stream to an analog signal of 2^M levels. The analog output filter is a low-pass filter that produces a clean signal.

5.4.8 Successive-Approximation or Cyclic or Algorithmic DAC

The block-level diagram of the successive-approximation, cyclic, or algorithm DAC is presented in Fig. 5.29 [22, 41, 83]. The DAC has the following blocks: summing amplifier, divide-by-two amplifier, sample-and-hold amplifier, and two switches. The cyclic DAC successively constructs the output signal during each cycle of the digital input. In essence, the digital input bits are processes during each

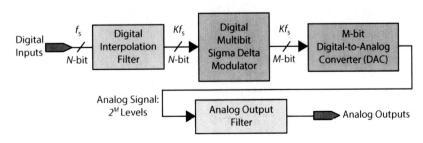

FIGURE 5.28 Architecture of the sigma-delta DAC [41].

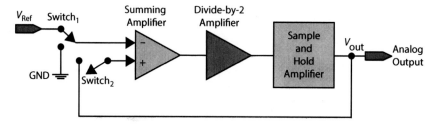

Figure 5.29 Architecture of the successive approximation or cyclic or algorithmic DAC [41, 83].

(a) Black-box representation of multiplying DAC

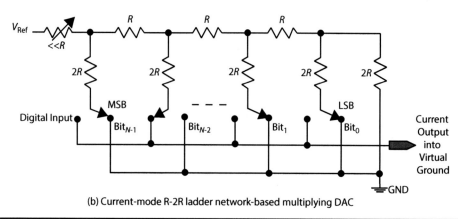

(b) Current-mode R-2R ladder network-based multiplying DAC

Figure 5.30 Architecture of the multiplying DAC [41].

cycle until all the input bits are processed. Switch$_2$ is always closed, whereas the Switch$_1$ is connected to the reference voltage (V_{ref}), if the digital input bit is "1" and Switch$_1$ is connected to ground if the digital input bit is "0." The signal from Switch$_1$ and Switch$_2$ is summed by the summing amplifier and divided by two by the divide-by-two amplifier; i.e., during each cycle, the previous conversion voltage is added with either reference voltage (V_{ref}) or "0" and then divided by two. This cyclic DAC that performs serial conversion is hardly used in the basic form presented as it is slow. The cyclic DAC, however, is extremely simple in terms of circuit design.

5.4.9 Multiplying DAC

The architecture of the multiplying DAC is presented in Fig. 5.30 [17, 41]. The key feature is that in a multiplying DAC, the reference voltage is a variable analog voltage instead of fixed voltage as in a typical DAC. The output of the multiplying DAC is proportional to both the analog input and digital input as depicted in Fig. 5.30(a) [41]. While many forms of multiplying DAC are possible, a simple implementation is presented in Fig. 5.30(b) [41]. The implementation in Fig. 5.30(b) is a current-mode ladder network and MOS switches that allow use of positive, negative, and AC voltage as V_{ref}.

5.4.10 Pipeline DAC

The schematic representation of a pipeline DAC is shown in Fig. 5.31 [71]. The pipeline DAC consists of N similar stages. Each stage is a 1-bit cyclic DAC (presented in a subsection before). The analog output voltage of the pipeline DAC is

$$V_{out} = -\left(\sum_{N}^{i=1}(-1)^{b_b(i)}\left(\frac{1}{2^i}\right)\right)V_{ref} + \left(\sum_{N}^{i=1}\Delta V_b(i)\left(\frac{1}{2^i}\right)\right) \quad (5.3)$$

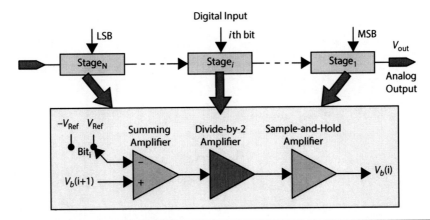

Figure 5.31 Architecture of the pipeline DAC [71].

where $b_b(i)$ is the ith digital input bit. $\Delta V_b(i)$ is the voltage across the ith stage of the pipeline DAC that processes the ith bit. $\Delta V_b(i) = (V_b(i) - V_b(i+1))$, where $V_b(i+1)$ is the voltage at the input and $V_b(i)$ is the voltage at the output of the ith stage of the pipeline DAC. The pipeline DAC can provide high throughput and resolution from simple implementations. The pipeline DAC does not need resistor, and hence is best suited for low-power and compact SoC applications.

5.5 Characteristics for Data Converters

Many terminologies are used to express different characteristics and attributes of data converters (both ADCs and DACs) [4, 14, 23, 29, 32, 41, 54, 62]. In addition, various figures-of-merit (FOM) are used to characterize the quality of data converters (both ADCs and DACs). Many of the characteristics or terminologies are the same for both the ADCs and DACs. However, many characteristics or terminologies are also specific to either ADCs or DACs. Some characteristics or terminologies may have the same name, but different definitions for ADCs and DACs. The characteristics of the data converters may be static characteristics or dynamic characteristics. The static characteristics are the characteristics that do not change over time. The dynamic characteristics are the characteristics that change over time. The static characteristics are easy to estimate but are important only for low-frequency operations. The dynamic characteristics are difficult to estimate but are essential for medium-frequency and high-frequency operations. It is difficult to provide a comprehensive list of all the terminologies. This section presents selected characteristics, terminologies, or FOMs for ADCs and DACs. It may be noted that for both the ADCs and DACs the transfer characteristics are a simple straight-line expression of the following form:

$$D = GA + K \quad (5.4)$$

where D is the digital code, A is the analog signal, G and K are constants. G is like the slope of the transfer characteristic that may correspond to gain and K may correspond to offset.

5.5.1 Characteristics for ADC

A selected characteristic of the ADC is depicted in Fig. 5.32 for the purpose of illustration [14, 41]. It considers a 3-bit ADC as an example. It shows three different transfer functions of the ADC: the infinite resolution characteristic, the ideal characteristic, and the actual characteristic.

1. *Resolution:* The resolution of an ADC is the number of bits used to represent the signal. The higher the resolution, the more accurate is the replication of the analog input signal. In other words, the resolution of the ADC is the smallest analog input voltage change that is distinguishable by an ADC.

2. *Effective Number of Bits (ENOB):* The ENOB of an ADC is the representation of the accuracy of an ADC for a given analog signal at specific frequency and sampling rate [4, 20, 41, 42].

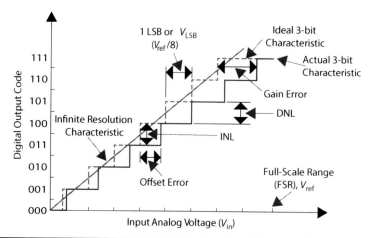

Figure 5.32 Selected characteristics of the ADC [14, 41].

Thus, ENOB is one of the important dynamic performances of the ADC. For a full-scale (FS) sinusoidal input signal, the ENOB of an ADC is calculated as follows [4, 20, 41, 42]:

$$\text{ENOB} = \left(\frac{\text{SINAD}_{\text{actual}} - 1.76\,\text{dB}}{6.02}\right) \tag{5.5}$$

where $\text{SINAD}_{\text{actual}}$ is the actual signal-to-noise-and-distortion ratio (SINAD or SNDR) to be discussed later. The above expression assumes that the input analog signal is of FS. However, this may not be the case in real life, as the value of SINAD reduces when the level of the input signal reduces. Thus, there is a need to correct the expression for accurate estimation of the ENOB. A more accurate expression for calculation of the ENOB is the following [42]:

$$\text{ENOB} = \left(\frac{\text{SINAD}_{\text{measured}} - 1.76\,\text{dB} + 20\log\left(\frac{A_{\text{FS}}}{A_{\text{in}}}\right)}{6.02}\right) \tag{5.6}$$

where $\text{SINAD}_{\text{measured}}$ is the measured SINAD, A_{FS} is the full-scale amplitude, and A_{in} is the input amplitude.

3. *Dynamic Range (DR)*: The DR of an ADC is the range of the input signal amplitude for which the ADC can produce a positive SNR [4, 18, 67]. Simplistically, it is the range of signal amplitudes that the ADC can resolve. The DR is the ratio of FS range (FSR) to the smallest voltage difference that can be resolved (i.e., 1 LSB or V_{LSB}). Thus, it is defined as follows [14]:

$$\text{DR} = \left(\frac{\text{FSR}}{V_{\text{LSB}}}\right) \tag{5.7}$$

$$= \left(\frac{\text{FSR}}{\frac{\text{FSR}}{2^N}}\right) \tag{5.8}$$

$$= 2^N \tag{5.9}$$

The DR in terms of decibels is calculated as follows [14]:

$$\mathrm{DR} = 6.02N \text{ dB} \tag{5.10}$$

4. *Acquisition Time* (T_{acq}): The acquisition time of an ADC is the time difference between the release of the hold state and the instant at which the voltage on the sampling capacitor is within 1 LSB of the input value [4, 14]. Simplistically, it is the time needed by the ADC to acquire the analog input voltage. The acquisition time is calculated as follows:

$$T_{acq} = \ln(2^N) + R_{source} C_{sample} \tag{5.11}$$

where N is the number of resolution bits, R_{source} is the source resistance, and C_{sample} is the sampling capacitance.

5. *Settling Time* (T_{stl}): The settling time of an ADC is the time needed for the voltage at the sampling capacitor to be within 1 LSB [4]. Simplistically, the settling time of the ADC is the time required by the ADC to settle to the final voltage value within an accuracy tolerance [14].

6. *Sampling Rate or Sampling Frequency*: The sampling rate, sampling frequency, or throughput rate is the rate at which an ADC sample acquires the analog input signal. The maximum sampling rate is calculated as follows [4, 14]:

$$f_{s,max} = \left(\frac{1}{T_s}\right) = \left(\frac{1}{T_{acq} + T_{stl}}\right) \tag{5.12}$$

where T_s is sampling time. Sampling time is the sum of acquisition time and the settling time and is presented as follows [14]:

$$T_s = T_{acq} + T_{stl} \tag{5.13}$$

7. *Integral Nonlinearity (INL)*: The INL is sometimes referred to as "relative accuracy" of the ADC. INL error of an ADC is the maximum difference between the actual finite resolution characteristic and ideal finite resolution characteristic measured as a percentage of LSB [2, 14, 41]. The idea of INL is depicted in Fig. 5.32. Simplistically, INL error is the deviation of the actual transfer function of the ADC from a straight line. These two approaches for the INL quantification are as follows: "best straight-line INL" and "end-point INL." The best straight-line approach is widely used as it produces results that are more accurate and it quantifies the INL as follows [2, 30, 32]:

$$\mathrm{INL} = \left|\left(\frac{V_D - V_{zero}}{V_{LSB,ideal}}\right) - D\right| \mathrm{LSB} \tag{5.14}$$

where $0 < D < 2^N - 1$. V_D is the analog voltage value represented by the digital output code D. N is the resolutions of the ADC resolution. V_{zero} is the minimum analog input signal corresponding to an all-zero output code. $V_{LSB,ideal}$ is the ideal voltage spacing for two adjacent output codes. In general, the voltage corresponding to the LSB is called V_{LSB} that is calculated as follows [32]:

$$V_{LSB} = \left(\frac{\text{Input Voltage Range}}{2^N}\right) \tag{5.15}$$

$$= \left(\frac{V_{end} - V_{start}}{2^N}\right) \tag{5.16}$$

A practical approach of estimating INL is using the histogram approach [7, 32]. In this equally spaced approach, voltages are created between the input range (V_{start}, V_{end}). At each of these values of the voltage, the ADC generates a code. The resultant codes for each of the ADC conversions are stored. Once all the conversions are completed for the entire input voltage steps, the INL is then calculated using the following expression [32]:

$$\text{INL}[i] = \text{code}_{width}[i] + \text{INL}[i-1] - 1 \tag{5.17}$$

where code_{width} contains the code-width calculations performed using the following expression [32]:

$$\text{code}_{width}[i] = \left(\frac{\text{bucket}[i]}{\text{hits} \times (2^N - 2)} \right) \tag{5.18}$$

where bucket holds the number of code hits for each code and hits is the total hits between codes 1 and $(2^N - 2)$.

8. *Differential Nonlinearity (DNL)*: The DNL error of the ADC is difference between two adjacent voltage levels corresponding to two successive output digital codes [2, 14, 41]. In an ideal scenario of the ADC, the analog input voltage levels that generate any two successive output digital codes should differ by 1 LSB; any deviation from 1 LSB is a DNL error. Thus, the DNL error of the ADC signifies the difference between the actual step width and the ideal value of 1 LSB (or V_{LSB}). The idea of DNL is depicted in Fig. 5.32. The DNL of an ADC is calculated using the following expression [2, 30, 32]:

$$\text{DNL} = \left| \left(\frac{V_{D+1} - V_D}{V_{LSB,ideal}} \right) - 1 \right| \text{LSB} \tag{5.19}$$

where $0 < D < 2^N - 1$. V_D is the analog voltage corresponding to the digital output code D. V_{D+1} is the analog voltage corresponding to the digital output code $(D + 1)$. $V_{LSB,ideal}$ is the ideal voltage difference for two adjacent digital codes. The DNL of an ideal ADC is 0 LSB. Similar to the case of INL, a practical approach of estimating DNL is using the histogram approach [7, 32]. Following the steps as presented for the INL, once all the conversions are completed for the entire input voltage steps, then the DNL is calculated using the following expression [32]:

$$\text{DNL}[i] = \text{code}_{width}[i] - 1 \tag{5.20}$$

where code_{width} contains the code-width calculations as in the case of INL calculations. A DNL error specification of less than or equal to 1 LSB guarantees a monotonic transfer function of the ADC with no missing output codes [2, 30, 32]. The higher values of the DNL limit the performances of the ADC in terms of SNR and spurious-free DR (SFDR).

9. *SNR*: The SNR is also called SNR without harmonics. The SNR for the ADC is the ratio of the amplitude of the target signal to the amplitude of the noise signals at any point of time [4, 14, 67]. For a sinusoid input signal, the theoretical maximum SNR is the ratio of the FS analog input to the RMS quantization error and is quantified in the following manner [1, 4, 14, 67]:

$$\text{SNR}_{max} = \left(\frac{\sqrt{6}}{2} \right) 2^N \tag{5.21}$$

In terms of decibels, the maximum SNR is calculated as follows [4, 14, 67]:

$$\text{SNR}_{max} = 6.02N + 1.76 \text{ dB} \tag{5.22}$$

10. *SNDR or SINAD*: The SNR is also known as SINAD or S/(N + D). It is the ratio of the RMS value of the sinusoid signal to the RMS value of the noise plus distortion without the sinusoid signal [1, 18, 30, 67]. The sinusoid signal is the input signal for an ADC or the reconstructed output signal for a DAC. The RMS value of the noise plus distortion includes all spectral components up to the Nyquist frequency; however, it excludes the fundamental and the DC offset. The SINAD, SNDR, or S/(N + D) of an ADC is calculated by the following expression [1, 30]:

$$\text{SINAD} = 20 \times \log_{10} \left(\frac{A_{\text{RMS,signal}}}{A_{\text{RMS,noise+harmonics}}} \right) \text{dB} \qquad (5.23)$$

where $A_{\text{RMS,signal}}$ is the RMS value of the fundamental amplitude of the signal. $A_{\text{RMS,noise+harmonics}}$ is the RMS value of the noise that includes the significant harmonics, e.g., the second to the fifth highest amplitudes. SINAD is a good metric for the overall dynamic performance of an ADC or DAC because it includes all components that constitute noise and distortion.

11. *Quantization Noise*: The quantization noise of an ADC is the $\pm \frac{1}{2}$ LSB uncertainty between its infinite resolution characteristic (which is shown as a straight line in Fig. 5.32) and its actual characteristic (which is shown as a solid staircase in Fig. 5.32) [4, 14, 41]. The quantization error of an ADC is the difference between the actual analog input value and the digital representation of that value. In an ADC, the analog input signal can have any values; however, the digital output of the ADC has quantized values. Thus, there is a difference between the actual analog input value and the exact value of the digital output. This difference is the quantization uncertainty or quantization error of the ADC. In an ADC when input is an AC signal the quantization error translates to the quantization noise.

12. *Offset Error*: The offset error is also called "zero-scale" error. The offset error of an ADC is the ideal finite resolution characteristic (which is shown as a dashed staircase in Fig. 5.32) and actual finite resolution characteristic (which is shown as a solid staircase in Fig. 5.32) [4, 14, 41]. The offset error of an ADC is indicated in Fig. 5.32. In terms of the generic transfer characteristic presented before, the offset error is the difference between the actual value of offset constant K and its ideal value.

13. *Gain Error*: The gain error of the ADC is the difference between its ideal finite resolution characteristic and the actual finite resolution characteristic for an FS input signal [4, 14, 41]. The gain error of an ADC is indicated in Fig. 5.32. In terms of the generic transfer characteristic presented before, the gain error is the difference between the actual value of gain constant G and its ideal value.

14. *Monotonicity*: An ADC is monotonic in nature if its digital output code always increases as the input analog signal increases. In other words, a monotonic ADC has all vertical jumps positive in its transfer characteristic. DNL is used to detect monotonicity. If the DNL error is less than ± 1 LSB, then the ADC is monotonic. The three transfer characteristics, presented in Fig. 5.32, are monotonic in nature.

15. *SFDR*: The SFDR of an ADC is the ratio of the RMS amplitude of the fundamental component to the RMS value of the largest distortion component [4, 14, 41]. SFDR is specified in terms of decibels to the relative carrier or dBc. For a sinusoid input signal, the SFDR can be calculated using the following expression [1, 4, 41]:

$$\text{SFDR} = 20 \times \log_{10} \left(\frac{A_{\text{RMS,signal}}}{A_{\text{RMS,HD-max}}} \right) \text{dBc} \qquad (5.24)$$

where $A_{\text{RMS,signal}}$ is the amplitude of the averaged discrete Fourier transform (DFT) value at the fundamental frequency. $A_{\text{RMS,HD-max}}$ is the amplitude of the averaged DFT value of the largest-amplitude harmonic. This definition of SFDR holds good for DAC and the SFDR calculations for a DAC can also be performed in a similar manner.

16. *Total Harmonic Distortion (THD)*: In general, the *i*th harmonic distortion in a signal indicates the power of the *i*th harmonic. The THD in an ADC is the sum of the powers of all the *i*th harmonic distortions [1, 4, 41, 56]. For an ADC, the THD is the ratio of the RMS sum of the selected harmonics of the input signal to the fundamental RMS amplitude. The THD of an ADC can thus be calculated using the following expression [1, 4, 41]:

$$\text{THD} = 20 \times \log_{10}\left(\frac{\sqrt{\sum_{i=2}^{N} A_{\text{RMS,HD}-i}^{2}}}{A_{\text{RMS,signal}}}\right) \text{dBc} \quad (5.25)$$

where $A_{\text{RMS,signal}}$ is the RMS amplitude of the fundamental frequency component. $A_{\text{RMS,HD}-i}$ are the RMS amplitudes of the second to *N*th order harmonics. In a similar manner, the THD can be calculated and used for the DACs.

17. *Nanoscale ADC FoM*: As evident from the above, there are many characteristics to specify the ADCs. However, a FoM that combines the important characteristics in a single form and can be used for ADC characterization is the following [49]:

$$\text{ADC}_{\text{FoM}} = \left(\frac{\text{Power}_{\text{ADC}}}{2^{\text{ENOB}} f_s}\right) \text{pJ/conversion step} \quad (5.26)$$

where $\text{Power}_{\text{ADC}}$ is the power dissipation of the ADC. ENOB is the effective number of bits of the ADC. f_s is the sampling frequency of the ADC. The lower the ADC_{FoM}, the better is the ADC.

5.5.2 Characteristics for DAC

Selected characteristics of the DAC are depicted in Fig. 5.33 for the purpose of illustration [14, 41]. It considers a 3-bit DAC as an example. It shows three different transfer functions of the DAC: the infinite resolution characteristic, the ideal characteristic, and the actual characteristic. The horizontal

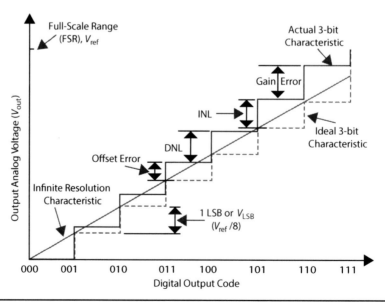

Figure 5.33 Selected characteristics of the DAC [14, 41].

and vertical axes of the Fig. 5.33 are swapped with Fig. 5.32, i.e., the horizontal input axis shows digital codes and the vertical output axis shows analog output voltage.

1. *Resolution:* Resolution of DAC has similar definition as that of the ADC [4, 14, 23]. The resolution of a DAC is the number of bits in the input digital code. The resolution of the DAC also represents the number of voltage levels that can be generated by the DAC. For example, an N-bit DAC can produce 2^N voltage levels at the output. A higher resolution DAC produces smaller step sizes in the analog voltage output.

2. *ENOB:* The ENOB of DAC has similar definition as that of an ADC and has not been repeated for brevity.

3. *Conversion Speed:* The conversion speed of a DAC is the time taken by the DAC to provide an analog output when the digital input code is changed [14].

4. *Average Analog Output Voltage:* The average output voltage of the DAC is the analog voltage due to the average on-time of the input bit stream. If the time frame is divided into 2^N parts for an N-bit DAC, then on-time is the number of parts in the time frame for which the input bitstream is "1." Under this assumption, the average output voltage of the DAC is calculated as follows [23]:

$$V_{out,avg} = \left(\frac{T_{on} FS}{2^N} \right) \tag{5.27}$$

where T_{on} is the on-time of the DAC and N is the resolution.

5. *Maximum Sampling Rate:* The maximum sampling rate of the DAC is the maximum speed at which the DAC can operate and still produce correct output signal. It can be calculated in a similar manner as that of an ADC [14].

6. *FS and FSR:* If the DAC is designed to represent in the range of (0, V_{ref}), the smallest digital code should produce 0 V and the largest digital code should produce the FS voltage V_{ref} V. The FS of a DAC is the difference of the output voltage when all inputs are "1" and output voltage when all inputs are "0." Thus, for an N-bit DAC, it can be calculated as follows [14, 23]:

$$FS = \left(V_{ref} - \frac{V_{ref}}{2^N} \right) - 0 \tag{5.28}$$

$$= V_{ref} \left(1 - \frac{1}{2^N} \right) \tag{5.29}$$

The related terminology, called FSR of a DAC is defined as follows [14, 23]:

$$FSR = \lim_{N \to \infty} (FS) \tag{5.30}$$

$$= \lim_{N \to \infty} \left(V_{ref} \left(1 - \frac{1}{2^N} \right) \right) \tag{5.31}$$

$$= V_{ref}. \tag{5.32}$$

7. *Offset Error:* The offset error of a DAC is the difference between its actual finite resolution characteristic and the ideal finite resolution characteristic measured at any vertical jump as depicted in Fig. 5.33 [4, 14]. In other words, the offset error of the DAC is the analog output response when the digital input code is all zeros.

8. *Gain Error:* The gain error of the DAC is the difference between the slope of its actual finite resolution characteristic and the corresponding ideal finite resolution characteristic measured at the right-most vertical jump as depicted in Fig. 5.33 [4, 14].

9. *INL:* The INL error of a DAC is the maximum difference between its actual finite resolution characteristic and the ideal finite resolution characteristic as depicted in Fig. 5.33 [4, 14, 51].

The INL of a DAC quantifies deviation of the transfer characteristic of a DAC from ideal characteristic, which is similar to the case of the ADCs.

10. *DNL:* The DNL error of a DAC is the difference between the ideal characteristic and the actual output characteristic for successive DAC codes. It is measured as the separation between adjacent voltage levels as shown in Fig. 5.33 [4, 14, 51]. Similar to an ADC, the DNL of a DAC signifies the deviation of analog values of two adjacent codes from the ideal 1 LSB.

11. *DR:* The DR of a DAC has a definition similar to that of an ADC and, for brevity, has not been repeated [14].

12. *Monotonicity:* The monotonicity is the ability of the DAC to produce output analog voltage in the direction that the digital input code changes [4, 36]. The DAC has monotonic characteristic if the analog voltage output always increases as the digital input increases. The DAC is monotonic if the DNL is less than $\pm$ 1 LSB.

13. *SNR:* The SNR of a DAC has a definition similar to that of an ADC and, for brevity, has not been repeated.

14. *SINAD:* The SINAD of a DAC has a definition similar to that of an ADC and, for brevity, has not been repeated. The approach used for the SINAD calculation of the ADCs can be used for SINAD calculation of the DACs.

5.6 A 90-nm CMOS-Based Flash ADC

As a specific example of flash ADC, a 90-nm CMOS-based design is discussed in this section. This includes circuit-level to physical-level designs and characterizations [29, 30, 31, 32, 45]. The high-level block diagram representation of an N-bit ADC is provided in Fig. 5.34 [90]. As shown in the block-diagram, the input is an analog voltage and output is an N-bit digital bitstream. The N-bit ADC needs a comparator bank consisting of (2^N-1) comparators. The nominal switching point of each of the comparators is determined by appropriately sizing the transistors in these comparators. The comparator bank generates a thermometer or unary code. The position of the 1-0 transition represents the analog input and is determined by the thermometer encoder circuit, which consists of "1 of N" code generators. The "1 of N" code generators produce a "1 of N" code. The $((2^N-1) \times N)$ NOR read-only memory (ROM) converts the thermometer code to a binary code.

5.6.1 Comparator Bank

Comparator bank is the first block responsible in the analog to digital conversion process that generates thermometer or unary code. The design of comparator is important in the flash ADC to achieve high-speed operations. The comparators can be implemented in many ways including differential comparator-based circuits that need resistor ladder for input voltage [33, 47]. The typical comparators that sample the input voltage in parallel have higher power consumption and reduced speed as the resolution increases. For power-efficient, compact implementation, and high-speed purposes, the comparators in the flash ADC are designed by using the threshold inverting quantization (TIQ) technique [47, 91]. The TIQ comparator has the advantage of high speed and simplicity and eliminates the need for complex high-gain differential input voltage comparators and associated resistor ladder circuit [47, 75].

The TIQ comparator is a pure inverter circuit particularly suitable for SoC applications as it can be implemented using digital CMOS technology. The TIQ comparator sets its switching threshold voltage $V_{switching}$ internally as the built-in reference voltage, based on its transistor sizes. The

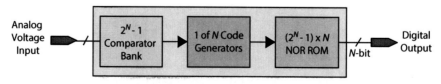

FIGURE 5.34 Block diagram of the N-bit flash ADC.

comparators in a conventional flash ADC are identical in size. In the TIQ-based comparator bank, all comparators have unique sizes. An N-bit ADC requires (2^N-1) different comparators in the comparator bank with different $V_{switching}$ value for each. As a specific example, for a 6-bit ADC, a total of 63 comparators are needed and each comparator has a different size of transistors. The switching voltage is generally calculated by using the following expression [29, 31, 32]:

$$V_{switching} = \left(\frac{(V_{DD} - |V_{Thp}|)\sqrt{\left(\frac{\mu_p W_p}{\mu_n W_n}\right)} + V_{Thn}}{1 + \sqrt{\left(\frac{\mu_p W_p}{\mu_n W_n}\right)}} \right) \quad (5.33)$$

where W_p is PMOS transistor width, W_n is NMOS transistor width, V_{DD} is supply voltage, V_{Thn} is NMOS threshold voltage, V_{Thp} is PMOS threshold voltage, μ_n is electron mobility, and μ_p is hole mobility. However, short-channel transistor may not accurately follow the square-law model used in deriving the above expression and other methods including experimental methods can be used to calculate $V_{switching}$. A more accurate expression to determine the $V_{switching}$ for short-channel devices is given by [29, 31, 32]:

$$V_{switching} = V_{DD} \times \left(\frac{R_n}{R_n + R_p} \right) \quad (5.34)$$

where R_n and R_p are the effective switching resistances for NMOS and PMOS short-channel transistors, respectively. During the design process, transistor length L is kept constant throughout the design and only W is varied. The width of the PMOS transistor (W_p) in the TIQ comparator is obtained as the switching resistance of a transistor depends on the width (W) of the transistor. Another reason for varying the width is that the channel length of the transistor more effectively controls the performance, e.g., frequency $\propto \left(\frac{1}{L^2}\right)$.

In the design of the comparators in the comparator bank, a good input voltage is determined with the following expression:

$$\text{Input Voltage Range} = V_{DD} - (V_{Thn} + |V_{Thp}|) \quad (5.35)$$

As a specific example, the TIQ comparator is obtained in order to achieve the 63 different switching voltages for a 6-bit ADC. In this specific case for a 90-nm CMOS technology, the input voltage range is varied from 493 mV (V_{start}) to 557 mV (V_{end}) [29, 31, 32]. The voltage for the LSB value V_{LSB} is calculated as follows:

$$V_{LSB} = \left(\frac{\text{Input Voltage Range}}{2^N} \right) \quad (5.36)$$

$$= \left(\frac{V_{end} - V_{start}}{2^N} \right) \quad (5.37)$$

The V_{LSB} value is calculated to be 1 mV for the specific technology and supply voltage. The V_{LSB} signifies the minimum step that the ADC can distinguish. The sizes of NMOS transistors in the comparator were kept constant at 240 nm/120 nm. For determining the sizes of PMOS transistors, an experimental approach is used. In this experimental approach, a DC parametric sweep is used in the analog simulator in which the input DC voltage was varied from 0 to 1.2 V in steps of 1 mV. During the analog simulations for 90-nm CMOS technology, as a specific example, L_p was kept at 120 nm, while W_p was varied from 240 nm ($W_p = W_n$) to 448 nm to obtain 63 linear quantization voltage levels. The TIQ comparator consists of four cascaded inverters as shown in Fig. 5.35. The inverters 1 and 2 form the baseline comparators, while inverters 3 and 4 provide increased gain and sharper switching.

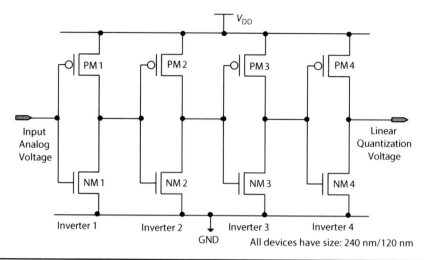

Figure 5.35 The TIQ comparator [29, 31, 32].

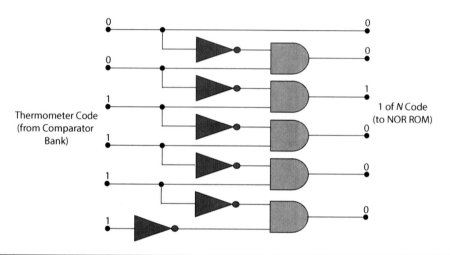

Figure 5.36 Logic-level diagram of a 1 of N code generator [24].

5.6.2 1 of N Code Generator

The output of the comparator bank discussed above is a thermometer code. At this point, a "1 of N" code generator is used to convert this thermometer code to a "1 of N" code. A single slice of "1 of N" code generator consists of AND gates as combinations of an inverter followed by a NAND gate as shown in Fig. 5.36 [24, 32]. One of the two inputs to the AND gate is fed from the TIQ comparator output. The other input to the AND gate is the inverted output from the next-level comparator. As a specific example, this "1 of N" code generator considers a 6-bit thermometer-code "001111" as input and converts it to "001000." The output from each of the AND gates is fed to the input of the NOR ROM.

5.6.3 NOR ROM

An N-bit flash ADC needs a $(2^N-1) \times N$ NOR ROM. The input to the NOR ROM is a "1 of N" code from the "1 of N" code generator. The NOR ROM converts this "1 of N" code to a binary code that is essentially the output of the flash ADC. As a specific example, the circuit of a 7×3 NOR ROM is shown in Fig. 5.37 [29, 32]. Similarly, there are 63-word lines and 6-bit lines, and a 63×6 NOR ROM for a 6-bit flash ADC. Simplistically speaking, the NOR ROM effectively maps an N-bit input into an arbitrary M-bit output. The NOR ROM consists of a grid of word lines (which is the address input) and bit lines (which is the data output) and are selectively joined together through the transistor switches. The NOR ROM consists of PMOS pull-up and NMOS pull-down devices. For a 90-nm

Electronic Signal Converter Circuits 241

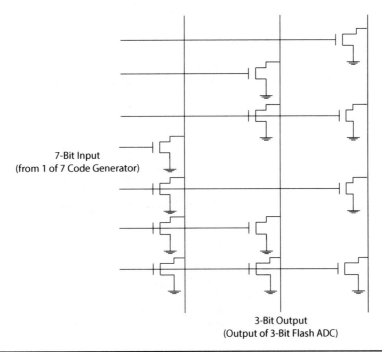

FIGURE 5.37 Circuit diagram of a 7 × 3 NOR ROM [29, 32].

CMOS, the PMOS and NMOS sizes have been taken as 135 nm/180 nm and 180 nm/180 nm, respectively. The sizing of transistors is important for the NOR ROM circuit design. In this design, sizing is performed such that $W_p < W_n$ for obtaining a good voltage swing at the output so that the pull-up transistor (PMOS) is narrow enough for the pull-down transistors (NMOS) to safely pull down the output [29, 32].

5.6.4 Physical Design and Characterization of 90-nm ADC

The complete design of a 6-bit ADC is performed at circuit-level and physical-level along with thorough characterization [29, 31, 32]. For the complete design of the flash ADC, the blocks such as thermometer bank, 1 of N code generator, and NOR ROM are individually designed as discussed in the previous subsections. Then, they are assembled for overall design of the flash ADC. At the end, buffers are placed at the output to obtain symmetrical waveforms with equalized rise and fall times. For a 90-nm CMOS technology, the buffer consists of two cascaded inverters having NMOS sizes of 240 nm/120 nm and PMOS sizes of 480 nm/120 nm. For illustration, a complete 3-bit flash ADC logic-level is shown in Fig. 5.38 [29, 31, 32].

As a specific case study, the physical design of the 6-bit flash ADC has been performed using a generic 90-nm salicide "1.2V/2.5V 1 Poly 9 Metal" process design kit (PDK). The use of a digital CMOS process for the physical design demonstrates the SoC readiness of the flash ADC. The three major blocks of the flash ADC, namely the comparator bank, the "1 of N" code generator, and the NOR ROM, have been laid out column-wise. The overall physical design of the 90-nm CMOS flash ADC is shown in Fig. 5.39. The power and ground routing of the flash ADC consist of wide vertical bars to ensure minimal electromigration risk as well as low IR drop. In addition, the generous use of contacts to the power and ground buses ensures the small IR drop. The summary of the characteristics of the 90-nm CMOS flash ADC is presented in Table 5.2 [29, 31, 32]. More discussions on the post-layout simulations and characterization are presented in subsequent subsections.

5.6.5 Post-Layout Simulation and Characterization

The functional simulation of the flash ADC using the post-layout parasitic-aware netlist is shown in Fig. 5.40. The transient analysis of the flash ADC is performed using a linearly varying ramp input signal that covers the FSR of the ADC. Output digital codes from 0 to 63 are obtained correctly, with no missing codes. A maximum sampling speed of 1 GS/s is observed. As expected, the least significant bit (bit 0) toggles the fastest. Successive bits toggle at half the frequency of the previous one.

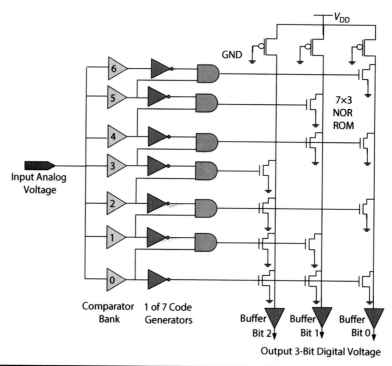

FIGURE 5.38 Complete logic-level diagram for a 3-bit flash ADC [29, 31, 32].

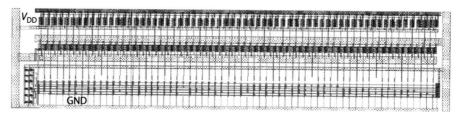

FIGURE 5.39 Complete physical design of the 90-nm flash ADC [29, 32].

Flash ADC Characteristics	Specific Values
Technology	90-nm CMOS 1P 9M
Resolution	6 bit
Supply Voltage (V_{DD})	1.2 V
Sampling Rate	1 GS/s
INL	0.344 LSB
DNL	0.459 LSB
SNDR	31.7 dB @ (f_{in} = 1 MHz)
Peak Power	5.794 mW @ 1.2 V
Average Power	3.875 mW @ 1.2 V
Input Voltage Range	493–557 mV
V_{LSB}	1 mV

TABLE 5.2 Flash ADC Characteristics with Nominal Supply and Threshold Voltages for 90-nm CMOS [29, 31, 32]

Electronic Signal Converter Circuits 243

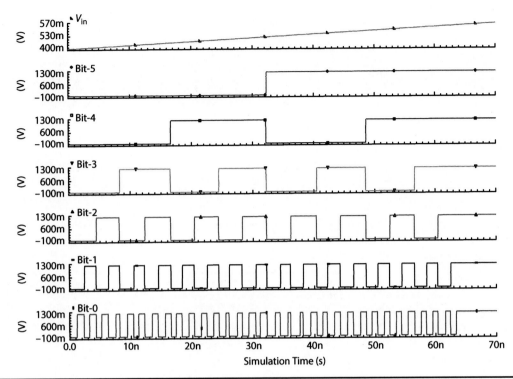

FIGURE 5.40 Functional simulation of the 6-bit ADC operating at 1 GS/s [29, 32].

All 64 codes of the flash ADC are covered as the input ramp traverses the FSR. The flash ADC thus works as desired. The flash ADC has been characterized for selected static and dynamic performances.

5.6.5.1 Static Characterization

The INL and DNL tests have been performed as discussed in the previous sections with nominal supply and threshold voltages to confirm satisfactory results of the flash ADC. The histogram test is used for linear INL and DNL characterization of the flash ADC as discussed before. A mixed-signal simulation environment is set up in which the INL and DNL calculating blocks are implemented in Verilog-A. The Verilog-A block generates a voltage V_{out}, which is sequentially set to 4096 equally spaced voltages between V_{start} and V_{end}, which is the input voltage range for flash ADC conversion. At each different value of V_{out}, a clock pulse is generated causing the flash ADC to convert this V_{out} value. The resultant code of each conversion is stored. When all the conversions have been performed, the INL and DNL are calculated. The total hits between codes 1 and 62 are denoted as hits. The maximum INL obtained is 0.344 LSB; low-INL ensures a low distortion flash ADC. The INL plot of the flash ADC is shown in Fig. 5.41. The maximum DNL is 0.459 LSB; a DNL < 1 LSB also ensures a monotonic transfer function for the flash ADC. The DNL plot of the flash ADC is shown in Fig. 5.42.

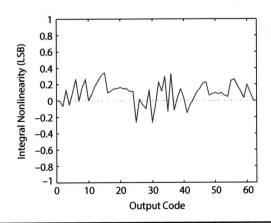

FIGURE 5.41 INL plot of the 90-nm CMOS-based flash ADC [29, 32].

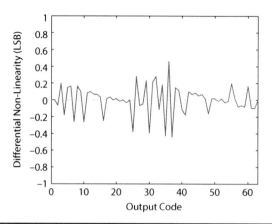

Figure 5.42 DNL plot of the ADC [29, 32].

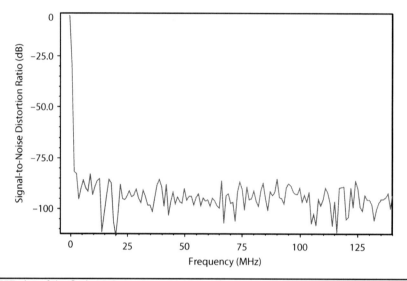

Figure 5.43 FFT plot of the flash ADC at a sinusoidal input frequency f_{in} of 1 MHz for SNDR calculation [29, 32].

5.6.5.2 Dynamic Characterization

The SNDR has been estimated as a specific dynamic performance of the flash ADC. SNDR is a good indicator of the overall dynamic performance of the ADC because it includes all the components that make up noise and distortion as discussed in the previous section. As a specific case study, a 1 MHz sinusoidal input has been fed to the ADC, and the FFT of the output has been obtained. The SNDR is calculated from this fast fourier transformation (FFT) plot that is shown in Fig. 5.43. The SNDR of the 90-nm CMOS-based flash ADC is 31.7 dB.

5.6.5.3 Power Analysis

The power dissipation of the 90-nm CMOS flash ADC circuit has been analyzed. As a specific example, the power analysis of the ADC was performed with a capacitive load of 100 fF. This capacitive load is a reasonable load for a 90-nm CMOS technology [57]. The 90-nm CMOS flash ADC consumes a peak power of 5.794 mW and an average power of 3.875 mW, which satisfies the lower bound on the power dissipation [77]. To provide the insight of the power-hungry components of the flash ADC, the component-wise power consumption is presented in Table 5.3 [29, 31, 32]. It can be observed from the table that the comparator bank consumes maximum power. A possible power reduction scheme in the flash ADC can be to turn off the unused TIQ comparators. The instantaneous power plot of the flash ADC is shown in Fig. 5.44. The instantaneous power plot has a parabolic nature with peak power at the middle of the conversion process because most of the comparators are turned ON at the middle voltage that is $\left| \frac{\text{Input Voltage Range}}{2} \right|$.

Component	Average Power (mW)
Comparator Bank	3.68125 (95%)
1 of N code generators	0.03875 (1%)
NOR ROM	0.155 (4%)
Total	3.875

TABLE 5.3 Component-wise Power Consumption of ADC [29, 31, 32]

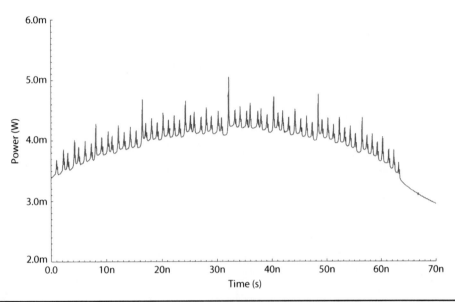

FIGURE 5.44 Instantaneous power plot of ADC with a load capacitance of 100 fF [29, 32].

5.7 A 45-nm CMOS-Based Flash ADC

The design and characterization of the flash ADC presented in the above section have been discussed for 45-nm technology [29, 30, 32]. The block diagram of the ADC remains the same as described earlier in the previous section. However, the sizes of the transistors and the power supply voltage are selected to suit 45-nm CMOS technology instead of 90 nm of the previous section.

5.7.1 Comparator Bank

The methodology followed for choosing the sizes of the transistors in the comparator is the same as for 90-nm technology. Specifically, the TIQ technique is used. The size of a PMOS transistor is made smaller than the NMOS transistors, i.e., $\left|\frac{W_p}{W_n}\right| < 1$, for many transistors in order to ensure that the power dissipation of the flash ADC is as low as possible [29, 30, 32]. The channel length of transistors has been adjusted accordingly. However, a suitable power supply of 0.7 V is adopted for the 45-nm CMOS-based flash ADC design. For the 45-nm CMOS flash ADC, the input voltage range (V_{start}, V_{end}) is: (296.3 mV, 328.3 mV). For this flash ADC, the 1 LSB voltage level VL_{SB} is calculated to be 500 μV. To simplify the sizing process, the following simplification is adopted. The length of the PMOS transistors L_p was kept constant at 90 nm. The size of PMOS transistors $\left|\frac{W_n}{L_n}\right|$ was kept constant at 90 nm/90 nm. The width of the PMOS transistors W_p was varied from 51 to 163 nm to obtain the 64-quantization levels that are 500 μV apart. The results of transistor sizing are presented in Table 5.4.

5.7.2 1 of N Code Generator

Even though there has been some discussion of 1-of-N code generator in the previous section, it is revisited here for more insight. As typically is the case, the 1-of-N code generator converts the thermometer code to a 1-of-N code that is then converted to binary in NOR ROM. The thermometer

Switching Voltage ($V_{switching}$)	PMOS Sizing $\left(\frac{W_p}{L_p}\right)$	NMOS Sizing $\left(\frac{W_n}{L_n}\right)$
296.3 mV (V_{start})	51 nm/90 nm	90 nm/90 nm
327.8 mV (V_{end})	163 nm/90 nm	90 nm/90 nm

TABLE 5.4 Comparator Transistor Sizes for Input Voltage Range at 45-nm Technology [29, 30, 32]

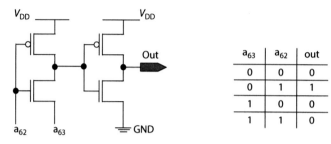

(a) Circuit of a single "01" generator cell   (b) Truth table of the single "01" generator cell

FIGURE 5.45 The circuit of a single cell "01" generator [91].

code is converted to the 1-of-N code by using the "01" generators. The circuit of a single cell "01" generator is shown in Fig. 5.45(a) [91]. The truth table of the single cell "01" generator is presented in Fig. 5.45(b) [91]. For a 6-bit flash ADC, 63 of these "01" generator cells are used in parallel to generate 1-out-of-63 code for the NOR ROM.

5.7.3 NOR ROM

The basic structure of NOR ROM remains the same as the 90-nm CMOS flash design in the previous section. As the specific example, the 6-bit flash ADC has NOR ROM of size 63 × 6. For a good voltage swing, the PMOS sizes were made smaller than the NMOS sizes, i.e., $W_p < W_n$. For the 45-nm CMOS technology, the following sizes are selected meeting the baseline performances: $\left(\frac{W_p}{L_p}\right) = 75$ nm/90 nm and $\left(\frac{W_n}{L_n}\right) = 90$ nm/90 nm [29, 30, 32].

5.7.4 Functional Simulation and Characterization

The functional simulation of the 45-nm flash ADC is shown in Fig. 5.46 [29, 30, 32]. The top signal in the figure is the input ramp covering FS of the flash ADC. The 6-bit output signals of the flash ADC are presented. It is observed that the output codes from 0 to 63 are obtained from the flash ADC with no missing codes. A maximum sampling speed of 100 MS/s has been obtained for this flash ADC design using 45-nm CMOS technology. This speed is sufficient for many applications including digital video [11, 29].

The 45-nm CMOS-based flash ADC has been characterized for INL, DNL, SNDR or SINAD, and power dissipation. In the power dissipation estimation of the flash ADC, both the peak power dissipation and the average power dissipation can be considered. The histogram approach is adopted for calculation of the INL and DNL of the flash ADC. The INL plot of the 45-nm CMOS flash ADC is presented in Fig. 5.47 [29, 30, 32]. The flash ADC for the 45-nm CMOS technology has maximum INL of 0.46 LSB. The DNL plot of the 45-nm CMOS flash ADC is presented in Fig. 5.48 [29, 30, 32]. The flash ADC for the 45-nm CMOS technology has maximum DNL of 0.7 LSB. The calculation of the SINAD of the flash ADC is performed using FFT test. The SINAD plot of the 45-nm CMOS-based flash ADC is shown in Fig. 5.49 [29, 30, 32]. The flash ADC is observed to have a SINAD of 31.9 dB for a 1-MHz sinusoid input signal. The plot of the instantaneous power dissipation for the 45-nm flash ADC at no load condition is shown in Fig. 5.50 [29, 30, 32]. The flash ADC is observed to have a peak power dissipation of 45.42 μW and average power dissipation of 8.8 μW. The complete summary of the characteristics of the 45-nm flash ADC is presented in Table 5.5 [29, 30, 32].

Electronic Signal Converter Circuits

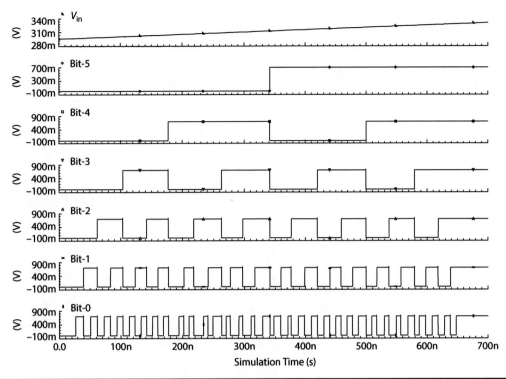

FIGURE 5.46 Functional simulation of the ADC at 45-nm CMOS node [29, 30, 32].

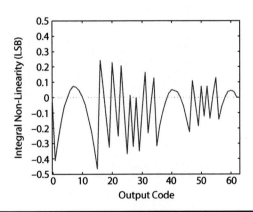

FIGURE 5.47 INL plot of the ADC at 45-nm CMOS node [29, 30, 32].

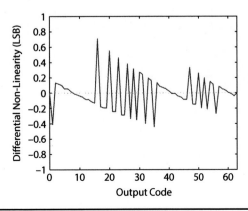

FIGURE 5.48 DNL plot of the ADC at 45-nm CMOS node [29, 30, 32].

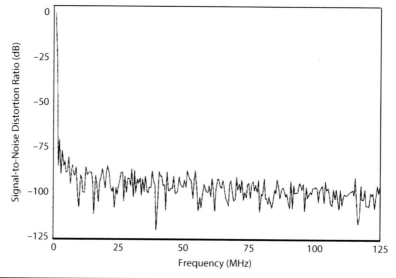

FIGURE 5.49 SNDR plot of ADC at 45-nm CMOS node [29, 30, 32].

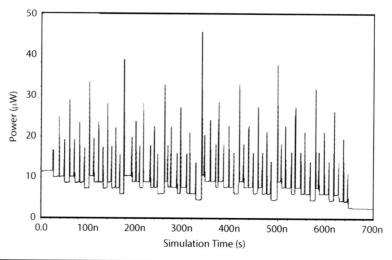

FIGURE 5.50 Instantaneous power plot of the ADC at 45-nm CMOS node [29, 30, 32].

Flash ADC Characteristics	Specific Values
Technology	45 nm
Resolution	6 bit
Supply Voltage (V_{DD})	0.7 V
Sampling Rate	100 MS/s
INL	0.46 LSB
DNL	0.7 LSB
SNDR	31.9 dB @ (f_{in} = 1 MHz)
Peak Power	45.42 µW @ 0.7 V
Average Power	8.8 µW @ 0.7 V
Input Voltage Range	296.3–328.3 mV
V_{LSB}	500 µV

TABLE 5.5 Flash ADC Characteristics for 45-nm CMOS Technology [29, 30, 32]

5.8 Single-Electron-Based ADC

5.8.1 Single-Electron Circuitry-Based ADC

Single-electron circuitry-based ADC designs are being explored for efficient analog to digital conversion. The single-electron circuitry consists of conducting isolated islands, tunnel junctions, and capacitors [43]. In the single-electron circuitry-based ADC, the digital output signal is produced directly without the use of reference voltages and auxiliary blocks including comparators, quantizers, digital counters, and PWMs [43]. These blocks make traditional ADCs power hungry, large in size, and slow in terms of conversion operation. The schematic representation of a 3-bit single-electron circuitry-based ADC is shown in Fig. 5.51 [43]. The circuit of the single-electron circuitry-based ADC consists of the following: (1) *Five Islands*: IS_1, IS_2, IS_3, IS_4, and IS_5; (2) *Five Tunnel Junctions*: TJ_1, TJ_2, TJ_3, TJ_4, and TJ_5; (3) *Five Capacitors*: C_{in}, C_1, C_2, C_3, and C_4. The islands IS_2, IS_3, IS_4, and IS_5 are connected to the ground through the capacitors C_1, C_2, C_3, and C_4, respectively. The input analog signal is applied to the ADC circuit through the capacitor C_{in}. The output signal of this 3-bit ADC circuit is taken from islands IS_2, IS_3, and IS_5 as depicted in Fig. 5.51.

5.8.2 Single-Electron Transistor-Based ADC

The circuit of a single-electron transistor (SET)-based 4-bit ADC is shown in Fig. 5.52 [12]. For the 4-bit ADC, there are four similar stages, as depicted in the figure. Each stage consists of a voltage ramp circuit and

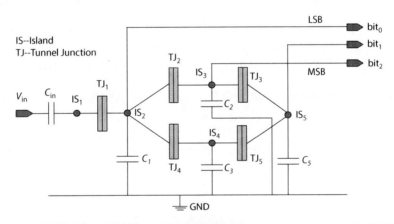

FIGURE 5.51 Circuit of the single-electron circuitry-based 3-bit ADC [43].

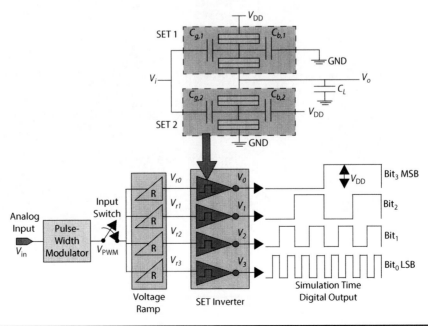

FIGURE 5.52 Circuit of the SET-based 4-bit ADC [12].

a SET-based inverter. In this ADC, the analog voltage signal is fed to a PWM. The PWM signal is applied to a switch. The output signal from the switch goes to a voltage ramp. The output of each voltage ramp rises from the input offset voltage V_{off} for the duration of the pulse that is the switch ON time. Each of the voltage ramp circuits has its own speed and voltage slope. SET inverter is an important component for each bit stage of the ADC [12, 34, 79]. The periodic output nature of the SET inverters is crucial for this ADC. At the end, the voltage produced by the ADC circuit is either 0 V or V_{DD}. The ADC design can be similarly extended for any N-bit version by simply adding N number of stages.

5.9 Organic Thin-Film Transistor-Based ADCs

5.9.1 Organic Thin-Film Transistor VCO-Based ADC

An organic thin-film transistor (OTFT) voltage-controlled oscillator (VCO)-based ADC is presented in Fig. 5.53 [64]. At the architecture level, it consists of an OTFT-based VCO and a ripple counter. The OTFT-based VCO produces the square waveform of frequency proportional to the input analog signal. The ripple counter quantizes the square waveform to produce digital output signal. The OTFT-based VCO consists of 7-inverting stages and 1-noninverting delay stage. The delay of the noninverting-delay stage is tunable. As the delay of the noninverting-delay stage is much larger than the delay of the inverting stages, the oscillation frequency of the VCO is mostly affected by it. In its operation, the output of the noninverting-delay stage is a saw tooth waveform. The inverting stages convert this saw tooth waveform to a large duty cycle square wave VCO_{out} [64]. The noninverting-delay stage consists of P-type OTFTs that are normally ON. The PM1 and PM2 OTFTs operate like transconductors [64, 65]. A linear characteristic of transconductors ensures a linear voltage-frequency characteristic of the VCO. The PM3 OTFT is a fast source follower. The characteristics of the OTFT-based ADC are presented in Table 5.6.

5.9.2 Complementary Organic Thin-Film Transistor-Based Successive-Approximation ADC

The block diagram of a SA-ADC with components made from OTFTs is shown in Fig. 5.54 [86, 87]. The SA-ADC contains the blocks like the sample and hold amplifier, the comparator, the SAR, and the DAC. The two blocks—the comparator and the DAC—are designed using OTFTs [86, 87]. Selected characteristics of the OTFT-based SA-ADC are presented in Table 5.7 [87].

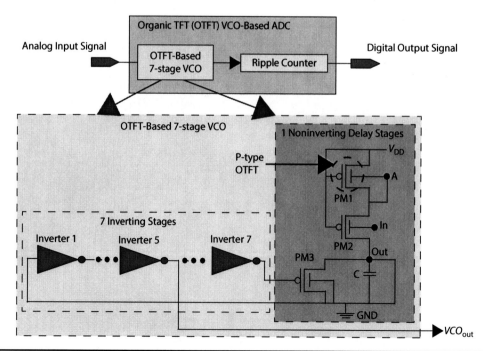

Figure 5.53 OTFT VCO-based ADC [64].

OTFT-Based ADC Parameters	Actual Values
Resolution	6 Bits
INL	1 LSB
DNL	0.6 LSB
$ENOB_{noise}$	7.7 Bits

TABLE 5.6 Characteristics of the OTFT-Based ADC [64]

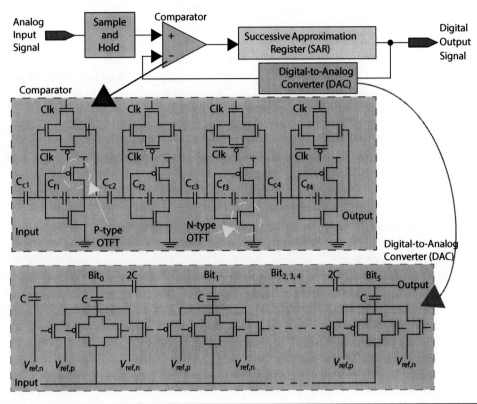

FIGURE 5.54 Complementary OTFT-based SA-ADC [86, 87].

OTFT-Based SA-ADC Parameters	Actual Values
Resolution	6 bits
Maximum INL	0.6 LSB (at 10 Hz) or −3 LSB (at 100 Hz)
Maximum DNL	−0.6 LSB (at 10 Hz) or 1.5 LSB (at 100 Hz)
Power Dissipation	3.6 μW

TABLE 5.7 Characteristics of the OTFT-Based SA- ADC [87]

The comparator design using OTFT is also presented in Fig. 5.54 [87]. It consists of a bank of inverters and transmission gates. The inverters provide self-biasing and tolerance to mismatches and process variability. The transmission gate provides a two-phase operation for the comparator: the reset phase and compare phase. In the reset phase, the transmission gates are ON to connect the input and output of the inverters to establish the DC operating points. In the compare phase, the transmission gates are OFF to allow the cascaded inverters to act. The error is essentially amplified by the inverters to

252 Chapter Five

compare the DAC code and the sampled input. As shown in the figure, C-2C DAC has been designed using OTFT transistors [86, 87]. The OTFT process offers no parasitic capacitance, unlike the classic CMOS technology processes.

5.10 Sigma-Delta Modulator-Based ADC

In general, an analog signal is encoded into a digital signal or even a higher-resolution digital signal is encoded into a lower-resolution digital signal using the sigma-delta ($\Sigma-\Delta$) modulation. The sigma-delta modulation technique can be used by both the ADCs and DACs in their architecture realization [67]. In the current section sigma-delta ($\Sigma-\Delta$) modulation based ADC is discussed and sigma-delta ($\Sigma-\Delta$) modulation based DAC will be discussed in a later section.

5.10.1 Broad Prospective

For a broad perspective of a sigma-delta modulator, a high-level presentation is provided in Fig. 5.55 [13, 16, 74]. The modulator samples the input analog signal at a higher frequency. The modulator then converts the input analog signal to a PWM signal that also includes noise. The decimator filters the noise present in the output signal from the modulator. The output signal of the modulator is a 1-bit output signal, whereas the output signal of the decimator is an N-bit digital signal. N is the resolution of the ADC that is dictated by the oversampling ratio. The order of the modulator dictates the order of the decimator. The order of the decimator is typically one or more than the order of the modulator.

5.10.2 Architecture Overview

A block-level representation of a sigma-delta modulator-based ADC architecture is shown in Fig. 5.56 [13, 16, 18, 67, 94]. It represents a first-order sigma-delta modulator ADC. Higher order sigma-delta modulator ADC can be constructed in a similar manner. For example, a second-order sigma-delta modulator ADC can be made using two modulator stages connected together in cascade with one DAC. In general, a sigma-delta modulator can be either continuous-time or discrete-time depending on the sampling stage [94]. The block-diagram of continuous-time sigma-delta modulator-based ADC is shown in Fig. 5.56(a) [13, 16, 18, 67, 94]. The block-diagram of discrete-time sigma-delta modulator-based ADC is shown in Fig. 5.56(b) [13, 16, 18, 67, 94]. The continuous-time sigma-delta modulator samples the analog signal inside the loop just before the quantizer. The discrete-time sigma-delta modulator samples the analog input signal outside the loop at the beginning, right after the anti-aliasing amplifier.

The components used in the architecture thus can be different based on the use of either a continuous-time sigma-delta modulator or discrete-time sigma-delta modulator. For example, an analog integrator is used in a continuous-time sigma-delta modulator as shown in Fig. 5.56(a). On the other hand, a discrete-time integrator is used in a discrete-time sigma-delta modulator as shown in Fig. 5.56(b). The different blocks of the sigma-delta modulator-based ADC architecture include the following: summing amplifier, discrete-time integrator, quantizer, 1-bit DAC, gain amplifier, low-pass filter, and downsampler. The individual block of the sigma-delta modulator-based ADC will be discussed in the following subsections. The simulation of an ideal sigma-delta modulator-based ADC is shown in Fig. 5.57 [13]. The top part of Fig. 5.57 shows the input clock signal to the sigma-delta modulator-based ADC. The bottom part of Fig. 5.57 shows the input and output data signal to the

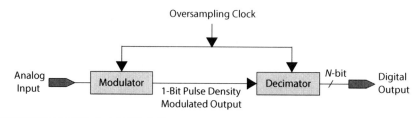

FIGURE 5.55 A broad perspective of a sigma-delta-modulator-based ADC [13, 16, 74].

Electronic Signal Converter Circuits 253

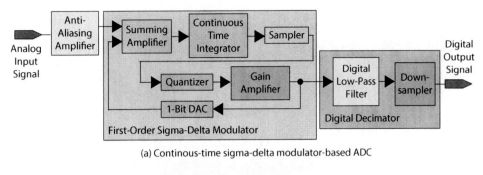

(a) Continuous-time sigma-delta modulator-based ADC

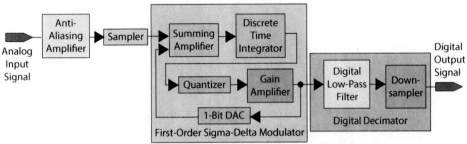

(b) Discrete-time sigma-delta modulator-based ADC

FIGURE 5.56 Block diagram of a sigma-delta modulator-based ADC [13, 16, 18, 67, 94].

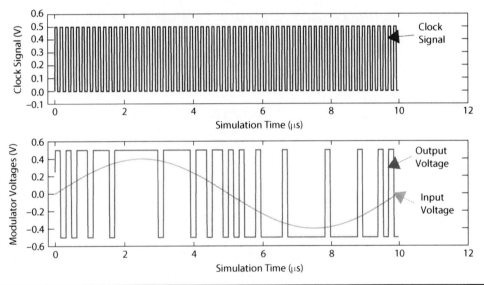

FIGURE 5.57 Simulation of an ideal sigma-delta modulator-based ADC architecture [13].

sigma-delta modulator-based ADC. The input to the sigma-delta modulator-based ADC is a sinusoid signal as shown in the figure. The output signal pulses are also shown in the same figure.

The question arises whether to design a continuous-time or a discrete-time sigma-delta modulator [70, 94]. Designing in the continuous-time domain can allow easy fine tuning of the design; however, it may be difficult to implement the continuous-time modulators. For example, all the design variables obtained in the design process in a continuous-time domain may not be easily implemented in practice. The continuous-time domain design process has been a classic approach. On the other hand, the discrete-time domain design allows easy selection of topologies and implementations, but accuracy can be compromised. At present, the design tools are quite mature and it is possible to perform the design in one domain and translate it to the other domain. However, the approach that is getting popular is to perform the design in the discrete-time domain using the discrete-time tools and obtain the desired target specifications. Then, perform mapping of the discrete-time domain design to the continuous-time domain design.

5.10.3 Architecture Components

5.10.3.1 Summing Amplifier

The summing amplifier in a sigma-delta modulator-based ADC considers an input voltage and an output voltage of the DAC. The summing amplifier calculates the difference between the two input signals. The error signal of the two is then amplified by the summing amplifier. The simulation of an ideal summing amplifier block is presented in Fig. 5.58 [13]. The design of the summing amplifier is based on OP-AMP. It can be a simple resistive ladder-based OP-AMPs similar to the ones discussed in DAC architectures. However, CMOS transistor-based efficient summing amplifiers are preferred design options for efficient sigma-delta modulator-based ADC implementations [78]. For a discrete-time sigma-delta modulator, the operation performed is the following:

$$V_{sa}[n] = V_{in}[n] - V_{FB}[n] \tag{5.38}$$

where $V_{sa}[n]$ is the discrete-time voltage generated from the summing amplifier, $V_{in}[n]$ is the discrete-time voltage input to the summing amplifier that is essentially generated from the sampler, and $V_{FB}[n]$ is the discrete-time voltage feedback voltage signal generated from the 1-bit DAC.

5.10.3.2 Integrator

The integrator in a sigma-delta modulator-based ADC checks the difference between two signals, the input clock signal and the transitions voltage clock signal. If the difference is positive, then the signal propagates through it. The output voltage produced by the integrator is the sum of the input voltages. The simulation of an ideal integrator block is presented in Fig. 5.59 [13]. For a discrete-time sigma-delta modulator, the operation the integrator performs is the following function:

$$V_{int}[n] = V_{sa}[n-1] + V_{int}[n-1] \tag{5.39}$$

where $V_{int}[n]$ is the discrete-time voltage generated from the discrete-time integrator. However, in general, a digital integrator can be of various types such as infinite-impulse response (IIR) and finite-impulse response (FIR). The digital integrator can be forward rectangular integrator or backward rectangular integrator. The above expression assumes a forward digital integrator.

5.10.3.3 Quantizer

The inputs for the quantizer block are the integrator's output voltage and a reference voltage. The quantizer block checks whether the difference between the two voltages is positive or negative. The quantizer blocks an output of level +1 if the difference is positive and produces an output of

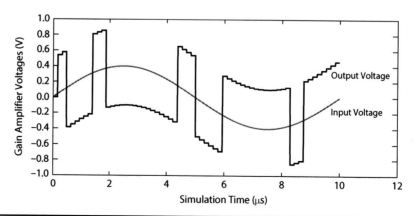

Figure 5.58 Simulation of an ideal summing amplifier [13].

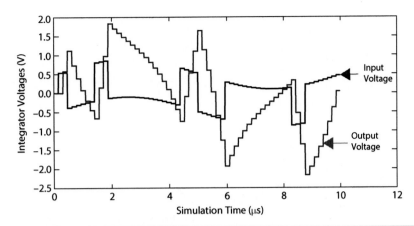

FIGURE 5.59 Simulation of an ideal integrator [13].

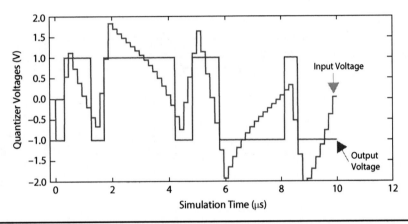

FIGURE 5.60 Simulation of an ideal quantizer [13].

level -1 if the difference is negative. The simulation of an ideal quantizer is presented in Fig. 5.60 [13]. The operation performed by the quantizer is expressed as

$$V_{\text{qnt}}[n] = \begin{cases} 1 & (V_{\text{int}}[n] - V_{\text{off}}) \geq 0 \\ -1 & (V_{\text{int}}[n] - V_{\text{off}}) < 0 \end{cases} \qquad (5.40)$$

where $V_{\text{qnt}}[n]$ is the discrete-time output signal from the quantizer. V_{off} is the threshold or offset voltage of the quantizer that also may be 0.

5.10.3.4 1-Bit Digital-to-Analog Converter

The input to the 1-bit DAC used in the feedback is the output from the quantizer. The output produced by it is an analog voltage that goes to the summing amplifier as a feedback signal. The simulation results of an ideal 1-bit DAC are presented in Fig. 5.61 [13].

5.10.3.5 Gain Amplifier

The input to the gain amplifier is the discrete-time and discrete-value (digital) signal from the quantizer. The output produced by the gain amplifier is the product of a specified gain of the input digital voltage. The simulation of an ideal gain amplifier of the sigma-delta ADC is shown in Fig. 5.62 [13]. The operation in essence is

$$V_{\text{gain}}[n] = K_{\text{gain}} V_{\text{qnt}}[n] \qquad (5.41)$$

where K_{gain} is gain of the gain amplifier of the sigma-delta ADC.

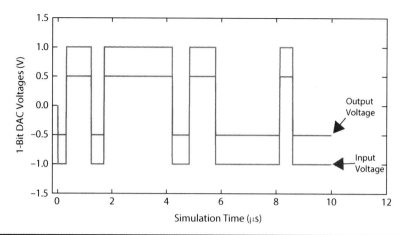

Figure 5.61 Simulation of an ideal 1-bit DAC [13].

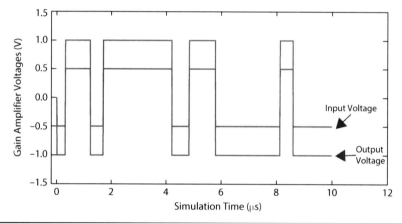

Figure 5.62 Simulation of an ideal gain amplifier [13].

5.10.3.6 Low-Pass Digital Filter

The low-pass digital filter is present after the gain amplifier of the sigma-delta modulator-based ADC. The low-pass digital filter removes all the frequency components above the target bandwidth of the ADC. It can be as simple as an averaging digital filter. A typical sigma-delta modulator-based ADC uses sinc filter as an averaging filter [19].

5.10.3.7 Downsampler

The downsampler is at the output of the sigma-delta modulator-based ADC. Both the input and output signals of this block are digital in nature. The operation of the downsampler is to bring the signal back to the Nyquist rate as input signal in a sigma-delta modulator-based ADC is oversampled to avoid aliasing.

5.11 Sigma-Delta Modulator-Based Digital-to-Analog Converter

The input to the sigma-delta-based DAC is an N-bit digital signal and the output produced is an output analog signal that is as accurate as the input signal. Both the sigma-delta-based ADC and sigma-delta-based DAC have similarity in terms of their block and operations. Simplistically speaking, the blocks and operation that are realized using analog signal processing circuits in a sigma-delta-based ADC

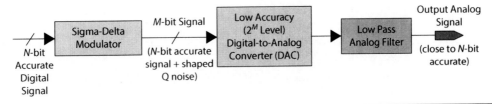

FIGURE 5.63 Sigma-delta modulator-based DAC [23, 35, 40].

are realized using digital signal processing circuits in a sigma-delta-based DAC, and vice versa. The high-level representation of a sigma-delta modulator-based DAC is shown in Fig. 5.63 [23, 35, 40]. It consists of a sigma-delta modulator, a low-accuracy DAC, and a low-pass analog filter. The sigma-delta modulator quantizes the N-bit high-precision input digital signal to an M-bit signal. The M-bit signal is thus effectively the N-bit accurate signal along with the quantization noise introduced by the modulator. The low-accuracy DAC converts the M-bit signal to analog voltage. The converted analog output signal from the low-accuracy DAC contains high-frequency noise as well. The low-pass analog filter block present at the final stage of the sigma-delta-based DAC removes the noise and produces a clean output analog signal close to the N-bit digital signal input.

5.12 Single Electron Transistor-Based Digital-to-Analog Converter

An SET-based DAC is shown in Fig. 5.64 [46]. It contains the following distinct components: SET-based CMOS inverters, classical-transistor-based CMOS inverters, amplifier, and capacitance network [46, 61, 86]. In this DAC, the input digital signal (N-bit) goes through SET-CMOS inverters and classical-CMOS inverters. The capacitance network performs weighted summation functions to generate analog voltage. The generated voltage from the capacitance network is amplified by the amplifier to obtain the final analog signal output V_{out}.

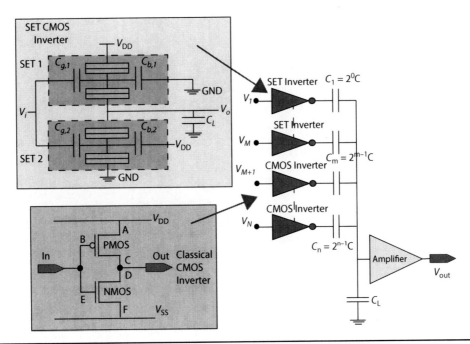

FIGURE 5.64 SET-based DAC [46].

5.13 Questions

5.1 Consider a real-life audio, image, or video-processing system. Discuss the signal processing flow and the role of ADCs and DACs in it.

5.2 Consider a ramp-compare ADC. Perform its simulation using a high-level language or platform.

5.3 Consider an integrating ADC. Perform its simulation using a high-level language or platform.

5.4 Consider a pipeline ADC. Discuss its principle of operation in a small paragraph with the help of block diagrams.

5.5 Consider a time-interleaved ADC. Perform its simulation using a high-level language or platform. The built-in blocks can be used from an available library.

5.6 Consider a folding ADC. Discuss its principle of operation in a small paragraph with the help of block diagrams.

5.7 Consider a tracking ADC. Perform its simulation using a high-level language or platform. The built-in blocks can be used from an available library.

5.8 Consider a binary-weighted DAC. Perform its transistor-level design using a schematic editor. Perform its SPICE simulation for a nanoscale CMOS technology. The existing library components can be used if available.

5.9 Consider an R-2R ladder network DAC. Perform its transistor-level design using a schematic editor. Perform its SPICE simulation for a nanoscale CMOS technology. The existing library components can be used if available.

5.10 Consider a segmented DAC. Perform its simulation using a high-level language or platform. The built-in blocks can be used from an available library.

5.11 Consider an interpolating DAC. Perform its simulation using a high-level language or platform. The built-in blocks can be used from an available library.

5.12 Consider a cyclic DAC. Perform its simulation using a high-level language or platform. The built-in blocks can be used from an available library.

5.13 Consider a pipeline DAC. Perform its simulation using a high-level language or platform. The built-in blocks can be used from an available library.

5.14 Discuss five important characteristics of the ADC. Discuss how each one of them can be characterized.

5.15 Discuss five important characteristics of the DAC. Discuss how each one of them can be characterized.

5.16 Consider a flash ADC. Discuss its principle of operation in a small paragraph with the help of block diagrams.

5.17 Consider a comparator design of a flash ADC. Perform its transistor-level design using a schematic editor. Perform its SPICE simulation for a nanoscale CMOS technology.

5.18 Perform the above comparator design of a flash ADC for a double-gate CMOS technology. Perform its SPICE simulation and characterization.

5.19 Consider a 1 of N generator design of a flash ADC. Perform its transistor-level design using a schematic editor. Perform its SPICE simulation for the following three nanoscale CMOS technology: 45-nm classical CMOS, 32-nm high-k CMOS, and double-gate.

5.20 Consider overall design of a flash ADC. Perform its simulation using a high-level language or platform or transistor-level schematic. The built-in blocks can be used from an available library.

5.21 Consider the design of the basic building block of digital circuits, the inverter. Perform its simulation and characterization for a SET-based CMOS technology.

5.22 Consider the overall design of a SET-based ADC. Perform its simulation using a high-level language or platform or transistor-level schematic. The built-in blocks can be used from an available library.

5.23 Consider overall design of an OTFT-based ADC. Perform its simulation using a high-level language or platform or transistor-level schematic. The built-in blocks can be used from an available library.

5.24 Consider overall design of a sigma-delta modulator-based ADC. Perform its simulation using a high-level language or platform. The built-in blocks can be used from an available library.

5.25 Consider overall design of a sigma-delta modulator-based DAC. Perform its simulation using a high-level language or platform. The built-in blocks can be used from an available library.

5.26 Consider overall design of a SET-based DAC. Perform its simulation using a high-level language or platform or transistor-level schematic. The built-in blocks can be used from an available library.

5.14 References

1. Defining and Testing Dynamic Parameters in High Speed ADCs, Part 1. http://www.maxim-ic.com/appnotes.cfm/an_pk/728 (2001). Accessed on 02 July 2013.
2. INL/DNL Measurements for High-Speed Analog-to-Digital Converters (ADCs). http://www.maximintegrated.com/app-notes/index.mvp/id/283 (2001). Accessed on 15 July 2013.
3. Understanding Pipelined ADCs. http://www.maximintegrated.com/app-notes/index.mvp/id/1023 (2001). Accessed on 07 July 2013.
4. ADC and DAC Glossary. http://www.maxim-ic.com/app-notes/index.mvp/id/641 (2002). Accessed on 02 July 2013.
5. Understanding Integrating ADCs. http://www.maximintegrated.com/app-notes/index.mvp/id/1041 (2002). Accessed on 07 July 2013.
6. A Simple ADC Comparison Matrix. http://www.maximintegrated.com/app-notes/index.mvp/id/2094 (2003). Accessed on 10 July 2013.
7. Histogram Testing Determines DNL and INL Errors. http://www.maxim-ic.com/appnotes.cfm/an_pk/2085 (2003). Accessed on 02 July 2013.
8. RIM's Camera BlackBerry Patent: Not What You Think! (2006). URL http://www.blackberrycool.com/2006/07/07/001980/
9. Electronics Tutorial about Summing Operational Amplifiers. http://www.electronics-tutorials.ws/opamp/opamp_4.html (2013). Accessed on 12 July 2013.
10. Adams, R., Nguyen, K.Q.: A 113-dB SNR Oversampling DAC With Segmented Noise-Shaped Scrambling. *IEEE Journal of Solid-State Circuits* **33**(12), 1871–1878 (1998). DOI 10.1109/4.735526.
11. Adeniran, O., Demosthenous, A., Clifton, C., Atungsiri, S., Soin, R.: A CMOS Low-Power ADC for DVB-T and DVB-H systems. In: *Proceedings of the International Symposium on Circuits and Systems*, pp. 209–212 (2004).
12. Ahn, S.J., Kim, D.M.: Asynchronous Analogue-to-Digital Converter for Single-Electron Circuits. *Electronics Letters* **34**(2), 172–173 (1998). DOI 10.1049/el:19980157.
13. Ale, A.K.: *Comparison and Evaluation of Existing Analog Circuit Simulators Using a Sigma-delta Modulator*. Master's thesis, Department of Computer Science and Engineering, University of North Texas, Denton (2006). URL http://books.google.com/books?id=c7aouAAACAAJ.
14. Allen, P.E., Holberg, D.R.: *CMOS Analog Circuit Design*, 3rd edn. Oxford University Press, USA (2011).
15. Allen, R.L., Mills, D.: *Signal Analysis: Time, Frequency, Scale, and Structure*. chap. Signals: Analog, Discrete, and Digital, pp. 1–100. Wiley-IEEE Press, Piscataway, NJ (2004).
16. Anantha, R.R.: *A Programmable CMOS Decimator for Sigma Delta Analog-to-Digital Converter and Charge Pump Circuits*. Master's thesis, Department of Electrical and Computer Engineering, Louisiana State University, Baton Rouge, LA (2005). Accessed on 06 July 2013.
17. Austerlitz, H.: *Data Acquisition Techniques Using PCs*. Academic Press, San Diego, CA, USA (2002). URL http://books.google.com/books?id=gK4DxMaqmYYC.
18. Aziz, P.M., Sorensen, H.V., van der Spiegel, J.: An Overview of Sigma-Delta Converters. *IEEE Signal Processing Magazine* **13**(1), 61–84 (1996). DOI 10.1109/79.482138.
19. Baker, B.: How delta-sigma ADCs Work, Part 2. *Analog Applications Journal* pp. 5–7 (2011). Accessed on 01 August 2013.
20. Belega, D., Dallet, D., Petri, D.: Estimation of the Effective Number of Bits of ADCs using the Interpolated DFT Method. In: *Proceedings of the IEEE Instrumentation and Measurement Technology Conference*, pp. 30–35 (2010). DOI 10.1109/IMTC.2010.5488200.
21. Black, B.: Analog-to-Digital Converter Architectures and Choices for System Design. *Analog Dialogue* **33**(8), 1–4 (1999). Accessed on 03 July 2013.
22. Chen, P., Liu, T.C.: Switching Schemes for Reducing Capacitor Mismatch Sensitivity of Quasi-Passive Cyclic DAC. *IEEE Transactions on Circuits and Systems II: Express Briefs* **56**(1), 26–30 (2009). DOI 10.1109/TCSII.2008.2010158.

23. Cheung, H., Raj, S.: Implementation of 12-bit Delta-Sigma DAC with MSC12xx Controller. *Analog Applications Journal* pp. 27–33 (2005).
24. Choudhury, J., Massiha, G.H.: Efficient Encoding Scheme for Ultra-Fast Flash ADC. In: *Proceedings of the 5th Topical Meeting on Silicon Monolithic Integrated Circuits in RF Systems*, pp. 290–293 (2004).
25. Delagnes, E., Breton, D., Lugiez, F., Rahmanifard, R.: A Low Power Multi-Channel Single Ramp ADC with Up to 3.2 GHz Virtual Clock. *IEEE Transactions on Nuclear Science* **54**(5), 1735–1742 (2007). DOI 10.1109/TNS.2007.906170.
26. Efstathiou, K.A., Karadimas, D.S.: An R-2R Ladder-Based Architecture for High Linearity DACs. In: *Proceedings of the IEEE Instrumentation and Measurement Technology Conference*, pp. 1–5 (2007). DOI 10.1109/IMTC.2007.379448.
27. Fossum, E.R.: CMOS Image Sensors: Electronic Camera-On-A-Chip. *IEEE Transactions on Electron Devices* **44**(10), 1689–1698 (1997).
28. Fusayasu, T.: A Fast Integrating ADC Using Precise Time-To-Digital Conversion. In: *IEEE Nuclear Science Symposium Conference Record*, vol. 1, pp. 302–304 (2007). DOI 10.1109/NSSMIC.2007.4436335.
29. Ghai, D.: *Variability Aware Low-Power Techniques for Nanoscale Mixed-Signal Circuits*. Ph.D. thesis, Department of Computer Science and Engineering, University of North Texas, Denton (2009).
30. Ghai, D., Mohanty, S.P., Kougianos, E.: A 45nm Flash Analog to Digital Converter for Low Voltage High Speed System on Chips. In: *Proceedings of the 13th NASA Symposium on VLSI Design*, p. 3.1 (2007).
31. Ghai, D., Mohanty, S.P., Kougianos, E.: A Process and Supply Variation Tolerant Nano-CMOS Low Voltage, High Speed, A/D Converter for System-on-Chip. In: *Proceedings of the Great Lakes Symposium on VLSI*, pp. 47–52 (2008).
32. Ghai, D., Mohanty, S.P., Kougianos, E.: A Variability Tolerant System-on-Chip Ready Nano-CMOS Analog-to-Digital Converter (ADC). *Taylor & Francis International Journal of Electronics (IJE)* **97**(4), 421–440 (2010).
33. Halim, I.S.A., Hassan, S.L.M., Akbar, N.D.B.M., Rahim, A.A.A.: Comparative Study of Comparator and Encoder in a 4-it Flash ADC Using 0.18 μ CMOS Technology. In: *Proceedings of the IEEE Symposium on Computer Applications and Industrial Electronics (ISCAIE)*, pp. 35–38 (2012). DOI 10.1109/ISCAIE.2012.6482064.
34. Heij, C.P., Hadley, P., Mooij, J.E.: Single-Electron Inverter. *Applied Physics Letters* **78**(8), 1140–1142 (2001). DOI 10.1063/1.1345822. URL http://link.aip.org/link/?APL/78/1140/1.
35. Hongqin, L., Xiumin, S.: 18 bit Delta-Sigma DAC in 0.18 μm CMOS Process. In: *Proceedings of the 2nd International Conference on Information Science and Engineering (ICISE)*, pp. 3376–3379 (2010). DOI 10.1109/ICISE.2010.5691093.
36. Horsky, P.: A 16 Bit+Sign Monotonic Precise Current DAC for Sensor Applications. In: *Proceedings of the Design, Automation and Test in Europe Conference and Exhibition*, vol. 3, pp. 34–38 (2004). DOI 10.1109/DATE.2004.1269195.
37. Huang, C.C., Wang, C.Y., Wu, J.T.: A CMOS 6-Bit 16-GS/s Time-Interleaved ADC Using Digital Background Calibration Techniques. *IEEE Journal of Solid-State Circuits* **46**(4), 848–858 (2011). DOI 10.1109/JSSC.2011.2109511.
38. Jiang, H., Olleta, B., Chen, D., Geiger, R.: A Segmented Thermometer Coded DAC with Deterministic Dynamic Element Matching for High Resolution ADC Test. In: *IEEE International Symposium on Circuits and Systems (ISCAS)*, pp. 784–787 (2005). DOI 10.1109/ISCAS.2005.1464705.
39. Julsereewong, A., Pongswatd, S., Riewruja, V., Sasaki, H., Fujimoto, K., Shi, Y.: Accurate Dual Slope Analog-to-Digital Converter Using Bootstrap Circuit. In: *Proceedings of the Fourth International Conference on Innovative Computing, Information and Control (ICICIC)*, pp. 1361–1364 (2009). DOI 10.1109/ICICIC.2009.65.
40. Kamath, A.S., Chattopadhyay, B.: A Wide Output Range, Mismatch Tolerant Sigma Delta DAC for Digital PLL in 90 nm CMOS. In: *Proceedings of the IEEE International Symposium on Circuits and Systems*, pp. 69–72 (2012). DOI 10.1109/ISCAS.2012.6272127.
41. Kester, W. (ed.): *Analog-Digital Conversion*. Analog Devices, Inc. (2004).
42. Kester, W.: Understand SINAD, ENOB, SNR, THD, THD + N, and SFDR so You Don't Get Lost in the Noise Floor. http://www.analog.com/static/imported-files/tutorials/MT-003.pdf (2009). Accessed on 15 July 2013.
43. Kiziroglou, M.E., Karafyllidis, I.: Design And Simulation of a Nanoelectronic Single-Electron Analog to Digital Converter. *Microelectronics Journal* **34**(9), 785–789 (2003). DOI 10.1016/S0026-2692(03)00153-8. URL http://www.sciencedirect.com/science/article/pii/S0026269203001538.
44. Kosonocky, S., Xiao, P.: *Digital Signal Processing Handbook*, chap. Analog-to-Digital Conversion Architectures, pp. 5.1–5.9. CRC Press LLC, Boca Raton, FL (1997).
45. Lad, K., Bhat, M.S.: A 1-V 1-GS/s 6-bit Low-Power Flash ADC in 90-nm CMOS with 15.75 mW Power Consumption. In: *Proceedings of the International Conference on Computer Communication and Informatics (ICCCI)*, pp. 1–4 (2013). DOI 10.1109/ICCCI.2013.6466320.
46. Le, J.Y., Jiang, J.F., Cai, Q.Y.: Design of Hybrid SET-CMOS D/A Converter. In: *Proceedings of the 4th International Conference on ASIC*, pp. 299–302 (2001). DOI 10.1109/ICASIC.2001.982558.
47. Lee, D., Yoo, J., Choi, K.: Design Method and Automation of Comparator Generation for Flash A/D Converter. In: *Proceedings of the International Symposium on Quality Electronics Design (ISQED)*, pp. 138–142 (2002).

48. Leung, B.: Pipelined Multi-Bit Oversampled Digital to Analog Converters With Capacitor Averaging. In: *Proceedings of the 34th Midwest Symposium on Circuits and Systems*, pp. 336–339 (1991). DOI 10.1109/MWSCAS.1991.252212.
49. Lewyn, L., Ytterdal, T., Wulff, C., Martin, K.: Analog Circuit Design in Nanoscale CMOS Technologies. *Proceedings of the IEEE* 97(10), 1687–1714 (2009). DOI 10.1109/JPROC.2009.2024663.
50. Li, J., Zhang, J., Shen, B., Zeng, X., Guo, Y., Tang, T.: A 10bit 30MSPS CMOS A/D Converter for High Performance Video Applications. In: *Proceedings of IEEE European Solid-State Circuits Conference*, pp. 523–526 (2005).
51. Lin, C.W., Lin, S.F.: A BIST scheme for testing DAC. In: *9th International Conference on Electrical Engineering/ Electronics, Computer, Telecommunications and Information Technology (ECTI-CON), 2012*, pp. 1–4 (2012). DOI 10.1109/ECTICon.2012.6254328.
52. Lin, S.M., Li, D.U., Chen, W.T.: 1 V 1.25 GS/s 8 mW D/A Converters for MB-OFDM UWB Transceivers. In: *Proceedings of the IEEE International Conference on Ultra-Wideband*, pp. 453–456 (2007). DOI 10.1109/ICUWB.2007.4380987.
53. Luo, C., McClellan, J.H.: Compressive Sampling with a Successive Approximation ADC Architecture. In: *Proceedings of the IEEE International Conference on Acoustics, Speech and Signal Processing (ICASSP)*, pp. 3920–3923 (2011). DOI 10.1109/ICASSP.2011.5947209.
54. Maloberti, F.: *Data Converters*. Springer (2007).
55. Marche, D., Savaria, Y.: Modeling R-2R Segmented-Ladder DACs. *IEEE Transactions on Circuits and Systems I: Regular Papers* 57(1), 31–43 (2010). DOI 10.1109/TCSI.2009.2019396.
56. *MobileReference: Electronics Quick Study Guide for Smartphones and Mobile Devices*. Mobi Study Guides. MobileReference.com (2007). URL http://books.google.com/books?id=AG6H633VIAcC.
57. Mohanty, S.P., Vadlamudi, S.T., Kougianos, E.: A Universal Voltage Level Converter for Multi-V_{dd} Based Low-Power Nano-CMOS Systems-on-Chips (SoCs). In: *Proceedings of the 13th NASA Symposium on VLSI Design*, p. 2.2 (2007).
58. Mortezapour, S., Lee, E.K.F.: A Reconfigurable Pipelined Data Converter. In: *Proceedings of the IEEE International Symposium on Circuits and Systems*, vol. 4, pp. 314–317 (2001). DOI 10.1109/ISCAS.2001.922235.
59. Okada, S., Matsuda, Y., Yamada, T., Kobayashi, A.: System On a Chip for Digital Still Camera. *IEEE Transactions on Consumer Electronics* 45(3), 1689–1698 (1999).
60. Ozalevli, E., Dinc, H., Lo, H.J., Hasler, P.: Design of A Binary-Weighted Resistor DAC Using Tunable Linearized Floating-Gate CMOS Resistors. In: *Proceedings of the IEEE Custom Integrated Circuits Conference*, pp. 149–152 (2006). DOI 10.1109/CICC.2006.320867.
61. Perry, J.C.: *Digital to Analog Converter Design Using Single Electron Transistors*. Master's thesis, Computer Engineering, Virginia Polytechnic Institute and State University, Blacksburg, Virginia (2005). Accessed on 03 July 2013.
62. Plassche, R.J.V.D.: CMOS Integrated Analog-to-Digital and Digital-to-Analog Converters. Springer; 2003, ISBN: 1-4020-7500-6.
63. Priemer, R.: *Introductory Signal Processing*. World Scientific (1991).
64. Raiteri, D., van Lieshout, P., van Roermund, A., Cantatore, E.: An Organic VCO-Based ADC for Quasi-Static Signals Achieving 1LSB INL at 6b Resolution. In: *Proceedings of the IEEE International Solid-State Circuits Conference Digest of Technical Papers (ISSCC)*, pp. 108–109 (2013). DOI 10.1109/ISSCC.2013.6487658.
65. Raiteri, D., Torricelli, F., Cantatore, E., Van Roermund, A.H.M.: A Tunable Transconductor for Analog Amplification and Filtering Based on Double-Gate Organic TFTs. In: *Proceedings of the European Solid-State Circuits Conference*, pp. 415–418 (2011). DOI 10.1109/ESSCIRC.2011.6044995.
66. Rathore, T.S., Jain, A.: Abundance of Ladder Digital-to-Analog Converters. *IEEE Transactions on Instrumentation and Measurement* 50(5), 1445–1449 (2001). DOI 10.1109/19.963222.
67. de la Rosa, J.M.: Sigma-Delta Modulators: Tutorial Overview, Design Guide, and State-of-the-Art Survey. *IEEE Transactions on Circuits and Systems I: Regular Papers* 58(1), 1–21 (2011). DOI 10.1109/TCSI.2010.2097652.
68. Sandler, M.B.: Digital-to-Analogue Conversion Using Pulse Width Modulation. *Electronics Communication Engineering Journal* 5(6), 339–348 (1993).
69. Sarkar, S., Banerjee, S.: An 8-bit 1.8 V 500 MSPS CMOS Segmented Current Steering DAC. In: *Proceedings of the IEEE Computer Society Annual Symposium on VLSI*, pp. 268–273 (2009). DOI 10.1109/ISVLSI.2009.12.
70. Schreier, R., Temes, G.C.: Understanding Delta-Sigma Data Converters. Wiley-IEEE Press; 2004. ISBN: 0471465852.
71. Shaber, M.U.: *Pipeline DA-converter Design and Implementation*. Master's thesis, Department of Electronic, Computers and Software Systems (ECS), School for Information and Communication Technology (ICT), Royal Institute of Technology (KTH) (2005). Accessed on 13 July 2013.
72. Shaker, M.O., Bayoumi, M.A.: A 6-bit 130-MS/s Low-Power Tracking ADC in 90 nm CMOS. In: *Proceedings of the 53rd IEEE International Midwest Symposium on Circuits and Systems (MWSCAS)*, pp. 304–307 (2010). DOI 10.1109/MWSCAS.2010.5548814.
73. Shaker, M.O., Gosh, S., Bayoumi, M.A.: A 1-GS/s 6-bit Flash ADC in 90 nm CMOS. In: *Proceedings of the 52nd IEEE International Midwest Symposium on Circuits and Systems*, pp. 144–147 (2009). DOI 10.1109/MWSCAS.2009.5236133.

74. Srivastava, A., Anantha, R.R.: A Programmable Oversampling Sigma-Delta Analog-to-Digital Converter. In: *Proceedings of the 48th Midwest Symposium on Circuits and Systems*, pp. 539–542 (2005). DOI 10.1109/MWSCAS.2005.1594157.
75. Stanoeva, M., Popov, A.: Investigation of a Parallel Resistorless ADC. In: *Proceedings of the International Conference on Computer Systems and Technologies*, pp. v.8-1–v.8-5 (2005).
76. Stewart, R.W.: An Overview of Sigma Delta ADCs and DAC Devices. In: *IEE Colloquium on Oversampling and Sigma-Delta Strategies for DSP*, pp. 1/1–1/9 (1995). DOI 10.1049/ic:19951371.
77. Svensson, C., Andersson, S., Bogner, P.: On the power consumption of analog to digital converters. In: *Proceedings of the 24th Norchip Conference*, pp. 49–52 (2006).
78. Titus, A.H., Gopalan, A.: A Differential Summing Amplifier for Analog VLSI Systems. In: *Proceedings of the IEEE International Symposium on Circuits and Systems*, vol. 4, pp. IV-57–IV-60 (2002). DOI 10.1109/ISCAS.2002.1010387.
79. Tucker, J.R.: Complementary Digital Logic Based on The "Coulomb Blockade." *Journal of Applied Physics* **72**(9), 4399–4413 (1992). DOI 10.1063/1.352206.
80. van Valburg, J., van de Plassche, R.: An 8b 650MHz Folding ADC. In: *Digest of Technical Papers 39th IEEE International Solid-State Circuits Conference*, pp. 30–31 (1992). DOI 10.1109/ISSCC.1992.200395.
81. van Valburg, J., van de Plassche, R.J.: An 8-b 650-MHz Folding ADC. *IEEE Journal of Solid-State Circuits* **27**(12), 1662–1666 (1992). DOI 10.1109/4.173091.
82. Vogel, C.: The Impact of Combined Channel Mismatch Effects in Time-interleaved ADCs. *IEEE Transactions on Instrumentation and Measurement* **54**(1), 415–427 (2005). DOI 10.1109/TIM.2004.834046.
83. Wang, S., Ahmad, M.O., Bhattacharrya, B.B.: A Novel Cyclic D/A Converter. In: *Proceedings of the IEEE International Symposium on Circuits and Systems*, pp. 1224–1227 (1993). DOI 10.1109/ISCAS.1993.393949.
84. Weaver, S., Hershberg, B., Hanumolu, P.K., Moon, U.K.: A Multiplexer-Based Digital Passive Linear Counter (PLINCO). In: *Proceedings of the 16th IEEE International Conference on Electronics, Circuits, and Systems*, pp. 607–610 (2009). DOI 10.1109/ICECS.2009.5410852.
85. Wietsma, T.A., Minsker, B.S.: Adaptive Sampling of Streaming Signals. In: *Proceedings of the IEEE 8th International Conference on E-Science (e-Science)*, pp. 1–7 (2012). DOI 10.1109/eScience.2012.6404475.
86. Xiong, W., Guo, Y., Zschieschang, U., Klauk, H., Murmann, B.: A 3-V, 6-Bit C-2C Digital-to-Analog Converter Using Complementary Organic Thin-Film Transistors on Glass. *IEEE Journal of Solid-State Circuits* **45**(7), 1380–1388 (2010). DOI 10.1109/JSSC.2010.2048083.
87. Xiong, W., Zschieschang, U., Klauk, H., Murmann, B.: A 3V 6b Successive-Approximation ADC using Complementary Organic Thin-Film Transistors on Glass. In: *Proceedings of the IEEE International Solid-State Circuits Conference Digest of Technical Papers (ISSCC)*, pp. 134–135 (2010). DOI 10.1109/ISSCC.2010.5434017.
88. Yang, J., Wu, X., Zhao, J.: A RC Reconstruction Filter for a 16-bit Audio Delta-Sigma DAC. In: *Proceedings of the IEEE 11th International Conference on Solid-State and Integrated Circuit Technology (ICSICT)*, pp. 1–3 (2012). DOI 10.1109/ICSICT.2012.6467729.
89. Yang, Y., Sculley, T., Abraham, J.: A Single-Die 124 dB Stereo Audio Delta-Sigma ADC with 111 dB THD. *IEEE Journal of Solid-State Circuits* **43**(7), 1657–1665 (2008). DOI 10.1109/JSSC.2008.923731.
90. Yoo, J., Choi, K., Ghaznavi, J.: A 0.07 μm CMOS Flash Analog to Digital Converter for High Speed and Low Voltage Applications. In: *Proceedings of the Great Lakes Symposium on VLSI*, pp. 56–59 (2003).
91. Yoo, J., Choi, K., Tangel, A.: A 1-GSPS CMOS Flash A/D Converter for System-on-Chip Applications. In: *Proceedings of the IEEE Computer Society Workshop on VLSI*, pp. 135–139 (2001).
92. Yuan, J., Fung, S.W., Chan, K.Y., Xu, R.: An Interpolation-Based Calibration Architecture for Pipeline ADC With Nonlinear Error. *IEEE Transactions on Instrumentation and Measurement* **61**(1), 17–25 (2012). DOI 10.1109/TIM.2011.2161026.
93. Yuan, J., Svensson, C.: A 10-bit 5-MS/s Successive Approximation ADC Cell Used in a 70-MS/s ADC Array in 1.2-μm CMOS. *IEEE Journal of Solid-State Circuits* **29**(8), 866–872 (1994). DOI 10.1109/4.297689.
94. Zheng, G., Mohanty, S.P., Kougianos, E.: Design and Modeling of a Continuous-Time Delta-Sigma Modulator for Biopotential Signal Acquisition: Simulink Vs. Verilog-AMS Perspective. In: *Proceedings of the Third International Conference on Computing Communication Networking Technologies (ICCCNT)*, pp. 1–6 (2012). DOI 10.1109/ICCCNT.2012.6396103.

CHAPTER 6
Sensor Circuits and Systems

6.1 Introduction

Energy appears in the universe in various forms including heat, mechanical, chemical (battery), and acoustics. The total energy of the universe is always constant as stated in the basic principles of energy conversion. However, the energy in various forms is always in a continuous state of transformation or conversion from one form to another. The broad term "transducers" covers all the devices needed for such energy conversion. Formally, the transducer is a device that converts one form of energy to another [3, 4, 93]. A list showing selected different transducers is presented in Fig. 6.1.

The transducers can measure and/or sense attributes such as light, temperature, force, speed, and sound [3, 4]. Simplistically, the transducers that convert nonelectrical energy to electrical energy are called "sensors." The transducers that convert electrical energy to mechanical energy are called "actuators." In this chapter, the discussion will be limited to sensors, i.e., transducers giving an electronic signal as the output. Numerous types of sensors are designed and deployed in day-to-day applications as any physical parameters essentially can be sensed [43]. Therefore, it is difficult to provide a comprehensive list. However, selected types of different sensors are presented in Fig. 6.2.

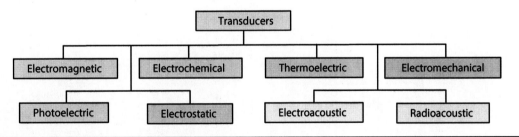

Figure 6.1 Different types of transducers.

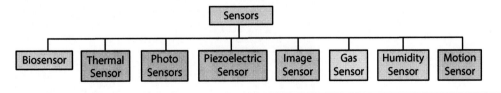

Figure 6.2 Different types of sensors.

6.2 Nanoelectronics-Based Biosensors

The biosensor is essentially a chemical sensing device in which a biologically derived recognition unit is coupled to a transducer unit to recognize a specific chemical [82]. For example, the sensors used in applications such as glucose monitoring, pollution monitoring, and deoxyribonucleic acid (DNA) detection are biosensors. In this section, two biosensors that are implemented using memristor and tunnel field-effect transistors (TFETs) are discussed.

6.2.1 Spintronic-Memristor-Based Biosensors

The architecture of a spintronic-memristor-based bionsensor is represented in Fig. 6.3 [22, 127]. The biosensor consists of the following distinct components: (1) 4×4 DNA spots, (2) frequency divider, (3) summing amplifiers, (4) 4-to-1 multiplexer, and (5) 2-bit counter. Each of the DNA spots has an array of $N \times N$ DNA cells. Each of the DNA cell contains a spintronic-memristor-based magnetic sensor that is the most crucial element of this biosensor. The architecture uses both frequency division multiplexing (FDM) and time division multiplexing (TDM) to have higher throughputs. Four DNA spots share one physical link with the FDM of four different frequencies. The different frequencies are generated with the help of the frequency divider. The four FDM channels are then connected to the 4-to-1 multiplexer for TDM. The output of TDM is a digital signal that is calibrated for the purpose of measurement.

The spintronic-memristor-based magnetic sensor senses the change in the magnetic field. The change is due to the additional magnetic field owing to the existence of the bound magnetic nanoparticles. The magnetic nanoparticles are the nanotags of the unknown DNA fragments or targets that are attached to the DNA target using a binding technique. The unknown DNA fragments or targets are to be recognized by the biosensor. For the detection of the unknown/target DNA, the target DNAs are captured by the complimentary DNA probes to form the hybridized DNA as shown in the Fig. 6.3 [22, 127]. The magnetic field in the magnetic nanoparticles changes due to the application of the external magnetic field. This change is sensed by the DNA cells to detect the density and the distribution of target DNAs. For sensing purposes, a current density (just above the critical

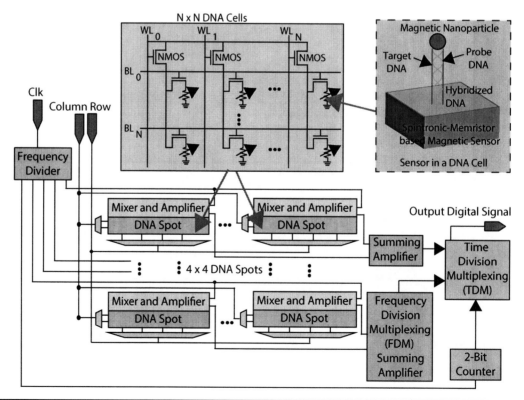

Figure 6.3 Architecture of a spintronic-memristor-based biosensor [22].

current density) is applied to the spintronic memristor for a specific duration. In such excitation of the magnetic nanoparticles, the domain wall of the magnetic field moves to different locations. The amplitude of the magnetic field generated by the magnetic nanoparticles is approximately proportional to the number of the magnetic nanoparticles.

The change in the resistance of the spintronic memristor is a function of the time and the number of magnetic nanoparticles. The memristance or resistance of the spintronic memristor is calculated as follows [22, 128]:

$$M(t) = M_{High} - \left(\frac{\left(M_{High} - M_{Low} \right) L_X(t)}{L_M} \right) \quad (6.1)$$

where L_M is the length of the memristor device, $L_X(t)$ is the domain wall position in the spintronic memristor, M_{Low} is the memristance when $L_X(t) = 0$, and M_{High} is the memristance when $L_X(t) = L_M$. The domain wall moves when the current-density applied is above the critical current density. The velocity of the domain wall is proportional to the applied current density of the spintronic memristor. The amplitude of the magnetic field in the sensors is thus sensed as a change in the domain-wall location or memristance of the spintronic memristor. As is the basic principle of the memristor, the memristance is stored in it even after the sensed magnetic field is removed [22].

6.2.2 Tunnel-FET-Based Biosensors

A PNPN double-gate TFET that can be used as a biosensor is shown in Fig. 6.4 [86, 87]. The superior behavior of the TFETs in the subthreshold region makes them accurate biosensors [114]. The different forms of band-to-band tunneling current mechanisms in the TFETs are the key characteristic [114]. The source is a heavily doped P^+ structure. The drain is a heavily doped N^+ structure. The channel is of P material. A heavily doped N^+ pocket is present between P^+ source and P channel. The gate length is L_{gate} as shown in the figure. There are two gates, one at the top and one at the bottom. A portion of the gate has typical high-κ oxide of dielectric constant $\varepsilon_{ox,2}$ under it. The nanogap cavity of length L_{gap} is the place in the TFET that can contain biomolecules [59, 63]. There is a sacrificial oxide layer of dielectric constant $\varepsilon_{ox,1}$ inside the cavity. The biomolecules change the electrical properties of the TFET such as threshold voltage, current flow, or conductance [63, 114]. These changes can be calibrated for sensing through the biosensor.

6.2.3 Graphene-FET-Based Biosensors

The schematic representation of a graphene FET (GFET)-based biosensor is shown in Fig. 6.5 [68, 70, 79, 96]. In addition to the typical silicon and oxide materials, the FET contains a single-layer of graphene. The particular example shown in the figure is an aptamer-modified electrolyte-gated GFET.

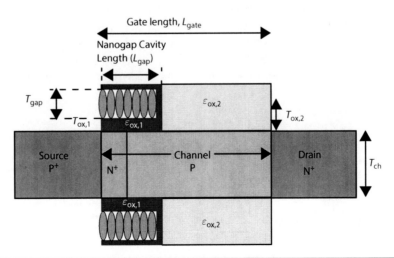

Figure 6.4 TFET-based biosensor [86, 87].

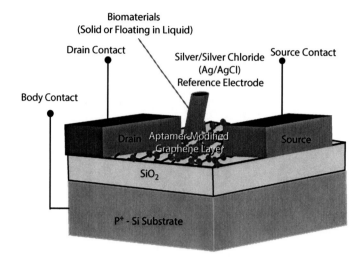

FIGURE 6.5 Aptamer-modified electrolyte-gated GFET-based biosensor [70, 96].

The aptamers, such as DNA or RNA, are molecules that can bind to a specific target molecule, thus facilitating sensing of specific materials. The silver/silver chloride (Ag/AgCl) that is the reference electrode provides top-gate voltage. The transconductance of the GFET changes as per the following relation [96]:

$$\left(\frac{\delta I_{ds}}{\delta V_{gs}}\right) = \mu C_{gate} V_{DD} \quad (6.2)$$

This change can be translated to a change in the other electrical characteristic of the GFET that is then calibrated for measurement purposes.

6.3 Thermal Sensors for Mixed-Signal Circuits and Systems

Sensing temperature is probably the most frequent sensing that is performed every moment for very diverse applications, for example, body temperature measurement, room temperature monitoring, refrigerator temperature, or even on-chip temperature sensing for microprocessor safety. Thus, the thermal sensors that are also known as temperature sensors are required all the time. An overview of different types of thermal sensors is presented in Fig. 6.6 [4]. The sensors are classified on the basis of the physical usage and the structure. The contact-type temperature sensors need to be in physical contact with a target object for temperature measurement purposes. Methods, such as convection and radiation, are used for temperature measurement in the non-contact-type temperature sensors. Based on the structure of the thermal sensor, it can be resistive, electromechanical, or electronic. The resistive thermal sensors are the simplest type. This section will focus on detailed discussions of the electronic sensors that can be used on the chips for temperature sensing [100, 102]. Such a thermal sensor is used to protect the circuits and system from high temperatures in which they are

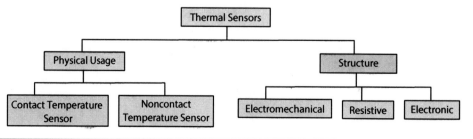

FIGURE 6.6 Different types of generic thermal sensors.

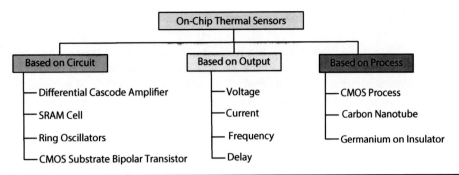

FIGURE 6.7 Different types of on-chip thermal sensors.

deployed. As empirically observed, a 1°C decrease in the on-chip temperature can reduce the failure of the circuit by 2% to 3% [108, 120]. As a result, the lifetime of the circuits and system will be significantly improved. The circuits or systems when overheated may function incorrectly or may even be destroyed without a good thermal solution such as an efficient heat sinker or a self-protection mechanism.

In general, the electronic temperature sensors can be either off-chip or on-chip types. The on-chip temperature sensors are preferred over the off-chip temperature sensors due to their fast response, low cost, low power, and low overhead. The different types of on-chip temperature sensors are shown in Fig. 6.7 [108]. They are divided into three groups based on circuit type, output produced, and manufacturing processes used in building them.

6.3.1 Performance Metrics for Thermal Sensors

Various types of on-chip thermal sensor designs have been discussed in this chapter [35, 99, 102, 106, 119, 144]. The design objectives for these on-chip thermal sensors include robust performance and accurate temperature measurement. Selected performance metrics that can be used to characterize the thermal sensors have been discussed in this subsection [53, 102, 108]:

1. *Temperature Resolution:* The temperature resolution of a thermal sensor is the minimum change in the temperature that can be measured or sensed by it. The smaller the resolution, the more stringent is the constraint on the sensing device and the measuring circuitry.

2. *Temperature Accuracy:* The temperature accuracy of a thermal sensor is the closeness of the thermal sensor measurement with respect to the actual temperature. It is desired that the thermal sensor reports temperature as close as possible to the actual temperature of the object or environment.

3. *Temperature Range:* The temperature range of a thermal sensor is the difference between the minimum and maximum temperature values that it can accurately sense and measurement.

4. *Sensing Linearity:* The relationship between the output signal of the thermal sensor and the input temperature range is expected to be linear. Linearity is an important requirement of the thermal sensors and is needed in making a decision if a specific circuit technology can be used to design a thermal sensor.

5. *Sensor Calibration:* The thermal sensor calibration involves the translation of the sensed signal to temperature so that it can be used for measurement. Accurate calibration of the thermal sensors is an important task and needs to account for the inherent variations of the circuit and systems of the sensors as well as the environmental variations.

6. *Power Dissipation:* Power dissipation of the thermal sensor is the total power consumed by it. The active power consumption of the thermal sensor and the leakage power dissipation contribute to the total power dissipation of the thermal sensor. The power overhead of the thermal sensor is required to be as low as possible to ensure that it is not a major fraction of the total power dissipation of a circuit and system in which it is used.

7. *Self-Heating*: Self-heating or joule-heating of the thermal sensor is the heat produced by it due to the resistive effect of the thermal sensors through which the current flows. The leakage power dissipation may add to the self-heating of the thermal sensors. The joule-heating or self-heating of the thermal sensors needs to be as little as possible to ensure that the measurement is not skewed. Otherwise, accuracy of the thermal sensors will be affected as calibration of the thermal sensors will become more difficult task.

8. *Silicon Area:* The silicon area is the area occupied by the physical design of a thermal sensor. The silicon area restricts the number of thermal sensors that can be used on a host circuit or system.

6.3.2 A Concrete Example: A 45-nm CMOS Ring Oscillator-Based Thermal Sensor

As a concrete example, a 45-nm complementary metal oxide semiconductor (CMOS) technology-based thermal sensor design is discussed in this subsection [99, 100, 102]. It is based on a ring oscillator made of inverters of odd number of stages. The strategy is to utilize the temperature sensitivity of the oscillating frequency for thermal sensing. In the thermal sensor design, the output frequency of the ring oscillator is proportional to the absolute temperature. This output frequency when used with a constant pulse generator and a bias calibrator, the count resulted between the pulse is proportional to the temperature change, thus forming an effective thermal sensor. The 45-nm thermal sensor that uses a ring oscillator as a major component for thermal sensing is shown in Fig. 6.8 [76, 99, 100, 102]. The thermal sensor also uses a 10-bit binary counter and a 10-bit register for accurately expressing the output signal. The output frequency of the ring oscillator is very sensitive to ambient temperature, and thus the output frequency changes in response to the effect from the ambient temperature. The binary counter is used to count the frequency difference between the ring oscillator output and the system clock. The count is stored in the 10-bit register and calibrated to measure the temperature change.

The schematic representation of the 15-stage ring oscillator that is used in this thermal sensor is shown in Fig. 6.9. The ring oscillator consists of a cascade of an odd number of inverters that have been connected in a closed loop leading to an unstable state that creates the oscillations. The ring oscillator shown in Fig. 6.9 has 15 stages. However, the first inverter is replaced with a NAND gate to

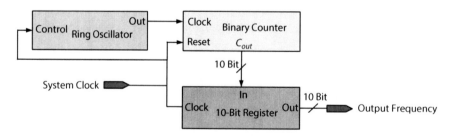

Figure 6.8 Block diagram of the thermal sensor system.

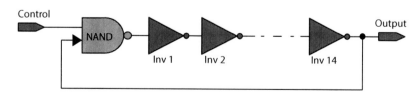

Figure 6.9 Block diagram of the ring oscillator used in 45-nm CMOS thermal sensor.

enable the ring oscillator operation. The oscillation frequency of the ring oscillator is calculated using the following expression [76, 99, 100, 102]:

$$f_{osc} = \frac{1}{N_{stage}(T_{pd,LH} + T_{pd,HL})} \quad (6.3)$$

where N_{stage} is the number of stages used in the ring oscillator and $T_{pd,LH}$ and $T_{pd,HL}$ are the low-to-high and high-to-low propagation delays, respectively. The propagation delays can be expressed as follows [76, 102]:

$$T_{pd,LH} = \frac{-2C_L V_{Th,p}}{\kappa_p(V_{DD} - V_{Th,p})^2} + \frac{C_L}{\kappa_p(V_{DD} - V_{Th,p})}\ln\left(\frac{1.5V_{DD} + 2V_{Th,p}}{0.5V_{DD}}\right) \quad (6.4)$$

$$T_{pd,LH} = \frac{2C_L V_{Th,n}}{\kappa_n(V_{DD} - V_{Th,n})^2} + \frac{C_L}{\kappa_p(V_{DD} - V_{Th,n})}\ln\left(\frac{1.5V_{DD} + 2V_{Th,n}}{0.5V_{DD}}\right) \quad (6.5)$$

where C_L is the capacitive load and κ_n and κ_p are the transconductance values of NMOS and PMOS, respectively. The transconductance can be calculated using the following expression [102]:

$$\kappa_{n/p} = \mu_{n/p} C_{ox}\left(\frac{W}{L}\right)_{n/p} \quad (6.6)$$

The threshold voltage V_{Th} and mobility of the MOS transistors are the factors that are most sensitive to temperature fluctuations. The dependency on temperature can be presented as follows [18, 102]:

$$V_{Th}(T) = V_{Th}(T_0) + \alpha_{V_{Th}}(T - T_0), \alpha_{V_{Th}} = -0.5 - 3.0 \text{ mV/}^\circ K \quad (6.7)$$

$$\mu(T) = \mu_0\left(\frac{T}{T_0}\right)^{\alpha_\mu}, \alpha_\mu = -1.2 - 2.0 \quad (6.8)$$

An increase in the temperature leads to an increase in the propagation delay of the inverter stages that translates to a decrease in oscillating frequency of the ring oscillator. The 10-bit binary counter that is used in the thermal sensor design is shown in Fig. 6.10. The counter consists of JK flip-flops and is a traditional counter design. The 10-bit register that is used in the thermal sensor is shown in Fig. 6.11. The register is used to store the value from the counter. The register is also implemented by using the JK flip-flops.

The physical design of the thermal sensor performed using a 45-nm CMOS is shown in Fig. 6.12 [99, 100, 102]. The thermal sensor design is characterized and it is observed that the thermal sensor can measure temperature in the range of 0°C to 100°C. The performance and accuracy of the physical design are degraded as compared to the schematic design of the thermal sensor. This is expected due to parasitic effects present in the physical design. The characterization of the schematic and physical design of the thermal sensor is shown in Table 6.1. The physical design of the thermal sensor shows a 29% increase in power dissipation and a 44% decrease in sensitivity as compared to

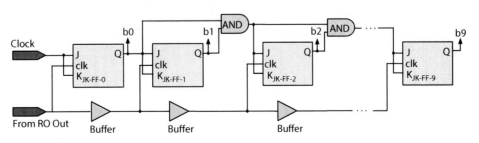

FIGURE 6.10 Block diagram of the 10-bit binary counter.

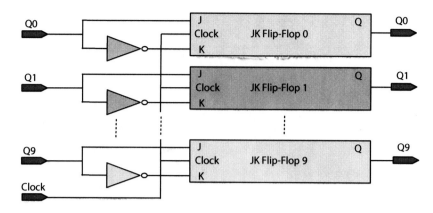

FIGURE 6.11 Block diagram of the 10-bit register.

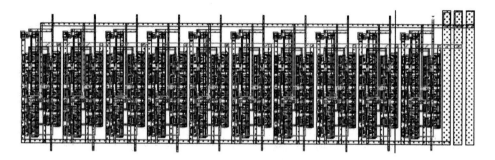

FIGURE 6.12 Physical design of a baseline circuit of a 45-nm CMOS thermal sensor [99, 102].

Different Design	Average Power (P_{TS})	Sensitivity (T_{TS})	Area (μm^2)
Schematic Design	293.1 μW	16.88 MHz/°C	–
Physical Design	379.4 μW	9.42 MHz/°C	1221.37
% Change	+29%	−44%	

TABLE 6.1 Characterization of the 45-nm CMOS Thermal Sensor [99, 102]

schematic design. The frequency versus temperature characteristic of the thermal sensor is presented in Fig. 6.13. This thermal sensor exhibits a linear dependence of oscillation frequency on the junction temperature as shown in Fig. 6.13.

The frequency versus temperature characteristics of the ring oscillators indicate that the oscillating frequency of the oscillator decreases as the temperature increases. For the 45-nm CMOS schematic design, the oscillating frequency is 5.9 GHz at 0°C and 4.2 GHz at 100°C. The effective resolution is calculated by dividing the temperature range by the number count 100°C/1024 bit that gives a 0.097°C/bit resolution. This is under the assumption that 6 GHz is the maximum clock rate for the ring oscillator and a 10-bit counter with 1024 maximum count. The frequency range of the ring oscillator is also severely degraded by the parasitics. In particular, the range degrades to 2.98 GHz from 3.87 GHz. The resolution is specified in terms of GHz per °C to reflect the degrading effect of parasitics in the physical design. A 47.8% change is observed in the frequency/temperature resolution between the schematic and physical design of the thermal sensor. In the thermal sensor design, the temperature measurements can be taken by using two different methods. The first uses the count versus temperature characteristic. The characteristic showing count versus temperature for the 45-nm CMOS technology is shown in Fig. 6.14. This count is then interpreted using a calibration table. The second approach uses a formula that is generated from the linear data fitting [99, 102]:

$$\text{Temperature} = -0.5167 \times \text{Count} + 395.3 \tag{6.9}$$

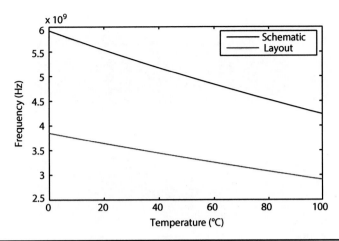

Figure 6.13 Ring oscillator frequency response versus temperature for 45-nm CMOS technology [99, 102].

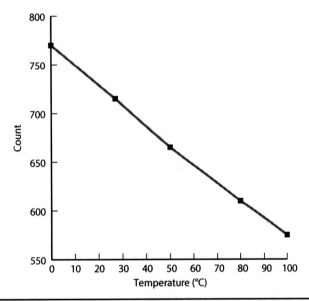

Figure 6.14 Count versus temperature characteristic of the 45-nm CMOS thermal sensor [99, 102].

The linear data fitting has a coefficient of determination (R^2) value of 0.9978, and hence is close to actual data. This equation serves as a predictive function that enables a direct temperature reading from the count value.

6.3.3 A Concrete Example: Spintronic-Memristor Temperature Sensor

The circuit diagram of a spintronic-memristor-based thermal sensor is shown in Fig. 6.15 [12, 128]. The thermal sensor consists of three distinct portions: reference current generator, driving circuit, and mode-control switch [12, 128]. The reference current generator has a modified NMOS Widlar current mirror consisting of NM1, NM2, NM3, and R_1 [12, 39, 128]. It also has an inverse PMOS Widlar current mirror consisting of PM1 and PM2. The driving circuit consists of two memristors, $Memristor_1$ and $Memristor_2$, and PM3. Two memristors are used in a series connection fashion instead of one memristor to increase the output voltage range with minimum switching time. The third part of the thermal sensor is a mode control switch made with transistor NM4. In this circuit, the PMOS current mirror consisting of PM2 and PM3 copies the reference current to the driving memristors. The mode control switch has two modes—either detect or reset based on its ON or OFF states. The mode control switch determines the driving pulse width and helps in temperature detection. A negative power supply V_{SS} can be used to reset the status of the memristors that reverses the current flow. The memristors

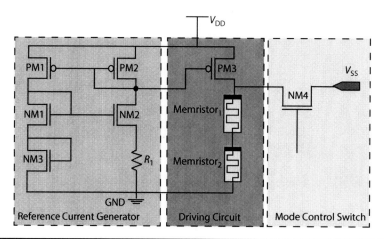

Figure 6.15 Circuit of a spintronic-memristor-based thermal sensor [12, 128].

initially provide maximum memristance as their domain walls are at the upper boundary. In the detection mode, the domain walls of the memristors are pushed toward the minimum memristance value by the constant current driving. At this time, the output voltage is also fed to the thermal management units. On the completion of the detection, the memristors are reset to the initial state.

In general, the thermal sensing of the spintronics memristor is controlled by two methods: the constant voltage driving and the constant current driving [12, 128]. The memristance or resistance of the spintronic memristor is calculated as follows [12, 22, 128]:

$$M(t) = m_{High} L_X(t) + m_{Low}(L_M - L_X(t)) \tag{6.10}$$

where L_M is the length of the memristor device. $L_X(t)$ is the domain-wall position in the spintronic memristor. M_{Low} is the memristance when $L_X(t) = 0$. M_{High} is the memristance when $L_X(t) = L_M$. m_{Low} is M_{Low} per unit length and m_{High} is M_{High} per unit length. When a constant voltage is applied at the two terminals of the memristor, the memristance decreases and the current increases. Effectively a positive feedback is established and domain-wall movement speeds up to result in further reduction in memristance. If $v_D(t)$ is the domain wall velocity, then it is $\left(\frac{dL_x(t)}{dt}\right)$. For constant voltage driving, the following expression holds true [12, 22, 128]:

$$(m_{High} - m_{Low}) v_D(t) = \left(\frac{V}{I^2(t)}\right)\left(\frac{dI(t)}{dt}\right) \tag{6.11}$$

where V is the constant voltage and $I(t)$ is the time-varying current. For constant current driving, the following expression holds true [12, 22, 128]:

$$-(m_{High} - m_{Low}) v_D(t) = \left(\frac{dV(t)}{I\,dt}\right) \tag{6.12}$$

where I is constant current and $V(t)$ is the time-varying voltage. The voltage across the memristor linearly decreases, which provides a good linearity curve for temperature sensing.

6.4 Solar Cells

Renewable energy, such as solar energy, wind energy, and geothermal energy, originate from resources that replenish continuously. Abundance of solar energy has always attracted attention for its usage in daily life. At the application level, the energy flow from the sun to end users is depicted in Fig. 6.16 [5, 54]. The solar panel converts sunlight to electrical energy. A battery stores the electrical energy for use when needed. A charger may be used to charge the battery. The power inverter converts the direct current (DC) that is stored in the battery to alternating current (AC) for use at home in conjunction

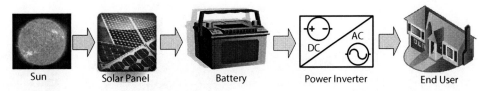

FIGURE 6.16 The solar energy to the end users.

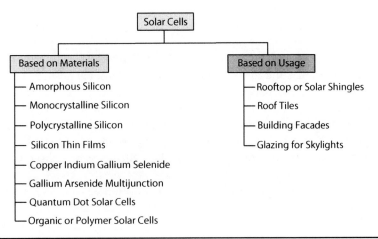

FIGURE 6.17 Different types of solar cells.

with typical AC energy. The solar cell is a solid-state device that converts light energy to electrical energy by photovoltaic effect. Thus, the solar cell is also called photovoltaic cell or photoelectric cell. The solar panel is the collection of multiple solar cells to generate the amount of solar power or solar energy needed for different applications [61]. While the photovoltaic effect was demonstrated by A. E. Becquerel in 1839, the first practical photovoltaic cell was developed by D. Chapin, C. S. Fuller, and G. Pearson in 1954. Now the solar cells are everywhere. Solar cells are even used in watches, computer key boards, and calculators to provide them with the solar power for long and battery-less operations.

The solar cells are built from several different materials and are also used for different applications. Selected different types of solar cells are presented in Fig. 6.17 [62, 130]. Based on the materials used to build the solar cells, the types include: amorphous silicon, crystalline silicon, silicon thin films, quantum dot solar cells, and organic polymer solar cells. In addition, the zinc oxide (ZnO) nanowires have been explored as promising material for solar cells [27, 28, 130]. Based on the usage, the solar cells include: rooftop or solar shingles, roof tiles, building facades, and glazing for skylights.

6.4.1 Operation and Performance of Cells

6.4.1.1 Operation of Solar Cells

The schematic representation of a solar cell with important layer is provided in Fig. 6.18 [62]. It has two semiconductor layers, the top N-type layer and bottom P-type layer. The two contacts are the front contact and back contact from which solar energy can be tapped for use. The working principle of this solar cell can be simplistically explained with a diode-like behavior. When the sunlight hits the solar cell, the input photons are absorbed by the semiconductor material. If the energy of photon is higher than the semiconductor material band gap value, then an electron-hole pair is generated and electrons and holes move in the semiconductor material in the following fashion. The energy of absorbed photon is transferred to the electrons in the semiconductor crystal lattice. On acquiring sufficient energy, the valence band electrons move to the conduction band and freely move in the semiconductor material. The movement of the electrons creates the missing covalent bond in the

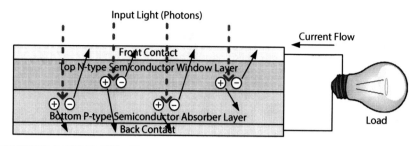

Figure 6.18 Basic operation of a solar cell.

lattice of the semiconductor material that is called hole. The electrons from the neighboring atoms move into this hole, thus making new holes in their original positions. In this way, the holes also move in the semiconductor material. All the electrons move in one direction and the holes move in the opposite direction. The solar cell effectively converts the solar energy into DC electric energy.

The above process of creation of the carriers in the solar cell is essentially the "Photovoltaic Effect." The photovoltaic effect is the creation of voltage or electric current in the semiconductor materials when exposed to light energy. The photovoltaic effect was first observed by A. E. Becquerel in 1839. It may be noted that a similar process called "photoelectric effect" is different from the photovoltaic effect. In the photoelectric effect, the electrons are emitted from solids, liquids, or gases upon exposure to light into the vacuum. The photovoltaic effect thus differs by the fact that the excited electrons pass directly from one material to another without going through the difficult step of passing through the vacuum in between.

6.4.1.2 Efficiency of Solar Cells

The efficiency of energy conversion in a solar cell has many aspects including the conductive efficiency, reflectance efficiency, thermodynamic efficiency, and carrier separation efficiency. The overall efficiency of the solar cell is the product of each of these individual efficiencies. The typical application scenario in which a solar cell system provides energy to a battery system is presented in Fig. 6.19 [132]. The steady-state efficiency of the solar cell system that transfers solar light to electrochemical energy in a battery system is expressed as follows [132]:

$$\eta_{SS} = \eta_{PV}\eta_{EC}\eta_{co} \qquad (6.13)$$

where η_{PV} is the efficiency of the photovoltaic cell, η_{EC} is the electrochemical process efficiency, and η_{co} is the efficiency of the coupling system such as a charger. The efficiency the of solar cell is affected by the voltage output, temperature coefficients, and shadow angles. The following expression describes the efficiency of energy conversion in the solar cells [91, 117]:

$$\eta_{PV} = \left(\frac{\alpha_{FF}V_{oc}I_{sc}}{P_{solar}}\right) \qquad (6.14)$$

where P_{solar} is the incident solar power, α_{FF} is the electronic fill factor, V_{oc} is the open circuit voltage of the solar cell, and I_{sc} is the short circuit current of the solar cell. The electric fill factor α_{FF} can be calculated from the empirical formulas [117, 132]. The typical values of α_{FF} are as follows: 0.7 for

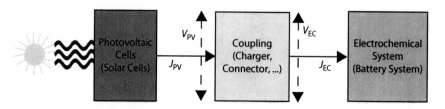

Figure 6.19 Solar energy to electrochemical energy conversion [132].

commercial grade, 0.4–0.65 for grade B cells, and 0.4–0.7 for amorphous solar or thin-film cells. In the open circuit scenario, there is no current flow and the potential is calculated using the following expression [91]:

$$V_{oc} = \left(\frac{k_B T_0}{q}\right) \ln\left(\frac{\phi_A}{\phi_{E,0}}\right) \tag{6.15}$$

where k_B is the Boltzmann constant, T_0 is the temperature, q is the electronic charge, ϕ_A is the absorbed photon rate, and $\phi_{E,0}$ is the emitted photon rate for zero potential. It may be noted that the photon rate or photo flux is the number of photons per second per unit area. In the short-circuit scenario, the potential is zero, and the short circuit current can be calculated as follows [91]:

$$I_{sc} = q\phi_A \tag{6.16}$$

The conversion efficiency is one of the critical challenges of solar cell design and implementation. It is observed that the silicon-based solar cells have maximum conversion efficiency of 22% [37]. The steady state of the solar cell system with energy transfer to the battery is calculated as follows [132]:

$$\eta_{SS} = \left(\frac{\phi_{therm} I_{op}}{P_{solar}}\right) \tag{6.17}$$

where ϕ_{therm} is the thermodynamic potential and I_{op} is the operating current density.

6.4.2 Selected Solar Cell Designs

6.4.2.1 Thin-Film Solar Cell

In this subsection, the structure of selected solar cells has been discussed to get insight into solar cells, in particular, thin-film solar cells (TFSCs). The cross-section of selected TFSC, or thin-film photovoltaic cell (TFPV), is depicted in Fig. 6.20 [62, 104, 125]. The TFSCs are of many different types based on the photovoltaic material used in them: (1) amorphous silicon (a-Si) solar cells, (2) cadmium telluride (CdTe) solar cells, (3) copper indium gallium selenide (CIS or CIGS) solar cells, and (4) dye-sensitized solar cell (DSC) and other organic solar cells.

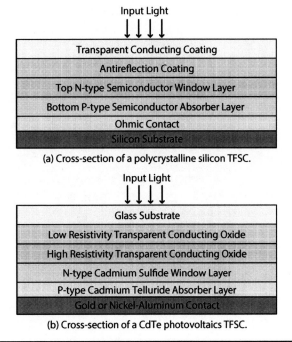

FIGURE 6.20 Cross-section of a silicon-based solar cell.

As a specific example of TFSC, the cross-section of a polycrystalline silicon TFSC is presented in Fig. 6.20(a). At the bottom, it has a silicon substrate and an ohmic contact. The two semiconductor layers are the bottom P-type semiconductor absorber layer and the top N-type semiconductor window layer. An antireflection coating is present at the top of the semiconductor layers. The topmost layer is the transparent conducting coating. The cross-section of a CdTe silicon TFSC is presented in Fig. 6.20(b). The bottom-most layer is the gold or nickel-aluminum contact. The photovoltaic material layers are the P-type CdTe absorber layer and N-type cadmium sulfide window layer. Two transparent conducting oxide layers of high-resistivity and low-resistivity are present at the top of photovoltaic materials.

6.4.2.2 Organic Solar Cell

The structure of an organic solar cell is presented in Fig. 6.21 [6, 17, 133]. The glass substrate is the top layer through which the incident light passes. The indium tin oxide (ITO) layer was patterned on the glass substrate by etching with zinc and acidum hydrochloricum. A layer of poly(3,4-ethylenedioxylenethiophene)-poly(stylenesulfonic) (PEDOT:PSS) is deposited on ITO by the spin coating process. The next layer is made of poly-(2-methoxy-5-(3',7'-dimethyloctyloxy)-1,4-phenylenevinylene)) (MDMO-PPV) and [6,6]-phenyl-C61-butyric acid methyl ester (PCBM). These are the two layers of organic polymer used for energy conversion in the solar cell. The last layer is aluminum contact. The organic-polymer solar cells have attractive features including lightweight, mechanical flexibility, low cost, and simple fabrication process. However, the conversion efficiency is lower than silicon solar cells.

6.4.3 Solar Cell Models for Circuit Simulations

In this subsection, equivalent circuit models for solar cells are discussed. These models can be used in SPICE and other CAD tools for simulation and design exploration of solar cells, solar panels, and solar cell-based systems. Two equivalent models of the organic solar cell are [17, 75, 94]: (1) the single-diode model and (2) the two-diode model. The single-diode model of the solar cell is presented in Fig. 6.22(a) [17, 75, 94]. The double-diode model for the solar cell is presented in Fig. 6.22(b) [17, 75, 94].

The single-diode model of the solar cell presented in Fig. 6.22(a) uses one diode, one current source, and two resistors to model the solar cell effectively [17, 75]. The series resistor is R_{se}. The series resistance depends on the resistivity of the semiconductor substrate, the metal electrodes, and the

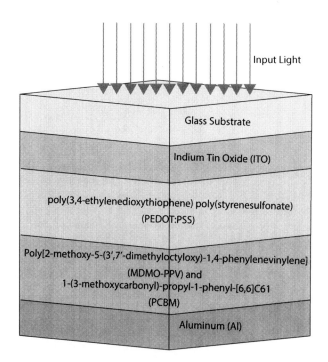

Figure 6.21 Structure of an organic solar cell [6, 17].

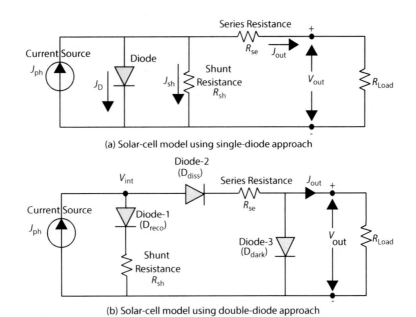

Figure 6.22 Circuit models for the solar cells [17, 75, 94].

semiconductor/metal interface. In an organic solar cell, the series resistance is affected by the resistivity of the organic materials, the metal electrodes, and the organic/metal interface. The shunt resistance is R_{sh}. The series resistance (R_{se}) and shunt resistance (R_{sh}) are related as follows [17]:

$$R_{se} = \left(\frac{J_{ph}}{J_{sc}} - 1\right) R_{sh} \tag{6.18}$$

where J_{sc} is the short-circuit current density. The diode provides a current of density J_D. This model represents the output current density as follows [17]:

$$J_{out} = J_{ph} - \left(\frac{V_{out} + R_{se} J_{out}}{R_{sh}}\right) - J_0 \left(\exp\left(\frac{V_{out} + R_{se} J_{out}}{v_{therm} N_i}\right) - 1\right) \tag{6.19}$$

where J_{ph} is the photo-current density, J_0 is the saturation current density of the diode under reverse bias condition. N_i is called the ideality factor, and v_{therm} is the thermal voltage. Based on the various assumptions, further simple analytical models are available to represent the currents [45].

The double-diode model shown in Fig. 6.22(b) has multiple diodes, one current source, and two resistors [17, 75]. Diode-1 is called D_{reco}, which models the loss due to polaron-pair recombination. Diode-2 is called D_{diss}, which that models the polaron-pair dissociation and collection of free carriers by electrodes. Diode-3 is called D_{dark}, which that models the current-voltage characteristics under the dark or no-light condition. R_{se} represents the loss due to the polaron recombination. R_{sh} represents the charge extraction to the electrodes. A simplified double-diode model is shown in Fig. 6.23 [45, 47, 132]. In this double-diode model, the output current density is the difference between the generated photocurrent and the recombination currents. In this double-diode model, the output current is calculated in the following manner [45, 47, 132]:

$$J_{out} = J_{ph} - J_{0,1} \exp\left(\frac{V_{out} + R_{se} J_{out}}{v_{therm} N_{i,1}}\right)$$

$$- J_{0,2} \exp\left(\frac{V_{out} + R_{se} J_{out}}{v_{therm} N_{i,2}}\right) - \left(\frac{V_{out} + R_{se} J_{out}}{R_{sh}}\right) \tag{6.20}$$

where $J_{0,1}$ and $J_{0,2}$ are the saturation current density of the diode under reverse bias condition for Diode-1 and Diode-2, respectively.

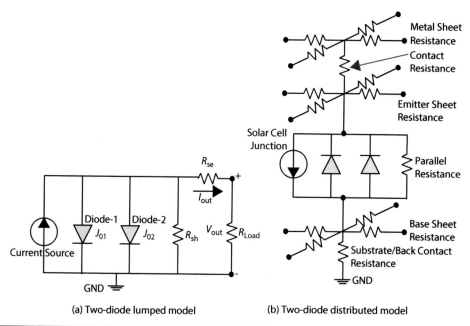

Figure 6.23 Circuit models for the solar cells [45, 47].

The diode-based lumped models have many simplistic assumptions as follows [45, 47]: (1) photon generation is linearly dependent on concentration; (2) the illumination and dark characteristics follow the superposition principle; (3) the neutral region recombination, the depletion region recombination, and the perimeter recombination current are considered; (4) all the ohmic losses in the solar cell are combined together; and (5) all the shunt leakage current through the PN junctions are modeled as a single shunt resistance. To overcome the limitations of the lumped models, the distributed models are explored. The two-diode distributed model is shown in Fig. 6.23(b) [45, 47]. As shown in the figure, there are many different resistances, metal-sheet resistance, contact resistance, emitter-sheet resistance, base-sheet resistance, and substrate resistance.

6.5 Piezoelectric Sensors

The piezoelectric sensors are sensors that directly convert stress (or strain) to electricity (electric potential) for measurement. This effect in piezoelectric sensors is called "piezoelectric effect" [48]. The piezoelectric effect is the interaction between the mechanical state and the electrical state in the crystalline materials. The effect is reversible in the sense that the crystalline materials generate electricity when stress is applied and at the same time strain is generated when electrical field is applied. In the piezoelectric effect, the electrical displacement is calculated using the following expression [66]:

$$D = \gamma_{i,j} S_j + \varepsilon_0 \varepsilon_{Si,k} E_k \tag{6.21}$$

where $\gamma_{i,j}$ is the strain piezoelectric coefficient, S_j is the mechanical strain, ε_0 is the permittivity of the free space, $\varepsilon_{Si,k}$ is the permittivity at constant strain S, and E_k is the electric field. The subscripts $i, j,$ and k denote the direction to which the physical properties of the crystalline material are related; $i = 1, 2, 3, j = 1, 2, 3, 4, 5, 6,$ and $k = 1, 2, 3$. The piezoelectric sensors have very wide applications and are used for the measurement of the force, torque, pressure, acoustic-emission, acceleration, and strain. It may be noted that similar sensors called "piezoresistive sensors" use change in the resistivity for measurement [66]. The piezoelectric sensors can detect stress or strain in all directions and simultaneously. On the other hand, the piezoresistive sensors sense stress/strain in a single direction.

As a specific example, a ZnO nanowire (ZnO-NW)–based piezotronic transistor is shown in Fig. 6.24 [130]. The fundamental principle of the piezotronic transistor is that application of strain creates a piezoelectric potential at the interface region in the semiconductor, which then controls the carrier transport. Therefore, the need for insulated gate to control the transistor, as is the case in

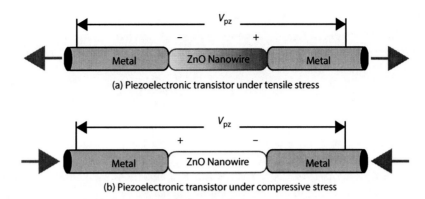

FIGURE 6.24 Basic concept of piezoelectric transistors [130].

a CMOS transistor, is eliminated. In effect, the piezotronic transistor in Fig. 6.24 has only two terminals, source and drain. When the ZnO-NW piezotronic transistor is subjected to strain, both the piezotronic effect and piezoresistive effect are originated. The carrier transport is affected by both these effects. The polarity of the piezopotential (V_{pz}) depends on the type of stress or strain. The piezotronic transistor for tensile stress/strain is shown in Fig. 6.24(a) [130]. The piezotronic transistor for compressive stress/strain is shown in Fig. 6.24(b) [130].

The practical piezoelectric sensor circuit with amplifier blocks is presented in Fig. 6.25 [66]. The piezoelectric sensor with an inverting OP-AMP is presented in Fig. 6.25(a) [66]. The feedback resistor (R_{FB}) helps in amplifying the voltage signal (V_{pz}) generated by the piezoelectric device. The capacitor (C_{FB}) stabilizes the OP-AMP circuit that otherwise is prone to oscillations if not compensated appropriately. The output voltage resulting from this piezoelectric sensor circuit that is used for measurement after calibration is the following:

$$V_{out} = -\left(\frac{R_{FB}}{R_{pz}}\right) V_{pz} \tag{6.22}$$

where R_{pz} is the effective resistance presented by the piezoelectric device at the input of the OP-AMP. The piezoelectric sensor with a noninverting OP-AMP is presented in Fig. 6.25(b) [66].

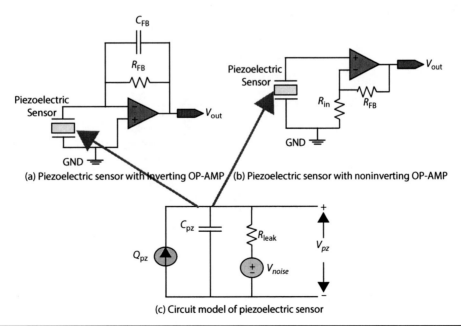

FIGURE 6.25 Schematic diagram of piezoelectric sensors [66].

The noninverting OP-AMP has larger input impedance than that of an inverting OP-AMP and thus is a better amplifier for the systems with small currents. The output voltage resulting from this piezoelectric sensor circuit that is used for measurement after calibration is

$$V_{out} = \left(1 + \frac{R_{FB}}{R_{in}}\right) V_{pz} \qquad (6.23)$$

For the circuit-level simulation of piezoelectric sensor-based circuits and systems, a circuit model for piezoelectric device or sensor is needed. A simple lumped model for piezoelectric device or sensor is presented in Fig. 6.25(c) [66]. The piezoelectric device is represented as a charge source (C_{pz}) in parallel with a capacitor (Q_{pz}). The resistor R_{leak} and the noise voltage source V_{noise} together represent the leakage current path for sensor-generated charges.

6.6 Image Sensors

The image sensors are sensors that convert an optical picture into an electrical signal. The electrical signal after the digitization becomes a digital image that is stored in memory. The image sensors play the most important roles in image sensing and capturing applications. The CMOS-based active-pixel sensors (APS) are omnipresent in various systems such as webcams, still digital cameras, video digital cameras, mobile phone cameras, smart toys, automobiles, cinematography, and robotics [85, 136]. As a typical example of an image sensor-based system, the digital camera system is depicted in Fig. 6.26 [85, 105]. The system uses an array of image sensors that is the key component of the digital camera.

The signal flow in a typical image sensor is depicted in Fig. 6.27 [46]. The incident photons of the light signal are converted to photocurrent by the photodetector of an image sensor. The examples of photodetector include photodiode and photogate. The photocurrent is integrated at the sensing node by the capacitance of the photodetector. Two types of integration are possible, linear integration and logarithmic integration. At the end of the integration period, the sensed charge in the sensing node capacitance may go through two alternative processes [46]: (1) in charge-coupled device (CCD)–based image sensors or passive-pixel sensors (PPS) in which the sensed charge is directly read out and then converted to voltage or (2) APS, in which the sensed charge is directly converted to photovoltage and then read out. The photovoltage may go through further processing such as filtering and amplification before the sensed voltage is actually used in the rest of the system that hosts the sensor. For example, in the APS and the digital-pixel sensors (DPS), the postprocessing may include amplification and analog-to-digital conversion (ADC).

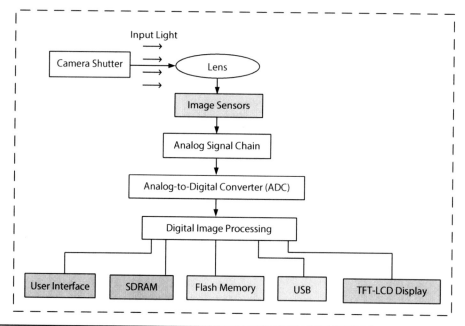

FIGURE 6.26 Image sensors deployed in a camera [85, 105].

Sensor Circuits and Systems

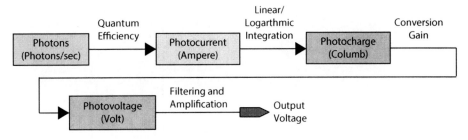

FIGURE 6.27 Image sensors deployed in a camera [46].

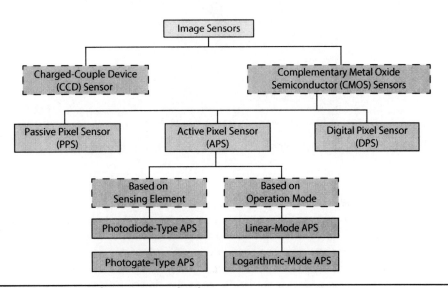

FIGURE 6.28 Different types of image sensors [105].

6.6.1 Types of Image Sensors

The different types of image sensors are listed in Fig. 6.28 [41, 71, 105]. Two basic types are the CCD sensors and the CMOS sensors. The CCD sensors were invented by W. Boyle and G. E. Smith in 1969 and a real-life image sensor was developed by C. H. Sequin, E. J. Zimany, M. F. Tompsett, and E. N. Fuls in 1976 [116]. Since then, the CCD image sensors have been part of many applications including fax machines, cameras, camcorders, digital cameras, and telescopes. Even though CCD sensors produce high-quality images, the high complexity and the high-power dissipation of CCD sensor-based circuits have led to exploration of alternative sensor technology [42, 85, 136]. The CMOS sensors include the following three types: PPS, APS, and DPS. The PPS, the first type of image sensor, was invented in 1967. A PPS does not have its own amplifier, which is essentially a switch working with the photodiode to read out the charge from the photodiode. The PPS had issues of fixed-pattern noise (FPN) and scalability. Although the APS was invented during the 1960s before CCD sensors, the FPN was holding it behind the CCD [41, 42]. Nevertheless, the advancement of CMOS technology in the 1990s again made CMOS sensors more popular than the CCD. The APS reduces the noise issues associated with the PPS. In the APS, the circuitry at each of the pixels determines the noise level and cancels it; thus, the name "active." The advantages of the APS over the other image sensors include the following [105]: (1) high sensitivity, (2) high fill factor of 50% to 70%, (3) computability due to integration with SoC, (4) lower cost, (5) low noise, (6) high readout speed, and (7) low-power dissipation. An APS is one of two types based on the actual light-sensing element in the pixel: the photodiode-type APS or the photogate-type APS. APSs also come in two types based on the mode of operation to integrate the sensed charge to voltage signal: the linear-mode APS and logarithmic-mode APS. Importantly, the CMOS-based image sensors are suitable for monolithic integration that allows integration of more functionality on the sensor array to build low-cost, low-power camera-on-a-chip. This had led to the design of a variety of CMOS sensors including the smart sensors and the secure sensors that will be discussed in the rest of the section. The DPS produces a digital signal straightaway through the pixel using a built-in ADC.

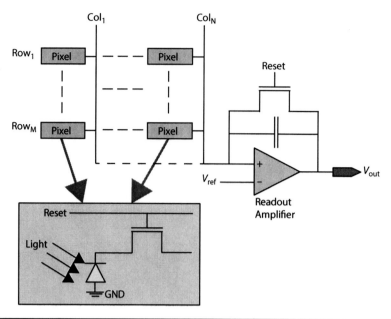

Figure 6.29 Architecture of a PPS array [105].

6.6.1.1 Passive Pixel Sensor

The architecture of the PPS with details of a single pixel is shown in Fig. 6.29 [71, 105]. The PSS working principle can be thought of as analogous to that of a single-transistor DRAM cell. Each pixel in the PPS array consists of a photodiode and a transistor [44, 78, 105]. The readout amplifier or charger integration amplifier is present in each column of the PPS array. The readout amplifier ensures constant voltage on each bus of the PPS array. The PPS has three modes of operation: (1) reset mode, (2) integration mode, and (3) readout mode. In the reset mode, the circuit is reset to ensure that the photodiode senses the next signal truthfully. The integration mode follows a reset mode. In the integration mode, the transistor is switched ON and the charge from the photodiode accumulates. The charges are accumulated for a period called "integration time." In the readout mode, the accumulated charge is read out as final output voltage of the pixel. PPSs have a high fill factor, lower area, low fabrication cost, and high quantum efficiency and are useful for small array sizes. The PPS has a larger capacitive load, slow operation, and FPN due to the individual readout amplifiers. The PPS is not suitable for large pixel arrays.

6.6.1.2 Active Pixel Sensor

The architecture of an APS array is depicted in Fig. 6.30 [16, 71, 105]. It consists of an array of $M \times N$ pixels that are individually one APS each. The APS array system has row decoder and column decoder that are used in addressing of the individual pixels. The row decoder is used to read the pixels in a row-by-row fashion from the APS array. The decoded signal from each row is readout to the parallel column first that are then amplified using the column amplifiers. The amplified signal then goes through the multiplexer and ADC to generate the final digital signal. The light-sensitive element of the APS may be a photodiode or a photogate depending on whether the APS is photodiode-APS or photogate-APS. The low-power dissipation, low dark current, low noise, and compact pixel size make the APS attractive for every day applications like digital cameras, digital single-lens reflex (DSLR) cameras, webcams, and smart phones [53, 57, 105].

The circuit topologies of the APS are presented in Fig. 6.31 [42, 71, 105, 136]. A single photodiode-based APS is presented in Fig. 6.31(a) [42, 71, 105, 136]. The photodiode-based APS or photodiode-type APS consists of one photodiode and three transistors per pixel. The photodiode is essentially a reverse-biased PN junction. The photodiode is for sensing light and the three transistors constitute the readout circuit. The three transistors are the following: reset transistor, source-follower transistor (or buffer transistor), and row-select transistor. When the reset transistor is OFF, the input light signal on the photodiode is converted to charge and then to voltage by the

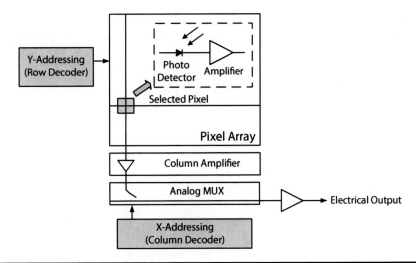

Figure 6.30 Architecture of an APS array [105].

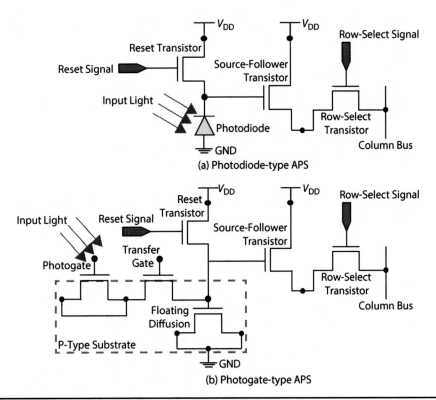

Figure 6.31 Photodiode and photogate-based APS [42, 64, 105, 136].

sensing node capacitor of the photodiode. When the reset transistor is ON, the photodiode is reset for new reading or conversion of light. Thus, effectively, the reset transistor controls the integration time. The source-follower transistor receives the sensed voltage from the photodiode as its gate input voltage. When row-select transistor is ON, the voltage is transferred to the column bus. The photodiode-type APS has a high fill factor, low-power dissipation, high sensitivity, and high quantum efficiency compared to the photogate-based APS.

A single photogate-based or photogate-type APS is presented in Fig. 6.31(b) [42, 64, 71, 105, 136]. In this circuit, the photodiode of the photodiode-type APS is replaced by the photogate. The photogate is essentially a MOS capacitor. There is also a transfer gate in addition to the photogate. Five transistors per pixel are used in the photogate-type APS. However, the photogate and transfer gate are overlapped using a double poly process [42]. In the photogate-based APS, the signal is integrated under the photogate. For readout of the signal, the output floating diffusion is resent and the resultant voltage is received by the source follower transistor. The charge is transferred by pulsing the transfer gate. The photogate-type APS has low output capacitance, low noise, and allows multiple integration as compared to photodiode-type APS.

The APS discussed in the above paragraphs operates in linear mode. In the linear-mode, the APS is reset by the reset transistor, and when the reset transistor is OFF, the integration of the charge on photodiode occurs. The linear-mode APS is shown in Fig. 6.32(a), which is the same as Fig. 6.31(a); however, it is shown again for a comparative perspective. The voltage sensed in the linear mode of APS operation is the following [92, 105]:

$$V_{\text{linear, APS}} = \left(\frac{1.2 \eta_{\text{QE}} \lambda_{\text{light}} P_{\text{light}}}{C_{\text{photo,A}}} \right) t_{\text{int}} \quad (6.24)$$

where η_{QE} is the quantum efficiency, λ_{light} is the wavelength of light in μm, P_{light} is the incident power in W, $C_{\text{photo,A}}$ is the per unit capacitance of the photodiode in $\left[\frac{F}{cm^2}\right]$, and t_{int} is the integration time. In the linear-mode APS, the output voltage is directly proportional to the incident light intensity of the photodiode. The linear-mode APS has high signal-to-noise and large output voltage signal. However, when nonlinear output signal of the APS is needed, a logarithmic-mode APS as shown in Fig. 6.32(b) may be used [42, 105]. In the logarithmic-mode APS, the input gate voltage of the reset transistor is the supply voltage V_{DD} instead of reset signal as is the case in linear-mode APS. As the photodiode does not get reset, the dark current in the photodiode is also converted to the output voltage. The output voltage of the logarithmic-mode APS is the following [15, 105]:

$$V_{\text{logarithmic, APS}} = \left(\frac{k_B T}{q} \right) \ln \left(\frac{I_{\text{photo}} + I_{\text{dark}}}{I_{\text{dark}}} \right) \quad (6.25)$$

where I_{photo} is the photocurrent, T is temperature, and I_{dark} is the dark current. The logarithmic-mode APS provides a high dynamic range of operation. However, the logarithmic-mode APS has slow response to low intensity and low signal-to-noise ratio (SNR).

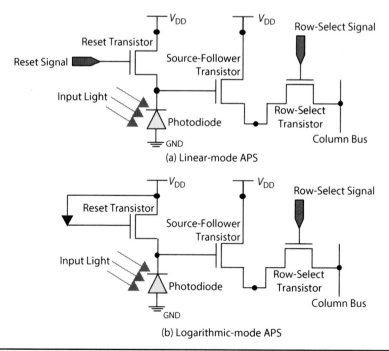

FIGURE 6.32 Linear-mode and logarithmic-mode APS [42, 105, 136].

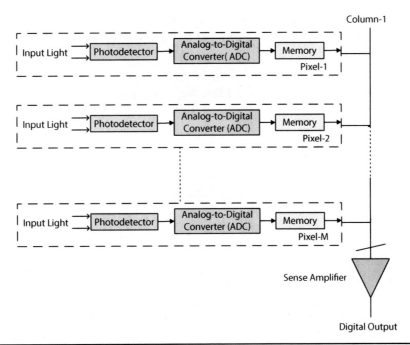

Figure 6.33 Schematic representation of the DPS [105].

6.6.1.3 Digital Pixel Sensor

DPS is the most recently developed CMOS image sensor. The schematic representation of a DPS array presenting one column is shown in Fig. 6.33 [65, 71, 105]. In a similar manner, N number of columns can be conceptualized. Each pixel in the DPS array consists of a photodetector, ADC, and a digital memory. The memory stores the digital data temporarily before output signal being read out through the digital bit lines. In the DPS array, any pixel can be accessed from digital memory at random. The readout columns are eliminated in a DPS. The DPS has several advantages over the analog (PPS/APS) CMOS sensors. DPS is more adaptive to CMOS technology scaling. The readout speed for the DPS is high, i.e., faster readout. The readout noise and FPN are low as the readout columns are eliminated. However, the drawback of the DPS is that the number of transistors per pixel in a DPS is more than that of the PPS and APS. This drawback leads to an increase in the pixel size of the DPS array that results in lower fill factors.

6.6.2 Characteristics of the Image Sensors

Selected characteristics and performance metrics of the image sensors are discussed in this subsection [51, 52, 85, 134].

1. *Quantum Efficiency:* The quantum efficiency of an image sensor is the rate at which the electron-hole pairs are generated per incident photon [95, 105]. The number of electrons generated in a time period is the ratio of the current and the electronic charge [95]:

$$\text{Number of Electrons in a Time Period} = \left(\frac{I_{\text{photo}}}{q}\right) \quad (6.26)$$

The number of photon incident on the image sensors can be expressed as the ratio of the power of the incident light and the energy of the photon [95]:

$$\text{Number of Photons in a Time Period} = \left(\frac{P_{\text{light}}}{\left(\frac{hc}{\lambda_{\text{light}}}\right)}\right) \quad (6.27)$$

where h is Planck's constant, c is velocity of light, λ_{light} is the wavelength of the incident light, and P_{light} is the power of the incident light. Thus, the quantum efficiency of an image sensor is given by the expression [95, 105]:

$$\eta_{QE} = \left(\frac{\text{Number of Electrons in a Time Period}}{\text{Number of Photons in a Time Period}} \right) \tag{6.28}$$

$$= \left(\frac{\left(\frac{I_{photo}}{q} \right)}{\left(\frac{P_{light}}{\frac{hc}{\lambda_{light}}} \right)} \right) \tag{6.29}$$

2. *Fill Factor:* The fill factor of an image sensor is the percentage of a photosite in a specific image sensor that is sensitive to light [1, 2, 105]. The CMOS sensors have an image-sensing element and need auxiliary circuits for operations such as readout, filtering, and amplification. The image-sensing element or photodetector such as photodiode and photogate may be covered by this circuitry. The actual percentage of the pixels that senses the input light signal is the fill factor of the image sensor. If A_{pixel} is the physical area of a pixel and A_{photo} is the area of the photodetector in the pixel, then the fill factor is calculated as follows [85]:

$$\text{Fill Factor}_{img} = \left(\frac{A_{photo}}{A_{pixel}} \right) \times 100\% \tag{6.30}$$

The CCDs can have fill factor up to 100%; the fill factor of the CMOS sensor is less. For example, if auxiliary circuits cover 30% of each pixel, then the image sensor is said to have a fill factor of 70%.

3. *Capture Time:* The capture time (t_{cap}) of the image sensor is defined as the time delay from the 50% level of the input photocurrent (I_{photo}) swing to the 50% level of the output voltage (V_{out}) [33].

4. *Output Voltage Swing:* The output voltage swing (V_{swing}) of the image sensors is defined as the maximum swing achieved by its output voltage [33]. It is an important characteristic of the image sensors as it affects the dynamic range.

5. *Dynamic Range:* The dynamic range (DR_{img}) of the image sensor is the ratio of the largest nonsaturating input signal or input signal swing of the image sensor and the smallest detectable input signal of the image sensor. It can be simplistically defined as follows [71, 85]:

$$DR_{img} = 20 \times \log_{10} \left(\frac{I_{photo,max}}{I_{photo,min}} \right) dB \tag{6.31}$$

where $I_{photo,max}$ is the largest nonsaturating photocurrent of the image sensor and $I_{photo,min}$ is the smallest detectable photocurrent of the image sensor. The dynamic range is calculated using the following expression [52, 71, 140, 141]:

$$DR_{img} = 20 \times \log_{10} \left[\frac{\left(\frac{q \times Q_{max}}{t_{int}} \right) - I_{dark}}{\left(\frac{q}{t_{int}} \right) \sqrt{\sigma_{total}^2 + \left(\frac{I_{dark} \times t_{int}}{q} \right)}} \right] dB \tag{6.32}$$

where σ_{total}^2 is the variance of noise due to readout of the image sensor and expressed in terms of $electron^2$, t_{int} is the integration period of the image sensor, and I_{dark} is the dark current of the photodetector. The maximum charge (Q_{max}) sensed by the photodetector on the incident of light is the following:

$$Q_{max} = \left(\frac{C_{photo} \times V_{swing}}{q} \right) \tag{6.33}$$

where, C_{photo} is the capacitance of the photodiode.

6. *Conversion Gain:* The two following conversions can happen in a typical image sensor [13]: the charge-to-voltage conversion $\left(\frac{V}{e^-}\right)$ and the ADC $\left(\frac{DN}{V}\right)$; where DN is a digital number. The conversion gain of the image sensor can be the charge-to-voltage conversion $\left(\frac{V}{e^-}\right)$. In other words, conversion gain (CG_{img}) is the amount of voltage change caused by one electron at the charge detection node of the image sensor [85]. The conversion gain is defined as follow [85]:

$$CG_{img} = \left(\frac{q}{C_{photo}} \right) \tag{6.34}$$

The conversion gain (CG_{img}) of the image sensor is also defined as the number of electrons represented by each digital number $\left(\frac{e^-}{DN}\right)$ [13].

7. *Photo Current:* The photodetectors such as the photodiode and the photogate generate photo-current (I_{photo}) from the radiant power (photons/s) [71]. For example, in a photodiode-APS, the photodiode is a reverse-biased PN junction. The photocurrent in the photodiode-APS-based image sensor has the following forms [38, 71]: (1) the current resulted due to carrier generation in depletion charge. All the carriers generated in this region are swept away by a strong electric field. (2) The current resulted due to holes generated in N-type quasi-neutral region. Some of these holes may also diffuse to the depletion region and get collected. (3) The current resulted due to electrons generated in P-type. Therefore, the total photocurrent in the photodiode-APS is [71]:

$$I_{photo} = I_{photo,dep} + I_{photo,h} + I_{photo,e} \tag{6.35}$$

where $I_{photo,dep}$ current in the depletion region of the photodiode, $I_{photo,h}$ is the current due to holes in N-type quasi-neutral region, and $I_{photo,e}$ current due to electrons in P-type region.

8. *Dark Current:* The dark current (I_{dark}) is the current in the photocurrent under no illumination or light. The dark current of the image sensors is the leakage current at the integration node of the APS circuit [71]. The dark current (I_{dark}) is not induced by the photogeneration in the photodetectors but originated due to nonideality in the silicon, as is the leakage in the typical CMOS circuit. Many different processes cause the dark current including the following [38]: (1) thermal generation of carriers, (2) band-to-band tunneling, (3) trap-assisted tunneling, and (4) impact ionization. The dark current can be quantified as the sum of surface-dependent component and sidewall-dependent components in the following manner [38]:

$$I_{dark} = J_{surface} A_{junction} + J_{sidewall} PT_{junction} \tag{6.36}$$

$$= J_{junction,EFF} A_{junction} \tag{6.37}$$

where $J_{surface}$ is the current density of the surface expressed in $\left(\frac{Amp}{m^2}\right)$, $A_{junction}$ is the area of the junction in (m^2), $J_{sidewall}$ is the current density of the surface expressed in $\left(\frac{Amp}{m}\right)$. $PT_{junction}$ is the perimeter of the junction in (m). $J_{junction,EFF}$ is the effective current density of the junction in $\left(\frac{Amp}{m^2}\right)$. The dark current limits the dynamic range of the APS due to the reduction in the signal swing and increase in the noise. In an APS array, the dark current varies from one pixel to another and causes FPN.

9. *Responsivity:* The responsivity (R_{img}) of the image sensors is defined as the change in the output voltage due to the photon collection [137]. It can be calculated from the ratio of the photocurrent and the power of the incident light as follows [85]:

$$R_{img} = \left(\frac{I_{photo}}{P_{light}}\right) \tag{6.38}$$

The responsivity can also be calculated from quantum efficiency and input light information using the following expression [85]:

$$R_{img} = \left(\frac{\eta_{QE} q \lambda_{light}}{hc}\right) \tag{6.39}$$

In general, the responsivity is affected by the quantum efficiency, capacitance of the photodetector, and photoelectron collection efficiency [137].

10. *Fixed-Pattern or Spatial Noise:* The noise in the image sensors can originate from the photodetectors as well as from the associated circuitry. The noise in the image sensors can be spatial noise and temporal noise. The spatial noise is also known as the FPN and discussed here [137]; temporal noise will be discussed next. The FPN is the spatial or pixel-to-pixel variations in the pixel outputs under uniform illuminations [46, 71]. The variations are originated from the mismatches in the device and interconnects across the image sensor array. The pixel output variations may cause two types of FPN: (1) *Offset FPN:* the offset FPN is caused by the readout devices. It is independent of pixel signal. The offset FPN due to variations in the dark current variation is called dark signal nonuniformity (DSNU). (2) *Gain FPN:* the other type of FPN, the gain FPN or photo response nonuniformity (PRNU), increases with signal level. In CMOS image sensors, the FPN is more than the CCD sensors due to the presence of the active circuitry. The significant FPN originates at the column amplifiers. Overall, the FPN can be a major issue as it may lead to visually objectionable streaks in the images or videos.

11. *Temporal Noise:* In addition to the above spatial noise or FPN, the image sensors can have temporal noise. The temporal noise may include the following [46, 137]: (1) shot noise from the photodetectors; (2) white noise and flicker noise from the source-follower transistors; (3) reset noise due to resetting of the floating diffusion capacitance; and (4) quantization noise due to sampling in the sample-and-hold capacitors. The temporal noise is a serious nonideality in the image sensors as it affects the signal fidelity. The temporal noise does not vary from pixel-to-pixel, but from frame-to-frame [46, 137].

All the above noises, i.e., spatial and temporal noises, have a significant effect on the signal of the image sensors [46]. The effects can be presented in terms of changes in the charges at different points of signal flow. The output charge including the noise right after reset operation can be expressed as follows [46]:

$$Q_{reset} = Q_{noise,reset} + Q_{noise,read,reset} + Q_{FPN} \tag{6.40}$$

where Q_{reset} is the effective charge after reset operation, $Q_{noise,reset}$ is the reset noise, $Q_{noise,read,reset}$ is the readout circuit noise at the reset node, and Q_{FPN} is the offset FPN. The output charge including the noise right after integration operation can be expressed as follows [46]:

$$Q_{int} = (I_{photo} + I_{dark})t_{int} + Q_{noise,shot} + Q_{noise,reset} + Q_{noise,read,int} \\ + Q_{FPN} + Q_{DSNU} + Q_{PRNU} \tag{6.41}$$

where Q_{int} is the effective charge after integration operation, $Q_{noise,shot}$ is the integrated shot noise, $Q_{noise,read,int}$ is the readout circuit noise at the integration node, Q_{DSNU} is offset FPN due to the DSNU, and Q_{PRNU} is the gain FPN due to PRNU. The overall signal charge is the difference between the above two Q_{int} and Q_{reset}. Thus, the overall charge from the pixel is the following that shows that some noises are canceled between Q_{int} and Q_{reset} [46]:

$$Q_{pixel} = Q_{int} - Q_{reset} \quad (6.42)$$

$$Q_{pixel} = (I_{photo} + I_{dark})t_{int} + Q_{noise,shot} - Q_{noise,read,reset}$$
$$+ Q_{noise,read,int} + Q_{DSNU} + Q_{PRNU} \quad (6.43)$$

12. *Signal-to-Noise Ratio:* The SNR of the image sensors is defined as the ratio of the input signal power and the average input referred to as noise power. In terms of the image sensor characteristics, the SNR can be calculated as follows [58]:

$$SNR_{img} = 20\log_{10}\left(\frac{I_{photo}t_{exp}}{qQ_{noise,temp}}\right) dB \quad (6.44)$$

where I_{photo} is the photocurrent, t_{exp} is the exposure time, and $Q_{noise,temp}$ is the overall temporal noise in terms of number of electrons in the image sensors including all noise components.

13. *Pixel Size:* The pixel size is the area occupied by the image sensor when built in actual silicon [21]. For compact electronic applications, smaller image sensors are desired. The comparative area between the photodetector and the associated circuitry has an impact on the fill factor of the image sensor. Larger fill factor means the photodetector is comparatively bigger than the associated circuitry. For a comparative perspective between two photodetectors, a larger area photodetector can sense more light. The large pixels can provide higher dynamic range as well as higher SNR. On the other hand, the small pixels can have higher spatial resolution and small die size.

14. *Resolution:* The image sensors essentially sample signals in spatial as well temporal domain [46]. The spatial resolution is the number of independent pixels per unit length, such as, lines per inch or linepairs per millimeter (lp/mm). A related characteristic is pixel resolution that is described as pixels per inch (PPI). For example, the spatial resolution of the computer monitors is approximately 72 to 100 lines per inch which corresponds to the pixel resolutions of 72 to 100 PPI. The spatial resolution measures how closely lines can be resolved in an image. It is the spatial resolution not the number of pixels that determines the clarity of the images. The spatial resolution is dictated by the Nyquist sampling theorem. At the spatial resolution below the Nyquist rate, the pixel response may be affected by low-pass filtering in the optics, spatial integration of the photocurrent, and crosstalk between pixels. The spatial frequencies that are above the Nyquist rate can cause aliasing and cannot be recovered after sampling [46].

15. *Power Dissipation:* This is the power dissipation in the image sensors that is needed to perform the operations or even when image sensor is ON [51, 52]. In nano-CMOS, implementation of the image sensor includes the dynamic power and leakage power dissipation.

6.6.3 A Concrete Example: 32-nm CMOS APS Design

In this subsection, a specific design of a photodiode-based APS and its characterization is discussed for 32-nm CMOS technology [51, 52].

6.6.3.1 The Baseline Design of APS

The schematic representation of a photodiode-based APS is shown in Fig. 6.34 [14, 46, 52]. This is a single-pixel design. It contains one photodiode and three transistors. The three transistors of the circuit are (1) PM1, reset transistor; (2) NM1, source-follower transistor; and (3) NM2, access transistor. In this APS, a PMOS transistor has been used as the reset transistor as it results in a higher output voltage swing as compared to a conventional APS [14, 118]. In the APS design, the transistor sizes are chosen carefully to achieve the following [14, 118]: (1) sufficient current, (2) sufficient source follower gain, and (3) isolation of source-follower output from the pixel output. In addition, the transistor sizes selected should be as small as possible for the maximum photodiode

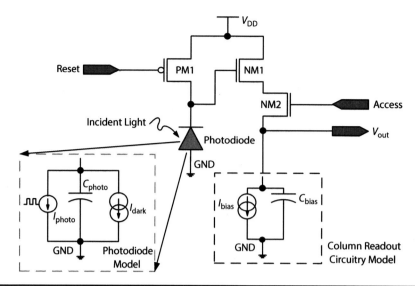

FIGURE 6.34 Model of a single APS for circuit simulation [51, 52].

APS Transistor Name	Size (W : L) for 32-nm CMOS
PM1	160 nm : 32 nm
NM1	320 nm : 32 nm
NM2	240 nm : 32 nm

TABLE 6.2 Transistor Sizes in the APS for 32-nm CMOS [51, 52]

to pixel ratio when considering the physical design in silicon as it has impact on the fill factor of the image sensor. The sizes chosen for the transistors of the APS are presented in Table 6.2.

The APS schematic presented in Fig. 6.34 also contains models used for simulation of the circuit [51, 52]. For the simulation of the APS using analog simulators, the photodiode is modeled as a pulsed current source representing the photocurrent (I_{photo} = 100–350 nA) in parallel with a capacitor representing the diode capacitance (C_{photo} = 20 fF) and a DC current source representing the dark current (I_{dark} = 2 fA) [51, 52, 140]. The biasing circuitry has the following values: I_{bias} = 500 nA and C_{bias} = 1 pF. The values are selected to be consistent with the 32-nm CMOS technology node. A higher bias current (I_{bias}) ensures a smaller readout time.

In a real-life application, a typical two-dimensional array of $M \times N$ pixels is organized into M rows and N columns. As a specific example, the block diagram of an 8×8 APS array implemented by using 64 single pixels is shown in Fig. 6.35 [51, 52]. In the APS array, the pixels in a specific row share reset lines so that a whole row is reset at a time. The row select lines of each pixel in a row are also connected together. The outputs of each pixel in any given column are connected together. The competition for the output line does not happen as only one row is selected at a given time. The amplifiers are used typically on a column basis. The APS array is accessed pixel-wise.

6.6.3.2 Characterization of the 32-nm CMOS APS

The functional simulation results of the APS array are shown in Fig. 6.36 for a high illumination photocurrent [51, 52]. The simulation result is obtained from transient analysis of the APS array in an analog simulator.

The characterization of the APS array with baseline size is now discussed. The APS array has been characterized for the following figures of merit or attributes: (1) average power dissipation P_{APS} that includes leakage as well; (2) capture time of the APS t_{cap}; (3) output voltage swing V_{swing} of the APS; and (4) dynamic range DR_{img} of the APS. An output voltage swing of 428 mV is observed [51, 52]. This value is 47.6% of V_{DD}, which is in the acceptable range. The DR_{img} of the baseline APS for 32-nm CMOS technology is calculated to be 59.47 dB. The capture time is calculated as the time delay from

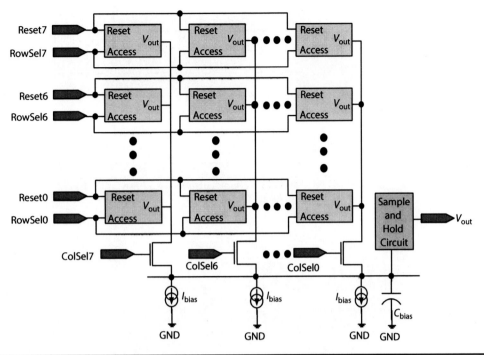

FIGURE 6.35 An 8 × 8 APS array constructed by using a collection of APS [51, 52].

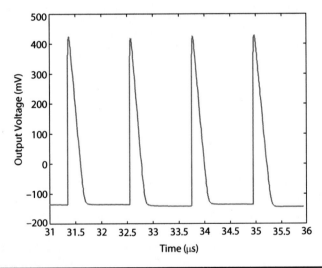

FIGURE 6.36 Circuit simulation of the 8 × 8 APS array [51, 52].

the 50% level of the input swing (I_{photo}) to the 50% level of the output voltage (V_{out}). For the measurement of capture time (t_{cap}) of the APS array, the pixel at the middle of the APS array is considered as it suffers the maximum loading. Thus, it gives the worst-case capture time of the APS array. The baseline APS array has a t_{cap} of 5.65 μs for 32-nm CMOS. The characterization results of the APS array are shown in Table 6.3 [51, 52].

6.6.4 Smart Image Sensors

In a bigger context, a "smart camera" is a camera that combines sensing, processing, and communication of video in a single platform [109]. It consists of four blocks as follows: (1) image sensors, (2) sensing unit, (3) processing unit, and (4) communication unit. The image sensors of the smart camera are any basic image sensors or pixel arrays that only sense the input data. The sensing unit reads data from image sensors, performs preprocessing such as white balance, and controls the image sensor

APS Parameters	Actual Values
Technology	32-nm CMOS
V_{DD}	0.9 V
P_{APS}	16.32 μW
t_{cap}	5.65 μs
V_{swing}	428 mV
DR_{img}	59.47 dB

TABLE 6.3 Characterization of 32-nm CMOS-Based APS Array [51, 52]

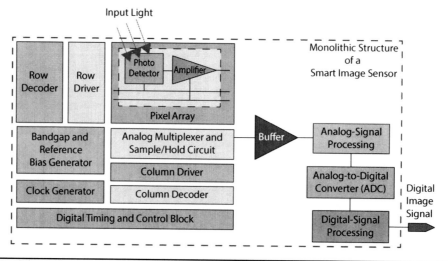

FIGURE 6.37 Schematic diagram of smart image sensor [40, 89, 115].

parameters. The processing unit performs processing of the image data. The communication unit provides interfaces such as Ethernet and USB. In the same spirit, the smart image sensors are image sensors with a variety of additional capabilities [10, 20, 74, 77]. The smart image sensors perform both analog and digital signal processing (DSP) on the same substrate as the photoelement and its interface in a monolithic fashion. CMOS-based APS has made it possible to fabricate smart image sensor monolithically to achieve low-voltage and low-power objectives.

A specific example of a smart image sensor is presented in Fig. 6.37 [40, 89, 115]. The main components of the smart sensor are the following: (1) pixel array, (2) scanning or addressing circuitry, (3) analog multiplexer and sample/hold analog front-end circuitry, (4) analog signal processing unit, (5) ADC, (6) DSP unit, (7) clock generator, (8) digital timing and control block, and (9) band gap and reference bias generator. The pixel array consists of photodetectors (photodiode) and amplifiers. The scanning and addressing circuitry builds digital memory-style random access reading of pixels in the APS array. Analog multiplexer and sample/hold constitute the analog front-end circuitry of the smart image sensor that processes all the pixels in a selected row simultaneously and samples onto sample/hold circuits at the end of their respective columns. The analog signal processing unit may perform analog processing when the smart image sensor is used in applications such as neuromorphic vision. The ADC converts analog signal to digital signal for further processing in digital domain. The DSP unit can perform a wide range of digital processing in the smart image sensor. The clock generator provides a base clock to the different components of the smart image sensor. The digital timing and control block control the operation of the complete smart image sensor. For example, it produces the appropriate sequencing of the row/column addresses, controls the ADC timing, and synchronizes the pixels, with the analog signal processing unit as well as DSP unit. The band gap and reference generator produces the on-chip analog voltage and current references for other units such as amplifiers, ADCs, and clock generators.

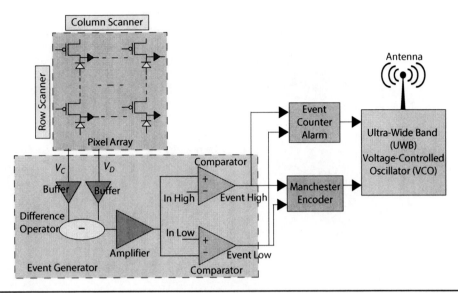

FIGURE 6.38 Schematic diagram of wireless image sensor [19, 20].

A smart sensor that has a temporal-difference image process as well as wireless communication capability is presented in Fig. 6.38 [19, 20]. This smart sensor has the following units: (1) pixel array, (2) row/column scanner, (3) event generator, (4) event counter alarm, (5) Manchester encoder, and (6) ultra-wide band (UWB) voltage-controlled oscillator (VCO). The pixel array is a typical photodiode-based APS array but each pixel has a capacitor, and hence the APS array can store the current frame as a reference image. The row/column scanner is used for random access of the APS array. The event generator consists of difference operator, amplifier, and comparators. The event counter alarm is made of a 12-bit counter that counts the number of events per frame and generates an alarm signal when the count exceeds a programmable threshold. In this smart sensor, the pixel-array receives the incident light signal and converts to voltage and also stores the voltage. When a specific pixel is accessed for readout, both the integration voltage from the photodiode and the previous voltage stored in the pixel capacitor are readout. The event generator calculates the difference between the two voltages and compares it with a positive and negative threshold. A digital event is generated if the difference exceeds the thresholds. The event bitstream and the clock are encoded by the Manchester encoder. The encoded digital signal from the Manchester encoder is converted into an impulse sequence in the UWB transmitter for wireless transmission. This is particularly useful in wireless sensor networks.

6.6.5 Secure Image Sensors

Secure digital cameras (SDC) are the cameras with built-in capabilities for the protection of the images that they capture [73, 80, 90]. However, in the SDC, the module, such as image sensor, encryption, and watermarking, are independent modules working together in the system-on-a-chip (SoC) architecture of the SDC. The secure image sensors are APS or DPS with integrated encryption or watermarking functionalities [73, 80]. The secure image sensors embed the copyright information right at the source to provide maximum security at lowest cost. The schematic representation of a secure image sensor is provided in Fig. 6.39 [90]. The secure image sensor consists of the following units: (1) APS array, (2) ADC, (3) linear feedback shift register (LFSR), and (4) Watermark Adder. In this secure image sensor, the output signal from the APS is converted to digital signal by ADCs. One ADC is deployed for each column of the pixels in the APS array. The LFSR circuit that is essentially the watermark creation unit generates a unique bit stream. The LFSR considers a sensor characteristic aware key as input. The input key is unique to the APS and it is a private key to ensure the false verification in the watermarking process. The watermark adder considers the digital signal from the ADC and the unique watermark from the LFSR and adds them together to create the watermarked output signal from the secure image sensor.

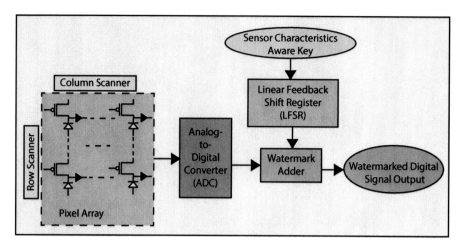

Figure 6.39 Schematic diagram of secure image sensor [90].

6.7 Nanoelectronics-Based Gas Sensors

CNTFETs have the following interesting properties that make them the most suitable for integrated amperometric sensors [23]: (1) high sensitivity to analytes, (2) fast response times, and (3) large surface area-to-volume ratios. The characteristics of CNT such as electrical resistance, thermoelectric power, and local density of states are extremely sensitive to oxygen or air [33]. Similarly, certain adsorbates such as the NH_3 (an electron donor) and the NO_2 (an electron acceptor) may lead to decrease and increase in conductance of the CNT [67]. The CNTs have been explored to be used as sensors including cantilevers and conductors for diverse applications [88]. For example, a simple conductance type CNT-based sensor has been built for the detection of streptavidin and mouse antibody [11] and also a DNA detection sensor using CNTFET [122].

6.7.1 CNTFET-Based Gas Sensor

The schematic representation of a specific CNTFET-based gas sensor is provided in Fig. 6.40 [121]. In this CNTFET-based sensor, the sensing element is a single-walled carbon nanotube (SWCNT). There are two CMOS inverters one each at the either end of the SWCNT. The change in the conductance of the SWCNT is used for calibration of the gas sensor. It is observed that conductance of a metallic SWCNT can increase by 50% with 10 ppm adsorption of thionyl chloride ($SOCl_2$) [69, 121, 135]. The signal generated from the inverter goes to a signal converter interface circuit and then to an ADC. The ADC generates a digital signal that is then used for the measurement purposes.

6.7.2 CNTFET-Based Chemical Sensor

The system architecture of a CNTFET-based chemical sensor is presented in Fig. 6.41 [24, 25, 26]. The different components of the CNTFET-based chemical sensor are: (1) CNT sensor array, (2) dynamic range current steering DAC, (3) successive approximation register (SAR) ADC,

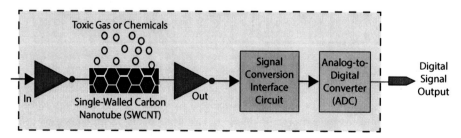

Figure 6.40 CNTFET-based gas sensor [121].

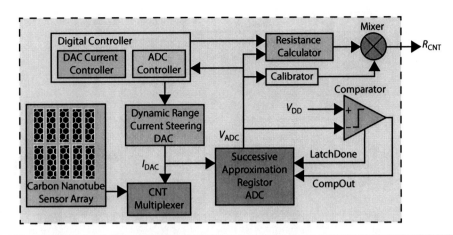

FIGURE 6.41   CNTFET-based chemical sensor [24, 25, 26].

(4) digital controller, (5) comparator, (6) resistance calculator, (7) calibrator, and (8) mixer. The key working principle of the CNT-based chemical sensor is that when a predetermined current source is applied to the CNT sensors, a voltage develops across the CNT sensors. The current and voltage is then converted to a resistance that is calibrated for the measurement purposes. The resistance resolution for the sensor is the following [24, 25, 26]:

$$R_{\text{CNT,resolution}} = \left(\frac{V_{\text{measured}}}{I_{\text{DAC}}} \right) \tag{6.45}$$

where I_{DAC} is the input current and V_{measured} is the measured voltage. The current steering DAC essentially works as the variable current source that can adjust the input current I_{DAC}. The sensed voltage is digitized through ADC. A SAR ADC is used as it can be selectively powered OFF to reduce power consumption. The digital controller adaptively controls the DAC current as the resistance changes as well as controls the ADC operation. The DAC current controls the output current of the DAC for next measurement based on the present resistance value. The resistance calculator uses the voltage from the ADC and current from the DAC controller to calculate the resistance as $\left(\frac{V_{\text{ADC}}}{I_{\text{DAC}}} \right)$. The calibrator is needed to reduce the linearity errors due to ADC and DAC. The calibrator measures the current with off-chip known as reference resistors and stores the ratio of the desired and measured current in a lookup table. The calibrator then multiplies the ratio from lookup table and resistance from the resistance calculator to obtain the output calibration resistance R_{CNT}.

6.8   Body Sensors

The concept of the wireless body sensor network (WBSN) is depicted in Fig. 6.42 [129, 143]. WBSN is being used in many applications, including medical application for vital sign monitoring, the diagnose assistant, and the drug delivery [81]. The body node, slave node, or sensing node consists of one or more of the sensors connected to or implanted in the body. The body node is connected to a base node or master node that may be a simple computing device such as a smart phone or tablet. From the base node, the information then goes to the Internet cloud for remote use. The body node can perform various tasks including biomedical information acquisition, signal processing, data storage, communications, drug delivery, and nerve stimulation.

The architecture of the body node or sensing node is presented in Fig. 6.43 [129, 143]. The body node consists of the following distinct units:

1. *Sensor and Stimulator:* This unit contains sensors and devices for biomedical signal sensing and stimulation. This unit converts a biological signal to an electrical signal. The ADC is used to convert the sensed signal to a digital format and the interface unit is used to communicate with the rest of the modules in the body node.

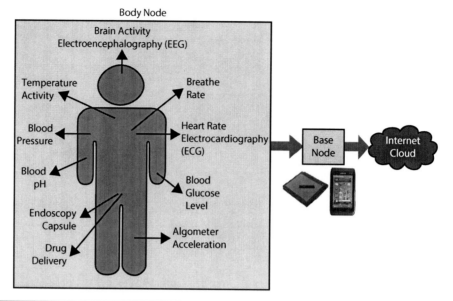

FIGURE 6.42 WBSN–basic concept [129, 143].

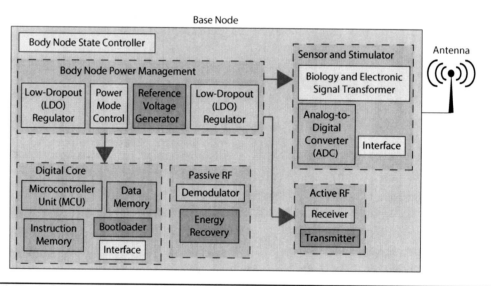

FIGURE 6.43 Architecture of a body node of the WBSN [129, 143].

2. *Active Radiofrequency (RF):* The active RF block of the body node is the active bidirection link that has both a receiver and a transmitter.

3. *Passive Radiofrequency:* The passive RF unit has ON-OFF keying demodulator and energy recovery/harvesting units. The passive RF unit provides the work-on-demand capability to the body node. The passive RF is activated when it gets modulated command from the base node or master node. The energy recovery block converts the RF energy received from the base node to a DC supply for the demodulator.

4. *Digital Core:* The digital core has multiple working units including the microcontroller unit, data memory, instruction memory, bootloader, and multimode transducer interface. The digital core processes signals in the body node.

5. *Body Node Power Management Unit:* The body node power management unit consists of low-dropout (LDO) regulator, power-mode control logic, and reference voltage generator. The power management unit generates right voltage supply for different block in the base node.

6. *Body Node State Controller:* The body node state controller is used for changing the states of the body node for its energy-efficient operations.

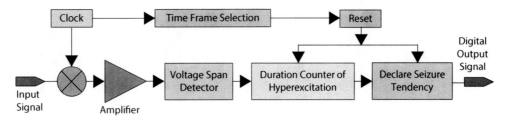

FIGURE 6.44 Architecture of an implantable epileptic seizure-onset detector [111, 112, 113].

6.9 Epileptic Seizure Sensors

Seizure refers to abnormal excessive neuronal activity in the brain. The seizure can be either epileptic or nonepileptic. The epileptic seizure is caused by abnormal and hypersynchronous discharge of cortical neurons [72, 111, 123]. The epileptic seizure sensor or epileptic seizure detector can detect the occurrence of the seizure from the neural signal. The start of the seizure is characterized by low-amplitude fast activity in the electroencephalography (EEG). The architecture of an implantable real-time epileptic seizure detector is presented in Fig. 6.44 [56, 111, 112, 113]. In this architecture, the input signals are divided into different time frames. The input signals are monitored over a time period. In a time frame, if a satisfied amount of seizure activities is detected then the seizure tendency is declared.

The epileptic seizure detector has three components: the modulation component, the amplification component, and the detection of seizure onset component. The modulation component ensures that the low-frequency noise from the instrumentation does not affect the EEG signal. The amplification component consists of the amplifier and voltage span detector (VSD). The detection component consists of a duration count detector (DCD) and seizure tendency declaration. If V_{EEG} is the input neural signal, then the modulated EEG signal that goes as input to the amplifier is the following:

$$V_{\text{mod}}[n] = \sum_{n=1}^{T_f/T_{\text{mod}}} ((-1)^n V_{EEG}(nT_{\text{mod}})) \tag{6.46}$$

where T_f is a time frame, F_{mod} is the modulated signal frequency, and $T_{\text{mod}} = \left(\dfrac{1}{F_{\text{mod}}}\right)$. The amplified modulated seizure signal is $V_{\text{mod,Amp}}$. The detected hyperexcitation signal that is the output of the voltage span detector is the following [112, 113]:

$$V_{\text{VSD}}[n] = \begin{cases} 1 & \text{When } V_{\text{mod,Amp}} \text{ within a threshold} \\ 0 & \text{Otherwise} \end{cases} \tag{6.47}$$

The seizure prediction signal is the following [112, 113]:

$$V_{\text{SZ}}[n] = \begin{cases} 1 \text{ (Seizure)} & \text{When } T_{SZ} > \eta_T \\ 0 \text{ (No Seizure)} & \text{Otherwise} \end{cases} \tag{6.48}$$

where the duration of the detected hyperexcitation is T_{SZ} in a time frame T_f and η_T is the hyperexcitation time threshold.

6.10 Humidity Sensors

The humidity sensors sense the moisture content in the surroundings in which they are placed. The instrument called "hygrometer" measures atmospheric humidity. The types of humidity sensors are quite diverse based on principles of sensing used including mechanical, gravimetric, thermal, resistive, and capacitive [97]. In this section, two selective electronic humidity sensors will be used.

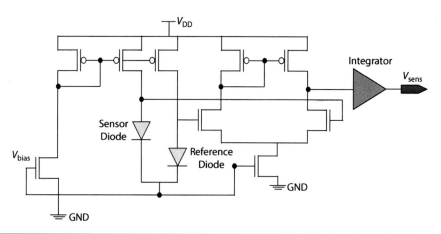

FIGURE 6.45 Schematic diagram of a diode-based humidity sensor [97].

6.10.1 A Diode-Based Humidity Sensor

The schematic representation of a diode-based humidity sensor is presented in Fig. 6.45 [97]. The circuit consists of two diodes, one sensor diode and another reference diode. In addition, there is on-chip readout circuitry. The readout circuitry contains a differential transconductance amplifier and an integrator. The diodes are given biased currents using the three transistors connected between the diode and the supply V_{DD}. In this humidity sensor, the sensor diode is exposed to the surroundings for sensing purposes. However, the reference diode is sealed to isolate it from the surroundings. When moisture in the surroundings increases, the temperature of the sensor diode decreases, which results in the increase in its thermal conductance. The output voltage of the sensor diode increases due to its negative temperature sensitivity. The temperature sensitivity of the PN junction sensor diode can be expressed by the following [97]:

$$\left(\frac{\delta V_D}{\delta T}\right) = \left(\frac{V_D}{T}\right) - v_{therm}\left(\frac{1}{I_{sat}}\frac{\delta I_{sat}}{\delta T}\right) \tag{6.49}$$

where V_D is the diode voltage, v_{therm} is the thermal voltage, and I_{sat} is the saturation current of the sensor diode. At the same time, the output voltage of the reference diode does not change as it is sealed. A differential transconductance amplifier converts the difference between the output voltages of the sensor diode and reference diode to a current. The output current of the differential transconductance amplifier is integrated by the integrator to provide the output voltage of the humidity sensor V_{sens} that is then calibrated for measurement purposes.

6.10.2 A CMOS Device–Based Humidity Sensor

The schematic diagram of a CMOS device–based humidity sensor is presented in Fig. 6.46 [32]. The humidity sensor has the following three components:

1. *CMOS Device Capacitive Humidity Sensor:* The CMOS device capacitive humidity sensor can be built using either: (1) a moisture-sensitive film sandwiched between two parallel plates, or (2) a moisture-sensitive film deposited on top of interdigitated electrodes. The first structure is highly sensitive but slow and the second structure is less sensitive but fast and easy to implement.

2. *Data Acquisition Unit:* The data acquisition unit consists of a charging unit, a programmable current source, and a divider. The current source is used to generate a current of $I_{control}$. This current is used to charge or discharge the capacitive humidity sensor with capacitance C_{sensor}. The capacitance generates a square waveform whose frequency is inversely proportional to the humidity.

3. *Interface Unit:* The interface unit consists of a monostable vibrator, a power-on-reset (PoR) block, a LDO voltage regulator, and an NMOS switch. The monostable vibrator is used to

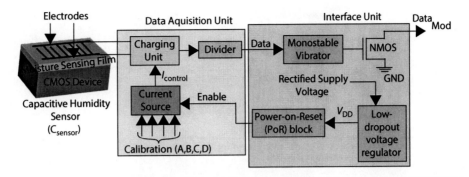

FIGURE 6.46 Schematic diagram of a CMOS-based humidity sensor [32].

modulate the data signal. The monostable vibrator generates a short pulse at the positive edge of the data. The short pulse is used to activate the NMOS at the final stage of the sensor that generates the modulated output signal Data$_{Mod}$. The LDO voltage regulator receives a rectified supply voltage (rectifier not shown in the picture) and converts it to the required voltage V_{DD} for the PoR block. The PoR block generates "Enable" signal to control the current source in the data acquisition unit.

The capacitance of the humidity sensor is calculated as follows [32]:

$$C_{sensor} = \varepsilon_0 \varepsilon_{hum} \theta_E + C_{para} \tag{6.50}$$

where ε_0 is the permittivity of the vacuum, ε_{wet} is the permittivity of the polyimide film with absorbed moisture, θ_E represents the geometry of the electrodes in the CMOS device sensor, and C_{para} is the parasitic capacitance. The output signal frequency that is generated by the charging circuit is the following:

$$f_{charge} = \left(\frac{I_{charge}}{2 V_{hyst} C_{sensor}} \right) \tag{6.51}$$

where I_{charge} is the charging current and V_{hyst} is the Schmitt triggers hysteresis.

6.11 Motion Sensors

The schematic representation of a photodetector-based motion sensor is provided in Fig. 6.47 [138, 139]. Each cell in this motion sensor consists of three photodetectors, a spatial edge detector, and a motion measurement block. The photodetectors, photodetector-1, photodetector-2, and photodetector-3, generate the signals S_1, S_2, and S_3, respectively. The spatial edge detector receives the S_1, S_2, and S_3 signals and generates output signals D_1 and D_2. The binary signal D_1 represents whether a spatial edge exists between the photodetector-1 and photodetector-2. The output signal D_1 is binary and is calculated as follows [138, 139]:

$$D_1 = \begin{cases} 1 & \text{If} \left(\frac{S_2}{S_1} \right) > \eta_S \\ 0 & \text{Otherwise} \end{cases} \tag{6.52}$$

where η_S is a threshold value of the ratio. The binary signal D_2 represents whether a spatial edge exists between the photodetector-2 and photodetector-3. The output signal D_2 is binary and is calculated as follows [138, 139]:

$$D_2 = \begin{cases} 1 & \text{If} \left(\frac{S_3}{S_2} \right) > \eta_S \\ 0 & \text{Otherwise} \end{cases} \tag{6.53}$$

300 Chapter Six

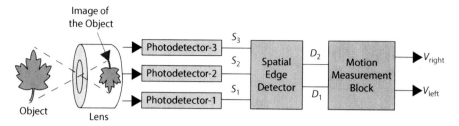

FIGURE 6.47 Schematic diagram of a photodetector-based motion sensor [138, 139].

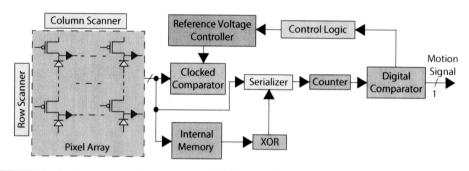

FIGURE 6.48 Block diagram of an APS-based motion sensor [142].

The motion measurement block receives the binary signals D_1 and D_2. It detects a motion in the object based on D_1 and D_2 signals. For example, a spatial edge of the target object is estimated as moving from left to right if the signals D_1 and D_2 have changed in the following order: (1) D_1 rises while D_2 is low, (2) D_2 rises before D_1 falls, (3) D_1 falls before D_2 falls, and (4) D_2 falls before D_1 rises again. Similarly, a spatial edge of the target object is estimated as moving from right to left if the signals D_1 and D_2 have changed in the following order: (1) D_2 rises while D_1 is low, (2) D_1 rises before D_2 falls, (3) D_2 falls before D_1 falls, and (4) D_1 falls before D_2 rises again.

The schematic diagram of a CMOS image sensor-based motion detector is presented in Fig. 6.48 [142]. The motion sensor consists of a pixel array, an internal memory, a clocked comparator, a reference voltage controller, a serializer, a counter, an XOR unit, and a digital comparator. The APS array samples a target object. The clocked comparator performs a comparison of the APS array output and initial reference voltage. The clocked comparator thus generates a quantized 1-bit image data. The digital comparator and the control logic decide whether the reference voltage is to be increased or decreased. The reference voltage controller is used to perform the change in the reference voltage of the clocked comparator. The internal memory stores frame of the image for a fixed reference voltage of the clocked comparator. The XOR unit is used to identify change in each pixel data. The serializer is used to serialize the digital image data. The counter is used to count the number of bright pixels in a frame by counting the number of "1s" in the serialized digital image data. When the number of "1s" exceeds the threshold value, the digital comparator generates a motion detection signal.

6.12 Sense Amplifiers

The sense amplifiers are circuits that are used to sense, amplify, or refresh the values of a bit stored in a memory cell [98, 103, 124]. The sense amplifier is used in volatile memory such as static random access memory (SRAM) and dynamic random access memory (DRAM). As the name suggests, the primary use of the sense amplifiers is to sense and amplify the voltage levels from the bitlines such that the voltage has proper voltage levels to be interpreted accurately [9, 36]. In addition, the sense amplifiers in a DRAM perform memory refresh in which data is immediately written back after the readout to make up the charge destruction due to the readout. Many variants of the sense amplifier

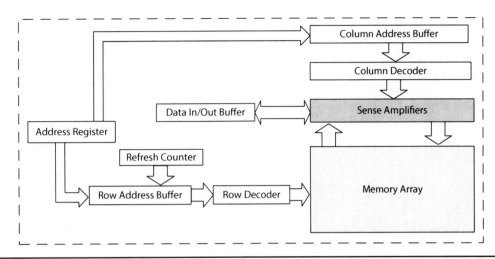

Figure 6.49 Deployment of sense amplifier in memory [98, 103].

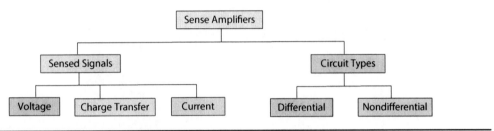

Figure 6.50 Different types of sense amplifiers.

have been proposed to increase the boost performance and increase yield [30, 31, 55]. The deployment of sense amplifier in memory architecture is presented in Fig. 6.49 [98, 103].

6.12.1 Types of Sense Amplifiers

The sense amplifiers are of various types based on their topology and signals. An overview of a selected types of sense amplifiers is presented in Fig. 6.50 [30, 84, 98, 131]. Based on the signal being sensed or amplified by the sense amplifier, different categories for sense amplifiers include the following [30, 84, 98, 126, 131]: (1) voltage-mode sense amplifier, (2) charge-transfer mode sense amplifier, and (3) current-mode sense amplifier. The voltage-mode sense amplifiers amplify a small differential voltage signal. The charge transfer sense amplifiers perform charge redistribution between high bitline capacitance and low output capacitance [107]. The voltage-mode sense amplifiers are of two types [83]: (1) current latch voltage-mode sense amplifier, and (2) voltage latch voltage-mode sense amplifier. The current-mode sense amplifiers amplify a differential current signal. The voltage-mode sense amplifiers are commonly used in DRAM [84, 131]. The current-mode sense amplifiers are more suitable for large memory due to their lower input capacitance. In addition, the lower voltage swings result in a faster sensing operation [29]. Based on the circuit types, the sense amplifiers can be differential or nondifferential sense amplifiers. The topology of the sense amplifiers could range from a basic gated flip-flop circuit to a full-latch cross-coupled sense amplifier based on their topology [60, 84].

The transistor-level schematic diagram of a full-latch cross-coupled sense amplifier is presented in Fig. 6.51 [29, 98, 110, 131]. There are two inverters that are cross coupled together to provide a positive feedback structure, which essentially forms the core of the sense amplifier. The sense amplifier circuit has essentially a differential coupling with two bitlines. During a read or refresh operation, the precharged bitlines are amplified to a full logic "1" or "0" values. During the amplification process, charging and discharging of the bitlines are required in this sense amplifier. The bitline capacitances of the memory make it difficult for the bitlines to accumulate the minimum voltage for fast charging and discharging. Thus, the amplification process in this sense amplifier can be slow for large-size memory.

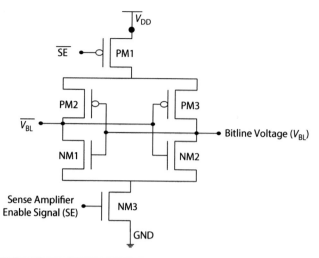

Figure 6.51 The full-latch cross-coupled sense amplifier circuit.

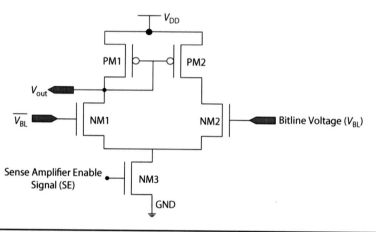

Figure 6.52 The current-mirror sense amplifier circuit.

The transistor-level circuit diagram of the current-mirror sense amplifier is provided in Fig. 6.52 [98]. This sense amplifier circuit consists of an NMOS differential amplifier and a PMOS current mirror as its load. The PMOS current mirror provides a stable bias current to work faithfully across a large voltage swing. In this sense amplifier, the transistors NM1 and NM2 receive the signals from BL and $\overline{BL}$ as gate inputs. Both the bitlines are precharged to a specific voltage. The sense amplifier is activated by raising the SE signal when there is sufficient voltage change on the bitline. One of the disadvantages of the current-mirror sense amplifier is that it requires a high current supply.

6.12.2 Performance Metrics for the Sense Amplifiers

The quality of the sense amplifier is determined by different performance and robustness metrics. For example, the sense delay is one metric that dictates the performance. Similarly, the robustness is determined by the minimum voltage difference between two bitlines called input offset voltage that can be accurately sensed by it. In general, the characteristics of figure-of-merit of the sense amplifier may include the following [83, 98, 101, 103]: (1) precharge and voltage equalization time, (2) sense delay, (3) sense margin, (4) power dissipation, and (5) silicon area. These characteristics are briefly discussed in the rest of this subsection.

1. *Precharge and Voltage Equalization Time:* The precharge and voltage equalization time is the time required by the sense amplifiers to charge both the bitlines BL and $\overline{BL}$ to $\left(\frac{V_{DD}}{2}\right)$ [98]. The bitlines are precharged to $\left(\frac{V_{DD}}{2}\right)$ to reduce the voltage swing of the sensing operation of the

sense amplifiers. The capacitance of the bitlines BL and $\overline{\text{BL}}$ have impact on the precharge and voltage equalization time. The precharge and voltage equalization time is an important factor affecting the overall speed of the sense amplifiers.

2. *Sense Delay:* Sense delay is the time required by the sense amplifiers to share a sufficient amount of voltage to produce the minimal voltage change that can be detected [98]. In a typical scenario, the voltage difference that appears on the bitlines is quite small. Thus, there is a need for some minimal signal that must accumulate for accurate operation of the sense amplifiers. There is a time delay between the time of wordline activation and sense amplifier activation to allow the minimal signal to appear for accurate sensing. Sense delay is an important characteristic of the sense amplifier as it constitutes a significant portion of the memory access time [49, 83]. The sense delay of a sense amplifier is affected by the supply voltage. A higher supply voltage results in a smaller sense delay but leads to higher power dissipation [7].

3. *Sense Margin:* The sense margin is the minimum amount of voltage difference that can be accurately sensed by the sense amplifiers [50, 55, 98]. The sense amplifier detects the change on the bitlines to be amplified to a logic "1" or a "0" value. For the sensing operation to work accurately, there must be sufficient voltage change on both BL and $\overline{\text{BL}}$. The direction of charge flow depends on the value in the memory cell when reading a value from a memory cell. When reading a value of logic "1," the charge flows from the storage capacitor to the bitline capacitor. The charge flows in the reverse direction when reading a value of logic "0." A positive voltage gain on the bitline indicates a logic "1" value read. On the other hand, a negative charge gain indicates a logic "0" value read. The resulting voltage shared is expressed as follows [98, 103]:

$$\Delta V = \frac{C_{\text{cell}}}{C_{\text{cell}} + C_{\text{BL}}} \left(V_{\text{cell}} - \frac{V_{\text{DD}}}{2} \right) \tag{6.54}$$

where C_{cell} and C_{BL} are the memory cell and bitline capacitances, respectively, V_{cell} is voltage stored in the memory cell, and V_{DD} is the supply voltage. Usually $C_{\text{BL}} \gg C_{\text{cell}}$, and hence the above expression can be simplified as follows [98, 103]:

$$\Delta V \cong \frac{C_{\text{cell}}}{C_{\text{BL}}} \left(V_{\text{cell}} - \frac{V_{\text{DD}}}{2} \right) \tag{6.55}$$

When the bit cell value is 1, $V_{\text{cell}} = V_{\text{DD}} - V_{\text{Th}}$ and the shared voltage is the following [98, 103]:

$$\Delta V(1) \cong \frac{C_{\text{cell}}}{C_{\text{BL}}} \left(\frac{V_{\text{DD}}}{2} - V_{\text{Th}} \right) \tag{6.56}$$

When the bit cell value is 0, $V_{\text{cell}} = 0$ and the shared voltage is the following [98, 103]:

$$\Delta V(0) \cong -\frac{C_{\text{cell}}}{C_{\text{BL}}} \left(\frac{V_{\text{DD}}}{2} \right) \tag{6.57}$$

Typically, ΔV is very small. As it is evident from the above expressions, the sense margin is different based on whether the cell data value sensed is a "1" (high) or "0" (low). It is higher when a logic "1" (high) is being read by the sense amplifiers. The sense margin is an important characteristic for the operation of the sense amplifier.

4. *Power Dissipation:* The power dissipation in a sense amplifier when designed using nano-CMOS technology has capacitive switching, subthreshold leakage, and gate-oxide leakage power dissipation as forms of power dissipation [8, 98]. Therefore, circuit-level methods available for estimation of power dissipation can be used [51]. The power dissipation in the sense amplifiers can be captured from the power dissipation of each transistor in the circuit as the following manner [98]:

$$P_{\text{SA}} = \sum_{\text{Over All Devices}} (V_{\text{DD}} * I_{\text{ds}} + V_{\text{gs}} * I_{\text{gate}}) \tag{6.58}$$

where V_{DD} is the supply voltage and current, I_{ds} is the drain current, V_{gs} is the input voltage at the gate, and I_{gate} is the gate current. The sense amplifiers are critical for sensing in nonvolatile memory and these memories such as SRAM/DRAM are heavily used in the computers. Thus, power dissipation in the sense amplifiers contribute to the power dissipation of the computers and embedded systems. Thus, low-power dissipation of the sense amplifier is highly desired.

5. *Silicon Area:* Silicon area occupied by the sense amplifier circuit is one of the important design criteria. The silicon area of the sense amplifier needs to be as small as possible so that the size of the memory does not increase. In particular, for the on-chip memory, area is a critical constraint.

6.12.3 A Concrete Example: 45-nm CMOS Clamped Bitline Sense Amplifier

The cross-coupled latch sense amplifier circuit schematic is presented in Fig. 6.53 [98, 101, 103]. It also shows the transistor sizes for a baseline sense amplifier design for 45-nm CMOS technology. The transistors NM1–PM2 and NM2–PM3 form the two cross-coupled inverters that are the core of the sense amplifier. These two cross-coupled inverters provide a positive feedback and are connected to the bitlines BL and $\overline{BL}$. The NMOS transistors NM5–NM6 form the precharge circuit that is used to precharge and equalize the bitlines to $\left(\frac{V_{DD}}{2}\right)$. The devices PM1 and NM3 are used to activate or deactivate the sense amplifier circuit. The physical design of the sense amplifier for 45-nm CMOS technology is presented in Fig. 6.54 [98].

The operation of the sense amplifier can be explained in the following simple steps.

1. At the beginning of the operation of the sense amplifier, the precharge circuit is activated and turned ON by the PRE signal. This activation drives both bitlines BL and $\overline{BL}$ to a voltage value of $\left(\frac{V_{DD}}{2}\right)$. The precharge circuit is then turned OFF and the bitlines BL and $\overline{BL}$ are left floating at $\left(\frac{V_{DD}}{2}\right)$.

2. The wordline is raised too high to turn ON the access transistors NM7 and NM8 in the sense amplifier circuit. A voltage difference appears on both bitlines BL and $\overline{BL}$. The voltage V_{BL} is higher than the voltage $\overline{V}_{BL}$, when a logic "1" is read.

3. The cross-coupled inverters are turned ON when the signal SE and $\overline{SE}$ are enabled. As the sense amplifier is now turned ON, the signal difference is detected by it. The signal is then amplified by the sense amplifier to a full swing of "1" logic value. The value on the bitline is written back to the memory cell as the wordline is still raised.

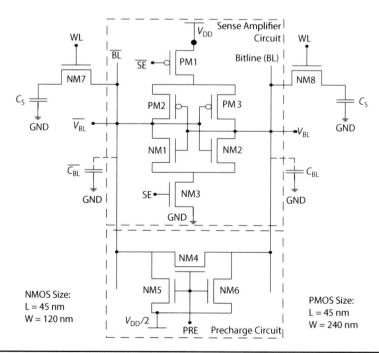

Figure 6.53 Circuit diagram of a full-latch cross-coupled sense amplifier with baseline sizes for a 45-nm CMOS technology node [98].

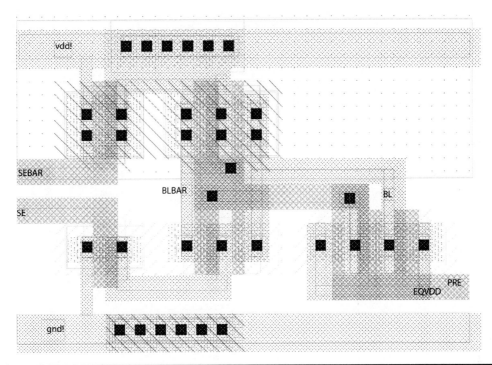

FIGURE 6.54 Physical design of a full-latch cross-coupled sense amplifier with baseline sizes for a 45-nm CMOS technology node [98].

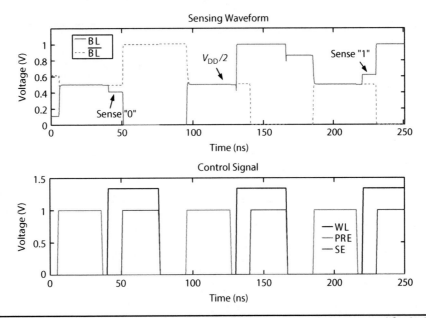

FIGURE 6.55 Simulation of the baseline design of 45-nm full-latch cross-coupled sense amplifier [98].

The function simulation of the 45-nm CMOS for the operating states of the DRAM is shown in Fig. 6.55 [98, 101, 103]. At the bottom of Fig. 6.55 the control signals SE, PRE, and WL are shown. The simulation shows both the bitlines BL and $\overline{BL}$. The read operations of logic "0" and logic "1" are shown in the simulation waveforms. The simulation is shown for a limited time window. Using an analog simulator, the sense amplifier circuit is characterized for various figures of merit. A selected set of the characteristics is presented in Table 6.4 [98, 101, 103].

The schematic circuit diagram of another variant of the full-latch sense amplifier is shown in Fig. 6.56 [98]. The sizes for a baseline sense amplifier design for 45-nm CMOS technology are also shown in the figure. As evident from the figures, this sense amplifier has different topology as

Characteristic	Precharge Time	Power Dissipation	Sense Delay	Sense Margin
Values	7.5 ns	153.1 nW	4.4 ns	43.4 mV

TABLE 6.4 Characterization of the Baseline Design of 45-nm Full-Latch Cross-Coupled Sense Amplifier [98]

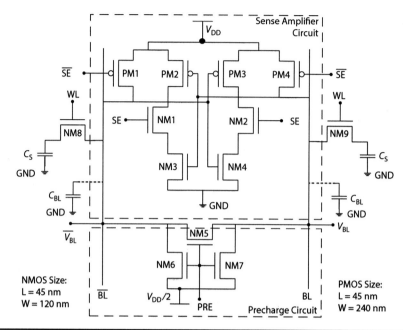

FIGURE 6.56 Circuit diagram of another variant of a full latch sense amplifier for 45-nm CMOS [98].

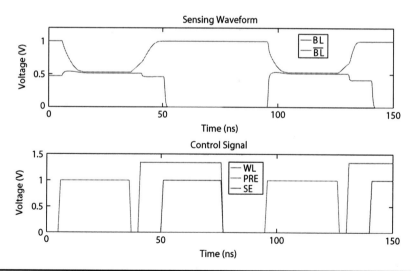

FIGURE 6.57 Simulation waveforms of the alternative bit latch sense amplifier. Another variant of a full latch sense amplifier for 45-nm CMOS [98].

compared to the previous one in Fig. 6.53. The sense amplifier has NM1 and NM2 transistors for enabling in one region. At the same time, the sense amplifier has PM1 and PM4 transistors for enabling in another region. The precharge circuit of this variant of the sense amplifier is the same as the previous sense amplifier topology. The simulation results for this variant of the sense amplifier for 45nm CMOS technology is shown in Fig. 6.57 [98]. The simulation waveforms show both the bitlines BL and $\overline{BL}$ of this sense amplifier for a limited time duration. The bottom figure of the simulation shows various control signals, SE, PRE, and WL.

6.13 Questions

6.1 Perform system-level behavioral simulation of the spintronic-memristor-based biosensor. Tools such as Simulink or Simscape can be used. Otherwise, Verilog-A-based simulation can be performed.

6.2 Explain the working principle of a TFET-based biosensor.

6.3 Explain the working principle of a GFET-based biosensor.

6.4 Perform system-level behavioral simulation of the ring-oscillator-based thermal sensor. Tools such as Simulink or Simscape can be used. Otherwise, hardware description language such as Verilog or VHDL-based simulation can be used.

6.5 Perform logic-level design and simulation of the ring-oscillator-based thermal sensor.

6.6 Perform transistor-level design and simulation of the ring-oscillator-based thermal sensor.

6.7 Perform transistor-level design and simulation of the spintronic-memristor-based thermal sensor. For the spintronic-memristor device, Verilog-A model or SPICE model can be used.

6.8 Explain the working principle of a solar cell.

6.9 Model the solar cells using Verilog-A and simulate in an analog simulation platform.

6.10 Explain the working principle of a piezoelectric sensor.

6.11 Simulate a piezoelectric sensor using Verilog-A or SPICE models.

6.12 Briefly explain different types of image sensors.

6.13 Simulate a PPS using Verilog-A or SPICE models.

6.14 Simulate a photodiode-based APS using Verilog-A or SPICE models.

6.15 Simulate a photogate-based APS using Verilog-A or SPICE models.

6.16 Simulate a logarithmic-mode photodiode-based APS using Verilog-A or SPICE models.

6.17 Perform system-level behavioral simulation of an APS array. Tools such as Simulink or Simscape can be used. Otherwise, hardware description language such as Verilog- or VHDL-based simulation can be used.

6.18 Simulate a DPS using Verilog-A or SPICE models.

6.19 Perform system-level behavioral simulation of a DPS array. Tools such as Simulink or Simscape can be used. Otherwise, hardware description language such as Verilog- or VHDL-based simulation can be used.

6.20 Perform system-level behavioral simulation of a secure image sensor. Tools such as Simulink or Simscape can be used. Otherwise, hardware description language such as Verilog- or VHDL-based simulation can be used.

6.21 Perform system-level behavioral simulation of a CNTFET-based gas sensor. Tools such as Simulink or Simscape can be used. Otherwise, hardware description language such as Verilog- or VHDL-based simulation can be used.

6.22 Perform system-level behavioral simulation of an epileptic seizure detector. Tools such as Simulink or Simscape can be used. Otherwise, hardware description language such as Verilog- or VHDL-based simulation can be used.

6.23 Simulate a diode-based humidity sensor using Verilog-A or SPICE models.

6.24 Perform system-level behavioral simulation of a CMOS-based humidity sensor. Tools such as Simulink or Simscape can be used. Otherwise, hardware description language such as Verilog- or VHDL-based simulation can be used.

6.25 Perform system-level behavioral simulation of an APS-based motion sensor. Tools such as Simulink or Simscape can be used. Otherwise, hardware description language such as Verilog- or VHDL-based simulation can be used.

6.26 Simulate a sense amplifier circuit using Verilog-A or SPICE models.

6.14 References

1. A Short Course in Digital Photography: Chapter 2 The Foundations of Digital Imaging. http://btc.montana.edu/ceres/malcolm/cd/universe/assets/multimedia/chapter02.pdf. Accessed on 26th August 2013.
2. CMOS Fundamentals. http://www.siliconimaging.com/cmos_fundamentals.htm. Accessed on 26th August 2013.
3. Instrumentation-Electronics. http://www.instrumentationtoday.com/.
4. Sensors and Transducers. http://www.electronics-tutorials.ws/io/io_1.html. Accessed on 05th August 2013.
5. Solar Electricity Basics. http://homepower.com/basics/solar/ (2012). Accessed on 19th August 2013.
6. Alem, S., de Bettignies, R., Nunzi, J.M., Cariou, M.: Efficient Polymer-Based Interpenetrated Network Photovoltaic Cells, *Applied Physics Letters* **84**(12), 2178–2180, Url http://link.aip.org/link/?APL/84/2178/1. DOI 10.1063/1.1669065.
7. Almazan, S.P.R., Zarsuela, J.V., Alarcon, A.P.B.L.P.: A Study on the Effect of Varying Voltage Supply on the Performance of Voltage Sense Amplifiers for 1-Transistor DRAM Memories. In: *Proceedings of the IEEE International Conference on Semiconductor Electronics*, pp. 108–112 (2008).
8. Almazan, S.P.R., Zarsuela, J.V., Ballesil, A.P., Alarcon, L.P.: A Study on the Effect of Varying Voltage Supply on the Performance of Voltage Sense Amplifiers For 1-Transistor DRAM Memories. In: *Proceedings of the IEEE International Conference on Semiconductor Electronics*, pp. 108–112 (2008). DOI 10.1109/SMELEC.2008.4770287.
9. Amrutur, B.S.: *Design and Analysis of Fast Low Power SRAMs*. Dissertation, Electrical Engineering, Stanford University (1999). Accessed on 06th September 2013.
10. Artyomov, E., Yadid-Pecht, O.: Adaptive Multiple-Resolution CMOS Active Pixel Sensor. *IEEE Transactions on Circuits and Systems* **53**(10), 2178–2186 (2006).
11. Atashbar, M., Bejcek, B., Singamaneni, S., Santucci, S.: Carbon Nanotube Based Biosensors. In: *Proceedings of IEEE Sensors*, pp. 1048–1051 (2004). DOI 10.1109/ICSENS.2004.1426354.
12. Bi, X., Zhang, C., Li, H., Chen, Y., Pino, R.: Spintronic Memristor Based Temperature Sensor Design with CMOS Current Reference. In: *Proceedings of the Design, Automation Test in Europe Conference Exhibition (DATE)*, pp. 1301–1306 (2012). DOI 10.1109/DATE.2012.6176693.
13. Bohndiek, S.E., Blue, A., Clark, A., Prydderch, M., Turchetta, R., Royle, G., Speller, R.: Comparison of methods for estimating the conversion gain of cmos active pixel sensors. *Sensors Journal, IEEE* **8**(10), 1734–1744 (2008). DOI 10.1109/JSEN.2008.2004296.
14. Casadei, B., Hu, Y., Dufaza, C., Martin, L.: Model for Electrical Simulation of Photogate Active Pixel Sensor. In: *Proceedings of the 16th International Conference on Microelectronics*, pp. 189–193 (2004).
15. Chamberlain, S.G., Lee, J.P.Y.: A Novel Wide Dynamic Range Silicon Photodetector and Linear Imaging Array. *IEEE Transactions on Electron Devices* **31**(2), 175–182 (1984). DOI 10.1109/T-ED.1984.21498.
16. Chapman, G., Audet, Y.: Creating 35 mm Camera Active Pixel Sensors. In: *Proceedings of the International Symposium on Defect and Fault Tolerance in VLSI Systems*, pp. 22–30 (1999). DOI 10.1109/DFTVS.1999.802865.
17. Cheknane, A., Hilal, H.S., Djeffal, F., Benyoucef, B., Charles, J.P.: An Equivalent Circuit Approach to Organic Solar Cell Modelling. *Microelectronics Journal* **39**(10), 1173–1180 (2008). DOI http://dx.doi.org/10.1016/j.mejo.2008.01.053.
18. Chen, P., Chen, C.C., Tsai, C.C., Lu, W.F.: A Time-to-Digital-Converter-Based CMOS Smart Temperature Sensor. Solid-State Circuits, IEEE Journal **40**(8), 1642–1648 (2005).
19. Chen, S., Tang, W., Culurciello, E.: A 64×64 Pixels UWB Wireless Temporal-Difference Digital Image Sensor. In: *Proceedings of IEEE International Symposium on Circuits and Systems (ISCAS)*, pp. 1404–1407 (2010). DOI 10.1109/ISCAS.2010.5537283.
20. Chen, S., Tang, W., Zhang, X., Culurciello, E.: A 64×64 Pixels UWB Wireless Temporal-Difference Digital Image Sensor. *IEEE Transactions on Very Large Scale Integration (VLSI) Systems* **20**(12), 2232–2240 (2012). DOI 10.1109/TVLSI.2011.2172470.
21. Chen, T., Catrysse, P.B., El Gamal, A., Wandell, B.A.: How Small Should Pixel Size Be?, Proceedings of the SPIE, 3965, pp. 451–459, (2000). DOI 10.1117/12.385463. URL http://dx.doi.org/10.1117/12.385463.
22. Chen, Y., Wang, X., Sun, Z., Li, H.: The Application of Spintronic Devices in Magnetic Bio-Sensing. In: *Proceedings of the 2nd Asia Symposium on Quality Electronic Design (ASQED)*, pp. 230–234 (2010). DOI 10.1109/ASQED.2010.5548244.
23. Chin, M., Kilpatrick, S.: Differential Amplifier Circuits Based on Carbon Nanotube Field Effect Transistors (CNTFETs). Tech. Rep. ARL-TR-5151, Army Research Laboratory (2010). Accessed on 10th August 2013.
24. Cho, T.S.: *An Energy Efficient CMOS Interface to Carbon Nanotube Sensor Arrays*. Master's thesis, Department of Electrical Engineering and Computer Science, Massachusetts Institute of Technology, Cambridge, MA (2007). Accessed on 10th August 2013.
25. Cho, T.S., Lee, K.J., Kong, J., Chandrakasan, A.: A Low Power Carbon Nanotube Chemical Sensor System. In: *Proceedings of the IEEE Custom Integrated Circuits Conference*, pp. 181–184 (2007). DOI 10.1109/CICC.2007.4405708.
26. Cho, T.S., Lee, K.J., Kong, J., Chandrakasan, A.P.: The Design of a Low Power Carbon Nanotube Chemical Sensor System. In: *Proceedings of the 45th ACM/IEEE Design Automation Conference*, pp. 84–89 (2008).
27. Choppali, U., Kougianos, E., Mohanty, S.P., Gorman, B.P.: Polymeric precursor derived nanocrystalline ZnO thin films using EDTA as chelating agent. *Solar Energy Materials and Solar Cells* **94**(12), 2351–2357 (2010).

DOI http://dx.doi.org/10.1016/j.solmat.2010.08.012. URL http://www.sciencedirect.com/science/article/pii/S0927024810004691.
28. Choppali, U., Kougianos, E., Mohanty, S.P., Gorman, B.P.: Maskless Deposition of ZnO Films. *Solar Energy Materials and Solar Cells* **95**(3), 870–876 (2011). DOI http://dx.doi.org/10.1016/j.solmat.2010.11.004. URL http://www.sciencedirect.com/science/article/pii/S0927024810006471.
29. Choudhary, A.: *Process Variation Tolerant Self Compensation Sense Amplifier Design*. Thesis, University of Massachusetts Amherst (2008).
30. Choudhary, A., Kundu, S.: A Process Variation Tolerant Self-Compensating Sense Amplifier Design. In: *Proceedings of the IEEE Computer Society Annual Symposium on VLSI*, pp. 263–267 (2009).
31. Chow, H.C., Hsieh, C.L.: A 0.5V High Speed DRAM Charge Transfer Sense Amplifier. In: *Proceedings of the 50th Midwest Symposium on Circuits and Systems*, pp. 1293–1296 (2007).
32. Cirmirakis, D., Demosthenous, A., Saeidi, N., Donaldson, N.: Humidity-to-Frequency Sensor in CMOS Technology With Wireless Readout. *IEEE Sensors Journal* **13**(3), 900–908 (2013). DOI 10.1109/JSEN.2012.2217376.
33. Collins, P.G., Bradley, K., Ishigami, M., Zettl, A.: Extreme Oxygen Sensitivity of Electronic Properties of Carbon Nanotubes. *Science* **287**(5459), 1801–1804 (2000). DOI 10.1126/science.287.5459.1801. URL http://www.sciencemag.org/content/287/5459/1801.abstract.
34. Datta, B.: *On-Chip Thermal Sensing in Deep Sub-Micron CMOS*. Master's thesis, Electrical and Computer Engineering, University of Massachusetts Amherst, Amherst, MA (2007).
35. Datta, B., Burleson, W.: Low-Power and Robust On-Chip Thermal Sensing Using Differential Ring Oscillators. In: *Proceedings of the 50th Midwest Symposium on Circuits and Systems*, pp. 29–32 (2007). DOI 10.1109/MWSCAS.2007.4488534.
36. Davis, B.T.: *Modern DRAM Architectures*. Dissertation, Computer Science and Engineering, The University of Michigan (2001). Accessed on 06th September 2013.
37. Dutta, A.K.: Prospects of Nanotechnology for High-Efficiency Solar Cells. In: *Proceedings of the 7th International Conference on Electrical Computer Engineering (ICECE)*, pp. 347–350 (2012). DOI 10.1109/ICECE.2012.6471558.
38. Feruglio, S., Hanna, V.F., Alquie, G., Vasilescu, G.: Dark Current and Signal-to-Noise Ratio in BDJ Image Sensors. *IEEE Transactions on Instrumentation and Measurement* **55**(6), 1892–1903 (2006). DOI 10.1109/TIM.2006.884291.
39. Fiori, F., Crovetti, P.S.: A New Compact Temperature-Compensated CMOS Current Reference. *IEEE Transactions on Circuits and Systems II: Express Briefs* **52**(11), 724–728 (2005). DOI 10.1109/TCSII.2005.852529.
40. Fish, A., Yadid-Pecht, O.: Low-Power "Smart" CMOS Image Sensors. In: *Proceedings of the IEEE International Symposium on Circuits and Systems*, pp. 1408–1411 (2008). DOI 10.1109/ISCAS.2008.4541691.
41. Fossum, E.R.: CMOS Image Sensors: Electronic Camera on a Chip. In: *Proceedings of the International Electron Devices Meeting*, pp. 17–25 (1995). DOI 10.1109/IEDM.1995.497174.
42. Fossum, E.R.: CMOS Image Sensors: Electronic Camera on a Chip. *IEEE Transactions on Electron Devices* **44**, 1689–16,984 (1997).
43. Fraden, J.: *Handbook of Modern Sensors: Physics, Designs, and Applications*. Springer (2010). URL http://books.google.com/books?id=W0Emv9dAJ1kC.
44. Fujimori, I.L., Wang, C.C., Sodini, C.G.: A 256×256 CMOS Differential Passive Pixel Imager with FPN Reduction Techniques. In: *Proceedings of the IEEE International Solid-State Circuits Conference*, pp. 106–107 (2000). DOI 10.1109/ISSCC.2000.839711.
45. Galiana, B., Algora, C., Rey-Stolle, I., Vara, I.G.: A 3-D Model for Concentrator Solar Cells Based on Distributed Circuit Units. *IEEE Transactions on Electron Devices* **52**(12), 2552–2558 (2005). DOI 10.1109/TED.2005.859620.
46. Gamal, A.E., Eltoukhy, H.: CMOS Image Sensors. *IEEE Circuits and Devices Magazine* **21**(3), 6–20 (2005).
47. Garcia, I., Algora, C., Rey-Stolle, I., Galiana, B.: Study of Non-Uniform Light Profiles on High Concentration III-V Solar Cells Using Quasi-3D Distributed Models. In: *Proceedings of the 33rd IEEE Photovoltaic Specialists Conference*, pp. 1–6 (2008). DOI 10.1109/PVSC.2008.4922908.
48. Gautschi, G.: *Piezoelectric Sensorics: Force, Strain, Pressure, Acceleration and Acoustic Emission Sensors, Materials and Amplifiers*. Engineering Online Library. Springer (2002). URL http://books.google.com/books?id=-nYFSLcmc-cC.
49. Geib, H., Raab, W., Schmitt-Landsiedel, D.: Block-Decoded Sense-Amplifier Driver for High-Speed Sensing in DRAM's. *IEEE Journal of Solid-State Circuits* **27**(9), 1286–1288 (1992).
50. Geib, H., Weber, W., Wohlrab, E., Risch, L.: Experimental Investigation of the Minimum Signal for Reliable Operation of DRAM Sense Amplifiers. *IEEE Jounal of Solid-State Circuits* **27**(7), 1028–1035 (1992).
51. Ghai, D.: *Variability Aware Low-Power Techniques for Nanoscale Mixed-Signal Circuits*. Ph.D. dissertation, Department of Computer Science and Engineering, University of North Texas, Denton (2009).
52. Ghai, D., Mohanty, S.P., Kougianos, E.: Variability-Aware Optimization of Nano-CMOS Active Pixel Sensors Using Design and Analysis of Monte Carlo Experiments. In: *Proceedings of the 10th International Symposium on Quality of Electronic Design*, pp. 172–178 (2009).
53. Tian, H., Fowler, B., Gamal, A.G.: Analysis of Temporal Noise in CMOS Photodiode Active Pixel Sensor. *IEEE Journal of Solid-State Circuits* **36**, 92–101 (2001).

54. Hidaka, Y., Kawahara, K.: Modeling of A Hybrid System of Photovoltaic and Fuel Cell for Operational Strategy in Residential Use. In: *Proceedings of the 47th International Universities Power Engineering Conference (UPEC)*, pp. 1–6 (2012). DOI 10.1109/UPEC.2012.6398416.
55. Hong, S., Kim, S., Wee, J.K., Lee, S.: Low-Voltage DRAM Sensing Scheme with Offset-Cancellation Sense Amplifier. *IEEE Journal of Solid-State Circuits* **37**(10), 1356–1360 (2002).
56. Hou, K.C., Chang, C.W., Chiou, J.C., Huang, Y.H., Shaw, F.Z.: Wireless and Batteryless Biomedical Microsystem for Neural Recording and Epilepsy Suppression Based on Brain Focal Cooling. *IET Nanobiotechnology* **5**(4), 143–147 (2011). DOI 10.1049/iet-nbt.2011.0017.
57. I. Shcherback, O.Y.P.: Photoresponse Analysis and Pixel Shape Optimization for Cmos Active Pixel Sensors. *IEEE Transactions on Electron Devices* **50**, 12–18 (2003).
58. Ignjatovic, Z., Maricic, D., Bocko, M.F.: Low Power, High Dynamic Range CMOS Image Sensor Employing Pixel-Level Oversampling $\Sigma\Delta$ Analog-to-Digital Conversion. *Sensors Journal, IEEE* **12**(4), 737–746 (2012). DOI 10.1109/JSEN.2011.2158818.
59. Im, H., Huang, X.J., Gu, B., Choi, Y.K.: A Dielectric-Modulated Field-Effect Transistor for Biosensing. *Nature Nanotechnology* **2**(7), 430–434 (2007). DOI http://dx.doi.org/10.1038/nnano.2007.180.
60. Jiang, T., Chiang, P.Y.: Sense Amplifier Power and Delay Characterization for Operation Under Low-V_{dd} and Low-voltage Clock Swing. *Proceedings of the IEEE International Symposium on Circuits and Systems*, pp. 181–184 (2009).
61. Jianping, S.: An Optimum Layout Scheme For Photovoltaic Cell Arrays Using PVSYST. In: *Proceedings of the International Conference on Mechatronic Science, Electric Engineering and Computer (MEC)*, pp. 243–245 (2011). DOI 10.1109/MEC.2011.6025446.
62. Khatri, I., Bao, J., Kishi, N., Soga, T.: Similar Device Architectures for Inverted Organic Solar Cell and Laminated Solid-State Dye-Sensitized Solar Cells. *ISRN Electronics* **2012**(180787), 1–11 (2012). DOI http://dx.doi.org/10.5402/2012/180787.
63. Kim, C.H., Jung, C., Lee, K.B., Park, H.G., Choi, Y.K.: Label-free DNA Detection with a Nanogap Embedded Complementary Metal Oxide Semiconductor. *IOP Nanotechnology* **22**(13), 135,502 (2011). DOI doi:10.1088/0957-4484/22/13/135502.
64. Kleinfelder, S., Bieser, F., Chen, Y., Gareus, R., Matis, H.S., Oldenburg, M., Retiere, F., Ritter, H.G., Wieman, H.H., Yamamoto, E.: Novel Integrated CMOS Sensor Circuits. *IEEE Transactions on Nuclear Science* **51**(5), 2328–2336 (2004). DOI 10.1109/TNS.2004.836150.
65. Kleinfelder, S., Lim, S., Liu, X., El Gamal, A.: A 10000 Frames/s CMOS Digital Pixel Sensor. *IEEE Journal of Solid-State Circuits* **36**(12), 2049–2059 (2001). DOI 10.1109/4.972156.
66. Kon, S., Horowitz, R.: A High-Resolution MEMS Piezoelectric Strain Sensor for Structural Vibration Detection. *IEEE Sensors Journal* **8**(12), 2027–2035 (2008). DOI 10.1109/JSEN.2008.2006708.
67. Kong, J., Franklin, N.R., Zhou, C., Chapline, M.G., Peng, S., Cho, K., Dai, H.: Nanotube Molecular Wires as Chemical Sensors. *Science* **287**(5453), 622–625 (2000). DOI 10.1126/science.287.5453.622. URL http://www.sciencemag.org/content/287/5453/622.abstract.
68. Kuila, T., Bose, S., Khanra, P., Mishra, A.K., Kim, N.H., Lee, J.H.: Recent Advances in Graphene-Based Biosensors. *Biosensors and Bioelectronics* **26**(12), 4637–4648 (2011). DOI http://dx.doi.org/10.1016/j.bios.2011.05.039. URL http://www.sciencedirect.com/science/article/pii/S0956566311003368.
69. Lee, C.Y., Baik, S., Zhang, J., Masel, R.I., Strano, M.S.: Charge Transfer From Metallic Single-Walled Carbon Nanotube Sensor Arrays. *The Journal of Physical Chemistry B* **110**(23), 11,055–11,061 (2006). URL http://www.ncbi.nlm.nih.gov/pubmed/16771365.
70. Liu, S., Guo, X.: Carbon Nanomaterials Field-Effect-Transistor-Based Biosensors. *NPG Asia Materials* **4**(8), 1–10 (2012). DOI http://dx.doi.org/10.1038/am.2012.42.
71. Liu, X.: *CMOS Image Sensors Dynamic Range and SNR Enhancement Via Statistical Signal Processing*. Dissertation, Stanford University (2002). Accessed on 26th August 2013.
72. Ludvig, N., Medveczky, G., French, J.A., Carlson, C., Devinsky, O., Kuzniecky, R.I.: Evolution and Prospects for Intracranial Pharmacotherapy for Refractory Epilepsies: The Subdural Hybrid Neuroprosthesis. *Epilepsy Research and Treatment* **2010**(725696), 1–10 (2009). DOI http://dx.doi.org/10.1155/2010/725696.
73. Lukac, R., Plataniotis, K.N.: Secure Single-Sensor Digital Camera. *Electronics Letters* **42**(11), 627–629 (2006). DOI 10.1049/iel:20060604.
74. M. Beiderman, M., Tam, T., Fish, A., Jullien, G.A., Yadid-Pecht, O.: A Low Noise CMOS Image Sensor with an Emission Filter for Fluorescence Applications. In: *Proceedings of the International Conference on Circuits and Systems*, pp. 1100–1103 (2008).
75. Mazhari, B.: An Improved Solar Cell Circuit Model For Organic Solar Cells. *Solar Energy Materials and Solar Cells* **90**(7–8), 1021–1033 (2006). DOI http://dx.doi.org/10.1016/j.solmat.2005.05.017. URL http://www.sciencedirect.com/science/article/pii/S0927024805001832.
76. Meng, T., Xu, C.: A Cross-Coupled-Structure-Based Temperature Sensor with Reduced Process Variation Sensitivity. *Journal of Semiconductors* **30**(4), 1642–1648 (2009).
77. Milirud, V., Fleshel, L., Zhang, W., Julien, G., Yadid-Pecht, O.: A Wide Dynamic Range CMOS Active Pixel Sensor with Frame Difference. In: *Proceedings of the International Conference on Circuits and Systems*, pp. 588–591 (2008).

78. Mizuno, S., Fujita, K., Yamamoto, H., Mukozaka, N., Toyoda, H.: A 256 × 256 Compact CMOS Image Sensor With On-Chip Motion Detection Function. *IEEE Journal of Solid-State Circuits* **38**(6), 1072–1075 (2003). DOI 10.1109/JSSC.2003.811988.
79. Mohanty, N., Berry, V.: Graphene-Based Single-Bacterium Resolution Biodevice and DNA Transistor: Interfacing Graphene Derivatives with Nanoscale and Microscale Biocomponents. *Nano Letters* **8**(12), 4469–4476 (2008). DOI 10.1021/nl802412n. URL http://pubs.acs.org/doi/abs/10.1021/nl802412n.
80. Mohanty, S.P.: A Secure Digital Camera Architecture for Integrated Real-Time Digital Rights Management. *Journal of Systems Architecture – Embedded Systems Design* **55**(10-12), 468–480 (2009).
81. Mohanty, S.P., Ghai, D., Kougianos, E., Joshi, B.: A Universal Level Converter Towards the Realization of Energy Efficient Implantable Drug Delivery Nano-Electro-Mechanical-Systems. In: *Proceedings of the 10th International Symposium on Quality of Electronic Design*, pp. 673–679 (2009).
82. Mohanty, S.P., Kougianos, E.: Biosensors: A Tutorial Review. *IEEE Potentials* 25(2), 35–40.
83. Mukhopadhyay, S., Mahmoodi, H., Roy, K.: A Novel High-Performance and Robust Sense Amplifier Using Independent Gate Control in Sub-50-nm Double-Gate MOSFET. *IEEE Transactions on Very Large Scale Integration (VLSI) Systems* **14**(2), 183–192 (2006). DOI 10.1109/TVLSI.2005.863743.
84. Mukhopadhyay, S., Raychowdhury, A., Mahmoodi, H., Roy, K.: Leakage Current Based Stabilization Scheme for Robust Sense-Amplifier Design for Yield Enhancement in Nano-scale SRAM. In: *Proceedings of the 14th Asian Test Symposium*, pp. 176–181 (2005).
85. Nakamura, J. (ed.). *Image Sensors and Signal Processing for Digital Still Cameras*. CRC Press, Boca Raton, FL, USA (2006).
86. Narang, R., Reddy, K.V.S., Saxena, M., Gupta, R.S., Gupta, M.: A Dielectric-Modulated Tunnel-FET-Based Biosensor for Label-Free Detection: Analytical Modeling Study and Sensitivity Analysis. *IEEE Transactions on Electron Devices* **59**(10), 2809–2817 (2012). DOI 10.1109/TED.2012.2208115.
87. Narang, R., Saxena, M., Gupta, R.S., Gupta, M.: Dielectric Modulated Tunnel Field-Effect Transistor – A Biomolecule Sensor. *IEEE Electron Device Letters* **33**(2), 266–268 (2012). DOI 10.1109/LED.2011.2174024.
88. Narayanan, A., Dan, Y., Deshpande, V., Lello, N.D., Evoy, S., Raman, S.: Dielectrophoretic Integration of Nanodevices With CMOS VLSI Circuitry. *IEEE Transactions on Nanotechnology* **5**(2), 101–109 (2006). DOI 10.1109/TNANO.2006.869679.
89. Narisawa, S., Masuda, K., Hamamoto, T.: High Speed Digital Smart Image Sensor with Image Compression Function. In: *Proceedings of IEEE Asia-Pacific Conference on Advanced System Integrated Circuits*, pp. 128–131 (2004). DOI 10.1109/APASIC.2004.1349426.
90. Nelson, G.R., Jullien, G.A., Yadid-Pecht, O.: CMOS Image Sensor with Watermarking Capabilities. In: *Proceedings of IEEE International Symposium on Circuits and Systems*, pp. 5326–5329 (2005). DOI 10.1109/ISCAS.2005.1465838.
91. Niv, A., Gharghi, M., Gladden, C., Abrams, Z., Zhang, X.: A New Analysis for Solar Cell Efficiency: Rigorous Electromagnetic Approach. In: *Proceedings of the 37th IEEE Photovoltaic Specialists Conference (PVSC)*, pp. 002,095–002,099 (2011). DOI 10.1109/PVSC.2011.6186366.
92. Noble, P.J.W.: Self-Scanned Silicon Image Detector Arrays. *IEEE Transactions on Electron Devices* **15**(4), 202–209 (1968). DOI 10.1109/T-ED.1968.16167.
93. Norton, H.N.: *Handbook of Transducers*. Prentice Hall (1989).
94. Nunzi, J.M.: Organic materials and devices for photovoltaic applications. In: J. Marshall, D. Dimova-Malinovska (eds.) *Photovoltaic and Photoactive Materials – Properties, Technology and Applications*, NATO Science Series, vol. 80, pp. 197–224. Springer Netherlands (2002). DOI 10.1007/978-94-010-0632-3_11. URL http://dx.doi.org/10.1007/978-94-010-0632-3_11.
95. Odiot, F., Bonnouvrier, J., Augier, C., Raynor, J.M.: Test Structures For Quantum Efficiency Characterization For Silicon Image Sensors. In: *Proceedings of the International Conference on Microelectronic Test Structures*, pp. 3–33 (2003). DOI 10.1109/ICMTS.2003.1197366.
96. Ohno, Y., Maehashi, K., Matsumoto, K.: Graphene Field-Effect Transistors for Label-Free Biological Sensors. In: *IEEE Sensors*, pp. 903–906 (2010). DOI 10.1109/ICSENS.2010.5690880.
97. Okcan, B., Akin, T.: A Low-Power Robust Humidity Sensor in a Standard CMOS Process. *IEEE Transactions on Electron Devices* **54**(11), 3071–3078 (2007). DOI 10.1109/TED.2007.907165.
98. Okobiah, O.: *Exploring Process-Variation Tolerant Design of Nanoscale Sense Amplifier Circuits*. Master's thesis, Department of Computer Science and Engineering, University of North Texas, Denton, TX (2010).
99. Okobiah, O., Mohanty, S.P., Kougianos, E.: Geostatistical-Inspired Metamodeling and Optimization of Nano-CMOS Circuits. In: *Proceedings of the 11th IEEE Computer Society Annual Symposium on VLSI* (2012) pp. 326–331.
100. Okobiah, O., Mohanty, S.P., Kougianos, E.: Geostatistical-Inspired Fast Layout Optimization of a Nano-CMOS Thermal Sensor. *IET Circuits, Devices & Systems* **7**(5), pp. 253–262 (2013). DOI 10.1049/iet-cds.2012.0358.
101. Okobiah, O., Mohanty, S.P., Kougianos, E., Garitselov, O.: Kriging-Assisted Ultra-Fast Simulated-Annealing Optimization of a Clamped Bitline Sense Amplifier. In: *Proceedings of the 25th International Conference on VLSI Design*, pp. 310–315 (2012).
102. Okobiah, O., Mohanty, S.P., Kougianos, E., Garitselov, O., Zheng, G.: Stochastic Gradient Descent Optimization for Low Power Nano-CMOS Thermal Sensor Design. In: *Proceedings of the 11th IEEE Computer Society Annual Symposium on VLSI* (2012).

103. Okobiah, O., Mohanty, S.P., Kougianos, E., Poolakkaparambil, M.: Towards Robust Nano-CMOS Sense Amplifier Design: A Dual-Threshold Versus Dual-Oxide Perspective. In: *Proceedings of the 21st ACM Great Lakes Symposium on VLSI*, pp. 145–150 (2011).
104. Palakodety, A.: *A Survey of Thin-Film Solar Photovoltaic Industry & Technologies*. Master's thesis, System Design and Managment Program, Massachusetts Institute of Technology, Boston, MA (2007). Accessed on 20th August 2013.
105. Palakodety, A.: *CMOS Active Pixel Sensors for Digital Cameras: Current State-of-the-Art*. Master's thesis, Department of Computer Science and Engineering, University of North Texas, Denton, TX (2007).
106. Park, S., Min, C., Cho, S.H.: A 95 nW Ring Oscillator-based Temperature Sensor for RFID Tags in 0.13 μm CMOS. In: *Proceedings of the IEEE International Symposium on Circuits and Systems*, pp. 1153–1156 (2009). DOI 10.1109/ISCAS.2009.5117965.
107. Patil, S., Wieckowski, M., Margala, M.: A Self-Biased Charge-Transfer Sense Amplifier. In: *Proceedings of the IEEE International Symposium on Circuits and Systems*, pp. 3030–3033 (2007). DOI 10.1109/ISCAS.2007.377985.
108. Remarsu, S.: *On Process Variation Tolerant Low Cost Thermal Sensor Design*. Master's thesis, Electrical and Computer Engineering, University of Massachusetts Amherst, Amherst, MA (2011).
109. Rinner, B., Wolf, W.: An Introduction to Distributed Smart Cameras. *Proceedings of the IEEE* **96**(10), 1565–1575 (2008). DOI 10.1109/JPROC.2008.928742.
110. Rodrigues, S., Bhat, M.S.: Impact of Process Variation Induced Transistor Mismatch on Sense Amplifier Performance. In: *Proceedings of the International Conference on Advanced Computing and Communications*, pp. 497–502 (2006).
111. Salam, M.T., Nguyen, D.K., Sawan, M.: A Low-Power Implantable Device for Epileptic Seizure Detection and Neurostimulation. In: *Proceedings of the IEEE Biomedical Circuits and Systems Conference (BioCAS)*, pp. 154–157 (2010). DOI 10.1109/BIOCAS.2010.5709594.
112. Salam, M.T., Sawan, M., Hamoui, A., Nguyen, D.K.: Low-Power CMOS-Based Epileptic Seizure Onset Detector. In: *Proceedings of the Joint IEEE North-East Workshop on Circuits and Systems and TAISA Conference*, pp. 1–4 (2009). DOI 10.1109/NEWCAS.2009.5290426.
113. Salam, M.T., Sawan, M., Nguyen, D.K.: A Novel Low-Power-Implantable Epileptic Seizure-Onset Detector. *IEEE Transactions on Biomedical Circuits and Systems* **5**(6), 568–578 (2011). DOI 10.1109/TBCAS.2011.2157153.
114. Sarkar, D., Banerjee, K.: Fundamental Limitations of Conventional-FET Biosensors: Quantum-Mechanical-Tunneling to The Rescue. In: *Proceedings of the 70th Annual Device Research Conference (DRC)*, pp. 83–84 (2012). DOI 10.1109/DRC.2012.6256950.
115. Schanz, M., Brockherde, W., Hauschild, R., Hosticka, B.J., Schwarz, M.: Smart CMOS Image Sensor Arrays. *IEEE Transactions on Electron Devices* **44**(10), 1699–1705 (1997). DOI 10.1109/16.628825.
116. Sequin, C.H., Zimany E.J., Jr., Tompsett, M.F., Fuls, E.N.: All-solid-state Camera for the 525-line Television Format. *IEEE Journal of Solid-State Circuits* **11**(1), 115–121 (1976). DOI 10.1109/JSSC.1976.1050685.
117. Servaites, J.D., Ratner, M.A., Marks, T.J.: Practical Efficiency Limits in Organic Photovoltaic Cells: Functional Dependence of Fill Factor and External Quantum Efficiency. *Applied Physics Letters* **95**(16), 163302 (2009). DOI 10.1063/1.3243986. URL http://link.aip.org/link/?APL/95/163302/1.
118. Shen, C., Xu, C., Wei, Q., Huang, W.R., Chan, M.: Low Voltage CMOS Active Pixel Sensor Design Methodology With Device Scaling Considerations. In: *Proceedings of the IEEE Hong Kong Electron Devices Meeting*, pp. 21–24 (2001).
119. Shenghua, Z., Nanjian, W.: A Novel Ultra Low Power Temperature Sensor for UHF RFID Tag Chip. In: *Proceedings of the IEEE Asian Solid-State Circuits Conference*, pp. 464–467 (2007). DOI 10.1109/ASSCC.2007.4425731.
120. Shih, Y.H., Hwu, J.G.: An On-Chip Temperature Sensor by Utilizing a MOS Tunneling Diode. *IEEE Electron Device Letters* **22**(6), 299–301 (2001). DOI 10.1109/55.924848.
121. Soundararajan, R., Srivastava, A., Xu, Y.: A Programmable Second Order Oversampling CMOS Sigma-Delta Analog-To-Digital Converter For Low-Power Sensor Interface Electronics. In: *Proceedings of the SPIE*, vol. 7646, pp. 76,460P–76,460P–11 (2010). DOI http://dx.doi.org/10.1117/12.847651.
122. Tang, X., Bansaruntip, S., Nakayama, N., Yenilmez, E., Chang, Y.l., Wang, Q.: Carbon Nanotube DNA Sensor and Sensing Mechanism. *Nano Letters* **6**(8), 1632–1636 (2006). DOI 10.1021/nl060613v. URL http://pubs.acs.org/doi/abs/10.1021/nl060613v.
123. Tetzlaff, R., Senger, V.: The Seizure Prediction Problem in Epilepsy: Cellular Nonlinear Networks. *IEEE Circuits and Systems Magazine* **12**(4), 8–20 (2012). DOI 10.1109/MCAS.2012.2221519.
124. Tsiatouhas, Y., Chrisanthopoulos, A., Kamoulakos, G., Haniotakis, T.: New Memory Sense Amplifier Designs in CMOS Technology. In: *Proceedings of the 7th IEEE International Conference on Electronics, Circuits and Systems*, vol. 1, pp. 19–22 vol.1 (2000). DOI 10.1109/ICECS.2000.911469.
125. Ullal, H.S., Zweibel, K., von Roedern, B.G.: Polycrystalline Thin-Film Photovoltaic Technologies: From The Laboratory to Commercialization. In: *Proceedings of the Twenty-Eighth IEEE Photovoltaic Specialists Conference*, pp. 418–423 (2000). DOI 10.1109/PVSC.2000.915857.
126. Wang, D.T.: *Modern DRAM Memory Systems: Performance Analysis and a High Performance, Power-Constrained DRAM Scheduling Algorithm*. Ph.D. thesis, University of Maryland, College Park (2005). Accessed on 23rd June 2013.
127. Wang, S.X., Li, G.: Advances in Giant Magnetoresistance Biosensors with Magnetic Nanoparticle Tags: Review and Outlook. *IEEE Transactions on Magnetics* **44**(7), 1687–1702 (2008). DOI 10.1109/TMAG.2008.920962.

128. Wang, X., Chen, Y., Gu, Y., Li, H.: Spintronic Memristor Temperature Sensor. *IEEE Electron Device Letters* **31**(1), 20–22 (2010). DOI 10.1109/EDL.2009.2035643.
129. Wang, Z., Zhang, X., Chen, X., Zhang, L., Jiang, H.: An Energy-Efficient ASIC With Real-Time Work-On-Demand For Wireless Body Sensor Network. In: *Proceedings of the IEEE International Conference on Electron Devices and Solid-State Circuits*, pp. 1–6 (2008). DOI 10.1109/EDSSC.2008.4760681.
130. Wang, Z.L.: From Nanogenerators To Piezotronics – A Decade-Long Study Of ZnO Nanostructures. *MRS Bulletin* **37**, 814–827 (2012). DOI 10.1557/mrs.2012.186. URL http://journals.cambridge.org/article_S0883769412001868.
131. Wicht, B., Nirschl, T., Schmitt-Landsiedel, D.: Yield and Speed Optimization of a Latch-type Voltage Sense Amplifier. *IEEE Journal of Solid-State Circuits* **39**(7), 1148–1158 (2004).
132. Winkler, M.T., Cox, C.R., Nocera, D.G., Buonassisi, T.: Modeling Integrated Photovoltaic–Electrochemical Devices Using Steady-state Equivalent Circuits. *Proceedings of the National Academy of Sciences* **110**(12), E1076–E1082 (2013). DOI 10.1073/pnas.1301532110. URL http://www.pnas.org/content/110/12/E1076.abstract.
133. Wu, T.Y., Dong, H.R., Huang, M.H., Chen, D.Z.: Evolution of Technology Fronts in Organic Solar Cells. In: *Proceedings of the Technology Management for Emerging Technologies (PICMET)*, pp. 2917–2924 (2012).
134. Xu, C., Ki, W.H., Chan, M.: A Low-Voltage CMOS Complementary Active Pixel Sensor (CAPS) Fabricated Using a 0.25 μm CMOS Technology. *IEEE Electron Device Letters* **23**(7), 398–400 (2002).
135. Xu, Y., Srivastava, A.: Transient Behavior of Integrated Carbon Nanotube Field Effect Transistor Circuits and Bio-Sensing Applications. In: *Proceedings of the SPIE*, vol. 7291, pp. 72,910I–72,910I–11 (2009). DOI http://dx.doi.org/10.1117/12.815392.
136. Yadid-Pecht, O., Etienne-Cummings, R.: *CMOS Imagers: From Phototransduction to Image Processing*. Kluwer Academic Publishers (2004).
137. Yadid-Pecht, O., Mansoorian, K., Fossum, E.R., Pain, B.: Optimization of noise and responsivity in cmos active pixel sensors for detection of ultralow-light levels. Proceedings of the SPIE, 3019, pp. 125–136 (1997). DOI 10.1117/12.275185. URL http://dx.doi.org/10.1117/12.275185.
138. Yamada, K., Soga, M.: A Compact Integrated Visual Motion Sensor for ITS Applications. In: *Proceedings of the IEEE Intelligent Vehicles Symposium*, pp. 650–655 (2000). DOI 10.1109/IVS.2000.898422.
139. Yamada, K., Soga, M.: A Compact Integrated Visual Motion Sensor For ITS Applications. *IEEE Transactions on Intelligent Transportation Systems* **4**(1), 35–42 (2003). DOI 10.1109/TITS.2002.808418.
140. Yang, D., Gamal, A.E.: Comparative Analysis of SNR for Image Sensors with Enhanced Dynamic Range. In: *Proceedings of the SPIE Electronic Imaging Conference*, pp. 197–211 (1999).
141. Yang, D.X.D., Gamal, A.E., Fowler, B., Tian, H.: A 640×512 CMOS Image Sensor with Ultrawide Dynamic Range Floating-Point Pixel-Level ADC. *IEEE Journal of Solid-State Circuits* **34**(12), 1821–1834 (1999). DOI 10.1109/4.808907.
142. Yang, S.H., Kim, K.-B., Kim, E.J., Baek, K.-H., Kim, S.: An Ultra Low Power CMOS Motion Detector. *IEEE Transactions on Consumer Electronics* **55**(4), 2425–2430 (2009). DOI 10.1109/TCE.2009.5373819.
143. Zhang, X., Jiang, H., Zhang, L., Zhang, C., Wang, Z., Chen, X.: An Energy-Efficient ASIC for Wireless Body Sensor Networks in Medical Applications. *IEEE Transactions on Biomedical Circuits and Systems* **4**(1), 11–18 (2010). DOI 10.1109/TBCAS.2009.2031627.
144. Zhang, Y., Srivastava, A.: Accurate Temperature Estimation Using Noisy Thermal Sensors. In: *Proceedings of the 46th ACM/IEEE Design Automation Conference*, pp. 472–477 (2009).

CHAPTER 7
Memory in the AMS-SoCs

7.1 Introduction

Memory is the key component of any computing platform. It performs its primary function of storing data, instructions, firmwires, system software, and application software. In addition, it temporarily stores data and instructions during the execution of an application or program. Depending on the usage of the memory, the memory can be of diverse types and forms. Selected types of memory are presented in Fig. 7.1 [76]. The objective of any computing platform design is to provide a large amount of memory to the users or programmers with a minimal cost. The cost, speed, and power dissipation of memory have affected the growth of VLSI technology and consumer electronics.

Based on the speed, cost, and capacity of the memory, the typical hierarchy is presented in Fig. 7.2 [93, 110]. The fastest, high bandwidth, and costly memory is closer to the microprocessor or central processing unit (CPU). The large capacity and slow memory are placed farthest from the processor. As depicted in Fig. 7.2, there are various types of memory in a typical computing system: register, cache, main memory, disk cache, solid-state drive, magnetic disk (hard drives), optical disk, and magnetic tapes. The registers in the microprocessor are the fastest. The size and number of registers depend on the architecture of the microprocessor. This is followed by multiple levels of cache with each element being static random-access memory (SRAM). The levels of cache vary depending on the microprocessors and dictate the speed and cost of the microprocessors. The main memory of the dynamic random-access memory (DRAM) is the first level of memory to which a user has direct access. It is slower than cache, faster than the storage memory, and its particular type in a computing system also depends on the motherboard. The solid-state (or flash) drive and magnetic disks are large

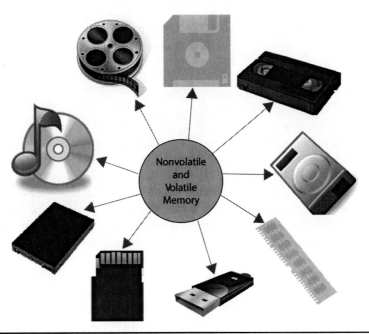

FIGURE 7.1 Different types of memory in various computing systems.

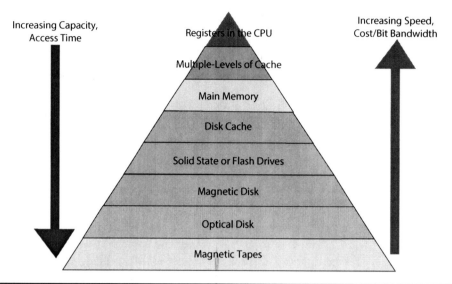

Figure 7.2 The memory hierarchy in a typical computing system.

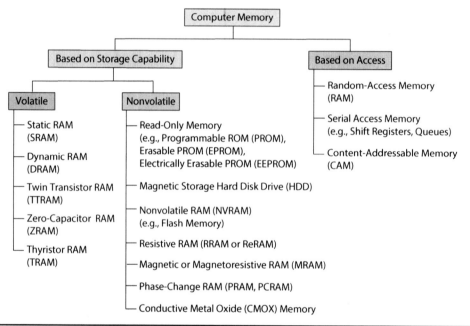

Figure 7.3 Different types of computer memory.

memory to store the large volume of data. The optical disks, magnetic tapes along with flash drive are used to store data while being portable to move data from one place to other.

Various types of computer memory can be classified as depicted in Fig. 7.3 [16, 93, 110]. Based on the storage capability, it can be either volatile or nonvolatile. The volatile memory such as SRAM stores data as long as a power supply is available to it. Many varieties of volatile memory have emerged in the current literature for improving power dissipation, performance, and process variation resilience. The nonvolatile memory such as read-only memory (ROM) or magnetic drive stores data permanently even when there is no power supply. Various nonvolatile memories such as flash memory and resistive memory are available or also investigated for high-density, high-capacity storage. Based on access types, the memory can be random-access memory (RAM), serial/sequential-access memory (SAM), and content-addressable memory (CAM). The RAM allows access to data from any memory location directly. The data is accessed sequentially by sequential-access memory in SAM. The examples of SAM include shift-registers, queues, hard disks, optical discs, and magnetic tapes.

In the CAM, the data is accessed from look-up table (LUT) in a single clock cycle [71, 97]. Detailed design and working principles of selected memory technologies will be discussed in the rest of the chapter.

7.2 Static Random-Access Memory

SRAM is high-speed volatile memory that forms the cache for the processors [93, 98]. The name "static" comes from the fact that a bit is stored electronically in its cell without needing periodic refreshing. It does not store data permanently as it is volatile but provides a high-speed buffering for processors when accessing data from low-speed storage. An SRAM circuit provides nondestructive writing ability, read access, and storage as long as power is supplied to it. In the modern system-on-a-chip (SoC), cache memory occupies a significant portion of the overall area [74, 98].

7.2.1 SRAM Array

The schematic representation of an SRAM array is provided in Fig. 7.4 [5, 58]. In general, it consists of M rows and N columns of SRAM cells. The SRAM array has a row decoder that receives a row address. The column decoder receives column address. The two buffers, input buffer and output buffer, are used for writing and reading data to the SRAM array. The sense amplifiers are collection of many sense amplifiers that are used to amplify the stored signals so that the accurate interpretation of signals as "1" or "0" is possible at the output buffer. The bitline bias circuitry provides bias for each column of the SRAM cells. The biasing circuitry that biases the lines between 0 to V_{DD} essentially improves write/read speed of SRAM array as well as the voltage swing. The SRAM is shown with a six-transistor SRAM (6T-SRAM) cell as an example. However, similar SRAM array architecture is possible for other SRAM cell topologies. The M×N array containing the classical double-ended 6T-SRAM cells has M rows with M lines and N columns with 2×N lines. Similar array is possible using any SRAM topologies, for example, single-ended seven-transistor SRAM (7T-SRAM; see Sec. 7.2.6.2) or double-ended 10-transistor (10T-SRAM; see Sec. 7.2.9). For the single-ended 7T-SRAM, the M×N array may have M rowlines and N columnlines. For the double-ended 10T-SRAM, the M×N array may have 4×M rowlines and 2×N columnlines.

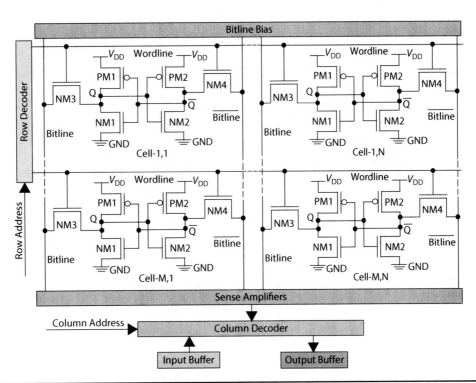

FIGURE 7.4 Schematic representation of SRAM array architecture.

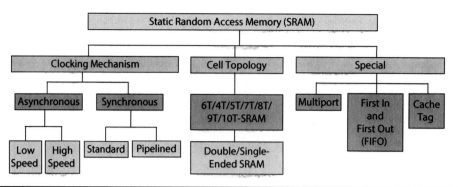

FIGURE 7.5 Different types of the SRAM.

7.2.2 Different Types of SRAM

The SRAMs are of various types as shown in Fig. 7.5 [5]. Based on the clocking mechanism, it can be either asynchronous or synchronous SRAM. The asynchronous SRAM is controlled by three clocks—chip select (CS), output enable (OE), and write enable (WE)—and is not operated on the basis of a processor clock. The asynchronous SRAM includes low-speed and high-speed variants. The low-speed asynchronous SRAMs are used in high-density hard drives. The high-speed asynchronous SRAMs are used as local storage in telecommunications systems. The synchronous SRAM has write or read operations synchronized with the microprocessor clock and can operate at a much higher speed than the asynchronous SRAM. The synchronous SRAM is preferred to be used as cache in high-performance microprocessors. The synchronous SRAM can be standard (unpipelined) or pipelined. The pipelined variant of the synchronous SRAM is cheaper than the standard for the same performance specifications. Based on the cell topology, the SRAM is of various types including: 6T-SRAM, eight-transistor SRAM (8T-SRAM), or 10-SRAM. The topology also classifies as single- or double-ended SRAM based on the number of bitlines. The single-ended SRAM has only bitline. The double-ended SRAM has both bitline and $\overline{\text{bitline}}$. Two bitlines lead to larger bitline capacitances, delays, and power dissipation. The special types of SRAM include multiport SRAM, first-in and first-out (FIFO) SRAM, and cache tag SRAM. The multiport SRAM with two or four ports can provide very fast operations. Both synchronous and asynchronous FIFO SRAMs are possible and are used for temporary storage for timing enhancement of nonsynchronized events, for example, between computer and local area network (LAN).

7.2.3 Traditional Six-Transistor SRAM

The 6T-SRAM is the most widely used SRAM to build cache in the computing systems [63, 98]. This is because six-transistor topology provides very good tradeoff among various characteristics including stability, area, leakage power, and speed. In this subsection, two different 6T-SRAMs will be discussed. The 6T-SRAM using classical transistors is discussed first, even though it is widely understood for the purpose of completeness and to serve as the baseline for the rest of the memory discussions. The discussion of a 6T-SRAM that is implemented using FinFET is also included in this subsection.

7.2.3.1 Classical Six-Transistor SRAM

The topology of a classical 6T-SRAM is presented in Fig. 7.6 [63, 98]. The SRAM cell stores the binary signal in the form of two states, state "1" and state "0." The binary signal is stored in the two cross-coupled inverters made of four transistors. There are two additional transistors to control access to the cross-coupled inverters for writing and reading operations. The schematic design shown in Fig. 7.6 has six transistors that are named as follows: (1) *driver transistors* (NM1 and NM2), (2) *load transistors* (PM1 and PM2), and (3) *access transistors* (NM3 and NM4). In the design of 6T-SRAM, the transistors NM1 and NM2 are sized wider than the transistors NM3 and NM4 to achieve read stability. At the same time, the transistors NM3 and NM4 are sized wider than PM1 and PM2 to achieve successful write. In a typical scenario, the minimum-sized transistors do not provide good stability and functionality.

Memory in the AMS-SoCs 319

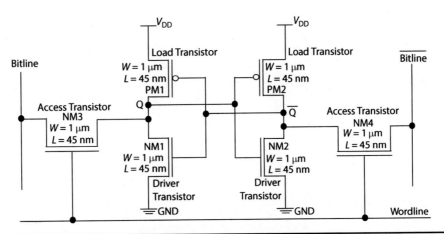

FIGURE 7.6 Transistor-level schematic of a classical 6T-SRAM with baseline sizes for a 45-nm CMOS [63, 98].

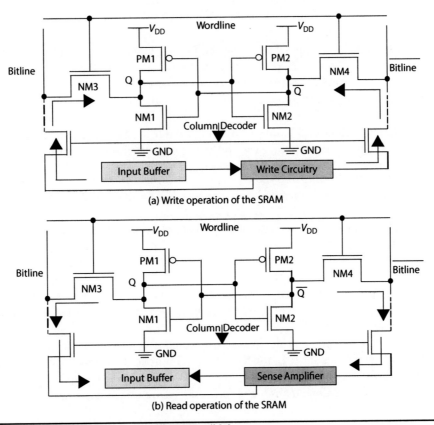

FIGURE 7.7 Write and read operations in a 6T-SRAM cell [5].

The write and read operations of the 6T-SRAM are depicted in Fig. 7.7 [5]. The access transistors are ON during write or read operations. The write operation is presented in 7.7(a). The write circuitry containing stronger transistors forces the data that are placed in the input buffer to the cross-coupled transistors. The read operation is presented in 7.7(b). The sense amplifier and voltage comparator read the voltage signal from the Bitline and $\overline{\text{Bitline}}$ and recognize the data stored in the cross-coupled inverters.

The simulation of a classical 6T-SRAM with baseline transistor sizes is shown in Fig. 7.8 [63, 98]. The SRAM cell can have three different states, "write," "read," and "hold" or "idle." The write cycle is initiated by applying appropriate values to be written to the Bitline. To write "0" to the SRAM, "0" is applied to Bitline and "1" is applied to $\overline{\text{Bitline}}$. Similarly, to write "1" to the SRAM, "1" is applied

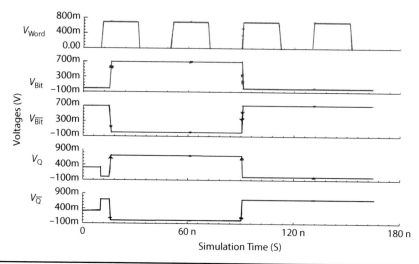

Figure 7.8 Simulation of the classical 6T-SRAM with baseline sizes for a 45-nm CMOS [63, 98].

to Bitline and "0" is applied to $\overline{\text{Bitline}}$. At this point, when the Wordline is made "1," the value applied at the Bitline is stored. The read cycle of the SRAM starts with the application of high signal to Wordline that turns ON both the access transistor NM3 and NM4. Now, the voltage values stored in Q (V_Q) and $\overline{Q}$ ($V_{\overline{Q}}$) are transferred to the Bitline (V_{Bitline}) and $\overline{\text{Bitline}}$ ($V_{\overline{\text{Bitline}}}$). For the hold state of the SRAM, the access transistors NM3 and NM4 are applied low signal through Wordline to switch them OFF. In essence, the Bitline and $\overline{\text{Bitline}}$ are disconnected from the cross-coupled inverters. The cross-coupled inverters continue to reinforce each other as long as they are disconnected from Bitline and $\overline{\text{Bitline}}$.

7.2.3.2 FinFET-Based 6T-SRAM

Many applications that need ultra-low power dissipation can have circuits operating in subthreshold mode. The stability of the classical 6T-SRAM degrades as the voltage scales down. In particular for the subthreshold mode and nanoscale process variations, the classical 6T-SRAM has very inadequate stability. The FinFET-based SRAM cells are becoming popular for low static power dissipation and compatibility with existing logic processes. The schematic diagram of a FinFET-based 6T-SRAM is shown in Fig. 7.9 [34, 36, 49, 77]. The topology of this 6T-SRAM is same as the previous classical transistor-based 6T-SRAM with NMOS replaced by an N-type FinFET and PMOS replaced by P-type FinFET, respectively. The FinFET-based SRAM can provide low-power dissipation and short access time.

6T-SRAM designs are also available in which tri-gate devices have been used [43, 88]. The tri-gate technology reduces short-channel effects and improves the subthreshold slope while the multiple fins enhance the driving current. The tri-gate FET (TFET)-based 6T-SRAM has further enhanced stability against voltage scaling. It has higher process variation tolerance and noise margin.

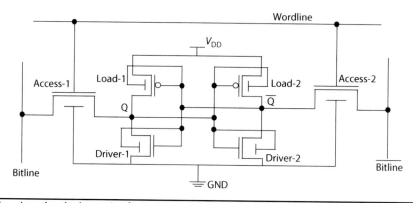

Figure 7.9 Transistor-level schematic of a FinFET-based 6T-SRAM [34, 36, 77].

7.2.4 Four-Transistor SRAM

The 6T-SRAM has been the most widely used SRAM. However, in the embedded systems, it has been a reason for large area overhead. The schematic diagram of a four-transistor SRAM (4T-SRAM) is shown in Fig. 7.10 [67, 113]. The 4T-SRAM topology consists of two NMOS as driver transistors and two PMOS transistors as access transistors. The access transistors in OFF state also act as load transistors. The pair of nodes formed by this specific combination of the transistors allows storing of full-swing voltage signal during the writing operation. The Bitlines are precharged to supply voltage in a stand-by cycle. The 4T-SRAM is smaller than 6T-SRAM but has comparable cell current and maximum Bitline swing and provides high-speed write and read operations. However, the 4T-SRAM has poor stability during low-voltage operations.

The schematic design of a 4T-SRAM that uses FinFET access transistors is shown in Fig. 7.11 [36]. The two driver transistors are still classical NMOS. The two access transistors are P-type FinFET. The storage nodes are cross-coupled with the back-gate of the access transistor. This type of connection provides high compensation current to node Q. This 4T-SRAM has better noise margin and higher β ratio than classical 4T-SRAM. FinFET-based 4T-SRAM has lower area compared to classical 6T-SRAM.

7.2.5 Five-Transistor SRAM

The schematic design of a single-ended five-transistor SRAM (5T-SRAM) is shown in Fig. 7.12 [22, 64, 65]. In comparison to the 6T-SRAM, it has only one access transistor. It has one bitline (Bitline), but $\overline{\text{Bitline}}$ is no longer present. Thus, 5T-SRAM can be much more area efficient as compared to a 6T-SRAM. However, the 5T-SRAM has lower write margin as compared to a 6T-SRAM and needs special mechanism to enhance the write margin. The lower write margin of the 5T-SRAM as compared to the 6T-SRAM is because writing "1" through the access transistor is difficult. In this 5T-SRAM,

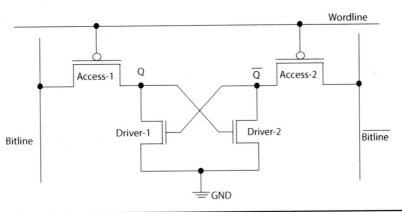

FIGURE 7.10 Transistor-level schematic of a 4T-SRAM [113].

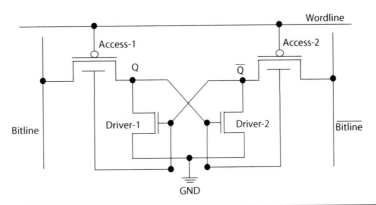

FIGURE 7.11 Transistor-level schematic of a FinFET-based 4T-SRAM [36].

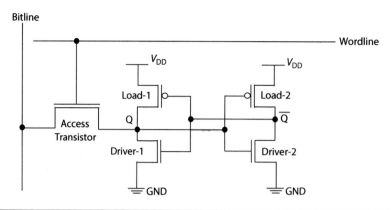

Figure 7.12 Transistor-level schematic of a 5T-SRAM [22, 64, 65].

writing "0" is not an issue as the NMOS access transistor passes strong "0." However, writing "1" is an issue as access transistor cannot pass a strong "1" and there is a ratioed fight between Driver-1 and access transistor. This issue can be resolved by using "write assist" methods. The write assist methods include Wordline boosting and cell supply voltage collapsing. In the collapsing method, the cell supply voltage is restored to full power rail before the end of Wordline pulse to ensure the completion of the write "1" operation. Overall, the 5T-SRAM has a smaller area, lower bitline leakage, and comparable read/write performance despite lower static noise margin (SNM).

7.2.6 Seven-Transistor SRAM

The quest for a robust SRAM design for low-power consuming, low leakage, process variation tolerance, and high noise margin has led to different SRAM topologies. In this subsection, two variants of seven-transistor SRAM (7T-SRAM) have been discussed. One is a double-ended 7T-SRAM with Bitline and $\overline{\text{Bitline}}$ and the other one is a single-ended 7T-SRAM with only one bitline (Bitline).

7.2.6.1 Double-Ended

The classical 6T-SRAM has stability issues at nanoscale technology due to reduced supply voltage and threshold voltage. In addition, the classical 6T-SRAM is easily affected by external noise as there are direct access paths from bitline to storage node that can destroy data during the read operations. A 7T-SRAM that can overcome the above two shortcomings is now discussed. The transistor-level schematic of a 7T-SRAM with two bitlines, Bitline and $\overline{\text{Bitline}}$, is shown in Fig. 7.13 [105]. This 7T-SRAM has two separate data access mechanisms as follows: one is for write operation and the

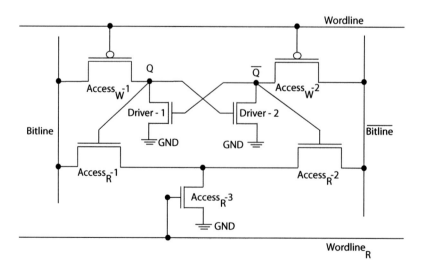

Figure 7.13 Transistor-level schematic of a 7T-SRAM [105].

other is for the read operation. The driver transistors are Driver-1 and Driver-2, the write access transistors are two PMOS devices, $Access_W - 1$ and $Access_W - 2$, and the read access transistors are three NMOS devices, $Access_R - 1$, $Access_R - 2$, and $Access_R - 3$.

During the writing operations for writing "1," the Bitline and $\overline{Bitline}$ are charged and discharged, respectively. The Wordline is set low that turns ON the write access transistors, $Access_W - 1$ and $Access_W - 2$. The node Q is charged to high by Bitline through $Access_W - 1$, and the node $\overline{Q}$ is discharged to low by $\overline{Bitline}$ through $Access_W - 2$. On the other hand, during the writing operations for writing "0," the Bitline and $\overline{Bitline}$ are discharged and charged, respectively. Similar steps are followed to discharge node Q and charge node $\overline{Q}$. The read operation of the 7T-SRAM cell is very different from the 6T-SRAM and 4T-SRAM. For the reading operation, the Wordline is set high that turns OFF the write access transistors $Access_W - 1$ and $Access_W - 2$. When $Wordline_R$ is made high, the access transistor $Access_R - 3$ is ON. If Q has "1" stored, then Bitline is discharged through $Access_R - 1$ and $Access_R - 3$. If Q has "0" stored, then $\overline{Bitline}$ is discharged through $Access_R - 2$ and $Access_R - 3$.

As evident from the above discussions, this double-ended 7T-SRAM has no direct path through bitlines to the data storage nodes and, hence, has higher endurance against external noise signals. The data destruction does not happen during the read operation. The double-ended 7T-SRAM has higher SNM than the 4T-SRAM but lower than the 6T-SRAM.

7.2.6.2 Single-Ended

The embedded systems that have a dominating memory component in the overall system are typically operated at lower voltages to reduce energy dissipation [92, 98]. For ultra-low power-embedded systems, particularly when the supply voltage is lower than the threshold voltage, building high-density cache is challenging. The classical 6T-SRAM cannot meet the high-density and high-yield requirements of modern circuit designs [92, 98, 99, 101, 102]. To address the above problems, the schematic design of a single-ended 7T-SRAM is shown in Fig. 7.14 [92, 98, 99, 101, 102]. The single-ended 7T-SRAM topology consists of two driver transistors (Driver-1 and Driver-2), two load transistors (Load-1 and Load-2), one access transistor (access transistor), and one transmission gate. There is only one bitline (Bitline) as compared to the classical 6T-SRAM for both write and read operations. The two inverters are connected back to back in a closed loop fashion to store 1-bit information. The transmission gate is essentially used to make or break the loop.

The simulation waveforms of a 45-nm CMOS technology-based single-ended 7T-SRAM are shown in Fig. 7.15 [98, 99, 101, 102]. Simplistically speaking, the Wordline is made high for write and

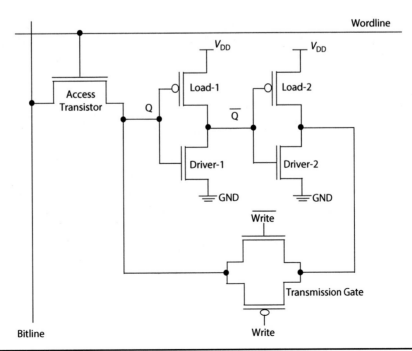

FIGURE 7.14 Transistor-level schematic of a single-ended 7T-SRAM [92, 98, 99, 101, 102].

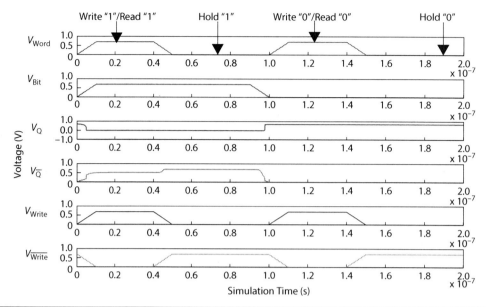

Figure 7.15 Functional simulation of a single-ended 7T-SRAM [98, 99, 101, 102].

read operations in the 7T-SRAM. The Wordline is made low for hold operation of this 7T-SRAM and the transmission gate provides a strong feedback to store the 1-bit information in the cross-coupled inverters. For write operation, the Wordline is asserted to high and Write signal made "1" to disconnect the feedback between the inverters. The signal (either "0" or "1") at the Bitline is then written to node Q through the access transistor. After write operation, the feedback connection is restored by applying write signal "0" to the transmission gate. The full voltage swing is stored with the help of node Q and $\overline{Q}$. For read operation, the Bitline is precharged to V_{DD}. The Wordline is asserted to voltage "1." The Bitline voltage remains at the precharge level upon read access. The access transistor and Driver-1 transistor form a voltage divider whose output is connected to the input of the Driver-2 and Load-2 inverter. During the hold operation, the Wordline is made low and Write is made "0" to make the cross-coupled inverters to store the 1-bit information.

The single-ended 7T-SRAM has many advantages over the 6T-SRAM [92, 98, 99, 101, 102, 105]. The single-ended 7T-SRAM circuit works well in the ultra-low voltage regions and allows subthreshold operation, which is unlike the classical 6T-SRAM circuit. The 7T-SRAM has better read and write stability as compared to the classical 6T-SRAM. The 7T-SRAM has better process variation tolerance as compared to the classical 6T-SRAM, and hence more suitable topology for nanoscale technology.

7.2.7 Eight-Transistor SRAM

The transistor-level schematic of an eight-transistor SRAM (8T-SRAM) is shown in Fig. 7.16 [23, 59, 106]. The 8T-SRAM can be thought of to have two distinct parts, a 6T-SRAM and a read buffer. There are two additional lines as compared to a 6T-SRAM, Bitline$_{RD}$ and Wordline$_{RD}$. The read buffer is made up of two transistors, read buffer transistor (RBT), RBT-1 and RBT-2. The read buffer isolates the data-retention structure during the read accesses buffer. In this 8T-SRAM, the write operation and read operations are isolated. The access transistors (Access-1 and Access-2) are used to perform write operation to the storage node. The write operation is thus same as in the case of a classical 6T-SRAM. The read operation is performed by transistors RBT-1 and RBT-2 in the read buffer. The read operation is quite different from the classical 6T-SRAM. For read operation, Wordline$_{RD}$ is enabled and data is read at Bitline$_{RD}$. The isolated read port in this 8T-SRAM enhances read noise margin (RNM) and more process variation tolerance. The isolation of write and read operations allows design optimization of the cell independently for both the operations. The additional wordline and bitline can increase the metal density of the memory chip. It can result in increased interconnect delay and capacitive coupling. It will lead to increased area penalty. The characterization results of a baseline 8T-SRAM are presented in Table 7.1 [59].

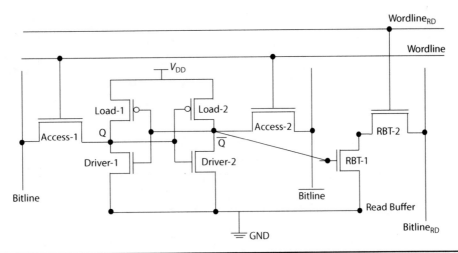

FIGURE 7.16 Schematic diagram of an 8T-SRAM designed using classical CMOS FET [59].

Characteristics	For 45 nm CMOS	For 32 nm CMOS
Average Power	5.81 nW	5.35 nW
SNM	281.44 mV	197.78 mV

TABLE 7.1 Characterization of the Baseline 8T-SRAM Cell [59]

For ultra-low voltage operation, the tunneling FET (TFET) is attractive due to steep subthreshold slope [24, 84]. However, the large crossover current may result in the inverters due to the broad soft transition region and delayed saturation characteristics that leads to degradation in hold SNM as well as read SNM. TFET-based 6T-SRAM, 7T-SRAM, and 8T-SRAM are available [24]. A TFET-based 8T-SRAM schematic design is presented in Fig. 7.17 [55]. It consists of N-type TFET driver transistors, Driver-1 and Driver-2 and P-type TFET load transistors, Load-1 and Load-2. The access transistors are completely different from classical 6T-SRAM. Four N-type TFETs have been used as access transistors, two each for Bitline and $\overline{\text{Bitline}}$. For Bitline, the access transistors are Access-1 and Access-2 and for $\overline{\text{Bitline}}$, the access transistors are Access-3 and Access-4. There are two wordlines, Wordline-1 and Wordline-2, as compared to one wordline of classical 6T-SRAM. It may be noted that in the operation point of view, the use of Bitline and $\overline{\text{Bitline}}$ may not be the same as that of a 6T-SRAM. For write operation, the $\overline{\text{Bitline}}$ is discharged to ground. Then, either Wordline-1 or Wordline-2 is set to V_{DD} depending on what value is to be written. The read operation is different from the classical

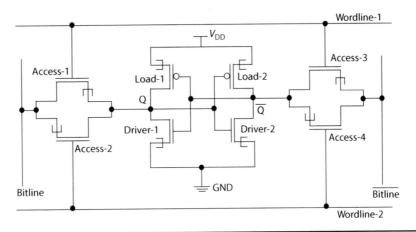

FIGURE 7.17 Schematic diagram of an 8T-SRAM designed using TFET [55].

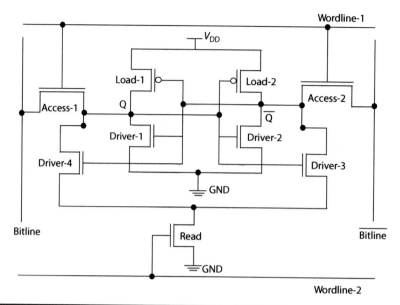

Figure 7.18 Transistor-level schematic of a 9T-SRAM [78].

6T-SRAM in which only Bitline is involved. For reading of "0," the Bitline is discharged. The Bitline retains its value of V_{DD} when "1" is stored in the cell. The TFET has its own limitations due to unidirectional behavior. As a TFET is a unidirectional device, it cannot conduct current in opposite directions that are needed by the access transistors during the write and read operation modes. The reverse behavior of the TFET in which the current is high for negative V_{DS} regardless of V_{GS} can be an issue for write operation [55, 93].

7.2.8 Nine-Transistor SRAM

The nine-transistor SRAM (9T-SRAM) is explored as a topology to overcome the issues faced by classical 6T-SRAM when operating at low voltage. The schematic diagram of a 9T-SRAM is presented in Fig. 7.18 [78]. The cross-coupled inverters consisting of Driver-1, Driver-2, Load-1, and Load-2 are typical, as is the case in 6T-SRAM. The access transistors Access-1 and Access-2 are also similar to the 6T-SRAM. The additional transistors in the 9T-SRAM are Driver-3, Driver-4, and Read. There are two wordlines, Wordline-1 and Wordline-2, as compared to the classical 6T-SRAM that has a single wordline. The write operation in the 9T-SRAM is similar to that of 6T-SRAM but using two wordlines. Similar to the 6T-SRAM for writing to Q and $\overline{Q}$ appropriate values are placed in Bitline and $\overline{\text{Bitline}}$. The enabling of Wordline-1 with disabling of Wordline-2 writes values to Q and $\overline{Q}$. For read operation, Wordline-2 is activated; this makes Read transistor ON. The transistors Driver-2 and Driver-3 provide stronger pull-down than 6T-SRAM, which improves RNM. The 9T-SRAM is also process variation tolerant compared to the 6T-SRAM. However, 9T-SRAM has larger area overhead as compared to the classical 6T-SRAM.

7.2.9 Ten-Transistor SRAM

The schematic representation of a 10T-SRAM is shown in Fig. 7.19 [58, 91, 98, 100]. The core storage portion of the 10T-SRAM circuit is the set of two inverters. The two inverters are connected back to back in a closed loop fashion to store the 1-bit information. There are three transmission gates TG_W, TG_R, and TG_H that are used as compared to two access transistors of a classical 6T-SRAM.

The write and read operations of the 10T-SRAM are quite different from the classical 6T-SRAM. For write operation, the write signal is made high. The transmission gate TG_W connects Bitline to node Q and forces node Q to the same level as Bitline. When the Write signal is made low, the loop is closed between the two inverters through the transmission gate TG_H. For read operation, read signal is made high. This enables the transmission gate TG_R and data is made available at the $\overline{\text{Bitline}}$. It may be noted that the Bitline in the case of 10T-SRAM is not the same as the 6T-SRAM. The Wordline of

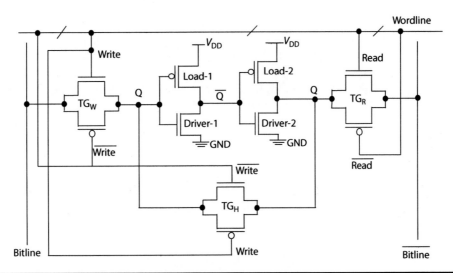

FIGURE 7.19 Transistor-level schematic of a 10T-SRAM [58, 91, 98, 100].

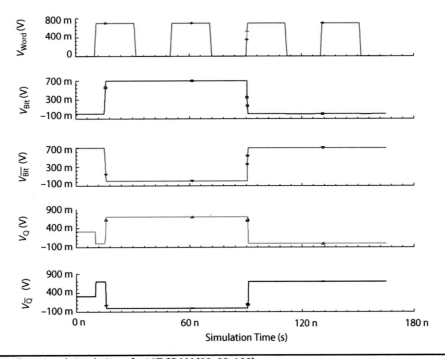

FIGURE 7.20 Functional simulation of a 10T-SRAM [58, 98, 100].

10T-SRAM is also not the same as in the case of classical 6T-SRAM. In the case of 10T-SRAM, the Wordline provides Write and Read signals. Selected waveforms of the 10T-SRAM simulation are shown in Fig. 7.20 [58, 98, 100]. It shows the Wordline, Q, $\overline{Q}$, Bitline, and $\overline{\text{Bitline}}$ voltage signals.

The use of transmission gates provides full voltage swing during write and read operations. The 10T-SRAM does not need sense amplifier and precharging circuitry. The elimination of sense amplifier and precharging circuitry achieves low-power SRAM. The 10T-SRAM is process variation tolerant as compared to classical 6T-SRAM. The 10T-SRAM is more stable and has more overhead as compared to classical 6T-SRAM.

7.2.10 Performance Metrics of SRAM

The characteristics of the SRAM are quantified by many performance metrics or figures-of-merit (FoM) [36, 77, 98]. This subsection will discuss selected performance metrics.

1. *SNM:* The SNM is a standard metric to measure the stability in SRAMs. The stability of the SRAM is the immunity of the SRAM to flip during its various operations. Smaller SNM means lower stability and it limits the reliability of storage data. SNM is sensitive to transistor parameter changes with aging, voltage fluctuations, and ionizing radiations [59]. Thus, SNM may decline with technology scaling even in defect-free SRAMs. The SNM of a SRAM cell is analytically expressed as follows [58, 85, 98]:

$$\text{SNM} = V_{\text{Th}} - \left(\frac{1}{k+1}\right) \times \left[\left(\frac{V_{\text{DD}} - \left(\frac{2r+1}{r+1}\right)V_{\text{Th}}}{1 + \left(\frac{r}{k(r+1)}\right)} \right) - \left(\frac{V_{\text{DD}} - 2V_{\text{th}}}{1 + \left(\frac{r}{q}k\right) + \sqrt{\left(\frac{r}{q}\right)\left(1 + 2k + \frac{r}{q}k^2\right)}} \right) \right] \quad (7.1)$$

In the above expression, r is the "cell ratio" that is the ratio of driver and access transistors $\left|\frac{\beta_d}{\beta_a}\right|$, β_d is the aspect ratio of driver transistor $\left|\frac{W}{L}\right|$, β_a is the aspect ratio of the access transistor $\left|\frac{W}{L}\right|$, and q is the "pull-up ratio" that is the ratio of sizes of load and access transistor $\left|\frac{\beta_l}{\beta_a}\right|$. k is defined as follows: $\left|\frac{r}{r+1}\right|\left|\sqrt{\frac{r+1}{r+1-V_s^2/V_r^2}} - 1\right|$ with V_s as $(V_{\text{DD}} - V_{\text{Th}})$ and V_r as $\left|V_s - \left|\frac{r}{r+1}V_{\text{Th}}\right|\right|$. From the above expression, it is evident that the SNM depends on the threshold voltage of the transistors and the sizes of the transistors. The SNM is measured as the maximum DC noise voltage (V_N) that can be tolerated by the SRAM without reliably storing the data. The setup for measurement of the SNM in a 6T-SRAM with an illustrative butterfly curve is presented in Fig. 7.21 [21, 59]. In Fig 7.21(a), noise sources with voltage (V_N) have been introduced at each of the internal nodes. In essence, the objective of SNM measurement experiment is to find the maximum V_N for which the SRAM safely holds the data. The graphical approach for calculating SNM is

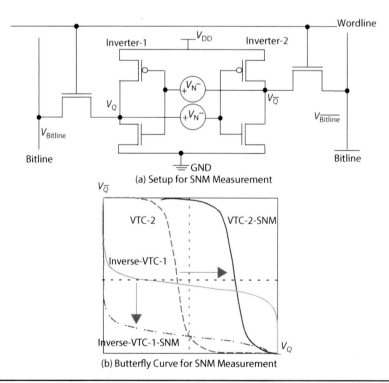

FIGURE 7.21 Measurement of SNM in SRAM using butterfly curve [21, 59].

presented in Fig 7.21(b). It consists of voltage transfer characteristics (VTCs) of two inverters for different voltage conditions. VTC-2 is the VTC for inverter-2. Inverse-VTC-1 is the inverse voltage transfer characteristic for inverter-1. The resulting two-lobed curve is called the "butterfly curve." SNM is quantified as the length of the largest square that can be fitted inside the lobes of the butterfly curve [12, 21, 59]. To understand why this calculation method works, consider the other curves in Fig 7.21(b) when the V_N is increased from 0. The VTC of inverter-2 moves to the right (VTC-2-SNM) and the inverse VTC of inverter-1 moves downward (Inverse-VTC-1-SNM). Once the values of V_N reach a maximum value, the curves meet at only two points beyond which the V_N will flip the SRAM cell.

The SNM measured in the above fashion to analyze stability of SRAM is essentially static (voltage) noise margin and has the following drawbacks: (1) it cannot be measured with automatic in-line testers. The static current noise margin (SINM) still has to be calculated by mathematical manipulation of the measured data after determining the butterfly curves of the SRAM. (2) It cannot generate statistical information on SRAM failures. (3) The butterfly curves delimit a maximal square side of maximum and 0.5 V_{DD} as an asymptotic limit for the SNM. Therefore, the supply voltage V_{DD} scaling limits the stability of the SRAM. The N-curve of SRAM that combines voltage and current information allows overcoming the above limitations. The experimental setup for N-curve measurement and an illustrative N-curve is presented in Fig. 7.22 [13, 35, 109]. The experimental setup is shown in Fig. 7.22(a) [13, 35, 109]. In this setup for N-curve measurement, Wordline, Bitline, and Bitline are clamped to V_{DD}. A voltage source V_{src} is applied at node Q and the current supplied by it is denoted as I_{src}. A voltage sweep is applied for V_{src} and the corresponding current values I_{src} are measured. An illustrative N-curve is shown in Fig. 7.22(b) [13, 35, 109]. The read margin is shown as SNM in the plot. The write margin is shown as write trip voltage (WTV). The positive peak current of the N-curve is the SINM that is the maximum current that can be injected in the SRAM to change its state. The negative peak current of the

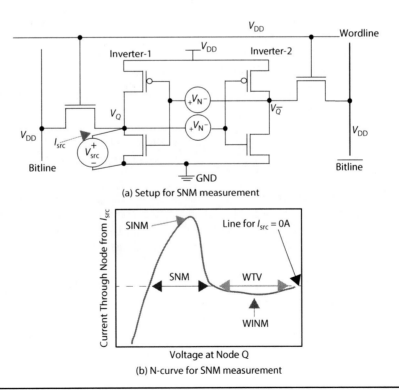

FIGURE 7.22 Measurement of SNM in SRAM using N-Curve [13, 35, 109].

N-curve is the write current noise margin (WINM) that is the amount of current needed to write the SRAM with both bitlines charged to V_{DD}.

2. *Hold Noise Margin (HNM)*: HNM is the SNM when the SRAM is in stand-by or hold operation mode [13, 21, 36]. During the hold operation, the wordlines are not active and the cross-coupled inverters are isolated from the bitlines to store the 1-bit data. The SNM measurements similar to the above that is performed over the equivalent circuit that is obtained for hold operation mode is essentially HNM. HNM is a very important metric for stability of SRAM to store the 1-bit data. Maintaining high HNM for nanoscale technology is very challenging as the typical trend is to reduce the supply voltage for low static power dissipation.

3. *RNM*: RNM is the SNM during the read operation of the SRAM. HNM is important for SRAM for storing the bits. Stability during the read operation for the SRAM is also significant as evident from the following discussions [13, 21, 36]. During read operation, when the wordline is "1," the bitlines are precharged to "1." When the SRAM node has "0," the node voltage gets pulled upward through the access transistor due to the voltage-dividing effect across drive and access transistor. This increase in the node voltage degrades the SNM during read operation.

4. *Write Noise Margin (WNM)*: The WNM is the minimum bitline voltage needed to flip the state of the SRAM [13, 74]. This is a statistical measure of the writeability of the SRAM. If the worst-case WNM is lower than the ground voltage, then the SRAM is not writeable.

5. *Dynamic Noise Margin (DNM)*: The SNM discussed above is the maximum DC voltage amplitude that SRAM can withstand for an infinitely long time without flipping its state. However, the noise in the form of short-time pulses can be of much higher amplitude than the SNM but may not affect the state of SRAM [32, 52, 108, 114]. The DNM essentially captures the amplitude of this pulse signal noise. The conventional SNM metric cannot provide accurate stability performance of nanoscale SRAM design with write/read assist circuits and shrinking access time. There is no strong correlation between the SNM and DNM of the SRAM, and hence it is difficult to determine whether a better DNM essentially indicates a better SNM. DNM takes into account spectral and time-dependent properties of the signal noise patterns. Similar to the SNM, there can be DNM for write, hold, and read operations of the SRAM.

6. *SRAM Access Time*: The access time is the specified time for which wordline is asserted high for write or read access to the cross-coupled inverters [36]. The significance of the metric comes from the fact that if the write or read operations cannot be successfully carried out before the wordline is made low, then access failure occurs.

7. *SRAM Cell Ratio*: The cell ratio of an SRAM is defined as the ratio of the sizes of the pull-down to the access transistors [13]:

$$\text{Cell Ratio} = \left(\frac{\text{Size of Pull-Down Transistor}}{\text{Size of Access Transistor}} \right) \quad (7.2)$$

The cell ratio of the SRAM decides the strength of "0" storage at the node to pull the precharged bitline to ground. A larger cell ratio indicates a larger pull-down network that ensures storage of a strong "0" robust from read operation noise originated from precharged bitline. The WNMs and HNMs of the SRAM are not affected much by the changes in its cell ratio.

8. *SRAM Pull-Up Ratio*: The pull-up ratio of an SRAM is defined as the ratio of the sizes of the pull-up to the access transistors [13]:

$$\text{Pull-Up Ratio} = \left(\frac{\text{Size of Pull-Up Transistor}}{\text{Size of Access Transistor}} \right) \quad (7.3)$$

The pull-up ratio of the SRAM also has major impact on the all noise margins. A good pull-up ratio ensures stability of "1" storage in the SRAM. A larger pull-up ratio makes it difficult to write or pull the node to ground which leads to a decrease in the write margin and WTV while increasing the read margin.

9. *SRAM Power Dissipation:* SRAM power dissipation is the power consumed by the SRAM during its write, hold, and read operations [36, 98]. The SRAM power dissipation includes capacitive switching power, subthreshold leakage, and gate-oxide leakage. In the embedded systems, power dissipated in the SRAM is a large portion of the total power dissipation of the system. The stand-by power or leakage power dissipation is very critical. A strong driving current reduces access time of the SRAM while increasing its power dissipation.

10. *SRAM Propagation Delay:* One way of defining propagation delay is the total time between the appearance of the row address at the address-decoder and the arrival of the stored data from memory at the output latch during the read operation [19, 36]. The propagation delay of SRAM depends on the interconnect delay and the column height. Larger device sizes reduce delay of SRAM but increase the power dissipation.

11. *SRAM Silicon Area:* SRAM silicon area is an important metric for applications that has area constraints. Larger device sizes provide large noise margin to guarantee functionality. However, larger device sizes increase SRAM area and lower memory density [36].

7.2.11 Characterization of Specific SRAMs

In this subsection, power and SNM characterization are discussed for selected SRAM cells.

7.2.11.1 6T-SRAM

The current flow path for write, hold, and read operations of a 6T-SRAM is shown in Fig. 7.23 [63, 98]. The figure shows the paths for active drain current (I_{dyn}), subthreshold leakage (I_{subth}), and gate-oxide leakage current (I_{ox}) for these states. The transient simulation of the SRAM for 45-nm CMOS showing various components of current is shown Fig. 7.24 [63, 98].

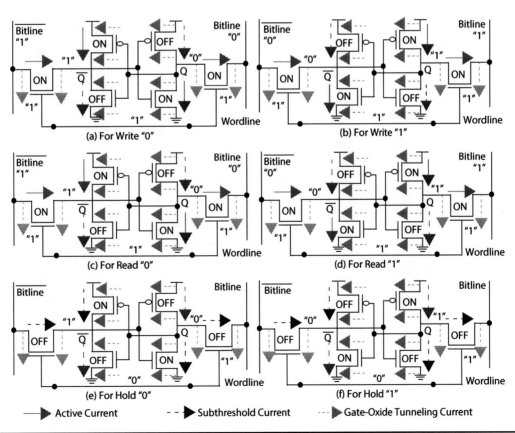

FIGURE 7.23 Paths of different current during write, read, and hold operation of a 6T-SRAM [63, 98].

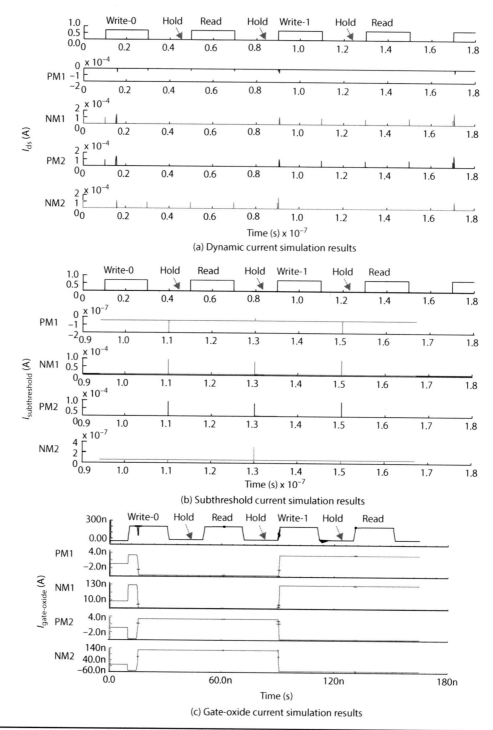

Figure 7.24 Analysis of different currents during write, read, and hold operation of a 6T-SRAM [63, 98].

For write "0" operation of SRAM, the current flow paths are presented in Fig. 7.23(a) [63, 98]. The ON transistors have (I_{ds}) and (I_{ox}) current components. The OFF transistors have (I_{subth}) and (I_{gate}) current components. The Bitline high makes PM1 and NM1 leaky. The access transistor NM3 is more leaky as both $\overline{\text{Bitline}}$ and Wordline are high compared to other access transistor NM4. For write "1" operation of SRAM, the situation is more or less symmetrical to that of write "0." The flow paths for various current components are depicted in Fig. 7.23(b) [63, 98]. The average power dissipation for the write operation of the SRAM can be expressed as follows [98]:

$$\text{Power}_{\text{Write,SRAM}} = \left(\frac{1}{T_{\text{Write}}}\right)\int_0^{T_{\text{Write}}} p(t)dt \quad (7.4)$$

$$= \left(\frac{V_{\text{DD}}}{T_{\text{Write}}}\right)\int_0^{T_{\text{Write}}} (I_{\text{ox}} + I_{\text{subth}} + I_{\text{dyn}})dt \quad (7.5)$$

For read "0" operation of the SRAM, the current flow paths are depicted in Fig. 7.23(c) [63, 98]. For read "1" operation of the SRAM, the current flow paths are depicted in Fig. 7.23(d) [63, 98]. The ON and OFF states of the transistors during read "0" operation are similar to the write "0" operation. However, the current actual values and durations may be different from write "0." The average power dissipation for the read operation of the SRAM can be expressed as follows [98]:

$$\text{Power}_{\text{Read,SRAM}} = \left(\frac{1}{T_{\text{Read}}}\right)\int_0^{T_{\text{Read}}} p(t)dt \quad (7.6)$$

$$= \left(\frac{V_{\text{DD}}}{T_{\text{Read}}}\right)\int_0^{T_{\text{Read}}} (I_{\text{ox}} + I_{\text{subth}} + I_{\text{dyn}})dt \quad (7.7)$$

For hold "0" operation of the SRAM, the current flow paths are presented in Fig. 7.23(e) [63, 98]. For hold "1" operation of the SRAM, the current flow paths are presented in Fig. 7.23(f) [63, 98]. In the hold operation of the SRAM, the two cross-coupled inverters continue to be active and have various forms of current flow in them even if they are disconnected from any external circuit. In the hold operation of the SRAM, the access transistors are OFF in this condition. The access transistors thus have subthreshold and gate-oxide leakage. The average power dissipation for the hold operation of the SRAM can be expressed as follows [98]:

$$\text{Power}_{\text{Hold,SRAM}} = \left(\frac{1}{T_{\text{Hold}}}\right)\int_0^{T_{\text{Hold}}} p(t)dt \quad (7.8)$$

$$= \left(\frac{V_{\text{DD}}}{T_{\text{Hold}}}\right)\int_0^{T_{\text{Hold}}} (I_{\text{ox}} + I_{\text{subth}})dt \quad (7.9)$$

The average power dissipation data for the simulation of a 45-nm CMOS technology is presented in Table 7.2 [63, 98]. It presents data for write "0," write "1," read "0," read "1," hold "0," and hold "0" operations of the SRAM. The results suggest that the write "1" operation has maximum power dissipation and the hold "0" operation has minimum power dissipation. The characterization of another 6T-SRAM for baseline sizes for 45- and 32-nm CMOS technology is presented in

6T-SRAM Operations	Average Power Value
Write 0	123.6 nW
Write 1	226.8 nW
Read 0	93.5 nW
Read 1	92.7 nW
Hold 0	96.1 nW
Hold 1	82.4 nW

TABLE 7.2 Average Power Dissipation in Write, Read, and Hold Operations of a 45-nm CMOS 6T-SRAM Cell [63, 98]

Characteristics	Data For 45 nm CMOS	Data For 32 nm CMOS
Average Power	5.70 nW	5.29 nW
SNM	141.94 mV	76.28 mV

TABLE 7.3 Characterization of the Baseline 6T-SRAM Cell for 45- and 32-nm CMOS [59]

Table 7.3 [59] that included SNM as well. The power results of the 6T-SRAM for the above two tables may not be identical as the sizes of the two designs are quite different. The sizes of the transistors affect I_{ox}, I_{subth}, and I_{dyn}.

7.2.11.2 7T-SRAM

The current paths in a 7T-SRAM for various operation modes are shown in Fig. 7.25 [98, 99, 101, 102]. The transient simulation results for various forms of current and total current are shown in Fig. 7.26 [98, 99, 101, 102].

For write "0" operation of the 7T-SRAM, the current paths are shown in Fig. 7.25(a) [98, 99, 101, 102]. In this situation, the access transistor is ON and has gate-oxide leakage and active drain current. The transistors in the transmission gate are OFF and have subthreshold leakage current and gate-oxide leakage current. In the cross-coupled transistors, some transistors are ON and some are OFF. While the ON transistors contribute to gate-oxide leakage and active drain current, the OFF transistors contribute to gate-oxide leakage and subthreshold leakage current. For write "1" operation of the 7T-SRAM, the current paths are shown in Fig. 7.25(b) [98, 99, 101, 102]. In this mode of operation, the state of transistors in the cross-coupled inverters has reversed as compared to the write "0" operation case. The state of access transistor as well as the transmission gate remains the same as that of the write "0" operation state. For read "0" operation of the 7T-SRAM, the current paths are

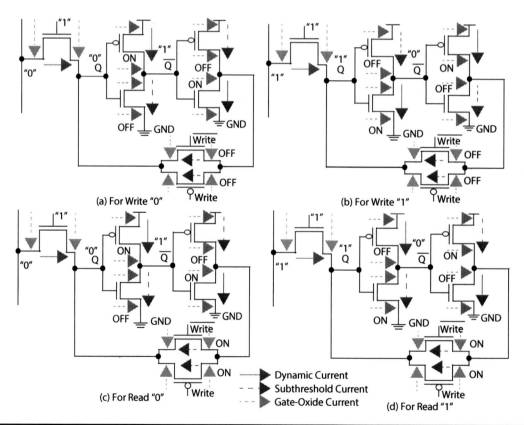

FIGURE 7.25 Paths of different current during write, read, and hold operation of a 7T-SRAM [98, 99, 101, 102].

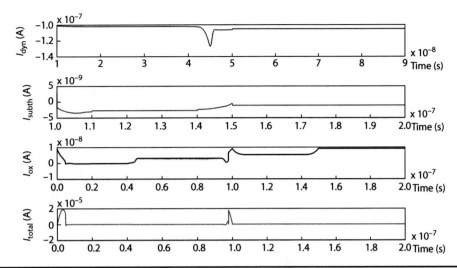

FIGURE 7.26 Analysis of different currents during write, read, and hold operation of a 7T-SRAM [98, 99, 101, 102].

Current Components	Average Value (nA)
Dynamic Current	155.1
Subthreshold Current	101.0
Gate-oxide Current	97.4
Total Current	162.3

TABLE 7.4 Various Currents in a 45-nm CMOS Baseline 7T-SRAM [98, 99, 101, 102]

shown in Fig. 7.25(c) [98, 99, 101, 102]. The transistors in the transmission gate are ON as compared to the write "0" operation. The state of all the other transistors is same as the write "0" operation. For read "1" operation of the 7T-SRAM, the current paths are shown in Fig. 7.25(d) [98, 99, 101, 102]. The read "1" state is similar to the write "1" state other than the transmission gate transistors that are ON. For the hold operation of the 7T-SRAM, either storing a "0" or "1" in the current flow can be similarly explained.

The instantaneous behavior of different current components for the 7T-SRAM is presented in Fig. 7.26 [98, 99, 101, 102]. Each of the components includes the current from different transistors for their different states during the 7T-SRAM operations. The average values of the various forms of current is shown in Table 7.4 [98, 99, 101, 102]. The results show that I_{dyn} is large due to the transition of bitlines from high to low or low to high. The total current has the maximum average current because along with the above-mentioned current components, it consists of several origins, reverse-biased currents, short circuit current.

SNM is another important characteristic of any SRAM. The setup for measurement of SNM in a 7T-SRAM is shown in Fig. 7.27(a) [98, 99, 101, 102]. The noise voltage source (V_N) is introduced at the two nodes for the measurement of the SNM. The resulting butterfly curve for a 45-nm CMOS technology-based implementation of the 7T-SRAM is presented in Fig. 7.27(b) [98, 99, 101, 102]. The average power dissipation of the 7T-SRAM along with the SNM is presented in Table 7.5 [98, 99, 101, 102].

7.2.11.3 10T-SRAM

The paths of different current components in the high-κ/metal-gate-based 10T-SRAM are shown in Fig. 7.28 [58, 98, 100]. The states of various transistors and their current contributions can be explained in a similar manner as in the case of 6T-SRAM and 7T-SRAM discussed previously. The gate-oxide leakage component is completely vanished due to the use of high-κ/metal-gate CMOS

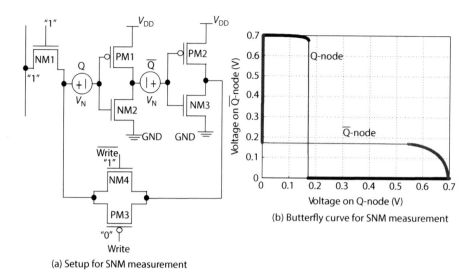

Figure 7.27 Measurement of SNM of a 7T-SRAM using butterfly curve [98, 99, 101, 102].

SRAM Characteristics	Actual Values
Average Power Dissipation	203.6 nW
SNM	170.0 mV

Table 7.5 Power and SNM for 45-nm CMOS-Based Baseline 7T-SRAM Cell [98, 99, 101, 102]

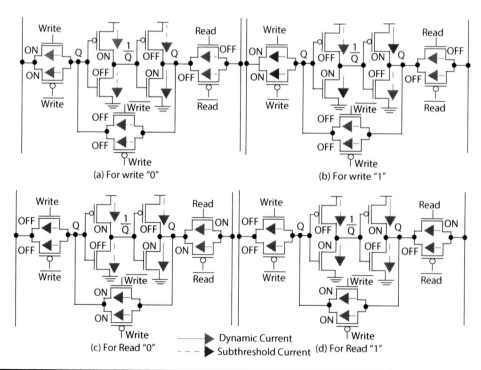

Figure 7.28 Paths of different currents during write, read, and hold operation of a high-κ/metal-gate-based 10T-SRAM [58, 98, 100].

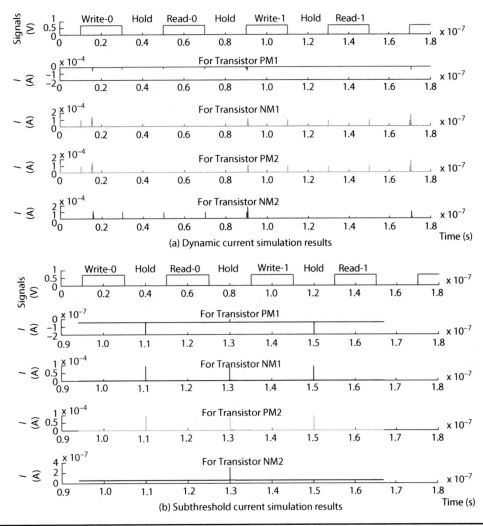

Figure 7.29 Analysis of different currents during write, read, and hold operation of a high-κ/metal-gate-based 10T-SRAM [58, 98, 100].

technology. The dynamic current component for various individual transistors indicating various operation modes is shown in Fig. 7.29(a) [58, 98, 100]. The subthreshold leakage current component for individual transistors indicating various operation modes is shown in Fig. 7.29(b) [58, 98, 100].

The power and access time values of the 10T-SRAM are presented in Table 7.6 [58, 98, 100]. The subthreshold leakage is maximum for write "0" and minimum for read "0." The active drain-to-source current is maximum for write "1" and minimum for write "0." The average power is maximum for write "1" and minimum for read "0." The access time is maximum for read "1" and minimum for write "0." The setup for SNM measurement in the 10T-SRAM is shown in Fig. 7.30(a) [58, 98, 100].

Operation	I_{subth} (nA)	I_{dyn} (nA)	Average Power (nW)	Access Time (ns)
Write "0"	202.41	67.85	189.20	0.02
Write "1"	201.5	897.5	769.3	5.64
Read "0"	71.90	141.10	149.11	14.56
Read "1"	118.00	129.30	173.14	15.56

Table 7.6 Characteristics during Read and Write Operations for 32-nm High-κ/Metal-Gate-Based 10T-SRAM Cell [58, 98, 100]

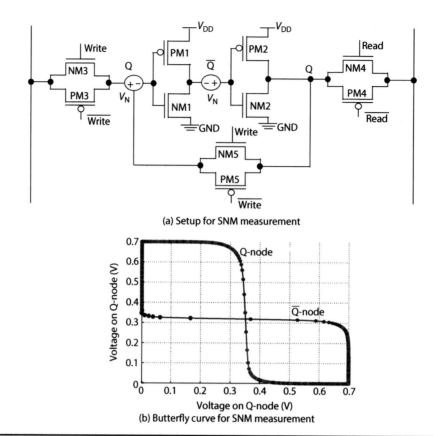

Figure 7.30 Measurement of SNM a high-κ/metal-gate-based 10T-SRAM using butterfly curve [58, 98, 100].

SRAM Characteristics	Experimental Value
Average Power Dissipation	2.27 μW
SNM	271 mV

Table 7.7 Experimental Results of a Baseline 32-nm High-κ/Metal-Gate-Based 10T-SRAM Cell [58, 98, 100]

The butterfly curve for 32-nm high-κ/metal-gate-based 10T-SRAM with 0.7 V supply voltage is presented in Fig. 7.30(b) [58, 98, 100]. The power and SNM results obtained for the baseline design of 32-nm high-κ/metal-gate-based 10T-SRAM with 0.7 V are presented in Table 7.7 [58, 98, 100]. The average dissipation is the power dissipation average value for several operations including write "0," write "1," read "0," and read "1."

7.3 Dynamic Random-Access Memory

DRAM is everywhere in almost all types of computing systems because it is cheap and relatively fast and possesses high-density volatile storage [45, 69, 115]. It can be a smart phone, embedded systems, media players, desktop computers, or printers. DRAM provides volatile memory to the processor for computing that has a good trade-off of high speed, low cost, and capacity. DRAM plays a crucial role in the memory hierarchy of computer systems that essentially dictates that "maximum possible and fast memory available to the users at a minimal possible cost." DRAM is in the middle of memory hierarchy in a computing system that consists of cache (made of SRAM), main memory (which is DRAM), and hard drive (magnetic memory or solid-state memory) [69]. DRAM is of different sizes depending on the computing system, e.g., few MB to several GB is a typical practice. In the VLSI

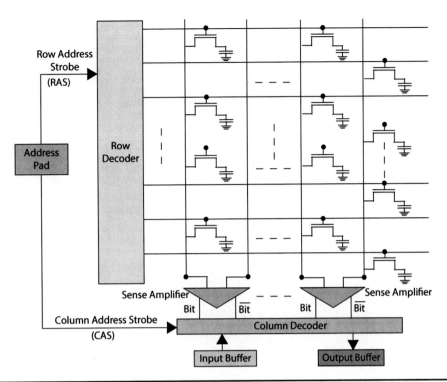

Figure 7.31 Schematic representation of DRAM array architecture.

trend, the DRAM is one of the bottlenecks in the computing, as when the processor operation is in the GHz range, the DRAM is in the MHz range. DRAM is slower than SRAM, but DRAM is cheaper than SRAM and has higher density as compared to SRAM.

7.3.1 DRAM Array

The DRAM array structure is depicted in Fig. 7.31 with one-transistor and one-capacitor (1T-1C) cell [1, 69]. It has an array of cells that can store 1-bit data, a set of sense amplifiers, address pad, row decoder, column decoder, input buffer, and output buffer. The gates of each cells are connected to the rowlines. The DRAM cells are written by placing charge in the capacitors of each cell. For write or read operation in the DRAM, the row and column addresses are made available on the address pad. In the first step, the row addresses are validated. In the second step, the column addresses are validated. During write cycle, the transistor in a DRAM cell is made ON through the rowline and a charge is made available from the columnline for storage in the capacitor in the DRAM cell. The rowline is then discharged to switch OFF transistor of the DRAM cell to isolate the stored charge in the DRAM cell capacitor.

7.3.2 Different Types of DRAM

Many different types of DRAM have been introduced for different purposes including to reduce the access time, speed up data flow, increase memory density, reduce area, and reduce power dissipation. Selected types of DRAM are presented in Fig. 7.32 [1, 69, 107]. The DRAMs are classified in many ways including the cell topology, data rate, and timing. Selected types of DRAMs are discussed in the next subsection.

Based on the topology, one-transistor to four-transistor DRAM are developed. DRAM has many different topologies depending on the circuit elements used; however, one-transistor and one-capacitor topology has been used extensively. Single- and double-data rate (DDR) DRAM is available based on different types of data rate. The different types of DRAM based on the modes of operation include fast page mode DRAM (FPM-DRAM), extended data out DRAM (EDO-DRAM), and burst EDO-DRAM (BEDO-DRAM). Audio DRAM (ADRAM) is DRAM with some defective cells that result from any normal DRAM fabrication that cannot be used in regular application but can be used

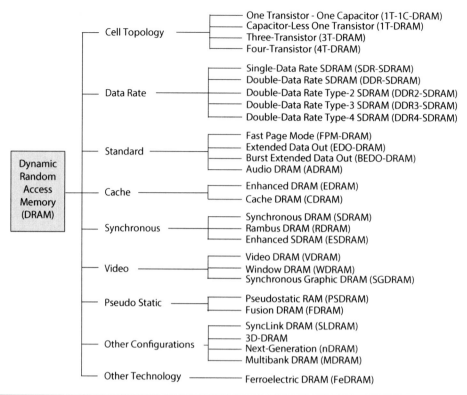

FIGURE 7.32 Different types of the DRAM [1, 69, 107].

for audio applications that have tolerance errors [1]. Thus, ADRAM is cheaper and has downgraded specifications as compared to the regular DRAM.

The high-speed cache DRAM (CDRAM) is of two types: enhanced DRAM (EDRAM) and CDRAM. The EDRAM has standard DRAM, integrated SRAM, and control. The CDRAM is integrated DRAM and SRAM cache memory on the same chip. The synchronous DRAM (SDRAM) has write and read cycles synchronized with the processor clock. The enhanced SDRAM (ESDRAM) is a DRAM that has SDRAM and cache SRAM on the same chip [1]. The video or graphic DRAMs are DRAMs with features specific to video and graphic-processing operations and performance requirements. The pseudostatic DRAM (PSDRAM) has classical PSDRAM and fusion DRAM (FDRAM). The PSDRAM, also referred to as pseudostatic SRAM (PSRAM), has storage mechanism of DRAM but on-chip circuitry of SRAM for speed and cost tradeoffs. FDRAM is also combination of DRAM and SRAM for speed and cost tradeoffs. The other configurations of the DRAM include SyncLink DRAM (SLDRAM), 3D-DRAM, next-generation DRAM (nDRAM), and multibank DRAM (MDRAM). The ferroelectric DRAM (FeDRAM) uses capacitors of ferroelectric materials such as lead zirconium titanate (PZT) and provides nonvolatile storage without requiring refreshing.

7.3.3 Selected DRAM Designs Based on Topology

The DRAM has four different topologies as follows [17, 53, 54, 69, 95]: (1) one-transistor and one-capacitor DRAM (1T-1C-DRAM), (2) one-transistor capacitor-less DRAM (1T-DRAM), (3) three-transistor DRAM (3T-DRAM), and (4) four-transistor DRAM (4T-DRAM). The 1T-1C-DRAM and 3T-DRAM are more commonly used DRAM topologies. The "dynamic" term comes from the fact that the DRAM needs refreshing operation as the charge in the stored capacitor slowly depletes. For example, the typical refresh rates may be 5 to 10 ms.

7.3.3.1 One Transistor-One Capacitor DRAM

The 1T-1C-DRAM is the classic DRAM cell that has been well-studied and used in real-life circuits and systems. The 1T-1C-DRAM became very popular because of its capacity for high-packing density. The transistor-level schematic of the 1T-1C-DRAM is shown in Fig. 7.33 [69]. It requires only one

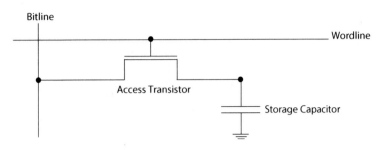

FIGURE 7.33 Transistor-level schematic of a 1T-1C-DRAM Cell [69].

transistor (access transistor) and one capacitor (storage capacitor). The Wordline and Bitline are used for access and storage of 1-bit data. The 1T-1C-DRAM uses a single rowline and a single columnline and has considerably less area. In order to write to the 1T-1C-DRAM, the Wordline is set "high" that switches the transistor ON. Then, the value of the Bitline is set to "1" or "0" depending on what needs to be written. Depending on that value, the storage capacitor is charged to discharged to write a logic "1" or "0," respectively. To read the 1T-1C-DRAM, the Bitline is precharged to a predetermined voltage level $\left(\text{e.g.,}\left(\frac{V_{DD}}{2}\right)\right)$ and then the Wordline is set "high." In this situation, charge sharing occurs between the storage capacitor and the Bitline. Depending on the values stored, the charge flow can happen from storage capacitor to Bitline or from Bitline to storage capacitor [69, 95]. If the stored value is "1," then the charge flow is from storage capacitor to Bitline. On the other hand, when the stored value is "0," then the charge flow is from Bitline to storage capacitor. The shared charge in the Bitline is small, and hence a sense amplifier is needed to detect the voltage change. The charge sharing makes the read operation in 1T-1C-DRAM destructive. The charge stored in the capacitor may be modified or corrupted during charge sharing between storage capacitor to Bitline. If "1" is stored in the storage capacitor, then the charge after sharing may not be strong enough to be considered "1." Similarly, if "0" is stored in the storage capacitor, then the charge after sharing may be too high to be considered "0." The sense amplifier helps to refresh the value of the 1T-1C-DRAM after read operation. The value read from the 1T-1C-DRAM is written back to maintain the signal. A read operation can be performed to refresh the charge in the storage capacitor that depletes its charge over time.

7.3.3.2 Capacitor-Less One-Transistor DRAM

The 1T-1C-DRAM has major limitations [68, 79]: (1) size of storage capacitor, (2) integration of storage capacitor, (3) leakage current, and (4) parasitic resistance. The main issue is the capacitance integration with area constraints for 100-nm CMOS technology and beyond. The storage capacitor does not scale with technology trend, as it needs to be large enough to store a detectable amount of charge [60].

A simpler DRAM is a DRAM without storage capacitor that gives rise to the concept of 1T-DRAM. This is also known as zero-capacitor RAM (ZRAM or Z-RAM) [68, 79]. The 1T-DRAM typically exploits the floating body effect of the silicon-on-insulator (SOI) transistor technology. The excess or lack of majority carriers in an SOI transistor represents two states for storage of 1-bit data [89, 96]. For example, excess holes in the floating body that lower V_{Th} represent storage of "1." Lack of holes in the floating body that increases V_{Th} represents storage of "0." When the body voltage of the SOI transistor is modified, it changes the current that in turn, changes the state. The 1T-DRAM cell uses only three signal lines, a single channel, and does not need the complex capacitor fabrication steps as needed in the case of a 1T-1C-DRAM.

The structure of triple-gate transistor-based 1T-DRAM is shown in Fig. 7.34 [18, 60]. It can also be made from single- or double-gate FinFET using SOI technology. The TGFET-based 1T-DRAM or Z-RAM, shown in Fig. 7.34, uses back gate for floating body storage node [18]. Gate-1 and Gate-2 are connected together to apply a voltage V_{gf}. Additionally, a voltage V_{gb} can be applied at the back gate. Application of a negative voltage at the back gate, i.e., $-V_{gb}$, creates an accumulation layer at the back interface. At this condition, the TGFET exhibits a partially depleted behavior that can be used as an 1T-DRAM cell to store 1-bit of information. In the Z-RAM, excess of majority carrier is

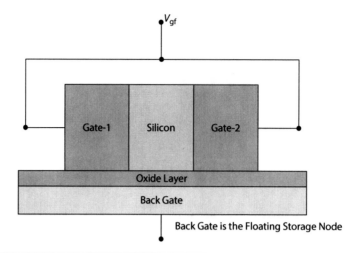

FIGURE 7.34 1T-DRAM or ZRAM or Z-RAM using a tri-gate FinFET [18, 60].

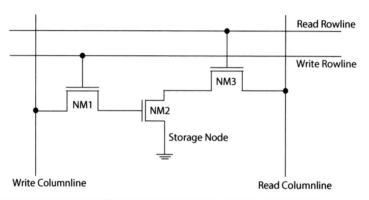

FIGURE 7.35 Transistor-level schematic of a 3T-DRAM cell [69].

state "1" and deficiency of majority carrier is state "0." The channel current in the TGFET is measured for sensing purposes.

7.3.3.3 Three-Transistor DRAM

The transistor-level schematic of 3T-DRAM is shown in Fig. 7.35 [31, 69]. 3T-DRAM consists of three different transistors. It has two pairs of rowlines and columnlines. There are write and read rowlines (Write Rowline and Read Rowline) as well as write and read columnlines (Write Columnline and Read Columnline). The transistor NM2 is used for storage purposes. The transistors NM1 and NM3 are used for access control purposes. The data to be stored is stored as charge on the gate capacitance of transistor NM2. For write operation, data is made available in the Write Columnline. At this point, the Write Rowline is made "high" to switch ON transistor NM1. The data from the Write Columnline charges or discharges the input capacitance of transistor NM2. The Write Rowline is disabled to switch transistor OFF after the input capacitance of transistor NM2 is sufficiently charged or discharged. For read operation in the 3T-DRAM, the Read Columnline is precharged to a predetermined voltage. The Read Rowline is now made "high" to switch ON transistor NM3. The Read Columnline is unchanged or discharged based on the data value of input capacitance of transistor NM2. When the stored value is "0," transistor NM2 is OFF and the Read Columnline remains unchanged at precharged value. When the stored value is "1," transistor NM2 is ON and Read Columnline discharges to the ground. Thus, the read operation is destructive as in 1T-1C-DRAM. As in the case of 1T-1C-DRAM, the stored charge in 3T-DRAM also depletes over time. Thus, charge refreshing is needed as in the case of 1T-1C-DRAM. The disadvantage of the 3T-DRAM is that it uses three transistors, two pairs of rowlines, and two pairs of columnlines. Thus, making the layout area of 3T-DRAM much larger than the 1T-1C-DRAM. Hence, 3T-DRAM did not become a popular candidate for use.

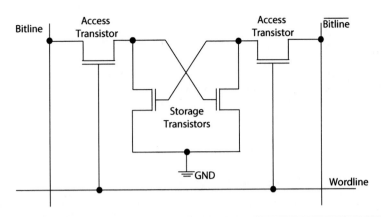

Figure 7.36 Transistor-level schematic of a 4T-DRAM cell [31, 41].

7.3.3.4 Four-Transistor DRAM

The transistor-level schematic of a 4T-DRAM is shown in Fig. 7.36 [31, 41]. Due to its topology and nature of operation, it is also called quasi-static RAM [41]. Topology-wise, it seems to be same as the 4T-SRAM shown in Fig 7.10, with access transistors being NMOS instead of PMOS. As compared to a 6T-SRAM, this 4T-DRAM does not have PMOS load transistors that connect to the V_{DD} and make up the stored charge decay over time. The silicon area of 4T-DRAM is smaller than the classical 6T-SRAM, but larger than the 1T-1C-DRAM. The 4T-DRAM naturally decays over time and needs to be automatically refreshed through the use of sense amplifier from the precharged bitlines whenever accessed. When 4T-DRAM is accessed the internal Q node is refreshed to voltage "high" without the need for a read-write cycle as in the case of 1T-DRAM.

7.3.4 DRAMs Based on Modes of Operation

A processor that does not get data in the cache, accesses the main memory (made from DRAM) to obtain the data that is in the next lower level in the memory hierarchy. The time that processor takes to fetch the first word so that processor stall due to cache miss is ended has severe impact on the system performance. Many different techniques have been explored to improve DRAM access [44, 62, 69]. Many locations of the DRAM array may be accessed concurrently to access a page of data. Depending on how a DRAM is accessed through the column address decoder, different modes of operation are possible.

7.3.4.1 Fast Page Mode DRAM

In conventional DRAM after the address is decoded for a DRAM cell access, the page is opened for access, which is again opened for each address decoding. In the "fast page mode (FPM)" DRAM (FPM-DRAM), the row or a page is kept open for multiple column reads [29, 69]. Multiple DRAM cells in the same row or page are accessed successively. FPM modes enhance the access time in which the time needed to open a page for successive reads in the same page is eliminated. The different variations of the FPM include "static column mode" and "nibble mode." In the static column-mode DRAM, the required column address is changed without strobing in a new access. In the nibble-mode DRAM, four serial bits are accessed for a selected bit.

7.3.4.2 Extended Data Out DRAM

The EDO-DRAM is a modification of the FPM-DRAM [62, 69]. The EDO-DRAM is like a FPM-DRAM with a latch at the output buffer. Thus, the valid data remains on the bus of the EDO-DRAM until the next address is strobed. Thus, the EDO-DRAM allows a new read operation to be started while the data of the current read operation is latched to the output buffer. The EDO-DRAM increases the valid output data time and effectively decreases the cycle time. The access time of the EDO-DRAM is approximately 10% to 15% lower than the access times of the FPM-DRAM.

7.3.4.3 Burst Extended Data Out DRAM

The BEDO-DRAM is based on additional design improvements over the EDO-DRAM [62, 69]. The BEDO-DRAM processes four read and write cycles in one "burst." The BEDO-DRAM has an additional

2-bit internal address counter to identify the next memory address to be accessed. In the BEDO-DRAM, only the first memory address to be accessed is decoded by the address decoder and the remaining addresses are generated by the internal address counter. This arrangement reduces the number of clock cycles required to access the address bits. In the BEDO-DRAM, a pipelined stage is used instead of the output latch as in the EDO-DRAM. The BEDO-DRAM increases the access time slightly, however, improves the memory bandwidth. Hence, the BEDO-DRAM is also known as the pipelined BEDO-DRAM.

7.3.5 Synchronous DRAMs

7.3.5.1 Classical Synchronous DRAM

In an asynchronous DRAM, a write or read cycle begins whenever appropriate signals (e.g., row-address strobe and column-address strobe) are available. In the SDRAM, a clock signal is used in addition to the signals in the asynchronous DRAM to determine when the write or read cycle begins [29, 62, 69]. In the SDRAM, the write operation and read operation are executed at the edge of the clock signal, which is the processor master clock. The SDRAM thus improves the access time by making it a multiple of the processor clock time that is not a case in the asynchronous DRAM [29, 69]. The SDRAM supports the memory access mode of BEDO-DRAM. The SDRAM has many different variants, for example, DDR. DDR improves the performance by performing multiple write and read operations in one cycle at both the rising and falling edges of the clock. Different versions of the DDR include DDR type-2 (DDR2) and DDR type-3 (DDR3). While DDR performs two operations in one cycle, the DDR2 and DDR3 perform four and eight write and read operations per cycle.

7.3.5.2 Direct Rambus DRAM

The direct Rambus DRAM (DRDRAM) also known as Rambus DRAM (RDRAM) is a type of SDRAM [1, 7, 28, 30, 40]. The core components such as the banks and the sense amplifiers of the DRDRAM are same as any DRAM, but the architecture and interface are different. The DRDRAM is an interleaved memory system in which the memory addresses are allocated to each memory bank in turn and is integrated onto a single memory chip. While the access time or latency does not improve much by DRDRAM, the bandwidth of memory is significantly higher. The high cost of DRDRAM is the main factor of not making this memory popular in computing systems. In the case of the SDRAM, dual in-line memory module (DIMM) holds a series of SDRAMs, which then fits into the DIMM slot of the motherboard. The analogous of DIMM is Rambus in-line memory module (RIMM) of the DRDRAM. RIMM of the DRDRAM is quite expensive as compared to the DIMM of the SDRAM.

7.3.6 Video or Graphics DRAM

Video or graphics DRAM is DRAM designed with video processing and display applications as the target. The key point is high-speed processing and high bandwidth data transfer to maintain high-throughput requirements. The three different types of video DRAM (VDRAM) are discussed in this subsection [1, 9].

7.3.6.1 Video DRAM

VDRAM is used generically to refer "video RAM" or "video memory," which is any memory in the video subsystem. However, the VDRAM discussed here is not any generic video RAM. The VDRAM, also called dual-port DRAM (DP-DRAM), has special features to support video systems [1, 9]. A VDRAM has dual ports that provide two access ports to support write and read simultaneously. A VDRAM has separate serial and parallel interfaces.

7.3.6.2 Window DRAM

The window DRAM (WDRAM) is a dual-port VDRAM with many added features [1, 10]. The WDRAM is more geared towards the graphics cards that has better performance but lower cost as compared to the VDRAM. Overall, the WDRAM can provide 25% more bandwidth than VDRAM.

7.3.6.3 Synchronous Graphics DRAM

The synchronous graphics DRAMs (SGDRAMs) are SDRAMs with capabilities specific to graphics processing including block write mode, masked write mode, and wide bus [1, 8]. However, the SGDRAM

Memory in the AMS-SoCs **345**

is single-ported as compared to the double-port VDRAM or WDRAM. SGDRAM provides high bandwidth data transfer through much higher speed operations as compared to VDRAM or WDRAM.

7.3.7 Ferroelectric DRAM

The FeDRAM is a DRAM that has storage capacitors made from ferroelectric material [1, 14, 75, 86]. The ferroelectric material is typically a ceramic film of PZT. The FeDRAM functions like a DRAM when supply is applied; however, it does not need refreshing as required for a typical DRAM. The FeDRAM stores the data after the power supply is removed and hence is nonvolatile in nature. It may be noted that ferroelectric memory is different from ferromagnetic memory. The ferromagnetic memory or core memory uses small magnetic cores for storage [86]. The ferromagnetic memory was too bulky, power hungry, and costly compared with semiconductor memory and hence became obsolete. On the other hand, the ferroelectric memory does not use iron (the ferromagnetic material). The ferroelectric memory is a semiconductor memory but the hysteresis loop characteristics of ferroelectric capacitor is similar to the ferromagnetic core. The ferroelectric material such as PZT has two stable polarization states that are reversed by applying a proper electric field [86]. Two cell topologies for FeDRAM cell are presented in Fig. 7.37 [1, 14, 75, 86]. One topology uses a single transistor and another topology uses two transistors.

The 1T-1C nonvolatile FeDRAM cell is presented in Fig. 7.37(a) [1, 86]. It has wordline and bitline as typical in the case of a standard DRAM. However, there is a plateline that is not present in a classical 1T-1C DRAM. The function of the access transistor is also similar to the case of a classical 1T-1C DRAM. The access transistor controls the access to the capacitor as typical in a classical 1T-1C DRAM. However, additionally, the access transistor eliminates the need for a square-like hysteresis loop of the ferroelectric capacitor. In general, the presence of the access transistor in series with the ferroelectric capacitor compensates for the softness of the hysteresis loop characteristics of the ferroelectric capacitor while blocking the noise signals from the other neighboring FeDRAM cells. When the access transistor is ON, the ferroelectric capacitor is connected to the bitline, and write or read operations can be performed with the help of the plateline. When the access transistor is OFF, the ferroelectric capacitor is disconnected from the bitline, and hence continues storing the data without any external disturbances.

A nonvolatile FeDRAM cell that uses two transistors for better fault tolerance is presented in Fig. 7.37(b) [75]. As compared to the 1T-1C nonvolatile FeDRAM cell, there is an additional transistor called the shunt transistor. The shunt transistor does not consume additional area as it is built under

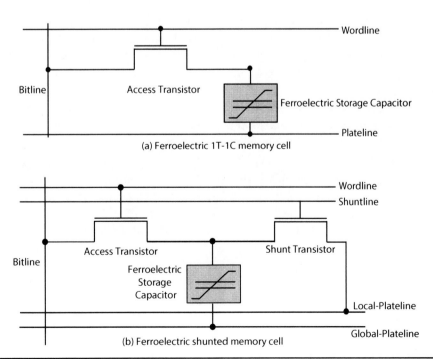

Figure 7.37 Transistor-level schematic of FeDRAM [1, 14, 75, 86].

the ferroelectric capacitor. The shunt transistor connects the common node of the access transistor and the ferroelectric capacitor to local plateline. The gate of the shunt transistor is connected to the shuntline. The local plateline is periodically connected to the global plateline when the shunt transistor is ON through high voltage at the shuntline. For the read operation, the shunt transistor is made ON and then the access transistor is made ON, which then connects the common node to the bitline. At this point, a destructive read of the FeDRAM cell happens. The precharge voltage level of the bitline is connected to the internal node, which develops a voltage across the ferroelectric capacitor. The voltage across the ferroelectric capacitor may keep it at the same polarization state or switch it to the opposite polarization state. A sense amplifier drives the bitline to the opposite voltage level when the switched voltage is sensed, thus restoring the original state of the ferroelectric capacitor.

7.3.8 Characteristics of DRAM

The DRAM has many characteristics to describe its capacity and performance [6, 107]. DRAM capacity is described that is representative of the number of cells in the memory array that can be used to describe bits of data. The typical capacity in modern computing systems is of gigabyte range. A higher capacity is desired for heavy-duty computing systems. The power dissipation in the DRAM is a major issue particularly when used in portable electronic systems. The power dissipation is the overall power dissipation due to switching activity power as well as standby power. The power dissipation may be different for write, read, and hold operations as in the case of SRAM discussed in the previous subsection. In the rest of this subsection, the focus will be on the timing-related characteristics of the DRAM. The important timing metrics of the DRAM include row-precharge time, row-access time, and column-access time. A DRAM is being rated using four timing characteristics as follows: column address strobe (select) latency (tCL), row address strobe (RAS) to column address strobe (CAS) delay (tRCD), row precharge time (tRP), and row address strobe (RAS) active time (tRAS). It is typically in the following order: tCL-tRCD-tRP-tRAS; in some cases, a different order is also used: tCL-tRCD-tRAS-tRP. These timing delays or latencies affect the speed of DRAM and are expressed in terms of clock cycles. The lower these timing values, are faster the performance of the DRAM is.

1. *Column Address Strobe or Column Address Select (CAS) Latency (tCL):* The tCL of the DRAM is the time between when the column address is sent to the DRAM and when the data response begins—in other words, the time between sending a DRAM reading command and the time to act on the command. The DRAM takes tCL time to read the first bit of data from a DRAM with the correct row already open.

2. *RAS to Column Address Delay (tRCD):* The RAS to CAS delay (tRCD) of the DRAM is the time between opening a row of DRAM and accessing the columns within it—in other words, the time for issuing an active command and write or read commands. The time to read the first bit from DRAM without an active row is the sum of tCL and tRCD.

3. *Row Precharge Time (tRP):* The tRP of the DRAM is the time required between issuing the precharge command and opening the next row—in other words, the time between active commands and the write or read of the next bank on the DRAM module. The time to read the first bit from a DRAM with the wrong row open is the sum of tCL, tRCD, and tRP.

4. *RAS Active Time (tRAS):* The tRAS of the DRAM is the time between a bank active command and issuing a precharge command—in other words, the time between a row being activated by precharge command and deactivated. tRAS is the time required to internally refresh the row. Typically, tRAS is calculated as (tRCD + 2 tCL); which is (tRCD + tCL) for a SDRAM module. Typically, tRAS is the sum of tRCD and write recovery time (tWR).

5. *Write Recovery Time:* The tWR of the DRAM is time difference between the last write command to a row and its precharging.

6. *Row Cycle Time (tRC):* The tRC is the sum of tRAS and tRP, i.e., tRC = tRAS + tRP [6]. This is also known as random access cycle time, which determines the minimum amount of time required between two random accesses.

7. *Fast Page Cycle Time (tPC):* The tPC is the time required to perform a write or read operation from a DRAM location on a page and the recovery time needed to prepare for the next operation.

8. *Random Access Time (tRAC):* The tRAC is the time required to read any random DRAM cell. This is the maximum time required to select a row and a column address of the DRAM, sense the charge on the selected DRAM cell, and send information to the data output [6].

9. *Column Access Time or Page Access Time (tCAC):* The tCAC is the time required to get data at the output [6]. This is the maximum time required for selection of column address, sensing the stored data, and then transferring data to the output.

10. *Retention Time:* The retention time of a DRAM is the amount of time for which the DRAM can safely store the data without being refreshed [51].

11. *Read-Modify-Write Cycle Time:* The read-modify-write cycle time is the time needed to select a row address, select a column address, read data from the DRAM cell, write new data to the same DRAM cell, complete the write operation, and prepare for the next DRAM operations [6].

12. *DRAM Bandwidth:* The DRAM bandwidth is the rate at which data can be written to or read from the DRAM [90]. The bandwidth of the DRAM is expressed in terms of bytes per second. The bandwidth of the DRAM is calculated as follows:

$$\begin{aligned} \text{DRAM Total Bandwidth} = &\ \text{Base Clock Frequency} \times \text{Number of Lines per Clock} \\ & \times \text{Interface/Bus Width} \times \text{Number of Interfaces} \end{aligned} \quad (7.10)$$

The number of lines per clock of the DRAM is two for DDR DRAM. The DRAM interface/bus width is typically 64 bit. The number of interfaces is two for dual-channel mode interfaces or bus.

7.4 Twin-Transistor Random-Access Memory

The transistor-level schematic of a capacitor-less twin transistor RAM (TTRAM) is presented in Fig. 7.38 [15, 61]. The TTRAM is built by two serial transistors. One is called storage transistor and the other is called access transistor. Each one of the two transistors has floating body structure. They are isolated from each other by full trench isolation. The floating body of the storage transistor is the storage node. The TTRAM stores data "0" as low voltage and "1" as high voltage of the storage node and it distinguishes the two states as current differences due to threshold voltage difference. The floating body voltage of the storage transistor changes the state of the storage data. The access transistor controls the write and read operation of the TTRAM. During the write operation, the access transistor controls the purge node. During the read operation, the access transistor limits the excessive current of the storage node.

The TTRAM has four types of operations as follows [15, 61]: (1) write operation, (2) hold operation, (3) read operation, and (4) refresh operation. For the write operation, the Bitline is discharged to 0 V and the Wordline is precharged to $\left(\frac{V_{DD}}{2}\right)$. The storage node is written as 0 V. The charge line coupling with the gate ensures storage of the written data. Both the floating nodes,

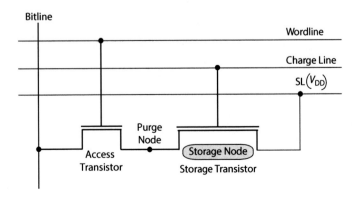

Figure 7.38 Transistor-level schematic of a capacitor-less TTRAM [15, 61].

the purge node and storage node, are raised to V_{DD} by capacitive coupling. In essence, two storage states are realized in TTRAM. The hold state of the TRAM is similar to the "restore operation" of the classical DRAM. For the read operation of the TTRAM, the charge line is asserted to V_{DD} and the read current flows to the Bitline. The data is identified as "1" or "0" as a difference of the read current through the serial transistors. The refresh operation of the TTRAM cell is performed by using a write operation right after the read operation.

All the operations of the TTRAM cell are possible by using $\left|\frac{V_{DD}}{2}\right|$ and V_{DD}, and complicated voltage control is not required. The size of a TTRAM cell is smaller than a classical 1T-1C DRAM cell. TTRAM is quite useful for large memories for high-speed embedded applications. The TTRAM has two additional lines as compared to the classical 1T-1C DRAM. These additional lines may make the design difficult and lead to additional parasitics as compared to 1T-1C DRAM. However, the dynamic power dissipation of TTRAM is reduced as the bitline voltage is lowered to the $\left|\frac{V_{DD}}{2}\right|$.

7.5 Thyristor Random-Access Memory

The thyristor RAM (TRAM) is a volatile memory that is a tradeoff between classical SRAM and DRAM [37, 66, 80, 82]. SRAM provides high performance but is larger in size, in other words, low-density. DRAM has low performance but is smaller in size, in other words high-density. TRAM provides performance and density tradeoffs. The data storage in DRAM is leaky, which requires regular refreshing. The read operation is destructive due to the use of passive capacitor without any internal gain. The DRAM requires write-back operation for every read and refreshes operation. Thus, the use of passive capacitor in DRAM limits the performance of the DRAM. The TRAM uses a bistable storage mechanism to provide nondestructive read operations and is self-refreshing. A thin capacitively coupled thyristor (TCCT) that provides negative differential resistance (NDR) is the key element. The bistable storage mechanism is due to gain resulting from the cross-coupled PNP-NPN bipolar junction transistors (BJTs). The transistor-level schematic of a TRAM cell is presented in Fig. 7.39 [37, 66, 80, 82]. TRAM has an additional wordline as compared to a 1T-1C DRAM cell.

The TRAM cell, shown in Fig. 7.39, has two elements, a TCCT and an access transistor [37, 80, 82]. The equivalent model of the TCCT is a pair of PNP and NPN BJTs that provide bistable characteristics. The gate of the TCCT provides capacitive coupling to the base of NPN BJT for high-speed switching of TCCT during write operation. For write or read operations, the Wordline-1 is asserted high. For write operation, the Wordline-2 is asserted to V_{DD}. At this point, the Bitline is biased at V_{DD} for "1" and a low voltage for "0". For read operation, the Wordline-2 is charged to a low voltage and the Bitline is precharged to ground voltage. If the TRAM has "1" stored, then TCCT is ON and the Bitline is pulled up by TCCT. On the other hand, if the TRAM has "0" stored, then TCCT is OFF and the Bitline stays at ground voltage.

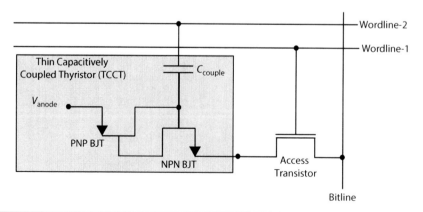

Figure 7.39 Transistor-level schematic of a TRAM [37, 66, 80, 82].

7.6 Read-Only Memory

The ROM is nonvolatile memory that retains the stored data even after the power supply is disconnected [4]. The data in ROM cannot be modified or is modified with difficulty. The ROM is used to store bootstrapping firmware and control firmware needed in various electronic components including TFT screen, hard drives, printers, graphics cards, and optical drives. The various types of ROM depending on programmable capabilities are discussed in the following subsections.

7.6.1 Programmable Read-Only Memory

The programmable ROM (PROM) is a digital memory in which the data storage is hardcoded by a fuse or antifuse [4, 104]. PROM also thus stores data permanently like a typical ROM, but the programming is applied much later. Typically, PROM is fabricated with blank information but is programmed or written at wafer, final test, or system depending on the technology used. A PROM programmer is used to program or write a PROM that uses a high voltage to permanently destroy or create internal links in the chip depending on fuses or antifuses technology. The PROM is also known as field PROM (FPROM) or one-time programmable nonvolatile memory (OTP NVM). The schematic representation of a PROM array is presented in Fig. 7.40 [4, 104]. At the array-level of the PROM, two different types of ROM architectures, NOR and NAND, will be discussed in a later part of the section. PROMs are used in many electronic systems including mobile phones, radiofrequency identification (RFID) tags, and video game consoles.

7.6.2 Erasable Programmable Read-Only Memory

The erasable PROM (EPROM) is a nonvolatile memory that can be programmed/written more often as compared to PROM or ROM that are programmed/written only once [2, 4]. EPROM is erased by exposing it to strong ultraviolet (UV) light typically for longer than 10 minutes and then rewritten by a method that needs much higher voltage than the typical supply voltage (V_{DD}). Repeated exposure to UV light causes some wear out of the EPROM, but the endurance of most EPROM chips is more than 1000 cycles of programming. The basic building block of the EPROM is the floating gate transistor as presented in Fig. 7.41. The figure also depicts the NAND and NOR architectures of the EPROM array.

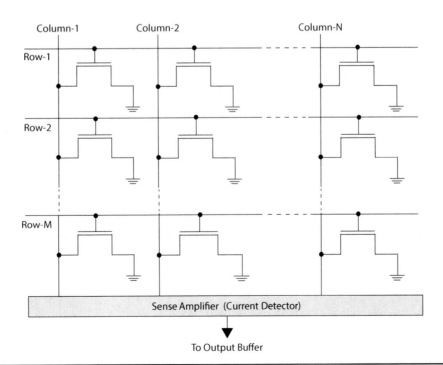

FIGURE 7.40 Schematic representation of PROM [4].

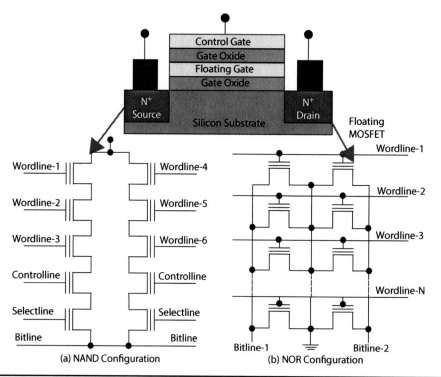

Figure 7.41 Schematic representation of EPROM [4].

The floating gate transistor, shown in Fig. 7.41, has an additional gate as compared with the classical MOS transistor [4]. As is typical for any MOS transistor, the floating gate transistor has a silicon substrate with source and drain regions. A layer of oxide is deposited over the silicon substrate. At the top of this oxide layer, a conductive silicon or aluminum gate electrode is deposited, which is the floating gate. A thicker oxide is deposited at the top of the floating gate. In essence, the floating gate is not electrically connected to other parts of the circuit and is completely insulated by the surrounding layers of oxide. The last layer in the gate structure is the control gate electrode. Hot electron injection is used to program or write to the floating gate.

For the purpose of writing or programming, high voltage is applied to gate and drain connection of a specific floating gate transistor. As a result, a large amount of electrons are released from the source that causes a large drain current flow. The large voltage at the gate attracts electrons that then penetrate the thin gate oxide and are stored on the floating gate. EPROM needs V_{DD} for read operations but a second power supply (V_{PP}) of much higher voltage for writing or programming operations. For example, if V_{DD} is of the order of 5 V then V_{PP} is of the range of 20 V. The writing or programming of the EPROM has the following steps [2]: (1) The V_{DD} and V_{PP} are set to their "program mode" levels. (2) The address and input data are applied at the address and data pins. (3) The required programming pulses are applied using a programming algorithm.

The EPROM can have two architectures, NAND and NOR, as shown in 7.41(a) and (b), respectively. The NAND architecture is very compact. However, as the transistors are in series, the sense amplifier notes a weaker signal. The weaker signal leads to slowing down of the NAND architecture-based EPROM array. The NOR architecture of EPROM does not have the signal strength issue such as NAND architecture. However, NOR EPROM uses one metal to diffusion contact per two cells that makes it larger compared to NAND EPROM. NOR architecture is typically more popular as compared with NAND architecture.

7.6.3 Electrically Erasable Programmable Read-Only Memory

The electrically erasable PROM (EEPROM) is a nonvolatile memory such as EPROM, PROM, or basic ROM [4, 70]. EEPROM is also known as E^2PROM or double-e PROM. EEPROM is similar to EPROM, but EEPROM allows its entire contents or selected banks to be electrically erased and

ROM Characteristic	EPROM	EEPROM
Storing Mechanism	Hot Electron Injection	Fowler-Nordheim Tunneling
Erasing Mechanism	UV Light	Electrical Signal
Basic Element	1 Transistor Cell	2 Transistor Cell
Oxide Layer	Thicker	Thinner
Area	Smaller	Larger
Programming Voltage (V_{PP})	External	Internal
Current during Programming (I_{PP})	Yes	No

TABLE 7.8 A Comparative Perspective of EPROM and EEPROM [4]

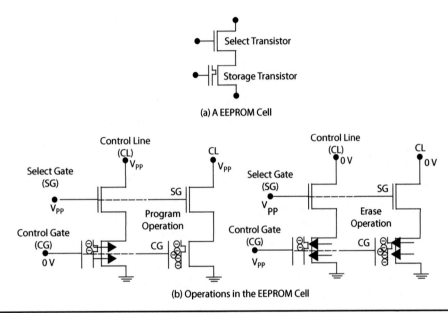

FIGURE 7.42 Schematic representation of EEPROM [4].

programmed. The writing of EEPROM is quite slow as erase and writing operations are performed on a byte basis. EEPROM that uses Fowler-Nordheim tunneling for storage needs much higher voltage than the usual supply voltage for writing. A comparative perspective of EPROM and EEPROM is presented in Table 7.8 [4].

The schematic representation of an EEPROM cell along with its programing/writing and erase operations is shown in Fig. 7.42 [4]. An EEPROM cell, as presented in Fig. 7.42(a), is composed of two transistors [4, 70]. The storage transistor has a floating gate that is similar to the EPROM storage transistor. The storage transistor will trap electrons for storage purposes. The other transistor, called access or select transistor, is needed for the erase operation of the EEPROM cell. The programming/writing as well as erase operations of the EEPROM are shown in Fig. 7.42(b). For write/programming operation of the EEPROM, control line (CL) and select gate (SG) are applied with a programming voltage (V_{PP}). For erase operation of the EEPROM, control line (CL) is applied with a 0 V and SG is applied with a programming voltage (V_{PP}). The erase operation of EEPROM happens when the electrons are trapped in the floating cell. On the other hand, the erase operation in EPROM occurs when the electrons are removed from the floating gate.

7.7 Flash Memory

The flash memory is a type of a nonvolatile RAM (NVRAM) that is a mix of EPROM and EEPROM [3, 27, 39, 45, 70, 73, 76]. An EPROM cell that consists of one transistor is programmed using channel hot electron injection and erased through UV light. An EEPROM cell that consists of two transistors

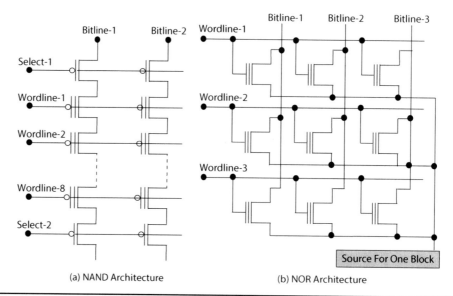

FIGURE 7.43 Schematic representation of flash memory architecture: NAND versus NOR [3].

is written/programmed and erased using Fowler-Nordheim tunneling. The flash memory is written/programmed and erased electrically but is composed of single transistors. The name "flash" signifies that a large chunk of memory can be erased at a time [3, 73]. Thus, flash memory is high performance, high endurance, and low area. Flash memory is thus in the process of replacing various types of memory such as EPROM, EEPROM, DRAM, and hard disks.

The flash memory has the following four architectures: (1) NAND, (2) NOR, (3) divided bitline NOR (DINOR), and (4) AND [3, 39, 45, 76]. The alternative flash memory array architectures are developed as a tradeoff between its silicon area and performance. The NAND and NOR architectures for the flash memory are shown in Fig. 7.43 [3]. NAND architecture has a compact design. NOR architecture has a larger area. DINOR and AND architectures are also smaller than the NOR architecture. NAND architecture is currently the most popular among the four due to high scalability, highly regular array, shallow source-drain junctions, self-aligned shallow trench isolation, and highly efficient contacts [38, 39, 50, 76, 103].

The flash memory cell consists of one floating gate transistor as in the case of EPROM cell. However, there are differences in the transistors of the flash memory and EPROM in terms of technology and geometry. In the flash memory, the gate oxide between the silicon substrate and the floating gate is thinner as compared to EPROM while similar to the tunneling oxide of the EEPROM. The source and drain diffusion areas are different in the floating gate transistor of the flash memory as compared with the EPROM. In general, the floating gate transistor of the flash memory has a larger transistor, thinner floating gate electrode, and thinner oxide.

7.8 Resistive Random-Access Memory

Resistive RAM (RRAM or ReRAM) is a nonvolatile memory technology that has potential to provide high-density memory at the cheapest possible cost per byte [11, 42]. It is a potential alternative for DRAM, SRAM, and NAND flash. The schematic representation of a RRAM array organization is shown in Fig. 7.44 [42]. The ReRAM can simply change resistance state with the applied voltage at its terminals and is more reliable and faster than other NVM options. In the following subsection, two types of resistive memory have been discussed.

7.8.1 Nonvolatile Resistive RAM for Storage

The schematic representation of a resistive memory cell is depicted in Fig. 7.45 [87, 94]. It has a ReRAM cell, an access transistor, a bitline, a wordline, and a sense amplifier. The diagram also presents

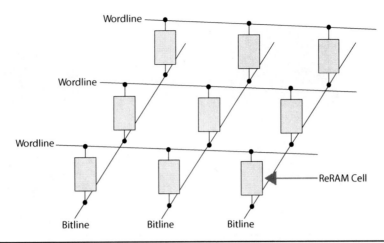

Figure 7.44 Schematic representation of RRAM array organization [42].

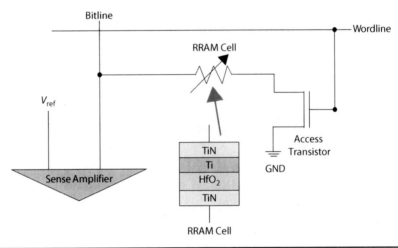

Figure 7.45 Schematic representation of RRAM or ReRAM [87, 94].

a specific resistive device. As a specific example, the ReRAM cell shown is a hafnium dioxide (HfO_2)-based resistive memory. The electrode of the ReRAM cell is made from titanium nitride (TiN). Titanium (Ti) is present below the top electrode. The Ti film has the capability to absorb oxygen atoms from buried HfO_2 [47]. The absorption of O_2 atoms by the Ti film changes its resistance. The change in the resistance is caused by the trapping and detrapping of process in the HfO_2. The traps are typically oxygen vacancies induced by the oxygen gettering of reactive Ti film. The Ti/TiN bilayer-based ReRAM cell has better performance than other similar structures.

7.8.2 Conductive Metal-Oxide Memory

The conductive metal oxide (CMOX) is a nonvolatile resistive memory (Fig. 7.46) [25, 56, 111]. The CMOX memory stores data in terms of moving ions rather than electric charges. It has two electrodes, a top electrode and a bottom electrode. The CMOX memory cell is formed by a combination of CMOX and insulating metal oxide adjacent to each other. The insulating metal oxide functions as a variable height tunnel barrier. The resistance change in the CMOX memory is formed by the exchange of the oxygen ions between the CMOX and insulating metal oxide. The ion movement even at room temperature is facilitated by an increase of the ion mobility under high-electric field during program and erase operations of the CMOX memory cell. The barrier height varies due to the changes in the charge in the barrier as oxygen moves in and out of the tunnel barrier.

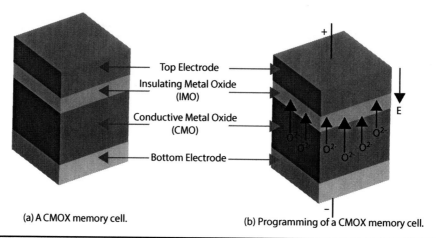

FIGURE 7.46 Operation of a CMOX memory cell [25, 56, 111].

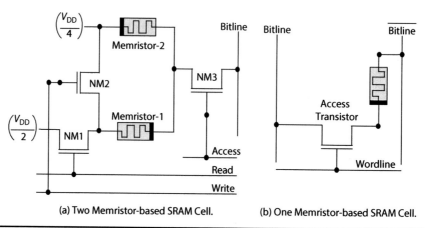

FIGURE 7.47 Schematic representation of memristor-based nonvolatile SRAM [57, 81, 83].

7.8.3 Memristor-Based Nonvolatile SRAM

The circuit level schematic of a memristor-based SRAM is presented in Fig. 7.47 [57, 81, 83]. Two different topologies are presented. A two memristor-based SRAM cell is presented in Fig. 7.47(a) [83]. On the other hand, one memristor-based SRAM cell is presented in Fig. 7.47(b) [57, 81]. The two memristor-based SRAM uses three transistors. The one memristor-based SRAM uses only one transistor for accessing purposes. Thus, the two memristor-based SRAM is very bulky compared with the one memristor-based SRAM.

The two memristor-based SRAM, shown in Fig. 7.47(a), uses the two memristors, Memristor-1 and Memristor-2, as storage elements. There are three transistors. Two transistors, NM1 and NM2, are pass transistors that help to connect the two memristors in a specific fashion. The third transistor, NM3, is used as an access transistor that is used to isolate the storage cell from neighboring cells. The two memristors are connected in parallel but in opposite polarity for write operations. For read operations, the two memristors are connected in series. The gate input of NM1 transistor is read. The gate input of NM2 transistor is write. The gate input of NM3 transistor is access. The access signal is the signal after performing the OR operation of read and write signals. For write operation, read is made low, write is made high, and access is made high. This connects Memristor-1 and Memristor-2 in parallel but with opposite polarity. For read operation, read is made high, write is made low, and access is made high. This connects the Memristor-1 and Memristor-2 in series.

The one memristor-based SRAM shown in Fig. 7.47(b) uses one memristor and one access transistor. It has one wordline and two bitlines, Bitline and $\overline{\text{Bitline}}$. In this cell, when Wordline is made high for read/write operation, the access transistor is ON and current can flow. On the other hand, when Wordline is made low the SRAM stays in its hold state and no current flows. As no power supply is needed to maintain the state of the memristor, the cell is a nonvolatile memory. For write operation, the Bitline and $\overline{\text{Bitline}}$ are supplied with data and its complement. The write current that flows through the memristor changes the state of the device. For read operation, a voltage sense amplifier is used that precharges the $\overline{\text{Bitline}}$ to high. The Bitline node is pulled down to low. At this point, when the Wordline is made high, the $\overline{\text{Bitline}}$ discharges through the memristor. At the end of read operation, the sense amplifier latches the read data to the output. The memory cell has reduced impact on state drift and has higher stability. This is achieved by restricting the read current to only go from the $\overline{\text{Bitline}}$ to Bitline, i.e., the OFF switching direction. The ON to OFF state transition of the memristor is slower than the OFF to ON state transition. The effect of the read operation is minimal when the memristor is in the OFF state.

7.9 Magnetic or Magnetoresistive Random-Access Memory

The magnetic/magnetoresistive RAM (MRAM) is a nonvolatile memory that stores data using magnetic polarization rather than electric charge [20, 72, 94, 112]. The MRAM uses tunneling resistance that depends on the directions of the magnetization of the ferromagnetic electrodes [20, 72, 94]. The key advantages of the MRAM include simple interfaces, compact size, wide-range operating temperature, superior soft error rate, and superior reliability [20, 72, 94]. The MRAM has been under development since the 1990s and has potential to replace many types of memory including SRAM, flash, and EEPROM.

A magnetic tunneling junction (MTJ) is the storage device in an MRAM. An MTJ consists of two electrodes that are ferromagnetic layers. The two layers sandwich a tunneling oxide layer. One of the ferromagnetic layers is called fixed layer or pinned layer as it is magnetically pinned so that its magnetization is fixed. The other ferromagnetic layer is called free layer as its magnetization is changed for storage purposes [48, 94]. MRAM can be either the field switching type or spin-transfer torque (STT) type depending on the writing technique (or magnetization of free layer) used in it. The switching-type MRAM uses external current-induced magnetic field for switching direction of the magnetization of the free layer during the write operation. For high-density MRAM, the field strength required for switching magnetization increases that leads to high-power dissipation. Thus, the field switching type MRAM that uses magnetic fields for controlling the free layer magnetization is not scalable. On the other hand, the STT-MRAM uses the direct injection of spin polarization current by MTJ to switch the magnetization of the free layer, thus requiring lesser write current. The STT-MRAM has simpler structure, low-power dissipation, and higher scalability and, hence, is a better choice for MRAM design.

The schematic representation of an MTJ that uses STT-MRAM is presented in Fig. 7.48 [48, 94]. There are two states in the MTJ depending on directions of the magnetization in the layers. The relative magnetization of the pinned layer and the fixed layer is sensed as resistance of the MTJ. In one scenario, the pinned/fixed layer and free layer are aligned in parallel to each other as depicted in Fig. 7.48(a). This is called low-resistance state with a resistance ($R_{\text{MTJ,P}}$). In another

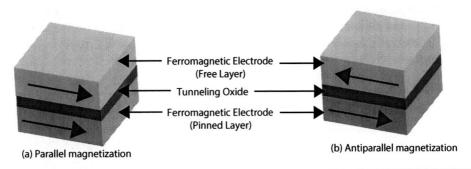

FIGURE 7.48 Operation of a STT-MRAM cell [48, 94].

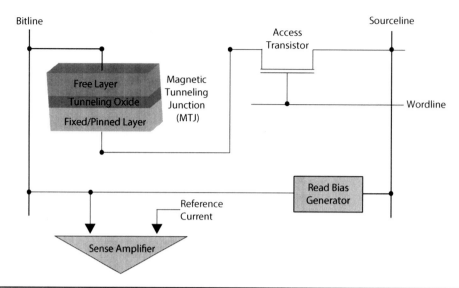

Figure 7.49 Transistor-level schematic of a STT-MRAM [48].

scenario, the pinned/fixed layer and free layer are aligned in antiparallel to each other as depicted in Fig. 7.48(b). This is called high-resistance state with a resistance ($R_{MTJ,AP}$). The tunnel magnetoresistance (TMR) defines the sensing margin between data 1 and data 0. The TMR ratio is defined as follows:

$$\text{TMR ratio} = \left(\frac{R_{MTJ,AP} - R_{MTJ,P}}{R_{MTJ,P}} \right) \tag{7.11}$$

TMR ratio of the MTJ is strongly affected by bias voltage and it reduces with the increase in the bias voltage. MTJ with high TMR ratio is preferred due to the enhanced noise margin that is essentially ($R_{MTJ,AP} - R_{MTJ,P}$).

The transistor-level schematic of a STT-MRAM cell that includes a MTJ that uses STT is presented in Fig. 7.49 [48]. It represents a one-transistor-one-MTJ (1T-1M) configuration that includes the following: (1) a spintronic-based MTJ, (2) an access transistor, (3) a read bias generator, (4) a sense amplifier, (5) a Bitline, (6) a Wordline, and (7) a Sourceline. The read operation of the STT-MRAM can be performed in two ways as follows: the parallel direction reading and the antiparallel direction reading [48]. In the parallel direction read operation of the STT-MRAM, a small bias is applied at Bitline and the Sourceline is grounded. In this condition, when the Wordline is activated, current flows from Bitline to Sourceline. For the antiparallel direction read operation of the STT-MRAM, the polarity of the voltages applied to the Bitline and Sourceline are swapped, i.e., a small bias is applied at Sourceline and the Bitline is grounded. The current then flows from Sourceline to Bitline. The write operation of the STT-MRAM needs much higher current than the read operation and, hence, is more power consuming than the read operation. A bidirectional current flow is required for write operation due to the hysteresis characteristic of the MTJ. For example, for writing a "0," the Bitline is precharged to V_{DD} and the Sourceline is discharged to ground. When the current of the STT-MRAM cell is above the switching threshold current of the MTJ, the magnetization direction of the free layer is changed from antiparallel to parallel.

7.10 Phase-Change RAM

The phase-change RAM (PRAM or PCRAM) is a nonvolatile memory that stores data using phase-change materials as compared to electrical charges of classical SRAM and DRAM technology [26, 33, 45, 94]. The basic operational structure of a PCRAM that shows phase-change of the

Memory in the AMS-SoCs 357

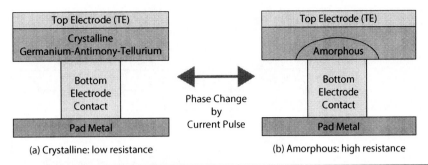

FIGURE 7.50 Operation of a PRAM or PCRAM cell [94].

materials is depicted in Fig. 7.50 [46, 94]. The electrodes are made using highly conducting titanium aluminum nitride (TiAlN) layers. A cylindrical resistive metal nitride electrode is deposited on the lower electrode that forms the bottom electrode contact, thereby, essentially forming a heater. The phase-change material that actually serves as the storage element is the germanium-antimony-tellurium ($Ge_2Sb_2Te_5$ or GST) chalcogenide alloy.

The PCRAM uses two phases of the GST alloy for the storage purposes. The amorphous phase provides high resistance and is the reset state. On the other hand, the crystalline phase offers low resistance and is the set state. The phase of the GST material is dependent on the joule heating and cooling rate. The joule heat is generated by an electrical current. The resistance ratio between the reset and set states is in the range of 100 to 10,000, which provides a large margin for sensing. The melting of the chalcogenide alloy by using joule heating and then quenching the molten state forms stable amorphous state. For a stable crystalline phase, enough heat is applied and time is given to rearrange the chalcogenide alloy atoms. The key characteristics of the PCRAM are the following [26, 33]:

1. A high resistance in the amorphous state. This is the OFF state or reset state.
2. A low resistance in the crystalline state. This is the ON state or set state.
3. A change in the conductivity occurs when the input voltage exceeds a threshold voltage and essentially a negative differential conductivity is exhibited.
4. The crystallization input pulse needed is of lower amplitude and longer duration than the amorphization input pulse. It may be noted that crystallization is the conversion of an amorphous material into a crystalline, whereas amorphization is the conversion of a crystalline material into an amorphous one.

For read operation of the PCRAM, a small voltage is applied across the electrodes. There is a large difference between the resistances of the reset state and the set state of the PCRAM. This leads to change in the current that is sensed for data. The read voltage is just sufficient to provide a measurable current while it remains low enough to avoid any write disturbance [33]. There are two kinds of write operations in the PCRAM: reset and set. The set operation switches the GST into the crystalline phase. During set operation, the GST is crystallized by heating it above its crystallization temperature, but below the melting temperature. The reset operation switches the GST into the amorphous phase. For reset operation, the GST is melted and quenched to make amorphous GST. The set and reset operations are controlled by electrical current that generate the required joule heat. For set operation, the moderate power but longer duration pulses are applied, whereas for reset operation high-power pulses are applied.

At present, various types of nonvolatile memories are available as discussed in the previous sections including flash, ReRAM, MRAM, and PCRAM. The memory operations that sense resistance change can have better design options than the charge-based memories such as DRAM or SRAM. The charge-based memory can suffer from charge loss, leakage current, and hot carrier issues that are eliminated in the resistance sensing-based memory. PCRAM has

Characteristics	NAND-Flash	ReRAM	MRAM	PCRAM
Cell Size	$\approx 5F^2$	$\approx 8F^2$	$\approx 8F^2$	$\approx 4F^2$
Density	xxGigabit		xxMegabit	xGigabit
Latency	20-200 μs	$\approx$ 50 ns–1 μs	$\approx$ 50 ns	$\approx$ 200 ns
Bandwidth	$\approx$100 MBps	$\approx$100 MBps	$\approx$1 GBps	$\approx$100 MBps
Endurance	$>10^5$	$>10^5$	$>10^{15}$	$>10^6$–10^{12}
Retention	>10 years	>10 years	>10 years	>10 years
Application	Data Storage	Data Storage	Working NVRAM	Code/Buffer Memory

TABLE 7.9 A Comparative Perspective of Selected Nonvolatile Memory [33, 46, 94]

many other interesting characteristics such as compact size, better scalability, faster access, reasonably high write speed, good retention time, excellent endurance, and erase-less programming. A comparative perspective of various nonvolatile memory is presented in Table 7.9 [33, 46, 94].

7.11 Questions

7.1 Discuss the various types of memories that are encountered in day-to-day applications.

7.2 Discuss the memory hierarchy in a typical computing system.

7.3 Discuss various types of volatile memories along with the advantages and disadvantages of each.

7.4 Discuss various types of nonvolatile memories along with the advantages and disadvantages of each.

7.5 Discuss various types of memories based on access type.

7.6 Discuss various types of SRAM.

7.7 Perform functional simulation of three different topologies of SRAM cells, for example, 6T-SRAM, 4T-SRAM, and single-ended 7T-SRAM, using a nanoscale CMOS technology.

7.8 Consider a working SRAM cell from the above. Now perform design of a small SRAM array using this cell and verify functionality of the array using simulations.

7.9 Consider a specific topology of an SRAM cell, for example, 6T-SRAM, 4T-SRAM, or single-ended 7T-SRAM. Perform functional simulations of this SRAM cell using three different nanoscale CMOS technology.

7.10 Perform the design and simulation of the FinFET-based 8T-SRAM.

7.11 Perform the design and simulation of the TFET-based 8T-SRAM.

7.12 Discuss five different performance metrics of SRAM.

7.13 Consider a specific topology of an SRAM cell, for example, 6T-SRAM, 4T-SRAM, or single-ended 7T-SRAM. Characterize its SNM using a butterfly curve for a specific CMOS technology.

7.14 Consider a specific topology of an SRAM cell, for example, 6T-SRAM, 4T-SRAM, or single-ended 7T-SRAM. Characterize its SNM using an N-curve for a specific CMOS technology.

7.15 Consider a TFET-based 8T-SRAM cell. Characterize its SNM using a butterfly curve and N-curve.

7.16 Consider a specific topology of an SRAM cell, for example, 6T-SRAM, 4T-SRAM, or single-ended 7T-SRAM. Characterize its power dissipation for write, read, and hold states for a specific CMOS technology.

7.17 Discuss the differences between SRAM and DRAM at the cell and architecture level.

7.18 Discuss different types of DRAM.

7.19 Perform functional simulation of three types of DRAM topologies.

7.20 Discuss various characteristics of the DRAM.

7.21 Consider the design of a TTRAM. Perform the functional simulation of a cell and a small array using it.

7.22 Consider the design of a TRAM. Perform the functional simulation of a cell and a small array based on it.

7.23 Discuss the different types of ROM with advantages and disadvantages of each.

7.24 Consider the design of a flash memory. Perform the functional simulation of a cell and a small array based on it.

7.25 Consider the design of a ReRAM. Perform the functional simulation of a cell and a small array based on it.

7.26 Consider the design of an MRAM. Perform the functional simulation of a cell and a small array based on it.

7.27 Consider the design of a PRAM. Perform the functional simulation of a cell and a small array based on it.

7.12 References

1. DRAM Technology. http://smithsonianchips.si.edu/ice/cd/MEMORY97/SEC07.PDF. Accessed on 26th October 2013.
2. Erasable Programmable Read-Only Memory (EPROM). http://www.siliconfareast.com/eprom.htm. Accessed on 10th November 2013.
3. Flash Technology. http://smithsonianchips.si.edu/ice/cd/MEM96/SEC10.pdf. Accessed on 17th September 2013.
4. ROM, EPROM, & EEPROM Technology. http://smithsonianchips.si.edu/ice/cd/MEM96/SEC09.pdf. Accessed on 17th September 2013.
5. SRAM Technology. http://smithsonianchips.si.edu/ice/cd/MEM96/SEC08.pdf. Accessed on 26th October 2013.
6. Understanding DRAM Specifications. http://www.cs.albany.edu/~sdc/CSI404/dramperf.pdf (1996). Accessed on 28th October 2013.
7. Direct Rambus DRAM (DRDRAM). http://www.pcguide.com/ref/ram/techDRDRAM-c.html (2001). Accessed on 13th October 2013.
8. Synchronous Graphics RAM (SGRAM). http://www.pcguide.com/ref/video/techSGRAM-c.html (2001). Accessed on 02nd November 2013.
9. Video RAM (VRAM). http://www.pcguide.com/ref/video/techVRAM-c.html (2001). Accessed on 02nd November 2013.
10. Window RAM (WRAM). http://www.pcguide.com/ref/video/techWRAM-c.html (2001). Accessed on 02nd November 2013.
11. Akinaga, H., Shima, H.: Resistive Random Access Memory (ReRAM) Based on Metal Oxides. *Proceedings of the IEEE* **98**(12), 2237–2251 (2010). DOI 10.1109/JPROC.2010.2070830.
12. Amelifard, B., Fallah, F., Pedram, M.: Reducing the Sub-threshold and Gate-tunneling Leakage of SRAM Cells using Dual-V_t and Dual-T_{ox} Assignment. In: *Proceedings of the Design Automation and Test in Europe*, pp. 1–6 (2006).
13. Arandilla, C.D.C., Alvarez, A.B., Roque, C.R.K.: Static Noise Margin of 6T SRAM Cell in 90-nm CMOS. In: *Proceedings of the UKSim 13th International Conference on Computer Modelling and Simulation*, pp. 534–539 (2011). DOI 10.1109/UKSIM.2011.108.
14. Paz de Araujo, C., McMillan, L., Joshi, V., Solayappan, N., Lim, M., Arita, K., Moriwaki, N., Hirano, H., Baba, T., Shimada, Y., Sumi, T., Fujii, E., Otsuki, T.: The Future of Ferroelectric Memories. In: *Digest of Technical Papers IEEE International Solid-State Circuits Conference*, pp. 268–269 (2000). DOI 10.1109/ISSCC.2000.839779.
15. Arimoto, K., Morishita, F., Hayashi, I., Dosaka, K., Shimano, H., Ipposhi, T.: A High-Density Scalable Twin Transistor RAM (TTRAM) with Verify Control for SOI Platform Memory IPs. *IEEE Journal of Solid-State Circuits* **42**(11), 2611–2619 (2007). DOI 10.1109/JSSC.2007.907185.
16. Atwood, G., Chae, S.I., Shim, S.S.Y.: Next-Generation Memory. *Computer* **46**(8), 21–22 (2013). DOI 10.1109/MC.2013.285.
17. Balwant Raj, A.S., Singh, G.: Analysis of Power Dissipation in DRAM Cells Design for NanoScale Memories. *International Journal of Information Technology and Knowledge Management* **2**, 371–374 (2009).
18. Bassin, C., Fazan, P., Xiong, W., Cleavelin, C.R., Schulz, T., Schruefer, K., Gostkowski, M., Patruno, P., Maleville, C., Nagoga, M., Okhonin, S.: Retention Characteristics of Zero-Capacitor Ram (Z-RAM) Cell Based On FinFET and Tri-Gate Devices. In: *Proceedings of the IEEE International SOI Conference*, pp. 203–204 (2005). DOI 10.1109/SOI.2005.1563588.
19. Blomster, K.A.: *Schemes For Reducing Power and Delay in SRAMs*. Master's thesis, School of Electrical Engineering and Computer Science, Washington State University (2006).
20. Butler, W.H., Zhang, X.G., Schulthess, T.C., MacLaren, J.M.: Spin-Dependent Tunneling Conductance of Fe|MgO|Fe Sandwiches. *Physical Review B* **63**, 054, 416 (2001). DOI 10.1103/PhysRevB.63.054416. URL http://link.aps.org/doi/10.1103/PhysRevB.63.054416.

21. Calhoun, B.H., Chandrakasan, A.P.: Static Noise Margin Variation for Sub-threshold SRAM in 65-nm CMOS. *IEEE Journal of Solid-State Circuits* **41**(7), 1673–1679 (2006). DOI 10.1109/JSSC.2006.873215.
22. Carlson, I., Andersson, S., Natarajan, S., Alvandpour, A.: A High Density, Low Leakage, 5T SRAM for Embedded Caches. In: *Proceeding of the 30th European Solid-State Circuits Conference*, pp. 215–218 (2004). DOI 10.1109/ESSCIR.2004.1356656.
23. Chang, L., Montoye, R.K., Nakamura, Y., Batson, K.A., Eickemeyer, R.J., Dennard, R.H., Haensch, W., Jamsek, D.: An 8T-SRAM for Variability Tolerance and Low-Voltage Operation in High-Performance Caches. *IEEE Journal of Solid-State Circuits* **43**(4), 956–963 (2008). DOI 10.1109/JSSC.2007.917509.
24. Chen, Y.N., Fan, M.L., Hu, V.H., Su, P., Chuang, C.T.: Design and Analysis of Robust Tunneling FET SRAM. *IEEE Transactions on Electron Devices* **60**(3), 1092–1098 (2013). DOI 10.1109/TED.2013.2239297.
25. Chevallier, C.J., Siau, C.H., Lim, S.F., Namala, S., Matsuoka, M., Bateman, B., Rinerson, D.: A 0.13 μm 64 Mb Multi-Layered Conductive Metal-Oxide Memory. In: *Digest of Technical Papers IEEE International Solid-State Circuits Conference*, pp. 260–261 (2010). DOI 10.1109/ISSCC.2010.5433945.
26. Cobley, R.A., Wright, C.D.: Parameterized SPICE Model for a Phase-Change RAM Device. *IEEE Transactions on Electron Devices* **53**(1), 112–118 (2006). DOI 10.1109/TED.2005.860642.
27. Crippa, L., Micheloni, R., Motta, I., Sangalli, M.: Nonvolatile Memories: NOR vs. NAND Architectures. In: R. Micheloni, G. Campardo, P. Olivo (eds.) *Memories in Wireless Systems, Signals and Communication Technology*, pp. 29–53. Springer Berlin Heidelberg (2008). DOI 10.1007/978-3-540-79078-5_2. URL http://dx.doi.org/10.1007/978-3-540-79078-5_2.
28. Crisp, R.: Direct RAMbus Technology: The New Main Memory Standard. *IEEE Micro* **17**(6), 18–28 (1997). DOI 10.1109/40.641593.
29. Cuppu, V., Jacob, B., Davis, B., Mudge, T.: A Performance Comparison of Contemporary DRAM Architectures. In: *Proceedings of the 26th International Symposium on Computer Architecture*, pp. 222–233 (1999).
30. DeMone, P.: Direct Rambus Memory. http://www.realworldtech.com/rambus-basics/ (1999). Accessed on 13th October 2013.
31. Diodato, P.W., Clemens, J.T., Troutman, W.W., Lindenberger, W.S.: A Reusable Embedded DRAM Macrocell. In: *Proceedings of the IEEE Custom Integrated Circuits Conference*, pp. 337–340 (1997). DOI 10.1109/CICC.1997.606642.
32. Dong, W., Li, P., Huang, G.M.: SRAM dynamic stability: Theory, Variability and Analysis. In: *Proceedings of IEEE/ACM International Conference on Computer-Aided Design*, pp. 378–385 (2008). DOI 10.1109/ICCAD.2008.4681601.
33. Dong, X., Jouppi, N., Xie, Y.: PCRAMsim: System-Level Performance, Energy, and Area Modeling for Phase-Change RAM. In: *Digest of Technical Papers IEEE/ACM International Conference on Computer-Aided Design*, pp. 269–275 (2009).
34. Fan, M.L., Wu, Y.S., Hu, V.H., Su, P., Chuang, C.T.: Investigation of Static Noise Margin of FinFET SRAM Cells in Sub-Threshold Region. In: *Proceedings of the IEEE International SOI Conference*, pp. 1–2 (2009). DOI 10.1109/SOI.2009.5318785.
35. Grossar, E., Stucchi, M., Maex, K., Dehaene, W.: Read Stability and Write-Ability Analysis of SRAM Cells for Nanometer Technologies. *IEEE Journal of Solid-State Circuits* **41**(11), 2577–2588 (2006). DOI 10.1109/JSSC.2006.883344.
36. Guo, Z., Balasubramanian, S., Zlatanovici, R., King, T.J., Nikolic, B.: FinFET-based SRAM Design. In: *Proceedings of the International Symposium on Low Power Electronics and Design*, pp. 2–7 (2005). DOI 10.1109/LPE.2005.195476.
37. Gupta, R., Nemati, F., Robins, S., Yang, K., Gopalakrishnan, V., Sundarraj, J., Chopra, R., Roy, R., Cho, H.J., Maszara, W.P., Mohapatra, N.R., Wuu, J., Weiss, D., Nakib, S.: 32 nm High-Density High-Speed T-RAM Embedded Memory Technology. In: *Proceedings of the IEEE International Electron Devices Meeting (IEDM)*, pp. 12.1.1–12.1.4 (2010). DOI 10.1109/IEDM.2010.5703345.
38. Haque Chowdhury, M.A., Kimy, K.H.: A Survey of Flash Memory Design and Implementation of Database in Flash Memory. In: *Proceedings of the 3rd International Conference on Intelligent System and Knowledge Engineering*, vol. 1, pp. 1256–1259 (2008). DOI 10.1109/ISKE.2008.4731123.
39. Harari, E.: Flash Memory – The Great Disruptor! In: *IEEE International Solid-State Circuits Conference – Digest of Technical Papers (ISSCC)*, pp. 10–15 (2012). DOI 10.1109/ISSCC.2012.6176930.
40. Hong, S.I., McKee, S.A., Salinas, M.H., Klenke, R.H., Aylor, J.H., Wulf, W.A.: Access Order and Effective Bandwidth for Streams on a Direct Rambus Memory. In: *Proceedings of Fifth International Symposium On High-Performance Computer Architecture*, pp. 80–89 (1999). DOI 10.1109/HPCA.1999.744337.
41. Hu, Z., Juang, P., Diodato, P., Kaxiras, S., Skadron, K., Martonosi, M., Clark, D.W.: Managing Leakage For Transient Data: Decay and Quasi-Static 4T Memory Cells. In: *Proceedings of the International Symposium on Low Power Electronics and Design*, pp. 52–55 (2002). DOI 10.1109/LPE.2002.146708.
42. Jung, M., Shalf, J., Kandemir, M.: Design of a Large-scale Storage-class RRAM System. In: *Proceedings of the 27th International ACM Conference on International Conference on Supercomputing, ICS'13*, pp. 103–114. ACM, New York, NY, USA (2013). DOI 10.1145/2464996.2465004. URL http://doi.acm.org/10.1145/2464996.2465004.
43. Karl, E., Wang, Y., Ng, Y.G., Guo, Z., Hamzaoglu, F., Bhattacharya, U., Zhang, K., Mistry, K., Bohr, M.: A 4.6 GHz 162 Mb SRAM Design in 22 nm Tri-Gate CMOS Technology with Integrated Active VMIN-Enhancing Assist Circuitry. In: *Proceedings of the IEEE International Solid-State Circuits Conference Digest of Technical Papers (ISSCC)*, pp. 230–232 (2012). DOI 10.1109/ISSCC.2012.6176988.

44. Keeth, B., Baker, R.J., Johnson, B., Lin, F.: DRAM Circuit Design: Fundamental and High-Speed Topics. IEEE Press Series on Microelectronic Systems. Wiley (2008). URL http://books.google.com/books?id=TgW3LTubREQC.
45. Kim, K.: Future Memory Technology: Challenges and Opportunities. In: *Proceedings of the International Symposium on VLSI Technology, Systems and Applications*, pp. 5–9 (2008). DOI 10.1109/VTSA.2008.4530774.
46. Lai, S., Lowrey, T.: OUM – A 180 nm Nonvolatile Memory Cell Element Technology for Stand Alone and Embedded Applications. In: *Technical Digest International Electron Devices Meeting*, pp. 36.5.1–36.5.4 (2001). DOI 10.1109/IEDM.2001.979636.
47. Lee, H.Y., Chen, P.S., Wu, T.Y., Chen, Y.S., Wang, C.C., Tzeng, P.J., Lin, C.H., Chen, F., Lien, C.H., Tsai, M.J.: Low Power and High Speed Bipolar Switching With a Thin Reactive Ti Buffer Layer in Robust HfO_2 based RRAM. In: *Proceedings of the IEEE International Electron Devices Meeting*, pp. 1–4 (2008). DOI 10.1109/IEDM.2008.4796677.
48. Li, J., Ndai, P., Goel, A., Salahuddin, S., Roy, K.: Design Paradigm for Robust Spin-Torque Transfer Magnetic RAM (STT MRAM) From Circuit/Architecture Perspective. *IEEE Transactions on Very Large Scale Integration Systems (TVLSI)* **18**(12), 1710–1723 (2010). DOI 10.1109/TVLSI.2009.2027907.
49. Li, Y., Lu, C.S.: Characteristic Comparison of SRAM Cells with 20 nm Planar MOSFET, Omega FinFET and Nanowire FinFET. In: *Proceedings of the Sixth IEEE Conference on Nanotechnology*, pp. 339–342 (2006). DOI 10.1109/NANO.2006.247646.
50. Li, Y., Quader, K.N.: NAND Flash Memory: Challenges and Opportunities. *Computer* **46**(8), 23–29 (2013). DOI 10.1109/MC.2013.190.
51. Liu, J., Jaiyen, B., Kim, Y., Wilkerson, C., Mutlu, O.: An Experimental Study of Data Retention Behavior in Modern DRAM Devices: Implications For Retention Time Profiling Mechanisms. In: *Proceedings of the 40th Annual International Symposium on Computer Architecture*, pp. 60–71. ACM, New York, NY, USA (2013). DOI 10.1145/2485922.2485928. URL http://doi.acm.org/10.1145/2485922.2485928.
52. Lohstroh, J.: Static and Dynamic Noise Margins of Logic Circuits. *IEEE Journal of Solid-State Circuits* **14**(3), 591–598 (1979). DOI 10.1109/JSSC.1979.1051221.
53. Luk, W.K., Cai, J., Dennard, R.H., Immediato, M.J., Kosonocky, S.V.: A 3-Transistor DRAM Cell with Gated Diode for Enhanced Speed and Retention Time. *Digest of Technical Papers Symposium on VLSI Circuits*, pp. 184–185 (2006).
54. Luk, W.K., Dennard, R.H.: A Novel Dynamic Memory Cell with Internal Voltage Gain. *IEEE Journal of Solid-State Circuits* **40**(4), 884–894 (2005).
55. Makosiej, A., Kashyap, R.K., Vladimirescu, A., Amara, A., Anghel, C.: A 32 nm Tunnel FET SRAM For Ultra Low Leakage. In: *Proceedings of the IEEE International Symposium on Circuits and Systems (ISCAS)*, pp. 2517–2520 (2012). DOI 10.1109/ISCAS.2012.6271814.
56. Meyer, R., Schloss, L., Brewer, J., Lambertson, R., Kinney, W., Sanchez, J., Rinerson, D.: Oxide Dual-Layer Memory Element For Scalable Non-Volatile Cross-Point Memory Technology. In: *Proceedings of the 9th Annual Non-Volatile Memory Technology Symposium*, pp. 1–5 (2008). DOI 10.1109/NVMT.2008.4731194.
57. Mohammad, B., Homouz, D., Elgabra, H.: Robust Hybrid Memristor-CMOS Memory: Modeling and Design. *IEEE Transactions on Very Large Scale Integration Systems (TVLSI)* **21**(11), 2069–2079 (2013). DOI 10.1109/TVLSI.2012.2227519.
58. Mohanty, S.P., Kougianos, E.: DOE-ILP Assisted Conjugate-Gradient Optimization of High-*k*/Metal-Gate Nano-CMOS SRAM. *IET Computers & Digital Techniques* **6**(4), 240–248 (2012).
59. Mohanty, S.P., Singh, J., Kougianos, E., Pradhan, D.K.: Statistical DOE-ILP Based Power-Performance-Process (P3) Optimization of Nano-CMOS SRAM. *Elsevier The VLSI Integration Journal* **45**(1), 33–45 (2012).
60. Moore, S.K.: Masters of Memory. *IEEE Spectrum* **44**(1), 45–49 (2007). DOI 10.1109/MSPEC.2007.273045.
61. Morishita, F., Noda, H., Gyohten, T., Okamoto, M., Ipposhi, T., Maegawa, S., Dosaka, K., Arimoto, K.: A Capacitorless Twin-Transistor Random Access Memory (TTRAM) on SOI. In: *Proceedings of the IEEE Custom Integrated Circuits Conference*, pp. 435–438 (2005). DOI 10.1109/CICC.2005.1568699.
62. Mormann, A.: Burst and Latency Requirements Drive EDO and BEDO DRAM Standards. In: *Digest of Papers – Technologies for the Information Superhighway*, pp. 356–359 (1996). DOI 10.1109/CMPCON.1996.501795.
63. Mukherjee, V., Mohanty, S.P., Kougianos, E., Allawadhi, R., Velagapudi, R.: Gate Leakage Current Analysis in READ/WRITE/ IDLE States of a SRAM Cell. In: *Proceedings of the IEEE Region 5 Conference*, pp. 196–200 (2006). DOI 10.1109/TPSD.2006.5507432.
64. Nalam, S., Calhoun, B.H.: Asymmetric Sizing in a 45 nm 5T SRAM to Improve Read Stability over 6T. In: *Proceeding of the IEEE Custom Integrated Circuits Conference*, pp. 709–712 (2009). DOI 10.1109/CICC.2009.5280733.
65. Nalam, S., Calhoun, B.H.: 5T SRAM With Asymmetric Sizing for Improved Read Stability. *IEEE Journal of Solid-State Circuits* **46**(10), 2431–2442 (2011). DOI 10.1109/JSSC.2011.2160812.
66. Nemati, F., Plummer, J.D.: A Novel Thyristor-Based SRAM Cell (T-RAM) for High-Speed, Low-Voltage, Giga-Scale Memories. In: *Technical Digest International Electron Devices Meeting (IEDM)*, pp. 283–286 (1999). DOI 10.1109/IEDM.1999.824152.
67. Noda, K., Matsui, K., Imai, K., Inoue, K., Tokashiki, K., Kawamoto, H., Yoshida, K., Takeda, K., Nakamura, N., Kimura, T., Toyoshima, H., Koishikawa, Y., Maruyama, S., Saitoh, T., Tanigawa, T.: A 1.9-μm^2 Loadless CMOS Four-Transistor Sram Cell In A 0.18-μm Logic Technology. In: *Technical Digest of International Electron Devices Meeting (IEDM)*, pp. 643–646 (1998). DOI 10.1109/IEDM.1998.746440.

68. Okhonin, S., Nagoga, M., Sallese, J.M., Fazan, P.: A Capacitor-Less 1T-DRAM Cell. *IEEE Electron Device Letters* **23**(2), 85–87 (2002). DOI 10.1109/55.981314.
69. Okobiah, O.: *Exploring Process-Variation Tolerant Design of Nanoscale Sense Amplifier Circuits*. Master's thesis, Department of Computer Science and Engineering, University of North Texas, Denton, TX (2010).
70. Owen, W.H., Tchon, W.E.: E²PROM Product Issues and Technology Trends. In: *Proceedings of the VLSI and Microelectronic Applications in Intelligent Peripherals and their Interconnection Networks*, pp. 1/17–1/19 (1989). DOI 10.1109/CMPEUR.1989.93334.
71. Pagiamtzis, K., Sheikholeslami, A.: Content-Addressable Memory (CAM) Circuits and Architectures: A Tutorial and Survey. *IEEE Journal of Solid-State Circuits* **41**(3), 712–727 (2006). DOI 10.1109/JSSC.2005.864128.
72. Parkin, S.S.P., Kaiser, C., Panchula, A., Rice, P.M., Hughes, B., Samant, M., Yang, S.H.: Giant Tunnelling Magnetoresistance at Room Temperature with MgO (100) Tunnel Barriers. *Nature Materials* **3**, 862–867 (2004). DOI 10.1038/nmat1256.
73. Pavan, P., Bez, R., Olivo, P., Zanoni, E.: Flash Memory Cells – An Overview. *Proceedings of the IEEE* **85**(8), 1248–1271 (1997). DOI 10.1109/5.622505.
74. Pavlov, A., Sachdev, M.: *CMOS SRAM Circuit Design and Parametric Test in Nano-Scaled Technologies: Process-Aware SRAM Design and Test*. Springer Science and Business Media B.V. (2008).
75. Philpy, S.T., Kamp, D.A., Derbenwick, G.F.: Nonvolatile and SDRAM Ferroelectric Memories For Aerospace Applications. In: *Proceedings of the IEEE Aerospace Conference*, vol. 4, pp. 2294–2299 (2004). DOI 10.1109/AERO.2004.1368023.
76. Quader, K.N.: Flash Memory at a Cross-Road: Challenges & Opportunities. In: *Proceedings of the 4th IEEE International Memory Workshop*, pp. 1–4 (2012). DOI 10.1109/IMW.2012.6213639.
77. Raj, B., Saxena, A.K., Dasgupta, S.: Nanoscale FinFET Based SRAM Cell Design: Analysis of Performance Metric, Process Variation, Underlapped FinFET, and Temperature Effect. *IEEE Circuits and Systems Magazine* **11**(3), 38–50 (2011). DOI 10.1109/MCAS.2011.942068.
78. Reddy, G.K., Jainwal, K., Singh, J., Mohanty, S.P.: Process Variation Tolerant 9T SRAM Bitcell Design. In: *Proceedings of the 13th International Symposium on Quality Electronic Design*, pp. 493–497 (2012)
79. Rodriguez, N., Cristoloveanu, S., Gamiz, F.: Novel Capacitorless 1T-DRAM Cell for 22-nm Node Compatible with Bulk and SOI Substrates. *IEEE Transactions on Electron Devices* **58**(8), 2371–2377 (2011). DOI 10.1109/TED.2011.2147788.
80. Roy, R., Nemati, F., Young, K., Bateman, B., Chopra, R., Jung, S.O., Show, C., Cho, H.J.: Thyristor-Based Volatile Memory in Nano-Scale CMOS. In: *Digest of Technical Papers IEEE International Solid-State Circuits Conference*, pp. 2612–2621 (2006). DOI 10.1109/ISSCC.2006.1696327.
81. Sakode, V., Lombardi, F., Han, J.: Cell Design and Comparative Evaluation of A Novel 1T Memristor-Based Memory. In: *Proceedings of the IEEE/ACM International Symposium on Nanoscale Architectures (NANOARCH)*, pp. 152–159 (2012).
82. Salling, C., Yang, K.J., Gupta, R., Hayes, D., Tamayo, J., Gopalakrishnan, V., Robins, S.: Reliability of Thyristor-Based Memory Cells. In: *Proceedings of the IEEE International Reliability Physics Symposium*, pp. 253–259 (2009). DOI 10.1109/IRPS.2009.5173259.
83. Sarwar, S.S., Saqueb, S.A.N., Quaiyum, F., Rashid, A.B.M.H.U.: Memristor-Based Nonvolatile Random Access Memory: Hybrid Architecture for Low Power Compact Memory Design. *IEEE Access* **1**, 29–34 (2013). DOI 10.1109/ACCESS.2013.2259891.
84. Seabaugh, A.: The Tunneling Transistor. *IEEE Spectrum* **50**(10), 35–62 (2013). DOI 10.1109/MSPEC.2013.6607013.
85. Seevinck, E., List, F.J., Lohstroh, J.: Static Noise Margin Analysis of MOS SRAM Cells. *IEEE Journal of Solid-State Circuits* **22**(5), 748–754 (1987).
86. Sheikholeslami, A., Gulak, P.G.: A Survey of Circuit Innovations in Ferroelectric Random-Access Memories. *Proceedings of the IEEE* **88**(5), 667–689 (2000). DOI 10.1109/5.849164.
87. Sheu, S.S., Cheng, K.H., Chang, M.F., Chiang, P.C., Lin, W.P., Lee, H.Y., Chen, P.S., Chen, Y.S., Wu, T.Y., Chen, F., Su, K.L., Kao, M.J., Tsai, M.J.: Fast-Write Resistive RAM (RRAM) for Embedded Applications. *IEEE Design & Test of Computers* **28**(1), 64–71 (2011). DOI 10.1109/MDT.2010.96.
88. Shin, C., Nikolic, B., Liu, T.J.K., Tsai, C.H., Wu, M.H., Chang, C.F., Liu, T.J.K., Kao, C.Y., Lin, G.S., Chiu, K.L., Fu, C.S., tzung Tsai, C., Liang, C.W.: Tri-Gate Bulk CMOS Technology for Improved SRAM Scalability. In: *Proceedings of the European Solid-State Device Research Conference (ESSDERC)*, pp. 142–145 (2010). DOI 10.1109/ESSDERC.2010.5618437.
89. Shino, T., Higashi, T., Fujita, K., Ohsawa, T., Minami, Y., Yamada, T., Morikado, M., Nakajima, H., Inoh, K., Hamamoto, T., Nitayama, A.: Highly Scalable FBC (Floating Body Cell) With 25 nm Box Structure for Embedded DRAM Applications. In: *Digest of Technical Papers Symposium on VLSI Technology*, pp. 132–133 (2004). DOI 10.1109/VLSIT.2004.1345435.
90. Shiva, S.G.: e-Study Guide for: Computer Organization, Design, and Architecture. Content Technology Inc. (2012). URL http://books.google.com/books?id=Jayapuotyg8C.
91. Singh, J., Mathew, J., Mohanty, S.P., Pradhan, D.K.: A Nano-CMOS Process Variation Induced Read Failure Tolerant SRAM Cell. In: *Proceedings of the IEEE International Symposium on Circuits and Systems*, pp. 3334–3337 (2008). DOI 10.1109/ISCAS.2008.4542172.

92. Singh, J., Mathew, J., Pradhan, D.K., Mohanty, S.P.: A Subthreshold Single Ended I/O SRAM Cell Design for Nanometer CMOS Technologies. In: *Proceedings of the IEEE International SoC Conference (SoCC)*, pp. 243–246 (2008).
93. Singh, J., Mohanty, S.P., Pradhan, D.K.: *Robust SRAM Designs and Analysis*. Springer Science and Business Media (2012).
94. Song, Y.J., Jeong, G., Baek, I.G., Choi, J.: What Lies Ahead for Resistance-Based Memory Technologies? *Computer* **46**(8), 30–36 (2013). DOI 10.1109/MC.2013.221.
95. Stein, K.U., Sihling, A., Doering, E.: Storage Array and Sense/refresh Circuit for Single-Transistor Memory Cells. *IEEE Journal of Solid-State Circuits* **7**(5), 336–340 (1972).
96. Tanaka, T., Yoshida, E., Miyashita, T.: Scalability Study on a Capacitorless 1T-DRAM: From Single-Gate PD-SOI To Double-Gate FinDRAM. In: *Technical Digest of IEEE International Electron Devices Meeting (IEDM)*, pp. 919–922 (2004). DOI 10.1109/IEDM.2004.1419332.
97. Tarigopula, S.: *A CAM Based High-Performance Classifier-Scheduler for a Video Network Processor*. Master's thesis, Department of Computer Science and Engineering, University of North Texas, Denton, TX (2007).
98. Thakral, G.: *Process-Voltage-Temperature Aware Nanoscale Circuit Optimization*. Ph.D. thesis, Computer Science and Engineering, University of North Texas, Denton (December 2010).
99. Thakral, G., Mohanty, S.P., Ghai, D., Pradhan, D.K.: A Combined DOE-ILP Based Power and Read Stability Optimization in Nano-CMOS SRAM. In: *Proceedings of the 23rd International Conference on VLSI Design (VLSID)*, pp. 45–50 (2010).
100. Thakral, G., Mohanty, S.P., Ghai, D., Pradhan, D.K.: A DOE-ILP Assisted Conjugate-Gradient Based Power and Stability Optimization in High-*k* Nano-CMOS SRAM. In: *Proceedings of the 20th ACM Great Lakes Symposium on VLSI (GLSVLSI)*, pp. 323–328 (2010).
101. Thakral, G., Mohanty, S.P., Ghai, D., Pradhan, D.K.: P3 (power-performance-process) optimization of nano-CMOS SRAM using statistical DOE-ILP. In: *Proceedings of the 11th International Symposium on Quality of Electronic Design (ISQED)*, pp. 176–183 (2010).
102. Thakral, G., Mohanty, S.P., Pradhan, D.K., Kougianos, E.: DOE-ILP Based Simultaneous Power and Read Stability Optimization in Nano-CMOS SRAM. *Journal Low Power Electronics* **6**(3), 390–400 (2010).
103. Thean, V.Y.A., Leburton, J.P.: Flash Memory: Towards Single-Electronics. *IEEE Potentials* **21**(4), 35–41 (2002). DOI 10.1109/MP.2002.1044216.
104. Thornton, C.: Overview Programmable Read-Only-Memories. In: *Digest of Technical Papers IEEE International Solid-State Circuits Conference*, pp. 180–181 (1977). DOI 10.1109/ISSCC.1977.1155709.
105. Tseng, Y.H., Zhang, Y., Okamura, L., Yoshihara, T.: A New 7-Transistor SRAM Cell Design with High Read Stability. In: *Proceedings of the International Conference on Electronic Devices, Systems and Applications (ICEDSA)*, pp. 43–47 (2010). DOI 10.1109/ICEDSA.2010.5503104.
106. Verma, N., Chandrakasan, A.P.: A 256 kb 65 nm 8T Subthreshold SRAM Employing Sense-Amplifier Redundancy. *IEEE Journal of Solid-State Circuits* **43**(1), 141–149 (2008). DOI 10.1109/JSSC.2007.908005.
107. Wang, D.T.: *Modern DRAM Memory Systems: Performance Analysis and Scheduling Algorithm*. Ph.D. thesis, Department of Electrical and Computer Engineering, University of Maryland, College Park, MD, USA (2005). Accessed on 14th September 2013.
108. Wang, J., Nalam, S., Calhoun, B.: Analyzing Static and Dynamic Write Margin For Nanometer SRAMs. In: *Proceedings of the ACM/IEEE International Symposium on Low Power Electronics and Design (ISLPED)*, pp. 129–134 (2008). DOI 10.1145/1393921.1393954.
109. Wann, C., Wong, R., Frank, D.J., Mann, R., Ko, S.B., Croce, P., Lea, D., Hoyniak, D., Lee, Y.M., Toomey, J., Weybright, M., Sudijono, J.: SRAM Cell Design For Stability Methodology. In: *IEEE VLSI-TSA International Symposium on VLSI Technology (VLSI-TSA-Tech)*, pp. 21–22 (2005). DOI 10.1109/VTSA.2005.1497065.
110. Weste, N.H.E., Harris, D.M.: *CMOS VLSI Design: A Circuits and Systems Perspective*, Fourth Edition, Addison-Wesley (2011).
111. Williams, T.: 1 Terabit on a Chip ? New Memory Technology Rises to Challenge NAND Flash. http://www.rtcmagazine.com/articles/view/102293 (2011). Accessed on 24th September 2013.
112. Xu, W., Zhang, T., Chen, Y.: Design of Spin-Torque Transfer Magnetoresistive RAM and CAM/TCAM with High Sensing and Search Speed. *IEEE Transactions on Very Large Scale Integration Systems (TVLSI)* **18**(1), 66–74 (2010). DOI 10.1109/TVLSI.2008.2007735.
113. Yamaoka, M., Osada, K., Tsuchiya, R., Horiuchi, M., Kimura, S., Kawahara, T.: Low Power SRAM Menu For SoC Application Using Yin-Yang-Feedback Memory Cell Technology. In: *Digest of Technical Papers Symposium on VLSI Circuits*, pp. 288–291 (2004). DOI 10.1109/VLSIC.2004.1346590.
114. Zhang, B., Arapostathis, A., Nassif, S., Orshansky, M.: Analytical Modeling of SRAM Dynamic Stability. In: *Proceedings of the IEEE/ACM International Conference on Computer-Aided Design (ICCAD)*, pp. 315–322 (2006). DOI 10.1109/ICCAD.2006.320052.
115. Zhang, W., Li, T.: Characterizing and Mitigating the Impact of Process Variations on Phase Change Based Memory Systems. In: *Proceedings of the 42nd Annual IEEE/ACM International Symposium on Microarchitecture*, pp. 2–13 (2009).

CHAPTER 8
Mixed-Signal Circuit and System Design Flow

8.1 Introduction

A typical analog/mixed-signal system-on-a-chip (AMS-SoC) has a variety of components including digital processors, analog circuitry, radiofrequency (RF) circuitry, true mixed-signal circuitry integrated together to achieve cost and performance tradeoffs [62, 67, 90, 118]. In addition, there is significant software presence in the AMS-SoC in the form of firmware, system software (operating system), and application software. The design of power supply components, which are battery packs and possibly accompanied by solar panels, involves different design cycles. This chapter will focus on the hardware components involving analog, RF, digital, and mixed-signal circuitry.

In the hardware components of the AMS-SoC, the digital circuitry performs most of the back-end processing of the data. The mixed-signal, analog, and RF components are present for front-end processing including communications and interfacing. In a typical design style, the digital components contain many more transistors than the mixed-signal, analog, and RF components. Of course, the digital designs of well-defined abstraction, such as from system to physical level, provide design flexibility to design engineers through the divide-and-conquer approach. However, at the last phase of the AMS-SoC design, all types of components (digital, mixed-signal, analog, and RF) are physical designs or layouts. At the circuit and the layout levels, the designs go through analog SPICE simulations for verification and characterization. The complexity of the AMS-SoC hardware component designs has increased multifold for many reasons [90]:

1. Integration of digital, analog, and mixed-signal functions along with the embedded software needs codesign for overall AMS-SoC optimization.
2. New signal processing algorithms and their corresponding architectures provide much serious challenges in terms of total AMS-SoC power dissipation and performance requirements.
3. The transistor count has increased tremendously to support the various functionalities of the AMS-SoC.
4. The rapid change of the process technologies demands consideration of different technology parameters during the design cycle for simulation and design space exploration.

As a result, the AMS-SoC design cycle is long and error prone. The overall design needs diverse skills, such as analog design, digital design, layout engineering, and design verification. The digital verification is becoming increasingly important and complex. The need for analog verification has also emerged [64]. To make the situation worse, the time-to-market has reduced significantly for such highly complex AMS-SoCs. In such a situation, computer-aided design (CAD) environments, including design and verification flows, are more important than ever to produce error-free, affordable, and functional AMS-SoCs on time.

8.2 AMS-SoC: A Complete Design Perspective

A real-life AMS-SoC realizes application-specific systems, such as mobile phones, have a variety of diverse components including digital components, analog circuitry, power supply component, and various types of software. Design and subsequent manufacturing of all these diverse

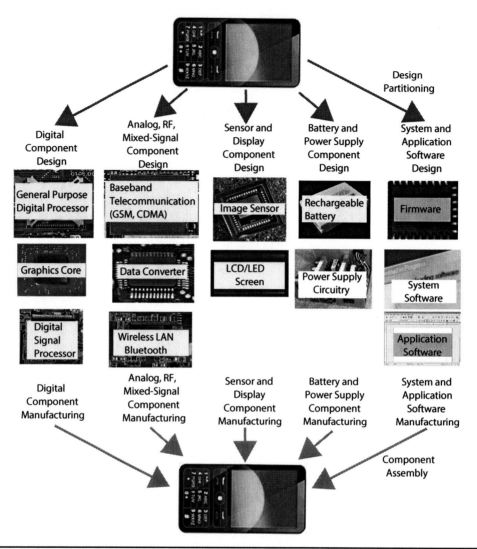

FIGURE 8.1 Broad perspective of the design cycle for the AMS-SoC.

components are necessary to build the overall system. A broad overview of this process is depicted in Fig. 8.1.

All the components have been grouped into five different categories. This is based on different types of skill sets needed to perform design of these components. Of course at a finer level of granularity, analog, RF, and mixed-signal components can be of three different categories. Similarly, the design of the battery and the design of the power supply component can be of two different categories.

1. *Digital Component Design:* The digital components of an AMS-SoC may include a general-purpose digital core, a digital signal processing (DSP) core, graphics core, and video codec core. These cores are classic digital designs covering general-purpose computing and special-purpose computing. The circuits perform integer and/or floating-point arithmetic. The general-purpose microprocessor that is typically a reduced instruction set computer (RISC) involves design flow of any typical general-purpose microprocessor. It may follow custom and semi-custom design flow. However, in the current trend of fabless design houses, the design of general-purpose microprocessor follows semi-custom approach. In this approach, a softcore of the general-purpose processor is obtained from a third party vendor with specific set of instruction architecture. The softcore refers to the hardware description language (HDL) implementation of the processor. This softcore is then further optimized and physical design is created for another third party manufacturing process technology. A similar approach of

intellectual property core design can be followed for the design of other digital cores to meet the short time-to-market demand. A standard cell design approach can be followed for the design of application-specific digital cores. In the standard cell, the design house develops synthesizable HDLs of the digital circuit whose physical design is then developed following register transfer level (RTL), logic synthesis, and/or physical synthesis approaches. The standard cells consist of a library of target process technology node.

2. *Analog, RF, and Mixed-Signal Component Design:* The analog, RF, and mixed-signal components are important components for the AMS-SoC with target applications such as mobile phones and tablets. The design of analog components and mixed-signal components (such as data converters) follow classical analog design flow involving custom design approach of circuit and layout levels of the design. However, the design of RF circuits may need different skills, methodologies, and tool sets. At the same time, behavioral simulation can be performed for first-level design verification of the components.

3. *Sensor and Display Component Design:* The design of various sensors in electronic systems needs different design skills, methodologies, and tool sets. Any type of sensor design is possible based on target electronic application. Various types of sensors have been discussed in the Sensor chapter of this book. For the design of sensors, a first-level verification using behavioral simulation can be performed. A custom design approach is a preferred approach for this type of components.

4. *Battery and Power Supply Component Design:* Design of the power supply system, including battery, battery charger, and power supply management, needs a different skill set from the engineers. It may involve a completely different design approach for the power source battery that may be rechargeable on its own. Some electronic systems may not be natively battery operated, for example, a desktop computer or digital TV. In general, a power management unit may be there in all types of electronic systems. The power management circuit design may be like other analog design. However, it may be quite different when it comes to the design of the power supply unit. The battery charging circuit design needs different design skill sets. These may be more inclined toward the power electronics design methodologies and tools.

5. *System and Application Software Design:* In the current-generation electronic systems, some levels of software are always present. The software may be of three types: firmware, system software (or operating system), or application software. The system software may also include some form of power management software for any dynamic power management of electronic systems. The design of software involves completely different skills and tools than the design of other components of the AMS-SoC.

Once all the components are designed, they are sent to different manufacturers or software developers. The designed hardware components are taped-out to various manufacturers. The batteries are manufactured by different manufacturers. The components that are not designed in house can also be acquired as commercial off the shelf (COTS). The components after manufacturing, acquisition, or development are assembled to build the complete AMS-SoC for the target electronic system. The manufacturing and development of the different AMS-SoC components can have the following distinct categories:

1. *Digital Component Manufacturing:* Manufacturing of the digital components is done in the most advanced fabrication facilities. The process technology used in the digital circuits evolves at a much faster pace than the other process technologies. Digital circuits are manufactured using single-gate high-κ or multigate devices to obtain low-leakage circuits.

2. *Analog, RF, and Mixed-Signal Component Manufacturing:* The analog, RF, and mixed-signal components are manufactured in different fabrication facilities with a different process technology node than the digital circuits.

3. *Sensor and Display Component Manufacturing:* The manufacturing of the sensors and display components may involve completely different types of fabrication plants. Sometimes micro-electro-mechanical systems (MEMS) technology may be used to build selected sensors.

4. *Battery and Power Supply Component Manufacturing*: The manufacturing of battery and power supply components is done at different facilities. The manufacturing of a battery is entirely different from all other components of the typical AMS-SoC.

5. *System and Application Software Development*: Firmwire, operating system, and the application software are developed quite differently from the fabrication of the hardware components of the AMS-SoC. The firmwire and operating systems are more specific to the architectures of the AMS-SoC components. The applications software comes in a wide variety that may be developed to work across different hardware and platforms.

The design of the hardware components in the AMS-SoC can have the high-level design flow depicted in Fig. 8.2 [65]. This depicts the situation where digital, analog, RF, and mixed-signal circuits are designed and optimized as a single unified component. Individual circuit design flows will be discussed separately in detail in the subsequent sections. Its overall design flow has three different levels. At the system level, the AMS-SoC is described as behavioral models. The system is partitioned into different components and the system requirements are derived. For each of the blocks, the interfaces are defined and the system is verified. For the true mixed-signal circuits as well as mixed-signal system modeling, simulation, and designs-space exploration, three distinct languages such as Verilog-AMS, VHDL-AMS, and SystemC-AMS, are used. Detailed discussions of various HDLs along with simulation methodologies are covered in future chapters. However, the use of digital languages, analog languages, as well as SPICE descriptions is possible for mixed-signal co-simulation [179]. At the circuit-level, design and/or synthesis of different types of circuits are performed. The RF and analog (along with true mixed-signal, e.g., ADC) blocks are designed. The digital circuit is obtained following various synthesis techniques. Each of the blocks of the AMS-SoC is simulated and verified at the circuit level for functionality and specifications. At the physical level of the AMS-SoC, the architecture of the AMS-SoC is converted to a floor plan. The blocks of the AMS-SoC are then placed and routed. At this point, the circuit netlist of the overall AMS-SoC is extracted. Depending on the level of accuracy needed for further simulation and characterization, resistances, capacitances, inductances, and/or mutual capacitances of the devices and/or wires are included in the netlist. At this level, functional verification, timing analysis, and power analysis is performed. The physical verification may include layout versus schematic (LVS) verification and design rule check (DRC). The

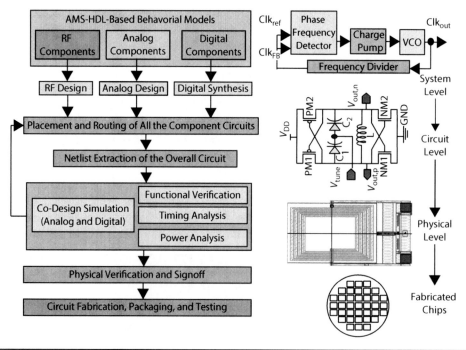

Figure 8.2 Design flow for the hardware components of the AMS-SoC.

AMS-SoC characterization is also performed for silicon-accurate characteristics of the AMS-SoC accounting of the parasitic effects. The characterization at this level is quite slow as compared to the circuit level and system level. When the design closure or signoff is met, the physical design of the AMS-SoC is taped-out [113]. From the taped-out design, manufacturers make masks that are then used for circuit fabrication. The packaged and tested integrated circuits (ICs) go to the market.

8.3 Integrated Circuit Design Flow: Top-Down versus Bottom-Up

With the nanoelectronics technology becoming a reality and increased circuit and system complexity, the debate of top-down versus bottom-up design is never ending [96, 115, 192]. The choice between top-down and bottom-up depends on the complexity of the circuit and the nature of design tasks involved. However, in reality, the overall design process of a chip follows a combination of top-down and bottom-up flows. For the circuits and systems with well-defined abstractions, top-down design flow can offer a fast and result-oriented design approach. For example, in large digital ICs with large transistor counts, top-down is preferred due to abstractions such as architecture level and logic level. However, the architecture-level library and logic-level library creations follow bottom-up approaches. The bottom-up is a good approach for small ICs with small transistor counts. For example, in analog circuits, the design is typically flat schematic-level design and flat layout-level design. There may be small level of hierarchical design. At the same time, it is a recent trend to perform system-level behavioral simulation for analog design using languages such as Verilog-AMS, VHDL-AMS, and Simulink®.

As a rule of thumb, top-down is preferred for large and complex ICs, whereas bottom-up is preferred for smaller ICs. The key aspect to control the design is the divide and conquer approach. Divide and conquer involves dividing the overall design into small manageable components so that the design time of each component can be controlled. Then, when the components are ready, they are put together to build the overall IC or system. The two types of design flows for ICs are presented in Fig. 8.3 [96, 115, 192]. The top-down, fully-automatic, no-custom approach heavily relies on automatic synthesis of the IC at different levels of design abstractions. The top-down flow is presented in 8.3(a) [96, 115, 192]. This is used too much for the digital ICs. Of course, various libraries that are themselves created using the bottom-up approach may be used in the design flow. The application-specific

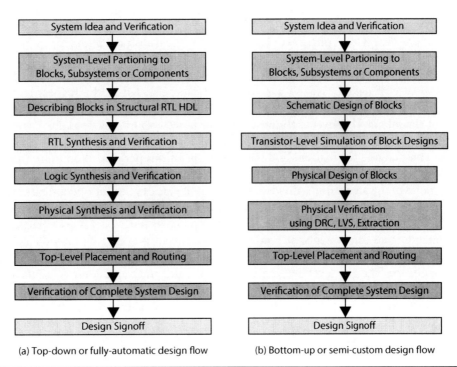

FIGURE 8.3 Integrated circuit design flow: top-down versus bottom-up [96, 115, 192].

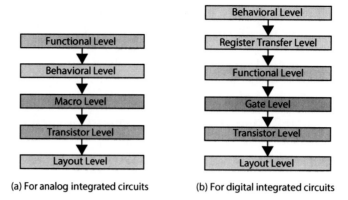

Figure 8.4 Abstraction levels for analog and digital integrated circuits for top-down design flow [71, 184].

integrated circuits (ASICs) such as DSP chips may be designed using this approach due to shorter time-to-market need. In the current design trend, this flow can easily adopt big cells such as intellectual property cores from a third party. The intellectual property cores that are megacells can be part of system-level or architecture-level blocks. The bottom-up or semi-custom design flow is depicted in Fig. 8.3(b) [96, 115, 131, 192]. This is good for analog, mixed-signal, RF, and small digital ICs. More detailed discussions on various design flows will be presented in this chapter.

In order to facilitate the design of complex ICs and systems, various levels of abstractions are available for both the analog and digital circuits [71, 184]. The various levels of abstractions are depicted in Fig. 8.4 [71, 184]. The various level of abstractions for analog IC design are shown in Fig. 8.4(a) [71, 184]. The functional level of the analog circuit involves describing the circuit in terms of mathematical functions. It may not satisfy conservation laws like Kirchoff current law (KCL) or Kirchoff voltage law (KVL) on the interconnecting nodes. At the behavioral level of the analog circuits, the mathematical functions are substituted by high-level blocks such as linear transfer functions, analog-to-digital converters (ADCs), and operational amplifiers (OP-AMPs). At the behavioral level, the conservation laws are enforced. The macro level of the analog ICs describes the circuits in term of the circuit elements such as resistors, capacitors, and linear/nonlinear controlled sources. The transistor level of the analog ICs describes the circuit in term of actual elementary components. This level is also called circuit level as well as schematic level. Sometimes logical level is also used to differentiate it from physical or layout level. The final abstraction called layout level (or physical level) is the representation of the IC in terms of manufacturing geometry and layers.

The various levels of abstractions in a digital IC are shown in Fig. 8.4(b) [71, 184]. However, in the case of digital systems, a system level can be added as the highest level of abstraction. The behavioral level of the digital IC describes the design in terms of signal transfers in the form of algorithms. This level of abstraction is thus termed algorithmic level, for example, video compression algorithms of a video compression chip. The RTL describes the design in terms of flow of digital signals among the registers and the logical operations performed on those signals. The RTL level is also known as the architecture level. The signals or data are in binary. At RTL, the execution time frame of the algorithm is divided into different clock cycles for operations scheduling and resource allocation at unique time stamps or clock cycles in the next level of abstraction. At the functional level, the operations are scheduled for a number of functional blocks (i.e., datapath components, architecture components, or hardware resource), such as adders, multipliers, and registers. At this level, the control signals are also determined for operation execution at the proper time stamp and sequence. The gate level (i.e., logic level) of the design abstraction represents the design in the form of various logic gates by replacing the architecture-level components with their logic gate realizations. At the transistor-level of design abstraction, the logic gates are implemented using transistors. The final level of the abstraction is the layout or physical design level that is eventually taped-out to the manufacturer for the fabrication of the digital integrated circuits.

The shorter design cycle and complexity of the circuits and systems are handled by the use of well-defined abstractions [71, 85, 131, 205, 210]. At any well-defined abstraction, the performance of the overall circuit or system needs to be evaluated at that abstraction level by using information from next lower level of abstraction that builds it. The information is provided in the form of models and libraries. The format of variables, types of models, types of libraries, and granularity of models and libraries depend on the level of abstraction. The models can include compact model, macromodel, and metamodel. Each of these models can calculate the output response based on input variables. Similarly, the libraries may be logic library, architecture library, or even intellectual property cores, depending on abstraction levels. The macromodels, metamodels, and libraries are made by the designer. The designers use information from a process design kit (PDK) provided by the manufacturers. However, the compact or device models also come from the manufacturers as a part of PDK. All these will be discussed in detail in the later part of this chapter or in next chapter.

8.4 Analog Circuit Design Flow

The standard design flow that can be used for the design of analog as well as big-Analog and small digital mixed-signal circuits is presented in Fig. 8.5 [33, 90, 91, 92]. At the very beginning of the design cycle, the behavioral simulations are performed to verify the functions and limited specifications at the very high broad level. As the first step of the circuit realization, the transistor-level schematic design is performed for a specific topology of the target design. At this stage, the transistors are sized for a specific technology node. The transistor-level design at this stage can be functionally verified as well as characterized for all the design specifications. Of course at this level, the characterization values may not be as accurate as the actual design built on the silicon because the SPICE netlist may not contain the parasitics. After the transistor-level schematic design passes a satisfactory level, the layout engineers perform the physical design of the circuit. The layout design is performed for a specific process technology as consistent with the transistor-level design phase and target foundry. This is a quite time-consuming phase that at the current design style follows a custom layout design approach. The layout engineer performs a DRC to ensure that the layout is clean to meet all the fabrication requirements. At this point, LVS verification is performed by the layout engineer to confirm that the physical design corresponds to the original transistor-level schematic design of the circuit. The LVS step involves a parasitic extraction substep so that the parasitic extraction phase has been shown before the LVS. At this phase, parasitic extraction is performed to obtain a netlist that contains the devices and their connections as well as the parasitics originated from the devices and the wires/interconnects. The extracted netlist is used in postlayout

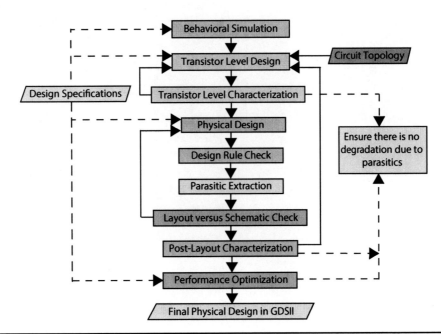

Figure 8.5 Flowchart showing the standard analog and mixed-signal circuit design flow.

characterization of the circuit. The characterization at this phase is close to what can be seen in actual silicon of the circuit. In a typical scenario, it is observed that the circuit characteristics may deviate quite a bit from the target specifications. This may lead to multiple back and forth iterations between the circuit designer and the layout engineer. It may be noted that these iterations may not be incremental. They may be quite time consuming and lead to an increase in the design cycle time as well as nonrecurrent engineering (NRE) cost of the chip. This is because the designers have to wait for another LVS-clean layout to get accurate parasitics and corresponding netlist. A typical method to avoid such time and effort-intensive iterations is to overdesign the circuit to compensate for unknown parasitics. Of course, the overdesign may be pessimistic and may lead to costly solutions. Circuit designers need to perform optimization of the circuit for various costs, characteristics, or figure-of-merits (FoMs) of the circuits including power, frequency, and phase noise.

The above design flow is the classical and most widely used design flow. However, it does not have all the steps needed to account for emerging challenges that appear due to the arrival of nanoelectronic technology. An advanced design flow is presented in Fig. 8.6 to accurately handle the nanoelectronic-based analog circuits. An additional step called variability analysis has been introduced in this flow. At this phase of the design cycle, the impact of process variation (P), voltage variation (V), and/or thermal variation (T) can be analyzed. For example, accurate statistical models for various device parameters can be used to study the distribution of the circuit characteristics. This can, in turn, be used for yield analysis of the circuit as well. The cost optimization in the advanced design flow can be classical as in the previous flow of Fig. 8.5. At the same time, statistical cost optimization can be performed to ensure the chip yield is not affected by the PVT variations [87, 145]. Thus, nanoelectronics-based circuit design needs greater skills from the circuit designer as well layout engineer. It increases the design cycle time and can have negative impact on the circuit yield. As a real-life illustration, the design life cycle of a nanoelectronic-based LC-VCO is depicted in Fig. 8.7 [82, 85, 147, 210].

8.4.1 Behavioral Simulation

The transistor count, details of device models, and complexity of the analog and/or mixed-signal circuits are ever increasing as the nanoelectronic process technology is evolving at a much rapid pace. At the same time, designers have diverse interests in terms of time-domain analysis, frequency-domain

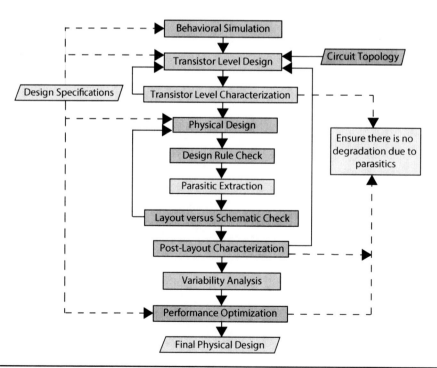

Figure 8.6 Flowchart showing the nanoelectronic-based analog and mixed-signal circuit design flow.

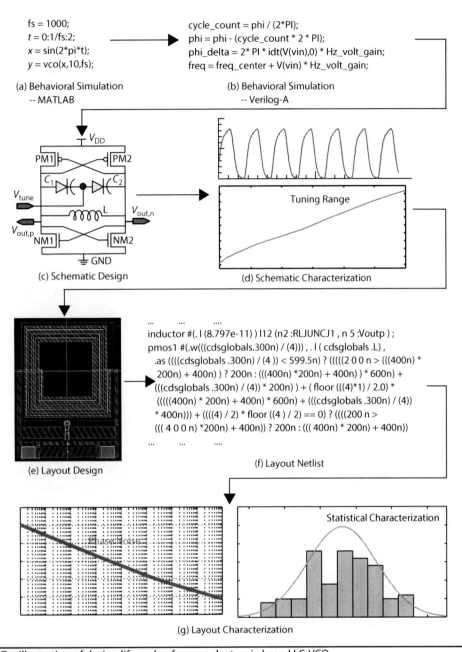

FIGURE 8.7 Illustration of design life cycle of a nanoelectronic-based LC-VCO.

analysis, continuous-time analysis, and discrete-time analysis. Behavioral modeling, behavioral simulation, and analog verification offer ways to abstract the features of the analog and/or mixed-signal circuits so that complex designs can be verified before going to longer design cycle phases [32, 37, 206]. Such modeling, simulation, and analysis may not be always possible through SPICE. The behavioral simulation performs the functional behavioral simulation of a complete system, components, circuits, or even single transistors. The analog SPICE modeling and representation uses the laws of physics to represent each circuit elements. The behavioral approach oversimplifies the modeling and complexity of the analog and/or mixed-signal circuits. Thus, structure or topology may not be quite as important at this phase. For example, LC-VCO and ring-oscillator may have similar behavior simulations. Thus, behavioral simulation provides a fast and flexible way for analog and/or mixed-signal circuit design. The behavioral modeling and simulations can be performed using many different languages and platforms including Verilog-A, Verilog-AMS, VHDL-AMS, SystemVerilog, SystemC-AMS, Saber®, MATLAB®, Simulink®, and Simscape®.

The behavioral model calibration is used to increase the accuracy of behavioral simulation. For the behavioral model calibration, the behavioral models are first generated using generic parameters. The generic parameters are then adjusted so that the characteristics obtained from the behavioral simulation are consistent with the characteristics obtained from the transistor-level simulations. To essentially achieve this, an approach called metamodeling has been used in which the transistor-accurate or silicon-accurate metamodels are integrated in the behavioral models [211, 212]. The design and verification are two separate tasks in the analog and mixed-signal circuit and system design [40, 64]. The circuit designers, layout engineers, and analog verification engineers in the present design and process technology trend work together to reduce the overall design cycle and cost. The behavioral modeling and simulation help in the analog verification process. The analog verification is not simple functional verification. The analog verification process uses the behavioral models of the analog blocks as well as the comprehensive self-checking testbench. The testbench is applied to both the behavioral models and the transistor-level schematic design.

8.4.2 Transistor-Level Design or Schematic Capture

The transistor-level design, schematic design, schematic capture, or logical design of an analog and/or mixed-signal circuit involve designing the circuit using basic elements such as transistors [91, 163]. The schematic capture is a visual way to understand, capture, and verify the circuit design. Appropriate topology selection and transistor sizing to meet performance specifications are important tasks at the transistor level design. At this point, a specific topology may be used for a specific design under consideration. It may involve a flat schematic capture of the overall design. It may also involve hierarchical schematic capture for large complex designs. In the case of hierarchical schematic capture, the circuit designer may use the "symbol creation" approach. In symbol creation, custom symbols or icons of different blocks or modules are created for easy reuse. The schematic design in addition to the basic elements such as transistors may include power supply, ground rail, interconnections, and input/output pins. All this ensures that a transistor-level netlist can be created for SPICE simulation and characterization. The generation of complete and accurate schematic design is important for the rest of the phases in the design cycle. During the schematic design, some sort of transistor sizing is performed to ensure that the designs meet the target specifications. Schematic editor is the schematic capture tool that is provided as a point tool or as a bundle with other tools, particularly in the framework of SPICE.

8.4.3 Transistor-Level Simulation and Characterization

The transistor-level characterization involves simulation and characterization of the schematic design of the circuit. This is the initial simulation phase of the design flow. This simulation is necessary for the basic purpose of detecting the design errors that might have been introduced during the schematic capture phase. A functional simulation is performed to verify that the schematic design has the functionalities for which it was designed. The schematic design can be characterized for various specifications of the circuit. Based on the simulation and characterization results, the designers may modify the schematic if needed. The modification may include the resizing of the devices as well as even selection of different topology of the circuit. A netlist of the schematic design is extracted for the transistor-level simulation and characterization. This netlist contains the circuit information, but may not contain the detailed information such as parasitics. The netlist is used through a SPICE or other analog simulators for simulation and characterization. Compact models are used in the analog simulators for a specific technology node. The schematic editors, such as Virtuoso Schematic Editor from Cadence®, Pyxis Schematic from Mentor Graphics®, and gschem from gEDA, can be used for transistor-level design.

8.4.4 Physical Design or Layout Design

The physical design or layout design phase follows the transistor-level design phase once the circuit is designed at the circuit level while meeting the target specifications [15, 121, 125]. The physical design essentially converts the elements of the circuits such as devices and interconnects into geometric representation of shapes such that the circuit can be manufactured using corresponding layers of materials. The physical design flow can be discussed in a top-down or a bottom-up perspective.

The analog physical design flow involves the following steps [15, 121, 125]: (1) hierarchical floor planning, (2) top-level (or block-level) assembly with a router, (3) chip finishing, (4) physical verification signoff with DRC and LVS, and (5) parasitic extraction. The hierarchical floor planning is top-down flow, whereas the top-level (or block-level) assembly is bottom-up flow. The floor planning is the positioning of the blocks or subblocks in the overall system or block so that the area and performance tradeoff is achieved. The floor-planning process uses approaches such as connectivity analysis, area estimation, and pin optimization to achieve efficient implementation. Routing deals with interconnecting various components and elements of the chip. The top-level place and route tool considers the layout of subblocks and assembles them one level higher in hierarchy iteratively till the complete layout is ready. Analog routing techniques such as critical signal, differential signal, shielded signal, and supply voltage routing are deployed to obtain better chip design. Chip finishing tasks such as metal density check, antenna check, metal filling, and guard rings are applied to complete the design process. The DRC, LVS, and parasitic extraction steps will be discussed in later subsections. Tools such as Virtuoso from Cadence®, L-Edit from Tanner EDA®, Pyxis Layout from Mentor Graphics®, and Magic are used for physical design. PDKs are used for layout design along with rule files to ensure manufacturability.

It may be noted that the physical design or layout design is performed using layout editors. The layout editor that may be part of an EDA tool bundle or an independent tool may use a proprietary or internal format to represent the layout design. However, the final layout or physical design, which needs to be taped-out to the manufacturer or building masks, is sent in a specific format or layout database called GDSII (which is often pronounced as GDS two). GDS stands for graphic database system. GDSII database files are the de facto industry standard for data exchange of IC layout artwork for last several decades. GDSII stream format is essentially a binary file format that represents the planar geometric shapes, text labels, and other information about the physical design in hierarchical form. With the current process technology and complex ICs, the sizes of GDSII files become too large—for example, tens of gigabytes—due to limited compressibility [66]. Another format, called open artwork system interchange standard (OASIS®), is a popular replacement of GDSII. The advantage of OASIS® is to provide better compressibility of layout data.

8.4.5 Design Rule Check

The design verification signoff includes various aspects including DRC, LVS, electrical rule check (ERC), and antenna check [40, 46, 121, 123]. In essence, DRC verifies whether the physical design can be manufactured by the fabrication plant using a mask set with a target yield. The DRC may check many aspects of the physical design including active-to-active spacing, well-to-well spacing, metal-to-metal spacing, minimum channel length, minimum metal width, and metal fill density. The DRC ensures that the physical design of a chip meets the geometric design rules and interconnect restrictions imposed by the target manufacturing process technology. The design rules are specific to a particular manufacturing process technology and corresponding mask set. The design rules may include a single layer rules, two layer rules, minimum area rule, and antenna rules. The single layer rules may include width rules that specify the minimum width of any shape in the layout. It may include a spacing rule that specifies the minimum distance between two adjacent objects. The two layer rules specify the relationships between two adjacent layers. For example, the enclosure rules suggest that the specific object types such as contact and via may be covered with additional margin by a metal layer. The minimum area rule dictates the minimal allowable area of the layout. The antenna rules dictate the ratio of metal area to gate area and are specified for each interconnect layer. Various tools such as Assura from Cadence®, Calibre from Mentor Graphics®, Hercules from Synopsys®, and HiPer Verify from Tanner EDA® can perform DRC.

For the nanoelectronic circuits, the concepts of DRC and design for manufacturability (DFM) have close connections [46]. However, the DRC and DFM need to be checked together to enhance yield while considering the time-to-market requirements. DFM can be considered a subset of requirements in the big set of DRC requirements. The DRC deals with much rigid criteria dealing with functionalities of the devices and the layout. The DFM rules are largely intended to address the die yield over the manufacturing process corners. For example, the DRC rules can be expanded to address DFM rules by introducing new rules at the lower level and higher level. At the lower level, the DFM rules such as scatter bars for placement optimization, enhancement of photoresist contrast

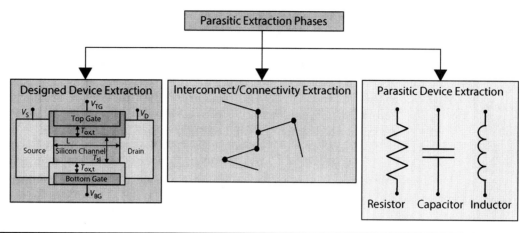

FIGURE 8.8 The three phases of parasitic extraction.

by avoiding slopes and side lobes, and reduction of jogs and intersections by restricting minimum jog sizes can be introduced. At the higher level, the DFM rules such as forbidden pitches and orientations, local and global pattern density, spacing from small to big features, and hierarchy management can be introduced in the DRC.

8.4.6 Parasitic (RCLK) Extraction

The parasitic extraction is the calculation of the parasitic effects of both the devices and interconnects in the physical design of an IC [92, 101, 116, 121, 123]. The parasitic extraction tool generates the netlist of the physical design of the IC. The objective of parasitic extraction is to create an analog model of the IC such that silicon accurate simulation, characterization, optimization, verification, and design closure can be performed. As the process technology shrinks very aggressively, the parasitic extraction becomes very important for functionality, verification, and design closure. Various tools such as Star-RCXT from Synopsys®, Calibre xACT3D from Mentor Graphics®, and QRC from Cadence® can perform parasitic extractions from the physical design or layout of the circuits.

The parasitic extraction process from the layout or physical design of the ICs can be logically divided into the following three phases [101, 116, 123, 133]: (1) designed device extraction, (2) interconnect extraction, and (3) parasitic device extraction. The three phases are depicted in Fig. 8.8. Of course, the above three phases are interrelated as various device extractions may change the connectivity of the IC. For example, the resistors irrespective of whether they are designed element of the circuit or a parasitic from the devices or interconnects need to be grouped into nets using multiple electrical nodes.

The designed devices are explicitly designed by the engineers and are made a part of the IC by the layout engineer. The designed device extraction may involve following multiple steps [123]: (1) the existence of the device is first recognized, (2) the exact form of the device is extracted, (3) the exact form of the device is verified for correctness, (4) an instance of the device is added to the circuit database, (5) each device is given a unique identification, (6) the terminal connections of each device are identified, (7) if a designed device creates a discontinuity of wire, then the device is removed from the interconnect layer before the interconnect analysis, and (8) the parameters of the device are measured if needed. The elements of the ICs are connected by the interconnects and contacts that essentially join the elements together. The interconnect or connectivity extraction is crucial to accurately define the IC. The netlist to be extracted must contain the nodes, connections, and function names so that it is useful for accurate analysis of the circuit using analog simulators.

At this point, parasitic devices such as resistance, capacitance, inductance, and mutual inductance are extracted [116, 123, 133]. The parasitic resistor is extracted to represent resistances associated with the substrate, interconnect, and package. The parasitic resistance extraction is performed using finite difference, finite element, boundary element, or Green function methods. The parasitic capacitance extraction uses 3D effects to meet the accuracy of nanoelectronic circuits. The 3D capacitance extraction tools divide the full-chip extraction process into three

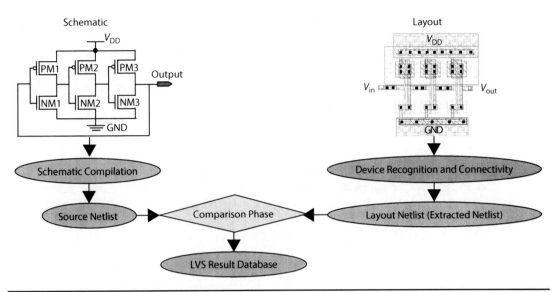

FIGURE 8.9 The mechanism of LVS [168].

major phases: (1) the technology precharacterization phase, which involves simulation of test structures using 2D/3D field solvers and presented data using empirical formulas or look-up tables (LUTs). (2) The second phase is the geometric parameter extraction. (3) The third phase is the calculation of the capacitances from the geometric parameters. The parasitic inductance extraction is more complicated than the parasitic resistance or parasitic capacitance extraction, as the parasitic inductance depends on the current flow through the wire as well as return current paths.

8.4.7 Layout versus Schematic Verification

DRC checks whether the physical design conforms to the design rules of a target manufacturing process technology. However, DRC does not check whether the physical design performs the intended functionalities of the IC for which is designed. The LVS is meant to deal with these functionality requirements [129, 168]. LVS is very critical in the physical design flow. Assura from Cadence®, Hercules LVS from Synopsys®, and Calibre from Mentor Graphics®, are the industry standard LVS tools. The generic steps of LVS verification process are depicted in Fig. 8.9 [168].

The LVS verification is performed by an EDA tool using LVS rule deck. The overall LVS process has two distinct phases: extraction and comparison. In the extraction phase, the layout database is compiled into basic elements such as transistors, diodes, or capacitors. The connectivity information among these elements is also extracted. The extraction process is guided by the LVS rule deck that is a set of code containing the layer definitions. The electrically connected regions in the layout called nets are recognized from the layout shapes through analysis between layout shapes in the layers. Each of the electrical nets is given a unique node number for identification purposes. This step may include a reduction phase. In the reduction phase, the extracted elements are combined into series/parallel combinations before generating the netlist of the layout database. In a similar fashion, the source or schematic netlist is generated. The second phase of the LVS verification is the comparison phase that also uses the LVS rule deck. At this phase, the electrical circuits from the layout netlist and the schematic netlist are compared. At this point, a one-to-one correspondence between the elements of the layout netlist and the schematic netlist is established. A layout is declared LVS clean if both layout and source netlist match electrically as well as in terms of the elements.

8.4.8 Electrical Rule Check

The physical design verification involves ERC in addition to the DRC and LVS. The ERC involves checking of the electrical connections of the netlist of the IC that are considered critical [17, 49, 100]. For example, the ERC may include checking of substrate and well areas for proper spacings or contacts. ERC ensures correctness of power and ground connections as well as checking of unconnected inputs and shorted outputs. While DRC and LVS have been very important for physical verifications, ERC is needed for complex systems with multiple power domains. ERC ensures the robustness of the IC for both schematic and layout levels using electronic design rules. The geometrical design rules that are used in DRC ensure that the IC is manufactured correctly. The electrical design rules that are used in ERC ensure that the IC works properly. ERC may include checking floating nets, floating devices, floating pins, floating well, net area ration of antenna rules, and susceptibility to electrostatic discharge (ESD) damage. Physical verfication system (PVS) programmable-ERC (PVS-PERC) from Cadence® and Calibre PERC from Mentor graphics® can perform ERC for nanoelectronic circuits.

8.4.9 Physical Design Characterization

The physical design characterization is the process for checking that the ICs meet all the target characteristics or performance specifications within the acceptable error margin [110, 167]. The postlayout (or physical design) simulation and characterization follow the physical design verification phase. The physical design verification mainly deals with the functional verification as well as DRC, ERC, and LVS to ensure manufacturability and robust function. On the other hand, the physical design characterization or postlayout characterization deals with analysis of the IC for its quality or how good it is. It may involve analysis of different characteristics, properties, performance metrics, FOMs, or quality metrics. For silicon-accurate characterization of the IC, it can be performed over the full-blown parasitic-extracted netlist. Various sets of EDA tools and other computation frameworks are used for analog/RF IC characterization.

The analog/RF ICs have multidimensional design space as depicted in Fig. 8.10 [167]. It may be noted that this multidimensional design space makes it difficult to use digital CMOS process technology for analog designs, as digital process is optimized for one tradeoff power and speed. In analog/RF design, every two axes that represent two analog/RF integrated circuit characteristics in the design space can trade with each other. The tradeoffs can be performed if they can be characterized for a circuit for a specific process technology. The analog/RF integrated circuit characterization is also difficult due to the lack of modeling as well as inaccurate modeling. The various analog attributes matching properties of transistors, self-resonance frequency of inductors, or nonlinearity in transistor characteristics are difficult to model. As depicted in Fig. 8.11, the analog/RF integrated circuits can be characterized for many different properties including the following [167]: (1) DC behavior, (2) AC behavior, (3) linearity, (4) matching, (5) noise, and (6) temperature dependence. Very accurate and exhaustive characterization is too slow even for small-size analog integrated circuits when performed over parasitic-extracted netlist [110, 167]. Additionally, for nanoelectronic analog integrated circuits with smaller feature sizes, the characterization is further challenging. The statistical characterization needed for process variation analysis will be discussed in the next subsection.

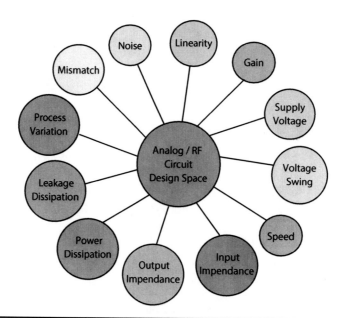

Figure 8.10 Design space of analog and RF integrated circuits [167].

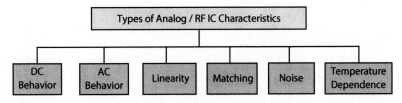

FIGURE 8.11 Different types of characteristics/properties of analog/RF integrated circuits [167].

8.4.10 Variability Analysis

The three major sources of variability in nanoelectronic analog/mixed-signal integrated circuit are process variation, voltage variation, and thermal variations. The variation types are depicted in Fig. 8.12 [88, 109]. The process variations originate from the variations in the sophisticated manufacturing process conditions of nanoelectronic integrated circuits. The voltage variations that are originated from interconnect resistance and inductance lead to noise in the supply voltage. The thermal variations are the variations of the temperature across the chip due to many reasons including different levels of power dissipation and leakage. The nanoelectronic variations are the major cause of yield loss in nanoelectronic integrated circuits. It makes designer tasks difficult, as design closure needs more iterations. Statistical design optimizations need to be performed by the designers instead of traditional nominal value-based design optimization, in other words, a shift from deterministic design to probability design. The overall effects of the PVT variations are the reduced reliability and the increased cost of integrated circuits.

The traditional corner-based analysis approach only checks extreme conditions and is a too pessimistic approach for nanoscale integrated circuits. The corner-based analysis approach cannot model intra-die and inter-die variations in these circuits [110]. The corner-based approach will have its importance but statistical analysis is needed to quantify the impact of process variations on the performances of the nanoscale integrated circuits as shown in Fig. 8.13 [110]. The variability in

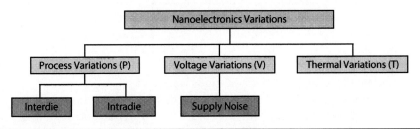

FIGURE 8.12 Process (P), voltage (V), and temperature (T) variation in integrated circuits.

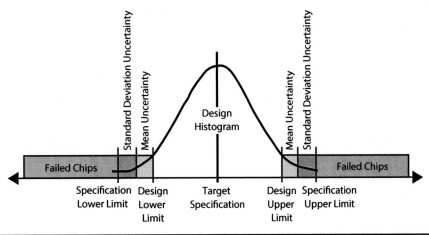

FIGURE 8.13 Statistical design margins [110].

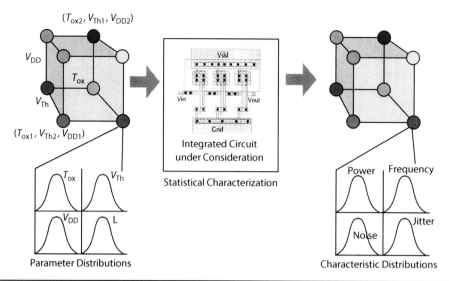

FIGURE 8.14 Analysis of process (P), voltage (V), and temperature (T) variations in integrated circuits.

integrated circuit performance metrics, such as power, noise, and frequency, can be characterized by the probability density function (PDF). For this purposes, the PVT variations are statistically incorporated to obtain the PDFs of the performance metrics in the following manner:

$$\widehat{\text{FoM}} = f(P, V, T) \tag{8.1}$$

where $\widehat{\text{FoM}}$ is a target performance metric such as power or noise of an analog IC. The process variation in essence includes the parametric variations of the device. Appropriate statistical distributions of different device parameters are used as inputs in the characterization process. A parasitic-aware netlist is used as the representation of the integrated circuit. The basic idea of statistical characterization is depicted in Fig. 8.14 [70, 80, 110, 143, 145]. There are many challenges of statistical characterization to capture variability. A realistic and accurate evaluation of FoM statistical distributions accommodating PVT variations while considering any device parameter correlations is needed. The relation between FoM variations that are caused by specific PVT variations needs to be established for speeding up the characterization process. The determination of accurate PDFs for each PVT variations is needed.

The existing techniques are mainly based on Monte Carlo analysis, derivatives of Monte Carlo, and parametric density estimation [80, 127, 143, 145, 154]. The classical Monte Carlo analysis that is helpful to generate sample points of PVT variations is quite computationally expensive and often slow to run. For a reasonable size circuit such as PLL with full-blown parasitics included in the simulations, it may be a matter of a few days to weeks. The Monte Carlo-based techniques have many advantages as well [88]. Monte Carlo techniques can provide accurate estimation of integrated circuit FoMs. Monte Carlo analysis does not need a closed form function relating the FoMs with the input parameters that would have been quite cumbersome. The input parameters during Monte Carlo analysis can assume more realistic values that lead to faithful estimations of FoM distributions. Various techniques are deployed to reduce the number of simulations needed for the statistical analysis to achieve tradeoff between computation time and accuracy [80, 88, 145]. For example, a design-of-experiments-assisted Monte Carlo (DOE-MC) analysis speeds up the process significantly without much compromise in accuracy [145]. The metamodel-based techniques are also being used as a tradeoff between accuracy and computation time [80, 126, 135, 156, 207]. The various forms of metamodel can be a polynomial or a posynomial function, a symbolic model, a Kriging model, or the neural networks (NNs). The existing EDA tools have limited capabilities for PVT variation analysis mostly that are based on Monte Carlo.

8.4.11 Performance Optimization

The integrated circuit optimization is a key step to ensure that the design meets the target specification. Otherwise, the integrated circuit may be classified as meeting a lower set of specifications or not even meeting the specifications. Overall, this leads to loss of yield and revenue. A manual iteration of design closure is quite slow and can have longer design cycles and nonrecurrent cost. It is desired that the design optimization be as automatic as possible as opposed to some form of netlist at the transistor level or layout level. The layout-level optimization with netlist containing all the parasitics can be most accurate. However, some level of optimization can be achieved at the transistor-level with netlist not containing all the parasitic information. A minimal-level of optimization can be performed at the behavioral level and system level; however, this may not be accurate as the circuit-level details are not captured at this stage of design flow. The optimization is performed for a target FoM such as power, noise, or frequency with constraints and is expressed as follows:

$$\begin{aligned} \text{Minimize:} \quad & \text{FoM} \\ \text{Subjected to:} \quad & \text{Constraints} \end{aligned} \quad (8.2)$$

An example of such an optimization is the minimization of noise of a PLL design for an operating frequency constraint. Another example, power dissipation is minimized for a PLL for a given operating frequency as a part of low-power design. The design parameters are the variables of optimization that may include the size of appropriate transistors as well as supply voltage level. As the number of design variables increases, the size of the design space increases and optimization slows. Selection of right number of design variables is important to reduce the search space during design exploration. It is also important to choose right optimization algorithm for fast design exploration and convergence.

The deterministic optimization may work well for technologies with larger feature sizes but, may not be sufficient for nanoelectronics-based integrated circuits in which process variations are prominent. In such cases, statistical design optimization of the following form is needed:

$$\begin{aligned} \text{Minimize:} \quad & \widehat{\text{FoM}} \\ \text{Subjected to:} \quad & \text{Constraints} \end{aligned} \quad (8.3)$$

The statistical optimization for performance and yield tradeoffs in integrated circuits is depicted in Fig. 8.15. A $(\mu + \sigma)$ quantification of $\widehat{\text{FoM}}$ that relates to "A" in Fig. 8.15 can provide a yield of 68.2%. This is the most optimistic optimization as redundancy provided in the design is the lowest. A $(\mu + 2\sigma)$ quantification of $\widehat{\text{FoM}}$ that relates to "B" in Fig. 8.15 can provide a yield of 95.4%. This is a pessimistic optimization as designs will have more redundancy than that corresponding to "A." Similarly, optimization with larger values like $(\mu + 3\sigma)$ is further pessimistic optimizations with 99.7% yield. Similar to the statistical analysis, the statistical optimizations can be performed on actual circuits (i.e., netlists) or metamodels. The optimization over metamodels will be faster by an order of $1000\times$ [83].

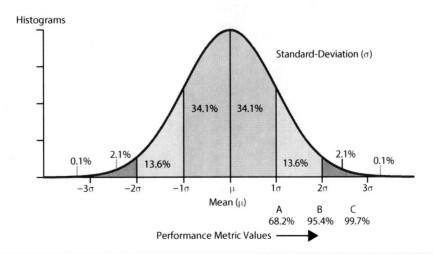

Figure 8.15 Statistical optimization for design tradeoffs.

8.5 Digital Circuit Design Flow

In the above section on the analog and mixed-signal integrated circuits, mostly two (transistor and physical) or three (system-level/behavioral, transistor, and physical) levels of abstraction can be observed. The design process is mostly full-custom design. It works well for the analog integrated circuits as the number of elements (such as transistors) involved in the design is small, in the order of thousands at present. However, the digital integrated circuits that are the main workhorses for most of the computing systems have large number of devices in them, in the order of billions as present. The digital integrated circuit design is thus very complex and time consuming. Design of full-custom digital chips is time consuming and may not be cost effective. However, the digital design comes with a built-in advantage, i.e., "abstraction." Due to well-developed theory, the digital design has abstractions such as system level, architecture level, logic level, transistor level, and physical level. For the digital design engineers, the abstraction helps to provide a black-box view of the next lower level of abstraction to keep the design process in control at a specific level. The digital design tools accordingly are well developed for design exploration at a specific or more than one design abstractions.

The overall digital design flow is presented in Fig. 8.16. In general, the digital design has two distinct design phases: (1) front-end design (via front-end design flow using a set of tools) and

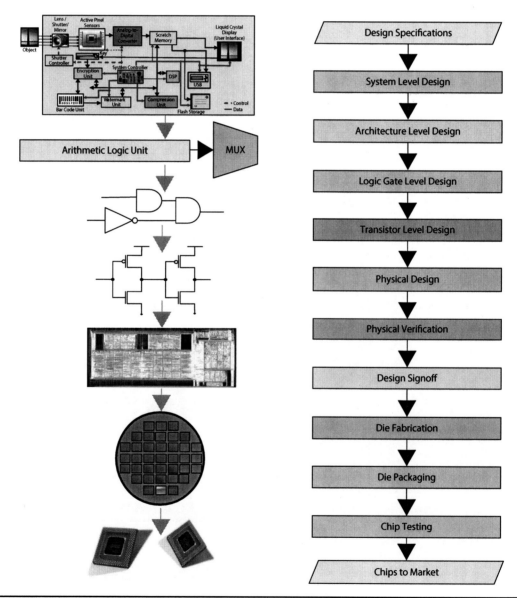

Figure 8.16 Standard digital circuit design flow.

(2) back-end design (via back-end design flow another set of tools). The front-end design includes digital design using HDLs and the corresponding design verification using simulation and various verification techniques. The back-end design is the physical design process along with library creation and characterization. Depending on what step or set of steps are performed, the design can be full-custom, semi-custom, or no-custom (or automatic) [79, 190]. The full-custom design uses the custom design approach with custom layout and the design can take months to complete. The semi-custom design uses standard-cell library with manual place and route but all remaining steps are automatic using EDA tools. The semi-custom design may take weeks for completion. The automatic or no-custom design uses EDA tools to carry every step automatically including gate sizing and place and route, and can complete the design in days. This is fast design approach that can produce low-cost ASICs to meet shortest time-to-market need. As a real-life illustration, the design life cycle of a digital ASIC is depicted in Fig. 8.17 [45, 141, 144, 148, 151].

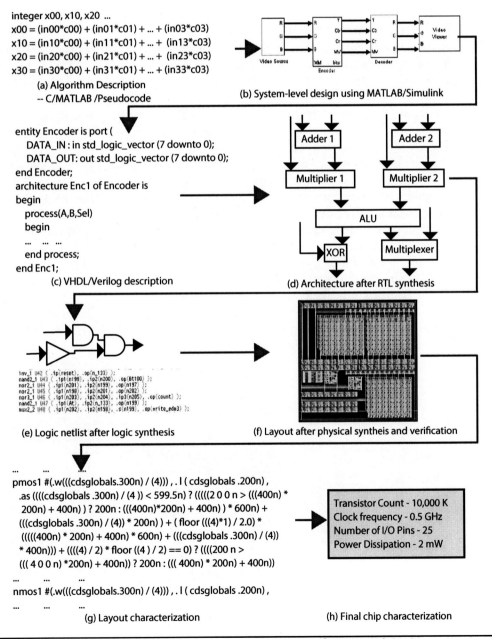

FIGURE 8.17 Illustration of design life cycle of a digital application-specific integrated circuit.

8.5.1 System-Level Design

System-level design is the design at the highest level of digital system or integrated circuit abstraction [42, 43, 180, 186]. With SoC becoming increasingly popular for low-power and high-performance applications such as tablet and smart mobile phones, the electronic system level (ESL) design and verification are becoming increasingly popular for system design. The ESL is a collection of complementary methodologies that can perform system-level design, verification, and debugging of embedded systems. The embedded system design can involve implementations using custom SoC, system-on-FPGA, system-on board, and entire multi-board systems for power and performance specifications. Thus, ESL provides design exploration techniques for selecting and refining architecture of embedded systems at the system-level. ESL design is the ability to assemble an embedded system using its constituent heterogeneous components [180, 186]. For example, for a smart mobile phone or a tablet, the components may include digital general-purpose core, video processor, image processor, RF chip, analog chip, battery, and operating system. ESL can use many languages and frameworks including SystemC, SystemVerilog, SystemC-AMS, Systems Modeling Language (SysML), MATLAB®, Simulink®, and Simscape®.

8.5.2 Architecture-Level Design

Architecture-level design of the digital ICs involves design exploration using architectural components to meet the specifications such as power and throughput [141, 150, 158, 159]. For example, an application-specific DSP can be designed using components of adders and multipliers. A datapath component library, architecture library, or RTL library consisting of components, such as adder, subtractor, multiplier, divider, comparator, multiplexer, and register, is used for architecture exploration. The datapath components may be of different bitwidths, e.g., 16 bits, 32 bits, or 64 bits. The library may contain integer units and/or floating-point units. The objective is to identify a target architecture consisting of datapath and control containing a specific number of different types of resources to meet the target specification. For example, an architecture containing 5 adders and 10 multipliers can perform its operations in 10 clock cycles. The architecture level is one level lower and closer to the actual design as compared to the system-level in the hierarchy of design abstractions. The architecture-level design decisions are more accurate compared to system-level decisions and provide reasonable degrees of freedom for alternative architecture and consequently hardware realization. The architecture-level design is alternatively called algorithm-level, high-level, or RTL design.

The automatic process of architecture exploration is called high-level synthesis. It is also known as behavioral synthesis, RTL synthesis, and algorithmic synthesis. The architecture-level synthesis generates RTL description or RTL netlist of a target digital integrated circuit from its corresponding algorithm description as depicted in Fig. 8.18 [141, 146, 149]. For example, for a

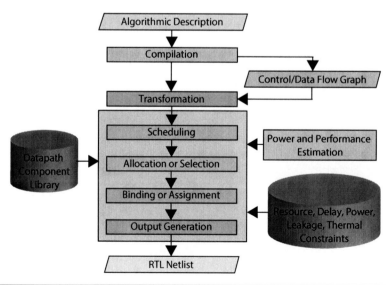

Figure 8.18 The architecture-level synthesis process [141, 146, 149].

video codec chip design, the input to the RTL synthesis process is the algorithm description of the video compression and decompression. The algorithm description can be in the form of many languages including C, MATLAB®, behavioral VHDL, or behavioral Verilog. It may be noted that behavioral synthesis is different from behavioral simulations. Behavioral synthesis results in a RTL structure, whereas the behavioral simulation is the only simulation of the algorithm to verify its functionality. The major tasks of the high-level synthesis include the following: (1) compilation, (2) transformation, (3) scheduling, (4) allocation or selection, and (5) assignment or binding. Compilation and transformation steps transform the input algorithm or behavioral description to a control/data-flow graph representation. The transformation phase may perform some technology-independent optimizations over the control data flow graph (CDFG) to result in an optimal CDFG representation of the input description. The scheduling task time stamps various operations in the CDFG for execution at the specific one or more number of clock cycles. Allocation decides the number and type of different resources to be used from the datapath component library. Obviously, there can be tradeoff between the number of components (more represents larger area) and number of clock cycles (more represents longer delay). The binding phase decides which operation to execute using available resources, affecting the above tradeoffs. In general, the three steps are related to each other in the sense that one decision may affect the other two decisions. Various optimizations for target specifications can be performed at these steps. For example, for low-power high-level synthesis to obtain energy-efficient chips, technology-dependent power optimization can be performed. Various EDA tools such as BlueSpec Compiler from Bluespec®, Catapult C from Calypto Design Systems®, C-to-Silicon from Cadence Design Systems®, and Synphony C Compiler from Synopsys® are currently available to perform high-level synthesis of digital integrated circuits [137, 149].

8.5.3 Logic-Level Design

The logic-level design deals with realization of the IC using various logic gates such as NAND, NOR and NOT. In a classical approach, one can start from a truth-table description of a digital design, draw its K-maps, and optimize the logic through K-map. Afterward derive the expression as sum of products (SOP) or product of sums (POS) and obtain logic level design for the SOP or POS. A typical OR-AND-NOT network can be obtained for the optimized Boolean expressions obtained from the K-map optimization. However, this approach of logic-level design can work for very small designs only, as the K-map optimization is not scalable due to its exponential (i.e., 2^n) complexity with respect to the number of bits.

In the automatic logic-level, design process is called logic-synthesis. The logic synthesis process assumes RTL netlist (expressed in VHDL or Verilog) as input and generates logic netlist (also expressed in VHDL or Verilog) as output as depicted in Fig. 8.19 [107, 128, 152]. The output logic netlist can be a NAND-NAND network, NOR-NOR network, or AND-OR-NOT network subjected to library and specification constraints. The various steps in logic-synthesis include parsing, elaboration, analysis/translation, technology-independent optimization, and technology-dependent optimization. The logic-synthesis process uses a standard cell library consisting various logic cells such as NAND gates, NOR gates, AND gates, OR gates, NOT gates, and flip-flops. It may include the cells with various fan-in configurations such as 2-input and 3-input. The design decisions may involve selection of type of network (e.g., NAND-NAND versus AND-OR-NOT network), logic gate fan-in, and design constraints dictated by the specifications. It has been observed that NAND-NAND network has lower leakage and high performance as compared to AND-OR-NOT network for nano-CMOS circuits [152, 153]. Similarly, use of higher fan-in logic gates may lead to lesser number of gates, but logic gates with higher fan-in have longer delay due to higher capacitances. In the technology-independent logic optimization, simplification and restructuring of the logic network or delay restructuring may be performed. In the technology-dependent logic optimization step, technology mapping, such as techniques using standard cell library, is performed subject to design constraints. For example, minimal power, minimal delay, and/or minimal area can be objectives of optimization at this step. The current EDA tools such as Design Compiler from Synopsys® and Encounter RTL Compiler from Cadence® can be used for logic synthesis.

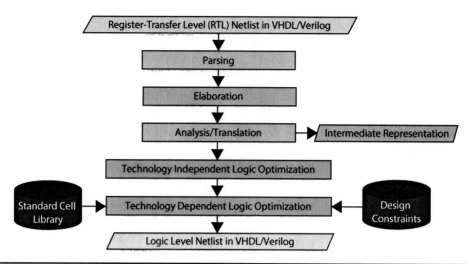

FIGURE 8.19 The logic synthesis process [128, 152].

8.5.4 Transistor-Level Design

Transistor-level design deals with different devices directly. It is very detailed and complex as compared to the higher levels of the abstractions such as architecture level as well as logic level. However, the transistor level is less complex than the layout level or physical-design level. Although historically, NMOS or PMOS only circuit implementations had been performed, CMOS technology has been the winner for last several decades because of low-leakage, high reliability, low signal noise, and high-packing density as compared to NMOS or PMOS only. Pass transistor technology has been explored to reduce the area footprint of the CMOS, but higher signal noise has slowed its usage. The multigate FET is adopted to further reduce the leakage of classical CMOS as well increase ON and OFF switching speed. To reduce standby leakage, a very good OFF has been achieved by multigate FET-based CMOS technology. The new technologies, such as spin-majority gate, spin-wave devices, tunnel field-effect transistors (TFETs), and graphene nanoribbon TFETs (GNR-TFETs), are being explored for classical nano-CMOS alternatives [122].

The digital designs when performed using the fully automatic approach may not need transistor-level design. In the fully automatic or semi-custom design approach when the layout is created automatically using silicon compilation from the physical-design of standard cells, then transistor-level design is bypassed. However, transistor-level digital cell optimization may lead to efficient overall chip designs [106, 170]. Various schematic editors or schematic capture tools, including Virtuoso Schematic Editor from Cadence, Pyxis Schematic from Mentor Graphics, Galaxy Custom Designer from Synopsys, and gschem from gEDA, can be used to perform transistor-level or schematic-level design. The transistor-level design may still use SPICE simulation that is performed over the transistor-level netlist extracted from the schematic design. The transistor-level netlist may not contain any parasitics and may not be as accurate as actual silicon-based implementation. However, it is useful in providing the designer a very good guideline of the target design under consideration. The compact models of transistors are used in SPICE simulations. Another approach is to use Verilog-A, Verilog-AMS, and VHDL-AMS modeling and simulation is also an option through EDA tools [94].

8.5.5 Physical Design

The flow that is used to generate physical design or layout of a digital chip is the back-end design flow of the digital integrated circuits. The steps that are followed to create a layout or physical design are presented in Fig. 8.20 [34, 68, 69, 114, 130]. The physical design or layout generation of the digital integrated circuits is semi-custom or almost fully automated [79, 138]. The EDA tools such as the following can perform physical design or layout synthesis of the digital integrated: (1) design compiler (DC) and IC compiler from Synopsys®, (2) encounter digital implementation (EDI) System from Cadence®, and (3) olympus-SoC from Mentor Graphics®.

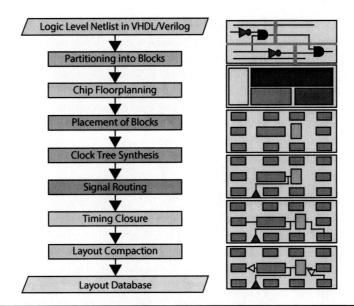

FIGURE 8.20 The physical design or synthesis flow for digital circuits [68, 114].

The first step of the physical design is the partitioning of the chip into different blocks that can be connected to make the complete chip. This step is needed as the typical chip is quite large in terms of the number of transistors and it is not possible to handle the complete chip at a time due to limited computational capabilities. In the floor-planning step of the physical design, the different blocks are identified to be placed for minimal chip area and spaces are allocated for each of the blocks. It involves the estimation of die area, arranging, and choosing the location of the blocks. Placement refers to the placing of the blocks in exact locations as per the floor-planning specifications to meet interconnect length, routing congestion, timing, and power requirements. The pin placement is also performed at this stage. In the synchronous digital circuit, the task of the clock tree synthesis step is to the ensure that clock signals are properly distributed from a clock source such as phase-locked loop (PLL) to the sequential elements such as registers with minimal skews and delays. The signal routing performs the completion of the interconnection among the blocks. It may be noted that the quality of the block placement becomes evident after the signal routing is performed. Timing closure is the process of modifying the placement and routing to meet its timing requirements in terms of shortest and longest path delay. The layout compaction is the process of compression of the layout from all possible directions reducing the chip area.

8.5.6 Physical Verification

The biggest challenge for the design engineers and the layout engineers is to verify and deliver a layout design that is free of DRC violations while meeting their tape-out schedule. The physical verification is the checking of the layout IC against certain criteria to ensure that it is manufacturable as well as that it meets the target specifications [166, 200]. The physical verification involves processes such as DRC, LVS, ERC, and antenna checks. This verification steps are performed on the parasitic-extracted netlist of the layout or physical design. The steps are similar to the analog chip, however, these may be more complex and time consuming due to very large transistor count as compared to analog. The EDA tools such as PVS from Cadence®, Calibre from Mentor Graphics®, and IC Validator from Synposys® can be used for physical verification. The importance of the verification is evident from the fact that more than 50% of design effort for most of the chips is devoted to it, and hence significantly contributes to the NRE cost of the chip.

8.5.7 Design Signoff

The design signoff is the last step of the chip design process before it is taped-out for mask generation and subsequent fabrication steps [18, 47]. The signoff step is more complex for nanoelectronics-based integrated circuits as additional effects that cannot be ignored have to be taken into account. The turnaround time for signoff is crucial, as it is the end of the design cycle. The signoff may involve a variety of steps including the following: (1) DRC, (2) LVS, (3) formal verification, (4) voltage drop analysis, (5) signal integrity analysis, (6) static time analysis, (7) electromigration lifetime checks, and (8) process variation analysis. During the signoff, a small number of errors can be fixed using engineering change-order (ECO) fashion by streaming the design back into the place and route, repairing the error, and performing another round of physical verification [47]. Tempus and Voltus from Cadence® and PrimeTime Suite from Synopsys® are EDA tools that can be used for design signoff.

8.5.8 Engineering Change Order

One important task in the nanoelectronic-based circuits is the ECO that has not been shown in the design flow that are presented. ECO though not much critical in older technologies is becoming increasingly important due to the current-generation complex integrated circuits [102, 165]. ECO is the fixing of the functions errors and/or performance deficiencies that are found on the integrated circuit designs at a very late stage of the design cycle. ECO can be performed at any stage of the design cycle, for example, prelayout, postlayout, postmask, or post-silicon. In the current-generation complex chips with shorter time-to-market, design errors as well performance efficiencies can emerge at any phase of their design cycle and manufacturing process. The key aspect is to have low-cost and faster solutions through ECO with minimal perturbations for the design process and incremental repair. ECO thus avoids redoing of full design or synthesis cycle to save time and NRE cost. Simple examples of ECO include gate sizing and buffer insertion to fix timing violations. Various EDA tools including PrimeTime from Synopsys® and Encounter Conformal ECO Designer from Cadence® have capabilities to facilitate ECO. Sometime ECOs called metal-mask ECOs or post-mask ECOs may be needed at a very late stage of chip design and manufacturing cycle, for example, after masks have been made [140]. ECOs at this stage may involve modification of a few layers, such as metal layers, which is way less costly than rebuilding the design from scratch.

8.5.9 Circuit Fabrication, Packaging, and Testing

IC manufacturing, packaging, and testing are not part of the design process, however, is presented here to provide a complete life cycle of chip to the readers. Rather, these are steps in manufacturing or fabrication process. However, it has been briefly presented to provide a clear picture of the overall process of obtaining a chip at hand, starting from its algorithmic concepts; for example, a video codec chip from video compress algorithm. At the end of the design signoff, the compacted database of the physical design or a layout of a chip is taped-out to the fabrication plants. The first step to translate the layout to geometry is to build the masks for manufacturing [14, 155]. The integrated circuit manufacturing is the most complex process resulting into a complex product present on the earth. The fabrication or manufacturing plants are the cleanest environments in the planet. This clearly justifies the name of the environment that is maintained for manufacturing research as the "clean room." The cleanroom has extremely minimal environmental pollution and is classified as class 1, 10, 100, 1000, 10,000, and 100,000 depending on the number of particles per cubic feet.

The integrated circuit manufacturing may takes hundreds of steps involving the following chemical, mechanical, or electrical processes: (1) slicing, (2) spin coating, (3) exposing, (4) etching, (5) cleaning, (6) deposition, (7) oxidation, (8) chemical vapor deposition, (9) ion implantation, (10) diffusion, (11) electrolysis, (12) metalization, (13) sputtering, and (14) dicing. The fabrication process starts from sand or quartz that is plentifully available on the earth. The sand or quartz after melting at a very high temperature and purified in multiple steps becomes electronic grade silicon in the form of a silicon ingot. The ingot that has a purity of 99.9999% when sliced becomes silicon wafers. The ICs are printed on these wafers that are analogous to printing photos in films on glossy papers. The steps followed in the manufacturing are unique based on a process recipe of a semiconductor house. The photolithography that essentially means "writing on stone" is the primary process involved. The principle can be simplistically explained in the following manner:

1. Grow oxide on the surface as a protection or passivation layer.
2. Coat the oxide-coated surface with photoresist.
3. Expose the photoresist with UV light using a mask.
4. Strip off the exposed and melted photoresist.
5. Etch the exposed oxide with hydrofluoric acid (HF).
6. Strip off the remaining unexposed and unmelted photoresist using a mixture of acids called piranha etch.
7. Pattern the surface that is now exposed in some area using processes such as ion implantation, CVD, and diffusion, while unexposed areas remain protected by the oxide protection or passivation layer.
8. Strip off the oxide protection or passivation layer using HF to obtain the desired patterned surface.

Steps similar to the above are followed for formation of the different layers of integrated circuit. All the dies present on the wafer are printed one after another. The dicing process gives an individual die that is essentially a silicon substrate printed with an integrated circuit.

Once all the dies in the wafer are printed, the wire bonding is necessary so that the individual pins with reasonable size can be made available for the access of the IC. The wire bonding thus making interconnections between an integrated circuit and its packaging [38]. Gold, copper, and aluminum wires can be used in the chips through binding methods like ball, wedge, and compliant. In a typical approach, the wires are attached using some combination of heat, pressure, and ultrasonic energy that leads to a weld. The important characteristics of the bonding wires include low-resistance, robust to external exposures, and good adhesiveness so that it does not form parasitics at the point of the contact. Packing involves covering the complete integrated circuit using a material so that the integrated circuit is protected and the heat generated from inside is smoothly transferred to the surroundings so that the chip does not melt itself. The packaging materials need to have a good balance of electrical, mechanical, and thermal properties while being low cost [39, 189]. The packaging materials need to have low parasitics, they need to be mechanically robust and reliable, and from thermal characteristic point of view it should be able to transfer heat.

In general, testing is a very important task in the design cycle of the integrated circuits. No chip is sent to market without going through rigorous testing. The testing involves checking of design errors as well as checking of manufacturing errors [173]. A class of equipments called automated test equipment (ATE) is used to perform testing and fault diagnosis of fabricated integrated circuits. Class testing is also performed in which the chips are tested for their key characteristics including the power dissipation and maximum operating frequency. After the class testing, the integrated circuits with the same capabilities are grouped.

8.6 Analog and Mixed-Signal Circuit Design Flow

This section discusses different techniques for mixed-signal integrated circuit design. It also discusses different techniques for automatic synthesis of mixed-signal circuits containing analog and digital portions. The mixed-signal design flows are not as mature like digital design flows. At the same time, unified mixed-design flow handling of both the analog and digital components of the mixed-signal circuit and system is very difficult due to the diverse nature of the analog and digital components. The EDA tools such as AMS Designer from Cadence®, Quest ADMS from Mentor Graphics®, Harmony from Silvaco®, Discovery AMS from Synopsys®, and Hiper Simulation AMS from Tanner® can be used for mixed-signal design.

8.6.1 Mixed-Signal Design Flow

To design AMS systems (i.e., analog and digital components not tightly coupled), it is straight-forward to think that analog and digital components can be designed independently and then

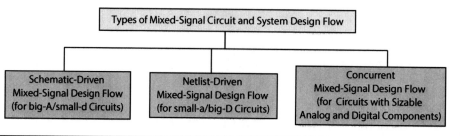

Figure 8.21 Different types of mixed-signal circuit and system design flow [15, 28, 136].

stitched together to build the AMS systems. Such independent analog and digital portion designs may not necessarily lead to overall optimal AMS system design. At the same time, some AMS circuits such as ADCs and PLLs may be natively AMS circuits needing closely coupled design of all the components to meet the target specifications. Unified AMS design flow to design AMS circuits and systems in a combined flow is desired for fast and optimal design exploration.

Three different types of AMS design flows are the following, as presented in Fig. 8.21 [15, 28, 44, 136, 184]:

1. *Schematic-Driven Mixed-Signal Design Flow:* This design flow is preferred for the design of the big-A/small-d mixed-signal circuits or systems. In such circuits and systems, the analog portion is the largest and digital portion is much smaller as compared to the analog portion. In the schematic-driven flow, a custom layout tool performs the overall design process including floor planning, chip assembly, and overall circuits or system integration. The digital portions that are either custom designed or synthesized using a digital synthesis tool are imported by the custom layout tool.

2. *Netlist-Driven Mixed-Signal Design Flow:* This design flow is preferred for the design of the small-a/big-D mixed-signal circuits or systems. In such circuits and systems, the digital portion is the largest and the analog portion is much smaller than the digital portion. In the netlist-driven flow, a digital design tool performs the overall design process including floor planning, chip assembly, and overall circuits or system integration. The digital portions are synthesized using the digital design or synthesis tool. The analog portions that are custom designed using a custom layout tool are imported by the digital design tool.

3. *Concurrent Mixed-Signal Design Flow:* The current generation of mixed-signal circuit and systems need design flow in which the analog and digital blocks are designed concurrently and overall optimal AMS design is obtained. In essence, the concurrent flow has the analog portion designed using a custom layout tool and the digital portion is synthesized using the digital synthesis tool. The floor planning, chip assembly, and ECOs are performed concurrently in a collaborative fashion by the analog and digital tools. Thus, the concurrent mixed-signal design flow represents a much more integrated approach for AMS circuit and system design.

A simplified concurrent mixed-signal design flow that highlights some of the important tasks is presented in Fig. 8.22 [15, 28, 44, 136, 184]. The mixed-signal system idea can be first verified at the system level using a language or framework such as Simulink®, Simscape®, Verilog-AMS, VHDL-AMS, or SystemC-AMS. At this point, the verification of functionalities is important not the characteristics of the system. In the next step, behavioral models can be used for architecture exploration. The behavioral model can be incorporated with mathematical equations, functional descriptions, algorithms, LUTs, or metamodels to behaviorally describe the input or output relation and the characteristics of circuits and systems [210]. The analog design engineer and digital design engineers work in close coordination with system design engineer for possible architectures of analog and digital portions to obtain an efficient mixed-signal system. The mixed-signal circuit and system is then partitioned into analog and digital subsystems. The specification of analog subsystem as well as digital subsystem is now defined. Now the analog and digital subsystems are ready for typical design flow using their respective top-down approaches. Each of the analog and digital subsystems can then

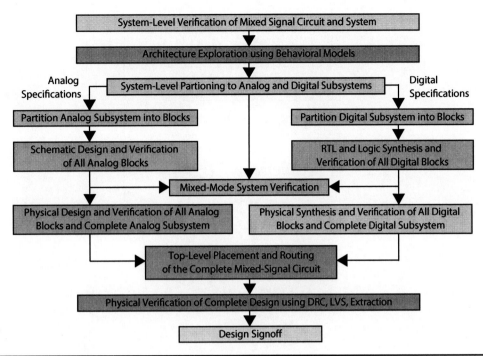

FIGURE 8.22 A simplified concurrent design flow for mixed-signal circuits [44, 184].

be hierarchically partitioned into smaller blocks to follow the "divide and conquer" approach for analog and digital subsystem design. The schematic design of analog blocks is performed. The digital blocks go through RTL and logic synthesis to obtain logic level netlist. At this point, a mixed-mode system verification can be performed to ensure that the overall system specification is not violated. The analog design engineer and digital design engineer need to work in coordination with system design engineer to make this possible. The physical design of analog block and by combining them, analog subsystem, can be made using a layout editor. The physical synthesis of digital blocks can be performed using a physical synthesis tool. Then, the entire blockwise layout can be assembled using a layout editor. Otherwise, even a physical synthesis tool can handle physical synthesis of the subsystem. The choice depends on the layout quality and layout engineering effort. The layouts of analog and digital subsystem are now individually verified for correctness as well as analog and digital specifications, respectively. The layouts of analog and digital subsystem are now assembled to obtain overall layout of the mixed-signal system through top-level placement and routing. Physical verification of the overall layout is performed as usual using DRC, LVS, and parasitic extraction.

8.6.2 Analog and/or Mixed-Signal Circuit Synthesis Techniques

Fully automated synthesis flow for mixed-signal circuits is not mature as compared with the digital circuits [208]. One reason is that the analog circuit-automated synthesis flow is not well developed. The automated synthesis of mixed-signal circuits is more difficult in a unified flow where analog and digital components are synthesized in a tightly coupled fashion. However, research is in full swing to make fully automated synthesis flow available for AMS circuits and systems. A top-down flow for mixed-signal synthesis is presented in Fig. 8.23 [48, 50, 62, 76, 90, 134, 175, 196, 197]. The flow consists of four main tasks: (1) system specification, (2) architecture synthesis, (3) constraint transformation, and (4) design verification.

The AMS synthesis framework considers Verilog-AMS or VHDL-AMS description of the target AMS circuit and system along with its specifications. The architecture synthesis step essentially performs architecture generation and selection. The architecture synthesis searches for optimal topology based on various design considerations. The constraint transformation phase derives all the building blocks (such as OP-AMP) to obtain an optimal topology. Although the automatic architecture synthesis has been a manual process, many automatic methods are now available in the

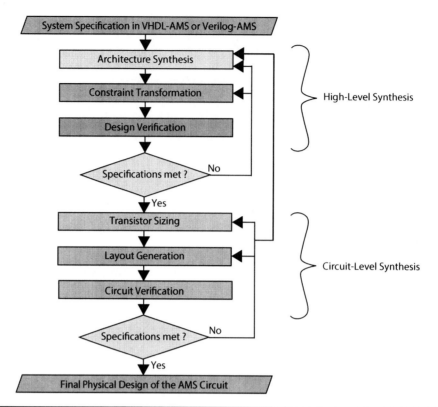

FIGURE 8.23 Analog and mixed-signal circuit synthesis flow [50, 76, 90, 134, 196, 197].

current literature. High-level design verification is performed to verify the architecture obtained. In case the specification is not met, the process is repeated with another topology. The circuit-level synthesis starts once a specific architecture is identified. The transistor sizing is performed for all transistors so that the building blocks meet the target specifications. Many automatic sizing tools are available to perform this task. The layout synthesis steps find a placement and routing solution for the sizes of the transistors obtained in the previous step so that the specifications are met. It may be noted that the transistor sizing (or circuit synthesis) and layout synthesis are two independent steps as they have different information to deal with. For example, the transistor sizing step is not aware of the parasitics that appear in the layout. The layout synthesis step need to minimize the silicon area while satisfying analog constraints such as matching and symmetry during placement and routing. Automatic layout synthesis is much difficult problem than the automatic transistor sizing. The layout of the AMS circuit is verified using DRC and LVS, and if the specifications are not met, then both the transistor sizing and layout generation steps can be repeated.

An overview of different automatic synthesis techniques applicable for both analog and mixed-signal circuits is presented in Fig. 8.24 [50, 78, 119, 120]. The existing analog synthesis tools perform two primary tasks, topology selection and transistor sizing [50, 78]. Some of the existing tools perform circuit-level synthesis and some perform system-level synthesis. Tools such as AMGIE, ANACONDA, ARCHGEN, ARIADNE, AZTECA, CADICS, CATALYST, GPCAD, HiFaDiCC, IDAC, KANSYS, MIDAS, OASYS, OPASYN, and VASS have been introduced for text-based design of analog circuits [36, 52, 72, 74, 76, 97, 99, 105, 111, 112, 117, 132, 161, 162, 176, 195, 202]. Similarly, many tools have been presented for mixed-signal synthesis. The MxSICO [55, 56, 57, 58] compiler aims synthesis of high performance mixed-signal circuits. Few tools for automatic synthesis of mixed-signal circuits from their VHDL-AMS descriptions have been presented in refs. [73, 75, 201]. The tool DAISY produces optimal $\Sigma - \Delta$ modulator topologies with building blocks for specified accuracy and bandwidth [81]. The tools such as MCAST and ADMS have been introduced for compiling VHDL-AMS or Verilog-AMS to generate device code in C for SPICE [124, 203].

The various automated approaches can be classified in four types based on the optimization methods used: knowledge based, equation based, simulation based, and model or learning-strategy

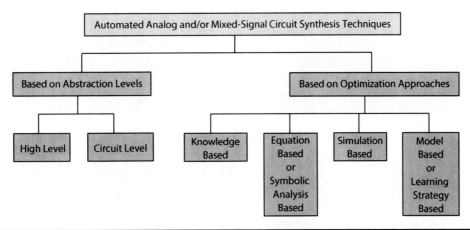

Figure 8.24 Different types of analog and/or mixed-signal circuit synthesis techniques.

based [119, 120, 169, 199]. The knowledge-based approaches use a design plan that is derived from expert knowledge to perform systematic design [104, 191]. The design plan may include descriptions of how to size the circuit components through design equations and design strategy so the target specifications are met. However, this approach does not guarantee optimal solutions. In the equation-based approaches, various analytic design equations are used to characterize the performance of the AMS circuit [54, 139]. The analytic equations can be automatically derived by using various symbolic analysis tools available. In essence, the symbolic analysis is the method of generating closed-form analytic expressions of different circuit characteristics in which the elements of the circuits are represented by symbols [89, 185]. The symbolic analysis method is complementary to numerical analysis method in which the variables and the circuit elements are represented by numbers instead of symbols. The equation-based approaches have drawbacks that new equations have to be derived each time a new topology is picked and not all characteristics of the AMS circuits can be expressed as analytical equations. The simulation-based approaches directly use analog simulators such as SPICE for sizing [63, 87, 198]. While any type of AMS circuit can be handled by simulation-based approaches, the drawback is the computational expensiveness and slow convergence. In the model or learning-strategy-based approaches, the simulations are performed over macromodels or metamodels instead of actual netlist of the AMS circuit [51, 84, 157, 169]. The models are automatically generated to describe the behavior of the AMS circuits based on distributions of variations using an analog simulator. The simulations over the macromodels and metamodels are faster, and hence the optimization takes way less time than the simulation-based approaches.

8.7 Design Flow Using Commercial Electronic Design Automation Tools

The history of EDA can be traced back to 1960s [177, 178]. However, mature EDA tools started emerging from 1980s. Since then, several EDA tool vendors have been established and have contributed to the fastest growing field of VLSI. This section discusses various commercial electronic design automation (EDA) tools. These tools are also known as electronic CAD (ECAD) or simply CAD. First, a quick introduction to selected commercial EDA tools has been provided. Then, various digital design, analog design, and mixed-signal design flows have been discussed using these tools. No specific order or preferences are followed in the discussions. The objective is to provide readers the knowledge of tools that they can use to perform design exploration using flows presented in the previous sections.

8.7.1 Selected Commercial EDA Tools

The EDA industry has matured significantly. This has resulted in reduction in the design cycle time and NRE costs of the chips. Selected commercial EDA vendors include: Cadence Design Systems®, Jasper Design Automation®, Mentor Graphics®, Microwind®, Silvaco International®, Synopsys®, and Tanner EDA®[20, 24, 25, 26, 27, 29, 30]. Most of these EDA tools run under Unix® and Linux. Many of these EDA tools run on Windows® and/or Mac OS. In addition, some EDA vendors provide

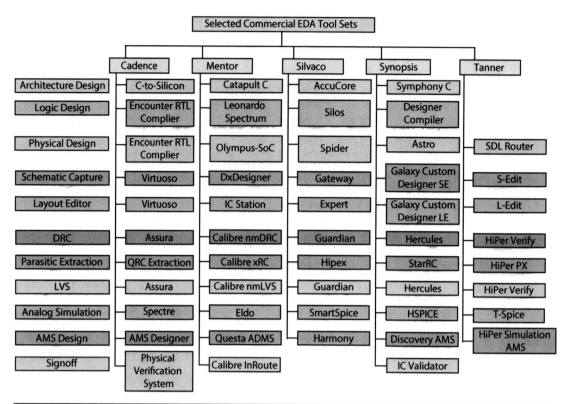

FIGURE 8.25 A selected commercial EDA tools [20, 24, 25, 26, 27, 29, 30].

different sets of tools for different platforms. The advantages of commercial EDA tools include [103, 118]: (1) engineering of quality-integrated circuits and systems, (2) getting exposure to real world experiences, (3) possibility of common user interface for all tools when bundled, (4) better product support may be available, and (5) better training support is possible. However, the disadvantages of commercial EDA tools include [103, 118]: (1) needs a steep learning curve for using the tools, (2) requires high-end computational facility, (3) requires dedicated system administration support, and (4) requires expensive annual license fee. Selected EDA tool sets are presented in Fig. 8.25. It shows selected tools that each commercial vendor provides. It may be noted that the EDA industry is fast evolving to cope with the fabrication process and their technology scaling. Thus, new tools may emerge, existing tools may be changed or even discontinued. Therefore, this should be considered as an overall guideline while expecting changes. It is a matter of choice of the design engineers to opt for a tool depending on the budget, computing platforms, and experiences.

Cadence Design Systems® have a large collection of tools for system design, functional verification, logic design, digital implementation, custom IC design, and silicon signoff, and verification [20]. Jasper Design Automation® has JasperGold® Apps for formal verification, low-power verification, connectivity verification, RTL development, and security path verification [24]. Mentor Graphics® has tool sets for electronic system design, IC design, functional verification, system modeling, and silicon test and yield analysis [25]. Microwind® has tools for schematic editor, layout editor, and nonvolatile floating gate memory simulation [26, 187, 188]. Silvaco International® has many tools for AMS and RF design, custom-integrated circuit design, interconnect modeling, and digital-integrated circuit design [27]. Synopsys® has a large collection of tools for system-level design, analog, digital, and mixed-signal verification, digital synthesis, custom integrated circuit design, and design signoff [29]. Tanner EDA® provides many tools for front-end design, back-end design, physical verification, and parasitic extraction [30].

8.7.2 For Analog Design

Most of the above EDA tool sets can perform analog design. However, two different tool sets have been randomly picked to explain analog design flow using them. Quite simplified analog design flows

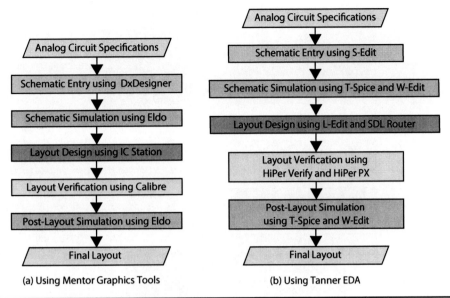

FIGURE 8.26 Simplified analog design flows using selected commercial EDA tools.

using selected commercial EDA tools are presented in Fig. 8.26. Of course, design process goes through various iterations and feedbacks depending on the need to make a successful design closure.

A simplified analog design flow using selected tools from Mentor Graphics® is presented in Fig. 8.26(a). The flow is presented in a schematic-based design flow. The schematic design of the analog circuit is performed using DxDesigner. The Eldo simulator performs the simulation and characterization of the schematic design to ensure that the specifications are met. IC Station can then be used to perform custom layout design of the analog circuit. The physical design or layout verification can be performed using Calibre through DRC and LVS checks to ensure accuracy of the layout. At this point, parasitic extraction is performed by Calibre. The postlayout parasitic accurate simulations and characterizations are performed by Eldo. The clean layout is finally obtained in GDSII format.

The schematic-driven analog design flow solution from Tanner EDA® is presented in Fig. 8.26(b) [16]. The schematic design of the analog circuit is performed using S-Edit. T-Spice and W-Edit tools can be used to perform schematic-level simulations of the design. L-Edit and schematic-driven-layout (SDL) Router tools can be used for physical design once the design meets specifications at the schematic level. Layout verification is performed using HiPer Verify tool and HiPer PX can be used for parasitic extraction of the layout. The postlayout simulations and characterization with parasitics were also performed using T-Spice and W-Edit tools.

A simplified design flow using selected tools available from Cadence Design Systems® is presented in Fig. 8.27 [21, 82, 147]. This is also a schematic-driven analog design flow. For a given set of analog design specifications, the Virtuoso schematic editor is used to perform schematic level design. The Spectre analog simulator is used to perform transistor-level simulations and characterizations. The Virtuoso Layout Suite is used at this point for custom layout of the circuit. The layout design goes through DRC/LVS using Assura. The postlayout simulations and characterizations are performed using Spectre after parasitic extraction is performed using the QRC extraction tool. The tools that are provided as a bundle can be driven through graphical user interface (GUI). However, for serious characterization, such as statistical process variation, ocean script can be used to control the tools and parameters. Sometimes multiple tool sets are needed for design optimization and other intensive design tasks. For example, tools from Cadence® and MATLAB® can be used together for design optimization. Ocean script is a preferred mechanism instead of GUI.

8.7.3 For Digital Design

The digital EDA tools are quite advanced and are available to perform automatic synthesis at almost all abstraction levels. The digital synthesis and simulations tools are available from most the EDA

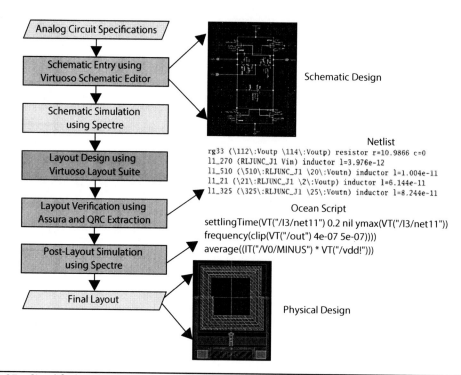

FIGURE 8.27 Simplified analog design flows using selected commercial EDA tools.

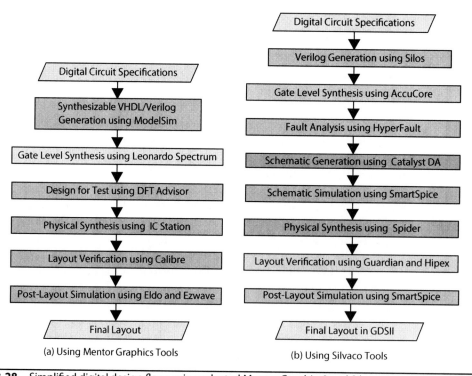

FIGURE 8.28 Simplified digital design flows using selected Mentor Graphics® and Silvaco International® EDA tools.

vendors. Selected tool sets have been presented at random to discuss the digital design flows. The simplified digital design flows have been presented in Fig. 8.28.

A simplified view of digital design flow using the selected Mentor Graphics® EDA tool is presented in Fig. 8.28(a) [59]. The inputs to the design flow are digital integrated circuit specifications. VHDL or Verilog digital HDLs are used to describe the digital design at the highest abstraction level. Synthesizable VHDL/Verilog descriptions of the digital circuit are generated from ModelSim.

A simplified digital design flow using selected tools from Silvaco International® is presented in Fig. 8.28(b) [23]. The input is a Verilog description of the digital circuit. The Verilog description needs to be in a structural format not a behavioral format so that it is synthesizable. The Silos tool can be used to simulate the Verilog to ensure that specifications are met at that level. The gate-level synthesis can be performed using AccuCore. Logic level fault analysis is possible using HyperFault. The schematic generation is performed at this point using Catalyst DA that converts the structural Verilog netlist into equivalent SPICE format netlist. This ensures that the circuit is fully verified at the schematic level before spending effort at the next lower levels of circuit abstraction. SmartSpice can be used for simulation and characterization. Once the circuit is found meeting the specifications at the transistor level, Spider tool can be used to perform physical synthesis of the digital circuit. DRC and LVS can be performed using Guardian tool. The parasitic extraction is performed using Hipex. SmartSpice can again be used to perform postlayout simulation accounting for the parasitics.

A simplified digital circuit design flow using Cadence® and Synopsys® tools is presented in Fig. 8.29 [45, 61, 96, 148]. The design of digital circuits starts with a VHDL and Verilog description of the chip. Incisive Enterprise Simulator from Cadence® or VCS from Synopsys® can be used to perform behavioral simulations of the digital circuit. Logic-level structural netlist can be obtained by using Design Compiler from Synopsys®. The logic netlist is obtained in terms of the logic cells available in the standard cell library. For example, the NOR, AND, MUX like cells as present in the Verilog netlist are actually the names of standard cells. For the physical design, the placement and routing tools from Cadence® can be used once the logic netlist is found clean in the previous stage. IC Compiler from Synopsys® can also be used. The DRC/LVS verifications of the physical design at this stage of the design flow are possible with Assura from Cadence® and Hercules from Synopsys®. For postlayout simulations purposes, the parasitic extractions can be performed using QRC Extraction from Cadence® or StarRC from Synopsys®. Spectre from Cadence® or HSPICE from Synopsys® can be used for postlayout simulations. These are industry's "gold standard" and the circuit that passes through these is as accurate as silicon data.

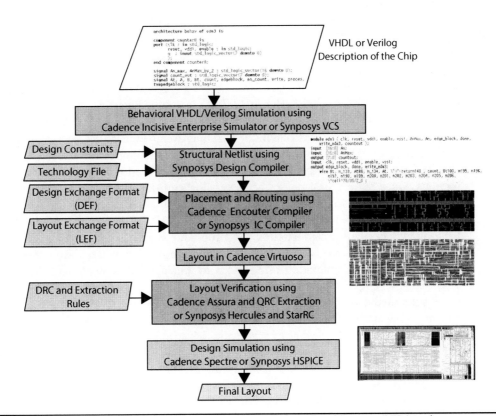

Figure 8.29 Simplified digital design flows using selected Cadence® and Synopsys® EDA tools.

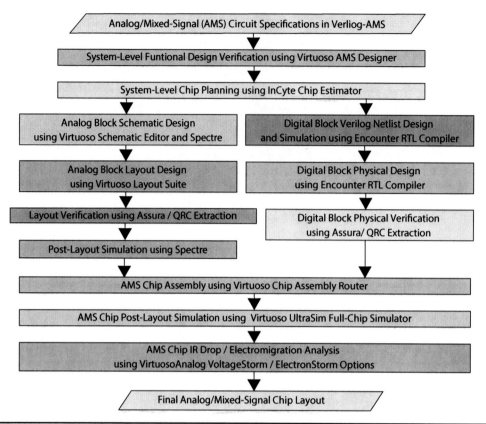

Figure 8.30 A simplified AMS design flow using selected Cadence® EDA tools [15, 28, 136].

8.7.4 For Mixed-Signal System Design

A simplified (AMS) design flow using selected Cadence® EDA tools is depicted in Fig. 8.30 [15, 28, 136]. This is essentially a concurrent AMS design flow in which design of analog and digital circuits takes place in parallel. At the same time, analog and digital designers communicate to ensure the overall AMS meets the specifications. The flow therefore can handle small-a/big-D, big-A/small-d, as well as analog and digital portions of comparable sizes each. The AMS circuit or system is first described using Verilog-AMS. Virtuoso AMS Designer tool is used for system-level design verification of AMS circuit or system. System-level chip planning can be performed using InCyte Chip Estimator. The analog portion of the AMS circuit or system is designed using schematic-drive flow as presented in the analog design flow in previous sections. The digital portion of the AMS circuit or system is designed using automatic digital synthesis flow as presented in the digital design flow in previous sections. Once the physical design of the analog and digital portions is obtained, Virtuoso Chip Assembly Router is used to assemble the overall AMS circuit or system. The postlayout simulations of the AMS chip can be performed using Virtuoso UltraSim Full-Chip Simulator. With the nanoelectronic technology-based integrated circuits, IR drop and electromigration are becoming increasingly important for signal integrity as well as reliability point of view. The layout can be checked for IR drop as well as electromigration using Virtuoso Analog VoltageStorm and ElectronStorm options. The final AMS chip layout is obtained in GDSII format.

A simplified AMS design flows using selected Tanner EDA® tools is depicted in Fig. 8.31 [19, 31]. This is a concurrent AMS design flow. However, both the analog and digital portions of the AMS circuit or system is performed using schematic-driven approach. The analog design flow is schematic-driven flow using tools from Tanner EDA®, as presented in the previous section. However, a different set of tools is needed for the synthesis of the digital portion of the AMS circuit or system. Behavioral simulation of the digital portion is possible through Aldec Riviera-PRO® tool. The overall AMS chip can be analyzed for functionality as well as characteristics using T-Spice, Aldec Riviera-PRO®, and Hiper Simulation AMS before physical design of analog and digital circuit is performed. The digital

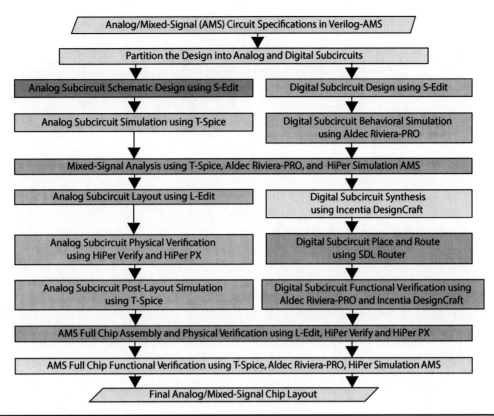

FIGURE 8.31 A simplified AMS design flows using Tanner EDA® tools [19, 31].

synthesis is performed using Incentia DesignCraft®. The place and route of the digital portion is performed using SDL Router. The digital circuit is verified using Aldec Riviera-PRO® and Incentia DesignCraft®. The AMS full chip assembly and physical verification is performed using L-Edit and HiPer. The physical design of the AMS chip is finally verified using T-Spice, Aldec Riviera-PRO®, and Hiper Simulation AMS before obtaining the final layout in GDSII format.

8.8 Design Flow Using Free or Open-Source EDA Tools

The commercial tools discussed in the above sections are important in producing quality industry standard designs. However, the cost factor including license and operational is an issue in many situations. In the institutes with low operational budget, commercial tools are difficult to acquire and maintain. In such a situation, free and/or open-source EDA tools are critical. In other situation, even an institute with huge budget can have a good mix of free/open-source and commercial tools to reduce the licensing fees. In such a situation, most of the design and verification phases are undertaken using the free/open-source tools. However, last phases of the design cycle goes through industry gold standard EDA tools. In this section, several open-source and/or free EDA tools have been discussed [108, 118].

8.8.1 Selected Free or Open-Source EDA Tools

Selected open-source and/or free EDA tools are presented in Fig. 8.32. The selection of a specific open-source and/or free tool or a set of tools is a matter of design, layout, and verification engineer's choice. The free EDA tools are the EDA tools that are made available without any fee. The source code may not be available. The open-source EDA tools are free and their source codes are available publicly. The tools are made available either under GNU General Public License (GPL) or Berkeley Open-Source License [12, 13].

The Alliance VLSI system consists of over 30 individual tools to cover complete design flow of integrated circuits [1, 108, 118, 160]. Specifically, the Alliance tool set has the following capabilities: (1) it can perform hierarchical physical design along with DRC and LVS; (2) it allows graphical entry

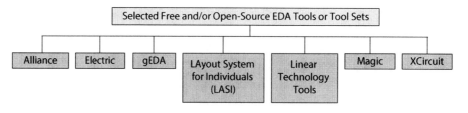

Figure 8.32 A selected free and/or open-source EDA tools.

of finite-state machine (FSM) along with its minimization and synthesis; (3) it presents VHDL simulation capabilities including customized C modules with waveform viewer; (4) it has VHDL to RTL synthesis and optimization capabilities; (5) it has standard cell placement and routing capabilities; and (6) it provides routing for pads capabilities. The Alliance tool set supports the standard VLSI description formats such as CIF, EDIF, GDSII, SPICE, and VHDL. The tool set consists of both design and validation tools. Alliance presents a set of λ-based CMOS libraries including standard cells, memories, and pads. In addition, scalable CMOS libraries are also available [160]. A brief description of selected Alliance tools follows [1]:

1. **asimut:** This is a VHDL logic simulation tool that can validate the input digital circuit behavior.
2. **boog:** This is the tool to map on a cell library logic synthesis tool.
3. **boom:** This is the logic optimizer and logic synthesis tool.
4. **Cougar:** This is a layout extraction tool.
5. **dreal:** This is a real layout-editing tool that supports rectangles in micron.
6. **graal:** This is an hierarchical symbolic layout-editing tool.
7. **loon:** This is a gate-level netlist optimization tool.
8. **Lvx:** This tool compares logical versus extracted netlist for LVS verification purposes.
9. **nero:** This is a tool for place and route using the over cell router.
10. **ocp:** This tool can be used to assemble the different parts of the circuit after placement.
11. **Ring:** This is a core-to-pads automatic routing tool that provides some indications on the pad placement for a given netlist.
12. **s2r:** This tool transforms the symbolic layout used in the tool set to a foundry compliant layout before the tape-out.
13. **syf:** This is the synthesis tool for the control flow described in the form of a finite state machine.
14. **vasy:** This tool converts most common VHDL descriptions of digital circuits to the restricted Alliance subsets.

The Electric™ VLSI Design System is a fully integrated open-source comprehensive EDA tool system that can perform various design tasks [108, 118, 171, 172]. Electric™ can handle formats such as VHDL, CIF, and GDSII. Electric™ can perform editing at layout, schematic, and architecture levels. It can perform tasks such as simulations, routing physical design, and DRC. Electric™ performs top-down while enforcing consistency of connections, and hence is a layout-constrained system. The difference between Electric™ and Alliance/Magic is that Electric™ uses connectivity for the circuits, whereas Alliance/Magic uses both connectivity and geometry. Electric™ considers circuit as set of nodes and arcs in which the nodes are electric elements and arcs are connections. This has multiple advantages to speed up and simplify the design process. For example, circuit schematic and layout can be performed in one interface. LVS can be run on layouts and then DRC can be performed without worrying about LVS match. However, Electric™ is fully integrated which indicates that establishing a flow with other individual tools is difficult.

The gEDA project has a set of tools for various tasks of design such as schematic capture, analog and digital simulation, and netlist generation into over 20 netlist formats [2]. gschem is the schematic capture tool for graphical input of components and circuits; gnetlist is the netlist generation tool. It takes gEDA/gaf .sch (schematic) files and the required .sym (symbol) files as inputs and converts them into netlists. Icarus Verilog is a Verilog simulation and synthesis tool from gEDA. GTKWave is a fully featured GTK+-based wave viewer for Unix®, Windows®, and Mac OSX. GTKWave can read and display LXT, LXT2, VZT, FST, and GHW files as well as standard Verilog VCD/EVCD files.

LAyout System for Individuals (LASI) is a versatile tool for integrated circuit as well as MEMS design that runs in Windows® platform [4]. LASI has a drawing editor and several "utility" tools. The tools include CIF, DXF, and GDS format converters, DRC tool, and matrix routing tool. LASI also has a tool to convert both schematic and layout design to SPICE netlist. However, LASI does not have a SPICE simulator and third party SPICE needs to be used for simulation purposes.

A set of analog design and simulation tools is available from Linear Technology [22]. The analog simulator tool, LTspice IV, is a high-performance SPICE simulator. It has a schematic capture tool and waveform viewer. All three are integrated within one user interface and run in Windows® platform.

Magic VLSI layout tool is one of the oldest tools written for integrated circuit layout design [5, 108, 118]. It runs in Unix® or Linux platforms. However, Magic also runs in Windows® through Cygwin application software. Magic contains many rules about circuits to assist design of circuits. Magic has built-in tools for DRC, routing, and circuit extraction.

XCircuit is a schematic capture tool that runs in Unix® or Windows® through Cygwin application software [10]. It produces circuit netlist through schematic capture. XCircuit considers the circuits as inherently hierarchical and produces hierarchical SPICE netlists. It also allows exporting of the circuit design using high-quality postscript code that can be used for publications.

8.8.2 For Analog Design

A schematic-driven analog circuit design flow using open-source/free tools is presented in Fig. 8.33 [118]. The schematic-driven analog design flow starts with a schematic entry using XCircuit [10, 118]. Transistor-level netlist is generated using XCircuit as well. The netlist at this stage can be manually edited for required analog simulations in the next phase of the design flow. The transistor-level simulation and characterization of the schematic design is performed using ngSPICE and GTKWave [3, 6]. Once the schematic design passes the functional verification and specifications, the layout design can be performed using Magic [5]. The layout verification using DRC can be performed using Magic. Parasitic extraction of the layout is also performed using Magic. LVS verification of the analog

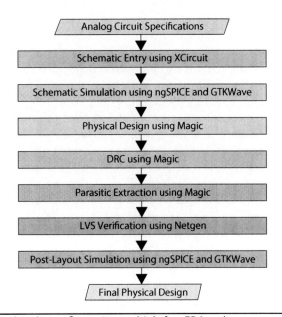

Figure 8.33 A simplified analog design flow using multiple free EDA tools.

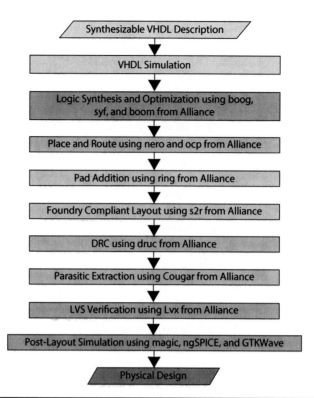

Figure 8.34 A simplified digital design flow using free EDA tools.

circuit can be performed using Netgen [7]. Postlayout simulation and characterization of the physical design accounting of all parasitics are performed using ngSPICE and GTKWave. Similarly, other tools can be used in different stages of the design flow.

8.8.3 For Digital Design

A digital design flow using open-source/free tools is presented in Fig. 8.34 [1, 108, 118, 160]. The overall digital design flow can be logically divided into five groups: (1) capture and simulation of the behavioral view, (2) capture and validation of the structural view, (3) logic synthesis, (4) physical design, and (5) verification and characterization. The complete flow is presented using a majority of the tools from Alliance tool set; however, it is possible to use other tools for individual tasks as well. At the beginning, the digital design is described through VHDL using the synthesizable subset. This step can take multiple substeps depending on the complexity of the design and a hierarchical mechanism can be used. ASIMUT tools from Alliance can be used to simulate the VDHL description. ASIMUT ensures that the VHDL contains the synthesizable subset and the design is functionally correct. Logic synthesis and optimization is performed using boog, syf, and boom tools from Alliance. For the physical design, place and route can be performed using nero and ocp tools. The addition of pads can be done using ring. The foundry compliance layout that can be taped-out can be obtained using s2r. DRC of the layout is performed using druc. Parasitic extraction of the layout is performed using Cougar. LVS verification of the physical design can be performed using lvx. The postlayout simulation accounting for all parasitics can be performed using Magic, ngSPICE, and GTKWave [3, 6]. In essence, as postlayout simulation cannot be performed due to lack of analog simulator in Alliance tool set, the layout is taken out of the Alliance and to Magic and then ngSPICE and GTKWave are used for postlayout simulations and characterizations of the layout.

8.8.4 For Mixed-Signal Design

A mixed-signal circuit and system design flow using open-source and/or free tools is presented in Fig. 8.35. At the first phase of the mixed-signal design flow, the AMS-SoC is partitioned to analog and digital portions so that analog and digital design can run concurrently. The design of analog portions

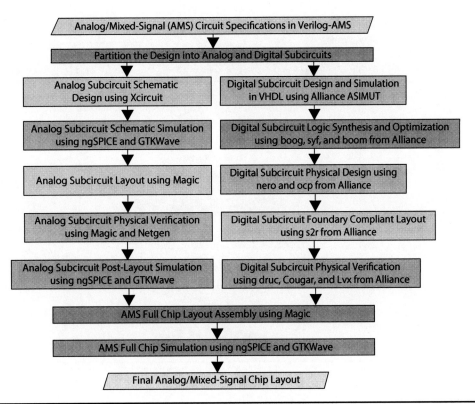

FIGURE 8.35 A simplified mixed-signal design flow using free EDA tools.

is schematic-driven and can be performed using open-source/free tools from the previous subsection. The synthesis of the digital portion can be performed mostly automatically using Alliance tools as presented in the previous subsection. Once the DRC/LVS clean layouts of analog and digital portions are obtained, then Magic can be used to assemble them. Full-chip simulation can be performed using ngSPICE and GTKWave tools before being finalized.

8.9 Comprehensive Design Flows

In the flows discussed in the above sections, simplified design flow is presented without showing design iterations that may happen during the design cycle of a chip. A real-life chip design cycle needs multiple iterations at each level for the design closure. At every level, these iterations may be part of design optimization over actual values of circuit/system characteristics. It may involve statistical design optimization over the PDFs of the circuit/system characteristics. At same time, there may be back and forth iterations from lower and higher levels of abstraction to ensure that the final chip meets the target specifications. This section discusses such design flows for analog as well as digital integrated circuit designs.

8.9.1 For Analog/Mixed-Signal Circuits and Systems

To provide a comparative perspective, a conventional design flow for AMS circuit and system that is typically used is presented in Fig. 8.36 [65, 204]. In this conventional design flow, schematic design and characterization are performed after any possible system-level behavioral simulation. This may be an automatic or manual process involving topology selection, sizing, and simulations. This may take M iterations until the schematic design meets the specifications. Then physical design or layout design of the integrated circuit is performed. This is conventionally performed by the layout engineers manually. Thus, it is a time-consuming step in the design flow. The DRC, parasitic extraction, and layout characterization are performed to verify the layout is clean and the specifications are met. Overall, N manual iterations on layout design are necessary to achieve parasitic closure [65, 204]. N can be any number depending on the skill-set of the design and layout engineers. The design

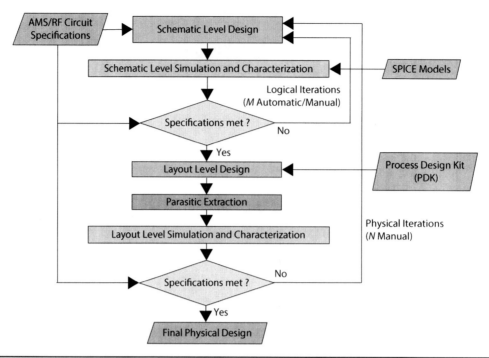

Figure 8.36 The conventional and slow AMS and RF integrated circuit design flow [86, 204].

iterations affect the design cycle and nonrecurrent cost of the chip. In addition, the manual design steps are often error prone that further affect the design cycle and overall cost of the chip. The convergence of the design cycle loop is difficult to predict as it depends on the skill of design and layout engineers, tool sets, and complexity of the integrated circuits. The design cycle convergence is also affected by the fact that the physical design parasitics extracted from previous design iteration is used in the schematic sizing of the current iteration. The handling of emerging challenges of nanoelectronics such as process variations is quite difficult in this design flow.

A fast AMS and RF design flow is presented in Fig. 8.37 [86, 87, 174, 204, 209]. This design flow starts with an initial schematic design for a specific topology and baseline size. Upon the testing of functional verification, a netlist of the schematic design is extracted. This netlist is parameterized over the design variables, e.g., width of devices. Automatic optimization of the parameterized schematic netlist is performed using search algorithms to obtain a schematic optimal design of the AMS/RF circuits. This may need M iterations over schematic netlist through SPICE. Once the optimal schematic design that meets all AMS/RF specifications is obtained, an initial layout design is performed. This is time consuming and manual layout, but only a one-time step. However, some techniques of layout optimization place and route refinement or floor-planning refinement can be applied at this point [204]. The physical design of the AMS/RF circuit goes through DRC, LVS, and parasitic (RLCK) extraction is performed for typical layout verification. The parasitic-aware netlist of the layout is simulated and characterized for the specifications using SPICE. If the specifications are not met, which is typically the case, a parasitic parameterized netlist is then created with the AMS/RF circuit design variables as parameters [86, 87]. The parasitic-aware netlist contains all the parasitics associated with the initial physical design that may be originated from the devices as well as interconnects. The devices that are selected to be varied are parameterized in terms of design variables like their length/width. The parasitics are captured based on the PDK specific to a technology. Automatic optimization of the parameterized parasitic-aware netlist is performed using search algorithms to obtain a layout optimal design of the AMS/RF circuits. This may need N iterations over parasitic-aware netlist through SPICE. It may be noted that the iterations over schematic netlist are faster than over the layout netlist. However, the number of iterations over parasitic-aware netlist may be smaller than over the schematic netlist, i.e., $N < M$. At the last phase of the design process, one final layout design may be performed manually again. This is the second layout step in the overall design flow. During the automatic optimization at the schematic as well as layout level, various FoMs of the AMS/RF circuit

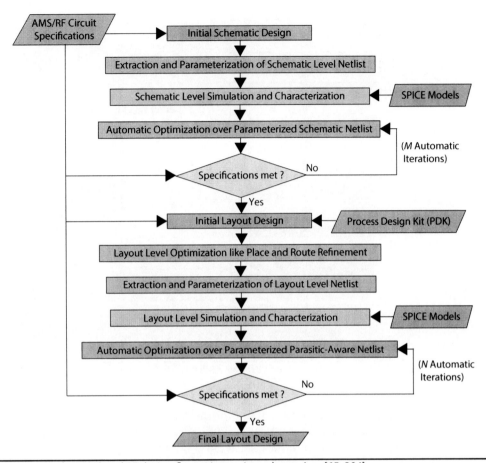

FIGURE 8.37 The fast AMS and RF design flow using various alternatives [65, 204].

and system can be considered, e.g., power, parasitic, noise, and variability based on target objectives that depend on applications. These will be discussed, in future chapters. The key aspect of this design is that it performs optimization over parasitic-aware netlist, and hence is silicon accurate. The flow needs only two manual layouts as follows: (1) one initial layout and (2) one final layout that is some modified version of the initial layout, thus needs lesser effort that the initial layout. Thus, the need for multiple layout iterations, which is shown in Fig. 8.36 is eliminated. This significantly reduced the design cycle time and NRE cost of the integrated circuit. The speed up of the overall design flow comes from the following: (1) the use of more automatic iterations instead of manual iterations in the flow, and (2) use of fast automatic optimization algorithms.

An ultra-fast design exploration flow for AMS/RF circuits is presented in Fig. 8.38 [82, 142, 210]. Three options are depicted in the figure. The option-1 is the slowest option, which has been discussed in Fig. 8.36 and time taken for design closure through manual iterations is unpredictable and error prone as discussed before. The option-2, which has been discussed in Fig. 8.37 is faster compared to the option-1. In option-2, the automatic design exploration is performed over the parasitic-aware netlist of the initial layout as discussed before. However, option-2 can be time consuming as well when the iterations need to invoke SPICE engine. For example, in a PLL, the convergence of optimization cycle in a parasitic-aware netlist within the SPICE can be of a few days' time [43]. For a specific timing scenario, an exhaustive search in a 21-parameter design space requires 10^{21} SPICE simulations [85]. If one iteration takes takes roughly 5 minutes, then the time for worst-case scenario is 5×10^{21} minutes. Running an optimization time that converged in 100 iterations is still a task of ≈ 7 days [85]. The mitigation of time problem is to perform optimization over metamodels instead of SPICE netlist as depicted in option-3 in design flow in Fig. 8.38. The metamodel is a mathematical description of the FoMs of the circuits and systems in terms of the design parameters. The metamodels provide design engineers with a simple, less computationally expensive (as it eliminates repetitive calling of SPICE engines during optimization iterations) and language-independent model that is sufficiently

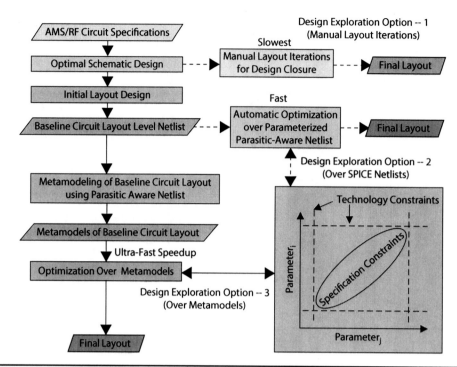

FIGURE 8.38 The metamodel assisted ultra-fast AMS and RF design flow [82, 142, 210].

accurate to produce an optimized result for the complex parametric problem. This flow can reduce design cycle time, NRE cost of the chip, and significant contribution to DFC.

The metamodel-assisted ultra-fast design flow, which is marked as option-3 in Fig. 8.38, starts with schematic design. Once the schematic design is optimal or meeting the target specifications, then an initial layout design is performed manually. This is the only major manual step in the overall ultra-fast design flow. Once the layout is DRC/LVS clean, then parasitic extraction is performed using an EDA tool to obtain a parasitic-aware netlist of the physical design. This netlist is parameterized for design variables of the integrated circuits [85]. Metamodels are then created using the parameterized parasitic-aware netlist for each of the target FoMs of the integrated circuits. Many different types of metamodels including simple polynomials, ANNs, or Kriging can be considered based on complexity of the circuits [142, 157]. Detailed discussions of different types of metamodels, metamodeling approaches, and metamodeling-assisted design optimization flow will be presented later in future chapters. It may be noted that even at the schematic-level design, the metamodeling approach can be followed. However, design exploration using schematic-level netlist that does not include much parasitic information may not be that slow. However, for larger circuits with a larger number of transistors metamodeling-assisted schematic optimization can be performed. Once the parasitic-aware metamodels are obtained from the layout-netlist, the automatic optimizations can be performed using algorithms. The automatic optimizations over metamodels are very fast and can be implemented in any high-level language and performed outside EAD tools. Once the algorithms converge to a solution, the initial layout is manually modified to obtain a final layout. This manual step is much less time consuming than the initial layout phase. In summary, the speed up in the ultra-fast design flow comes from the following: (1) the use of more automatic iterations instead of manual iterations in the flow. (2) The automatic iteration over metamodels instead of SPICE netlists. The iterations invoke a tool outside EDA framework instead of SPICE engine. (3) The use of fast optimization algorithms. It may be stated that more sophisticated optimization algorithms, which possibly are very slow when run over SPICE netlists, can be used as they run over metamodels.

8.9.2 For Digital Circuits and Systems

A comprehensive view of overall design flow for a digital circuit or system covering all the levels of abstraction is presented in Fig. 8.39 [141, 149]. At the system-level design, the system-level library that contains intellectual property cores, caches, and routers is used for design exploration [164].

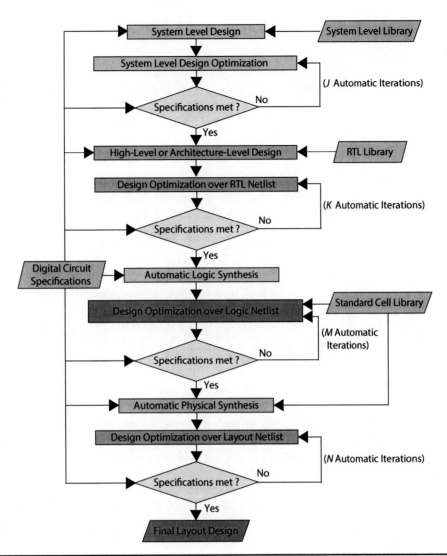

FIGURE 8.39 The comprehensive digital design flow using libraries and multiple iterations at every abstraction level [141, 149].

Automatic tools perform iterative system-level design exploration involving intellectual property configuration and logical partitioning using the system-level library and models. The high-level, architecture-level, or RTL design exploration also go through many iterations before meeting the target specifications [143, 149]. RTL, architecture-level, or datapath component library along with their RTL models are used for design exploration using automatic tools. The target architecture consisting of datapath and control is generated after it meets the specifications. The automatic logic synthesis that has a very advanced set of EDA tools performs multiple iterations before converging to an optimal logic description of the digital circuit or system meeting all the specifications [35]. For automatic physical synthesis, or silicon compilation of digital circuits, many mature EDA tools exist. The EDA tools can automatically perform physical synthesis tasks such as placement and routing with many iterations [34]. In general, the physical design optimization can take many iterations and these iterations are quite slow at this abstraction level. At each level of design abstraction, various FoMs such as power, leakage, throughput, and area can be optimized for the digital circuits or systems.

8.10 Process Design Kit and Libraries

The PDK is the first thing that is created when a process technology is introduced by a manufacturer [181, 182, 183, 193, 194]. PDKs communicate process technology information as well as possible design flows of different manufacturers or semiconductor houses to the design and layout engineers

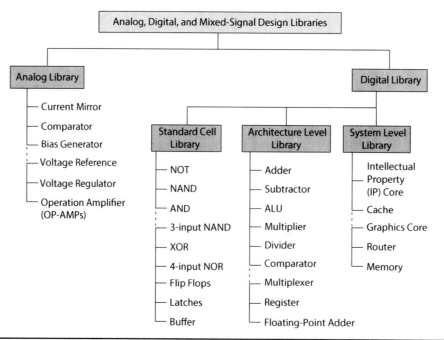

Figure 8.40 Libraries for analog, digital, mixed-signal circuit, and system design.

of the design houses. The PDKs are natives of manufacturing plants, and hence are foundry specific. At the same time, PDKs drive the evaluation engines of device parameters, rules, and models, and are EDA-vendor specific as well. The PDKs can be considered as the DNA of the entire semiconductor industry as they are the basic building blocks of the integrated circuits. A PDK contains the descriptions of the basic building elements of the process technology such as transistors and contacts that are expressed algorithmically as PCells so that they are adjusted depending on their parameters. For example, to obtain larger contact area, additional contact openings are created based on the design rules. The PDK also includes the physical parameters, electrical parameters, constraints, design rules, devices models, and layout information. All design libraries depend on the PDKs, and thus eventually PDKs affect the manufacturing capability and yields.

A broad overview of the various different libraries used in the AMS and digital circuit design is presented in Fig. 8.40. In the current trend of low-power and high-performance application design such as smart phones, the AMS-SoC is the implementation technology and needs these libraries during different phases of the design. The analog library may contain cells such as current mirrors, OP-AMPs, and bias generators [93]. On the other hand, the digital integrated circuits have different libraries depending on the level of the abstractions. It may be standard cell library (or logic cell library), architecture-level library (or RTL library), and system-level library as briefly summarized in Fig. 8.40 with selected examples for each libraries.

A typical standard-cell library contains two main components [53, 98, 160]: (1) a library database consisting of a number of views including layout, schematic, symbol, and other logical or simulation views and (2) a timing abstract needed to provide functional definitions, timing, power, and noise information for each cell. The standard cell library may contain logic cells such as inverters, AND, OR, and flip flops that are optimized for semi-custom designs [53, 98, 160]. The standard cell library contains cell layouts, cell SPICE models, parasitic extraction models, and design rules. Logic synthesis and physical synthesis processes may use standard cell library during design flow for the physical design of the integrated circuits. The architecture-level library, also known as RTL library or datapath library, contains bigger cells than the standard cell library [77, 149]. The RTL library may contain various cells such as adders, subtractors, arithmetic logic unit (ALU), and multipliers. Their integer and floating-point designs may be present. These are much bigger cells as compared to standard cells. The RTL library is used during RTL synthesis or high-level synthesis for architecture generation. For the ESL design, the system-level library is needed [60, 164]. These are the largest cells as compared to RTL or standard cells. In the current trend of AMS-SoC design with pressing time-to-market complete cores from third party may be part of this library.

8.11 EDA Tool Installation

In the above sections, many design methodologies are presented for analog, digital, and AMS circuit and/or system design. Many design flows are presented using commercial as well as open-source/free tools. The next point is how to install these tools in computing platform depending on the availability of the resources and need of the design and layout engineers. This section may not be necessarily dealing with the science and engineering aspects of the integrated circuit design. However, it, deals with the computational platforms that are needed to host the EDA tool so that they can be used for analog circuits, digital circuits and systems, and AMS circuits and systems. This may not be important when system administration and technical support are easily available. However, in many situations that is not the case. This section attempts to quickly present some discussions on how the EDA tools can be installed to build a reasonable EDA infrastructure to be used by a sizable number of designers, researchers, and students [41, 59, 95, 118]. Three different computing platforms for installation of EDA tools are presented in Fig. 8.41. Appropriate mix of hardware and software provides these platforms. The choice of these platforms depends on the choice of EDA tools.

8.11.1 Client-Server Platform

The client-server platform models for EDA installation are presented in Fig. 8.42 [59, 118]. The server can be a 64-bit multiple core server-class microprocessor with 24 to 64 GB RAM and 8 or 16 TB hard drive storage to allow multiple users running simulations at the same time. A 64-bit enterprise class linux-operating system for accessing the multicore processor and large RAM is the choice [8]. Each of the computers on client side can be low-end servers or PCs. The operating system for each client computer can be either Linux® or Windows®. The clients are x86-based PCs that are configured as workstations. The performance requirements are much less than the server. A minimal powerful processor and hardware configurations are required, e.g., 4 to 8 GB of RAM and 1 to 2 TB hard disk storage are sufficient for clients. In the case of Windows®-based clients, free X server, such as Xming, or free virtual network software, such as TightVNC, can be used to access the server [9, 11]. The main

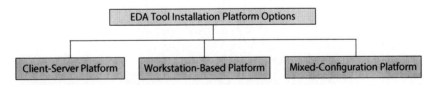

FIGURE 8.41 Different options to install EDA tools.

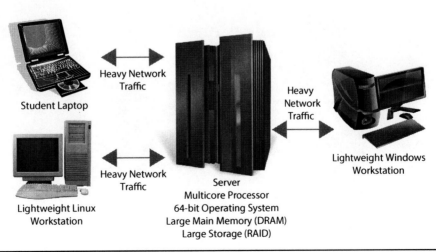

FIGURE 8.42 The client-server platform for EDA tool installation [59, 118].

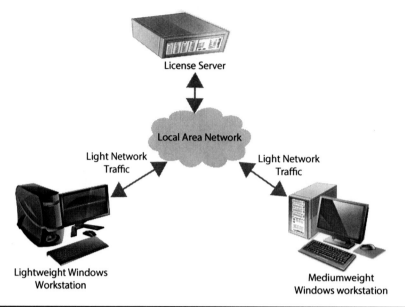

FIGURE 8.43 The workstation-based platform EDA tool installation [118].

advantage of a client-server platform is that only one machine, i.e., the server, needs to be custom configured in terms of operating system and software installations. A further advantage is that centralized accounting management is possible on the server with a network login service.

8.11.2 Workstation-Based Platform

Another platform for EDA tool installation known as a workstation-based platform is depicted in Fig. 8.43 [118]. In this platform, the use of a heavy-duty high-end server is eliminated. Each of the computing terminals is a mid-level workstation with reasonable computational capabilities. For example, a mid-level powerful processor and hardware configurations are required, e.g., quad-core processor, 8 GB of RAM and 1 TB hard disk storage are sufficient for clients. Each of the workstations will run the EDA tools. However, a centralized license server is also needed. The license server can be of minimal hardware capabilities, as it will not execute any EDA tools. The individual workstations can have dual-boot with Linux® or Windows® operating systems to allow use of all types of EDA tools.

8.11.3 Mixed-Configuration Platform

As a third alternative, the mixed-configuration platform EDA installation is depicted in Fig. 8.44 [118]. This can be considered as a combination of the two above platforms discussed in the previous

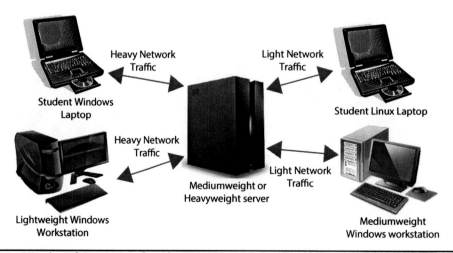

FIGURE 8.44 Mixed-configuration platform for EDA tool installation [118].

subsections. This is the most flexible configuration due to availability of a high-end server as well as individual workstations. In the mixed-configuration platform, the server handles authentication and storage but the execution of the tools can be performed either locally on the workstations or remotely on the server. The home directories of the individual users can be mounted upon login to the local workstation Network File Server (NFS). If the workstations are used as dump terminals, then the home directories can be on the server. This computing platform reduces the hardware requirements on the server at the expense of additional configuration and maintenance of the workstations. The workstation can be of the same hardware capabilities in the case of workstation-based platform in the previous subsection. The dump terminals can be with minimum hardware and software as in server-client platform in the previous subsection.

8.12 Questions

8.1 Briefly discuss the big design perspective of a smart mobile system using AMS-SoC technology.

8.2 Briefly discuss the top-down versus bottom-up design flow of the integrated circuits.

8.3 Briefly discuss the different levels of abstractions for the analog integrated circuits.

8.4 Briefly discuss the different levels of abstractions for the digital integrated circuits.

8.5 Briefly discuss the different broad steps of analog integrated circuit design. Identify at least one EDA tool that can be used at each of the steps.

8.6 Briefly discuss the different broad steps of digital integrated circuit design. Identify at least one EDA tool that can be used at each of the steps.

8.7 Briefly discuss schematic-driven, netlist-driven, and concurrent design flows of mixed-signal circuits and systems.

8.8 Do an exhaustive literature search. Identify the commercial EDA tools that can be used in Linux® platform.

8.9 Do an exhaustive literature search. Identify the commercial EDA tools that can be used in Windows® platform.

8.10 Do an exhaustive literature search. Identify the open-source/free EDA tools that can be used in Linux® platform.

8.11 Do an exhaustive literature search. Identify the open-source/free EDA tools that can be used in Windows® platform.

8.12 Do an exhaustive literature search. Identify the commercial/open-source/free EDA tools that can be used in Mac OS X platform.

8.13 Consider analog and digital circuit design flows. Comment on the time spent for each step of the design flows.

8.14 Briefly discuss PDK and analog cell library.

8.15 Briefly discuss digital library for different abstraction levels.

8.16 Discuss the parasitic extraction and LVS processes of physical design.

8.17 Consider schematic-level design of a ring oscillator and LC-VCO using a commercial and a free schematic capture and analog simulation tool. Discuss the results and your experiences.

8.18 Consider physical design including DRC/LVS/parasitic extraction of a ring oscillator using a commercial layout editor and analog simulation tool. Discuss the results and your experiences. Comment on the results as compared with the schematic level design.

8.19 Consider physical design including DRC/LVS/parasitic extraction of a ring oscillator using a free layout editor and analog simulation tool. Discuss the results and your experiences. Comment on the results as compared with the schematic-level design.

8.20 Consider the schematic and layout netlists of the ring oscillator design. Perform Monte Carlo simulations using an analog simulation tool for its power dissipation and center frequency. Discuss the statistical results. Comment on the results with LVS-level design.

8.21 Perform Monte Carlo simulations using an analog simulation tool for the power dissipation and center frequency of the ring oscillator layout netlist for two different temperatures. Tabulate the statistical results for the two temperatures. Comment on the results for the two temperatures.

8.22 Discuss the phases of high-level synthesis. Identify commercial and free tools available for this purpose.

8.23 Discuss the phases of logic synthesis. Identify commercial and free tools available for this purpose.

8.24 Discuss the phases of physical synthesis. Identify commercial and free tools available for this purpose.

8.13 References

1. Alliance: A Complete CAD System for VLSI Design. http://www-soc.lip6.fr/en/recherche/cian/alliance/, ftp://ftp.ece.lsu.edu/pub/koppel/alliance/overview.pdf. Accessed on 04th February 2014.
2. gEDA project. http://www.geda-project.org/. Accessed on 04th February 2014.
3. GTKWave Waveform Viewer. http://gtkwave.sourceforge.net/. Accessed on 10th February 2014.
4. LAyout System for Individuals (LASI). http://lasihomesite.com/index.htm. Accessed on 10th February 2014.
5. Magic VLSI Layout Tool. http://opencircuitdesign.com/magic/. Accessed on 11th February 2014.
6. Mixed – Mode Mixed – Level Circuit simulator. http://ngspice.sourceforge.net/, http://ngspice.sourceforge.net/docs/ngspice-manual.pdf. Accessed on 10th February 2014.
7. Netgen. http://opencircuitdesign.com/netgen/. Accessed on 11th February 2014.
8. The Community Enterprise Operating System. http://www.centos.org/. Accessed on 15th February 2014.
9. TightVNC software. http://www.tightvnc.com/. Accessed on 15th February 2014.
10. XCircuit. http://opencircuitdesign.com/xcircuit/. Accessed on 11th February 2014.
11. Xming X Server. http://www.straightrunning.com/XmingNotes/. Accessed on 15th February 2014.
12. Open Source Initiative. http://opensource.org/ (1998). Accessed on 10th February 2014.
13. GNU General Public License. https://www.gnu.org/copyleft/gpl.html (2007). Accessed on 10th February 2014.
14. AMAZING PICS: How a Chip Is Made. APC Magazine (2009). Accessed on 28th January 2014.
15. Cadence Analog/Mixed-signal Design Methodology. http://www.cadence.com/rl/Resources/overview/ams_methodology_ov.pdf (2009). Accessed on 07th January 2014.
16. New Approach to Accelerating Analog Layout Surpasses Full Custom and Traditional Automation Methodologies. http://www.tannereda.com/images/pdfs/Whitepapers/hiper_devgen_whitepaper.pdf (2010). Accessed on 2nd February 2014.
17. Pre and Post Layout Electrical Rule Checking using PVS Programmable ERC (PERC). http://s218101435.online-home.us/cadence/jan12/Pre_and_Post_Layout_Electrical_Rule_Checking_using_PVS_Programmable.pdf (2011). Accessed on 16th January 2014.
18. The Fastest Path to Design Signoff. Cadence Design Systems (2011). Accessed on 28th January 2014.
19. Driving A/MS Innovation: An EDA Ecosystem Approach. http://www.soccentral.com/results.asp?CatID=488 EntryID=38596 (2012). Accessed on 04th February 2014.
20. Cadence Design Systems, Inc. http://www.cadence.com/ (2014). Accessed on 30th January 2014.
21. Custom IC Design. http://www.cadence.com/products/cic/pages/default.aspx (2014). Accessed on 02nd February 2014.
22. Design Simulation and Device Models. http://www.linear.com/designtools/software/ (2014). Accessed on 10th February 2014.
23. Digital CAD. http://www.silvaco.com/products/digital_cad.html (2014). Accessed on 02nd February 2014.
24. Jasper Design Automation, Inc. http://jasper-da.com/ (2014). Accessed on 30th January 2014.
25. Mentor Graphics Corporation. http://www.mentor.com/ (2014). Accessed on 30th January 2014.
26. Microwind. http://www.microwind.net/ (2014). Accessed on 30th January 2014.
27. Silvaco, Inc. http://www.silvaco.com/ (2014). Accessed on 30th January 2014.
28. SMIC-Cadence Analog Mixed-Signal Reference Flow 1.1. http://www.smics.com/eng/design/reference_flows06.php (2014). Accessed on 04th February 2014.
29. Synopsys, Inc. http://www.synopsys.com/ (2014). Accessed on 30th January 2014.
30. Tanner EDA. http://www.tannereda.com/ (2014). Accessed on 30th January 2014.
31. Tanner EDA Mixed-Signal Design Flow. http://www.tannereda.com/ams (2014). Accessed on 04th February 2014.
32. Abidi, A.A.: Behavioral Modeling of Analog and Mixed Signal IC's: Case Studies of Analog Circuit Simulation Beyond SPICE. In: *Proceedings of the IEEE Conference on Custom Integrated Circuits*, pp. 443–450 (2001). DOI 10.1109/CICC.2001.929819.

33. Agarwal, A.: *Algorithms for Layout-Aware and Performance Model Driven Synthesis of Analog Circuits*. Ph.D. thesis, Department of Computer Science & Engineering, University of Cincinnati, Ohio (2005). URL http://rave.ohiolink.edu/etdc/view?acc_num=ucin1132259454.
34. Alpert, C.J., Karandikar, S.K., Li, Z., Nam, G.J., Quay, S.T., Ren, H., Sze, C.N., Villarrubia, P.G., Yildiz, M.C.: Techniques for Fast Physical Synthesis. *Proceedings of the IEEE* **95**(3), 573–599 (2007). DOI 10.1109/JPROC.2006.890096.
35. Amaru, L., Gaillardon, P.E., De Micheli, G.: BDS-MAJ: A BDD Based Logic Synthesis Tool Exploiting Majority Logic Decomposition. In: *Proceedings of the 50th ACM/EDAC/IEEE Design Automation Conference*, pp. 1–6 (2013).
36. Antao, B., Brodersen, A.: ARCHGEN: Automated Synthesis of Analog Systems. *IEEE Transactions on Very Large Scale Integration (VLSI) Systems* **3**(2), 231–244 (1995).
37. Antao, B.A.A., Brodersen, A.J.: Behavioral Simulation for Analog System Design Verification. *IEEE Transactions on Very Large Scale Integration (VLSI) Systems* **3**(3), 417–429 (1995). DOI 10.1109/92.406999.
38. Appelt, B.K., Tseng, A., Lai, Y.S.: Copper Wire Bonding Experiences From a Manufacturing Perspective. In: *Proceedings of the 18th European Microelectronics and Packaging Conference*, pp. 1–4 (2011).
39. Appelt, B.K., Tseng, A., Uegaki, S., Essig, K.: Thin Packaging – What Is Next? In: *Proceedings of the 2nd IEEE CPMT Symposium Japan*, pp. 1–4 (2012). DOI 10.1109/ICSJ.2012.6523400.
40. Arafa, A., Wagieh, H., Fathy, R., Ferguson, J., Morgan, D., Anis, M.H., Dessouky, M.: Schematic-Driven Physical Verification: Fully Automated Solution For Analog IC Design. In: *Proceedings of the IEEE International SOC Conference (SOCC)*, pp. 260–264 (2012). DOI 10.1109/SOCC.2012.6398358.
41. Armenti, C.: Managing EDA Tools in a Global Design Environment – Part 1. http://blog.zuken.com/index.php/2014/02/managing-eda-tools-in-a-global-design-environment-part-1/ (2014). Accessed on 04th February 2014.
42. Atienza, D., Bobba, S.K., Poli, M., De Micheli, G., Benini, L.: System-Level Design for Nano-Electronics. In: *Proceedings of the 14th IEEE International Conference on Electronics, Circuits and Systems*, pp. 747–751 (2007). DOI 10.1109/ICECS.2007.4511099.
43. Bailey, B., Martin, G.: *ESL Models and their Application: Electronic System Level Design and Verification in Practice*. Embedded Systems. Springer (2009). URL http://books.google.com/books?id=5ILek8YFmx4C.
44. Bakeer, H.G., Shaheen, O., Eissa, H.M., Dessouky, M.: Analog, Digital and Mixed-Signal Design Flows. In: *Proceedings of the 2nd International Design and Test Workshop*, pp. 247–252 (2007). DOI 10.1109/IDT.2007.4437470.
45. Balakrishnan, K.: *A VLSI System for Digital Watermarking in Images*. Master's thesis, Department of Computer Science and Engineering, University of South Florida, Tampa, FL (2003).
46. Balasinski, A.: Question: DRC or DfM? Answer: FMEA and ROI. In: *Proceedings of the 7th International Symposium on Quality Electronic Design*, pp. 788–794 (2006). DOI 10.1109/ISQED.2006.110.
47. Bali, S.: "In-Design" Physical Verification is "On-Time" Physical Verification. EDN Network (2009). Accessed on 28th January 2014.
48. Balkir, S., Dundar, G., Ogrenci, A.S.: *Analog VLSI Design Automation*. CRC Press (2003).
49. Bard, K., Dewey, B., Hsu, M.T., Mitchell, T., Moody, K., Rao, V., Rose, R., Soreff, J., Washburn, S.: Transistor-Level Tools for High-End Processor Custom Circuit Design at IBM. *Proceedings of the IEEE* **95**(3), 530–554 (2007). DOI 10.1109/JPROC.2006.889385.
50. Barros, M., Barros, M.F.M., Guilherme, J., Horta, N.: *Analog Circuits and Systems Optimization Based on Evolutionary Computation Techniques*. Studies in Computational Intelligence. Springer (2010). URL http://books.google.com/books?id=r_VfPhGq8wsC.
51. Barros, M., Guilherme, J., Horta, N.: Analog Circuits Optimization Based on Evolutionary Computation Techniques. *Integration, the VLSI Journal* **43**(1), 136–155 (2010). DOI http://dx.doi.org/10.1016/j.vlsi.2009.09.001. URL http://www.sciencedirect.com/science/article/pii/S0167926009000406.
52. Beenker, G., Conway, J., Schrooten, G., Slenter, A.: *Analog CAD for Consumer ICs*, pp. 347–367. Kluwer Academic Publishers (1993).
53. Bekiaris, D., Papanikolaou, A., Stamelos, G., Soudris, D., Economakos, G., Pekmestzi, K.: A Standard-Cell Library Suite For Deep-Deep Sub-Micron CMOS Technologies. In: *Proceedings of the 6th International Conference on Design and Technology of Integrated Systems in Nanoscale Era (DTIS)*, pp. 1–6 (2011). DOI 10.1109/DTIS.2011.5941445.
54. Bennour, S., Sallem, A., Kotti, M., Gaddour, E., Fakhfakh, M., Loulou, M.: Application of the PSO technique to the Optimization of CMOS Operational Transconductance Amplifiers. In: *Proceedings of the 5th International Conference on Design and Technology of Integrated Systems in Nanoscale Era (DTIS)*, pp. 1–5 (2010). DOI 10.1109/DTIS.2010.5487582.
55. Berkcan, E.: MxSICO: A Mixed Analog Digital Compiler: Application to Oversampled A/D Converters. In: *Proceedings of the IEEE Custom Integrated Circuits Conference*, p. 14.9.1 (1990).
56. Berkcan, E.: MxSICO: A Silicon Compiler for Mixed Analog Digital Circuits. In: *Proceedings of the IEEE International Conference on Computer Design*, pp. 33–36 (1990).
57. Berkcan, E., Currin, B.: Module Compilation for Analog and Mixed Analog Digital Circuits. In: *Proceedings of the IEEE International Symposium on Circuits and Systems*, pp. 831–834 (1990).

58. Berkcan, E., Yassa, F.: Towards Mixed Analog/Digital Design Automation: A Review. In: *Proceedings of the IEEE International Symposium on Circuits and Systems*, pp. 809–815 (1990).
59. Bhatti, M.K., Minhas, A.A., Najam-ul Islam, M., Bhatti, M.A., Ul Haque, Z., Khan, S.A.: Curriculum Design Using Mentor Graphics Higher Education Program (HEP) for ASIC Designing from Synthesizable HDL To GDSII. In: *Proceedings of IEEE International Conference on Teaching, Assessment and Learning for Engineering*, pp. W1D-1–W1D-6 (2012). DOI 10.1109/TALE.2012.6360406.
60. Bocchi, M., Brunelli, C., De Bartolomeis, C., Magagni, L., Campi, F.: A System Level IP Integration Methodology For Fast SoC Design. In: *Proceedings of the International Symposium on System-on-Chip*, pp. 127–130 (2003). DOI 10.1109/ISSOC.2003.1267734.
61. Brunvand, E.: *Digital VLSI Chip Design with Cadence and Synopsys CAD Tools*. ADDISON WESLEY Publishing Company Incorporated (2010). URL http://books.google.com/books?id=YippPgAACAAJ.
62. Carley, L.R., Gielen, G.G.E., Rutenbar, R.A., Sansen, W.M.C.: Synthesis Tools for Mixed-signal ICs: Progress on Frontend and Backend Strategies. In: *Proceedings of the Design Automation Conference*, pp. 298–303 (1996). DOI 10.1109/DAC.1996.545590.
63. Castro-Lopez, R., Guerra, O., Roca, E., Fernandez, F.V.: An Integrated Layout-Synthesis Approach for Analog ICs. *IEEE Transactions on Computer-Aided Design of Integrated Circuits and Systems* **27**(7), 1179–1189 (2008). DOI 10.1109/TCAD.2008.923417.
64. Chang, H., Kundert, K.: Verification of Complex Analog and RF IC Designs. *Proceedings of the IEEE* **95**(3), 622–639 (2007). DOI 10.1109/JPROC.2006.889384.
65. Chen, F.: RF/RF-SoC Overview and Challenges. http://www.vlsi.uwindsor.ca/presentations/chen1w_seminar_1.pdf (2004). Accessed on 24th November 2013.
66. Chen, Y., Kahng, A.B., Robins, G., Zelikovsky, A., Zheng, Y.: Evaluation of The New OASIS Format For Layout Fill Compression. In: *Proceedings of the 11th IEEE International Conference on Electronics, Circuits and Systems*, pp. 377–382 (2004). DOI 10.1109/ICECS.2004.1399697.
67. Chou, E.Y., Sheu, B.: System-on-a-Chip Design for Modern Communications. *IEEE Circuits and Devices Magazine* **17**(6), 12–17 (2001).
68. Coudert, O.: *Logic Synthesis and Verification*. chap. Logical and Physical Design: A Flow Perspective, pp. 167–196. Kluwer Academic Publishers, Norwell, MA, USA (2002). Accessed on 17th January 2014.
69. Coudert, O.: Timing and Design Closure in Physical Design Flows. In: *Proceedings of the International Symposium on Quality Electronic Design*, pp. 511–516 (2002). DOI 10.1109/ISQED.2002.996796.
70. De Jonghe, D., Maricau, E., Gielen, G., McConaghy, T., Tasić, B., Stratigopoulos, H.: Advances in Variation-Aware Modeling, Verification, and Testing of Analog ICs. In: *Proceedings of the Design, Automation Test in Europe Conference Exhibition*, pp. 1615–1620 (2012). DOI 10.1109/DATE.2012.6176730.
71. De Smedt, B., Gielen, G.: Models For Systematic Design and Verification of Frequency Synthesizers. *IEEE Transactions on Circuits and Systems II: Analog and Digital Signal Processing* **46**(10), 1301–1308 (1999). DOI 10.1109/82.799680.
72. Degrauwe, M.G.R., Nys, O., Dijkstra, E., Rijmenants, J., Bitz, S., Goffart, B.L.A.G., Vittoz, E.A., Cserveny, S., Meixenberger, C., van der Stappen, G., Oguey, H.J.: IDAC: An Interactive Design Tool for Analog CMOS Circuits. *IEEE Journal of Solid-State Circuits* **22**(6), 1106–1116 (1987).
73. Dhanwada, N.R., Vemuri, R.: Constraint Allocation in Analog System Synthesis. In: *Proceedings of the International Conference on VLSI Design*, pp. 253–258 (1998).
74. Doboli, A., Vemuri, R.: A VHDL-AMS Compiler and Architecture Generator for Behavioral Synthesis of Analog Systems. In: *Proceedings of the Design Automationa and Test in Europe (DATE) Conference*, pp. 338–345 (1999).
75. Doboli, A., Vemuri, R.: Behavioral Modeling for High-Level Synthesis of Analog and Mixed-Signal Systems from VHDL-AMS. *IEEE Transactions on CAD of Integrated Circuits* **22**(11), 1504–1520 (2003). DOI 10.1109/TCAD.2003.818302.
76. Doboli, A., Vemuri, R.: Exploration-Based High-Level Synthesis of Linear Analog Systems Operating at Low/Medium Frequencies. *IEEE Transactions on Computer-Aided Design of Integrated Circuits and Systems* **22**(11), 1556–1568 (2003). DOI 10.1109/TCAD.2003.818374.
77. Economakos, G., Xydis, S.: Optimized Reconfigurable RTL Components for Performance Improvements During High-Level Synthesis. In: *Proceedings of the 12th Euromicro Conference on Digital System Design, Architectures, Methods and Tools*, pp. 164–171 (2009). DOI 10.1109/DSD.2009.193.
78. Eeckelaert, T., Schoofs, R., Gielen, G., Steyaert, M., Sansen, W.: An Efficient Methodology for Hierarchical Synthesis of Mixed-signal Systems with Fully Integrated Building Block Topology Selection. In: *Proceedings of the Conference on Design, Automation and Test in Europe*, pp. 81–86 (2007).
79. Eleyan, N., Lin, K., Kamal, M., Mohammad, B., Bassett, P.: Semi-Custom Design Flow: Leveraging Place and Route Tools in Custom Circuit Design. In: *Proceedings of the IEEE International Conference on IC Design and Technology*, pp. 143–147 (2009). http://www2.dac.com/46th/proceedings/slides/02U_2.pdf (2009). Accessed on 07th January 2014.

80. Filiol, H., O'Connor, I., Morche, D.: A New Approach For Variability Analysis of Analog ICs. In: *Proceedings of the Joint IEEE North-East Workshop on Circuits and Systems and TAISA Conference*, pp. 1–4 (2009). DOI 10.1109/NEWCAS.2009.5290422.
81. Francken, K., Vancorenland, P., Gielen, G.: DAISY: A Simulation-Based High-Level Synthesis Tool for $\Delta\Sigma$ Modulators. In: *Proceedings of International Conference on Computer Aided Design*, pp. 188–192 (2000).
82. Garitselov, O.: *Metamodeling-Based Fast Optimization Of Nanoscale AMS-SoCs*. Ph.D. thesis, Computer Science and Engineering, University of North Texas, Denton, 76203, TX, USA., Denton, TX 76207 (2012).
83. Garitselov, O., Mohanty, S.P., Kougianos, E.: Fast Optimization of Nano-CMOS Mixed-signal Circuits Through Accurate Metamodeling. In: *Proceedings of the 12th International Symposium on Quality Electronic Design (ISQED)*, pp. 1–6 (2011).
84. Garitselov, O., Mohanty, S.P., Kougianos, E.: A Comparative Study of Metamodels for Fast and Accurate Simulation of Nano-CMOS Circuits. *IEEE Transactions on Semiconductor Manufacturing* **25**(1), 26–36 (2012). DOI 10.1109/TSM.2011.2173957.
85. Garitselov, O., Mohanty, S.P., Kougianos, E.: Accurate Polynomial Metamodeling-Based Ultra-Fast Bee Colony Optimization of A Nano-CMOS Phase-Locked Loop. *Journal of Low Power Electronics* **8**(3), 317–328 (2012).
86. Ghai, D.: *Variability Aware Low-Power Techniques for Nanoscale Mixed-Signal Circuits*. Ph.D. thesis, Computer Science and Engineering, University of North Texas, Denton, TX 76203, USA (2009).
87. Ghai, D., Mohanty, S., Kougianos, E.: Design of Parasitic and Process-Variation Aware Nano-CMOS RF Circuits: A VCO Case Study. *IEEE Transactions on Very Large Scale Integration (VLSI) Systems* **17**(9), 1339–1342 (2009). DOI 10.1109/TVLSI.2008.2002046.
88. Ghai, D., Mohanty, S.P., Kougianos, E., Patra, P.: A PVT Aware Accurate Statistical Logic Library For High-κ Metal-gate Nano-CMOS. In: *Proceedings of the 10th International Symposium on Quality of Electronic Design*, pp. 47–54 (2009).
89. Gielen, G., Wambacq, P., Sansen, W.M.: Symbolic Analysis Methods and Applications For Analog Circuits: A Tutorial Overview. *Proceedings of the IEEE* **82**(2), 287–304 (1994). DOI 10.1109/5.265355.
90. Gielen, G.G.E., Rutenbar, R.A.: Computer-Aided Design of Analog and Mixed-Signal Integrated Circuits. *Proceedings of the IEEE* **88**(12), 1825–1854 (2000). DOI 10.1109/5.899053.
91. Goering, R.: Cadence Design Tools Tutorial. http://www.vlsi.wpi.edu/cds/. Accessed on 07th December 2013.
92. Goering, R.: How Parasitic-Aware Design Flow Improves Custom/Analog Productivity. http://www.cadence.com/community/blogs/ii/archive/2011/03/14/how-parasitic-aware-design-improves-custom-analog-productivity.aspx. Accessed on 27th July 2012.
93. Gopinath, M.: *The Characterization and Model Optimization of an Analog Integrated Circuit Standard Cell Library*. Master's thesis, Department of Electrical and Computer Engineering, University of Louisville, Louisville, Kentucky (2004).
94. Grabinski, W., Schreurs, D.: *Transistor Level Modeling for Analog/RF IC Design*. Springer (2006). URL http://books.google.com/books?id=8WsV2YcyzUEC.
95. Gray, J.L.: Managing EDA Tool Installations. http://www.coolverification.com/2005/08/managing_eda_to.html (2005). Accessed on 04th February 2014.
96. Grimblatt, V.: Tutorial: Digital IC Design. http://www.sase.com.ar/2012/files/2012/09/IC-Design-Flow-SASE-2012.pdf (2012). Accessed on 31st January 2014.
97. Gupta, S.K., Hasan, M.M.: KANSYS: A CAD Tool for Analog Circuit Synthesis. In: *Proceedings of the International Conference on VLSI Design*, pp. 333–334 (1996).
98. Ha, D.S.: Cell Libraries to Support VLSI Research and Education. http://www.vtvt.ece.vt.edu/vlsidesign/cell.php (2009). Accessed on 14th February 2014.
99. Harjani, R., Rutenbar, R., Carley, L.: OASYS: A Framework for Analog Circuit Synthesis. *IEEE Transactions on Computer-Aided Design of Integrated Circuits and Systems* **8**(12), 1247–1266 (1993).
100. Hegazy, H.: Programmable Electrical Rule Checking. http://www.eetimes.com/document.asp?doc_id=1276155 (2008). Accessed on 07th January 2014.
101. Heimlich, M.: Circuit Extraction Techniques Provide Faster Interconnect Modeling and Analysis. *High Frequency Electronics* **6**(6), 42–50 (2007).
102. Ho, K.H., Jiang, J.R., Chang, Y.W.: TRECO: Dynamic Technology Remapping for Timing Engineering Change Orders. *IEEE Transactions on Computer-Aided Design of Integrated Circuits and Systems* **31**(11), 1723–1733 (2012). DOI 10.1109/TCAD.2012.2201480.
103. Hodson, R.F., Doughty, D.C.: Using Commercial EDA Software in Computer Engineering. In: *Proceedings of the ASEE Annual Conference. American Society for Engineering Education*, (1997). Accessed on 28th January 2014.
104. Horta, N.: Analogue and Mixed-Signal Systems Topologies Exploration Using Symbolic Methods. *Analog Integrated Circuits and Signal Processing* **31**(2), 161–176 (2002). DOI 10.1023/A:1015098112015. URL http://dx.doi.org/10.1023/A.
105. Horta, N., Franca, J., Leme, C.: *Automated High Level Synthesis of Data Conversion Systems*. Peregrinus, London (1991).

106. Puhan, J., Fajfar, I., Tuma, T., Bürmen, A.: Transistor Level Optimisation of Digital Cells. *Electrotechnical Review* **78**(1-2), 31–35 (2011).
107. Jiang, J.H.R., Devadas, S.: Logic Synthesis in a Nutshell (2008). Accessed on 23rd January 2014.
108. Jin, L., Liu, C., Anan, M.: Open-Source VLSI CAD Tools: A Comparative Study. In: *Proceedings of American Society for Engineering Education*, Illinois – Indiana Section Conference (2010). URL http://ilin.asee.org/Conference2010/Papers/A1_Liu_Anan.pdf.
109. Johansson, A.: *Investigation of Typical 0.13 µm CMOS Technology Timing Effects in a Complex Digital System on Chip*. Master's thesis, Division of Electronics Systems, Department of Electrical Engineering, Linkoping University (2004).
110. Johnson, K.: Characterize Nanometer Analog/RF Circuits. *Chip Design Magazine* pp. 21–22 (2010). Accessed on 17th January 2014.
111. Jusuf, G., Gray, P.R., Sangiovanni-Vincentelli, A.L.: CADICS – Cyclic Analog-to-Digital Converter Synthesis. In: *Proceedings of the International Conference on Computer Aided Design*, pp. 286–289 (1990).
112. Jusuf, G., Gray, P.R., Sangiovanni-Vincentelli, A.L.: A Performance-Driven Analog-to-Digital Converter Module Generator. In: *Proceedings of the IEEE International Symposium on Circuits and Systems*, pp. 2160–2163 (1992).
113. Kahng, A.B.: The Future of Signoff. *IEEE Design Test of Computers* **28**(3), 86–89 (2011). DOI 10.1109/MDT.2011.66.
114. Kahng, A.B., Lienig, J., Markov, I.L., Hu, J.: *VLSI Physical Design: From Graph Partitioning to Timing Closure*. Springer (2011). URL http://books.google.com/books?id=DWUGHyFVpboC.
115. Kang, S.M., Leblebici, Y.: *CMOS Digital Integrated Circuits*. Tata McGraw-Hill Education (2003).
116. Kao, W.H., Lo, C.Y., Basel, M., Singh, R.: Parasitic Extraction: Current State of the Art and Future Trends. *Proceedings of the IEEE* **89**(5), 729–739 (2001). DOI 10.1109/5.929651.
117. Koh, H.Y., Sequin, C.H., Gray, P.R.: OPASYN: A Compiler for CMOS Operational Amplifiers. *IEEE Transactions on Computer-Aided Design of Integrated Circuits and Systems* **9**(2), 113–125 (1990).
118. Kougianos, E., Mohanty, S., Patra, P.: Digital Nano-CMOS VLSI Design Courses in Electrical and Computer Engineering through Open-Source/Free Tools. In: *Proceedings of the International Symposium on Electronic System Design*, pp. 265–270 (2010). DOI 10.1109/ISED.2010.57.
119. Kubar, M.: *Novel Optimization Tool for Analog Integrated Circuits Design*. Ph.D. thesis, Electrical Engineering and Information Technology, Czech Technical University, Prague, Czech Republic (2013). 24 Feb 2014.
120. Kubar, M., Jakovenko, J.: A Powerful Optimization Tool for Analog Integrated Circuits Design. *Radioengineering* **22**(3), 921–931 (2013).
121. Lampaert, K., Gielen, G., Sansen, W.M.C.: *Analog Layout Generation Performance and Manufacturability*. Kluwer international series in engineering and computer science: Analog circuits and signal processing. Springer (1999).
122. LaPedus, M.: What's after CMOS? Semiconductor Engineering (2014). Accessed on 24th January 2014.
123. Lavagno, L., Martin, G., Scheffer, L.: *Electronic Design Automation for Integrated Circuits Handbook – 2 Volume Set*. CRC Press, Inc., Boca Raton, FL, USA (2006).
124. Lemaitre, L., McAndrew, C.C., Hamm, S.: ADMS – Automatic Device Model Synthesizer. In: *Proceedings of the IEEE Custom Integrated Circuits Conference*, pp. 27–30 (2002).
125. Lewyn, L.L.: Physical Design and Reliability Issues in Nanoscale Analog CMOS Technologies. In: *Proceedings of the NORCHIP*, pp. 1–10 (2009). DOI 10.1109/NORCHP.2009.5397862.
126. Li, X., Gopalakrishnan, P., Xu, Y., Pileggi, L.T.: Robust Analog/RF Circuit Design With Projection-Based Posynomial Modeling. In: *Proceedings of the International Conference on Computer Aided Design*, pp. 855–862 (2004). DOI 10.1109/ICCAD.2004.1382694.
127. Li, X., Le, J., Gopalakrishnan, P., Pileggi, L.T.: Asymptotic Probability Extraction for Nonnormal Performance Distributions. *IEEE Transactions on Computer-Aided Design of Integrated Circuits and Systems* **26**(1), 16–37 (2007). DOI 10.1109/TCAD.2006.882593.
128. Lin, M.B.: Synthesis. In: *Digital System Designs and Practices Using Verilog HDL and FPGAs*. John Wiley (2008). Accessed on 23th January 2014.
129. Luo, T.C., Leong, E., Chao, M.C.T., Fisher, P.A., Chang, W.: Mask Versus Schematic – An Enhanced Design-Verification Flow for First Silicon Success. In: *Proceedings of the IEEE International Test Conference*, pp. 1–9 (2010). DOI 10.1109/TEST.2010.5699238.
130. Mair, H., Xiu, L.: An ASIC Design Flow of Deep Submicron Succeeds on First Pass. In: *Proceedings of 5th International Conference on Solid-State and Integrated Circuit Technology*, pp. 352–355 (1998). DOI 10.1109/ICSICT.1998.785894.
131. Manganaro, G., Kwak, S.U., Cho, S., Pulincherry, A.: A Behavioral Modeling Approach to the Design of a Low Jitter Clock Source. *IEEE Transactions on Circuits and Systems II: Analog and Digital Signal Processing* **50**(11), 804–814 (2003). DOI 10.1109/TCSII.2003.819134.
132. del Mar Hershenson, M., Boyd, S.P., Lee, T.H.: GPCAD: A Tool for CMOS OP-AMP Synthesis. In: *Proceedings of the International Conference on Computer Aided Design*, pp. 296–303 (1998).
133. Marques, N.A., Kamon, M., Silveira, L.M., White, J.K.: Generating Compact, Guaranteed Passive Reduced-Order Models of 3-D RLC Interconnects. *IEEE Transactions on Advanced Packaging* **27**(4), 569–580 (2004). DOI 10.1109/TADVP.2004.831867.

134. Martens, E., Gielen, G.: Top-Down Heterogeneous Synthesis of Analog and Mixed-Signal Systems. In: *Proceedings of the Design, Automation and Test in Europe*, pp. 1–6 (2006). DOI 10.1109/DATE.2006.244137.
135. McConaghy, T., Eeckelaert, T., Gielen, G.: CAFFEINE: Template-Free Symbolic Model Generation of Analog Circuits via Canonical Form Functions and Genetic Programming. In: *Proceedings of the International Conference on Design, Automation and Test in Europe*, pp. 1082–1087 (2005), doi=10.1109/DATE.2005.89, issn=1530-1591.
136. McCrorie, P., Nizic, M.: Solutions for Mixed-Signal IP, IC, and SoC Implementation. http://www.cadence.com/rl/Resources/white_papers/ms_implementation_wp.pdf (2014). Accessed on 04th February 2014.
137. Meeus, W., Van Beeck, K., Goedemé, T., Meel, J., Stroobandt, D.: An Overview of Today's High-level Synthesis Tools. *Design Automation for Embedded Systems* **16**(3), 31–51 (2012). DOI 10.1007/s10617-012-9096-8. URL http://dx.doi.org/10.1007/s10617-012-9096-8.
138. Mehrotra, A., van Ginneken, L., Trivedi, Y.: Design Flow and Methodology for 50M Gate ASIC. In: *Proceedings of the Asia and South Pacific Design Automation Conference*, pp. 640–647 (2003). DOI 10.1109/ASPDAC.2003.1195102.
139. Meng, K.H., Pan, P.C., Chen, H.M.: Integrated Hierarchical Synthesis of Analog/RF Circuits with Accurate Performance Mapping. In: *Proceedings of the 12th International Symposium on Quality Electronic Design (ISQED)*, pp. 1–8 (2011). DOI 10.1109/ISQED.2011.5770817.
140. Modi, N.A., Marek-Sadowska, M.: ECO-Map: Technology Remapping for Post-Mask ECO using Simulated Annealing. In: *Proceedings of the IEEE International Conference on Computer Design*, pp. 652–657 (2008). DOI 10.1109/ICCD.2008.4751930.
141. Mohanty, S.P.: *Energy and Transient Power Minimization during Behavioral Synthesis*. Ph.D. thesis, Department of Computer Science and Engineering, University of South Florida, Tampa, USA (2003).
142. Mohanty, S.P.: Ultra-Fast Design Exploration of Nanoscale Circuits through Metamodeling). http://www.cse.unt.edu/smohanty/Presentations/2012/Mohanty_SRC-TxACE_Talk_2012-04-27.pdf (2012). Accessed on 13th February 2014.
143. Mohanty, S.P., Gomathisankaran, M., Kougianos, E.: Variability-Aware Architecture Level Optimization Techniques for Robust Nanoscale Chip Design. *Computers & Electrical Engineering* **40**(1), 168–193 (2014). DOI http://dx.doi.org/10.1016/j.compeleceng.2013.11.026. URL http://www.sciencedirect.com/science/article/pii/S004579061300308X. 40th-year commemorative issue.
144. Mohanty, S.P., Kougianos, E.: Real-Time Perceptual Watermarking Architectures for Video Broadcasting. *Journal of Systems and Software* **84**(5), 724–738 (2011).
145. Mohanty, S.P., Kougianos, E.: Incorporating Manufacturing Process Variation Awareness in Fast Design Optimization of Nanoscale CMOS VCOs. *IEEE Transactions on Semiconductor Manufacturing* **27**(1), 22–31 (2014). DOI 10.1109/TSM.2013.2291112.
146. Mohanty, S.P., Kougianos, E., Pradhan, D.K.: Simultaneous Scheduling and Binding for Low Gate Leakage Nano-Complementary Metal-Oxide-Semiconductor Datapath Circuit Behavioural Synthesis. *IET Computers & Digital Techniques* **2**(2), 118–131 (2008).
147. Mohanty, S.P., Kougnianos, E.: Fast Design Exploration of Nanoscale Circuits: Designer's Guide. http://www.cse.unt.edu/smohanty/Projects/CNS_0854182/CNS_0854182_Designer_Guide.pdf (2012). Accessed on 04th February 2014.
148. Mohanty, S.P., Ranganathan, N., Balakrishnan, K.: A Dual Voltage-Frequency VLSI Chip for Image Watermarking in DCT Domain. *IEEE Transactions on Circuits and Systems II: Express Briefs* **53**(5), 394–398 (2006). DOI 10.1109/TCSII.2006.870216.
149. Mohanty, S.P., Ranganathan, N., Kougianos, E., Patra, P.: *Low-Power High-Level Synthesis for Nanoscale CMOS Circuits*. Springer (2008). 0387764739 and 978-0387764733.
150. Mohanty, S.P., Ranganathan, N., Krishna, V.: Datapath Scheduling using Dynamic Frequency Clocking. In: *Proceedings of the IEEE Computer Society Annual Symposium on VLSI*, pp. 58–63 (2002).
151. Mohanty, S.P., Ranganathan, N., Namballa, R.: A VLSI Architecture for Visible Watermarking in a Secure Still Digital Camera (S^2DC) Design. *IEEE Transactions on Very Large Scale Integration Systems (TVLSI)* **13**(8), 1002–1012 (2005).
152. Mukherjee, V.: *A Dual Dielectric Approach for Performance Aware Reduction of Gate Leakage in Combinational Circuits*. Master's thesis, Department of Computer Science and Engineering, University of North Texas, Denton, TX, USA (2006).
153. Mukherjee, V., Mohanty, S.P., Kougianos, E.: A Dual Dielectric Approach for Performance Aware Gate Tunneling Reduction in Combinational Circuits. In: *Proceedings of the 23rd IEEE International Conference on Computer Design*, pp. 431–437 (2005).
154. Mukhopadhyay, S.: A Generic Method For Variability Analysis of Nanoscale Circuits. In: *Proceedings of the International Conference on Integrated Circuit Design and Technology*, pp. 285–288 (2009). DOI 10.1109/ICICDT.2008.4567297.
155. Nishi, Y., Doering, R.: *Handbook of Semiconductor Manufacturing Technology*. Taylor & Francis (2012). URL http://books.google.com/books?id=PsVVKz_hjBgC.

156. Okobiah, O., Mohanty, S., Kougianos, E.: Geostatistics Inspired Fast Layout Optimization Of Nanoscale CMOS Phase Locked Loop. In: *Proceedings of the 14th International Symposium on Quality Electronic Design*, pp. 546–551 (2013). DOI 10.1109/ISQED.2013.6523664.
157. Okobiah, O., Mohanty, S.P., Kougianos, E.: Geostatistical-Inspired Fast Layout Optimization of a Nano-CMOS Thermal Sensor. *IET Circuits, Devices Systems* 7(5), 253–262 (2013). DOI 10.1049/iet-cds.2012.0358.
158. Pandey, S., Glesner, M., Muhlhauser, M.: Architecture Level Design Space Exploration and Mapping of Hardware. In: *Proceedings of the International Symposium on Signals, Circuits and Systems*, pp. 553–556 (2005). DOI 10.1109/ISSCS.2005.1511300.
159. Peixoto, H.P., Jacome, M.F.: Algorithm and architecture-level design space exploration using hierarchical data flows. In: *Proceedings of the IEEE International Conference on Application-Specific Systems, Architectures and Processors*, pp. 272–282 (1997). DOI 10.1109/ASAP.1997.606833.
160. Petley, G.: VLSI and ASIC Technology Standard Cell Library Design. http://www.vlsitechnology.org/. Accessed on 10th February 2014.
161. Phelps, R., Krasnicki, M., Rutenbar, R.A., Carley, L.R., Hellums, J.R.: Anaconda: Simulation-Based Synthesis of Analog Circuits Via Stochastic Pattern Search. *IEEE Transactions on Computer-Aided Design of Integrated Circuits and Systems* 19(6), 703–717 (2000). DOI 10.1109/43.848091.
162. Van der Plas, G., Debyser, G., Leyn, F., Lampaert, K., Vandenbussche, J., Gielen, G.G.E., Sansen, W., Veselinovic, P., Leenarts, D.: AMGIE-A synthesis environment for CMOS analog integrated circuits. *IEEE Transactions on Computer-Aided Design of Integrated Circuits and Systems* 20(9), 1037–1058 (2001). DOI 10.1109/43.945301.
163. Pratt, G., Jarrett, J.: *Top-Down Design Methods Bring Back The Useful Schematic Diagram*. http://electronicdesign.com/boards/top-down-design-methods-bring-back-useful-schematic-diagram (2001). Accessed on 08th December 2013.
164. Priyadarshi, S., Hu, J., Choi, W.H., Melamed, S., Chen, X., Davis, W.R., Franzon, P.D.: Pathfinder 3D: A Flow for System-Level Design Space Exploration. In: *Proceedings IEEE International 3D Systems Integration Conference (3DIC)*, pp. 1–8 (2012). DOI 10.1109/3DIC.2012.6262961.
165. Rangarajan, S., Chakrabarti, P., Sahais, S., Datta, A., Subramanya, A.: Tutorial T3B: Engineering Change Order (ECO) Phase Challenges and Methodologies for High Performance Design. In: *Proceedings of the 27th International Conference on VLSI Design*, pp. 7–8 (2014). DOI 10.1109/VLSID.2014.118.
166. Rashinkar, P., Paterson, P., Singh, L.: *System-on-a-Chip Verification: Methodology and Techniques*. Springer (2001). URL http://books.google.com/books?id=MpzE-kclSMUC.
167. Razavi, B.: CMOS Technology Characterization For Analog and RF Design. *IEEE Journal of Solid-State Circuits* 34(3), 268–276 (1999). DOI 10.1109/4.748177.
168. Agarwal, R., Kumar, S., Sinha, S., Gotra, V.: An Insight into Layout Versus Schematic. http://www.edn.com/design/integrated-circuit-design/4418390/An-insight-into-layout-versus-schematic (2013). Accessed on 07th January 2014.
169. Rocha, F.A.E., Martins, R.M.F., Lourenco, N.C.C., Horta, N.C.G.: State-of-the-Art on Automatic Analog IC Sizing. In: *Electronic Design Automation of Analog ICs combining Gradient Models with Multi-Objective Evolutionary Algorithms, SpringerBriefs in Applied Sciences and Technology*, pp. 7–22. Springer International Publishing (2014). DOI 10.1007/978-3-319-02189-8_2.
170. Roy, R., Bhattacharya, D., Boppana, V.: Transistor-Level Optimization of Digital Designs with Flex Cells. *Computer* 38(2), 53–61 (2005). DOI http://doi.ieeecomputersociety.org/10.1109/MC.2005.74.
171. Rubin, S.M.: Electric VLSI System. http://www.staticfreesoft.com/index.html. Accessed on 04th February 2014.
172. Rubin, S.M.: *Computer Aids for VLSI Design*. R. L. Ranch Press (2009). URL http://books.google.com/books?id=kP0uPwAACAAJ.
173. Ruparel, K.N., Chin, C., Fitzgerald, J.: A Vertically Integrated Test Methodology Based on JTAG IEEE 1149.1 Standard Interface. In: *Proceedings of the Fourth Annual IEEE International ASIC Conference and Exhibit*, pp. P11-4.1–4.4 (1991). DOI 10.1109/ASIC.1991.242912.
174. Rutenbar, R.A.: Design Automation for Analog: The Next Generation of Tool Challenges. In: *Proceedings of the IEEE/ACM International Conference on Computer-Aided Design*, pp. 458–460 (2006). DOI 10.1109/ICCAD.2006.320157.
175. Rutenbar, R.A., Gielen, G.G.E., Antao, B.A.: *Computer-Aided Design of Analog Integrated Circuits and Systems*. Wiley-IEEE Press (2002).
176. Sabiro, S.G., Sen, P., Tawfik, M.S.: HiFADiCC: A Prototype Framework of a Highly Flexible Analog to Digital Converters Silicon Compiler. In: *Proceedings of the IEEE International Symposium on Circuits and Systems*, pp. 1114–1117 (1990).
177. Sangiovanni-Vincentelli, A.: The Tides of EDA. *IEEE Design Test of Computers* 20(6), 59–75 (2003). DOI 10.1109/MDT.2003.1246165.
178. Sangiovanni-Vincentelli, A.: Corsi e ricorsi: The eda story. *Solid-State Circuits Magazine, IEEE* 2(3), 6–25 (2010). DOI 10.1109/MSSC.2010.937693.
179. Sarivisetti, G.: *Design and Optimization of Components in a 45 nm CMOS Phase Locked Loop*. Master's thesis, Dept. of Computer Science and Egniniering, University of North Texas, Denton, TX, USA (2006).
180. Schaumont, P., Verbauwhede, I.: A Component-Based Design Environment for ESL Design. *IEEE Design Test of Computers* 23(5), 338–347 (2006). DOI 10.1109/MDT.2006.110.

181. Schulz, S.: Understanding Process Design Kit Standards: Part I. http://www.si2.org/?page=1129 (2010). Accessed on 04th February 2014.
182. Schulz, S.: Understanding Process Design Kit Standards: Part II. http://www.si2.org/?page=1136 (2010). Accessed on 04th February 2014.
183. Schulz, S.: Understanding Process Design Kit Standards: Part III. http://chipdesignmag.com/bayer/2010/03/15/understanding-process-design-kit-standards-part-iii/ (2010). Accessed on 04th February 2014.
184. Shariat-Yazdi, R.: *Mixed Signal Design Flow: A Mixed Signal PLL Case Study*. Master's thesis, Electrical & Computer Engineering, Waterloo, Ontario, Canada (2001).
185. Shi, C.J.R., Tan, X.D.: Canonical Symbolic Analysis of Large Analog Circuits With Determinant Decision Diagrams. *IEEE Transactions on Computer-Aided Design of Integrated Circuits and Systems* **19**(1), 1–18 (2000). DOI 10.1109/43.822616.
186. Shukla, S.K., Pixley, C., Smith, G.: Guest Editors' Introduction: The True State of the Art of ESL Design. *IEEE Design Test of Computers* **23**(5), 335–337 (2006). DOI 10.1109/MDT.2006.121.
187. Sicard, E., Bendhia, S.D.: *Advanced CMOS Cell Design*. Professional Engineering. Mcgraw-hill (2007). URL http://books.google.com/books?id=LPyQb6xPulQC.
188. Sicard, E., Dhia, S.B.: *Basics of CMOS Cell Design*. Tata McGraw-Hill professional. Tata McGraw-Hill (2005). URL http://books.google.com/books?id=iwzvoWrU9cIC.
189. Singer, A.T., Wzorek, J.F.: Cost Issues For Chip Scale Packaging. In: *Proceedings of the Twenty-First IEEE/CPMT International Electronics Manufacturing Technology Symposium*, pp. 224–228 (1997). DOI 10.1109/IEMT.1997.626922.
190. Smith, M.J.S.: *Application-Specific Integrated Circuits*. VLSI Systems Series. Addison-Wesley Publishing Company Incorporated (1997). URL http://books.google.com/books?id=im63OgAACAAJ.
191. Somani, A., Chakrabarti, P.P., Patra, A.: An Evolutionary Algorithm-Based Approach to Automated Design of Analog and RF Circuits Using Adaptive Normalized Cost Functions. *IEEE Transactions on Evolutionary Computation* **11**(3), 336–353 (2007). DOI 10.1109/TEVC.2006.882434.
192. Stan, M., Cabe, A., Ghosh, S., Qi, Z.: Teaching Top-Down ASIC/SoC Design vs Bottom-Up Custom VLSI. In: *Proceedings of IEEE International Conference on Microelectronic Systems Education*, pp. 89–90 (2007). DOI 10.1109/MSE.2007.84.
193. Stine, J.E., Castellanos, I., Wood, M., Henson, J., Love, F., Davis, W.R., Franzon, P.D., Bucher, M., Basavarajaiah, S., Oh, J., Jenkal, R.: FreePDK: An Open-Source Variation-Aware Design Kit. In: *Proceedings of the IEEE International Conference on Microelectronic Systems Education*, pp. 173–174 (2007). DOI 10.1109/MSE.2007.44.
194. Stine, J.E., Chen, J., Castellanos, I., Sundararajan, G., Qayam, M., Kumar, P., Remington, J., Sohoni, S.: FreePDK v2.0: Transitioning VLSI Education Towards Nanometer Variation-Aware Designs. In: *Proceedings of the IEEE International Conference on Microelectronic Systems Education*, pp. 100–103 (2009). DOI 10.1109/MSE.2009.5270820.
195. Swings, K., Sansen, W.: ARIADNE: A Constraint-Based Approach to Computer-Aided Synthesis and Modeling of Analog Integrated Circuits. *Analog Integrated Circuits and Signal Processing*, Kluwer Publications **3**, 197–215 (1993).
196. Tang, H., Doboli, A.: High-Level Synthesis of $\Delta\Sigma$ Modulator Topologies Optimized for Complexity, Sensitivity, and Power Consumption. *IEEE Transactions on Computer-Aided Design of Integrated Circuits and Systems* **25**(3), 597–607 (2006). DOI 10.1109/TCAD.2005.854633.
197. Tang, H., Zhang, H., Doboli, A.: *Languages for system specification*. chap. Towards High-level Analog and Mixed-signal Synthesis from VHDL-AMS Specifications: A Case Study for a Sigma-delta Analog-digital Converter, pp. 201–216. Kluwer Academic Publishers, Norwell, MA, USA (2004).
198. Thakker, R.A., Baghini, M.S., Patil, M.B.: Low-Power Low-Voltage Analog Circuit Design Using Hierarchical Particle Swarm Optimization. In: *22nd International Conference on VLSI Design*, pp. 427–432 (2009). DOI 10.1109/VLSI.Design.2009.14.
199. Tlelo-Cuautle, E., Guerra-Gomez, I., Duarte-Villasenor, M.A., de la Fraga, L.G., Flores-Becerra, G., Reyes-Salgado, G., Reyes-Garcia, C.A., Rodriguez-Gomez, G.: Applications of Evolutionary Algorithms in the Design Automation of Analog Integrated Circuits. *Journal of Applied Sciences* **10**(17), 1859–1872 (2010). DOI 10.3923/jas.2010.1859.1872.
200. Velayutham, E.: Accelerating Physical Verification with an In-Design Flow. Synopsys (2009). Accessed on 28th January 2014.
201. Vemuri, R., Dhanwada, N., Nunez, A., Campisi, P.: VASE: VHDL-AMS Synthesis Environment - Tools for Synthesis of Mixed-Signal Systems from VHDL-AMS. In: *Proceedings of the Analog and Mixed-Signal Applications Conference*, pp. 1C:77–1C:84 (1997).
202. Vital, J.C., Franca, J.E.: Synthesis of High-Speed A/D Converter Architectures with Flexible Functional Simulation Capabilities. In: *Proceedings of the IEEE International Symposium on Circuits and Systems (ISCAS)*, pp. 2156–2159 (1992).
203. Wan, B., Hu, B.P., Zhou, L., Shi, C.J.R.: MCAST: An Abstract-Syntax-Tree Based Model Compiler for Circuit Simulation. In: *Proceedings of the IEEE Custom Integrated Circuits Conference*, pp. 249–252 (2003).

204. Wang, X., McCracken, S., Dengi, A., Takinami, K., Tsukizawa, T., Miyahara, Y.: A Novel Parasitic-Aware Synthesis and Verification Flow for RFIC Design. In: *Proceedings of the 36th European Microwave Conference*, pp. 664–667 (2006). DOI 10.1109/EUMC.2006.281498.
205. Wei, Y., Doboli, A.: Structural Macromodeling of Analog Circuits through Model Decoupling and Transformation. *IEEE Transactions on Computer-Aided Design of Integrated Circuits and Systems* **27**(4), 712–725 (2008). DOI 10.1109/TCAD.2008.917575.
206. Wong, W., Gao, X., Wang, Y., Vishwanathan, S.: Overview of Mixed Signal Methodology for Digital Full-chip Design/Verification. In: *Proceedings 7th International Conference on Solid-State and Integrated Circuits Technology*, vol. 2, pp. 1421–1424 (2004). DOI 10.1109/ICSICT.2004.1436852.
207. Yu, G., Li, P.: Yield-Aware Analog Integrated Circuit Optimization Using Geostatistics Motivated Performance Modeling. In: *Proceedings of the International Conference on Computer Aided Design*, pp. 464–469 (2007). DOI 10.1109/ICCAD.2007.4397308.
208. Zeng, K., Huss, S.A.: Structure Synthesis of Analog and Mixed-Signal Circuits Using Partition Techniques. In: *Proceedings of the 7th International Symposium on Quality Electronic Design*, pp. 225–230 (2006). DOI 10.1109/ISQED.2006.125.
209. Zhang, G., Dengi, A., Rohrer, R.A., Rutenbar, R.A., Carley, L.R.: A Synthesis Flow Toward Fast Parasitic Closure for Radio-Frequency Integrated Circuits. In: *Proceedings of the 41st Design Automation Conference*, pp. 155–158 (2004).
210. Zheng, G.: *Layout-Accurate Ultra-Fast System-Level Design Exploration Through Verilog-AMS*. Ph.D. thesis, Computer Science and Engineering, University Of North Texas, Denton, 76203, TX, USA., Denton, TX 76207 (2013).
211. Zheng, G., Mohanty, S.P., Kougianos, E., Okobiah, O.: iVAMS: Intelligent Metamodel-Integrated Verilog-AMS for Circuit-Accurate System-Level Mixed-Signal Design Exploration. In: *Proceedings of the 24th IEEE International Conference on Application-specific Systems, Architectures and Processors (ASAP)*, pp. 75–78 (2013). DOI 10.1109/ASAP.2013.6567553.
212. Zheng, G., Mohanty, S.P., Kougianos, E., Okobiah, O.: Polynomial Metamodel Integrated Verilog-AMS for Memristor-Based Mixed-Signal System Design. In: *Proceedings of the 56th IEEE International Midwest Symposium on Circuits & Systems (MWSCAS)*, pp. 916–919 (2013).

CHAPTER 9
Mixed-Signal Circuit and System Simulation

9.1 Introduction

Breadboarding that involves prototyping of a system using a breadboard with connector holes is not a feasible option for an integrated circuit. The current-generation equivalent of breadboarding for integrating circuits is simulation [248]. The simulation of integrated circuits can help the overall design process during the initial design phase, debugging phase, and diagnostic phase [176]. It is the determination of the desired signals using computers [218]. In general, it may involve simulation for many aspects including the current signals, the voltage signals, the timing information, as well as power dissipation information. The simple concept of circuit and/or system is depicted in Fig. 9.1. The complete integrated circuit or overall system is made of many different components or elements. The models of these components or elements are constructed in different forms including mathematical expression, look-up-table (LUT), or plots that a computer can understand. The circuits or systems as well as the models are described in various languages that the computer can understand with various simulation frameworks. The simulation engine that uses the models and circuit/system descriptions solves the circuit/system for specific input and setup conditions. The results obtained from the simulation are viewed as plain text data and/or graphical waveform viewers. This chapter discusses the simulation engines, various forms of circuit/system descriptions, and models in detail.

Simulation is essential as it provides insight into a circuit and/or a system before it is actually built [189, 218, 278]. Various possible forms of simulations ensure that the circuit or system design is analyzed, characterized, and verified before proceeding to the next step in the design cycle. Hence simulations stop propagation of any design errors to the next level of design abstraction that may be difficult or costly to correct at a later stage. Thus, the design cycle time nonrecurrent engineering (NRE) cost, and the overall chip cost are reduced. Different forms of simulations such as ultra-fast system-level simulations, fast switch-level simulations, or slow SPICE simulations are used in various phases of the design flow for verification as well as characterization purposes. Fast and accurate simulations are needed for analog/mixed-signal system-on-a-chip (AMS-SoC) design simulation and verification. Different languages and corresponding compliers and simulation tools are needed for various levels of simulations. The digital languages like SystemC, VHDL, and SystemVerilog are used for discrete event simulations. These digital languages are not designed for the modeling and simulation of analog or continuous-time circuits or systems. Languages like Verilog-A are used for analog device and circuit simulations. Languages like Verilog-AMS, VHDL-AMS, and SystemC-AMS

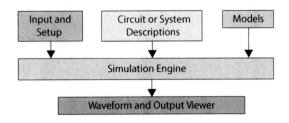

FIGURE 9.1 The concept of integrated circuit simulation.

are used for true mixed-signal circuit and system simulations [229, 263]. System-level tools like Simulink®, Simscape®, and Ptolemy II are used for functional simulation of continuous-time or mixed-signal systems [184]. In this chapter several modeling and simulation frameworks are discussed to provide a complete perspective of analog simulations, digital simulations, and mixed-signal simulations.

9.2 Simulation Types and Languages for Circuits and Systems

Integrated circuit and system simulation has many aspects and can happen in various forms and ways. Different types of simulations are shown in Fig. 9.2. Simulations need many languages and frameworks. They can be carried out at different levels of design abstractions. Based on the types of signal, a simulation can be analog, digital, or mixed-signal. Based on the models used for circuit and system description, the simulation can be behavioral, structural, or a combination of behavioral and structural. Different simulations are required for various design task stages such as initial design, post-layout, design optimization, and design verification. There are hundreds of hardware description languages (HDLs) as well as hardware verification languages (HVLs) for design and verification of circuits and systems. It is quite difficult to list all of these languages. Some of these languages are listed in Fig. 9.2 and are discussed in this chapter. The languages can be proprietary from commercial vendors or open standards. It may be noted that in addition to the simulation types shown in Fig. 9.2, other forms of simulation such as Monte Carlo analysis and statistical analysis are often used for circuit and system characterization, which are discussed in subsequent chapters.

9.2.1 Simulations Based on Abstraction Levels

Many forms of simulation are needed at different levels of design abstraction. The abstraction levels are different for analog circuits and digital circuits. However, the digital circuits and/or systems that are way large and complex in terms of the number of circuit elements are abstracted to system-level, architecture-level, logic-level, switch-level, circuit-level, layout-level, and technology-level in accordance with the divide-and-conquer design approach. At the system-level, the frameworks and languages like MATLAB®, Simulink®, Simscape®, and SystemC are used [92, 98, 99, 102, 244]. The frameworks and languages like MATLAB®, Simulink®, Simscape®, SystemC-AMS, Verilog-AMS, VHDL-AMS, and MAST HDL are used at the system- or architecture-level simulation of analog circuits as well

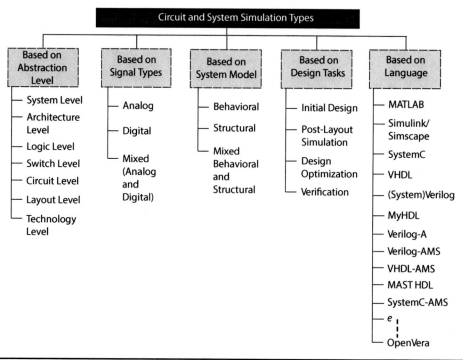

FIGURE 9.2 Different types of simulations in circuit and system design.

as AMS-SoCs [31, 49, 53, 92, 160, 293]. The architectures of digital integrated circuits are simulated using languages like VHDL, SystemVerilog, and MyHDL [29, 68, 103]. These digital HDLs can also be used for logic-level or gate-level simulation of the digital integrated circuits [276]. In general, each of these languages for analog, digital, or mixed-signal modeling needs associated compilers or simulators for the simulation purposes. The switch-level simulations of digital integrated circuits use simplistic transistor models for fast simulations [21, 192]. For example, the transistors are modeled as switched linear resistors and all the capacitors are assumed to be connected to the ground. The circuit-level simulation of the analog as well as digital circuits is performed using SPICE or SPICE-like analog simulators [179]. The analog simulators are used for layout-level simulation of an integrated circuit that contains parasitic information at this level. At the technology level, the simulations performed using the manufacturing processes are called technology computer-aided design (TCAD) [143].

9.2.2 Simulations Based on Signal Types

As presented in Fig. 9.2, three types of simulations are possible based on the signal types: analog, digital, and mixed-mode with analog and digital. The analog simulation is a continuous-time and continuous-amplitude simulation of integrated circuits using analog simulators like SPICE [216]. The Verilog-A models and corresponding simulators can also be used for analog simulations. It is interesting to note that simulators are run on digital computers but are still called analog simulations! Digital simulations differ from analog simulations in many ways [213, 216]: the digital simulations need not satisfy the Kirchhoff's conservation laws (KVL or KCL). They determine whether a change in the logic state of a node has taken place and then propagates this change to the connecting elements. Such a change in the logic state is called an event. On the occurrence of an event the digital simulations examine only the elements that are affected by the event. Thus, the digital simulations do not require to iteratively solve the voltage or current matrix. They do not solve the behavior of the entire circuit at a time. Therefore, in general, the digital simulations are way faster than the analog simulations. The languages like VHDL and SystemVerilog are among the popular languages for digital circuit modeling. Mixed-signal simulations are performed either using analog and digital simulators in parallel or using an unified mixed-signal simulator [132]. Various languages like Verilog-AMS, VHDL-AMS, MAST HDL, SystemC-AMS, and Simulink® are used for mixed-signal simulation purposes [31, 49, 53, 70, 104, 247, 258].

9.2.3 Simulations Based on System Models

For the overall simulation of a circuit or system, behavioral, structural, or a combination of behavioral and structural models can be used [129, 224, 272]. In the behavioral modeling, an integrated circuit or system is described in terms of its operations with respect to time, like an algorithm that performs the overall task. On the other hand, the structural modeling involves connecting different components of the integrated circuit together to obtain the final design of it. For a comparative perspective, Algorithm 9.1 presents partial models for a 2-bit register in VHDL [129]. The behavioral architecture model has process statements with assignment and wait statements. The structural architecture model has signal declarations, component instances, port maps, and wait statements. The behavioral model is difficult to synthesize automatically to the next level of design abstraction. A subset of behavioral models can be automatically synthesized using an available behavioral or high-level synthesis tool. The structural modeling connects components hierarchically to form a netlist and is easier to manage and reuse. The structural model can be easily synthesized automatically. In practice a combination of behavioral and structural models is used in hierarchical fashion for the simulation of large circuits and systems modeling.

9.2.4 Simulations Based on Design Tasks

The types of simulations needed for various design tasks of the integrated circuits and the tools used for these purposes are different [155, 156, 196, 255]. The initial design phase for analog and digital circuit or system can be behavioral simulation through MATLAB® and Simulink® like frameworks. However, the initial structural design of analog circuits can start from a topology selection and schematic design in a schematic editor. At this stage SPICE can be used for simulation and functional verification of the circuit. Another method, called symbolic analysis, that studies the behavior of the

ALGORITHM 9.1 Behavioral versus Structural Modeling: A 2-bit Register Simulation using VHDL [129]

```
        Behavioral Modeling                          Structural Modeling

architecture Behave of Register_2b is     architecture Structure of Register_2b is
begin                                        component D_Latch
   storage : process                            port (D, clk : in bit; Q : out bit);
   variable stored_D0, stored_D1 : bit;      end component;
   begin                                     signal int_clk : bit;
      if clk = '1' then                   begin
         stored_D0 := D0;                    bit0 : D_Latch
         stored_D1 := D1;                    port map (D0, int_clk, Q0);
      end if;                                bit1 : D_Latch
      Q0 <= stored_D0 after 2 ns;            port map (D1, int_clk, Q1);
      Q1 <= stored_D1 after 2 ns;         end Structure;
      wait on D0, D1, clk;
   end process storage;
end Behave;
```

integrated circuits in terms of symbolic parameters can also be used. Although symbolic analysis can be used for many design aspects including topology selection and behavioral modeling, it has not been widely used due to exponential complexity with the circuit size. The structural design and verification can start with register-transfer level (RTL) modeling and simulation. The languages like SystemVerilog and VHDL can be used at this stage. The post-layout simulation of the integrated circuit over its netlist can be performed using a SPICE [155, 203, 220, 286]. However, for large circuits with many transistors, the simulation using SPICE is slow and may even be infeasible due to heavy demand for the computational capacity during the simulations. The post-layout simulations can be made significantly faster using techniques like macromodeling and metamodeling. For analog design optimization, simulations through SPICE can be used for iterative optimizations. At the same time, faster and efficient simulations for optimization over macromodels, metamodels, and other methods can be performed [155, 203, 220, 286]. For analog and mixed-signal integrated circuits, the simulations can be performed using languages like Verilog-A and Verilog-AMS. Different types of simulations are performed for optimization of digital design at a specific level of abstraction, such as system level or logic level. In addition, behavioral and structural modelings are possible at each of the design abstractions. So the selection of a language for a digital design simulation depends on the abstraction and type of model for each of the elements; it may be SystemC, SystemVerilog, VHDL, and MyHDL [68, 102, 103, 111, 112]. Design verification may need different languages. Analog design verification can be performed using languages like Simulink®, Verilog-AMS, VHDL-AMS, and SystemC-AMS [131, 163]. Verification of small analog circuits can always be performed at SPICE level itself. Macromodeling-based simulations can be performed for analog design verification. The digital integrated circuit verification can be performed using languages like SystemVerilog, e, OpenVera, and SystemC [32, 103, 167].

9.2.5 Simulation Languages

Various types of simulations are possible using different languages available for integrated circuit and AMS-SoC design [121, 133, 136, 256, 292]. At a very high level, the languages can be categorized into different types based on the type of signal, circuit, or system and also based on their capabilities for hardware design and verification as depicted in Fig. 9.3. Based on signal, circuit, and system types, the languages for hardware can be analog, digital, or mixed-signal. Verilog-A is an example of pure analog language [223]. The digital languages include SystemVerilog, VHDL, and MyHDL. The mixed-signal languages include Verilog-AMS and VHDL-AMS. Based on the capabilities, the hardware languages can be categorized into the following types: hardware description language (HDL), hardware verification language (HVL), and hardware description and verification language (HDVL). Examples of HDL are SystemVerilog and VHDL. Languages such as e and OpenVera are HVLs. HDVL, which is essentially (HDL+HVL), includes SystemVerilog and SystemC.

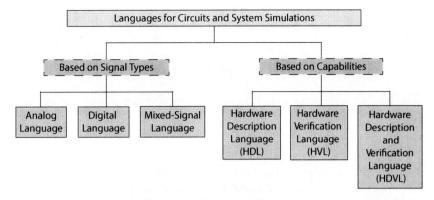

FIGURE 9.3 Categories of languages for circuit and system simulation.

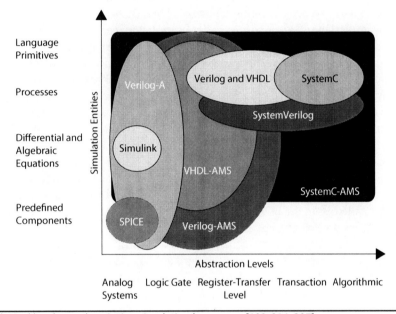

FIGURE 9.4 Selected hardware description simulation languages [108, 214, 227].

The scope and capabilities of selected languages are depicted in Fig. 9.4 [104, 108, 180, 214, 227]. On the horizontal axis, different abstraction levels such as analog systems, logic level, RTL, transaction, and algorithmic are presented. On the vertical axis, different simulation entities such as predefined components, differential equations, algebraic equations, processes, and language primitives are presented. It is evident that two different languages may have some common features, but they also have additional features to justify their need in hardware design and verification. Verilog-AMS supports most of the simulation entities and allows analog and multilevel digital abstractions. SystemC-AMS claims to have maximum simulation entities and supports most of the analog as well as digital abstractions.

9.3 Behavioral Simulation using MATLAB®

MATLAB® is a high-level language and interactive framework presented for numerical computation, visualization, and programming. A design engineer can analyze result data, develop algorithms, create models, and simulate applications. MATLAB® and similar frameworks that can be used for textual-based behavioral simulation of circuits and systems are presented in Fig. 9.5. MATLAB®, which stands for matrix laboratory, is a commercial simulator [92]. Other simulators such as GNU Octave, Scilab, and SciPy are open-source and/or free alternatives of MATLAB® [87, 95, 96]. GNU Octave is a high-level interpreted language that performs numerical computations of linear and nonlinear problems

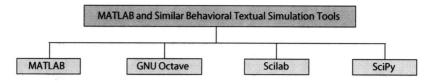

Figure 9.5 MATLAB® and selected similar behavioral textual simulation frameworks.

and is distributed under GNU General Public License (GPL) [75]. Scilab is distributed under GPL compatible license, which has several mathematical functions that facilitate advanced data structures as well as graphical functions. SciPy is a Python framework that provides a large set of mathematical algorithms and functions for numerical computation. For brevity, the rest of the discussion is on MATLAB®; however, a similar discussion can be made for similar tools.

There is no doubt that the language, tools, and built-in functions of MATLAB® can be used to explore multiple approaches and reach alternative solutions way faster than with spreadsheets or traditional C like programming languages [161, 207]. The question arises why MATLAB® for the hardware designers when so many hardware languages are available. MATLAB® satisfies multiple needs at various stages of the design cycle. At the initial stage, behavioral simulations of analog as well as digital designs can be performed for the first proof of the concept. During the design optimization, the large set of algorithms that are available in MATLAB® can be used by the design engineers for the optimization purposes without writing the algorithms from scratch. This will save time for the design and verification engineers and in turn will lead to reduced design cycle time. MATLAB® can be used to generate random input data needed for process variation analysis using a large set of statistical tools. It can be used for generating metamodels [151, 220], and also for experimental result analysis and display. In addition, graphical user interface (GUI) can be developed using MATLAB® for easy usage of the electronic design automation (EDA) and non-EDA tools seamlessly during the complete design cycle [202]. MATLAB® and EDA tools can work together in many ways [151, 220]. MATLAB® can be called from an EDA tool when needed or EDA tools can be called from MATLAB® when needed. Scripts provided by EDA tools, for example, ocean scripts, MATLAB® programs, or third-party scripting languages, can be used for close coordination of MATLAB® and EDA tools. MATLAB® can be used for rapid prototyping of designs as VHDL-AMS and VHDL descriptions can be generated from it [159, 247]. The large collection of filters and digital signal processing (DSP) blocks will come in handy for this purpose. However, in general, MATLAB® cannot replace HDLs, HVLs, or HDVLs but can complement them.

9.3.1 System- or Architecture-Level Simulations

In this section, system- or architecture-level modeling using MATLAB® is discussed with selected examples. First analog or mixed-signal circuit or system is discussed. Then a digital circuit example is discussed.

9.3.1.1 Analog/Mixed-Signal (AMS) Design Modeling

In this section, the design of a continuous-time sigma-delta modulator (CT-SDM) analog-to-digital (ADC) system is presented [258, 259, 282, 286, 287]. The design of a third-order CT-SDM with feedforward loop filter (LPF) is discussed as a specific case study [286, 287]. The system-level design flow for the CT-SDM along with the simulation frameworks for each of the steps is shown in Fig. 9.6 [286, 287]. The design and simulation frameworks for discrete-time sigma-delta modulator (DT-SDM) are quite mature compared to the CT-SDM design [246, 286, 287]. Thus, a practical way of designing a good CT-SDM is to first obtain a system-level DT-SDM design with desired performance using the widely available tools. Then map this DT-SDM design to a CT-SDM topology to obtain the target CT-SDM [226]. This approach is presented in the flow in Fig. 9.6.

The system-level design of DT-SDM begins with specific parameters of SDM, such as order, oversampling ratio (OSR), out-of-band gain (OBG) using the MATLAB® toolbox [246, 286, 287]. The system-level design parameters and specifications for each building block of the DT-SDM are obtained taking into consideration the signal-to-noise ratio (SNR) and stability. The resulting design of the DT-SDM is then converted to a CT-SDM implementation. At this stage of the design process, necessary

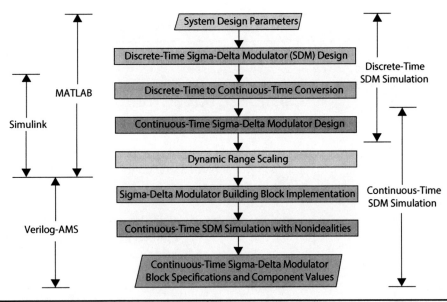

FIGURE 9.6 A system-level sigma-delta modulator design flow [286, 287].

dynamic range scaling is performed to ensure that the outputs of all stages of the SDM are within the range. Simulations are performed throughout the design process to estimate the design specification, in order to ensure that the target requirements are met. Critical nonidealities are modeled and simulated, which leads to reasonable specifications for each modulator building block. The behavioral modeling of the resultant CT-SDM design is now performed using Verilog-AMS. Realistic nonidealities for each of the building blocks of the CT-SDM are included in the models, and simulations are performed to verify the design through mixed-signal behavioral simulations.

The system block diagram representation of a DT-SDM is presented in Fig. 9.7 [286, 287]. A typical sigma-delta modulator consists of an LPF, an analog-to-digital converter (ADC), and a feedback digital-to-analog converter (DAC). The z-domain transfer function of the discrete-time LPF is $L(z)$. The ADC converts its input discrete-time sampled signal $y[n]$ to digital output signal $v[n]$. The DAC presents as the feedback converts the digital signal $v[n]$ back to discrete-time signal so that it can be subtracted from the sampled input of $u(t)$ to calculate the error signal. The segment of the MATLAB® program that has been used in the design, simulation, and analysis of DT-SDM is presented in Algorithm 9.2 [245, 286, 287]. The program needs the sigma-delta toolbox from MATLAB® central [245].

A single-bit ADC (aka quantizer) has been used in this design demonstration for simplicity and low-power dissipation even though multibit ADC can provide better noise-shaping, lower jitter sensitivity, and better stability. The OBG is another high-level design factor of the SDM. OBG determines the gain of the signal with a frequency component at sampling frequency f_s. A higher OBG leads to lower in-band noise at the cost of higher jitter noise and increases instability. The use of

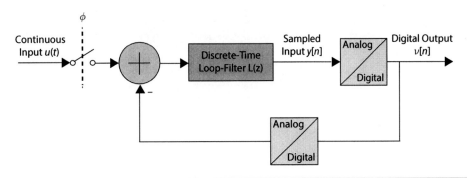

FIGURE 9.7 System-level block diagram of a DT-SDM [286, 287].

ALGORITHM 9.2 High-Level Overview of a MATLAB® Model for DT-SDM [245, 286, 287]

```
% Design Parameter Initialization and Simulation Setup
order = 3; osr = 128; nlev = 2; obg = 1.3;
f0 = 0; opt = 0; form = 'CIFF'; msa = 0.8;
xlim_MIN = 0.1;              % Minimum State Amplitude
xlim_MAX = 0.8;              % Maximum State Amplitude
td = 0.2;                    % Excess Loop Delay
modname = ['mod' num2str(order) '_' form '_opt' num2str(opt)];
sin_AMP = 0.85;              % Input Sinewave Amplitude (<= 1)
freq_in = (8/20)/(2*osr);    % Input Frequency
time_sim = [];               % Simulation Time
N_short = 5;                 % Number of input signal period to simulate (small)
N_long = 2^5;                % Number of input signal period to simulate (large)
% Design Discrete Time (DT) Noise Transfer Function (NTF)
NTF_DT = synthesizeNTF(order, osr, opt, obg);
loop_filter = 1 / NTF_DT - 1;
show_dttf(NTF_DT, modname); % Show DT NTF Pole/Zero and Spectrum Plots
time_sim = N_long / frequency_in - 1;
u_dt = sin_AMP * sin(2*pi * [0 : time_sim] * freq_in);
v_dt = sim_dtdsm(NTF_DT, nlev, form, u_dt);    % Simulate DT-DSM
dsm_spec(v_dt, time_sim, osr, modname, {'DT'}); % Spectrum Analaysis
% Discrete Time (DT) to Continous Time (CT) Mapping
Ns = 30;   % Number of Samples for the Coefficient Fitting
a = dsm_dt2ct(NTF_DT, order, td, Ns);
```

functions like `synthesizeNTF` in the sigma-delta toolbox simplifies the DT-SDM design tasks. The determination of the noise transfer function (NTF) taking into consideration all system-level specifications is the starting step of DT-SDM design. The resultant NTF of the DT-SDM is the following for specific constraints [286, 287]:

$$\mathrm{NTF}_{\mathrm{DT}}(z) = \left(\frac{(z-1)^3}{(z-0.7701)(z^2 - 1.708z + 0.7684)} \right) \tag{9.1}$$

The NTF of the DT-SDM is evaluated in z-plane and frequency domain as shown in Fig. 9.8(a) and (b), respectively. The output power spectrum density (PSD) of the DT-SDM is shown in Fig. 9.9 [286, 287]. The plot shows that the DT-SDM design results in a signal-to-noise ration $\mathrm{SNR}_{\mathrm{DT}}$ of 100.5 dB,

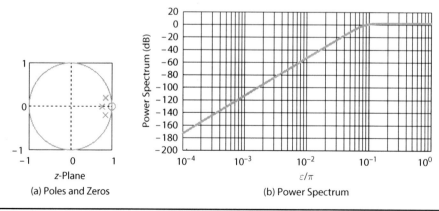

(a) Poles and Zeros (b) Power Spectrum

FIGURE 9.8 Synthesized NTF power spectrum and its poles and zeros in z-domain [286, 287].

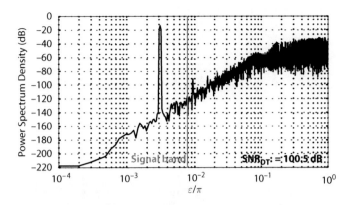

FIGURE 9.9 Output power spectral density (PSD) of the DT-SDM from MATLAB® simulations [286, 287].

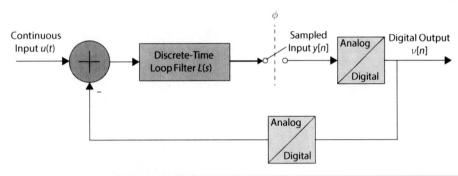

FIGURE 9.10 System-level block diagram of a CT-SDM [286, 287].

which is sufficient for a low-power ADC requirement. The z-domain transfer function of the LPF is the following [286, 287]:

$$\text{LPF}_{\text{DT}}(z) = \left(\frac{0.513(z^2 - 1.756z + 0.783)}{(z-1)^3} \right) \quad (9.2)$$

To obtain the CT-SDM design, first a CT-SDM architecture is selected. Then the coefficients of the LPF are calculated for the selected CT-SDM architecture. The system-level architecture of the CT-SDM is shown in Fig. 9.10 [286, 287]. The major difference between DT-SDM and CT-SDM is that the DT-SDM samples the analog input signal outside the loop whereas the CT-SDM samples the signal at the output of the LPF. The sampling in the SDM is controlled by a clock signal ϕ with sampling frequency f_s. A CT-SDM has an implicit anti-aliasing filter (AAF). Typically the power dissipation of a CT-SDM is lesser than that of a DT-SDM with a similar structure [154, 286].

9.3.1.2 Digital Design Modeling – Watermarking Chip

MATLAB® is particularly important for digital design of signal processing applications. The DSP chip design engineers can take particular advantage of the rich set of functions available in the toolboxes of MATLAB®. The MATLAB® simulation ensures the correctness of the target DSP hardware. It allows a full-blown verification before more design effort and corresponding dollars toward NRE cost of the chip are spent on the hardware design. Once the design is verified, the rest of the digital design can be performed by following automatic or semi-custom design steps. As a specific example, the design of watermarking is discussed in this section [197]. Digital watermarking has been used for digital rights management (DRM) and in particular for copyright protection. Digital watermarking embeds watermark in images or videos and detects it for copyright proof whenever needed.

A high-level block diagram view of the watermarking digital design is depicted in Fig. 9.11 [130, 199, 204]. In this specific example, the watermarking chip considers an input image or video and a watermark image. Various signal processing operations are performed over the host image or video

430 Chapter Nine

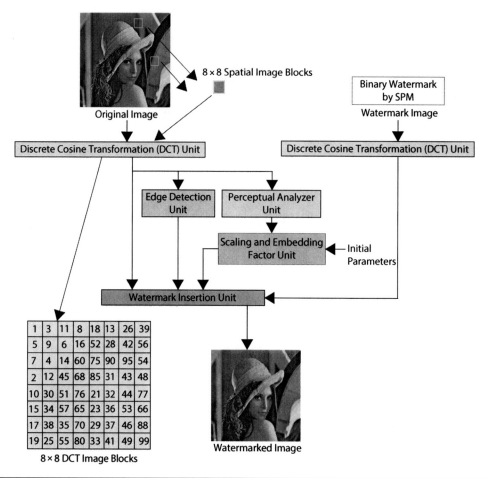

Figure 9.11 MATLAB®-Based modeling for digital design.

for quality assessment as well as to determine how to embed the watermark. The overall architecture is divided into the following five different units: discrete cosine transformation (DCT) unit, edge detection unit, perceptual analyzer unit, scaling and embedding factor unit, and watermark insertion unit. DCT units calculate the DCT coefficients of the image considering one 8 × 8 block at a time. The edge detection unit is used to determine the edge blocks in the host image so that these blocks can be minimally altered during watermarking in order to preserve the quality of the host image. The perceptual analysis unit performs perceptual analysis of the host image using the DCT coefficients to determine how much watermark can be added in each of the host image blocks so that the overall quality is not perceptually degraded. The scaling and embedding factor unit calculates the scaling and embedding factors for each image block. The watermark insertion unit inserts the DCT coefficients of the watermark into the original image. The built-in functions of MATLAB® such as DCT are used for the simulations. The overall chip architecture, which is divided into different components, can be designed using various MATLAB® functions. The overall chip can be hierarchically modeled and simulated after modeling the individual components shown in Fig. 9.11. A high-level view of a MATLAB®-based model that is used for overall chip simulation is presented in Algorithm 9.3 [130, 199, 204]. The watermarked image obtained out of the chip is also shown.

9.3.2 Circuit-Level Simulations

The use of MATLAB® for circuit level analysis is not common or obvious as it is not an EDA tool. The SPICE engine with a schematic editor is much more straightforward for this purpose. The circuit is typically captured in the schematic editor and solved using the SPICE engine for various analysis purposes. However, MATLAB® as a language and framework is quite powerful in terms of numerical solvers and a huge set of optimization algorithms. SPICE was not intended for optimization. In this section the use of MATLAB® for analysis of small circuits is presented. Both linear and nonlinear

ALGORITHM 9.3 High-Level Overview of a MATLAB® Model for the Watermarking Chip [130, 199, 204]

```
% High-Level View of MATLAB Modeling of a Watermarking Chip
InTMP = imread('test-image.jpg');
[imrows imcols dimension] = size(InTMP);
InIMG = rgb2gray(InTMP);
WmTMP = imread('watermark-image.jpg');
[wmrows wmcols dimension] = size(WmTMP);
WmIMG = rgb2gray(WmTMP);
No-Of-Blocks = floow(imrows*imcols/wmrows*wmcols);
InIMG-DCT = DCT-Module(InIMG);
WmIMG-DCT = DCT-Module(WmIMG);
[Scale-Coeff Embed-Coeff] = Perceptual-Analysis-Unit(InIMG-DCT, No-Of-Blocks);
Edge-Block-Index = Edge-Detection-Unit(InIMG-DCT, No-Of-Blocks);
OutIMG = Visible-Insertion-Unit(InIMG-DCT, WmIMG-DCT, Scale-Coeff,
                                Embed-Coeff, Edge-Block-Index);
```

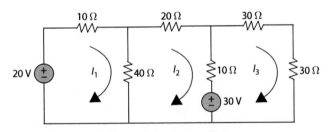

Figure 9.12 A simple resistive circuit for MATLAB®-based simulation [116].

ALGORITHM 9.4 MATLAB® Model for Simple Resistive Circuit-Based Simulation [116]

```
% Z is the impedance (resistance) matrix of the circuit
Z = [50  -40    0;
    -40   70  -10;
      0  -10   70];
% V is the voltage matrix of the circuit
V = [20  -40   30]';
% Solution of loop currents I using simple matrix operations
I = inv(Z)*V;
% Power supplied by the two sources
PowerC1 = I(1)*20;
PowerC2 = (I(3) - I(2))*30;
% Power consumed by a specific resistor, e.g. first resistor
PowerD1 = I(1)^2*10;
```

analysis can be performed. Both direct current (DC) and alternating current (AC) analysis can be performed using MATLAB®. However, the handling of more complicated circuits needs a compiler from SPICE netlist to KVL/KCL formulation! This is done automatically inside SPICE. As a specific example, a small resistive linear circuit is presented in Fig. 9.12 [116]. The loop equations are used and voltage, impedance, and current matrices are formulated. The voltage and impedance matrices need to be solved for current matrix. The steps of circuit analysis are presented in Algorithm 9.4 [116]. The current matrix is a simple one step in MATLAB®. However, when MATLAB® and EDA work together, they are extremely powerful for any complicated analysis and optimization [151, 220].

ALGORITHM 9.5 A Simple MATLAB® Model for a Thin-Film Memristor [153, 198, 233, 291]

```
% Parameter Definitions
% V                    Voltage Vector
% Time_Sim             Time Vector
% Default Values of Parameters
L_M = 10*10^-9;     % Device Length in nm
mu_v = 10^-14;      % Dopant drift mobility of the device material
M_OFF = 38*10^3;    % OFF (maximum) memristance of the device in ohm
M_ON = 0.1*10^3;    % ON (minimum) memristance of the device in ohm
M_init = 2*10^3;    % Initial memristance of the device in ohm
% Memristor Resistance Calculation
k = mu_v * M_ON / L_M^2;
M_diff = M_OFF - M_ON;
M(1) = M_init;
A(1) = 0;
j = 1;
for i = 2:length(Time_Sim)
    A(i) = A(j) + 0.5*(V(i) + V(j)) * (Time_Sim(i) - Time_Sim(j));
    M_tmp(i) = sqrt( M_tmp(1)^2 + 2 * k * M_diff * A(i) );
    j = i;
end
% Saturation Condition
M_tmp(M_tmp > M_OFF) = M_OFF;
M_tmp(M_tmp < M_ON)  = M_ON;
```

9.3.3 Device-Level Simulations

This section discusses the use of MATLAB® for device-level simulations. The device-level MATLAB® model can be used in hierarchical simulations of any system with a function call. The model can be moduled in Simulink®- or Simscape®-based simulation of systems [25, 106, 153, 285]. As a specific case study, the modeling of a thin-film memristor drive is considered [153, 198, 233, 242, 291]. The MATLAB® model for the TiO_2 thin-film memristor is presented in Algorithm 9.5 [153, 198, 233, 291]. The default values of various parameters are chosen as follows [153]: device length L_M = 10 nm, OFF state memristor resistance M_{OFF} = 38 kΩ, ON state memristor resistance M_{ON} = 100 Ω, initial state memristor resistance M_{init} = 2 kΩ, and mobility $\mu_v = 10^{-14} m^2 s^{-1} V^{-1}$.

The memristance characteristic of the memristor with respect to time is presented in Fig. 9.13(a) [153, 198] for a sinusoidal input signal. At the initial stage the input voltage is 0 V and the resistance is M_{init}. Along with time, the value of memristance changes in proportion to the value of the sinusoidal input voltage. More or less the memristance follows the voltage pattern in time. The memristance versus voltage characteristic is presented in Fig. 9.13(b) [153, 198]. This characteristic is like combining the memristance versus time and voltage versus time together. At the initial state the input voltage across the memristor is 0 V, the current is 0 A, and the memristance is M_{init} [198, 233]. The memristance value depends on the sign and amplitude of the input voltage $v_{in}(t)$ in the following manner: memristance [M_{init}, M_{OFF}] for $v_{in}(t) < 0$ and (M_{ON}, M_{init}] for $v_{in}(t) > 0$. This can be explained by the fact that the current in the memristor follows voltage applied across the memristor while memristance keeps increasing as long as voltage is positive. At the same time, when the applied voltage falls back to 0 V, the memristance is maximum M_{OFF}. The shape of memristance versus applied voltage characteristic is a sin^2 () function [198, 233]. The current versus voltage ($I-V$) characteristic of the memristor is shown in Fig. 9.13(c) [153, 198]. The $I-V$ characteristic shows the pinched hysteresis effect. The change in the slope of the $I-V$ characteristic shows a change between different memristance states. The memristance is positive when the applied voltage increases; on the other hand the memristance is negative when the applied voltage decreases. The symmetrical applied voltage leads to double-loop hysteresis in $I-V$ characteristic, which will collapse to a straight line for high frequencies.

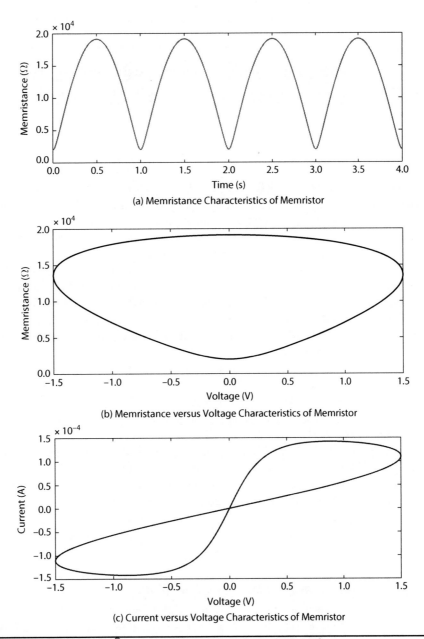

Figure 9.13 Simulation of MATLAB® model of the memristor device [153, 198].

9.4 Simulink®-or Simscape®-Based Simulations

The MATLAB® language and framework discussed in the previous section allows text-based behavioral modeling of devices and systems. This serves as a very good first-level verification before the design progresses to more detailed phases. A selected set of tools for graphical/visual simulation of analog as well as digital systems is presented in Fig. 9.14 [80, 90, 91, 98, 99, 101, 217]. Such graphical simulation of systems provides better insight into different system components, visual perspective of the system data, and control flow and promotes hierarchical design methodology to handle large and complex systems. While a quick introduction is provided, the simulation examples are based on Simulink® or Simscape® for brevity.

Simulink® is a framework that allows designers of embedded systems to model, block-diagram-based simulations for a fast system-level verification [99, 100]. It supports system-level simulation, system-level design verification, and automatic code generation from the system-level description. It is interesting to note that Simulink® now runs on single-board computers like BeagleBoard and

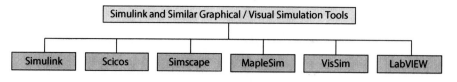

Figure 9.14 Simulink® and selected similar graphical/visual simulations frameworks.

Raspberry Pi, which indicates its lower computational requirements [86, 94]. Simulink® is a commercial software.

Scicos or ScicosLab is a freely available simulator that can be used for graphical modeling and simulation of dynamical systems simulators [80]. It is similar to Simulink® in terms of capabilities and features. ScicosLab can be used for simulation of physical and biological systems along with usual signal processing systems. It in essence can simulate continuous-time as well as discrete-time systems under the same framework. It is possible to generate code from Scicos system model.

Simscape® is a much more versatile system-level simulator as compared to Simulink® [98]. It is also a commercial software. Simscape® can model and simulate multidomain physical systems including electrical and mechanical. It has large and diverse building blocks that can be combined to graphically prototype desired systems. Simscape® model library can be extended by addition of custom-made, MATLAB®-based, text-based models. Code generation is possible from Simscape®. Simulation of hardware-in-the-loop (HIL) system is also possible using Simscape® [201]. A similar simulator called MapleSim is a modeling environment for simulating complex multidomain physical systems including HIL [91, 217]. It may not be explicitly geared toward electronic systems; however, it has sensitivity analysis and optimization capabilities and can talk to Simulink® or LabVIEW by exporting code. VisSim™ is another simulator that can perform modeling and simulation of a variety of systems [101, 217].

LabVIEW is also a graphical simulation software with signal acquisition and processing capabilities [90, 217]. However, it is more geared toward measurement, input/output, and control hardware simulations. Typically, LabVIEW is popular as a software interface for external measuring instruments. The tools like MATLAB®/Simulink® are mostly used for signal processing and corresponding system simulations, although they can also control external hardwares. However, LabVIEW can talk to MATLAB®/Simulink® and together in a complementary fashion can serve the need of the system design engineer.

9.4.1 System- or Architecture-Level Simulations

Simulink® provides a large collection of building blocks for modeling and simulation of complex electronic systems as depicted in Fig. 9.15 [79, 84, 100, 280]. This list essentially presents a quick selection of the building blocks actually available. For example, both continuous-time and discrete-time integrators are present. Various subsystems like input and output ports are available. A large collection of electronic sources is available. Various math operations are presented. In addition, any custom block can be designed by the system design engineers whenever necessary. The results can be displayed or written to a file. VHDL and Verilog models of the system can be generated from Simulink® using a tool called HDL Coder™ that can speed up the hardware design process [88]. Simulink® has additional capability for RF system simulation for gain, noise, and harmonics analysis [97]. This section provides a few examples of Simulink®/Simscape® modeling and simulation at different levels.

9.4.1.1 AMS Design Modeling – Sigma-Delta Modulator (SDM)

The CT-SDM simulation using Simulink® is now discussed as per the simulation flow presented in Fig. 9.6. The CT-SDM obtained using MATLAB® is shown in Fig. 9.10. This section resumes from the same design. The s-domain structural model of the CT-SDM is shown in Fig. 9.16 [258, 259, 286, 287]. It contains a third-order cascade of integrators with feedforward (CIFF) LPF.

The Simulink® model of the CT-SDM is presented in Fig. 9.17 [286, 287]. The coefficients of the LPF can be obtained from time-domain simulations [286, 287]. The dynamic range scaling can be

Mixed-Signal Circuit and System Simulation 435

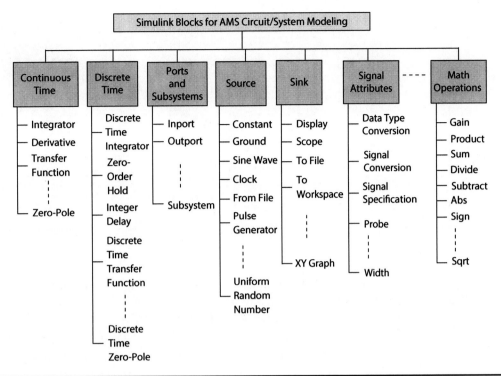

FIGURE 9.15 Selected blocks of Simulink® for analog and/or mixed-signal system modeling [79, 84].

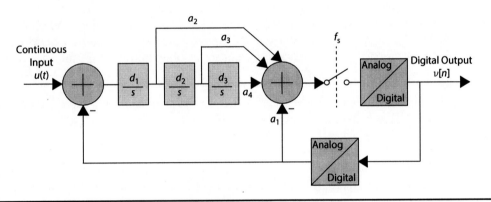

FIGURE 9.16 s-Domain model of the CT-SDM [286, 287].

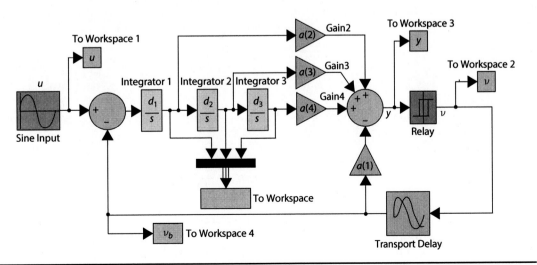

FIGURE 9.17 Simulink® model of the CT-SDM [286, 287].

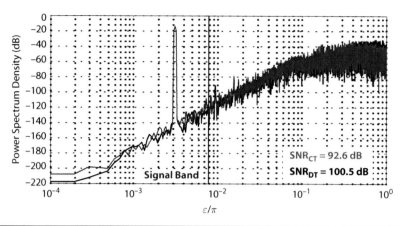

Figure 9.18 Output spectra comparison of DT-SDM and CT-SDM [286, 287].

performed through the use of Simulink® to ensure that the outputs of the integrator are within proper range. It is needed in the real-life circuits as all the voltages have an upper bound of the power supply voltage. The coefficients of the LPF are the following:

$$\begin{aligned}[a(1), a(2), a(3), a(4)] &= [0.0948, 1.2090, 0.5760, 0.8716] \\ [d(1), d(2), d(3)] &= [0.4017, 0.5022, 0.0787]\end{aligned} \quad (9.3)$$

Simulink® simulations can be performed to generate CT-SDM without any nonidealities. Simulink® simulations can also be performed to generate PSD of both DT-SDM and CT-SDM. The results are shown in Fig. 9.18 [286, 287]. The PSD plots of both DT-SDM and CT-SDM match very well. The difference in SNR of the two designs is due to the limitations in simulation accuracy.

9.4.1.2 AMS Design Modeling – Memristor-Based Wien Oscillator

The Simulink® simulator is now presented for the simulation of a memristor-based programmable oscillator [106, 153, 228]. The programmability of the oscillator is achieved by the variable memristance of memristor(s). The schematic design of a Wien oscillator, which is used as a specific case study circuit, is shown in Fig. 9.19(a) [106, 153]. Typically, a Wien oscillator is used to generate signals with frequency from 20 Hz to 20 kHz and is deployed for the generation of audio signals. It comprises of four resistors (R_1, R_2, R_3, R_4) and two capacitors (C_1, C_2). The Wien oscillator works without use of any external oscillation source. The combination of negative and positive feedback with an operational amplifier (OP-AMP) drives the oscillator into an unstable state and hence sustained oscillation is achieved in the overall circuit. A typical Wien oscillator has constant resistances and capacitances to obtain constant amplitude and the frequency of the oscillation. The condition for sustained oscillation of the Wien oscillator is given by the following expression [106, 153]:

$$\frac{C_2}{C_1} + \frac{R_1}{R_2} = \frac{R_3}{R_4} \quad (9.4)$$

The frequency oscillation of the Wien oscillator is calculated using the following expression [106, 153]:

$$f_c = \left(\frac{1}{2\pi\sqrt{R_1 R_2 C_1 C_2}}\right) \quad (9.5)$$

If the resistors of the Wien oscillator are replaced by the memristors, then the oscillation frequency can be varied by controlling the memristance. The initial condition of the memristor determines the frequency of the oscillator if the memristance is not intentionally changed during normal oscillation. It is possible to replace more than one of the resistor(s) with memristor(s) to achieve programmability.

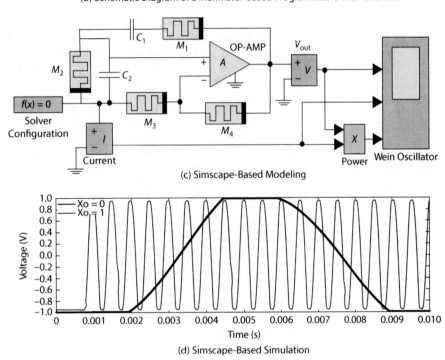

FIGURE 9.19 Simulink®/Simscape® simulation of a memristor-based Wien oscillator [106, 153].

The Wien oscillator in which all resistors are replaced by memristors is shown in Fig. 9.19(b) [106, 153]. The condition for oscillation of this Wien oscillator is the following [106, 153]:

$$1 + \frac{M_1}{M_2} = \frac{M_3}{M_4} \tag{9.6}$$

All memristors were replaced with the same rating but R_3 was replaced with $2M$ to achieve the condition for oscillation. The frequency of oscillation of this Wien oscillator is the following [106, 153]:

$$f_c = \frac{1}{2\pi C \sqrt{M_1 M_2}} \tag{9.7}$$

The functionality and characteristics are verified using Simulink®/Simscape® simulations. The graphical Simulink®/Simscape® model is presented in Fig. 9.19(c) [106, 153]. The oscillator output

Average Power (for $M = M_{OFF}$)	Average Power (for $M = M_{ON}$)	Operating Frequency (for $M = M_{OFF}$)	Operating Frequency (for $M = M_{ON}$)
9.3 μW	1.4 mW	44 Hz	12.4 kHz

TABLE 9.1 Memristor-Based Wien Oscillator Characterization using Simulink®/Simscape® [106, 153]

waveform that has been obtained from Simulink®/Simscape® is shown in Fig. 9.19(d) [106, 153]. The same simulation framework can also be used to characterize the oscillator, e.g., average power dissipation and operating frequency. The memristor-based Wien oscillator characterization results are presented in Table 9.1 [106, 153].

9.4.1.3 Digital Design Modeling – Watermarking Chip

The modeling and simulation of complex digital circuits or systems can be performed using Simulink® before the design progresses to the next level of abstraction for fully automatic or semi-custom design. As a specific case study of digital circuit design, digital watermarking chip of video is considered in this section [178, 201]. The compressed domain video watermarking algorithm flow is presented in Fig. 9.20(a) [130, 178]. The compressed domain video watermarking does not consider the video frames as independent like still images do. The video frames are rather correlated with each other in temporal mode as inter frames (predicted picture P or bi-directionally predicted picture B) and are predicted from intra frame I. The watermark when embedded in intra frame I also appears in inter frames (P or B) even though watermark is not embedded in P/B frames. The drift compensation is performed for stable and clear watermark even in the presence of fast-moving objects in the video. The watermark is embedded into Y color space only for a monochrome video. On the other hand, the watermark is embedded in all Y, C_b, and C_r components in the color space. The architecture for the algorithm in Fig. 9.20(a), which can be realized as a hardware following field programmable gate array (FPGA), standard cell integrated circuit design using semi-custom design approach, is presented in Fig. 9.20(b) [130, 178]. The architecture has three different paths to embed watermark and to perform drift compensation for overall stable watermark in the video. In path A, the watermark is embedded into all I/B/P frames, which gives rise to two watermarks (one stable and one drifted) in B/P frames as one is acquired by these two from frame I. In path B, watermark is embedded in I frames only and B/P frames get one watermark (drifted) from frame I. The subtraction of results from B and C cancels the drifted watermark. In path C, no watermark is embedded in any frames, which effectively cancels the encoding noise in the drift compensation output. The overall architecture has the following 14 units: watermark embedding IBP, watermark embedding I, frame buffer, DCT, inverse DCT (IDCT), motion estimation, motion compensation, quantization, inverse quantization, zigzag scanning, inverse zigzag scanning, entropy coding, inverse entropy coding, and controller. Each unit performs the typical operations as in the video compression algorithm and are not discussed further for brevity.

The partial Simulink® models for the graphical simulation of the overall architecture is presented in Fig. 9.21 [130, 178]. As evident from the figure, a hierarchical modeling approach is followed in this Simulink®-based simulation of the compressed domain video watermarking chip. Simulink® provides several image and video processing functions and modules for rapid system modeling. The functions such as DCT and IDCT can be used directly. In addition, the sum of the absolute difference (SAD) used for motion estimation, block processing, and split, and delay (for buffer) can be easily implemented. Units such as quantization, zigzag scanning, and entropy encoding are designed based on the available primitives. It is observed from the Simulink® system-level simulation that for a specific input condition and DVD-resolution video, 43 frames per second could be processed and the resulting video has peak-signal-to-noise ratio (PSNR) of approximately 30 dB.

9.4.2 Circuit-Level Simulations

In this section, a popular oscillator LC-VCO is used to demonstrate Simscape® for circuit-level simulation [175]. The schematic diagram of a double-gate graphene FET (GFET)-based LC-VCO is presented in Fig. 9.22(a) [175]. GFET-based RF circuits such as oscillators, frequency multipliers, and mixers can have stable higher frequency operations due to the high carrier mobility of graphene

Mixed-Signal Circuit and System Simulation 439

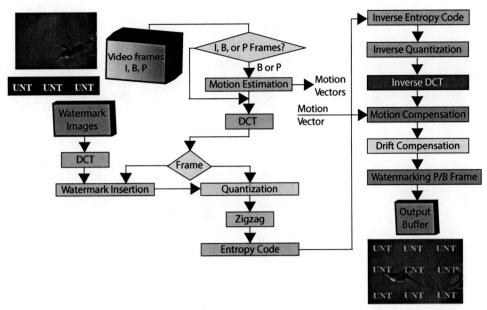

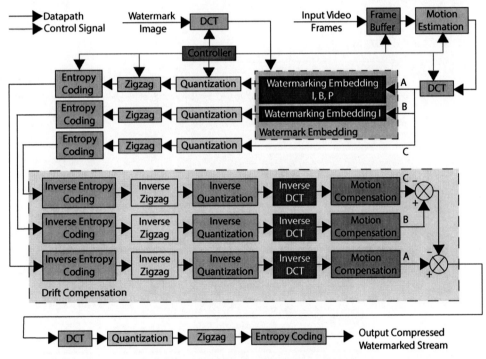

FIGURE 9.20 Compressed domain digital watermarking of video [130, 178].

[175, 235]. As in the LC-VCO topology, it has cross-coupled structure with double-gate GFETs as active devices to compensate for the losses in the LC tank. The varactors are used as variable tuning capacitors using tuning voltage V_{tune}. The inductance value is determined for a specific center frequency after proper operating points of the GFETs are selected.

The Simscape® graphical model for double-gate GFET based LC-VCO simulation is presented in Fig. 9.22(b). It uses Simscape®-based GFET device model, which is discussed in the next section. The P-type GFETs are arranged in a cross-coupled topology and operated close to the saturation region. The source-to-drain conductance is reduced in order to achieve intrinsic voltage gain greater than one. The back gate voltage is set to obtain the appropriate charge neutral point and desired bias point.

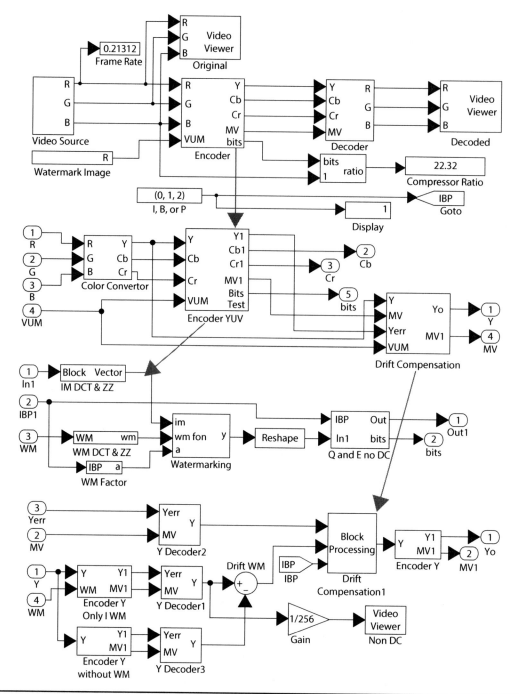

Figure 9.21 Simulink model of the compressed domain digital watermarking of video [130, 178].

In the Simscape® model, ode14x (extrapolation) solver is used [42]. The ode14x solver uses a combination of Newton's method and extrapolation from the current value to compute the state of the model at the next time step. The ode14x solver is more accurate for a specific step size; however, it has a higher computational need. The output voltage that results from Simscape® simulation is shown in Fig. 9.22(c). A selected set of characterization data is presented in Table 9.2.

9.4.3 Device-Level Simulations

Simscape®-based device-level modeling is discussed in this section. As an example, two emerging devices, memristor and graphene FET (GFET), are considered.

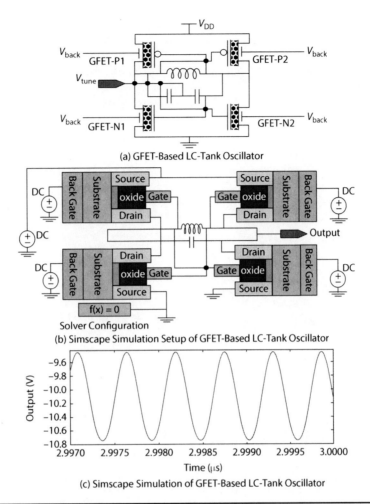

FIGURE 9.22 Simulink®-based simulation of a graphene FET (GFET)-based LC-VCO.

Circuit or Device Parameters	Estimated Values
Center Frequency f_c	1.8 GHz
Tank Voltage Swing $V_{tank}\,(p-p)$	1.286 V
Bias Current I_{bias}	4.6 mA
Total Tank Conductance g_{tank}	0.134 mΩ^{-1}
Transconductance of the active device g_{active}	4.68 mΩ^{-1}

TABLE 9.2 GFET-Based LC-VCO Characterization using Simulink® Simulation

9.4.3.1 Memristor Device

The device-level modeling of the memristor can be performed through Simulink® and/or Simscape® [106, 153, 268, 285]. Memristor modeling using Simulink® is available [268, 285]. The heart of the device model is the window function that models the nonlinearities of the charge carrier transport in the memristor [106, 126, 153, 268, 285]. The nature of the window function dictates the accuracy of the models. In the Simulink® model, the dynamics of the memristor is described in integral format [126, 285]. This is despite the fact that the original dynamics of the memristor were described in differential equation format. The physical signal through the memristor is first converted to a Simulink® signal. This is followed by integration and windowing operation and then the resulted signal is converted back to the physical domain. This hybrid approach of changing back and forth from one domain to another limits the efficiency of the model. A Simscape® model that is native to Simscape® as it is completely

442 Chapter Nine

ALGORITHM 9.6 A Simscape®-Based Simple Memristor Model [106, 153]

```
% A Memristor Model using Simscape Language
component memristor<foundation.electrical.branch
    parameters
        Muv = {1e-14,'m^2/s/V'};   % Mobility
        Lrat0 = {0,'1'};           % Initial Effective and Physical Length Ratio
        Lm = {5e-9, 'm'};          % Memristor Length
        Mon = {100,'Ohm'};         % ON State Memristance
        Moff = {36e3, 'Ohm'};      % OFF State Memristance
    end
    variables
        Lrat = {1,'1'};
        Mtmp = {1e3,'Ohm'};
    end
    function setup
        Lrat = Lrat0;
    end
    equations
        let
            az = Muv * Mon / Lm^2;
        in
            if(Lrat <= 0 && v <= 0)||(Lrat >= 1 && v >= 0)
                Lrat.der == 0;
            else
                Lrat.der == az * i;
            end
        end
        Mtmp == Mon * Lrat + Moff * (1 - Lrat);
        v == i * Mtmp;
    end
end
```

---

described in Simscape® language is presented in Algorithm 9.6 [106, 153]. In this model, the dynamics of memristor is realized in a differential equation form, which is the original mathematical equation described in [274]. The first line with `component memristor` declares memristor as a component. Simscape® language has an important feature of subclassing to develop a component model as an extension of a base model [46, 65]. In addition, the physical domain definitions of Simscape® language ensures that the new custom components are compatible. In the first line, `foundation.electrical.branch` makes `component memristor` a part of branch of electrical foundation.

Simulink®/Simscape® graphical simulation setup is presented in Fig. 9.23 [106, 153]. The memristor has been simulated with specific device parameter values and an external sinusoidal voltage source of amplitude 2 V. A selected memristor characteristic resulting from the simulation is shown in Fig. 9.24 [106, 153]. The memristor resistance (memristance) with respect to time simulation is shown in Fig. 9.24(a) [106, 153]. As observed in a typical memristor, the memristance increases with positive applied voltage and decreases with negative applied voltage. Initially there is no change in memristance as it is essentially M_{off} and memristance cannot be increased above physical maximum. The I–V characteristic of the memristor is shown in Fig. 9.24(b) [106, 153]. It is a typical double-loop characteristic as expected in the case of a memristor.

9.4.3.2 Graphene FET (GFET) Device

In this section Simscape® device-level modeling for a graphene FET (GFET) is presented. This Simscape® GFET model is based on the VHDL-AMS model presented in [260]. The textual Simscape®

Mixed-Signal Circuit and System Simulation 443

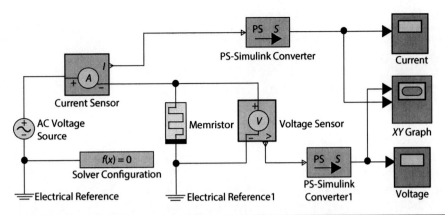

FIGURE 9.23 Simulink®/Simscape®-based simulation of a memristor device [106, 153].

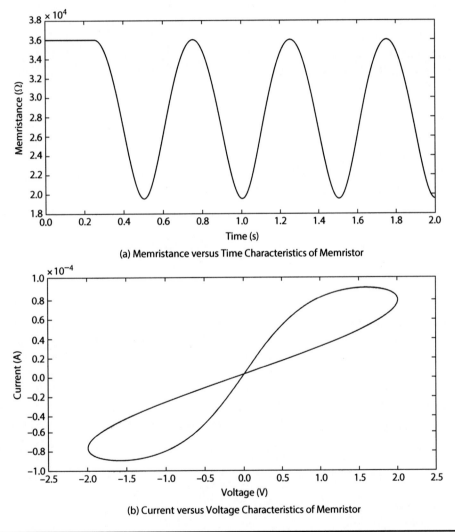

(a) Memristance versus Time Characteristics of Memristor

(b) Current versus Voltage Characteristics of Memristor

FIGURE 9.24 Memristor device characteristic from Simulink®/Simscape®-based simulation [106, 153].

model for GFET, which is completely written in Simscape® language, is presented in Algorithm 9.7. This model is fully compatible with Simscape® and in turn can be used in graphical system-level simulation of GFET-based circuits like amplifier and oscillators. The I–V characteristic of the GFET drain current versus drain-to-source voltage is shown in Fig. 9.25. The device parameters selected are based on published data [193, 277]. In this specific simulation, top-gate voltages of -0.8 V, -1.3 V,

ALGORITHM 9.7 A Simscape® Model for Graphene FET (GFET)

```
% A Graphene FET (GFET) Model using Simscape Language
component GFET
    nodes
    p1 = foundation.electrical.electrical; % Source
    p2 = foundation.electrical.electrical; % Top Gate
    p3 = foundation.electrical.electrical; % Drain
    p4 = foundation.electrical.electrical; % Bottom Gate
    end
    parameters
        Rs = {800, 'Ohm'}; mu = {700,'cm^2/s/V' }; Ec = {4.5e5,'V/m'};
        Cgio = {0.8072,'1'}; Hsub = {285.0e-9,'m'}; tox = {15e-9, 'm'};
        L = {1e-6, 'm'}; W = {2.1e-6, 'm'}; ntop = {2.1209e16,'cm^-3'};
        Vgs0 = {1.45, 'V'}; Vbs0 = {2.7, 'V'}; k = {16, '1'};
        k_sub = {3.9, '1'}; pi = {3.1415926, '1'}; q = {1.60e-19, 'c'};
        eps0= {8.854187817e-12, 'F/m'}; h= {6.62606876e-34, 'Js'}; vf={1e6,'m/s'};
    end
    variables
        Vds = {1,'V'}; Ids={1,'A'}; Vgs= {1,'V'}; Vbs= {1,'V'};
        Ctop ={1,'F/cm^2'}; Cback ={1,'F/cm^2'}; Vo ={1,'V'};
        Vg0 ={1,'V'}; Vc ={1,'V'}; Rc ={1,'Ohm'}; Gamma ={1,'1'};
        Vdsat ={1,'V'}; Io={1,'A'};
    end
    function setup
        across(Vds, p3.v, p1.v); through(Ids, p3.i, p1.i);
        across(Vgs, p2.v, p1.v); across(Vbs, p4.v, p1.v);
    end
    equations
        let
            Cq = q^2*(ntop/pi)^0.5/(vf*(h/(2*pi))); Ce = Cgio*eps0*k/tox;
        in
            Ctop == Cq*Ce/(Cq+Ce);
        end
        Cback == eps0*k_sub/Hsub; Vo == Vgs0 + (Cback/Ctop)*(Vbs0 - Vbs);
        Vg0 == Vgs - Vo; Vc == Ec*L; Rc == 1.0/((W/L)*mu*Ctop*Vc); Gamma == Rs/Rc;
        Vdsat == (2.0*Gamma*Vg0/(1.0+Gamma)+ (1.0-Gamma)/(1.0+Gamma)^2.0
                * (Vc-(Vc^2.0-2.0*(1.0+Gamma)*Vc*Vg0)^0.5) );
        Io == 2.0*(W/L)*mu*Vc*Ctop*(Vgs-Vo-Vds/2.0);
        if (Vds > Vdsat)
            Ids == 1.0/4.0/Rs *(Vds-Vc+Io*Rs +((Vds-Vc+Io*Rs)^2.0-4.0*Io*Rs*Vds)^0.5);
        else
            if (Vds <= Vdsat)
                Ids == (Gamma/Rs/(1.0+Gamma)^2.0* (-Vc+(1.0+Gamma)*Vg0+(Vc^2.0-2.0
                   *(1.0+Gamma)*Vc*Vg0)^0.5)+ mu*Cback*abs(Vbs-Vbs0)
                   *Vds*W/L/10.0*(Vds/Vdsat -1.0)^2.0);
            else
                Ids == 0.0;
            end
        end
    end
end
```

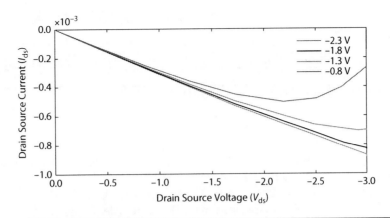

Figure 9.25 Simulink®/Simscape®-based simulation of a GFET.

-1.8 V, -2.3 V, and -2.8 V were used. The drain-to-source voltage V_{ds} varied from 0 to -3 V. The top-gate capacitance C_{top} of the transistor was 759.5 nF/cm² and the back-gate capacitance was 12.12 nF/cm². The trends of the $I-V$ characteristics results are consistent with those presented for similar transistors in the literature [193, 260].

9.5 Circuit-Level and/or Device-Level Analog Simulations

A thorough discussion of simulation programs with integrated circuit emphasis (SPICE) is presented in this section. SPICE background and historical perspectives are presented. Different types of SPICE tools are described. An insightful view of how SPICE really performs the simulation of large circuits is discussed. A brief discussion is provided on SPICE features and capabilities. In short, this section is a combined perspective of design engineers and SPICE developers. It is not an overstatement to say that every integrated circuit available goes through SPICE [109, 176, 212, 232, 270]. There is no foreseeable alternative to it even after five decades of development.

SPICE is a general-purpose circuit simulation program that accepts a description of a circuit and performs several forms of accurate and detailed simulations using different models and setups. The basic idea of SPICE is depicted in Fig. 9.26 [109, 176, 212, 232, 270]. The circuit and system description targeted for SPICE is called SPICE netlist. The overall SPICE file containing SPICE netlist, simulation and analysis options is called a SPICE deck. The individual circuit elements in a SPICE deck is called a SPICE card or line. The physical description of different elements of the netlist is provided in the form of SPICE models. The text file netlist contains the circuit elements such as transistors and capacitors and their connections. Then it translates the netlist into a set of linear and nonlinear equations to be solved. The general set of equations typically results from nonlinear differential algebraic equations (DAEs) that are solved using implicit integration methods such as Newton's

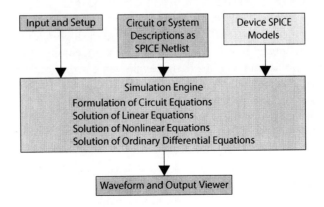

Figure 9.26 A simple concept of SPICE simulation [212].

446 Chapter Nine

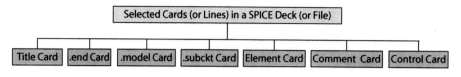

FIGURE 9.27 Different cards in a SPICE deck [170, 209, 212].

methods. Thus, in general, SPICE is an analog simulator that uses digital numerical routines to predict the behaviors of the circuits before manufacturing. Any simulation results obtained out of SPICE must follow commonsense VLSI to be acceptable!

Different types of SPICE cards are presented in Fig. 9.27 [170, 176, 209, 212, 216]. The "Title Card" is always the first line or start line of the SPICE deck. The ".end Card" is the end of SPICE deck. The ".model Card" specifies different models of elements used in the SPICE deck. A ".subckt Card" specifies a subcircuit that is used in the circuit description. The "Element Card" gives the element name and the number of nodes to which that element is attached. The first letter of the name specifies wheather the element is a MOSFET, BJT, or diode. The "Comment Card" is included to improve explanation of the program as in any computer program writing. It starts with a symbol "*" and is ignored by the SPICE. The "Control Card" specifies the type of analysis as well as the output to be generated, for example, .tran, .dc, .ac, .tf, .four, .sens, .noise, .disto, .temp. The control cards such as .plot or .print are used to define plot and tabulation with respect to the independent variable. The ".option Card" is used to set parameter values that control simulations, and there are many SPICE options that can be used by the design engineers depending on the need.

9.5.1 SPICE Analog Simulation Background

The long history of analog simulator starts from the program called "Computer Analysis of Non-Linear Circuits Excluding Radiation (CANCER)" in the year 1971. CANCER was a simple program that could perform DC, AC, and transient analysis only [166, 210, 211]. In the year 1972, the first version of "Simulation Programs with Integrated Circuit Emphasis (SPICE)" called SPICE 1 came to public. In 1975, the second and improved version of SPICE called SPICE 2 was released. In 1985, the SPICE was thoroughly rewritten in C and the third version SPICE 3 came to public domain. Due to the public domain availability of the SPICE program with liberal license, the SPICE program has proliferated to a variety of SPICE programs from many different sources. However, SPICE was never intended to simulate very large circuits and was never designed with scalability as an objective for the current-generation nanoelectronic circuits or system-on-a-chip (SoC) with analog, digital, and software components! A large number of SPICE and SPICE-like analog simulators are available. It is an impossible task to find, execute, and discuss them in detail. However, an attempt has been made to discuss a selected few. A selected set of SPICE and SPICE-like analog simulators is listed in Fig. 9.28 [157, 170, 212]. This is no way a complete list. Some of them run in Unix®/Linux, some of them in Windows®, and some in Mac OS. Some of them are from commercial vendors and some of them are open-source and/or free SPICE tools. A designer can choose one of these depending on the needs, target design, and available budget.

The selected SPICE and SPICE-like simulators are grouped into five different categories as follows: (1) commercial SPICE simulators, (2) free and/or open-source SPICE simulators, (3) fast SPICE simulators, (4) analog-fast SPICE simulators, and (5) high-speed SPICE simulators. The commercial simulators are mostly based on the original SPICE 3 engine. In general, these are very accurate and often used as golden standard by design engineers. The free and/or open-source SPICE simulators are made available to the public without any charge. These are primarily based on original SPICE 3. Some of them have limited capabilities, but some are as powerful as their commercial counterpart. However, the most important factor is the free usage by the public. These are also slow as the original SPICE was not intended to simulate large and complex chips that are available at present. In fact, a full-chip SPICE simulation is still difficult to impossible.

The emergence of nanoelectronic technology, the significant increase in the transistor count, and the emergence of SoC are beyond the ability of SPICE as it was never designed for it [73, 236]. In fact, the accurate SPICE engine has been overutilized due to lack of any alternatives. If the integrated circuit netlist has N number of nodes when its matrix is populated in the SPICE engine, then, depending on

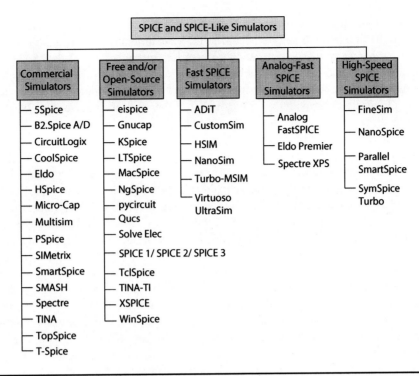

FIGURE 9.28 A selected set of different SPICE and SPICE-like simulators.

the circuit topology, the time complexity of the SPICE solver is $O(N^{1.5})$ or $O(N^2)$. In addition, significant time is spent in the evaluation of complex equation-based device models. The accurate SPICE thus can support approximately 100,000 active elements. However, the current integrated circuits can have 100 million to 1 billion active devices and hence accurate SPICE will take forever to simulate such an integrated circuit. Verification of large-scale digital and memory circuits is very slow and almost impossible. The functional simulation of full-chip AMS-SoC is also infeasible using accurate SPICE.

Several different techniques including the following have been attempted to overcome the speed and scalability limitations of the accurate SPICE [33, 69]. Look-up table (LUT) model has been explored to overcome the speed limitation caused by the use of highly accurate models. In the LUT method, charge and current are evaluated as functions of nodes and stored as LUTs. Subsequently, the LUTs are used instead of the model equations; thus achieving significant speedup. The circuit partitioning approach, in which different time steps are applied to different parts of the subcircuits, is explored during time analysis. In essence, the simulation becomes event-driven when a particular voltage or current threshold (called event) only triggers the reevaluation of the subcircuits. Thus, event-driven and multi-time step algorithms have been explored as an option to speed up the solution and convergence. Hierarchical simulation can result in reduction in the computational resources including memory. In addition, parallel computation can speed up the simulation if one or more steps of simulation can be run in parallel. Hardware acceleration of SPICE has also been attempted to improve simulation time. Implementations of one or more of these feature lead to Fast SPICE (including flat Fast SPICE and hierarchical Fast SPICE), Analog-Fast SPICE, and high-speed SPICE. A high-level discussion of all these different types of SPICE and SPICE-like simulators will be presented with more detailed discussion of selected SPICE or SPICE like simulators for brevity.

9.5.2 Commercial Accurate Analog Circuit Simulators

A few selected commercial accurate analog simulators are described in this section. They are mostly based on original SPICE 3. However, a few of them are not based on SPICE 3 and have been developed from scratch. This is in no way a complete list and many more might be available in the market. It is left to the designers to identify the appropriate analog simulator for the design being undertaken.

5Spice has integrated schematic entry and waveform viewing capability [110]. It can define and store a large number of analyses and simulation results. B2.Spice A/D can perform continuous transient simulation, parametric sweep, and Monte Carlo analysis [4]. It allows live altering of parameters even when the simulation is running. It has virtual instruments like oscilloscope, ammeter, and voltmeter for a visual perceptive of the simulation process. CircuitLogix has simulation capability for analog, digital, and mixed-signal circuits and a built-in schematic capture tool [6]. CoolSpice is based on NgSpice, which includes a schematics capture tool, a text editor for manual netlist editing, and a plotter [7].

Eldo® Classic provides an extensive set of device models including BSIM and HiSIM [10]. It can simulate any design containing up to 300,000 active elements. It also supports a variety of macromodels to simulate designs that may not be described using the available basic elements. Eldo® has facility for RC reduction, optimization, and enhanced Monte Carlo analysis. It can use distributed computing to increase the speed of the simulation process. It provides library encryption for protection of the design.

HSPICE® is a highly accurate gold standard circuit simulator [17, 77]. It performs accurate simulation, cell characterization, and signal integrity analysis of very large netlists. Fast Monte Carlo is available for statistical simulation. It has flexible mechanism for defining process variation effects using variation blocks while accounting variations in both device and interconnects. Hot carrier injection (HCI) and negative bias temperature instability (NBTI) device aging effects are easily analyzed.

Micro-Cap has integrated analog and digital simulators that support analog as well as digital behavioral simulations using 32,000 components in the library [26]. It can perform Monte Carlo analysis as well as optimize design. Multisim can analyze analog, digital, and power electronics using 26,000+ devices from its library [28]. PSpice® is a complete analog and mixed-signal circuit simulator containing 4,500 parameterized device models [34, 271]. It can perform Monte Carlo analysis, sensitivity analysis, and component optimization while detecting stress in components. PSpice® can work with Simulink® for co-simulation of larger integrated circuits and systems.

SIMetrix has integrated SPICE simulator, schematic editor, and waveform viewer [38]. It can perform analog simulation as well as logic-level digital simulation (Verilog) and uses upto four cores to increase the speed of the simulation. It supports Verilog-A-based analog models. SmartSpice™ can perform simulation of analog as well as AMS circuits and has Verilog-A-based behavioral simulation capability [39]. SmartSpice™ has a large set of calibrated SPICE models for traditional MOSFET/BJT technologies as well as emerging FinFET, TFT, and SOI technologies. SMASH is a single kernel mixed-signal and multidomain simulator [40]. It supports SPICE, SystemVerilog, VHDL, Verilog-A, Verilog-AMS, and VHDL-AMS languages.

Spectre® is one of the gold standard analog simulators [43, 179]. It was not developed from the original SPICE 3 unlike a typical SPICE simulator. It supports analog as well as RF simulation while providing high accuracy and fast convergence. It can handle post-layout simulation of large netlists. It can perform high-speed Monte Carlo analysis to characterize yield and analyze mismatch effects while supporting Verilog-A. It works with Simulink® for co-simulation of large circuits and systems.

TINA™ is built based on original SPICE 3 and XSPICE for analog, digital, and mixed-signal simulations [56]. It supports HDLs such as HDL, Verilog, Verilog-A, and Verilog AMS. TopSpice is a full-fledged, mixed-mode, mixed-signal circuit simulator [57]. It has model libraries containing over 30,000 analog and digital components. It supports analog behavioral and frequency-domain modeling as well as logic simulations. T-Spice™ supports SPICE as well as Verilog-A behavioral simulations and optimization using Levenberg–Marquardt nonlinear optimizer [51].

9.5.3 Free and/or Open-Source Accurate SPICE

Several SPICE and SPICE-like analog simulators, which were discussed in the last section, are available from various commercial vendors. Many of these are gold standard simulators. However, commercial simulators are sometimes quite expensive. So, design houses with small budget may not be able to buy their licenses. Moreover, to reduce design cost, a mix of commercial analog simulators and free simulators can be used in which limited number of commercial simulator licenses are acquired with the use of a large number of free simulators. A large set of free and/or open-source analog simulators are available for the use of designers. It is quite a difficult task to cover all these. A selected few of free and/or open-source SPICE simulators are: eispice, Gnucap, KSpice, LTSpice, MacSpice, NgSpice, pycircuit, Quite universal circuit simulator (Qucs), Solve Elec, TclSpice, TINA-TI™, XSPICE, and Winspice [24, 30, 36, 41, 52, 55, 66, 67, 71, 89, 144, 169, 273]. Some of these are briefly discussed in the rest of this section.

Gnucap is an open-source circuit analysis package that is not based on original SPICE 3 [71, 139]. Unlike any typical SPICE, Gnucap is a true mixed-mode simulation with both analog and digital simulations taking place through a single engine. While Gnucap is fully interactive, it can also be executed in batch mode. In batch mode, it is very much compatible with SPICE. In Gnucap the digital device netlist is as digital without converting to analog between gates. Thus, the digital circuits are simulated faster than that happens in a typical SPICE. Gnucap also has a simple behavioral modeling language.

NgSpice is a mixed-signal and mixed-level circuit simulator that is based on three open-source software simulators: SPICE 3, Cider, and Xspice [5, 30, 109, 216]. These three simulators are seamlessly integrated to provide a much powerful simulator that takes advantage of their features. Verilog models can be integrated to NgSpice. NgSpice has a very good facility for statistical circuit analysis, random number generation, and RF analysis. NgSpice can perform circuit optimization using selected optimizers. NgSpice has reasonable technology CAD (TCAD) capabilities through "general-purpose semiconducor simulator" (GSS) or "genius device simulator" that allows NgSpice to be used for device and circuit co-simulation [13, 14]. NgSpice is equipped with tcl scripting capability and several instances of NgSpice can be executed in parallel on a multicore computer.

pycircuit is a Python based circuit analysis package [169]. It provides post-processing facility as well as interfacing options for other simulators. pycircuit can perform symbolic analysis as well as be used for developing new circuit analysis methods. It can be used for performing complex modified nodal analysis (MNA). pycircuit can be used for post-processing of the data generated by simulators Spectre® and Eldo®.

TINA-TI™ is the complete version of TINA™ (which is based on SPICE 3) integrated with a library of macromodels from Texas Instruments (TI) Incorporated [55]. It has a schematic editor. TINA-TI™ also supports multicore processor execution for faster simulations.

XSPICE is an open-source circuit simulator that is based on SPICE 3 [67, 109, 138]. It provides the ability to easily add new models to it. XSPICE performs simulation of the digital functions using event-driven algorithm of SPICE core, and supports many more event-driven data types. XSPICE has a library containing more than 40 functional units, including summers, multipliers, digital gates, and digital storage elements, and a generalized digital state-machine.

9.5.4 Fast SPICE

SPICE and SPICE-like simulators presented in the preceding section are no doubt quite accurate. However, accurate simulation of current nanoelectronic gigascale complexity circuit or SoCs is a quite time-consuming and even infeasible task [275]. For the full-chip simulation of the circuits and systems, there has always been a need for faster SPICE engines. The EDA industry has addressed this issue by introducing fast SPICE simulators. A comparative perspective of SPICE and fast SPICE is presented in Table 9.3 [73, 236]. The fast SPICE can be either flat fast SPICE or hierarchical fast SPICE [69, 73]. Flat fast SPICE simulators can handle 10 million transistors and hierarchical fast SPICE can handle more than 1 billion transistors. Various speedup techniques are integrated in the fast SPICE simulators so

Simulation Features	SPICE	Fast SPICE
Models	Compact Models	Simplified Models
Matrix Structure	Single Matrix for Entire Circuit	Partitioned Matrix
Convergence	Each Time Global Convergence	Event Driven, No Global Convergence
Analysis Support	Transient, AC, Noise, RF	Transient Analysis
Transistor Count Support	50K	10M
Computation Time	Long	Short
Memory Requirement	High	Low
Accuracy	High	Low
Circuit Application	Analog, Mixed-Signal	Mixed-Signal, Digital, Memory

TABLE 9.3 Comparative Perspective of SPICE and Fast SPICE [73, 236]

Fast SPICE Methods	Effect on Computation Time and Memory Usage
Partitioning	Reduces size of the matrices. Reduces memory requirement.
Event-Driven or Multirate	Reduces events. Only active partitions are solved. 2 – 100 × speedup.
Hierarchical isomorphic method	Reduces memory requirement. Identical partitions shared. 2 – 10,000 × speedup.
Analog table, digital representative model	LUT reduces model evaluation time. 5 – 20 × speedup.
Parasitic reduction	Reduces number of elements. Reduces matrix size by 5 – 20 × in layout netlists.
Parasitic stitching	Generates hierarchical postlayout view from flat netlist. 4 × speedup.

TABLE 9.4 Effect of Fast SPICE Methods on Computation Time and Memory Usage [73]

that full-chip simulation is possible. The objective has been to reduce the computation time (aka processor execution time or CPU time) as well as reduce memory requirement during the computation. The effect of selected fast SPICE methods on computation time and memory usage is presented in Table 9.4 [73]. Fast SPICE simulators that make full-chip simulation a reality include the following: ADiT™, CustomSim™, HSIM™, Turbo-MSIM™, and Virtuoso® UltraSim [2, 8, 16, 58, 64].

9.5.5 Analog-Fast SPICE

Analog-fast SPICE is a SPICE simulator between accurate SPICE and fast SPICE in terms of simulation time and accuracy. An analog-fast SPICE simulator is much faster than accurate SPICE and slower than fast SPICE. From the accuracy point of view, the analog-fast SPICE is very close to accurate SPICE. So, simplistically speaking, the analog-fast SPICE is a faster and near-SPICE accurate analog simulator. A careful selection of fast SPICE and high-speed methods are deployed in analog-fast SPICE to achieve this. In some cases analog-fast SPICE may use a single matrix without partitioning as in accurate SPICE and improve simulation speed by using multithreading, multicore techniques. Some analog-fast SPICE may use partitioning and solve partitioned matrix. Speed up to 20× can be achieved by using analog-fast SPICE that can handle more than 10 million transistors. A selected set of examples of analog-fast SPICE include the following: Analog FastSPICE™ (AFS), Eldo® Premier, and Spectre® eXtensive Partitioning Simulator (Spectre® XPS) [3, 11, 44].

9.5.6 High-Speed SPICE

The high-speed SPICE simulators deploy techniques including multicore simulation, multimachine simulation, parallelization, modified mathematical methods, and hardware acceleration to enhance simulation speed [48, 134, 164]. In the process, speed as high as 100× and circuits as large as 100 million transistors could be handled. A selected set of examples of high-speed SPICE include the following: FineSim™, NanoSpice™, Parallel SmartSpice™, SymSpice Turbo [12, 15, 48, 69].

9.5.7 Different Types of Analysis using SPICE

SPICE and SPICE-like simulators are essentially general-purpose analog simulators [45, 147, 168, 179, 195, 216, 281]. These simulators can perform various types analysis over the SPICE netlist of the circuit. Selected different types of SPICE analysis are presented in Fig. 9.29.

9.5.7.1 DC Analysis

In the DC analysis, the SPICE finds the DC operating point of the circuit for input DC voltages. In this situation the capacitors are assumed to be open circuits and inductors are assumed to be short circuits. The DC analysis is performed using .DC, .OP, and .TF control lines.

9.5.7.2 AC Small-Signal Analysis

The AC small-signal analysis studies the circuit for sinusoidal supply voltages. AC analysis is for the analog nodes only and the output response is determined for one frequency or a set of frequencies. In

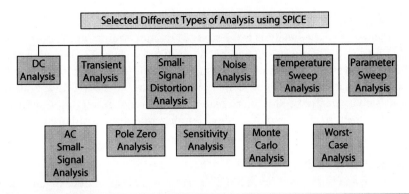

FIGURE 9.29 Different types of analysis using SPICE.

the AC small-signal analysis, the SPICE determines the DC operating point of the circuit and then finds linearized, small-signal models for all the nonlinear devices of the circuit. The linear circuit is then analyzed for a range of frequencies. The AC analysis is performed using .AC control line.

9.5.7.3 Transient Analysis

The transient analysis of SPICE is the computation of the transient output variables as a function of time in a specific time interval. It is like an extension of DC analysis of SPICE to the time domain. The SPICE determines the initial conditions of the circuit using a DC analysis. Once the DC operating point is obtained, the time-dependent aspects of the circuit are reintroduced and the circuit is incrementally solved for time-varying behavior. The transient analysis in SPICE is performed through a .TRAN control line.

9.5.7.4 Pole-Zero Analysis

In general, a circuit (or network) is described by its network transfer function (NTF). The transfer function of a linear time-invariant (LTI) circuit or system has the following format:

$$\text{NTF}(s) = \frac{N(s)}{D(s)} = \left(\frac{a_0 s^m + a_1 s^{m-1} + \cdots + a_m}{b_0 s^n + a_1 s^{n-1} + \cdots + a_n} \right) \tag{9.8}$$

In factorized form the NTF can be expressed in the following manner:

$$\text{NTF}(s) = \left(\frac{a_0}{b_0} \right) \left(\frac{(s+z_1)(s+z_2)\cdots(s+z_i)\ldots(s+z_m)}{(s+p_1)(s+p_2)\cdots(s+p_i)\ldots(s+p_m)} \right) \tag{9.9}$$

The roots z_i of the numerator of the NTF are called zeros of the NTF. The roots p_i of the denominator of the NTF are called poles of the NTF, which are also called the natural frequencies of the circuit or system. The dynamic behavior of the circuit or system depends on the location of the poles and zeros on the network function curve. SPICE has the capability to compute the poles and zeros in the small-signal AC transfer function of the circuit through pole-zero analysis. SPICE first computes the DC operating point of the netlist and then determines the linearized, small-signal models of all the nonlinear devices in the netlist. This linearized circuit is then used to determine the poles and zeros of the NTF. Two types of NTFs are allowed in SPICE: (1) output voltage over input voltage and (2) output voltage over input current. The pole-zero analysis in SPICE is performed through a .PZ control line.

9.5.7.5 Small-Signal Distortion Analysis

In the small-signal distortion analysis of the circuit netlist, SPICE calculates the steady-state harmonic and intermodulation products for small input signal magnitudes. SPICE finds the distortion characteristics of the circuit in an AC small-signal, sinusoidal, steady-state analysis through the use of .DISTO control line.

9.5.7.6 Sensitivity Analysis

In general, sensitivity analysis is the study of the uncertainty in the output of a system with respect to different sources of uncertainty in the inputs of the system. The sensitivity analysis of circuits through SPICE involves calculation of the sensitivity of output variables with respect to the circuit variables and model parameters. The sensitivity analysis can be either DC operating-point sensitivity or AC small-signal sensitivity. SPICE calculates the difference in an output variable (e.g., node voltage) by changing each parameter of each device independently. Sensitivity analysis of circuits is performed by SPICE using the .SENS control line.

9.5.7.7 Noise Analysis

The noise analysis operation of SPICE performs analysis of device-generated noise for a circuit. The noise contribution of each device to the output voltage is calculated in this analysis by SPICE. Sensitivity analysis of circuits is performed by SPICE using the .NOISE control line.

9.5.7.8 Monte Carlo Analysis

In general, the Monte Carlo analysis varies parameters within their specified limits with each new run to present a useful perspective of actual circuit performance when run many times. Typically, SPICE varies component values or model parameter using a random distribution (e.g., Gaussian) of certain mean and variance. Different SPICE analyses such as DC, AC, and transient simulation are run for a specified number of times using the randomly generated parameters. The response obtained from these runs follows a statistical distribution as well.

9.5.7.9 Temperature Sweep Analysis

SPICE allows simulation of a circuit for different temperatures so that the effect of the temperature on circuit performance metrics can be determined. In SPICE, .TEMP control line sets the operating temperature of the circuit for the entire simulation.

9.5.7.10 Worst-Case Analysis

Worst-case analysis is performed by SPICE to identify the extreme operating condition of an integrated circuit. In the worst-case analysis, SPICE varies one component at a time to calculate the sensitivity of an output response for each component of the circuit. A final run is then performed with all the component parameters to generate the worst-case output response.

9.5.7.11 Parameter Sweep Analysis

The parameter sweep analysis of SPICE allows design engineers to simulate the circuit netlist across a range of values of a parameter. It is the same as simulating the netlist multiple times. However, parameter sweep achieves that using one setup. The statements such as .PARAM, .STEP. are used along with .AC, .TRAN for specific analysis.

9.5.8 SPICE-Based Simulation Examples

Now simple examples are presented for easier understanding of SPICE-based simulation. First nanoscale CMOS-based circuits and then memristor modeling are presented.

9.5.8.1 CMOS Nanoelectronics

A simple 45-nm CMOS-based ring oscillator is shown in Fig. 9.30(a) [203]. This is a baseline design with nominal sizes chosen for a 45-nm CMOS technology. A nominal device length of 45 nm is used for all transistors. The NMOS are sized for a width of 10 times the length, making it 450 nm. The NMOS are sized for a width of 20 times the length, making it 900 nm. The corresponding circuit-level SPICE netlist is presented in Fig. 9.30(b). The netlist is extracted from a schematic editor and does not contain any parasitics. The first line has the voltage supply that says source VDD is connected between node VDD and node 0 and has a value of 1 V. The next six lines are for transistors M1 to M6. For example, for the first transistor M1: drain, gate, source, and body of M1 are connected at node N002, node N001, node 0, and node 0, respectively; M1 is a NMOS with length 45 nm and width 450 nm. This is how other lines for transistors can be interpreted. The .model lines specify the model, e.g., NMOS name is an NMOS type. The .lib control line includes library. The .tran control line directs transient simulation for 1 ns. Similarly, any other analysis using SPICE can be performed. The .ic control line sets

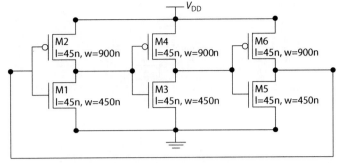

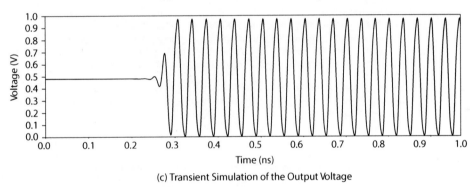

(a) Schematic-Level Design

```
VDD VDD 0 1
M1 N002 N001 0 0 NMOS l=45n w=450n
M2 N002 N001 VDD VDD PMOS l=45n w=900n
M3 N003 N002 0 0 NMOS l=45n w=450n
M4 N003 N002 VDD VDD PMOS l=45n w=900n
M5 N001 N003 0 0 NMOS l=45n w=450n
M6 N001 N003 VDD VDD PMOS l=45n w=900n
.model NMOS NMOS
.model PMOS PMOS
.lib \...\standard.mos
.tran 1n
.ic v(osc)=1
.include 45nm_bulk.pm.txt
.end
```

(b) SPICE Deck of the Schematic Design

(c) Transient Simulation of the Output Voltage

Figure 9.30 SPICE analysis: 45-nm CMOS ring oscillator example.

initial conditions for transient analysis. The .include control line includes an external model file for the devices. The transient simulation result for the output node is shown in Fig. 9.30(c) [203].

As a second example of SPICE simulation, a 45-nm CMOS-based sense amplifier circuit is considered. This circuit is more complex than the previous example. The schematic design has been presented and is shown in Fig. 9.31(a) [219, 222]. For a 45-nm CMOS technology, the baseline sizes are chosen as follows. For NMOS, the length is 45 nm and width 120 nm. For PMOS, the length is 45 nm and width 240 nm. The extracted netlist is shown in Fig. 9.31(b). The netlist uses an important aspect of SPICE modeling and simulation, "subckt." The .subckt control line of SPICE specifies subcircuit definitions in the netlist. This allows use of the subcircuit multiple times in one and more designs and facilitates design reuseability and hierarchy. As a specific example of this circuit simulation, the voltage signals at the bitline and $\overline{\text{bitline}}$ have been probed. The voltage signals are shown in Fig. 9.31(c) [219, 222].

9.5.8.2 Memristor Nanoelectronics

As a third example, a SPICE model of a memristor device is presented in this section: Algorithm 9.8 [105, 125, 126, 153]. The core analytical model is more or less the same as the models presented when

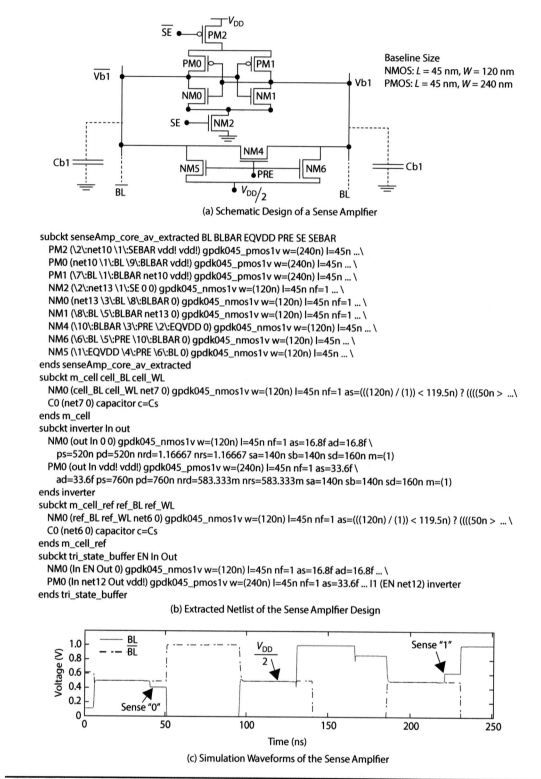

FIGURE 9.31 SPICE analysis: 45-nm CMOS sense amplifier example [219, 222].

MATLAB®- or Simulink®- based models for memristor were discussed in the previous sections. In the SPICE model, the first few lines that start with "*" are comments. It is a good idea to have the SPICE program well documented and commenting is a way to do that as is the case in any programming language. The SPICE device is modeled as a subcircuit with .subckt statement in the SPICE file. During SPICE simulation the parameters like ON state memristance Mon, OFF state memristance

ALGORITHM 9.8 A Simple SPICE Model for a Thin-Film Memristor [125, 126, 153]

```
* Mon  -   ON state memristances of the memristor
* Moff -   OFF state memristances of the memristor
* Minit - Initial memristances at T=0
* Lm   -   Memristor length
* Muv  -   Dopant mobility
* p    -   A parameter in function for modeling nonlinear dopant drift
* Lme  -   Length of the doped area between 0 to Lm
* Lrat -   Lme over Lm Ratio
.subckt memristor Plus Minus PARAMS:
+ Mon=100 Moff=36K Minit=1K Lm=10n Muv=14F p=1
Grat 0 Lrat value={I(Emem)*Muv*Mon/Lm^2*f(V(Lrat),p)}
Crat Lrat 0 1 IC={(Moff-Minit)/(Moff-Mon)}
Raux Lrat 0 1T
Emem plus aux value={-I(Emem)*V(Lrat)*(Moff-Mon)}
Moff aux minus {Moff}
Eflux flux 0 value={SDT(V(plus,minus))}
Echarge charge 0 value={SDT(I(Emem))}
.func f(Lrat,p)={1-(2*Lrat-1)^(2*p)}
.func f(Lrat,i,p)={1-(Lrat-stp(-i))^(2*p)}
.end memristor
```

Moff, initial memristance Minit, memristor length Lm, the dopant mobility Muv, and parameter p of the function that models nonlinear dopant drift in the memristor should be used to invoke the model for simulation using this SPICE model. The simulation results for an input sinusoidal signal of frequency 1 Hz are presented in Fig. 9.32 [153]. The characteristics are consistent with those obtained from MATLAB®- or Simulink®-based models.

9.5.9 Inside of SPICE

Having learned about many forms of SPICE and SPICE-like simulators, the obvious questions arises, "What is inside a SPICE engine?" "How does SPICE really work?" This section addresses these questions. Even though SPICE is extensively used for VLSI design and simulation, the inside of SPICE is really math and in particular numerical routine. Any results obtained from SPICE should match common sense VLSI to be acceptable by the design and verification engineers!

9.5.9.1 SPICE Elements

In order to support modeling of integrated circuits, SPICE provides several elements that can be used as components to build the integrated circuits. A selected set of SPICE elements are presented in Fig. 9.33 [168, 209, 216, 231, 249]. The elements can be either linear (e.g., resistor) and nonlinear (e.g., diode) or passive (e.g., resistor) and active (e.g., transistor). The passive elements such as resistors can be easily described in the SPICE. However, the semiconductor devices need several parameters for their accurate description. The .model line in SPICE specifies the semiconductor device type. Overall, the SPICE elements can be classified into eight different categories as follows:

1. *Primitive Elements:* These are basic electronic elements such as resistors and capacitors that can be explicitly used in the circuits. They can also originate as parasitic effects from the integrated circuits.

2. *Primitive Sources:* These are various types of independent and dependent voltage and current sources. The two independent sources are voltage and current sources. In addition, four types of dependent sources are presented: current-controlled current source (CCCS),

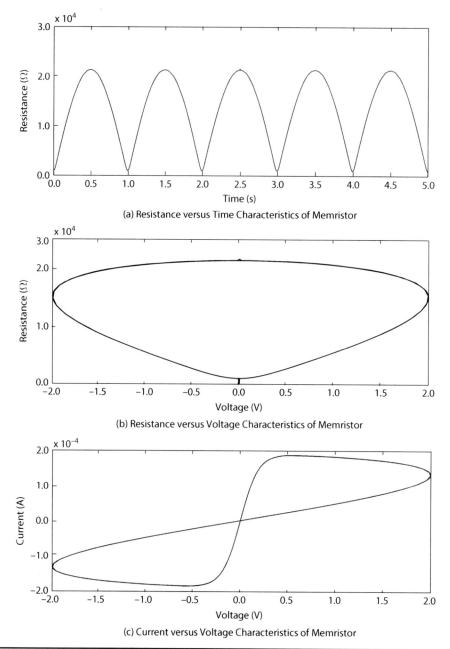

Figure 9.32 Simulation plot from the SPICE model of a thin-film memristor [153].

current-controlled voltage source (CCVS), voltage-controlled current source (VCCS), and voltage-controlled voltage source (VCVS), depending on the different effect signal and cause signal types. There is also an arbitrary source (ASRC) to specify any source, which works using various functions in a behavioral fashion.

3. *Transformers:* The transformer and mutual inductor may be similar in terms of behavior, but they can be different in terms of voltage conversion.

4. *Ideal Elements:* The ideal elements in SPICE include amplifier, isolator, circulator, gyrator, attenuator, phase shifter, and coupler, for simulation of a large variety of circuits and systems.

5. *Transmission Lines:* Lossless transmission lines, lossy transmission lines, and uniform distributed RC lines are available in SPICE to model lines that deliver signal from one node to another.

Mixed-Signal Circuit and System Simulation 457

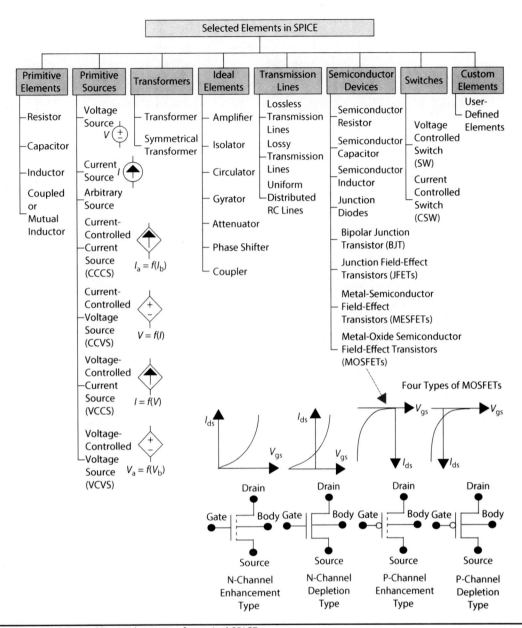

FIGURE 9.33 Selected basic elements of a typical SPICE.

6. *Semiconductor Devices:* Several semiconductor devices are available in SPICE, including MOSFET, as presented in the Fig. 9.33. The semiconductor resistor, semiconductor capacitor, and semiconductor inductor are different from resistor, capacitor, and inductor, respectively. For example, the resistance of a semiconductor resistor made from a semiconductor can be calculated from process and geometry information.

7. *Switches:* Switches such as the voltage-controlled switch (WS) and current-controlled switch (CSW) are available in SPICE to effectively model switches.

8. *Custom Elements:* Custom elements can be modeled using SPICE language and can be used in SPICE simulations. For example, the memristor device model presented in the preceding section is not a native SPICE model, but after being described in SPICE language can be a part of SPICE as an element. Similarly, as new devices emerge due to the rapid development of semiconductor technology and other related technologies, they can be easily simulated using SPICE.

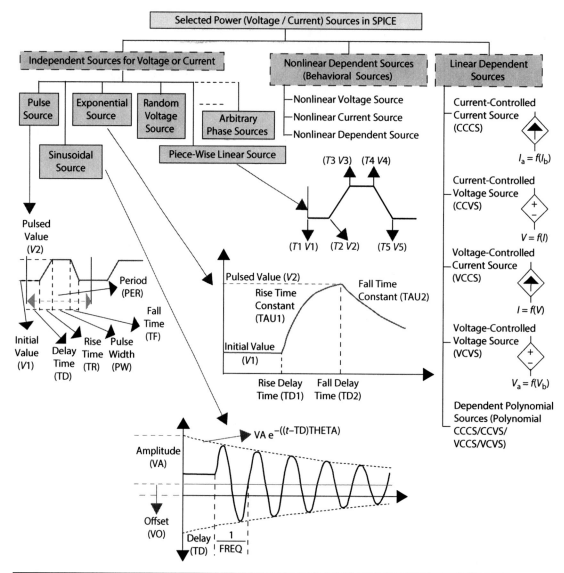

FIGURE 9.34 Different current or voltage sources in a typical SPICE.

The source available in the SPICE as elements are used to model voltage and current signals. A selected set of sources in the SPICE are presented in Fig. 9.34 [209, 216, 249]. At a high level the sources can be either independent or dependent. They can also be voltage or current sources. The dependent sources can be linear or nonlinear depending on the relation between cause (input parameter) and effect (output parameter). The independent sources irrespective of voltage or current signals can be of various types, including pulse source, sinusoidal source, exponential source, random-voltage source, and piece-wise linear source. In general, the pulse signal is specified as PWL(T1 V1 < T2 V2 T3 V3 ... >) with Ti being the time when the new value starts and Vi the pulse value. The value of the source at intermediate values of time is calculated through the use of linear interpolation on the input values by the SPICE engine. A general sine wave is described in the following manner [216]:

$$V_{\sin}(t) = \begin{cases} \text{VO} & \text{if } 0 \leq t < \text{TD} \\ \text{VO} + \text{VA} e^{(-(t-\text{TD})\text{THETA})} \sin(2\pi \text{FREQ}(t-\text{TD})) & \text{if } \text{TD} \leq t < \text{TSTOP} \end{cases} \quad (9.10)$$

where VO is offset, VA is amplitude, FREQ is frequency, TD is delay, and THETA is damping factor of the signal, which may be voltage. Similar expression can be used for current signal. A general exponential signal is described in the following manner [216]:

$$V_{\exp}(t) = \begin{cases} V1 & \text{if } 0 \leq t < \text{TD1} \\ V1 + V21\left(1 - e^{-\left(\frac{t-\text{TD1}}{\text{TAU1}}\right)}\right) & \text{if } \text{TD1} \leq t < \text{TD2} \\ V1 + V21\left(1 - e^{-\left(\frac{t-\text{TD1}}{\text{TAU1}}\right)}\right) + V12\left(1 - e^{-\left(\frac{t-\text{TD2}}{\text{TAU2}}\right)}\right) & \text{if } \text{TD2} \leq t < \text{TSTOP} \end{cases} \quad (9.11)$$

where V1 is initial value, V2 is pulsed value, TD1 is rise delay time, TAU1 is rise time constant, TD2 is fall delay time, and TAU2 is the fall time constant of the exponential signal. A current signal can also be described in a similar manner.

9.5.9.2 SPICE Engine

A large variety of integrated circuits can be described in SPICE using the components discussed in the previous section. SPICE has been typically treated as a black box that provides solutions to a variety of integrated circuits. This may be due to the fact that the choice of solver algorithms and tolerances and the majority of the integrated circuits simulate reasonably using the default settings [122, 176, 209, 216, 236, 266]. The circuit elements can be linear as well as nonlinear and a variety of governing equations describe their behavior. It can be said that various device models, numerical integration methods that can solve DAEs, and robust nonlinear and linear solvers form the core of a SPICE engine. Simplistically speaking, SPICE is a numerical solver for integration and an ordinary differential equation (ODE) solver. It may be noted that SPICE is not a solver for partial differential equations (PDEs).

The SPICE engine was originally designed in 1970s. Most of the current SPICE engines are no doubt based on this original. However, since then a lot of modifications have been made in terms of software architecture. It is difficult to accurately present the software architecture of the SPICE engine. The schematic view is depicted in Fig. 9.35, which gives an idea of the engine to the reader [77, 176, 209, 216]. However, this may not be fully complete and unique. Not all the available SPICE engines have the architecture with all the modules present. Each may have different capabilities, support different models, and support different analysis.

The SPICE engine may have a large number of modules for specific tasks. However, not all of them are invoked for all types of analyses. Rather a selected set of modules are used in a specific sequence

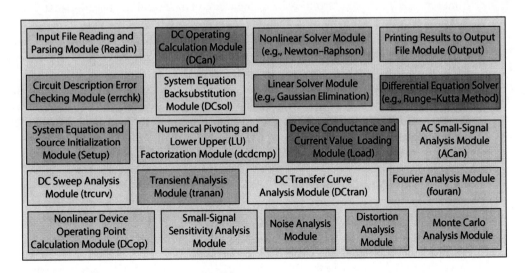

FIGURE 9.35 Architecture of the SPICE engine showing selected typical modules.

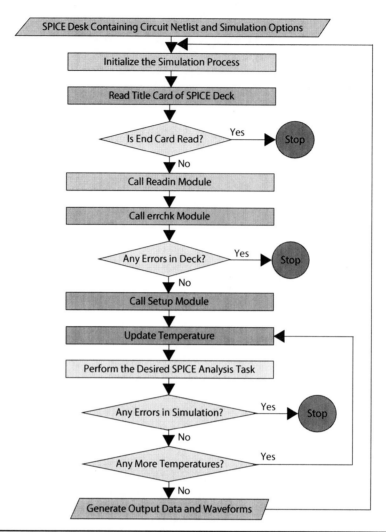

Figure 9.36 A simplified and generic flow for simulation using the modules of a SPICE engine [209].

based on the need of the analysis, for example, DC analysis or AC small-signal analysis. The generic flow of SPICE analysis is presented in Fig. 9.36 [209]. However, different flows that are specific to different analysis are discussed in the next section. The input netlist can be created manually or automatically from schematic editor or layout tools. A single circuit or several can be simulated using one SPICE run. A typical sequence is that the SPICE engine reads the cards of the SPICE deck till a .END card is reached.

The current integrated circuits are made of millions to billions of nanoscale transistors. The full-chip simulation of these gigascale complexity nanoscale integrated circuits is a difficult to impossible task. As discussed in the previous section, fast SPICE or analog-fast SPICE can handle full-chip simulations. However, there can be loss in accuracy when fast SPICE is used while analog-fast SPICE can provide near-SPICE accuracy although it can handle a lesser number of transistors than fast SPICE. Parallel and high-performance engines have been explored to provide high-speed SPICE simulation while maintaining original accuracy. A parallel SPICE engine with selected modules is depicted in Fig. 9.37 [15]. This solves the complete matrix using accurate SPICE models with high memory efficiency.

9.5.10 SPICE Simulation Flow

This section discusses the basics of SPICE simulation, with specific examples for selected types of SPICE analysis, such as DC and transient analysis.

9.5.10.1 SPICE Simulation : Theoretical Perspective

Once the SPICE deck is available to the SPICE engine, the SPICE engine builds a data flow graph abstract data type using a matrix data structure [77, 122, 171, 172, 195, 209, 236, 266, 281]. This data

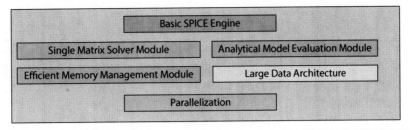

FIGURE 9.37 Simple architecture of a parallel SPICE engine [15].

flow graph represented in the form of a matrix essentially provides information on the topology of the netlist at this point. For example, it contains the information about the nodes at the SPICE elements (e.g., resistor, diode, or transistor) between the nodes. While formulating the equation for any kind of analysis through SPICE, the two conservation laws—Kirchhoff's current law (KCL) and Kirchhoff's voltage law (KVL)—need to be satisfied [168]. The KCL states that the algebraic sum of all the currents flowing through any specific node of the circuit is zero. This is charge conservation law. For any node (aka vertex) n out of total N nodes in the circuit, KCL is the following:

$$\sum_{j=1}^{J} I_{n,j} = 0 \qquad (9.12)$$

where j is any link (aka broach or edge) out of total J for the node n. Any current $I_{n,j}$ is positive if it is flowing into n or negative if it is flowing out of n. Similarly, KVL states that the algebraic sum of all the potential differences in any specific mesh (or loop) of the circuit is zero. This is energy conservation law. For any mesh m out of total M meshes in the circuit, KVL is the following:

$$\sum_{k=1}^{K} V_{m,k} = 0 \qquad (9.13)$$

where k is any link (aka broach or edge) out of total K for the mesh m. Any voltage $V_{m,k}$ is positive if it is a potential rise or negative if it is a potential drop. So the question is whether to adopt KCL or KVL. In theory it depends on the circuit topology. KCL is better if the number of nodes of the circuit is less than the number of loops, which results in a lesser number of node equations to solve. On the other hand, KVL is better if the number of loops of the circuits is less than the number of nodes, which results in a lesser number of loop equations to solve. However, in practical situations with any circuit topology, the number of nodes and the number of loops are comparable. So either KCL (i.e., charge conservation) or KVL (i.e., energy conservation) can be used. It is observed that most of the SPICE/SPICE-like engines use KCL and solve the circuit using (modified) nodal analysis, and few SPICE/SPICE-like engines use KVL and solve the circuit using mesh analysis. It may be noted that essentially mesh is a loop in the circuit that does not contain any other loop.

For any given circuit with N number of nodes, a total of N charge conservation equations can be derived using KCL as presented above for each node n. The equations are then modified by using the branch constitutive equations (BCEs) that represent the current–voltage characteristic of each of the branch elements. For the purpose of solving the equations for the unknown using the MNA, the equations are represented in the following matrix format [168, 171, 195]:

$$[A][X] = [B] \qquad (9.14)$$

where $[A]$, $[X]$, and $[B]$ are matrices of different dimensions. In the above equation, $[A]$ is the matrix of conductances, $[B]$ is the matrix of known currents and voltages, and $[X]$ is the matrix of unknown voltages and branch currents. In essence, this is a representation of DAEs of the given circuit. It may be noted that the DAEs are different from ODEs. For a circuit with N nodes and M independent voltage sources, the matrix $[A]$ has a dimension of $(N + M) \times (N + M)$. The dimension of matrix $[X]$ is $(N + M) \times 1$. The dimension of matrix $[B]$ is also $(N + M) \times 1$. Using computer-aided numerical methods, SPICE solves the above set of equations for unknown $[X]$ of the following format:

$$[X] = [A]^{-1}[B] \qquad (9.15)$$

The above can be stated as "equation-formulation phase" in SPICE simulation. This approach of MNA is a sparse matrix technique and requires less memory.

The above SPICE formulations may contain SPICE circuit linear BCEs for elements like resistors and capacitors as well as nonlinear BCEs for elements like diodes and transistors while satisfying the conservation constraints at the different circuit nodes [171, 172, 266]. It may be noted that whenever an element is present between two or more nodes in the matrix, the current values need to be calculated for that element using its model. This is performed by using the SPICE (compact) models and this process of SPICE is called "model-evaluation phase." In essence, the model-evaluation phase of SPICE calculates the currents and conductances through different elements of the circuit and then updates the corresponding entries in the matrix with these values. This is performed for every element using its SPICE model. Model evaluation is performed multiple times during the numerical iterations undertaken by SPICE for solving the unknowns. It is a fact that the SPICE engines spend quite a bit of time during the overall simulation in model evaluation. That is the specific reason why fast SPICE used LUT instead of compact models so that the model evaluation is not needed.

The numerical solution of SPICE may involve linear equation solution, nonlinear solution, and numerical integration depending on the type of elements present in the circuit as well as the type of analysis needed [77, 122, 171, 172, 195, 209, 236, 266, 281]. Each of these solutions can be obtained through several algorithms and the choice depends on the speed and accuracy of the algorithm. There are many aspects to be considered when choosing algorithms for this purpose, e.g., accuracy, speed, and memory requirements. The linear equation solutions can be obtained through numerical techniques such as the Gaussian elimination method that relies on lower upper (LU) factorization and pivoting. The nonlinear equation solution involves calculating the small-signal linear approximations for the nonlinear elements and then solving the resulting linear equations till it reaches the fixed point. For nonlinear equation solutions, the SPICE must search for an operating point using Newton–Raphson (NR) iterations that may require repeated model evaluations with corresponding updates in the matrix. SPICE may use techniques such as the Euler method and trapezoidal method to perform numerical integration. This phase of SPICE is called the "matrix-solution phase." This is accompanied by "iteration control phase" in which the following can be effectively executed during the simulation: (1) linearization iterations for nonlinear devices, (2) adaptive time-stepping for time-varying devices, and (3) convergence condition evaluation.

9.5.10.2 SPICE Simulation : Specific Analysis Flows

The SPICE engine follows specific flow and uses selected engines based on the type of analysis that needs to be performed. The SPICE simulation flow chart for DC analysis is presented in Fig. 9.38 [77, 168]. The flow is applicable for the analysis of a large variety of circuits including both linear and nonlinear circuits. If the circuit is linear, then DC analysis is the simplest and fastest. The overall matrix is solved once using linear solution techniques such as the Gaussian elimination algorithm. If the circuit is nonlinear, then the NR algorithm is used. The convergence for nonlinear circuits is improved by the use of fall-back convergence helpers as well as application of nodesets to force NR iteration in a specific direction.

The SPICE flow for transient analysis is presented in Fig. 9.39 [77, 209]. For each time step, the time step is incremented and numerical integration is performed to approximate the time-derivative components in the model equations. The NR iterations are then used to solve the circuit at each time step. Each of the NR iterations involves linearization of the model equations. This is followed by formulation of the linearized circuit equations. The linearized set of equations are then solved using the Gaussian elimination algorithm. The solution for this time step is stored after the convergence of linear solution. The above NR iteration is repeated for the next time step. The transient analysis stops and the results are saved/displayed after all the time steps are covered. It may be noted that for the linear circuits, the inner NR iteration loop is not required as the solution needs only one iteration.

Mixed-Signal Circuit and System Simulation 463

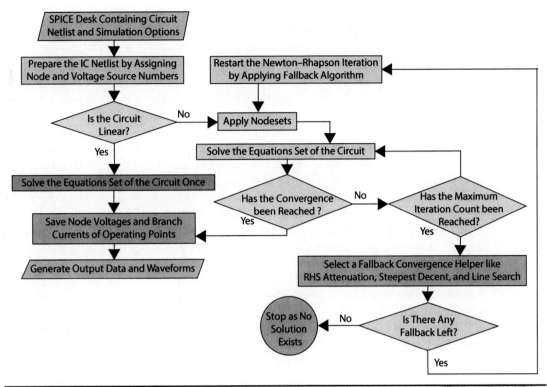

FIGURE 9.38 SPICE simulation DC analysis flow [168].

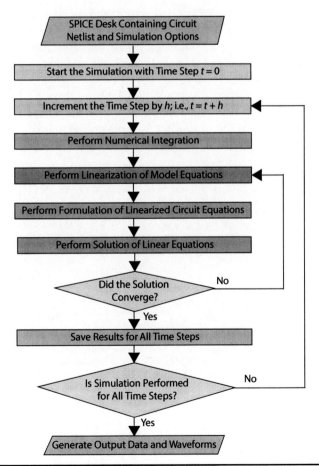

FIGURE 9.39 SPICE simulation transient analysis flow [209].

9.6 Verilog-A-Based Analog Simulation

While SPICE has been in the mainstream of analog simulation, Verilog-A hardware description language (Verilog-A HDL or simply Verilog-A) is emerging as an industry standard for modeling language for analog integrated circuits and systems [141, 194, 208, 252, 257]. In 1996, Verilog-A was initiated as the analog extension to Verilog HDL [223]. However, it was later integrated with Verilog-AMS HDL and has been made available as a subset of Verilog-AMS HDL [104, 180]. Verilog-A (through Verilog-AMS) provides a new dimension in modeling and simulation of analog circuits and systems. It allows analog design engineers to behaviorally model the analog components or elements and use them at a higher level of design abstraction to design and simulate larger circuits and systems without worrying about the insides of these lower-level modules [104]. Verilog-A has the following important features: (1) Verilog-A behavioral model descriptions are part of separate analog blocks. (2) Verilog-A models are compatible with Verilog-AMS for mixed-signal modeling. (3) Verilog-A based modules can be parametrized within specified range. (4) Branches in Verilog-A models can be named for easy usage. (5) Verilog-A modeling can take advantage of operators, variables, and signals for system modeling. (6) Operators like derivative, integral, and trigonometric functions are supported in Verilog-A. (7) Verilog-A can access SPICE primitives within the language itself. (8) Verilog-A behavioral descriptions can be used in multiple disciplines including electrical, mechanical, and thermodynamic.

The question arises as to why use Verilog-A when SPICE is so successful and has been widely used over the past several decades. SPICE language and netlist format is quite complex to describe large circuits and systems [59, 141, 257]. It is not easy to directly write circuits or system descriptions using SPICE language as it is not a high-level language. However, Verilog-A is a high-level language that allows design engineers to model circuits and systems. A comparative perspective of Verilog-A modeling with respect to SPICE modeling is presented in Fig. 9.40 [59]. Hierarchical modeling through Verilog-A is easier and selected components of circuits or systems of a netlist can be replaced by Verilog-A to reduce modeling efforts. SPICE simulations always have to solve a large number of equations, particularly when active elements are present in the circuit, and large matrices, which can be reduced through Verilog-A. However, as of now Verilog-A-based simulation seems to be slower than SPICE simulations [141, 208].

One can ask why Verilog-A when Verilog HDL has been around and well accepted. Verilog HDL, which was among the first HDLs developed, was meant for modeling and simulation of digital integrated circuit only. On the other hand, Verilog-A HDL was introduced for modeling and

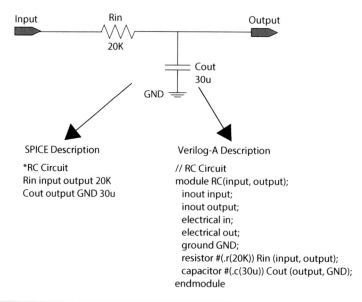

Figure 9.40 Verilog-A versus SPICE simulation [59].

ALGORITHM 9.9 Verilog-A versus Verilog Modeling [194, 252]

```
          Verilog Template                        Verilog-A Template

module Test-Verilog(<signal list>);    module Test-Verilog-A(<signal list>);
   <signal_declarations>                  <signal_declarations>
   <parameter_definitions>                <parameter_definitions>
   <local_structure>                      <local_structure>
   always begin                           analog begin
      <behavioral_definitions>               <behavioral_definitions>
   end                                    end
endmodule;                             endmodule;
```

simulation of analog integrated circuits. The two languages are entirely different: Verilog-A is for analog circuits and Verilog for digital circuits [194, 252]. The digital simulator that supports Verilog will not accept Verilog-A models for simulations. Of course, Verilog-AMS simulators can support both of them. For a comparative perspective, Verilog-A modeling is presented with respect to Verilog in Algorithm 9.9 [194].

Yet another point, if Verilog-A is for behavioral modeling, then why cannot well-established MATLAB® language be used? MATLAB® is well established for DSP and communication systems, but it does not contain features for hardware modeling [107]. Even Simulink®, which has features for hardware modeling, needs more effort for circuit-level modeling. For example, when modeling logic gates and flip-flops, delay blocks need to be inserted manually to obtain realistic frequency of operation. Modeling of circuit-level loading effects such as RC-delay and fall/rise time is not simple using Simulink®. Verilog-A language, which has mode circuit-level and device-level features, is suited for analog modeling.

Many of the current SPICE engines can be used for Verilog-A-based simulation. For example, a selected set of simulators for Verilog-A-based simulations include the following: SIMetrix, SMASH, Spectre®, TINA™. In addition, Verilog-A-based simulations can be performed using the same simulators that are used for Verilog-AMS-based simulations.

9.6.1 Verilog-A-Based Circuit-Level Simulations

Few examples of Verilog-A-based modeling for circuit-level simulation are now presented. Three widely used circuits are considered: VCO, GFET based LC-VCO, and low-pass filter.

9.6.1.1 Voltage-Controlled Oscillator (VCO)

As a specific example of Verilog-A modeling, the simulation of a VCO is discussed. The Verilog-A code for a VCO is presented in Algorithm 9.10 [104, 113, 243]. The lines starting with // character are comments that are not executed in the simulation but are used as a better programming practice. Then the three statements of this Verilog-A model that start with accent grave ` character are compiler directives that are used to include disciplines and constants. The `define globally defines constants for easy use repeatedly and while facilitating easy modification at a single place when required. The module defines VCO_Behavioral module with two ports, Vin and Vout, which are input and output, respectively. Verilog-A also allows inout port. Vin and Vout are electrical ports. Parameter types and default values have been specified in several lines. The behavior of the module is described within analog begin-end. The model presents a VCO with a center frequency f_c. The instantaneous frequency is calculated based on center frequency, gain and input voltage. The idtmod (instantaneous_frequency, 0, 1) line performs circular integration and returns a value in the range [0,1]. The phase of the VCO is enforced to $[0, 2\pi]$. Sine wave is used through the trigonometric function sin to describe output of the VCO. As this is behavioral modeling both ring oscillator and LC-VCO can be represented

ALGORITHM 9.10 A Verilog-A Model for a Voltage-Controlled Oscillator (VCO) [104, 243]

```
// Verilog-A Model for a Voltage Controlled Oscillator (VCO)
'include "discipline.h"
'include "constants.h"
'define PI   3.1416
module VCO_Behavioral(Vin, Vout);
input Vin;
output Vout;
electrical Vin, Vout;
parameter real amp = 0.35;
parameter real center_frequency = 394.25M;
parameter real VCO_gain = 563.2M;
parameter integer steps_per_period = 32;
real VCO_phase;
real instanteous_frequency;
analog begin
   instanteous_frequency = center_frequency + VCO_gain * V(Vin);
   $bound_step (1.0 / (steps_per_period * instanteous_frequency));
   VCO_phase = idtmod(instanteous_frequency, 0, 1);
   V(Vout) <+ amp * sin (2 * 'PI * VCO_phase) + 0.35;
end
endmodule
```

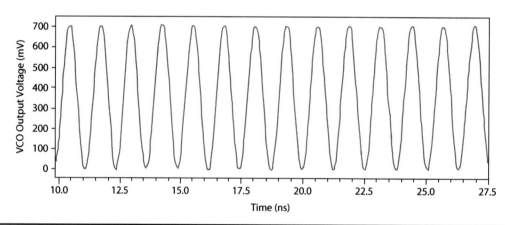

FIGURE 9.41 Verilog-A-based simulation of a voltage-controlled oscillator (VCO) [243].

in this fashion. The waveforms obtained after simulating the Verilog-A model of VCO in an analog simulator is presented in Fig. 9.41 [243].

9.6.1.2 Graphene FET-Based LC-VCO

The schematic diagram of a graphene FET (GFET)-based LC-tank oscillator is presented in Fig. 9.42(a) [175]. This is the same GFET LC-VCO topology that is discussed in the previous section on Simulink®-based simulation. However, in this section Verilog-A-based modeling has been used for its simulation. The circuit-level simulation of the LC-VCO uses Verilog-A-based device model, which is discussed in the next section. Schematic-level design and simulation are performed in an analog simulator in which the GFET used Verilog-A models. The output voltage simulation data is presented in Fig. 9.42(b) [175]. The frequency versus tuning voltage characteristics is shown in Fig. 9.42(c) [175]. The tuning characteristics show that it is linear in the operating voltage range. The characterization of the GFET VCO obtained from its Verilog-A-based simulations is presented in Table 9.5 [175].

Mixed-Signal Circuit and System Simulation

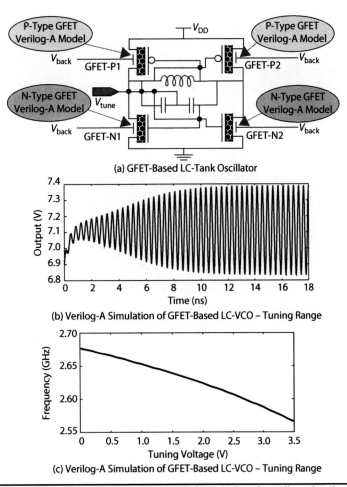

Figure 9.42 Verilog-A simulation of a graphene FET (GFET)-based LC-tank oscillator [175].

GFET LC-VCO Parameters	Simulation Values
Center Frequency f_c	2.64 GHz
Tank Voltage Swing $V_{tank_{p-p}}$	0.55 V
Bias Current I_{bias}	0.88 mA
Tuning Range	4.2%
Phase Noise (at 1 MHz offset)	−92.66 dBc/Hz

Table 9.5 Characterization of the GFET VCO Simulated using Verilog-A Models [175]

9.6.1.3 Low-Pass Filter (LPF)

The circuit-level Verilog-A-based simulation of an LPF is now discussed. A simple RC circuit-based LPF with a resistor and a capacitor is considered. However, Verilog-A model of LPF with inductor could also be presented [113]. The Verilog-A model of the LPF is presented in Algorithm 9.11 [104, 243]. The `module` defines LPF_1st-Order_Behavioral module with electrical input and output ports Vin and Vout, respectively. In this Verilog-A model, the value for the bandwidth of the filter is 788 MHz. The values for the resistance and capacitance are calculated for this bandwidth. The `ddt` operator in the Verilog-A model calculates the time derivative of its argument Cin*Vout. The waveforms obtained after simulating the low-pass RC filter are shown in Fig. 9.43 [243].

ALGORITHM 9.11 A Verilog-A Model for a Low-Pass Filter (LPF) [104, 243]

```
// Verilog-A Model for a Low Pass Filter (LPF)
`include "discipline.h"
`include "constants.h"
`define PI      3.1416
module LPF_1st-Order_Behavioral(Vin, Vout);
input Vin;
output Vout;
electrical Vin, Vout;
parameter real bandwidth = 788M;
real Rin;
real Cin;
analog begin
   @(initial_step("tran","ac","dc")) begin
      Rin = 1K;
      Cin = 1/(2*`PI*r*bandwidth);
   end
   V(Vout, Vin) <+ R*I(Vout, Vin);
   I(Vout) <+ ddt(Cin*V(Vout));
end
endmodule
```

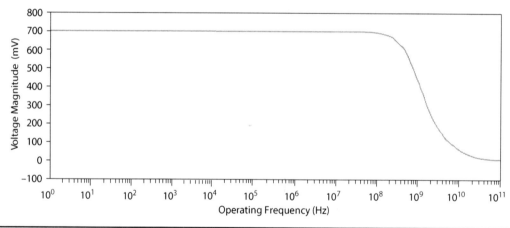

FIGURE 9.43 Verilog-A-based simulation of a voltage-controlled oscillator (VCO) [104, 243].

9.6.2 Verilog-A-Based Device-Level Simulations

In this section device-level simulation of selected nanoelectronic devices is discussed. Two models for memristor and one model for GFET are considered. These device models can be used for hierarchical circuit-level and system-level simulations.

9.6.2.1 Verilog-A-Based Modeling for Memristor-1

The Verilog-A model of the thin-film memristor whose Simscape® model is presented in the previous section is presented in Algorithm 9.12 [137, 146, 153, 233, 290]. The `include in the first line includes mathematical and physical constants from Verilog-AMS as Verilog-A is a subset of Verilog-AMS. The memristor is defined as a `module` with designer given `inout` ports pos and neg. The `electrical` keyword declares pos and neg are of electrical discipline or nature in Verilog-A. In the statements that begin with `parameter`, many different device parameters have been declared

ALGORITHM 9.12 A Simple Verilog-A Model for a Thin-Film Memristor [153, 233, 290]

```
// Verilog-A Model for a Thin-Film Memristor
'include "constants.vams"
'include "disciplines.vams"
module memristor (pos, neg) ;
inout pos, neg;
electrical pos, neg;
parameter real Muv = 10e-15;    // Dopant Mobility
parameter real Lm = 10e-9;      // Memristor Length
parameter real Mon = 100;       // ON State Memristance
parameter real Moff = 38e3;     // OFF State Memristance
parameter real Minit = 5e3;     // Initial Memristance
real k, r1, r2, Mmem;
analog begin
   k = 2 * Muv * Mon * (Moff - Mon) / pow(Lm,2) ;
   r1 = pow(Minit,2) + k * idt( V(pos,neg), 0 );
   r2 = min( pow(Moff,2), max(r1, pow(Mon,2)) );
   Mmem = sqrt(r2);
   V(p,n) <+ Mmem * I(pos,neg);
end
endmodule
```

and default values have been assigned. The memristor module description is provided after `analog begin`. Different functions like `pow`, `min`, `max`, and `sqrt` have been used in the behavioral model description. The `idt` is the time-integral operator that computes the time integral of `V(pos, neg)`. The device is simulated in an analog simulator for the following parameters: Lm =10 nm, Mon = 100 Ω, Moff = 38 kΩ, Minit = 5 kΩ, and Muv = $10^{-14} m^2 s^{-1} V^{-1}$ for a sinusoidal input signal of frequency 1 Hz. The memristance versus voltage characteristics are shown in Fig. 9.44(a) [153] and the I–V characteristics are presented in Fig. 9.44(b) [153]. The results match with the results obtained from memristor models discussed in the previous sections.

9.6.2.2 Verilog-A-Based Modeling for Memristor-2

This section discusses another memristor model using Verilog-A. The coupled variable-resistor model of a titanium dioxide (TiO_2) thin-film memristor is expressed as follows [238, 250, 289]:

$$V_M = [M_{on} L_{rat} + M_{off}(1 - L_{rat})] I_M \qquad (9.16)$$

$$\left(\frac{dx}{dt}\right) = \mu_v \left(\frac{M_{on}}{L_m^2}\right) I_M \qquad (9.17)$$

In the above model $L_{rat} \in [0, 1]$ is an internal state variable, M_{off} is the OFF state memristance, M_{on} is the ON state memristance, μ_v is the dopant mobility of the semiconductor, and L_m is the length of the memristor. The boundary conditions of the memristors, in general, present major limitations in the memristor modeling [289]. An issue called "hard-switching" or "terminal-state" problem needs to be addressed [230, 289]. When the state variable L_{rat} is pushed to its boundaries (0 or 1) due to external signal source, its value should remain unchanged (0 or 1) until the applied signal changes its polarity. The window functions as used in previous memristor models may prevent the state variable from returning from its boundaries even after the polarity of the applied signal changes [230, 289]. In addition, the nonlinearity generated by the window function does not conform to the characteristics of the already fabricated memristors. Moreover, further inaccuracy to the models is caused by the

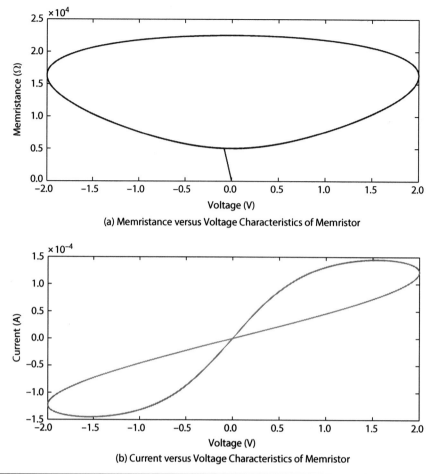

FIGURE 9.44 Simulation of Verliog-A model of the thin-film memristor [153].

noninclusion of an explicit threshold. To examine threshold-type behavior, a coupled variable-resistor model is now introduced [251, 289]:

$$\left(\frac{dx}{dt}\right) = \begin{cases} \mu_v \left(\frac{V_p}{L_m^2}\right) exp\left(\frac{M_{on}}{V_p} I_M\right) & \text{if } V_M \geq V_p \\ \mu_v \left(\frac{V_n}{L_m^2}\right) exp\left(\frac{M_{on}}{V_n} I_M\right) & \text{if } V_M \leq V_n \\ \mu_v \left(\frac{R_{on}}{L_m^2}\right) I_M & \text{otherwise} \end{cases} \quad (9.18)$$

In the above model, V_p and V_n are two different threshold values for different polarities of V_M. The above analytical model is presented as a Verilog-A model as shown in Algorithm 9.13 [289]. Time-domain simulation of the memristor is presented for the following parameters: $M_{on} = 1$ kΩ, $M_{off} = 10$ kΩ, $\mu_v = 10^{-14}$m^{2}s^{-1}V^{-1}, $V_p = 1.7$ V, $V_n = -1.7$ V with an input sinusoidal voltage of 2 V amplitude and 40 Hz frequency. In Fig. 9.45(a) voltage and current are shown [289]. The left side y-axis has input voltage in V and right side y-axis shows current in mA. The I-V characteristic of the memristor is shown in Fig. 9.45(b) for three different frequencies of the input sinusoidal voltage [289].

ALGORITHM 9.13 Verilog-A-Based Coupled Variable Resistor Model for a Thin-Film Memristor [289]

```
// Verilog-A based Coupled Variable Resistor Model for A Thin Film Memristor
`include "constants.vams"
`include "disciplines.vams"
module memristor (pos, neg);
parameter real Lrat0 = 0.5;        // Initial State
parameter real Muv = 3e-18;        // Dopant Mobility
parameter real d = 40e-9;          // Thickness
parameter real Moff = 10e3;        // OFF State Memrisistance
parameter real Mon = 1e3;          // ON State Memrisistance
parameter real Vp = 1.7;           // Positive Threshold
parameter real Vn = -1.7;          // Negative Threshold
real Lrat, Lrata, Mm, dxdt, fz, fp, fn, integ;
inout pos, neg;
electrical pos, neg;
branch (pos, neg) mem;
analog begin
   @(initial_step) begin
      fz = Muv * Mon / (d**2);
      fp = Muv * Vp / (d**2);
      fn = Muv * Vn / (d**2);
      Lrata = Lrat0;
   end
   @(cross(V(mem), 0) or cross(V(mem)-Vp, 0) or cross(V(mem)-Vn, 0));
   if (V(mem) >= Vp)
      dxdt = fp * limexp(I(mem) * Mon / Vp);
   else if (V(mem) <= Vn)
      dxdt = fn * limexp(I(mem) * Mon / Vn);
   else dxdt = fz * I(mem);
   if ((integ == -1 && V(mem) > 0)||(integ == 1 && V(mem) < 0)) integ = 0;
   @( cross(Lrat,-1) ) begin
      integ = -1;
      Lrata = 0;
   end
   @( cross(Lrat-1,1) ) begin
      integ = 1;
      Lrata = 1;
   end
   Lrat = idt(dxdt, Lrata, integ, 1e-12);
   rm = Mon * Lrat + Moff * (1 - Lrat);
   I(mem) <+ V(mem) / Mm;
end
endmodule
```

9.6.2.3 Verilog-A-Based Modeling for Graphene FET (GFET) Device

As a third example of Verilog-A-based simulation, graphene FET (GFET) device is considered in this section. The same GFET whose Simscape® model is presented in the previous section is considered. The Verilog-A models are based on the VHDL-AMS models presented in [260]. The Verilog-A model for a double-gate GFET is presented in Algorithm 9.14 [175, 260]. The Verilog-A model of GFET is simulated using an analog simulator. The results match well with the results obtained through other models. As example, the I–V characteristic is presented in Fig. 9.46 [175].

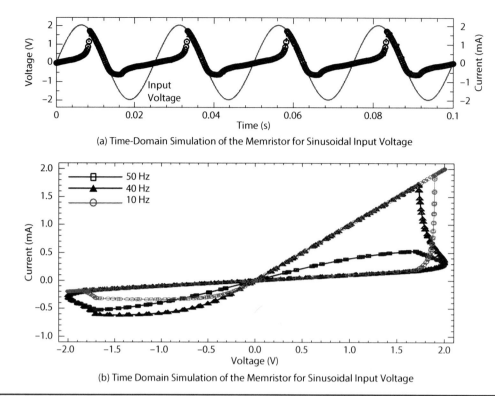

Figure 9.45 Simulation plot of coupled variable resistor-based Verliog-A model of the memristor [289].

Algorithm 9.14 A Verilog-A Model for a Graphene FET (GFET) [175]

```
// Verilog-A Model for a Graphene FET (GFET)
'include "disciplines.vams"
'include "constants.vams"
module graphene(drain, gate, back_gate, source);
parameter real Rs = 800.0; parameter real mu = 700.0e-4; parameter real Ec = 15e5;
parameter real Cq = 0.02; parameter real Cgio = 0.8072;
parameter real Hsub = 285.0e-9;   // Substate thickness
parameter real tox = 15.0e-9;     // Top gate dielectric thickness
parameter real L = 3.0e-6;        // Gate lenght
parameter real W = 2.1e-6;        // Channel Width
parameter real ntop = 2.1209e16;  // carrier concentration
parameter real Vgs0 = 1.45;       // Gate-to-Source voltage at the Dirac point
parameter real Vbs0 = 2.7;        // Bulk-to-Source voltage at the Dirac point
parameter real k = 16.0;          // Top gate dielectric
parameter real k_sub = 3.9;       // Back gate substrate dielectric
inout drain, gate, back_gate, source;
electrical drain, gate, back_gate, source; electrical drainint, sourceint;
branch(drainint, sourceint) Ids; branch(drain, source) Vds; branch(gate, source) Vgs;
branch(gate, drain) Vgd; branch (back_gate, source) Vbs;
branch(back_gate, sourceint) Vbsint; branch(gate, sourceint) Vgsint;
branch(gate, drain) Vgdint; branch (back_gate, drainint) Vbdint;
branch (drain, drainint) Resistor1; branch (source, sourceint) Resistor2;
parameter real c1=('P_EPS0 * k * Cgio * W * L/ tox) from [0:inf);
parameter real c2=('P_EPS0 * k_sub * W * L *0.5/Hsub) from [0:inf);
parameter real r=800.0; real ids;
analog function real Fgfet;
    input Vds, Vgs, Vbs, Rs, mu, Ec, Cgio, Hsub, tox, L, W, ntop, Vgs0, Vbs0, k, k_sub;
```

```
    real Vds, Vgs, Vbs, Rs, mu, Ec, Cgio, Hsub, tox, L, W, ntop, Vgs0, Vbs0, k, k_sub;
    real Ce, Ctop, Cback, Vc, Ids, Rc, Gamma, Vo, Vg0, Vdsat, Io, Cq;
    begin
        Cq = pow((ntop/'M_PI),0.5)*pow('P_Q,2)/1.0E6/('P_H/(2*'M_PI));
        Ce = Cgio*'P_EPS0*k/tox; Ctop = Cq*Ce/(Cq+Ce); Cback = 'P_EPS0*k_sub/Hsub;
        Vo = Vgs0 + (Cback/Ctop)*(Vbs0 - Vbs); Vg0 = Vgs - Vo;
        Vc = Ec*L; Rc = 1.0/((W/L)*mu*Ctop*Vc); Gamma = Rs/Rc;
        Vdsat = 2.0*Gamma*Vg0/(1.0+Gamma)+((1.0- Gamma)/pow((1.0+Gamma),2.0))
              *(Vc-sqrt(pow(Vc,2.0)-2.0*(1.0+Gamma)*Vc*Vg0));
        Io = 2.0*(W/L)*mu*Vc*Ctop*(Vgs-Vo-Vds/2.0);
        if (Vds < Vdsat)
            Ids = -1.0/4.0/Rs*(Vds-Vc+Io*Rs+sqrt(pow((Vds-Vc+Io*Rs),2.0)-4.0*Io*Rs*Vds));
        else if (Vds >= Vdsat)
            Ids = -(Gamma/Rs/pow((1.0+Gamma),2.0)*(- Vc+(1.0+Gamma)*Vg0
                 +sqrt(pow(Vc,2.0)-2.0*(1.0+Gamma)*Vc*Vg0))-mu*Cback*abs(Vbs-Vbs0)
                 *Vds*W/L/10.0*pow((Vds/Vdsat -1.0),2.0));
        else
            Ids = 0.0;
        Fgfet = Ids;
    end
endfunction
analog begin
    ids = Fgfet(V(Vds),V(Vgs),V(Vbs),Rs,mu,Ec,Cgio,Hsub,tox,L,W,ntop,Vgs0,Vbs0,k,k_sub);
    I(Ids) <+ ids; I(gate,source) <+ c1*ddt(V(Vgs)); I(gate,drain) <+ c1*ddt(V(Vgd));
    I(back_gate,sourceint)<+c2*ddt(V(Vbsint)); I(back_gate,drainint)<+c2*ddt(V(Vbdint));
    if (r > 1e12*Resistor1.flow.abstol) begin
        I(Resistor1) <+ V(Resistor1)/Rs;
    end
    else begin
        V(Resistor1) <+ Rs*I(Resistor1);
    if (r > 1e12*Resistor2.flow.abstol) begin
      I(Resistor2) <+ V(Resistor2)/Rs;
    end
    else begin
        V(Resistor2) <+ Rs*I(Resistor2);
    end
endmodule
```

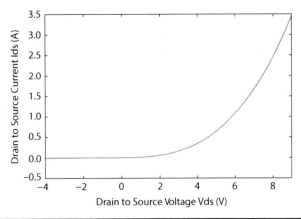

FIGURE 9.46 Simulation plot of Verilog-A model of the Graphene FET (GFET) [175].

9.7 Simulations of Digital Circuits or Systems

The digital components in an AMS-SoC are much more complex in terms of transistor count. In addition, the digital fabrication technology scales at a much faster rate. However, digital design has the advantage of more mature EDA tools as well as HDLs. In this section, several digital HDLs, such as (System)Verilog, VHDL, MyHDL, and SystemC, are briefly introduced [117, 213, 239]. These languages can model digital circuits and systems, and can be used for digital simulations. Some can be used for ASIC simulations, some for system simulations, and some for simulation in FPGA platforms. Each of these HDLs has its own style and heredity. A designer can make a decision to use one of them for his/her design need based on various aspects including design budget, complexity, and computing platform. The best way to learn these languages is to learn by simulating diverse types of design examples. In addition, there are digital verification languages which are out of scope of this book. The question arises why HDLs when C/C++ is there. No doubt the general-purpose languages like C/C++ are used almost everywhere but HDLs are needed for the following reasons. The language C/C++ do not have the capability for hardware data types, concurrency, notion of time, event sequencing, tristate bus value with "Z". C/C++ can handle sequential instructions, the concurrency of instructions is not possible. C/C++ can lead to 0 or 1 numbers, but there is no facility to handle tristate "Z". C/C++ do not have any feature to arbitrary width bit vector which is typically needed in hardware modeling. C/C++ has no features to natively support processor, memory, I/O devices.

9.7.1 SystemVerilog-Based Simulation

In this section the HDL SystemVerilog is briefly discussed [103, 117, 237, 254, 294]. Verilog is one of the oldest HDLs, which was developed in the 1980s. Verilog since then has had a standard language and has been widely used in industry. An HDL called Superlog that existed around early 2000 was the starting point of SystemVerilog. The HVL called Vera (aka OpenVera™) was merged with it to provide verification capabilities. Eventually Verilog was merged with this. Overall a new and powerful language called "SystemVerilog" originated, which is an HDVL, i.e., HDL+HVL. SystemVerilog gives both design and verification capabilities to hardware designers in a single platform.

The distinct features of SystemVerilog are the following [103, 117, 237, 254, 294]: All features of design language Verilog-2005 along with additional RTL capabilties are covered by SystemVerilog. The verification features of the SystemVerilog are supported by Java-like extensive object-oriented programming capabilities. SystemVerilog has C-like data type including int, typedef, and enum. In addition, dynamic data types including array, classes, struct are supported in it. All data types are predefined in the language itself and all data types can have a bit-level representation. SystemVerilog supports many control flow features including foreach, continue, and break. It can be used for the multiple levels of digital design abstractions like algorithm, RTL, and logic level. The tools that can be used for (System)Verilog simulation include: Incisive Enterprise Simulator, ModelSim®, SMASH, Super-FinSim, and Synopsys® VCS [18, 27, 40, 47, 60].

As a specific example, the SystemVerilog model of an 8-bit up counter is presented in Fig. 9.47 [61, 103, 254]. The first line with // character is commented till the end of the line and anything after

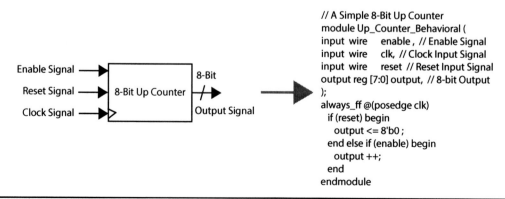

Figure 9.47 SystemVerilog-based implementation of a 8-bit up counter [61, 103, 254].

that is ignored by the compiler. The `module` keyword declared a user given name Up_Counter_Behavioral as module. It has several `input` and `output` ports. The `wire` keyword declares net or connection elements. The `reg` keyword declares a variable that can hold a value; but this variable is not a hardware register. The `output` port named output can hold 8-bit binary. The behavioral description of the module is included in `always_ff` block. This block runs whenever a signal in the sensitivity list (which are specified within bracket) changes its value. The statements within the `always` block are executed sequentially just like C/C++. It may be noted that SystemVerilog has `always_ff`, `always_comb`, and `always_latch` blocks in addition to `always` block to explicitly specify logic. For example, `always_ff` block indicates a sequential logic. The `posedge` clk indicated positive edge, i.e., transition from 0 to 1 in clk. Similarly, there can be negedge edge, i.e., transition from 1 to 0 in clk. The keyword `endmodule` suggests the end of the module.

9.7.2 VHDL-Based Simulation

The second digital HDL, called very high-speed integrated circuit (VHSIC) hardware description language (VHDL), is introduced in this section [68, 78, 114, 117]. The development of VHDL started way back in 1981. VHDL was developed based on the Ada programming language and is more verbose than Verilog due to the corresponding language requirements [117]. For a comparative perspective, modeling of a simple 4-bit module that can perform addition or subtraction based on an input signal is presented in Fig. 9.48 [61]. Both SystemVerilog modeling and corresponding VHDL modeling are presented for the same module. The `entity` keyword is used to declare the entity instead of `module` keyword SystemVerilog. In VHDL, a set of ports is declared for the entity using keyword `port`. The description of the entity in the VHDL is presented within an `architecture` body as compared to `always` keyword in SystemVerilog. The architecture is described within a `process` body behaviorally in the VHDL model. In a similar manner the (System)Verilog and HVDL can be compared to understand the two languages. In general, VHDL is a strongly typed language and hence

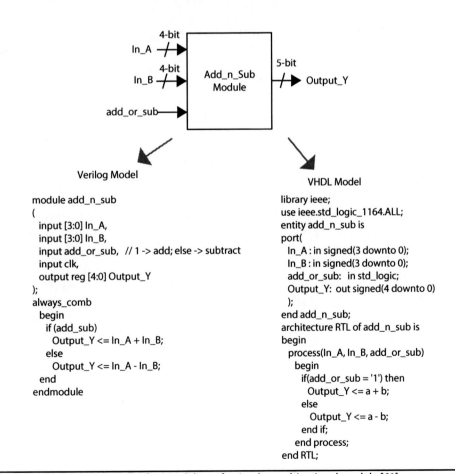

FIGURE 9.48 VHDL versus SystemVerilog modeling of a simple combinational module [61].

requires more lines of code to explicitly convert from one data type to another data type (e.g., from bit-vector to integer) [117]. The VHDL models can be simulated using various tools as follows: Incisive Enterprise Simulator, ISE Simulator (ISim), ModelSim®, Quartus® II Simulator, SMASH, Super-FinSim, Synopsys® VCS [18, 19, 22, 27, 40, 47, 60].

VHDL can be used to model systems at different levels of digital abstraction such as behavioral level, RTL, and logic level [78, 114, 240]. VHDL supports hierarchical modeling and simulation of the system. It supports modeling of digital system using either structural or behavioral descriptions at different levels of abstraction. In structural modeling, predefined components in VHDL are instantiated and connected together to build the complete system. In behavioral modeling, the overall system uses the behavior of the system without worrying about the structure or the components of the system. The behavior of the system can be presented in one of the two ways. One is to describe the system using concurrent signal assignment statements. The other way is to describe the behavior of the system by means of high-level language constructs using processes. VHDL supports many types of objects, statements, packages, libraries, and operators. A VHDL library contains packages, entities, architectures, and configurations. A VHDL package contains functions, procedures, type declarations, variable declarations, and component declarations. VHDL allows a variety of ports including input port in, output port out, and bidirectional port inout. The standard types of VHDL include boolean type, integer type, real type, and bit_vector type. The boolean operators of VHDL include not, nand, shift-right logical (srl), and shift-left arithmetic (sla). The six comparison operators of VHDL include equality (=), inequality (/=), greater-than-or-equal (>=), and less-than (<). The arithmetic operators of VHDL include the following: addition (+), subtraction (-), division (/), and absolute value (abs).

As an example, in order to go over the VHDL modeling, the schematic design of a 32-bit ALU is presented in Fig. 9.49. The VHDL model of this 32-bit ALU is presented in Algorithm 9.15 [81]. The first line of this VHDL starting with -- character is comment line till the end of the line. The library keyword includes the IEEE library to the VHDL model. The use IEEE.STD_LOGIC _1164.all includes the package STD_LOGIC_1164 from the IEEE library. It gives access to all the entities and data type within the STD_LOGIC_1164 package. The use IEEE.NUMERIC_STD.ALL includes the package for the signed, unsigned types and arithmetic operations to the VHDL model. The entity ALU_32_Behavioral is declares a designer's chosen name "ALU_32_Behavioral" as entity. The port defines external view or pins of the entity. The port declaration contains name and types of more than one port. In this VHDL model, four input ports and one output port are defined. These are of different data types such as signed and unsigned. The architecture Behavioral of ALU_32_Behavioral is declares a designer's chosen name "Behavioral" as architecture. The signal keyword declares the items are signals and all signals need to have initial values when the simulation of the VHDL model begins. The lines such as "Register1 <= In1" are signal assignments. The line process (Clk) describes the architecture in a behavioral fashion, which is sensitive to the sensitivity list that contains only Clk in this particular VHDL model. A process contains several statements that are executed in a sequential order just like a C program. The edge function rising_edge( ) finds rising edge or 0 to 1 transition of Clk. Similarly, the edge function falling_edge( ) can find falling edge or 1 to 0 transition of a signal. The behavioral description is based on if and case statements. The process is completed with end keyword. The last line end Behavioral ends the description of architecture Behavioral.

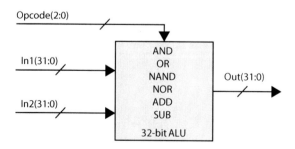

Figure 9.49 Block diagram of a simple ALU.

ALGORITHM 9.15 VHDL-Based Implementation of the 32-bit ALU shown in Fig. 9.49 [81]

```vhdl
-- Simple VHDL Model of a 32-bit ALU
library IEEE;
use IEEE.STD_LOGIC_1164.ALL;
use IEEE.NUMERIC_STD.ALL;
entity ALU_32_Behavioral is
port(Clk : in std_logic; -- clock signal
    In1, In2 : in signed(31 downto 0); -- two input operands of ALU
    Opcode : in unsigned(2 downto 0);   -- for ALU operations
    Out : out signed(31 downto 0)       -- output of ALU
    );
end ALU_32_Behavioral;
architecture Behavioral of ALU_32_Behavioral is
signal Register1, Register2, Register3 : signed(31 downto 0) := (others => '0');
begin
Register1 <= In1;
Register2 <= In2;
R <= Register3;
process(Clk)
begin
    if(rising_edge(Clk)) then -- Operate at the positive clock edge
        case Opcode is
            when "000" =>
                Register3 <= Register1 + Register2;   -- addition
            when "001" =>
                Register3 <= Register1 - Register2;   -- subtraction
            when "010" =>
                Register3 <= not Register1;           -- NOT
            when "011" =>
                Register3 <= Register1 nand Register2; -- NAND
            when "100" =>
                Register3 <= Register1 nor Register2;  -- NOR
            when "101" =>
                Register3 <= Register1 and Register2;  -- AND
            when "110" =>
                Register3 <= Register1 or Register2;   -- OR
            when "111" =>
                Register3 <= Register1 xor Register2;  -- XOR
            when others =>
                NULL;
        end case;
    end if;
end process;
end Behavioral;
```

9.7.3 MyHDL-Based Simulation

A new language called MyHDL is briefly discussed in this section. MyHDL is a high-level HDVL [29, 112, 140, 264]. MyHDL is an open-source and free language that has been developed based on Python object-oriented programming language. The key idea of MyHDL is to utilize the Python program to model hardware concurrency. A number of useful features of Python, including linear algebra, math function, control functions, array support, and object-oriented programming, make MyHDL a

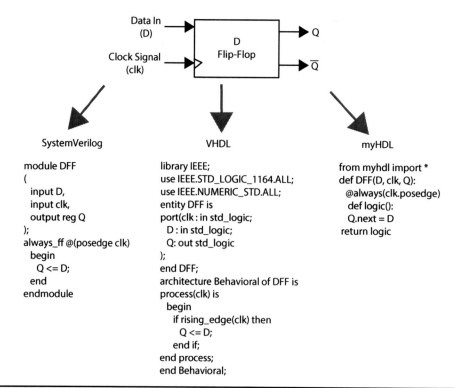

FIGURE 9.50 MyHDL versus SystemVerilog versus VHDL modeling of a simple sequential element [29, 140].

powerful language. In MyHDL, a hardware module is modeled as a function that returns generators. The generators of MyHDL are like always blocks of SystemVerilog or process blocks of VHDL. MyHDL has a class feature to model traditional hardware descriptions. For example, MyHDL provides a signal class to support communications and a class to support bit operations. MyHDL model can be converted to Verilog or VHDL, if needed. MyHDL can also be used as HVL for Verilog designs. MyHDL can also be used for co-simulation with traditional HDL simulators. The built-in simulator of MyHDL runs on top of the Python interpreter that performs simulation of the MyHDL model. This simulator also supports waveform viewing on the completion of the simulation. For a comparative perspective of various HDL modeling, a simple D flip-flop is presented in Fig. 9.50 [29, 140].

9.7.4 SystemC-Based Simulation

SystemC is a class library in C++, which has been developed for system-level design and simulation [124, 127, 128, 191, 206, 239]. SystemC was first introduced in 1999 and has become a standardized modeling language like many HDLs. SystemC allows modeling of SoCs containing both hardware and software components while facilitating intellectual property exchange. These features are unique to SystemC and are not present in HDLs. However, SystemC and an HDL (e.g., SystemVerilog) can provide complementary mechanisms for system design and simulation [121]. A system-level designer may want a language that is flexible in order to present the complex interactions of hardware and software components. A hardware designer may prefer HDLs that can represent more abstract system-level constructs. In summary, both system designers and hardware engineers have been looking for a more powerful framework that can facilitate presilicon hardware simulation and software validation to meet the short time-to-market demands. A comparative perspective of SystemC modeling is presented along with SystemVerilog and VHDL modeling for a D flip-flop in Fig. 9.51 [102].

SystemC presents a comprehensive set of constructs and extensions to C++ [54, 102, 124, 239], thus providing features to represent hardware and software as well as various functionalities. It provides a set of modules that are system components as well as interface components for SoC modeling. SystemC has constructs for structural description, concurrency, communication, and synchronization. A module of SystemC has processes, ports, and channels. A SystemC module

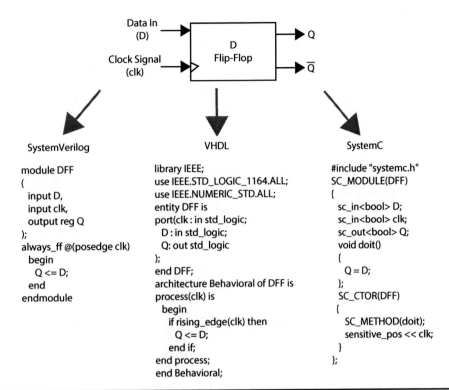

FIGURE 9.51 SystemC, SystemVerilog, and VHDL modeling comparison of a simple sequential element [102].

similar to a module of SystemVerilog or entity/architecture pair of VHDL. It represents the basic building block for hierarchical system modeling. A process in a module accesses the interface of a channel using a port (i.e., C++ object). A process that inherits concurrency essentially provides the behavioral description of a module. The behavior of a channel is accessed by a set of publicly visible functions (aka methods in C++) called interfaces. The channels of the SystemC are communication mechanisms just like wires and signals of HDLs. SystemC in essence supports hierarchical modeling through the use of modules. It supports both C/C++ native types as well as explicit SystemC types. It has a large set of data types to facilitate a large variety of digital system modeling. The SystemC types include arbitrary size signed/unsigned integers, two values ('0', '1'), four values ('0', '1', 'Z', 'X'), logic vector, and types for systems modeling. SystemC supports arithmetic operators similar to HDLs including, bitwise operators, arithmetic operators, rotational operators, and assignment operators.

A specific example of SystemC is now presented for a 8-bit synchronous up counter, as shown in Fig. 9.52 [54, 253]. The first line starting with // character makes the statement a comment statement to the end of the line and is ignored by the simulator during simulation. The second line with `#include` includes "systemc.h" header file to the SystemC module. This is essential for any SystemC module. `SC_MODULE` declares a user given Syn_up_Counter as SystemC module. For hierarchical modeling point of view it can be said that `SC_MODULE` creates a SystemC module class object named Syn_up_Counter. `sc_in` defines input signal ports. `sc_out` defines output signal ports. `sc_in_clk` defines clk as clock signal. It may be noted that clock gives real-time notion to SystemC modeling. `bool` is like boolean type of VHDL. `sc_uint<N>`, in general, defines N-bit unsigned integer vector in the example that 8-bit used. The `void` keyword defines a user-named function increment_counter C++. The function is described as C++ like using if-else statements. `SC_CTOR` is the constructor for Syn_up_Counter module. When the `SC_MODULE` is instantiated, it is created and at that time the constructor `SC_CTOR` is called. The ports, signals, and variables are initialized during the construction phase. `SC_METHOD` registers the function increment_counter with the simulation kernel. `sensitive` keyword declares the sensitivity list. { } start-end functions as well as modules are similar to begin-end in HDLs.

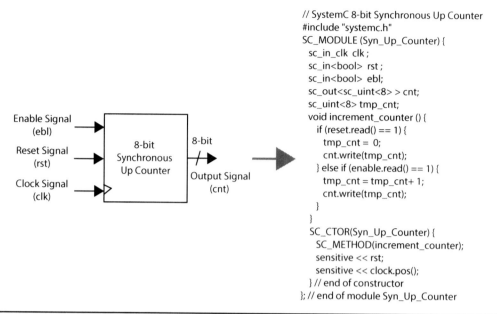

FIGURE 9.52 SystemC-based implementation of the 8-bit synchronous up counter [253].

9.8 Mixed-Signal HDL-Based Simulation

So far in the last several sections, many different languages and frameworks have been discussed to model and simulate analog circuits as well as digital circuits and systems. The modeling and simulations are performed at different levels of design abstraction using various components as needed at these abstraction levels. The simulations can be based on behavioral or structural modeling. For analog circuit-level modeling, SPICE no doubt has the best possible accuracy. However, it is quite computational intensive and slow. SPICE was never designed for large circuits, and scalability for large nanoelectronic circuits has been an issue due to the computation demand. SPICE only supports ODEs solving, and cannot solve PDEs. Verilog-A-based behavioral modeling and simulation can address some of the issues of SPICE. For digital circuits and systems, SystemVerilog, VHDL, MyHDL, and SystemC have been discussed for their discrete event simulations. For high-level unified simulation AMS circuits and systems, analog/mixed-signal hardware description languages (AMSHDLs) have been developed [225, 229, 261, 267]. The major AMSHDLs include: Verilog-AMS, VHDL-AMS, OpenMAST™, SystemC-AMS [31, 76, 104, 214]. In addition, as suggested in the previous sections, Simulink® can be used to model AMS systems as many supporting libraries and functions are available. A comparative perspective of AMSHDLs and SPICE with respect to selected AMS simulation characteristics is presented in Table 9.6 [225]. AMSHDLs allow design engineers to describe AMS circuits and systems at different levels of abstraction. Modeling and simulation of the entire mixed-signal circuit and system can be performed using a single AMSHDL to verify the nonlinear behavior of these complex circuits and systems.

Simulation Characteristics	Verilog-AMS	VHDL-AMS	SystemC-AMS	SPICE
Accuracy	Medium – High	Medium – High	Medium – High	High
Design Abstraction Level	Any Level	Any Level	Any Level	Circuit/Layout Level
Modeling Effort	Less to Medium	Less to Medium	Less	High
Simulation Run Time	Less	Less	Medium – High	High
Tools Required	AMS Simulator	AMS Simulator	AMS Libraries with GCC/ Simulators	SPICE Engine

TABLE 9.6 Comparative Perspective of AMSHDLs and SPICE [225]

9.8.1 Verilog-AMS-Based Simulation

Verilog-AMS is a language that is a superset of Verilog, Verilog-A, and their mixed-signal extensions. Verilog-AMS provides facility to build and use modules for circuit and system design exploration [104, 150, 180, 194, 227]. The modules that are encapsulated at the system level are described in either behavioral or structural forms. The behavioral description is presented using the constituent mathematical relations among the interface ports. Verilog-AMS allows both conservative and signal-flow descriptions through the concepts of nets, signals, ports, nodes, and branches. A contiguous, hierarchical collection of nets makes a signal. A signal is analog if all the nets in it are of analog domain. Similarly, the signal is digital if all the nets in it are of discrete domain. A signal that has nets of both analog and digital domain is a mixed signal. A port with two analog connections is an analog port, a port with two digital connections is a digital port, and a port whose connections are both analog and digital is a mixed port. The modules connect through ports and nets at a certain point that is called a node. A path of flow between two nodes through a component is called a branch. These nodes and branches can then be used to formulate KCLs and KVLs for simulation purposes. Verilog-AMS has many unique features including [104, 180, 227]: (1) Initial, always, and analog procedural blocks can appear in the same module. (2) Signals of both analog and digital types can be present in the same module. (3) Signal values can be written from any context even outside of an analog procedure. (4) For mixed-signal ports using hierarchical connections of analog and digital ports, user-defined connection modules are automatically included for signal conversions.

At a higher-level perspective, it can be said that Verilog-AMS supports all the features, types, and operations of Verilog and Verilog-A. The line commenting using // character and block commenting using /* ... */ is possible in Verilog-AMS, just as in C++, to improve documentation of the program. Similar to Verilog-A, the accent grave ` character declares the compiler directives such as `include, `define, `default_discipline, and `default_transition. Similar to Verilog, integer, real, and parameter data types are supported in Verilog-AMS. A data type called wreal is introduced to support real value nets. Another data type called net_discipline supports analog nets and disciplines of all nets and regs. The arithmetic and logical operators of Verilog-AMS are similar to C++ programming. Many bitwise operations such as bitwise negation, bitwise AND, and bitwise equivalence are supported in Verilog-AMS. Standard mathematical functions of Verilog-AMS include min, pow, and sqrt. Verilog-AMS has many trigonometric and hyperbolic functions including sin, acos, and tanh. In Verilog-AMS, the ddt(...) operator calculates the time derivative of its argument(s), the idt(...) operator calculates the time integral of its argument, and the idtmod(...) operator performs the circular integration. Overall the large set of features, data types, and operators can be used for multidiscipline and multiple domain system modeling. A selected set of frameworks that can be used for Verilog-AMS simulations include Discovery AMS, Virtuoso® AMS Designer, Questa® ADMS [9, 35, 63]. In this section a selected set of examples are presented for demonstration of Verilog-AMS modeling.

9.8.1.1 Verilog-AMS Simulation Example: Continuous-Time Sigma-Delta Modulator

Verilog-AMS modeling and simulation of continuous-time sigma-delta modulator (CT-SDM) is discussed in this section. This is functionally the same CT-SDM that was simulated using Simulink® in a previous section in this chapter. The schematic diagram of the CT-SDM is presented in Fig. 9.53 [286, 287]. The CT-SDM has been implemented using the following components [286, 287]: (1) three active RC-integrators, (2) one summing amplifier, and (3) one clocked quantizer. The major component of all the integrators and summing amplifier is the operational amplifier (OP-AMP) as evident from the schematic diagram. There are a total of four OP-AMPs. The transfer function of an ideal OP-AMP can be expressed as follows [286, 287]:

$$\frac{V_{\text{out}}(s)}{V_{\text{in}}(s)} = -\left(\frac{1}{sRC}\right) \tag{9.19}$$

where R and C are the input resistance and forward capacitance of the OP-AMP. The transfer function of an OP-AMP with a finite gain-bandwidth product (GBP) has the following form:

$$\frac{V_{\text{out}}(s)}{V_{\text{in}}(s)} = -\left(\frac{1}{s(RC + \frac{1}{\omega_{\text{ug}}}) + s^2 \frac{RC}{\omega_{\text{ug}}}}\right) \tag{9.20}$$

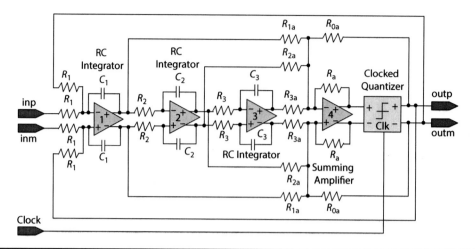

FIGURE 9.53 A CT-SDM realized using several active-RC integrators [286, 287].

where R and C are the input resistance and forward capacitance of the OP-AMP. ω_{ug} is the unity-gain frequency and is calculated as follows [286, 287]:

$$\omega_{ug} = 2\pi \text{GBP} \tag{9.21}$$

$$= 2\pi G_{DC} f_{co} \tag{9.22}$$

In the above expression, GBP is gain-bandwidth product. G_{DC} is the DC gain of the OP-AMP. f_{co} is the cutoff or −3 dB frequency also called bandwidth of the OP-AMP. This is the frequency at which the voltage is 0.707 of its intended value and the corresponding gain is −3 dB.

The Verilog-AMS model for the CT-SDM is shown in Algorithm 9.16 [286, 287]. The first line with // character is a comment till the end of line and is ignored by the simulator. The second line with compiler directive `include includes the standard definition package constants.vams. The next line

---

ALGORITHM 9.16 Verilog-AMS Model for an Active-RC Integrator in the CT-SDM [286, 287]

```
// Verilog-AMS Model for a Integrator in the CT-SDM
`include "constants.vams"
`include "disciplines.vams"
module Active_RC_Integrator (inp, inm, fbp, fbm, outp, outm);
   parameter real Gdc0 = 128.0 from (0 : inf); // OP-AMP Open-Loop DC Gain
   parameter real fco = 5e6 from (0:inf);      // OP-AMP 3dB Bandwidth
   parameter real R = 1.0 from (0 : inf);      // Input Resistor
   parameter real C = 1.0 from (0 : inf);      // Forward Capacitor
   parameter real Vdd = 1.0 from (0:10);
   parameter real Vgnd = 0.0;
   input inp, inm, fbp, fbm;
   output outp, outm;
   electrical inp, inm;
   electrical outp, outm, fbp, fbm, outd;
   real Vcm, Vout;
   real d [0:2];
   real wug;                                    // Unity-Gain Frequency
   analog begin
      @(initial_step)
      begin
         wug = Gdc0 * 2 * `M_PI * fco;
```

```
            d[0] = 0;
            d[1] = R * C + 1 / wug;
            d[2] = R * C / wug;
            Vcm = (Vdd - Vgnd) / 2;
        end
        V(outd) <+ laplace_nd( (V(inp, inm) + V(fbp, fbm)), {1}, d );
        V(outp) <+ Vcm + V(outd) / 2;
        V(outm) <+ Vcm - V(outd) / 2;
    end
endmodule
```

with compiler directive `include includes another standard definition package disciplines.vams. The module keyword declares a designer given name Active_RC_Integrator as Verilog-AMS module with a set of input and output ports within (...). The parameter keyword declares various parameters of different types, which are like constants and not supposed to be changed during run time. The input keyword defines input ports and the output keyword defines output ports. The electrical keyword assigns the ports as electrical physical type and associates voltage and currents with them. The real keyword declares the variables to be real type variables. The behavioral description of the module is presented within the analog keyword. The laplace_nd function calculates Laplace transfer function expressed in terms of numerator and denominator polynomial coefficients.

The plot of PSD for various GBPs of the CT-SDM is presented in Fig. 9.54 [286, 287]. The GBP of the CT-SDM is infinite when the SNR_{Ideal} is 97.6 dB. When GBP is very low, the SNR denoted as $SNR_{GBP(Lo)}$ is 81.7 dB. For optimal GBP, the SNR denoted as $SNR_{GBP(Optimal)}$ is 96.9 dB. The specification of the CT-SDM for this optimal design is presented in Table 9.7 [286, 287]. It may be noted that there are minor differences in the simulation results of the Verilog-AMS model versus the Simulink®-based simulations discussed in a preceding section.

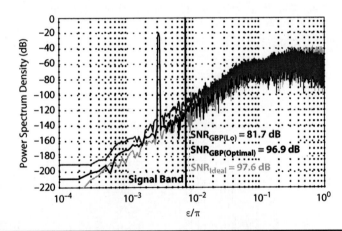

Figure 9.54 Power spectrum density (PSD) of the CT-SDM with different gain-bandwidth-product (GBP) [286, 287].

CT-SDM Components	Specific Values of Components
Resistors (Ω)	R_1 = 100 k, R_2 = 100 k, R_3 = 100 k, R_{1a} = 1.188 k, R_{2a} = 1.857 k, R_{3a} = 947, R_{0a} = 10.546 k, R_a = 1 k
Capacitors (F)	C_1 = 3.384 p, C_2 = 5.225 p, C_3 = 32.16 p
OP-AMP 1	G_{DC} = 128, f_{co} = 12 kHz
OP-AMP 2	G_{DC} = 128, f_{co} = 12 kHz
OP-AMP 3	G_{DC} = 128, f_{co} = 12 kHz
OP-AMP 4	G_{DC} = 128, f_{co} = 80 kHz

Table 9.7 Specifications for the Components of the CT-SDM for $SNR_{GBP(Optimal)}$ [286, 287]

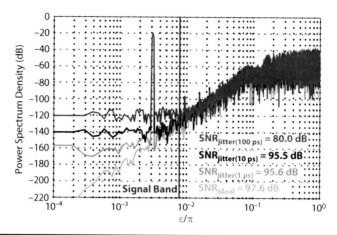

Figure 9.55 Power spectrum density (PSD) of the CT-SDM with different RMS jitter [286, 287].

The plot of PSD for various clock jitters of the CT-SDM is presented in Fig. 9.55 [286, 287]. The clock jitter is estimated using the `rdist_normal` function in Verilog-AMS. The PSDs of the CT-DSM for different RMS jitter values (e.g., 1 ps, 10 ps, and 100 ps) are shown in the figure. The time-domain simulation of the Verilog-AMS model showing input and output voltages of the CT-SDM for the optimal design with parameters of Table 9.7 is presented in Fig. 9.56 [286, 287].

9.8.1.2 Verilog-AMS Simulation Example: PLL Components

Detailed design of charge pump-based phase-locked loop (CPPLL) has been presented in an earlier chapter. However, the objective of considering CPPLL again in this chapter is to perform its Verilog-AMS using behavioral description of individual components then hierarchical system-level modeling and simulation. The block-level schematic representation of the CPPLL is presented in Fig. 9.57 [119, 123, 187, 286, 288]. It contains the typical building blocks such as phase-frequency detector (PFD), charge pump (CP), LPF, and VCO. The simulation of this CPPLL system can be performed using full SPICE, pure Verilog-A, and AMSHDL. Pure SPICE circuit-level simulation is accurate and slow and can be performed at the circuit level. Verilog-A-based simulation model is a complete analog-based system. In a mixed-signal simulation of this system, some components can be modeled as analog components and some components can be modeled as digital components. A mixed analog and digital modeling that provides a fast design verification with given accuracy requirements without depending on analog-only simulation is slow and has high computational requirements [180, 286, 288].

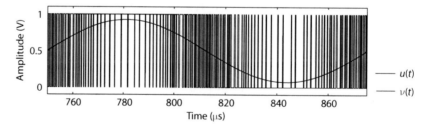

Figure 9.56 The time-domain simulation result of the Verilog-AMS model [286, 287].

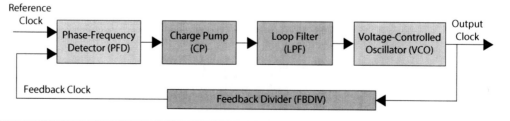

Figure 9.57 Block diagram of a charge-pump-based phase-locked loop (CPPLL) [187, 286].

The PFD can be modeled as an analog block or digital block and can be called analog PFD and digital PFD, respectively. Either of the two modules can be integrated in the overall CPPLL system for its mixed-signal simulation and verification [243, 286]. Analog PFD can be modeled using Verilog-A as presented in Algorithm 9.17 [72, 243]. This analog behavioral model essentially determines two signals V_{UP} and V_{Down} needed, respectively, for increasing or decreasing frequency of oscillation of the VCO. The simulation of the Verilog-A model for an analog PFD is shown in Fig. 9.58 [243]. The Verilog model of a digital PFD is shown in Algorithm 9.18 [286, 288]. This Verilog-based behavioral model calculates Up and Down signals as needed to increase or decrease oscillation frequency.

Algorithm 9.17   Verilog-A Model for an Analog Phase-Frequency Detector (PFD) of a PLL [72, 243]

```
// Verilog-A based Model for an Analog Phase-Frequency Detector (PFD)
'include "discipline.h"
'include "constants.h"
'define behind 0
'define same 1
'define ahead 2
module Analog_PFD(vin_if, vin_lo, sigout_Up, sigout_Down);
input vin_if, vin_lo;
output sigout_Up, sigout_Down;
electrical vin_if, vin_lo, sigout_Up, sigout_Down;
parameter real vlogic_high = 0.7;
parameter real vlogic_low  = 0;
parameter real vtrans = 0.35;
parameter real t_del = 20n from [0:inf);
parameter real t_rise = 0.1n from (0:inf);
parameter real t_fall = 0.1n from (0:inf);
integer tpos_on_if, tpos_on_lo;
real sigout_Up_val, sigout_Down_val;
integer state;
   analog begin
      @ ( initial_step ) begin
      sigout_Up_val = 0;
      sigout_Down_val = 0;
      state = 'same;
      end
   tpos_on_if = 0;
      @ ( cross(V(vin_if) - vtrans, +1) )
         tpos_on_if = 1;
         tpos_on_lo = 0;
      @ ( cross(V(vin_lo) - vtrans, +1) )
         tpos_on_lo = 1;
         if (tpos_on_if && tpos_on_lo) begin
         state = 'same;
      end else if (tpos_on_if) begin
         if (state == 'behind) begin
            state = 'same;
         end else if (state == 'same) begin
            state = 'ahead;
         end
      end else if (tpos_on_lo) begin
         if (state == 'ahead) begin
            state = 'same;
         end else if (state == 'same) begin
```

```
                    state = 'behind;
                end
            end
            if (tpos_on_if || tpos_on_lo) begin
                if (state == 'ahead) begin
                    sigout_Up_val = vlogic_high;
                    sigout_Down_val = vlogic_low;
                end else if (state == 'same) begin
                    sigout_Up_val = vlogic_low;
                    sigout_Down_val = vlogic_low;
                end else if (state == 'behind) begin
                    sigout_Up_val = vlogic_low;
                    sigout_Down_val = vlogic_high;
                end
            end
            V(sigout_Up) <+ transition(sigout_Up_val, t_del, t_rise, t_fall);
            V(sigout_Down) <+ transition(sigout_Down_val, t_del, t_rise, t_fall);
        end
endmodule
```

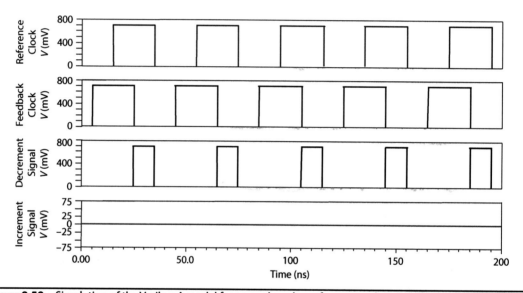

FIGURE 9.58 Simulation of the Verilog-A model for an analog phase-frequency detector (PFD) [243].

ALGORITHM 9.18 Verilog Model for a Digital Phase-Frequency Detector (PFD) of a PLL [286, 288]

```
// Verilog based Model for a Digital Phase-Frequency Detector (PFD)
'timescale 10ps / 1ps
module Digital_PFD(q_Up, q_Down, ref, clk);
    output q_Up, q_Down;
    input clk, ref;
    reg q_Up, q_Down;
    wire reset;
    assign #1 reset = q_Up && q_Down;
    always @(posedge ref or posedge reset)
        begin
            if (reset) q_Up <= 1'b0;
            else q_Up <= 1'b1;
```

```
            end
    always @(posedge clk or posedge reset)
        begin
            if (reset) q_Down <= 1'b0;
            else q_Down <= 1'b1;
        end
endmodule
```

In a PLL with digital PFD, the CP considers the digital signals from the PFD. The CP generates analog signal that is fed to LPF. Thus, CP is a very good example for Verilog-AMS-based mixed-signal modeling. Verilog-AMS model for the CP of the PLL is shown in Algorithm 9.19 [286, 288]. In this model, the output current of the CP is calculated using behavioral models. The branch contribution operator <+ changes the output current.

The Verilog-A model of an LPF that can be used as LPF of this CPPLL is presented in the Verilog-A simulation section of this chapter [104, 113, 243]. The same LPF model can be used and has not been presented again to avoid duplication. On the other hand, the LPF can be modeled using SPICE netlist as well. A simple LPF can be made from three passive components: R_1, C_1, and C_2 [286].

Many different types of VCO designs including LC-VCO and ring oscillator (RO) have been discussed in the PLL design chapter of this book. Also, in the Verilog-A section of this chapter, Verilog-A-based behavioral model and simulation are presented. This Verilog-A model of the VCO presented in the previous section can be used in the simulation of the CPPLL in the current section [72, 113, 286, 288].

The frequency divider or feedback divider (FIBDIV) can be implemented using either an analog circuit or a digital circuit. Accordingly Verilog-A or Verilog model can be used for the simulation. Verilog-AMS Model for an analog frequency divider of the PLL is shown in Algorithm 9.20 [113, 165]. Essentially, the feedback loop through the frequency divider indirectly makes the reference frequency identical with the VCO output frequency. The idtmod operator in the analog description block performs circular integration. The Verilog model of a digital frequency divider is shown in Algorithm 9.21 [205, 283]. An arrangement of flip-flops is a straightforward method for integer-N

ALGORITHM 9.19 Verilog-AMS Model for a Charge Pump of the PLL [286, 288]

```
// Verilog-AMS based Model for a Charge-Pump of the PLL
'include "disciplines.vams"
'timescale 10ps / 1ps
module Charge_Pump (Up, Down, CP_out);
    parameter real cur = 50u;       // Output current in A
    input Up, Down;
    electrical CP_out;
    real iout;
    parameter real Vdd = 1.2;
    parameter real Vgnd = 0;
    analog begin
        @(initial_step) iout = 0.0;
        if (Down && !Up && (V(CP_out) > Vgnd))
            iout = - cur;
        else if (!Down && Up && (V(CP_out) < Vdd))
            iout = cur;
        else
            iout = 0;
        I(CP_out) <+ -transition(iout, 0.0, 2p, 2p);
    end
endmodule
```

488 Chapter Nine

ALGORITHM 9.20 Verilog-AMS Model for an Analog Frequency Divider of the PLL [113, 165]

```
// Verilog-AMS Model for a Analog Frequency-Divider
`include "disciplines.vams"
`include "constants.vams"
module Analog_Frequency_Divider(FD_in, FD_out);
inout FD_in, FD_out;
electrical FD_in, FD_out;
parameter real FB_div = 100;
parameter real FD_gain = 0.40;
parameter real fc = 4.9e9;
analog
   V(FD_out) <+ sin(`M_TWO_PI * idtmod(fc/FD_div + FD_gain * V(FD_in), 0, 1, 0));
endmodule
```

ALGORITHM 9.21 Verilog Model for a Digital Frequency Divider of the PLL [205, 283]

```
// Verilog Model for a Digital Frequency-Divider
module Digital_Frequency_Divider(clk, reset, FD_out);
input clk, reset;
output FD_out;
reg FD_out;
wire clk, reset;
parameter period = 10;
parameter half_period = period/2;
reg [3:0] count;
always
   @(posedge clk) begin
   if(reset) begin
      count = 0;
      FD_out <= 0;
   end
   else begin
      if(count == period-1) begin
         countvalue = 0;
         FD_out <= 0;
      end
      else  count = count + 1;
      if(count == half_period)
         FD_out <= 1;
      end
   end
endmodule
```

---

division [164, 283]. For power-of-2 integer division, a simple binary counter can be used, clocked by the input signal. The least-significant output bit alternates at 1/2 the rate of the input clock, the next bit at 1/4 the rate, and the third bit at 1/8 the rate.

9.8.1.3 Simulink® versus Verilog-AMS

A brief comparative perspective of Simulink® modeling with respect to Verilog-AMS modeling is presented in this section. The overview of the comparison is presented in Table 9.8 [286, 287]. For the CT-SDM design and simulation presented in the current section and in the Simulink® section before, it is observed that the simulation using Simulink® takes longer time. In particular, the Simulink®-based simulation needs more than twice of the simulation time as compared to the Verilog-AMS simulation

Modeling Framework	Simulation Time	Simulation Accuracy	Modeling Effort	Integrability	Nonideal Characteristic Analysis
Simulink® Modeling	125.8 s	Good	Easy	Difficult	Difficult
Verilog-AMS Modeling	57.0 s	Better	Difficult	Easy	Easy

TABLE 9.8 Simulation Time for Simulink® and Verilog-AMS Modeling [286, 287]

for the same CT-SDM design. This may be due to the fact that the relative tolerance is set to be two times smaller than that of the Verilog-AMS simulations. However, tolerance adjustment in Simulink® for comparable accuracy with Verilog-AMS is not easy [286, 287]. The modeling effort in Simulink® is less as it has a large set of libraries with many fundamental building blocks. In addition, when the mixed-signal design is modified, it is also easier in Simulink® than in Verilog-AMS. The simulation of nonideal characteristic analysis such as clock jitter is easier in Verilog-AMS than in Simulink®. Moreover, Verilog-AMS models are easier to integrate with the actual implementation of circuit and systems to perform co-simulations.

9.8.2 VHDL-AMS-Based Simulation

VHDL-AMS is another unified language like Verilog-AMS for behavioral modeling of analog, digital, as well as mixed-signal circuits and systems [1, 62, 76, 115, 135, 181, 227]. VHDL-AMS language is quite similar to Verilog-AMS in concept and goals; however, it uses different semantics, coding techniques, and modeling style than does the Verilog-AMS. A comparison of Verilog-AMS and VHDL-AMS is presented in Table 9.9 [227]. VHDL-AMS is a superset of VHDL 1076-1993 and VHDL 1076.1-1999 supporting description and the simulation of analog and mixed-signal circuits and systems [68, 70]. VHDL-AMS has the following new features to support such modeling and simulations:

1. It has a new simulation model supporting continuous behavior.
2. It has continuous models based on the DAEs.
3. It has a dedicated analog kernel to solve these DAEs.
4. It has capabilities to handle initial conditions, piecewise-defined behavior, as well as discontinuities.
5. It has capabilities for the optimization of the set of DAEs being solved.
6. It has extended structural semantics like conservative semantics to model physical systems in terms of KCL/KVL, nonconservative semantics for abstract models through signal-flow descriptions, and mixed-signal interfaces.
7. It has mixed-signal semantics for unified time modeling for a consistent synchronization of continuous and event-driven behavior, mixed-signal initialization and simulation cycle, and mixed-signal descriptions of behavior.
8. It has frequency-domain support for small-signal frequency and noise analysis.

Language Features	Verilog-AMS	VHDL-AMS
Language Inheritance	C like (case sensitive)	Ada like (case insensitive)
Analog Subset	Yes (with some support of SPICE primitives)	No
Port Structure	Continuous ports are of inout mode	Continuous ports are modeless
Objects	Variables, wires, registers	Terminals, quantities, signals, variables, constants
Conservative Formulation	Nodal formulation enforced	No specific graph enforced
Solvability for Simulation	No solvability checks	Solvability check done at design unit level

TABLE 9.9 Verilog-AMS versus VHDL-AMS Modeling [227]

VHDL-AMS supports hierarchical modeling of circuits and systems at various levels of design abstractions just similar to Verilog-AMS [20, 76, 115, 269]. In VHDL-AMS, the system descriptions can be presented either in continuous-time or in discrete-time form. The descriptions can be in either conservative or nonconservative sets. VHDL-AMS models typically comprise two distinct portions such as an entity and an architecture. The entity specifies the interface of the module through the mechanism of ports. The architecture of the module specifies the implementation semantics to describe the behavioral style or function of the model. The system architecture description can be in one style or in a combination of more than one modeling styles, like behavioral, structural, or behavioral/structural. VHDL-AMS provides six classes of "objects" such as constants, terminals, quantities, variables, signals, and files. Various types of terminals called "natures" in VHDL-AMS include electrical, thermal, fluidic, and translational. VHDL-AMS has three types of quantities, such as free quantity, branch quantity, and source quantity. VHDL-AMS has significant capability to solve DAEs due to availability of extensive numerical solution methods. However, the PDEs are not explicitly supported in VHDL-AMS. The methods and algorithms for the solution of the DAEs are entirely up to the implementation of the tool that simulates the system. A selected set of frameworks that can be used for Verilog-AMS simulations include Virtuoso®, AMS Designer, Questa® ADMS [35, 63]. In this section a selected set of examples is presented for demonstration of VHDL-AMS modeling and simulations.

As an example VHDL-AMS modeling and simulation of a phase-locked loop (PLL) is considered. This provides a close comparative perspective of VHDL-AMS model with respect to Verilog-AMS model of the previous section. The schematic diagram of the PLL is presented in Fig. 9.59 [119, 123, 185]. It has a lesser number of blocks than the example in the previous section. However, a detailed structure of CP and LPF is also presented. A high-level view of the VHDL-AMS model of this PLL is shown in Algorithm 9.22 [173, 185]. As in VHDL, – character comments a line in VHDL-AMS model [70, 76, 115]. The `entity` keyword defines a designer given name like CP_PLL in this case as an entity, module, or component. The `port` keyword specifies designer given names as ports of a specified type like input or output and port nature like electrical or mechanical. The `architecture High_Level_PLL of CP_PLL is` declares High_Level_PLL as a specific architecture of CP_PLL, which is to be defined in a later part of this model. The `signal` keyword declares designer given names like Up, Down as signals. The `quantity` keyword declares designer given name V_c as `voltage` quantity. The `terminal` keyword declares designer given name t_VCO as `electrical` terminals. The architecture description is presented in `begin ... end`. In this model four different entities such as phase detector (PD), CP, LPF, and VCO are presented. The `generic map` keyword is used for parameter assignment. The `port map` keyword is used for terminal assignment.

In the above VHDL-AMS model, a high-level presentation CPPLL is provided. Each of the entities is instantiated and mapped using `port map`. In the hierarchical modeling, each of these components needs to be described in VHDL-AMS just similar to the Verilog-A or Verilog presented in the above sections, of course in VHDL-AMS or VHDL formats. As a specific example, the VHDL-AMS model of the VCO component of the PLL is presented in Algorithm 9.23 [174, 265]. In this VHDL-AMS model, the VCO is described by a differential equation in which its frequency is related to the time derivative of its phase. The `library` includes IEEE, library to the VHDL-AMS model. The `use IEEE.math_real.all`

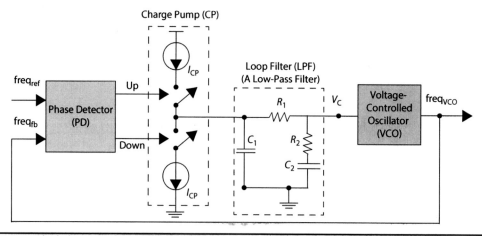

Figure 9.59 Block diagram of a simplified charge-pump-based phase-locked loop (CPPLL) [185].

Mixed-Signal Circuit and System Simulation 491

ALGORITHM 9.22 High-Level VHDL-AMS Behavioral Model a CPPLL [173, 185]

```
-- VHDL-AMS Behavioral Model of a Simple CPPLL
entity CP_PLL is
    generic (freq_center, freq_slope, I_CP, R1, R2, C1 , C2 : real);
    port (freq_ref : in bit);
end entity CP_PLL;
architecture High_Level_PLL of CP_PLL is
signal Up, Down, freq_VCO : bit;
quantity V_c : voltage;
terminal t_VCO, t_top, t_CP : electrical;
begin
    -- Phase Detector
    PD: entity work.PD_Behave
    port map (ref_in => freq_ref, VCO_out => freq_VCO, Up => Up, Down => Down);
    -- Charge Pump
    CP: entity work.CP_Behave
    generic map (I_CP => I_CP)
    port map (Switch_Up => Up, Switch_Down => Down, T_top => t_top,
              T_CP => t_CP, ref => GND);
    DC: entity work.dc
    generic map (amp => 5.0)
    port map (p => t_top, m => GND);
    -- Loop Filter
    LPF: entity work.Low_Pass_Filter
    generic map (R1 =>RI,  R2 =>R2, Cl=> Cl, C2 => C2)
    port map (t1 => t_VCO, t2 => t_CP, ref => GND);
    -- Voltage Contolled Oscillator
    VCO: entity work.VCO_Behave
    generic map (freq_center => freq_center, freq_slope => freq_slope)
    port map (t_in => t_VCO, V_out => V_c, ref => GND);
    freq_VCO <= '1' when V_c'above(0.0)
       else '0' when not V_c'above(0.0);
end architecture High_Level_PLL;
```

includes the math_real package of IEEE, which contains many math constants to be used in the VHDL-AMS model. The use IEEE.electrical_systems.all includes the electrical_systems package from the IEEE library. This package declares many electrical types including charge, resistance, capacitance, inductance, and flux. The entity defines VCO as a component with various ports. VCO_Behavioral is the architecture name of the VCO module as specified by the architecture keyword. The constant keyword declares designed given name PI as a constant of real type. The quantity V_out across I_out through out_terminal to GND statement defines the branch quantities V_out and I_out for the two terminal ports out_terminal and GND. V_out refers to the across and I_out refers to the through aspect of the two terminal ports. In general, a branch quantity is an analog object that is used for conservative energy systems. The sig_phase `above(2*PI) returns 1 or 0 depending on whether the phase is greater than threshold level 2*PI. In essence, 'above attribute is used to convert continuous quantity into a discontinuous quantity by detecting a threshold. The time derivative of the sig_phase, sig_phase `dot is calculated as the maximum of calculated frequency and a designer defined specific value of 500 MHz. In the last statement the output voltage V_out is calculated as a sin function of sig_phase.

9.8.3 OpenMAST™-Based Simulation

The third AMSHDL, called OpenMAST™, is discussed in this section [132, 241]. The history of OpenMAST™ goes back to 1985 when the proprietary mixed-signal language called MAST was introduced [31, 53]. MAST was an integral part of the simulator called Saber® [37, 136]. OpenMAST™

ALGORITHM 9.23 VHDL-AMS Model of the VCO of the CPPLL [174, 265]

```
-- VHDL-AMS Model of the VCO of the CPPLL
library IEEE;
use IEEE.math_real.all;
use IEEE.electrical_systems.all;
entity VCO is
   generic(
   freq_center: real := 1.2E6;
   freq_slope: real := 0.4E6;
   V_c: voltage := 0.0
   );
   port(quantity V_in: in voltage; terminal out_terminal, GND: electrical);
end entity VCO;
-- Architecture Description
architecture VCO_Behavioral of VCO is
constant PI: real := 3.1416;
quantity sig_phase : real;
quantity V_out across I_out through out_terminal to GND;
begin
   if domain = quiescent_domain use
      sig_phase == 0.0;
   elsif sig_phase'above(2*PI) use
      sig_phase == 0.0;
   else
      sig_phase'dot == 2*PI*realmax(0.5E6, freq_center+(V_in-V_c)*freq_slope);
   end use;
   break on sig_phase'above(2*PI);
   V_out == 2.5*(1.0+sin(sig_phase));
end architecture VCO_Behavioral;
```

is the open-source delivery of the MAST proprietary language [31, 53]. One of the objectives of making MAST open source is to facilitate model interoperability between OpenMAST™ and Verilog-AMS or VHDL-AMS. Through OpenMAST™, a design engineer has access to tons of original models of MAST.

OpenMAST™ is similar to Verilog-AMS or VHDL-AMS languages that support hierarchical description and simulation of analog, digital, and mixed-signal circuits and systems [31, 136]. It can model the circuits and systems at various abstraction levels just like other AMSHDLs. OpenMAST™ allows both conservative and nonconservative semantics in the modeling of AMS circuits and systems. It combines C-like programming constructs with strengths of familiar mathematical expressions. Moreover, OpenMAST™ has additional constructs targeted for statistical, parametric, robust, and reliability design verification for developing a wide range of systems. A high-level comparison of OpenMAST™ with respect to VHDL-AMS is presented in Table 9.10 [162].

Language Concepts	VHDL-AMS	OpenMAST™
Analog Threshold Detection	Above Quantity Attribute	Threshold Function
Digital and Analog Simulators Synchronization	Break Instruction	Schedule Next Time Function
Digital Event Detection	Sensitivity List Wait Instruction	Event On Function
Digital Signals Assignment	<=	Schedule Event Function

TABLE 9.10 High-Level Comparison of OpenMAST™ versus VHDL-AMS [162]

Mixed-Signal Circuit and System Simulation 493

The terminologies of OpenMAST™ and Verilog-AMS or VHDL-AMS are different [31, 136, 162]. For example, a behavioral model in OpenMAST™ is a model whose behavior is described by using the features of the language, not by connecting the existing models together. The behavioral models may also include existing models; however the existing models are not included as a replacement of the language-based functionality. The macromodel of OpenMAST™ is a model that is created by connecting the existing models and is equivalent to a structural model of Verilog-AMS or VHDL-AMS. The test circuit or netlist of OpenMAST™ is like a test bench of Verilog-AMS or VHDL-AMS.

For a comparative perspective of OpenMAST™ and VHDL-AMS, modeling of a small circuit is presented in Fig. 9.60 [136]. OpenMAST™ models are divided into two main portions, the template header (similar to `entity` of VHDL-AMS) and the template body (similar to `architecture` of VHDL-AMS). The # character is used for commenting a line till the end of the line, which is ignored by the simulators. The first noncommented line is called header, which starts with `template` and includes template name, pin names, and input arguments. In the presented example, the `template` keyword defines the designer named design entity (e.g., VCVS_R_in), which is similar to `module` of Verilog-AMS or `entity` of VHDL-AMS. It has connection points V_P, V_M, P, and M. The input arguments are VCVS_gain and R_in. The header declarations follow the header, the pin types are defined, and the input types are specified. For example, the `electrical` keyword defines V_P, V_M, P, and M as electrical type. The `number` keyword defines VCVS_gain and R_in as numbers. The template body is described within { } typically like a C++, rather than `begin` ... `end` of VHDL-AMS. The `branch` keyword defines the branch variables of the OpenMAST™ model, which are similar to `quantity` keyword of VHDL-AMS. The `equation` keyword is the beginning of the description of the OpenMAST™. In the `equation` section the OpenMAST™ template is described just similar to architecture description of the entity in VHDL-AMS. It is observed that there are no; characters at the end of every line in OpenMAST™ as in VHDL-AMS.

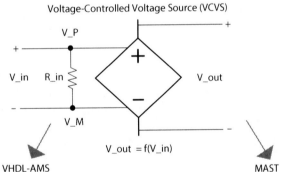

FIGURE 9.60 OpenMAST™ and VHDL-AMS modeling comparison of a simple circuit element [136].

9.8.4 SystemC-AMS-Based Simulation

The fourth AMSHDL called SystemC-AMS, is briefly discussed in this section. Generally speaking SystemC-AMS cannot be considered as a replacement for Verilog-AMS or VHDL-AMS, but rather as a complement to them with many complementary features including software component modeling [50, 118, 120, 158, 262]. The history of SystemC-AMS, while it can be traced back to 2000, really started in 2010 with the development of SystemC-AMS 1.0 standard. The constructs and semantics of the SystemC-AMS are defined as C++ class library just like SystemC. The SystemC-AMS supports design and simulation of true heterogeneous AMS-SoC [49, 50, 108, 120, 158, 262, 284]. It can perform simulation, architectural exploration, and refinement along with software integration of AMS-SoCs. The AMS-SoC may contain analog components, digital components, software components, mixed-signal components, or RF components as a part of overall design.

SystemC-AMS presents facilities for both continuous-time and discrete-time descriptions [49, 50, 108, 120, 158, 234, 262, 284]. In the SystemC-AMS continuous-time modeling, the behaviors can be described in terms of mathematical equations that can include time-domain derivatives of any order, such as ODEs or DAEs. However, PDEs are not accepted in SystemC-AMS modeling. The discrete-time modeling in SystemC-AMS represents the signals as sequences of values that are defined at discrete-time intervals, as is a typical case. SystemC-AMS supports both conservative models (i.e., following KCL/KVL) and nonconservative models. SystemC-AMS has the following required modeling formalisms to support the behavioral modeling at different levels of abstraction:

1. *Electrical Linear Networks (ELN)*: SystemC-AMS has predefined linear network primitives such as resistors and capacitors, which can be used as macromodels for describing the continuous-time behaviors.

2. *Linear Signal Flow (LSF)*: SystemC-AMS has a set of primitive modules such as addition, multiplication, and integration, or delay to describe continuous-time behaviors.

3. *Timed Data Flow (TDF)*: SystemC-AMS has discrete-time modeling and simulation capabilities through time-data flow (TDF) that allow use of discrete-event kernel without the overhead.

For a simple understanding of SystemC-AMS, a high-level skeleton is presented in Algorithm 9.24. [50, 120, 158, 262]. The comment character is // and the description is presented within { }, just like C++. For SystemC-AMS model, the first line with #include adds the header file <systemc-ams.h> to bring classes, symbols, and macro definitions to the scope of the program just like <systemc.h> of SystemC. Sometimes a stripped-down version called <systemc-ams> (i.e., without .h) can be used as header file. A module in SystemC-AMS is defined using the SCA_MODULE macro for a designer given name. It may be noted that the SCA_prefix distinguishes SystemC-AMS from SystemC that uses SC_prefix. In general, during hierarchical modeling using SystemC-AMS, it is possible to use SCA_MODULE for continuous-time primitives and SC_MODULE for discrete-time primitives. An SCA_MODULE can register at most four member functions: a

ALGORITHM 9.24 SystemC-AMS Module Template [50, 120, 158, 262]

```
// SystemC-AMS Module Template
#include <systemc-ams.h>
SCA_MODULE(Some_Module_Name)
// Different Ports, Member Functions, and Internal Data ...
   SCA_CTOR(Some_Module_Name)
   {
     // Constructor for SDF Synchronization
     SCA_SDF_ATTR(Some_Attribute_Function)
     SCA_SDF_INIT(Some_Initialization_Function)
     SCA_SDF_POST(Some_Postprocessing_Function)
   }
};
```

ALGORITHM 9.25 SystemC-AMS Model for VCO [188, 279]

```
// SystemC-AMS Model for VCO
SCA_SDF_MODULE(VCO_Behavioral)
{
   double freq_center;
   double freq_slope;
   double V_c_at_fc;
   void (init)
   {
      mul_fac = 2.0 * M_PI * 1e-12;
      deg_bias = 36 * M_PI / 180;
   }
   void sig_proc
   {
      Control_V = V_in - V_c_at_fc;
      f_VCO = (freq_center - freq_slope * Control_V);
      theta += mul_fac * f_VCO;
      for (int count_cycles = 0; count_cycles <= 19; count_cycles ++)
         sin_wave[count_cycles] = sin(theta + count_cycles * deg_bias);
   }
}
```

mandatory signal processing function and three optional functions, i.e., an attribute function, an initialization function, and a post-processing function. Thus, discrete-event processes are not allowed in an SCA_MODULE. The required signal-processing function essentially describes the continuous-time behavior of the module in the form of procedural assignments. The attribute function of an SCA_MODULE defines the values of the attributes for the static data flow simulation. Once after elaboration is done, the initialization function is forked before the simulation process starts for the SystemC-AMS model. The post-processing function of an SCA_MODULE is executed once just before the completion of simulations.

As a specific example, the SystemC-AMS model of the VCO component of a CPPLL is presented in Algorithm 9.25 [188, 279]. This model essentially represents a VCO model that can generate 20 cycles of the same phase at a frequency of 500 MHz. A designer given name VCO_Behavioral is declared as an AMS module through the use of SCA_MODULE. The double keyword declares the variable double data type. Inside the init function, variables like mul_fac and deg_bias are defined. The sig_proc function presents the behavioral description of the VCO_Behavioral SystemC-AMS-based VCO module. It is a simple sin function calculation for 20 cycles.

9.9 Mixed-Mode Circuit-Level Simulations

In the preceding section, mixed-signal circuits and systems have been presented with emphasis on system-level behavioral modeling. In the current section the mixed-mode and/or mixed-signal simulation frameworks are presented with circuit level as emphasis. First, a brief discussion is presented on whether there is a real need for mixed-signal simulations. Then different mixed-mode simulators are discussed. For the rest of the discussion, it is important to clarify the following terms [215, 216]: (1) mixed-signal/mixed-mode simulators and (2) mixed-level simulators. The mixed-signal/mixed-mode simulators can perform simulation of both analog and digital circuits. Mixed-level simulators allow simulation of different parts of circuit/system using different levels. For example, the critical parts are simulated at circuit level and the noncritical parts are simulated at functional level. Some simulators are both mixed-mode and mixed-level. However, the terms "mixed-signal," "mixed-level," and "mixed-mode" have been used interchangeably in the existing literature. In this text, mixed-signal/mixed-mode means mixed analog and digital simulation. A selected set of mixed-mode simulators currently available include: B2.Spice A/D, Discovery AMS, Gnucap, IsSpice4,

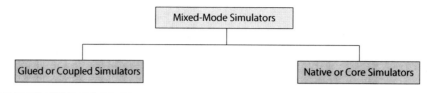

FIGURE 9.61 Different mixed-mode simulators for integrated circuits and systems.

Micro-Cap, NgSpice, Questa® ADMS, SIMetrix, SMASH, TINA™, TopSpice, and Virtuoso® UltraSim Full-Chip Simulator [4, 9, 23, 26, 30, 35, 38, 40, 56, 57, 64, 71].

The current-generation circuits, or AMS-SoCs, are a combination of analog and digital circuits or subsystems. To simulate such circuits or systems, an accurate mix of analog and digital simulation techniques is required. This is essentially the purpose of mixed-signal/mixed-mode simulators [215, 216]. A comparative mixed-mode simulator taxonomy is presented in Fig. 9.61 [139, 215, 216]. The mixed-mode simulators are of two types based on how the simulator has been implemented: (1) glued or coupled mixed-mode simulators and (2) native or core mixed-mode simulators. The glued or coupled mixed-mode simulators are implemented using two kernels or executables: one is analog and the other is digital in nature. On the other hand, the native or core mixed-mode simulators implement both the analog and the digital simulation algorithms in the same executable.

9.9.1 Nanoelectronics Analog versus Mixed-Signal Simulation: A Comparative Perspective

The question arises as to why cannot design and simulation be performed using an only analog or only digital simulator [177, 200, 243]. The AMS-SoC have heterogeneous mix of analog components, digital components, and software components. Generally speaking, in terms of hardware analog and digital components, one can imagine the following three possible options: small-a/big-D, big-A/small-d, and comparable-A/comparable-D. This is based on the size of the analog and digital portions present in the AMS-SoC. The small-a/big-D AMS-SoC can be discussed with a set of analog "islands" in a digital "sea." Simulation of this AMS-SoC using an accurate analog/SPICE simulator is impossible. Simulation using digital simulator due to the digital "sea" can still have inaccuracy issues from an overall AMS-SoC perspective. The big-A/small-d AMS-SoC can be discussed with a set of digital "islands" in an analog "sea." Again this AMS-SoC can not be simulated using an accurate analog/SPICE simulator due to high complexity with analog "sea." This AMS-SoC when simulated using digital simulators will be highly inaccurate from overall AMS-SoC perspective due to only digital "islands" situation. For comparable-A/comparable-D AMS-SoC with comparable analog and digital portions, either of the above situations can be used but accuracy will be an issue.

As a specific example of accuracy issue, a system with partial components of PLL is presented in Fig. 9.62 [177, 200, 243]. Specifically, a VCO has been considered along with a frequency divider. In one case, an analog component of the VCO along with a digital frequency divider is considered for the mixed-signal simulations as shown in Fig. 9.62(a) [177, 200, 243]. For mixed-signal and mixed-domain simulation purposes, the VCO is designed as the circuit level and the frequency divider is modeled as a digital circuit using Verilog. In this circuit, the analog clock output of the VCO goes as input to the digital frequency divider. The overall output is the digital clock, with half the frequency as that of the VCO output being obtained, as shown in Fig. 9.63(a) [177, 200, 243]. To facilitate the communication among the analog and digital components, interface elements predefined in the mixed-signal design and simulation framework need to be used for this kind of mixed-mode simulations. Specifically, in this case study, an analog-to-digital interface is created between the analog VCO and the digital frequency divider.

In the other case, an analog component of the VCO along with an analog frequency divider is considered for the analog signal simulations, as shown in Fig. 9.62(b) [177, 200, 243]. The analog simulation of the overall circuit can be performed using transistor-level design of both the components using accurate SPICE/analog simulators. In this case the analog output of the VCO is provided as input to the analog frequency divider. The overall output is the analog signal, with half the frequency as that of the VCO analog output being obtained, as shown in Fig. 9.63(b) [177, 200, 243]. The frequencies of signals at the output of the VCO, digital frequency divider, and analog frequency divider are observed to be the following: f_{VCO} = 717.9 MHz, $f_{FD,D}$ = 394.0 MHz, $f_{FD,A}$ = 357.9 MHz,

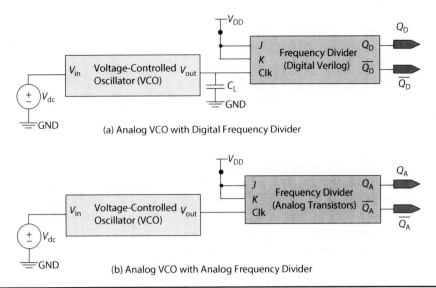

FIGURE 9.62 Block diagram of the VCO along with an analog frequency divider and a digital frequency divider [177, 200, 243].

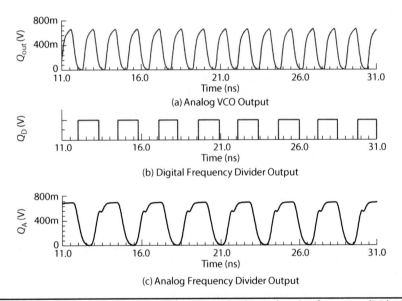

FIGURE 9.63 Output waveforms of the VCO, digital frequency divider, and analog frequency divider [177, 200, 243].

respectively. From these simulation results it is obvious that the output frequency of the digital frequency divider is about 36.1 MHz higher than the output frequency of the analog frequency divider. In other words, there is a frequency discrepancy of approximately 10% between the two types of simulations. This kind of discrepancy will increase further for more complex circuits and other characteristics.

9.9.2 Mixed-Mode with Individual Analog and Digital Engine

The glued or coupled engine-based mixed-mode simulator uses two individual engines: one for analog simulation and another for digital simulation [139, 215, 216]. The information between the two engines of the glued mixed-mode simulators is exchanged based on task prioritization as well as synchronization algorithms. The design engineers using the glued mixed-mode simulators have the option of choosing either loosely coupled simulators or tightly coupled simulators. The glued mixed-mode simulators apply more than one algorithm to the circuit to be simulated. The circuit is partitioned into parts to which each algorithm is applied during the simulation. The algorithms run in either of the engines. The glued mixed-mode simulator allows the component models developed for separate executables to be used without modification. However, the glued mixed-mode simulators have reduced speed and accuracy due to communication constraints and synchronization needs of the two executables.

9.9.3 Mixed-Mode with Unified Analog and Digital Engine

The native or core engine-based mixed-mode simulator uses one engine for either analog or digital simulation [139, 215, 216]. In other words, native or core engine-based simulators provide both analog- and discrete-event-based simulations in the same executable. By necessity, native mixed-mode simulators use either an analog core or a digital core in their simulation algorithms with extensions to support the other type. For example, a native mixed-mode simulator may use an analog simulator with an extension for digital simulations, but in the same executable. In another option, it may use a digital simulator with an extension for analog simulations, but in the same executable. In other words, both types of algorithms needed for analog and digital simulations are executed in one simulation engine. Thus, native mixed-mode simulators do not need any synchronization mechanisms and hence can be faster than the glued mixed-mode simulators.

9.10 Models for Circuit Simulations

In general, model is a relation of the behavior of an entity of a circuit or system to its design or process parameters. The relation is represented in the form of mathematical formulas, LUTs, or graphs. Models of different components are needed at different levels of abstraction to simulate the circuit or system at that level of abstraction. For example, for the architecture-level simulation of the systems, the models of the architecture components are needed. Similarly, for the logic-level simulations, models for logic gates are needed. In this section, the device models needed for the use in SPICE or similar analog simulators are discussed. A compact model or SPICE model is the mathematical description of the electrical behavior of a semiconductor device that is suitable for simulation of integrated circuits and systems [148, 149, 182, 183, 216]. A selected set of compact or SPICE models available for the use by design engineers is presented in Fig. 9.64 [183, 216]. These are briefly discussed in the rest of this section. A broad overview of the compact model generation flow is also discussed to give a quick understanding of the modeling effort.

9.10.1 Compact Model Generation Flow

When discussing model development it is important to distinguish two scenarios as follows: (1) Creating a new model for a new device that does not already exist in SPICE library at all. This is for a completely new device with or without fabrication data. (2) Calibrating (i.e., fitting model parameters) a new device to the existing SPICE model. This is for an already existing device with access to fabrication data or partial process information. The first approach needs more effort and can be done by the manufacturers. The second approach is a typically faster alternative. The first approach is discussed in this section to present a comprehensive view to the reader for in-depth understanding. The overall flow for device model generation is presented in Fig. 9.65 [74, 142, 152, 190]. The overall modeling mechanism can be categorized into the following levels: (1) accurate modeling, (2) compact modeling, and (3) metamodeling.

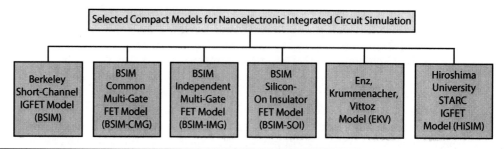

Figure 9.64 Different compact models for integrated circuit simulations.

Mixed-Signal Circuit and System Simulation   499

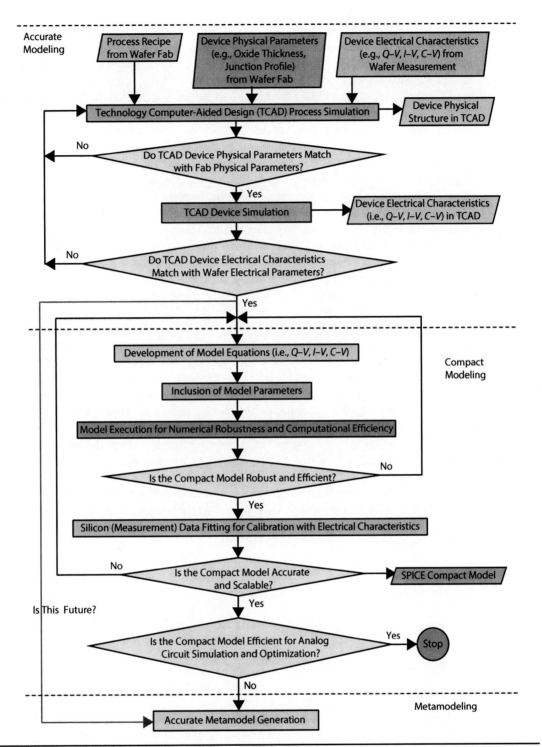

FIGURE 9.65 An illustrative compact model generation flow.

The accurate modeling at the detailed or lowest level involves technology computer-aided design (TCAD) process simulations. The process recipe from wafer fab is the primary input to the TCAD simulations. The wafer fab can also provide information on the devices physical parameters, such as oxide thickness and junction profile. Device electrical characteristics, such as Q-V, I-V, and C-V, can be obtained from wafer measurements. The TCAD process simulations generate device physical structure. The TCAD physical structure is compared with the fab physical parameters. If there is a match, then the next phase of simulation involved is TCAD device simulation. The TCAD device simulation generates various electrical characteristics of the TCAD device, including Q-V, I-V, and

C–V. At this point TCAD electrical characteristics are compared with the device characteristics obtained from the wafer measurements. The simulation iterations stop if a good match of physical and electrical parameters is obtained from the TCAD simulations. In essence, at this point a "comprehensive device model" is obtained from the TCAD simulations. The comprehensive device model is the most accurate device model and, in theory, can be used for simulation purposes, but it is too slow for practical usage for circuit-level simulations.

The next phase in the overall modeling process is compact modeling. At this phase "compact" mathematical equations are developed to capture the physics of the device. Then physics- and technology-related model parameters are included to the model. In the TCAD device in the above step, the parameter-like junction depth and doping profile are very complex. Thus, to make the models compact and yet accurate, physics- and technology-related parameters are added to the model. The model is then executed to verify its numerical robustness and computational efficiency. The compact model should not have any convergence issues and should be computationally efficient to facilitate circuit-level simulation with minimal overhead. In case of any issues some of the physics-based equations may be modified. In the next step, the actual silicon data fitting is performed for calibration of the compact model electrical characteristics. In case of any issues, the compact model equation may need modification, or new model parameters may even be introduced. Once the compact model is verified to be accurate and scalable, the compact modeling stops.

It is a fact that, with the available compact models, which use 100s of parameters to improve accuracy, and the analog simulation engine, simulation of the full-blown circuits and systems is difficult to impossible. To mitigate the problem, fast SPICE and parallel SPICE have been developed, as discussed in the prior sections. However, these are still not adequate for large AMS-SoCs. Thus, if the above compact models are not found to be efficient for large analog circuit simulation and optimization, then accurate metamodeling can be generated [152, 221]. The metamodels of the devices can be generated from the compact models generated above. In another way, the metamodels can be generated from the accurate TCAD device models from the phase before the compact modeling phase. The latter approach may result in more accurate metamodels than generated from the compact models.

For the nanoelectronic technology, corner-based modeling is not performed satisfactorily due to several factors. The corner models ignore the correlations in the process parameters. The corner-based models cannot distinguish between intra-die and inter-die variations. In the corner models, the simultaneous instance of process corners may lead to unrealistic process combinations. For the analog circuits, it may be difficult to identify the worst-case conditions using corner-based models. The process-dependent compact SPICE models or statistical compact SPICE models are developed for efficient simulation of nanoelectronic circuits [74, 93].

9.10.2 Types of Compact Models

A selected set of compact SPICE models are discussed in this section. These models can be used for circuit- or device-level simulation of integrated circuits.

9.10.2.1 BSIM Compact Model

The Berkeley short-channel insulated-gate field-effect transistor (IGFET) models (BSIMs) are quite popular and widely used [82, 85, 216]. The successful history of BSIM started around early 1990s with the introduction of BSIM1 (level 4) model, which is a polynomial equations-based model with less emphasis on physics. BSIM4 model (levels 14 $\rightarrow$ 54) has been a very widely used SPICE compact model. It is a physics-based, accurate, scalable, and predictive MOSFET model. The latest model in this series is BSIM6.0 compact SPICE model for MOSFET.

9.10.2.2 BSIM Common Multi-Gate (BSIM-CMG) FET Model

The multi-gate MOSFETs can be divided into two main categories as follows: common multi-gate (CMG) MOSFETs and independent multi-gate (IMG) MOSFETs. In a CMG MOSFET, all gates are ON at a time. The gate stacks have identical materials, dielectric thicknesses, and work functions. BSIM common multi-gate (BSIM-CMG) FET models are compact SPICE models available for the simulation of circuits that are designed using them as basic elements [186].

9.10.2.3 BSIM Independent Multi-Gate (BSIM-IMG) FET Model

In an IMG MOSFET, the gates are independent. The gate stacks have different materials, dielectric thicknesses, and work functions. BSIM independent multi-gate (BSIM-IMG) FET models are compact SPICE models available for the simulation of circuits that are designed using them as basic elements [186].

9.10.2.4 BSIM Silicon-On-Insulator (BSIM-SOI) FET Model

BSIMSOI (levels 10 → 58) models are compact SPICE models for silicon-on-insulator-based FET [82, 216]. The BSIMSOI compact models include partially depleted and fully depleted devices.

9.10.2.5 Enz, Krummenacher, Vittoz (EKV) Compact Model

The Enz, Krummenacher, Vittoz (EKV) compact model started in 1995 and is named after the original developers [145, 216]. The EKV compact SPICE models are mathematical models that can support analog circuit design and simulation. The EKV model is accurate for subthreshold region of operation. It captures many effects that are prominent in nanoscale technology.

9.10.2.6 HiSIM Models

Another compact SPICE model is Hi-Roshima University STARC IGFET model (HiSIM) [83, 216]. HiSIM implementation is available in two models: surface-potential-based MOSFET model (HiSIM2) and surface-potential-based HV/LD-MOSFET model (HiSIMHV). HiSIM-SOI is also available for SOI FETs.

9.10.3 Automatic Device Model Synthesizer (ADMS)

The development, implementation, and maintenance of compact SPICE models for analog circuit simulators are very labor intensive. In order to make implementation of the compact SPICE model simple, efficient, and robust, automatic device model synthesizer (ADMS) was introduced [182, 183]. ADMS simplifies compact SPICE model development, implementation, and maintenance as well as distribution and sharing. The key idea of ADMS is to describe the compact model in a high-level language like Verilog-AMS and then automatically generate the low-level model from it.

9.11 Questions

9.1 What is the concept of circuit simulation? Discuss different forms of simulation based on abstraction levels of digital circuits.

9.2 Discuss different types of simulations based on the tasks of the integrated circuit design.

9.3 How is hardware description language (HDL) different from hardware verification language (HVL).

9.4 Discuss two languages or frameworks for system-level simulation.

9.5 Discuss the difference between behavioral and structural simulation.

9.6 Simulate an example analog block using MATLAB® language and framework.

9.7 Using MATLAB® language and framework, read a JPEG image and perform its DCT coefficient calculation.

9.8 Consider a resistive, RC, or RCL circuit with DC sources. Write its mesh equations and perform current analysis using MATLAB®.

9.9 Simulate a MATLAB® of a memristor to verify its characteristics.

9.10 Consider an example analog integrated circuit. Perform its modeling and simulation using Simulink®.

9.11 Using Simulink® language and framework, read a JPEG image and perform its DCT coefficient calculation.

9.12 Simulate a Simulink® of a memristor to verify its characteristics.

9.13 Consider an example memristor-based analog integrated circuit. Perform its modeling and simulation in Simulink® using the memristor model from the above.

9.14 Simulate a Simulink® of a graphene FET to verify its characteristics.

9.15 Consider an example graphene FET-based analog integrated circuit. Perform its modeling and simulation in Simulink® using the graphene FET model from the above.

9.16 What is SPICE? Discuss five different SPICE simulators available at present?

9.17 Briefly discuss different types of analysis that SPICE can perform.

9.18 Consider a simple ring oscillator and perform its simulation using SPICE for the latest device models available.

9.19 Briefly discuss how a SPICE engine performs simulation.

9.20 Select an analog circuit of your choice. Perform its simulations using Verilog-A.

9.21 Select a digital circuit of your choice. Perform its simulations using SystemVerilog.

9.22 Select a digital circuit of your choice. Perform its simulations using VHDL.

9.23 Select a digital circuit of your choice. Perform its simulations using MyDL.

9.24 Select a digital system component of your choice. Perform its simulations using SystemC.

9.25 Select an analog circuit of your choice. Perform its simulations using Verilog-AMS.

9.26 Select an analog circuit of your choice. Perform its simulations using VHDL-AMS.

9.27 Select an analog circuit of your choice. Perform its simulations using SystemC-AMS.

9.28 Discuss why mixed-mode simulation is necessary as compared with pure analog or digital simulations.

9.12 References

1. Analog and Mixed-Signal Modeling Using the VHDL-AMS Language. http://tams-www.informatik.uni-hamburg.de/vhdl/doc/P1076.1/tutdac99.pdf. Accessed on 05th April 2014.
2. Analog Digital Turbo Simulator (ADiT™). http://www.mentor.com/products/ic_nanometer_design/analog-mixed-signal-verification/adit/. Accessed on 26th March 2014.
3. Analog FastSPICE™ (AFS). http://www.berkeley-da.com/prod/datasheets/Berkeley_DA_Platform_DS.pdf. Accessed on 27th March 2014.
4. B2.Spice A/D Circuit Simulation Software. http://www.5spice.com/. Accessed on 25th March 2014.
5. Cider Mixed-Level Circuit Simulator. http://www-cad.eecs.berkeley.edu/Software/cider.html. Accessed on 26th March 2014.
6. CircuitLogix. https://www.circuitlogix.com/. Accessed on 25th March 2014.
7. CoolSpice™. http://coolcadelectronics.com/coolspice/. Accessed on 25th March 2014.
8. CustomSim™. http://www.synopsys.com/Tools/Verification/AMSVerification/CircuitSimulation/Pages/CustomSim-ds.aspx. Accessed on 26th March 2014.
9. Discovery AMS. http://www.synopsys.com/Tools/Verification/AMSVerification/AnalogDigital/AMS/Pages/default.aspx. Accessed on 03rd April 2014.
10. Eldo® Classic. http://www.mentor.com/products/ic_nanometer_design/analog-mixed-signal-verification/eldo/. Accessed on 25th March 2014.
11. Eldo® Premier. http://www.mentor.com/products/ic_nanometer_design/analog-mixed-signal-verification/eldo-premier/. Accessed on 27th March 2014.
12. FineSim™. http://www.synopsys.com/Tools/Verification/AMSVerification/Pages/finesim-ds.aspx. Accessed on 26th March 2014.
13. General-purpose Semiconducor Simulator (GSS) TCAD for NgSpice. http://ngspice.sourceforge.net/gss.html. Accessed on 26th March 2014.
14. Genius Device Simulator. http://www.cogenda.com/. Accessed on 26th March 2014.
15. Giga-scale Parallel SPICE Simulator- NanoSpice™. http://www.proplussolutions.com/en/pro2/Giga-scale-Parallel-SPICE-Simulator—NanoSpice.html. Accessed on 09th March 2014.
16. HSIM™. http://www.synopsys.com/Tools/Verification/AMSVerification/CircuitSimulation/HSIM/Pages/default.aspx. Accessed on 26th March 2014.
17. HSPICE®. http://www.synopsys.com/Tools/Verification/AMSVerification/CircuitSimulation/HSPICE/Documents/hspice_ds.pdf. Accessed on 25th March 2014.
18. Incisive Enterprise Simulator. http://www.cadence.com/rl/Resources/datasheets/incisive_enterprise_specman.pdf. Accessed on 31st March 2014.

19. Introduction to Quartus® II Software. http://www.altera.com/literature/manual/quartus2_introduction.pdf. Accessed on 31st March 2014.
20. Introduction to VHDL-AMS. http://www.denverpels.org/Downloads/Denver_PELS_20071113_Cooper_VHDL-AMS.pdf. Accessed on 06th April 2014.
21. IRSIM Version 9.7 Switch-level Simulator. http://opencircuitdesign.com/irsim/. Accessed on 20th March 2014.
22. ISE Simulator (ISim). http://www.xilinx.com/tools/isim.htm. Accessed on 31st March 2014.
23. IsSpice4. http://www.intusoft.com/products/ICAP4Professional.htm. Accessed on 07th April 2014.
24. LTSpice. http://www.linear.com/designtools/software/. Accessed on 26th March 2014.
25. MATLAB® Function. http://www.mathworks.com/help/simulink/slref/matlabfunction.html. Accessed on 23rd March 2014.
26. Micro-Cap. http://www.spectrum-soft.com/down/ug11.pdf. Accessed on 25th March 2014.
27. ModelSim™. http://www.mentor.com/products/fpga/model. Accessed on 31st March 2014.
28. Multisim. http://www.ni.com/multisim/. Accessed on 25th March 2014.
29. MyHDL - From Python to Silicon! http://www.myhdl.org/. Accessed on 13th March 2014.
30. Ngspice. http://ngspice.sourceforge.net/. Accessed on 26th March 2014.
31. OpenMAST™. http://www.openmast.org/. Accessed on 16th March 2014.
32. OpenVera®. http://www.open-vera.com/. Accessed on 20th March 2014.
33. Parallel SmartSpice: Fast and Accurate Circuit Simulation Finally Available. http://www.silvaco.com/tech_lib_TCAD/simulationstandard/1997/may/a1/a1.html. Accessed on 25th March 2014.
34. PSpice®. http://www.cadence.com/products/orcad/pspice_simulation/pages/default.aspx. Accessed on 26th March 2014.
35. Questa® ADMS. http://www.mentor.com/products/ic_nanometer_design/analog-mixed-signal-verification/advance-ms/. Accessed on 03rd April 2014.
36. Quite Universal Circuit Simulator (Qucs). http://qucs.sourceforge.net/. Accessed on 26th March 2014.
37. Saber®. http://www.synopsys.com/Systems/Saber/Pages/default.aspx. Accessed on 26th March 2014.
38. SIMetrix. http://www.simetrix.co.uk/site/simetrix-classic.html. Accessed on 26th March 2014.
39. SmartSpice™. http://www.silvaco.com/products/analog_mixed_signal/smartspice.html. Accessed on 26th March 2014.
40. SMASH - Mixed-Signal Simulator. http://www.dolphin.fr/medal/downloads/pdf/smash/Brochure_SMASH.pdf. Accessed on 26th March 2014.
41. Solve Elec. http://www.physicsbox.com/indexsolveelec2en.html. Accessed on 26th March 2014.
42. Solver Pane. http://www.mathworks.com/help/simulink/gui/solver-pane.html. Accessed on 24th March 2014.
43. Spectre® Circuit Simulator. http://www.cadence.com/products/cic/spectre_circuit/pages/default.aspx. Accessed on 26th March 2014.
44. Spectre® eXtensive Partitioning Simulator (Spectre® XPS). http://www.cadence.com/products/cic/Spectre_eXtensive_Partitioning_Simulator/pages/default.aspx. Accessed on 26th March 2014.
45. SPICE Analysis Fundamentals. http://www.ni.com/white-paper/12794/en/. Accessed on 27th March 2014.
46. Subclassing and Inheritance. http://www.mathworks.com/help/physmod/simscape/lang/subclassing-and-inheritance.html. Accessed on 25th March 2014.
47. Super-FinSim. http://www.fintronic.com/product_description.html#Super-FinSim. Accessed on 31st March 2014.
48. SymSpice Turbo. http://www.symica.com/technology/fast-spice-technology. Accessed on 27th March 2014.
49. SystemC-AMS. http://www.systemc-ams.org/. Accessed on 17th March 2014.
50. SystemC AMS extensions User's Guide. http://kona.ee.pitt.edu/socvlsi/lib/exe/fetch.php?media=osci_systemc_ams_users_guide.pdf. Accessed on 18th March 2014.
51. T-Spice™. http://www.penzar.com/topspice/topspice.htm. Accessed on 26th March 2014.
52. TclSpice. http://qucs.sourceforge.net/. Accessed on 26th March 2014.
53. The Analog, Mixed-Technology and Mixed-Signal HDL for Saber. http://www.synopsys.com/Systems/Saber/Pages/MAST.aspx. Accessed on 26th March 2014.
54. The Guide to SystemC. https://www.doulos.com/knowhow/systemc/. Accessed on 02nd April 2014.
55. TINA-TI™. http://www.ti.com/tool/tina-ti. Accessed on 26th March 2014.
56. TINA™. http://www.tina.com/. Accessed on 26th March 2014.
57. TopSpice. http://www.penzar.com/topspice/topspice.htm. Accessed on 26th March 2014.
58. Turbo-MSIM. http://www.legenddesign.com/products/tbmsim.shtml. Accessed on 26th March 2014.
59. Using Verilog-A to Simplify a SPICE Netlist. http://www.silvaco.com/content/appNotes/analog/1-004_Verilog-A.pdf. Accessed on 12th March 2014.
60. VCS: Functional Verification Choice of Leading SoC Design Teams. http://www.synopsys.com/Tools/Verification/FunctionalVerification/Documents/vcs-ds.pdf. Accessed on 31st March 2014.
61. Verilog HDL: Behavioral Counter. http://www.altera.com/support/examples/exm-index.html. Accessed on 13th March 2014.
62. VHDL-AMS Introduction. http://www.dolphin.fr/projects/macros/dissemination/pdf/tutorial_vhdl_ams.pdf. Accessed on 05th April 2014.
63. Virtuoso® AMS Designer. http://www.cadence.com/products/cic/ams_designer/pages/default.aspx. Accessed on 03rd April 2014.

64. Virtuoso® UltraSim Full-Chip Simulator. http://www.cadence.com/products/cic/UltraSim_fullchip/pages/default.aspx. Accessed on 26th March 2014.
65. What Is the Simscape Language? http://www.mathworks.com/help/physmod/simscape/lang/what-is-the-simscape-language.html. Accessed on 25th March 2014.
66. Winspice. http://www.winspice.com/. Accessed on 26th March 2014.
67. XSPICE. http://users.ece.gatech.edu/mrichard/Xspice/. Accessed on 26th March 2014.
68. IEEE Standard VHDL Language Reference Manual. ANSI/IEEE Std 1076-1993 (1994). DOI 10.1109/IEEESTD.1994.121433.
69. Parallel SmartSpice: Fast and Accurate Circuit Simulation Finally Available. *Simulation Standard* **8**(5), 1–11 (1997). Accessed on 25th March 2014.
70. IEEE Standard VHDL Analog and Mixed-Signal Extensions. IEEE Std 1076.1-1999 (1999). DOI 10.1109/IEEESTD.1999.90578.
71. *The Gnu Circuit Analysis Package, User's manual.* http://www.gnu.org/software/gnucap/man/index.html (2002). Accessed on 26th March 2014.
72. Behavioral Modeling of PLL Using Verilog-A. http://www.silvaco.com/tech_lib_TCAD/simulationstandard/2003/jul/a2/july2003_a2.pdf (2003). Accessed on 15th March 2014.
73. Using Hierarchy and Isomorphism to Accelerate Circuit Simulation. Tech. rep., Cadence Design Systems, Inc., San Jose, CA 95134 USA (2004). Accessed on 06th March 2014.
74. Paramos: A Process-Dependent Compact SPICE Model Extractor. http://www.synopsys.com/Tools/TCAD/CapsuleModule/news_dec06.pdf (2006). Accessed on 08th April 2014.
75. GNU General Public License (GPL). http://www.gnu.org/copyleft/gpl.html (2007). Accessed on 21st March 2014.
76. IEEE Standard VHDL Analog and Mixed-Signal Extensions. IEEE Std 1076.1-2007 (Revision of IEEE Std 1076.1-1999) pp. c1–328 (2007). DOI 10.1109/IEEESTD.2007.4384309.
77. *HSPICE® User Guide: Simulation and Analysis.* http://cseweb.ucsd.edu/classes/wi10/cse241a/assign/hspice_sa.pdf (2008). Accessed on 11th March 2014.
78. IEEE Standard VHDL Language Reference Manual - Redline. IEEE Std 1076-2008 (Revision of IEEE Std 1076-2002) - Redline pp. 1–620 (2009).
79. Engineering Education and Research Using MATLAB, chap. Mixed-Signal Circuits Modelling and Simulations Using MATLAB. InTech (2011). DOI 10.5772/21556. Accessed on 28th February 2014.
80. Scicos: Block Diagram Modeler/Simulator. http://www.scicos.org/ (2011). Accessed on 23rd March 2014.
81. VHDL Code for a Simple ALU. http://vhdlguru.blogspot.com/2011/06/vhdl-code-for-simple-alu.html (2011). Accessed on 13th March 2014.
82. BSIM Group. http://www-device.eecs.berkeley.edu/bsim/ (2012). Accessed on 08th April 2014.
83. HiSIM: Hi-Roshima University STARC IGFET Model. http://www.hisim.hiroshima-u.ac.jp/ (2012). Accessed on 08th April 2014.
84. Block Libraries. http://www.mathworks.com/help/simulink/block-libraries.html (2013). Accessed on 01st March 2014.
85. BSIM6.0 MOSFET Compact Model. http://www-device.eecs.berkeley.edu/bsim/Files/BSIM6/BSIM6.0.0/BSIM6.0.0_Technical_manual.pdf (2013). Accessed on 08th April 2014.
86. BeagleBone. http://beagleboard.org/ (2014). Accessed on 23rd March 2014.
87. GNU Octave. https://www.gnu.org/software/octave/ (2014). Accessed on 21st March 2014.
88. HDL Coder™. http://www.mathworks.com/products/hdl-coder/ (2014). Accessed on 23rd March 2014.
89. KSpice. http://embedded.eecs.berkeley.edu/pubs/downloads/spice/kspice.html (2014). Accessed on 26th March 2014.
90. LabVIEW System Design Software. http://www.ni.com/labview/ (2014). Accessed on 23rd March 2014.
91. MapleSim. http://www.maplesoft.com/products/maplesim/ (2014). Accessed on 23rd March 2014.
92. MATLAB®: The Language of Technical Computing. http://www.mathworks.com/products/matlab/ (2014). Accessed on 20th March 2014.
93. Mystic Statistical Compact Model Extractor. http://www.goldstandardsimulations.com/products/mystic/ (2014). Accessed on 08th April 2014.
94. Raspberry Pi. http://www.raspberrypi.org/ (2014). Accessed on 23rd March 2014.
95. Scilab. http://www.scilab.org/ (2014). Accessed on 21st March 2014.
96. SciPy. http://www.scipy.org/ (2014). Accessed on 21st March 2014.
97. SimRF™. http://www.mathworks.com/products/simrf/ (2014). Accessed on 23rd March 2014.
98. Simscape®: Model and Simulate Multidomain Physical Systems. http://www.mathworks.com/products/simscape/ (2014). Accessed on 20th March 2014.
99. Simulink®: Simulation and Model-Based Design. http://www.mathworks.com/products/simulink/ (2014). Accessed on 20th March 2014.
100. Simulink® User's Guide. http://www.mathworks.com/help/releases/R2014a/pdf_doc/simulink/sl_using.pdf (2014). Accessed on 23rd March 2014.
101. VisSim™. http://www.vissim.com/ (2014). Accessed on 23rd March 2014.
102. Accelera International, Inc.: SystemC Version 2.0 Users Guide (2002). Accessed on 14th March 2014.
103. Accelera International, Inc.: SystemVerilog 3.1a Language Reference Manual (2004). Accessed on 13th March 2014.

104. Accelera International, Inc.: Verilog-AMS Language Reference Manual: Analog & Mixed-Signal Extensions to Verilog-HDL (2009). Accessed on 12th March 2014.
105. Adzmi, A., Nasrudin, A., Abdullah, W., Herman, S.: Memristor Spice model for designing analog circuit. In: *Proceedings of the IEEE Student Conference on Research and Development (SCOReD)*, pp. 78–83 (2012). DOI 10.1109/SCOReD.2012.6518615.
106. Agu, E., Mohanty, S.P., Kougianos, E., Gautam, M.: Simscape Based Design Flow for Memristor Based Programmable Oscillators. In: *Proceedings of the 23rd ACM/IEEE Great Lakes Symposium on VLSI*, pp. 223–224 (2014).
107. Ahmed, S.I., Kwasniewski, T.A.: A Multiple-Rotating-Clock-Phase Architecture for Digital Data Recovery Circuits using Verilog-A. In: *Proceedings of the 2005 IEEE International Behavioral Modeling and Simulation Workshop*, pp. 112–117 (2005).
108. Al-Junaid, H., Kazmierski, T.: Analogue and Mixed-Signal Extension to SystemC. *IEE Proceedings – Circuits, Devices and Systems*, pp. 682–690 (2005). DOI doi:10.1049/ip-cds:20045204.
109. Ale, A.K.: *Comparison and Evaluation of Existing Analog Circuit Simulators Using a Sigma-Delta Modulator*. Master's thesis, Computer Science and Egnineering, University of North Texas, Denton, TX, USA (2006).
110. Andresen, R.P.: 5Spice Analysis Software. http://www.5spice.com/. Accessed on 25th March 2014.
111. Dillon, T., Paatela, J., Dannoritzer, G., Hussong, S.: Accelerating Algorithm Implementation in FPGA/ASIC Using Python. http://www.ll.mit.edu/HPEC/agendas/proc07/Day2/12_Dillon_Poster.pdf. Accessed on 21st March 2014.
112. Dillon, T., Paatela, J., Dannoritzer, G., Hussong, S.: Accelerating Algorithm Implementation in FPGA/ASIC Using Python. In: *Proceedings of the High Performance Embedded Computing Workshop (2007)*. Accessed on 13 Mar 2014.
113. Ashari, Z.M., Nordin, A.N., Ibrahimy, M.I.: Design of a 5GHz Phase-Locked Loop. In: *Proceedings of the IEEE Regional Symposium on Micro and Nanoelectronics (RSM)*, pp. 167–171 (2011). DOI 10.1109/RSM.2011.6088316.
114. Ashenden, P.J.: *The Student's Guide to VHDL*. Morgan Kaufmann. Elsevier Science & Technology Books (2008).
115. Ashenden, P.J., Peterson, G.D., Teegarden, D.A.: *The System Designer's Guide to VHDL-AMS: Analog, Mixed-Signal, and Mixed-Technology Modeling*. Systems on Silicon. Elsevier Science (2002).
116. Attia, J.O.: *Electronics and Circuit Analysis Using MATLAB*, second edition. MATLAB Series. Taylor & Francis (2004).
117. Bailey, S.: Comparison of VHDL, Verilog and SystemVerilog. http://www.fpga.com.cn/advance/vhdl_14919.pdf. Accessed on 13th March 2014.
118. Banerjee, A., Sur, B.: *Systemc and Systemc-AMS in Practice*. Springer Verlag (2013).
119. Banerjee, D.: *PLL Performance, Simulation and Design*. Dog Ear Publishing (2006).
120. Barnasconi, M.: Introduction to SystemC AMS. http://www.iscug.in/sites/default/files/ISCUG-2013/ppt/day1/1_4_3_Introduction_SystemC_AMS_March2013.pdf (2013). Accessed on 18th March 2014.
121. Berman, V.: A Tale of Two Languages: SystemC and SystemVerilog. http://chipdesignmag.com/display.php?articleId=116. Accessed on 14th March 2014.
122. Bernstein, H.: Understanding the SPICE Simulation Engine. *DESIGN SOLUTIONS* 1, 9–10 (2003). Accessed on 10th March 2014.
123. Berny, A.D., Meyer, R.G., Niknejad, A.: *Analysis and Design of Wideband LC VCOs*. Ph.D. thesis, EECS Department, University of California, Berkeley (2006). URL http://www.eecs.berkeley.edu/Pubs/TechRpts/2006/EECS-2006-50.html.
124. Bhasker, J.: *A SystemC Primer*. Star Galaxy Publishing (2004).
125. Biolek, D., Biolek, Z., Biolkova, V.: SPICE Modeling of Memristive, Memcapacitative and Meminductive Systems. In: *Proceedings of the European Conference on Circuit Theory and Design*, pp. 249–252 (2009).
126. Biolek, Z., Biolek, D., Biolkova, V.: SPICE Model of Memristor with Nonlinear Dopant Drift. *Radioengineering* **18**(2), 210–214 (2009).
127. Bjørnsen, J., Ytterdal, T.: Behavioral Modeling and Simulation of High Speed Analog-to-Digital Converters using SystemC. In: *Proceedings of the International Symposium on Circuits and Systems*, pp. 906–909 (2003).
128. Brown, A.D., Zwolinski, M.: The Continuous-Discrete Interface – What Does This Really Mean? In: *Proceedings of the International Symposium on Circuits and Systems*, pp. 894–897 (2003).
129. Brown, S.D., Brown, S., Vranesic, Z.: *Fundamentals of Digital Logic with VHDL Design*. McGraw-Hill Series in Electrical and Computer Engineering Series. McGraw-Hill Higher Education (2004).
130. Cai, W.: *FPGA Prototyping of A Watermarking Algorithm*. Master's thesis, Electrical Egnineering Technology, University of North Texas, Denton, TX, USA (2006).
131. Chang, H., Kundert, K.: Verification of Complex Analog and RF IC Designs. *Proceedings of the IEEE* **95**(3), 622–639 (2007). DOI 10.1109/JPROC.2006.889384.
132. Chaudhary, V., Francis, M., Huang, X., Mantooth, H.A.: Paragon – A Mixed-Signal Behavioral Modeling Environment. In: *Proceedings of the International Conference on Communications, Circuits and Systems and West Sino Expositions*, pp. 1315–1321 (2002). DOI 10.1109/ICCCAS.2002.1179024.
133. Chen, G.: *A Short Historical Survey of Functional Hardware Languages*. *ISRN Electronics* 2012 (271836), 11 pages (2012). DOI 10.5402/2012/271836. Accessed on 21st march 2014.

134. Chen, X., Wang, Y., Yang, H.: An Adaptive LU Factorization Algorithm for Parallel Circuit Simulation. In: *Proceedings of the 17th Asia and South Pacific Design Automation Conference (ASP-DAC)*, pp. 359–364 (2012). DOI 10.1109/ASPDAC.2012.6164974.
135. Christen, E., Bakalar, K.: VHDL-AMS – A Hardware Description Language for Analog and Mixed-Signal Applications. *IEEE Transactions on Circuits and Systems – II: Analog and Digital Signal Processing* **46**(10), 1263–1272 (1999).
136. Cooper, R.S.: *The Designer's Guide to Analog & Mixed-Signal Modeling: Illustrated with VHDL-AMS and MAST*. http://www.openmast.com (2004). Accessed on 16th March 2014.
137. da Costa, H.J.B., de Assis Brito Filho, F., de Araujo do Nascimento, P.I.: Memristor Behavioural Modeling and Simulations Using Verilog-AMS. In: *Proceedings of the IEEE Third Latin American Symposium on Circuits and Systems (LASCAS)*, pp. 1–4 (2012). DOI 10.1109/LASCAS.2012.6180334.
138. Cox III, F.L., Kuhn, W.B., Murray, J.P., Tynor, S.D.: Code-Level Modeling in XSPICE. In: *Proceedings of the IEEE International Symposium on Circuits and Systems*, vol. 2, pp. 871–874 (1992). DOI 10.1109/ISCAS.1992.230083.
139. Davis, A.T.: *Implicit Mixed-Mode Simulation of VLSI Circuits*. Ph.D. thesis, Department of Electrical Engineering, College of Engineering and Applied Science, University of Rochester, Rochester, New York (1991). Accessed on 18th March 2014.
140. Decaluwe, J.: *MyHDL Manual (2013)*. Accessed on 12th March 2014.
141. Depeyrot, G., Poullet, F., Dumas, B.: Verilog-A Compact Model Coding Whitepaper. In: *Proceedings of the Nanotech*, pp. 821–824. Anaheim, CA, USA (2010). Accessed on 29th March 2014.
142. Dunga, M.V.: *Nanoscale CMOS Modeling*. Ph.D. thesis, EECS Department, University of California, Berkeley (2008). Accessed on 19th March 2014.
143. Dutton, R.W., Strojwas, A.J.: Perspectives on Technology and Technology-Driven CAD. *IEEE Transactions on Computer-Aided Design of Integrated Circuits and Systems* **19**(12), 1544–1560 (2000). DOI 10.1109/43.898831.
144. Eidsness, C.: eispice. http://www.thedigitalmachine.net/eispice.html. Accessed on 26th March 2014.
145. Enz, C., Krummenacher, F., Vittoz, E.: An Analytical MOS Transistor Model Valid in All Regions of Operation and Dedicated to Low-voltage and Low-Current Applications. *Journal on Analog Integrated Circuits and Signal Processsing* **8**, 83–114 (1995).
146. Eshraghian, K., Kavehei, O., Cho, K.R., Chappell, J.M., Iqbal, A., Al-Sarawi, S.F., Abbott, D.: Memristive Device Fundamentals and Modeling: Applications to Circuits and Systems Simulation. *Proceedings of the IEEE* **100**(6), 1991–2007 (2012). DOI 10.1109/JPROC.2012.2188770.
147. Fitzpatrick, D.: Analog Design and Simulation Using OrCAD Capture and PSpice. Newnes (2011).
148. Foty, D.: Mosfet Modeling for Circuit Simulation. *IEEE Circuits and Devices Magazine* **14**(4), 26–31 (1998). DOI 10.1109/101.708477.
149. Foty, D., Foty, D.P.: MOSFET Modeling with SPICE: Principles and Practice. Prentice Hall series in innovative technology. Prentice Hall PTR (1997).
150. Frey, P., O'Riordan, D.: Verilog-AMS: Mixed-Signal Simulation and Cross Domain Connect Modules. In: *IEEE/ACM International Workshop on Behavioral Modeling and Simulation*, pp. 103–108 (2000).
151. Garitselov, O.: *Metamodeling-Based Fast Optimization Of Nanoscale AMS-SoCs*. Ph.D. thesis, Computer Science and Engineering, University Of North Texas, Denton, 76203, TX, USA (2012).
152. Garitselov, O., Mohanty, S., Kougianos, E.: A Comparative Study of Metamodels for Fast and Accurate Simulation of Nano-CMOS Circuits. *IEEE Transactions on Semiconductor Manufacturing* **25**(1), 26–36 (2012). DOI 10.1109/TSM.2011.2173957.
153. Gautam, M.: *Exploring Memristor Based Analog Design in Simscape*. Master's thesis, Computer Science and Egnineering, University of North Texas, Denton, TX, USA (2006).
154. Gerfers, F., Ortmanns, M.: *Continuous-Time Sigma-Delta A/D Conversion: Fundamentals, Performance Limits and Robust Implementations*, 1 edn. Springer (2005).
155. Ghai, D.: *Variability Aware Low-Power Techniques for Nanoscale Mixed-Signal Circuits*. Ph.D. thesis, Computer Science and Engineering, University of North Texas, Denton, TX 76203, USA. (2009).
156. Gielen, G., Wambacq, P., Sansen, W.M.: Symbolic Analysis Methods and Applications for Analog Circuits: A Tutorial Overview. *Proceedings of the IEEE* **82**(2), 287–304 (1994). DOI 10.1109/5.265355.
157. Gopinath, A.: What's New in SPICE. *Electronics For You*, pp. 107–109 (2013). Accessed on 25th March 2014.
158. de Graaf, A.: SystemC-AMS Analog & Mixed-Signal System Design. http://ens.ewi.tudelft.nl/Education/courses/et4351/SystemC-AMS-2011v2.pdf (2011). Accessed on 18th March 2014.
159. Grout, I.A., Keane, K.: A MATLAB to VHDL Conversion Toolbox for Digital Control. In: *Proceedings of the IFAC Symposium on Computer Aided Control Systems Design* (2000).
160. Gu, B., Gullapalli, K.K., Hamm, S., Mulvaney, B.: Implementing Nonlinear Oscillator Macromodels Using Verilog-AMS for Accurate Prediction of Injection Locking Behaviors of Oscillators. In: *Proceedings of the IEEE International Behavioral Modeling and Simulation Workshop*, pp. 43–47 (2005).
161. Guan, X.H., Zhang, M.M., Zheng, Y.: MATLAB Simulation in Signals & Systems Using MATLAB at Different Levels. In: *Proceedings of the First International Workshop on Education Technology and Computer Science*, pp. 952–955 (2009). DOI 10.1109/ETCS.2009.476.
162. Guihal, D., Andrieux, L., Esteve, D., Cazarre, A.: VHDL-AMS Model Creation. In: *Proceedings of the International Conference Mixed Design of Integrated Circuits and System*, pp. 549–554 (2006). DOI 10.1109/MIXDES.2006.1706640.

163. Gupta, S., Krogh, B., Rutenbar, R.A.: Towards Formal Verification of Analog Designs. In: *Proceedings of the IEEE/ACM International Conference on Computer Aided Design*, pp. 210–217 (2004). DOI 10.1109/ICCAD.2004.1382573.
164. Han, L., Zhao, X., Feng, Z.: TinySPICE: A Parallel SPICE Simulator on GPU for Massively Repeated Small Circuit Simulations. In: *Proceedings of the Design Automation Conference (DAC)*, pp. 1–8 (2013).
165. Harasymiv, I., Dietrich, M., Knochel, U.: Fast Mixed-Mode PLL Simulation Using Behavioral Baseband Models Of Voltage-Controlled Oscillators and Frequency Dividers. In: *Proceedings of the XIth International Workshop on Symbolic and Numerical Methods, Modeling and Applications to Circuit Design (SM2ACD)*, pp. 1–6 (2010). DOI 10.1109/SM2ACD.2010.5672294.
166. Hewlett, J.D., Wilamowski, B.M.: SPICE as a Fast and Stable Tool for Simulating a Wide Range of Dynamic Systems. *International Journal of Engineering Education* **27**(2), 217–224 (2011).
167. Iman, S., Joshi, S.: *The e-Hardware Verification Language*. Springer (2004).
168. Jahn, S., Margraf, M., Habchi, V., Jacob, R.: Quite Universal Circuit Simulator (Qucs). http://www.thedigitalmachine.net/reference/QUCS_White_Paper.pdf (2005). Accessed on 09th March 2014.
169. Johansson, H., Frisk, J., Drejfert, A.: pycircuit. http://docs.pycircuit.org/. Accessed on 26th March 2014.
170. Johns, D.A.: The SPICE Circuit Simulator. http://www.eecg.toronto.edu/~johns/spice/spice_manual.html. Accessed on 05th March 2014.
171. Kapre, N.: *SPICE² – A Spatial Parallel Architecture for Accelerating the SPICE Circuit Simulator*. Ph.D. thesis, California Institute of Technology, Pasadena, California (2010). Last Accessed on 04th March 2014.
172. Kapre, N., DeHon, A.: Accelerating SPICE Model-Evaluation Using FPGAs. In: *Proceedings of the 17th IEEE Symposium on Field Programmable Custom Computing Machines*, pp. 37–44 (2009). DOI 10.1109/FCCM.2009.14.
173. Kazmierski, T.: Phase-Locked Loop Frequency Multiplier. http://www.syssim.ecs.soton.ac.uk/vhdl-ams/examples/pll.htm (1997). Accessed on 15th March 2014.
174. Kazmierski, T.: VCO with Phase Integration. http://www.syssim.ecs.soton.ac.uk/vhdl-ams/examples/vco.htm (1997). Accessed on 15th March 2014.
175. Khan, A., Mohanty, S.P., Kougianos, E.: Statistical Process Variation Analysis of a Graphene FET based LC-VCO for WLAN Applications. In: *Proceedings of the 15th IEEE International Symposium on Quality Electronic Design*, pp. 569–574 (2014).
176. Kielkowski, R.M.: *Inside SPICE: Overcoming the Obstacles of Circuit Simulation. v. 1*. McGraw-Hill (1994).
177. Kougianos, E., Mohanty, S.P.: Impact of Gate-Oxide Tunneling on Mixed-Signal Design and Simulation of a Nano-CMOS VCO. *Microelectronics Journal* **40**(1), 95–103 (2009).
178. Kougianos, E., Mohanty, S.P., Pradhan, D.K.: Simulink Based Architecture Prototyping of Compressed Domain MPEG-4 Watermarking. In: *Proceedings of the 12th IEEE International Conference on Information Technology (ICIT)*, pp. 10–16 (2009).
179. Kundert, K.: *The Designer's Guide to SPICE and Spectre®*. The Designer's Guide Book Series. Springer (1995).
180. Kundert, K., Zinke, O.: *The Designer's Guide to Verilog-AMS*. The Designer's Guide Book Series. Springer (2004).
181. Lallement, C., Francois Pecheux, A.V., Pregaldiny, F.: *Transistor Level Modeling for Analog/RF IC Design*, chap. Compact Modeling of the MOSFET in VHDL-AMS. Springer (2006).
182. Lemaitre, L., Grabinski, W., Mcandrew, C.: Compact Device Modeling Using Verilog-AMS and ADMS. *Electron Technology – Internet Journal* **35**(3), 1–5 (2003).
183. Lemaitre, L., McAndrew, C.: An Open-source Software Tool for Compact Modeling Applications. *IEEE Circuits and Devices Magazine* **20**(2), 6–41 (2004). DOI 10.1109/MCD.2004.1276164.
184. Li, R., Zhou, R., Li, G., He, W., Zhang, X., Koo, T.: A Prototype of Model-Based Design Tool and Its Application in the Development Process of Electronic Control Unit. In: *Proceedings of the IEEE 35th Annual Computer Software and Applications Conference Workshops (COMPSACW)*, pp. 236–242 (2011). DOI 10.1109/COMPSACW.2011.50.
185. Liyi, X., Bin, L., Yizheng, Y., Guoyong, H., Jinjun, G., Peng, Z.: A Mixed-Signal Simulator for VHDL-AMS. In: *Proceedings of the Asia and South Pacific Design Automation Conference*, pp. 287–294 (2001).
186. Lu, D.: *Compact Models for Future Generation CMOS*. Ph.D. thesis, EECS Department, University of California, Berkeley (2011). Accessed on 19th March 2014.
187. Lu, J.: Analysis and Design Of 5GHz Phase Locked Loops. In: *Proceedings of the 7th International Conference on Solid-State and Integrated Circuits Technology*, pp. 1488–1491 (2004). DOI 10.1109/ICSICT.2004.1436886.
188. Ma, K., Van Leuken, R., Vidojkovic, M., Romme, J., Rampu, S., Pflug, H., Huang, L., Dolmans, G.: A Fast and Accurate SystemC-AMS Model for PLL. In: *Proceedings of the 18th International Conference Mixed Design of Integrated Circuits and Systems*, pp. 411–416 (2011).
189. Maas, S.: Historical Trends and Evolution of Circuit-Simulation Technology. In: *Proceedings of the IEEE MTT-S International Microwave Symposium*, pp. 968–971 (2010). DOI 10.1109/MWSYM.2010.5518025.
190. Maiti, T.K., Maiti, C.K.: *Nanowires*, chap. Technology CAD of Nanowire FinFETs. 978-953-7619-79-4. InTech (2010). DOI 10.5772/39522. Accessed on 19th March 2014.
191. Martin, G.: SystemC and the Future of Design Languages: Opportunities for Users and Research. In: *Proceedings of the 16th Symposium on Integrated Circuits and Systems*, pp. 61–62 (2003).
192. McDonald, C.B., Bryant, R.E.: CMOS Circuit Verification with Symbolic Switch-Level Timing Simulation. *IEEE Transactions on Computer-Aided Design of Integrated Circuits and Systems* **20**(3), 458–474 (2001). DOI 10.1109/43.913762.

193. Meric, I., Han, M.Y., Young, A.F., Ozyilmaz, B., Kim, P., Shepard, K.L.: Current Saturation in Zero-Bandgap, Top-Gated Graphene Field Effect Transistors. *Nature Nanotechnology* **3**, 654–659 (2008).
194. Miller, I.: Verilog-A and Verilog-AMS Provides a New Dimension in Modeling and Simulation. In: *Proceedings of the Third IEEE International Caracas Conference on Devices, Circuits and Systems*, pp. C49/1–C49/6 (2000).
195. Mohanty, S.P.: SPICE. http://www.cse.unt.edu/~smohanty/Teaching/2010Fall_AVS/AVS7_SPICE.pdf. Accessed on 27th March 2014.
196. Mohanty, S.P.: *Energy and Transient Power Minimization during Behavioral Synthesis*. Ph.D. thesis, Department of Computer Science and Engineering, University of South Florida, Tampa, USA (2003).
197. Mohanty, S.P.: ISWAR: An Imaging System with Watermarking and Attack Resilience. CoRR abs/1205.4489 (2012).
198. Mohanty, S.P.: Memristor: From Basics to Deployment. *IEEE Potentials*, pp. 34–39 (2013).
199. Mohanty, S.P., Bhargava, B.K.: Invisible Watermarking Based on Creation and Robust Insertion-Extraction of Image Adaptive Watermarks. *ACM Transactions on Multimedia Computing, Communications, and Applications* **5**(2), (2008).
200. Mohanty, S.P., Kougianos, E.: Impact of Gate Leakage on Mixed Signal Design and Simulation of Nano-CMOS Circuits. In: *Proceedings of the 13th NASA Symposium on VLSI Design*, vol. paper # 2.4, 6 pages (2007).
201. Mohanty, S.P., Kougianos, E.: Real-Time Perceptual Watermarking Architectures for Video Broadcasting. *Journal of Systems and Software* **84**(5), 724–738 (2011).
202. Mohanty, S.P., Kougianos, E.: *Models, Methods, and Tools for Complex Chip Design*, Selected Contributions from FDL 2012, chap. Polynomial Metamodel-Based Fast Optimization of Nanoscale PLL Components. 978-3-319-01417-3. Springer (2014).
203. Mohanty, S.P., Kougianos, E.: Polynomial Metamodel Based Fast Optimization of Nano-CMOS Oscillator Circuits. *Analog Integrated Circuits and Signal Processing* **79**(3), 437–453 (2014). DOI 10.1007/s10470-014-0284-2. URL http://dx.doi.org/10.1007/s10470-014-0284-2.
204. Mohanty, S.P., Ranganathan, N., Balakrishnan, K.: A Dual Voltage-Frequency VLSI Chip for Image Watermarking in DCT Domain. *IEEE Transactions on Circuits and Systems II* **53**(5), 394–398 (2006). DOI 10.1109/TCSII.2006.870216.
205. Mohanty, S.P., Ranganathan, N., Krishna, V.: Datapath Scheduling Using Dynamic Frequency Clocking. In: *Proceedings of the IEEE Computer Society Annual Symposium on VLSI*, pp. 65–70 (2002).
206. Moondanos, J.: SystemC Tutorial. http://embedded.eecs.berkeley.edu/research/hsc/class/ee249/lectures/l10-SystemC.pdf. Accessed on 14th March 2014.
207. Moore, H.: *MATLAB® for Engineers. Always Learning*. Pearson (2013).
208. Mysore, O.: *Compact Modeling of Circuits and Devices in Verilog-A*. Master's thesis, Department of Electrical Engineering and Computer Science, Massachusetts Institute of Technology, MA, USA (2012).
209. Nagel, L.W.: SPICE2: A Computer Program to Simulate Semiconductor Circuits. Tech. Rep. ERL-M520, Electronics Research Laboratory, College of Engineering, University of California, Berkeley, CA 94720, USA (1975). Accessed on 07th March 2014.
210. Nagel, L.W.: The Life of SPICE. http://bear.ces.cwru.edu/eecs_cad/cad_spice_history_nagel.html (1996). Accessed on 25th March 2014.
211. Nagel, L.W.: Is It Time for SPICE4? In: *Numerical Aspects of Circuit Device Modeling Workshop* (2004). Accessed on 05th March 2014.
212. Nagel, L.W., Pederson, D.O.: Simulation Program with Integrated Circuit Emphasis (SPICE). Tech. Rep. ERL-M382, Electronics Research Laboratory, College of Engineering, University of California, Berkeley, CA 94720, USA (1973). Accessed on 05th March 2014.
213. Nance, R.E.: A Histroy of Discrete Event Simulation Programming Languages. Tech. Rep. TR 93-21, Department of Computer Science, Virginia Polytechnic Institute and State University, Blacksburg, Virginia 24061, USA (1993). Accessed on 03rd March 2014.
214. Narayanan, R., Abbasi, N., Zaki, M., Al Sammane, G., Tahar, S.: On the Simulation Performance of Contemporary AMS Hardware Description Languages. In: *Proceedings of the International Conference on Microelectronics*, pp. 361–364 (2008). DOI 10.1109/ICM.2008.5393509.
215. Neira, H.G.: Electronics Mixed-Mode Simulation. Tech. Rep. ARFSD-TR-93009, U.S. Army Armament Research, Development and Engineering Center, Fire Support Armaments Center, Picatinny Arsenal, New Jersey (1993). Accessed on 18th March 2014.
216. Nenzi, P., Vogt, H.: *Ngspice Users Manual: Version 26plus*. Tech. rep. (2014). Accessed on 08th March 2014.
217. Nikam, S.D.: *A Comparison of Software Engines for Simulation of Closed-Loop Control Systems*. Master's thesis, Department of Electrical and Computer Engineering, New Jersey Institute of Technology, NJ, USA (2010). Accessed on 23rd March 2014.
218. Ogrodzki, J.: *Circuit Simulation Methods and Algorithms. Electronic Engineering Systems*. Taylor & Francis (1994).
219. Okobiah, O.: *Exploring Process-Variation Tolerant Design of Nanoscale Sense Amplifier Circuits*. Master's thesis, Department of Computer Science and Engineering, University of North Texas, Denton, TX (2010).
220. Okobiah, O.: *Geostatistical Inspired Metamodeling and Optimization of Nanoscale Analog Circuits*. Ph.D. thesis, Computer Science and Engineering, University Of North Texas, Denton, 76203, TX, USA (2014).
221. Okobiah, O., Mohanty, S.P., Kougianos, E.: Geostatistical-Inspired Fast Layout Optimization of a Nano-CMOS Thermal Sensor. *IET Circuits, Devices Systems* **7**(5), 253–262 (2013). DOI 10.1049/iet-cds.2012.0358.

222. Okobiah, O., Mohanty, S.P., Kougianos, E., Poolakkaparambil, M.: Towards Robust Nano-CMOS Sense Amplifier Design: A Dual-Threshold versus Dual-Oxide Perspective. In: *Proceedings of the 21st ACM Great Lakes Symposium on VLSI*, pp. 145–150 (2011).
223. Open Verilog International: *Verilog-A Language Reference Manual : Analog Extensions to Verilog HDL* (1996). Accessed on 21st Mar 2014.
224. Ostroumov, S., Tsiopoulos, L., Sere, K., Plosila, J.: Generation of Structural VHDL Code with Library Components from Formal Event-B Models. In: *Proceedings of the Euromicro Conference on Digital System Design*, pp. 111–118 (2013). DOI 10.1109/DSD.2013.20.
225. Padmaraju, N.: Analog and Mixed Signal Modeling Approaches. http://www.design-reuse.com/articles/22773/analog-mixed-signal-modeling.html. Accessed on 02nd April 2014.
226. Pavan, S.: Systematic Design Centering of Continuous Time Oversampling Converters. *IEEE Transactions on Circuits and Systems II: Express Briefs* **57**(3), 158–162 (2010). DOI 10.1109/TCSII.2010.2041814.
227. Pêcheux, F., Lallement, C., Vachoux, A.: VHDL-AMS and Verilog-AMS as Alternative Hardware Description Languages for Efficient Modeling of Multidiscipline Systems. *IEEE Transactions on Computer-Aided Design of Circuits and Systems* **24**(2), 204–225 (2005).
228. Pershin, Y.V., Di Ventra, M.: Practical Approach to Programmable Analog Circuits with Memristors. *IEEE Transactions on Circuits and Systems I* **57**(8), 1857–1864 (2010).
229. Popescu, G., Goldgeisser, L.: Mixed Signal Aspects Of Behavioral Modeling and Simulation. In: *Proceedings of the IEEE International Symposium on Circuits and Systems*, pp. V-628–V-631 (2004). DOI 10.1109/ISCAS.2004.1329886.
230. Prodromakis, T., Peh, B.P., Papavassiliou, C., Toumazou, C.: A Versatile Memristor Model with Nonlinear Dopant Kinetics. *IEEE Transactions on Electron Devices* **58**(9), 3099–3105 (2011).
231. Quarles, T., Newton, A.R., Pederson, D.O., Sangiovanni-Vincentelli, A.: *SPICE3 Version 3f3 Users Manual.* http://www.physics.ucdavis.edu/Classes/Physics116/SPICE_cp.pdf (1993). Accessed on 11th March 2014.
232. Quarles, T.L.: Analysis of Performance and Convergence Issues for Circuit Simulation. Tech. Rep. UCB/ERL M89/42, Electronics Research Laboratory, College of Engineering (1989). Accessed on 04th March 2014.
233. Radwan, A.G., Zidan, M.A., Salama, K.N.: On the Mathematical Modeling of Memristors. In: *Proceedings of the International Conference on Microelectronics*, pp. 284–287 (2010). DOI 10.1109/ICM.2010.5696139.
234. Rahman, A.B.A.: *Modelling of Mixed Physical-Domain System.* Master's thesis, Faculty of Engineering, Science and Mathematics, School of Electronics and Computer Science, UK (2010). Accessed on 07th April 2014.
235. Ramon, M.E., Parrish, K.N., Chowdhury, S.F., Magnuson, C.W., Movva, H.C.P., Ruoff, R.S., Banerjee, S.K., Akinwande, D.: Three-Gigahertz Graphene Frequency Doubler on Quartz Operating Beyond the Transit Frequency. *IEEE Transactions on Nanotechnology* **11**(5), 877–883 (2012). DOI 10.1109/TNANO.2012.2203826.
236. Rewieński, M.: A Perspective on Fast-SPICE Simulation Technology. In: P. Li, L.M. Silveira, P. Feldmann (eds.) *Simulation and Verification of Electronic and Biological Systems*, pp. 23–42. Springer Netherlands (2011). DOI 10.1007/978-94-007-0149-6_2. URL http://dx.doi.org/10.1007/978-94-007-0149-6_2.
237. Rich, D.I.: The Evolution of SystemVverilog. *IEEE Design Test of Computers* **20**(4), 82–84 (2003). DOI 10.1109/MDT.2003.1214355.
238. Rose, G.S., Rajendran, J., Manem, H., Karri, R., Pino, R.E.: Leveraging Memristive Systems in the Construction of Digital Logic Circuits. *Proceedings of the IEEE* **100**(6), 2033–2049 (2012).
239. Rowen, C.: *Engineering the Complex SOC: Fast, Flexible Design with Configurable Processors.* Pearson Education (2008).
240. Rushton, A.: *VHDL for Logic Synthesis.* Wiley (2011).
241. Sagesaka, H., Irii, H., Asai, H.: SPADE : Analog/Digital Mixed Signal Simulator With Analog Hardware Description Language. In: *Proceedings of the IEEE International Conference on Electronics, Circuits and Systems*, pp. 517–520 (1998). DOI 10.1109/ICECS.1998.813375.
242. Sah, M.P., Kim, H., Chua, L.O.: Brains Are Made of Memristors. *IEEE Circuits and Systems Magazine* **14**(1), 12–36 (2014). DOI 10.1109/MCAS.2013.2296414.
243. Sarivisetti, G.: *Design and Optimization of Components in a 45 nm CMOS Phase Locked Loop.* Master's thesis, Computer Science and Egineering, University of North Texas, Denton, TX, USA (2006).
244. Schneider, T., Mades, J., Windisch, A., Glesner, M., Monjau, D., Ecker, W.: A System-Level Simulation Environment for System-on-Chip Design. In: *Proceedings of the 13th Annual IEEE International ASIC/SOC Conference*, pp. 58–62 (2000).
245. Schreier, R.: Delta Sigma Toolbox. http://www.mathworks.com/matlabcentral/fileexchange/19-delta-sigma-toolbox (2011). Code covered by the BSD License, Accessed on 01st March 2014.
246. Schreier, R., Temes, G.C.: *Understanding Delta-Sigma Data Converters*, 1st ed. Wiley-IEEE Press (2004).
247. da Silva, A.C.R., Grout, I., Ryan, J., O'Shea, T.: Generating VHDL-AMS Models of Digital-to-Analogue Converters From MATLAB/SIMULINK. In: *Proceedings of the International Conference on Thermal, Mechanical and Multi-Physics Simulation Experiments in Microelectronics and Micro-Systems*, pp. 1–7 (2007). DOI 10.1109/ESIME.2007.360029.
248. Smith, M.J.S.: *Application Specific Integrated Circuits.* VLSI Systems Series. Addison-Wesley (1997). URL http://books.google.com/books?id=3hxTAAAAMAAJ.

249. Steer, M.B.: *SPICE: Users Guide and Reference.* http://www.freeda.org/doc/SPICE/spice.pdf (2007). Accessed on 11th March 2014.
250. Strukov, D.B., Snider, G.S., Stewart, D.R., Williams, R.S.: The Missing Memristor Found. *Nature* **453**(7191), 80–83 (2008).
251. Strukov, D.B., Williams, R.S.: Exponential Ionic Drift: Fast Switching and Low Volatility of Thin-Film Memristors. *Applied Physics A: Materials Science & Processing* **94**(3), 515–519 (2009).
252. Suzuki, K., Nishio, A., Kamo, A., Watanabe, T., Asai, H.: An Application of Verilog-A to Meodeling of Back Propagation Algorithm in Neural Networks. In: *Proceedings of the 43rd IEEE Midwest Symposium on Circuits and Systems*, pp. 1336–1339 (2000).
253. Tala, D.K.: SystemC Tutorial. http://www.asic-world.com/systemc/tutorial.html (2014). Accessed on 13th March 2014.
254. Tala, D.K.: SystemVerilog Tutorial. http://www.asic-world.com/systemverilog/tutorial.html (2014). Accessed on 13th March 2014.
255. Tan, X.D., Shi, C.J.R.: Hierarchical Symbolic Analysis of Analog Integrated Circuits via Determinant Decision Diagrams. *IEEE Transactions On Computer Aided Design of Integrated Circuits and Systems* **19**(4), 401–412 (2006). DOI 10.1109/43.838990. URL http://dx.doi.org/10.1109/43.838990.
256. Trofimov, M., Mosin, S.: The Realization of Algorithmic Description on VHDL-AMS. In: *Proceedings of the International Conference on Modern Problems of Radio Engineering, Telecommunications and Computer Science*, pp. 350–352 (2004).
257. Troyanovsky, B., O'Halloran, P., Mierzwinski, M.: *Transistor Level Modeling for Analog/RF IC Design*, chap. Compact Modeling In Verilog-A. Springer (2006).
258. Tsuboi, K., Okumura, N.: A Next-Generation Workflow for System-Level Design of Mixed-Signal Integrated Circuits. http://www.mathworks.com/tagteam/75539_92085v00_a-next-generation-workflow-for-system-level-design.pdf. Accessed on 28th Feb 2014.
259. Uehara, J.: Epson Toyocom Designs and Verifies Mixed-Signal Integrated Circuit in Two Months. http://www.mathworks.com/company/user_stories/Epson-Toyocom-Designs-and-Verifies-Mixed-Signal-Integrated-Circuit-in-Two-Months.html. Accessed on 28th February 2014.
260. Umoh, I.J., Kazmierski, T.J.: VHDL-AMS Model of a Dual Gate Graphene FET, In: *Proceedings of the Forum on Specification and Design Languages*, pp. 1–5 (2011).
261. Vachoux, A., Bergé, J., Levia, O., Rouillard, J.: *Analog and Mixed-Signal Hardware Description Language. Current Issues in Electronic Modeling.* Kluwer Academic Publishers (1997).
262. Vachoux, A., Grimm, C., Einwich, K.: Towards Analog and Mixed-Signal SoC Design with SystemC-AMS. In: *Proceedings of the IEEE International Conference on Field-Programmable Technology*, pp. 97–102 (2004). DOI 10.1109/DELTA.2004.10008.
263. Vachoux, A., Grimm, C., Kakerow, R., Meise, C.: Embedded Mixed-Signal Systems: New Challenges for Modeling and Simulation. In: *Proceedings of the IEEE International Symposium on Circuits and Systems*, 4 pp. (2006). DOI 10.1109/ISCAS.2006.1692754.
264. Villar, J.I., Juan, J., Bellido, M.J., Viejo, J., Guerrero, D., Decaluwe, J.: Python as a Hardware Description Language: A Case Study. In: *Proceedings of the VII Southern Conference on Programmable Logic*, pp. 117–122 (2011). DOI 10.1109/SPL.2011.5782635.
265. Vishak, C., Sunil, G.: Phase Locked Loop : Mixed Signal Design Flow. http://www2.ece.ohio-state.edu/~bibyk/ece822/VishakandSunilNonThesis.pdf. Accessed on 15th March 2014.
266. Vladimirescu, A.: *THE SPICE BOOK.* J. Wiley (1994).
267. Vladimirescu, A.: SPICE-The Fourth Decade Analog and Mixed-signal Simulation - A State of the Art. In: *Proceedings International Semiconductor Conference*, pp. 39–44 (1999). DOI 10.1109/SMICND.1999.810383.
268. Walsh, A., Carley, R., Feely, O., Ascoli, A.: Memristor Circuit Investigation through a New Tutorial Toolbox. In: *Proceedings of the European Conference on Circuit Theory and Design (ECCTD)*, pp. 1–4 (2013). DOI 10.1109/ECCTD.2013.6662261.
269. Wang, L., Kazmierski, T.J.: VHDL-AMS Based Genetic Optimization of Mixed-Physical-Domain Systems in Automotive Applications. *Simulation* **85**(10), 661–670 (2009).
270. Warwick, C.: Everything You Always Wanted to Know about SPICE – But Were Afraid to Ask. *The EMC Journal* **82** (2009). Accessed on 04th March 2014.
271. Weiqiang, Z., Peimin, W.: PSpice System Simulation Application in Electronic Circuit Design. In: *Proceedings of the 32nd Chinese Control Conference*, pp. 8634–8636 (2013).
272. Wenqing, C., Donglin, S., Derong, C., Chaoxian, Z.: A Behavioral Simulation Method to Predict and Estimate EMi Characteristics of Electronic System. In: *Proceedings of the Asia-Pacific Symposium on Electromagnetic Compatibility*, pp. 742–745 (2008). DOI 10.1109/APEMC.2008.4559982.
273. Williams, C.D.H.: MacSpice. http://www.macspice.com/. Accessed on 26th March 2014.
274. Williams, R.: How We Found The Missing Memristor. *IEEE Spectrum* **45**(12), 28–35 (2008).
275. Wong, W., Gao, X., Wang, Y., Vishwanathan, S.: Overview of Mixed Signal Methodology for Digital Full-Chip Design/Verification. In: *Proceedings of the 7th International Conference on Solid-State and Integrated Circuits Technology*, pp. 1421–1424 (2004). DOI 10.1109/ICSICT.2004.1436852.

276. Woods, S., Casinovi, G.: Multiple-Level Logic Simulation Algorithm. *IEE Proceedings – Computers and Digital Techniques* **148**(3), 129–137 (2001). DOI 10.1049/ip-cdt:20010485.
277. Xia, F., Farmer, D.B., Lin, Y.M., Avouris, P.: Graphene Field-Effect Transistors with High On/Off Current Ratio and Large Transport Band Gap at Room Temperature. *Nano Letters* **10**(2), 715–718 (2010). DOI 10.1021/nl9039636. URL http://pubs.acs.org/doi/abs/10.1021/nl9039636.
278. Xiu, L.: *VLSI Circuit Design Methodology Demystified: A Conceptual Taxonomy*. Wiley (2007). URL http://books.google.com/books?id=69hbACmcZrAC.
279. Xu, T., Arriens, H.L., Van Leuken, R., de Graaf, A.: A Precise SystemC-AMS Model for Charge Pump Phase Lock Loop with Multiphase Outputs. In: *Proceedings of the IEEE 8th International Conference on ASIC*, pp. 50–53 (2009). DOI 10.1109/ASICON.2009.5351608.
280. Xue, D., Chen, Y.Q.: *System Simulation Techniques with MATLAB® and Simulink®*. Wiley (2013).
281. Yang, Y., Zhou, S.: A Circuit Simulation Experimental System Based On Re-development SPICE. In: *Proceedings of the 4th International Congress on Image and Signal Processing*, pp. 2531–2535 (2011). DOI 10.1109/CISP.2011.6100711.
282. Yerra, T.N.: Design of a Second-Order Delta-Sigma Modulator for Use in Biomedical Signal Acquisition. Master's thesis, Department of Electrical and Computer Engineering, Southern Illinois University Edwardsville Edwardsville, Illinois, USA (2009). Accessed on 28th February 2014.
283. Yoder, S.: Timers, Frequency Divider Examples. http://www3.nd.edu/~cse/2013sp/20221/handouts/L15%20Verilog%20Sequential%20Logic.pdf (2013). Accessed on 15th March 2014.
284. Zaidi, Y., Grimm, C., Haase, J.: Fast and Unified SystemC AMS - HDL Simulation. In: *Proceedings of the Forum on Specification Design Languages*, pp. 1–6 (2009).
285. Zaplatilek, K.: Memristor Modeling in MATLAB & Simulink. In: *Proceedings of the 5th European Conference on European Computing Conference*, pp. 62–67 (2011).
286. Zheng, G.: *Layout-Accurate Ultra-Fast System-Level Design Exploration through Verilog-AMS*. Ph.D. thesis, Computer Science and Engineering, University Of North Texas, Denton, 76203, TX, USA., Denton, TX 76207 (2013).
287. Zheng, G., Mohanty, S.P., Kougianos, E.: Design and Modeling of a Continuous-Time Delta-Sigma Modulator for Biopotential Signal Acquisition: Simulink vs Verilog-AMS Perspective. In: *Proceedings of the 3rd International Conference on Computing, Communication and Networking Technologies* (2012).
288. Zheng, G., Mohanty, S.P., Kougianos, E., Garitselov, O.: Verilog-AMS-PAM: Verilog-AMS integrated with Parasitic-Aware Metamodels for Ultra-Fast and Layout-Accurate Mixed-Signal Design Exploration. In: *Proceedings of the ACM Great Lakes Symposium on VLSI*, pp. 351–356 (2012).
289. Zheng, G., Mohanty, S.P., Kougianos, E., Okobiah, O.: Polynomial Metamodel Integrated Verilog-AMS for Memristor-Based Mixed-Signal System Design. In: *Proceedings of the IEEE 56th International Midwest Symposium on Circuits and Systems (MWSCAS)*, pp. 916–919 (2013).
290. Zidan, M.A., Omran, H., Radwan, A.G., Salama, K.N.: Memristor Model. http://sensors.kaust.edu.sa/tools/memristor-model (2011). Code covered by the BSD License, Accessed on 21st February 2014.
291. Zidan, M.A., Radwan, A.G., Salama, K.N.: Memristor Model. http://www.mathworks.com/matlabcentral/fileexchange/31530-memristor-model (2011). Code covered by the BSD License, Accessed on 21st Feb 2014.
292. Zorzi, M., Franzè, F., Speciale, N.: Construction of VHDL-AMS Simulator in MATLAB. In: *Proceedings of the International Workshop on Behavioral Modeling and Simulation*, pp. 113–117 (2003).
293. Zorzi, M., Franzè, F., Speciale, N., Masetti, G.: A Tool for the Integration of New VHDL-AMS Models in SPICE. In: *Proceedings of the 2004 International Symposium on Circuits and Systems*, pp. IV–637–IV–640 (2004).
294. Zwolinski, M.: *Digital System Design with SystemVerilog*. Prentice Hall Signal Integrity Library. Pearson Education (2009).

CHAPTER 10

Power-, Parasitic-, and Thermal-Aware AMS-SoC Design Methodologies

10.1 Introduction

In the chapter Design for eXcellence (DFX), a large set of issues that are encountered during the design of nanoelectronic-based circuits and systems was presented. Addressing one or more of these DFX issues is a challenge and demands more efforts from system, design, and layout engineers. The scope of this chapter is the detailed discussion of a selected subset from DFX that was discussed in a previous chapter. In particular, power, parasitics, and thermal issues have been discussed. In the case of power, static and dynamic power dissipation have been presented along with their effects and techniques to handle them at various levels of design abstraction. The parasitics such as resistors, capacitors, and inductors that arise from the active devices as well as passive interconnects can be the origin of many problems such as performance and power dissipation. The parasitics can be distributed and lumped; the distributed ones are much more difficult to handle. The origin and modeling of parasitics and methods to handle them during design flow have been presented in this chapter. Thereafter, the thermal or temperature issue of the circuits and systems are discussed. The thermal issues are due to high on-chip heating due to high power dissipation or may be high ambient temperature. The on-chip thermal issues may arise from the high power dissipation. Although the power dissipation issue and thermal issues have some commonalities, they are different issues and different approaches may be used to solve them. Therefore, this chapter discusses the thermal issues as a different issue from power dissipation.

10.2 Power Dissipation: Key Design Constraint

The research in low-power has been one of the primary focuses of VLSI design for the last several decades [27, 32, 40, 55, 119, 162, 175, 198]. However, it remains one of the major issues along with the additional emerging issues in the nanoelectronics era. The explosive growth of portable systems with serious computing capabilities has been a major driving factor of the low-power design. No doubt, with technology scaling and smaller feature size, the devices are operated at low supply voltages. Hence, it can be stated that the power dissipation of individual transistors per technology generation has declined. However, the numbers of transistors that are packed in the same die, i.e., integration density or packing density of the chips has increased. It is estimated that with each generation, feature size has scaled down by 0.7, integration density has increased by $2\times$, cost of computing has decreased by $2\times$, while die size has a minor increase of 14%. Thus, the power consumption of the mainstream chip has increased. In the recent years, low-power systems-on-a-chip (SoCs) have been in great demand over pure hardware mainstream microprocessors as they operate with limited battery life. Of course, in the process, the performance has been compromised. One might think why not have high-performance battery as a solution! First, the chemical technology that is used in a battery has its own limitations. The size of the battery needed to provide higher current would not be helpful in making portable small electronic systems. In general, the need for high performance and yet portable small

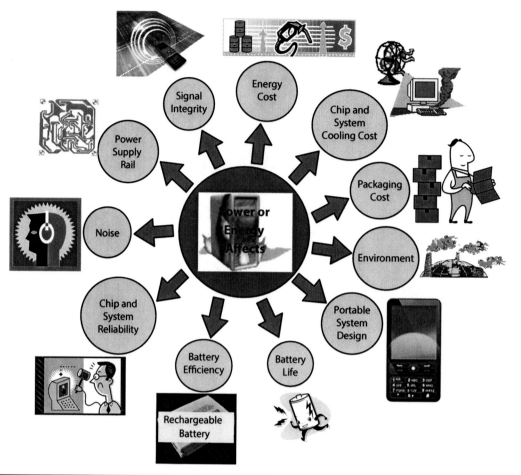

Figure 10.1 Effects of power dissipation [162].

electronic systems and many other factors, which are to be discussed in this section, has kept the power dissipation (including leakage dissipation) as the key constraint for the design engineers.

10.2.1 The Effects of High-Power Dissipation

The amount of power consumption or dissipation of a circuit or systems has profound effect on its various performances as well as cost. A brief overview of various aspects of circuit and system design and operation is depicted in Fig. 10.1 [107, 162, 192]. The high-power dissipation has direct impact on the energy cost of the circuit or system operation. High-power dissipation translates to higher cost of computing and operational budget. Both the power-critical and energy-critical systems can have heat dissipation needs. The chip and system cooling costs become high due to higher power or energy dissipation, which, in turn, increases heat dissipation requirements. In many cases, for example the circuits and system used in the gaming platforms, the cooling cost can easily outperform the cost of the chip itself. High-power dissipation circuits need costlier packaging for high-rate heat dissipation of the chip while at the same time protecting the circuits underneath that it is supposed to protect. All of these aspects are macroscopic aspects of the power dissipation and can then add to the environmental issues due to higher energy dissipation. As presented in the DFX chapter, appliances, electronics, and lighting in which electronic circuits and/or systems are integral parts that contribute to 34.6% of the total energy consumption.

At the microscopic level, the power dissipation issue provides many challenges for the circuit and system design engineers. For example, signal integrity, which is the quality of the electrical signal (i.e., simply speaking, how close it is to the expected value) can be affected by higher power dissipation. In a worst-case scenario, for example, the voltage values that represent logic "1" in a digital circuit may be not be high enough to be treated as logic "1." The signal integrity issues are challenging for various aspects of circuit and system design including on-chip interconnects, chip packaging, printed circuit board (PCB), as well as at higher levels the inter-system communications. The higher power dissipation that translates to higher currents may need bigger power supply rails to sustain that current flow. This may lead to various associated issues such as larger area, larger current drop, and higher Joule heating. Higher power dissipation does affect noise of the signal in the integrated circuit (IC) and may need a larger noise budget. The various fixed and proportional noise sources such as cross-talk, interference, offset, and supply noise are affected by the amount of power and current flow in the IC. The chip and system reliability is affected by higher power dissipation as well as higher temperatures originating from this high-power dissipation. In the current electronic system growth, portable electronic systems such as smart mobile phones, tablets, and notebooks are omnipresent and have profound impact on the society. These systems heavily depend on battery as their energy source. The other energy-critical systems include medical applications such as cardiac pacemaker and other implantable systems. The electronic systems that are used in military and sensor networks are also energy-critical systems. Battery life, which is essentially how long the battery will last before losing all its charge, is directly affected by the amount and rate of power (i.e., current) dissipation. The battery efficiency that refers to the energy conversion between electrical and chemical may be affected by the power dissipation, in particular, the fluctuation in the power dissipation profile. The overall design of the portable electronic system, including form factor, battery type, and processing capabilities, is affected by the power dissipation.

10.2.2 Power Dissipation Sources

The power or energy dissipation in nanoelectronic circuits and systems occurs when the hardware is executed, when the hardware is just supplied power but not doing useful work (called leakage), software is executed on the hardware, or information is stored in the hardware. However, in most of the cases, at the device level, various forms of current flow are responsible for this power or energy dissipation. In the prior chapter discussing design issues (DFX), various forms of power dissipation were presented. For a quick reference, the various sources of current flow in a nano-CMOS transistor are listed in Fig. 10.2 [162, 204]. Based on the process technology and feature size, the quantitative significance of these components differs. The modeling and origin of these current components have been discussed in the same chapter. The current or power components have been classified as static and dynamic types as presented in Fig. 10.2. The static power dissipation takes place when the active devices do not do any useful work but are connected to the supply voltage. On the other hand, the dynamic power consumption is due to useful work performed by the active devices in the circuits and systems. For a clear perspective, the dynamic and static power breakdown for selected bulk nano-CMOS technology is presented in Table 10.1 [88]. However, for below 70-nm CMOS technology, the leakage aspects have been aggravated due to emergence of gate-oxide leakage as a major leakage component [3, 96]. Moreover, the power dissipation components for high-κ/metal-gate MOSFET technology may not have gate-oxide leakage component. For the FinFET and multiple-gate FET technology, the standby leakage components may have insignificant values [61]. In terms of the power dissipation density, the trends are depicted in Fig. 10.3 [6]. It clearly distinguishes between the dynamic power dissipation and leakage power dissipation components. Observation has shown that beyond 45-nm technology node, the leakage power density component

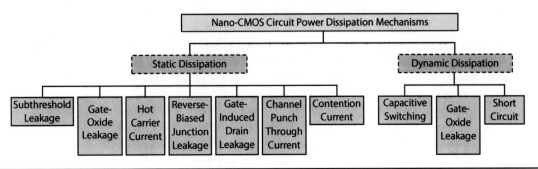

FIGURE 10.2 Various forms of power dissipation in a nano-CMOS circuit.

Technology Node	Dynamic Power Dissipation	Static Power Dissipation
350 nm	85%–90%	5%–10%
180 nm	80%–85%	10%–15%
90 nm	60%–80%	10%–20%
70 nm	30%–50%	30%–50%

TABLE 10.1 Power Dissipation Breakdown for Different Bulk Nano-CMOS Technology [88]

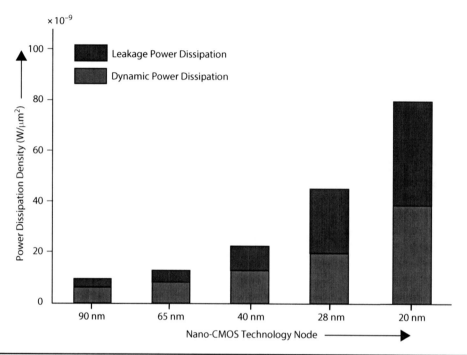

FIGURE 10.3 Power dissipation trend in a nano-CMOS circuit with dynamic versus leakage power components [6].

matches the dynamic power density component. Moreover, beyond the 28-nm technology node, the leakage power density exceeds the dynamic power density. Therefore, the low-power techniques need to take into account the target technology, circuit topology, architecture, as well as applications for effective solutions.

10.2.3 Power or Energy Dissipation Metrics

The low-power design is very difficult due to various aspects such as a large variety of custom circuits, application domains, technology limitations, diverse forms of current flow and their interrelations, and yet high-performance requirements. A simple differentiation can be made between applications as power critical and energy critical. The power-critical applications, for example a desktop PC, can operate at low power for requirements such as low-cooling cost and low-heat dissipation. However, low-energy consumption may not be a critical need as it is plugged to wall and has an ample supply of power. On the other hand, many applications, particularly portable electronics, implantable medical devices, military devices, wireless sensor networks (WSN; in particular, deployed at inaccessible locations), are energy-critical applications. In addition, the performance requirements may be imposed on any of the electronic applications. At the same time, maximum power consumption of a circuit or system and rate of power dissipation are important for various design needs [141, 156, 161, 162]. Thus, for the design engineers, different power or energy dissipation metrics are necessary to take care of them during the early phases of the design cycle.

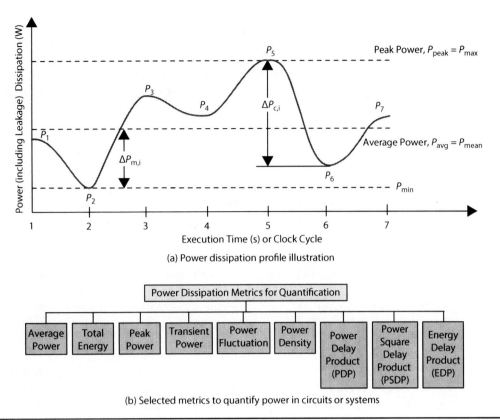

FIGURE 10.4 Different power dissipation metrics to quantify its profile [141, 162].

An illustrative power dissipation profile is depicted in Fig. 10.4(a) [141, 162]. It represents the overall power dissipation of a circuit or system accounting for all the forms such as leakage and capacitive switching power dissipation at different instance of time. For a digital IC, it can be for every clock cycle. Several forms of power, energy, power density, and implicit forms of power/energy and performance metrics have been presented in Fig. 10.4(b) [141, 156, 161, 162]. The average power dissipation P_{avg} is the mean or average power over an extended period of power profile. The total energy E is the total energy consumed by the circuit or system during its overall operation period. These metrics can be minimized to increase battery lifetime, to enhance noise margin, to reduce energy costs, and to increase system reliability. The peak power dissipation is the maximum power dissipation of the circuit and system over a period of time. The peak power metric needs to be minimized to maintain supply voltage levels, to increase reliability, to allow the use of smaller heat sinks, and to reduce the packaging cost. The power fluctuation and transient power represent the changes in the power dissipation rate from time to time or cycle to cycle. The quantification of power fluctuation is possible in two ways. One measure of power dissipation change $\Delta P_{m,i}$ is the power dissipation difference at any instance or clock cycle from the mean power dissipation. The other measure of power dissipation change $\Delta P_{c,i}$ is the power dissipation difference at any instance or clock cycle from the power dissipation of a previous time instance or clock cycle. The average or mean of either $\Delta P_{m,i}$ or $\Delta P_{c,i}$ quantifies the overall power fluctuation or transience of the circuit and system during its period of operation. The power fluctuation or transience metrics can be minimized to increase battery efficiency, to increase reliability, to reduce power supply noise, and to reduce crosstalk and electromagnetic noise. The power density is the power consumption per unit area and has impact on the reliability. The power or energy metrics with implicit time or performance measures include the following: power-delay-product (PDP), power-square-delay-product (PSDP), and energy-delay-product (EDP). These metrics can be minimized to achieve power or energy efficiency of the circuits and systems while optimizing delay (a measure of performance).

10.2.4 Energy/Power Dissipation: Application Perspectives

In this subsection, power dissipation for specific applications has been discussed to understand the nature of power dissipation in different operation scenarios for different electronic systems [30, 108, 127, 180]. The most widely used electronic system, the mobile phone, which is an excellent example of AMS-SoC, is discussed first. Thereafter, sensor network application and energy dissipation scenario are discussed.

10.2.4.1 Mobile Phone

A mobile phone spends maximum amount of time in waiting for calls. As compared to the waiting periods (T_{OFF}), the active talk times (T_{ON}) are typically shorter [92, 107, 162, 192]. During the waiting period, the mobile phone turns ON for short bursts for bookkeeping and base-station communication. In this situation, the ratio between the times $\left(\frac{T_{ON}}{T_{OFF}+T_{ON}}\right)$ is small and essentially the communications are in the form of short bursts. In the smart mobile phones, the active time may be spent in various digital signal processing (DSP) applications. In this scenario, passive leakage and active leakage power dissipation occurs along with a minimal dynamic power dissipation. During the talk period, both the active power dissipation and leakage power dissipation occur in the mobile phone.

A smart mobile phone consists of many heterogeneous components including analog baseband, DSP, and memory [4, 30, 88]. The total power dissipation of the smart mobile phone may vary depending on the operations that it is performing at any point of time. In addition, the proportion of the power dissipation components in the total power dissipation varies during the different operations. For specific examples, four different operation scenarios of the smart phone such as the following have been presented: (1) idle mode operation, (2) Global System for Mobile Communications (GSM) call operation, (3) video playback operation, and (4) WiFi email communication operation. For specific component specification of the smart mobile phone, the power dissipation for the above four operations is presented in Table 10.2 [30]. This provides a very good perspective to the reader. Of course, for other component specifications and different smart phones, this data will vary. The power dissipation in the idle mode operation is the lowest, whereas it is highest for the GSM call operation.

The component-wise breakdown of the power dissipation of the above smart mobile phone is presented in Fig. 10.5 [30, 108, 180]. Selected components of the smart mobile phone such as GSM module, microprocessor (a.k.a. CPU), random access memory (RAM), WiFi module, graphics module, liquid crystal display (LCD), audio, and other modules are analyzed. The proportion of the power dissipation in these components varies significantly in the four modes of the operation. In the ideal state of a smart mobile phone, power dissipation breakdown is shown in Fig. 10.5(a) [30]. In this idle mode operation, graphics, GSM, and LCD module consume a large portion of the total power dissipation. In the idle mode operation, if the applications are running in background and not suspended, then overall power dissipation that is largely static in nature can be high. The power dissipation during the GSM call operation is shown in Fig. 10.5(b) [30]. In this mode of operation, the GSM module consumes as much as 80% of total power dissipation. In other words, most of the power dissipation is in radiofrequency (RF)-ICs. The power dissipation distribution in the video playback operation is shown in Fig. 10.5(c) [30]. In this case, the graphics and CPU consume the maximum portion of the power. In other words, the digital ICs consume maximum portions of the total power. The power dissipation breakdown during email communication over WiFi of the smart

Different Operations of a Smart Mobile	Power Dissipation
Idle mode operation	269 mW
GSM call operation	1054 mW
Video playback operation	453 mW
WiFi email communication operation	432 mW

TABLE 10.2 Power Dissipation during Different Operations of a Smart Mobile Phone [30]

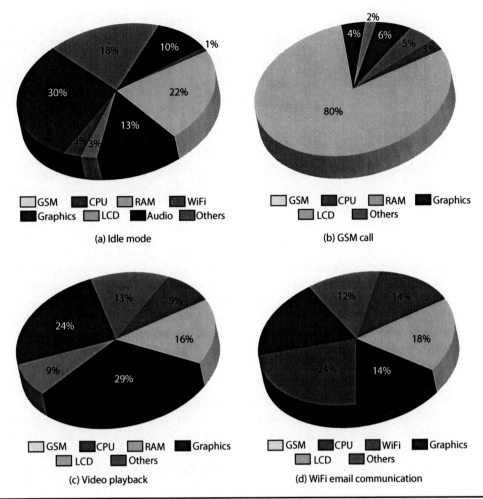

FIGURE 10.5 Power breakdown for a smart mobile phone for its different states of its operations [30].

mobile phone is shown in Fig. 10.5(d) [30]. During this operation, WiFi, graphics, and GSM modules consumed a large portion of the total power. In this case, analog, RF, and digital ICs consume power.

10.2.4.2 Wireless Sensor Network

The WSN is an important application in which energy criticality is of utmost importance [22, 78, 223]. The WSNs have wide-spread application in environmental monitoring, bio-warfare early warning, military, structure monitoring, medicine, and inventory management. The application domain is mostly energy critical and somewhat power critical due to lack of access and space constraints. The maintenance and/or replacement of a battery in the WSNs can be difficult as well as expensive. The design and architecture of the WSNs is influenced by various factors including operating conditions, energy/power dissipation, fault tolerance, scalability, and transmission media for maximum lifetime, high flexibility, and minimal maintenance [217]. In essence, a WSN that consists of data acquisition network, data distribution network, and management center is used to sense and monitor the physical world from a remote location. The WSN in which the communications take place through the Internet or satellite is depicted in Fig. 10.6 [13, 66, 184, 217].

In a WSN, a larger number of sensor nodes are present to gather information of real-world physical world. Each of the sensor nodes may consist of a combination of sensor or sensing module, power module, communication module, processor module, and memory [13, 22, 184, 217]. The sensing module is a device that monitors the environment and converts the sensed information to an analog signal. The analog-to-digital converter (ADC) converts the analog response from the sensor to digital signal. For on-site processing of sensor data, a DSP may be present. Such a DSP

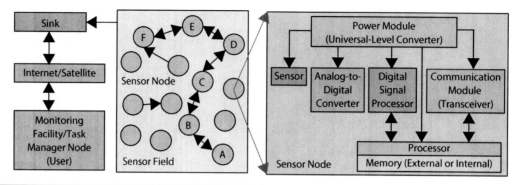

FIGURE 10.6 Deployment of the low-power sensors in a WSN.

can postprocess the acquired data on-site in a digital form before transmission. The digital data without postprocessing in DSP can also be transmitted. One possible implementation of a power supply module is the use of universal voltage-level converter (ULC) for energy-efficient design [66]. The sensor nodes can have active, sleep, and test modes for efficient operations [22]. A power module with ULC very well meets that need for energy efficiency through the three modes of sensor node operation.

10.2.5 Limits to Low-Power Design

For the last several decades, the energy-efficient and/or low-power design research has been undertaken. Such effort has definitely been a success as demonstrated by the emergence of reasonable computing-capable mobile systems such as smart mobile phones, tablets, and notebooks that are omnipresent in the society. However, low-power design research needs to go far ahead before we can dream of recharging such portable systems, least frequently, may be once in a week or month; not on a daily basis that is the current case.

In general, there are fundamental limits for the design engineer that prevents pushing the boundaries of the design space [24, 58, 226]. In an analog IC, there is a need for minimal power consumption for maintaining energy above the fundamental noise to maintain the required signal-to-noise ratio (SNR). The minimum power consumption of an analog IC at a specific temperature is dictated by the required frequency/bandwidth and SNR. The minimum power consumption of the analog circuit is also proportional to the ratio between the power supply voltage and the peak-to-peak signal amplitude. Therefore, the analog circuit should be designed for maximum voltage swing. The minimum power consumption that is required for analog circuits in order to handle rail-to-rail signal voltages is given by the following expression [24, 58, 226]:

$$P_{minimum,analog} = 8k_B T SNR f \tag{10.1}$$

where k_B is the Boltzmann constant, T is the temperature, and f is the operating frequency. Thus, based on the above discussions, it can be stated that the supply voltage in the analog ICs handles the frequency/bandwidth and frequency specifications; however, it may not directly affect its minimum power consumption.

The minimum power consumption in the digital ICs is expressed as follows [24, 58, 226]:

$$P_{minimum,digital} = N_{gates} E_{transition} f \tag{10.2}$$

where N_{gates} is the number of logic gates transitions when used by a basic operation in a clock cycle. Each of the transitions consumes an energy of $E_{transition}$. The signal bandwidth or frequency is f. If there are N_{bit} number of digital bits, then the following expression holds true [226]:

$$N_{gates} \approx N_{bit}^\alpha \approx (\log SNR)^\alpha \tag{10.3}$$

where is α is a real-number parameter. Thus, the power dissipation of the digital IC has weak dependency on the SNR. The power dissipation of the digital circuit can also be expressed as follows [24, 163]:

$$P_{\text{minimum,digital}} = N_{\text{gates}} C_{\text{sw,min}} V_{\text{DD}}^2 f \tag{10.4}$$

where $C_{\text{sw,min}}$ is the minimum gate capacitance that is switched. This is dependent on the feature size. At the same time, the supply voltage is dictated by the propagation delay time of the digital ICs that are expressed as follows [162]:

$$D_{\text{prop}} = \gamma_{\text{tech}} \frac{V_{\text{DD}}}{(V_{\text{DD}} - V_{\text{Th}})^\alpha} \tag{10.5}$$

where V_{Th} is the threshold voltage, α is a technology-dependent factor, and γ_{tech} is a constant.

In addition, a number of obstacles including technological limits and psychological limits that the design engineers need to encounter while incorporating the low-power features in the realization of ICs and systems [24, 226]. For example, a practical technological limit is that capacitors increase the power necessary to achieve a specific frequency/bandwidth. The power spent in the biasing circuitry and clock networks is practical limit of the circuits.

10.3 Different Energy or Power Reduction Techniques for AMS-SoC

This section will briefly outline various techniques for low-power or energy-efficient design of AMS-SoC. Detailed discussions of selected techniques will be presented in a subsequent section. There has been a need for low-power or energy-efficient design for various reasons. This has been a topic for serious attentions from the worldwide community in both industry and academia for last several decades. There have been many successful solutions that have been widely applied in industry. For example, multiple supply voltage or variable voltage for dynamic power reduction is well deployed in various real-life designs. Similarly, multiple threshold voltage and/or variable threshold voltage are widely accepted techniques for leakage reduction. A typical AMS-SoC may have analog hardware, digital hardware, mixed-signal hardware, RF hardware, various types of memory, and software. Separate research is needed and has been undertaken for each of these due to the diverse nature and specifications for each one of them. For example, for analog circuits many low-voltage low-power techniques have been explored [33, 205, 206, 219]. The low-power or energy-efficient digital design got tremendous attention as digital circuits have been the main workhorse for the maximum amount of signal processing that takes place in the AMS-SoCs [86, 162]. The memory circuit that has different forms in terms of cache, DRAM, or permanent storage, has lots of impact on the cost, form factors, and power budget [125, 211]. Similarly, low-power software has also been a serious research topic [138, 181].

10.3.1 AMS-SoC Energy or Power Reduction Techniques: An Overview

Many of the existing techniques that are used for the overall AMS-SoC or for the different components of it are listed in Fig. 10.7 [46, 65, 92, 107, 132, 141, 162, 173, 208, 245]. The existing literature may have many such techniques, but it is an extremely difficult task to find all the techniques and include them in the list. The existing techniques can be grouped into different categories in many different ways. However, this text classifies them into presilicon and postsilicon techniques [143]. The detailed discussions of selected techniques will be presented later in subsequent sections.

The presilicon techniques, a.k.a. structural techniques or static techniques, are developed during the design phase and are also incorporated in the circuit and system during the design phase [143]. The presilicon techniques once incorporated stay in the circuit and system forever and there is no scope to change or tune it later. Selected presilicon techniques include multiple voltage islands, multiple threshold devices, multioxide devices, capacitance minimization by custom design, and microarchitecture parallelism. In the multiple voltage island, the chip or system is partitioned into various voltage islands consisting of blocks or subsystems. Then, the different blocks and subsystems are provided different supply voltages for power/energy-efficient operations. This technique has been very well accepted and incorporated in many different industrial designs. In the multiple threshold

522 Chapter Ten

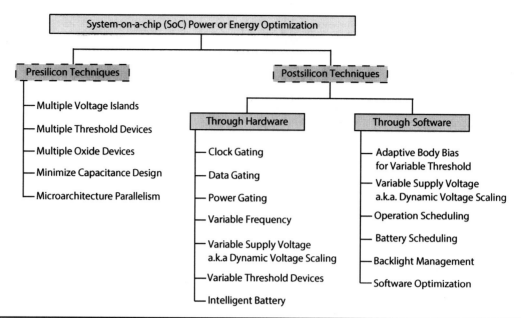

FIGURE 10.7 Different power or energy reduction techniques for AMS-SoCs.

devices technique, transistors of different threshold voltages are deployed for subthreshold leakage optimization along with performance tradeoffs. This technique has also seen widespread applications. In the multiple oxide devices technique, transistors of different oxide thicknesses are deployed for gate-oxide as well as subthreshold leakage optimization along with performance tradeoffs. Custom design including appropriate transistor sizing is used to reduce capacitance of the circuits that can, in turn, reduce dynamic power dissipation and enhance performance. The architecture parallelism that uses resources in parallel so that they can operate at slower clock to reduce power dissipation while maintaining throughput or overall performance has widespread usage.

The postsilicon techniques, a.k.a. dynamic techniques, are the techniques deployed after the circuit or system is fabricated [143]. However, the design decisions for these techniques are made during the design phase, appropriate hardware and software support are incorporated during the manufacturing phase, and the operation is tuned during runtime, using these design decisions. The postsilicon techniques can be hardware based and can be software-based implementations. Selected different postsilicon techniques include clock gating, data gating, power gating, variable supply voltage [a.k.a. dynamic voltage scaling (DVS)], variable threshold devices, and intelligent battery. The software-based postsilicon techniques include adaptive body bias (ABB) for variable threshold, variable supply voltage (a.k.a. DVS), operation scheduling, battery scheduling, backlight management, and software optimization.

The DVS or variable voltage scaling is the technique in which the supply voltage is varied or scaled from one time to another based on the load. Typically, it is accompanied by variable frequency to achieve the maximum possible power or energy reduction while maintaining performance level. The DVS can be realized in various different forms presented in Fig. 10.8 [141, 162, 183, 245]. A variable voltage processor may include a special instruction in order to control the power consumption of the processor. Based on the instruction in the operating system (OS) or the application program, the supply voltage and clock frequency are changed to ensure correct execution of the instructions.

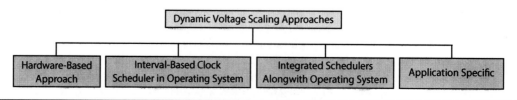

FIGURE 10.8 Four approaches for DVS [141, 162, 183].

In the hardware-based approach, there is now interaction of exchange between power dissipation and overall performance of the microprocessor or the corresponding system. In the interval-based clock scheduler in the OS, the load information is only used and clock is scheduled based on that. In the third alternative, an integrated schedule is used in the OS and all the OS statistics are at the disposal to take scheduling decisions with close interaction between hardware, OS, and application program. The fourth approach for DVS is the application-specific approach in which the complete knowledge of microprocessor load is used based on which the current hardware usages by the applications and future usages are predicted for better power and performance tradeoffs. Overall, dynamic/variable supply voltage operation is possible with hardware and software.

In the variable threshold device approach, the threshold voltage of transistors is changed using various mechanisms to control the threshold voltage while maintaining the performance [201, 245]. A widely used technique for changing threshold voltage is ABB [5, 133, 201, 230]. Both hardware- and software-based methods are available for deployment of ABB in various AMS-SoCs. ABB can be used at the chip level in which all the modules in the die have a single body voltage and ABB is applied to the whole chip. The ABB can also be used at the module level in which each module can have its own and different body voltage so that ABB is applied to each module. In general, ABB can be either forward body bias (FBB) or reverse body bias (RBB). In RBB, either voltage of the N-well is increased relative to power rail voltage (V_{DD}) or voltage of the substrate is lowered relative to ground (GND). Thus, RBB effectively increases the threshold voltage (V_{Th}) and reduces subthreshold leakage. On the other hand, opposite happens in FBB, i.e., FBB effectively decreases the threshold voltage (V_{Th}), which increases the operating frequency in active mode and leads to higher subthreshold leakage.

The use of intelligent battery for portable AMS-SoC is an attractive option for power savings [132, 154]. The intelligent battery uses an array of cells that are scheduled depending on the operational requirements to provide proper voltage and current output. Hardware integrated in the battery pack can perform such scheduling without the need for the OS, thus resulting in a much faster approach.

Better operation scheduling can be performed to reduce the power dissipation of the system [173]. The interaction of OS and application software can be used to predict future workloads accordingly to perform operation scheduling. The workload can be accordingly balanced to meet the deadlines for the performance while saving power. Software- or OS-based battery scheduling is a good power-saving method. Backlight management of the LCD display can also save lots of energy in applications such as mobile phones and tablet. Software in itself does not consume power or energy as it is a dependent entity of the system. The software uses hardware that in turn consumes power or energy. However, better software means less resource consumption and less power or energy dissipation of the system. Therefore, the software optimization can be an important way for energy/power optimization.

10.3.2 Analog Circuit Power Optimization: An Overview

Before moving on to specific details of low-power analog design, a quick overview of selected techniques is presented in this subsection. A selected set of power reduction techniques that can be used in analog IC design is presented in Fig. 10.9 [63, 65, 94, 111, 174, 245]. One type of ABB called FBB is used to control the threshold voltage in analog IC for the purpose of leakage reduction [174, 245]. This achieves variable threshold or dynamic threshold for the individual active devices. This

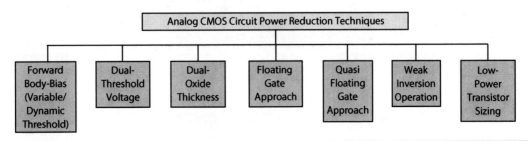

FIGURE 10.9 Different power or energy reduction techniques for analog circuits.

is essentially a postsilicon method. The static version of this is dual-threshold voltage option in which two types of devices are used, one with nominal threshold voltage and the other with higher threshold voltage for subthreshold leakage and performance tradeoffs. The dual-oxide thickness technique can be used for gate-oxide leakage and subthreshold leakage reductions and is presilicon or static technique. The floating-gate approaches that are used in EEPROMs and flash memories can be used for low-power design of analog circuits [94, 174]. The floating-gate essentially shifts the threshold voltage and helps in subthreshold leakage reduction. The quasi-floating-gate MOSFET are advanced versions of the floating-gate MOSFET in which initial trapped charge is not an issue and the area overhead is less. The weak-inversion approach is used to reduce power consumption in the analog components of CMOS image sensors [63]. Proper sizing of transistors can be performed during design optimization time with an power dissipation minimization objective [65, 111].

A fast and automatic design optimization flow that results in low-power analog as well as mixed-signal circuits is presented in Fig. 10.10 [60, 65, 69]. In this flow, as the first step, the initial schematic design is performed in a schematic capture tool. The schematic-level netlist is then

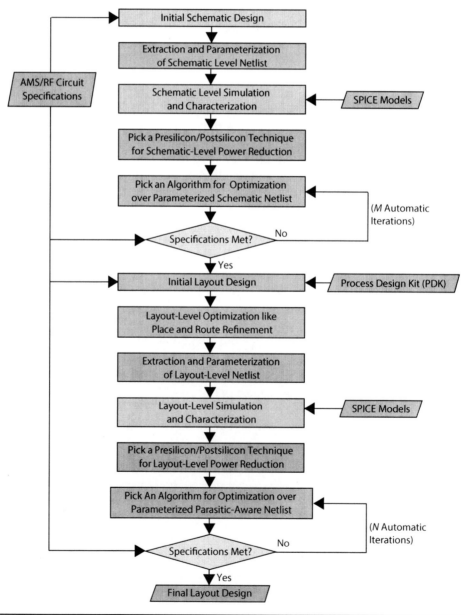

Figure 10.10 Single iteration automatic power-aware design flow.

extracted and parameterized for the design variables. These variables will then be used as tuning parameters during the automatic optimization. A suitable presilicon/postsilicon technique can be picked for power optimization based on the target power objective and applications. At this point, an algorithm can be picked to automatically perform optimization by performing design space exploration through an analog simulator. The schematic-level optimization ends once a suitable topology and optimized device sizes are obtained. At this point, layout design of the schematic design is performed and is called initial layout design. Place and route refinement such as layout-level optimization can be performed over the initial layout design. The layout-level netlist is extracted from the resulting layout design that has much complex elements as compared to the schematic-level netlist in the above stages. The layout-level netlist is parametrized for design variables that are to be tuned for layout-level optimization. The number of design variables may be different from that of the schematic-level optimization. A suitable presilicon/postsilicon technique can be picked for layout-level power optimization. This technique should be the same as the schematic-level presilicon/postsilicon technique as final design will have the same technique. An algorithm is now used for automatic design exploration through the tuning of design variables in conjunction with an analog simulator. The algorithm picked at layout-level can be the same as or different from the schematic level. However, it must be noted that the simulations at the layout level during algorithm iterations using an analog simulator are much slower than the schematic level. Thus, it may be a good idea to pick an algorithm that takes minimal iterations to converge. Of course, one advantage that already exists is that the schematic design is already power optimal due to optimization at that phase. When the algorithm convergence occurs, the final layout is performed using the final design variables that are obtained from the algorithm. As can be observed, the above design optimization flow needs only two manual layout steps, which are called one manual iteration. All the remaining steps are automatic steps.

10.3.3 Digital SoC Power or Energy Optimization Procedures: An Overview

As compared to the analog or mixed-signal circuits, the digital circuits or SoCs have more abstractions. This provides the opportunity for the divide-and-conquer approach to control the design of large circuits and SoCs. A typical procedure covering all levels of abstractions for a digital SoC is presented in Fig. 10.11 [88, 107, 137, 141]. It can be assumed that this design optimization flow is also valid for digital hardware portions of an AMS-SoC. The input for this design optimization flow is a target application for which a SoC architecture is desired. At the system-level, the SoC descriptions containing different hardware and software components are obtained through system-level design

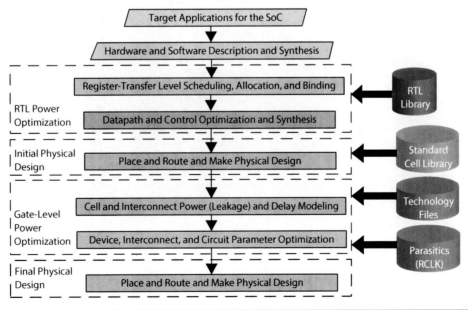

FIGURE 10.11 A typical procedure of power and energy optimization in a digital SoC.

and synthesis. At this stage, the digital hardware components are considered for power optimization through the subsequent phases of the proposed flow. At the register-transfer level (RTL), various tasks such as scheduling, allocation, and binding are performed. Various energy and power optimization tasks can be performed at the RTL [141, 162]. The RTL library needs to have appropriate low-power datapath components for such low-power or energy-efficient RTL optimization. For example, multiple voltage library and multiple threshold voltage library can be used. An initial physical design can be performed using standard cell library for a first-hand silicon realization of the hardware. The standard cell library contains physical design of various logic cells. Using technology files and parasitics (RCLK) information, the power and delay modeling is performed for the logic cells and interconnects. Logic-level optimization for low power is performed through tuning of the device, interconnect, and circuit parameters. At the final phase of the digital hardware design, final physical design is obtained through placement and routing. It may be noted that in the above low-power digital design flow, the ICs are described in terms of a digital HDL at various levels of design abstraction. However, the HDLs do not have explicit features to describe power connectivity, voltage levels, or threshold voltages for low-power design exploration at these abstractions [3, 40]. Thus, task of design engineers for low-power/energy design is still involved.

Various low-power and/or energy-efficient techniques covering different levels of abstractions for a digital SoC are presented in Fig. 10.12 [41, 124, 162, 171, 178, 185, 230, 241]. The basic elements or building blocks at each of the levels of the digital circuit abstractions are different. They present different levels of granularity for the design engineers at the different levels. At the higher level of abstractions, the optimization possibilities are higher and the granularity is bigger due to the use of large basic elements or building blocks. At the same time, design iteration time for low-power optimization takes minimum time. The accuracy of the results at the higher level is less than the lower levels of abstractions. The optimization possibilities are lower at the lower level

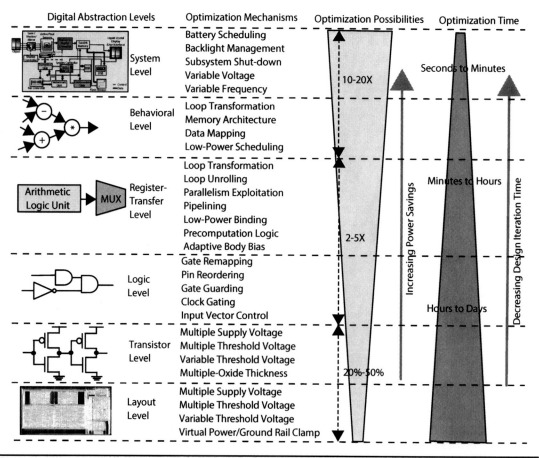

FIGURE 10.12 Power reduction at different abstractions of a digital SoC.

of circuit abstraction. The granularity of basic elements or building blocks is finer. However, the iterations of design optimization take a longer time. In the case of large digital circuits, the power optimization at the physical level can be even infeasible if an analog/SPICE simulator is used in the iteration loop.

10.4 Presilicon Power Reduction Techniques

The presilicon power/energy reduction techniques are the static or structural techniques that are decided during the design time and implemented at the design time, with no future change/tuning made by the designers or users after fabrication. This section briefly discusses some presilicon techniques. Later in this section, concrete and detailed discussions of selected analog and digital designs are presented.

10.4.1 Brief Discussion

A large number of presilicon, static, or structural techniques for power/energy dissipation reduction exist. Selected presilicon techniques are illustrated in Fig. 10.13 [65, 141, 162].

The widely used "multiple supply voltage" technique assigns higher supply voltages to the blocks or components in the critical path and reduces supply voltages to other blocks for power/energy and delay tradeoffs [3, 40, 141, 162]. Consider an application scenario: a microprocessor may operate at a 1 GHz frequency, but a universal serial bus (USB) block may just require some 100 MHz as per the USB protocol [3, 40]. Thus, the USB block can be operated at a lower frequency and correspondingly lower supply voltage. Thus, overall dynamic power and/or energy dissipation

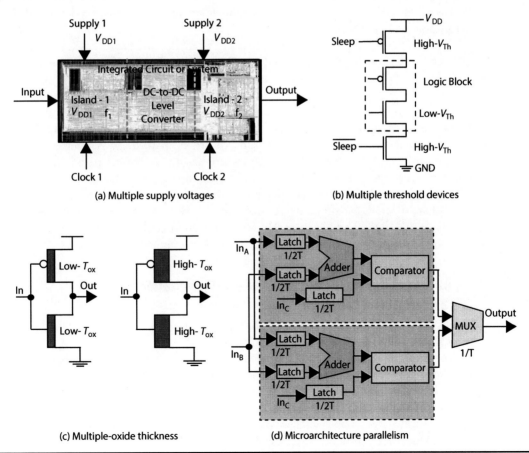

Figure 10.13 Brief concepts of selected presilicon (a.k.a. static or structural) power reduction techniques [65, 141, 162].

can be significantly reduced. As a specific example, "dual supply voltage"-based static design time solution for dynamic power reduction is shown in Fig. 10.13(a) [3, 40, 65, 141, 162]. In this particular case, the chip is partitioned into two different islands and each of which operate at different supply voltage and frequency [159]. The islands talk to each other through a large number of DC-to-DC voltage-level converters as the signal strengths representing logic "1" in the two islands are different. The use of DC-to-DC voltage-level converters is the area overhead of the multiple supply voltage techniques, but a large power reduction comes out of this technique making it an attractive option. Specific discussions on the level converters and this technique will be presented later in this section.

The "multiple threshold voltage"-based technique, which is a static design time solution for subthreshold leakage power reduction, is shown in Fig. 10.13(b) [3, 65, 141, 149, 162, 231]. It shows a specific example of the multiple threshold voltage-based technique called dual-threshold voltage-based technique. In this technique, the devices of two threshold voltages, i.e., low-V_{Th} and high-V_{Th} are used [73, 91, 197, 229]. The semiconductor houses may provide dual-threshold CMOS (DTCMOS) library with low-V_{Th} and high-V_{Th} for power and performance tradeoffs. For example, a designer can use low-V_{Th} library components for small delay, but high leakage in the critical path, or high-V_{Th} library components for low leakage, but high-delay in the off critical path. Gating can be effectively implemented in the context of this technique. When not in use, the logic block can be completely disconnected from the supply using high-V_{Th} devices. Thus, reducing the subthreshold leakage completely during standby operation when the transistors in the logic block are supposed to be OFF.

The gate-oxide leakage static solution using "multiple-oxide thickness" devices is shown in Fig. 10.13(c) [65, 141, 149, 162]. In these techniques, the devices with low-T_{ox} and high-T_{ox} are used for gate-oxide leakage and delay tradeoffs [110, 166]. For example, the transistors, logic gates, or module in the critical path can have devices with low-T_{ox} and rest all high-T_{ox}. As an optimization task, dual-oxide thickness assignment can be performed at the architecture level, logic level, and transistor level. At the end, everything is implemented at the transistor level and then layout level before going for chip fabrication.

Architecture-level parallelism to reduce power dissipation is presented in Fig. 10.13(d) [65, 141, 162]. In this technique, two identical datapaths are used that perform the same computation but at a reduced speed [59, 162]. For example, each of the datapaths operates at half of the original frequency, i.e., double the execution time while maintaining the original operation rate or throughput. The slowing down of the datapaths can be further operated at lower voltage due to lower operating frequency for multifold reduction of dynamic power. Let us consider a specific example of an architecture in which the timing constraint or critical path delay is 3 nanoseconds [3, 40]. If the parallelism is incorporated and the architecture has a new critical path delay of 1.7 nanoseconds, then there is a time slack of 1.3 nanoseconds. Thus, the supply voltage can be reduced to a value small enough to make the critical path delay of parallel architecture 3 nanoseconds again. The overall parallel architecture has the same performance but very low power/energy dissipation and a larger area than the original architecture. However, this architecture-level parallelism approach can have large area overhead as compared with the multiple supply voltage technique.

10.4.2 Dual-Threshold-Based Circuit-Level Optimization of a Universal Level Converter

In this subsection, the transistor-level power/energy optimization of a ULC a.k.a. universal voltage-level shifter (ULS) is considered using dual-V_{Th} presilicon technique [66, 146, 154, 165, 224]. The voltage-level converters are absolutely needed in the circuits and system using multiple supply voltage technique for dynamic power reduction. An ULC or ULS can additionally help to reduce the subthreshold leakage if used at appropriate places in the circuit, e.g., between the power rail and pull-up network of the static CMOS circuit [154]. A ULC has an input voltage signal called V_{in}, two control signals S_1 and S_0, two supply voltages V_{DDh} and V_{DDl}, and results in an output voltage signal V_{out}. It may be noted that for the multiple supply voltage-based designs, it is possible to use other types of voltage converters, for example, single supply voltage-based differential cascode voltage switch (DCVS) as shown in Fig. 10.24 [20, 141, 186]. In a ULC/ULS, the control signals decide the specific function that has to be performed. The four possible operations a ULC/ULS can perform for power management of the AMS-SoC are:

1. *Level-up conversion:* This operation converts a low-voltage signal to a high-voltage level signal.
2. *Level-down conversion:* This operation converts a high-voltage signal to a low-voltage level signal.
3. *Signal blocking:* This operation blocks the input signal from appearing at the other side. This operation can be used to shut-off the unused blocks of a circuit in the standby mode to reduce the standby leakage of these blocks.
4. *Signal passing:* This operation passes the input signal to the output side without doing any operation on the signal.

The ULC can be programmed for any of these four functionalities depending on the type of requirement for effective power and energy management. However, all the supporting operations may not be needed every time and one or a combination of two operations can be sufficient. For example, signal blocking and step-down conversion are needed for management dynamic power/energy dissipation of a AMS-SoC. Similarly, signal blocking and level-up conversion are needed to reduce short-circuit power and leakage power in an AMS-SoC. Energy- as well as area-efficient design of the ULC is necessary as ULC brings some overhead to the AMS-SoC in which it is used for power/energy management. A design flow for energy-efficient design of the ULC is presented in Algorithm 10.1 [66, 154].

A 32-transistor design of the ULC with nominal sizes for a 32-nm CMOS technology is shown in Fig. 10.14(a) [66, 154]. This ULC design is obtained by stitching subcircuits of level-up converter, level-

---

ALGORITHM 10.1 Power and Delay Optimal Universal Voltage-Level Shifter (ULC) Design Methodology [66, 154]

1: Perform the design and simulation of the level-up conversion subcircuit of the ULC.
2: Perform the design and simulation of the level-down conversion subcircuit of the ULC.
3: Perform the design and simulation of the pass and block subcircuit of the ULC.
4: Stitch the partial subcircuits obtained in the above steps to obtain the design of the complete ULC circuit.
5: Eliminate the gate-oxide leakage power of the ULC circuit by using high-κ/metal-gate nano-CMOS technology.
6: Perform the functional simulation for the verification of the ULC for different functionality and programmability.
7: Perform the reduce transistor design of the ULC by eliminating any redundancy in the circuit.
8: Perform the functional simulation for the verification of the reduced-transistor ULC for different functionality and programmability.
9: Obtain the netlist of the ULC circuit and parameterize the netlist for the design variable, e.g. transistor width.
10: Perform ranking of the individual transistors of the ULC circuit in the order of total power (including subthreshold leakage) dissipation.
11: Perform identification of the power-hungry transistors in the ULC circuit which collectively dissipate a designer-defined percentage of total power of the ULC circuit.
12: Assign high-V_{Th} to the power-hungry transistors of the ULC circuit to reduce its subthreshold leakage power dissipation.
13: Extract and parameterize the transistor-level netlist of this ULC circuit.
14: Use an optimization algorithm go through the netlist to determine the optimal width for all the transistors of the ULC circuit.
15: Assign the new width obtained from the optimization algorithm to all the transistors of the ULC circuit.
16: Perform the power, parametric, and load characterization of the final ULC circuit.

down converter, and pass/block. By eliminating the redundant transistors, a reduced 28-transistor ULC circuit is constructed that is shown in Fig. 10.14(b) [66, 154]. A further reduced 22-transistor ULC circuit is shown in Fig. 10.14(c) [66, 154]. This ULC circuit has two output nodes instead of one as in the case of 28-transistor ULC design. The choice of the use of 28-transistor versus 22-transistor depends on the target applications. The single-output 28-transistor ULC has more flexible programmability; however, it has more energy and area overhead. It is more suitable for field-programmable gate array

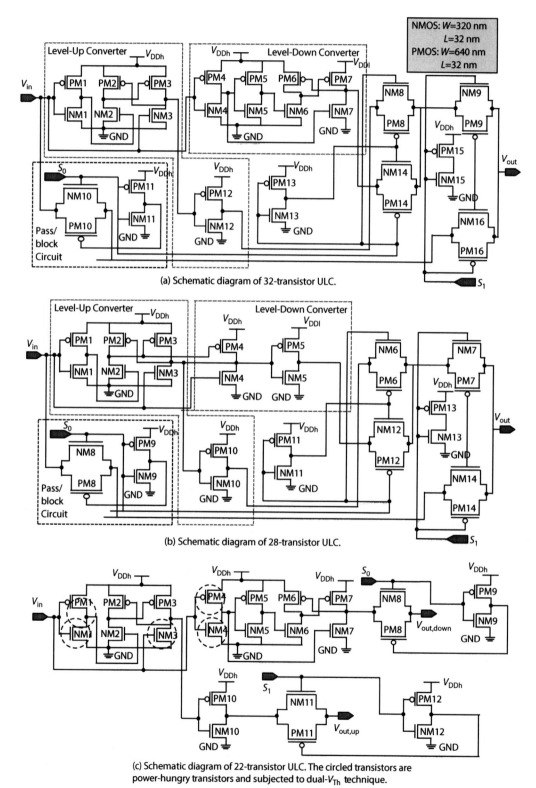

Figure 10.14 Dual-V_{Th} based optimization of ULC circuit [66, 154].

(FPGA) environments. On the other hand, the two-output 22-transistor has less flexible programmability; however, it has lesser energy and area overhead. It is more suitable for application-specific ICs (ASICs). The functional simulation of the 22-transistor ULC is shown in Fig. 10.15 [66, 154]. It shows that the input voltage signal is either up-converted or down-converted based on the control signals.

Any of the above circuit realizations of the ULC can be considered for energy optimization using the dual-V_{Th} technique. The 22-transistor-based ULC is considered for the optimization. For the purpose of energy/power optimization, the power-hungry transistors of the ULC circuit are identified and are assigned higher V_{Th} values. Power-hungry NMOS are assigned 20% higher V_{Th} and power-hungry PMOS are assigned 50% higher V_{Th} as compared with the nominal values specified for the technology node [145, 154]. The power-hungry transistors are marked as dashed-circles in Fig. 10.14 [66, 154]. This dual-V_{Th} assignment reduces the power/energy consumption considerably; however, it is observed that the propagation delay of the ULC circuit increased. Thus, using it could become a performance overhead for the host AMS-SoC. Hence, the geometry of transistors is also explored for optimization, in which the widths of all the transistors in the level-up and level-down converters subcircuits are considered for sizing. In general, the sizing of device parameters and determination of appropriate value of V_{Th} can be considered during optimization [66, 70, 154]. However, for simplicity, for experimentally selected value of V_{Th}, the sizing of W can be considered while keeping length at the technology-defined nominal value. An algorithm can be used to automatically perform this over the ULC netlist parameterized for W. Specific examples of such algorithms, such as conjugate gradient and simulated annealing, are presented in future sections of this chapter. The characterization of the optimal 22-transistor ULC is presented in Table 10.3 [66, 154].

10.4.3 Dual-Oxide-Based Logic-Level Optimization of Digital Circuits

In this subsection, logic-level optimization of digital IC has been considered using dual-oxide presilicon technique for gate-oxide leakage optimization [170, 171]. The dual-oxide-based logic-level netlist optimization flow is presented in Fig. 10.16 [170, 171]. The flow considered the logic-level netlist as an input. This netlist is generated following design and synthesis steps outlined in digital design and synthesis flows in design flow chapter from a given higher level description of a digital IC. At the logic level, a combination circuit can be modeled as a weighted directed acyclic graph $G_{logic}(V,E)$ [170, 171]. In the G_{logic}, the node set V is composed of the primary inputs (PIs), the primary outputs (POs), and the combinational logic elements. Any node $v_i \in V$. The edges $e_{i,j} \in E$ represent the interconnections between any two nodes v_i and v_j. At the first phase of the logic-level optimization flow, various technology-independent logic-level optimization, including local-Boolean optimization, selective collapsing, and algebraic decomposition, can be performed over the logic-level netlist. Thus, generating an intermediate technology-independent optimal logic-level netlist for the digital IC. Logic-level technology mapping is now performed for the resulted logic-level netlist. Technology mapping and optimization including tree covering for gate selection and load buffering for fanout-tree construction is performed at this phase of the flow. The design flow uses a logic cell library that has standard cells of multiple-oxide thickness technology.

At this phase of the logic-level optimization flow, the dual-dielectric-based optimization can be considered for leakage and delay tradeoffs. Based on the observation of the logic-level data in the standard cell library, the logic-level netlist is converted to a NAND netlist [170, 171]. This is because the NAND is observed to have minimal gate leakage as well as delay as compared to other basic logic cells. At the logic level, the gate-oxide leakage problem can be the following: given a weighted directed acyclic graph $G_{logic}(V,E)$, the objective is to determine the logic gates that can take transistors of higher dielectric thickness such that the total gate-oxide leakage is minimal for a given delay constraint. In the G_{logic}, the weights of the nodes can be associated with the delay in the combinational elements. Moreover, as the dual-dielectric assignment exploration is performed at the technology mapping phase of the logic

ULC Characteristics	Estimated Values
ULC Average Power Dissipation P_{ULC}	5.0 µW
ULC Propagation Delay D_{ULC}	1.6 ns

TABLE 10.3 Characteristics of the Optimal ULC for 32-nm High-κ CMOS [66, 154]

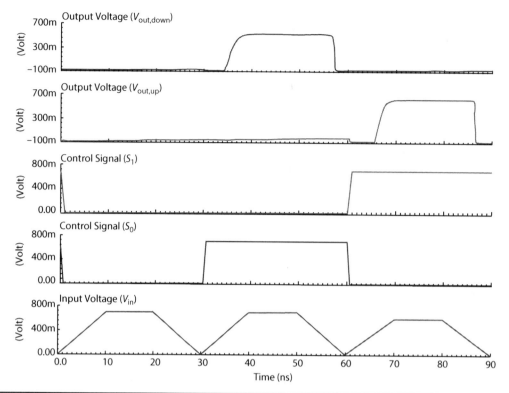

FIGURE 10.15 Functional simulation of ULC circuit [66, 154].

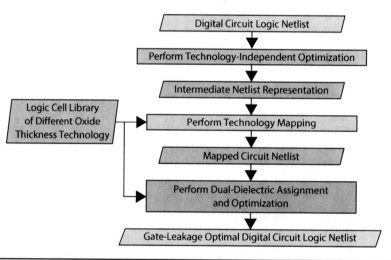

FIGURE 10.16 Dual-oxide-based logic-level leakage optimization flow [170, 171].

design and as the exact layout information is not available, the interconnecting edges can be modeled as constant delay values. The gate-oxide leakage and delay tradeoff at the logic levels can be formulated as an optimization problem as follows: let us assume V be the set of all vertices in the G_{logic} representing the digital IC at the logic level. Further assume V_{CP} as the set of all vertices in the critical path from the PIs to the POs. The power- and performance-driven two-dimensional problem can thus be formulated as follows:

$$\text{Minimize:} \sum_{v_i \in V} P_{ox}(v_i) \quad \forall \quad v_i \in V \qquad (10.6)$$

$$\text{Subjected To:} \sum_{v_i \in V_{CP}} D_i(v_i) \leq D_{CP} \quad \forall \quad v_i \in V_{CP} \qquad (10.7)$$

Power-, Parasitic-, and Thermal-Aware AMS-SoC Design Methodologies

In the above formulation, $P_{ox}(v_i)$ is the gate-oxide leakage of the sample node v_i of the G_{logic}. In the above formulation, the constraints in the second row ensures that the total delay $D_i(v_i)$ in a given path in the G_{logic} is less than the critical path delay D_{CP}. Any algorithm such as simulated annealing or tabu search can be used for this optimization task.

The logic-level gate leakage optimization is demonstrated in Fig. 10.17 [170, 171]. The original G_{logic} for a digital circuit logic-level benchmark is shown in Fig. 10.17(a) [170, 171]. The technology-independent

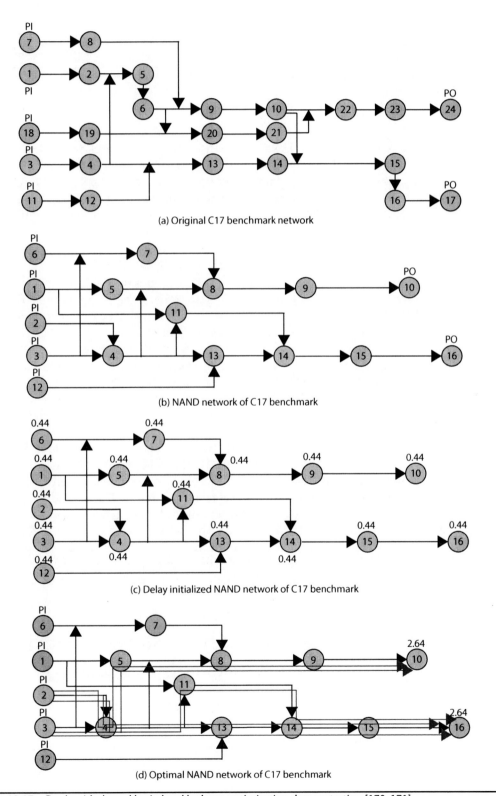

FIGURE 10.17 Dual-oxide-based logic-level leakage optimization demonstration [170, 171].

Standard Benchmark Circuits	Number of Logic Gates	Number of Nodes in NAND Network	Critical Path Delay (in ps)	Gate Leakage for SiO₂ with Thickness (nA)	Gate Leakage with Dual-Dielectric Assignment (nA)	Percentage Reduction (%)
C432	160	398	3.8	3949.4	253.2	93.5
C499	202	503	2.0	5708.5	590.4	89.6
C880	383	576	6.1	6537.0	337.8	94.8
C2670	1193	1764	24.6	17863.3	1560.6	91.2
C3540	1669	1792	18.2	34637.1	2215.7	93.6

TABLE 10.4 Gate-Oxide Leakage Optimization Results for Selected Digital Circuits for 45-nm CMOS [170, 171]

optimal G_{logic} containing the NAND gates only is shown in Fig. 10.17(b) [170, 171]. The optimization algorithm then initialized the logic-level netlist with the nominal cells from the standard library. The G_{logic} with NAND gates and initial delay assignment is shown in Fig. 10.17(c) [170, 171]. After the optimization algorithm converged to a solution, the optimal G_{logic} with NAND gates is shown in Fig. 10.17(d) [170, 171]. The experimental results for selected benchmarks of the digital circuits are presented in Table 10.4 [170, 171].

10.4.4 Dual-Oxide-Based RTL Optimization of Digital Circuits

In this subsection, RTL optimization of gate-oxide leakage is presented using dual-oxide static solution for power optimization. The overall RTL gate-oxide leakage optimization flow is presented in Fig. 10.18 [44, 63, 129, 162, 166, 167]. The input to the RTL optimization flow is the digital HDL description of the digital IC. The digital HDL description is then compiled in a sequencing data flow graph (DFG) or control DFG (CDFG) [162, 166, 167]. At this level, the DFG representing the behavioral representation of the digital circuit is $G_{RTL}(V,E)$; each vertex in V is an operation and each edge in E represents a dependency among the vertices. The optimization flow then performs scheduling as soon as possible (ASAP) for the given resource constraints. This is followed as late as possible (ALAP) by scheduling for the given resource constraints. The mobility graph that gives degree of freedom for each of the vertices can now be constructed from the ASAP and ALAP schedules.

At this phase of the design flow, a simulated annealing algorithm is used for gate-oxide leakage optimization during the high-level synthesis (HLS) or RTL optimization tasks [63, 129, 162, 166, 167]. The simulated annealing algorithm performs operation scheduling as well as resource allocation and binding while optimizing the RTL cost to obtain a scheduled DFG denoted as $G_{RTL,S}(V,E)$. This can be formally stated as an optimization problem in the following manner: Let V be the set of all vertices and V_{CP} be the set of vertices in the critical path from the source vertex of the DFG to the sink vertex. $FU_{k,i}$ denotes a functional unit, datapath resource, or datapath component of type k and is made up of transistors of dielectric thickness level i. Let us assume c is a clock cycle or a time stamp in the total number of clock cycles N_{DFG} in the $G_{RTL,S}(V,E)$. The optimization problem can then be stated as [143]:

$$\text{Minimize:} \quad P_{ox,DFG} \quad (10.8)$$

$$\text{Subjected To: Allocated } (FU_{k,i}) \leq \text{Available } (FU_{k,i}), \forall c \in N_{DFG} \quad (10.9)$$

$$\text{Subjected To:} \quad D_{CP,DFG} \leq D_{Constraint,DFG} \quad (10.10)$$

The objective function ensures that the total gate leakage for the datapath in the target architecture is minimized. The constraints ensure that the total allocation of the number of functional units allocated does not exceed that available for any clock cycle in the N_{DFG} number of total clock cycles that are obtained as the maximum of the ASAP and ALAP schedules above.

In order to minimize the gate leakage, the simulate annealing algorithm ensures that every vertex in the $G_{RTL,S}(V,E)$ is scheduled in such a way that less leaky functional units (which are slow) can be

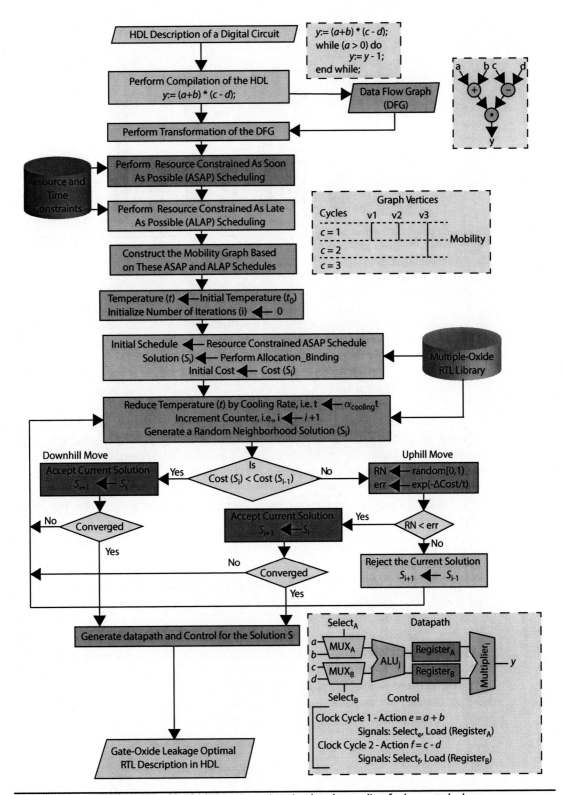

FIGURE 10.18 Dual-oxide-based RTL optimization using simulated annealing for low-gate leakage.

assigned to noncritical vertices so that the total delay in the critical path is not affected. The simulated annealing algorithm starts with initial temperature. The initial schedule is assumed to be the resources constrained scheduled from an above step [166, 167]. As the initial solution (S_i), Allocate_Binding is performed in which low-T_{ox} (or nominal oxide thickness for a specific technology) functional units are assigned from the multiple-oxide RTL library. The total gate-oxide leakage is determined as the sum of gate-oxide leakages of all the allocated functional units that is Cost(S_i) The delay constraint

536 Chapter Ten

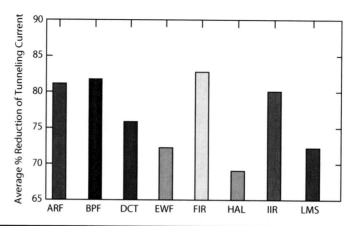

Figure 10.19 Average % reduction in gate leakage for different digital circuits through dual-oxide technique [166, 167].

$D_{\text{constraint,DFG}}$ is calculated from this initial assignment. The algorithm during each iteration evaluates a neighborhood solution by making random transitions from the current solution. If this solution has less gate-oxide leakage than the current solution for the overall DFG, the neighborhood solution is made the current solution. In generating a neighborhood solution, the algorithms randomly select a vertex and check if a less leaky high-T_{ox} functional unit can be assigned in all possible clock cycles while satisfying a time constraint.

The experimental results for a selected DSP benchmarks circuits are shown in Fig. 10.19 [166, 167]. In particular, digital ICs for auto-regressive filter (ARF), band-pass filter (BPF), discrete cosine transformation (DCT), elliptic wave filter (EWF), finite impulse response (FIR) filter, HAL differential equation solver, infinite impulse response (IIR) filter, and linear mean square (LMS) are explored. The results take into account the gate-oxide leakage and propagation delay of functional units, interconnect units, and storage units present in the datapath circuit. The critical path delay of the circuit is estimated as the sum of the delays of the vertices in the longest path of the $G_{\text{RTLS}}(V,E)$ for single cycle case and number of control steps times the slowest delay resource for multicycling-chaining case.

10.5 Hardware-Based Postsilicon Power Reduction Techniques

The postsilicon power/energy optimization techniques are designed during the design cycle of the circuit and system, but tuned during the operation of the circuit and system to get the power/energy efficiency. The postsilicon techniques can be implemented using hardware and/or software mechanisms. The running of the postsilicon techniques at the top of circuit or system operations is a continuous overhead for the host. However, these techniques are quite attractive, particularly, for system-level power management. In this section, selected hardware-based postsilicon reductions have been discussed. First overview is presented and then selected few are discussed in detail at various levels of circuit abstraction.

10.5.1 Brief Discussion

Selected postsilicon techniques have been presented in Fig. 10.20 [92, 106, 107, 132, 162, 188, 201]. Each of these selected techniques will be briefly discussed in this subsection.

In general, the "gating" can be defined as the process of selection of a portion of signal between specified time intervals or between specified amplitude limits. The "data gating" technique has been illustrated in Fig. 10.20(a) [188]. The "data gating" involves the prevention of the portions of the logic blocks from toggling when the results from it are not used. One way of implementing it is by using "operator isolation" as shown in this figure. In this case, the multiplier will not be activated if the results from the multiplier will be selected to propagate through the multiplexer. Thus, power/energy reduction occurs as the multiplier is not executed. A drawback of this technique is that the additional logic needed to implement this technique can lead to large area overhead for the host ICs.

Power-, Parasitic-, and Thermal-Aware AMS-SoC Design Methodologies 537

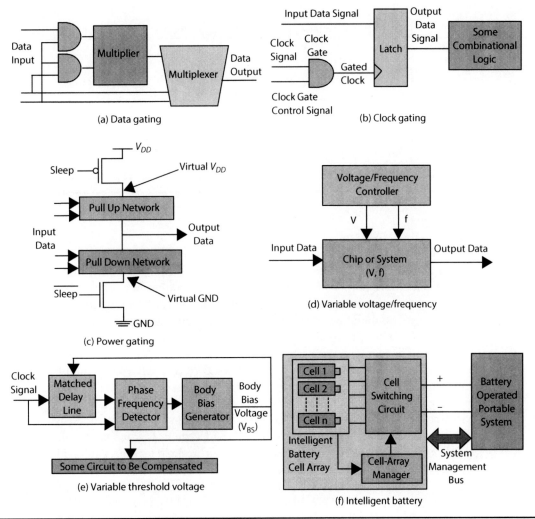

FIGURE 10.20 Brief concepts of postsilicon power reduction techniques [92, 107, 132, 162, 188].

The "clock gating" technique is illustrated in Fig. 10.20(b) [3, 128, 162, 185, 188]. The clock gating technique involves the prevention of clocking of the register, latch, or flip-flop when the input data signal has not changed [128, 162, 188]. In essence, the clock gating reduces unnecessary switching of these sequential blocks and the effective switching capacitance ($C_{switching}$) of these blocks and clock circuity. Thus, the power/energy reduction can happen at a linear scale as the dynamic power and energy dissipation is linearly proportional to the switching capacitance. The technique has some area overhead due to the need for additional logic gates for the implementation. However, clock gating is not quite effective for overall power reduction when the leakage component is high [201].

The "power gating" technique is illustrated in Fig. 10.20(c) [3, 39, 79, 106, 122, 188, 201]. The power-gating techniques involve cutting off the power and/or ground rails from the logic blocks when the logic block is not in use, thus completely saving the power/energy dissipation during the standby mode of the logic blocks. There are many variants of power-gating techniques based on structural and control aspects of the techniques. The basic power gating technique can be of three types [122]: (1) PMOS gating, (2) NMOS gating, and (3) dual gating. For example, the figure shows dual-type power gating in which both PMOS and NMOS gatings have been used. It is possible to have PMOS gating and NMOS gating to achieve power gating in which only PMOS power-gate with "Sleep" control or NMOS power-gate with "$\overline{\text{Sleep}}$" control is used, respectively. The power-gating technique can also lead to area overhead due to additional transistors. This dynamic or postsilicon technique can be immensely useful for a battery-driven mobile system and save battery life significantly.

The "variable supply voltage," "DVS," or "variable frequency" technique has been illustrated in Fig. 10.20(d) [135, 141, 162, 201]. A straightforward thinking from dynamic power dissipation and

Tuning Options	Operating Frequency (f)	Supply Voltage (V_{DD})	Power Dissipation (P_{dyn})	Energy Dissipation (E_{dyn})
1 - Normal Operation	f_{max}	V_{DD}	P_{dyn}	E_{dyn}
2 - Low Power	$\left(\frac{f_{max}}{2}\right)$ - 50% Delay Penalty	V_{DD}	$\left(\frac{P_{dyn}}{2}\right)$ - 50% Reduction	E_{dyn} - 0% Reduction
3 - Low Energy/Power	$\left(\frac{f_{max}}{2}\right)$ - 50% Delay Penalty	$\left(\frac{V_{DD}}{2}\right)$	$\left(\frac{P_{dyn}}{8}\right)$ - 87.5% Reduction	$\left(\frac{E_{dyn}}{4}\right)$ - 75% Reduction
4 - Infeasible	f_{max}	$\left(\frac{V_{DD}}{2}\right)$	$\left(\frac{P_{dyn}}{4}\right)$ - 75% Reduction	$\left(\frac{E_{dyn}}{4}\right)$ - 75% Reduction

TABLE 10.5 Variable Frequency versus Variable Voltage [141, 162]

dynamic energy dissipation is that variable voltage can reduce dynamic power/energy significantly due to its square dependence, i.e., $P_{dyn} \propto V_{DD}^2$ as well as $E_{dyn} \propto V_{DD}^2$. Thus, reduction of the supply voltage can reduce both of them. In addition, the operating frequency is linearly dependent on the dynamic power dissipation and does not affect the dynamic energy dissipation. Thus, voltage and frequency in combination have multifold impact on the dynamic power dissipation as follows: $P_{dyn} \propto V_{DD}^2 f$. Some specific instances of voltage and frequency scaling are presented in Table 10.5 [141, 162]. Therefore, voltage scaling and frequency scaling are implemented together for maximum efficiency. An example of practical approach for DVS is supply voltage hopping in which the voltage and frequency are changed based on the workload. This is a very effective method for power and/or energy reduction of circuits and systems operating in real time. This technique needs efficient voltage and frequency controller circuitry so that the voltage and/or frequency changing is possible with minimal performance penalty on the circuit and system. It may also be noted that supply voltage (V_{DD}) reduction is also effective to reduce the subthreshold leakage, gate-oxide leakage, and junction leakage, as these leakages are also directly/indirectly affected by V_{DD}.

The "variable threshold voltage," "threshold voltage scaling," or "dynamic threshold voltage" technique is illustrated in Fig. 10.20(e) [92, 106, 107, 135, 162, 188, 201]. In essence, in this technique, the threshold voltage of the low-threshold voltage devices is varied by applying a variable body-bias voltage (V_{BS}). An increase in the threshold voltage (V_{Th}) decreases the subthreshold leakage. However, the delay is affected and hence tradeoff is necessary between leakage power and propagation delay. An ABB generator is used to generate such body bias voltages. The ABB generators work with leakage sensors and other auxiliary circuitry, such as matched delay line and phase-frequency detector, for effective control of the V_{Th}.

The concept of "intelligent battery" or "smart battery" is illustrated in Fig. 10.20(f) [98, 131, 132, 227]. The overall battery consists of an array of cells, a cell switching circuit, and a cell-array manager. Each of the cells is the same and has some discrete voltage levels much lower than the overall battery output voltage specifications. The cell switching circuit connects the cell array into the main terminal of the battery. The cell-array manager decides the reconfiguration of the individual cells based on the need of the loading system (e.g., battery-operated portable system) on the battery. The system management bus is the port for information exchange between the smart battery and the loading system. The energy efficiency or lifetime of the battery is enhanced by this intelligent battery in various ways. The dynamic reconfiguration of cells to match load requirements eliminates the need for DC-to-DC voltage converters and voltage regulators that themselves have energy/power overheads. The intelligent battery can be preconfigured to a specific current/voltage delivery pattern to exactly meet the requirements load thereby saving energy. The intelligent battery takes advantage of the charge recovery effects of the cells that provide energy advantages over a single monolithic battery.

10.5.2 Dynamic or Variable Frequency Clocking for Power Reduction

In this subsection, variable frequency 'or dynamic frequency' postsilicon approach for dynamic power reduction has been presented [29, 38, 141, 155, 157, 162, 164]. In this approach, all the functional units are clocked by a single clock line. The clock frequency is varied during the runtime based on the functional units active in that cycle. The design decisions for the variable clocking can be taken at

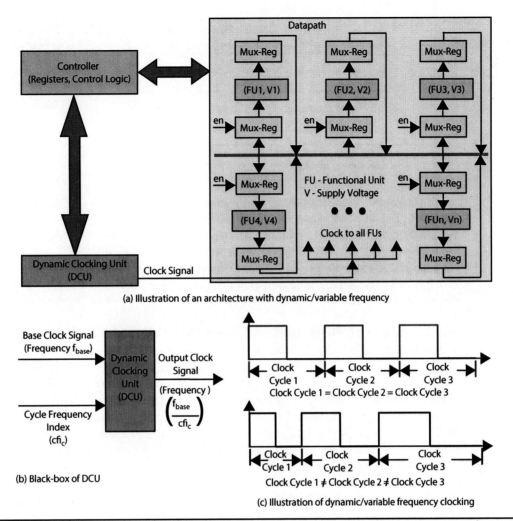

FIGURE 10.21 Dynamic or variable frequency clocking for power reduction [29, 141, 155, 157, 162, 164].

the system level or architecture level. The clocking unit and controller make the datapath operate based on the clocking mechanism.

The key idea and working principles of this technique are presented in Fig. 10.21 [29, 141, 157, 162, 164]. The architecture that can operate at a variable or dynamic frequency is depicted in Fig. 10.21(a) [29, 141, 155, 157, 162, 164]. The architecture consists of the datapath, controller, and a dynamic clocking unit (DCU). In general, the datapath consists of N_{FU} number of functional units that are resources such as adders and subtractors. Each of the functional units has registers and multiplexers. Each of the functional units performs single cycle operations. The controller activates the functional units that need to be executed during each clock cycle and the functional units that are not active are disabled. The operating frequency is changed dynamically and along with appropriate supply voltage assignment can be performed; thus, providing savings in both dynamic power and energy consumption. The DCU has been depicted in Fig. 10.21(b) [29, 141, 155, 157, 162, 164]. The inputs to the DCU are the base clock signal (f_{base}) and the cycle frequency index (cfi_c). A set of values of cfi_c is stored in a register in the controller. The DCU generates an output clock signal of frequency $\left(\frac{f_{base}}{cfi_c}\right)$. The base frequency is typically the maximum or multiple of maximum operating frequency of any of the functional units. A specific value of cfi_c is loaded from the controller for a clock cycle. The dynamic clock frequency is illustrated in Fig. 10.21(c) [29, 141, 157, 162, 164].

For specific experimental conditions, the dynamic power and energy reductions possible from the dynamic clocking along with the voltage scaling are presented in Table 10.6 [156, 162]. The results are presented for selected DSP IC designs that have significant real-life applications in various multimedia consuming devices, mobile phones, and media players. The power and/or energy savings possible from

Digital Circuits	Average Power Reduction	Average Energy Reduction	Delay Penalty
BPF	64.9%	43.5%	30%
DCT	58.7	43.6	20
EWF	47.4	43.0	10
Filter FIR	52.2	41.5	20

TABLE 10.6 Power and Energy Reduction through Dynamic/Variable Frequency Clocking [156, 162]

the use of dynamic frequency/voltage are quite significant. However, this postsilicon technique has various overheads. The use of additional DCU is an overhead. DC-to-DC voltage-level converters are required as the data is transferred among the functional units operating at different supply voltages. The use of DC-to-DC voltage-level converters when multiple voltage is used along with the frequency is also an overhead. It is a fact that in a real-life scenario, the clock frequency cannot be changed quite as frequently otherwise data inconsistency in sequential elements, such as registers, will be an issue. Thus, at what intervals frequency needs to be changed is an important decision at the circuit and system design time.

10.5.3 Adaptive Voltage Scaling for Power and Energy Reduction

A postsilicon technique called "adaptive voltage scaling (AVS)" is presented in this subsection [8, 9]. The AVS is a continuous, closed-loop, and real-time power and/or energy management technique. The AVS technique provides optimum supply voltage to the digital SoC or digital ASIC for the highest possible power and/or energy savings while compensating for process and temperature variations. In terms of the working principle, this is like DVS or variable supply voltage. However, there are differences between the DVS and AVS. The DVS technique does not have any mechanism to compensate for process or temperature variations. As a consequence, the scaled/reduced supply voltage needs to be high enough to maintain functionalities in the presence of the process and temperature variations. Unless both frequency and voltages are not scaled in a coordinated fashion, the DVS may not result in both power and/or energy reductions. The AVS with a closed-loop mechanism not only saves power and energy, but also compensates for the process and temperature variations.

The key concept of the AVS is depicted in Fig. 10.22 [8,9]. The energy/power-efficient ASIC or SoC needs to have two additional units to effectively deploy AVS technique: the hardware power monitor (HPM) and advanced power controller (APC). The HPM units monitor the silicon characteristics based on the process and temperature variations. The APC decides whether the voltage change should be made. The energy management unit (EMU) is external to the digital SoC/ASIC. If needed, the APC sends commands to the EMU through an open-standard PowerWise® Interface (PWI) [8, 9]. EMU is used to change the supply voltage of the digital SoC/ASIC according to the needs.

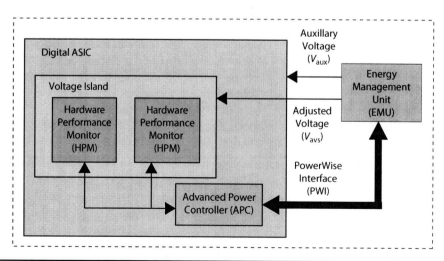

FIGURE 10.22 AVS for power and energy reduction [8, 9].

10.6 Dynamic Power Reduction Techniques

In this section, the techniques that can be used for the reduction of dynamic power or energy are discussed. First, the techniques discussed in brief and then a detailed discussion of specific examples is presented for better understanding of the techniques.

10.6.1 Brief Discussion

A high-level overview of selected techniques that can be used for dynamic power/energy reduction is presented in Fig. 10.23 [141, 162, 220]. The dynamic power/energy dissipation has three components, as presented in the figure: the capacitive switching power or energy consumption, the short-circuit power/energy consumption, and the transient gate-oxide leakage dissipation that is particularly important for nano-CMOS technology using SiO_2 as gate dielectric [148, 220].

For the reduction of the capacitive-switching power, various static and dynamic approaches can be used. The static approaches are structural or presilicon approaches, such as multiple frequency, multiple supply voltage, and physical capacitance reduction that can be used. The multiple supply voltage technique that can reduce both capacitive-switching power and energy has already been discussed in the past and some specific detailed discussions will be presented in the later part of this section. The reduction of physical capacitance can be done by reducing the device sizes. At the circuit-level, this can be performed for small to medium size IC only. It may be noted that the reduced size transistors will have reduced the current drive that will make the circuit slow. The dynamic approaches for capacitive-switching power reduction include variable frequency clock f, variable supply voltage (V_{DD}), and switching capacitance ($\alpha_{sw}C_L = C_{sw}$) reduction. No switching means no power loss! The switching activity reduction is essentially achieved by the techniques such as data gating and clock gating that have been discussed in the previous section. It may be noted that the switching activity is affected by the logic function being performed. The switching activity is user

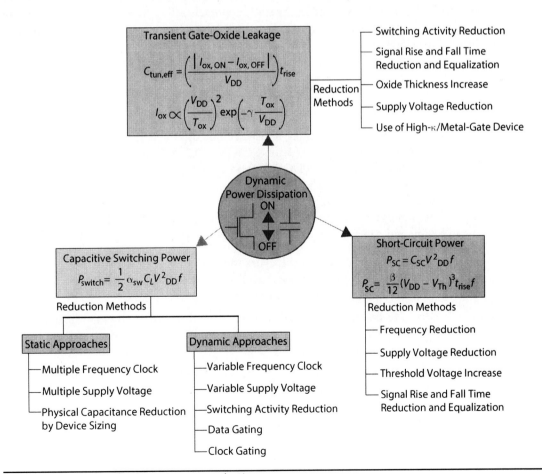

FIGURE 10.23 Options for dynamic power reduction.

dependent and it is difficult for a designer to handle. The spatial and temporal correlations in the switching activity are also difficult for the designers to handle.

The "short-circuit power" dissipation happens due to the direct path between power rail and ground rail when both the pull-up and pull-down network are ON. Based on the expressions from the short-circuit power dissipation, it can be deduced that short-circuit power dissipation reduction can be achieved in the following ways: (1) frequency reduction, (2) supply-voltage reduction, (3) threshold-voltage reduction, and (4) signal rise/fall time reduction and equalization. The frequency reduction and supply-voltage reduction can be performed in the same way as the supply-voltage scaling of the capacitive-switching power reduction mechanism. Thus operating frequency and/or supply-voltage scaling for capacitive-switching power reduction can reduce the short-circuit power and overall dynamic power of the IC. The threshold voltage increase for short-circuit power/energy reduction can be done using techniques such as multiple threshold voltage or variable threshold voltage just as used for subthreshold leakage reduction. Equalization of the rise time and fall time of the signal and their reductions can decrease the short-circuit power/energy dissipation as time for which the direct path is created between supply and ground reduces [11, 15, 225].

The transient gate-oxide leakage that exists when the transistor makes transitions between ON and OFF states increases the dynamic power dissipation [99, 148, 220]. The transient gate-oxide leakage can be quantified as a capacitance $C_{tun,eff}$. This capacitance poses additional loading effects on the logic gates [148]:

$$C_{L^*} = C_L + C_{tun,eff} \qquad (10.11)$$

where C_L is the intrinsic load capacitance used in the capacitive-switching power calculation. This increases effective switching capacitance to the following $\alpha_{sw}C_{L^*}$. Just like the short-circuit power/energy reduction, equalization and reduction of rise and fall time of the signal are possible options to maintain low transient gate-oxide leakage. Use of transistors of high gate-oxide can reduce transient gate-oxide leakage as discussed in the static approaches. Supply voltage scaling is also an option, just similar to capacitive-switching power reduction. The use of high-κ/metal-gate-based CMOS technology eliminates the gate-oxide leakage [154].

10.6.2 Dual-Voltage and Dual-Frequency-Based Circuit-Level Technique

As a concrete example of static dual-voltage/frequency-based dynamic power reduction, design of a digital IC, specifically a watermarking chip, has been discussed in this subsection. The dual-voltage/frequency-based low-power technique has been incorporated at the circuit-level. The big picture of the design optimization process is depicted in Fig. 10.24 [20, 141, 158, 159].

The overall architecture of a DCT domain watermarking chip that can insert both invisible and visible watermarks to the host image is shown in Fig. 10.24(a) [20, 141, 158, 159]. The overall architecture has been logically partitioned into two different portions as invisible watermarking and visible watermarking for easy understanding. The digital watermarking chip architecture has the following units to perform the watermarking operations:

1. *DCT Unit*: The DCT unit calculates the DCT coefficients of the host image for analysis and watermarking in the DCT domain of the image. It may also calculate DCT coefficients of the watermark image depending on the type of the watermark.

2. *Edge Detection Unit*: The edge detection unit identifies whether any host image block contains an edge that has more perceptual information and needs to be preserved during the watermarking process.

3. *Perceptual Analyzer Unit*: The perceptual analyzer unit calculates the perceptual statistics of the host image blocks, e.g., mean gray level. It then determines the perceptual significant region of the host image using these statistics.

4. *Scaling and Embedding Factor Unit*: The scaling and embedding factor unit determines the scaling and embedding factors for visible watermark insertion based on the statistics from the perceptual analyzer unit. It considers user-specified initial parameters to change the watermarking strength.

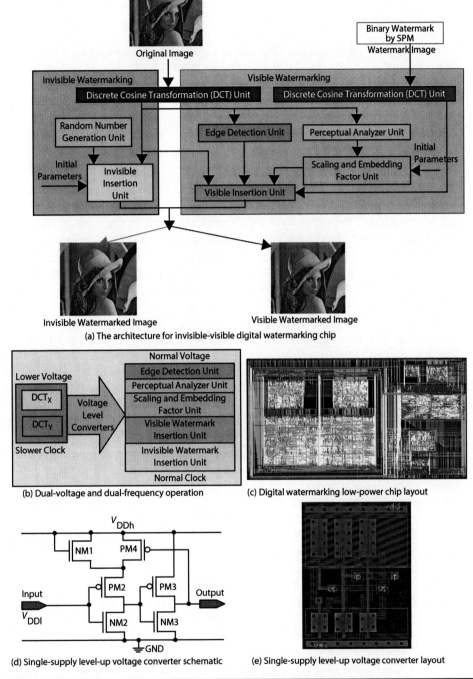

FIGURE 10.24 Dual-voltage and dual-frequency-based design for dynamic power reduction using digital watermarking chip example [20, 141, 158, 159].

5. *Random Number Generator Unit*: The random number generator unit generates random numbers needed in the invisible watermarking process; to be used as watermarks in this case.

6. *Invisible Insertion Unit*: The invisible insertion unit is used to insert the invisible watermark in the host image. This unit considers initial parameters to tune the invisible watermarking strength.

7. *Visible Insertion Unit*: The visible insertion unit is used to insert the visible watermark in the host image.

The overall architecture is a decentralized controller architecture in which each of the units in the architecture has its own controller. In addition there is an overall centralized controller. Each of the arithmetic units in the overall architecture performs fixed-point operations. In this architecture, although two different DCT units have been shown, it is possible to have the architecture with only one DCT unit and run it multiple times on different data to process different images, e.g., host and watermark images.

The architectural perspective uses dual voltage and dual frequency for the minimization of dynamic power/energy consumption as shown in Fig. 10.24(b) [20, 141, 158, 159]. This architecture is obtained based on the critical path delay analysis of each of the units of the overall architecture. Other configurations of this architecture with the dual-voltage and dual-frequency are possible. The DCT unit has slower clock and the other units have normal clock. In a specific case, the DCT unit processes 4 pixels and the other unit processes 1 pixel at a time. Thus, it is logical to operate the slower clock at a $\left(\frac{1}{4}\right)$ frequency of that of the normal clock. In consistence with the operating frequency, two different supply voltages have been used in this chip. The DCT unit has a supply voltage much lower than the normal supply voltage of the other units. In addition to the dual clock operation, the local clocks are automatically generated to trigger the operations of some units of the architecture. These local clocks are generated from the integrated local controllers of each of the units of the architecture. It may be noted that this type of clock generation within the chip essentially indirectly implements the clock-gating technique used for switching power reduction.

The overall architecture is typically first modeled and verified at various levels as discussed in the digital design flow in a previous chapter. For example, at the RTL modeling and verification using a digital HDL/HDVL can be performed. For complex chips, this standard cell synthesis approach is a quicker way to obtain the physical design of the chip. The physical design that has been automatically generated using digital HDL and standard cell synthesis was shown in Fig. 10.24(c) [20, 141, 158, 159]. The voltage-level converters that are needed to stitch the low-voltage units and normal-voltage units can be made part of the standard cell library. The layout characterization using a Fast-SPICE can provide reasonable accuracy. A SPICE characterization can be very slow due to large transistor count. The characteristics of the dual-voltage and dual-frequency-based low-power watermarking chip are presented in Table 10.7 [20, 141, 158, 159]. As is evident from the results, a reduction of total power of 84.2% is possible by this dual-voltage and dual-frequency technique.

One of the important design considerations for this dual-voltage chip is the use of DC-to-DC voltage-level converters as it can provide large area overhead for the chip. It is possible to use many types of the level converters including ULC that has been discussed in a previous section [14]. However, as a specific example, a DCVS-based single supply DC-to-DC voltage-level converter has been used in this design. The schematic design of a single supply level converter that is fast and energy-efficient is shown in Fig. 10.24(d) [20, 141, 186]. The physical design of this DC-to-DC voltage-level converters has been presented in Fig. 10.24(e) [20, 141, 158, 159].

10.6.3 Multiple Supply Voltage-Based RTL Technique

As another specific example of dynamic power reduction, multiple supply voltage RTL or architecture-level power optimization is presented in this subsection. The optimization is presented in the context of the steps in the HLS of digital ICs.

Watermarking Chip Characteristics	Estimated Values
Process Technology	TSMC 0.25 μ
Chip Silicon Area	16.2 sq mm
Dual Clock Frequencies	280 MHz and 70 MHz
Dual Supply Voltages	2.5V and 1.5V
Number of Transistors	1.4 million
Power Dissipation for Single-Voltage and Single-Frequency	1.9 mW
Power Dissipation for Dual-Voltage and Dual-Frequency	0.3 mW
Power Reduction	84.2%

TABLE 10.7 Dual-Voltage and Dual-Frequency-Based Watermarking Chip Characteristics [20, 141, 158, 159]

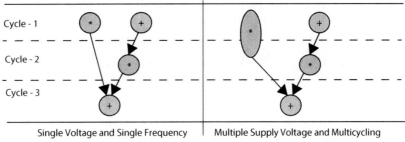

FIGURE 10.25 Multiple supply voltage-based energy and power reduction [141, 160, 162].

To understand the advantage of multiple supply voltage, the behavioral description in the form of a DFG of a hypothetical digital IC is shown in Fig. 10.25 [141, 160, 162]. In a single-voltage single-cycle mode, the operations of the DFG are executed in the functional units that operate at a single supply voltage as shown in Fig. 10.25(a). The clock cycle delay is determined by the slowest functional unit that is a multiplier in this specific case. For the same clock cycle delay, if a multiplication operation is executed in a multiplier operating slowly and supplied with a lower voltage, then it can complete the execution in two clock cycles. The DFG obtained is then the multicycle DFG as shown in Fig. 10.25(a). The datapath is a multiple supply voltage and multicycle datapath. The power/energy saving is a result of this low-operating voltage functional unit. Further use of low-voltage operating adder/ALU is possible based on the resource availability/constraints and timing constraints. A specific set of operating parameters and energy reduction possibilities is presented in Fig. 10.25(b) [141, 160, 162]. The total energy consumption and critical path delay of the datapath in this case is the following:

$$E_{\text{RTL,MV}} = E_{\text{mult}} + \left(\frac{E_{\text{mult}-}}{2}\right) + E_{\text{add}-} + \left(\frac{E_{\text{mult}-}}{2}\right) + E_{\text{add}-} \tag{10.12}$$

$$T_{\text{pd,RTL}} = 4 T_{\text{pd,mult}} \tag{10.13}$$

In the above expression, $E_{\text{mult}-}$ is the energy consumption of the multiplier operating at supply voltage V^-, $E_{\text{add}-}$ is the energy consumption of the adder operating at supply voltage V^-, and $T_{\text{pd,mult}}$ is the delay of the multiplier operating at supply voltage V. $E_{\text{RTL,MV}}$ is much less than the energy consumption of single cycle and single voltage operation. In many cases, energy/power reductions are possible without a delay penalty by the use of multiple voltage operations. However, the multiple voltage multiple cycle datapath can have area penalty and time penalty based on the constraints.

Now the question is how to scale and automate the above idea for bigger and real-life circuits, This subsection discusses that in the context of scheduling step of HLS. In particular, optimization of EDP has been considered during the operation scheduling step of the HLS. Let us assume that the behavioral datapath is represented in the form of a sequencing DFG. The total EDP of the DFG for multiple supply voltage operation can be modeled as follows [141, 160, 162]:

$$\text{EDP}_{\text{RTL,MV}} = \sum_{c=1}^{N_{\text{DFG}}} \text{EDP}_c = \sum_{c=1}^{N_{\text{DFG}}} \left(\frac{1}{f}\right) \sum_{i=1}^{N_{\text{FU},c}} \alpha_{i,c} C_{i,c} V_{i,c}^2 = \sum_{c=1}^{N_{\text{DFG}}} \left(\frac{1}{f}\right) \sum_{i=1}^{N_{\text{FU},c}} \alpha_{i,c} C_{i,c} V_{i,c}^2 \quad (10.14)$$

In the above expression, N_{DFG} is the number of clock cycles, $N_{\text{FU},c}$ is the number of functional units active in a clock cycle c, $\alpha_{i,c}$ is the switching activity of the functional unit i active in clock cycle c, and $V_{i,c}$ is the operating voltage of the functional unit i active in clock cycle c.

In the context of HLS steps, the above cost function can be minimized using an optimization algorithm. As a specific example, integer linear programming (ILP) formulation and its solutions are now presented. The ILP formulation has an objective function and various constraints. The objective function is to be minimized for power and/or delay reduction. The various constraints need to be satisfied to maintain the operation precedences as well as resource constraints.

Objective Function: One objective can be to minimize the EDP of the whole DFG over all control steps as follows:

$$\text{Minimize}: EDP_{\text{RTL,MV}} \quad (10.15)$$

The objective can be energy, average power as well. The EDP objective can be rewritten as the following:

$$\text{Minimize}: \sum_l \sum_i \sum_{L_V} x_{i,L_V,l,(l+L_{i,L_V}-1)} EDP(i, L_V) \quad (10.16)$$

where $x_{i,L_V,l,m}$ is the decision variable of ILP that takes the value of "1" if vertex v_i uses the functional unit FU_{k,L_V} and scheduled in control steps $l \to m$, FU_{k,L_V} is a functional unit of type k and operating at voltage level L_V, and $EDP(i,L_V)$ is the EDP of a functional unit used by vertex v_i for its execution that is operating at voltage level L_V.

Uniqueness Constraints: The uniqueness constraints of ILP formulation ensure that each vertex v_i is scheduled in the appropriate clock cycle within the mobility range $(C_S[i], C_L[i])$ and is assigned the specific supply voltage. The mobility range is obtained from the ASAP and ALAP schedules. Any operation may be executed with more than one clock cycle, depending on the supply voltage of the corresponding function units that it needs. The uniqueness constraints are represented as follows:

$$\sum_{L_V} \sum_{l=C_S[i]}^{C_S[i]+C_L[i]+1-L_{i,L_V}} x_{i,L_V,l,(l+L_{i,L_V}-1)} = 1, \quad \forall i, 1 \le i \le N_{\text{DFG}} \quad (10.17)$$

It may be noted that when an operation of the DFG is scheduled to use a normal-voltage operating functional unit, then it is scheduled in one unique clock cycle. On the other hand, when it is scheduled to use a lower-voltage operating functional units, then it may need more than one clock cycle for completion. Thus, for lower voltage vertices, the mobility is restricted.

Precedence Constraints: The precedence constraints guarantee that for a vertex v_j, all its predecessors are scheduled in earlier clock cycles and its successors are scheduled in later clock cycles. These constraints are needed to maintain the data dependency. These constraints should also account for the multicycling operations. These constraints are expressed as follows:

$$\sum_{L_V} \sum_{l=C_S[i]}^{C_L[i]} (l + L_{i,L_V} - 1) x_{i,L_V,l,(l+L_{i,L_V}-1)} - \sum_{L_V} \sum_{l=C_S[j]}^{C_L[j]} l x_{j,L_V,l,(l+L_{j,L_V}-1)} \le -1, \quad \forall i, j, v_i \in \text{Pred}_{v_j} \quad (10.18)$$

Resource Constraints: The resource constraints ensure that each of the clock cycles of the DFG contains no more than $FU_{k,v}$ operations of type k operating at voltage level L_V. This can be expressed as follows:

$$\sum_i \sum_l x_{i,L_V,l,(l+L_{i,L_V}-1)} \le N_{FU,k,L_V}, \quad \forall L_V, \forall l, 1 \le l \le N_{\text{DFG}} \quad (10.19)$$

where N_{FU,k,L_V} is the maximum number of functional units of type FU_{k,L_V}.

Power-, Parasitic-, and Thermal-Aware AMS-SoC Design Methodologies

The steps of solving the above ILP formulation in the context of HLS synthesis task is similar to the RTL optimization flow in the previous section. It is presented in the current context as pseudocode in Algorithm 10.2 [141, 160, 162]. For ILP formulation and modeling, many languages and frameworks can be used. The solutions steps have been demonstrated with the help of a small DFG as shown in Fig. 10.26 [141, 160, 162]. The ASAP is shown in Fig. 10.26(a). The ALAP is shown in Fig. 10.26(b). The mobility graph constructed from the ASAP and ALAP schedules is

---

Algorithm 10.2 ILP-Based Scheduling for EDP Minimization [141, 160, 162]

1: Determine the as-soon-as-possible (ASAP) of the unscheduled sequencing DFG.
2: Determine the as-late-as-possible (ALAP) of the unscheduled sequencing DFG.
3: Determine the mobility graph for each of the vertices in the DFG.
4: Construct the integer linear programming (ILP) formulations for the DFG.
5: Model the ILP formulation using a modeling language or framework.
6: Solve the ILP formulations using a ILP solver for given constraints.
7: Find the scheduled sequencing DFG with resulting resource allocation and binding.
8: Estimate the energy, propagation delay, and EDP of the scheduled DFG.

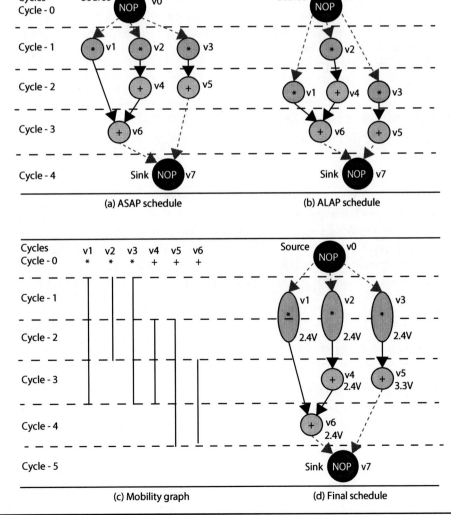

Figure 10.26 Multiple supply voltage-based scheduling for energy and power reduction [141, 160, 162].

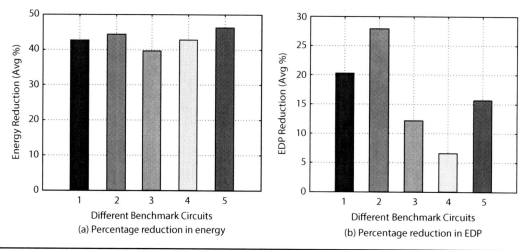

FIGURE 10.27 Results for multiple supply voltage-based scheduling for energy power reduction [141, 160, 162].

presented in Fig. 10.26(c). The final schedule along with the voltage assignments that resulted after solving the ILP formulations for a specific constraint are shown in Fig. 10.26(d). Based on the resulted scheduling, allocation, and binding, target architecture containing datapath and control can be generated. The characterization results of the target architecture are shown in Fig. 10.27. The energy reductions for a set of five DSP benchmark ICs have been presented. The energy reductions are more than 40% for all the digital ICs as shown in Fig. 10.27(a). The EDP reductions for the same digital ICs have been presented in Fig. 10.27(b). The EDP reductions have been in the range of approximately 8% to 28%, thus, suggesting that the energy reductions are obtained without any performance or propagation delay penalty.

10.7 Subthreshold Leakage Reduction Techniques

Selected different presilicon and postsilicon techniques that can be used for reduction for subthreshold leakage have been discussed in this section. First, high-level brief overview has been presented and then detailed discussions of selected examples have been presented.

10.7.1 Brief Discussion

The subthreshold leakage current was discussed in the chapter on design issues for DFX; however, it is being quickly revisited here. In the OFF state, many current components are present in a transistor. The two prominent components are current due to weak inversion and current due to drain-induced barrier lowering (DIBL), thus, leading to the following expression for the current in the OFF state [35, 89]:

$$I_{\text{subth}} = e^{1.8} \mu_0 \left(\frac{\varepsilon_{\text{ox}}}{T_{\text{ox}}}\right)\left(\frac{W}{L}\right) v_{\text{therm}}^2 \exp\left(\frac{V_{\text{gs}} - V_{\text{Th0}} + \gamma_{\text{body}} V_{\text{bs}} + \eta_{\text{DIBL}} V_{\text{ds}}}{\tau v_{\text{therm}}}\right)\left(1 - \exp\left(\frac{-V_{\text{ds}}}{v_{\text{therm}}}\right)\right) \quad (10.20)$$

where the terms have same meaning as when they were used in the chapter on design issues for DFX. η_{DIBL} is the DIBL coefficient. Driven by this expression, a high-level overview of selected possible techniques useful for subthreshold leakage is presented in Fig. 10.28 [35, 77, 89, 91, 136, 218].

Various presilicon (static) and postsilicon (dynamic) techniques can be used for subthreshold leakage as listed in the figure. Various static approaches include multiple-threshold voltage devices, multiple-oxide thickness devices, nonminimum channel length device, stacked devices, supply voltage reduction, low-temperature operation, and multiple-gate device. The multiple-threshold CMOS (MTCMOS) voltage or (DTCMOS) voltage-based static approach has been discussed in a

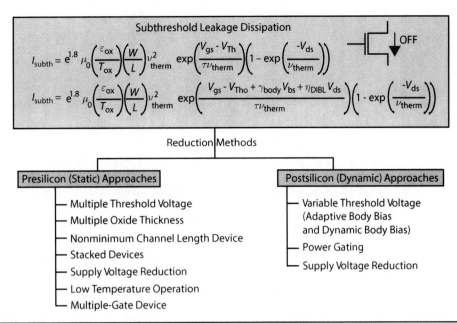

FIGURE 10.28 Different techniques for subthreshold leakage reduction.

previous section. The use of multiple-oxide thickness for subthreshold leakage reduction is not straightforward. Let us consider the relationship of gate-oxide thickness and threshold voltage [153]:

$$V_{Th} = V_{fb} + 2\phi_F + \left(\frac{T_{ox}}{\varepsilon_{ox}}\right)\sqrt{2q\varepsilon_{Si}N_{sub}(2\phi_F + V_{bs})} \qquad (10.21)$$

where V_{fb} is the flat-band voltage, ϕ_F is the Fermi-level, T_{ox} is the gate-oxide thickness, ε_{ox} is the gate dielectric permittivity, q is the electronic charge, ε_{sub} is the silicon permittivity, N_{sub} is the substrate doping concentration, and V_{bc} is the body bias. Thus, it is evident from the above expression that V_{Th} and T_{ox} have a linear relationship [153, 212]. In other words, an increase in T_{ox} will increase V_{Th}. As higher V_{Th} leads to the reduction of the subthreshold leakage, a higher T_{ox} will lead to reduction of subthreshold leakage as well.

The use of "nonminimum channel length" devices can reduce subthreshold leakage [35, 212]. For short-channel MOSFET, the threshold voltage increases almost linearly with respect to the physical channel length or drawn channel length that can be expressed as follows [35]:

$$\Delta V_{Th0} = V_{Th0}\left(\frac{\Delta L_{phy}}{L_{phy}}\right) \qquad (10.22)$$

Thus, an increase in the channel length leads to an increase in threshold voltage [212]. This, in turn, results in the reduction in the subthreshold leakage. The "stacked devices technique" involves stacking or series connections of OFF devices for the reduction of subthreshold leakage [91]. Two series connection OFF devices have low leakage current as compared to one OFF device due to the self-reverse biasing effects. Supply voltage reduction through supply voltage scaling just similar to the case of dynamic power reduction can be used for subthreshold leakage reduction. Low-temperature operations are also useful for subthreshold leakage current reduction as the thermal voltage (v_{therm}) has strong dependence on the subthreshold leakage current as evident from the expression.

The various postsilicon or dynamic approaches for subthreshold leakage reduction include variable threshold voltage, power gating, and supply voltage reduction. The ABB or dynamic body bias approaches deploying RBB can increase threshold voltage and reduce leakage as discussed in a previous section. Supply voltage reduction, such as DVS, can be effective for subthreshold leakage reduction. The useful multigate MOSFET is a practical approach for leakage reduction that is adopted

Subthreshold Leakage Reduction Technique	Percentage Reduction
Dual-Threshold Voltage	75%
Dual-Oxide Thickness	80%
Nonminimum Channel Length	85%
Stacked Devices	90%
Supply Voltage Reduction	50%
Variable Threshold Voltage	55%
Multiple-Gate Devices	90%

TABLE 10.8 Comparison of Techniques for Subthreshold Leakage Reduction [26, 35, 162]

by industry for high-end microprocessor design [26, 207]. A comparative perspective of the subthreshold leakage reduction techniques is provided in Table 10.8 [35]. This is just a comparative perspective that should not be treated for fair comparison.

10.7.2 Dual-Threshold-Based Circuit-Level Optimization of Nano-CMOS SRAM

In this subsection, as a specific detailed example of subthreshold leakage reduction technique, the transistor-level power optimization of a static RAM (SRAM) using dual-V_{Th} approach is considered [220, 221, 222]. For the SRAMs, the static noise margin (SNM) is an important characteristic to maintain its stability. Thus, power and SNM tradeoff is considered during power optimization of the SRAM. Two different design optimization flows have been shown in Fig. 10.29 [220, 221, 222]. The input to the design flows is a baseline SRAM design that is obtained by using nominal sizes for a specific technology using a schematic capture tool. For the power and SNM tradeoffs, it is important to identify appropriate transistors for high-V_{Th} assignment such that power and SNM tradeoff is possible. For this purpose, combined Design of Experiments (DOE) and ILP algorithms are used. The combined DOE-ILP approach reduces the optimization search space and convergence solutions much faster due to the use of DOE while maintaining the accuracy and optimality of the ILP.

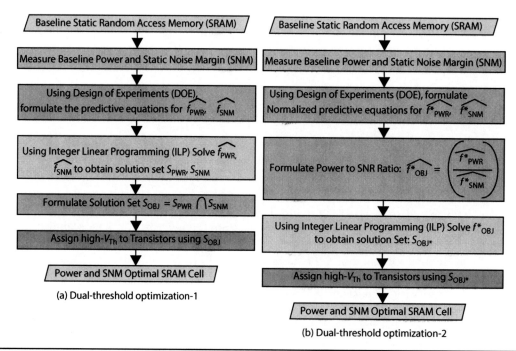

FIGURE 10.29 Design optimization flow for dual-threshold voltage-based nano-CMOS SRAM [220, 221, 222].

In the first, optimization flow called dual-threshold optimization-1 is presented in Fig. 10.29(a) [220, 221, 222]. In this flow, the predictive equations for total power dissipation ($\widehat{f_{PWR}}$), and SNM ($\widehat{f_{SNM}}$) are formulated for each of the transistors using DOE approach. The predictive equations are linear and ILP can be used to solve the predictive equations. The ILP formulation for power dissipation minimization of a 45-nm CMOS-based 7-transistor SRAM has the following form [220, 221, 222]:

$$\text{Minimize:} \quad \widehat{f_{PWR}}(=118.2 - 5.9x_1 - 28.9x_2 - 23.1x_3 - 10.9x_4 - 10.6x_5 - 12.1x_6 + 6.4x_7) \quad (10.23)$$

$$\text{Subjected To:} \; 0 \leq x_1 \leq 1, 0 \leq x_2 \leq 1, 0 \leq x_3 \leq 1, 0 \leq x_4 \leq 1, 0 \leq x_5 \leq 1, 0 \leq x_6 \leq 1, 0 \leq x_7 \leq 1$$

$$f_{SNM} > \tau_{SNM} \quad (10.24)$$

where the constraints "1" and "0" represent coded values for high-V_{Th} and low-V_{Th} states. τ_{SNM} is the SNM of the baseline design of the SRAM. The optimal solution of this ILP is the following solution set: $S_{PWR} = [x_1 = 1, x_2 = 1, x_3 = 1, x_4 = 1, x_5 = 1, x_6 = 1, x_7 = 0]$. The ILP formulation for read SNM maximization of a 45-nm CMOS-based 7-transistor SRAM has the following form [220, 221, 222]:

$$\text{Maximize:} \quad \widehat{f_{SNM}}(=156.6 - 44.0x_1 + 58.7x_2 - 53.9x_3 - 6.4x_4 + 32.5x_5 + 19.3x_6 - 19.6x_7) \quad (10.25)$$

$$\text{Subjected To:} \; 0 \leq x_1 \leq 1, 0 \leq x_2 \leq 1, 0 \leq x_3 \leq 1, 0 \leq x_4 \leq 1, 0 \leq x_5 \leq 1, 0 \leq x_6 \leq 1, 0 \leq x_7 \leq 1$$

$$f_{PWR} < \tau_{PWR} \quad (10.26)$$

where τ_{PWR} is the power consumption of the baseline design of the SRAM. The optimal solution of this ILP is the following solution set: $S_{SNM} = [x_1 = 0, x_2 = 1, x_3 = 0, x_4 = 0, x_5 = 1, x_6 = 1, x_7 = 0]$. To obtain the power optimal SRAM while maximizing the SNM, the overall solution set S_{OBJ} is obtained as the intersection of the two solutions as $S_{PWR} \cap S_{SNM}$; $S_{OBJ} = [x_1 = 0, x_2 = 1, x_3 = 0, x_4 = 0, x_5 = 1, x_6 = 1, x_7 = 0]$. The high-$V_{Th}$ values are assigned to these transistors for the power dissipation and SNM optimality of the SRAM circuit.

In the second, optimization flow called dual-threshold optimization-2 is presented in Fig. 10.29(b) [220, 221, 222]. In this flow, the predictive equations for normalized total power dissipation ($\widehat{f_{PWR}}^*$), and normalized SNM ($\widehat{f_{SNM}}^*$) are formulated for each of the transistors using DOE approach. The normalization of these two entirely different dimensional characteristics, i.e., Watt and Volt, allows formulation of a combined objective function that is to be minimized for the power dissipation and SNM optimality of the SRAM circuit. The combined objective function called $\widehat{f_{OBJ}}^*$ is formed as the ratio of $\widehat{f_{PWR}}^*$ and $\widehat{f_{SNM}}^*$. The ILP formulation for simultaneous power dissipation and read SNM maximization of a 45-nm CMOS-based 7-transistor is the following [220, 221, 222]:

$$\text{Maximize:} \quad \widehat{f_{OBJ}}^*(=0.18x_3 - 0.02x_4 + 0.11x_5 + 0.06x_6 - 0.06x_7) \quad (10.27)$$

$$\text{Subjected To:} \; 0 \leq x_1 \leq 1, 0 \leq x_2 \leq 1, 0 \leq x_3 \leq 1, 0 \leq x_4 \leq 1, 0 \leq x_5 \leq 1, 0 \leq x_6 \leq 1, 0 \leq x_7 \leq 1$$

$$f_{PWR} < \tau_{PWR}, f_{SNM} > \tau_{SNM} \quad (10.28)$$

The optimal solution that is obtained from this ILP formulation is the following: $S_{OBJ}^* = [x_1 = 0, x_2 = 1, x_3 = 0, x_4 = 0, x_5 = 1, x_6 = 1, x_7 = 1]$ and high-V_{Th} values can be assigned to the transistors accordingly.

The dual-V_{Th}-based SRAM design that is the result of the above design optimization flow-1 is presented in Fig. 10.30(a) [220, 221, 222]. At the same time, the dual-V_{Th}-based SRAM design that is the result of the above design optimization flow-2 is presented in Fig. 10.30(b) [220, 221, 222]. As is evident from the figures, the resulting solutions are not the same. One of the second configuration has

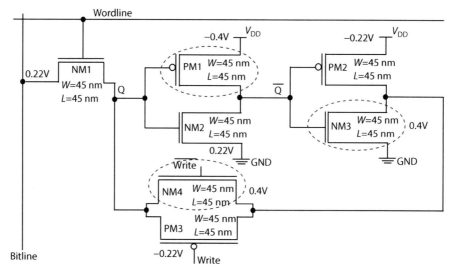

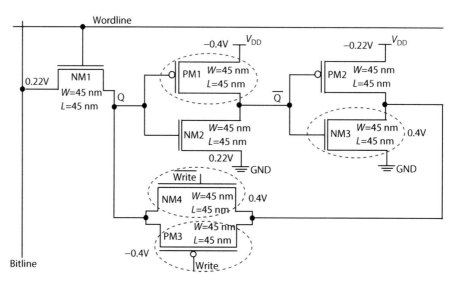

FIGURE 10.30 Dual-threshold voltage-based nano-CMOS SRAM topologies for 45-nm CMOS [220, 221, 222].

SRAM Design Alternative	SRAM Characteristics	Estimated Values	Changes
SRAM Design From Baseline Schematic	P_{SRAM}	203.6 nW	–
	SNM_{SRAM}	170.0 mV	–
SRAM Design From Optimization-1	P_{SRAM}	113.6 nW	44.2 % Decrease
	SNM_{SRAM}	303.3 mV	43.9 % Increase
SRAM Design From Optimization-2	P_{SRAM}	100.5 nW	50.6 % Decrease
	SNM_{SRAM}	303.3 mV	43.9 % Increase

TABLE 10.9 Characterization of Different 45-nm CMOS-Based 7T-SRAM Designs [220, 221, 222]

an additional transistor compared with the first configuration for high-V_{Th} assignment. The resulting two SRAM designs along with the baseline SRAM design are characterized for power dissipation and SNM. The characterization results for the three SRAM designs are presented in Table 10.9 [220, 221, 222]. It is observed that the decrease in the SRAM power dissipation is much more in the second SRAM design than in the first SRAM design for the same increase in the SNM. Thus, the use of the second design optimization flow is more beneficial.

10.8 Gate-Oxide Leakage Reduction Techniques

This section discusses different techniques that can be used for gate-oxide leakage reduction. First, a high-level overview of many possible techniques is presented. This is followed by a detailed discussion of specific examples at various levels of design abstraction for analog and digital ICs.

10.8.1 Brief Discussion

A detailed discussion including origin and modeling of gate-oxide leakage has been presented in the previous chapter on design issues for DFX. A simple explanation is that for sub-90-nm CMOS technology with an ultra-thin gate-oxide thickness (e.g., 0.7–1.2 nm), the carrier quantum mechanically tunnel through the gate-oxide layer [147, 162]. The probability of tunneling of a carrier is a function of the barrier height that is the voltage drop across gate oxide and the barrier thickness that is oxide thickness T_{ox}. The gate-oxide leakage current is present in the ON, OFF, and transition states of the nanoscale MOSFET. The various techniques that can be used for the reduction of gate-oxide leakage are presented in Fig. 10.31 [100, 109, 162, 167, 171].

It is observed that a small difference in the oxide thickness can lead to an order of magnitude change in the gate-oxide leakage current [120, 162, 209, 212, 235]. For example, an increase of gate-oxide thickness from 1.2 to 1.8 nm can decrease the gate-oxide leakage current by 1000×. Thus, an increase of oxide thickness is an attractive option for gate-oxide leakage reduction. However, increase of oxide thickness is contrary to technology scaling that dictates a decrease in oxide thickness. In addition, the increase in oxide thickness leads to an increase of threshold voltage and critical path delay. The multiple/dual oxide thickness technology is an attractive technique for gate-oxide leakage and propagation delay tradeoffs. A dual-oxide technique explores the usage of two values of oxide thickness, e.g., high-oxide thickness (T_{oxH}) and low oxide thickness (T_{oxL}), for gate-oxide leakage and performance tradeoffs. In this technique, low-oxide thickness logic gate or functional units are assigned to the critical path and high-oxide thickness logic gate or functional units are assigned to the noncritical path. The optimization exploration using dual-oxide thickness can be performed at the circuit-level, logic-level, or RTL. This is essentially a presilicon, static, or structural technique. The increase in T_{ox} leads to an increase in threshold voltage (V_{Th}) and reduction in the subthreshold leakage as well as discussed in the previous section.

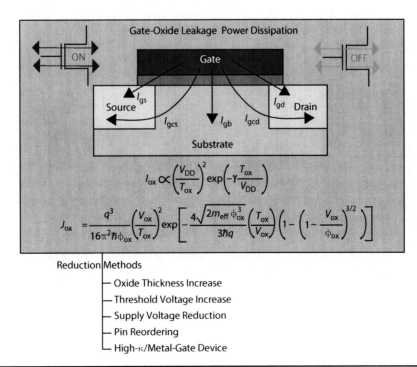

FIGURE 10.31 Different techniques for gate-oxide leakage reduction.

Dual-Oxide Technology	Dual-Threshold Voltage Technology	Dual-Channel Length Technology
One parameter varies	Several parameters vary	One parameter varies
Less area overhead	More area overhead	Less area overhead
Gate/subthreshold leakage affected	Subthreshold/gate leakage affected	Subthreshold/gate leakage affected
Higher leakage reduction	Higher leakage reduction	Medium leakage reduction
Dynamic power reduces	Dynamic power unaffected	Dynamic power increases
Less sensitive to process variation	More sensitive to process variation	Less sensitive to process variation
High process complexity	Low process complexity	Medium process complexity

TABLE 10.10 Comparative Perspective of Dual-Oxide and Dual-Threshold Technology [153, 212]

The increase in threshold voltage (V_{Th}) leads to reduction in the subthreshold leakage as discussed in the previous section. At the same time, an increase in V_{Th} may lead to a drop of voltage across the oxide V_{ox}. This in turn decreases the gate-oxide current density and gate-oxide leakage current. The threshold voltage increase can be static approach such as dual-threshold voltage (i.e., dual-V_{Th} or DTCMOS) or postsilicon approach such as variable-threshold voltage. Similarly, the increase in channel length increases the threshold voltage for short-channel devices. This in turn reduces the subthreshold leakage and gate-oxide leakage as evident from the above discussions. Thus, multiple-channel length CMOS (MLCMOS) or dual-channel length CMOS (DLCMOS) is a possible static or structural approach for leakage reduction. A comparative perspective of the dual-oxide, dual-threshold, and dual-channel length technologies is presented in Table 10.10 [153, 212].

The supply voltage reduction can decrease the gate-oxide leakage current [147, 152, 168]. Hence, supply voltage scaling through static and dynamic approaches can be helpful in gate-oxide leakage optimization. During the analysis of the leakage, it is observed that the magnitude of the subthreshold leakage current depends on the number of OFF transistors in the circuit. On the other hand, the magnitude of the gate-oxide leakage current depends on the position of the ON/OFF transistors as well as on their states (i.e., ON or OFF). For example, for a 2-input NAND gate, the magnitude of gate-oxide leakage current for input "10" and can be quite different from that of input "01" [148]. Thus, "pin ordering" in which the position of input pins are changed without changing the function of the overall stack of the transistors can reduce gate leakage [109, 209]. In fact, a combined approach involving pin order and "state assignment" can be quite effective in reducing the total standby leakage current of the circuit. The use of high-κ/metal-gate process technology for the nanoelectronic circuits can completely eliminate the issue of gate-oxide leakage [17, 101].

10.8.2 Dual-Oxide-Based Circuit-Level Optimization of a Current-Starved VCO

In this subsection, dual-oxide technique-based circuit-level optimization of a current-starved voltage-controlled oscillator (VCO) is presented [65, 70]. The proposed design optimization flow is presented in Fig. 10.32 [65, 70]. The design method for a current-starved VCO was presented in the chapter on phase-locked loop (PLL). In this flow, the baseline schematic design of VCO is performed that can be optimized at the schematic level as well. Once the specifications have been met at the schematic-level, the baseline physical design of the VCO is performed. Typically, the physical design is subjected to design rule check (DRC) and layout versus schematic (LVS) verification to ensure that a correct physical design has been obtained. At this stage, a netlist of the VCO can be extracted for the analysis of the VCO using an analog simulator. It is a designer's decision to include one or more types of parasitics in the netlist or even no parasitics. The inclusion of layout parasitics is better for accuracy, which will be the matter for subsequent discussions. A worst-case process analysis of the physical design of the VCO is performed for the target VCO performances, such as center frequency and phase noise.

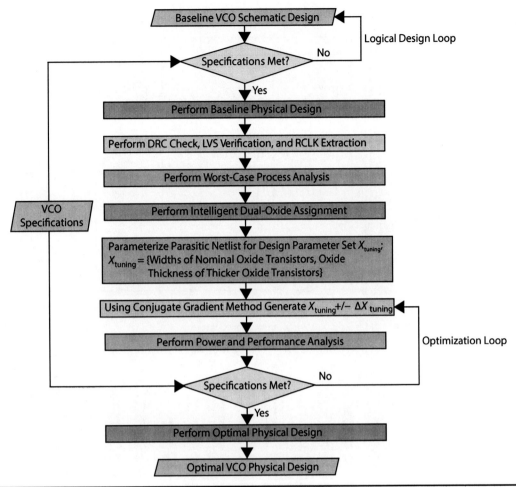

FIGURE 10.32 Design optimization flow for dual-oxide thickness-based low-power VCO [65, 70].

At this phase of the flow, dual-oxide transistors are explored for power and performance tradeoffs. For the purpose of assignment of dual-oxide thickness transistors, transient analysis is performed on the extracted netlist of the VCO. The average power dissipation of each of the transistors is monitored and the transistors are ranked in terms of their average power dissipation values. The number of transistors to be considered for the higher-oxide thickness is a tradeoff decision as well. For a 90-nm CMOS current-starved VCO, the transistors that are selected for higher-oxide thickness are circled in Fig. 10.33(a) [65, 70]. The input-stage transistors consume 48% of the total average power dissipation of the VCO. The output buffer stage transistors consumed 11.5% of the total average power dissipation of the VCO. A different design optimization decision can be just to consider the input-stage transistors. Yet another design choice can be to consider all the transistors and determine values of oxide thickness. In any case, there is a need for optimization to determine proper value of higher-oxide thickness for design tradeoffs.

For the purpose of optimization of the VCO, the VCO netlist is parameterized in terms of design parameter that will be tuned during the optimization process. This design parameter set selected for tuning is denoted as X_{tuning}. As a specific example, the width of all the transistors of the VCO and the oxide thickness of the input-stage transistors (T_{oxnin}, T_{oxpin}) are considered for tuning; in other words, to determine their specific values through optimization. Any automated algorithm can be used to perform this optimization over the parameterized netlist. As a specific example, a conjugate gradient-based optimization algorithm has been presented in this flow; however, any other algorithms can be used. The automated algorithm returns final values of the VCO tuning parameters after its convergence. A second physical design of the VCO is now performed that can be easily done by modifying the baseline physical design of the VCO.

The automated algorithm used for the optimization is based on the conjugate-gradient method. The conjugate gradient method has been used due to its low memory requirements and faster

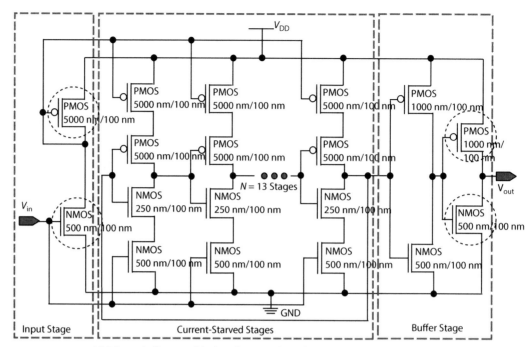

(a) Schematic diagram of dual-oxide 90-nm CMOS-based current-starved VCO.

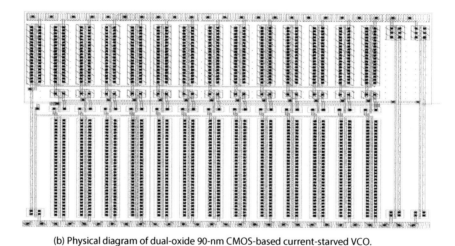

(b) Physical diagram of dual-oxide 90-nm CMOS-based current-starved VCO.

FIGURE 10.33 Design of dual-oxide thickness 90-nm CMOS-based low-power VCO [65, 70].

convergence [65, 74]. For the low-power VCO using the conjugate gradient method, the optimization problem is formulated as follows [65, 70, 74, 115, 240]:

$$\text{Minimize:} \quad P_{\text{VCO}}(X_{\text{tuning}}) \quad (10.29)$$

$$\text{Subjected To:} \; X_{\text{lower}} \leq X_{\text{tuning}} \leq X_{\text{upper}} \text{ and } f_c \geq f_{\text{spec}} \quad (10.30)$$

where P_{VCO} is the total power dissipation of the VCO. X_{lower} and X_{upper} are the lower and upper bounds of the tuning design variable X_{tuning}, respectively. f_{spec} is the frequency specification of the VCO. The detailed steps of the conjugate-gradient-based automatic iteration has been presented in Algorithm 10.3 [65, 70]. The range of VCO tuning parameter values has been presented in Table 10.11 [65, 70]. It also presents the optimal tuning variables that resulted from the conjugate-gradient method. The VCO design that resulted out of the optimal design flow has been shown in Fig. 10.33(a) [65, 70]. The final physical design of the 90-nm CMOS-based current-starved VCO is shown in Fig. 10.33(b) [65, 70]. The characterization results of this 90-nm CMOS VCO is presented in Table 10.12 [65, 70].

ALGORITHM 10.3 Conjugate Gradient-based Optimization of a Nano-CMOS VCO [65, 70]

1: **Input:** Parameterized VCO netlist, Target Specifications $[P_{VCO}, f_c]$, Maximum Number of Iterations N_{cnt},
2:         Tuning parameter set $X_{tuning} = [W_n, W_p, W_{ncs}, W_{pcs}, T_{oxnin}, T_{oxpin}]$, Tuning parameter limits $[X_{lower}, X_{upper}]$.
3: **Output:** Optimized tuning parameters $X_{optimal}$.
4: Perform initial simulations for lower and upper limits of the tuning parameters to analyze the feasibility of values of tuning parameters for the given specifications of the VCO.
5: Consider the lower limit of the tuning variable X_{lower} as the initial parameter set.
6: Initialize the optimal tuning parameter set as the lower limit of the tuning variable, $X_{optimal} \leftarrow X_{lower}$.
7: Initialize the iteration counter, $n_{cnt} \leftarrow 0$.
8: **while** ($n_{cnt} < N_{cnt}$) **do**
9:     Use the finite difference perturbation to generate new set of parameters X_{tuning^*} in the range from ($X_{tuning} - \Delta X_{tuning}$) to ($X_{tuning} + \Delta X_{tuning}$), based on design error margin by simultaneously varying design parameters as:
        (a) ($W - \Delta W$) to ($W + \Delta W$) for all nominal oxide thickness transistors.
        (b) ($T_{ox} - \Delta T_{ox}$) to ($T_{ox} + \Delta T_{ox}$) for thicker-oxide transistors.
10: Perform simulation to compute the VCO characteristics $[P_{VCO}, f_c]$.
11: Increment iteration counter, $n_{cnt} \leftarrow n_{cnt} + 1$.
12: Update the optimal tuning parameter as $X_{optimal} \leftarrow X_{tuning^*}$.
13: **end while**
14: **return** Optimal tuning parameter of the VCO, $X_{optimal}$.

Design Parameters (X_{tuning})	Lower Bound (X_{lower})	Upper Bound (X_{upper})	Optimal Values ($X_{optimal}$)
Inverter NMOS Width W_n	200 nm	500 nm	210 nm
Inverter PMOS Width W_p	400 nm	1 μm	415 nm
Current-Starved Circuitry NMOS Width W_{ncs}	1 μm	5 μm	8.5 μm
Current-Starved Circuitry PMOS Width W_{pcs}	5 μm	10 μm	5 μm
Input Stage NMOS Oxide Thickness T_{oxnin}	2.33 nm	5 nm	3.54 nm
Input Stage PMOS Oxide Thickness T_{oxpin}	2.48 nm	5 nm	5 nm

TABLE 10.11 Tuning of the VCO Design Parameters [65, 70]

VCO Design Characteristics	Estimated Values
Technology	90 nm CMOS 1P 9M
Supply Voltage	1.2 V
Center Frequency for Nominal Process	2.3 GHz
Center Frequency for Worst-Case Process	1.98 GHz
Tuning Parameter Set	6 ($W_n, W_p, W_{ncs}, W_{pcs}, T_{oxnin}, T_{oxpin}$)
Number of Objectives	2 (P_{VCO} = Minimum and $f_c \geq 2$ GHz)
Power Dissipation for Baseline Design	212 μW
Power Dissipation for Dual-Oxide Optimal Design	158 μW
Power Dissipation Reduction	25 %
Process and Supply Voltage Variations	V_{Th} (+10%), T_{ox} (+10%), V_{DD} (−10%)

TABLE 10.12 Characteristics of a Dual-Oxide 90-nm CMOS-Based Current-Starved VCO [65, 70]

10.8.3 Dual-Oxide-Based RTL Optimization of Digital ICs

A dual-oxide-based RTL optimization of digital IC is discussed in this subsection [147, 152, 153, 168]. In a previous section that discussed the presilicon techniques for power reduction, a similar RTL optimization flow has been presented that used Simulated Annealing as the core algorithm. In this section, the dual-gate oxide leakage optimization has been presented using ILP as the optimization algorithm. The overall RTL optimization flow has been presented in Fig. 10.34 [142, 153, 162]. The high-level steps, such as compilation, transformation, datapath scheduling, resource allocation, operation binding, connection allocation, and architecture generation, are similar to steps shown in the previous section. So for brevity, the dual-oxide technique-based gate-oxide leakage optimization using ILP is discussed in the rest of the subsection.

In this specific RTL optimization problem, the tradeoffs between gate-oxide leakage and critical path delay has been considered. For the unified gate-oxide leakage and delay optimization, a combined metric called leakage delay product (LDP) has been formulated that implicitly captures both leakage and delay characteristics of the target digital IC. The gate-oxide leakage optimization problem during HLS can be formalized as follows: for an unscheduled DFG $G_{RTL,U}(V,E)$ that represents the behavior of

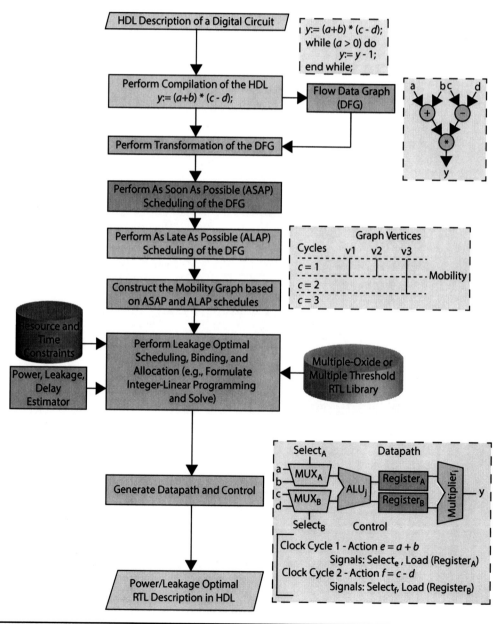

Figure 10.34 Dual-oxide or dual-threshold RTL optimization flow [153].

the circuit, it is required to determine the scheduled SDFG $G_{\text{RTLS}}(V,E)$ with appropriate resource binding such that the LDP is minimized for given resource constraints. Let us assume that V is the set of all vertices and V_{CP} is the set of vertices in the critical path from the source vertex of the DFG to the sink vertex. $FU_{k,t}$ denotes functional units of type k made up of transistors of a specific T_{ox}. t is called technology index that represents high-T_{ox} or low-T_{ox} for this dual-oxide technique. The total number of clock cycles in a DFG is N_{DFG} with any cycle denoted as c. The RTL optimization problem can then be stated as follows [142, 153]:

$$\text{Minimize: } \text{LDP}_{\text{DFG}} \tag{10.31}$$

$$\text{Subjected To: Allocated}(FU_{k,t}) \leq \text{Available}(FU_{k,t}), \quad \forall c \in N_{\text{DFG}} \tag{10.32}$$

In the above formulation, the objective function ensures minimization of gate-oxide leakage and delay simultaneously. The constraints ensure that the total allocation of the ith functional units of type k and made up of transistors of technology index t is less than the total number of same functional units available. The LDP of the DFG is estimated as the sum for all clock cycles using the following expression [153]:

$$\text{LDP}_{\text{DFG}} = \sum_{c=1}^{N_{\text{DFG}}} \text{LDP}_c = \sum_{c=1}^{N_{\text{DFG}}} \sum_{\forall v_{i,c}} P_{\text{ox}}(v_{i,c}) D_c \tag{10.33}$$

where, $v_{i,c}$ is a vertex v_i scheduled in cycle c of propagation delay D_c, and $P_{\text{ox}}(v_{i,c})$ is the oxide leakage of the corresponding functional unit active due to the execution of the same vertex.

The ILP formulation for LDP minimization through dual-oxide at RTL is mathematically similar to the EDP minimization through dual-voltage technique at RTL that has been discussed in a previous section on dynamic power reduction. Let us assume the following notation for ILP formulations: $N_{\text{FU},k,t}$ is the maximum number of $FU_{k,t}$ type functional units, $C_S[i]$ is the ASAP time stamp for the vertex v_i, $C_L[i]$ is ALAP time stamp for the vertex v_i, $\text{LDP}(i,t)$ is the LDP of $FU_{k,t}$ used by vertex v_i, $x_{i,t,c}$ is the decision variable that takes the value of 1 if v_i is using $FU_{k,t}$ and scheduled in clock cycle c, and $L_{i,t}$ is the latency in number of cycles for v_i using $FU_{k,t}$. The ILP formulation needs to minimize the LDP while satisfying the resource constraints and data dependency ensuring that every vertex is scheduled in uniquely allowable clock cycle. The formulations are presented for single-cycle scenario that can be easily modified for multicycling through $L_{i,t}$. For a multicycle operation, when a vertex uses a nominal-T_{ox} functional unit, then it is scheduled in one unique clock cycle. On the other hand, when a vertex uses a high-T_{ox} functional unit, then it may need more than one clock cycle $L_{i,t}$ for completion, thus restricting the mobility of that vertex.

Objective Function: The objective is to minimize the LDP of the whole DFG over all control steps. This can be expressed using decision variable as follows:

$$\text{Minimize: } \sum_c \sum_i \sum_t x_{i,t,c} \text{LDP}(i,t) \tag{10.34}$$

Uniqueness Constraints: The uniqueness constraints ensure that each vertex v_i is scheduled in the appropriate control step within the mobility range $(C_S[i], C_L[i])$ being assigned the resource $FU_{k,t}$. A vertex may be operated with more than one clock cycle sometimes, depending on the propagation delay of a functional unit. The uniqueness constraints are represented as follows:

$$\sum_c \sum_t x_{i,t,c} = 1, \quad \forall i, 1 \leq i \leq V \tag{10.35}$$

Precedence Constraints: The precedence constraints make sure that for a vertex v_i, all its predecessors are scheduled in earlier clock cycles and its successors are scheduled in later clock cycles. The precedence constraints are modeled as follows:

$$\sum_t \sum_{d=C_S[i]}^{C_L[i]} dx_{i,t,d} - \sum_t \sum_{e=C_S[j]}^{C_L[j]} ex_{j,t,e} \leq -1, \quad \forall i,j, v_i \in \text{Pred}_{v_j} \tag{10.36}$$

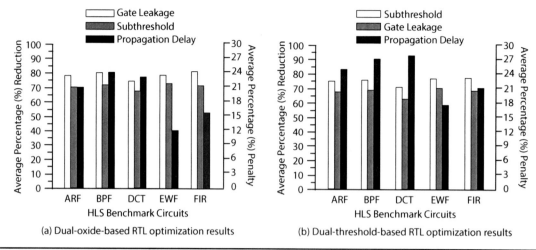

FIGURE 10.35 Dual-oxide or dual-threshold RTL optimization results [153].

Resource Constraints: The resource constraints ensure that each clock cycle needs functional units not exceeding available number of functional units. The resource can be enforced as follows:

$$\sum_{i \in \text{FU}_{k,t}} x_{i,t,c} \leq N_{FU,k,t}, \quad \forall t, \forall c, 1 \leq c \leq N_{DFG} \tag{10.37}$$

The above ILP formulation can be solved in the context of HLS by following the steps that are similar to the steps followed during ILP solutions for dual-supply voltage technique for dynamic power reduction in a previous section. For brevity, the steps are not again presented in this section. The results obtained for various DSP benchmark circuits are presented in Fig. 10.35(a) [153]. The reduction in the gate-oxide leakage and the accompanying subthreshold leakage are significant for all the benchmark digital circuits. The delay penalty as a result of the optimization is presented in the same figure for the benchmark circuits. The above optimization flow, ILP formulation, and corresponding solution can be similarly performed using dual-threshold voltage technique for subthreshold leakage reduction. The results from such an approach are presented in Fig. 10.35(b) [153]. As evident from the figure, the subthreshold leakage reduction is quite significant for all the benchmark digital circuits. The accompanied gate-oxide leakage is also significant for all the circuits. There is some delay penalty as a result of the optimization process.

10.9 Parasitics: Brief Overview

The parasitics that arise in the ICs, which are not actually designed by the designers or built by the manufacturers, are the single most important challenge for the design engineers that consume a large portion of the design cycle time. All the real-life nanoelectronic IC elements such as transistors, diodes, and even metal wires have significant parasitics that create discrepancies between the actual and ideal characteristics of the ICs. In a previous chapter on the issues for DFX, parasitics were briefly introduced. In an IC, the parasitics arise from both active elements as well as the interconnects. The parasitics arise due to fact that neither the active elements nor the interconnects are perfect due to their structure limitations, fabrication process limitations, as well as the building materials limitations. A quick overview of different parasitics are presented in Fig. 10.36 [16, 43, 65, 69, 144, 176]. The parasitics in the ICs can arise from different active and passive elements. The parasitics elements that originate can be either passive or active in nature. The different parasitic elements can be spread across the element, for example, resistance of the metal wire or silicon substrate. At the same time, many different parasitics can be lumped in nature, for example, the gate capacitance of the MOSFET. What is crucial for the design engineers is that the exact and accurate prediction of the AMS-SoC component characteristics becomes challenging

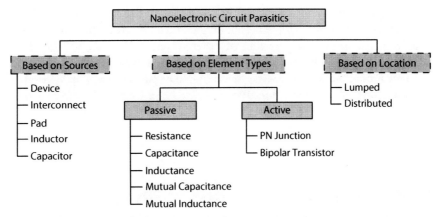

(a) Different types of parasitics in a nanoelectronic IC.

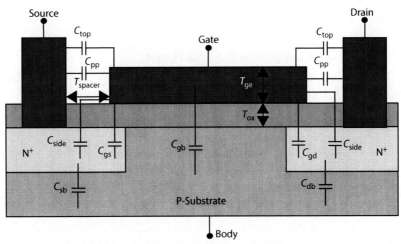

(b) Different parasitics capacitor components in a nanoscale MOSFET.

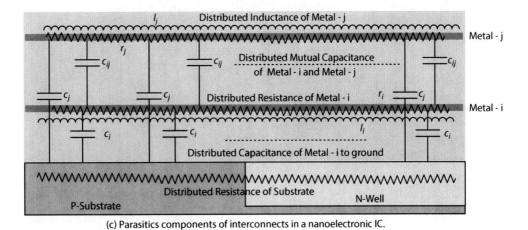

(c) Parasitics components of interconnects in a nanoelectronic IC.

FIGURE 10.36 Brief overview of parasitics in nanoelectronic integrated circuits.

irrespective of digital, analog, and RF components in the presence of these large and complex networks of parasitics.

The parasitics have profound impact on various characteristics of the ICs such as dynamic power, noise, and signal propagation delay. There is a huge gap between the ideal and nonideal (parasitics inclusive) characteristics of the ICs. The design decisions taken without accounting for the parasitics are inaccurate, as fabricated ICs will not match the design closure time specifications. For various reasons including design efficiency improvement, yield improvement, and cost reduction, it is essential to predict the effects of the parasitics such that appropriate compensations can be made during the design time. For the design engineers, many issues including the following arise: (1) how to accurately capture the parasitics of different basic elements; (2) how to model the parasitics of different AMS-SoC blocks; (3) how to model for the parasitics of a full-chip; and (4) how to model the parasitics at the different levels of the digital circuit abstractions. Various methods for extraction and modeling of the above parasitics are discussed in the following sections. This is followed by parasitic-aware design optimization techniques. Parasitic-aware design and synthesis flow is an approach to incorporate the parasitics effects during the early phase of the design cycle of ICs and AMS-SoCs [16, 43, 65, 69, 176].

10.10 The Effects of Parasitics on Integrated Circuits

The differences between layout and schematic design is first discussed in this section with selected examples that arise due to the emergence of the parasitics elements. Then, a broad overview is presented discussing various effects of parasitics on the IC characteristics.

10.10.1 Parasitics in Real-Life Example Circuits

To provide the difference between schematic- and layout-level designs of ICs, two small examples are presented in Fig. 10.37 [64, 150]. A 180-nm CMOS-based LC-VCO design is shown in Fig. 10.37(a) [64, 150]. As is shown, the schematic design that has no parasitics has five elements only, with two transistors that are the active elements, two capacitors and one inductor, which are explicitly designed passive elements. In the layout design, there are a total of 724 elements with two transistors as active elements. However, the number of capacitors, inductors, and resistors are significant in number in the layout design even though only two capacitors and one inductor were explicitly designed. All these parasitics will significantly alter the simulation time, design exploration time, and characteristics of the LC-VCO. Specifically, for this 180-nm CMOS-based LC-VCO, the simulation time for the layout is increased by 3-fold as compared to the schematic design. The oscillating frequency is decreased by 20% for this LC-VCO design. As a second example, a 45-nm CMOS-based ring oscillator design is shown in Fig. 10.37(b) [64, 150]. In the schematic design, there are six transistors, which are actually designed by the design engineers and should be there in the layout design. However, as is shown, the layout design has a total of 107 elements containing 82 capacitors and 19 resistors, along with the designer-designed 6 transistors. Thus, the simulation and design exploration will be quite involved due to some additional elements that appeared in the IC. Specifically, for this 45-nm CMOS ring oscillator, the simulation times increased by three times due to the emergence of the parasitics. The oscillating frequency of the layout is decreased by 40% as compared with the schematic design of this ring oscillator.

10.10.2 Effects of the Parasitics

The parasitics have severe effects on all the types of the circuits including digital, analog, and mixed-signal. The effects may be ignored at the low-frequency operation IC, however, for the high-frequency operation ICs, the effects are most severe and should not be ignored during design phase. An overview of the effects of the parasitics is shown in Fig. 10.38 [16, 43, 65, 69, 144].

The parasitics can increase the signal propagation delay in an IC. For example, the parasitics can be additive to the capacitance and resistance and cause larger RC delay of the circuit. The parasitics can increase power dissipation of the IC. For example, the parasitic capacitances can be additive to the load capacitance and in effect switching a larger capacitance, thus increasing the power dissipation. Moreover, the parasitic diodes increase the leakage through the junction leakage mechanism. Signal noise is created by parasitic resistances, capacitances, and inductances. Increase in power supply noise as well as ground noise occurs due to parasitic resistances, capacitances, and inductances. The parasitics such as mutual capacitances and mutual inductance can be sources for coupling and lead to noise called cross-talk. In essence, the cross-talk means a change in the signal in one wire that is manifested as a signal in a victim wire. These signal interferences can lead to unreliable operations. For example, in closely spaced computer buses and cables, the signals from

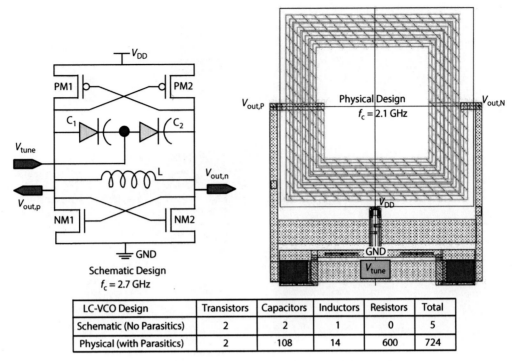

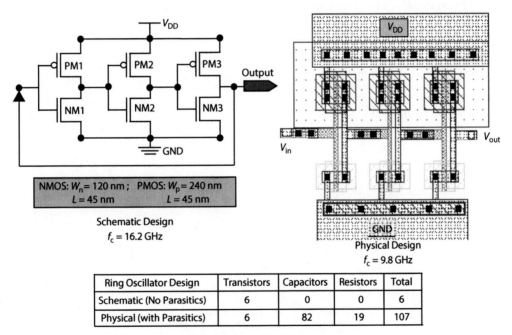

Figure 10.37 Parasitic elements in the designs of LC-VCO and ring oscillator [64, 150].

one block or IC can affect the signals of another block or IC, thus causing unreliability in the victim bus or cable.

The various circuit performances, such as power and propagation delay, are affected by the parasitics as discussed above. Many other circuit performances are specific to analog circuits. For example, in an amplifier circuit, it is possible that the parasitic capacitance between the output and input can act as a feedback path that will lead the circuit to oscillate at higher frequency. These

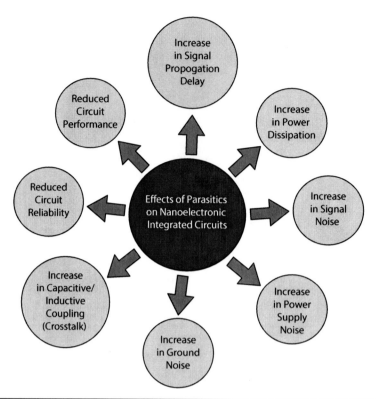

FIGURE 10.38 Various effects of parasitics in nanoelectronic circuits.

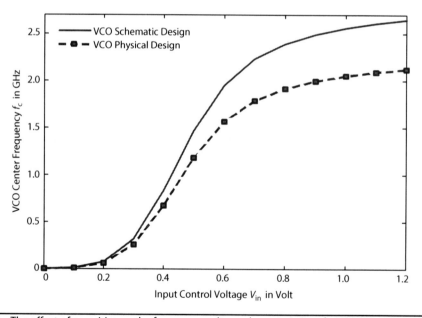

FIGURE 10.39 The effect of parasitics on the frequency-voltage characteristics of a VCO [65, 69].

unwanted oscillations of the amplifier circuit are called parasitic oscillations. Similarly, in the case of VCOs, the parasitics can affect its characteristics such as center frequency and phase noise. As a specific example, the frequency versus tuning voltage characteristics for a 90-nm CMOS-based current-starved VCO for its schematic and layout implementations are shown in Fig. 10.39 [65, 69]. As is evident from the characteristics, the layout design that contained RCLK parasitics has 25% degradation in the oscillating frequency.

10.11 Modeling and Extraction of Parasitics

The parasitics of ICs are originated from various sources, materials, and geometry. While the parasitics in some cases are in lumped form, these are in distributed form in others. Extraction and modeling of these parasitics are very complex tasks as would be evident from the discussions in this section. This section focuses on these aspects of the parasitics.

10.11.1 Signal Propagation: In a Real Wire

Before moving on to the details of parasitics modeling and extraction, it is important to comprehend the issues of signal propagation even in a metal wire or interconnect in an IC. Let us consider the hypothetical case illustrated in Fig. 10.40 [57, 130, 189]. It shows an inverter driver gate sending a signal through a nonideal metal interconnect to a receiver inverter gate. In many nanoelectronic circuits, the interconnect can be a long transmission line of maximum length X_L. It can have all the forms of parasitics that in a real-life scenario are distributed in nature. A simple question: If at time $t = 0$, a input signal is applied, then what is the voltage in the interconnect at any distance x from the origin/source $x = 0$ and at any time t, i.e., $V(x,t) = ?$ This is a difficult problem to analyze, as it involves both space x and time t variables. The accurate solution involves second order partial differential equations (PDEs).

In a more accurate representation of long interconnects, it can be a transmission line with all the distributed parasitics as depicted in Fig. 10.40(b) [57, 130, 189]. In this interconnect, the signal propagates as a wave by alternatively transferring energy from electric (in capacitors) to magnetic

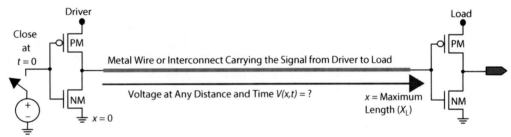

(a) Signal propagation from a driver to load through a transmission line

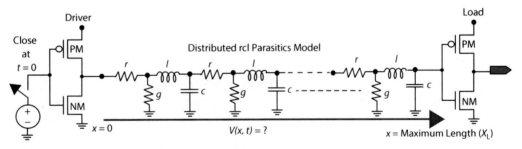

(b) Signal propagation from a driver to load through a transmission line with distributed rcl parasitics

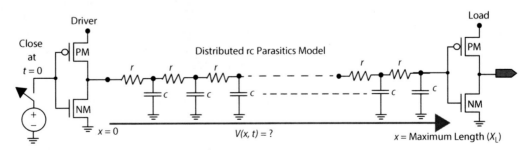

(c) Signal propagation from a driver to load through a transmission line with distributed rc parasitics

FIGURE 10.40 Signal propagation in a transmission line [10, 57, 130, 189].

field (in inductors). The governing expression for this wave is called the "wave propagation equation" that is expressed as follows [45, 189]:

$$\frac{\partial^2 v}{\partial x^2} = lc\frac{\partial^2 v}{\partial t^2} + rc\frac{\partial v}{\partial t} \qquad (10.38)$$

where r, c, and l are per unit parasitic resistance, capacitance, and inductance, respectively. The above equation needs to be solved to accurately determine $V(x,t)$.

In a simpler scenario, when the inductance of the interconnect can be neglected, the interconnect can be modeled as a transmission line as depicted in Fig. 10.40(c) [57, 130, 189, 237]. In this interconnect, the signal diffuses from the source to the destination. The governing expression for this diffusion is called the "diffusion equation" that is expressed as follows [189, 237]:

$$\frac{\partial^2 v}{\partial x^2} = rc\frac{\partial v}{\partial t} \qquad (10.39)$$

The above equation needs to be solved to accurately determine $V(x,t)$. This is simpler to solve in comparison to the wave propagation equation.

It is important to note that the analog simulators such as SPICE cannot solve the PDEs. Hence, solving the interconnects and the use of the above governing equations determined $V(x,t)$ are not an easy task. There are many methods available to solve such PDEs, but they may be quite slow. These are even slower and computationally intensive than the SPICE that solves the ordinary differential equations (ODEs) only. However, SPICE uses lumped models for the interconnects like Π and T models to model the interconnects with parasitics.

10.11.2 Parasitics Modeling and Simulation: The Key Aspects

To accurately represent the real wire or interconnects and then perform the IC simulation and design exploration, many key aspects need to be fulfilled. The key aspects of the parasitic modeling and simulations have been presented in Fig. 10.41 [199, 200]. Device parasitic extraction process calculates the parasitics of the active devices. These are mostly lumped in nature. The interconnect parasitics extraction estimates the parasitics of the wires. These are distributed in nature and hence the interconnect parasitics extraction process is much more complex than the device parasitics extraction. The parasitics modeling techniques deal with analysis of the signal propagation in the presence of parasitics. The model order reduction (MOR) involves development of simpler models so that IC simulations including parasitics can be performed with minimal usage of computational resources. All of these aspects will be discussed in the rest of this subsection.

10.11.3 Circuit (Device+Parasitic) Extraction Process

In general, for the purpose of simulations of the physical design (layout) of IC using analog/SPICE simulators, the electrical circuit extraction from the physical design is necessary [1, 194]. The "electrical circuit extraction" process measures the actual shapes and spacings of the lithography mask layers and predicts the electrical characteristics of the physical design. The first step of the electrical circuit extraction is called "device extraction" that finds the circuit elements such as transistors, diodes, and spiral inductors, which are explicitly designed by the design engineers. The next phase of the electrical circuit extraction process is called the "parasitics extraction" process that calculates the unwanted/byproducts of the devices and the interconnects. The parasitics extraction is the calculation of the parasitics associated with active devices as well as the wire interconnects of an IC [95, 199, 200]. The electrical circuit (device+parasitic) extraction process in essence creates a reasonably accurate

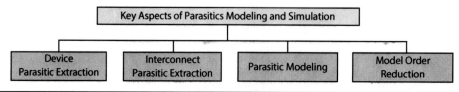

Figure 10.41 Key aspects of parasitics modeling and simulation.

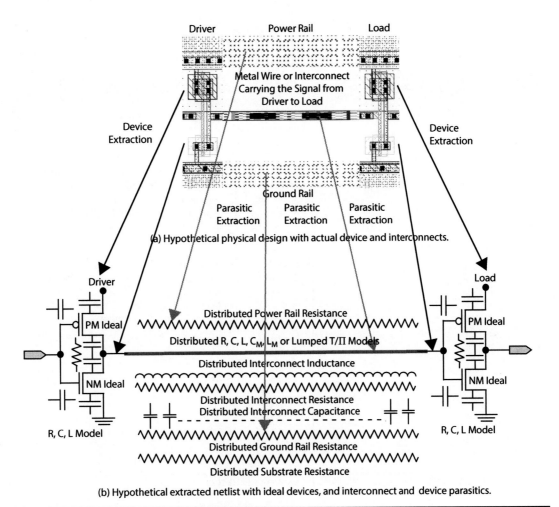

FIGURE 10.42 Illustration of device and parasitic extraction.

analog model of the overall (analog/digital/mixed-signal) IC to perform its simulation using analog/ SPICE simulators. The key concept of the electric circuit (device+parasitic) extraction process is depicted in Fig. 10.42. An inverter driver driving an inverter load through a long interconnect is presented in Fig. 10.42(a). In addition, there are power and ground rails as well as silicon substrate as shown in the case of any IC.

The electrical circuit extraction from the physical design is depicted in Fig. 10.42(b). The device extraction process extracts the active devices, i.e., NMOS and PMOS ideal transistors. This step is useful in LVS phase in the design flow. Then, the resistance and capacitance parasitics of these NMOS and PMOS devices are calculated. It may include inductance parasitics of these NMOS and PMOS devices for high-frequency operations. These parasitics are lumped values. The parasitics extractions for power line, ground line, interconnect, and substrate can be challenging. First, these need to be extracted as distributed parameters. Of course, for simplified usage, lumped parameter can be extracted as well. For the power line, distributed resistance is extracted. The power line can exhibit distributed capacitance and distributed inductance as well as mutual capacitances that can be effective based on operating conditions. In addition, there can be distributed mutual capacitances with respect to signal interconnects or even with the ground line. The ground line may exhibit the similar parasitics as the power line, but their effectiveness will depend on the potential differences and current paths. For example, for the ground line, capacitance with respect to the substrate may be ineffective as both may be at ground potential. The signal interconnect can have distributed resistance, capacitance, and inductance. It can have various mutual parasitics such as distributed mutual capacitance and distributed mutual inductance. The substrate can exhibit distributed parasitic resistances. Instead of distributed parasitics, lumped T or Π models are extracted to perform simulations using existing analog/SPICE simulators.

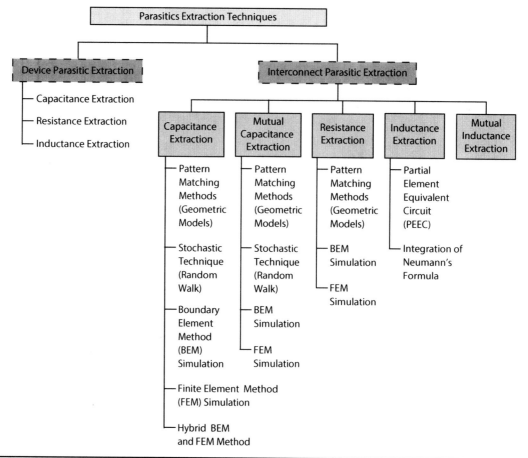

Figure 10.43 Different techniques for parasitics extraction.

10.11.4 Parasitics Extraction Techniques

The different parasitics extraction techniques are presented in Fig. 10.43 [199, 200]. The device parasitics extraction are done mostly using "pattern matching methods" or "geometric models" [1, 194, 238]. In this method, a rule-based extractor uses layout pattern matching (pattern recognition) algorithms and predefined models placed in a database to extract the parasitics.

The interconnect parasitics extractions may vary depending on if capacitance, mutual capacitance, resistance, inductance, or mutual-inductance is extracted. There are four different types of interconnect parasitics extractions, viz. [200]: (1) pattern matching methods or geometric model-based methods, (2) stochastic techniques (e.g, random walk), (3) numerical methods, and (4) analytical method. The pattern matching methods or geometric model-based methods extract parasitics based on predefined models in a database. The pattern matching methods are quite fast, but may have an inaccuracy issue for complex geometries with multiple layers. Hence, they may not perform accurate parasitics extraction for nanoelectronic ICs. The stochastic techniques such as random walk can be used for capacitance extraction. The stochastic technique has low memory requirement, and hence has capability for full-chip parasitic extraction.

The numerical methods include boundary element method (BEM) and finite element method (FEM) [200]. Both of these techniques have their own advantages and disadvantages. For example, BEM results in a small matrix, but is only efficient for regular stratified dielectrics. FEM results in a large sparse matrix and iterative solvers can be applied. To overcome that, a hybrid BEM and FEM method can be used. A fast numerical method is called "measured equation of invariance" that is geometry independent. An "analytical method" is used for capacitance calculation of simple structures such as an infinitely long straight line over a ground plane.

The inductance extraction can be used using the partial element equivalent circuit (PEEC) approach [199, 200]. PEEC approach is essentially an integral equation-based method. PEEC models are used to describe the discrete approximation of the electric field integral equation (EFIE) that represents the electromagnetic problem for structure such as interconnects and packaging. The solutions of the EFIEs lead to the extraction of the parasitics.

The inductance as well as mutual inductance can be calculated by solving the Neumann formula [200]. For the self-inductance, the Neumann formula can be reproduced to the following [54]:

$$L = \left(\frac{\mu_0}{4\pi}\right)\left(\oint \frac{d\mathbf{x}.d\mathbf{x}'}{|\mathbf{x}-\mathbf{x}'|}\right)_{|s-s'|>a/2} + \left(\frac{\mu_0}{4\pi}\right) K_J l + \cdots \quad (10.40)$$

where μ_0 is the permeability of the free space, $\oint$ is integral over closed path, $\mathbf{x}$ is a vector distance along the wire axis, s is a length along the wire axis, a is wire radius, K_J is a constant depending on the distribution of the current in wire cross section, and l is the length of the wire. For the mutual inductance, the Neumann formula gives the following [54]:

$$L_M = \left(\frac{\mu_0}{4\pi}\right)\left(\oint \frac{d\mathbf{x}_1.d\mathbf{x}_2}{|\mathbf{x}_1-\mathbf{x}_2|}\right) \quad (10.41)$$

In the above expression, there are two current loops. $\mathbf{x}_1$ and $\mathbf{x}_2$ are vectors distances in the direction of these current loops. Many numeric techniques can be used to solve the Neumann formula that will lead to the extraction of the parasitics [200]. For simple structures, analytical methods can be used to solve the formulation.

10.11.5 Parasitics Modeling

In a conventional parasitics modeling approach, the interconnect parasitics are extracted to form the approximate representation of the ICs [95, 199, 200]. These lumped models can then be evaluated in an analog/SPICE simulator for the purpose analysis. However, such a modeling approach that relies on lumped parameters may not always be accurate, for example, for high-frequency operations. In such a situation, distributed models as well as higher-dimensional modeling is needed. There are several ways of parasitic modeling as presented in Fig. 10.44 [95, 199, 200]. The time domain as well as frequency domain modeling of the parasitics can be performed through Maxwell equations. The one-dimensional (1D), distributed transmission line modeling is performed as *rlc* lossy transmission line, *rc* lossy transmission line, or *lc* lossless transmission line. Based on the accuracy need, the parasitics modeling may involve higher dimensions such as two-dimensional (2D), 2.5-dimensional (2.5D), and three-dimensional (3D) solutions. The following are four Maxwell equations in the differential form [199]:

$$\begin{aligned}
\nabla \cdot \mathbf{D} &= q & \text{(Gauss Law)} \\
\nabla \cdot \mathbf{B} &= 0 & \text{(Gauss Law for Magnetism)} \\
\nabla \times \mathbf{E} &= -s\mathbf{B} & \text{(Faraday Law of Induction)} \\
\nabla \times \mathbf{B} &= \mu \mathbf{J} + \mu\varepsilon s\mathbf{E} & \text{(Ampere Law)}
\end{aligned} \quad (10.42)$$

In the above expressions, s is either $\left|\frac{\partial}{\partial t}\right|$ for time-domain modeling or it is the Laplace transform variable. $\mathbf{D}$ is electric displacement, $\mathbf{B}$ is magnetic field, $\mathbf{E}$ is electric filed, $\mathbf{J}$ is current density, μ is

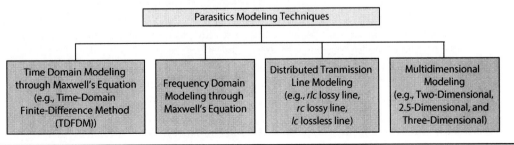

FIGURE 10.44 Different techniques for parasitics modeling.

permeability, and ε is permittivity. The complete set of Maxwell equations is considered for full-wave simulation. The time-domain solution of the Maxwell equations has issues of instability, boundary condition inaccuracy, and high computational overhead. The frequency domain solution of the Maxwell equations is easier to handle.

10.11.6 Parasitics Model Order Reduction
This subsection discusses the need and options for parasitics MOR. The widely used technique for RC-delay calculation "Elmore Delay model" has also been discussed with specific examples.

10.11.6.1 Different Methods for Model Order Reduction
For most accurate modeling, 3D full-wave solution of the PDEs is needed [199, 200]. This often needs a large algebraic system the solution of which is time consuming and highly computational resource intensive. Therefore, there is a need for approaches to reduce the complexity of such solution spaces. MOR that is applied for linear systems is a mechanism to generate models with lower complexity. The two key requirements are the following: (1) the reduced-order model should capture the input-output behavior of the original system in the desired frequency range with acceptable accuracy. (2) The reduced-order model should have efficient representations for both time-domain and frequency-domain solutions. Parasitic MOR approaches are of two types [49, 52, 134, 182, 199]: (1) physics-based MOR and (2) mathematical-based MOR. Some examples of the MOR are the following: the model reduction has been made using the Eigen subspaces of small dimension and a SPICE compatible $Y - \Delta$ method reduces model order while improving simulations stability [48, 62, 187].

10.11.6.2 Elmore Delay Model : Any Generic Network
A modeling approach that simplified the job of parasitics modeling and delay analysis is Elmore delay model [10, 57, 130, 189]. The Elmore delay model, even though proposed in 1948, is still used because of its simplicity. Elmore delay model only considers resistance and capacitance parasitics, but not inductor parasitics. It also does not consider mutual capacitance or mutual inductance parasitics. Thus, the use of Elmore delay model for the nanoelectronic IC has been a concern; however, it is still used for determining the upper bound of the delay, specifically the RC delay. Improvised versions of the Elmore delay called the transformed Elmore delay (TED) model and fitted Elmore delay (FED) are available as well [10].

To understand the use and advantage of Elmore delay model, a random network of logic gates is presented in Fig. 10.45(a). It represents a hypothetical scenario in which several inverters are present. These are connected with each other through metal interconnects. While in a real-life scenario, the interconnects have distributed resistance, capacitance, inductance, mutual capacitance, and mutual inductance, for simplicity lumped T or Π models can be used or higher order lumped T or Π models. Higher order lumped models such as T2 or Π2 models can be used to improve accuracy. In these, capacitance calculations, both the parallel component as well as fringing components can be used for the best possible accuracy. For the transistors, the R and C models with lumped values for gate, drain, source, and body terminals can be used.

The lumped RC model of the above circuit is presented in Fig. 10.45(b). In this RC network, any node capacitance C_i is a combination of capacitances from interconnects as well as gate, source, or drain capacitances from transistors depending on their terminal connective of different capacitance components. The resistance between two nodes is the lumped values of the interconnect resistances for that length. Any two random nodes have been marked as "Source" and "Destination" nodes. The objective is to calculate the RC delay between these two nodes. The calculation is not straightforward as the current path is affected not by just the path of our interest but also by any other paths that can come its way. The Elmore delay model makes the job easy for the design engineers. The steps for the calculation of the Elmore delay are presented in Fig. 10.46 [57, 130, 189]. By following these steps, the delay between the source and destination nodes of Fig. 10.45(b) can be calculated as follows. The "main path" from the source node to the destination node is R_{01}, R_{12}, R_{24}, R_{46}, and R_{67}. For the node-1, the path from source node is R_{01} and the "shared path" with the main path is R_{01}. Thus, the product of the node-1 capacitance and the shared resistances is $R_{01}C_1$. Similarly, for the node-2, the path from source node is R_{01}, R_{12} and the "shared path" with the main path is R_{01}, R_{12}. Thus, the product of the node-2 capacitance and the shared resistances is $(R_{01}+R_{12})C_2$. Similarly, for the node-3, the path

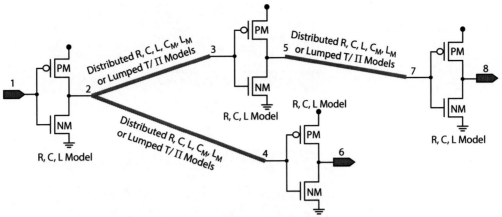

(a) A random network gates with wires (Interconnects) connecting them.

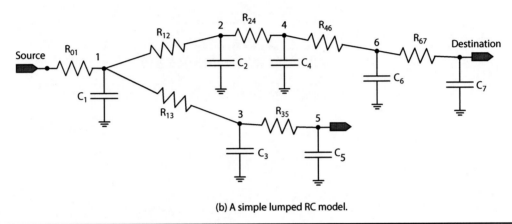

(b) A simple lumped RC model.

FIGURE 10.45 Illustration of Elmore delay modeling.

from source node is R_{01}, R_{13} and the "shared path" with the main path is R_{01}. Thus, the product of the node-3 capacitance and the shared resistances is $R_{01}C_3$. In the similar manner, node-4, node-5, node-6, and node-Destination can be covered. The overall RC delay between the source and destination node is the following:

$$D_{\text{RC,Source-Destination}} = R_{01}C_1 + (R_{01} + R_{12})C_2 + R_{01}C_3 + (R_{01} + R_{12} + R_{24})C_4 + R_{01}C_5 \\ + (R_{01} + R_{12} + R_{24} + R_{46})C_6 + (R_{01} + R_{12} + R_{24} + R_{46} + R_{67})C_7 \quad (10.43)$$

10.11.6.3 Elmore Delay Model: A RC-Chain Network of Lossy Long Transmission Line

Let us now revisit the lossy transmission line that is shown in Fig. 10.47(a) [57, 130, 189] and was also presented in another section before. It was discussed that the voltage $V(x,t)$ can be calculated by solving either the wave propagation equation for a distributed *rcl*-parasitics modeling or diffusion equation distributed *rc*-parasitics modeling. The second-order PDE solution needed can be complex, slow, and computational intensive. For the distributed *rc*-parasitics modeling, the solution of the diffusion equation will lead to the following RC-delay between $x = 0$ and $x = X_L$:

$$D_{\text{RC},X_L,\text{DE}} = \left(\frac{rcX_L^2}{2}\right) \quad (10.44)$$

The delay in the above lossy transmission line can be calculated using a lumped model presented in Fig. 10.47(b) [57, 130, 189]. The line has N number of nodes with individual node capacitance as

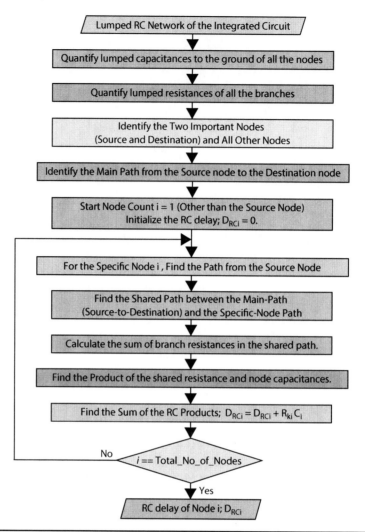

Figure 10.46 Steps for the Elmore delay calculation in a lumped RC network.

shown. The resistances between the nodes are also shown. Using the Elmore delay calculation steps, the RC delay of this lossy transmission line for the overall length can be calculated as follows:

$$D_{RC,X_L,\text{Elmore}} = R_1 C_1 + (R_1 + R_2)C_2 + \cdots + (R_1 + R_2 + \cdots + R_N)C_N \tag{10.45}$$

Let us assume the length of each segment to be equal with equal branch resistance of value $\left(\frac{R}{N}\right)$ each and equal node capacitance of value $\left(\frac{C}{N}\right)$. Under this assumption, the above expression can be rewritten as follows:

$$D_{RC,X_L,\text{Elmore}} = \frac{R}{N}\frac{C}{N} + \left(\frac{R}{N} + \frac{R}{N}\right)\frac{C}{N} + \cdots + \left(\frac{R}{N} + \frac{R}{N} + \cdots + \frac{R}{N}\right)\frac{C}{N} \tag{10.46}$$

$$= \frac{R}{N}\frac{C}{N} + 2\frac{R}{N}\frac{C}{N} + \cdots + N\frac{R}{N}\frac{C}{N} \tag{10.47}$$

$$= (1 + 2 + \cdots + N)\frac{R}{N}\frac{C}{N} \tag{10.48}$$

$$= \left(\frac{N(N+1)}{2}\right)\frac{R}{N}\frac{C}{N} \tag{10.49}$$

$$\approx \left(\frac{RC}{2}\right) \tag{10.50}$$

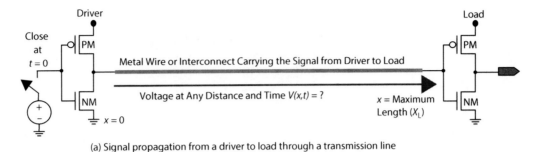

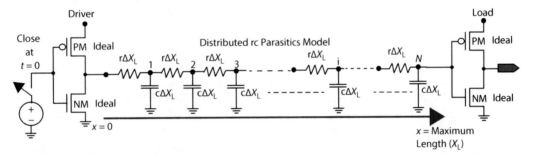

FIGURE 10.47 Elmore delay modeling for RC or rc chain lossy transmission line [57, 130, 189].

The above delay will be equal to the delay value of the $D_{RC,X_L,DE}$, which is obtained from the diffusion equation solution, if the overall length is assumed as one segment with a lumped resistance R and lumped capacitance C, in other words, $R = rX_L$ and $C = cX_L$.

As another approach, the delay in the above lossy transmission line can be calculated using a distributed rc chain as shown in Fig. 10.47(c) [57, 130, 189]. Let us assume that each of the nodes is equidistant from each other so that the overall X_L length is divided into equal-length segments of length $\left(\Delta X_L = \frac{X_L}{N}\right)$. The lumped resistance of each branch is $r\Delta X_L$. The capacitance of each node is $c\Delta X_L$. Using the Elmore delay, calculation steps, the RC delay of this lossy transmission line for the overall length can be calculated as follows:

$$D_{RC,X_L,\text{Elmore}} = (r\Delta X_L)(c\Delta X_L) + \big((r\Delta X_L) + (r\Delta X_L)\big)(c\Delta X_L) + \cdots \tag{10.51}$$

$$+ \big((r\Delta X_L) + (r\Delta X_L) + \cdots + (r\Delta X_L)\big)(c\Delta X_L) \tag{10.52}$$

$$= (r\Delta X_L)(c\Delta X_L) + 2(r\Delta X_L)(c\Delta X_L) + \cdots + N(r\Delta X_L)(c\Delta X_L) \tag{10.53}$$

$$= \big(1 + 2 + \cdots + N\big)(r\Delta X_L)(c\Delta X_L) \tag{10.54}$$

$$= \left(\frac{N(N+1)}{2}\right)(r\Delta X_L)(c\Delta X_L) \tag{10.55}$$

$$\approx \left(\frac{N^2}{2}\right)(r\Delta X_L)(c\Delta X_L) \tag{10.56}$$

$$\approx \left(\frac{1}{2}\right)(rN\Delta X_L)(cN\Delta X_L) \tag{10.57}$$

$$\approx \left(\frac{1}{2}\right)(rX_L)(cX_L) \tag{10.58}$$

$$\approx \left(\frac{rcX_L^2}{2}\right) \tag{10.59}$$

The above delay is equal to the delay value of the $D_{RC,X_L,DE}$, which is obtained from the diffusion equation solution. Thus, the Elmore delay could result in the same delay calculation as the diffusion equation for the lossy transmission line without inductance, mutual capacitance, and mutual inductance.

10.12 Design Flows for Parasitic-Aware Circuit Optimization

From the discussion in many of the above sections, it is evident that the parasitics are big issues and the design engineers need to deal with it for accurate design closure [21, 42, 43, 65, 102, 114, 144, 177]. The parasitic-aware design and/or synthesis flow is one of the attractive approaches to compensate the degradations that may surface in the ICs and affect the specifications. In this section, a selected parasitic-aware design example has been presented for detailed understanding.

10.12.1 Parasitic-Aware Analog Design Flow with Multilevel Optimizations

A parasitic-aware design flow with multilevel optimization processes is presented in Fig. 10.48 [12, 31, 43]. The flow relies on physical design optimization at multiple stages to minimize the parasitics effects. In this design optimization flow, the baseline design of the analog IC is formed for a given set of specifications. The schematic design is partitioned into subcircuit blocks. The partition process helps in deciding which devices can have physical designs together and which need special attention. The specialized blocks are subcircuits with these devices. The schematic design is then optimized through the help of an analog simulator. After the specifications are met at the schematic-level design, the placement and topological-placement optimization of specialized block are performed. The topological placement optimization determines the packing configuration so that performance is maximized while meeting layout constraints and design rules. To speed up the topological placement optimization, only a subset of all configurations that satisfy all analog constraints is searched using careful selection of the encoding scheme and move sets. If the specifications of the design are met, either manual or automatic routing of the best placement candidates is performed. At this stage, analog compaction optimization is performed through incremental block displacement. The incremental displacement of blocks in the analog IC helps in improving the accuracy of the parasitic estimation while avoiding large changes in extracted parasitics values during optimization iterations. It also makes sure that the optimized topological placement of the blocks remains unchanged. At this point, performance-driven routing optimization is performed using analog simulator in the loop. In this routing optimization phase, the parasitics-aware netlist is updated. During the routing optimization process, each interconnect is assigned a set of rerouting options and choice of widths. To determine any new routing candidates, the interconnects are rerouted sequentially in the order of criticality by matching several characteristics. The candidate netlists are updated efficiently using the interconnect sensitivity data. The end result of this multilevel optimization is a parasitic-optimal layout of an analog IC.

10.12.2 A Rapid Parasitic-Aware Design Flow for Analog Circuits

One of the key challenges in the analog design is the time for layout generation that occupies a main chunk of design cycle time. For example, a typical manual layout is a matter of 2 weeks. A fast parasitic-aware design flow that takes few hours for setup and run to provide the physical design of the analog ICs is presented in Fig. 10.49 [72].

In this design flow, the design process starts with schematic-level circuit design for a set of design constraints. Initial circuit simulations are performed to verify the design for the functions. At this

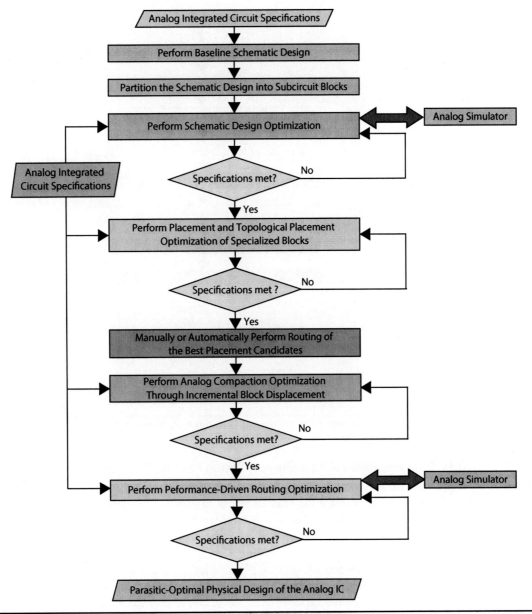

FIGURE 10.48 Parasitic-aware analog design flow with multilevel optimizations [31].

point, prelayout parasitic estimation and resimulation are performed. The prelayout extraction is a circuit extraction process in which actual layout is not created [18]. The process relies on the geometry information and physical property and provides an estimate of the parasitics without actually doing the layout. The prelayout extraction process therefore is much faster than the postlayout extraction and can be effectively used in the layout-aware circuit sizing processes. The circuit optimization using various techniques can be performed on the extracted netlist that has now prelayout parasitics information. The module creation phase is used to create one placeable object from multiple possible instances. Now device placement and net routing is performed to obtain the physical design. It may be noted that this physical design has a high possibility of meeting the specifications as the prelayout parasitic netlist has been optimized. In the in-design physical verification phase, the layout is verified including DRC and timing analysis. The postlayout parasitics extraction provides realistic parasitics as compared to the prelayout parasitics. At the same time, the postlayout and presilicon parasitics can be compared as a checking point for the layout accuracy. If these two do not match then the optimization can be performed again. At the end, a rapid analog design is obtained which is DRC and LVS correct and meets the signoff quality.

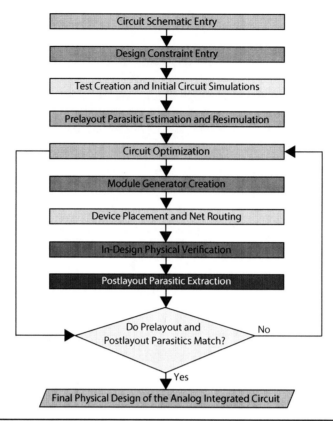

Figure 10.49 A parasitic-aware design flow for fast estimation of postlayout parasitics [72].

10.12.3 Single-Manual Iteration Fast Design Flow for Parasitic-Optimal VCO

An accurate parasitic-aware design flow has been presented in Fig. 10.50 [65, 150]. The objective of this design flow is to reduce the manual layout to exactly "2" while accommodating the full-blown parasitics in the design loop. This is called a single-manual iteration. Following the self-explanatory steps in Fig. 10.50, the baseline physical design is created. This is the "1st manual layout" step. The baseline layout is fully verified including DRC/LVS. A full-blown parasitics extraction is then performed. As compared to the prelayout parasitics of rapid design flow of the previous subsection, this step is slow but highly accurate. This baseline layout step will also have the additional advantage of making final layout easily as compared to the layout creation step in the rapid design flow of the previous subsection.

The extracted parasitics-aware netlist is parameterized for design variables (X_{tuning}) to be tuned for optimization of the design. As a specific example, the set of design variables used for optimization are the following: widths of NMOS devices in the inverter (W_n), widths of PMOS devices in the inverter (W_p), widths of NMOS devices in the current-starved circuitry (W_{ncs}), and widths of PMOS devices in the current-starved circuitry (W_{pcs}). In other words, the tuning variable set is the following: ($X_{tuning} = W_n, W_p, W_{ncs}, W_{pcs}$). At this point, any automated algorithm, such as simulated annealing, particle swarm optimization, can be used for the automatic iterations over the "parameterized parasitic-aware" netlist of the VCO. The conjugate-gradient-based optimization algorithm that was discussed in the previous section is also a good example and can be picked up for use in the context of this design optimization flow. The optimization problem can be formulated as follows:

$$\text{Maximize}: \quad f_c(X_{tuning}) \tag{10.60}$$

$$\text{Subjected To}: X_{lower} \leq X_{tuning} \leq X_{upper} \text{ and } f_c \geq f_{spec} \tag{10.61}$$

It may be noted that while the design optimization is presented for center frequency of the VCO, the design flow in principle can also be applied for optimization of other parameters including phase noise and jitter of the VCO [65, 75].

Power-, Parasitic-, and Thermal-Aware AMS-SoC Design Methodologies

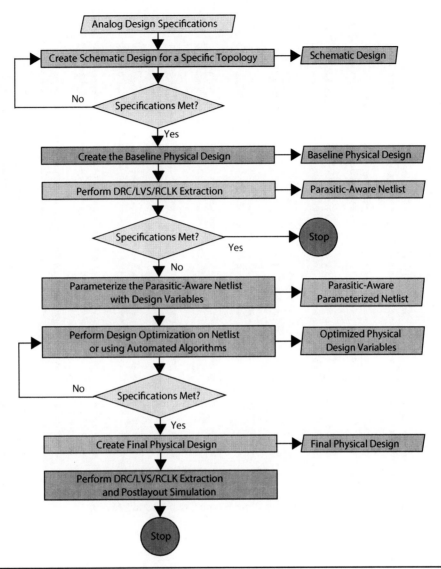

FIGURE 10.50 Proposed single-manual iteration parasitic-aware IC design flow [65, 150].

It may be noted that the frequency oscillation is affected by the channel length of the transistors, i.e., $f_c \propto \left(\dfrac{1}{L^2}\right)$. So even the channel length can be considered for optimization. As a specific example, for a 90-nm CMOS-based technology, the channel length is fixed at 100 nm and widths are tuned for optimization. The tuning of the design variables during the process of automatic iteration using conjugate-gradient algorithm is presented in Table 10.13 [65, 68, 69].

On the convergence of the conjugate-gradient optimization algorithm the optimal tuning variables (X_{optimal}) are fixed. The final physical design that was obtained using the optimal design variables is

Design Parameters (X_{tuning})	Lower Bound (X_{lower})	Upper Bound (X_{upper})	Optimal Values (X_{optimal})
W_n	200 nm	500 nm	220 nm
W_p	400 nm	1000 nm	440 nm
W_{ncs}	500 nm	2 μm	945 nm
W_{pcs}	5 μm	20 μm	9.45 μm

TABLE 10.13 Tuning of the Design Variables for Parasitic-Aware VCO Design [65, 68, 69]

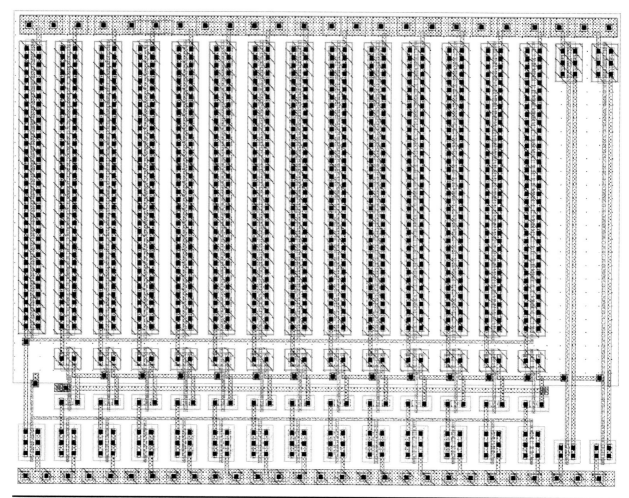

Figure 10.51 Physical design of the parasitic-optimal 90-nm CMOS VCO [65, 68, 69].

shown in Fig. 10.51 [65, 68, 69]. This is the "second manual physical design" of the VCO in the proposed design flow Fig. 10.50 [45, 65]. However, the second manual layout step is not time consuming as at this point a baseline layout is already available from a previous step of this design optimization flow. The frequency versus voltage transfer characteristic for the schematic and the parasitic-aware VCO is shown in Fig. 10.52 [65, 68, 69]. It is evident that the curve for the parasitic-aware VCO closely follows the baseline design in the desired tuning voltage range. It is also observed that the frequency versus input voltage characteristic has linear behavior in a desired voltage range. The physical design of the VCO has been characterized for selected characteristics and the selected characterstics have been presented in Table 10.14 [65, 68, 69].

10.12.4 Parasitic-Aware Low-Power Design of the ULC

In this subsection, detailed discussion of a parasitic-aware design is performed that is also targeted for low-power using dual-oxide technology. As a specific case study, a ULC design has been considered. The overall design flow using pseudocode is presented in Algorithm 10.4. The flow is similar to the single manual-iteration design flow presented in the previous subsection, but with additional steps for dual-oxide (DOXCMOS)-based low-power design. The key point is that the total power consumption (including leakage power) of the ULC circuit is reduced while not compromising the propagation delay. The power and delay optimization is performed by tuning the design parameters: T_{ox} and W. The ULC optimization problem can be formulated as follows:

$$\text{Minimize:} \quad P_{\text{ULC}}(X_{\text{tuning}}) \tag{10.62}$$

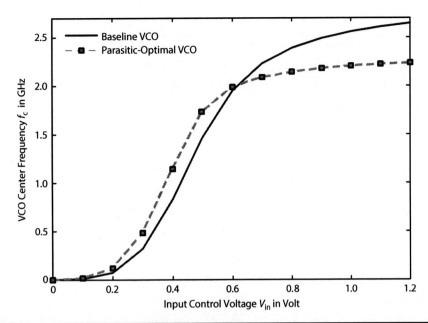

Figure 10.52 Frequency versus voltage transfer characteristics of the 90-nm CMOS VCO obtained from an accurate parasitic-aware optimization [65, 68, 69].

VCO Characteristics	Calculated Values
Process Technology	90-nm CMOS 1P 9M
Supply Voltage (V_{DD})	1.2 V
VCO Oscillation Frequency	1.98 GHz
VCO Phase Noise@10 MHz	−109.13 dBc/Hz
Number of Design Variables	4 (W_n, W_p, W_{ncs}, W_{pcs})
Number of objectives	1 ($f_0 \geq 2$ GHz)

Table 10.14 Measured Performance of the Parasitic-Aware Optimal 90-nm CMOS VCO [65, 68, 69]

$$\text{Minimize}: \quad D_{\text{Prop,ULC}}(X_{\text{tuning}}) \tag{10.63}$$

$$\text{Subjected To}: X_{\text{lower}} \leq X_{\text{tuning}} \leq X_{\text{upper}} \tag{10.64}$$

In the above expression, P_{ULC} is the power dissipation of the ULC circuit. The power dissipation includes dynamic power dissipation, subthreshold leakage power, and gate-oxide leakage power. $D_{\text{Prop,ULC}}$ is the propagation delay of the ULC circuit. The ULC design variables to be tuned during optimization is (X_{tuning}). The lower limit of the design variable is (X_{lower}). The upper limit of the design variable is (X_{upper}). In the automatic design optimization methodology, power consumption and propagation delay are estimated from analog simulations performed on the parameterized parasitics-aware netlist of the ULC. The schematic diagram of a 20-transistor ULC obtained with reduced number of transistors is shown with nominal transistor sizes for a 90-nm CMOS technology in Fig. 10.53 [67, 151].

Any algorithm can be picked based on choice to perform the automatic iterations over the parameterized parasitic-aware netlist of the ULC circuit. However, for simplicity and low-memory usage the conjugate-gradient-based method is chosen again and is presented in Algorithm 10.5 [67, 151]. This netlist of ULC is parameterized for W and T_{ox} of the various transistors. After the convergence of the algorithm for design engineer-specified error margin, the optimal values of the ULC tuning parameter are obtained. The schematic ULC design with nominal transistor sizes and power-hungry transistor identified has been shown in Fig. 10.53(a) [67, 151]. The constraints and

ALGORITHM 10.4 Parasitic-Aware Low-Power Design Flow for ULC [67, 151]

1: **Input:** Specification of the ULC circuits in terms of power and delay.
2: **Output:** Sizes of parasitic-aware, power and delay optimal ULC circuit.
3: Perform design and simulation of voltage-level up conversion circuitry.
4: Perform design and simulation of voltage-level down conversion circuitry.
5: Combine the different subcircuits to design the overall ULC circuit.
6: Perform functional simulations of ULC to verify voltage-level up, voltage-level down, block, and pass operations.
7: Perform transistor reduction in the ULC overall circuit by eliminating any redundant transistors.
8: Perform manual physical design of the baseline ULC circuit using the nominal values L, W, T_{ox}.
9: Perform ULC layout verification through DRC and LVS.
10: Perform parasitics extraction of the ULC physical design to obtain parasitics-aware netlist.
11: Perform characterization of the baseline physical design using its parasitic-aware netlist.
12: Parameterize the parasitics-aware netlist for tuning of design variables (X_{tuning}).
13: Rank the individual transistors of the ULC circuit according to the total power (including leakage) dissipation of the overall ULC circuit.
14: Identify the power-hungry transistors in the ULC circuit which collectively consume a designer-defined percentage of total power of the overall ULC circuit.
15: Use an automated algorithm to tune the design variables (X_{tuning}) such that their optimal values ($X_{optimal}$) are selected while meeting the specifications.
16: Assign high-T_{ox} to the power-hungry transistors of the ULC circuit and new W to all transistors of the ULC circuit.
17: Perform the manual final physical design of the parasitics-optimal low-power ULC with new transistor sizes.
18: Perform physical design verification of the parasitics-optimal low-power ULC.
19: Extract the parasitic-aware netlist of the parasitics-optimal low-power physical design of the ULC.
20: Perform characterization of the parasitics-optimal low-power physical design of the ULC.

ALGORITHM 10.5 Conjugate-Gradient Optimization for DOXCMOS-Based Power and Delay Aware Nano-CMOS ULC [67, 151]

1: **Input:** Parasitic-aware parameterized netlist of the baseline ULC, Target Objective Set $Cost_{target} \leftarrow [P_{ULC}, D_{Prop,ULC}]$, Termination Criterion S, Tuning Variable Set $X_{tuning} \leftarrow [T_{oxn}, T_{oxp}, W_{nup}, W_{pup}, W_{ndown}, W_{pdown}]$, Tuning parameter limits $[X_{lower}, C_{upper}]$, Stopping criterion ε_{target} (a designer defined error margin).
2: **Output:** Optimal tuning variable $X_{optimal}$ of the ULC circuit.
3: Assume the initial solution as the lower limit; $X_{optimal} \leftarrow X_{lower}$.
4: **while** ($X_{lower} < X_{tuning} < X_{upper}$) **do**
5:    Using conjugate gradients generate new design solutions $X_{tuning'}$ in the range from ($X_{tuning} - \Delta X_{tuning}$) to ($X_{tuning} + \Delta X_{tuning}$), by simultaneously varying the tuning parameters as:
   (a) ($T_{ox} - \Delta T_{ox}$) to ($T_{ox} + \Delta T_{ox}$) for power-hungry transistors and
   (b) ($W - \Delta W$) to ($W + \Delta W$) for all transistors of the ULC.
6:    Perform simulations to characterize ULC for $[P_{ULC}, D_{Prop,ULC}]$.
7:    Compute the error between target and current characteristics $\varepsilon_{current} \leftarrow Cost_{target} - Cost(X_{tuning'})$.
8:    **if** ($\varepsilon_{current} \leq \varepsilon_{target}$) **then**
9:      {i.e. Stopping criterion is in the design error margin.}
     Update the solution, i.e. $X_{optimal} \leftarrow X_{tuning'}$.
10:   **end if**
11: **end while**
12: **return** $X_{optimal}$.

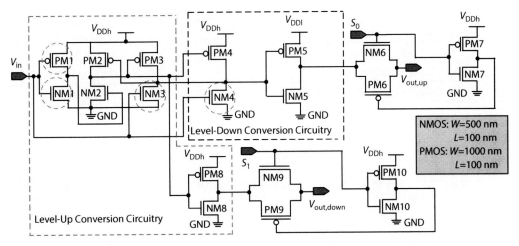
(a) 20-Transistor ULC schematic with high-oxide thickness transistors marked in circles.

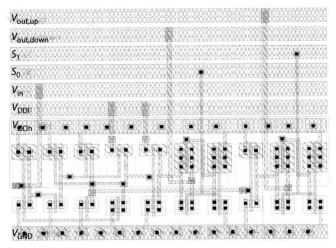
(b) Physical design of a dual-oxide thickness-based parasitic-optimal 20-transistor ULC.

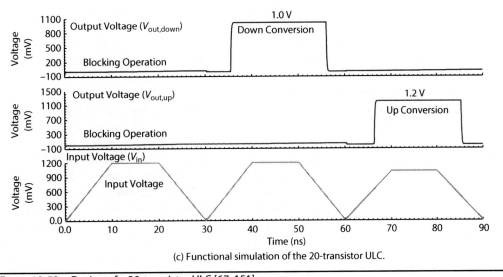
(c) Functional simulation of the 20-transistor ULC.

FIGURE 10.53 Design of a 20-transistor ULC [67, 151].

optimal values of the parameter tuned for ULC circuit optimization are shown in Table 10.15 [67, 151]. The final layout design that has been manually performed using the optimal tuning variables is shown in Fig. 10.53(b) [67, 151]. The functional simulation of the ULC that proves blocking, voltage-down conversion, and voltage-up conversion operations of the ULC is shown in Fig. 10.53(c) [67, 151]. The characterization results of the 90-nm CMOS-based physical design of the ULC is presented in Table 10.16 [67, 151].

ULC Design Parameters (X_{tuning})	Lower Limit (X_{lower})	Upper Limit (X_{upper})	Optimal Value ($X_{optimal}$)
NMOS Oxide Thickness T_{oxn}	2.6 nm	4.7 nm	2.6 nm
PMOS Oxide Thickness T_{oxp}	2.7 nm	5.0 nm	5.0 nm
Up Conversion Circuitry NMOS W_{nup}	120 nm	1 μm	365 nm
Up Conversion Circuitry PMOS W_{pup}	120 nm	1 μm	380 nm
Down Conversion Circuitry NMOS W_{ndown}	120 nm	1 μm	550 nm
Down Conversion Circuitry PMOS W_{pdown}	120 nm	1 μm	120 nm

TABLE 10.15 Tuning Range and Optimal Design Parameters for the 20-Transistor DOXCMOS ULC [67, 151]

ULC Circuits	Power Dissipation (P_{ULC})	Power Savings	Propagation Delay ($T_{d_{ULC}}$)	Delay Reduction	Silicon Area	Area Savings
24-Transistor Baseline ULC	97.72 μW	–	1058.0 ps	–	146.5 μm²	–
20-Transistor Baseline ULC	73.70 μW	25.0 %	894.0 ps	15.5 %	118.6 μm²	19.0 %
20-Transistor DOXCMOS ULC	12.26 μW	87.5 %	113.8 ps	87.3 %	115.7 μm²	21.0 %

TABLE 10.16 Characterization of 90-nm CMOS-Based ULC Circuits [67, 151]

10.13 Temperature or Thermal Issue: An Overview

In the previous sections, power dissipation aspects were discussed with many detailed examples. A brief introduction to the thermal issue was presented in the chapter on the issues for DFX. The high temperature of ICs has several effects on the various characteristics of the ICs and systems (Fig. 10.54) [179, 190]. A logical progress after power dissipation is the thermal/temperature behavior of the ICs. In the following sections, details of thermal characteristics of the ICs along with the techniques to model the thermal effects and techniques to reduce the effects of the temperature-related degradations will be discussed. The first thing that comes to mind is why thermal needs special attention when power dissipation that mainly causes it has already been discussed. The IC power metrics cannot capture the thermal behavior of the ICs due to various reasons [28, 215]. For example, average power dissipation will not capture hot spots developed in the circuits or systems. The lateral coupling will be captured by even the localized average power dissipation metrics. Moreover, the power density of different units in the circuits and systems may be quite different, and hence discussions in terms thermal/temperature may be more suitable. The analysis of the peak chip performance under highest permissible temperature limits may be possible only by using a combined electrical and thermal simulation of the ICs [179]. When discussing the thermal aspects of the ICs and systems, one can talk about two different aspects as follows: (1) on-chip temperature or thermal issues that arise due to the power dissipation of the ICs and systems and (2) the ambient or environmental temperature in which a chip or system operates.

The interrelationships among the power dissipation, temperature rise (due to heat dissipation that may be also due to power dissipation), and reliability are quite complex [28, 103]. A specific snapshot of such an interrelationship has been presented in Fig. 10.55 [28]. The power dissipation effects are a subset of the effects that were discussed in the power dissipation section before such as battery life. The heat dissipation that occurs due to the dynamic power dissipation as well as leakage power dissipation may increase the temperature that may have a very severe impact on performance characteristics as

Power-, Parasitic-, and Thermal-Aware AMS-SoC Design Methodologies

FIGURE 10.54 Effect of temperature on the circuits and system performances.

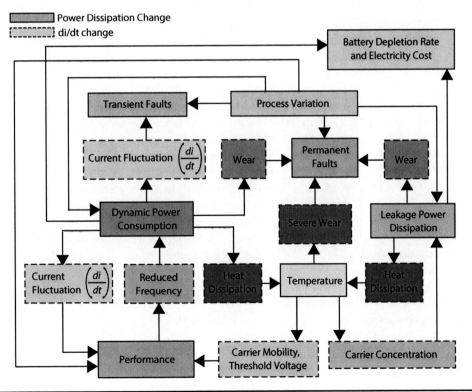

FIGURE 10.55 Power dissipation, performance, temperature, and reliability interrelations [28].

well as device properties such as mobility. Essentially, the power density is manifested as heat dissipation that leads to the increase in the temperatures in nanoelectronic ICs that having high-packing density drive, by smaller feature sizes [233]. The additional effects that results due to $\left|\frac{di}{dt}\right|$, i.e., change in current or change in power dissipation, have negative impact on the propagation delay and are even cause of permanent faults.

Appropriate modeling, optimization, and runtime support at different design stages and abstractions are necessary to manage power, temperature, and reliability of the circuits and systems. The low-power techniques do not necessarily address the power densities in the hot spots of the circuits and systems [93, 214]. The low-power techniques or power management techniques often exploit any timing slack for power-savings purposes, but do not do a good job in the absence of such slacks [39, 236, 243]. The "temperature-aware" or "thermal-aware" design flow fills the need. In the thermal or temperature-aware design flow, the design decisions are different phases and abstraction is taken based on heat dissipation and temperature as the explicit objective.

10.14 Thermal Modeling

This section first discusses the overall physical structure of the chip and corresponding heat dissipation mechanism available to ensure its smooth operations even at higher temperatures. Then, discussions on modeling of thermal elements have been provided to have quantification of heat transfer.

10.14.1 Heat Dissipation: Structure View

The common visual structure that is always found inside any typical computing platform is presented in Fig. 10.56(a) [76, 81, 82, 83, 105, 123, 126]. This structure consists of the actual IC packaged and sitting at the top of a motherboard. A heat sink is then placed on the top of that while being attached using a semisolid paste thermal compound. In the effect, the structure ensures that the chip is kept cool and the system seamlessly operates without thermal failures.

The elevation view of such a structure showing details of all the layers is schematically presented in Fig. 10.56(b) [76, 81, 82, 83, 105, 123, 126]. The lowest layer shown is the motherboard or PCB. A ceramic ball grid array (CBGA) joint connects the ceramic substrate with PCB. CBGA is a surface-mount packaging that attaches the chip to the ceramic substrate and provides more pins than a flat package or dual-in-line package (DIP). The controlled collapse chip connection (C4) pads and underfill connects the chip to the external circuitry. In essence, the solder pumps that are deposited on the chip pads are connected to the external circuitry through soldering. This is different from the wire bonding in which the chip is mounted upright and the wire is used to connect the chip pads to external circuitry. The underfill is an electrical-insulation adhesive used to fill the gaps between the chip, and underlaying mounting facilitates the heat transfer between them and also makes sure that the solder joints are not stressed due to thermal differences. The silicon substrate that hosts the actual IC is shown at the top of the interconnect layer. One can easily imagine that this is reverse vertical, i.e., 180° rotated view as on a silicon substrate first active elements (such as transistors) and then interconnects are fabricated following a process technology. The thermal interface material (TIM) a.k.a. thermal paste, thermal gel, thermal compound, or heat sink compound is a viscous fluid that closely connects the two dissimilar surfaces, the chip surface and the heat sink, effectively acting as a heat spreader. It improves the contact area as well as reduces the thermal resistance at the interface of chip and heat sink. The top large surface of the structure is the heat sink. The heat sink, made of copper or aluminum, is designed to maximize its surface area in contact with the cooling medium (such as air) surrounding it with fins and ducts. At the top of the heat sink, fans and water ducts can be present (not shown in the figure) for forced-air and forced-water cooling, respectively.

The heat dissipation on the above structure can happen through two primary paths as indicated in Fig. 10.56(b). The heat generation layer, in the middle, is essentially the logic representation of the silicon substrate or the interconnect layer. The primary path of the heat transfer is through the TIM heat spreader, and the heat sink. Both radiation and convection mechanisms are responsible for the heat transfer to the ambient air. The secondary path of heat transfer is through the ceramic substrate, CBGA joint, and PCB. Again radiation and convection mechanisms are drivers of heat dissipation. The thermal modeling is needed to analyze the heat dissipation rates.

Power-, Parasitic-, and Thermal-Aware AMS-SoC Design Methodologies

(a) Structure containing chip, package, and heat sink.

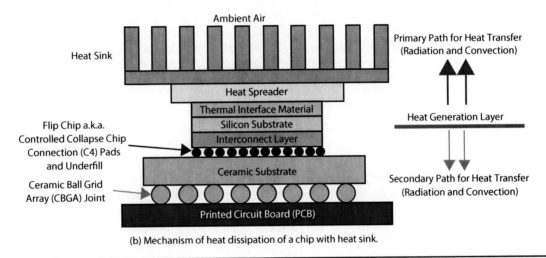

(b) Mechanism of heat dissipation of a chip with heat sink.

FIGURE 10.56 Paths for heat dissipation of a chip in a system [76, 81, 83, 105, 123, 126].

In the above chip and heat sink system, the silicon substrate and the interconnect layer generate the heat [23, 179]. In the substrate, the active and inactive devices dissipate power due to switching and leakage, respectively. The Joule heating in the interconnects is also the major contributing factor to the temperature rise as the interconnect is far away from the heat sink. In addition, the various parasitics present in the device and interconnects increase the power dissipation and aggravate the thermal conditions leading to further temperature rise. The operating temperature of the chip can be estimated using the following expression [179]:

$$\Theta_{chip} = \Theta_{ambient} + R_{therm}\left(\frac{P_{total}}{A_{chip}}\right) \qquad (10.65)$$

where Θ_{chip} is the average operating temperature of the chip, $\Theta_{ambient}$ is the ambient temperature, R_{therm} is the equivalent thermal resistance of the silicon substrate (Si), package, and heat sink, P_{total} is the total power dissipation of the chip, and A_{chip} is the overall area of the chip. The temperature of the metal wire can then be calculated using the following expression [23, 179]:

$$\Theta_{wire} = \Theta_{chip} + R_{elec,wire}I^2_{RMS,wire}R_{therm,wire} \qquad (10.66)$$

where the second component on the right side of the equation is the temperature rise of the metal interconnect due to the flow of current, i.e., Joule heating. $R_{elec,wire}$ is the electrical resistance of the interconnect, $I_{RMS,wire}$ is the root-mean-square value of the current flow in the interconnect, and $R_{therm,wire}$ is the thermal resistance of the interconnect line to the substrate.

Before moving on to the details of the thermal modeling, an important discussion on heat sink design is provided that should also serve as a motivation for the need for further thermal modeling [126, 243]. At the package design level, the maximum or peak sustainable power dissipation of an IC is quantified as thermal design power (TDP). It may be noted that this is not the peak power dissipation of the ICs as discussed in the context of power dissipation. This is because if the peak power dissipation happens for a time that is smaller than the thermal time constant, then the peak power may not affect the heat sink design. As a result of TDP, if the peak temperature is maintained above the ambient temperature, then the maximum thermal resistance of the heat sink is expressed by the following [126, 243]:

$$R_{therm,sink} = \left(\frac{\Theta_{peak}}{P_{peak,sust}}\right) \quad (10.67)$$

where $R_{therm,sink}$ is maximum thermal resistance of the heat sink, Θ_{peak} is maximum temperature, and $P_{peak,sust}$ is the maximum sustained power dissipation. The choice of the heat sink for effective heat dissipation is dictated by the value of $R_{therm,sink}$. This is analogous to the electrical resistance (R) that may dictate the choice of electrical conductors, e.g., copper is a better electrical carrier than aluminum.

10.14.2 Compact Thermal Modeling

Similar to the current flow and its corresponding parameter (e.g., resistance) modeling in the electrical system, thermal modeling is possible. Two types of thermal modeling are available as shown in Fig. 10.57(a) [76, 81, 82, 83, 105, 123, 126]. A high-level analogy of the electrical and thermal parameters is presented in Table 10.17 [76, 202, 214]. The compact models differ based on the basic elements used in the modeling. The basic compact thermal models are based on the resistors only. The dynamic compact thermal models also include thermal capacitors.

The compact thermal model of the chip and heat sink system of Fig. 10.57(b) is presented in Fig. 10.57(c) [76, 81, 82, 83, 105, 123, 126]. It is a resistance-based model that represents various thermal resistances to model the structures. The thermal resistance between two adjacent layers is determined by the common border length shared by these two layers. The various different thermal resistances are shown which model the thermal effects of the different material layers; e.g., interconnect layer, C4 pads and underfill, ceramic substrate, CBGA joint, and PCB. It also includes a thermal resistance due to convention path. The ground symbols at the bottom and top represent thermally ground, e.g., ambient air. The thermal resistance of a structure is proportional to its material thickness and inversely proportional to the cross-sectional area across which the heat is transferring; just analogous to the electrical resistance. Thus, the thermal resistance of a layer is quantified as follows [76, 214, 216]:

$$R_{therm} = \left(\frac{T_{mat} \rho_{therm}}{A_{cross}}\right) \quad (10.68)$$

where T_{mat} is the material thickness, ρ_{therm} is the thermal resistivity of the material per unit volume, and A_{cross} is the cross-sectional area. Based on this expression, various thermal resistances presented in the model can be calculated. The thermal resistance values will be different for each layer due to different material properties and dimensions. The resistance to represent the heat transfer due to the convection is calculated as follows [236]:

$$R_{convection} = \left(\frac{1}{h_{convection} A_{sink}}\right) \quad (10.69)$$

where $h_{convection}$ is the convection heat-transfer coefficient that is a function of rate of airflow and A_{sink} is the effective surface area of the heat sink.

The dynamic compact thermal model that includes the capacitance as additional elements is shown in Fig. 10.57(c) [76, 81, 82, 83, 105, 123, 126]. This is the RC-based dynamic compact thermal

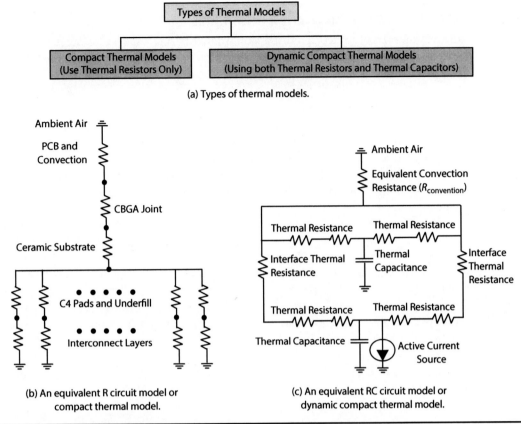

FIGURE 10.57 Thermal models for heat dissipation of a chip [76, 81, 83, 105, 123, 126].

Electrical Parameters	Measurement Units	Thermal Parameters	Measurement Units
I – Current Flow	Amperes	P – Heat Flow, Power	Watts
J – Current Density	Amperes/m²	ϕ_h – Heat Flux, Power Density	Watts/m²
V – Voltage	Volts	Θ_{Diff} – Temperature Difference	Kelvin or °C
ρ – Electrical Resistivity	Meters × Ω	ρ_{therm} – Thermal Resistivity	(Meter × Kelvin) per Watts
R – Electrical Resistance	Ω = Volt/Ampere	R_{therm} – Thermal Resistance	Kelvin/Watts
C – Electrical Capacitance	Farad = Ampere/Volt	C_{therm} – Thermal	Joules/Kelvin
$R \times C$ – Electrical RC Constant	Seconds	$R_{therm} \times C_{therm}$ – Thermal RC Constant	Seconds

TABLE 10.17 The Duality between Electrical and Thermal Parameters [76, 202, 214]

model of the chip and heat sink system of Fig. 10.56. It has different thermal resistances that can be calculated using the same equations that have been presented in the previous paragraph. Each of the material has a thermal capacitance to the thermal ground that is determined by the area of each material or each units in a material layer. The thermal capacitance is proportional to both thickness and area of the material and is calculated as follows [76, 113]:

$$C_{therm} = c_{therm} T_{mat} A_{cross} \qquad (10.70)$$

where c_{therm} is the thermal capacitance per unit volume of a material, T_{mat} is the material thickness, and A_{cross} is the area of cross-section of the materials. The capacitances present in the above models can be calculated for different materials or units based on the type of material and dimensions.

10.15 Thermal Analysis or Simulation Techniques

This section presents a detailed discussion on the thermal analysis principles. It discusses different thermal analysis techniques. Many of the subsections have been dedicated for comprehensive discussions of selected techniques.

10.15.1 Heat Transfer Basics

In general, many different forms of heat transfer are possible based on the material layers and the material between the layers. The three modes of the heat transfer are shown in Fig. 10.58(a) [81]. The basic "conduction mode" heat transfer happens when the media is solid. The heat transfer between a solid surface and fluid happens by the "convection mode." Radiation is the mode for heat transfer when the transfer is in the form of electromagnetic waves. This mode of heat transfer can happen in a vacuum.

The chip and heat sink heat transfer system is schematically presented in Fig. 10.58(b) to explain various forms of heat dissipation [81, 196]. For a clear demonstration of what is taking place in the chip, the structure is rotated 180° as compared to the structure shown in Fig. 10.56. In other words, the CPU fan is at the bottom and motherboard is at the top. The big structure at the bottom is the heat sink, which was at the top in Fig. 10.56. The IC of active area (WxL) is shown, which is dissipating power that is manifested in the form of heat and temperature rise. The heat generated in the IC transfers through the substrate and flow toward the heat sink. The heat transfer happens due to heat flux (φ_{therm}) which is rate of heat/thermal power transfer per unit area. In a 1D scenario, the heat transfer can be in one direction, i.e., toward the top of heat sink toward ambient air. In addition, in 2D scenario, the heat transfer can happen in the sides through the side heat flux, which is orthogonal to the top heat flux.

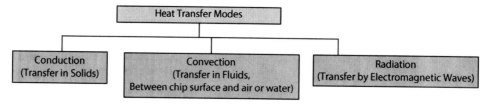

(a) Different modes of heat transfer.

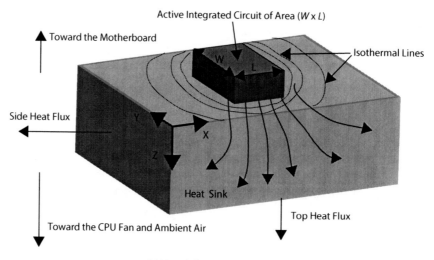

(b) Heat diffusion illustration.

Figure 10.58 Different modes of heat transfer [81, 196].

10.15.2 Thermal Analysis Basics

In a previous section, lumped compact thermal models have been presented. However, accurate thermal analysis needs distributed thermal modeling. The scenario is analogous to lumped-parasitics modeling versus distributed-parasitics modeling. In general, thermal analysis is the simulation of heat transfer through different materials from an entity that produces heat to any entity that receives heat [80, 81, 82, 191, 196, 202, 236]. In electrical analogy, this is similar to a driver gate driving a load gate through a transmission line. The thermal analysis can also be done through the similar PDEs that were used in signal propagation through lossy transmission line. The 3D solving of the heat diffusion equation describing the heat transfer in a chip is the most accurate way of thermal analysis and calculation of the temperature profile [179, 196, 228]. However, the analysis is most computational intensive. As an example, the simple 1D heat diffusion equation follows [179]:

$$\frac{\partial^2 \theta(x)}{\partial x^2} = -\rho_{\text{therm}} \phi_h(x, \theta(x)) \quad (10.71)$$

where $\theta(x)$ is the temperature at any distance x, ρ_{therm} is the thermal resistivity of the materials per unit volume, and $\varphi_h(x, \theta(x))$ is the temperature-dependent power density of heat source. For the interconnects, the power density is defined as [179]:

$$\phi_h(x, \theta(x)) = J_{\text{RMS}}^2 \rho_{\text{elec}}(x, \theta(x)) \quad (10.72)$$

where J_{RMS} is the RMS current density and $\rho_{\text{elec}}(x, \theta(x))$ is the temperature-dependent metal resistivity at any distance x.

The thermal analysis of simulation essentially involves solving these heat transfer equations [81, 179, 196]. The PDEs presenting the heat transfer equations can be solved for various boundary conditions that on discretization result in a system of equations. In a specific case, i.e., for constant thermal conductivity, the set of equations is a linear set of equations. In general, for the nanoelectronic ICs, the solution of the PDEs can be quite resource and time consuming. Therefore, similar to parasitics MOR, MOR techniques can be used to reduce the complexity [228]. The matrix resulting from the discretized linear system of equations can be solved using multigrid techniques to speedup the simulation.

10.15.3 Thermal Analysis Types

The thermal analysis techniques provide the design engineer with the ability to predict the thermal behavior of the ICs during the design stage and before the actual chip is fabricated. An overview of the various thermal analysis techniques is presented in Fig. 10.59 [76, 121, 243].

Based on the application of the thermal analysis, it can be system-level analysis, architecture-level analysis, cell-level analysis, or device-level analysis [28, 76, 236, 243]. Higher the level of analysis earlier is the access of thermal data for the design engineers, and hence enabling the thermal as a design parameter at an earlier phase of the nanoelectronic IC design cycle. Based on the circuit

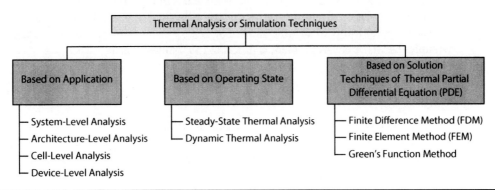

Figure 10.59 Different techniques for thermal simulations.

operating state, the thermal analysis can be either steady-state thermal analysis or dynamic thermal analysis [28, 236]. In the steady-state thermal analysis, the thermal analysis of the IC is performed at its steady-state operation for specific power and thermal-conductivity profiles, i.e., infinitely long operation state. On the other hand, the dynamic thermal analysis is performed at any time for specific initial temperature, power, heat capacity, and thermal-conductivity information. The steady-state thermal analysis is sufficient when the IC thermal profile converges before the subsequent changes; it is faster and has lesser computational overheads. On the other hand, the dynamic thermal analysis is needed when the power profile of the IC vary rapidly; however, it has higher computational overheads.

Based on the solution techniques used in solving the PDE representing the heat transfer equation, it can be the finite difference method (FDM), FEM, and Green function method [76, 243]. Both FDM and FEM numerical approaches rely on the discretization of the entire chip and formulate a linear system of equations relating the distribution of temperature and the distribution of power density. Due to the discretization of the entire chip, the FDM and FEM methods may have large computational overheads. On the other hand, the Green function method is a semi-analytical approach in which only the layers that are the source of power dissipation or whose temperature is of interest are analyzed. Thus, the Green function method is faster than the FDM and FEM.

10.15.4 A Runge-Kutta-Based Method

A Runge-Kutta-based method for thermal simulation is shown in Fig. 10.60 [7, 76, 81, 215, 216]. The simulation considers the geometry of the IC, physical properties of the metal layers, and map power density in the form of a matrix. The active IC is divided into several blocks. For each of the blocks, dynamic compact thermal models, i.e., RC-based models are created. During the simulation of each time step, differential equations for the RC circuit representing thermal model are formulated. These differential equations are solved using a fourth-order Runge-Kutta method with four iterations and as a result the new temperature of each block is calculated. The above steps of differential equation formulation and solutions are then repeated for all the blocks of the IC.

10.15.5 An Integrated Space-and-Time-Adaptive Chip Thermal Analysis Framework

The numerical simulations such as fourth-order Runge-Kutta method and improved Runge-Kutta-Fehlberg method have been widely used for thermal analysis [39, 76, 236]. However, these techniques are quite slow, which makes the full-chip analysis computational intensive and time consuming. To overcome this issue, an integrated space- and time-adaptive chip (ISAC) thermal analysis framework is presented in Fig. 10.61 [236]. It shows two routes: one for dynamic thermal analysis and another for steady-state thermal analysis. The simulation framework considers the power dissipation profile, heat capacity, and thermal conductivity profiles as inputs. All these parameters are needed for the dynamic thermal analysis; however, the chip/package heat-capacity profile information may not be needed in the steady-state thermal analysis. The simulation framework supports 3D arbitrary heterogeneous thermal conduction models for the purposes of accurate simulations. For example, a model may consist of active layer, silicon substrate, packaging material, heat spreader, and a heat sink in a forced-air ambient environment; similar to what is depicted in Fig. 10.56.

In the simulation framework, the dynamic thermal analysis is performed by partitioning the simulation period into small time steps. For $N_{element}$ number of discretized elements of the ICs, the dynamic thermal analysis problem is described as follows [236]:

$$\overline{C}_{therm} \overline{\Theta}^T(t) + \overline{\sigma}_{therm} \overline{\Theta}(t) = \overline{P}_{total} u(t) \tag{10.73}$$

where $\overline{C}_{therm}$ is a (N × N) thermal-capacitance matrix, $\overline{\Theta}$ is a (N × 1) temperature vector, $\overline{\sigma}_{therm}$ is a (N × N) thermal-conductivity matrix, $\overline{P}_{total}$ is a (N × 1) power dissipation vector, and $u(t)$ is a time-step function. A multigrid incremental solver is deployed to iteratively refine the discretization of the thermal element to produce a thermal profile. The local times of all elements are then advanced in

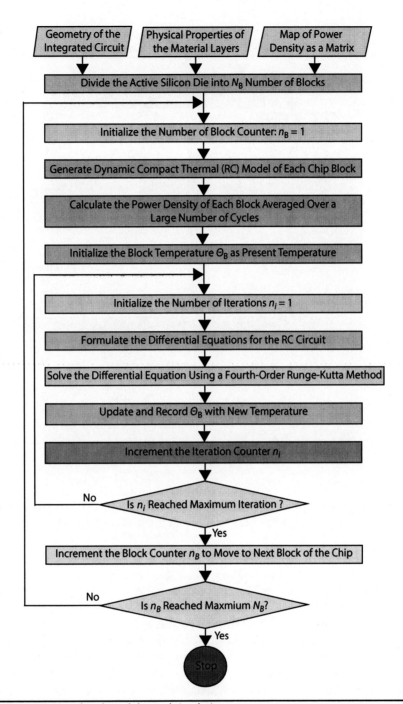

FIGURE 10.60 Runge-Kutta solver-based thermal simulation.

lock-step using transient temperature approximations yielded by the difference equations. The steady-state thermal analysis problem is described as follows [236]:

$$\bar{\sigma}\bar{\Theta}(t) = \bar{P}_{\text{total}} u(t) \tag{10.74}$$

For spatial-adaptation purposes, the simulation framework uses a data structure called "hybrid octree," which is a tree data structure in which each node has 2, 4, or 8 children and maintains spatial relationships among rectangular parallelepipeds for 3D analysis. It is a fact that the recalculation of the thermal-conductivity values after each iteration for all the discretized elements is computational

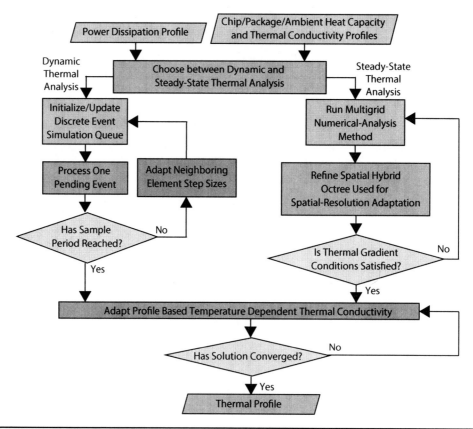

Figure 10.61 Simulation flow of ISAC thermal analysis framework [236].

intensive. For the purpose of accuracy and simulation time tradeoffs, a feedback loop is used at the end to find out the impact of the variations in thermal conductivity on temperature profile of the ICs.

10.15.6 A Fast Asynchronous Time Marching Technique

A dynamic thermal analysis method called "fast asynchronous time marching technique (FATA)" is presented in Fig. 10.62 [39]. It relies on the FDM to perform the simulation and allows different elements to have different step sizes and own local times. The inputs for the simulations algorithm are power dissipation profile, initial thermal profile, and thermal model parameters. The event queue that contains temperature update events sorted by their target times is initialized after initializing the step size of the elements. During each iteration, the event with the earliest target time is selected and temperature of the corresponding element is updated. The simulation flow at this point determines whether the thermal profile has reached quiescent state. If yes, then the local times of all the thermal elements are advanced to the design engineers specified end time of simulation. If no, then the next step size of the elements is calculated and the simulation flow reinserts the temperature update event into the event queue with a new target time. The iterations similar to the above repeats until the design engineers'-specified end time of simulation is reached.

10.15.7 Green Function-Based Method

The thermal analysis through the solution of PDE that describes that the heat transfer is typically very slow. MOR similar to parasitics MOR is one of the techniques that can be used to overcome this issue [228]. The Green function-based method is considered a faster method for thermal analysis and particularly useful for full-chip analysis [50, 71, 228, 243]. The Green function method for full-chip thermal analysis is presented in Fig. 10.63 [228, 243, 244].

This full-chip thermal simulation method makes use of the Green function, 2D DCT, and spectral domain computations for accurate and yet fast thermal analysis [243, 244]. It considers geometry of the ICs, physical properties of the material layers, and the map of power density matrix as inputs for

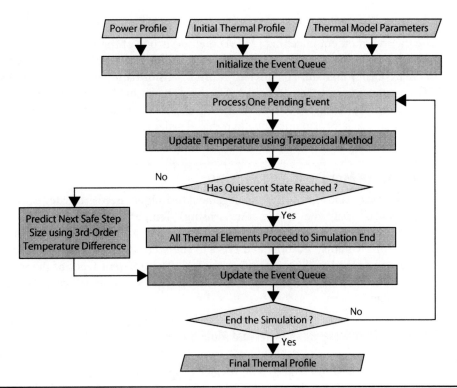

Figure 10.62 Simulation flow of FATA [39].

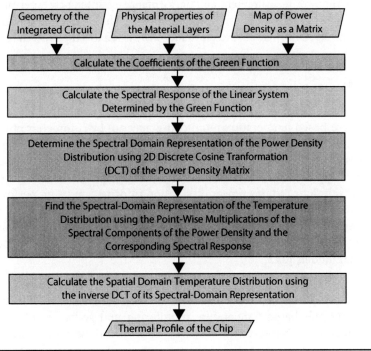

Figure 10.63 Green's function-based method for full-chip thermal simulation.

simulations. The simulation algorithm first calculates the coefficients of the Green function. One of the advantages of this Green function method-based simulation is that spectral responses of the linear system depend on the geometry of the IC and material properties; however, independent of the power dissipation and placement of the units. The calculation of the spectral responses needs to be made only once, but can be used as many times as needed during the thermal-aware design flows.

10.15.8 Thermal Moment Matching Method

The thermal moment matching (TMM) method is one of the MOR techniques that can be used to speedup the thermal analysis [116, 123]. The TMM approach applies fast moment matching techniques in the frequency domain, e.g., discrete Fourier transformation domain. The TMM only processes the DC components and disregards the AC components in the input power dissipation used in the thermal simulation and thus effectively implements a MOR method. However, the thermal profile estimation is an approximation due to reduced order processing.

10.16 Temperature Monitoring or Sensing

The dynamic or runtime thermal management techniques need estimated or measured temperatures, in particular, real-time temperature-sensing. Temperature-monitoring or temperature-sensing techniques essentially help in sensing thermal profile or temperature in various part of the circuits and systems. Number, location, and types of the thermal sensors and sensing techniques affect the accuracy and speed of the temperature sensing. The different types of the temperature-monitoring techniques are shown in Fig. 10.64 [76, 202]. Selected techniques will be discussed in detail in the rest of this section.

10.16.1 Hardware-Based Thermal Monitoring

Various types of on-chip thermal sensors such as based on circuit, based on output, and based on processes have been presented in a previous chapter on sensor circuits and systems. Those on-chip sensors in principle can be used for thermal monitoring and belong to this category of hardware-based thermal monitoring [19, 53, 202]. Many hardware-based temperature/thermal monitoring/sensing techniques are possible [47, 76]. For example, a hardware temperature-monitoring approach uses total power measurement and performance-counter-based per-unit power estimations [76, 87]. In another hardware-based thermal monitoring, the digital output current is produced using a ring oscillator [19, 202]. The generated square wave is fed to a digital counter for the purpose of calibration and measurement.

10.16.2 Software-Based Temperature Monitoring

For dynamic thermal management (DTM), particularly at the higher levels of abstractions, hardware-based temperature monitoring may not be completely effective. In such a situation, the software-based temperature monitoring is more useful. For example, a software can be a part of the system kernel and probe the hardware components to estimate temperature at runtime [232]. Various types of software-based techniques are possible for complex AMS-SoCs. A software-based technique obtains power dissipation information using performance counters and then an analytical model is used to estimate the temperature of the ICs [25]. A similar software-based temperature-monitoring technique measures power dissipation using performance counters at the granularity of functional units and temperature estimation is peformed using precalibrated thermal models [232].

10.16.3 Hybrid Hardware- and Software-Based Thermal Monitoring

Hardware-only-based temperature monitoring and software-only-based temperature monitoring have their own advantages and disadvantages when these are considered for use at the various levels

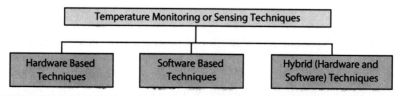

Figure 10.64 Different methods for temperature monitoring or sensing in circuits and systems.

of AMS-SoC abstraction, system-level, architecture-level, or device-level. A hybrid-based temperature-monitoring technique can provide better tradeoff of the two and reduce the overall cost while maintaining effectiveness of monitoring. For example, sensor-fusion-based techniques receive input information from hardware temperature sensors and effectively reduce the number of times the temperature values are calculated [76, 112].

10.17 Temperature Control or Management

In this section, the temperature control, thermal control, or thermal management techniques are discussed. First, basic principles of thermal management or temperature control are presented. Then, different types of temperature control techniques are discussed.

10.17.1 Basic Principle

The temperature control or thermal management techniques are methods to keep the temperature of a circuit and system under control so that it operates in a safe mode. In general, many different strategies can be used for effective temperature control as follows [126]: (1) reduction of the power dissipation of circuits and systems and, in particular, peak power dissipation. (2) Minimization of the impact of local hot spots by improving heat spreading. (3) Increase of the heat absorption capability of the thermal solutions. (4) Improvement of forced air or water-cooling system heat transfer capabilities. (5) Expansion of the thermal envelopes of circuits and systems. Temperature control techniques have many technical as well as cost challenges to be implemented in circuits and systems. Use of a specific technique is a decision based on many factors including cost, form factor, and system application. Design of low-cost and smaller size thermal solutions that meet cost constraints as well as form factor constraints is the key.

The basic principle of DTM or temperature control is depicted in Fig. 10.65 [126, 213, 215]. The overall thermal management system has the following units:

1. *Error Amplifier:* It first calculates the difference between the setpoint/target current ($i_{\text{setpoint}}(t)$) and measured current ($i_{\text{measured}}(t)$) and then produces an amplified current signal ($i_{\text{error}}(t)$).

2. *Thermal Controller:* The thermal controller timely enables the temperature sensors. In the process, it performs the steps required to enable the temperature sensors as well read back the temperature values from these sensors. The thermal controller generates the output signal ($\text{PID}_{\text{out}}(t)$). In principle, it is a proportional, integral, and derivative (PID) controller.

3. *Thermal Actuator:* The thermal actuator makes adjustment for the runtime or DTM.

4. *Structure Thermal Dynamics:* This represents the thermal behavior of the circuit and system that is being monitored.

5. *Temperature Sensor:* This is the temperature sensor or even a set of temperatures to measure the temperature and generate an output signal called measured current ($i_{\text{measured}}(t)$).

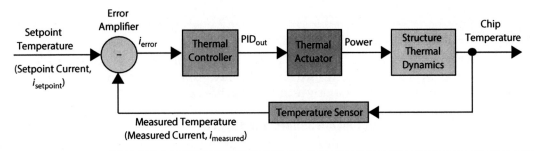

FIGURE 10.65 Basic principle of temperature control [213, 215].

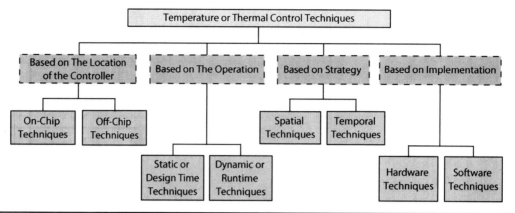

FIGURE 10.66 Different methods for thermal control in circuits and systems.

In effect, the overall temperature control is based on the PID control principle. The output control signal generated is the following [213]:

$$\text{PID}_{\text{out}} = K_P i_{\text{error}}(t) + K_I \int i_{\text{error}}(t) dt + K_D \frac{i_{\text{error}}(t)}{dt} \qquad (10.75)$$

where, $i_{\text{setpoint}}(t)$ is current value corresponding to the setpoint or target temperature of the circuit or system that needs to be maintained and $i_{\text{measured}}(t)$ is the current value corresponding to the measured temperature of the chip. The measured temperature is the chip temperature sensed by the temperature sensor that may be different from it depending on the accuracy of the temperature sensor. The tuning parameters of the PID controller are the following: K_P is the proportional gain, K_I is the integral gain, and K_D is the derivative/differential gain. In principle for the change in the current rate, P depends on the present error, I depends on the accumulation of past errors, and D is a prediction of future errors. The weighted sum of these three actions with tuning parameters is used for effective thermal control of the target circuit and system. The PID controller-based thermal monitoring is very effective and it maintains the circuit and system temperature within the setpoint temperature with a maximum of 0.2° overshoot.

10.17.2 Types

The different methods of the thermal control or temperature monitoring that are used for various circuits and systems are listed in Fig. 10.66 [172, 179]. The temperature control techniques may be either on-chip or off-chip depending on whether these are an integral part of the circuit or system that they monitor [179]. The off-chip techniques by necessity runtime or dynamic techniques are implemented at the system level or at the board level. However, the on-chip techniques can be either static or dynamic techniques. Thus, in general, the techniques are either static (i.e., design time techniques) or dynamic (i.e., runtime techniques). Based on the cooling strategy, the technique can be either spatial or temporal [172]. The spatial techniques explore minimization of the performance penalties by distributing the power dissipation units across larger area subjected to constraints. The temporal techniques explore the possibilities of slowing down or even stop the computation from time to time for cooling of the circuit and system. Based on the implementation, the thermal control techniques can be either hardware-based or software-based.

10.18 Thermal-Aware Circuit Optimization

As a specific insight of the thermal-aware design optimization, two case studies have been presented in this section. First, widely used SRAM has been considered. Then, as a second example, physical-design optimization of a widely used VCO has been presented.

10.18.1 A Thermal-Aware SRAM Optimization

As a specific example, thermal-aware design of a 7-transistor single-ended SRAM has been discussed in this subsection [140, 210, 220]. A thermal-aware design flow that performs circuit-level optimization of a SRAM is presented in Fig. 10.67(a) [140, 220]. The flow also incorporates process variations in the design flow that may or may not be part of the thermal-aware design flow. The design flow as typical begins with the baseline design of the SRAM. The baseline SRAM is characterized for selected characteristics, e.g., total power dissipation and SNM. At this step, effects of the ambient temperature on the characteristics of the SRAM can be analyzed. This can be performed using a temperature variable in an analog/SPICE simulator. The on-chip temperature

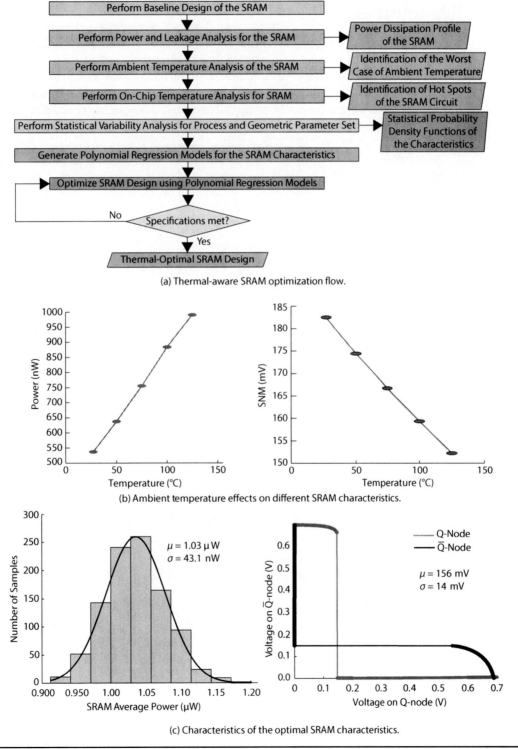

(a) Thermal-aware SRAM optimization flow.

(b) Ambient temperature effects on different SRAM characteristics.

(c) Characteristics of the optimal SRAM characteristics.

FIGURE 10.67 Thermal-aware SRAM optimization technique [140, 220].

analysis for the SRAM circuits can be performed using a thermal analysis tool. The on-chip temperature analysis and ambient temperature analysis can be performed in any order. A decision can be made if the design flow will accommodate the effects of both the ambient or on-chip or one of these [139, 220]. It is also a decision of a design engineer if the worst-case or average-case temperature values need to be considered. The design optimization for the worst-case temperature is a pessimistic design approach. On the other hand, the design optimization for the average-case temperature is an optimistic design approach. As a specific example, the worst-case ambient temperature scenario is considered during optimization in this design flow. The worst-case temperature is the temperature for which the SRAM has maximum power dissipation and minimum SNM. At this phase, the design flow process and voltage variation analysis can be performed. This is not necessarily needed in the thermal-aware flow, but demonstrates the important nanoelectronic challenge variability along with the thermal awareness [140, 220]. To reduce the iteration time during the optimization process, the design flow implements a technique of "polynomial regression model" in which such models for the SRAM characteristics are created. Thus, the optimization iterations are performed on the polynomial regression models instead of analog/SPICE simulators that take longer time and have computational overhead. This can be treated like a metamodel that is the scope of a future chapter. However, thermal-aware optimization iterations can be directly performed through an analog/SPICE simulator instead of these polynomial regression model.

As a specific example, the characteristics of SRAM that are considered for optimization are average power dissipation (accounting the leakage power) and SNM. For simultaneous minimization of power dissipation and maximization of SNM, a figure-of-merit (FoM) called power-to-SNM ratio (PSR) has been considered. The PSR is defined as follows:

$$\widehat{f}_{\text{PSR}} = \left(\frac{\widehat{f}_{\text{power}}}{\widehat{f}_{\text{SNM}}} \right) \tag{10.76}$$

where the $\wedge$ presents that the value of the SRAM characteristic has been normalized by division with the maximum value in its range. The minimization of PSR minimizes the average power dissipation and maximizes the SNM, one of the SRAM performance metrics. The SRAM characteristics and the FoM are analyzed for a range of different temperature, e.g., 25°C, 50°C, 75°C, 100°C, and 125°C. Selected SRAM characteristics with respect to the temperature have been presented in Fig. 10.67(b) [140, 220]. The increase in the temperature increases the power dissipation of the SRAM, one reason is the increase in the leakage with the rise in temperature. It is evident from the figure that the SNM of the SRAM is degraded as the temperature increases. To obtain the polynomial regression models, the surface plots are generated for the SRAM characteristic by tuning the target design variables. Any characteristics or FoM of SRAM is represented using the following quadratic polynomial models [140, 220]:

$$\widehat{f}_X = \sum_{i,j=1}^{N_T} \alpha_{ij} W_{n,i} W_{p,j} \tag{10.77}$$

where X is SRAM characteristics or FoM, e.g., PSR in this case, α_{ij} is the matrix of coefficients obtained during the polynomial regression, N_T is the number of tuning levels that also dictate the size of the matrix. For the minimization of f_{PSR}, the following are solved [140, 220]:

$$\frac{\partial f_X}{\partial W_n} = \frac{\partial f_X}{\partial W_p} = 0 \tag{10.78}$$

$$\frac{\partial^2 f_X}{\partial W_n^2}, \frac{\partial^2 f_X}{\partial W_p^2} \gtrless 0 \tag{10.79}$$

In the above expression, the > 0 criterion is used for minimization and the < 0 criterion is used for maximization, thus getting specific values for tuning variable for minimal f_{PSR}. With this size, the

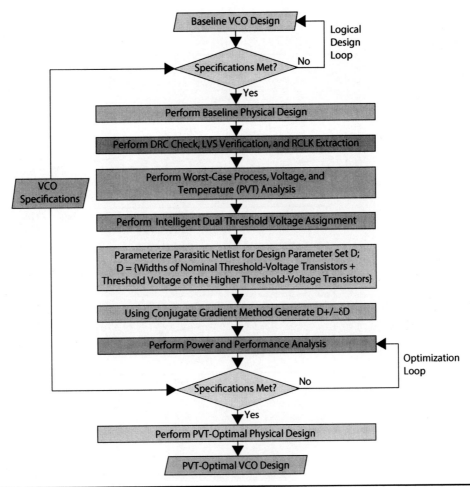

Figure 10.68 Thermal-aware VCO optimization technique [145].

thermal-optimal circuit of SRAM is designed. Selected characteristics of the thermal-optimal SRAM are presented in Fig. 10.67(c) [140, 220]. The process variation analysis of the average power dissipation of SRAM is presented that has low standard deviation. The butterfly curve has also been presented. The thermal-optimal SRAM for a 45-nm CMOS technology has power dissipation of 1.0 μW and SNM of 154 mV.

10.18.2 A Thermal-Aware VCO Optimization

In this subsection, thermal-aware optimization of a VCO has been presented. The overall thermal-aware design flow is presented in Fig. 10.68 [145]. The baseline schematic, as well as physical designs of the VCO, is created using typical approaches along with the DRC, LVS, and parasitics extractions steps. At this stage, thermal analysis of the baseline VCO design can be performed. Both on-chip and ambient temperature analysis can be performed as just similar to the previous subsection. Process and voltage variation analysis of the VCO can be performed to have variability along with the thermal-aware. Moreover, as an approach for low-power design along with the thermal awareness, a dual-threshold voltage technique can be incorporated. It may be noted that both the variability and low-power steps may not be explicitly needed in the thermal-aware design flow. As has been demonstrated in many VCO optimization flows in the previous sections, the tuning design variable X_{tuning} consists of widths of the nominal threshold voltage transistors and threshold voltage of higher threshold voltage transistors. Any automatic algorithm including the conjugate-gradient method that has been presented before can be used for automatic tuning the parameters X_{tuning}.

As a specific design example, the target specification is the center frequency of the VCO $f_{osc} \geq 2$ GHz. In this design flow, a worst-case process variation analysis of the physical design is performed for the

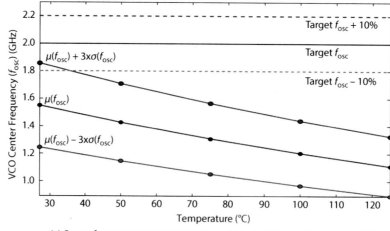

(a) Center frequency versus temperature characteristics for the baseline VCO.

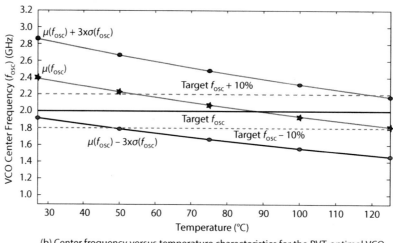

(b) Center frequency versus temperature characteristics for the PVT-optimal VCO.

FIGURE 10.69 Thermal-aware VCO optimization frequency characteristics [145].

center frequency. The VCO is analyzed for a selected set of ambient temperatures in the range of 27°C to 125°C, in particular at 27°C, 50°C, 75°C, 100°C, and 125°C. Worst-case process variation analysis can be performed for each of these temperatures using Monte-Carlo simulations. The behavior of the VCO center frequency for $\mu(f_{osc})$, $\mu(f_{osc}) + 3\sigma(f_{osc})$, and $\mu(f_{osc}) - 3\sigma(f_{osc})$ quantifications of the statistical distribution of the center frequency is shown in Fig. 10.69(a) [145]. The center frequency of the VCO decreases with the increase in the temperature.

For the power reduction purposes, dual-threshold voltage assignment approach is incorporated in the design flow [145]. The power-hungry transistors are identified and higher threshold voltages are assigned; i.e., V_{ThnH} for the power-hungry NMOS and V_{ThpH} for the power-hungry PMOS. The parasitic-inclusive VCO netlist is parameterized for the design variables X_{tuning} to be tuned for VCO optimization. A conjugate-gradient optimization algorithm is now used to tune X_{tuning} and evaluate the VCO characteristics and optimization. It uses an analog/SPICE simulator in the design optimization iterations. In the process, the optimal values of the X_{tuning} variables are obtained. The thermal-aware VCO design is performed for a 90-nm Salicide 1.2V/2.5V 1 Poly 9 Metal PDK. The resulting schematic design of thermal-aware VCO is shown in Fig. 10.70(a) and the physical design of the thermal-aware VCO is shown in Fig. 10.70(b) [145]. The frequency characteristics of the thermal-optimal VCO are shown in Fig. 10.69(b) [145]. The center frequency f_{osc} of the thermal-optimal VCO meets the target specifications of $f_{osc} \geq 2$ GHz with a 10% degradation at worst-case across the entire range of the target temperature range. The characterization results of the thermal-optimal 90-nm CMOS-based current-starved VCO is presented in Table 10.18 [145].

Power-, Parasitic-, and Thermal-Aware AMS-SoC Design Methodologies 601

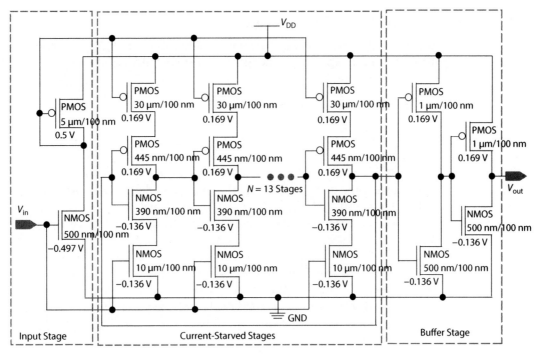

(a) Schematic diagram of thermal-optimal 90-nm CMOS-based current-starved VCO.

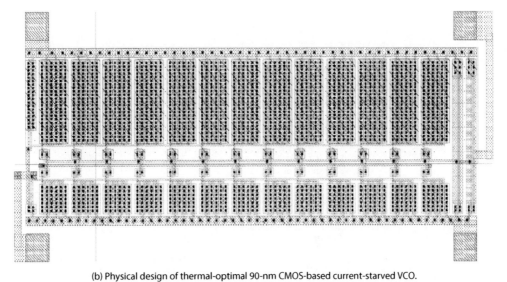

(b) Physical design of thermal-optimal 90-nm CMOS-based current-starved VCO.

FIGURE 10.70 Thermal-aware VCO optimization technique resulting the 90-nm CMOS-based circuits [145].

VCO Characteristics	Calculated Values
Process Technology	90nm CMOS 1P 9M
Supply Voltage (V_{DD})	1.2V
Nominal Center Frequency	2.4GHz
Worst-Case Center Frequency	1.8GHz
Phase Noise@10 MHz	−109.1dBc/Hz
Worst-Case PVT	V_{Th} (+10%), T_{ox} (+10%), V_{DD} (−10%), 125°C
Tuning Parameters	X_{tuning} (W_n, W_p, W_{ncs}, W_{pcs}, V_{ThnH}, V_{ThpL})
Silicon Area	547.7 μm^2 (58.3% penalty)

TABLE 10.18 Characteristics of the Thermal-Aware 90-nm CMOS VCO [145]

10.19 Thermal-Aware Digital Design Flows

In this section, selected techniques that are specifically suitable for digital ICs are discussed. Different techniques for various levels of digital design flow are briefly discussed.

10.19.1 Thermal-Aware Digital Synthesis

A thermal-aware design synthesis flow that has optimizations at selected phases of digital IC synthesis is shown in Fig. 10.71 [83, 195]. Specifically, thermal-aware optimizations are performed at logic-synthesis phase, block floor-planning phase, and cell placement and routing phase to obtain a thermal-aware physical design of digital ICs as a result of this flow, as presented in Fig. 10.71(a) [83, 195]. The design flow considers the three limits of the thermal-aware design, the host-spot limit, power limit, and reliability limit as depicted in Fig. 10.71(b) [83, 195]. The design flow estimates the following two thermal parameters—mean increase in junction temperature ($\Delta\Theta_{jun}$) and maximum and mean temperature difference ($\delta\Theta$)—and performs optimization so that the thermal behavior and timing constraints of the ICs are within the specified constraints. The parameter $\delta\Theta$ is an estimate of the intensity of the hot spots and hence can be used to identify the host spots in the ICs. During the optimizations at each level of the abstraction, if the thermal constraints are not met, then the iterations are continued. One thermal constraint is specified as follows [195]:

$$\Delta\Theta_{jun} + \delta\Theta < \Delta\Theta_{ref} \tag{10.80}$$

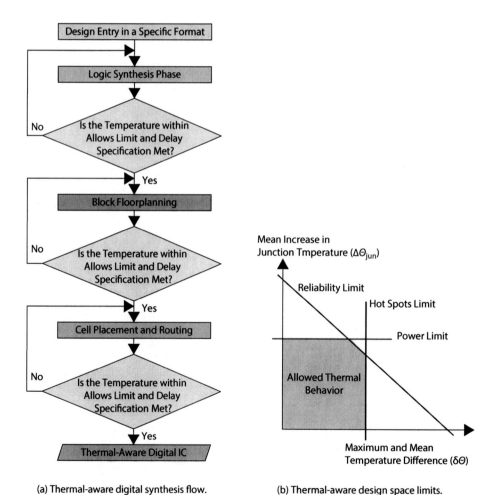

(a) Thermal-aware digital synthesis flow.   (b) Thermal-aware design space limits.

Figure 10.71 Thermal-aware digital synthesis for ICs [83, 195].

where $\Delta\Theta_{ref}$ is temperature increase over the ambient temperature and can be considered as a constraint in the design flow. Another constraint that puts limit on the maximum power dissipation of the ICs is the following [195]:

$$\Delta\Theta_{jun} < R_{therm} P_{IC,max} \qquad (10.81)$$

where $P_{IC,max}$ is the maximum allowable power dissipation of the IC and R_{therm} is the thermal resistance of the IC package. The above expression originates from the fact that self-heating is proportional to the total power dissipation. During the logic synthesis, the mean increase in the junction temperature is calculated as follows [90, 195]:

$$\Delta\Theta_{jun} = R_{therm} \sum_{i=1}^{N_{gates}} P_{gate,i} \qquad (10.82)$$

where $P_{gate,i}$ is the power dissipation of a logic gate with N_{gates} number of logic gates present in the logic-level netlist of the digital IC. The maximum and mean temperature difference during the logic synthesis is calculated as follows [195]:

$$\delta\Theta = \left(\frac{\alpha_{proc}}{\sigma_{therm,Si}\sqrt{A_{IC,total}}}\right)\left[\sum_{i=1}^{N_{gates}} P_{gate,i} \ln\left(\frac{P_{gate,i}}{A_{gate,i}}\right) - P_{IC,total} \ln\left(\frac{P_{IC,total}}{A_{IC,total}}\right)\right] \qquad (10.83)$$

where α_{proc} is a process technology-dependent parameter, $\sigma_{therm,Si}$ is the thermal conductivity of the silicon, $A_{IC,total}$ is the total area of the IC, and $P_{IC,total}$ is the total power dissipation of the IC. At the circuit-level of digital design abstraction, the IC is divided into ($L_{element} \times L_{element}$) size of elements. The temperature of an element is then calculated using the following expression [193, 195, 196]:

$$\Theta_{element} \approx \Theta_{ambient} + R_{therm} P_{IC,total} + 3.74\left(\frac{P_{element}}{L_{element}} + 0.283 \sum_{i=1}^{N_{neighbor}} \frac{P_{element,i}}{L_{element,i}}\right) \qquad (10.84)$$

where $\Theta_{ambient}$ is the ambient temperature, $P_{element}$ is the total power dissipation of the element, $N_{neighbor}$ is the number of neighboring elements that this specific element has, $L_{element,i}$ is the distance of neighboring element i from the current element under consideration, and $P_{element,i}$ is the total power dissipation of the neighboring element i. At the thermal-aware cell placement and routing phase, the distance between two cells, i.e., a current "element" and its neighbor element i needs to satisfy the following to be thermally safe [195]:

$$L_{element,i} > 2.8 L_{element}\left(\frac{P_{element,i}}{P_{element}}\right) \qquad (10.85)$$

10.19.2 Thermal-Aware Physical Design

The different steps of physical design of digital ICs along with the techniques for thermal-awareness options is presented in Fig. 10.72 [2]. The thermal analysis techniques can be used in the context of temperature-aware power analysis, temperature-aware IR drop analysis, as well as temperature-aware timing analysis. The thermal awareness can be applied at any stage of the design flow, e.g., temperature-aware partitioning, temperature-aware floor planning, temperature-aware placement, temperature-aware routing, temperature-aware sizing. Specific discussions of a selected few appear in the following subsection. In each of the steps, thermal analysis can be performed from the initial power estimate of the power-generating sources and the layout information for the ICs.

Thermal-aware or temperature-aware floor planning improves the overall temperature distribution of the IC by using mechanisms such as reordering the placement of blocks that have a significant impact on the temperatures. Thermal-aware floor-planning technique is available that can reduce the peak temperature of the IC [203]. This technique calculates the temperature and the

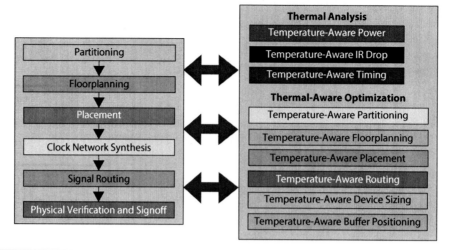

Figure 10.72 Thermal-aware physical design flow for digital ICs [2].

delay of interconnects for each clock cycle using a simulated annealing method. Another thermal-aware floor-planning solutions to reduce peak temperature is also available that is based on the genetic algorithms [84].

Thermal-aware placement techniques can be used for better distribution of the heat across the IC [36, 37]. A specific example of such techniques is a partition-driven placement approach for standard cell [36, 37]. This technique relies on a multigrid-like heuristic method to incorporate thermal awareness as placement constraints. In this technique, the thermal equation is simplified at each level of partitioning.

10.20 Thermal-Aware Register-Transfer-Level Optimization

In this section, selected techniques that consider thermal awareness or temperature constraint at the architecture-level are discussed. They include high-level synthesis or behavioral synthesis tasks for digital ICs.

A thermal-aware HLS approach that reduces the peak temperature is depicted in Fig. 10.73 [97, 104, 239]. This is an HLS flow that accounts for the impact of operation scheduling, resource binding, and floor planning on the power distribution of the units as well as overall temperature distribution. In this thermal-aware HLS flow, the DFG is simulated for profiling. The physical designs of RTL components are performed and their netlists are extracted. The profile information from the DFG are used as an input for these netlists and power models in the form of look-up tables are created. At this phase, a simulated annealing-based algorithm is used to perform simultaneous operation scheduling, resource binding, and floor planning while minimizing the power dissipation and peak temperature. The temperature profiles of each of the RTL components are calculated by the steady-state thermal analysis [236]. For the calculation of the temperature of the RTL

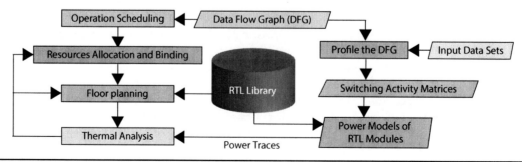

Figure 10.73 Overview of a thermal-aware high-level synthesis approach for digital circuits [104, 239].

components, floor plan and power dissipation for different inputs are considered in the thermal analysis tool.

Another HLS framework relies on a simultaneous operation scheduling and resource binding algorithm to reduce the peak temperature reached by the hottest resource [169]. The temperature-aware optimization is incorporated into both operation scheduling and resource binding stages of this HLS. In this thermal-aware HLS flow, first resource-constrained and time-constrained scheduling are performed that are followed by low-power resource binding. This forms the initial solution for the thermal awareness. Then, for the thermal-aware HLS, operation rescheduling and resource binding are performed to reduce the peak temperature. Thermal-aware floor planning is performed at the beginning of the thermal-aware HLS and the floor plan is not changed after that to simplify the HLS process. The temperatures of the RTL components are estimated using an architecture-level temperature analysis tool [83].

10.21 Thermal-Aware System-Level Design

In this section, different thermal-aware techniques that are specifically used for system-level design are discussed [51, 56, 117, 118]. The multiprocessor SoC (MPSoC) architectures are popular due to their low-power high-performance features particularly suitable for portable systems such as smart mobile phones and tablets. There are two different alternatives to design such MPSoC-based architectures as follows [234]: platform-based design and hardware-software cosynthesis. In the platform-based design, a predefined MPSoC platform is selected and the applications of the embedded system are mapped onto this specific platform. On the other hand, in the hardware-software cosynthesis approach, the MPSoC architecture is custom made for target specifications using a technology library containing various processing elements (PEs). Although hardware-software cosynthesis is outside of the scope of this book, for the sake of completeness to cover all the levels of abstraction, this section has been included without providing many details of the hardware-software cosynthesis.

A thermal-aware design flow for MPSoC architecture design is shown in Fig. 10.74 [85, 234]. In the first step, the target architecture of the MPSoC is partitioned into hardware and software components to meet the target specifications. In the MPSoC architecture exploration, the allocation and scheduling algorithms are used to determine the schedule of tasks on the PEs. These algorithms are very critical to produce good architecture solutions. The thermal-aware platform-based MPSoC design flow has been shown in Fig. 10.74(a) [85, 234]. In this design flow, the given task graph, the allocation and scheduling are performed for temperature constraints. The temperature profile of the

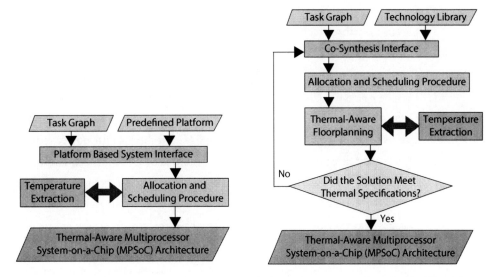

(a) Thermal-aware platform-based MPSoC design flow. (b) Thermal-aware co-synthesis flow for MPSoC design.

FIGURE 10.74 A thermal-aware design flow for MPSoC [85, 234].

target architecture is extracted using architecture-level temperature analysis tool [34, 215]. The thermal-aware hardware-software cosynthesis-based MPSoC design flow is shown in Fig. 10.74(b) [85, 234]. In this flow, the allocation and scheduling are performed first and then thermal-aware floor planning considers the placements of the PEs. The thermal-aware floor planning works in conjunction with a architecture-level temperature analysis tool to converge to a thermal-optimal MPSoC [34, 215].

The thermal control method for MPSoC can be based on control theory [242]. The key idea is to obtain thermal balancing of the thermal profile of the MPSoC to avoid any hotspots. The modeling of the thermal behavior of the MPSoC allows the design of an optimal frequency controller independent of the thermal profile of the IC. For a control theory perspective, thermal balancing is essentially the minimization of the thermal gradients on the MPSoC with scheduler requirements as input reference signal and the thermal profile of the IC as a feedback signal. An example of a PID control-based thermal control was presented in a previous section.

10.22 Questions

10.1 What are the different effects of high power dissipation in an nanoscale IC or system? Discuss any three of them in detail.

10.2 Consider a specific nanoscale technology. Discuss the major components of power dissipation for that technology-based IC.

10.3 Discuss different metrics to capture power dissipation profile of the ICs. Discuss one such metric that also accounts for performance along with power.

10.4 Discuss the power dissipation in a smart mobile phone for its different operating conditions.

10.5 Discuss any three presilicon and postsilicon power reduction techniques.

10.6 Discuss any three ways of scaling voltage for dynamic power reduction.

10.7 Discuss any three ways of reducing power dissipation in analog ICs.

10.8 Discuss an overview of a power-aware analog design flow.

10.9 Discuss an overview of a power-aware digital design flow.

10.10 Select any three levels of digital design abstraction and discuss three techniques for power reduction in each of these abstractions.

10.11 Discuss any three presilicon techniques for power reduction.

10.12 Discuss any three postsilicon techniques for power reduction.

10.13 Discuss the commonalities and differences between power and energy reduction.

10.14 Discuss any three techniques for capacitive-switching power reduction.

10.15 Discuss any three techniques for short-circuit power reduction.

10.16 Discuss any three techniques for transient gate-oxide leakage power reduction.

10.17 Discuss a specific DVS or dynamic frequency scaling technique for power reduction in detail.

10.18 Discuss a specific multiple supply-voltage technique for power reduction in detail.

10.19 Discuss a specific multiple supply-voltage HLS technique for power reduction in detail.

10.20 Discuss any three techniques for subthreshold leakage power reduction.

10.21 Discuss a specific subthreshold leakage power reduction technique in detail.

10.22 Discuss any three techniques for gate-oxide leakage power reduction.

10.23 Discuss a specific dual-oxide technique for analog circuit power reduction in detail.

10.24 Discuss a specific dual-oxide HLS technique for power reduction in detail.

10.25 What do you mean by parasitics in an IC? Discuss any three parasitics in a nanoelectronic transistor.

10.26 Discuss different parasitics in the interconnects of nanoelectronic circuits.

10.27 What are the effects of the parasitics on nanoelectronic ICs? Discuss any three in detail.

10.28 Discuss the circuit extraction process from the physical design of an IC.

10.29 Discuss the parasitics extraction process from the physical design of an IC.

10.30 Discuss any three techniques for parasitics extraction in an IC.

10.31 Discuss a specific example of parasitics modeling in an IC.

10.32 Why is parasitics model reduction needed? Discuss a specific example in detail.

10.33 Discuss the Elmore delay model with a specific example.

10.34 Consider signal propagation in a long transmission line. Consider its modeling as a lumped-RC network. Using Elmore delay model, calculate its delay at the farthest distance from the source end.

10.35 Consider signal propagation in a long transmission line. Consider its modeling as a distributed-rc network. Using Elmore delay model, calculate its delay at the farthest distance from the source end.

10.36 Discuss an overview of a parasitics-aware analog design flow.

10.37 Discuss a specific example of a parasitics-aware analog design.

10.38 Discuss any three effects of high temperature on an IC in detail.

10.39 Discuss how power, temperature, and reliability are interrelated.

10.40 Discuss the different structures responsible for heat dissipation from a chip.

10.41 Discuss the paths of heat dissipation from a chip in a computing system.

10.42 How are chip operating temperature and wire operating temperature calculated?

10.43 How does the design of a heat disk form the basis of power dissipation?

10.44 Discuss two types of lumped compact thermal models.

10.45 Discuss the analogy between electrical and thermal parameters with three examples.

10.46 Discuss how thermal resistance and capacitance are calculated.

10.47 Discuss different techniques for thermal/temperature analysis of circuits and systems.

10.48 Discuss the steady-state thermal analysis versus dynamic thermal analysis.

10.49 Discuss a specific example of thermal analysis technique in detail.

10.50 Discuss different techniques for temperature monitoring in circuits and systems.

10.51 Discuss different techniques for temperature control in circuits and systems.

10.52 Discuss the power-aware techniques versus the thermal-aware techniques.

10.53 Discuss a specific example of thermal-aware circuit optimization.

10.54 Discuss a specific example of thermal-aware digital logic-synthesis technique.

10.55 Discuss a specific example of thermal-aware HLS technique.

10.56 Discuss a specific example of thermal-aware system-level design technique.

10.23 References

1. Hipex Full-Chip Parasitic Extraction. http://www.silvaco.com/content/kbase/hipex_jpws.pdf. Accessed on 09th May 2014.
2. Thermally Aware Design Methodology. Tech. rep., Gradient Design Automation Inc. (2005). URL http://www.cdnusers.org/community/encounter/resources/resources_imp/route/dtp_cdnlive2005_1349_moynihan.pdf. Accessed on 14th August 2012.
3. Taking A Bite Out of Power: Techniques for Low-power-ASIC Design. http://edn.com/Home/PrintView?contentItemId=4314765 (2007). Accessed on 05th May 2014.
4. OMAP-Vox™ Single-Chip Solution for Affordable Multimedia-Rich Phones. http://www.ti.com/pdfs/wtbu/TI_omapv1035.pdf (2008). Accessed on 01st May 2014.

5. Body Effect and Body Biasing. http://www.suvolta.com/files/2513/0713/0696/BodyEffect_BodyBiasing_TechBrief_Final_6.3.2011.pdf (2011). Accessed on 18th April 2014.
6. Building Energy-Efficient ICs from the Ground Up. http://www.cadence.com/rl/Resources/white_papers/low_power_impl_wp.pdf (2011). Accessed on 25th may 2014.
7. HotSpot. http://lava.cs.virginia.edu/hotspot/ (2011). Accessed on 26th April 2014.
8. National Semiconductor PowerWise Adaptive Voltage Scaling Technology. http://www.ti.com/lit/wp/snvy007/snvy007.pdf (2012). Accessed on 17th April 2014.
9. PowerWise Adaptive Voltage Scaling (AVS). http://www.ti.com/ww/en/analog/power_management/powerwise-avs.shtml (2014). Accessed on 17th April 2014.
10. Abou-Seido, A.I., Nowak, B., Chu, C.: Fitted Elmore Delay: A Simple and Accurate Interconnect Delay Model. *IEEE Transactions on Very Large Scale Integration (VLSI) Systems* **12**(7), 691–696 (2004). DOI 10.1109/TVLSI.2004.830932.
11. Acar, E., Arunachalam, R., Nassif, S.R.: Predicting Short Circuit Power from Timing Models. In: *Proceedings of the Asia and South Pacific Design Automation Conference*, pp. 277–282 (2003). DOI 10.1109/ASPDAC.2003.1195029.
12. Agarwal, A.: *Algorithms for Layout-Aware and Performance Model Driven Synthesis of Analog Circuits*. Ph.D. thesis, Department of Computer Science & Engineering, University of Cincinnati, Ohio (2005). URL http://rave.ohiolink.edu/etdc/view?acc_num=ucin1132259454.
13. Akyildiz, I., Su, W., Sankarasubramaniam, Y., Cayirci, E.: A Survey on Sensor Networks. *IEEE Communications Magazine* **40**(8), 102–114 (2002). DOI 10.1109/MCOM.2002.1024422.
14. Ali, S., Tanner, S., Farine, P.A.: A Robust, Low Power, High Speed Voltage Level Shifter with Built-in Short Circuit Current Reduction. In: *Proceedings of the 20th European Conference on Circuit Theory and Design (ECCTD)*, pp. 142–145 (2011). DOI 10.1109/ECCTD.2011.6043302.
15. Alipour, S., Hidaji, B., Pour, A.S.: Circuit Level, Static Power, and Logic Level Power Analyses. In: *Proceedings of the IEEE International Conference on Electro/Information Technology (EIT)*, pp. 1–4 (2010). DOI 10.1109/EIT.2010.5612180.
16. Allstot, D.J., Choi, K., Park, J.: *Parasitic-Aware Optimization of CMOS RF Circuits*. Springer US (2003).
17. Auth, C.: 45 nm High-κ + Metal Gate Strain-enhanced CMOS Transistors. In: *Proceedings of the IEEE Custom Integrated Circuits Conference*, pp. 379–386 (2008). DOI 10.1109/CICC.2008.4672101.
18. Badaoui, R.F., Sampath, H., Agarwal, A., Vemuri, R.: A High Level Language for Pre-Layout Extraction in Parasite-Aware Analog Circuit Synthesis. In: *Proceedings of the 14th ACM Great Lakes Symposium on VLSI*, pp. 271–276 (2004).
19. Bakker, A., Huijsing, J.H.: *High-Accuracy CMOS Smart Temperature Sensors*. Series in Engineering and Computer Science. Springer (2000).
20. Balakrishnan, K.: *A Dual Voltage and Dual Frequency Low Power VLSI Implementation of DCT Domain Image Watermarking Schemes*. Master's thesis, Electrical Egnineering Technology, University of South Florida, FL, USA (2003).
21. Ballweber, B.M., Gupta, R., Allstot, D.J.: A fully integrated 0.5–5.5 GHz CMOS distributed amplifier. *IEEE Journal of Solid State Circuits* **35**(2), 231–239 (2000).
22. Balouchestani, M.: Low-Power Wireless Sensor Network with Compressed Sensing Theory. In: *Proceedings of the 4th Annual Caneus Fly by Wireless Workshop (FBW)*, pp. 1–4 (2011). DOI 10.1109/FBW.2011.5965565.
23. Banerjee, K., Mehrotra, A., Sangiovanni-Vincentelli, A., Hu, C.: On Thermal Effects in Deep Sub-Micron VLSI Interconnects. In: *Proceedings of the 36th Design Automation Conference*, pp. 885–891 (1999). DOI 10.1109/DAC.1999.782207.
24. Baschirotto, A., Chironi, V., Cocciolo, G., DAmico, S., De Matteis, M., Delizia, P.: Low Power Analog Design in Scaled Technologies. In: *Proceedings of the Topical Workshop on Electronics for Particle Physics*, pp. 103–110 (2009).
25. Bellosa, F., Weißel, A., Waitz, M., Kellner, S.: Event-Driven Energy Accounting for Dynamic Thermal Management. In: *Proceedings of the Workshop on Compilers and Operating Systems for Low Power* (2003).
26. Bohr, M., Mistry, K.: Intels Revolutionary 22 nm Transistor Technology. http://download.intel.com/newsroom/kits/22nm/pdfs/22nm-details_presentation.pdf (2011). Accessed on 06th May 2014.
27. Brodersen, R.W., Chandrakasan, A., Sheng, S.: Technologies For Personal Communications. In: *Digest of Technical Papers Symposium on VLSI Circuits*, pp. 5–9 (1991). DOI 10.1109/VLSIC.1991.760053.
28. Brooks, D., Dick, R.P., Joseph, R., Shang, L.: Power, Thermal, and Reliability Modeling in Nanometer-Scale Microprocessors. *IEEE Micro* **27**(3), 49–62 (2007). DOI 10.1109/MM.2007.58.
29. Brynjolfson, I., Zilic, Z.: Dynamic Clock Management for Low Power Applications in FPGAs. In: *Proceedings of the IEEE Custom Integrated Circuits Conference*, pp. 139–142 (2000).
30. Carroll, A., Heiser, G.: An Analysis Of Power Consumption in a Smartphone. In: *Proceedings of the 2010 USENIX Annual Technical Conference*, pp. 21–21 (2010).
31. Chan, H., Zilic, Z.: Parasitic-Aware Physical Design Optimization of Deep Sub-Micron Analog Circuits. In: *Proceedings of the 50th Midwest Symposium on Circuits and Systems*, pp. 1022–1025 (2007). DOI 10.1109/MWSCAS.2007.4488736.

32. Chandrakasan, A.P., Sheng, S., Brodersen, R.W.: Low-Power CMOS Digital Design. *IEEE Journal of Solid-State Circuits* **27**(4), 473–484 (1992). DOI 10.1109/4.126534.
33. Chang, L., Morton, S., Chang, K., Jin-Man, H., Malcovati, P., Stojanovic, V.: F2: VLSI Power-Management Techniques: Principles and Applications. In: *IEEE International Solid-State Circuits Conference Digest of Technical Papers (ISSCC)*, pp. 502–503 (2013). DOI 10.1109/ISSCC.2013.6487601.
34. Chaparro, P., Gonzalez, J., Gonzalez, A.: Thermal-aware Clustered Microarchitectures. In: *Proceedings of the IEEE International Conference on Computer Design*, pp. 48–53 (2004). DOI 10.1109/ICCD.2004.1347897.
35. Chatterjee, B., Sachdev, M., Hsu, S., Krishnamurthy, R., Borkar, S.: Effectiveness and Scaling Trends of Leakage Control Techniques for Sub-130 nm CMOS Technologies. In: *Proceedings of the 2003 International Symposium on Low Power Electronics and Design*, pp. 122–127 (2003). DOI 10.1109/LPE.2003.1231847.
36. Chaudhury, S.: A Tutorial and Survey on Thermal-Aware VLSI Design: Tools and Techniques. *International Journal of Recent Trends in Engineering* **2**(8), 18–21 (2009). Accessed on 28th April 2014.
37. Chen, G., Sapatnekar, S.: Partition-driven Standard Cell Thermal Placement. In: *Proceedings of the International Symposium on Physical Design*, pp. 75–80 (2003). DOI 10.1145/640000.640018.
38. Chen, S., Yao, Y., Yoshimura, T.: A Dynamic Programming Based Algorithm for Post-scheduling Frequency Assignment in Energy-efficient High-level Synthesis. In: *Proceedings of the 10th IEEE International Conference on Solid-State and Integrated Circuit Technology (ICSICT)*, pp. 797–799 (2010). DOI 10.1109/ICSICT.2010.5667427.
39. Chen, S.Y., Lin, R.B., Tung, H.H., Lin, K.W.: Power Gating Design for Standard-Cell-Like Structured ASICs. In: *Proceedings of the Design, Automation Test in Europe Conference Exhibition (DATE)*, pp. 514–519 (2010). DOI 10.1109/DATE.2010.5457152.
40. Chinnery, D., Keutzer, K.: *Closing the Power Gap between ASIC & Custom: Tools and Techniques for Low Power Design*. Springer (2008).
41. Choi, J., Cha, H.: System-Level Power Management for System-On-a-Chip-based Mobile Devices. *IET Computers Digital Techniques* **4**(5), 400–409 (2010). DOI 10.1049/iet-cdt.2008.0074.
42. Choi, K., Allstot, D.J.: Parasitic-Aware Design and Optimisation of RF Power Amplifiers. *Proceedings of the IEE Circuits, Devices and Systems* **149**(5/6), 369–375 (2002).
43. Choi, K., Allstot, D.J.: Parasitic-Aware Design and Optimization of a CMOS RF Power Amplifier. *IEEE Transactions on Circuits and Systems I: Regular Papers* **53**(1), 16–25 (2006). DOI 10.1109/TCSI.2005.854608.
44. Choi, K., Allstot, D.J., Kiaei, S.: Parasitic-Aware Synthesis of RF CMOS Switching Power Amplifiers. In: *Proceeddings of the IEEE International Symposium on Circuits and Systems*, pp. I–269–I–272 (2002). DOI 10.1109/ISCAS.2002.1009829.
45. Choudhary, A., Maheshwari, V., Singh, A., Kar, R.: Wave Propagation Based Analytical Delay and Cross Talk Noise Model For Distributed On-Chip RLCG Interconnects. In: *Proceedings of the IEEE International Conference on Semiconductor Electronics (ICSE)*, pp. 153–157 (2010). DOI 10.1109/SMELEC.2010.5549381.
46. Chung, E.Y.: *Software Approaches for Energy Efficient System Design*. Ph.D. thesis, Department of Electrical Engineering, Stanford University (2002).
47. Chung, S.W., Skadron, K.: Using On-Chip Event Counters for High-Resolution, Real-Time Temperature Measurements. In: *Proceedings of the Thermal and Thermomechanical Phenomena in Electronics Systems*, pp. 114–120 (2006).
48. du Cloux, R., de Graaf, W.J., Maas, G.P.J.F.M., van der Veeken, R.W.: EMC Simulations and Measurements. In: *Proceedings of International Ž rich Symposium on Electromagnetic Compatibility*, pp. 185–190, (1995).
49. Coen, G., Zutter, D.D.: Reduction of Circuit Complexity Using Tensor Analysis of Networks. In: *Proceedings of the International Ž rich Symposium on Electromagnetic Compatibility* (1997).
50. Costa, J., Chou, M., Silveira, L.: Efficient Techniques for Accurate Modeling and Simulation of Substrate Coupling in Mixed-signal IC's. *IEEE Transactions on Computer-Aided Design of Integrated Circuits and Systems* **18**(5), 597–607 (1999). DOI 10.1109/43.759076.
51. Cui, J., Maskell, D.L.: High Level Event Driven Thermal Estimation for Thermal Aware Task Allocation and Scheduling. In: *Proceedings of the 15th Asia and South Pacific Design Automation Conference*, pp. 793–798 (2010). DOI 10.1109/ASPDAC.2010.5419781.
52. Daniel, L., White, J.K.: Automatic Generation of Geometrically Parameterized Reduced Order Models for Integrated Spiral RF Inductors. In: *Proceedings of IEEE International Behavioral Modeling and Simulation Conference*, pp. 18–23 (2003).
53. Datta, B.: *On-Chip Thermal Sensing in Deep Sub-Micron CMOS*. Master's thesis, Electrical And Computer Engineering, University of Massachusetts Amherst, Amherst, MA (2007). Accessed on 27th Arpil 2014.
54. Dengler, R.: Self Inductance Of A Wire Loop As A Curve Integral. ArXiv e-prints (2012).
55. Dennington, B.: Low Power Design from Technology Challenge to Great Products. In: *Proceedings of the International Symposium on Low Power Electronics and Design*, pp. 213–213 (2006). DOI 10.1109/LPE.2006.4271838.
56. Donald, J., Martonosi, M.: Temperature-Aware Design Issues for SMT and CMP Architectures. In: *Proceedings of the Workshop on Complexity-Effective Design*, pp. 48–53 (2004).
57. Elmore, W.C.: The Transient Response of Damped Linear Networks with Particular Regard to Wideband Amplifiers. *Journal of Applied Physics* **19**(1), 55–63 (1948). DOI 10.1063/1.1697872.

58. Enz, C.C., Vittoz, E.A.: CMOS Low-Power Analog Circuit Design. In: *Proceedings of the Designing Low Power Digital Systems, Emerging Technologies*, pp. 79–133 (1996). DOI 10.1109/ETLPDS.1996.508872.
59. Finchelstein, D.F., Sze, V., Sinangil, M.E., Koken, Y., Chandrakasan, A.P.: A Low-Power 0.7-V H.264 720p Video Decoder. In: *Proceedings of the IEEE Asian Solid-State Circuits Conference (A-SSCC '08)*, pp. 173–176 (2008). DOI 10.1109/ASSCC.2008.4708756.
60. Fino, M.H., Coito, F.: On the Use of Compact Modeling for RF/analog Design Automation. In: *Proceedings of the 20th International Conference Mixed Design of Integrated Circuits and Systems (MIXDES)*, pp. 41–47 (2013).
61. Fulde, M., Schmitt-Landsiedel, D., Knoblinger, G.: Analog and RF Design Issues in High-κ Multi-gate CMOS Technologies. In: *Proceedings of the IEEE International Electron Devices Meeting (IEDM)*, pp. 1–1 (2009). DOI 10.1109/IEDM.2009.5424326.
62. Gallivan, K., Grimme, E., Dooren, P.V.: A Rational Lanczos Algorithm for Model Reduction. *Numerical Algorithms* **12**, 33–63 (1996).
63. Gao, W.: *Low Power Design Methodologies in Analog Blocks of CMOS Image Sensors*. Ph.D. thesis, Computer Science and Engineering Department, York University, Toronto, Ontario, Canada (2011). Accessed on 02nd May 2014.
64. Garitselov, O., Mohanty, S.P., Kougianos, E.: A Comparative Study of Metamodels for Fast and Accurate Simulation of Nano-CMOS Circuits. *IEEE Transactions on Semiconductor Manufacturing* **25**(1), 26–36 (2012).
65. Ghai, D.: *Variability Aware Low-Power Techniques for Nanoscale Mixed-Signal Circuits*. Ph.D. thesis, University of North Texas (2009).
66. Ghai, D., Mohanty, S., Kougianos, E.: A Dual Oxide CMOS Universal Voltage Converter for Power Management in Multi-V_{DD} SoCs. In: *Proceedings of the 9th International Symposium on Quality Electronic Design*, pp. 257–260 (2008).
67. Ghai, D., Mohanty, S.P., Kougianos, E.: A Dual Oxide CMOS Universal Voltage Converter for Power Management in Multi-V_{DD} SoCs. In: *Proceedings of the 9th International Symposium on Quality Electronic Design*, pp. 257–260 (2008). DOI 10.1109/ISQED.2008.4479735.
68. Ghai, D., Mohanty, S.P., Kougianos, E.: Parasitic Aware Process Variation Tolerant Voltage Controlled Oscillator (VCO) Design. In: *Proceedings of the 9th International Symposium on Quality of Electronic Design*, pp. 330–333 (2008).
69. Ghai, D., Mohanty, S.P., Kougianos, E.: Design of Parasitic and Process-Variation Aware Nano-CMOS RF Circuits: A VCO Case Study. *IEEE Transactions on Very Large Scale Integration (VLSI) Systems* **17**(9), 1339–1342 (2009). DOI 10.1109/TVLSI.2008.2002046.
70. Ghai, D., Mohanty, S.P., Kougianos, E.: Unified P4 (Power-Performance-Process-Parasitic) Fast Optimization of A Nano-CMOS VCO. In: *Proceedings of the 19th ACM Great Lakes Symposium on VLSI*, pp. 303–308 (2009).
71. Gharpurey, R., Meyer, R.: Modeling and Analysis of Substrate Coupling in Integrated Circuits. *IEEE Journal of Solid-State Circuits* **31**(3), 344–353 (1996). DOI 10.1109/4.494196.
72. Goering, R.: How Parasitic-Aware Design Flow Improves Custom/Analog Productivity. URL http://www.cadence.com/community/blogs/ii/archive/2011/03/14/how-parasitic-aware-design-improves-custom-analog-productivity.aspx. Accessed on 27 July 2012.
73. Gupta, P., Kahng, A.B., Sharma, P.: A Practical Transistor-Level Dual Threshold Voltage Assignment Methodology. In: *Proceedings of the Sixth International Symposium on Quality of Electronic Design*, pp. 421–426 (2005). DOI 10.1109/ISQED.2005.13.
74. Hager, W.W., Zhang, H.: Algorithm 851: CG-DESCENT, A Conjugate Gradient Method with Guaranteed Descent. *ACM Transactions on Mathematical Software* **32**(1), 113–137 (2006).
75. Hajimiri, A., Limotyrakis, S., Lee, T.H.: Jitter and Phase Noise in Ring Oscillators. *IEEE Journal of Solid State Circuits* **34**(6), 790–804 (1999).
76. Han, Y.: *Temperature Aware Techniques for Design, Simulation and Measurement in Microprocessors*. Ph.D. thesis, Electrical and Computer Engineering, University of Massachusetts Amherst (2007).
77. Hanson, S., Seok, M., Sylvester, D., Blaauw, D.: Nanometer Device Scaling in Subthreshold Circuits. In: *Proceedings of the 44th ACM/IEEE Design Automation Conference*, pp. 700–705 (2007).
78. Hempstead, M., Lyons, M.J., Brooks, D., Wei, G.Y.: Survey of Hardware Systems for Wireless Sensor Networks. *Journal Low Power Electronics* **4**(1), 11–20 (2008). URL http://dblp.uni-trier.de/db/journals/jolpe/jolpe4.html#HempsteadLBW08.
79. Henry, M.B., Nazhandali, L.: From Transistors to NEMS: Highly Efficient Power-Gating of CMOS Circuits. *ACM Journal on Emerging Technologies in Computing Systems (JETC)* **8**(1), 2:1–2:18 (2012). DOI 10.1145/2093145.2093147. URL http://doi.acm.org/10.1145/2093145.2093147.
80. Huang, P.Y., Lee, Y.M.: Full-Chip Thermal Analysis for the Early Design Stage via Generalized Integral Transforms. *IEEE Transactions on Very Large Scale Integration (VLSI) Systems* **17**(5), 613–626 (2009). DOI 10.1109/TVLSI.2008.2006043.
81. Huang, W.: *HotSpot - A Chip and Package Compact Thermal Modeling Methodology for VLSI Design*. Ph.D. thesis, Department of Electrical and Computer Engineering, University of Virginia, Charlottesville, Virginia (2007). Accessed on 26th April 2014.
82. Huang, W., Ghosh, S., Velusamy, S., Sankaranarayanan, K., Skadron, K., Stan, M.R.: Hotspot: A Compact Thermal Modeling Methodology for Early-stage VLSI Design. *IEEE Transactions on Very Large Scale Integration (VLSI) Systems* **14**(5), 501–513 (2006). DOI 10.1109/TVLSI.2006.876103.

83. Huang, W., Stan, M.R., Skadron, K., Sankaranarayanan, K., Ghosh, S., Velusamy, S.: Compact Thermal Modeling for Temperature-Aware Design. In: *Proceedings of the Design Automation Conference*, pp. 878–883 (2004).
84. Hung, W.L., Xie, Y., Vijaykrishnan, N., Addo-Quaye, C., Theocharides, T., Irwin, M.: Thermal-Aware Floorplanning Using Genetic Algorithms. In: *Proceedings of the Sixth International Symposium on Quality of Electronic Design*, pp. 634–639 (2005). DOI 10.1109/ISQED.2005.122.
85. Hung, W.L., Xie, Y., Vijaykrishnan, N., Kandemir, M., Irwin, M.J.: Thermal-aware Task Allocation and Scheduling for Embedded Systems. In: *Proceedings of the Design, Automation and Test in Europe*, pp. 898–899 (2005). DOI 10.1109/DATE.2005.310.
86. Hung, Y.C., Chen, J.C., Shieh, S.H., Tung, C.K.: A Survey of Low-voltage Low-power Technique and Challenge for CMOS Signal Processing Circuits. In: *Proceedings of the 12th International Symposium on Integrated Circuits*, pp. 554–557 (2009).
87. Isci, C., Martonosi, M.: Runtime Power Monitoring in High-end Processors: Methodology and Empirical Data. In: *Proceedings of the 36th Annual IEEE/ACM International Symposium on Microarchitecture*, pp. 93–104 (2003). DOI 10.1109/MICRO.2003.1253186.
88. Jian, H., Xubang, S.: The Design Methodology and Practice of Low Power SoC. In: *Proceedings of the International Conference on Embedded Software and Systems Symposia*, pp. 185–190 (2008). DOI 10.1109/ICESS.Symposia.2008.38.
89. Johnson, M.C., Somasekhar, D., Roy, K.: Models and Algorithms for Bounds on Leakage in CMOS Circuits. *IEEE Transactions on Computer-Aided Design of Integrated Circuits and Systems* **18**(6), 714–725 (1999). DOI 10.1109/43.766723.
90. Kanda, K., Nose, K., Kawaguchi, H., Sakurai, T.: Design Impact of Positive Temperature Dependence on Drain Current in Sub-1-V CMOS VLSIs. *IEEE Journal of Solid-State Circuits* **36**(10), 1559–1564 (2001). DOI 10.1109/4.953485.
91. Kao, J., Narendra, S., Chandrakasan, A.: Subthreshold Leakage Modeling and Reduction Techniques. In: *Proceedings of the IEEE/ACM International Conference on Computer Aided Design*, pp. 141–148 (2002). DOI 10.1109/ICCAD.2002.1167526.
92. Kao, J.T.: *Subthreshold Leakage Control Techniques for Low Power Digital Circuits*. Ph.D. thesis, Department of Electrical Engineering and Computer Science, Massachusetts Institute of Technology, MA, USA (2001). Accessed 15th April 2014.
93. Karn, T., Rawat, S., Kirkpatrick, D., Roy, R., Spirakis, G., Sherwani, N., Peterson, C.: EDA Challenges Facing Future Microprocessor Design. *IEEE Transactions on Computer-Aided Design of Integrated Circuits and Systems* **19**(12), 1498–1506 (2000). DOI 10.1109/43.898828.
94. Khateb, F., Dabbous, S.B.A., Vlassis, S.: A Survey of Non-conventional Techniques for Low-voltage Low-power Analog Circuit Design. *Radioengineering* **22**(2), 415–427 (2013). Accessed 12th April 2014.
95. Kim, H., Chen, C.C.P.: Be Careful of Self and Mutual Inductance Formulae. http://ccf.ee.ntu.edu.tw/ cchen/research/CompInduct9.pdf (2001). Accessed on 21st April 2014.
96. Kim, N.S., Austin, T., Baauw, D., Mudge, T., Flautner, K., Hu, J.S., Irwin, M.J., Kandemir, M., Narayanan, V.: Leakage Current: Moore's Law Meets Static Power. *Computer* **36**(12), 68–75 (2003). DOI 10.1109/MC.2003.1250885.
97. Kim, T., Lim, P.: Thermal-aware High-level Synthesis Based on Network Flow Method. In: *Proceedings of the 4th International Conference Hardware/Software Codesign and System Synthesis*, pp. 124–129 (2006). DOI 10.1145/1176254.1176285.
98. Kim, T., Qiao, W., Qu, L.: Series-Connected Reconfigurable Multicell Battery: A Novel Design Towards Smart Batteries. In: *Proceedings of the IEEE Energy Conversion Congress and Exposition (ECCE)*, pp. 4257–4263 (2010). DOI 10.1109/ECCE.2010.5617723.
99. Kougianos, E., Mohanty, S.P.: Effective Tunneling Capacitance: A New Metric oo Quantify Transient Gate Leakage Current. In: *Proceedings IEEE International Symposium on Circuits and Systems*, pp. 2937–2940 (2006). DOI 10.1109/ISCAS.2006.1693240.
100. Kougianos, E., Mohanty, S.P.: Metrics to Quantify Steady and Transient Gate Leakage in Nanoscale Transistors: NMOS vs. PMOS Perspective. In: *Proceedings of the 20th International Conference on VLSI Design*, pp. 195–200 (2007).
101. Kougianos, E., Mohanty, S.P.: A Comparative Study on Gate Leakage and Performance of High-κ Nano-CMOS Logic Gates. *International Journal of Electronics (IJE)* **97**(9), 985–1005 (2010).
102. Krasnicki, M.J., Phelps, R., Hellums, J.R., McClung, M., Rutenbar, R.A., Carley, L.R.: ASF: A Practical Simulation-Based Methodology for the Synthesis of Custom Analog Circuits. In: *Proceedings of the IEEE/ACM International Conference on Computer Aided Design*, pp. 350–357 (2001).
103. Krishnamoorthy, S., Venkatraman, V., Apanovich, Y., Burd, T., Daga, A.: Thermal-Aware Reliability Analysis of Nanometer Designs. In: *Proceedings of the 19th Conference on Electrical Performance of Electronic Packaging and Systems (EPEPS)*, pp. 277–280 (2010). DOI 10.1109/EPEPS.2010.5642793.
104. Krishnan, V., Katkoori, S.: TABS: Temperature-Aware Layout-Driven Behavioral Synthesis. *IEEE Transactions on Very Large Scale Integration (VLSI) Systems* **18**(12), 1649–1659 (2010). DOI 10.1109/TVLSI.2009.2026047.
105. Ku, J.C., Ismail, Y.: Area Optimization for Leakage Reduction and Thermal Stability in Nanometer-Scale Technologies. *IEEE Transactions on Computer-Aided Design of Integrated Circuits and Systems* **27**(2), 241–248 (2008). DOI 10.1109/TCAD.2007.913393.

106. Kursun, V., Friedman, E.G.: Domino Logic with Variable Threshold Voltage Keeper. *IEEE Transactions on Very Large Scale Integration (VLSI) Systems* **11**(6), 1080–1093 (2003). DOI 10.1109/TVLSI.2003.817515.
107. Kyung, C.M.: Various Low-Power SoC Design Techniques. http://vswww.kaist.ac.kr/course/ee877/lecture/lecture2. Low-Power_SoC_Design_Techniques.ppt (2006). Accessed on 15th July 2012.
108. Lambrechts, A., Raghavan, P., Leroy, A., Talavera, G., Aa, T., Jayapala, M., Catthoor, F., Verkest, D., Deconinck, G., Corporaal, H., Robert, F., Carrabina, J.: Power Breakdown Analysis for a Heterogeneous NoC Platform Running a Video Application. In: *Proceedings of a 16th IEEE International Conference on Application-Specific Systems, Architecture Processors*, pp. 179–184 (2005).
109. Lee, D., Blaauw, D., Sylvester, D.: Gate Oxide Leakage Current Analysis and Reduction for VLSI Circuits. *IEEE Transactions on Very Large Scale Integration (VLSI) Systems* **12**(2), 155–166 (2004). DOI 10.1109/TVLSI.2003.821553.
110. Lee, D., Blaauw, D., Sylvester, D.: Runtime Leakage Minimization through Probability-Aware Optimization. Very Large Scale Integration (VLSI) Systems. *IEEE Transactions on* **14**(10), 1075–1088 (2006). DOI 10.1109/TVLSI.2006.884149.
111. Lee, J., Kim, Y.B.: ASLIC: A Low Power CMOS Analog Circuit Design Automation. In: *Proceedingso the Sixth International Symposium on Quality of Electronic Design*, pp. 470–475 (2005). DOI 10.1109/ISQED.2005.23.
112. Lee, K.J., Skadron, K.: Using Performance Counters for Runtime Temperature Sensing in High-Performance Processors. In: *Proceedings of the 19th IEEE International Parallel and Distributed Processing Symposium*, April (2005). DOI 10.1109/IPDPS.2005.448.
113. Lee, S., Song, S., Au, V., Moran, K.P.: Constriction/Spreading Resistance Model for Electronics Packaging. In: *Proeedings of ASME/JSME Thermal Engineering Conference*, pp. 199–206 (1995).
114. Leenaerts, D., Gielen, G., Rutenbar, R.A.: CAD Solutions and Outstanding Challenges for Mixed-Signal and RF IC Design. In: *Proceedings of the IEEE/ACM International Conference on Computer Aided Design*, pp. 270–277 (2001).
115. Li, H., Fan, J., Qi, Z., Tan, S., Wu, L., Cai, Y., Hong, X.: Partitioning-Based Approach to Fast On-Chip Decoupling Capacitor Budgeting and Minimization. *IEEE Transactions on Computer-Aided Design of Integrated Circuits and Systems* **25**(11), 2402–2412 (2006). DOI 10.1109/TCAD.2006.870862.
116. Li, H., Liu, P., Qi, Z., Jin, L., Wu, W., Tan, S.X.D., Yang, J.: Efficient Thermal Simulation for Run-Time Temperature Tracking and Management. In: *Proceedings IEEE International Conference on Computer Design*, pp. 130–133 (2005). DOI 10.1109/ICCD.2005.46.
117. Li, Y., Brooks, D., Hu, Z., Skadron, K.: Evaluating the Thermal Efficiency of SMT and CMP Architectures. In: *Proceedings of the IBM Watson Conference on Interaction between Architecture, Circuits, and Compilers* (2004).
118. Li, Y., Lee, B., Brooks, D., Hu, Z., Skadron, K.: Impact of Thermal Constraints on Multi-Core Architectures. In: *Proceedings of the The Tenth Intersociety Conference on Thermal and Thermomechanical Phenomena in Electronics Systems*, pp. 132–139 (2006). DOI 10.1109/ITHERM.2006.1645333.
119. Lin, J.F.: Low-Power Pulse-Triggered Flip-Flop Design Based on a Signal Feed-through. *IEEE Transactions on Very Large Scale Integration (VLSI) Systems* **22**(1), 181–185 (2014). DOI 10.1109/TVLSI.2012.2232684.
120. Lin, Q., Ma, M., Vo, T., Fan, J., Wu, X., Li, R., Li, X.Y.: Design-for-Manufacture for Multigate Oxide CMOS Process. *IEEE Transactions on Semiconductor Manufacturing* **21**(1), 41–45 (2008). DOI 10.1109/TSM.2007.913190.
121. Lin, S.C., Mahajan, R., De, V., Banerjee, K.: Analysis and Implications of IC Cooling for Deep Nanometer Scale CMOS Technologies. In: *Technical Digest IEEE International Electron Devices Meeting*, pp. 1018–1021 (2005). DOI 10.1109/IEDM.2005.1609537.
122. Lin, T., Chong, K.S., Gwee, B.H., Chang, J.S.: Fine-Grained Power Gating for Leakage and Short-Circuit Power Reduction by Using Asynchronous-Logic. In: *Proceedings of the IEEE International Symposium on Conference on Circuits and Systems*, pp. 3162–3165 (2009). DOI 10.1109/ISCAS.2009.5118474.
123. Liu, P., Qi, Z., Li, H., Jin, L., Wu, W., Tan, S.D., Yang, J.: Fast Thermal Simulation for Architecture Level Dynamic Thermal Management. In: *Proceedings of the IEEE/ACM International Conference on Computer-Aided Design*, pp. 639–644 (2005). DOI 10.1109/ICCAD.2005.1560145.
124. Lu, Y.H., De Micheli, G.: Comparing System Level Power Management Policies. *IEEE Design Test of Computers* **18**(2), 10–19 (2001). DOI 10.1109/54.914592.
125. Macii, A., Benini, L., Poncino, M.: Memory Design Techniques for Low Energy Embedded Systems. Springer (2002).
126. Mahajan, R., Chiu, C-P., Chrysler, G.: Cooling a Microprocessor Chip. *Proceedings of the IEEE* **94**(8), 1476–1486 (2006). DOI 10.1109/JPROC.2006.879800.
127. Mahesri, A., Vardhan, V.: Power Consumption Breakdown on a Modern Laptop. In: *Proceedings of the 4th International Conference on Power-Aware Computer Systems*, pp. 165–180. Springer-Verlag, Berlin, Heidelberg (2005). DOI 10.1007/11574859_12. URL http://dx.doi.org/10.1007/11574859_12.
128. Mahmoodi, H., Tirumalashetty, V., Cooke, M., Roy, K.: Ultra Low-Power Clocking Scheme Using Energy Recovery and Clock Gating. *IEEE Transactions on Very Large Scale Integration (VLSI) Systems* **17**(1), 33–44 (2009). DOI 10.1109/TVLSI.2008.2008453.
129. Majizdadeh, V., Shoaei, O.: A Power Optimized Design Methodology for Low-distortion Sigma-delta-pipeline ADCs. In: *Proceedings of the 16th ACM Great Lakes Symposium on VLSI*, pp. 284–289. ACM, New York, NY, USA (2006). DOI 10.1145/1127908.1127974.

130. Mal, A.K., Dhar, A.S.: Modified Elmore Delay Model for VLSI Interconnect. In: *Proceedings of the 53rd IEEE International Midwest Symposium on Circuits and Systems (MWSCAS)*, pp. 793–796 (2010). DOI 10.1109/MWSCAS.2010.5548693.
131. Mandal, S.K., Bhojwani, P., Mohanty, S.P., Mahapatra, R.N.: IntellBatt: Towards Smarter Battery Design. In: *Proceedings of the 45th Design Automation Conference*, pp. 872–877 (2008).
132. Mandal, S.K., Mahapatra, R.N., Bhojwani, P., Mohanty, S.P.: IntellBatt: Toward a Smarter Battery. *IEEE Computer* **43**(3), 67–71 (2010).
133. Martin, S.M., Flautner, K., Mudge, T., Blaauw, D.: Combined Dynamic Voltage Scaling and Adaptive Body Biasing for Lower Power Microprocessors Under Dynamic Workloads. In: *Proceedings of the IEEE/ACM International Conference on Computer Aided Design*, pp. 721–725 (2002). DOI 10.1109/ICCAD.2002.1167611.
134. McCormick, S.P., Allen, J.: Waveform Moment Methods for Improved Interconnection Analysis. In: *Proceedings of Design Automation Conference* (1990).
135. Mehta, N., Amrutur, B.: Dynamic Supply and Threshold Voltage Scaling for CMOS Digital Circuits Using In-Situ Power Monitor. *IEEE Transactions on Very Large Scale Integration (VLSI) Systems* **20**(5), 892–901 (2012). DOI 10.1109/TVLSI.2011.2132765.
136. Meindl, J.D.: Low Power Microelectronics: Retrospect and Prospect. *Proceedings of the IEEE* **83**(4), 619–635 (1995). DOI 10.1109/5.371970.
137. Mengibar-Pozo, L., Lorenz, M., Lopez, C., Entrena, L.: Low-Power Design in Aerospace Circuits: A Case Study. *IEEE Aerospace and Electronic Systems Magazine* **28**(12), 46–52 (2013). DOI 10.1109/MAES.2013.6693668.
138. Mera, D.E., Santiago, N.G.: Low Power Software Techniques for Embedded Systems Running Real Time Operating Systems. In: *Proceedings of the 53rd IEEE International Midwest Symposium on Circuits and Systems (MWSCAS)*, pp. 1061–1064 (2010). DOI 10.1109/MWSCAS.2010.5548830.
139. Meterelliyoz, M., Kulkarni, J.P., Roy, K.: Thermal Analysis of 8-T SRAM for Nano-Scaled Technologies. In: *Proceeding of the 13th International Symposium on Low Power Electronics and Design*, pp. 123–128 (2008).
140. Mohanty, S., Kougianos, E.: PVT-tolerant 7-Transistor SRAM Optimization via Polynomial Regression. In: *Proceedings of the International Symposium on Electronic System Design (ISED)*, pp. 39–44 (2011). DOI 10.1109/ISED.2011.11.
141. Mohanty, S.P.: *Energy and Transient Power Minimization during Behavioral Synthesis*. Ph.D. thesis, Department of Computer Science and Engineering, University of South Florida, Tampa, USA (2003).
142. Mohanty, S.P.: ILP Based Gate Leakage Optimization Using DKCMOS Library during RTL Synthesis. In: *Proceedings of the 9th International Symposium on Quality Electronic Design*, pp. 174–177 (2008). DOI 10.1109/ISQED.2008.4479721.
143. Mohanty, S.P.: Unified Challenges in Nano-CMOS High-Level Synthesis. In: *Proceedings of the 22nd International Conference on VLSI Design*, pp. 531–531 (2009). DOI 10.1109/VLSI.Design.2009.124.
144. Mohanty, S.P.: Power, Parasitics, and Process-Variation (P3) Awareness in Mixed-Signal Design. *Journal Low Power Electronics* **8**(3), 259–260 (2012).
145. Mohanty, S.P., Ghai, D., Kougianos, E.: A P4VT (Power-Performance-Process-Parasitic-Voltage-Temperature) Aware Dual-V_{th} Nano-CMOS VCO. In: *Proceedings of the 23rd IEEE International Conference on VLSI Design (ICVD)*, pp. 99–104 (2010).
146. Mohanty, S.P., Ghai, D., Kougianos, E., Joshi, B.: A Universal Level Converter Towards the Realization of Energy Efficient Implantable Drug Delivery Nano-Electro-Mechanical-Systems. In: *Proceedings of the International Symposium on Quality Electronic Design*, pp. 673–679 (2009). DOI 10.1109/ISQED.2009.4810374.
147. Mohanty, S.P., Kougianos, E.: Modeling and Reduction of Gate Leakage during Behavioral Synthesis of NanoCMOS Circuits. In: *Proceedings of the 19th International Conference on VLSI Design*, pp. 83–88 (2006).
148. Mohanty, S.P., Kougianos, E.: Steady and Transient State Analysis of Gate Leakage Current in Nanoscale CMOS Logic Gates. In: *Proceedings of the IEEE International Conference on Computer Design*, pp. 210–215 (2006). DOI 10.1109/ICCD.2006.4380819.
149. Mohanty, S.P., Kougianos, E.: Simultaneous Power Fluctuation and Average Power Minimization during Nano-CMOS Behavioural Synthesis. In: *Proceedings of the 20th IEEE International Conference on VLSI Design*, pp. 577–582 (2007).
150. Mohanty, S.P., Kougianos, E.: Polynomial Metamodel Based Fast Optimization of Nano-CMOS Oscillator Circuits. *Analog Integrated Circuits and Signal Processing* **79**(3), 437–453 (2014). DOI 10.1007/s10470-014-0284-2. URL http://dx.doi.org/10.1007/s10470-014-0284-2.
151. Mohanty, S.P., Kougianos, E., Okobiah, O.: Optimal Design of a Dual-Oxide Nano-CMOS Universal Level Converter for Multi-V_{dd} SoCs. *Analog Integrated Circuits and Signal Processing* **72**(2), 451–467 (2012). DOI 10.1007/s10470-012-9887-7. URL http://dx.doi.org/10.1007/s10470-012-9887-7.
152. Mohanty, S.P., Mukherjee, V., Velagapudi, R.: Analytical Modeling and Reduction of Direct Tunneling Current during Behavioral Synthesis of Nanometer CMOS Circuits. In: *Proceedings of the 14th ACM/IEEE International Workshop on Logic and Synthesis (IWLS)*, pp. 249–256 (2005).
153. Mohanty, S.P., Panigrahi, B.K.: ILP Based Leakage Optimization During Nano-CMOS RTL Synthesis: A DOXCMOS Versus DTCMOS Perspective. In: *Proceedings of the World Congress on Nature Biologically Inspired Computing*, pp. 1367–1372 (2009). DOI 10.1109/NABIC.2009.5393744.

154. Mohanty, S.P., Pradhan, D.K.: ULS: A Dual-V_{th}/high-κ Nano-CMOS Universal Level Shifter for System-Level Power Management. *ACM Journal on Emerging Technologies in Computing Systems (JETC)* **6**(2), 8:1–8:26 (2010).
155. Mohanty, S.P., Ranganathan, N.: Energy Efficient Scheduling for Datapath Synthesis. In: *Proceedings of the 16th International Conference on VLSI Design*, pp. 446–451 (2003). DOI 10.1109/ICVD.2003.1183175.
156. Mohanty, S.P., Ranganathan, N.: A Framework For Energy and Transient Power Reduction During Behavioral Synthesis. *IEEE Transactions on VLSI Systems* **12**(6), 562–572 (2004).
157. Mohanty, S.P., Ranganathan, N.: Energy-efficient Datapath Scheduling Using Multiple Voltages and Dynamic Clocking. *ACM Transactions on Design Automation of Electronic Systems (TODAES)* **10**(2), 330–353 (2005). DOI 10.1145/1059876.1059883. URL http://doi.acm.org/10.1145/1059876.1059883.
158. Mohanty, S.P., Ranganathan, N., Balakrishnan, K.: Design of a Low Power Image Watermarking Encoder Using Dual Voltage and Frequency. In: *Proceedings of the 18th International Conference on VLSI Design*, pp. 153–158 (2005). DOI 10.1109/ICVD.2005.73.
159. Mohanty, S.P., Ranganathan, N., Balakrishnan, K.: A Dual Voltage-Frequency VLSI Chip for Image Watermarking in DCT Domain. *IEEE Transactions on Circuits and Systems II: Express Briefs* **53**(5), 394–398 (2006). DOI 10.1109/TCSII.2006.870216.
160. Mohanty, S.P., Ranganathan, N., Chappidi, S.K.: Transient Power Minimization through Datapath Scheduling in Multiple Supply Voltage Environment. In: *Proceedings of the 10th IEEE International Conference on Electronics, Circuits and Systems*, pp. 300–303 (2003). DOI 10.1109/ICECS.2003.1302036.
161. Mohanty, S.P., Ranganathan, N., Chappidi, S.K.: ILP Models for Simultaneous Energy and Transient Power Minimization During Behavioral Synthesis. *ACM Transactions on Design Automation Electronic Systems* **11**(1), 186–212 (2006).
162. Mohanty, S.P., Ranganathan, N., Kougianos, E., Patra, P.: Low-power High-level Synthesis for Nanoscale CMOS Circuits. Springer (2008). 0387764739 and 978-0387764733.
163. Mohanty, S.P., Ranganathan, N., Kougianos, E., Patra, P.: Low-power High-level Synthesis for Nanoscale CMOS Circuits. Springer Science+Business Media LLC, New York, NY 10013, USA (2008).
164. Mohanty, S.P., Ranganathan, N., Krishna, V.: Datapath Scheduling Using Dynamic Frequency Clocking. In: *Proceedings of the IEEE Computer Society Annual Symposium on VLSI*, pp. 58–63 (2002). DOI 10.1109/ISVLSI.2002.1016876.
165. Mohanty, S.P., Vadlamudi, S.T., Kougianos, E.: A Universal Voltage Level Converter for Multi-V_{dd} Based Low-Power Nano-CMOS Systems-on-Chips (SoCs). In: *Proceedings of the 13th NASA Symposium on VLSI Design*, p. 2.2 (2007).
166. Mohanty, S.P., Velagapudi, R., Kougianos, E.: Dual-K Versus Dual-T Technique for Gate Leakage Reduction: A Comparative Perspective. In: *Proceedings of the 7th International Symposium on Quality Electronic Design*, pp. 564–569 (2006). DOI 10.1109/ISQED.2006.52.
167. Mohanty, S.P., Velagapudi, R., Kougianos, E.: Physical-aware Simulated Annealing Optimization of Gate Leakage in Nanoscale Datapath Circuits. In: *Proceedings of the Conference on Design, Automation and Test in Europe*, pp. 1191–1196 (2006).
168. Mohanty, S.P., Velagapudi, R., Mukherjee, V., Li, H.: Reduction of Direct Tunneling Power Dissipation during Behavioral Synthesis of Nanometer CMOS Circuits. In: *Proceedings of the IEEE Computer Society Annual Symposium on VLSI (ISVLSI)*, pp. 248–249 (2005).
169. Mukherjee, R., Memik, S.O.: An Integrated Approach to Thermal Management in High-Level Synthesis. *IEEE Transactions on Very Large Scale Integration (VLSI) Systems* **14**(11), 1165–1174 (2006). DOI 10.1109/TVLSI.2006.886408.
170. Mukherjee, V.: *A Dual Dielectric Approach for Performance Aware Reduction of Gate Leakage in Combinational Circuits*. Master's thesis, Department of Computer Science and Engineering, University of North Texas, Denton, TX, USA (2006).
171. Mukherjee, V., Mohanty, S.P., Kougianos, E.: A Dual Dielectric Approach for Performance Aware Gate Tunneling Reduction in Combinational Circuits. In: *Proceedings of the IEEE International Conference on Computer Design: VLSI in Computers and Processors*, pp. 431–436 (2005). DOI 10.1109/ICCD.2005.5.
172. Naderlinger, A.: A Survey of Dynamic Thermal Management and Power Consumption Estimation. Software System Seminar, University of Salzburg, Austria, Summer, (2007).
173. Naik, K.: A Survey of Software Based Energy Saving Methodologies for Handheld Wireless Communication Devices. Tech. Rep. Technical Report No. 2010-13, Department of Electrical and Computer Engineering, University of Waterloo, Waterloo, Ontario, Canada, N2L3G1 (2010). URL https://ece.uwaterloo.ca/ snaik/energy.pdf.
174. Pan, J.: *CMOS Analog Integrated Circuit Design Techniques for Low-Powered Ubiquitous Device*. Ph.D. thesis, Graduate School of Information, Production and Systems, Waseda University, Tokyo, Japan (2008).
175. Panda, P.R., Silpa, B.V.N., Shrivastava, A., Gummidipudi, K.: *Power-efficient System Design*. Springer (2010).
176. Park, J., Choi, K., Allstot, D.J.: Parasitic-Aware Design and Optimization of a Fully Integrated CMOS Wideband Amplifier. In: *Proceedings of the Asia South Pacific Design Automation Conference*, pp. 904–907 (2003).
177. Park, J., Choi, K., Allstot, D.J.: Parasitic-aware RF Circuit Design and Optimization. *IEEE Transactions on Circuits and Systems I: Regular Papers* **51**(10), 1953–1966 (2004). DOI 10.1109/TCSI.2004.835691.
178. Pedram, M., Abdollahi, A.: IEE Proceedings. *Computers and Digital Techniques* **152**(3), 333–343 (2005). DOI 10.1049/ip-cdt:20045111.

179. Pedram, M., Nazarian, S.: Thermal Modeling, Analysis, and Management in VLSI Circuits: Principles and Methods. *Proceedings of the IEEE* **94**(8), 1487–1501 (2006). DOI 10.1109/JPROC.2006.879797.
180. Pering, T., Agarwal, Y., Gupta, R., Want, R.: CoolSpots: Reducing the Power Consumption of Wireless Mobile Devices with Multiple Radio Interfaces. In: *Proceedings of the 4th international conference on Mobile systems, applications and services*, pp. 220–232 (2006). DOI 10.1145/1134680.1134704. URL http://doi.acm.org/10.1145/1134680.1134704.
181. Peymandoust, A., Simunic, T., De Micheli, G.: Low Power Embedded Software Optimization Using Symbolic Algebra. In: *Proceedings of the Design, Automation and Test in Europe Conference and Exhibition*, pp. 1052–1058 (2002). DOI 10.1109/DATE.2002.998432.
182. Pillage, L.T., Rohrer, R.A.: Asymptotic Waveform Evaluation for Timing Analysis. *IEEE Transaction on Computer-Aided Design* **9**, 352–366 (1990).
183. Pouwelse, J., Langendoen, K., Sips, H.: Dynamic Voltage Scaling on a Low-power Microprocessor. In: *Proceedings of the 7th Annual International Conference on Mobile Computing and Networking*, pp. 251–259. ACM, New York, NY, USA (2001). DOI 10.1145/381677.381701. URL http://doi.acm.org/10.1145/381677.381701.
184. Puccinelli, D., Haenggi, M.: Wireless Sensor Networks: Applications and Challenges of Ubiquitous Sensing. *IEEE Circuits and Systems Magazine* **5**(3), 19–31 (2005). DOI 10.1109/MCAS.2005.1507522.
185. Punitha, A., Joseph, M.: Survey of Memory, Power and Temperature Optimization Techniques in High Level Synthesis. *International Journal of Recent Trends in Engineering* **2**(8), 22–26 (2009).
186. Puri, R., Stok, L., Cohn, J., Kung, D., Pan, D., Sylvester, D., Srivastava, A., Kulkarni, S.: Pushing ASIC Performance in a Power Envelope. In: *Proceedings of the Design Automation Conference*, pp. 788–793 (2003). DOI 10.1109/DAC.2003.1219126.
187. Qin, Z., Cheng, C.K.: Realizable Parasitic Reduction Using Generalized Y-Δ Transformation. In: *Proceedings of the Design Automation Conference*, pp. 220–225 (2003).
188. Rabaey, J.: Low Power Design Essentials. *Integrated Circuits and Systems*. Springer (2009).
189. Rabaey, J.M., Chandrakasan, A.P., Nikolić, B.: *Digital Integrated Circuits*, 2nd edn. Pearson Education (2003).
190. Rahimipour, S., Flayyih, W.N., El-Azhary, I., Shafie, S., Rokhani, F.Z.: A Survey of On-Chip Monitors. In: *Proceedings of the IEEE International Conference on Circuits and Systems (ICCAS)*, pp. 243–248 (2012). DOI 10.1109/ICCircuitsAndSystems.2012.6408286.
191. Ramalingam, A.: *Analysis Techniques for Nanometer Digital Integrated Circuits*. Ph.D. thesis, Electrical and Computer Engineering, The University of Texas at Austin, Austin, USA (2007). Accessed on 25th April 2014.
192. Ramirez-Angulo, J., Carvajal, R.G., Lopez-Martin, A.: Techniques for the Design of Low Voltage Power Efficient Analog and Mixed Signal Circuits. In: *Proceedings of the 22nd International Conference on VLSI Design*, pp. 26–27 (2009). DOI 10.1109/VLSI.Design.2009.112.
193. Rencz, M., Szekely, V., Poppe, A.: A Fast Algorithm for the Layout Based Electro-thermal Simulation. In: *Proceeding of the Design, Automation and Test in Europe Conference and Exhibition*, pp. 1032–1037 (2003). DOI 10.1109/DATE.2003.1253740.
194. Robertson, C.: Solving the Next Parasitic Extraction Challenge. http://www.techdesignforums.com/practice/technique/solving-the-next-parasitic-extraction-challenge/ (2010). Accessed on 09th May 2014.
195. Rossello, J.L., Bota, S., Rosales, M., Segura, J., de Les Illes Balears, U., de Mallorca, P., Keshavarzi, A.: Thermal-Aware Design Rules for Nanometer ICs. In: *Proceedings of the 11th International Workshop on Thermal Investigations of ICs and Systems* (2005).
196. Rossello, J.L., Canals, V., Bota, S.A., Keshavarzi, A., Segura, J.: A Fast Concurrent Power-Thermal Model for Sub-100 nm Digital ICs. In: *Proceedings of the Design, Automation and Test in Europe*, pp. 206–211 (2005). DOI 10.1109/DATE.2005.12.
197. Roy, K., Mukhopadhyay, S., Mahmoodi-Meimand, H.: Leakage Current Mechanisms and Leakage Reduction Techniques in Deep-submicrometer CMOS Circuits. *Proceedings of the IEEE* **91**(2), 305–327 (2003). DOI 10.1109/JPROC.2002.808156.
198. Roy, K., Prasad, S.: SYCLOP: Synthesis of CMOS Logic for Low Power Applications. In: *Proceedings of the International Conference on Computer Design: VLSI in Computers and Processors*, pp. 464–467 (1992). DOI 10.1109/ICCD.1992.276316.
199. Ruehli, A.E., Cangellaris, A.C.: Progress in the Methodologies for the Electrical Modeling of Interconnects and Electronic Packages. *Proceedings of the IEEE* **89**(5), 740–771 (2001).
200. Sabelka, R., Harlander, C., Selberherr, S.: The State of the Art in Interconnect Simulation. In: *Proceedings of the International Conference on Simulation of Semiconductor Processes and Devices*, pp. 6–11 (2000). DOI 10.1109/SISPAD.2000.871194.
201. Sakurai, T.: Low Power Digital Circuit Design. In: *Proceeding of the 34th European Solid-State Device Research Conference*, pp. 11–18 (2004). DOI 10.1109/ESSDER.2004.1356476.
202. Sankaranarayanan, K.: *Thermal Modeling and Management of Microprocessors*. Ph.D. thesis, Computer Science, School of Engineering and Applied Science, University of Virginia (2009). Accessed on 14th May 2013.
203. Sankaranarayanan, K., Velusamy, S., Stan, M., Skadron, K.: A Case for Thermal-Aware Floorplanning at the Microarchitectural Level. *Journal of Instruction-Level Parallelism* **7**, 8–16 (2005).

204. Semenov, O., Vassighi, A., Sachdev, M.: Impact of Technology Scaling on Thermal Behavior of Leakage Current in Sub-Quarter Micron MOSFETs: Perspective of Low Temperature Current Testing. *Microelectronics Journal* **33**(11), 985–994 (2002). DOI http://dx.doi.org/10.1016/S0026-2692(02)00071-X.
205. Serdijn, W.A.: Low-Voltage Low-Power Analog Integrated Circuits: A Special Issue of Analog Integrated Circuits and Signal Processing. *An International Journal* **8**(1) (1995). The Springer International Series in Engineering and Computer Science. Springer US (2012).
206. Serdijn, W.A., Mulder, J., Rocha, D., Marques, L.C.C.: Advances in Low-voltage Ultra-low-Power Analog Circuit Design. In: *Proceedings of the 8th IEEE International Conference on Electronics, Circuits and Systems*, vol. 3, pp. 1533–1536 (2001). DOI 10.1109/ICECS.2001.957507.
207. Shin, C., Sun, X., Liu, T.J.K.: Study of Random-Dopant-Fluctuation (RDF) Effects for the Trigate Bulk MOSFET. *IEEE Transactions on Electron Devices* **56**(7), 1538–1542 (2009).
208. Siddhu, L., Mishra, A., Singh, V.: Operand Isolation with Reduced Overhead for Low Power Datapath Design. In: *Proceedings of the 27th International Conference on VLSI Design*, pp. 483–488 (2014). DOI 10.1109/VLSID.2014.90.
209. Sill, F., Grassert, F., Timmermann, D.: Total Leakage Power Optimization with Improved Mixed Gates. In: *Proceedings of the 18th Symposium on Integrated Circuits and Systems Design*, pp. 154–159 (2005). DOI 10.1109/SBCCI.2005.4286849.
210. Singh, J., Mathew, J., Pradhan, D.K., Mohanty, S.P.: A Subthreshold Single Ended I/O SRAM Cell Design for Nanometer CMOS Technologies. In: *Proceedings of the IEEE International SOC Conference*, pp. 243–246 (2008). DOI 10.1109/SOCC.2008.4641520.
211. Singh, J., Mohanty, S.P., Pradhan, D.K.: *Robust SRAM Designs and Analysis*. Springer Science and Business Media (2012).
212. Sirisantana, N., Roy, K.: Low-Power Design Using Multiple Channel Lengths and Oxide Thicknesses. *IEEE Design Test of Computers* **21**(1), 56–63 (2004). DOI 10.1109/MDT.2004.1261850.
213. Skadron, K., Abdelzaher, T., Stan, M.R.: Control-Theoretic Techniques and Thermal-RC Modeling for Accurate and Localized Dynamic Thermal Management. In: *Proceedings of the Eighth International Symposium on High-Performance Computer Architecture*, pp. 17–28 (2002). DOI 10.1109/HPCA.2002.995695.
214. Skadron, K., Stan, M., Huang, W., Velusamy, S., Sankaranarayanan, K., Tarjan, D.: Temperature-aware Computer Systems: Opportunities and Challenges. *IEEE Micro* **23**(6), 52–61 (2003). DOI 10.1109/MM.2003.1261387.
215. Skadron, K., Stan, M., Huang, W., Velusamy, S., Sankaranarayanan, K., Tarjan, D.: Temperature-aware Microarchitecture. In: *Proceedings of the 30th Annual International Symposium on Computer Architecture*, pp. 2–13 (2003). DOI 10.1109/ISCA.2003.1206984.
216. Skadron, K., Stan, M.R., Sankaranarayanan, K., Huang, W., Velusamy, S., Tarjan, D.: Temperature-aware Microarchitecture: Modeling and Implementation. *ACM Transactions on Architecture and Code Optimization* **1**(1), 94–125 (2004). DOI 10.1145/980152.980157.
217. Soundararajan, R., Srivastava, A., Xu, Y.: A Programmable Second Order Oversampling CMOS Sigma-Delta Analog-to-digital Converter for Low-power Sensor Interface Electronics. In: *Proceedings of the SPIE*, vol. 7646, pp. 76, 460P–76,460P–11 (2010). DOI http://dx.doi.org/10.1117/12.847651.
218. Stan, M.R.: Optimal Voltages and Sizing for Low Power. In: *Proceedings of the 12th International Conference on VLSI Design*, pp. 428–433 (1999). DOI 10.1109/ICVD.1999.745193.
219. Takeuchi, K., Chang, K., Zhang, K., Yamauchi, T., Gastaldi, R.: F2: Ultra-low Voltage VLSIs for Energy Efficient Systems. In: *IEEE International Solid-State Circuits Conference Digest of Technical Papers (ISSCC)*, pp. 514–515 (2011). DOI 10.1109/ISSCC.2011.5746429.
220. Thakral, G.: *Process-voltage-temperature Aware Nanoscale Circuit Optimization*. Ph.D. thesis, Department of Computer Science and Engineering, University of North Texas, Denton (2010).
221. Thakral, G., Mohanty, S.P., Ghai, D., Pradhan, D.K.: A Combined DOE-ILP Based Power and Read Stability Optimization in Nano-CMOS SRAM. In: *Proceedings of the 23rd International Conference on VLSI Design*, pp. 45–50 (2010). DOI 10.1109/VLSI.Design.2010.14.
222. Thakral, G., Mohanty, S.P., Pradhan, D.K., Kougianos, E.: DOE-ILP Based Simultaneous Power and Read Stability Optimization in Nano-CMOS SRAM. *Journal of Low Power Electronics* **6**(3), 390–400 (2010).
223. Torfs, T., Sterken, T., Brebels, S., Santana, J., van den Hoven, R., Spiering, V., Bertsch, N., Trapani, D., Zonta, D.: Low Power Wireless Sensor Network for Building Monitoring. *IEEE Sensors Journal* **13**(3), 909–915 (2013). DOI 10.1109/JSEN.2012.2218680.
224. Vadlamudi, S.T.: *A Nano-CMOS Based Universal Voltage Level Converter for Multi-VDD SoCs*. Master's thesis, Department of Computer Science and Engineering, University of North Texas (2007).
225. Veendrick, H.J.M.: Short-circuit Dissipation of Static CMOS Circuitry and its Impact on the Design of Buffer Circuits. *IEEE Journal of Solid-State Circuits* **19**(4), 468–473 (1984). DOI 10.1109/JSSC.1984.1052168.
226. Venkataraman, H., Muntean, G.M.: *Green Mobile Devices and Networks: Energy Optimization and Scavenging Techniques*. Taylor & Francis (2012).
227. Visairo, H., Kumar, P.: A Reconfigurable Battery Pack for Improving Power Conversion Efficiency in Portable Devices. In: *Proceedings of the 7th International Caribbean Conference on Devices, Circuits and Systems*, pp. 1–6 (2008). DOI 10.1109/ICCDCS.2008.4542628.

228. Wang, B., Mazumder, P.: Fast Thermal Analysis for VLSI Circuits Via Semi-Analytical Green's Function in Multi-Layer Materials. In: *Proceedings of the IEEE International Symposium on Circuits and Systems*, pp. II-409–II-412 (2004). DOI 10.1109/ISCAS.2004.1329295.
229. Wang, C.C., Lee, P.M., Chen, K.L.: An SRAM Design Using Dual Threshold Voltage Transistors and Low-power Quenchers. *IEEE Journal of Solid-State Circuits* **38**(10), 1712–1720 (2003). DOI 10.1109/JSSC.2003.817254.
230. Wang, F., Wu, X., Xie, Y.: Variability-Driven Module Selection With Joint Design Time Optimization and Post-Silicon Tuning. In: *Proceedings of the Asia and South Pacific Design Automation Conference*, pp. 2–9 (2008). DOI 10.1109/ASPDAC.2008.4483963.
231. Wei, L., Chen, Z., Roy, K., Johnson, M.C., Ye, Y., De, V.K.: Design and Opimization of Dual-Threshold Circuits for Low-voltage Low-power Applications. *IEEE Transactions on VLSI Systems* **7**(1), 16–24 (1999).
232. Wu, W., Jin, L., Yang, J., Liu, P., Tan, S.X.D.: Efficient Power Modeling and Software Thermal Sensing for Runtime Temperature Monitoring. *ACM Transactions on Design Automation of Electronic Systems (TODAES)* **12**(3), 25:1–25:29 (2008). DOI 10.1145/1255456.1255462. URL http://doi.acm.org/10.1145/1255456.1255462.
233. Xie, Y., Hu, Z.: Tutorial 1 Thermal-Aware Design Techniques for Nanometer VLSI Chip. In: *Proceedings of the 6th International Conference on ASIC*, vol. 1, pp. 1–1 (2005). DOI 10.1109/ICASIC.2005.1611232.
234. Xie, Y., Hung, W.L.: Temperature-aware Task Allocation and Scheduling for Embedded Multiprocessor Systems-on-Chip (MPSoC) Design. *Journal of VLSI Signal Processing* **45**(3), 177–189 (2006). DOI 10.1007/s11265-006-9760-y. URL http://dx.doi.org/10.1007/s11265-006-9760-y.
235. Yang, S., Wang, H., Yang, Z.-J.: Low Leakage Dynamic Circuits with Dual Threshold Voltages and Dual Gate Oxide Thickness. In: *Proceedings of the 7th International Conference on ASIC*, pp. 70–73 (2007). DOI 10.1109/ICASIC.2007.4415569.
236. Yang, Y., Gu, Z., Zhu, C., Dick, R., Shang, L.: ISAC: Integrated Space-and-Time-Adaptive Chip-Package Thermal Analysis. *IEEE Transactions on Computer-Aided Design of Integrated Circuits and Systems* **26**(1), 86–99 (2007). DOI 10.1109/TCAD.2006.882589.
237. Yanzhu, Z., Dingyu, X.: Modeling and Simulating Transmission Lines Using Fractional Calculus. In: *Proceedings of the International Conference on Wireless Communications, Networking and Mobile Computing*, pp. 3115–3118 (2007). DOI 10.1109/WICOM.2007.773.
238. You, E., Varadadesikan, L., MacDonald, J., Xie, W.: A Practical Approach to Parasitic Extraction for Design of Multimillion-Transistor Integrated Circuits. In: *Proceedings of the Design Automation Conference*, pp. 69–74 (2000). DOI 10.1109/DAC.2000.855279.
239. Yu, J., Zhou, Q., Qu, G., Bian, J.: Behavioral Level Dual-Vth Design for Reduced Leakage Power with Thermal Awareness. In: *Proceedings of the Design, Automation Test in Europe Conference Exhibition (DATE)*, pp. 1261–1266 (2010). DOI 10.1109/DATE.2010.5457000.
240. Yuan, G., Wei, Z., Chen, H.: A Modified Conjugate Gradient Algorithm for Optimization Problems. In: *Proceedings of the International Conference on Multimedia Technology (ICMT)*, pp. 6175–6178 (2011). DOI 10.1109/ICMT.2011.6002381.
241. Yuan, L., Qu, G.: A Combined Gate Replacement and Input Vector Control Approach for Leakage Current Reduction. *IEEE Transactions on Very Large Scale Integration (VLSI) Systems* **14**(2), 173–182 (2006). DOI 10.1109/TVLSI.2005.863747.
242. Zanini, F., Atienza, D., De Micheli, G.: A Control Theory Approach for Thermal Balancing of MPSoC. In: *Proceedings of the Asia and South Pacific Design Automation Conference*, pp. 37–42 (2009). DOI 10.1109/ASPDAC.2009.4796438.
243. Zhan, Y., Kumar, S.V., Sapatnekar, S.S.: Thermally Aware Design. *Foundations and Trends in Electronic Design Automation* **2**(3), 255–370 (2008). DOI http://dx.doi.org/10.1561/1500000007. 25 April 2014.
244. Zhan, Y., Sapatnekar, S.S.: High-efficiency Green Function-based Thermal Simulation Algorithms. *IEEE Transactions on Computer-Aided Design of Integrated Circuits and Systems* **26**(9), 1661–1675 (2007). DOI 10.1109/TCAD.2007.895754.
245. Zhang, C.: *Techniques For Low Power Analog, Digital and Mixed Signal CMOS Integrated Circuit Design*. Ph.D. thesis, The Department of Electrical and Computer Engineering, Louisiana State University and Agricultural and Mechanical College, Baton Rouge, USA (2005). Accessed 12 April 2014.

CHAPTER 11
Variability-Aware AMS-SoC Design Methodologies

11.1 Introduction

Process variation has been discussed as a major issue in the chapter on design issues for DFX. Simplistically speaking, process variations lead to discrepancy between sizes intended during the design time and the sizes obtained during the manufacturing. For example, it is estimated that the variations in the channel length and threshold voltage of a MOSFET are as high as 30% in the case of a 65-nm CMOS process technology node [25, 28, 60]. The variations can be larger for smaller process technology nodes. As depicted in Fig. 11.1(a), no two chips on the same wafer have the same characteristics or even no two transistors in the same die are the same [4, 11, 48, 49, 50]. However, in a bigger perspective, the nanoelectronic parametric variations include process variations (P), voltage variations (V), thermal variations (T), and transistor aging that takes place in nanoelectronic integrated circuits (ICs) and systems [12, 13, 20, 64]. As presented in Fig. 11.1(b), the parameter variation can be either static or dynamic variations [12, 13]. The static parameter variations are caused by the variability in the manufacturing process; on the other hand, the dynamic parameter variations occur during the operation of the circuit and system due to the changes in the environmental and workload conditions. The process variations originate from many possible sources as presented in Fig. 11.1(b) [23, 50]. As a result, the nature of process variations is quite different at wafer level, reticle level, and local level. For the purpose of modeling and accurate statistical analysis, they can be classified in various ways as depicted in the figure. Many of these have been discussed in the chapter on design issues for DFX.

The process variations have significant negative impact on the ICs, system-on-chips (SoCs), as well as multiple core systems in terms of their energy dissipation and performance characteristics [4, 11, 12, 20, 48, 49, 50, 64]. For example, the process variations affect functionality of design. The variations in the channel length can affect current-carrying capability and delay. The variations in the threshold voltage can affect subthreshold leakage and delay. The process variations may change the characteristics of the circuits and system as compared with the design specification and hence may affect the yield. For example, not meeting the power dissipation or performance specifications even if the chip is fully functional is loss of yield in this competitive market. The design cycle is complicated and design engineers' skills are really tested due to the process variations. For example, the number of process variation sources leads to more corner cases needed for meaningful simulations of the designs. The design decisions may need to be made based on statistical distributions rather than the actual characteristic data. Moreover, all these directly or indirectly affect the cost of the designed and fabricated circuits and systems. The cost may increase due to the increase in the design cost as well as the reduction in the number of good chips resulting from the fabrications.

Variability tolerant design is necessary to produce robust circuits and systems with maximum possible yield and reduced cost. The important aspect is to incorporate variability awareness during the early stages of the design cycles such that the resulting chip is process variation tolerant. Such design flows for ICs are called "process-variation tolerant" or "process-variation aware" design flows. For a quick reference, such flows are presented in Fig. 11.2 for both analog and digital ICs [25, 28, 48, 49, 50]. The manufacturing process variation information for a specific manufacturing process needs to be available to the design engineers for use at different levels of design abstractions.

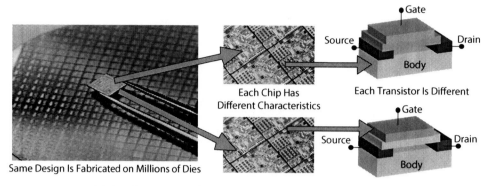

(a) Nanoelectronics variability makes each die/transistor different from each other.

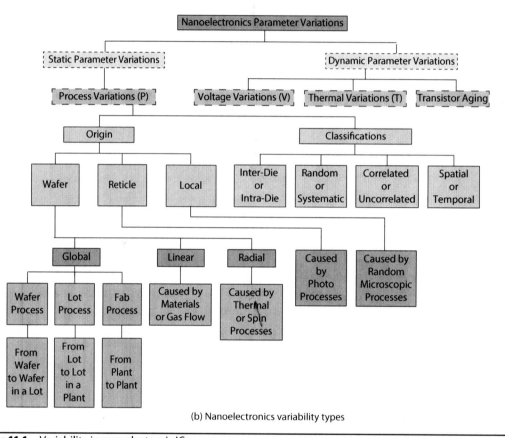

(b) Nanoelectronics variability types

Figure 11.1 Variability in nanoelectronic ICs.

For example, what kind of variation channel length follows? For the ICs, variability-aware analysis as well as variability-aware design optimization can be used. The variability-aware analysis techniques are needed for process-variation-aware characterizations. Statistical timing analysis (STA) and statistical power analysis are available. At the same time, variability-aware optimization techniques are needed for variability-aware design exploration that can iteratively use the statistical characterization data. The process variation-aware analog IC design flow uses process variation-aware analysis and optimization at the functional level, behavioral level, macro level, transistor level, and layout level as presented in Fig. 11.2(a) [25, 28, 48, 49, 50]. Several specific examples of such techniques are discussed in this chapter for analog ICs such as voltage-controlled oscillators (VCOs). The process variation-aware analog digital IC design flow uses process variation-aware analysis and optimization at behavioral level, register-transfer level (RTL), functional level, logic level, transistor level, and layout level as presented in Fig. 11.2(b) [25, 28, 48, 49, 50]. Several specific examples of such techniques are discussed in this chapter for digital ICs. For the digital ICs, selected examples are presented for various levels of design abstraction.

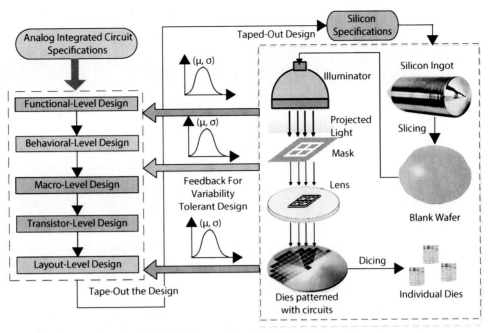

(a) Variability-aware analog IC design.

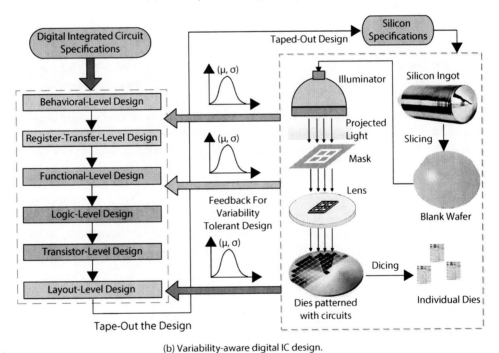

(b) Variability-aware digital IC design.

FIGURE 11.2 Variability-aware design of nanoelectronic ICs.

11.2 Methods for Variability Analysis

Variability analysis is the technique that considers the variations in the MOSFET device parameters and studies the effects of these variations on the target characteristics of the designs [11, 23, 28, 43, 54, 69, 77]. The important challenge is to accurately model the MOSFET device parameters. The other challenge is to accurately simulate the circuits and systems for these device parameters so that characteristics can be accurately captured. At the same time, the computational requirements for the simulations are key aspects of the simulations so that too much design cycle time is not consumed. This section discusses selected nanoelectronics variability analysis techniques. Specific case study circuits have been discussed for a clear understanding of the analysis and effect of process variations.

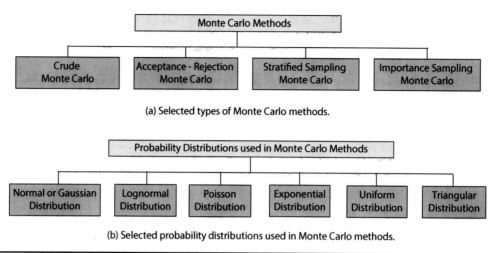

Figure 11.3 Selected Monte Carlo methods.

11.2.1 Monte Carlo Method

In this section, the most widely used technique, the Monte Carlo method, is discussed. First, a brief overview of the technique is presented along with the introduction to various types. Then, specific case studies are presented.

11.2.1.1 Brief Overview

The history of the Monte Carlo method can be traced back to Enrico Fermi when the technique was used but was not named [2, 6, 46, 66]. In 1940s, the modern Monte Carlo method was invented by Stanislaw Ulam. The Monte Carlo method solves deterministic problems using probability distribution of inputs. The Monte Carlo method is popular because of its anti-aliasing property and ability to provide an approximate answer to the problem, which is otherwise time consuming to solve, using deterministic approach. Although the Monte Carlo methods vary, they have some common features as follows [2, 6, 66]: (1) a domain for possible inputs is defined, (2) the inputs are randomly generated over the domain using probability distributions, (3) deterministic computations are performed on the inputs, and (4) the results are aggregated.

Selected Monte Carlo methods, such as crude Monte Carlo, acceptance-rejection Monte Carlo, stratified Monte Carlo, and importance sampling Monte Carlo, are listed in Fig. 11.3(a) [2, 23, 40, 41, 66]. For the purpose of discussion of the different Monte Carlo methods, let us consider any generic problem, for example, solution of an integral as follows [66]:

$$Y = \int_a^b f(x)dx \qquad (11.1)$$

In the crude Monte Carlo method, N random samples are taken between a and b. The function $f(x)$ is evaluated for each of these random samples. The integral is evaluated essentially as the mean of these N function values. The crude Monte Carlo computes this integral in the following manner [66]:

$$Y = \left(\frac{b-a}{N}\right)\sum_{i=1}^{N} f(x_i) \qquad (11.2)$$

In the acceptance-rejection Monte Carlo method (a.k.a. rejection Monte Carlo method), for any x in the interval (a,b), the upper limit of the function is calculated. Based on this upper limit, a rectangle that is large enough to cover the function in the entire interval is used. A random sample is taken within the rectangle, and if the sample is within the function value, then it is a successful sample. At the end of N samples, the integral Y is approximated as the area of the surrounding rectangle as follows [66]:

$$Y \approx \left(\frac{\text{Number of Successful Samples}}{\text{Total Number of Samples}}\right) M(b-a) \qquad (11.3)$$

where $M > 1$ is an appropriate bound on $f(x)/\text{Max}(f(x))$. In the stratified Monte Carlo method, the interval (a,b) is divided into subintervals. Then the crude Monte Carlo method is used in these subintervals. For example, let us assume the interval (a,b) is divided into two subintervals (a,c) and (c,b). Then, the integral under consideration can be rewritten as follows [66]:

$$Y = \int_a^b f(x)dx = \int_a^c f(x)dx + \int_c^b f(x)dx \qquad (11.4)$$

Then using the crude Monte Carlo method in these two subintervals, the integral is evaluated as follows:

$$Y = \left(\frac{c-a}{N}\right)\sum_{i=1}^N f(x_i) + \left(\frac{b-c}{N}\right)\sum_{i=1}^N f(x_i) \qquad (11.5)$$

In the important sampling Monte Carlo method, the areas of the function that are more important are sampled more. For this purpose, a probability density function (PDF) is introduced. Through this PDF, higher probability is assigned to the interval that needs to be sampled more than the other intervals. Thus, the integral is calculated as follows [66]:

$$Y = \int_a^b f(x)dx = \int_a^b \left(\frac{f(x)}{Pr(x)}\right) Pr(x)dx \qquad (11.6)$$

where Pr is a PDF.

The accuracy and speed of the Monte Carlo method can be affected by the number of samples used as well as the type of distributions used for sampling. Selected distributions used for sampling in the Monte Carlo simulation are shown in Fig. 11.3(b) [2, 66]. Some of the commonly used probability distributions or curves for the Monte Carlo methods are [2]: normal or Gaussian distribution, lognormal distribution, Poisson distribution, exponential distribution, uniform distribution, and triangular distribution. A design engineer can use these distributions to model the MOSFET device parameter variations and also to model the circuit or system characteristics. The choice will depend on his/her experience and manufacturing information.

11.2.1.2 Monte Carlo Method Example: Nano-CMOS Oscillator Circuits

Accurate design space sampling is one of the basic applications of the Monte Carlo analysis [25, 26]. Sampling of nanoelectronic IC design space is quite time intensive. The computational overhead of the design space sampling, in general, depends on many aspects including: (1) number of design or tuning parameters; (2) range of design or tuning parameters; (3) steps that tune the design parameters; (4) complexity of the ICs in terms of device counts, interconnects, and parasitics; and (5) speed of the analog simulators. The design space sampling for two VCO ICs is shown in Fig. 11.4 [25, 26]. The design space sampling is shown for two different design parameters, NMOS transistor channel width (W_n) and PMOS transistor channel width (W_p). In more complex examples, a greater number of design parameters can be included that will make the design space multidimensional and cannot be easily plotted on a 2D or 3D figure. The design space sampling is presented for one characteristic, i.e., center frequency, of the VCOs. Moreover, similar design space can be presented for other characteristics such as power and jitter. The design sampling is time consuming due to the computational overhead of the simulator. Then, the design sampling process is affected by the range of values, i.e., lower limit to upper limit, of the design parameters and the size of the increments within that range. Based on these facts, imagine as an example a 90-nm CMOS-based phase-locked loop (PLL) with 1000s of devices with 10 design parameters and their multiple simulations for design exploration. In general, N number of design variables with M steps for each need N^M analog simulations. Figure 11.4(a) shows the design space of 45-nm CMOS ring oscillator for 10,000 and 5000 samples that have been obtained using the Monte Carlo simulations [25, 26]. The Monte Carlo simulations picked W_n and W_p values at random using a distribution and simulated the 45-nm CMOS ring oscillator using an analog simulator. Figure 11.4(b) shows the design space of 180-nm CMOS LC-VCO for 10,000 and 5000 samples that have been obtained using the Monte Carlo simulations [25, 26]. The Monte Carlo simulations

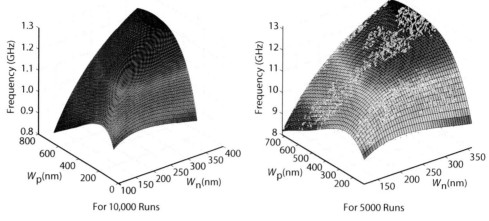

(a) For 45-nm CMOS ring oscillator circuit.

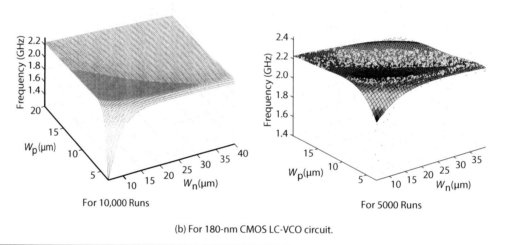

(b) For 180-nm CMOS LC-VCO circuit.

Figure 11.4 Sampling of design space of nano-CMOS oscillators using the Monte Carlo method [25, 26].

picked W_n and W_p values at random using a distribution and simulated the 180-nm CMOS LC-VCO using an analog simulator. In both the cases, only a few hours of computational time are needed for design sampling.

11.2.1.3 Monte Carlo Method Example: Nano-CMOS Sense Amplifier Circuits

The Monte Carlo method can be used for statistical process variation analysis of nanoelectronic ICs. As an example, a 45-nm CMOS-based sense amplifier IC design has been considered in this subsection as presented in Fig. 11.5 [61, 62]. The detailed design of this sense amplifier has been discussed in a previous chapter on sensor circuits and systems. The Monte Carlo method can be used at both the schematic and layout levels for statistical process variation analysis. Accordingly, either a schematic-level parasitic-free netlist or layout-level parasitic-aware netlist is extracted from a schematic capture tool or a layout editor tool. The other decisions taken by the design engineers include: (1) the parameters to be considered for process variations, (2) the sense amplifier characteristics that need to be analyzed, (3) the number of simulation runs to be performed, and (4) the statistical distributions to be used for modeling the device parameters. It is important to run a large number of simulations to obtain accurate results. It is also important to use accurate statistical models to accurately capture the device and design parameter variations. The overall idea of this Monte Carlo method-based statistical process has been depicted in Fig. 11.5(a) [61, 62].

In this example of a 45-nm nano-CMOS-based sense amplifier, all the parasitics are included in the layout-level netlist. The following design and device parameters are considered: L_n, L_p, V_{Thn}, V_{Thp}, V_{DD}, T_{oxn}, T_{oxp}, W_n, and W_p. The following four sense amplifier characteristics are analyzed for the process

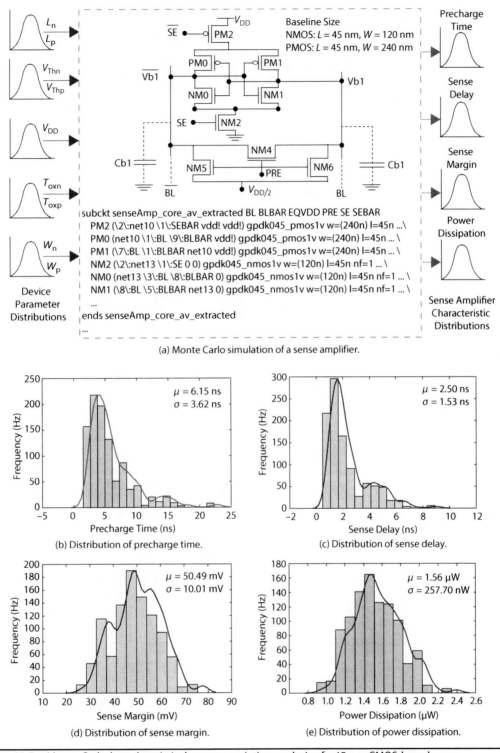

FIGURE 11.5 Monte Carlo-based statistical process variation analysis of a 45-nm CMOS-based sense amplifier [61, 62].

variation study: precharge time, sense delay, sense margin, and power dissipation. The standard distributions of the parameter are assumed to be Gaussian or normal. The standard deviation (σ) for the distributions is assumed to be 5% of the mean (μ). The Monte Carlo simulations are performed from 1000 runs. The statistical distributions of the characteristics of the sense amplifier are presented in Fig. 11.5(b–e) [61, 62]. The mean and standard deviations of the distributions of the characteristics of the sense amplifier are also shown in the figures.

11.2.1.4 Monte Carlo Method Example: Graphene LC-VCO Circuit

The graphene FET-based LC-VCO design has been presented in the chapter on mixed-signal circuit and system simulation using Simulink®/Simpscape® as well as Verilog-A-based models. In this section, the statistical process variation analysis of the GFET-based LC-VCO has been presented. The overall perspective of the graphene FET-based LC-VCO is depicted in Fig. 11.6 [42]. In contrast to the previous subsection on sense amplifier in which process variation analysis is performed over the SPICE netlist, in this case the analysis is performed over the Verilog-A-based design.

The following device-level parameters are considered for process variation for this GFET-based LC-VCO: length of N-type devices (L_n), length of P-type devices (L_p), mobility of N-type devices (μ_n), mobility of P-type devices (μ_p), drain to source resistance of N-type devices (R_{dsn}), oxide-thickness

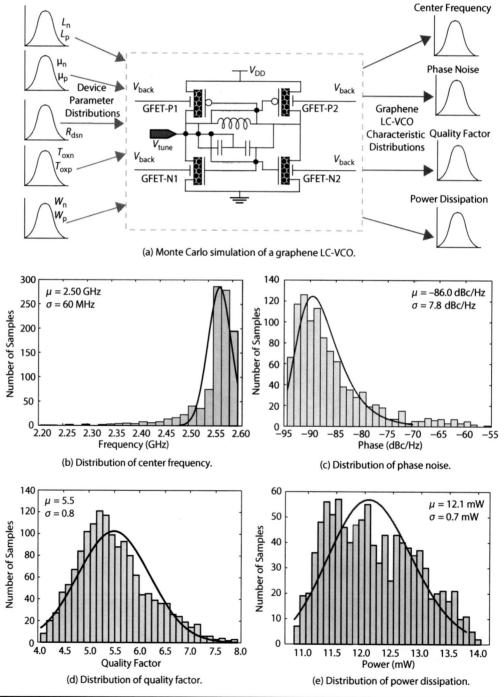

Figure 11.6 Monte Carlo-based statistical process variation analysis of a graphene LC-VCO [42].

of N-type devices (T_{oxn}), oxide-thickness of P-type devices (T_{oxp}), width of N-type devices (W_n), and width of P-type devices (W_n). The following four characteristics of this GFET-based LC-VCO are considered for statistical characterization: center frequency, phase noise, quality factor, and power dissipation. This is depicted in Fig. 11.6(a) [42]. Gaussian or normal distributions are used to model the statistical distributions of the device-level parameter in which standard deviations are assumed to be 10% of the mean. The Monte Carlo simulations are performed for 1000 runs. The resulting statistical distributions of the characteristics of the GFET LC-VCO are presented in Fig. 11.6(b–e) along with their mean values and standard deviation values [42]. It is observed that the distribution of the center frequency that is presented in Fig. 11.6(b) follows the extreme value distribution. The distribution of the phase noise that is presented in Fig. 11.6(c) is also an extreme value distribution. The statistical distribution of the quality factor has a Gaussian nature as shown in Fig. 11.6(d). The statistical distribution of the power dissipation is log-normal in nature as shown in Fig. 11.6(e).

11.2.1.5 Monte Carlo Method Example: Digital Logic Level

In this subsection, statistical process variation analysis of logic gates is discussed. The logic gates can then be used as building blocks to build bigger digital circuits. As a specific example, a 45-nm CMOS-based NAND logic gate is presented for two reasons [58]. First, using NAND gate, any digital IC can be designed as a universal gate. In addition, it observed that NAND gate has lower leakage and lower delay as compared to the other logic gates. Therefore, for low leakage and faster digital circuits, use of NAND is a better choice. The overall perspective of statistical process analysis of the 2-input NAND gate has been presented in Fig. 11.7 [50, 52]. The device parameters that are considered for the process variation analysis are: L_n, L_p, V_{Thn}, V_{Thp}, V_{DD}, T_{oxn}, T_{oxp}, W_n, and W_p. The parameters are assumed to be Gaussian distribution with standard deviations of 10% of their mean. The following four characteristics are analyzed for the effect of process variations: capacitive switching current, subthreshold current, gate-oxide leakage current, and propagation delay. The nominal sizes used for the NMOS and PMOS transistors of the NAND gate are shown the Fig. 11.7(a) [50, 52]. The paths of the various forms of the current are also shown in the figure. A supply voltage V_{DD} of 0.7 V is used for the 45-nm CMOS technology. The Monte Carlo simulations for 1000 runs are performed for each of the characteristics of the NAND gate. The statistical distributions of the characteristics of the NAND gate are shown in the Fig. 11.7(b–e) [50, 52]. The mean and standard deviations of each of the distribution characteristics have been presented. The distributions of the capacitive switching, subthreshold leakage, and gate-oxide leakage current are of log-normal type. The propagation delay follows Gaussian or normal distribution.

As a second example, 32-nm high-κ CMOS-based NAND gate process variation analysis is now discussed, as shown in Fig. 11.8 [35]. In this case, variations of 15 different device-level parameters are incorporated as follows: L_n, L_p, V_{Thn}, V_{Thp}, V_{DD}, NMOS gate-doping concentration (N_{gaten}), PMOS gate-doping concentration (N_{gatep}), NMOS channel-doping concentration (N_{chn}), PMOS channel-doping concentration (N_{chp}), NMOS source/drain-doping concentration (N_{sdn}), PMOS source/drain-doping concentration (N_{sdp}), T_{oxn}, T_{oxp}, W_n, and W_p. These parameters are not necessarily independent and may have correlations. The statistical variations of these parameters can be modeled as Gaussian distributions. The nominal values of these parameters are assumed as specified in the device model files and standard deviation is assumed as 10% of the mean. A total of 1000 runs are performed for the Monte Carlo simulations of each of the four target characteristics, capacitive switching current, subthreshold leakage current, gate-induced drain leakage (GIDL) current, and propagation delay. Due to the use of high-κ gate oxide, the gate-oxide leakage current is not present in this circuit. The overall statistical analysis perspective along with the nominal sizes for a 32-nm technology node is presented in Fig. 11.8(a) [35]. The statistical distributions of the characteristics of the NAND gate are shown in Fig. 11.8(b–e) [35]. It is observed that capacitive switching current, subthreshold leakage current, and GIDL current have log-normal distribution. The propagation delay has a Gaussian distribution.

The discussion of the above statistical analysis is now expanded to another axis with temperature as a variable. The overall picture is depicted in Fig. 11.9 [35]. For the statistical process variations, the same set of device level parameters has been used with a 32-nm high-κ CMOS technology node. The same nominal transistor sizes are used as in the above case, shown in Fig. 11.9(a) [35]. The process variations of the devices (P) and the supply voltage variations (V) are chosen as Gaussian

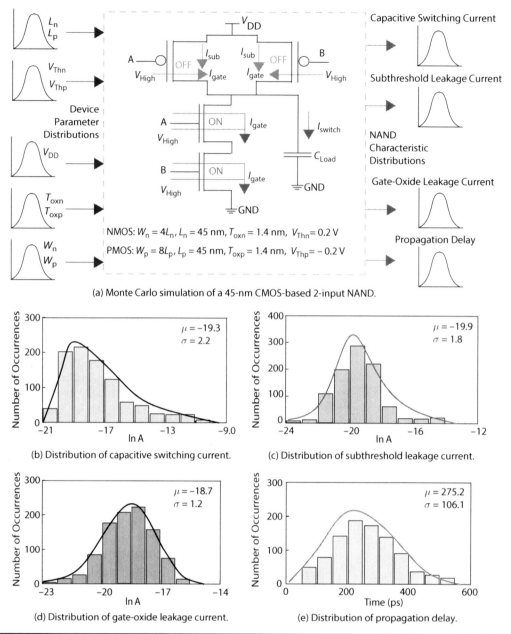

FIGURE 11.7 Monte Carlo-based statistical process variation analysis of a 2-input NAND gate for 45-nm CMOS [50, 52].

distributions with the standard deviation 10% of the mean. The changes in the operating temperature (T) that are due to self-heating, ambient heat, or a combination of the two are varied in the steps of the following: 0°C, +50°C, +100°C, and +125°C. The distribution of various NAND characteristics for the range of temperature is shown in Fig. 11.9(b–e) [35]. The capacitive switching current does not change much with the change in the temperature. The subthreshold leakage current shows an increase with the temperature. The GIDL current does not change much with the change in the temperature. The propagation delay is also affected by the temperature variation.

11.2.2 Design of Experiments Method

This subsection discusses an efficient mechanism for circuit and system analysis called design of experiments (DOE) method. The DOE is an efficient technique for planning experiments such that the resulting data can be efficiently analyzed to reach meaningful conclusions [1, 3, 8, 56]. The DOE is

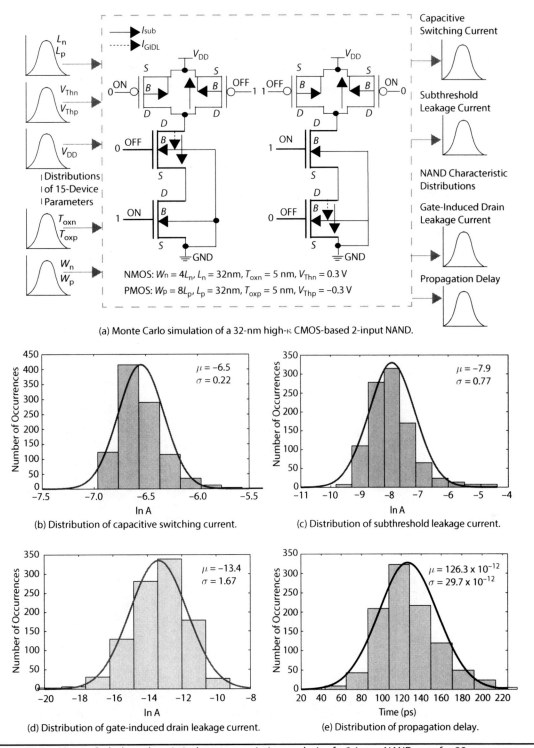

FIGURE 11.8 Monte Carlo-based statistical process variation analysis of a 2-input NAND gate for 32-nm high-κ CMOS [35].

also known as experimental design or even designed experiments. In general, DOE can be used for various tasks including screening, characterizing, modeling, comparison, and optimization.

11.2.2.1 Brief Overview

In general, the term "experiment" is the systematic procedure that is conducted under controlled conditions for various purposes including: (1) to determine any unknown effect, (2) to illustrate the known effects, and (3) to test the hypothesis. The experiments can be designed in various different ways.

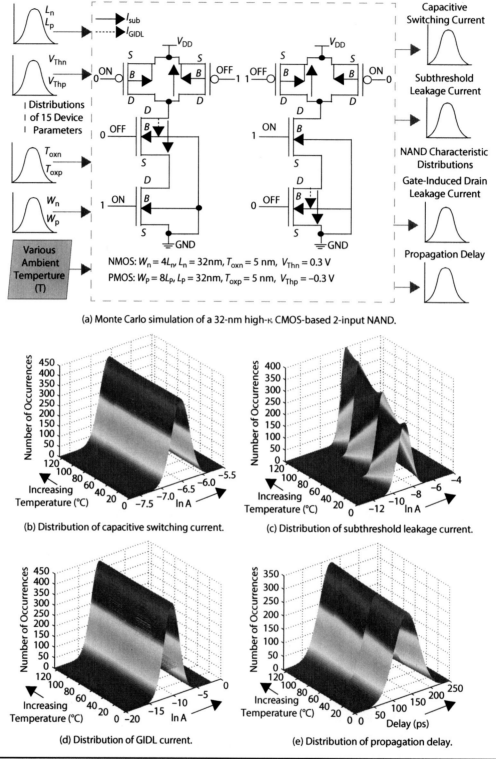

FIGURE 11.9 Monte Carlo-based statistical PVT analysis of a 2-input NAND gate for 32-nm high-κ CMOS [35].

The DOE is the systematic, rigorous, and efficient procedure applied at the data collection stage for generating valid and supportable engineering conclusions [1, 3, 5, 44, 56, 74]. DOE is an efficient technique in terms of number of runs, effort, and cost. The generic overview of the DOE is depicted in Fig. 11.10(a) [1, 3, 5, 44, 56, 72, 74]. As presented in the figure, there are several aspects of the DOE. The "process" is the entity that is being analyzed. In this specific case, it is an IC or a system. The "factors" are controlled inputs or causes. The "co-factors" are uncontrolled inputs or causes that can

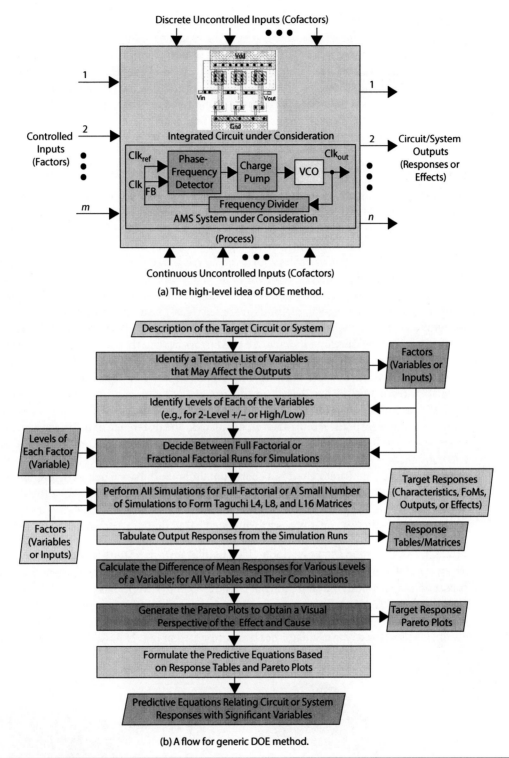

FIGURE 11.10 The general idea of the DOE method.

be either discrete uncontrolled inputs or continuous uncontrolled inputs. The factors can be of various "levels" or "settings." The "responses" or "effects" are the outputs of the process. The key idea of the DOE is to determine the factors and their settings for the best overall outcome for the critical-to-quality characteristics.

The generic flow for the DOE method has been presented in Fig. 11.10(b) [1, 3, 5, 44, 56, 72, 74]. Description of the target circuit or system is the input for the DOE method. Based on the type of circuit or system and the level of design abstraction, it can be described in the form of HDL, schematic

632 Chapter Eleven

design, or SPICE netlist. At the first step, a list of variables (or factors) that may affect target outputs (or responses) is identified. The design experiences can speed up this decision. A simple approach is to pick the variables that will be tuned for design optimization. For each of the variables, the levels or settings are now decided. For example, for a variable if two possible values are to be used, then the two levels are $+/-$ or high/low. For the purposes of simulations, the decision is needed between the full factorial or fractional factorial approaches. In the full-factorial experiment, each factor is tested at each of its levels using every possible combination with the other factors and their levels. In the full-factorial experiments, the following number of runs are needed:

$$\text{Number of Simulation Runs} = \text{Number of Levels}^{\text{Number of Factors}} \quad (11.7)$$

For example, for an experiment with m variables and two levels (high and low), the number of simulation runs needed is 2^m. Thus, full-factorial experiment is practical for a few factors, each with two to three levels, as the number of runs increases exponentially with the number of factors and number of levels. Thus, the full-factorial experiment, while exhaustive, may not be possible for cases in which a large number of variables with several levels may be involved. The practical approach is to use fractional factorial experiments in which only a subset of all possible combinations of the factors is evaluated. Thus, it is practical and efficient. An example of fractional factorial experiments is the Taguchi methods. The Taguchi methods are based on the assumption that interactions are not likely to be significant, and correspondingly the Taguchi matrices are derived from the full-factorial arrays. The Taguchi designs are developed to analyze the factors at two, three, four levels, as well as with mixed levels. The Taguchi designs use the L4, L8, and L16 matrices for two-level designs. While the L4 matrix is used up to three factors, the most popular are the L8 and L16 metrics that are used for up to 7 and 15 factors, respectively. Thus, based on the number of factors and number of levels, L4, L8, or L16 matrix is used in the experiment. Based on the above decisions, simulation runs are performed and the output responses from the simulations are tabulated. The difference of the mean responses for various levels of a variable is calculated from the response tables and the process is repeated for all variables and combinations. At this point, Pareto charts are generated for the visual perspective of the effect and cause. In general, the Pareto chart is a chart containing both bars (such as histograms) and a line plot in which the individual values are shown in descending order by bars while the cumulative total is represented by the line plot. One Pareto chart is needed for each of the target responses or characteristics. At this point, predictive equations are formulated based on the response tables and Pareto charts.

11.2.2.2 DOE Method Example: Nano-CMOS Oscillator Design Space Sampling

The use of DOE method for fast and accurate sampling of analog IC design space is discussed in this subsection. Case study circuits considered are 45-nm CMOS-based ring oscillator and 180-nm CMOS-based LC-VCO. Two variables considered are NMOS transistor width (W_n) and PMOS transistor width (W_p). DOE involved three levels of these design variables. The design space for the 45-nm CMOS-based ring oscillator has been presented in Fig. 11.11(a) [25, 26]. For the ring oscillator,

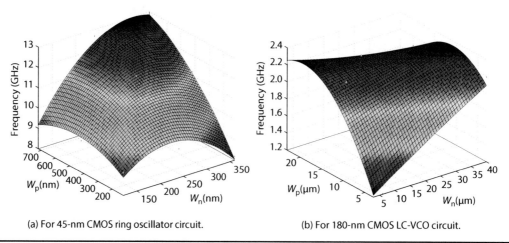

(a) For 45-nm CMOS ring oscillator circuit.   (b) For 180-nm CMOS LC-VCO circuit.

Figure 11.11 Sampling of design space of nano-CMOS oscillators using DOE method [25, 26].

the root mean square error (RMSE) for DOE-based sampling is approximately 7%. The design space for the 180-nm CMOS-based LC-VCO is presented in Fig. 11.11(b) [25, 26]. For the LC-VCO, the RMSE error for DOE-based sampling is 9%. It is observed that the DOE technique was not easily scalable and had larger error than other sampling techniques.

11.2.2.3 DOE Method Example: Nano-CMOS Oscillator Design Optimization

In this subsection, a detailed example of DOE method is presented using a 45-nm CMOS-based current-starved VCO. The overall perspective of the DOE for this design is presented in Fig. 11.12(a) [44, 51, 74]. The sizing in terms of aspect ratio $\left(\frac{W}{L}\right)$ of various transistors and the gate-oxide thicknesses (T_{ox}) of these transistors can be tuning parameters for design exploration. The higher values (T_{ox}) lead to reduction in the power dissipation of nano-CMOS circuits but also lead to increase in the delay; thus, needing optimization for proper sizing for tradeoffs. As a specific example, the following five are considered as the input or factor for the DOE method in this case of nano-CMOS-based current-starved VCO: (1) the gate-oxide thickness (T_{ox}), (2) the size or ratio of the PMOS devices of the inverter (k_{invp}), (3) the size or ratio of the NMOS devices of the inverter (k_{invn}), (4) the size or ratio of the current-starved PMOS devices (k_{csp}), and (5) the size or ratio of the current-starved PMOS devices (k_{csn}). As another example, individual channel length of the transistors and width of the transistors can be considered input or factors. The two outputs or responses of the DOE are the following: center frequency (f_{osc}) and average power dissipation (P_{VCO}). Similarly, additional characteristics of the CS-VCO can be considered as response. The flow for the DOE-based design optimization of the current-starved VCO is presented in Fig. 11.12(b) [44, 51, 74]. This flow is similar to the generic DOE flow presented in the previous subsection, however, customized for the current-starved VCO.

In this DOE-based design flow, five input variables are considered, each with two different levels. As a specific example, for 45-nm CMOS-based CS-VCO, the two levels of the different input variables are chosen as follows [44, 51, 74]: (1) T_{ox} - 1.4 and 1.7 nm, (2) k_{invp} - 5 and 10, (3) k_{csp} - 5 and 10, (4) k_{invn} - 1.72 and 3.44, and (5) k_{csn} - 1.72 and 3.44. In this example, five input factors with two levels each and two target responses are considered, so it is possible to use full factorial. However, for efficient experimentations, the Taguchi method is considered with Taguchi L8 design matrix. Simulations of the CS-VCO are performed for eight different cases using an analog simulator such as SPICE or Spectre®. The simulation results in the cause and effect form have been presented in Table 11.1. For visual perspective of the responses, Pareto charts are generated for each of the target responses, center frequency, and average power, as shown in Fig. 11.13 [44, 51, 74]. Based on the response table and Pareto charts, the following predictive linear equations are generated for the 45-nm CMOS-based CS-VCO [44, 51, 74]:

$$f_{osc} = 786.4 - 93.4 T_{ox} + 60.3 k_{invn} \tag{11.8}$$

$$P_{VCO} = 35.0 + 5.7 k_{csn} + 3.3 k_{csp} \tag{11.9}$$

From the above predictive equations, it is evident that oscillating frequency of the CS-VCO is maximum affected by the gate-oxide thickness as NMOS transistors of the inverters. At the same time, the average power dissipation of the CS-VCO is affected by the sizes of the current-starved transistors. The above predictive equations can be used for various nanoelectronic design tasks including optimization and process variations analysis. Such tasks will be much faster over these equations as compared to the circuit or layout netlist using the analog/SPICE simulator. The above predictive linear equations can be called polynomial metamodels of the first order that will be the topic of discussion in a future chapter.

11.2.3 Corner-Based Method

In this section, the corner-based method is discussed. First, generic discussion of the method is presented. This is then followed by a specific example. The corner-based method has been widely used in the past due to its simplicity and fast usage [28, 34, 43, 69]. It may be noted that the corner-based method may not be as accurate as traditional Monte-Carlo simulations, but it is very useful in getting a quick estimate of the impact of process variation. Although the traditional Monte-Carlo

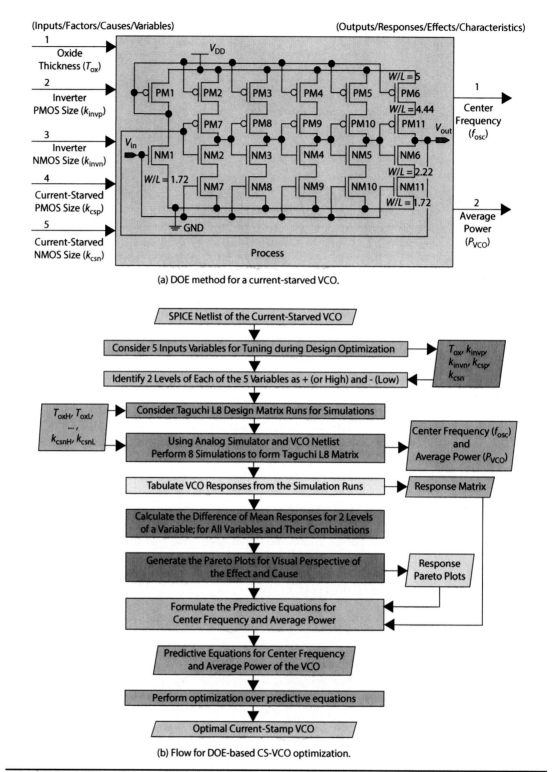

FIGURE 11.12 DOE method-based optimization of a current-starved VCO [44, 51, 74].

simulations can be accurate, they need 1000s of runs to provide such accuracy, hence can be quite time consuming.

11.2.3.1 Brief Overview

A specific example of the DOE technique is the process-corner method or corner-based method. The process corners are the extreme cases of parameter variations within which the ICs must be designed and manufactured in order to function correctly. While the term *process corner* is interchangeably

DOE Run	T_{ox} (nm)	For Inverter PMOS (k_{invp})	For Inverter NMOS (k_{invn})	For Current Starved PMOS (k_{csp})	For Current Starved NMOS (k_{csn})	Center Frequency (f_{osc}) (MHz)	Average Power (P_{VCO}) (μW)
1	1.4	5	1.72	10	3.44	787.9	46.0
2	1.4	5	3.44	5	1.72	925.0	29.6
3	1.4	10	1.72	10	1.72	813.8	32.0
4	1.4	10	3.44	5	3.44	992.5	38.3
5	1.7	5	1.72	5	3.44	630.6	32.9
6	1.7	5	3.44	10	1.72	692.1	29.6
7	1.7	10	1.72	5	1.72	672.3	26.2
8	1.7	10	3.44	10	3.44	777.2	45.7

TABLE 11.1 Tabulation of DOE Runs [44, 51, 74]

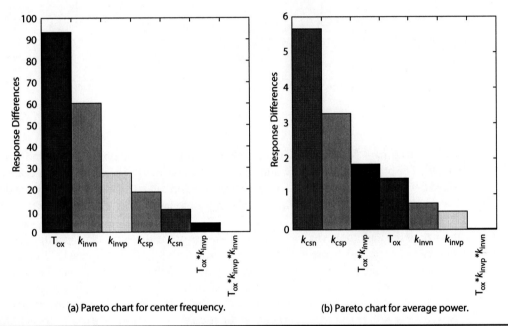

(a) Pareto chart for center frequency.   (b) Pareto chart for average power.

FIGURE 11.13 Pareto plots for DOE method-based optimization of a current-starved VCO [44, 51, 74].

used, the overall corner are of two broad types as presented in Fig. 11.14 [23, 25, 28, 48, 49]. The two types of corners are front-end of line (FEOL) corners and back-end of line (BEOL). The FEOL corners are the corners used at the front end of the design phases. BEOL corners are the corners used at the back end of the design and phases. FEOL corners affect the performances including leakage and delay of the ICs. BEOL corners affect the parasitics of the ICs. Of course, both the corners are related to each other. The corner analysis is more effective for digital ICs than the analog ICs due to the direct effect of process variations on the speed of transistor switching in the digital domain.

For CMOS ICs using two types of transistors, the five possible FEOL corners are shown in Fig. 11.14 [25, 28, 48, 49]. The five process corners that can be used for design analysis are the following:

1. *SS corner or Slow-Slow corner*: It is the corner in which NMOS is slow and PMOS is also slow.
2. *FF corner or Fast-Fast corner*: It is the corner in which NMOS is fast and PMOS is also fast.
3. *FS corner or Fast-Slow corner*: It is the corner in which NMOS is fast, but PMOS is slow.
4. *SF corner or Slow-Fast corner*: It is the corner in which NMOS is slow, but PMOS is fast.
5. *TT corner or Typical-Typical corner*: It is the corner in which NMOS is typical and PMOS is also typical.

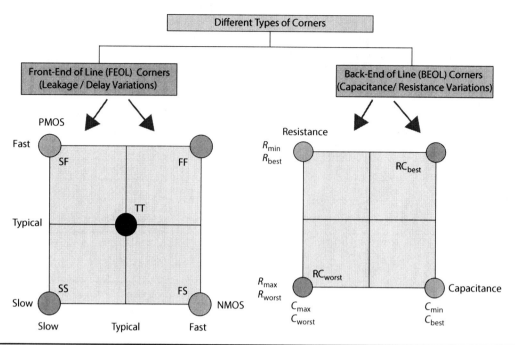

FIGURE 11.14 Process corners for CMOS ICs design and manufacturing [25, 28, 48, 49].

The three corners, SS, FF, and TT, are called "even corners," as both NMOS and PMOS are affected equally. The other two corners, FS and SF, are called "skewed corners," as there is imbalanced switching speed between the two types of devices. These are difficult corners to handle by the design engineers. For example, imbalanced switching speed can lead to large short-circuit currents.

The BEOL corners are based on the parasitics of the ICs. The capacitance values are affected by the size of the interconnect as well as the thickness of the dielectric. For example, maximum size and minimum separation between the interconnects is the C_{max} or C_{worst} case. Similarly, minimum size and maximum separation between the interconnects is the C_{min} or C_{best} case. The resistance values are affected by the size of the interconnect. For example, minimum size interconnects is the R_{max} or R_{worst} case. Similarly, maximum size interconnects is the R_{min} or R_{best} case. Accounting both resistances and capacitances RC_{worst} and RC_{best} corners can be used.

11.2.3.2 A Specific Example

In this subsection, the corner-based method is demonstrated for a 90-nm CMOS-based flash analog-to-digital converter (ADC) IC [28, 29, 34]. Process variations through the change of the threshold voltage, voltage variations, and temperature variations are analyzed for different figures of merits, such as integral nonlinearity (INL) and differential nonlinearity (DNL) of the ADC.

The corner-based method for the worst-case threshold voltage variations is depicted in Fig. 11.15 [28, 29, 34]. The variations in the threshold voltages of both the NMOS transistor and PMOS transistors are considered. As a specific example, the worst-case change in both V_{Thn} and V_{Thp} are assumed to be ±5% from their nominal values. Thus, four different corners as presented in the figure are analyzed. The results of the corner analysis for this case are presented in Table 11.2 [28, 29, 34]. The INL of the 90-nm CMOS-based flash ADC shows a variation from 0.2% to 10.5% and the DNL shows a variation from 2.2% to 5.7%. The INL and DNL values are always less than 0.5 LSB that indicates that the flash ADC is tolerant of process variations.

For the analysis of the ADC for supply voltage variations, the voltage is changed by ±10% of the nominal value. The ADC is simulated for the four corners. INL, DNL, and input voltage range values of the ADC have been recorded. The results are presented in Table 11.3 [28, 29, 34]. It is observed that the INL of the ADC changed in the range of −1% to +4%. At the same time, the DNL of the ADC varied +1.8% to +4.8%. The flash ADC design is supply voltage variation tolerant as the INL < 0.5 LSB as well as the DNL < 0.5 LSB.

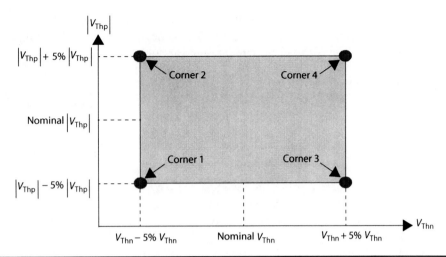

FIGURE 11.15 Worst-case process variation analysis of the ADC using a corner-based method [28, 29, 34].

Threshold Voltage Change	Input Voltage Range (mV)	V_{LSB} (mV)	INL (LSB)	DNL (LSB)
V_{Thp} (Nominal), V_{Thn} (Nominal)	493–557	1	0.34	0.46
V_{Thp} (−5%), V_{Thn} (−5%) (Corner 1)	491–556	1.016	0.34	0.48
V_{Thp} (+5%), V_{Thn} (−5%) (Corner 2)	501–566	1.016	0.38	0.48
V_{Thp} (−5%), V_{Thn} (+5%) (Corner 3)	500–564	1	0.36	0.48
V_{Thp} (+5%), V_{Thn} (+5%) (Corner 4)	495–557	0.969	0.33	0.46

TABLE 11.2 Corner-Based Analysis Results: Process Variations [28, 29, 34]

Supply Voltage (V_{DD})	Input Voltage Range (mV)	V_{LSB} (mV)	INL (LSB)	DNL (LSB)
1.08 V (−10%)	448–500	0.81	0.36	0.47
1.20 V (Nominal)	493–557	1	0.34	0.46
1.32 V (+10%)	537–614	1.20	0.34	0.48

TABLE 11.3 Corner-Based Analysis Results: Supply Voltage Variations [28, 29, 34]

Temperature	Input Voltage Range (mV)	V_{LSB} (mV)	INL (LSB)	DNL (LSB)
−40°C	484.5–544.5	0.937	0.30	1
27°C (Nominal)	493.0–557.0	1	0.34	0.46
90°C	503.5–570.5	1.047	0.67	1

TABLE 11.4 Corner-Based Analysis Results: Temperature Variations [28, 29, 34]

The flash ADC is now analyzed for temperature changes. As a specific example, the ADC is analyzed for an extensive temperature range of −40°C to 90°C. For three different temperatures, the values of input voltage range, V_{LSB}, INL and DNL, have been recorded. The simulation results for three different temperatures have been shown in Table 11.4 [28, 29, 34]. It is observed that the input voltage range of the flash ADC changes with the temperature, i.e. decreases at −40°C and increases at 90°C, which changes the V_{LSB} value. It is observed that the DNL values change significantly in the temperature range. At the same time, the INL values changed, but the change was less severe. Thus, the flash ADC is sensitive to the temperature variations.

11.2.4 Fast Monte Carlo Methods

In general, the sufficient coverage of the design space for a large number of parameters is a nondeterministic polynomial-time hard (NP hard) problem [32, 53, 63, 69]. Various techniques such as Monte Carlo, DOE, and corner-based have been presented in the previous subsections. Monte Carlo is the de facto approach used in the industry and most research, as it is very well understood. Monte Carlo techniques have many drawbacks including the following. (1) Remote points in the design space may not receive sufficient coverage. (2) Any estimate for the variability obtained from the simulations is global, as each point contributes equally to the statistical sample. (3) Monte Carlo techniques can be accurate if a large number of runs are used to attain the accuracy; however, they can be computationally highly expensive [68]. For example, a Monte Carlo simulation of 1000 runs over the parasitics-aware netlist of a 180-nm CMOS PLL takes 5 days in a state-of-the-art server [32, 53]. The DOE or corner-based techniques can be straightforward and computationally efficient. However, these techniques on their own cannot generate statistical variability information. Thus, different approaches that are needed to improve the speed at the same time generate accurate results as well as statistical information. For example, a two-level hybrid sampling approach as shown in Fig. 11.16 is a possible option. In essence, at the global level, a less accurate sampling approach, and at the local level more accurate sampling using Monte Carlo can meet the need. This subsection discusses some of these techniques with specific examples.

11.2.4.1 Design-of-Experiment-Assisted Monte Carlo Method

The flow of a statistical process variation analysis approach that uses DOE and Monte Carlo effectively is presented in Fig. 11.17 [32, 53]. It is called DOE-assisted Monte Carlo (DOE-MC) method. The key idea of the DOE-MC technique is to perform the Monte Carlo simulations for a very small number of runs for each trial of DOE. The flow assumes a netlist of the analog or mixed-signal IC as input. The netlist can be either obtained from the schematic design or layout design; of course, layout will be more accurate but will have more computational overhead. Identification for the parameters is then performed for which statistical variability effects will be analyzed as it can be one or multiple parameters. The netlist is then parameterized for these process and design parameters. Appropriate statistical distributions are identified for the selected process and design parameters. The number of DOE trials is then decided for full factorial or fractional factorial options. In general for n number of design and process parameters with two levels each, 2^n trials are needed for a full-factorial

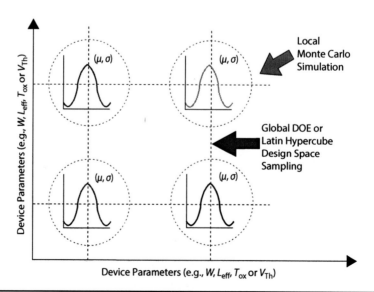

FIGURE 11.16 Illustration of hybrid sampling to speed up Monte Carlo simulations.

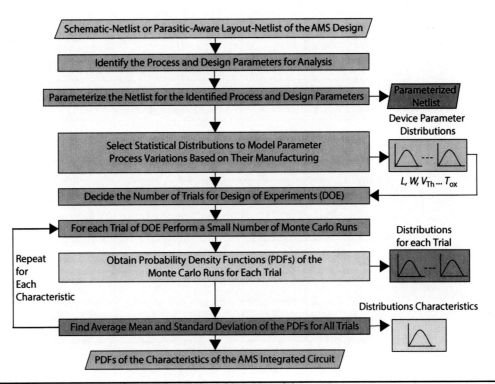

FIGURE 11.17 Flow of DOE-MC method [32, 53].

experiment. A fractional factorial will need fewer trials and is performed for large design with more variables. The Monte Carlo simulations for a small number of runs are performed for each of the DOE trials. In worst-case for full factorial DOE, the Monte Carlo simulations can generate 2^n number of intermediate $PDF_i(\mu_i, \sigma_i)$. The PDF of a target characteristic of FoM is calculated as the statistical average of these intermediate PDFs as follows:

$$PDF_{FoM}(\mu_{FoM}, \sigma_{FoM}) = \text{Statistical-Average-Over-}2^n(PDF_i(\mu_i, \sigma_i)) \quad (11.10)$$

For simplicity, under the assumption of statistical independence, it can be calculated as follows: μ_{FoM} is the average of μ_is and σ_{FoM} is the average of σ_is. Otherwise, more complex statistical approaches considering their correlations are needed for accurate estimations. The Monte Carlo simulations can be repeated or setup can be done under one Monte Carlo simulation for each of the target characteristics or FoMs of interest for the design engineers and their PDFs can be estimated.

For a 90-nm CMOS-based current-starved VCO, the following five processes and design parameters have been considered for statistical process variations [32, 53]: (a) supply voltage (V_{DD}), (b) NMOS threshold voltage (V_{Thn}), (c) PMOS threshold voltage (V_{Thp}), (d) NMOS gate-oxide thickness (T_{oxn}), and (e) PMOS gate-oxide thickness (T_{oxp}). As a specific example, two levels for each of the five parameters are assumed as follows:

1. *Level 1:* This is represented by "−." This is nominal value with −10%.
2. *Level 2:* This is represented by "+." This is nominal value with +10%.

The nominal values are the baseline values obtained from the 90-nm CMOS PDK. The choice of 10% is driven by the intension for 2σ–3σ coverage. A typical 90-nm CMOS process has a standard deviation of 3% to 5% of the mean; thus, a 10% value results in a 2σ–3σ coverage. The use of two levels is good enough for screening purposes, however, more levels may be needed for accurate modeling. DOE is performed for a full factorial of 2^5, i.e., 32 trials. The center frequency (f_{osc}) is assumed as the target characteristic in this specific example. The 32 trials are grouped into 16 runs according to the values of T_{oxp} as shown in Table 11.5 [32, 53]. Simulations of the CS-VCO netlist are performed for these 32 trials using a SPICE/analog simulator. From the above experiment, the

DOE Trials	V_{DD}	V_{Thn}	V_{Thp}	T_{oxn}	T_{oxp}
Run 1	+	−	−	−	−
	+	−	−	−	+
Run 2	+	−	+	−	−
	+	−	+	−	+
Run 3	+	+	−	−	−
	+	+	−	−	+
Run 4	−	−	−	−	−
	−	−	−	−	+
Run 5	+	+	+	−	−
	+	+	+	−	+
Run 6	−	−	+	−	−
	−	−	+	−	+
Run 7	−	+	−	−	−
	−	+	−	−	+
Run 8	−	+	+	−	−
	−	+	+	−	+
Run 9	+	−	−	+	−
	+	−	−	+	+
Run 10	+	−	+	+	−
	+	−	+	+	+
Run 11	−	−	−	+	−
	−	−	−	+	+
Run 12	−	−	+	+	−
	−	−	+	+	+
Run 13	+	+	−	+	−
	+	+	−	+	+
Run 14	+	+	−	+	−
	+	+	−	+	+
Run 15	−	+	−	+	−
	−	+	−	+	+
Run 16	−	+	+	+	−
	−	+	+	+	+

TABLE 11.5 Full Factorial Experiments with 32 Trials for the Five Parameters of a 90-nm CMOS Current-Starved VCO [32, 53]

worst-case process variation for f_{osc} is the corner in which V_{DD} is reduced by 10%, and all the process parameters (V_{Thn}, V_{Thp}, T_{oxn}, and T_{oxp}) are increased by 10%. For this specific case study of 90-nm CMOS CS-VCO, the worst-case degradation of f_{osc} is 43.5%.

In the above, full-factorial, DOE needed 32 trials. As a specific case, if a Monte Carlo simulation of five runs is performed for f_{osc} characterization, then there are 32 values of μ and σ for f_{osc}, one for every trial of the DOE. For the overall DOE-MC, the PDF of the f_{osc} is the average of these 32 PDFs. The results obtained from the DOE-MC that are PDFs for f_{osc} for different parameter variations are presented in Fig. 11.18 [32, 53]. The distribution of f_{osc} is observed to be Gaussian in all the cases with μ and σ presented in the figure. The speedup by DOE-MC versus a traditional Monte Carlo can be explained as follows: For five Monte Carlo runs for 32 trials, the analog/SPICE simulator is

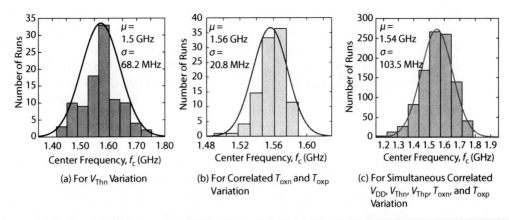

FIGURE 11.18 Statistical process variation analysis of a 90-nm CMOS-based current-starved VCO [32, 53].

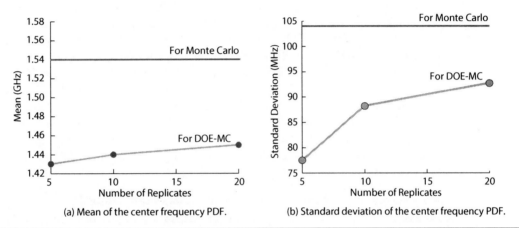

FIGURE 11.19 Accuracy versus speed tradeoffs in the DOE-MC method for center frequency of the VCO [32, 53].

Monte Carlo Runs Per Trial	Total Runs	% Error of μ	% Error of σ	Speedup Over Monte Carlo
05	160	7.47	25.1	6.2×
10	320	6.78	14.7	2.0×
20	640	5.78	10.3	1.6×

TABLE 11.6 Accuracy and Speedup Information the DOE-MC Approach [32, 53]

invoked $32 \times 5 = 160$ times. This is substantially less than 1000 runs needed for a traditional Monte Carlo. In a similar manner, DOE-MC is performed through 10 MC per trial as well 20 MC per trial. The μ and σ of f_{osc} PDF for DOE-MC are plotted in Fig. 11.19(a) and Fig. 11.19(b), respectively [32, 53]. The figures also have μ and σ of f_{osc} PDF obtained from traditional Monte Carlo of 1000 runs. The results are tabulated in Table 11.6 along with the errors in μ and σ of f_{osc} PDF with respect to the traditional Monte Carlo of 1000 runs [32, 53]. The speedup achieved due to the use of DOE-MC is also presented in the table. The speedup will be much higher if the traditional Monte Carlo needs more runs for example 5000 or 10,000 to provide good coverage of the CS-VCO design space. In a similar manner, other characteristics of the CS-VCO such as power dissipation and jitter can be analyzed.

11.2.4.2 Latin Hypercube Sampling-Based Monte Carlo Method

In a typical case, the Monte Carlo with random sampling is used. However, in general, random sampling, uniform sampling, and orthogonal sampling are possible options to use [22, 23, 25, 26, 41, 89].

In random sampling or brute force sampling, the new points are sampled without taking into account the previously generated sample points, and hence it is not necessary to know how many sample points are needed beforehand. The Latin hypercube sampling (LHS) is a type of stratified sampling. In LHS, the design space is divided in a specific manner to ensure that all portions of the range of an input variable are represented. In LHS, the number of total samples needed must be decided beforehand and based on that the range of each input variable is divided into overlapping intervals on the basis of equal probability, and hence this is a uniform sampling approach. In general, if N input variables are divided into M equally probable intervals, then the LHS needs to pick from $(M!)^{(N-1)}$ combinations. For example, two variables with four divisions of each gives 24 combinations in the design space and three variables with four divisions of each gives 576 combinations in the design space for the LHS. For each sample point, it needs to be remembered in which row and column the sample point was picked. LHS has lesser computational overhead and can easily handle multiple input variables and thus multidimensional design space. In the orthogonal sampling, the design space is divided into equally probable subspaces. All sample points are then picked simultaneously while ensuring that the total ensemble of sample points is an LHS and that each subspace is sampled with the same density. Thus, the orthogonal sampling has the requirement that the entire sample space must be sampled evenly. Although the orthogonal sampling is efficient, it is more difficult to implement as all the samples must be generated simultaneously.

As a specific application of LHS-MC, two nanoelectronic VCO circuits have been used as a case study, the 45-nm CMOS ring oscillator and the 180-nm CMOS LC-VCO as in the previous sections. The two variables picked are NMOS channel length (W_n) and PMOS channel length (W_p). For the sampled value of the W_n and W_p, the netlist is simulated using a analog/SPICE simulator for center frequency as the target characteristic. The design space generated using LHS-MC for the 45-nm CMOS ring oscillator is shown in Fig. 11.20(a) [25, 26]. The average RMSE is 21.6 MHz for a target center frequency of 9.8 GHz that is 0.2%. The design space generated using LHS-MC for the 180-nm CMOS LC-VCO is shown in Fig. 11.20(b) [25, 26]. The average RMSE is 32.9 MHz for a target center frequency of 2.1 GHz that is 1.57%.

11.2.4.3 Middle Latin Hypercube Sampling-Based Monte Carlo Method

The middle LHS is similar to the LHS. In the MLHS technique, the design space is partitioned similar to the LHS. However, unlike LHS, instead of random sampling from these partitions, the middle value is picked from each of the partitions. Thus, the name "middle" LHS. MLHS technique generates more uniform samples than the LHS technique. However, the main drawback of the MLHS technique is that it cannot sample the region of the design space that is close to the edges. The design space generated using MLHS-MC is shown in Fig. 11.21 for the same ring oscillator and LC-VCO as the previous subsection [25, 26]. The design space generated using MLHS-MC for the 45-nm CMOS ring oscillator is shown in Fig. 11.21(a) [25, 26]. In this case, the RMSE is 21.7 MHz for a center frequency of 9.8 GHz that is 0.22%. The design space generated using MLHS-MC for the 180-nm CMOS LC-VCO is shown in Fig. 11.21(b) [25, 26]. In this case, the RMSE is 37.3 MHz for a center frequency of 2.1 GHz that is 1.77%.

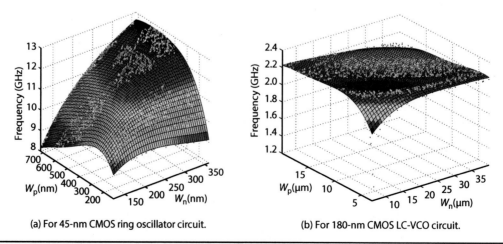

(a) For 45-nm CMOS ring oscillator circuit.   (b) For 180-nm CMOS LC-VCO circuit.

Figure 11.20 LHS of design space of nano-CMOS oscillators [25, 26].

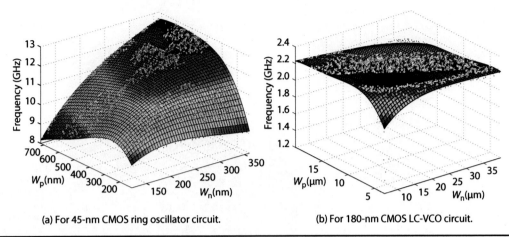

(a) For 45-nm CMOS ring oscillator circuit. (b) For 180-nm CMOS LC-VCO circuit.

FIGURE 11.21 Middle LHS of design space of nano-CMOS oscillators [25, 26].

11.3 Tool Setup for Statistical Analysis

In general, the methods used for statistical analysis, such as Monte Carlo, have major computational overheads during their executions. In addition, there are other requirements such as preprocessing and postprocessing in the context of such computations. For example, the generated data can be significantly large and proper tabulation or plotting may be needed to provide accurate prospective to the design engineers. The electronic design automation (EDA) tools may not necessarily have all the features needed. Therefore, a combination of tools and even the use of scripts are important for such analysis. Two options for tool usage for statistical analysis of nanoelectronic integrated circuits are presented in Fig. 11.22. These frameworks are useful for circuit and layout level of the analog as well as digital ICs as netlist is simulated. However, statistical analysis at the higher level of abstraction, such as logic-level, architecture-level, or system-level, may involve different flows, models, and frameworks.

The first option is the use of EDA-only framework. In this case, all the tasks are performed in the EDA tool only. The simulation engine is no doubt an analog simulator such as a SPICE or Spectre®. Some EDA tools may have graphics user interface (GUI) to easily provide parameters for the simulations. Some EDA tools may need to be fed these parameters through scripts in command mode simulations. The tools have options for process variation analysis, mismatch analysis, or both together. The process performs the statistical variations for all the transistors in the circuit. The mismatch performs an instance statistical variation. There may be options to select various sampling methods, e.g., random, LHS, or MHLS. Moreover, the statistical distributions needed to model the randomness in the parameters may be selected from various options such as Gaussian, uniform, or log normal. In addition, other parameters such as number of runs of Monte Carlo can be given as parameters to the EDA tool framework. A supplementary model file may be needed to specify the device-level variables that are to be statistically varied and to be included along with the simulation setup. This process variation file identifies the variables to be varied for analysis and sets the nominal values as well as statistical variance for each variable. The results for each simulation run can be saved to allow for postprocessing operations, such as histogram display or statistical distribution identification. One of the advantages of native EDA environment is that the process and mismatch and all selections can be easily done. However, the EDA tools may lack easy postprocessing features and the algorithms for optimization available in the EDA frameworks are limited.

Another option is the EDA–non-EDA framework. In this case, preprocessing and postprocessing are done through non-EDA tools using an analog/SPICE simulation engine in the backend, such as MATLAB®. The non-EDA tool is used to generate the sample of input design or process variables that need to be varied. Then, from the large collection of the statistical distributions available in the tool, the parameters are generated. A script is used to feed these parameters to an analog simulator in the EDA tool set for the simulation of the netlist for the generated parameters. The results generated

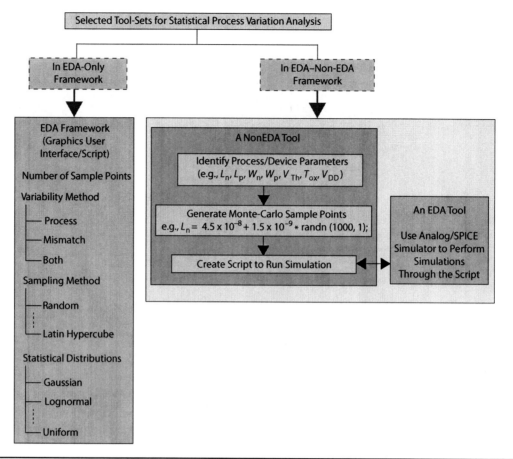

Figure 11.22 Tool setup for statistical process variation analysis.

from the analog simulator are saved in a file. The non-EDA tool through script is used to postprocess the results, for example, plotting the results as histograms. The non-EDA tools have more options for these as well as more distributions to fit for accurate estimation. In the EDA–non-EDA framework, the parameters of each instance would have to be treated as a unique variable for combined process and mismatch, thus making it more difficult to set up. The non-EDA tools on the other hand allow for more flexibility in creating the Monte Carlo sampling for a process variation analysis. In the non-EDA tools, the optimization can be performed easily through its large collection of optimization algorithms.

11.4 Methods for Variability-Aware Design Optimization

The process variations effects can be taken care of right at the early design phase of the ICs and systems. The techniques are static design time compensation to make the design variability aware. For the analog and mixed-signal ICs, such techniques are deployed during the schematic-level or layout-level design exploration. Additionally, for the digital ICs and systems, the design time techniques are incorporated during the design time at different levels of abstractions. In this section, generic discussion is presented for variability-aware design optimization or robust optimization. The subsequent section presents detailed examples with specific case studies.

11.4.1 Brief Concept

The device-level variations are variations in the parameters such as effective channel length (L_{eff}), threshold voltage (V_{Th}), and gate-oxide thickness (T_{ox}). As a specific example, the threshold voltage variations can be originated from various reasons including doping concentration variations and gate-oxide thickness variations [14, 70]. A large number of transistors, which are the same during the design time, may not have the same threshold voltage after they are fabricated. For a large number of

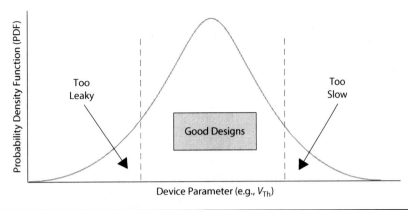

FIGURE 11.23 Process variation due to threshold voltage variation (V_{Th}) produces leaky or slow chips and robust optimization is needed for good designs.

transistors, the histograms of threshold voltage and number of transistors having specific value can present information of the V_{Th} variation. Such histograms of the V_{Th} are presented in Fig. 11.23 to understand the impact of the process variations. The threshold voltage when reduces to low values due to variations can make the transistors too leaky as the subthreshold leakage is affected by it. On the other hand, increases in the threshold voltage to higher values due to variations can make the transistors slow. The IC designs containing many of these transistors will not be considered good designs as they will not meet the target specification due to higher leakage or higher delay. Thus, design time or post design time techniques are needed to compensate the effects such that most of the transistors have V_{Th} within acceptable values to result in good designs.

The objective of the variability-aware design optimization is to perform optimization of the nanoelectronic circuit and system using approaches such as statistical so that the characteristics or FoMs of the circuits and system are minimal to maximize the yield. In the statistical optimization over the circuit and system characteristics, the mean and standard deviation are candidates for optimization as depicted in Fig. 11.24(a) [28, 33]. The baseline PDF has the mean of $\mu_{baseline}$ and standard deviation of $\sigma_{baseline}$. The statistical optimizations can be performed over the nanoelectronic circuit and system to shift mean and/or standard deviation. Two instances that can possibly result due to such statistical optimization are illustrated. The left of the baseline PDF is the new PDF with mean of $\mu_{minimized}$ and standard deviation of $\sigma_{minimized}$. In this case, the mean is shifted to the left from $\mu_{baseline}$ to $\mu_{minimized}$; which can lead to low power design or low noise design. The shifting of the standard deviation from $\sigma_{baseline}$ to $\sigma_{minimized}$ indicates reduction of design variability; which can result in higher yield. Thus, simultaneous reduction of $\mu_{minimized}$ and $\sigma_{minimized}$ can result in design like low-power robust designs or low-power high-yield designs. The shifting of $\mu_{baseline}$ to $\mu_{maxmized}$ can be used for high-performance designs. Simultaneous $\mu_{maximized}$ and $\sigma_{minimized}$ lead to high-performance robust designs or high-performance high-yield designs. The typical cost of statistical optimization is minimized ($\mu + \alpha\sigma$) as presented in Fig. 11.24(b), where α is a real number. For example, the minimization of ($\mu + \sigma$) can ensure yield of 68.2% that refers to 'A' in the figure. The minimization of ($\mu + 2\sigma$) can ensure a yield of 95.4% that refers to 'B' in the figure.

11.4.2 Variability-Aware Schematic Design Optimization Flow

A generic variability-aware schematic-level design optimization flow is shown in Fig. 11.25. This flow can be used for optimization of analog, mixed-signal, and digital ICs. However, it is more useful for analog and mixed-signal ICs as large real-life digital ICs are preferably optimized at higher levels of abstractions such as architecture-level and system-level. At the same time, the digital ICs can be automatically/semi-automatically synthesized by skipping the schematic-level design step. In this variability-aware design flow, following the typical steps presented, a parameterized schematic-level netlist is created. The schematic-level netlist is parameterized for the design and process variables whose variability effects need to be analyzed and then compensated in the statistical optimization flow. Using accurate distributions for the design and process parameters, the schematic-level netlist is

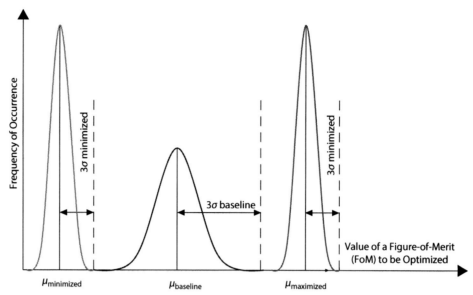

(a) Optimizing both mean (μ) and standard deviation (σ) of the target characteristics.

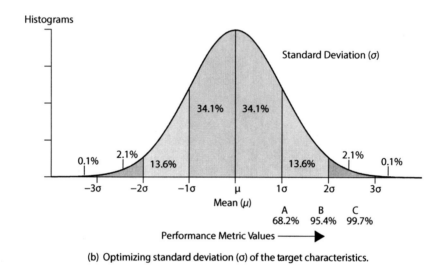

(b) Optimizing standard deviation (σ) of the target characteristics.

FIGURE 11.24 Objectives of the statistical optimization for variability awareness [28, 33].

characterized to obtain the statistical distributions of the target FoMs. For this purpose, any of the methods presented in the variability analysis technique section can be used. After this phase, optimization is performed to minimize ($\mu + \alpha\sigma$) of the target FoM subjected to specified constraints. The selection of α depends on the yield specifications as discussed in the previous subsection. The statistical characterization and optimization steps work together in the variability-aware design flow. A large variety of optimization algorithms can be used to perform automatic optimization. The examples showing these steps will be presented in later sections. Based on the optimal sizes and variables obtained from the optimization, the final design is performed. This final design meets the target specifications is variability aware as well.

11.4.3 Single Manual Layout Iteration Automatic Flow for Variability-Aware Optimization

A generic variability physical-design optimization flow that needs only two layout steps (i.e., single manual layout iteration) is presented in Fig. 11.26 [28, 31]. This flow can be used for the layout of analog, digital, and mixed-signal ICs as long as its complexity can be handled from the computation overhead point of view. Simplistically speaking, this flow may look similar to the schematic-level

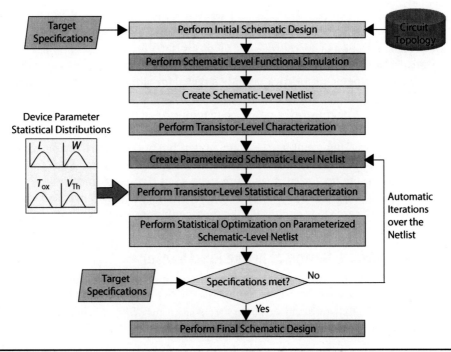

FIGURE 11.25 Proposed variability-aware schematic-level design optimization flow.

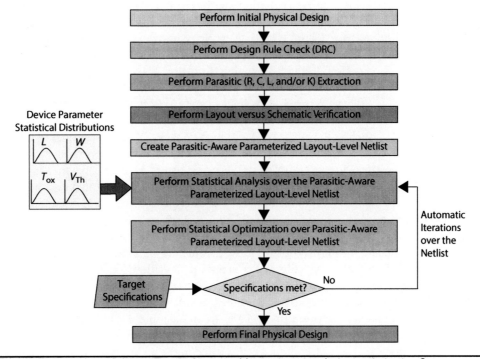

FIGURE 11.26 Proposed variability-aware single-manual-layout iteration design optimization flow.

flow of the previous subsection, but the optimization performed in this flow is over the physical design, not the schematic design. The optimization over layout is more computational intensive due to all the parasitics present which leads to long simulation time in the analog simulator. In this flow, following the standard design steps, the initial physical design is made. For an analog/mixed-signal design, it is a manual layout step (the first manual step), whereas for digital, it can be a semi-automatic or an automatic step. The design is verified through DRC and LVS as typical in an analog design flow. The parasitic extraction generates a parasitic-aware netlist of the initial physical design.

At this point, the process and design parameters that are to be studied for process variations and/or tuned for optimization are identified. The parasitic-aware netlist is then parameterized for these variables. Selection of appropriate distributions is made to capture the variations of these parameters. Statistical analysis of the parasitic-aware parameterized netlist is performed through an analog/SPICE simulator for the characteristics or FoMs targeted for optimization. Determination of the distributions of the target characteristics can be made by fitting known distributions with minimal errors. An optimization algorithm can be now used to minimize $(\mu + \alpha\sigma)$ of the target FoM subjected to specified constraints. Once the algorithm converges, the final physical design is performed using the new transistor sizes. This layout step is a manual step, but not necessarily a completely new physical design. The initial physical design can be modified with new transistor sizes to obtain the final physical design. Thus, the flow needs maximum two layout steps, one initial physical design and second the final physical design; thus, the name single manual-layout iteration. Detailed examples demonstrating this flow with case study circuits will be presented in the subsequent sections.

11.5 Variability-Aware Design of Active Pixel Sensor

A specific example of variability-aware schematic-level optimization is discussed in this section. Design optimization of a 32-nm CMOS-based active pixel sensor (APS) has been considered as a case study circuit. First statistical variability analysis of the APS and then optimization using a combined DOE and the Monte Carlo approach are presented.

11.5.1 Impact of Variability on APS Performance Metrics

The basic design of APS has been discussed in the chapter on sensor circuits and systems. In this subsection, statistical process variation analysis of that APS is presented. An APS array of size 8×8 constructed using 32-nm CMOS-based technology node for the purpose of experimentation [28, 33]. The baseline sizes for 32-nm CMOS technology node are picked to obtain a baseline design at the schematic level. The APS array has been subjected to simultaneous "intra-pixel mismatch" and "inter-pixel process variation" through an analog simulator. The effects on the FoMs or characteristics have been studied. The process parameters identified for mismatch and process variation are: (1) supply voltage (V_{DD}), (2) NMOS threshold voltage (V_{Thn}), (3) PMOS threshold voltage (V_{Thp}), (4) NMOS gate-oxide thickness (T_{oxn}), and (5) PMOS gate-oxide thickness (T_{oxp}). As an example, two values of supply voltage (V_{DD}) such as low-V_{DD} (V_{DDl}) and high-V_{DD} (V_{DDh}) and two values of oxide-thickness (T_{ox}) as low-T_{ox} (T_{oxL}) and high-T_{ox} (T_{oxH}) are used to obtain four design corners. In a traditional CMOS process, the gate oxides of NMOS and PMOS transistors are grown together and hence equal thickness T_{ox} with $T_{oxn} = T_{oxp}$ is assumed. In other words, T_{oxn} and T_{oxp} are scaled together, i.e., they are assigned either low-T_{ox} or high-T_{ox} value together. The example values for a 32-nm CMOS technology node are the following: $V_{DDl} = 0.7$ V, $V_{DDh} = 0.9$ V, $T_{oxL} = 1.65$ nm, and $T_{oxH} = 2.0$ nm.

For the "intra-pixel mismatch," the parameters are assumed to have a Gaussian or normal distribution and mean (μ) values of these parameters are taken as the baseline values specified in the 32-nm CMOS technology file and a standard deviation (σ) of 5%. For the "inter-pixel process variation," the parameters are also assumed to have a Gaussian or normal distribution and mean (μ) values of these parameters are taken as the baseline values specified in the 32-nm CMOS technology file and a standard deviation (σ) of 10%. As an example, Monte Carlo simulation is performed for 1000 runs. It is possible to perform Monte Carlo simulations of higher number of runs as well as use other statistical analysis techniques as discussed in the previous section. The results obtained from the Monte Carlo simulations have been plotted as histograms. The PDFs of the power dissipation of the APS (P_{APS}) for the four corners have been shown in Fig. 11.27 [28, 33]. It is observed that P_{APS} follows log-normal distribution. The mean and sigma for the distributions have been shown in the figure. It is also observed that the power dissipation is lowest with lowest mean when V_{DDl} and T_{oxH} corner are used. The variations in the power dissipation are lowest with lowest sigma when V_{DDh} and T_{oxH} corner are used. The PDFs of the output voltage swing of the APS (V_{swing}) for the four corners have been shown in Fig. 11.28 [28, 33]. V_{swing} is observed to have a Gaussian or normal distribution with mean and sigma as shown in the figure. As can be seen from the figures, the output voltage swing is highest with highest mean when V_{DDh} and T_{oxL} corner are used. The variations in the output voltage swing are lowest with lowest sigma also when V_{DDh} and T_{oxL} corner are used. As in a same corner, low power

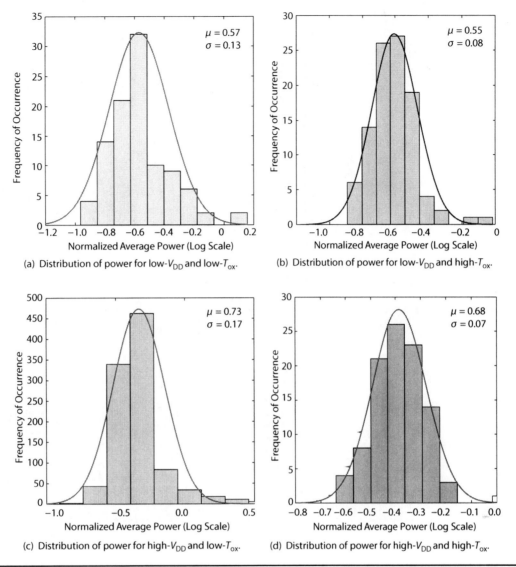

Figure 11.27 Distribution of average power P_{APS} for the four corner cases for two variables V_{DD} and T_{ox} and their statistical process variations for the 32-nm CMOS APS [28, 33].

dissipation, high output voltage swing, and their lowest variations are not obtained, there is need for optimization. The statistical optimization for power and output voltage swing tradeoffs is discussed in the next subsection.

11.5.2 Variability-Aware APS Optimization

The high-level view of the variability-aware design flow APS is shown in Fig. 11.29 [28, 33]. The baseline APS array of size $M \times N$ is assumed as an input to this flow. The target FoM or characteristics of the APS is then estimated. The parameters are then identified for the mismatch and process variation analysis purposes. Then, "intra-pixel mismatch" and "inter-pixel process variation" analysis are performed using a statistical approach as discussed in the previous subsection. The FoMs that need to be optimized are identified for cost function and constraints so that tradeoffs can be achieved during optimization. An algorithm performs variability-aware optimization to determine optimal design parameters. The final $M \times N$ APS array design is then performed using the optimal sizes and characterization is then performed.

A detailed flow of the variability-aware optimization approach is presented in Fig. 11.30 [28, 33]. First, the parameter to be tuned during optimization is identified and denoted as X_{APS}. Then, possible design corners are identified using different levels of these tuning variables. As a specific

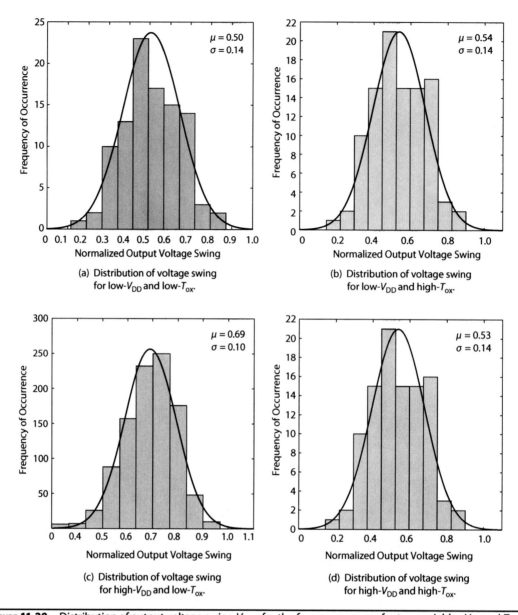

FIGURE 11.28 Distribution of output voltage swing V_{swing} for the four corner cases for two variables V_{DD} and T_{ox} and their statistical process variations for the 32-nm CMOS APS [28, 33].

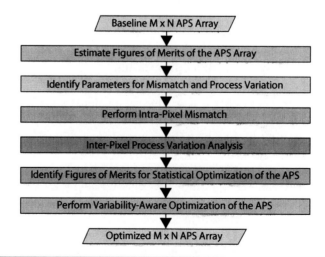

FIGURE 11.29 A design flow for process variation optimal design of nano-CMOS APS array [28, 33].

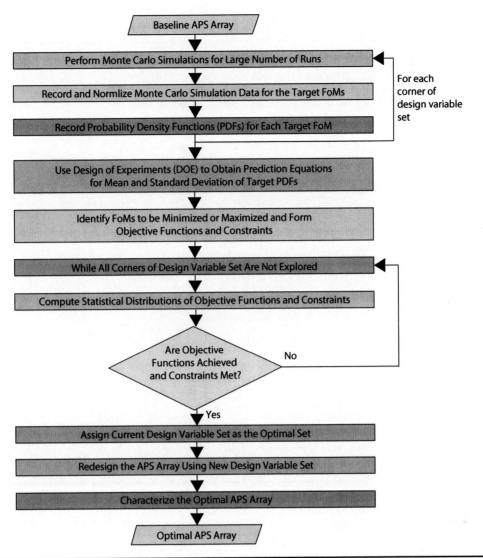

FIGURE 11.30 Flowchart of a variability-aware approach for APS optimization [28, 33].

example, if V_{DD} and T_{ox} are considered as variables for tuning and each has two levels, then there are four corners that are discussed in the previous subsection. The baseline APS array is then statistically characterized as discussed in the previous subsection accounting for intra-pixel mismatch and inter-pixel process variations to obtain the PDF for each of the target FoMs that are denoted as FoM_{APS}. Traditional Monte Carlo and other techniques can be used for this purpose. At this point, the DOE approach is used to obtain predictive equations for mean (μ) and standard deviation (σ) of the PDFs of the target FoMs. The optimization is performed to determine the optimal values of the tuning parameters. The steps followed to perform optimization over the algorithm for optimization of any given circuit are discussed in Algorithm 11.1. The algorithm takes advantage of the strengths of both Monte Carlo analysis and DOE techniques just like the DOE-MC discussed in a prior section. The algorithm returns optimal sizes of X_{APS} that are denoted as $X_{APS,optimal}$ upon the convergence. The optimal sizes are then used to obtain final APS design.

The steps of the variability-aware algorithm are presented with specific details. As a specific example, an 8×8 APS array has been implemented using 32-nm CMOS technology node. The baseline design used the baseline sizes and characterization results are presented in Table 11.7 [28, 33]. For the optimization purposes, the following variable set has been considered: $X_{APS} = \{V_{DD}, T_{ox}\}$ and the following are considered as target FoM: $FoM_{APS} = \{P_{APS}, V_{swing}\}$. In more comprehensive examples, a larger array size, more design variables, and more target FoMs can be considered. For the statistical process, variability analysis Gaussian distribution of the parameters such

ALGORITHM 11.1 Variability-Aware Optimization with APS Using Hybrid Monte Carlo and DOE [28, 33]

1: **Input:** Baseline APS Circuit, Objective Set FoM$_{APS}$, Design Variable Set X_{APS}.
2: **Output:** Design Variable Set $X_{APS,optimal}$.
3: **for** (Each corner of design variable set X_{APS}) **do**
4:    Perform Monte Carlo simulation for N runs.
5:    Record the Monte Carlo results for the target FoM$_{APS}$.
6:    Normalize the Monte Carlo results for the target FoM$_{APS}$.
7:    Tabulate statistical characteristics (mean and standard deviation) of FoM$_{APS}$ distributions.
8: **end for**
9: Obtain prediction equations for μ_{FoM} and σ_{FoM} using design of experiments approach.
10: Formulate the objective function as: Cost$_{FoM} \leftarrow \mu_{FoM} \pm 3\sigma_{FoM}$.
11: **while** (All corners of design variable set X_{APS} not explored) **do**
12:    Compute the objective function as: Cost$_{FoM}$.
13:    **if** (Cost$_{FoM}$ is within error margin) **then**
14:       Update the optimal parameter set; $X_{APS,optimal} \leftarrow X_{APS}$.
15:    **end if**
16: **end while**
17: **return** Optimal parameter set $X_{APS,optimal}$.

Different APS Circuits	Power Dissipation P_{APS} (μW)	Output Voltage Swing V_{swing} (μV)
Baseline APS Circuit	16.3	428
Optimal APS Circuit	12.9	325

TABLE 11.7 Selected Characteristics of the 32-nm CMOS-Based APS Array [28, 33]

as V_{DD}, V_{Thn}, V_{Thp}, T_{oxn}, and T_{oxp}, with mean as nominal value for 32-nm CMOS technology node and standard deviation (σ) equal to 10% of the mean (μ) is assumed. A Monte Carlo simulation of 1000 runs to statistically characterize the FoM$_{APS}$ produces the distributions just like the PDFs presented in the previous subsection, which is lognormal for P_{APS} and Gaussian for V_{swing}. At this stage, the algorithm is ready for process variation-aware optimization.

Now the algorithm performs DOE to obtain the predictive equations of the μ_{FoM} and σ_{FoM} in terms of tuning variables X_{APS}. As a specific example, two levels of design variables are considered. Thus, the possible values of the design variables are the following: V_{DDh}, V_{DDl}, T_{oxH}, and T_{oxL}. Although only four different scenarios are considered in this example, the situation is complex when other discrete sets of design variables with additional levels are considered. These values are assigned to the μ of optimization parameters for Monte Carlo 100 runs. The APS array is subjected to 5% intra-pixel mismatch and 10% inter-pixel process variation for each of the four combinations. The Monte Carlo results for FoM$_{APS}$ are obtained and normalized. The normalization process is the division of each value of the FoMs by their maximum values. The statistical distributions are shown in the previous subsection for this baseline APS design. The prediction equations to be explored for the μ_{FoM} and σ_{FoM} is of the following form [28, 33]:

$$Y_{response} = Y_{average} + \left(\frac{\Delta V_{DD}}{2}\right) V_{DD} + \left(\frac{\Delta T_{ox}}{2}\right) T_{ox} \tag{11.11}$$

where $Y_{response}$ is a specific response, $Y_{average}$ is the average of the responses, and $\left|\frac{\Delta V_{DD}}{2}\right|$ and $\left|\frac{\Delta T_{ox}}{2}\right|$ are the half effects of the design variables. Thus, a linear relationship between the design variables and the output responses is assumed to reduce the complexity considerably. The complexity would increase exponentially if a nonlinear relationship is assumed. For the 32-nm CMOS-based APS, the following prediction equations are obtained by using the DOE method on Monte Carlo simulations [28, 33]:

$$\mu_{P_{APS}} = 0.64 + 0.07 V_{DD} - 0.02 T_{ox} \quad (11.12)$$

$$\sigma_{P_{APS}} = 0.12 + 0.01 V_{DD} - 0.04 T_{ox} \quad (11.13)$$

$$\mu_{V_{swing}} = 0.61 + 0.09 V_{DD} + 0.01 T_{ox} \quad (11.14)$$

$$\sigma_{V_{swing}} = 0.12 - 0.02 V_{DD} + 0.002 T_{ox} \quad (11.15)$$

The predictive equations are very accurate with a maximum discrepancy of 1% between the experimental results and the results calculated with these predictive equations. Selected Pareto charts used in the experiments are shown in Fig. 11.31 [28, 33]. From the above predictive equations, it can be observed that $\mu_{P_{APS}}$ and $\sigma_{P_{APS}}$ are to be minimized for power minimization. On the other hand, $\mu_{V_{swing}}$ needs to be maximized for V_{swing} maximization and $\sigma_{V_{swing}}$ needs to be minimized for variation reduction. It is also observed that $\mu_{P_{APS}}$ and $\sigma_{P_{APS}}$ are perfectly correlated; in other words, optimizing the mean would also optimize the standard deviation. However, $\mu_{V_{swing}}$ and $\sigma_{V_{swing}}$ are not correlated. Hence, a combined effect of the mean and the standard deviation is needed. Thus, for the variability-aware optimization of the APS, the following is used as objective functions, $f_{FoM} = \mu_{FoM} \pm 3\sigma_{FoM}$. Using the predictive equations, the two objective functions are obtained as follows [28, 33]:

$$f_{P_{APS}} = \mu_{P_{APS}} + 3\sigma_{P_{APS}} = 0.98 + 0.10 V_{DD} - 0.12 T_{ox} \quad (11.16)$$

$$f_{V_{swing}} = \mu_{V_{swing}} - 3\sigma_{V_{swing}} = 0.24 + 0.14 V_{DD} + 0.01 T_{ox} \quad (11.17)$$

In general, after optimization of the above, the design corner that gives optimal values is called $X_{APS,optimal}$. In this specific case, $X_{APS} = \{V_{DDl}, T_{oxH}\}$ leads to the minimum value of $f_{P_{APS}}$ with acceptable value of $f_{V_{swing}}$. The characterization of the optimal APS is shown in Table 11.7 [28, 33]. Thus, a reduction of 21% P_{APS} is obtained for a 24% reduction penalty in V_{swing}. To evaluate variations in the characteristics of the APS, coefficient of variation is used that is defined as follows: $\left(c_v = \frac{\sigma}{\mu}\right)$. The optimal APS exhibits 1.5× decrease in the spread of coefficient of variation as compared to the baseline APS.

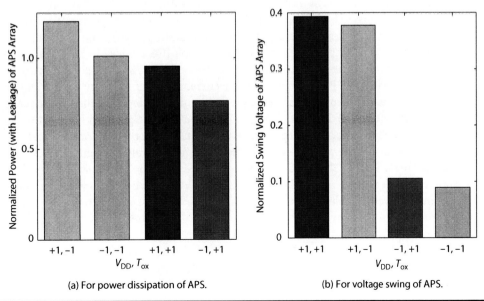

FIGURE 11.31 Pareto charts for power dissipation and output voltage swing for the 32-nm CMOS APS [28, 33].

11.6 Variability-Aware Design of Nanoscale VCO Circuits

The above section discussed a specific example of schematic-level process variation-aware optimization. In this section, more detailed level, the layout-level process variation analysis, and optimization are presented.

11.6.1 A Conjugate-Gradient-Based Optimization of a 90-nm CMOS Current-Starved VCO

In this subsection, optimization of a current-starved VCO is presented in the presence of process variations [28, 30, 31]. As a specific example, the 13-stage current-starve VCO that has been discussed in previous chapters has been considered again for 90-nm CMOS technology node. The statistical variations of the following device-level parameters are considered: V_{DD}, V_{Thn}, V_{Thp}, T_{oxn}, and T_{oxp}, for analysis of the CS-VCO. Statistical process variation analysis is performed over the 90-nm CMOS-based current-starved VCO using the Monte Carlo method. Of course, other statistical process variation analysis methods can be used. The device-level parameters are modeled using Gaussian distributions with $\sigma = 10\%$ of the mean (μ). Five different cases are presented as follows: only V_{DD} variation, only V_{Thn} variation, only V_{Thp} variation, simultaneous T_{oxn} and T_{oxp} variation, and simultaneous V_{DD}, V_{Thn}, V_{Thp}, T_{oxn} and T_{oxp} variation. In the case of simultaneous T_{oxn} and T_{oxp} variations, the parameters are modeled as correlated parameter with a correlation coefficient of 0.9 to be consistent with the fact that the gate oxide of NMOS and PMOS transistors are grown together during manufacturing. Selected resulting statistical distributions of VCO characteristics are shown in Fig. 11.32 [28, 30, 31]. The resulting PDFs following Gaussian distributions with the estimated mean and standard deviations have been reported in the figure. From the PDFs, it is observed that in only V_{DD} variation case, there is maximum reduction in the mean value of the center frequency. On the other hand, the mean value of the center frequency decreased significantly as well as the standard deviations increased significantly for the case when all the parameters are varied. This is the case when the process variations have maximum impact on the center frequency. The frequency versus voltage characteristics of the unoptimized 90-nm CMOS-based 13-stage current-starved VCO for a target oscillating frequency of 2 GHz are shown in Fig. 11.33 [28, 30, 31]. As evident from the figure, the schematic design has highest frequency as there are no parasitics in the design. The oscillating frequency for the physical design of the CS-VCO decreases by 25% as compared to the schematic design due to the presence of RCLK parasitics. For the worst-case process variations that is the case when V_{DD} is reduced by 10% and all the process parameters increased by 10%, the oscillating frequency of the CS-VCO reduced by 44%.

The aim of the variability-aware optimization of the CS-VCO is to improve the oscillating frequency specification as in the worst case it is reduced by 44%. The conjugate-gradient algorithm, which was discussed in the previous chapter, has been reused for process variation-aware optimization of the CS-VCO in this subsection. The parasitic-aware netlist of the unoptimized CS-VCO design is parameterized for the following: (1) widths of NMOS devices in the inverter (W_n), (2) widths of PMOS devices in the inverter (W_p), (3) widths of NMOS devices in the current-starved circuitry (W_{ncs}), (4) widths of PMOS devices in the current-starved circuitry (W_{pcs}), and (5) lengths of all the transistors as L. The objective is to achieve a minimum target oscillation frequency of 2 GHz with a stopping criterion of 2%. The conjugate-gradient-based algorithm iterates for discrete values of the tuning variables X_{tuning} until target oscillating frequency is obtained for the worst-case process variation environment and the corresponding values of design variables are denoted as $X_{optimal}$. The final physical design of the 90-nm CMOS CS-VCO is performed using the $X_{optimal}$ parameters. This process variation-aware physical design of the 90-nm CMOS CS-VCO has been shown in Fig. 11.34 [28, 30, 31]. The frequency versus voltage characteristics for the baseline schematic ideal design and the process-variation optimal physical design of the CS-VCO have been shown in Fig. 11.35 [28, 30, 31]. The process variation optimal CS-VCO is further characterized for various other characteristics and a selected few are summarized in Table 11.8 [28, 30, 31]. For a comparative perspective, the optimal CS-VCO is presented with respect to the unoptimized CS-VCO in Table 11.9 [28, 30, 31]. The oscillating frequency of the optimal CS-VCO is within a 4% margin of the objective ideal center frequency or oscillating frequency.

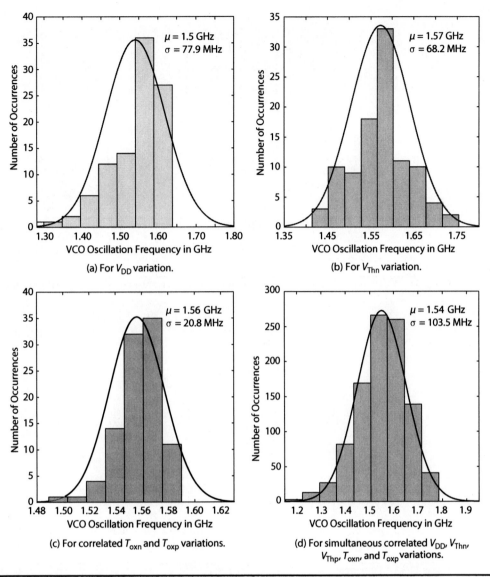

Figure 11.32 Statistical process variation analysis of the baseline or unoptimized 90-nm CMOS-based VCO [28, 30, 31].

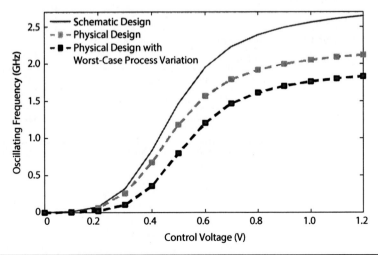

Figure 11.33 Frequency versus voltage characteristics of the 90-nm CMOS-based unoptimized current-starved VCO showing worst-case process variation scenario [28, 30, 31].

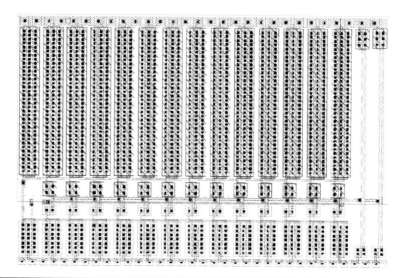

Figure 11.34 Physical design of process variation-aware current-starved VCO for 90-nm CMOS technology node [28, 30, 31].

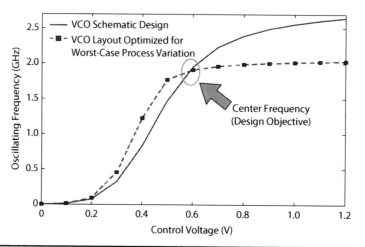

Figure 11.35 Frequency versus voltage characteristics of the variability-aware 90-nm CMOS-based VCO [28, 30, 31].

VCO Parameters and Characteristics	Specific Values
Process Technology	90-nm CMOS 1P 9M
Supply Voltage	1.2 V
Number of Design variables	5 (W_n, W_p, W_{ncs}, W_{pcs}, L)
Number of Objectives	1 (i.e., f_{osc} of 2 GHz)
Oscillation Frequency (Nominal Process)	2.5 GHz
Oscillation Frequency (Worst-Case Process Variation)	2 GHz
Phase Noise (For an offset frequency of 10 MHz)	−109.1 dBc/Hz

Table 11.8 Characteristics of the Process Variation-Aware 90-nm CMOS-Based VCO [28, 31]

VCO Characteristics	For Unoptimized Physical Design	Unoptimized Physical Design and Process variation	Optimized Physical Design and Process variation
Frequency	1.5 GHz	1.1 GHz	2 GHz
Discrepancy	25%	44%	4%

Table 11.9 Frequency Discrepancy and Worst-Case Process Values for a Target Oscillating Frequency of 2 GHz of the 90-nm CMOS-Based VCO [28, 30, 31]

11.6.2 A Particle Swarm Optimization Approach for a 90-nm Current-Starved VCO

In this subsection, variability-aware optimization of the 90-nm CMOS-based 13-stage current-starved VCO is presented that relies on particle swarm optimization (PSO) algorithm for automatic search of the sizes. A variability-aware low-power single-manual-layout iteration VCO physical design optimization flow is presented in Fig. 11.36 [53]. In the first step of this flow, the baseline physical design is first created using standard analog steps. Once the parasitic-aware netlist of the physical design is extracted, the VCO baseline physical design is characterized for the process variations. At this step, any of the statistical analysis techniques, including the exhaustive approach for accurate analysis, are not part of an iteration loop. If the baseline design does not meet the target specifications, which is the most likely situation, the next steps are followed. The design parameters of the CS-VCO to be tuned for optimization are now identified and denoted as X_{tuning}. The parasitic-aware netlist is then parameterized for these parameters for iterations during the optimizations. The VCO is now statistically analyzed using appropriate distributions of the device parameters for target characteristics of the VCO that are to be optimized in the following steps. For low-power design of the VCO, a power-prioritized dual-oxide assignment can be performed at this stage. Such a technique has been discussed in power-aware design chapter. The parameterized parasitic-aware netlist of the VCO

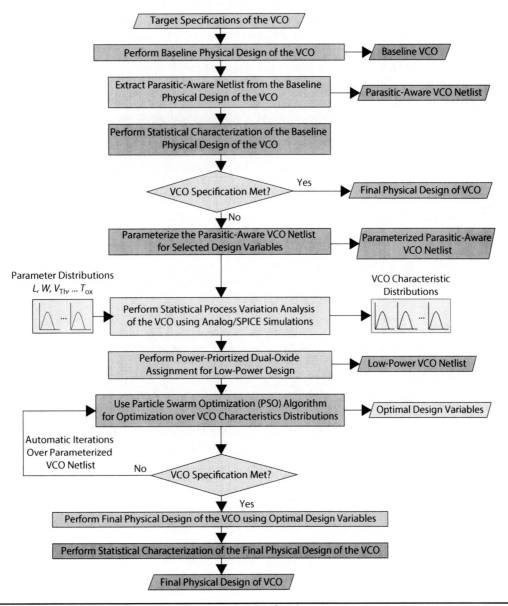

FIGURE 11.36 A PSO-based process variation-aware design flow for VCO [53].

then goes through a process variation-aware statistical optimization loop based on PSO to meet the target specifications. Once the parameter values for which the specifications are met are obtained, a final physical design of the VCO is created using these optimal parameter values denoted as $X_{optimal}$. This design flow also uses a single manual layout design iteration approach in which the layout is performed only twice.

As a case study circuit, the same 13-stage 90-nm CMOS baseline design of current-starved VCO that was the previous subsection has been used here again. Therefore, its statistical process variation is the same as presented in the previous subsection for the baseline unoptimized CS-VCO. The tuning parameters (X_{tuning}) for optimization are: widths of the inverter transistors (W_n, W_p) and current-starved transistors (W_{ncs}, W_{pcs}), and the oxide thicknesses of thick-oxide input-stage transistors (T_{oxnin}, T_{oxpin}). Thus, in this specific example, the tuning variable set is: $X_{tuning} = \{W_n, W_p, W_{ncs}, W_{pcs}, T_{oxnin}, T_{oxpin}\}$; i.e., of dimension $N_{dimension} = |X_{tuning}| = 6$. Each of these tuning variables $X_{tuning}[i]$ can be tuned in the range of ($X_{tuning,min}[i]$, $X_{tuning,max}[i]$); i.e., $X_{tuning}[i] \in [X_{tuning,min}[i], X_{tuning,max}[i]]$. For example, a specific variable W_n may vary between $W_{n,min}$ to $W_{n,max}$, i.e., $W_n \in [W_{n,min}, W_{n,max}]$. In a similar manner, other five tuning variables can be varied within their minimum and maximum allowable range.

As a specific example of automatic search algorithm, a PSO-based algorithm is used for process variation-aware optimization of the VCO [19, 53, 67, 75]. In 1995, this metaheuristic algorithm PSO was introduced; in essence, it uses a population (called swarm) of candidate solutions (called particles) to explore a search space. Being metaheuristic it does not make assumptions about the problem being optimized. PSO can search very large spaces of candidate solutions and is well suited for multidimensional nanoelectronic IC optimization with a large number of variables [22, 76]. The PSO algorithm deploys multiple particles at a time to obtain a solution based on the cost function. In this algorithm, the particle movement is calculated based on the local intelligence of each particle that is offset using global knowledge. Each particle location information holds a multidimensional location. The PSO algorithm starts at a random location of each parameter using each particle and with a random velocity. The steps of the PSO are shown in Algorithm 11.2. In the PSO algorithm, there are $N_{particle}$ number of particles. Each of these particles has N_{tuning} dimensions where N_{tuning} is the number of tuning variable X_{tuning} and it is 6 in this specific case. In general, $x_{tuning,i}[j]$ denotes a position of particle i in a dimension j, where $i = 1 \rightarrow N_{particle}$ and $j = 1 \rightarrow N_{tuning}$ and $x_{tuning,i}$ denotes a position for all dimensions. A particle i has a velocity v_i for all dimensions $j = 1 \rightarrow N_{tuning}$. The best-known position of particle i is $x_{tuning,i,best}$ the best-known position of the entire swarm is $x_{tuning,S,best}$. The "inertia weight" constant for velocity effect α_v, the "acceleration coefficient" constant for the particle β_p, and the "acceleration coefficient" constant for the swarm β_S are used by the design engineers to control the behavior and convergence of the PSO algorithm. The cost function in the algorithm is calculated using a Monte Carlo method around the various points of tuning parameter X_{tuning} in the following manner. First, a small number of Monte Carlo simulations are performed for $x_{tuning,i}$. The statistical distribution of the center frequency is obtained from it. The statistical distribution of the center frequency is then quantified as $\left(\mu_{f_c} + 3\sigma_{f_c}\right)$. It is a fact that the PSO algorithm is implemented as a cost minimization problem; however, in this specific example, the interest is frequency maximization. The cost at a position $x_{tuning,i}$ is calculated as follows: $-\left(\mu_{f_c} + 3\sigma_{f_c}\right)$. Similarly, the cost is calculated for $x_{tuning,i,best}$ and $x_{tuning,S,best}$.

The above variability-aware design flow is used along with the PSO algorithm to experiment for a target center frequency of 2 GHz. The algorithm provided the optimal tuning parameter $X_{optimal}$. The corresponding schematic-level design with different sizes of the transistors and gate-oxide thicknesses is shown in Fig. 11.37(a) [53]. Accordingly, the final physical design is performed and characterized. The oscillating frequency versus tuning voltage characteristics is shown in Fig. 11.37(b) [53]. For a comparative perspective, characteristics for both baseline schematic ideal VCO design and that of the variability-optimal VCO design have been shown. The oscillating frequency versus voltage characteristic for the variability-optimal VCO show acceptable linearity. The overall characterization results of the variability-optimal CS-VCO are presented in Table 11.10 [53]. The calculation of power dissipation of the 90-nm CMOS-based CS-VCO includes the subthreshold leakage power and the gate-oxide leakage power dissipation.

For the study of the process variation effects of the optimal CS-VCO, statistical analysis is performed for the following five different cases: only V_{DD} variation, only V_{Thn} variation, only V_{Thp}

ALGORITHM 11.2 PSO for Variability-Aware VCO [19, 27, 53]

1: **Input:** Parameterized CS-VCO Netlist, Target Specifications $[f_c]$, Maximum Number of Iterations N_{cnt},
2:            Tuning Parameter Set $X_{tuning} = [W_n, W_p, W_{ncs}, W_{pcs}, T_{oxnin}, T_{oxpin}]$, Tuning Parameter Limits $[X_{min}, X_{max}]$.
3: **Output:** Optimized tuning parameters $X_{optimal}$.
4: Initialize the number of particles or population size, $N_{particle} \leftarrow$ number of particles.
5: **for** (Each Particle i from 1 to $N_{particle}$) **do**
6:      Initialize the particle's position with a uniform distribution, $x_{tuning,i} \leftarrow U(X_{min}, X_{max})$.
7:      Initialize the particle's best known position to its initial position, $x_{tuning,i,best} \leftarrow x_{tuning,i}$.
8:      Initialize the entire swarm's best known position, $x_{tuning,S,best} \leftarrow$ Minimum $(x_{tuning,i,best})$.
9:      Initialize the particle's velocity $v_i \leftarrow U(-|X_{max} - X_{min}|, |X_{max} - X_{min}|)$.
10: **end for**
11: Initialize a counter as, cnt $\leftarrow 1$.
12: Initialize the inertia weight constant for velocity effect α_v.
13: Initialize the acceleration coefficient constant for the particle β_p.
14: Initialize the acceleration coefficient constant for the swarm β_S.
15: **while** (cnt $< N_{cnt}$) **do**
16:      **for** (Each Particle i from 1 to N_p) **do**
17:          Generate a random number from uniform distribution, $r_p \leftarrow U(0, 1)$.
18:          Generate another random number from uniform distribution, $r_S \leftarrow U(0, 1)$.
19:          Update the particle for all N_{tuning} dimensions as: $v_i \leftarrow \alpha_v v_i + \beta_p r_p (x_{tuning,i,best} - x_{tuning,i}) + \beta_S r_S (x_{tuning,S,best} - x_{tuning,i})$.
20:          Update the particle's position, $x_{tuning,i} \leftarrow x_{tuning,i} + v_i$.
21:          Calculate the cost for the particle's position, $x_{tuning,i}$.
22:          Calculate the cost for the particle's best known position, $x_{tuning,i,best}$.
23:          **if** (Cost $(x_{tuning,i}) <$ Cost $(x_{tuning,i,best}))$ **then**
24:              Update the particle's best known position, $x_{tuning,i,best} \leftarrow x_{tuning,i}$.
25:              Calculate the cost for the particle's best known position, $x_{tuning,i,best}$.
26:              Calculate the cost for the entire swarm's best known position, $x_{tuning,S,best}$.
27:              **if** (Cost $(x_{tuning,i,best}) <$ Cost $(x_{tuning,S,best}))$ **then**
28:                  Update the entire swarm's best known position, $x_{tuning,S,best} \leftarrow x_{tuning,i,best}$.
29:              **end if**
30:          **end if**
31:      **end for**
32:      Increment counter as, cnt $\leftarrow$ cnt $+ 1$.
33: **end while**
34: **return** Optimal tuning parameter $X_{optimal}$.

variation, simultaneous T_{oxn} and T_{oxp} variation, and simultaneous V_{DD}, V_{Thn}, V_{Thp}, T_{oxn} and T_{oxp} variation. The device-level parameters are modeled using Gaussian distributions with $\sigma = 10\%$ of the mean (μ). In the case of simultaneous T_{oxn} and T_{oxp} variations, the parameters are modeled as correlated parameter with a correlation coefficient of 0.9. For brevity, selected resulting statistical distributions of the variability-optimal VCO characteristics are shown in Fig. 11.38 [53]. A comparison of the parasitic-optimal and process variation optimal CS-VCO with respect to selected characteristics is

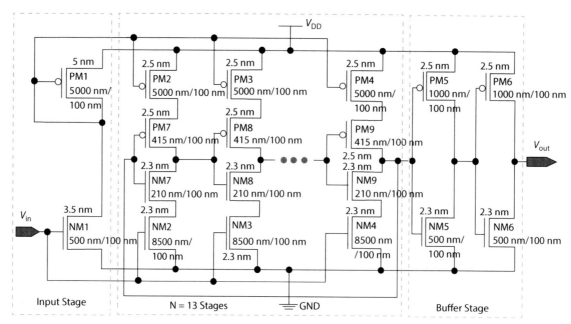

(a) Schematic design of a variability-optimal 90-nm CMOS current-starved VCO.

(b) Frequency-voltage characteristics of the baseline and variability-optimal VCO.

Figure 11.37 Process variation-aware VCO design: schematic and characteristics [53].

VCO Characteristics	Specific Values
Technology Node	90-nm CMOS 1P 9M
Supply Voltage	1.2 V
Center Frequency (Nominal Process)	2.3 GHz
Center Frequency (Worst-Case Process Variation)	1.98 GHz
Tuning Parameters X_{tuning}	$\{W_n, W_p, W_{ncs}, W_{pcs}, T_{oxn}, T_{oxp}\}$
Design Objectives	2 ($f_c \geq 2$ GHz, P_{VCO} = Minimum)
Power Dissipation	158 μW
Silicon Area	389 μm^2

Table 11.10 Characteristics of the Process Variation Optimal Current-Starved VCO [53]

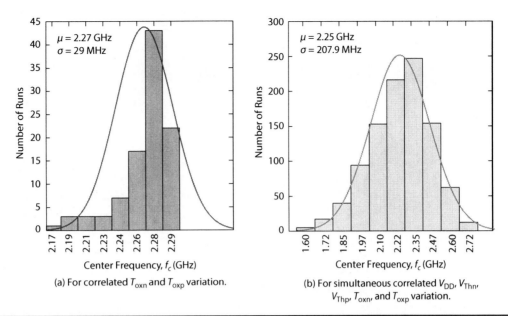

FIGURE 11.38 Statistical characterization of process variation-aware 90-nm CMOS-based current-starved VCO [53].

VCO Design	Power Dissipation		Center Frequency		Silicon Area		Coefficient of Variation	
	P_{VCO}	% Change	f_c	% Change	A_{VCO}	% Change	c_v (For 5 Parameters)	% Change
Parasitic-Optimal	461 μW	–	1.98 GHz	–	325 μm²	–	15.30%	–
Variability-Optimal	158 μW	−65.7%	2.30 GHz	+16.2%	389 μm²	+19.7%	9.24%	−1.66×

TABLE 11.11 Comparison of Characteristics of Baseline and Optimal 90-nm CMOS CS-VCO [53]

presented in Table 11.11 [53]. It is observed that power dissipation decreased and center frequency increased for the variability-optimal CS-VCO. For the process variation effect comparison purposes, the coefficient of variation that is defined as follows is calculated, $c_v = \left|\frac{\sigma}{\mu}\right|$. The table reports c_v for the case of simultaneous V_{DD}, V_{Thn}, V_{Thp}, T_{oxn} and T_{oxp} variations. It is observed that the c_v of the variability-optimal CS-VCO is 1.66 times lower than that of the parasitic-optimal CS-VCO. In other words, variability-optimal CS-VCO is 1.66 times less sensitive or more robust to the process variations than the parasitic-optimal CS-VCO.

11.6.3 Process Variation Tolerant LC-VCO Design

The above two subsections and two sections included several process variation-aware design techniques. All these are design time static methods that generate process variation-tolerant ICs. Other methods for process variation mitigation are the dynamic or run time techniques. In this subsection, a specific LC-VCO design is presented that mitigates the process variations effects on power dissipation and phase noise using adaptive voltage supply, and hence is a dynamic method.

A schematic design of process variation-tolerant LC-VCO is shown in Fig. 11.39 [7, 83]. The characterization results of a 90-nm CMOS based realization of this LC-VCO is shown in Table 11.12 [83]. The core LC-VCO circuit is a typical CMOS-based LC-VCO design that has been discussed in prior chapter on PLL. An adaptive linear voltage regulator is used to generate voltage of V_{reg} that is the supply for the LC-VCO instead of V_{DD} that is the typical case. In essence, the adaptive linear voltage regulator ensures that the current drawn by the LC-VCO is independent of the process parameters, thus mitigating the effects of the process variations. The adaptive linear voltage regulator has a reference voltage generator and a comparator. The reference voltage generator consists of two

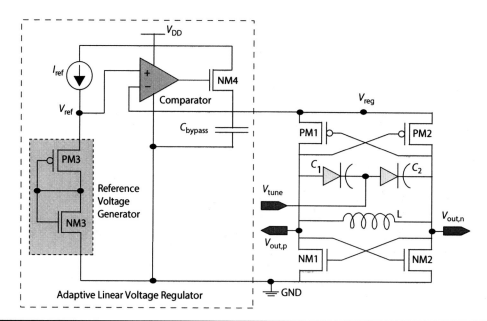

Figure 11.39 Process variation tolerant LC-VCO design using adaptive linear voltage regulator [83].

Technology Node	Supply Voltage	Center Frequency	Phase Noise	Current Dissipation
90-nm CMOS	1.0 V	$\mu = 1.83$ GHz and $\pm 3\sigma = 15.62$ MHz	-96.58 dBc/Hz @ 100 kHz	0.4 mA

Table 11.12 Characteristics of the Voltage-Adaptive LC-VCO [83]

MOSFETs that operate as diodes to behave as high impedance current mirror. Effectively in this circuit, for a reference current I_{ref}, a reference voltage is generated as follows [7, 83]:

$$V_{\text{ref}} = V_{\text{gs,N}} + V_{\text{gs,P}} \qquad (11.18)$$

where $V_{\text{gs,N}}$ is the voltage across the NM3 and $V_{\text{gs,P}}$ is the voltage across PM3. The generated reference voltage V_{ref} will track process/temperature variations for a reference current I_{ref}, as $V_{\text{gs,N}}$ and $V_{\text{gs,P}}$ are affected by the process and temperature variations.

11.7 Variability-Aware Design of the SRAM

The static-random access memory (SRAM) has widespread and important usage in various computing platforms. This section discusses variability-aware optimization of SRAM [36, 55, 57, 59, 84, 85]. As a specific case-study, a 7-transistor SRAM (7T-SRAM) has been considered and statistically optimized using a combination of DOE and integer linear programming (ILP) [55, 84, 85]. The design and characterization of 7T-SRAM were presented in a previous chapter on memory for AMS-SoCs.

The design flow steps of the variability-aware SRAM design are presented in Algorithm 11.3 [55, 84, 85]. As a specific case study, the 7T-SRAM shown in Fig. 11.40(a) is considered for optimization [55, 84, 85]. The key idea of this variability-aware optimization is performing four different minimizations as follows: (1) mean of the power dissipation of the SRAM (μ_{PWR}), (2) standard deviation of the power dissipation of the SRAM (σ_{PWR}), (3) mean of the read static noise margin of the SRAM (μ_{SNM}), and (4) standard deviation of the read static noise margin of the SRAM (σ_{SNM}). Then, forming a global solution that is the combination of these four solutions. The threshold voltage of each the transistors with two levels of each (V_{ThL}, V_{ThH}) are the factors for this DOE-assisted optimization. For the specific 7T-SRAM design, there are seven V_{Th} parameters, one for each of the transistors; leading to seven factors. The two levels of each of the factors are (V_{ThL}, V_{ThH}) or (0, 1). For speeding of the design process, Taguchi L8 matrix method of DOE is considered. For each of the eight

---

ALGORITHM 11.3 Statistical DOE-ILP Optimization of Nano-CMOS SRAM [55, 84, 85]

1: **Input:** Baseline SRAM Design.
2: **Output:** Variability-Aware Low-Power SRAM Design.
3: Estimate the power dissipation (PWR_{SRAM}) and performance (SNM_{SRAM}) of the baseline SRAM.
4: Consider threshold voltage (V_{Th}) of each transistor of the SRAM cell as factors for a DOE method with two levels.
5: Consider the statistical parameters for the power dissipation (μ_{PWR}, σ_{PWR}) and static noise margin (μ_{SNM}, σ_{SNM}) of the SRAM as 4 responses of the DOE method.
6: **for** (Each 1:8 Experiments of 2-Level Taguchi L8 Matrix) **do**
7:     Perform Monte Carlo simulations of N runs for power and SNM of the SRAM.
8:     Plot the histograms for the resulted Monte Carlo data for power and SNM of the SRAM.
9:     Fit known statistical distributions with minimal error to these histograms.
10:     Estimate mean and standard deviation for these statistical distributions.
11: **end for**
12: Formulate Taguchi L8 matrix and Pareto plots for the responses.
13: Formulate linear predictive equations for the 4 responses, (μ_{PWR}, σ_{PWR}, μ_{SNM}, σ_{SNM}).
14: Formulate integer linear programming (ILP) for μ_{PWR}, solve it and denote the solution set $S_{\mu PWR}$.
15: Formulate ILP for σ_{PWR}, solve it and denote the solution set $S_{\sigma PWR}$.
16: Formulate ILP for μ_{SNM}, solve it and denote the solution set $S_{\mu SNM}$.
17: Formulate ILP for σ_{SNM}, solve it and denote the solution set $S_{\sigma SNM}$.
18: Obtain the global set of transistor V_{Th}s which is the intersection of the above 4 sets as: $S_{obj} = S_{\mu PWR} \cap S_{\sigma PWR} \cap S_{\mu SNM} \cap S_{\sigma SNM}$.
19: Assign high threshold voltage (V_{ThH}) to transistors based on the global solution set S_{obj}.
20: Perform process variation characterization of SRAM using device parameters.

---

trails of this DOE, the Monte Carlo analysis with a small number of runs is performed to estimate the PDF for the power dissipation and read SNM of the 7T-SRAM design. In the DOE method, predicting equations are formulated for the four responses (μ_{PWR}, σ_{PWR}, μ_{SNM}, σ_{SNM}) based on the information from Taguchi matrix and Pareto chart. The four responses are then minimized using ILP approach for different constraints. The four solution sets obtained from these solutions are the following: $S_{\mu PWR}$, $S_{\sigma PWR}$, $S_{\mu SNM}$, and $S_{\sigma SNM}$. The solutions contain either V_{ThL} or V_{ThH} assignment for each of the seven transistors. The overall solution of the variability-aware design flow is the SRAM with the solution set meeting all the four solutions and is determined as follows: $S_{SRAM-PV} \leftarrow S_{\mu PWR} \cap S_{\sigma PWR} \cap S_{\mu SNM} \cap S_{\sigma SNM}$.

The specific details of the variability-aware design optimization flow is now presented. As a specific example, the Monte Carlo simulations of 100 runs are performed for each trial in the DOE. Hence, overall, 800 runs to generate eight distributions with eight means and eight standard deviations for each of the two characteristics (power and SNM). These are used to generate predictive equations for four responses corresponding to two characteristics. The following 12 process parameters are considered for the Monte Carlo simulations [47, 55, 91]: (1) NMOS access transistor channel length (L_{na}), (2) PMOS access transistor channel length (L_{pa}), (3) NMOS driver transistor channel length (L_{nd}), (4) PMOS load transistor channel length (L_{pl}), (5) NMOS access transistor channel width (W_{na}), (6) PMOS access transistor channel width (W_{pa}), (7) NMOS driver transistor channel width (W_{nd}), (8) PMOS load transistor channel width (W_{pl}) (nm), (9) NMOS gate-oxide thickness (T_{oxn}) (nm), (10) PMOS gate oxide thickness (T_{oxp}) (nm), (11) NMOS channel doping concentration (N_{chn}), and (12) PMOS channel doping concentration (N_{chp}). Each of these process parameters is considered to have a Gaussian distribution with mean taken as the nominal values specified in 45-nm CMOS and

664 Chapter Eleven

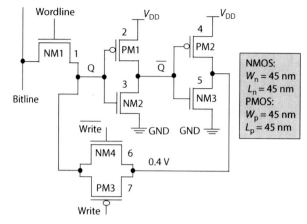

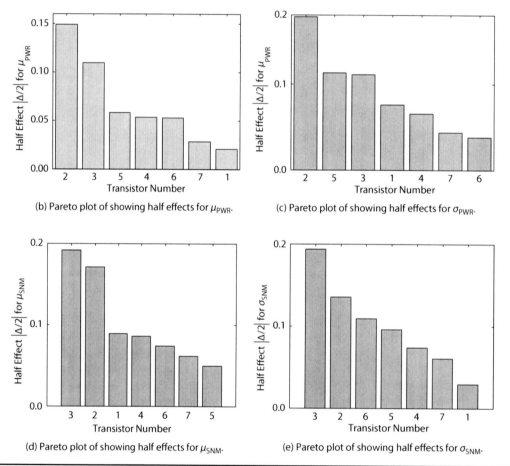

FIGURE 11.40 A 7T-SRAM and its Pareto plot for DOE-ILP optimization [55, 84, 85].

standard deviation as 10% of the mean. Among the above 12 parameters, some are independent and others are correlated, and hence a correlation coefficient of 0.9 is considered during the simulation. The four responses (μ_{PWR}, σ_{PWR}, μ_{SNM}, σ_{SNM}) are recorded. The half effects of the responses are recorded using the following expression:

$$\frac{\Delta(n)}{2} = \left(\frac{\text{avg}(1) - \text{avg}(0)}{2}\right) \qquad (11.19)$$

where $\left|\frac{\Delta(n)}{2}\right|$ is the half effect of the nth transistor, avg(1) is the average value of response when transistor n is on the V_{ThH} state (or 1 state), and avg(0) is the average value of response when transistor

n is on the V_{ThL} state (or 0 state). The responses are normalized so that the values with different units, such as power in W and SNM in V, can be handled together. The normalized predictive equations of the different responses can be obtained from the following:

$$Y_{\text{response}} = Y_{\text{average}} + \sum_{n=1}^{7}\left(\frac{\Delta(n)}{2} x_n\right) \tag{11.20}$$

$$= Y_{\text{average}} + \sum_{n=1}^{7}\left(\left(\frac{\text{avg}(1) - \text{avg}(0)}{2}\right) x_n\right) \tag{11.21}$$

where Y_{response} is the predicted response, Y_{average} is the average of the responses, $\left|\frac{\Delta(n)}{2}\right|$ is the half effect of the nth transistor. x_n is the V_{Th} state of the nth transistor and has a value of either 1 and 0.

In this specific 45-nm CMOS-based 7T-SRAM, the Pareto charts of the half effects of the transistors for the four responses are shown in Fig. 11.40(b–e) [55, 84, 85]. The predictive equation for mean of the power dissipation of the SRAM is the following:

$$\mu_{\text{PWR}} = 0.58 - 0.02x_1 - 0.15x_2 - 0.10x_3 - 0.05x_4 - 0.59x_5 - 0.05x_6 + 0.02x_7 \tag{11.22}$$

where x_1 represents the V_{Th} state of transistor 1, x_2 represents the V_{Th} state of transistor 2, and so on. For the minimization of μ_{PWR}, for low-power SRAM design, with μ_{SNM} as the constraint the following ILP formulation is obtained:

$$\begin{aligned}&\text{Minimize:} \quad \mu_{\text{PWR}}\\ &\text{Subjected To:} \quad \mu_{\text{SNM}} > \tau_{\text{SNM}} \text{ and } x_n \in \{0, 1\} \forall n\end{aligned} \tag{11.23}$$

In the above formulation, τ_{SNM} is a constraint on the SNM of the SRAM. Solving the ILP problem, the following optimal solution is obtained: $S_{\mu_{\text{PWR}}} = [x_1 = 1, x_2 = 1, x_3 = 1, x_4 = 1, x_5 = 1, x_6 = 1, x_7 = 0]$. This can also be interpreted as transistors 1, 2, 3, 4, 5, and 6 are the transistors with V_{ThH} and the transistor 7 is with V_{ThL}. Similarly, for the standard deviation of the average power dissipation of the SRAM, the following predictive equation is obtained:

$$\sigma_{\text{PWR}} = 0.61 + 0.07x_1 - 0.18x_2 - 0.11x_3 - 0.06x_4 - 0.11x_5 \tag{11.24}$$

The ILP formulation for the minimization of σ_{PWR} so that the variability in the SRAM power dissipation is minimal has the following form:

$$\begin{aligned}&\text{Minimize:} \quad \sigma_{\text{PWR}}\\ &\text{Subjected To:} \quad \mu_{\text{SNM}} > \tau_{\text{SNM}} \text{ and } x_n \in \{0, 1\} \forall n\end{aligned} \tag{11.25}$$

The solution of the ILP results in the following set: $S_{\sigma_{\text{PWR}}} = [x_1 = 0, x_2 = 1, x_3 = 1, x_4 = 1, x_5 = 1, x_6 = 1, x_7 = 0]$. In other words, the transistors 2, 3, 4, 5, and 6 are transistors with V_{ThH}, and transistors 1 and 7 are transistors with V_{ThL}.

The above two solution sets are solutions for either low μ_{PWR} or for low σ_{PWR}. In the next phase, optimization of read SNM is considered under power dissipation constraints. The predictive equation for μ_{SNM} of the SRAM is the following:

$$\mu_{\text{SNM}} = 0.45 - 0.09x_1 + 0.17x_2 - 0.19x_3 - 0.09x_4 + 0.05x_5 + 0.07x_6 - 0.06x_7 \tag{11.26}$$

For robust SRAM design, read SNM is maximized. The following ILP is formulated for the maximization of the read SNM:

$$\begin{aligned}&\text{Maximize:} \quad \mu_{\text{SNM}}\\ &\text{Subjected To:} \quad \mu_{\text{PWR}} < \tau_{\text{PWR}} \text{ and } x_n \in \{0, 1\} \forall n\end{aligned} \tag{11.27}$$

In the above expression, τ_{PWR} is the constraint on the power dissipation of the SRAM. The solution of the ILP results in the following set: $S_{\mu_{\text{SNM}}} = [x_1 = 0, x_2 = 1, x_3 = 0, x_4 = 0, x_5 = 1, x_6 = 1, x_7 = 0]$.

Thus, the transistors 2, 5, and 6 are transistors with V_{ThH}, and transistors 1, 3, 4, and 7 are transistors with V_{ThL}. The predictive equation for σ_{SNM} is the following:

$$\sigma_{SNM} = 0.35 + 0.03x_1 - 0.13x_2 + 0.19x_3 + 0.07x_4 - 0.09x_5 - 0.11x_6 + 0.06x_7 \quad (11.28)$$

The ILP formulation for the minimization σ_{SNM} that will lead to reduction in the variations of SNM has the following form:

$$\begin{aligned}\text{Minimize:} \quad & \sigma_{SNM} \\ \text{Subjected To:} \quad & \mu_{PWR} < \tau_{PWR} \text{ and } x_n \in \{0, 1\} \forall n\end{aligned} \quad (11.29)$$

Solving the ILP problem generates the following optimal solution for the SRAM that has minimal variations in the SNM: $S_{\sigma_{SNM}} = [x_1 = 0, x_2 = 1, x_3 = 0, x_4 = 0, x_5 = 1, x_6 = 1, x_7 = 0]$. This can also be interpreted as transistors 2, 5, and 6 are with V_{ThH}, and transistors 1, 3, 4, and 7 are with V_{ThL}.

The overall SRAM that will have lowest possible power dissipation, highest possible read SNM, and lowest possible statistical variations in these is the intersection of the above four sets. The following set is the overall solution for the variability-aware SRAM:

$$S_{SRAM-PV} = S_{\mu_{PWR}} \cap S_{\sigma_{PWR}} \cap S_{\mu_{SNM}} \cap S_{\sigma_{SNM}} \quad (11.30)$$

Thus, the solution low-power and high-stability SRAM that is the result of this variability-aware design flow is: $S_{SRAM-PV} = [x_1 = 0, x_2 = 1, x_3 = 0, x_4 = 0, x_5 = 1, x_6 = 1, x_7 = 0]$. Thus, the transistors 2, 5, and 6 are with V_{ThH} and the transistors 1, 3, 4, 7, with V_{ThL}. The schematic design of the SRAM is shown in Fig. 11.41(a) [55, 84, 85]. The characterization of this SRAM is performed. Selected

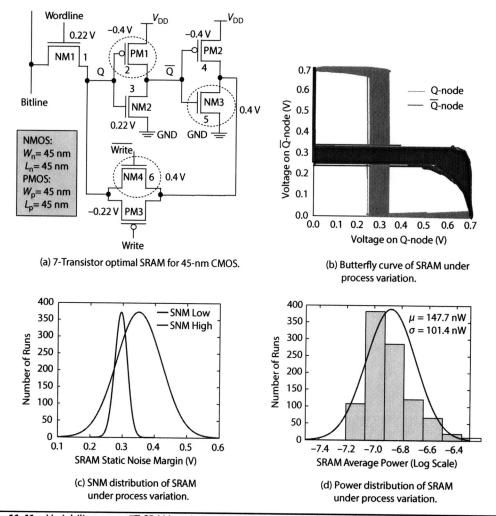

FIGURE 11.41 Variability-aware 7T-SRAM optimization and its statistical characterization [55, 84, 85].

SRAM Design	Power Dissipation (PWR$_{SRAM}$)		Static Noise Margin (SNM$_{SRAM}$)	
	Estimated Value	Change	Estimated Value	Change
Baseline	203.6 nW	–	170.0 mV	–
Variability-Optimal	113.6 nW	44.2%	303.3 mV	78.4%

TABLE 11.13 Statistical DOE-ILP Results for the 7T-SRAM Cell [55, 84, 85]

characteristics are presented in Table 11.13 in comparison to the baseline SRAM design [55, 84, 85]. The variability-aware 7T-SRAM has 44.2% lower power dissipation and 78.4% higher SNM as compared to the baseline 7T-SRAM. The statistical characterization of the 7T-SRAM is also performed. Selected results of this statistical characterization are shown in Fig. 11.41(b–d) [55, 84, 85]. The effect of the process variation on the butterfly curve is shown in Fig. 11.41(b) [55, 84, 85]. The distribution of the SNM that is obtained from the Monte Carlo simulations is presented in Fig. 11.41(c) [55, 84, 85]. The distributions for "SNM Low" and "SNM High" that are extracted from the Monte Carlo simulations are shown; however, "SNM Low" is treated as the actual SNM of the 7T-SRAM [21]. The distribution is Gaussian or normal form. The distribution of average power dissipation of the 7T-SRAM is shown in Fig. 11.41(d) that has a log-normal distribution [55, 84, 85].

11.8 Register-Transfer-Level Methods for Variability-Aware Digital Circuits

In the digital ICs, the process variation awareness has been accounted in various levels, such as system-level, architectural-level, logic-level, transistor-level, and layout-level. The RTL or architecture-level has been a popular abstraction for digital ICs for early optimization of the designs [48, 50, 54]. In this section, selected RTL or architecture-level process variation-aware design methodologies are presented. First, the generic idea of the variability-aware RTL optimization is introduced. Then, selected methods for variability-aware power RTL optimization as well as timing-aware RTL optimization are discussed.

11.8.1 Brief Overview

The different variability-aware RTL optimization techniques have been presented in Fig. 11.42 [48, 50, 54]. Both presilicon and postsilicon techniques are available to maximize the yield while accounting process variations at the RTL or architecture level [17, 48, 50, 54, 88, 90]. The presilicon/static techniques are design time techniques and no additional tuning is needed after the IC is fabricated to achieve variability tolerance. The postsilicon/dynamic/runtime techniques additionally need tuning or adaptation to be effective for process variation mitigation. Both statistical and parametric techniques are possible and a variety of techniques that perform yield optimization during scheduling, resource sharing, and module selection steps of RTL are available. In the statistical methods, the power and delay are estimated as statistical distributions and optimizations tasks are performed over them. In the parametric methods, feedback from the process recipes is taken into account at the architectural level and on-the-fly during the various RTL optimization tasks.

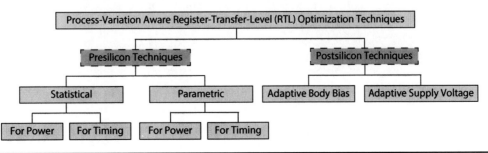

FIGURE 11.42 Various types of variability-aware RTL optimization techniques [48, 50, 54].

The postsilicon techniques use adaptive body biasing (ABB) and adaptive supply voltage such as dynamic mechanism to tune the fabricated ICs to make them variable tolerance. The techniques can perform variability-aware power and timing optimization.

In general, the parametric yield is defined as the probability of the designed hardware meeting a specified constraint [9, 17, 50]. Yield of an IC at the RTL or architecture-level is the probability of the design meeting the power and performance specification during the various RTL design tasks [48, 52, 54, 88]. In terms of power dissipation characteristic, the yield is defined as follows:

$$\text{Yield}_{\text{Power}} = \text{Probability (Power} \leq \text{Power}_{\text{Constraint}}) \quad (11.31)$$

The above is the power yield of an IC. The constraint of the power dissipation $\text{Power}_{\text{Constraint}}$ that is dictated by the power specifications of the IC can be calculated as a multiple of the average or peak power dissipation during the RTL optimization tasks. Similarly, in terms of a performance characteristic, such as delay, the yield is defined as follows:

$$\text{Yield}_{\text{Delay}} \text{ or } \text{Yield}_{\text{Timing}} = \text{Probability (Delay} \leq \text{Delay}_{\text{Constraint}}) \quad (11.32)$$

This is the delay, timing, or performance yield of the IC. Simplistically speaking, timing yield is the probability that an RTL functional unit can complete its execution in a given time period. The delay constraint $\text{Delay}_{\text{Constraint}}$ that is a constraint on the delay, timing, or performance of the chip is calculated as a multiple of the critical path delay of the design during the RTL optimization tasks. The power yield and performance yield can either be considered separately or together in a design optimization flow.

11.8.2 A Simulated-Annealing-Based Statistical Approach for RTL Optimization

A flow for variability-aware statistical RTL optimization is shown in Fig. 11.43 [48, 50, 54]. The input to the flow is a description of the digital IC in the form of an HDL. The overall design framework is divided into the following engines: (1) input generation engine, (2) datapath and control

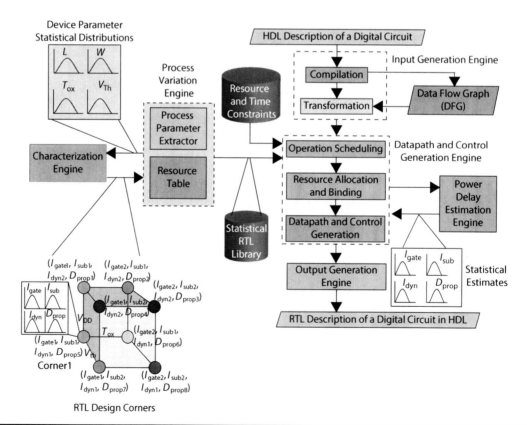

Figure 11.43 Flow of a variability-aware RTL statistical optimization [48, 50, 54].

generation engine, (3) characterization engine, (4) process variation engine, (5) power-delay estimation engine, and (6) output generation engine. The "input generation engine," as typical for RTL synthesis, generates a sequencing data flow graph (DFG) through compilation and transformation from the HDL description of the digital ICs. The "datapath and control generation engine" performs operation scheduling, resource allocation, and binding to generate optimal datapath and control using optimization over the statistical RTL library. The "characterization engine" generates the statistical power and performance models for various design corners while considering statistical models for various device-level parameters. The "process variation engine" performs a dual task. It extracts design and process parameters and maintains a database of statistical distributions for their modeling. It also consists of a resource table populated by the characterization engine. The "power delay estimation engine" estimates the statistical distribution for the intermediate DFGs with intermediate allocation and binding during the RTL optimization tasks. The "output generation engine" generates the variability-optimal RTL description containing datapath and control.

A detailed RTL or architecture-level statistical optimization flow that is based on a simulated annealing algorithm has been presented in Fig. 11.44 [48, 50, 54]. The input to this statistical optimization flow is a sequencing DFG. The mobility graph of each of the vertices in the DFG is constructed by following standard resource-constrained ASAP and ALAP schedules. The target delay is assumed as the product of the critical path delay and delay tradeoff factor. The simulated annealing algorithm is similar to the simulated annealing algorithm presented before in the chapter on power-aware design, but the flow has significant differences in terms of cost function calculation. As a specific example, the steps followed to calculate the statistical cost of the DFG is presented in Algorithm 11.4 [48, 50]. The statistical data for each of the power components and delay is estimated using random Monte Carlo simulation. Under the assumption of independent distributions, the computation of statistical summation can be simplified, otherwise correlations need to take into account. For example, summation of two correlated Gaussian distributions PDF$_A$ and PDF$_B$ is the Gaussian distributions PDF$_{\text{sum}}$ that has the following mean and standard deviation [88]:

$$\mu_{\text{sum}} = \mu_A + \mu_B \tag{11.33}$$

$$\sigma_{\text{sum}} = \sqrt{\sigma_A^2 + \sigma_B^2 - 2\rho\sigma_A\sigma_B} \tag{11.34}$$

In the above expression, PDF$_A$ has a mean of μ_A and standard deviation of σ_A, PDF$_B$ has a mean of μ_B and standard deviation of σ_B, and ρ is the correlation coefficient. Similarly, the maximum between PDF$_A$ and PDF$_B$ can be calculated through tightness probability and moment matching techniques [88]. The cost function calculated is essentially the power delay product (PDP) of the DFG for a specific schedule, allocation, and binding. The RTL statistical design flow can be used for digital ICs including autoregressive filter (ARF), bandpass filter (BPF), and discrete cosine transformation (DCT) for various resource and time constraints. For brevity, the results for selected RTL benchmarks are presented in Fig. 11.45 [48, 50]. In particular, results are presented for two RTL benchmarks, ARF and DCT for four different time constraints averaged over various resource constraints. The power reduction results are presented for different components of power dissipation as well as for total power dissipation. The percentage of power reduction for individual components as well as total has been very significant as evident from the charts.

11.8.3 A Taylor-Series Expansions Diagram-Based Approach for RTL Optimization

A variability-aware Taylor-series expansion diagram (TED)-based RTL statistical optimization flow is presented in Fig. 11.46 [9, 10, 50]. This flow is particularly suitable for variability-aware design exploration of digital signal processing (DSP) ICs or digital signal processor. The input to this flow is the DSP algorithm expressed in polynomial terms. Toward the first step of structural form of the algorithm, the polynomial is graphically represented as a TED. A DFG is generated from the TED of the input DSP polynomial. A heuristic algorithm is then used for simultaneous operation scheduling, and resource allocation and binding. It uses the resource and timing constraints, and statistical RTL library. Then, power and performance yield estimation is performed. Rescheduling of the DFG is performed to optimize timing yield. Finally, datapath and control generation of the DSP IC is performed that is expressed using RTL HDL.

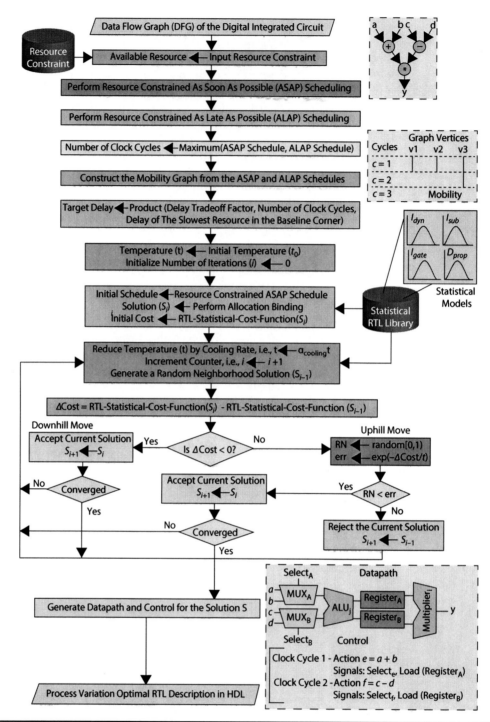

Figure 11.44 Simulated annealing-based statistical RTL optimization [48, 50, 54].

The TED as evident from the nomenclature is modeled on the well-known Taylor series expansion and is based on a nonbinary decomposition principle. TED is a graph-based data structure that gives a compact and efficient way to represent word-level computation in a canonical, factored form. As a specific example of DSP algorithm, let us consider a 4-tap finite-impulse response (FIR) filter. The polynomial representation of a 4-tap FIR filter is the following [9, 50]:

$$y_{\text{out}} = a_0 x_n + a_1 x_{n-1} + a_2 x_{n-2} + a_3 x_{n-3} \tag{11.35}$$

The TED of the above 4-tap FIR filter is shown in Fig. 11.47(a) [9, 10, 50]. TED can capture an entire class of structural solutions not just a single DFG. TED is converted into a DFG optimized for a

ALGORITHM 11.4 RTL-Statistical-Cost-Function to Calculate the Cost of a DFG with Scheduling, Allocation, and Binding [48, 50]

1: **Input:** Data Flow Graph (DFG) with some scheduling, allocation, and binding.
2: **Output:** Power and Delay Product Cost of the DFG.
3: Calculate the probability density function of dynamic power of each clock cycle,
$$P_{dyn,c}(\mu_{dyn,c}, \sigma_{dyn,c}) = \text{Statistical-Summation}_{\forall v \in c}(P_{dyn,FU_{k,i,v}}(\mu_{dyn,k,i}, \sigma_{dyn,k,i})).$$
4: Calculate the probability density function of subthreshold leakage power of each clock cycle,
$$P_{sub,c}(\mu_{sub,c}, \sigma_{sub,c}) = \text{Statistical-Summation}_{\forall v \in c}(P_{sub,FU_{k,i,v}}(\mu_{sub,k,i}, \sigma_{sub,k,i})).$$
5: Calculate the probability density function of gate-oxide leakage power of each clock cycle,
$$P_{ox,c}(\mu_{ox,c}, \sigma_{ox,c}) = \text{Statistical-Summation}_{\forall v \in c}(P_{ox,FU_{k,i,v}}(\mu_{ox,k,i}, \sigma_{ox,k,i})).$$
6: Calculate the probability density function of total power of each clock cycle,
$$P_{total,c}(\mu_{total,c}, \sigma_{total,c}) = \text{Statistical-Summation}(P_{dyn,c}(\mu_{dyn,c}, \sigma_{dyn,c}), P_{sub,c}(\mu_{sub,c}, \sigma_{sub,c}), P_{ox,c}(\mu_{ox,c}, \sigma_{ox,c})).$$
7: Calculate the probability density function of power dissipation of the overall DFG,
$$P_{total,DFG}(\mu_{P,DFG}, \sigma_{P,DFG}) = \text{Statistical-Summation}_{N_{cc}}(P_{total,c}(\mu_{total,c}, \sigma_{total,c})).$$
8: Calculate the power cost of the DFG using 3σ coverage, $\text{Cost-Power}_{P,DFG} = \mu_{P,DFG} + 3\sigma_{P,DFG}$.
9: Calculate the probability density function of clock cycle delay,
$$D_{prop,c}(\mu_{D,c}, \sigma_{D,c}) = \text{Statistical-Maximum}_{\forall v \in c}(D_{prop,FU_{k,i,v}}(\mu_{D,k,i}, \sigma_{D,k,i})).$$
10: Calculate the delay cost of the DFG using 3σ coverage, $\text{Cost-Delay}_{prop,c} = \mu_{D,c} + 3\sigma_{D,c}$.
11: Calculate the critical path delay for the DFG, $\text{Cost-Delay}_{D,DFG} = N_{cc}D_{prop,c}$.
12: Calculate the power delay cost for the overall DFG and corresponding datapath,
$$\text{Cost-PDP} = \text{Cost-Power}_{P,DFG} \times \text{Cost-Delay}_{D,DFG}.$$
13: **return** Cost-PDP.

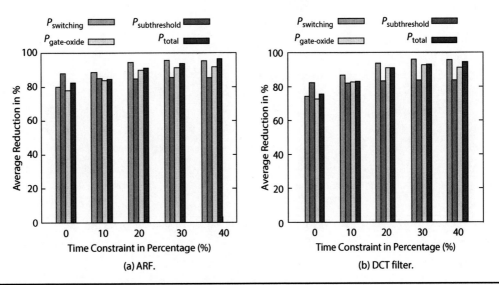

FIGURE 11.45 Simulated annealing-based statistical RTL optimization results [48, 50].

particular design objective by using decompositions. The resulted DFG for the 4-tap FIR filter is shown in Fig. 11.47(b) [9, 10, 50]. As shown in the design flow, the resulted DFG goes through statistical analysis and optimization to enhance power and timing yield.

The proposed flow uses a combination of low-T_{ox} and high-T_{ox} for low-power RTL optimization under timing yield constraint [9, 10, 50]. In the variation-aware design flow, STA is performed to

672 Chapter Eleven

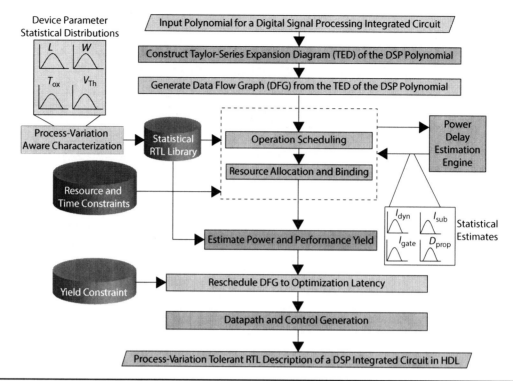

FIGURE 11.46 TED-based variability-aware RTL optimization flow [9, 10, 50].

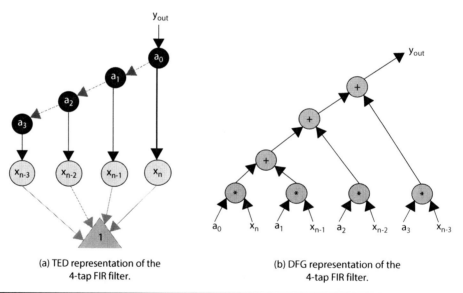

(a) TED representation of the 4-tap FIR filter.

(b) DFG representation of the 4-tap FIR filter.

FIGURE 11.47 TED versus DFG representation of digital IC behavior [9, 10, 50].

determine the critical and noncritical vertices in the DFG. Then, for power optimization purposes, low-T_{ox} is assigned to critical vertices and high-T_{ox} to noncritical vertices. At this point, the critical path delay and power yield are again calculated. The DFG is then iteratively searched to replace as many low-T_{ox} with high-T_{ox} nodes power dissipation as minimally as possible while satisfying delay yield and critical path delay. On average across the RTL benchmarks, the power yield is improved by 4.5% for the timing yield of 95% by the variability-aware design flow.

To understand the concept of power and timing yield, let us refer to the scheduling of the DFG of the 5-tap FIR filter shown in Fig. 11.48 [9, 10, 50]. For the 5-tap FIR filter, an example of scheduling

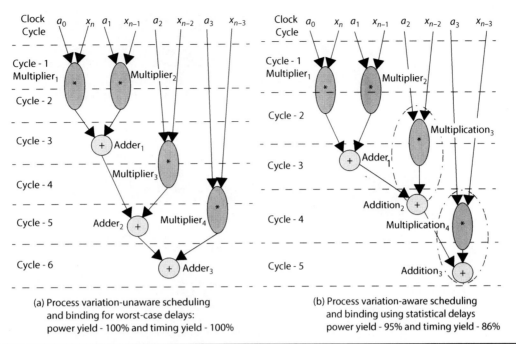

FIGURE 11.48 Power yield and timing yield illustration through a DFG [9, 10, 50].

and binding is shown in Fig. 11.48(a) that has critical latency of six clock cycles. This process variation-unaware solution is for power yield of 100% and timing yield of 100%. An example of process variation-aware solution is presented for power yield of 95% and timing yield of 86% in Fig. 11.48(b). Thus, one clock cycle is saved at the tradeoff of power and timing yield. In the variation-aware design flow, the process variation aware latency algorithm performs iterative searches on the scheduled DFG to reduce delay under timing yield constraints. In the first iteration, the DFG scheduling and binding is performed for 100% power yield and 100% timing yield. It then determines the mobility graph for each of the vertices in the DFG. For each of the possible movements of the operation in the DFG, the gain in delay is calculated in terms of clock cycles. For each movement of the operation, the timing yield is calculated, and if the timing yield is acceptable, then only the movement is allowed. This process continues in the algorithm until all the movements of the operations of the DFG in the mobility graph are checked or the designer-specified timing yield has reached. The variability-aware design flow, on average over all benchmark circuits, achieves a delay reduction of 12.5% for power yield of 96.1% and timing yield of 95%.

11.8.4 Variability-Aware RTL Timing Optimization

A flow for variability-aware RTL timing optimization is shown in Fig. 11.49 [37, 38, 39, 48, 50]. The underneath approach of this variability-aware design flow is a branch-and-bound algorithm for scheduling and binding. The algorithm uses a window-based search for faster convergence in which window is assumed as the maximum number of consecutive clock cycles satisfying resource constraints. The design flow assumes DFG of a digital IC as input description. As the initial condition, the timing slack is assigned as the submultiple of clock cycle time with a "timing constraint" less than one, for example 0.1. The initial latency of the DFG is assign as ∞. The operations that can be scheduled at earlier clock cycles are identified as frontier operations using certain rules. At this point, a window that is a subgraph in the DFG is identified based on resource constraints. Yield equivalent subgraphs are extracted from the window using a branch-and-bound approach. For these subgraphs, all possible scheduling and binding combinations are generated for given resource constraints. A solution is identified for minimal delay for specified yield constraint. For this solution, an intermediate structure with register and multiplexer is synthesized. If the yield constraint has not met then the timing slack is decreased and the above iterations are repeated. If the yield constraint is met and the latency is improved then the timing slack is increased and the above iterations are repeated.

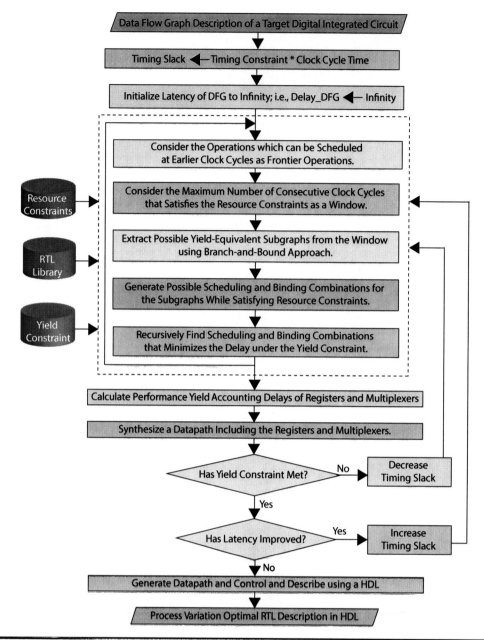

Figure 11.49 RTL optimization flow for timing variation awareness [37, 38, 39, 48].

Once the algorithm converges to a solution, i.e., schedule and binding, the variability-optimal RTL is generated for the datapath and control of the target digital IC.

The important aspect of the above variability-aware design flow is the delay and timing yield tradeoffs as depicted in Fig. 11.50 [37, 38, 39, 48, 50]. The distribution of the delay of a multiplier that is a Gaussian PDF is shown in Fig. 11.50(a) [37, 38, 39, 48, 50]. From the figure, it is observed that if the multiplier is executed for nine cycles, then 100% timing yield is achieved, and if the multiplier is scheduled for eight cycles, then the yield is reduced by 14.7% and becomes 85.3%. Similar discussions are possible for adder that is depicted in Fig. 11.50(b) [37, 38, 39, 48, 50]. However, to present a realistic scenario, the scheduling of a multiplication after the addition operation is presented in Fig. 11.50(c) [37, 38, 39, 48, 50]. As a result, an adder is active and then a multiplier is active to perform the operations. The corresponding statistical distribution is also shown that is also Gaussian. If the two operations are scheduled for 10 cycles, then 100% timing yield is achieved. On the other

Variability-Aware AMS-SoC Design Methodologies 675

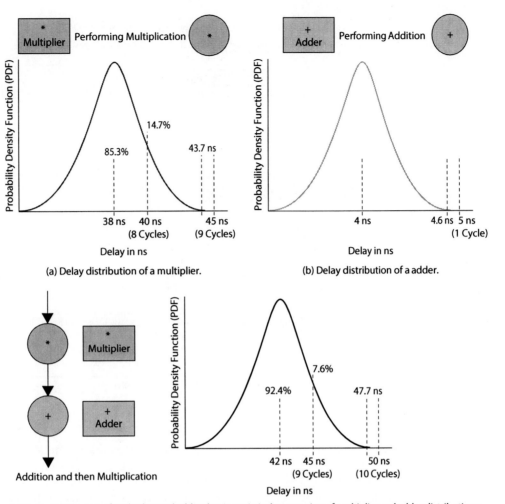

Figure 11.50 Tradeoffs between the clock cycles and performance yield during variability-aware RTL optimization [37, 38, 39, 48].

Benchmark Circuits	Yield Constraint	Yield Resulted	Yield Penalty	Delay Reduction
Average Over Benchmarks	90%	92.9%	7.1%	18.8%
Average Over Benchmarks	80%	88.1%	11.9%	20.2%

Table 11.14 Summary of Results for Timing Variation-Aware RTL Optimization [37, 38, 39, 48]

hand, a nine-cycle schedule is possible for a 7.6% tradeoff in the timing yield, i.e., to obtain a 92.4% of timing yield. The results of this variability-aware RTL timing optimization is presented in Table 11.14 [37, 38, 39, 48, 50]. For a yield constraint of 90%, on an average, a delay reduction of 18.8% is obtained with an yield loss of 7.1%. For a yield constraint of 80%, on an average, a delay reduction of 20.2% is obtained for a yield loss of 11.9%.

11.8.5 RTL Postsilicon Techniques for Variability Tolerance

The three RTL or architecture-level variability-aware methods presented in the above three subsections are essentially static methods. This subsection presents a postsilicon/dynamic method for variability mitigation based on ABB. The RTL design flow is shown in Fig. 11.51 that performs

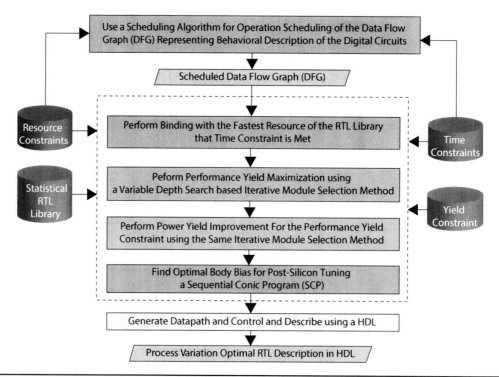

Figure 11.51 Flow for postsilicon tuning for variability-driven RTL design.

power yield improvement under performance yield and determines optimal body bias [17, 18, 50, 88]. In the first phase using a scheduling algorithm, a scheduled DFG is obtained. Then binding is performed with the fastest resource from the RTL library for specific time constraints. Using the iterative module selection method, the performance yield is maximized for the scheduled DFG with the fastest resource binding. In this design flow, the performance is estimated as using the following expression:

$$\text{Yield}_{\text{Timing}} = \text{Probability}\,(\text{Cost-Timing}_{\text{DFG}}\,(\text{Statistical-Maximum}_{\forall c \in N_{cc}} \quad (11.36)$$
$$(D_{\text{prop},c}\,(\mu_{D,c}, \sigma_{D,c}))) \le D_{\text{clk}})$$

where D_{clk} is the clock cycle time, $D_{\text{prop},c}\,(\mu_{D,c}, \sigma_{D,c})$ is the distribution of the delay of the functional unit scheduled in a clock cycle c that is calculated using STA. Then maximum delay is calculated using the tightness probability and moment matching techniques. The timing yield is then calculated. These calculations are similar to the methods adopted in the previous subsection on statistical variability-aware RTL optimization. The same algorithm is then used to optimize power yield for timing yield constraint. The distribution of power dissipation of the overall DFG and power yield is calculated as in the previous subsection on statistical variability-aware RTL optimization. In the last phase of the design flow, a sequential conic program (SCP) is used to determine the optimal V_{bs} to compensate the variability caused by the process variations. For this purpose, SCP minimizes the following:

$$\begin{aligned}\text{Minimize:} \quad & \text{Cost-Power}_{\text{DFG}}\left(P_{\text{total,DFG}}\left(\mu_{P,\text{DFG}}, \sigma_{P,\text{DFG}}\right)\right) \\ \text{Subjected To:} \quad & \text{Yield}_{\text{Timing}} > \tau_{\text{Timing-Yield}}\end{aligned} \quad (11.37)$$

In the above formulation, $\tau_{\text{Timing-Yield}}$ is the threshold in the timing yield as specified by the design engineer. The outcome of this optimization is a set of V_{bs} that will be used for postsilicon tuning, for example, using one V_{bs} for each functional unit. The experimental results over a set of RTL benchmark circuits are presented in Table 11.15 [50, 88].

Benchmark Circuits	Power Constraint	Yield$_{Power}$ for Timing Variation Awareness	Yield$_{Power}$ for Postsilicon Tuning	Yield$_{Power}$ Enhancement
Average Over Benchmarks	No	66%	88%	38%
Average Over Benchmarks	Yes	83%	92%	11%

TABLE 11.15 Summary of Results for Postsilicon Tuning for Variation-Aware RTL [50, 88]

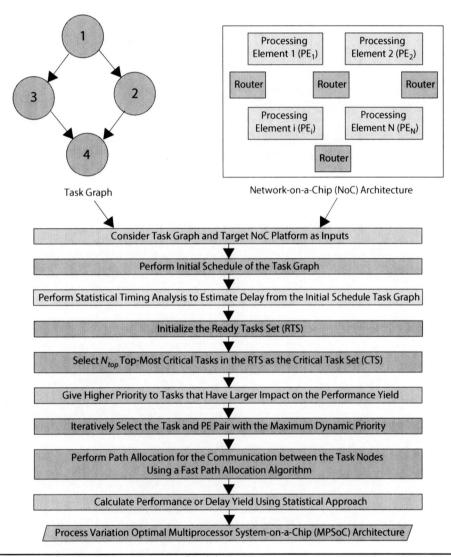

FIGURE 11.52 Flow for variation-aware system-level design.

11.9 System-Level Methods for Variability-Aware Digital Design

In this section, variability awareness at the highest level of abstraction, the system-level, is discussed for digital circuits or systems. A variability-aware system-level design flow is presented in Fig. 11.52 [24, 65, 87]. This system-level variability-aware flow considers a task graph and a target network-on-a-chip (NoC)-based multiprocessor SoC (MPSoC) architecture. A tasks graph is a directed acyclic graph $G(V,E)$ in which V is the set of vertices and E is the set of edges; similar to the DFG of architecture-level or RTL representation. Any vertex v_i in the set V of the task graph contains a task. Any edge or arc $e(i,j)$ represents the precedence constraint as well as communications between task v_i and task v_j. The weight

$w[e(i,j)]$ associated with an edge or arc $e(i,j)$ represents the amount of data that transfers from task v_i to task v_j. The MPSoC architecture consists of many components including the synschronous computational components, embedded processors (e.g., processing elements or PEs), hard intellectual property cores, and router. The MPSoC executes the task graph of a specific application.

The variability-aware scheduling of the task graph can be stated as follows: given a task graph and NoC-based MPSoC architecture, determine a mapping that maximizes the performance yield for specified performance constraints. In the variability-aware design flow, presented in Fig. 11.52, an initial schedule is performed at the beginning. A STA over the initial scheduled task graph is performed to estimate the delay. At this point, the design flow initializes the ready tasks set (RTS). A selected top task in the RTS is called a critical task set (CTS); critical tasks are the tasks that have higher probability to be on the critical path of the task graph. During the scheduling process, the tasks that have larger impact on the performance yield are given higher priority over other tasks. The performance or timing yield $Yield_{Timing}$ is calculated for the PEs using similar approach that is used in the previous section on RTL design flow. In the next phase of the design flow, the task-PE pair with maximum dynamic priority is selected iteratively. In the last phase of the design flow, path allocation is performed through the routers for the communication between the task nodes. The variability of the routers is incorporated while performing the routing path allocation. Once the solution is obtained, the performance yield analysis is again performed using a statistical method. In the above design flow, variability of both PE and router is performed assuming that delay have Gaussian distribution. The variability-aware system-level design flow resulted in yield of 90% to 100% for various embedded-system benchmarks.

11.10 An Adaptive Body Bias Method for Dynamic Process Variation Compensation

The ABB has already been discussed as a method for leakage reduction in a previous chapter on power reduction. In this section, a system is explained that uses ABB for process variation compensation. A combination of forward body bias (FBB) and reverse body bias (RBB) has been used for process variation compensation and performance tradeoff. The FBB and RBB effectively change the threshold voltage and performance tradeoffs as discussed in the power management section in the chapter on power reduction. The ABB technique for process variation compensation has been illustrated in Fig. 11.53 [73, 86].

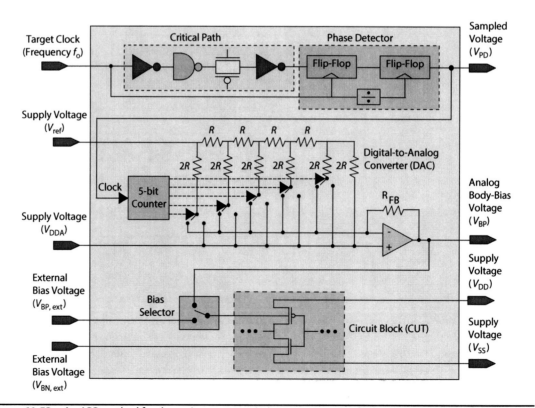

Figure 11.53 An ABB method for dynamic process variation compensation [86].

As presented in Fig. 11.53, the key idea is that the replica of the actual circuit is monitored for speed change and a circuit changes the body bias for process variation fluctuation compensation [73, 86]. The overall IC that is to be compensated is divided into multiple subsites or blocks and is controlled separately. A circuit block contains a complete ABB generator and control circuit. In addition, each block has a critical path circuit block. The critical path is a replica of the actual critical path of the circuit block. A signal of target clock frequency (f_o) is applied externally. A phase detector (PD) is used to sample the output signal of the critical path. In the process, the target clock period and critical path delay are compared by the PD. The output signal generated by the PD is the sample voltage (V_{PD}). The PD has two flip-flops in which the output signal of the first flip-flip is sampled by the second flip-flop that is clocked by a divided clock signal or a lower-frequency signal. In effect, the two flip-flops allow sufficient time for the ABB generator to stabilize as well as the critical path frequency adapts to the new body bias before the PD is updated again. The sampled voltage V_{PD} is used to clock a 5-bit digital counter whose value is essentially the desired body bias. The digital-to-analog converter (DAC) is used to convert the value of the digital counter to an analog voltage that is the body bias voltage (V_{BP}). Thus, V_{BP} is a function of target clock frequency as well as the supply voltage V_{ref} and V_{CCA}. The body bias voltage V_{BP} is used to bias the PMOS devices of the circuit block. The body bias voltage V_{BP} can be controlled by changing the supply voltage V_{ref} and V_{CCA} and by setting a counter control bit, thus, effectively implementing the ABB, including both FBB and RBB.

11.11 Parametric Variation Effect Mitigation in Clock Networks

In the synchronous ICs and systems, the clock network is very important. The clock network is a global interconnect that provides clock signal to all portions of the IC, for example, sequential elements such as flip-flops for synchronization [92, 93]. The clock network has strong impact on the performance of the ICs and systems in terms of overall power dissipation as well as speed of operation. For the clock network, robust operation in the presence of parametric variations is one of the important requirements. In this section, techniques for parameter variation mitigation in clock network is discussed.

One of the techniques for parameter variation effect mitigation in clock network is timing control using elastic clocking or dynamic clock period stretching [15, 16]. The key idea is the use of time-borrowing flip-flop (TBFF) and clock shifter to increase the clock frequency even at a high activation probability of critical paths. The concept of dynamic timing control is presented in Fig. 11.54(a) [15, 16]. The minimum clock period of an IC is limited by the maximum critical path delay, i.e., Clock Period $\geq T_{Clk,Max}$. If the clock period is reduced $0.8T_{Clk}$, then the paths whose delay is between $0.8T_{Clk}$ to T_{Clk} will have timing failures. To avoid this timing failure, time ($T_{Borrow} = 0.25T_{Clk,Min}$) is borrowed from the next stage of a pipeline through a TBFF as illustrated in the same figure. This time borrowing prevents the timing failure of the current pipeline stage, but leads to an increase in the path delay in the next pipeline stage and can be a cause for its timing failure. The timing failure of the next pipeline stage is then prevented by stretching the clock period by T_{Borrow} when timing borrowing by the previous stage is detected. Thus, when critical paths are not activated, the IC can operate at a clock period of $0.8T_{Clk}$ with the TBFF and clock shifter arrangement. The control mechanism for the dynamic clock stretching is presented in Fig. 11.54(b) [15, 16]. If the activation probability of the critical paths is 1, i.e., $Pr_c = 1$, the clock period stretching happens and the clock period is the maximum delay Clock Period $= T_{Clk,Max}$. On the other hand, if the activation probability of the critical paths is 0, i.e., $Pr_c = 0$, then the clock period stretching does not happen and the clock period is the minimum delay Clock Period $= T_{Clk,Min}$. In other words, the overall design operates at two different clock periods $T_{Clk,Max}$ and $T_{Clk,Min}$. The effective operating frequency of this IC can be calculated as follows [15, 16]:

$$f_{Clk,Eff} = \left(\frac{1}{Pr_c T_{Clk,Max} + (1 - Pr_c) T_{Clk,Min}} \right) \quad (11.38)$$

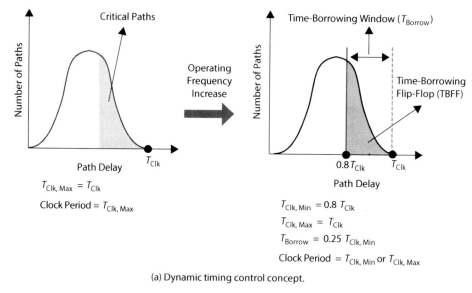

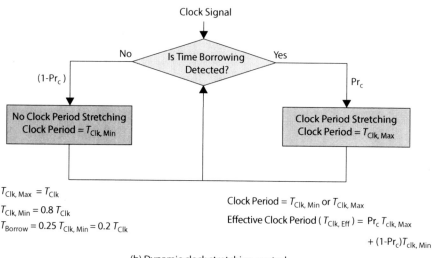

FIGURE 11.54 The concept of dynamic clock stretching [15, 16].

In the above expression, Pr_c is the activation probability of critical paths. The range of effective operating frequency is the following [15, 16]:

$$\left(\frac{1}{T_{\text{Clk,Max}}}\right) \leq f_{\text{Clk,Eff}} \leq \left(\frac{1}{T_{\text{Clk,Min}}}\right) \tag{11.39}$$

Another approach that uses clock stretching is now briefly discussed. The key idea of this approach is that a dynamic delay detection logic is used to identify the uncertainty in the delay due to parametric variations and then stretches the clock in the destination sequential elements. The delay detection and clock stretching logic (CSL) is used in the critical path and near-critical path cells only. In the presence of timing variations, the CSL delays the instant of the active clock edge trigger in the critical path on the destination sequential elements. The logic level design of the unit that performs detection of timing variations and then performs dynamic clock stretching is shown in Fig. 11.55 [45]. In essence, additional logic consisting of a D flip-flop, a delay-flag setting XOR, and a clock stretching buffer is used between the source and destination sequential

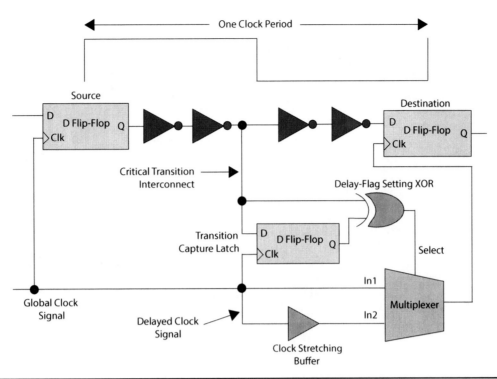

FIGURE 11.55 The logic-level design of the unit for dynamic clock stretching for dynamic mitigation for timing variation [45].

elements. In the absence of timing variations, the critical path operates normally in a case in which the clock signal passes through the In1 of the multiplexer to the destination sequential element. In the presence of timing variations, the two inputs to the delay-flag setting XOR will be different, thus generating an output of "1." In such a situation, select signal selects In2 signal from the multiplexer. Thus, through In2, the delayed clock signal is sent to the destination of the destination D flip-flop. Thus, the proposed unit dynamically selects between the normal clock and delayed clock. It stretches the clock for the destination sequential elements in the case of timing variations in the critical path.

11.12 Statistical Methods for Yield Analysis

In the above sections, the methods presented deal with power and/or timing optimizations. The statistical methods estimate the power dissipation or delay as PDF. Then, the methods perform optimization to ensure that power and timing are statistically optimized to meet constraints. In essence these optimization methods work for maximization of most of the circuit and systems meeting the constraints. A complementary question arises how to determine how many circuits/systems will fail? This question may not always be easy to answer. For example, let us consider chip yield of 99.99%, in other words, no more than one chip per 10,000 should fail; how to find that one chip out of the 10,000 chips. This is a rare event, extreme event, or statistically unlikely event [21, 68, 79, 80, 81, 82]. On an average 100 million simulations are needed for statistical simulation of a 5σ event. The traditional Monte Carlo simulation is not suitable for such a scenario due to long computational time. This section discusses the simulation of rare or extreme events and estimates the statistics of such events.

As a specific example of rare event analysis method, statistical blockade (SB) method has been discussed [78, 79, 80, 81, 82]. The concept of SB has been depicted in Fig. 11.56 with an SRAM circuit example [81, 82]. Even though the SRAM circuit is used as a case study circuit, the SB can be applied for other designs, e.g., arithmetic circuits [81, 82]. As presented in the figure, device-level

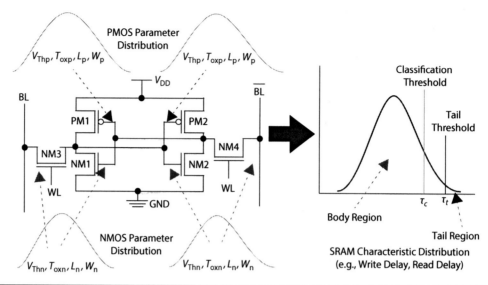

Figure 11.56 The concept of SB using SRAM circuit as a case study [79, 81, 82].

parameter variations can cause variations in the characteristics of the SRAM circuit. The PDF of the SRAM characteristic illustrated in the figures can be divided into body region and tail region. The traditional Monte Carlo simulations capture the body region more easily than the tail region. For such process variations, the SB method provides the ability to the design engineers to efficiently determine the failure probability of the SRAM circuits. An SB method addresses two issues for fast and accurate evaluation of the SRAM circuits. The tail threshold (τ_t) indicates the beginning of the tail region of the distribution. The classification threshold (τ_c) is a specific value of the SRAM performance metric to be used by a classifier. This is a relaxed boundary of the classifier that is used to block most unwanted sample points from simulation. The key idea of the SB method is the following. It is a fact that generating random Monte Carlo sample points is way faster and computationally cheaper than simulating the points using an analog/SPICE simulator. Thus, if the simulation points of the traditional Monte Carlo, which are unlikely to fall in the tail region, are not simulated (i.e., blocked) rather generated, then the overall simulation process will be faster. The decrease in the simulation time or speedup in the simulation is very significant as most of the analog/SPICE simulations are blocked and only the rare events that are very small in number are simulated using the analog/SPICE simulator. Thus, the name SB originated from blocking activity of the method.

The flow of an SB method is shown in Fig. 11.57 [79, 81, 82]. In this method, first the training samples are generated using the random variables that accurately model the device parameters. Using the accurate distribution of the random variables, a small number of Monte Carlo simulations are performed for the IC. A tail threshold (τ_t) is selected based on the simulation data of the IC. The minimal value of the classification threshold (τ_c) as well as the minimum number of initial sampling (N_{min}) required to ensure that all tail points are covered can be automatically determined by using available algorithms [81]. At this point, a classifier is trained using the simulation data and the specified classification threshold τ_c. An example of a classifier that has been tested for SB method is a support vector machine (SVM) classifier [79]. However, a large variety of classifiers that are available in the literature can be used for this purpose. It can be noted that the time consumed for training the classifier as well as the classification process is negligible as compared to the overall simulation time. At this phase of the design flow, larger samples points are generated using the random variables for the device parameters of the IC. The sample points that fall in the body region are discarded. The IC is simulated using an analog/SPICE simulator for the tail points only. The tail points for a specified tail threshold τ_t, called the true tail points, are obtained from the simulation data. At this point extreme value theory (EVT) is used to obtain a tail model of the tail distribution [71, 79]. The sample point in the tail region does not follow Gaussian distribution after the classification of the sample points is performed. Thus, there is a need to determine the generalized

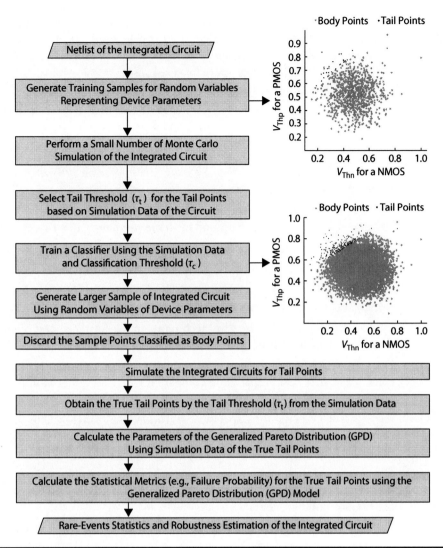

Figure 11.57 Flow for SB approach for circuit evaluation [79, 81, 82].

Pareto distribution (GPD). The cumulative distribution function (CDF) of the GPD is then considered as the tail model. The coefficients of the GPD such as the scale parameter (α_{GPD}) and the shape parameter (β_{GPD}) are calculated for the tail model. The error measures such as maximum likelihood estimation (MLE) are used for best fitting of the GPD.

11.13 Questions

11.1 Discuss different types of parameter variations possible in a nanoelectronic circuit or system.

11.2 Discuss different types of process variations possible in a nanoelectronic IC.

11.3 Discuss the concept of variability-aware design flow for analog and digital ICs.

11.4 Discuss the concept of dynamic compensation of parameter variations in nanoelectronic circuits and systems.

11.5 What is a Monte Carlo method? Discuss any three Monte Carlo methods.

11.6 Discuss any three selected statistical distributions or PDFs that can be used for nanoelectronic variability modeling.

11.7 Discuss the different device-level parameters that are affected by the nanoelectronic process variations. Discuss what specific statistical distributions should be used to model their variability.

11.8 Consider a three-stage ring oscillator circuit for a specific nanoelectronic technology. Extract its netlist from the schematic design. Perform Monte Carlo for 500, 1000, and 10,000 runs considering width of NMOS and PMOS as device parameters and center frequency as the characteristic. Plot the design space for three cases.

11.9 Consider the physical design of the above ring oscillator for a specific technology node, extract the netlist including all the parasitics. Repeat the above Monte Carlo experiments and plot the design space. Comment on the accuracy and simulation time for schematic versus layout designs.

11.10 Repeat the above for a LC-VCO circuit schematic design for a technology of your choice.

11.11 Repeat the above for the LC-VCO physical design for a technology of your choice.

11.12 Briefly discuss the DOE method.

11.13 Consider a three-stage ring oscillator design for a specific nanoelectronic technology. Use DOE to demonstrate its design exploration for center frequency in terms of NMOS and PMOS sizes.

11.14 Using DOE obtain predictive equations for the center frequency of the above ring oscillator.

11.15 Briefly discuss the corner method considering the FEOL corners.

11.16 Consider a LC-VCO design for a specific application. Using corner-based analysis, identify its worst-case variability impact assuming ± change in threshold voltage.

11.17 Discuss the DOE-MC. Use this technique to perform process variation analysis of the three-stage ring oscillator.

11.18 For a LC-VCO design for a specific application, use LHS-MC to sample 1000 points in its center frequency design space and compare it with traditional Monte Carlo in terms of speed and accuracy.

11.19 For a LC-VCO design for a specific application, use MLHS-MC to sample 1000 points in its center frequency design space and compare it with LHS-MC as well as traditional Monte Carlo in terms of speed and accuracy.

11.20 Consider an APS design for a specific nanoelectronic technology. Perform statistical variability analysis using a method for three of its characteristics. Discuss the nature of statistical distributions of these characteristics.

11.21 Consider the 13-stage current-starved VCO for a specific application. Perform its schematic-level and layout-level design for a specific technology node. Perform statistical process variation analysis for its center frequency and compare the schematic-level versus layout-level distributions obtained.

11.22 Implement the adaptive linear voltage regulator-based process variation tolerant LC-VCO at the schematic level for a specific nanoelectronic technology and perform study of the impact of process variation on its center frequency characteristics.

11.23 Consider a 6-transistor and an 8-transistor SRAM cell. Perform statistical process variation analysis for their power and SNM characteristics and compare the distributions.

11.24 Consider a 7T-SRAM design. Using the DOE approach, obtain the predictive equations for its power and SNM characteristics.

11.25 Do a survey of recent literature on variability-aware system-level optimization. Pick one of the articles and write a one-page summary of it.

11.26 Do a survey of recent literature on variability-aware RTL optimization. Pick one of the articles and write a one-page summary of it.

11.27 Do a survery of recent literature on variability-aware circuit-level optimization. Pick one article and write a one-page summary of it.

11.28 What is an ABB in nanoelectronic circuits and systems? Discuss with an example how it can be used for process variation compensation.

11.29 Discuss the concept of dynamic clock stretching. How can it be used for process variation mitigation?

11.30 Briefly discuss the method of SB for yield analysis in nanoelectronic circuits and systems.

11.14 References

1. Design of Experiments (DOE). https://www.moresteam.com/toolbox/design-of-experiments.cfm. Accessed on 27th May 2014.
2. Monte Carlo Analysis. http://www.tutorialspoint.com/management_concepts/monte_carlo_analysis.htm. Accessed on 26th May 2014.
3. NIST/SEMATECH e-Handbook of Statistical Methods. National Institute of Science and Technology (NIST) (2003). Accessed on 27th May 2014.
4. How a Chip Is Made. http://apcmag.com/picture-gallery-how-a-chip-is-made.htm (2009). Accessed on 25th May 2014.
5. Design of Experiments - Taguchi Experiments. http://www.qualitytrainingportal.com/resources/doe/taguchi_concepts.htm (2012). Accessed on 28th May 2014.
6. Anderson, H.L.: Metropolis, Monte Carlo, and the MANIAC. *Los Alamos Science* (LAUR-86-2600), 96–107 (1986). Accessed on 26th May 2014.
7. Ang, K.C.M., Chia, M.Y.W., Li, D.P.M.: A Process Compensation Technique for Integrated VCO. In: *Digest of Papers IEEE Radio Frequency Integrated Circuits (RFIC) Symposium*, pp. 591–594 (2004). DOI 10.1109/RFIC.2004.1320690.
8. Babanova, S., Artyushkova, K., Ulyanova, Y., Singhal, S., Atanassov, P.: Design of Experiments and Principal Component Analysis as Approaches for Enhancing Performance of Gas-diffusional Air-breathing Bilirubin Oxidase Cathode. *Journal of Power Sources* **245**, 389–397 (2014). DOI http://dx.doi.org/10.1016/j.jpowsour.2013.06.031.
9. Banerjee, S., Mathew, J., Mohanty, S.P., Pradhan, D.K., Ciesielski, M.J.: A Variation-Aware Taylor Expansion Diagram-based Approach for Nano-CMOS Register-Transfer Level Leakage Optimization. *Journal of Low Power Electronics* **7**(4), 471–481 (2011).
10. Banerjee, S., Mathew, J., Pradhan, D., Mohanty, S.P., Ciesielski, M.: Variation-aware TED-based Approach for Nano-CMOS RTL Leakage Optimization. In: *Proceedings of the 24th International Conference on VLSI Design (VLSID)*, pp. 304–309 (2011). DOI 10.1109/VLSID.2011.40.
11. Blaauw, D., Chopra, K., Srivastava, A., Scheffer, L.: Statistical Timing Analysis: From Basic Principles to State of the Art. *IEEE Transactions on Computer-Aided Design of Integrated Circuits and Systems* **27**(4), 589–607 (2008). DOI 10.1109/TCAD.2007.907047.
12. Bowman, K.A., Tokunaga, C., Tschanz, J.W., Karnik, T., De, V.K.: Adaptive and Resilient Circuits for Dynamic Variation Tolerance. *IEEE Design Test* **30**(6), 8–17 (2013). DOI 10.1109/MDAT.2013.2267958.
13. Bowman, K.A., Tschanz, J.W., Lu, S.L., Aseron, P.A., Khellah, M.M., Raychowdhury, A., Geuskens, B.M., Tokunaga, C., Wilkerson, C.B., Karnik, T., De, V.K.: A 45 nm Resilient Microprocessor Core for Dynamic Variation Tolerance. *IEEE Journal of Solid-State Circuits* **46**(1), 194–208 (2011). DOI 10.1109/JSSC.2010.2089657.
14. Bukhori, M.F., Brown, A.R., Roy, S., Asenov, A.: Simulation of Statistical Aspects of Reliability in Nano CMOS Transistors. In: *Proceedings of the IEEE International Integrated Reliability Workshop*, pp. 82–85 (2009). DOI 10.1109/IRWS.2009.5383028.
15. Chae, K., Lee, C.H., Mukhopadhyay, S.: Timing Error Prevention Using Elastic Clocking. In: *Proceedings of the IEEE International Conference on IC Design Technology (ICICDT)*, pp. 1–4 (2011). DOI 10.1109/ICICDT.2011.5783192.
16. Chae, K., Mukhopadhyay, S., Lee, C.H., Laskar, J.: A Dynamic Timing Control Technique Utilizing Time Borrowing and Clock Stretching. In: *Proceedings of the IEEE Custom Integrated Circuits Conference (CICC)*, pp. 1–4 (2010). DOI 10.1109/CICC.2010.5617392.
17. Chen, Y., Wang, Y., Xie, Y., Takach, A.: Parametric Yield-Driven Resource Binding in High-level Synthesis with Multi-V_{th}/V_{dd} Library and Device Sizing. *Journal of Electrical and Computer Engineering* (2012). DOI http://dx.doi.org/10.1155/2012/105250.
18. Chen, Y., Xie, Y., Wang, Y., Takach, A.: Parametric Yield Driven Resource Binding in Behavioral Synthesis With Multi-V_{th}/V_{dd} Library. In: *Proceedings of the 15th Asia and South Pacific Design Automation Conference (ASP-DAC)*, pp. 781–786 (2010). DOI 10.1109/ASPDAC.2010.5419783.
19. Cui, S., Weile, D.S.: Application of a Parallel Particle Swarm Optimization Scheme to the Design of Electromagnetic Absorbers. *IEEE Transactions on Antennas and Propagation* **53**(11), 3616–3624 (2005). DOI 10.1109/TAP.2005.858866.
20. De, V.: Keynote: Variation-Tolerant Adaptive and Resilient Designs in Nanoscale CMOS. In: *Proceedings of the IEEE 19th International Symposium on Asynchronous Circuits and Systems (ASYNC)*, pp. xv–xv (2013). DOI 10.1109/ASYNC.2013.35.
21. Doorn, T.S., ter Maten, E.J.W., Croon, J.A., Di Bucchianico, A., Wittich, O.: Importance Sampling Monte Carlo Simulations for Accurate Estimation of SRAM Yield. In: *Proceedings of the European Solid-State Circuits Conference*, pp. 230–233 (2008). DOI 10.1109/ESSCIRC.2008.4681834.
22. Fang, K.T., Li, R., Sudjianto, A.: *Design and Modeling for Computer Experiments*. Chapman and Hall/CRC, 23–25 Blades Court, London SW15 2NU, UK (2006).
23. Forzan, C., Pandini, D.: Statistical Static Timing Analysis: A Survey. *Integration, the VLSI Journal* **42**(3), 409–435 (2009). DOI http://dx.doi.org/10.1016/j.vlsi.2008.10.002.

24. Garg, S., Marculescu, D.: System-level Leakage Variability Mitigation for MPSoC Platforms Using Body-Bias Islands. *IEEE Transactions on Very Large Scale Integration (VLSI) Systems* **20**(12), 2289–2301 (2012). DOI 10.1109/TVLSI.2011.2171512.
25. Garitselov, O.: *Metamodeling-Based Fast Optimization of Nanoscale AMS-SoCs*. Ph.D. thesis, Department of Computer Science and Engineering, University of North Texas, Denton (2012).
26. Garitselov, O., Mohanty, S.P., Kougianos, E.: A Comparative Study of Metamodels for Fast and Accurate Simulation of Nano-CMOS Circuits. *IEEE Transactions on Semiconductor Manufacturing* **25**(1), 26–36 (2012). DOI 10.1109/TSM.2011.2173957.
27. Garitselov, O., Mohanty, S.P., Kougianos, E., Zheng, G.: Particle Swarm Optimization over Non-Polynomial Metamodels for Fast Process Variation Resilient Design of Nano-CMOS PLL. In: *Proceedings of the ACM Great Lakes Symposium on VLSI*, pp. 255–258 (2012).
28. Ghai, D.: *Variability Aware Low-Power Techniques for Nanoscale Mixed-signal Circuits*. Ph.D. thesis, University of North Texas, Denton (2009).
29. Ghai, D., Mohanty, S.P., Kougianos, E.: A Process and Supply Variation Tolerant Nano-CMOS Low Voltage, High Speed, A/D Converter for System-on-Chip. In: *Proceedings of the 18th ACM Great Lakes Symposium on VLSI*, pp. 47–52 (2008).
30. Ghai, D., Mohanty, S.P., Kougianos, E.: Parasitic Aware Process Variation Tolerant Voltage Controlled Oscillator (VCO) Design. In: *Proceedings of the 9th International Symposium on Quality of Electronic Design*, pp. 330–333 (2008).
31. Ghai, D., Mohanty, S.P., Kougianos, E.: Design of Parasitic and Process-variation Aware Nano-CMOS RF Circuits: A VCO Case Study. IEEE Transactions on VLSI Systems. **17**(9), 1339–1342 (2009).
32. Ghai, D., Mohanty, S.P., Kougianos, E.: Unified P4 (Power-Performance-Process-Parasitic) Fast Optimization of a Nano-CMOS VCO. In: *Proceedings of the 19th ACM Great Lakes Symposium on VLSI*, pp. 303–308 (2009).
33. Ghai, D., Mohanty, S.P., Kougianos, E.: Variability-aware Optimization of Nano-CMOS Active Pixel Sensors using Design and Analysis of Monte Carlo Experiments. In: *Proceedings of the 10th International Symposium on Quality of Electronic Design*, pp. 172–178 (2009).
34. Ghai, D., Mohanty, S.P., Kougianos, E.: A Variability Tolerant System-on-Chip Ready Nano-CMOS Analog-to-Digital Converter (ADC). *International Journal of Electronics (IJE)* **97**(4), 421–440 (2010).
35. Ghai, D., Mohanty, S.P., Kougianos, E., Patra, P.: A PVT Aware Accurate Statistical Logic Library for High-κ Metal-gate Nano-CMOS. In: *Proceedings of the 10th International Symposium on Quality of Electronic Design*, pp. 47–54 (2009).
36. Gupta, V., Anis, M.: Statistical Design of the 6T SRAM Bit Cell. *IEEE Transactions on Circuits and Systems I: Regular Papers* **57**(1), 93–104 (2010). DOI 10.1109/TCSI.2009.2016633.
37. Jung, J., Kim, T.: Timing Variation-aware High-level Synthesis. In: *Proceedings of the IEEE/ACM International Conference on Computer-Aided Design*, pp. 424–428 (2007). DOI 10.1109/ICCAD.2007.4397302.
38. Jung, J., Kim, T.: Timing Variation-aware High Level Synthesis: Current Results and Research Challenges. In: *Proceedings of the IEEE Asia Pacific Conference on Circuits and Systems*, pp. 1004–1007 (2008). DOI 10.1109/APCCAS.2008.4746194.
39. Jung, J., Kim, T.: Scheduling and Resource Binding Algorithm Considering Timing Variation. *IEEE Transactions on Very Large Scale Integration (VLSI) Systems* **19**(2), 205–216 (2011). DOI 10.1109/TVLSI.2009.2031676.
40. Kalos, M.H., Whitlock, P.A.: *Monte Carlo Methods*. Wiley (2008).
41. Keramat, M., Kielbasa, R.: Latin Hypercube Sampling Monte Carlo Estimation of Average Quality Index for Integrated Circuits. *Analog Integrated Circuits and Signal Processing* **14**(1/2), 131–142 (1997). DOI 10.1007/978-1-4615-6101-9_11. URL http://dx.doi.org/10.1007/978-1-4615-6101-9_11.
42. Khan, A., Mohanty, S.P., Kougianos, E.: Statistical Process Variation Analysis of a Graphene FET based LC-VCO for WLAN Applications. In: *Proceedings of the 15th IEEE International Symposium on Quality Electronic Design*, pp. 569–574 (2014).
43. Kocher, M., Rappitsch, G.: Statistical Methods for the Determination of Process Corners. In: *Proceedings of the International Symposium on Quality Electronic Design*, pp. 133–137 (2002). DOI 10.1109/ISQED.2002.996713.
44. Kougianos, E., Mohanty, S.P.: Impact of Gate-Oxide Tunneling on Mixed-Signal Design and Simulation of a Nano-CMOS VCO. *Microelectronics Journal* **40**(1), 95–103 (2009).
45. Mahalingam, V., Ranganathan, N., Hyman Jr., R.: Dynamic Clock Stretching for Variation Compensation in VLSI Circuit Design. *ACM Journal on Emerging Technologies in Computing Systems (JETC)* **8**(3), 16:1–16:13 (2012). DOI 10.1145/2287696.2287699.
46. Mascagni, M.: Monte Carlo Methods: Early History and The Basics. http://www.cs.fsu.edu/mascagni/MC_Basics.pdf (2011). Accessed on 11th June 2014.
47. Mizuno, T., Okamura, J., Toriumi, A.: Experimental Study of Threshold Voltage Fluctuation Due to Statistical Variation of Channel Dopant Number in MOSFETs. *IEEE Transactions on Electron Devices* **41**(11), 2216–2221 (1994).
48. Mohanty, S.P.: Unified Challenges in Nano-CMOS High-Level Synthesis. In: *Proceedings of the 22nd International Conference on VLSI Design*, pp. 531–531 (2009).

49. Mohanty, S.P.: DfX for Nanoelectronic Embedded Systems. In: *Keynote Address, International Conference on Control, Automation, Robotics and Embedded System* (2013). Accessed on 25th May 2014.
50. Mohanty, S.P., Gomathisankaran, M., Kougianos, E.: Variability-aware Architecture Level Optimization Techniques for Robust Nanoscale Chip Design. *Computers & Electrical Engineering* **40**(1), 168–193 (2014). DOI http://dx.doi.org/10.1016/j.compeleceng.2013.11.026.
51. Mohanty, S.P., Kougianos, E.: Impact of Gate Leakage on Mixed Signal Design and Simulation of Nano-CMOS Circuits. In: *Proceedings of the 13th NASA Symposium on VLSI Design*, vol. paper # 2.4, 6 pages (2007).
52. Mohanty, S.P., Kougianos, E.: Simultaneous Power Fluctuation and Average Power Minimization during Nano-CMOS Behavioral Synthesis. In: *Proceedings of the 20th International Conference on VLSI Design*, pp. 577–582 (2007).
53. Mohanty, S.P., Kougianos, E.: Incorporating Manufacturing Process Variation Awareness in Fast Design Optimization of Nanoscale CMOS VCOs. *IEEE Transactions on Semiconductor Manufacturing* **27**(1), 22–31 (2014). DOI 10.1109/TSM.2013.2291112.
54. Mohanty, S.P., Ranganathan, N., Kougianos, E., Patra, P.: *Low-power High-level Synthesis for Nanoscale CMOS Circuits*. Springer (2008).
55. Mohanty, S.P., Singh, J., Kougianos, E., Pradhan, D.K.: Statistical DOE-ILP Based Power-Performance-Process (P3) Optimization of Nano-CMOS SRAM. *Integration* **45**(1), 33–45 (2012).
56. Montgomery, D.C.: *Design and Analysis of Experiments*. John Wiley & Sons (2008).
57. Mostafa, H., Anis, M., Elmasry, M.: Adaptive Body Bias for Reducing the Impacts of NBTI and Process Variations on 6T SRAM Cells. *IEEE Transactions on Circuits and Systems I: Regular Papers* **58**(12), 2859–2871 (2011). DOI 10.1109/TCSI.2011.2158708.
58. Mukherjee, V., Mohanty, S.P., Kougianos, E.: A Dual Dielectric Approach for Performance Aware Gate Tunneling Reduction in Combinational Circuits. In: *Proceedings of the IEEE International Conference on Computer Design: VLSI in Computers and Processors*, pp. 431–436 (2005). DOI 10.1109/ICCD.2005.5.
59. Mukhopadhyay, S., Kim, K., Mahmoodi, H., Roy, K.: Design of a Process Variation Tolerant Self-Repairing SRAM for Yield Enhancement in Nanoscaled CMOS. *IEEE Journal of Solid-State Circuits* **42**(6), 1370–1382 (2007). DOI 10.1109/JSSC.2007.897161.
60. Nassif, S.R.: Modeling and Analysis of Manufacturing Variations. In: *Proceedings of the IEEE Conference on Custom Integrated Circuits*, pp. 223–228 (2001).
61. Okobiah, O.: *Exploring Process-variation Tolerant Design of Nanoscale Sense Amplifier Circuits*. Master's thesis, Department of Computer Science and Engineering, University of North Texas, Denton, TX (2010).
62. Okobiah, O., Mohanty, S.P., Kougianos, E., Poolakkaparambil, M.: Towards Robust Nano-CMOS Sense Amplifier Design: A Dual-threshold Versus Dual-oxide Perspective. In: *Proceedings of the 21st ACM Great Lakes Symposium on VLSI*, pp. 145–150 (2011).
63. Orshansky, M., Chen, J.C., Hu, C.: A Statistical Performance Simulation Methodology for VLSI Circuits. In: *Proceedings of the Design Automation Conference*, pp. 402–407 (1998).
64. Orshansky, M., Nassif, S., Boning, D.: *Design for Manufacturability and Statistical Design: A Constructive Approach. Integrated Circuits and Systems*. Springer (2007).
65. Palermo, G., Silvano, C., Zaccaria, V.: Variability-aware Robust Design Space Exploration of Chip Multiprocessor Architectures. In: *Proceedings of the Asia and South Pacific Design Automation Conference*, pp. 323–328 (2009). DOI 10.1109/ASPDAC.2009.4796501.
66. Pengelly, J.: Monte Carlo Methods. Tech. rep., Department of Computer Science, University of Otago, Dunedin, New Zealand (2002). Accessed on 26th May 2014.
67. Poli, R.: Analysis of the Publications on the Applications of Particle Swarm Optimisation. *Journal of Artificial Evolution and Applications* **2008** (685175), 10 pages (2008). DOI 10.1155/2008/685175.
68. Qazi, M., Tikekar, M., Dolecek, L., Shah, D., Chandrakasan, A.: Loop Flattening & Spherical Sampling: Highly Efficient Model Reduction Techniques for SRAM Yield Analysis. In: *Proceedings of the Design, Automation Test in Europe Conference Exhibition*, pp. 801–806 (2010).
69. Rappitsch, G., Seebacher, E., Kocher, M., Stadlober, E.: SPICE Modeling of Process Variation Using Location Depth Corner Models. *IEEE Transactions on Semiconductor Manufacturing* **17**(2), 201–213 (2004). DOI 10.1109/TSM.2004.826940.
70. Reid, D., Millar, C., Roy, G., Roy, S., Asenov, A.: Analysis of Threshold Voltage Distribution Due To Random Dopants: A 100000-Sample 3-D Simulation Study. *IEEE Transactions on Electron Devices* **56**(10), 2255–2263 (2009). DOI 10.1109/TED.2009.2027973.
71. Resnick, S.I.: *Extreme Values, Regular Variation and Point Processes*. Springer-Verlag, New York (1987).
72. Roy, R.K.: *A Primer on the Taguchi Method*, second edn. Society of Manufacturing Engineers (2010).
73. Sakurai, T.: Low Power Digital Circuit Design. In: *Proceeding of the 34th European Solid-State Device Research Conference*, pp. 11–18 (2004). DOI 10.1109/ESSDER.2004.1356476.
74. Sarivisetti, G.: *Design and Optimization of Components in a 45 nm CMOS Phase Locked Loop*. Master's thesis, Department of Computer Science and Egniineering, University of North Texas, Denton, TX, USA (2006).
75. Seo, J.H., Im, C.H., Heo, C.G., kwang Kim, J., Jung, H.K., Lee, C.G.: Multimodal Function Optimization Based on Particle Swarm Optimization. *IEEE Transactions on Magnetics* **42**(4), 1095–1098 (2006). DOI 10.1109/TMAG.2006.871568.

76. Shi, X., Yeo, K.S., Ma, J.G., Do, A.V., Li, E.: Scalable Model of On-Wafer Interconnects for High-Speed CMOS ICs. *IEEE Transactions on Advanced Packaging* **29**(4), 770–776 (2006). DOI 10.1109/TADVP.2006.884781.
77. Singh, J., Mathew, J., Mohanty, S.P., Pradhan, D.K.: Statistical Analysis of Steady State Leakage Currents in Nano-CMOS Devices. In: *Proceedings of the 25th IEEE Norchip Conference (NORCHIP)*, pp. 1–4 (2007).
78. Singhee, A., Rutenbar, R.A.: Statistical Blockade: A Novel Method for Very Fast Monte Carlo Simulation of Rare Circuit Events, and its Application. In: *Proceedings of the Design, Automation Test in Europe Conference Exhibition*, pp. 1–6 (2007). DOI 10.1109/DATE.2007.364490.
79. Singhee, A., Rutenbar, R.A.: Statistical Blockade: Very Fast Statistical Simulation and Modeling of Rare Circuit Events and Its Application to Memory Design. *IEEE Transactions on Computer-Aided Design of Integrated Circuits and Systems* **28**(8), 1176–1189 (2009). DOI 10.1109/TCAD.2009.2020721.
80. Singhee, A., Wang, J., Calhoun, B.H., Rutenbar, R.A.: Recursive Statistical Blockade: An Enhanced Technique for Rare Event Simulation with Application to SRAM Circuit Design. In: *Proceedings of the International Conference on VLSI Design*, pp. 131–136 (2008).
81. Sun, L., Mathew, J., Pradhan, D.K., Mohanty, S.P.: Algorithms for Rare Event Analysis in Nano-CMOS Circuits Using Statistical Blockade. In: *Proceedings of the International SoC Design Conference*, pp. 162–165 (2010). DOI 10.1109/SOCDC.2010.5682948.
82. Sun, L., Mathew, J., Pradhan, D.K., Mohanty, S.P.: Enhanced Statistical Blockade Approaches for Fast Robustness Estimation and Compensation of Nano-CMOS Circuits. Special Issue on Power, Parasitics, and Process-Variation (P3) Awareness in Mixed-Signal Design. *ASP Journal of Low Power Electronics* **8**(3), 261–269 (2012).
83. Tanguay, L.F., Sawan, M.: Process Variation Tolerant LC-VCO Dedicated to Ultra-Low Power Biomedical RF Circuits. In: *Proceedings of the 9th International Conference on Solid-State and Integrated-Circuit Technology*, pp. 1585–1588 (2008). DOI 10.1109/ICSICT.2008.4734869.
84. Thakral, G.: *Process-voltage-temperature Aware Nanoscale Circuit Optimization*. Ph.D. thesis, Department of Computer Science and Engineering, University of North Texas, Denton (2010).
85. Thakral, G., Mohanty, S., Ghai, D., Pradhan, D.: P3 (Power-Performance-Process) Optimization of Nano-CMOS SRAM Using Statistical DOE-ILP. In: *Proceedings of the 11th International Symposium on Quality Electronic Design (ISQED)*, pp. 176–183 (2010). DOI 10.1109/ISQED.2010.5450470.
86. Tschanz, J.W., Kao, J.T., Narendra, S.G., Nair, R., Antoniadis, D.A., Chandrakasan, A.P., De, V.: Adaptive Body Bias for Reducing Impacts of Die-to-Die and within-Die Parameter Variations on Microprocessor Frequency and Leakage. *IEEE Journal of Solid-State Circuits* **37**(11), 1396–1402 (2002). DOI 10.1109/JSSC.2002.803949.
87. Wang, F., Chen, Y., Nicopoulos, C., Wu, X., Xie, Y., Vijaykrishnan, N.: Variation-aware Task and Communication Mapping for MPSoC Architecture. *IEEE Transactions on Computer-Aided Design of Integrated Circuits and Systems* **30**(2), 295–307 (2011). DOI 10.1109/TCAD.2010.2077830.
88. Wang, F., Wu, X., Xie, Y.: Variability-driven Module Selection with Joint Design Time Optimization and Post-Silicon Tuning. In: *Proceedings of the Asia and South Pacific Design Automation Conference*, pp. 2–9 (2008). DOI 10.1109/ASPDAC.2008.4483963.
89. Wyss, G.D., Jorgensen, K.H.: *A User's Guide to LHS: Sandia's Latin Hypercube Sampling Software*. Tech. rep., Risk Assessment and Systems Modeling Department, Sandia National Laboratories, PO Box 5800, Albuquerque, NM 87185-0747 (1998). Accessed on 16th June 2014.
90. Xie, Y., Chen, Y.: Statistical High-Level Synthesis under Process Variability. *IEEE Design Test of Computers* **26**(4), 78–87 (2009). DOI 10.1109/MDT.2009.85.
91. Zhao, W., Cao, Y.: New Generation of Predictive Technology Model for sub-45 nm Design Exploration. In: *Proceedings of the International Symposium on Quality Electronic Design*, pp. 585–590 (2006).
92. Zhao, X., Tolbert, J., Liu, C., Mukhopadhyay, S., Lim, S.K.: Variation-aware Clock Network Design Methodology for Ultra-Low Voltage (ULV) Circuits. In: *Proceedings of the International Symposium on Low Power Electronics and Design (ISLPED)*, pp. 9–14 (2011). DOI 10.1109/ISLPED.2011.5993615.
93. Zhao, X., Tolbert, J., Mukhopadhyay, S., Lim, S.K.: Variation-aware Clock Network Design Methodology for Ultralow Voltage (ULV) Circuits. *IEEE Transactions on Computer-Aided Design of Integrated Circuits and Systems* **31**(8), 1222–1234 (2012). DOI 10.1109/TCAD.2012.2190825.

CHAPTER 12
Metamodel-Based Fast AMS-SoC Design Methodologies

12.1 Introduction

The design of various components of analog/mixed-signal system-on-a-chip (AMS-SoC) is quite complex and time intensive, specifically at the optimization and physical design stages [14, 16, 23, 49, 59, 83]. The parasitics present in the physical designs of AMS-SoC components that influence the circuits' characteristics have significant impact on the numerical simulations. In order to reduce the design effort, design cycle, nonrecurrent (NRE) cost, and overall the chip cost, fast and accurate AMS-SoC design optimization flow is needed. Design engineers can resort to various alternatives including the following:

1. Reduction of complexity of models (e.g., fast SPICE models, look-up tables) used in the circuit simulations.
2. Reduction of the simulation time by using fast numerical solvers.
3. Reduction of the design optimization time.
4. Reduction of the number of layout steps.

The fast SPICE models, look-up tables, reduced-order models, and macromodels speed up the simulation process. Macromodels are widely used in circuit simulations and verifications in the framework of EDA tools. Fast numerical solvers such as parallel solvers can speed up the simulation and design exploration as discussed in Chapter 9 on mixed-signal circuit and system simulation. Use of algorithms that can iterate and converge faster is always useful for speeding up the design process. The reduction of manual-layout steps, which are particularly used for mixed-signal and analog components and are time consuming, can speed up the design effort. A metamodel or surrogate (which is essentially a model of a model, i.e., a mathematical model of a SPICE model) is used to perform design exploration either outside the EDA tools or within high levels of abstractions in the EDA tools to significantly speed up the design exploration of AMS-SoC components. This chapter discusses selected metamodels, metamodeling techniques, metamodel-assisted analysis techniques, and metamodel-assisted design flows.

12.2 Metamodel: An Overview

In this section, various aspects of a metamodel are presented. The concept of metamodel is introduced. Different types of metamodels are briefly discussed. The important features of metamodels for accurate and efficient representation of the circuits or systems are discussed. Various techniques used for error or accuracy analysis of metamodels are also presented.

12.2.1 Concept

A high-level idea of metamodel and metamodeling is depicted in Fig. 12.1 [18, 32, 45, 68, 69, 80, 90]. The schematic or layout of an integrated circuit (IC) is represented as a SPICE netlist or SPICE

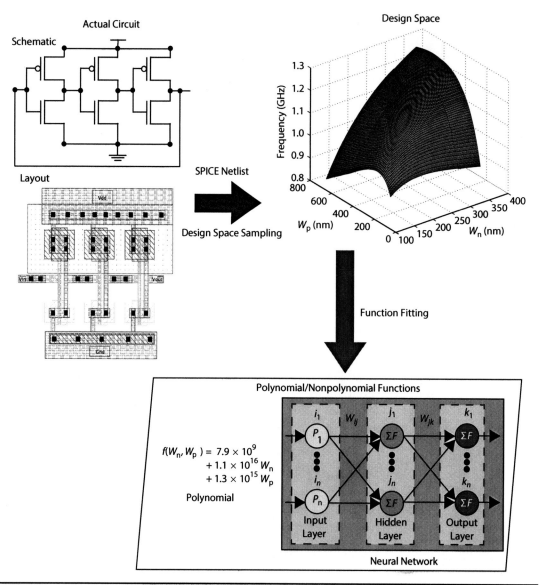

Figure 12.1 A high-level concept of metamodel [18, 45].

model. The SPICE netlist/model of the IC can be simulated and design exploration can be performed on it using an analog simulator or SPICE. This can be time consuming and computational intensive depending on the size and complexity of the IC. If the SPICE netlist is simulated at the sample points of design/circuit/process parameter for specific IC characteristics, then design space can be obtained. For example, in the figure, center frequency is characterized for different discrete sizes of the design variables W_n and W_p. As the step sizes of the discrete values W_n and W_p decrease, the number of analog simulations and the amount of time needed to generate the design space increase. Thus, design exploration through the traversal of the design space using an analog simulator or SPICE is quite slow. In another way, the function fitting based on a small number of analog simulations can generate metamodels. The metamodels can then be used to generate the values in the design space, which otherwise could have come from slow and computational-intensive analog/SPICE simulations. Various forms of metamodels can be generated based on the need of the design engineer. A simple example is a polynomial metamodel presented in Fig. 12.1. An example of a nonpolynomial metamodel is a neural network (NN). Once trained from an initial set of data, the NN can then be used to generate data for unknown parameters. In principle, many other mathematical functions can be used as metamodels. The metamodels can then be used for analysis or design exploration.

The metamodel is a "model of a model" or "surrogate model," that is, an approximation model used to replace the computationally expensive simulation model during the analysis, design, and optimization processes [18, 32, 45, 68, 69, 80, 90]. In general, the metamodels are mathematical functions or algorithms that relate the characteristics or figures-of-merit (FoMs) of ICs or systems in terms of design or tuning parameters. For the design variable x_{tuning} if the response is y_{response}, then any generic relation between them has the following form:

$$y_{\text{response}} = f(x_{\text{tuning}}) \tag{12.1}$$

The metamodel of the original response y_{response} is the following:

$$y^*_{\text{response}} = g(x_{\text{tuning}}) \tag{12.2}$$

where g is another function that may be different from the original function f. y^*_{response} is related to the original response as follows:

$$y_{\text{response}} = y^*_{\text{response}} + \varepsilon \tag{12.3}$$

where ε is the error originating from approximation and measurement and can also be random error. In addition to speedup in the simulation, analysis, and design exploration as discussed in the preceding paragraph, metamodels can provide a better understanding of the relationship between x and y. The metamodels can be easily integrated into the domain-dependent tools. They have been used in many different forms in the design and simulation of analog and digital ICs as well as circuit and system level [14, 18, 37, 41, 80].

12.2.2 Types

A selected set of different types of metamodels is presented in Fig. 12.2(a) [11, 32, 45, 49, 68, 69, 72, 80, 83]. These include polynomial regression function, Kriging methods, multivariate adaptive regression spline (MARS), radial basis function (RBF), Bayesian process regression (BPR), Gaussian process regression (GPR), and machine learning-based metamodels. A detailed discussion of these metamodels along with specific IC examples is presented in many of the following sections. The polynomial regression functions of response surface are simple metamodels that are represented as polynomials of specific orders [15, 18, 40, 45]. Kriging methods are of various forms including simple Kriging, ordinary Kriging, and universal Kriging as shown in Fig. 12.2. Kriging metamodels have typically a polynomial part and an additional departure to improve accuracy [27, 54, 81, 82]. MARS metamodels adaptively select a set of basis functions for approximating the response function using a forward/backward iterative approach. A MARS metamodel can have the following form [32, 74]:

$$y^*_{\text{response}} = \sum_{i=1}^{N} \alpha_i B_i(x) \tag{12.4}$$

where α_i is the coefficient of the expansion and B_i the basis functions. The RBF metamodel is a combination of radially symmetric function based on Euclidean distance or other distance functions to approximate the response functions. It can be expressed as follows [30, 32]:

$$y^*_{\text{response}} = \sum_{i=1}^{N} \alpha_i \phi_{\text{RBF}}(\|x - x_i\|) \tag{12.5}$$

where α_i is the weight coefficient of the expression and x_i the observed input. ϕ_{RBF} is a radial function that can be of many forms including Euclidean distance function, Gaussian function, inverse multiquadric, inverse quadratic, and polyharmonic spline. In the GPR, the interpolated/missing values are modeled by a Gaussian process; the regression function $f(x)$ instead of being a specific model such as polynomial is represented obliquely [13, 61]. Kriging is considered as a specific type of Gaussian process prediction [61]. The BPR is a completely different metamodeling. The other metamodelings are within the big category of either "frequentist" or "likelihoodist" as shown in Fig. 12.2(b) [6]. Machine learning-based metamodels are of many types including the following: NN, support vector machine (SVM), inductive learning, genetic programming, case-based learning, and analytical learning. These are widely used in many application domains. Artificial neural networks (ANNs)

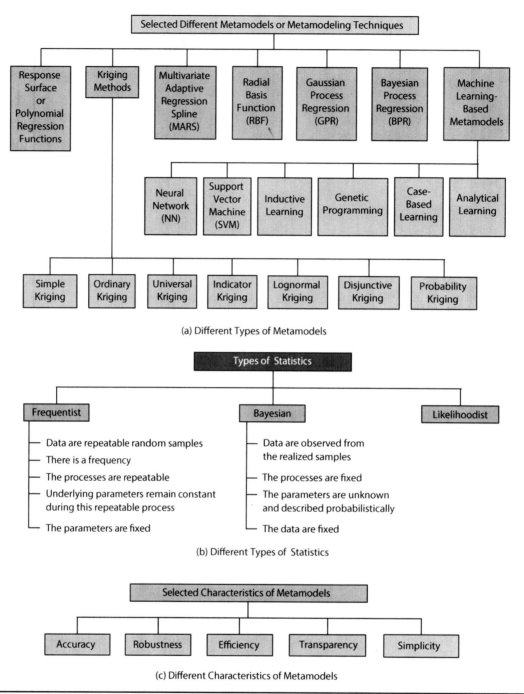

FIGURE 12.2 Different types of metamodels and their characteristics [6, 45, 49, 61].

are discussed with specific examples in the rest of this chapter [33, 75]. SVMs are also used as metamodels in many applications. An example approximating function for an SVM metamodel follows [4, 10, 35, 58]:

$$y^*_{\text{response}} = \text{sign}\left(\sum_{i=1}^{N} \alpha_i \exp\left(-\gamma_{\text{fit}} |x - x_i|^2\right) - \beta_{\text{con}}\right) \quad (12.6)$$

where α_i are weighting coefficients, x_i are input samples, β_{con} is a constant, and γ_{fit} is the fitting parameter.

A selected set of desired characteristics of the metamodel or metamodeling is presented in Fig. 12.2(c) [11, 32, 45, 49, 68, 69, 72, 80, 83]. The metamodel characteristics include accuracy, robustness, efficiency, transparency, and simplicity [32, 43]. The accuracy of metamodels is the capability of the metamodels to predict the responses of the IC and system over the entire design space. The robustness of metamodeling is its ability to achieve good accuracy for different problems. The efficiency of metamodels is the computational efforts required to construct the metamodels. The transparency characteristic of the metamodel refers to its capability of providing the information concerning contributions and variations of design variables and correlation among these variables. The conceptual simplicity characteristic of the metamodel refers to the easiness in its implementation. The simple metamodeling requires less manual input and can be easily adapted to different problems.

12.2.3 Generation Flow

A generic flow for the generation of metamodels or metamodeling method is depicted in Fig. 12.3(a) [19, 32, 45, 49, 68, 69, 80]. The SPICE netlist of an AMS-SoC component that is extracted from the schematic design or physical design is the input to this metamodeling flow. As the first step, the parameters or variables (e.g., width of NMOS W_n and width of PMOS W_p) for which the metamodeling will be performed are identified. The identification can be based on the experience of the design engineer or can be made using some initial analysis like design of experiments (DOE). The netlist is then parameterized for these design/tuning variables/parameters to obtain a parameterized netlist of the AMS-SoC components: target characteristics or figures-of-merit (FoMs) (e.g., center frequency, jitter, or power dissipation of a PLL or a VCO). Each of the characteristics of the AMS-SoC components will have a metamodel. At this phase, a sampling technique like Latin hypercube sampling (LHS) or middle Latin hypercube sampling (MLHS) is used to obtain discrete samples/values of the design/tuning variables/parameters. This sampling can be performed using a non-EDA tool such as MATLAB®, which has a large variety of options compared to the EDA tools. The number of sample points will affect the accuracy of metamodels and time of metamodeling. Of course, the larger the number of sample points is, the more accurate is the metamodel, but it can take more time during the simulations in the next step. As an example of sample input variable or design variable generator, let X_{tuning} be the set of design variables with $X_{tuning} = [x_1, x_2, ..., x_N]^T$, where any x_i is a design variable. The range of each of these variables is denoted as X_i and is divided into M contiguous intervals as follows [16, 19, 45 63]:

$$X_{tuning,matrix} = \begin{bmatrix} X_1 \\ X_2 \\ ... \\ X_N \end{bmatrix} = \begin{bmatrix} x_{11} & x_{12} & ... & x_{1M} \\ x_{21} & x_{22} & ... & x_{2M} \\ ... & ... & ... & ... \\ x_{N1} & x_{N2} & ... & x_{NM} \end{bmatrix} \quad (12.7)$$

In the above expression, N number of statistical distributions, i.e., one for each design variable, are used to generate M size sample for each of the variables.

Using an analog simulator or SPICE engine, the parameterized netlist is simulated for the target characteristics or the FoMs of the AMS-SoC components for the sampled values of the design parameters. The number of runs during the simulations is dictated by the accuracy and computational time of metamodeling. The number of simulations is related to the number of different types of analysis needed for FoM characterization. The number of simulations translates into the amount of time it takes to generate the metamodel that is calculated as follows [16, 19, 45]:

$$T_{Metamodel} = \left(\frac{N_s}{i}\right) \times \sum T_{s_i} \quad (12.8)$$

where T_{s_i} is average time to perform a separate simulation for each i analysis and N_s is the total number of simulations. For the simulations, a script is typically used to import the sample design variables to the analog simulator or SPICE. The response/FoM generated from the analog simulations is normalized to improve the accuracy of the training/fitting of the metamodels. The normalization

694 Chapter Twelve

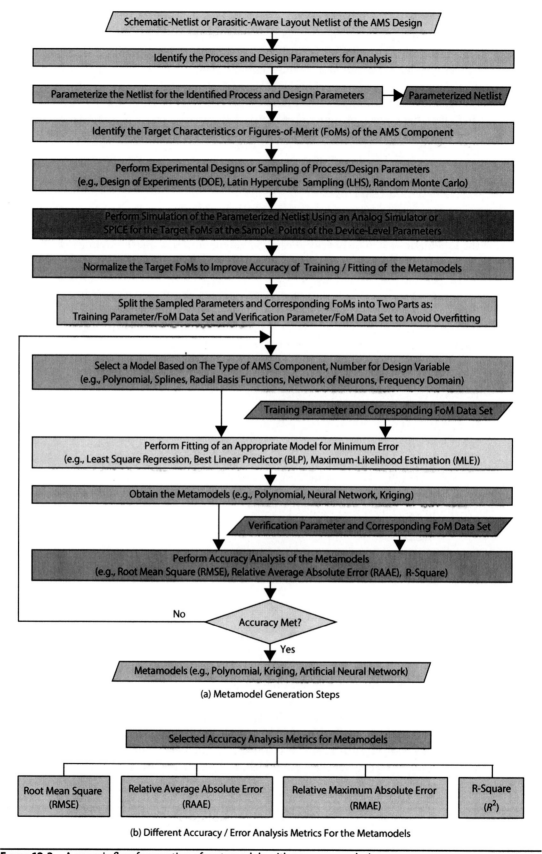

Figure 12.3 A generic flow for creation of metamodels with accuracy analysis.

reduces the dynamic range in the FoM data and reduces the complexity and improves accuracy of the metamodel fitting. An example of such process is the mean data centering. In the mean data centering, the FoM data is centered and scaled before it goes for regression or fitting in the next phase. For the ith FoM, FoM$_i$ with mean μ_i and standard deviation σ_i in its range FoM$_{ij}$, the normalized value is calculated as follows [16, 19, 45]:

$$\text{FoM}_{ij}^* = \left(\frac{\text{FoM}_i - \mu_i}{\sigma_i} \right) \tag{12.9}$$

The input variables and corresponding characterization data are divided into two sets: one large training set and another small verification set. At this phase, a metamodel, such as a polynomial or an NN, is selected for regression or fitting of the training parameter/FoM set. The verification of the metamodels is performed using the verification parameter/FoM set. Various accuracy analysis metrics are used for the verification to ensure minimal error between actual data from simulation and data generated from the metamodels.

A selected set of accuracy metrics that can be used for metamodel accuracy analysis is presented in Fig. 12.3(b) [19, 32, 44, 45, 49, 68, 69, 80]. The mean square error (MSE) and root mean square error (RMSE) are calculated as follows [32, 43]:

$$\text{MSE} = \left(\frac{1}{N} \right) \sum_{i=1}^{N} \left(\text{FoM}_i - \text{FoM}_i^* \right)^2 \tag{12.10}$$

$$\text{RMSE} = \sqrt{\left(\frac{1}{N} \right) \sum_{i=1}^{N} \left(\text{FoM}_i - \text{FoM}_i^* \right)^2} \tag{12.11}$$

where FoM$_i$ is the simulation data from verification parameter/FoM set and FoM$_i^*$ is the generated FoM from the metamodels for the same input parameters/variables as the verification set. The smaller the MSE/RMSE are, more accurate is the metamodel. The relative average absolute error (RAAE) is calculated as follows [32, 43]:

$$\text{RAAE} = \left(\frac{\sum_{i=1}^{N} |\text{FoM}_i - \text{FoM}_i^*|}{N \sigma_{\text{FoM}}} \right) \tag{12.12}$$

$$= \left(\frac{\sum_{i=1}^{N} |\text{FoM}_i - \text{FoM}_i^*|}{N \sqrt{\sum_{i=1}^{N} \left(\text{FoM}_i - \text{FoM}_i^* \right)^2}} \right) \tag{12.13}$$

where σ_{FoM} is the standard deviation in the FoM, which is calculated as follows:

$$\sigma_{\text{FoM}} = \sqrt{\sum_{i=1}^{N} \left(\text{FoM}_i - \text{FoM}_i^* \right)^2} \tag{12.14}$$

A smaller RAAE means that the metamodel is more accurate. The relative maximum absolute error (RMAE) is calculated as follows [32]:

$$\text{RMAE} = \left(\frac{\text{maximum}\left(|\text{FoM}_i - \text{FoM}_i^*| \big|_{\forall i = 1 \to N} \right)}{\sigma_{\text{FoM}}} \right) \tag{12.15}$$

$$= \left(\frac{\text{maximum}\left(|\text{FoM}_1 - \text{FoM}_1^*|, |\text{FoM}_2 - \text{FoM}_2^*|, \ldots, |\text{FoM}_N - \text{FoM}_N^*| \right)}{\sqrt{\sum_{i=1}^{N} \left(\text{FoM}_i - \text{FoM}_i^* \right)^2}} \right) \tag{12.16}$$

A smaller RMAE means that the metamodel is more accurate. The coefficient of determination R-square (R^2) is calculated as follows [32, 43]:

$$R^2 = \left(1 - \frac{\text{MSE}}{\text{Variance}_{\text{FoM}}}\right) \tag{12.17}$$

$$= \left(1 - \frac{\text{MSE}}{\sigma_{\text{FoM}}^2}\right) \tag{12.18}$$

$$= \left(1 - \frac{\left(\frac{1}{N}\right)\sum_{i=1}^{N}\left(\text{FoM}_i - \text{FoM}_i^*\right)^2}{\sum_{i=1}^{N}\left(\text{FoM}_i - \text{FoM}_i^*\right)^2}\right) \tag{12.19}$$

A larger R-square (R^2) means that the metamodel is more accurate. RAAE has a high correlation with MSE and hence RMAE and R-square may not be needed. A large RMAE indicates a large error in one region of the design space even though the overall accuracy indicated by RAAE and R-square can be very good. Thus, RMAE is not as important as RAAE or R-square [32].

12.2.4 Metamodel versus Macromodel

For a comparative perspective of metamodels, four types of IC and system modeling are presented in Fig. 12.4(a) [18, 45, 83]. The SPICE netlist and the symbolic representation are structural representations of the ICs or systems. The macromodels used in the IC and system simulation can be either structural or behavioral in nature. The metamodels are essentially behavioral models of the ICs or systems. In terms of simulation speed, the SPICE netlist is the slowest as analog simulators take the longest time due to the use of numerical routines with a large number of nodes and branches. The metamodels that are behavioral in nature can simulate faster. In terms of accuracy, SPICE simulations provide the highest possible accuracy; on the other hand, the metamodels have the lowest accuracy among the four options.

In the literature, in many cases metamodeling and macromodeling have been used interchangeably even though they are different. For a clear understanding, the differences between metamodel and macromodel are summarized in Fig. 12.4(b) [18, 45, 83]. As evident from the discussion in the preceding sections, which will be carried on in more detail in the following sections, metamodels relate the input variables with output responses. Thus, metamodel is a behavioral model. However, metamodels are developed using the netlist of an integrated circuit (which is structural information). In short, metamodels are models of models or surrogated models that are essentially mathematical or algorithmic behavioral models derived from SPICE structural models. The metamodels can be integrated into the hardware description languages (HDLs), such as Verilog-AMS and VHDL-AMS, for simulation in EDA tools. They can also be used for IC analysis and optimization outside EDA tools.

The typical macromodels are simply reduced-order (complexity) models and the IC representation can still be a structural netlist. Macromodels produce simpler circuits that are faster to simulate than the actual SPICE netlist. In such a case the same EDA tool (e.g., SPICE) is needed for its simulation. Thus, macromodels are faster than SPICE, but slower than the metamodels. The macromodels retain the circuit structure and provide insight into the ICs. However, the modeling effort needed is huge as the design engineers need to understand the effects of the basic building blocks on the overall circuit or system characteristics as well as their interactions. A quasi-proportional increase in the simulation time may happen with the increase in the size of the IC.

For a comparative perspective, the symbolic representation and analysis of the ICs is now briefly discussed [24, 29, 66]. In the symbolic representations, the ICs are represented in terms of symbolic parameters. The symbolic analysis at the circuit level calculates the characteristic of an IC, with the independent variables (such as time or frequency), the dependent variables (such as voltages and currents), and some or all of the circuit elements represented by symbols. It explicitly presents the circuit parameters that determine the circuit characteristics, which is quite different from the numerical or SPICE analysis. Symbolic simulators, thus, can provide more advantages over the numerical or

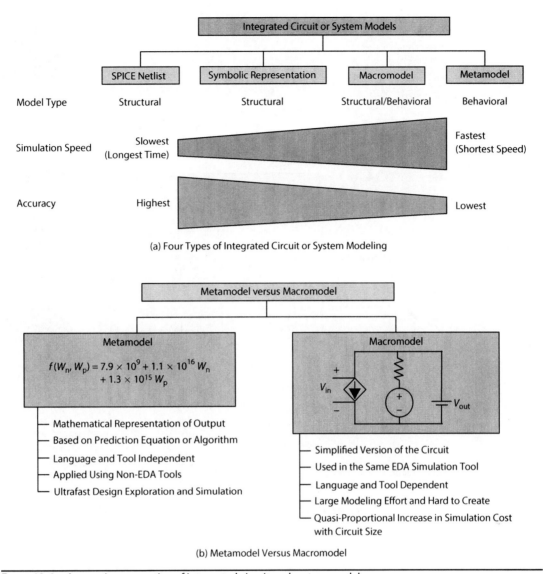

FIGURE 12.4 A generic perspective of integrated circuit and system models.

SPICE simulators in many design tasks, including topology selection, design space exploration, and behavioral model generation. However, symbolic analysis has not been widely used by design engineers as it is considered inefficient for larger circuits due to the exponential increase in the number of product terms in a symbolic expression.

12.3 Metamodel-Based Ultrafast Design Flow

Once the metamodels are generated from the SPICE netlist of the ICs, they can be used for various design and optimization tasks. The metamodel-assisted ultrafast design flow is presented in Fig. 12.5 [45, 49, 55]. A SPICE netlist is extracted from the baseline schematic or layout design of the AMS-SoC component whose metamodeling needs to be performed. The design space sampling is then performed for the AMS-SoC components. It consists of sampling the design parameters and simulating the SPICE netlist at those sample design parameters using an analog or SPICE tool. As discussed in the previous section, one or more metamodels are generated for different characteristics of the AMS-SoC components. The metamodel now becomes the representative of the AMS-SoC component in place of the SPICE netlist for various design tasks. An accurate metamodel can provide design engineers with a good understanding of the behavior of AMS-SoC components as their design spaces are traversed. For the Monte Carlo analysis over the metamodels, the device or process parameter values

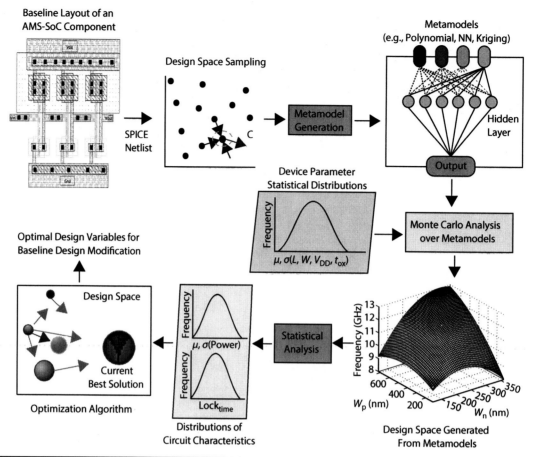

FIGURE 12.5 Generic perspective of metamodel-assisted ultrafast AMS design flow [45, 49, 55].

can be generated using statistical distributions and then for these specific parameter values, the metamodels can be used to generate values of the FoMs. This creates a design space from the metamodels as shown in Fig. 12.5. This is in contrast to generating FoM values from analog/SPICE simulations. It significantly speeds up the analysis process. The resulting characteristics can then be fitted with known statistical distributions to analyze the nature of the various FoMs of the AMS-SoC component. This fitting step in itself may take similar time irrespective of whether the data is generated (metamodels) or simulated (SPICE). Design optimization can now be performed using an algorithm over the metamodels. This is significantly faster than iteratively running analog/SPICE simulation in the optimization loop. Even statistical design optimization is possible using the statistical distributions of the FoMs obtained from the statistical analysis step. A large collection of algorithms and metamodels together can be used outside the EDA framework for ultrafast design exploration.

12.4 Polynomial-Based Metamodeling

The simplest form of metamodels is the polynomial metamodel [15, 18, 32, 40, 45, 68, 80]. In this section, a detailed discussion of polynomial metamodeling is presented. Various different representative case studies are used as specific examples of polynomial metamodeling. Then integration of the polynomial metamodels into various HDLs is discussed.

12.4.1 Theory

Polynomial metamodels can be of any order. As a specific example, a second-order polynomial model can be expressed as follows [18, 32, 45, 68, 80]:

$$y^*_{\text{response}} = \alpha_0 + \sum_{i=1}^{k}\alpha_i x_i + \sum_{i=1}^{k}\alpha_{ii} x_i^2 + \sum_{i=1}^{k}\sum_{j=1}^{k}\alpha_{ij} x_i x_j \qquad (12.20)$$

In the above expression, $y^*_{response}$ is the response that is characteristic or FoM of an AMS-SoC component, e.g., center frequency of a VCO. The x_i are input parameters that are design or tuning variables, e.g., $X_{tuning} = [x_1, x_2] = [W_n, W_p]$ is the vector of design or tuning variables. α_{ij} are the coefficients of the polynomial metamodel. These are determined by using a least-squares regression analysis to fit the response surface approximations to the available data. The polynomial metamodels are easy to use and quick convergence is possible when used in optimization. However, they are useful for small circuits with a smaller number of design variables. Modeling of the nonlinear characteristics as well as modeling for a large set of design variables using polynomial metamodels can be difficult.

12.4.2 Generation

The detailed steps for generation of a polynomial metamodel are presented in Fig. 12.6 [16, 44, 46]. As in the typical design flow, the parameterized SPICE netlist of the AMS-SoC component is created for the design/tuning variables. The number of sample points for the design variables is initialized. Sampling techniques like LHS and MLHS can be used for the sampling of the design variable range within a range [17, 39, 71]. As a specific example, let us consider a second-order polynomial with two design variables, $X_{tuning} = [x_1, x_2]$. A matrix X_p is generated to code each row, as a term in the polynomial power for each variable in which the value is 0 implies that the term is not present. As an example, the matrix for second-order code for two parameters is the following [16, 44, 46]:

$$X_p = \begin{bmatrix} 0 & 0 \\ 1 & 0 \\ 2 & 0 \\ 0 & 1 \\ 0 & 2 \\ 1 & 1 \end{bmatrix} \quad (12.21)$$

Then the design matrix template (DMT) for X_{tuning} using X_p code is the following [16, 44, 46]:

$$\text{DMT}(X_{tuning}, X_p) = [0, x_1, x_1^2, x_2, x_2^2, x_1 x_2] \quad (12.22)$$

The design parameter matrix is created by using LHS of x_1 and x_2 parameters, which has the following form [16, 44, 46]:

$$\text{DMT}(X_{tuning}, X_p) = \begin{bmatrix} 0 & \cdots & 0 \\ \text{lhs}(x_1)_1 & \cdots & \text{lhs}(x_1)_n \\ \text{lhs}(x_1)_1^2 & \cdots & \text{lhs}(x_1)_n^2 \\ \text{lhs}(x_2)_1 & \cdots & \text{lhs}(x_2)_n \\ \text{lhs}(x_2)_1^2 & \cdots & \text{lhs}(x_2)_n^2 \\ \text{lhs}(x_1)_1 \text{lhs}(x_2)_1 & \cdots & \text{lhs}(x_1)_n \text{lhs}(x_2)_n \end{bmatrix} \quad (12.23)$$

Similarly, the design parameter matrix, which is created by using MLHS of the x_1 and x_2 parameters, has the following form [16, 44, 46]:

$$\text{DMT}(X_{tuning}, X_p) = \begin{bmatrix} 0 & \cdots & 0 \\ \text{mlhs}(x_1)_1 & \cdots & \text{mlhs}(x_1)_n \\ \text{mlhs}(x_1)_1^2 & \cdots & \text{mlhs}(x_1)_n^2 \\ \text{mlhs}(x_2)_1 & \cdots & \text{mlhs}(x_2)_n \\ \text{mlhs}(x_2)_1^2 & \cdots & \text{mlhs}(x_2)_n^2 \\ \text{mlhs}(x_1)_1 \text{mlhs}(x_2)_1 & \cdots & \text{mlhs}(x_1)_n \text{mlhs}(x_2)_n \end{bmatrix} \quad (12.24)$$

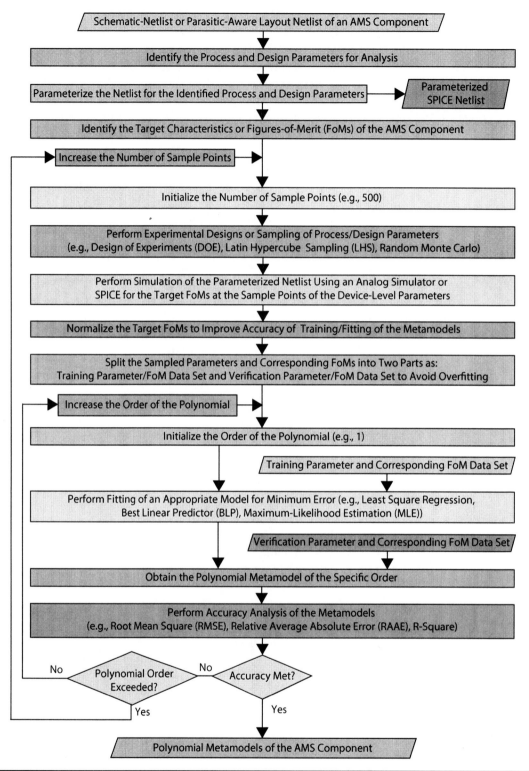

Figure 12.6 A flow for polynomial metamodel generation [16, 44, 46].

The SPICE netlist is now simulated at the LHS/MLHS sample points of the design variables to generate target FoMs or characteristics that need to be modeled. The data resulting out of the simulations is normalized to improve accuracy of the metamodels. The design parameters and corresponding FoMs are splitted to 1000 and 100 for training and verification purposes, respectively. At this phase of polynomial metamodel generation, stepwise regression is performed on the sample parameter/FoM data to create the polynomial of first to p order, e.g., $p = 2$ for the second-order or

$p = 6$ for the sixth-order polynomial [16, 42]. The resulting polynomial may not include all terms as the stepwise regression iteratively removes coefficients that are not statistically significant, resulting in metamodels with fewer coefficients without much loss in the accuracy. The function stepwise(), which is embedded in MATLAB®, attempts the different initial models and generates the coefficients for the best model for a specific order. The final outcome is a multivariate polynomial function of degree p in the N design variables $X_{\text{tuning}} = \{x_1, x_2\}$ for the predicted response y_{response}. The metamodels are verified using the verification parameter/FoM data through the use of various accuracy analysis metrics. If the polynomial metamodels do not pass the accuracy test, then either higher-order polynomials or more samples are tried as shown in Fig. 12.6.

12.4.3 Ring Oscillator

As a specific example, polynomial metamodeling of a 45-nm CMOS ring oscillator is presented in Fig. 12.7 [16, 17, 18, 45]. The two design variables are W_n for NMOS and W_p for PMOS. Thus, the polynomial metamodel for two parameters has the following form [14, 16, 18, 45, 48]:

$$y^*_{\text{response}} = \sum_{i,j=0}^{k} \left(\alpha_{ij} W_n^i W_p^j \right) \tag{12.25}$$

where y^*_{response} is the response being modeled, e.g., center frequency in the case of ring oscillator, W_n and W_p are variables, and α_{ij} are the coefficients determined by the polynomial regression, and k varies from 1 to 5. The polynomial metamodels generated for the center frequency of the ring oscillator are the following [16, 18, 45]:

Ring oscillator – Order 1:

$$f_{\text{osc}}(W_n, W_p) = 7.9 \times 10^9 + 1.1 \times 10^{16} W_n + 1.3 \times 10^{15} W_p \tag{12.26}$$

Ring oscillator – Order 2:

$$\begin{aligned} f_{\text{osc}}(W_n, W_p) = &\, 6.4 \times 10^9 + 2.2 \times 10^{16} W_n + 6.1 \times 10^{15} W_p - 5.0 \times 10^{22} W_n^2 \\ &+ 3.3 \times 10^{22} W_n W_p - 1.5 \times 10^{22} W_p^2 \end{aligned} \tag{12.27}$$

Ring oscillator – Order 3:

$$\begin{aligned} f_{\text{osc}}(W_n, W_p) = &\, 4.7 \times 10^9 + 3.9 \times 10^{16} W_n + 1.3 \times 10^{16} W_p - 1.7 \times 10^{23} W_n^2 \\ &+ 8.8 \times 10^{22} W_n W_p - 5.1 \times 10^{22} W_p^2 + 1.8 \times 10^{29} W_n^3 \\ &+ - 3.8 \times 10^{28} W_n^2 W_p - 4.8 \times 10^{28} W_n W_p^2 + 3.8 \times 10^{28} W_p^3 \end{aligned} \tag{12.28}$$

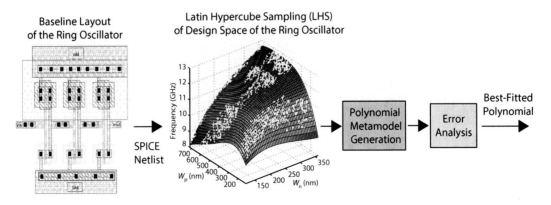

FIGURE 12.7 Polynomial metamodeling of a 45-nm CMOS-based ring oscillator [16, 18, 45].

12.4.4 LC-VCO

As another simple example, the polynomial metamodeling of a 180-nm CMOS-based LC-VCO is presented in Fig. 12.8 [16, 18, 45]. The steps presented in Sec. 12.4.2 on metamodel generation have been followed. The polynomial metamodels generated for center frequency of the 180-nm CMOS-based LC-VCO are the following [16, 18, 45]:

LC-VCO – Order 1:

$$f_{\text{osc}}(W_n, W_p) = 2.4 \times 10^9 - 3.5 \times 10^{12} W_n - 6.7 \times 10^{12} W_p \tag{12.29}$$

LC-VCO – Order 2:

$$\begin{aligned} f_{\text{osc}}(W_n, W_p) &= 2.3 \times 10^9 + 6.2 \times 10^{11} W_n + 1.5 \times 10^{11} W_p - 4.2 \times 10^{16} W_n^2 \\ &\quad - 2.0 \times 10^{17} W_n W_p - 1.0 \times 10^{17} W_p^2 \end{aligned} \tag{12.30}$$

LC-VCO – Order 3:

$$\begin{aligned} f_{\text{osc}}(W_n, W_p) &= 2.1 \times 10^9 + 1.6 \times 10^{13} W_n + 4.7 \times 10^{13} W_p - 4.7 \times 10^{17} W_n^2 \\ &\quad - 1.4 \times 10^{18} W_n W_p - 3.4 \times 10^{18} W_p^2 + 4.5 \times 10^{21} W_n^3 \\ &\quad + 9.7 \times 10^{21} W_n^2 W_p + 3.3 \times 10^{22} W_n W_p^2 + 7.2 \times 10^{22} W_p^3 \end{aligned} \tag{12.31}$$

Fig. 12.8 also shows the number of coefficients needed for the polynomial metamodels of different orders.

12.4.5 Verilog-AMS Integrated with Polynomial Metamodel for an OP-AMP

In this section, the polynomial metamodeling of a 90-nm CMOS-based operational amplifier (OP-AMP) is considered, with emphasis on integration of the polynomial metamodels with Verilog-AMS to generate Verilog-AMS-PoM. The overall flow is presented in Fig. 12.9 [83, 85]. This polynomial metamodel generation flow is, in principle, more or less the same as the previously discussed flows, but customized for OP-AMP. Another important difference is that two different types of polynomial metamodels are considered in this flow as compared to one type in the previous flow. The two groups of the polynomial metamodels are the following:

1. *Circuit Parameter Metamodel (CPM)*: This is a paradigm shift approach. The OP-AMP circuit parameters considered for polynomial metamodeling are open-loop DC gain (A_{DC}), transconductance of input stage (g_{mi}), maximum available positive current ($I_{\text{max+}}$), maximum available negative current ($I_{\text{max-}}$), function for poles ($P_{\text{OP-AMP}}$), and function for zeros ($Z_{\text{OP-AMP}}$).

2. *Performance Metric Metamodel (PMM)*: This is the same FoMs, characteristics, or performance metamodel as expressed in terms of device-level parameters. In the case of the OP-AMP, the performance metrics considered are power dissipation ($PD_{\text{OP-AMP}}$), bandwidth ($BW_{\text{OP-AMP}}$), phase margin ($PM_{\text{OP-AMP}}$), and slew rate ($SR_{\text{OP-AMP}}$).

The device parameters that are considered as design variables are the width and length of an NMOS and a PMOS, i.e., $X_{\text{tuning}} = \{L_n, L_p, W_n, W_p\}$.

The SPICE netlist of the 90-nm CMOS-based fully differential OP-AMP presented in Fig. 12.10 is used as the input for the polynomial metamodel generation flow of Fig. 12.9 [83, 85]. The LHS of the design variables $X_{\text{tuning}} = \{L_n, L_p, W_n, W_p\}$ is performed. The polynomial metamodel of the following format has been used for metamodeling both OP-AMP parameters and characteristics [83, 85]:

$$y^*_{\text{response}} = \sum_{i=0}^{N_B-1} \beta_i \prod_{j=0}^{N_D-1} x_j^{P_{ij}} \tag{12.32}$$

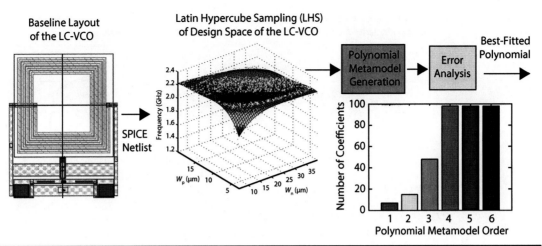

FIGURE 12.8 Polynomial metamodeling of a 180-nm CMOS-based LC-VCO [16, 18, 45].

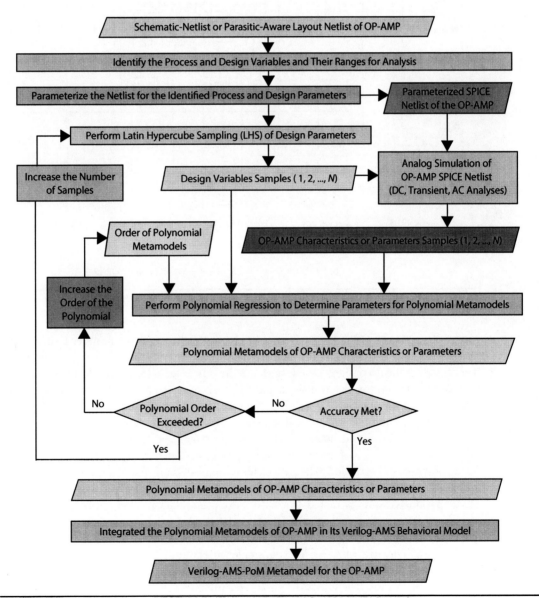

FIGURE 12.9 Polynomial metamodeling of an operational amplifier (OP-AMP) [83, 85].

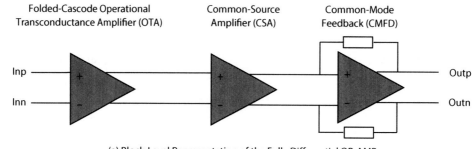
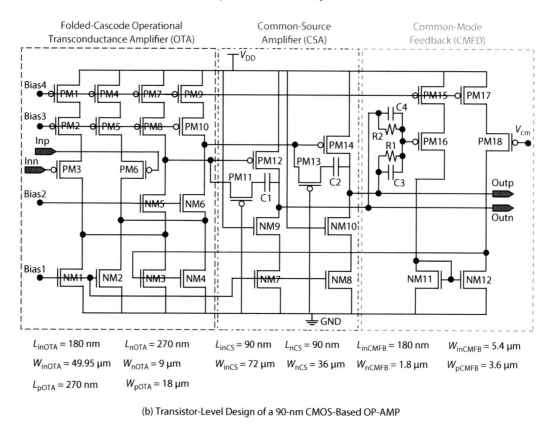

FIGURE 12.10 A 90-nm CMOS-based fully differential operation amplifier (OP-AMP) [83, 85].

where $y^*_{response}$ is any OP-AMP parameter or characteristic. N_B is the number of basis functions of this polynomial metamodel. β_i is the coefficient for the ith basis function. N_X is the number of design variables. x_j is the jth design variable and p_{ij} is the power term for the jth design variable in the ith basis function. Based on the DC, transient, and AC analysis of the parameterized netlist of the OP-AMP in analog/SPICE simulations, OP-AMP parameter and characteristics samples are obtained. In the first group of polynomial metamodeling, six OP-AMP parameters: open-loop DC gain (A_{DC}), transconductance of input stage (g_{mi}), maximum available positive current (I_{max+}), maximum available negative current (I_{max-}), function for poles (P_{OP-AMP}), and function for zeros (Z_{OP-AMP}), are modeled using polynomial regression. The parameter metamodels are very useful for design exploration of the OP-AMP circuit as evident from the following discussion. In a typical design exploration, the OP-AMP parameters are extracted from the SPICE simulations. Whenever the values of the design variables change, the analog/SPICE simulations need to be generated to extract the OP-AMP parameters again. Thus, the iterations during the design exploration OP-AMP circuit can be time consuming. The use of OP-AMP parameter metamodels can eliminate this need for repeated extraction of the OP-AMP parameters whenever values of design variables change, thus speeding up the design iterations. The OP-AMP parameter metamodels are effectively meta-macromodels as the OP-AMP parameters are essentially parameters of macromodels (consisting of transfer characteristic and small-signal transfer function) describing the OP-AMP. The OP-AMP performance metrics or characteristics considered for polynomial metamodeling are power dissipation (PD_{OP-AMP}), bandwidth (BW_{OP-AMP}), phase margin (PM_{OP-AMP}), and slew rate (SR_{OP-AMP}) using polynomial regression. In the polynomial regression for both OP-AMP parameter and characteristics, the objective is to determine β_i and p_{ij} for target accuracy for a specified number of samples and polynomial order.

The block-level representation of the fully differential OP-AMP is shown in Fig. 12.10(a) [3, 67, 83, 85]. A fully differential OP-AMP suppresses the common-mode noise and hence the even-order harmonic distortions. It has three distinct components: folded-cascode operational transconductance amplifier (OTA), common-source amplifier (CSA), and common-mode feedback (CMFB) circuit. This OP-AMP is a DC-coupled high-gain electronic voltage amplifier with differential inputs and differential outputs. The folded-cascode architecture has good self-compensation and good input common-mode range, and maintains the gain of a two-stage OP-AMP. In the typical application of this OP-AMP, the output is controlled by two feedback paths that determine the output voltage for any given input due to the high gain of this OP-AMP. The OTA without any output buffer can not drive resistive load or large capacitive load; it can only drive small capacitive load. However, it has smaller size and is simple in design. The folding of the cascode provides the configuration of a common-source stage driving a common-gate stage with complementary transistors to maintain the input and output voltages at the same level. The CSA as output stage of the OTA provides a low output resistance. The CMFB circuit regulates the output common-mode voltages. It has a very small impedance for the common-mode signals, but it is transparent for the differential signals.

The transistor-level schematic OP-AMP is shown in Fig. 12.10(b) [3, 67, 83, 85]. In the two-stage OP-AMP, the transistors M_{21} and M_{22} act as resistors to remove the zeros of the right-half plane. At the same time, the two-stage OP-AMP is compensated using Miller capacitors C_1 and C_2. A 90-nm CMOS process with 1 V supply voltage is used for the implementation of the OP-AMP. The characterization results of this 90-nm CMOS-based fully differential OP-AMP is presented in Table 12.1 [83, 85]. In a specific example, second-order polynomial metamodels are constructed using 200 samples while maintaining high accuracy without noticeably increasing the initialization time. For the OP-AMP design variables $X_{tuning} = \{L_n, L_p, W_n, W_p\}$, $N_D = 4$. As an example, the polynomial metamodel for the DC gain of the OP-AMP is the following [83, 85]:

$$A_{DC} = 4.5 \times 10^2 \cdot L_n^0 \cdot L_p^0 \cdot W_n^0 \cdot W_p^0 + 0.8 \times 10^9 \cdot L_n^0 \cdot L_p^1 \cdot W_n^0 \cdot W_p^0 \\ - 1.2 \times 10^{15} \cdot L_n^1 \cdot L_p^0 \cdot W_n^1 \cdot W_p^0 + 0.3 \times 10^{16} \cdot L_n^0 \cdot L_p^0 \cdot W_n^0 \cdot W_p^2 \quad (12.33)$$

where A_{DC} consists of four basis functions and is a second-order polynomial.

The concept of polynomial metamodel integrated Verilog-AMS (Verilog-AMS-PoM) is now presented. A high-level perspective of the Verilog-AMS-PoM is provided in Fig. 12.11 [83, 85]. The meta-macromodels or parameter metamodels of the OP-AMP are integrated with Verilog-AMS to obtain Verilog-AMS-PoM. Verilog-AMS-PoM incorporates circuit-level information in the system-level Verilog-AMS. Thus, the Verilog-AMS-PoM is fast like Verilog-AMS and accurate like transistor level. An example Verilog-AMS-PoM for the OP-AMP is provided in Algorithm 12.1 [83, 85]. The `initial` block of the Verilog-AMS-PoM reads the text files containing the coefficients β_i and exponentials p_{ij} for each OP-AMP parameter polynomial metamodel as well as the given design variable values. Verilog-AMS-PoM then calculates the OP-AMP circuit parameters: A_{DC}, g_{mi}, I_{max+},

OP-AMP Characteristics	Target Specifications	Values for the Baseline Design
Power Dissipation (PD_{OP-AMP}) (μW)	To Be Minimized	252.8
Bandwidth (BW_{OP-AMP}) (kHz)	40	58.0
Phase Margin (PM_{OP-AMP}) (degree)	65	92.5
Slew Rate (SR_{OP-AMP}) (mV/ns)	4	05.1
Open-Loop DC Gain (A_{DC}) (dB)	43	52.3

TABLE 12.1 Characterization Results for the 90-nm CMOS-Based Fully Differential OP-AMP [83, 85]

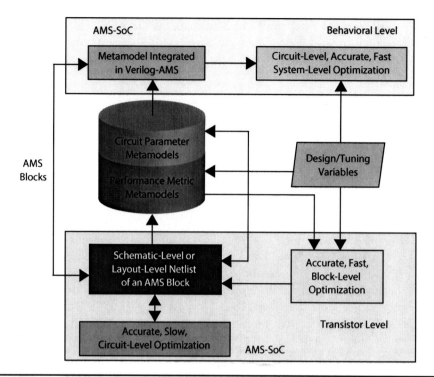

FIGURE 12.11 A high-level concept of the Verilog-AMS-PoM for circuit-accurate system-level behavioral modeling [83, 85].

ALGORITHM 12.1 Verilog-AMS-PoM for the Fully Differential Operational Amplifier (OP-AMP) [83, 85]

```
'include "constants.vams"
'include "disciplines.vams"
module OP-AMP_Verilog-AMS-PoM(outp, outn, inp, inn);
... ...
initial begin
// Reading the Design Variable Values
fd = $fopen({"OP-AMP_vars.csv", file_var}, "r");
... ...
readfile = $fscanf(fd, "%e", OP-AMP_var[i]);
... ...
$fclose(fd);
// Reading the Polynomial Metamodel Coefficients
fd_exParameter = $fopen({"exponetial_p_ij.csv"}, "r");
fd_coParameter = $fopen({"coefficient_beta_i.csv"}, "r");
... ...
for (k = 0; k < NB; k = k + 1) begin
   bf0 = 1.0;
   for (i = 0; i < ND; i = i + 1) begin
      readfile = $fscanf(fd_exParameter, "%e", expo);
      bf0 = bf0 * pow(OP-AMP_var[i], expo);
   end
   for (j = 0; j < N_Parameter; j = j + 1) begin
      readfile = $fscanf(fd_coParameter, "%e", coef);
      parameter[j] = parameter[j] + coef * bf0;
   end
end
... ...
```

```
$fclose(fd_expoParameter);
$fclose(fd_coefParameter);
// Obtain the OP-AMP Circuit Parameters
gmi = parameter[0];
... ...
end
analog begin
   vin = V(inp, inn);
   if (vin > vmax)
      I = Imaxp;
   else if (vin < vmin)
      I = Imaxn;
   else I = gm * vin;
   V(outd) <+ A0 * laplace_nd(I, num, den)/gm;
   ... ...
end
endmodule
```

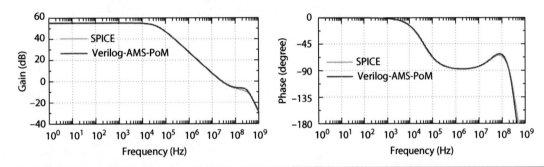

FIGURE 12.12 AC analysis of 90-nm CMOS-based OP-AMP using SPICE and Verilog-AMS [83, 85].

I_{max-}, P_{OP-AMP}, and Z_{OP-AMP}. The analog process in the Verilog-AMS-PoM implements the OP-AMP model. AC analysis of 90-nm CMOS-based OP-AMP using SPICE and Verilog-AMS-PoM is presented in Fig. 12.12 for a comparative perspective [83, 85]. As evident, Verilog-AMS-PoM at system level could match the accuracy of SPICE at circuit level. The polynomial metamodels can also be integrated with VHDL-AMS and other similar HDLs for system modeling and design exploration.

12.4.6 Verilog-AMS Integrated with Polynomial Metamodel for a Memristor Oscillator

The design of a memristor-based programmable Schmitt trigger oscillator is presented in Chapter 4 on PLL component circuit. In addition, Verilog-A-based coupled variable resistor model for a thin-film memristor is presented in Chapter 9 on mixed-signal circuit and system simulation. Verilog-AMS integrated with polynomial metamodel (Verilog-AMS-PoM) for a memristor-based programmable Schmitt trigger oscillator as an example is presented in this section. The memristor-based oscillator circuit is presented in Fig. 12.13 for a quick reference [87, 89].

The polynomial metamodel generation flow presented in the previous section can be customized for the memristor oscillator. This is shown in Fig. 12.14 [87, 89]. For the memristor oscillator, the device-level variables and circuit-level parameters selected are the following: ON state memristance (M_{on}), OFF state memristance (M_{off}), width of NM3/NM4 devices (W_n), width of PM7/PM8 devices (W_p), bias current (I_{bias}), set time (t_{set}), and memristor state (x_{state}) as presented in Table 12.2 with their ranges [87, 89]. The polynomial metamodels of the same format as those described in the previous section are used in this metamodeling as well, with N_B being the number of basis functions, N_D the number of design variables, β_i coefficients, and p_{ij} power terms. Specifically, second-order polynomial metamodels generate 500 LHS and the corresponding analog/SPICE simulation data. Polynomial metamodels for the following two memristor oscillator

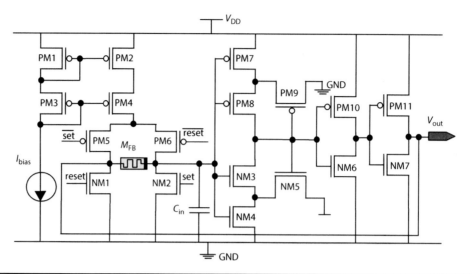

FIGURE 12.13 Memristor-based programmable Schmitt trigger oscillator circuit [87, 89].

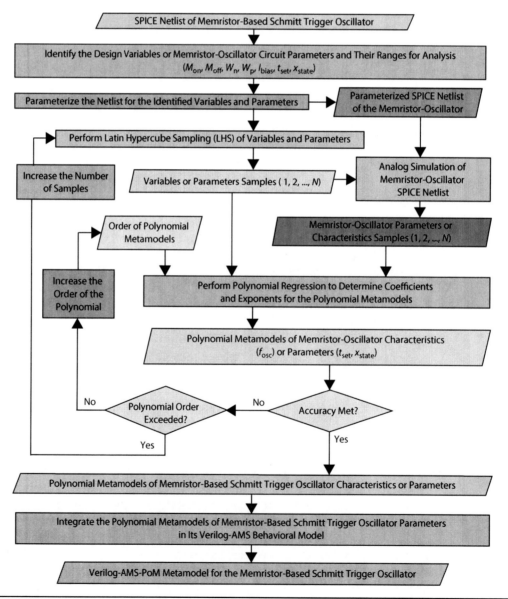

FIGURE 12.14 Polynomial metamodeling of a memristor-based Schmitt trigger oscillator [87, 89].

Memristor-Oscillator Parameters	Minimum Value	Maximum Value	Device
ON State Memristance M_{on}	8 kΩ	12 kΩ	M_{FB}
OFF State Memristance M_{off}	80 kΩ	120 kΩ	M_{FB}
NMOS Device Width W_n	0.8 μm	1.2 μm	NM3, NM4
PMOS Device Width W_p	1.6 μm	2.4 μm	PM7, PM8
Bias Current I_{bias}	80 μA	120 μA	-
Set Time t_{set}	0 ms	100 ms	-
Memristor State x_{state}	0	1	-

TABLE 12.2 Variables and Their Ranges for Polynomial Metamodeling of Memristor-Based Oscillator [87, 89]

parameters are generated: memristor state x_{state} and set time t_{set}. At the same time polynomial metamodels for the following memristor oscillator characteristic are generated: oscillating frequency f_{osc}. The x_{state} metamodel approximates the response surface of the memristor state x_{state}. This x_{state} metamodel is generated from the device variables and circuit parameter set $\{M_{on}, M_{off}, W_n, W_p, I_{bias}, t_{set}\}$. The t_{set} metamodel is generated from the device variables and circuit parameter set $\{M_{on}, M_{off}, W_n, W_p, I_{bias}, x_{state}\}$. The oscillating frequency f_{osc} metamodel is generated from the input variables set $\{M_{on}, M_{off}, W_n, W_p, I_{bias}, t_{set}\}$. The f_{osc} metamodel approximates the response surface of the output frequency of the memristor oscillator. For the verifications of the metamodels, again 500 SPICE simulation points are generated and various accuracy metrics such as RMSE, coefficient of determination (R^2), and RMAE are used. The polynomial metamodels are integrated with the Verilog-AMS model of the memristor oscillator to generate Verilog-AMS-PoM for the memristor-based Schmitt trigger oscillator, which is shown in Algorithm 12.2 [87, 89]. The transient analysis of the memristor oscillator is presented in Fig. 12.15(a) for both SPICE and Verilog-AMS-PoM and, as evident, both produce the same results [87, 89]. The oscillating frequency of the memristor oscillator obtained from both SPICE and Verilog-AMS-PoM simulations is shown in Fig. 12.15(b) [87, 89]. As evident from the figure, Verilog-AMS-PoM could match the SPICE with a minimal error.

ALGORITHM 12.2 Verilog-AMS-PoM for the Memristor-Based Schmitt Trigger Oscillator [87, 89]

```
'timescale 10ps / 1ps
'include "constants.vams"
'include "disciplines.vams"
module Memristor_Schmitt-Trigger_Oscillator_Verilog-AMS-PoM(out, reset, set);
parameter integer NB = 126; // Number of Basis Functions of the Polynomial
parameter real VDD = 1.0 from (0:10);
parameter real GND = 0.0;
input set, reset;
output out;
logic out, set, reset;
reg out;
real Xvar[0:4];  // Design Variables of the Schmitt Trigger Oscillator
real  bf0, expo, coef, f_osc;
integer readfile, fd, fd_expo, fd_coef, i , k;
initial  begin
out = 0;
// Reading the Design Variable Values
fd = $fopen("XVar_values.csv", "r");
i = 0; // Counter Initialization
while(! $feof(fd)) begin
   readfile = $fscanf(fd, "%e", Xvar[i]);
```

```
      i = i + 1;
   end
   $fclose(fd);
   // Reading the Polynomial Metamodel Coefficients
   fd_coeff = $fopen("coefficient_beta_i.csv", "r");
   fd_expo = $fopen("exponential_p_ij.csv", "r" );
   for(k = 0; k < NB; k = k + 1) begin
      bf0 = 1.0;
      for(i = 0; i < 5; i = i + 1) begin
         readfile = $fscanf(fd_expo, "%e", expo);
         bf0 = bf0 * pow(Xvar[i], expo);
      end
      readfile = $fscanf(fd_coef, "%e", coef);
      f_osc = f_osc + coef * bf0;
   end
   $fclose(fd_expo);
   $fclose(fd_coef);
end
always begin
   if(f_osc < 0) f_osc = 1M;
   else if(f_osc > 1G) f_osc  = 1G;
   #(0.5 / f_osc / 10p) out = ~out;
end
endmodule
```

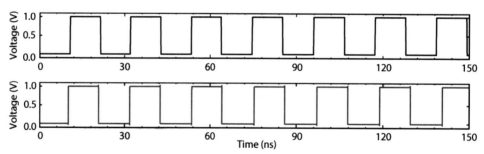

(a) Transient Analysis of the Memristor-Based Oscillator: SPICE versus Metamodel

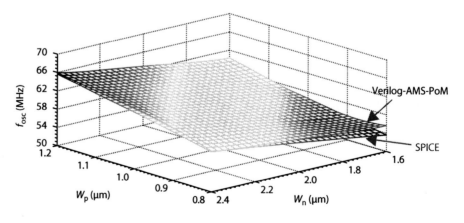

(b) Memristor-Based Oscillator Design Space: SPICE versus Metamodel

FIGURE 12.15 Verilog-AMS-PoM for memristor-based programmable Schmitt trigger oscillator [87, 89].

12.4.7 Verilog-AMS Integrated with Parasitic-Aware Metamodel

In the preceding two sections, the polynomial metamodeling is demonstrated for OP-AMP and memristor oscillator using their schematic-level designs. Then the concept of Verilog-AMS-PoM was introduced. In this section, the polynomial metamodels are generated from the physical design accounting the parasitic effects with LC-VCO as a case study design. Then the concept of Verilog-AMS integrated with parasitic-aware metamodel (Verilog-AMS-PAM) is introduced [83, 88]. A flow for polynomial metamodel generation for LC-VCO is presented in Fig. 12.16 [83, 88]. Based on all the discussions in this chapter, this figure can be considered as a self-explanatory flow. The parameterized parasitic-aware SPICE netlist is used for analog/SPICE simulations at the sample points of the design parameters obtained from LHS. The polynomial metamodel of the following form is used in this example [83, 88]:

$$y^*_{\text{response}} = \sum_{i=0}^{N_B-1} \beta_i W_p^{p_{1i}} W_n^{p_{2i}} V_{\text{tune}}^{p_{3i}} \quad (12.34)$$

where the three input variables for the metamodeling are W_n, W_p, and V_{tune}; N_B is the number of basis functions; and β_i is the coefficient for the basis functions. In order to generate the metamodels for a

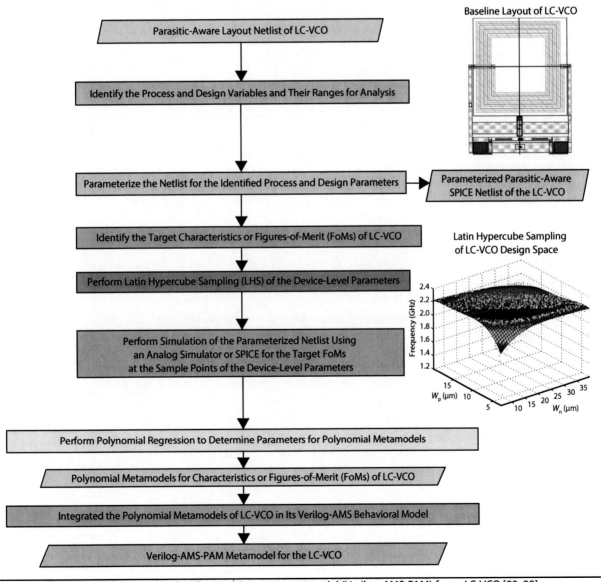

FIGURE 12.16 Verilog-AMS integrated with parasitic-aware metamodel (Verilog-AMS-PAM) for an LC-VCO [83, 88].

given LC-VCO design, for each basis function the coefficient β_i and the power terms p_{1i}, p_{2i}, and p_{3i} for each input variable need to be determined using polynomial regression for minimal error. The typical steps of polynomial metamodeling are used for this purpose. It may be noted that for the LC-VCO design, the inductor and capacitors are not parameterized because they have a first-order impact on the parasitics and varying them may lead to potential inaccuracy. For the LC-VCO design, the parameterization is performed by replacing the original transistor models in the netlist with parameterized ones. The target LC-VCO characteristics considered for metamodeling in this specific example are the power dissipation and center frequency of the LC-VCO.

The parasitic-aware polynomial metamodel integrated Verilog-AMS (Verilog-AMS-PAM) code for the LC-VCO is presented in Algorithm 12.3 [83, 88]. The part of the basis function that is related to the input variables W_p and W_n is constructed in the `initial` block. The remainder of the basis function is constructed in the `always` block as the third variable V_{tune} needs to be updated continuously during the simulation. The output signal of the Verilog-AMS-PAM module is implemented to be the digital logic type to reduce the computation overhead. In this specific example, the power and center frequency metamodels share the same input variables, while the coefficient β_i for the two metamodels is different. After these values are calculated, they are written into a text file that is read by the Verilog-AMS-PAM module of the LC-VCO. An example of such a text file is presented in Table 12.3 [83, 88]. These are values for the 180-nm CMOS-based LC-VCO with 100 LHS of design variables and corresponding SPICE simulation of the parasitic-aware SPICE netlist.

Algorithm 12.3 Verilog-AMS-PAM for the LC-VCO [83, 88]

```
'timescale 10ps / 1ps
'include "disciplines.vams"
module LC_VCO_Polynomial_Metamodel (VCO_out, VCO_in);
    parameter metafile = "LC_VCO_Polynomial_Metamodel.csv"; // LC-VCO Metamodel File
    parameter integer NB = 3; // Number of variables of the polynomial metamodel
    parameter powerfile = "LC_VCO_Metamodel_Exponent.csv"; // File to store exponents
    parameter real VDD = 1.2 from (0:10);
    parameter real GND = 0;
    parameter real W_n = 10u from [0.2u : 20u];
    parameter real W_p = 20u from [0.4u : 40u];
    input VCO_in;
    output VCO_out;
    electrical VCO_in;
    logic VCO_out;
    reg VCO_out;
    reg VCO_clk;    // Clock signal for strobing and generating results
    real V_in;
    integer i = 1;
    real Xvar[1:5];
    real bf[1:NB], bp[1:NB];
    real pvin[1:NB];
    real VCO_frequency, VCO_power;
    integer metafile, readfile, presult;
    initial begin   // Reading of the metamodels
        VCO_out = 0;   // Initialize vco digital output
        VCO_clk = 0;
        metafile = $fopen({"path_to_metamodel_file", metafile}, "r");
        while (!$feof(metaf)) begin
            readfile = $fscanf(metaf, "%e,%e,%e,%e,%e\n",
                                Xvar[1], Xvar[2], Xvar[3], Xvar[4], Xvar[5]);
            bf[i] = pow(wp, Xvar[1]) * pow(wn, Xvar[2]) * Xvar[4];
            bp[i] = pow(wp, Xvar[1]) * pow(wn, Xvar[2]) * Xvar[5];
```

```
          pvin[i] = Xvar[3];
          i = i + 1;
      end
      $fclose(metafile);
      metafile = $fopen({"path_to_save_power", powerfile}, "w");
  end
  always begin
      V_in = V(in);
      if (V_in > V_DD) V_in = V_DD;
      else if (vin < gnd) V_in = GND;
      VCO_frequency = 0;
      VCO_power = 0;
      for (i = 1; i <= NB; i = i + 1) begin
          VCO_frequency = VCO_frequency + bf[i] * pow(V_in, pvin[i]);
          VCO_power = VCO_power + bp[i] * pow(V_in, pvin[i]);
      end
      if (VCO_frequency < 0) VCO_frequency = 1G;
      else if (VCO_frequency > 10G) VCO_frequency = 10G;
       #(0.5 / freq / 10p)    // 10p is the unit time in the timescale
       VCO_out = ~VCO_out;
  end
  always #5 VCO_clk = ~VCO_clk;
  always @(posedge VCO_clk) begin
      $fstrobe(metafile, "%e, %e", $abstime, VCO_power);
  end
endmodule
```

	Polynomial Exponents			Polynomial Coefficients	
i	p_{1i}	p_{2i}	p_{3i}	$\beta_{i,f}$	$\beta_{i,p}$
0	0	0	0	2.11e + 09	1.38e − 05
1	1	0	0	−3.21e + 12	44.46e + 00
2	2	0	0	3.46e + 16	−2.80e + 05
3	0	1	0	6.87e + 12	39.73e + 00
4	1	1	0	−1.02e + 17	2.91e + 05
5	0	2	0	−2.07e + 17	−1.08e + 06
6	0	0	1	3.51e + 08	−8.27e − 04
7	1	0	1	−2.56e + 12	−31.28e + 00
8	0	1	1	−5.33e + 12	−11.39e + 00
9	0	0	2	0.00e + 00	1.04e − 03

TABLE 12.3 A Text File Storing Exponents and Coefficients for the LC-VCO Polynomial Metamodel [83, 88]

The quality of Verilog-AMS-PAM is examined by using the LC-VCO in a PLL system as presented in Fig. 12.17(a) [83, 88]. For a comparative perspective, schematic design, physical design, and Verilog-AMS forms of LC-VCO are used in this PLL. The center frequency versus tuning voltage characteristics for the three different cases is shown in Fig. 12.17(b) [83, 88]. As evident, the layout implementation and schematic implementation of the LC-VCO cause differences in the center frequency due to the presence of the parasitics. When the LC-VCO is implemented in Verilog-AMS-PAM, the PLL tuning characteristics closely follow if the LC-VCO used has a physical design.

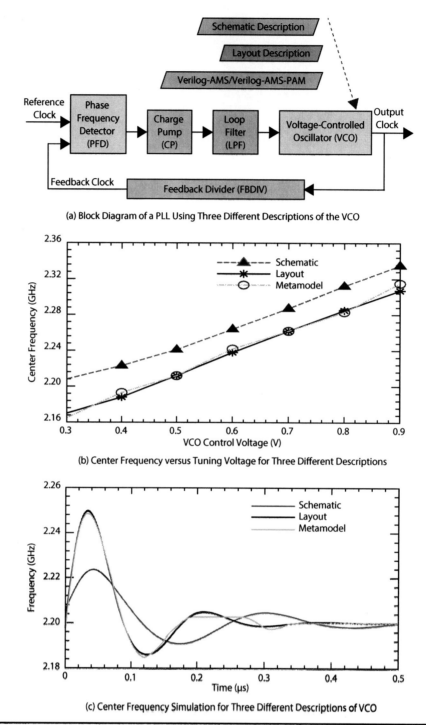

FIGURE 12.17 Verilog-AMS integrated with parasitic-aware metamodel (Verilog-AMS-PAM) for an LC-VCO [83, 88].

This Verilog-AMS-PAM is as accurate as the layout-aware SPICE netlist, i.e., physical design. The transient simulation for the center frequency of the PLL is presented in Fig. 12.17(c) for the three different implementations of the LC-VCO [83, 88]. As evident from the figure, the characteristics match for the Verilog-AMS-PAM and layout implementations of the LC-VCO.

12.5 Kriging-Based Metamodeling

After the discussion of polynomial metamodeling in the previous section, this section is dedicated to Kriging metamodeling [8, 27, 36, 49, 81, 82]. Kriging metamodeling is based on a set of geostatistical methods that are used for estimation of the surfaces based on limited data; thus,

Kriging metamodeling is also called geostatistics-inspired metamodeling. In 1950, the Kriging method was proposed by Georges Matheron, who named it after Daniel Krige who was a pioneer in geostatistical mining. This section presents brief theory of the Kriging methods with discussions on various types of Kriging methods. Generation of Kriging metamodel is presented, which is followed by specific case study designs.

12.5.1 Theory

The simulation data generated from the simulations may not be necessarily completely random. There may be correlations from one sample to another in the data generated. Thus, for highest accuracy the data can be modeled using a combination of polynomial model and departure. The general expression of a Kriging model has the following form [8, 32, 36, 49, 51, 68, 80]:

$$y^*_{\text{response}} = \text{Polynomial Metamodel} + \text{Departure} \tag{12.35}$$

$$y^*_{\text{response}} = \sum_{i=1}^{N_B} \lambda_i B_i(X) + Z(X) \tag{12.36}$$

In the above metamodel y^*_{response} is the predicted response. X is a set of input variables or design parameters. $\{B_i(X), i = 1, \ldots, N_B\}$ is a specific set of basis functions over the domain D_{N_B}; the basis functions are polynomial in nature. N_B is the number of basis functions. λ_i are fitting coefficients also known as weights. $Z(X)$ is the random error. Contrary to the other least-square-based approaches, the Kriging prediction method models $Z(X)$ to be a random process and not independent, unique to each weight and not distributed identically. The random process $Z(X)$ has mean μ_Z, variance σ_Z^2, and correlation function as follows [49, 51]:

$$R_Z(X_1, X_2) = \text{Corr}(Z(X_1), Z(X_2)) \tag{12.37}$$

where X_1 and X_2 are two samples of multidimensional variables or set X. This correlation function is called "variogram" in geophysics. The mean μ_Z, variance σ_Z^2, and correlation function (R_Z) are assumed known. The covariance structure of $Z(X)$ relates to the smoothness of the approximating surface. The key aspect of the Kriging method is that the observed data samples can be modeled as a random process with spatial autocorrelation [9, 49, 73]. The spatial autocorrelation between the observed data samples is explicitly modeled as weighted averages of nearby points using correlation function or variogram. The weights are unique to each predicted point and are a function of the distance between the points to be predicted and observed points. The weights are chosen so that the prediction error is minimized. Thus, simplistically speaking, the Kriging metamodels are a set of linear regression functions that minimize estimation variance from a predefined covariance model. The Kriging method is flexible due to the availability of a variety of correlation functions. The other advantage of the Kriging method is that it provides a basis for a stepwise algorithm to determine the important factors, and the same data can be used for screening and building the prediction model. The major disadvantage of the Kriging method is that metamodel generation can be quite time consuming.

The Kriging metamodeling has two different steps: examining the gathered data and estimating the data at those locations that have not been sampled (this step is essentially Kriging) [8, 36, 49]. The second step has many substeps such as calculating the empirical semivariogram, fitting a model to the empirical values, generating the matrices of Kriging equations, and solving them to obtain a predicted value and the error associated with it for each location. The estimation of the correlation between sampled points and a predicted point is performed using the semivariogram model. The semivariogram is calculated using the covariance and the correlation between the different sample values. It is then used with the points to be predicted. The empirical semivariogram is used to study the spatial trend of the design variables. However, in the actual point Kriging prediction, lag distances that are not available in the empirical semivariogram are required; hence, the empirical semivariograms are replaced with semivariogram models. The semivariogram models also ensure that the Kriging equations are solved due to non-negative definiteness. As presented in Fig. 12.18, the most common models used are the following: linear, spherical, Gaussian, exponential, and nugget [8, 32, 36, 49, 62, 68, 80]. Based on the nature of the observed data, one of these models could fit. The smoothness of the predicted points is affected by the use of this semivariogram model. A steeper model reduces the smoothness as it places more weight on closer neighbors. The most common semivariogram model used is the spherical model and is expressed as follows [49]:

$$\gamma(h) = C_0 + C\left[\frac{3h}{2a} - \frac{1}{2}\left(\frac{h}{a}\right)^3\right] \quad \text{for } 0 < h \leq a \tag{12.38}$$

where C_0, C, and a are shape parameters. The Gaussian model is expressed as follows [49]:

$$\gamma(h) = C_0\left[1 - \exp\left(-\frac{h^2}{r^2 a}\right)\right] \tag{12.39}$$

Based on the assumption of the distribution of the prediction error or noise $Z(X)$, mean μ_Z, and variance σ_Z^2, the Kriging methods are of various types, as shown in Fig. 12.18, including simple Kriging, ordinary Kriging, and universal Kriging [8, 32, 36, 49, 68, 80]. The simple Kriging method assumes a constant and known mean over the global domain. The ordinary Kriging method assumes a mean that is constant in the local domain of a predicted point. The universal Kriging method assumes the mean as a deterministic function. The simple Kriging method estimates the weights λ_i using the following expression [49, 51]:

$$\begin{pmatrix} \lambda_1 \\ \vdots \\ \lambda_N \end{pmatrix} = \Gamma^{-1} \begin{pmatrix} \gamma(x_1, x_0) \\ \vdots \\ \gamma(x_N, x_0) \end{pmatrix} \tag{12.40}$$

where N is the number of sample points of the input variable X, γ and Γ is the covariance matrix of the observed data points and is expressed by the following:

$$\Gamma = \begin{pmatrix} \gamma(x_1, x_1) & \cdots & \gamma(x_1, x_n) \\ \vdots & \ddots & \vdots \\ \gamma(x_n, x_1) & \cdots & \gamma(x_n, x_n) \end{pmatrix} \tag{12.41}$$

where the following expression is used for each correlation calculation:

$$\gamma(x_1, x_2) = E\left(|Z(x_1) - Z(x_2)|^2\right) \tag{12.42}$$

In the ordinary Kriging method, the weights are chosen to minimize the Kriging variance under the unbiasedness constraint that $E(\hat{Z}(X) - Z(X)) = 0$. Hence the weights are chosen so that the following expression is satisfied:

$$\sum_{j=1}^{N} \lambda_j = 1 \tag{12.43}$$

This above condition is not required for simple Kriging. The weights then for ordinary Kriging technique are calculated by the following expression [49, 57]:

$$\begin{pmatrix} \lambda_1 \\ \vdots \\ \lambda_N \\ \Lambda \end{pmatrix} = \Gamma^{-1} \begin{pmatrix} \gamma(x_1, x_0) \\ \vdots \\ \gamma(x_N, x_0) \\ 1 \end{pmatrix} \tag{12.44}$$

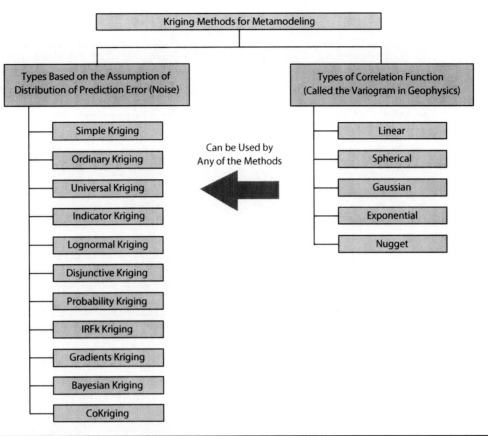

FIGURE 12.18   Different Kriging methods.

where Λ is a Lagrange multiplier. Γ is the covariance matrix of the observed data points and for ordinary Kriging is given as follows:

$$\Gamma = \begin{pmatrix} \gamma(x_1,x_1) & \cdots & \gamma(x_1,x_n) & 1 \\ \vdots & \ddots & \vdots & 1 \\ \gamma(x_n,x_1) & \cdots & \gamma(x_n,x_n) & 1 \\ 1 & 1 & 1 & 0 \end{pmatrix} \quad (12.45)$$

It may be noted that Fig. 12.18 indicates that any of the correlation functions or semivariogram models can be used in any of the Kriging methods. It is a design engineer's choice based on the characterization data.

12.5.2 Generation

The flow of generation of the Kriging metamodels for nanoelectronic circuits is presented in Fig. 12.19 [49, 54, 57]. The initial few steps such as design parameter and target FoM identification, SPICE netlist parameterization, parameter sampling, analog/SPICE simulation, simulation data normalization, and parameter/FoM data splitting for training/fitting and verifications are similar to those in the case of the polynomial metamodel generation flow that is presented in a previous section. In the next phase, for the Kriging metamodel generation, a correlation function or semivariogram model is selected from available options like linear, spherical functions. Now a specific Kriging function from the available options, such as simple Kriging and ordinary Kriging, is selected for regression. This involves calculation of weights from the variogram such that the error variance is minimal. The tools such as mGstat [1] and surrogate modeling (SUMO) can be used in the MATLAB® framework for the Kriging metamodeling [1, 26]. A script can be used to ordinate the analog/SPICE simulation and metamodeling. The verification parameter/FoM set is used at this

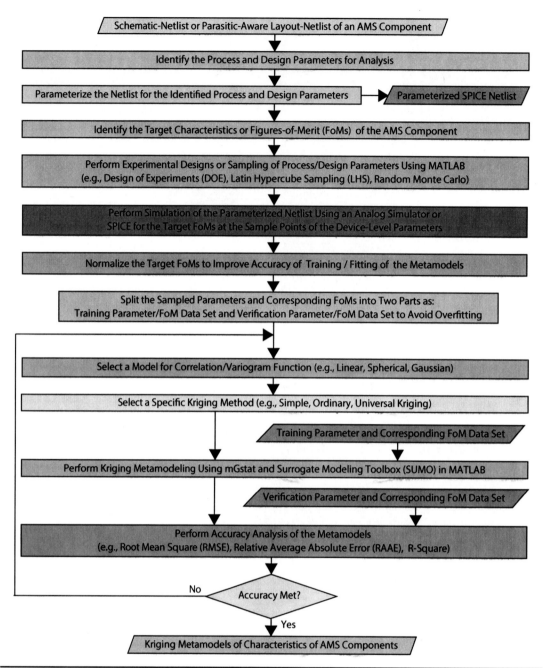

Figure 12.19 Steps for Kriging methods-based metamodel generation.

phase to validate the Kriging metamodels. A Kriging metamodel is generated for each of the target characteristic/FoM of the AMS-SoC component.

12.5.3 Simple Kriging Metamodeling of a Clamped Bitline Sense Amplifier

Simple Kriging metamodeling of 45-nm CMOS-based sense amplifier is presented in this section as a specific case study of Kriging metamodeling. The overall view is presented in Fig. 12.20 [49, 51, 54, 57]. Following the typical approaches, a parameterized netlist of the 45-nm CMOS-based clamped bitline sense amplifier is generated. There are 10 NMOS transistors and 2 PMOS transistors in this sense amplifier design and hence W_n and W_p are assumed to be design variables. The characteristics such as precharge time (T_{PC}), sense delay (T_{SD}), average power dissipation (P_{SA}), and sense margin (V_{SM}) are targets for which metamodels need to be generated. The validations of

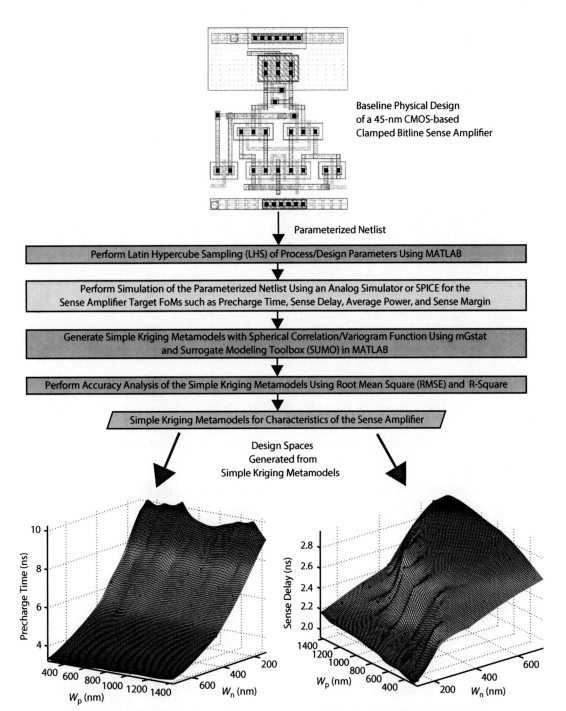

FIGURE 12.20 Simple Kriging metamodeling of a 45-nm CMOS-based clamped bitline sense amplifier [49, 51, 54, 57].

these Kriging metamodels are performed using accuracy metrics. As a specific example, 100 samples of W_n and W_p are picked using LHS. Analog/SPICE simulations are performed for characterizing the target FoMs at the 100 sample points. The Kriging surfaces are generated once the weights of Kriging metamodel are calculated. As a specific example, Kriging surfaces for precharge time (T_{PC}) and sense delay (T_{SD}) are shown in Fig. 12.20. For design explorations of the sense amplifier, these simple Kriging metamodels or simple Kriging surfaces can be used.

The accuracy analysis of the simple Kriging metamodels for the five target FoMs is presented in Table 12.4 [49, 51, 54, 57]. Specifically, RMSE and R^2 are reported for these five FoMs. The results are presented for two different sampling counts, one for 20 LHS and another for 100 LHS. For a

Characteristics	Precharge Time (T_{PC})		Sense Delay (T_{SD})		Average Power (P_{SA})		Sense Margin (V_{SM})	
Sample Count	20	100	20	100	20	100	20	100
RMSE	1.3×10^{-09}	2.7×10^{-10}	3.8×10^{-07}	5.2×10^{-08}	2.3×10^{-10}	1.2×10^{-11}	2.4×10^{-03}	2.8×10^{-04}
R^2	0.98	0.98	0.96	0.96	0.99	0.99	0.94	0.95

TABLE 12.4 Error Analysis for the Simple Kriging Metamodels for the 45-nm CMOS-Based Clamped Bitline Sense Amplifier [49, 51, 54, 57]

Sense Amplifier Characteristics	RMSE		R^2	
	For Kriging	For Polynomial	For Kriging	For Polynomial
Precharge Time (T_{PC})	2.7×10^{-10}	5.0×10^{-10}	0.98	0.93
Sense Delay (T_{SD})	5.2×10^{-08}	8.6×10^{-08}	0.97	0.90
Average Power (P_{SA})	1.2×10^{-11}	5.9×10^{-11}	0.99	0.96
Sense Margin (V_{SM})	2.8×10^{-04}	3.7×10^{-04}	0.95	0.89

TABLE 12.5 Comparison of Simple Kriging and Polynomial Metamodels for the 45-nm CMOS-Based Clamped Bitline Sense Amplifier [49, 51, 54, 57]

comparative perspective with polynomial metamodels, the accuracy metrics of Kriging and polynomial metamodels are presented side by side in Table 12.5 [49, 51, 54, 57]. It can be observed that Kriging metamodels consistently perform better than polynomial metamodels when analyzed with RMSE and R^2 as statistical metrics.

12.5.4 Ordinary Kriging Metamodeling of a Sense Amplifier

In this section another Kriging metamodeling is explored for a different circuit as compared to the previous section. Specifically, ordinary Kriging metamodels are generated for a 45-nm CMOS-based conventional sense amplifier circuit. The overall flow is presented in Fig. 12.21 [49, 50, 57]. The steps followed are similar to the steps used in the previous section. Same set of design variables, same spherical variogram functions, tools, and accuracy analysis metrics are used. The ordinary Kriging surface is plotted for the ordinary Kriging metamodels when weight coefficients are obtained. The accuracy analysis of the four ordinary metamodels of this sense amplifier circuit is presented in Table 12.6 [49, 50, 57]. The results are presented for two different sample counts of 100 and 200. It can be observed that the ordinary Kriging metamodels are quite accurate with very low RMSE values and average R^2 values in the range of 0.98–0.99. As expected the metamodels generated from 200 LHS sampling points are generally more accurate than the ones generated from 100 sampling points. A comparison of ordinary Kriging metamodels with simple Kriging metamodels from the previous section for same sample count of 100 indicates that simple Kriging metamodels are slightly more accurate for most of the FoMs of the sense amplifier.

12.5.5 Universal Kriging Metamodeling of a Phase-Locked Loop

In this section the third Kriging method called universal Kriging method is presented for an entirely different circuit, a 180-nm CMOS-based PLL, as compared to the previous section. The overall universal Kriging metamodel generation flow is shown in Fig. 12.22 [53]. A total of 21 parameters consisting of W_n and W_p of transistors present in four components of the PLL (i.e., PFD, CP-LPF, LC-VCO, and FBDIV) as shown in the figure are considered as design parameters. The following characteristics of the PLL are considered for universal Kriging metamodel generation: locking time, center frequency, and average power. A Gaussian variogram model is used in this specific Kriging metamodeling. A comparison with SPICE analysis data shows that universal Kriging metamodels are quite accurate, e.g., metamodel locking time has 0.33% error and metamodel for power dissipation has 0.7% error as compared to the SPICE data.

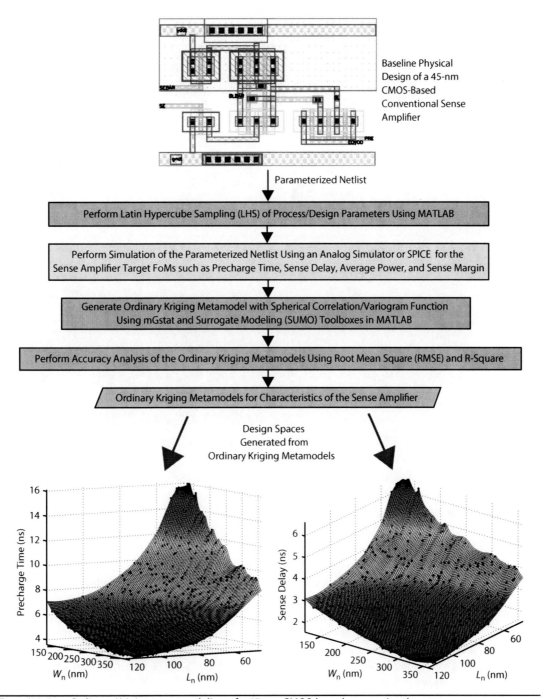

FIGURE 12.21 Ordinary Kriging metamodeling of a 45-nm CMOS-based conventional sense amplifier [49, 50, 57].

Characteristics	Precharge Time (T_{PC})		Sense Delay (T_{SD})		Average Power (P_{SA})		Sense Margin (V_{SM})	
Sample Count	100	200	100	200	100	200	100	200
RMSE	4.7×10^{-10}	2.3×10^{-10}	2.0×10^{-10}	1.1×10^{-10}	3.4×10^{-10}	3.3×10^{-11}	3.4×10^{-03}	1.8×10^{-04}
R^2	0.96	0.99	0.95	0.99	0.84	0.85	0.98	0.99

TABLE 12.6 Error Analysis for the Ordinary Kriging Metamodeling of a 45-nm CMOS-Based Conventional Sense Amplifier [49, 50, 57]

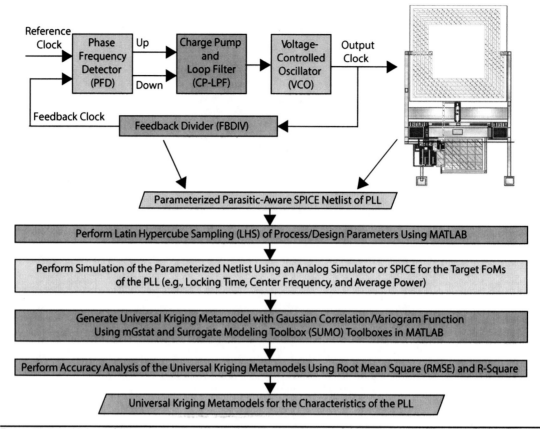

Figure 12.22 Universal Kriging metamodeling of a 180-nm CMOS-Based PLL [53].

12.6 Neural Network–Based Metamodeling

In this section the focus is on the NN-based metamodel. Both ANN and NN have been used interchangeably in the literature. This text will use NN. The NN computational models are inspired by the central nervous system, specifically the brain [5, 16, 31, 68, 69]. The history of NN can be traced back to 1943, when Warren McCulloch and Walter Pitts created the threshold logic computational model. This model eventually led to two fronts of NN research: one focused on the biological processes in the brain and the other focused on the application of NNs to artificial intelligence. In this section, a brief introduction on the NN model is presented. Then use of NN metamodels for specific nanoelectronic ICs is presented. The concept of intelligent Verilog-AMS (iVAMS) is introduced, which is similar to Verilog-AMS-PoM or Verilog-AMS-PAM presented in the polynomial metamodel discussions in Sec. 12.4.

12.6.1 Theory

The NN consists of neurons in various layers as presented in Fig. 12.23(a) [5, 31, 68, 69]. As evident from the figure, a single neuron consists of a weighted sum function or layer function of the following form:

$$z(X) = \sum_{i=1}^{N} \omega_i x_i + \omega_0 \tag{12.46}$$

where $X = \{x_1, x_2, \ldots, x_N\}$ are the input variables and ω_i called weights are regression coefficients. The neuron also includes an activation function (f). For the specific activation function shown in the figure, the output from the neuron is [31, 68, 69]:

$$y = f(Z) = \left(\frac{1}{1 + \exp(-\lambda z)}\right) \tag{12.47}$$

In the above function, λ is called the slope parameter. Using the weighted sum function in the above function, the output of the neuron can be expressed as follows:

$$y(X) = \left(\frac{1}{1 + \exp\left(-\lambda\left(\sum_{i=1}^{N}\omega_i x_i + \omega_0\right)\right)} \right) \tag{12.48}$$

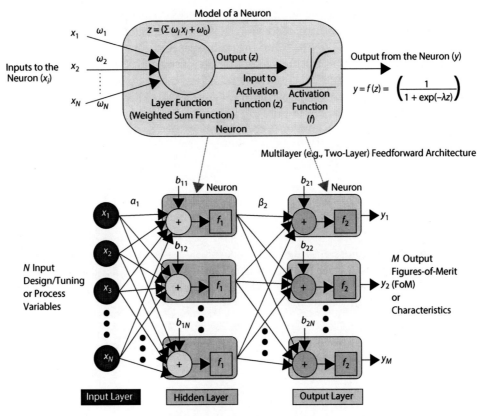

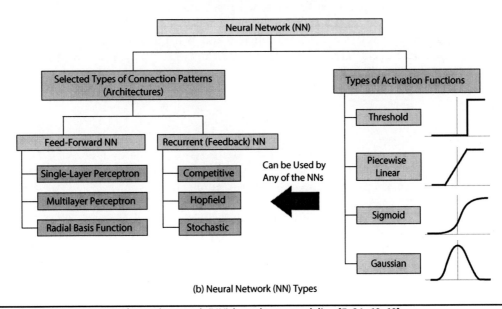

FIGURE 12.23 The concept of neural network (NN)-based metamodeling [5, 31, 68, 69].

There are two key aspects of building the NN computational model [31, 68, 69]:

1. *Specifying the NN architecture*: An NN computational model is created by assembling the neurons into an architecture.
2. *Training the NN architecture*: The NN architecture is trained by minimizing a least-square error so that it behaves just as the original training data was supposed to behave.

The above process is analogous to specifying the regression model and estimating the parameters of the regression mode from the reference/training data. An NN architecture is shown in Fig. 12.23(a) for the purpose of illustrations. A selected variety of NN computational models based on their architecture are shown in Fig. 12.23(b) [5, 31, 68, 69]. The figure also shows a selected set of activation functions such as threshold, piecewise linear, sigmoid, and Gaussian.

One of the most popular NN is the multilayer feed-forward NN [16, 20, 21]. It has an input layer, a nonlinear activation function in the hidden layer, and a linear activation function in the output layer. The multilayer NN needs to have at least one nonlinear activation function, otherwise the composition of linear functions becomes just another linear function. The multilayer NNs are very flexible and powerful due to their capability to model the nonlinear as well as the linear functions. The two common nonlinear activation functions are the following [14, 16]:

$$y(Z) = \left(\frac{1}{1+e^{-\lambda z_i}}\right) \tag{12.49}$$

$$y(Z) = \tanh(\lambda z_i) \tag{12.50}$$

In the above expressions i denotes a neuron in the hidden layer, z_j and y_j are its input and output, respectively, and λ is the neuron transfer function steepness. For f_1 a hyperbolic tangent function and f_2 a pure linear function, the NN architecture presented in Fig. 12.23(a) can be expressed mathematically as follows [83, 84]:

$$y_l = b_{2l} + \sum_{j=1}^{N} \left(\beta_{2,jl} \tanh\left(b_{1j} + \sum_{i=1}^{N} \alpha_{1,ij} x_i\right)\right) \tag{12.51}$$

where there are M number of outputs.

Another NN architecture called the radial NN is also a two-layer network. In this NN architecture, the first layer having radial base neurons calculates its weighted inputs with distance and its net input with a radial function. The second layer having linear neurons calculates its weighted input with a dot product function and its net inputs by combining its weighted inputs and biases. Both layers of radial NN have biases. The radial NN architecture is mathematically modeled as follows [16, 20, 21]:

$$y = \sum_{i=1}^{N} \omega_i \rho\left(\|x - c_i\|\right) \tag{12.52}$$

where c_i is the center vector of neuron i, x is the prediction point, ρ is the transfer function of the neuron, and ω_i are the weights of the linear neuron.

The NNs can operate in a parallel and distributed fashion that resembles the biological NNs. The NNs have great potential for parallelism as the computations of the components are independent of each other. It has been proven in the universal approximation theorem that an NN with one hidden layer can estimate any continuous function that maps to real numbers [16, 20, 21]. However, selection of appropriate NN architecture and training of the NN can be time consuming. The NN metamodels can be efficient for large and complex integrated circuits as compared to the polynomial metamodels.

12.6.2 Generation

The flow for NN metamodel generation is presented in Fig. 12.24 [16, 20, 21]. Starting from a SPICE netlist of the schematic or physical design, typical steps are followed to create parameterized SPICE netlist, sample the design/process parameters, and perform analog/SPICE simulations. After the simulation data is obtained, the parameter sample and FoM are split into two sets: training parameter/FoM data and verification parameter/FoM data. For example, 70%–80% for training and 20%–30% for verification. This step is important as training and verification data needs to be distinct so that the NN is not artificially inflated. Overfitting of the NN is the scenario in which the NNs behave worse after certain point of training when it is trained excessively to

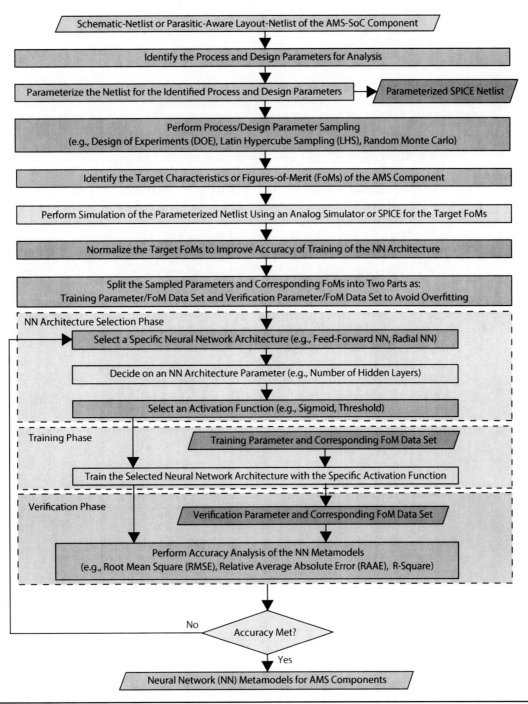

FIGURE 12.24 The flow for neural network (NN)-based metamodel generation [16, 20, 21].

more or less make zero error. This happens as excessive training or use of a large amount of neurons in the hidden layers of the architecture may make the NN memorize the training patterns and stop adjusting the weights. One method to avoid overfitting is to stop training at the point of optimal generalization. Another option is regularization that limits the complexity of the NN to avoid its learning of the peculiarities.

The rest of the NN metamodel generation has the following three distinct phases: (1) architecture selection phase, (2) training phase (or learning phase), and (3) verification phase. In the neural architecture selection phase, a specific architecture like feed-forward or radial is selected. In this phase the architecture parameters such as the number of hidden layers can be decided. Selection of an activation function from the available options like sigmoid and threshold is also part of the architecture selection phase. Once the specific architecture with all its features and parameters is selected, the NN training phase or learning phase starts. The training of NNs determines the weights of functions such that the output of the NN matches the FoM data. The minimal least-square criterion can be used for this purpose [16, 20, 21]:

$$E_{\text{NN}} = \sum_{i=1}^{N}(y_{\text{response},i} - y^*_{\text{response},i})^2 \qquad (12.53)$$

where $y_{\text{response},i}$ and $y^*_{\text{response},i}$ are the actual and predicted responses, respectively, at the ith training point out of a total of N. The NN verification phase follows the training phase. The aim of the verification phase is to verify the NN architecture obtained such that it passes the accuracy criteria. The verification phase uses verification parameter/FoM data set that is different from the training parameter/FoM data set. If the accuracy requirements are not met, the NN architecture is not an accurate representation of the IC and system. So, the three above phases of architecture: selection, training, and verification can be repeated with different options. However, the analog/SPICE simulations need not be performed again. Once the accuracy requirements are met for a specific architecture and activation function, the NN metamodel is obtained for use in IC and system analysis, verification, and optimization tasks in place of the SPICE netlist.

12.6.3 Neural Network Metamodel of PLL Components

As a specific example of NN metamodeling, a 180-nm CMOS-based PLL is considered in this section. The overall NN metamodeling for the 180-nm CMOS-based PLL is shown in Fig. 12.25 [20, 21]. The flow is essentially the generic NN metamodel generation flow customized for the PLL. The 180-nm CMOS-based PLL has a target specification as follows: center frequency = 2.7 GHz, power dissipation = 3.9 μW, and locking time = 8.5 μs. A total of 21 device-level design variables consisting of W_n and W_p of transistors present in the four components of the PLL (i.e., PFD, CP-LPF, LC-VCO, and FBDIV) are considered as input variables. The target FoMs of this integrated circuit that are considered for metamodeling are locking time, center frequency, and power dissipation.

In this specific metamodeling, LHS is used to sample the 21 design variables. As a specific example, 100 sample points of the design variables and corresponding analog/SPICE generated data are used for NN training. A two-layer feed-forward NN architecture is trained. Different alternative functions such as log-sigmoid and hyperbolic tangent sigmoid are used for activation functions f_1 and f_2 present in the two layers. A different number of neurons are used in the NN architecture. As a specific example, 30 sample points of the design variables and corresponding analog/SPICE generated data are used for NN verification. The results for NN metamodeling for center frequency and locking time of the PLL are shown in Table 12.7 [20, 21]. The results are presented for two different sets of activation functions as shown in the table. The two cases are reported with and without data normalization as well as with different numbers of neurons in the hidden layer. The comparison results for NN and polynomial metamodeling for the PLL are presented in Table 12.8 [20, 21]. It is observed that the NN metamodels have a 56% increase in accuracy over the polynomial metamodels.

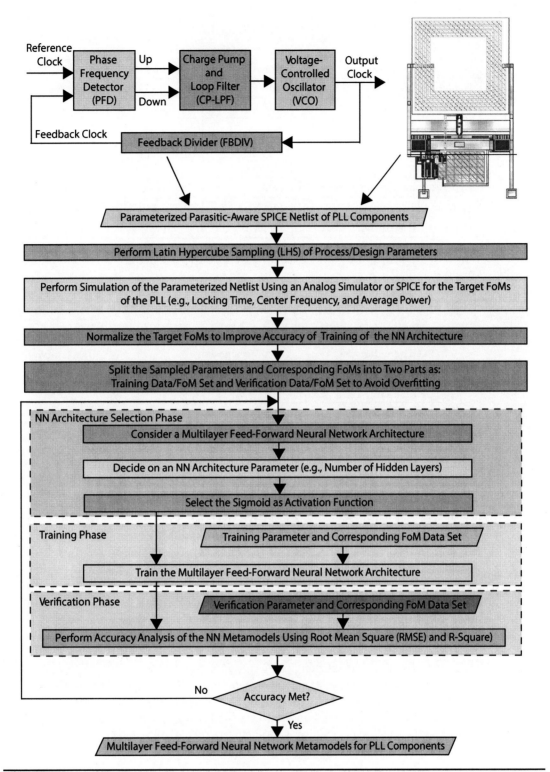

FIGURE 12.25 Neural network (NN) metamodeling of PLL components [20, 21].

12.6.4 Intelligent Verilog-AMS

In this section NN metamodeling is presented for 90-nm CMOS-based operational amplifier (OP-AMP). It is the same OP-AMP design whose polynomial metamodels are presented in a previous section. The NN metamodels are integrated with Verilog-AMS to generate what is called "intelligent Verilog-AMS (iVAMS)" for the OP-AMP circuit. Metamodeling of both the OP-AMP circuit

Functions (f1, f2)	Data Filtering	R²-Verification	RMSE	Neurons Count
For Center Frequency of the PLL				
Log-sigmoid, Linear	None	0.72	52.74 MHz	4
Hyperbolic Tangent Sigmoid, Linear	None	0.71	51.24 MHz	3
Log-sigmoid, Linear	Min-Max	0.66	48.90 MHz	9
Hyperbolic Tangent Sigmoid, Linear	Min-Max	0.70	53.65 MHz	1
For Locking Time of the PLL				
Log-sigmoid, Linear	None	0.87	1.30 μs	1
Hyperbolic Tangent Sigmoid, Linear	None	0.72	1.44 μs	9
Log-sigmoid, Linear	Min-Max	0.87	1.29 μs	1
Hyperbolic Tangent Sigmoid, Linear	Min-Max	0.94	1.12 μs	10

TABLE 12.7 Two-Layer Feed-Forward Neural Network (NN) Metamodeling of a 180-nm CMOS-Based PLL Component [20, 21]

PLL Characteristics	RMSE for Polynomial Metamodels	RMSE for Neural Network Metamodels
Center Frequency	78 MHz	48 MHz
Locking Time	1.9 μs	1.2 μs
Power Dissipation	2.6 mW	0.29 mW

TABLE 12.8 Metamodel Comparison: Polynomial versus Neural Network (NN) for a 180-nm CMOS-Based PLL [20, 21]

parameters and performances is considered, in other words, CPM and PMM are generated. The iVAMs generation through the creation of NN metamodels of the OP-AMP is shown in Fig. 12.26 [84, 86]. In this NN metamodeling of the OP-AMP, the following are considered as device-level variables, design variables, or input variables $X_{tuning} = \{L_n, L_p, W_n, W_p\}$. A total of eight variables are used by properly grouping the transistors in the OP-AMP circuit and using the same variables if more than one transistor are of same size. The following OP-AMP circuit parameters are considered for NN-based CPM: open-loop gain A_{DC}, transconductance of input stage g_{mi}, maximum available positive current I_{max+}, maximum available negative current I_{max-}. The following OP-AMP characteristics, performance metrics, or FoM are considered for NN-based PMM: power dissipation (PD_{OP-AMP}), bandwidth (BW_{OP-AMP}), phase margin (PM_{OP-AMP}), slew rate (SR_{OP-AMP}).

As a specific architecture, a multilayer feed-forward NN is considered in this metamodeling. The input layer has design variables $X_{tuning} = \{L_n, L_p, W_n, W_p\}$ and bias currents as input variables. The multilayer feed-forward NN architecture consisted of a hidden layer of four neurons with hyperbolic tangent function as the activation function f_1. The output layer of the NN architecture is a single neuron employing a linear activation function f_2. The output of this NN architecture can be expressed mathematically as follows [84, 86]:

$$y_{response} = b_{21} + \sum_{j=1}^{N} \omega_{2,j} \tanh\left(b_{1j} + \sum_{i=1}^{N} \omega_{1,ij} x_i\right) \qquad (12.54)$$

where $\omega_{1,ij} \in W_1$ and $\omega_{2,j} \in W_2$. W_1 is a matrix containing the weights of the connections from the design variables in the input layer to the neurons in the hidden layer. W_2 is a matrix containing the weights of the connections from the hidden layer to the output layer. The bias b_{ij} ($i = 1, 2$ and $j = 1, 2, ..., N$) are useful for additional control of each neuron. The NN architecture is trained using

Metamodel-Based Fast AMS-SoC Design Methodologies 729

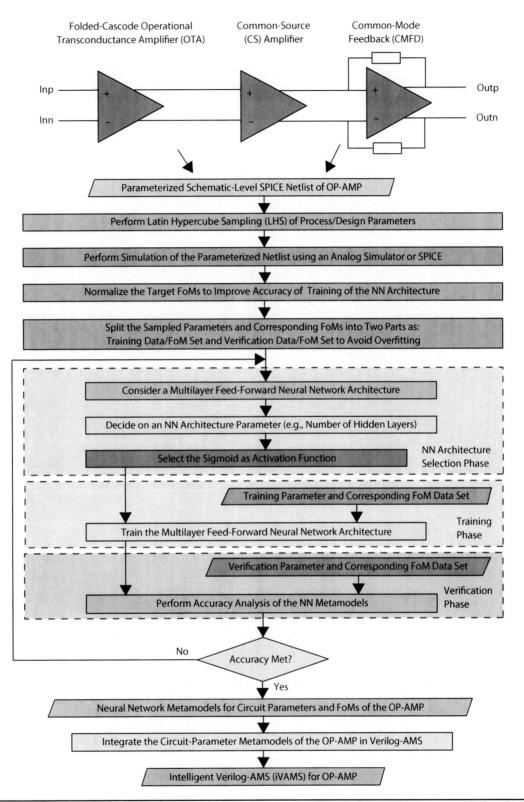

FIGURE 12.26 Neural network (NN) metamodel integrated intelligent Verilog-AMS (iVAMS) generation flow for an OP-AMP [84, 86].

ALGORITHM 12.4 *iVAMS Code for the 90-nm CMOS-Based OP-AMP [84, 86]*

```
function real OP-AMP_NN_metamodel;
integer weight_w1, weight_w2, bias_b1, bias_b2, i, j, readfile,...;
real weight_w, b, v, u;
// Read neural network metamodel weights and bias from
// text files weight_w1, weight_w2, bias_b1, and bias_b2.
begin
   weight_w1 = $fopen("weight_w1.txt", "r");
   weight_w2 = $fopen("weight_w2.txt", "r");
   bias_b1 = $fopen("bias_b1.txt", "r");
   bias_b2 = $fopen("bias_b2.txt", "r");
   v = 0.0;
   for (j = 0; j < nl; j = j + 1) begin
      u = 0.0;
      for (i = 0; i < size_x; i = i + 1) begin
         readfile = $fscanf(weight_w1, "%e", weight_w);
         u = u + weight_w * x[i];
      end
      readfile = $fscanf(weight_w2, "%e", weight_w);
      readfile = $fscanf(bias_b1, "%e", bias_b);
      v = v + weight_w * tanh(u + bias_b);
   end
   readfile = $fscanf(bias_b2, "%e", bias_b);
   OP-AMP_NN_metamodel = v + bias_b;
   $fclose(weight_w1);
   $fclose(weight_w2);
   $fclose(bias_b1);
   $fclose(bias_b2);
end
endfunction
```

500 samples with Bayesian regulation training. Similar to the case of the polynomial metamodeling in the previous section, the integration of the NN CPMs into the macromodel of the OP-AMP produces meta-macromodels, which when described in Verilog-AMS is called iVAMS. In addition, the iVAMS can include NN-based PMMs. An example iVAMS code of this OP-AMP is presented in Algorithm 12.4 [84, 86]. In the iVAMS, `initial` block of the OP-AMP iVAMS module, these weights and biases are read from the files, and the function `OP-AMP_NN_metamodel` computes the circuit parameter values. The computed circuit parameter values can be used in an `analog` process in the iVAMS module to realize the model. Then various analysis on the OP-AMP, for example, small-signal function, can be described using a function such as `laplace_nd`. As an example, AC analysis of 90-nm CMOS-based OP-AMP using SPICE and iVAMS is shown in Fig. 12.27 [84, 86]. As evident from the figure, the accuracy of iVAMS is closely matching with that of the SPICE simulations; however, iVAMS simulation is way faster than SPICE.

The accuracy results for various NN metamodels for the 90-nm CMOS-based OP-AMP are shown in Table 12.9 [84, 86]. Three accuracy analysis metrics such as coefficient of determination (R^2), RMAE, and RMSE are used. The values of R^2 in these cases have been close to 1, and the values of RMAE as well as RMSE are very small, which indicates that the NN metamodels of the OP-AMP are highly accurate. However, in case of lower accuracy, the architecture parameters can be changed to obtain a different NN metamodel.

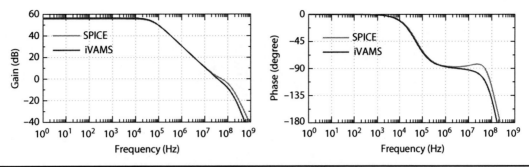

FIGURE 12.27 AC analysis of 90-nm CMOS-based OP-AMP using SPICE and iVAMS [84, 86].

Different OP-AMP Parameters and Characteristics	Accuracy Metric		
	R^2	RMAE	RMSE
For OP-AMP Circuit Parameter Metamodeling			
Open-loop DC Gain A_{DC}	0.96	1.32	41.93 V/V
Transconductance of Input Stage g_{mi}	0.99	0.11	1.77 μA/V
Maximum Available Positive Current I_{max+}	0.99	0.67	0.31 μA
Maximum Available Negative Current I_{max-}	0.99	0.49	0.26 μA
For OP-AMP Performance Metamodeling			
Power Dissipation (PD_{OP-AMP})	0.99	0.52	8.31 μW
Bandwidth (BW_{OP-AMP})	0.99	0.89	2.12 kHz
Phase Margin (PM_{OP-AMP})	0.90	2.16	4.99 degree
Slew Rate (SR_{OP-AMP})	0.99	0.48	0.29 mV/ns

TABLE 12.9 Accuracy Results for NN Metamodels for the 90-nm CMOS-Based OP-AMP [84, 86]

12.6.5 Kriging Bootstrapped Training for Neural Network Metamodeling

In the Kriging metamodeling, the correlations among the variations in the design and process parameters are taken into account in the calculation of the weight for each predicted point. However, the drawback of the Kriging metamodel is that the weight for each predicted point is unique and involves matrix manipulations that could become cumbersome for a multidimensional design space involving a large number of variables. On the other hand, the NN metamodels are accurate and robust for high-dimensional design space but do not effectively account for correlations in the process variations. Hence, one idea is to combine the Kriging and NN methods for robust, accurate, and efficient metamodeling of large nanoelectronic circuits. The flow of Kriging bootstrapped NN metamodeling for a 180-nm CMOS-based PLL is presented in Fig. 12.28 [49, 55, 56].

Based on the previous discussions on the Kriging metamodeling as well as NN metamodeling, it can be said that the steps in this flow are self-explanatory. In essence, the sample parameters/FoM data points are fed into a Kriging method that produces an intermediate set of sample data points (i.e., bootstrapped). This data is then used to train a specific NN architecture. Thus, instead of random Monte Carlo samples being used for the NN training, the Kriging-generated, correlated sample data points are used. A comparison of Kriging metamodeling, NN metamodeling, and Kriging-NN metamodeling is presented in Table 12.10 [49, 55, 56]. As evident, Kriging metamodeling is the slowest. While both NN metamodeling and Kriging-NN metamodeling take the same amount of time, Kriging-NN metamodel has higher accuracy than the NN metamodel.

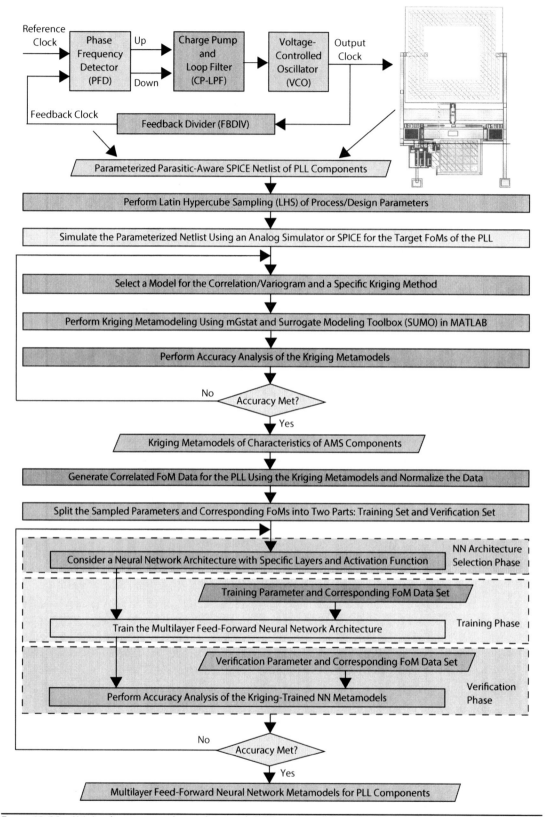

FIGURE 12.28 Kriging bootstrapped neural network (NN) metamodeling of a PLL [49, 55, 56].

Metamodels Types	Kriging Metamodeling	NN Metamodel	Kriging-NN Metamodeling
Simulation Time	468 s	19 s	19 s
Speedup	1	24.6 ×	24.6 ×

TABLE 12.10 Monte Carlo Time Analysis Comparison for Metamodels [49, 55, 56]

12.7 Ultrafast Process Variations Analysis Using Metamodels

Different types of metamodels and flow for the generations of such metamodels have been discussed in the previous sections of this chapter. The metamodels can be used in various design tasks such as analysis, verification, and optimization with or in place of SPICE netlists. As an example of application of the metamodels, process variation analysis is presented in this section.

12.7.1 Kriging-Metamodel-Based Process Variation Analysis of a PLL

The Kriging-metamodel-assisted statistical process variation analysis flow for a 180-nm CMOS-based PLL is shown in Fig. 12.29 [53]. The Kriging metamodel generation flow is similar to what has been presented in the previous section on Kriging metamodeling. The sampled device parameters that are generated from their statistical distributions are generated and analog/SPICE simulations are performed for those sampled points. In total, 21 design and process parameters have been used for this PLL. Gaussian distribution with standard deviation of 10% of the mean is used to generate the parameters. Using this parameter/FoM data, universal Kriging metamodels are generated with a Gaussian semivariogram model, which is then verified with a distinct set of parameter/FoM data. At this point, accurate universal Kriging metamodels of the PLL are available, which can be used for further analysis of the PLL in place of its SPICE netlist. The analysis over Kriging metamodels is ultrafast as compared to analysis over SPICE netlist, which is slow due to use of slow analog/SPICE engine. The device parameter samples generated from their distributions can be used in MATLAB® to generate responses of the FoMs whose Kriging metamodels are now available. As a specific example, the statistical distributions of two characteristics of PLL, such as power dissipation and locking time, are presented in the figure. For a broad perspective of accuracy, selected results are shown in Table 12.11 [53]. The statistical analysis for mean (μ) and standard deviation (σ) of the power dissipation and locking time of the PLL are shown. The accuracy of the results validates the proposed statistical models while reducing the amount of time required for the conventional Monte Carlo analysis in SPICE. For this case study PLL circuit that has 21 design and process parameters, even a high and low test case will require 2^{21} simulations. The simulation time for the random Monte Carlo analysis of 1000 run on the actual netlist in SPICE is about 5 days for this PLL, while the Kriging metamodel generation and process analysis using them takes only a few hours. Thus, the simulations time saving is very significant as it will translate to reduction in chip cost.

12.7.2 Neural Network Metamodel-Based Process Variation Analysis of a PLL

As a second example of metamodel application, in this section, an NN metamodel-assisted process variation analysis is presented. Specifically, an NN metamodel-assisted process variation analysis flow for a 180-nm CMOS PLL is presented in Fig. 12.30 [22]. The parameter with the same statistical distributions as the previous section is used in the current analysis. A 1000 run Monte Carlo simulation for statistical analysis over the SPICE netlist using statistical distributions of 35 variables takes roughly 5 days' time. Instead the Monte Carlo simulation over the NN metamodels using MATLAB® is a matter of a few hours. A selected set of probability density functions (PDFs) obtained from the Monte Carlo analysis on the NNs is shown in the figure, specifically center frequency and power dissipation. The statistical parameters of these PDFs are also presented in Table 12.12 [22]. As evident from the table, the results from NN metamodels match very closely those obtained from SPICE and have close to zero errors.

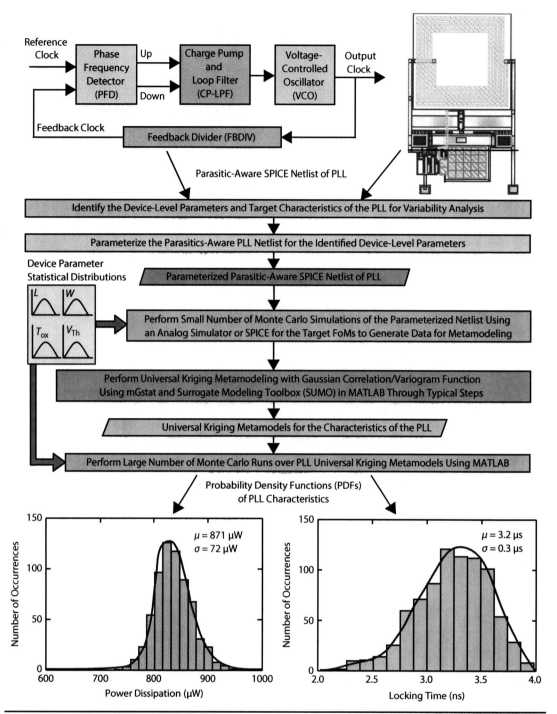

Figure 12.29 Universal Kriging-based process variation analysis of a PLL [53].

Characteristics of PLL	From SPICE		From Kriging Metamodels		Errors	
	Mean (μ)	Standard Deviation (σ)	Mean (μ)	Standard Deviation (σ)	Mean (μ)	Standard Deviation (σ)
Power Dissipation	877 μW	73 μW	871 μW	72 μW	0.7%	1.4%
Locking Time	3.24 μs	1.1 μs	3.23 μs	0.30 μs	0.3%	69.1%

Table 12.11 Kriging-Metamodel-Based Statistical Analysis of a 180-nm CMOS-Based PLL [53]

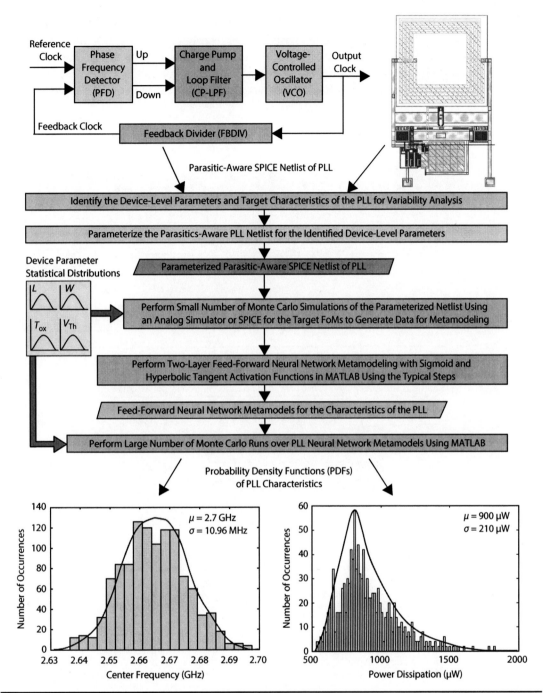

FIGURE 12.30 Neural network-based process variation analysis of a PLL [22].

Characteristics of PLL	From SPICE		From Neural Network		Errors	
	Mean (μ)	Standard Deviation (σ)	Mean (μ)	Standard Deviation (σ)	Mean (μ)	Standard Deviation (σ)
Center Frequency	2.7 GHz	10.95 MHz	2.7 GHz	10.96 MHz	0.0%	0.1%
Power Dissipation	900 μW	210 μW	900 μW	210 μW	0.1%	1.3%

TABLE 12.12 Two-Layer Feed-Forward Neural Network Metamodel-Based Statistical Analysis of a 180-nm CMOS-Based PLL [22]

12.7.3 Kriging-Trained Neural Network–Based Process Variation Analysis of a PLL

Kriging-trained NN metamodel-based process variation analysis of a 180-nm CMOS PLL is now discussed in this section. The overall self-explanatory flow for process variations analysis over Kriging-trained NN metamodel is presented in Fig. 12.31 [56]. For all the 21 device parameters, the samples are generated using Gaussian distributions just as was the case in the previous two sections. Selected

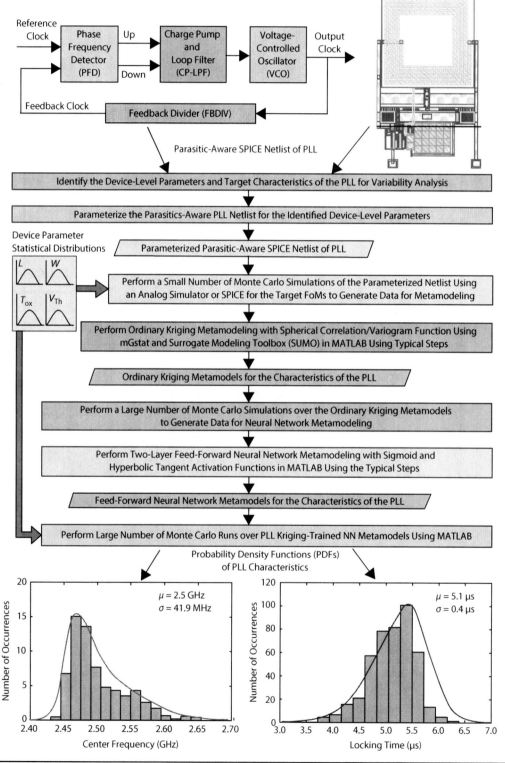

Figure 12.31 Kriging-trained neural network metamodel-based process variation analysis of a PLL [56].

Characteristics of PLL	From SPICE		From Kriging-Trained NN		Errors	
	Mean (μ)	Standard Deviation (σ)	Mean (μ)	Standard Deviation (σ)	Mean (μ)	Standard Deviation (σ)
Center Frequency	2.7 GHz	10.95 MHz	2.5 GHz	41.9 MHz	5.6%	283%
Locking Time	5.5 µs	0.7 µs	5.1 µs	0.4 µs	7.3%	38.9%

TABLE 12.13 Kriging-Trained Neural Network Metamodel-Based Statistical Analysis of a 180-nm CMOS-Based PLL [56]

results obtained from 1000 run Monte Carlo simulations over the Kriging-Bootstrapped NN metamodels are presented in the figure. The distributions of center frequency and locking time are specifically shown. For a comparative perspective, the results are also presented in Table 12.13 along with the SPICE simulation results for the same 1000 run Monte Carlo simulations [56].

12.8 Polynomial-Metamodel-Based Ultrafast Design Optimization

To demonstrate the use of the polynomial metamodel for ultrafast design optimization, three different case studies are presented in this section. The speedup in the design optimization comes from the fact that the design optimizations are performed over the metamodels using non-EDA tools instead of SPICE netlists, thus completely eliminating the use of slow analog simulation engines in the design optimization iteration loop.

12.8.1 Polynomial-Metamodel-Based Optimization of a Ring Oscillator

As the first demonstration of the ultrafast optimization over the polynomial metamodels, the 45-nm CMOS-based ring oscillator case study circuit, which was discussed in the previous section, is considered in this section. The polynomial metamodels generated in a previous section for their performance or characteristics are used in this section as well for brevity. The overall flow for the polynomial metamodel-assisted ultrafast layout optimization of the ring oscillator is presented in Fig. 12.32 [17, 45].

In the ultrafast layout optimization flow, the polynomial metamodels are generated following the typical steps discussed in the previous sections. The design variables or tuning variables are considered as $X_{\text{tuning}} = \{W_n, W_p\}$, which are the width of NMOS and PMOS transistors of the ring oscillator circuit. The performance metrics or characteristics of the ring oscillator circuit that are considered for metamodeling are the following: (1) center frequency, (2) power dissipation, and (3) power over frequency ratio (PFR). The objective of power minimization and center frequency maximization can be translated to PFR minimization. Once the polynomial metamodels are obtained, a DOE-assisted tabu search optimization (DOE-TSO) is used to search for minimal PFR for the range of values of X_{tuning}. The physical design of the ring oscillator is then modified based on the result of the DOE-TSO X_{opimal}, the optimal transistor sizes. The detailed steps of DOE-TSO are presented in Algorithm 12.5 [17, 45].

The tabu search is a metaheuristic algorithm that uses a more aggressive approach than most other search alternatives [25, 38, 47]. It skips inferior solutions other than the scenarios when it needs to skip out the local optimum. The tabu search algorithm effectively uses the entire search constrained space and applies the divide and conquer approach for speeding up the convergences. The classic and widely used tabu search has been further enhanced by the integration of the DOEs as presented in the DOE-TSO steps in Algorithm 12.5 [17, 45]. The DOE-TSO performs the DOE analysis of the ring oscillator design space. The design space is recursively divided into four adjacent subspaces and the analysis of each of these areas gives the optimal area. The algorithm then moves to that area by constraining to its boundaries and then conducts the same analysis of that subspace recursively. In the process the DOE-TSO algorithm is able to perform a detailed search in the region that the optimal results may exist. As a specific example, the FoM used in this DOE-TSO is PFR. PFR is calculated during each iteration of the algorithm using the polynomial metamodels instead of the computationally

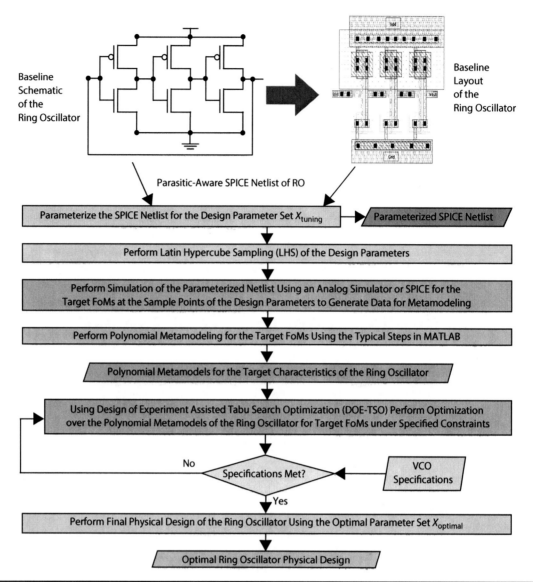

Figure 12.32 Polynomial metamodel-based ultrafast optimization of a ring oscillator [17, 45].

Algorithm 12.5 Design of Experiments (DOEs) Assisted Tabu Search Optimization of the Ring Oscillator over the Polynomial Metamodels [17, 45]

1: **Input:** Tuning parameters or design variables X_{tuning} with their ranges for the ring oscillator, Polynomial metamodels of the ring oscillator, Maximum Number of Iterations N_{cnt}.
2: **Output:** Optimized tunning parameters $X_{optimal}$ of the ring oscillator.
3: Initialize iteration counter $Counter \leftarrow 0$ for the Tabu search.
4: Perform design of experiments (DOE) analysis for the tuning parameters X_{tuning} using a center design with 9 points.
5: Generate initial solution for tuning parameters $X_{tuning,init}$ and assume as $X_{optimal}$.
6: Calculate the figure of merit (FoM) of the ring oscillator from the polynomial metamodels as follows:
   $FoM_{optimal} \leftarrow Polynomial\_Metamodel\_Functions\ (X_{optimal})$.
7: **while** ($Counter < N_{cnt}$) **do**
8:   Perform DOE analysis to generate a set of 5 points with resulting FOMs calculated from the polynomial metamodels.

9:   Generate the next feasible solution of tuning parameter size in the selected quadrant as X_{optimal^*}.
10: Increment the counter of the Tabu search, $Counter \leftarrow Counter + 1$.
11: **if** (X_{optimal^*} is not visited in the previous iterations) **then**
12:      Calculate the figure of merit (FoM) of the ring oscillator from the polynomial metamodels as follows:
         $\text{FoM}_{\text{optimal}^*} \leftarrow \text{Polynomial\_Metamodel\_Functions}(X_{\text{optimal}^*})$.
13:      **if** ($\text{FoM}_{\text{optimal}^*} < \text{FoM}_{\text{optimal}}$) **then**
14:          Update the optimal size with new optimal size as the FoM has improved as: $X_{\text{optimal}} \leftarrow X_{\text{optimal}^*}$.
15:      **else**
16:          Discard the solution X_{optimal^*} as it is an inferior solution as compared to the current solution.
17:      **end if**
18: **end if**
19: **end while**
20: **return** The optimal tuning parameter X_{optimal} of the ring oscillator.

Iteration Count	Target Frequency	Resulted Frequency	Accuracy	Simulation Time
30	9.0 GHz	9.4 GHz	4.4%	8.6 ms
07	9.5 GHz	9.4 GHz	0.9%	6.0 ms
12	10 GHz	9.9 GHz	0.7%	7.2 ms
24	10.5 GHz	10.5 GHz	0.3%	7.4 ms
10	11 GHz	11.1 GHz	0.8%	6.4 ms
19	11.5 GHz	11.4 GHz	0.7%	7.1 ms

TABLE 12.14 Results of DOE-TSO Algorithm for 10 GHz Frequency of 45-nm CMOS-Based Ring Oscillator [17, 45]

expensive analog/SPICE simulations. Other options are to use $-f_{\text{osc}}$ as FoM for minimization (as center frequency needs maximization) under the power as constraints. Even power dissipation as FoM for minimization with center frequency as the constraint can be used. The details of the DOE-TSO algorithm need to be modified slightly to accommodate these options. The results obtained from the DOE-TSO algorithm for 10 GHz frequency of 45-nm CMOS-based ring oscillator are presented in Table 12.14 [17, 45].

Table 12.14 shows the number of iterations used by the algorithm to search a target center frequency for the ring oscillator [17, 45]. For example, the search for a center frequency of 9.0 GHz could be made possible in 30 iterations. For the specific design of this 45-nm ring oscillator, 120-nm width is the minimum value for both W_n and W_p and 360 nm and 720 nm are the maximum values for W_n and W_p, respectively. The optimization is performed for a target center frequency within 5% accuracy. The simulation time for the convergence of this search is 8.6 ms. Similarly, the other rows of the table can be interpreted. On an average, DOE-TSO algorithm takes 46.8 s over SPICE netlist and 0.12 s over the polynomial metamodels, thus obtaining 387.5× speedup. For larger circuits like PLL the optimization over SPICE netlist is not an easy option.

12.8.2 Polynomial-Metamodel-Based Optimization of a PLL

As the second example of application of the polynomial metamodel for design optimization, the optimization of a PLL is presented in this section. The 180-nm CMOS-based PLL has a target center frequency specification of 2.6 GHz with minimum power dissipation and jitter. The overview of the polynomial metamodel-assisted ultrafast design optimization flow is depicted in Fig. 12.33 [16, 19]. The typical steps of polynomial metamodel generation are followed to generate the polynomial metamodels of various PLL characteristics based on their physical design. A total of 21 device-level

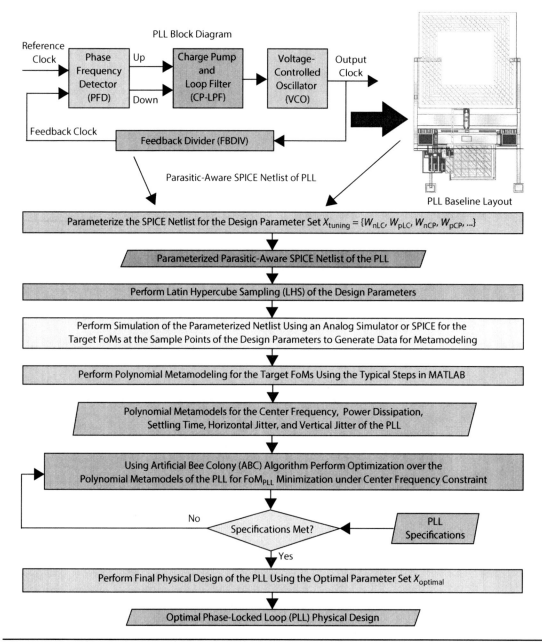

Figure 12.33 Polynomial metamodel-based ultrafast optimization of a PLL [16, 19].

design variables consisting of W_n and W_p of transistors present in the four components of the PLL (i.e., PFD, CP-LPF, LC-VCO, and FBDIV) are considered as input variables, i.e., $X_{tuning} = \{W_{nLC}, W_{pLC}, W_{nCP}, \ldots\}$. The target FoMs of this PLL that are considered for metamodeling are center frequency, power dissipation, locking time, horizontal jitter, and vertical jitter. The polynomial metamodels of various PLL characteristics are third order or fourth order and each has several number of coefficients, which are presented in Table 12.15 [16, 19]. The RMSE and R^2 values indicate that the polynomial metamodels are of very good accuracy.

The polynomial metamodel-assisted ultrafast design flow relies on an artificial bee colony (ABC) or bee colony optimization (BCO) algorithm [16, 19, 45]. The ABC algorithm is also a metaheuristic approach that mimics the behavior of honey bee swarms [2, 34, 64, 78]. The working principle of the ABC algorithm is that a swarm of artificial bees is generated and moved randomly. These artificial bees interact when they gather some nectar. The solution of the problem is obtained from the intensity

PLL Performances	Order of Polynomial Metamodels	Number of Coefficients in Metamodels	R^2 Values	RMSE Values
Center Frequency	3	48	0.91	33.40 MHz
Power Dissipation	3	56	0.99	0.16 mW
Locking Time	4	57	0.87	9.96 ns
Horizontal Jitter	3	34	0.84	36.37 ps
Vertical Jitter	3	41	0.80	1.60 mV

TABLE 12.15 Polynomial Metamodel Results for Selected Characteristics of the 18-nm CMOS-Based PLL Circuit [16, 19]

of the interactions of these artificial bees. The quality of the solution is related to the amount of the nectar. In the ABC algorithm, the bee colony is classified into following three types:

1. *Worker Bee or Employed Bee:* This type of bee stays on a food source to provide the neighborhood of the food source from its memory. These bees are roughly 50% of the total population.

2. *Onlooker Bee:* This type of bee collects the information regarding the food sources from the worker bees and selects one of the food sources to evaluate the nectar. These bees are also roughly 50% of the total population.

3. *Scout Bee:* This type of bee is responsible for finding new food sources and new nectar. The worker bee whose food source has been exhausted becomes a scout bee.

A solution to the problem is essentially the position of the food source. The number of worker bees is the number of solutions of the population.

A high-level overview of the ABC approach is presented in Algorithm 12.6 [34, 43]. The ABC algorithm starts with a random food source or initial solution for all worker bees and iterates the following steps while convergence requirements are met. Each worker bee goes to a food source and evaluates its nectar amount. Each onlooker bee watches the dance of worker bees and chooses one of their sources depending on the dances and evaluates its nectar amount. Each scout bee determines abandoned food sources and replaces them with new food sources. The actions of the artificial bees in the context of PLL optimization are depicted in Fig. 12.34 [16, 19]. The movement of the worker, onlooker, and scout bees is dependent on two factors: the probability of food source ($Prob_{food}$) and the FoM of the PLL. The worker bee continues its FoM evaluation as long as the current FoM exceeds the previous FoM, i.e., $FoM_{cur} > FoM_{pre}$. Otherwise the transition is made to an onlooker bee. The onlooker bee continues if $Prob_{food}$ is low; when $Prob_{food}$ is high and $FoM_{cur} > FoM_{pre}$, a transition is made to a worker bee; and when $Prob_{food}$ is high and $FoM_{cur} < FoM_{pre}$, a transition is made to a scout bee. The scout bee continues its evaluation as long as the next FoM does not exceed the current FoM, i.e., $FoM_{cur} < FoM_{pre}$. The transition from scout bee to worker bee is made when $FoM_{cur} > FoM_{pre}$. A detailed pseudocode of the ABC-based approach for PLL optimization is presented in Algorithm 12.7 [16, 19, 45].

ALGORITHM 12.6 Overview of the Artificial Bee Colony (ABC) Optimization [34, 43]

1: Initialize the bee population with the food sources.
2: **while** (requirements are met) **do**
3:    Each worker bee goes to a food source and evaluates its nectar amount.
4:    Each onlooker bee watches the dance of worker bees and chooses one of their sources depending on the dances and evaluates its nectar amount.
5:    Determine abandoned food sources and replace with the new food sources discovered by scout bees.
6:    Memorize the best food source determined so far.
7: **end while**

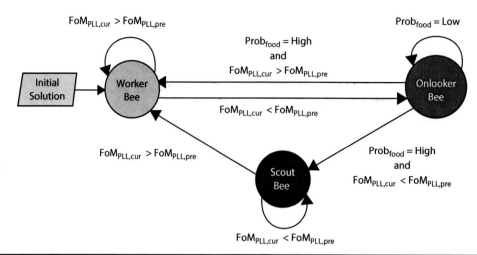

Figure 12.34 The states of the artificial bee colony (ABC) optimization of a PLL [16, 19].

Algorithm 12.7 Artificial Bee Colony (ABC) Optimization of a PLL over the Polynomial Metamodels [16, 19, 45]

1: **Input:** Baseline tuning parameters or design variables X_{tuning} with their ranges for the PLL, Polynomial metamodels of the PLL, Maximum Number of Iterations N_{cnt}.
2: **Output:** Optimized tunning parameters $X_{optimal}$ of the PLL.
3: Initialize a parameter called the Number of bees which is the Number of variables as: $N_{Bee} \leftarrow$ Size (X_{tuning}).
4: Initialize a matrix called Bee-Matrix $(3, N_{Bee})$ = [Workers, Onlookers, Scouts] using 0/1 entries.
5: Initialize the food sources which are sample points of the design variables.
6: Initialize the iteration counter as: $counter \leftarrow 1$
7: **while** ($counter \leq N_{cnt}$) **do**
8:    **for** (Each i from 1 to N_{Bee}) **do**
9:       **if** (Bee-Matrix(1,i) == 1) **then**
10:         Send a worker bee to a random known food source.
11:         Calculate FoM_{PLL} of the PLL using polynomial metamodels, $FoM_{PLL} \leftarrow$ Polynomial_Metamodels (Food Source).
12:         **if** (Calculated FoM_{PLL} is smaller than the Existing FoM_{PLL} of the PLL) **then**
13:             Update FoM_{PLL} of PLL and locations of the bees which are the design parameters.
14:         **else**
15:             Convert the worker bee to an onlooker bee and update Bee-Matrix accordingly.
16:         **end if**
17:       **else if** (Bee-Matrix(1,i) == 1) **then**
18:         Send onlooker bee for the food source i.e. random sample for a design parameter.
19:         Calculate the probability of the food source ($Prob_{food}$).
20:         **if** ($Prob_{food}$ == High) **then**
21:             Send the onlooker bee to a random food source i.e. random sample for a design parameter.
22:             Calculate FoM_{PLL} of the PLL using polynomial metamodels, $FoM_{PLL} \leftarrow$ Polynomial_Metamodels (Food Source).
23:             **if** (Calculated FoM is smaller than the Existing FoM of the PLL) **then**
24:                 Update FoM_{PLL} of PLL and locations of the bees which are the design parameters.

25:             Convert onlooker bee to worker bee and update Bee-Matrix accordingly.
26:         else
27:             Convert onlooker bee to scout bee and update Bee-Matrix accordingly.
28:         end if
29:     end if
30: else
31:     Send the scout bee to the random locations for each food sources.
32:     if (Calculated FoM$_{PLL}$ is smaller than the Existing FoM$_{PLL}$ of the PLL) then
33:         Update FoM$_{PLL}$ of PLL and locations of the bees which are the design parameters.
34:         Convert scout bee to worker bee and update Bee-Matrix accordingly.
35:     end if
36: end if
37: if (Calculated FoM$_{PLL}$ is smaller than the Existing FoM$_{PLL}$ of the PLL) then
38:     Update FoM$_{PLL}$ of PLL and locations of the bees which are the design parameters.
39: end if
40: end for
41: Increment iteration counter as: *counter* ← *counter* + 1.
42: **end while**
43: **return** The optimal locations of the bees i.e. design parameters X_{optimal} of the PLL.

In the steps of Algorithm 12.7, the initial food source is calculated using the following expression [2, 16, 19, 45]:

$$x_{j,k} = x_{\min,k} + \text{rand}(0,1)\left(x_{\max,k} - x_{\min,k}\right) \tag{12.55}$$

where x is any food source, j is any integer from 1 to the number of food sources (i.e., sample points), k is any integer from 1 to the number of design variables, and rand(0,1) generates a uniform random number between 0 and 1. The same expression can be used to generate any random food source during the iteration of the algorithm. An onlooker bee chooses a food source based on the probability associated with that food source, which is calculated using the following expression [2, 16, 19, 45]:

$$\text{Prob}_{\text{food},i} = \left(\frac{\text{fitness}_i}{\sum_{i=1}^{\text{\#of food sources}} \text{fitness}_i}\right) \tag{12.56}$$

where fitness$_i$ is the fitness value of the solution i which is proportional to the nectar amount of the food source in the position i. In the current PLL optimization, the following is used as FoM for minimization [16, 19, 45]:

$$\text{FoM}_{\text{PLL}} = -\left(\frac{1}{\text{Power}_{\text{PLL}} \times \text{Jitter}_{h,\text{PLL}} \times \text{Jitter}_{v,\text{PLL}}}\right) \tag{12.57}$$

where Power$_{\text{PLL}}$, Jitter$_{h,\text{PLL}}$, and Jitter$_{v,\text{PLL}}$ are the power, horizontal jitter, and vertical jitter of the PLL, respectively. The minimization of this FoM$_{\text{PLL}}$ leads to a PLL design that will have minimized power and jitter while able to achieve a lock. The optimization is performed under a constraint of locking

PLL Characteristics	For Baseline Design	For Optimal Design	Improvement
Power Dissipation	9.3 mW	0.9 mW	91%
Jitter Vertical	168.3 μV	3.3 nV	≈100%
Jitter Horizontal	189.0 ps	180.0 ps	5%

TABLE 12.16 Power and Jitter of the 180-nm CMOS PLL [16, 19]

time. The locking time is a characteristic of the PLL, which is an indicator whether the PLL will be able to lock within reasonable time. The optimization target is that the center frequency is within 0.5% of the specification of 2.6 GHz.

The results of the ABC-based optimization of the PLL are presented in Table 12.16 [16, 19, 43]. The optimal PLL has 91% lower power dissipation, 100% vertical jitter improvement, and 5% horizontal jitter improvement over the baseline PLL design. The final PLL layout has no area penalty, which is due to the layout configuration and the selected parameters that do not affect the final layout considerably. The time spent for the polynomial metamodel-based ABC optimization of the PLL is as follows: (1) LHS and its analog simulation takes 11 hours for sampling the design variables in 21 dimensions, (2) polynomial metamodel creation takes 1 minute, and (3) ABC for 100 iterations over the polynomial metamodel takes 5 minutes. Thus, the total time for overall optimization is 11 hours and 6 minutes. The same 100 iterations of ABC over the SPICE netlist in the framework of an analog/SPICE iteration would have taken 7.3 days in worst case; assuming 21 bees, 21 × 5 minute × 100 iteration ≈ 10,500 minute ≈ 7.3 days.

12.8.3 Polynomial-Metamodel-Based Optimization of an OP-AMP

As the third example of a polynomial metamodel for optimization, the optimization of a 90-nm CMOS-based OP-AMP using cuckoo search optimization (CSO) algorithm is presented in this section. The OP-AMP considered in this section is the same OP-AMP that was discussed in Sec. 12.4 on polynomial-metamodeling and correspondingly Verilog-AMS-PoM. The same polynomial metamodels and Verilog-AMS-PoM can be used in this section as well. The overall ultrafast design flow is shown in Fig. 12.35 [83, 85].

In the ultrafast flow, using the typical steps the polynomial metamodels and Verilog-AMS-PoM for the OP-AMP are generated as depicted in Fig. 12.35 [83, 85]. The width and length of an NMOS and a PMOS $X_{tuning} = \{L_n, L_p, W_n, W_p\}$, a total of 8, are considered design variables. The OP-AMP circuit parameters, such as open-loop DC gain (A_{DC}), transconductance of input stage (g_{mi}), maximum available positive current (I_{max+}), maximum available negative current (I_{max-}), function for poles (P_{OP-AMP}), and function for zeros (Z_{OP-AMP}), and the performance metrics, such as power dissipation (PD_{OP-AMP}), bandwidth (BW_{OP-AMP}), phase margin (PM_{OP-AMP}), and slew rate (SR_{OP-AMP}), are considered for metamodeling. Verilog-AMS meta-macromodel is then generated, which is called Verilog-AMS-PoM as discussed in a previous section.

The polynomial metamodels are used in a CSO algorithm to minimize the OP-AMP power dissipation with the gain, bandwidth, phase margin, and slew rate as constraints. The cuckoo search is a metaheuristic optimization algorithm that can handle optimization problems with high nonlinearity and complex constraints and can achieve better performance than particle swarm optimization (PSO) [77, 78, 83]. The following three rules are the basis of the CSO algorithm [77]:

1. Each cuckoo lays one egg at a time and places it in a randomly chosen host nest out of total N_{nest} nests.
2. The best nests with high quality eggs (i.e., solutions) will carry over to the next generations (i.e., iterations).
3. The number of available host nests is fixed and a host bird can discover an alien egg with a probability $Pr_{Alien} \in [0,1]$. In such a situation, the host bird can either throw the egg away or

abandon the nest so as to build a new nest in a new location. For simplicity, this can be modeled as a Pr_{Alien} fraction of the N_{nest} nests being replaced by new nests or with new random eggs (solutions) at new locations.

In the algorithm, a parameter in the set X_{tuning} is updated using the following expression [77, 78]:

$$X_{tuning,t+1} = X_{tuning,t} + \alpha \otimes \text{Levy}(\lambda) \quad (12.58)$$

$$X_{tuning,t+1} = X_{tuning,t} + \alpha \otimes t^{-\lambda} \quad (12.59)$$

In the above expression, $\alpha > 0$ is the step size, which is selected based on the problem of interest, and product $\otimes$ is the entrywise multiplication. The Lévy flights essentially provide a random walk while their random steps are picked from a Lévy distribution for large steps with an infinite variance and an infinite mean, hence the exponential function. The detailed steps of the polynomial metamodel-assisted CSO of the OP-AMP are shown in Algorithm 12.8 [83, 85].

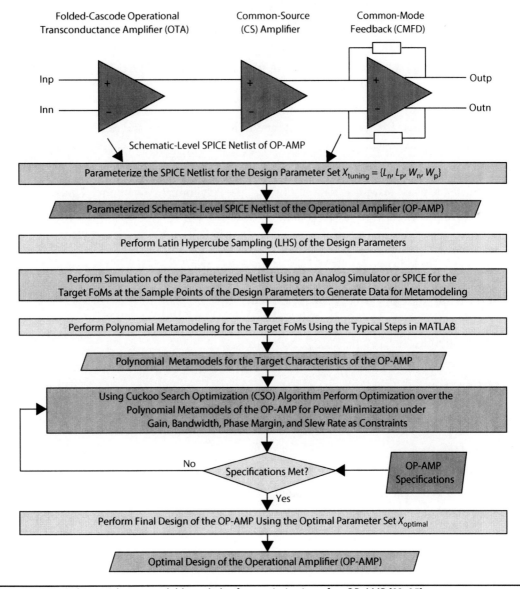

FIGURE 12.35 Polynomial metamodel-based ultrafast optimization of an OP-AMP [83, 85].

ALGORITHM 12.8 Cuckoo Search Optimization of an OP-AMP over Polynomial Metamodels [83, 85]

1: **Input:** Baseline tuning parameters or design variables X_{tuning} with their ranges for the OP-AMP, Polynomial metamodels of the OP-AMP, Maximum Number of Iterations N_{cnt}.
2: **Output:** Optimized tunning parameters $X_{optimal}$ of the OP-AMP.
3: Initialize the iteration counter as $Counter \leftarrow 0$;
4: Initialize a population of N_{nest} host nests for each variables in X_{tuning}, $x_{tuning,i}$ ($i = 1, 2, \ldots N_{sample}$) and call it $X_{tuning,nest}$.
5: Evaluate the power dissipation ($PD_{OP\text{-}AMP}$) values of the N_{nest} points of X_{tuning} using the polynomial metamodels as:
$$PD_{OP\text{-}AMP}(X_{tuning,nest}) \leftarrow \text{Polynomial\_Metamodel\_Functions}(X_{tuning,nest}).$$
6: Determine the minimum power dissipation from the various samples using polynomial metamodels of the OP-AMP, $PD_{OP\text{-}AMP,min} \leftarrow \text{Minimum}(PD_{OP\text{-}AMP}(X_{tuning,nest}))$.
7: Determine the minimum bandwidth from the various samples using polynomial metamodels of the OP-AMP, $BW_{OP\text{-}AMP,min} \leftarrow \text{Minimum}(BW_{OP\text{-}AMP}(X_{tuning,nest}))$.
8: Determine the minimum phase margin from the various samples using polynomial metamodels of the OP-AMP, $PM_{OP\text{-}AMP,min} \leftarrow \text{Minimum}(PM_{OP\text{-}AMP}(X_{tuning,nest}))$.
9: Determine the minimum slew rate from the various samples using polynomial metamodels of the OP-AMP, $SR_{OP\text{-}AMP,min} \leftarrow \text{Minimum}(SR_{OP\text{-}AMP}(X_{tuning,nest}))$.
10: Determine the minimum open-loop DC gain from the various samples using polynomial metamodels of the OP-AMP, $A_{DC,min} \leftarrow \text{Minimum}(A_{DC}(X_{tuning,nest}))$.
11: **while** ($PD_{OP\text{-}AMP,target} < PD_{OP\text{-}AMP,min}$) and ($Counter < N_{cnt}$) **do**
12:     Pick a new sample (cuckoo) from X_{tuning} randomly by Lévy flights and call it $X_{tuning,i}$.
13:     Evaluate the power dissipation as $PD_{OP\text{-}AMP}(X_{tuning,i}) \leftarrow \text{Polynomial\_Metamodel\_Functions}(X_{tuning,i})$.
14:     Evaluate the bandwidth as $BW_{OP\text{-}AMP}(X_{tuning,i}) \leftarrow \text{Polynomial\_Metamodel\_Functionst}(X_{tuning,i})$.
15:     Evaluate the phase margin as $PM_{OP\text{-}AMP}(X_{tuning,i}) \leftarrow \text{Polynomial\_Metamodel\_Functions}(X_{tuning,i})$.
16:     Evaluate the slew rate as $SR_{OP\text{-}AMP}(X_{tuning,i}) \leftarrow \text{Polynomial\_Metamodel\_Functions}(X_{tuning,i})$.
17:     Evaluate the open loop DC gain as $A_{DC}(X_{tuning,i}) \leftarrow \text{Polynomial\_Metamodel\_Functions}(X_{tuning,i})$.
18:     Pick another sample from X_{tuning} randomly by Lévy flights and call it $X_{tuning,j}$.
19:     Evaluate the power dissipation as $PD_{OP\text{-}AMP}(X_{tuning,j}) \leftarrow \text{Polynomial\_Meta model\_Functions}(X_{tuning,j})$.
20:     **if** ($PD_{OP\text{-}AMP}(X_{tuning,i}) < PD_{OP\text{-}AMP}(X_{tuning,j})$) **then**
21:         $\text{Constraint}_{BW} \leftarrow BW_{OP\text{-}AMP}(X_{tuning,i}) > BW_{OP\text{-}AMP,min}$.
22:         $\text{Constraint}_{PM} \leftarrow PM_{OP\text{-}AMP}(X_{tuning,i}) > PM_{OP\text{-}AMP,min}$.
23:         $\text{Constraint}_{SR} \leftarrow SR_{OP\text{-}AMP}(X_{tuning,i}) > SR_{OP\text{-}AMP,min}$.
24:         $\text{Constraint}_{DC} \leftarrow A_{DC}(X_{tuning,i}) > A_{DC,min}$.
25:         **if** (All OP-AMP constraints are met) **then**
26:             Replace $\mathbf{x}_j$ by $\mathbf{x}_i$.
27:         **end if**
28:     **end if**
29:     Abandon a fraction of worst sample points.
30:     Increment iteration counter, $counter \leftarrow counter + 1$.
31: **end while**
32: **return** The optimal tunning parameters $X_{optimal}$ of the OP-AMP.

OP-AMP Characteristics	Objectives and Constraints	From SPICE Netlist	From Polynomial Metamodels
Power Dissipation (μW)	≈65	65.5	68.1
Bandwidth (kHz)	>50	85.5	58.9
Phase Margin (degree)	>70	84.4	87.7
Slew Rate (mV/ns)	>05	8	7.1
Open Loop Gain (dB)	>43	52.8	56.4

TABLE 12.17 Results for the 90-nm CMOS-Based OP-AMP Optimization Using Cuckoo Search over Polynomial Metamodels [83, 85]

Algorithm Performance	Over SPICE Netlist	Over Polynomial Metamodels
Power Reduction	×3.71	×3.86
Number of Iterations	1200	1200
Computation Time	12.5 h	2.6 sec
Normalized Speed	1	×17120

TABLE 12.18 Comparison of Optimization Performance 90-nm CMOS-Based OP-AMP [83, 85]

In a specific execution of the cuckoo search algorithm, the initial number of solutions and maximum number of iterations are set at 10 and 1200, respectively. The power dissipation was targeted for 65 μW. The results for the cuckoo search algorithm are shown in Table 12.17 [83, 85]. The results are reported when the algorithm was executed over the SPICE netlist and over the polynomial metamodels. It may be noted that the results from the polynomial could match that from SPICE netlist in terms of accuracy. The power dissipation is close to the target specification. The other OP-AMP characteristics such as bandwidth, phase margin, slew rate, and open-loop gain exceeded the target value. Further performance comparisons are presented in Table 12.18 [83, 85]. It is evident from the results that polynomial metamodel-based optimization is 17,120× faster than the SPICE netlist. The accuracy due to the use of polynomial metamodels has not been compromised. Thus, metamodel-assisted techniques can provide a very good option for design engineers to reduce the design cycle time and design effort.

12.9 Neural Network Metamodel-Based Ultrafast Design Optimization

After learning the use of polynomial metamodels for design optimization, this section discusses use of NN metamodels for ultrafast design optimization. Two case study circuits are considered: schematic-level design of the 90-nm CMOS-based OP-AMP and physical design of 180-nm CMOS-based PLL. Different algorithms are also discussed for the automatic iterations of the optimization.

12.9.1 Neural Network Metamodel-Based Optimization of an OP-AMP

This section presents the first example of ultrafast design optimization flow using NN metamodels using a 90-nm CMOS OP-AMP as case study IC. This is the OP-AMP whose NN metamodels have been generated in a previous section and iVAMS models have also been presented for them. The NN metamodels and the 90-nm CMOS based OP-AMP discussed before are used through the overall ultrafast design flow presented in Fig. 12.36 [84, 86]. The objective is to maximize the slew rate and minimize the power dissipations while meeting the bandwidth, phase margin, and open-loop gain. An effective method used in this design flow is to determine the Pareto front (PF) that consists of a set of nondominated solutions for the optimization problem. The design engineers can then select one of the alternative designs from this solution set to implement. A multiobjective firefly algorithm (MOFA) is used as the core optimization engine.

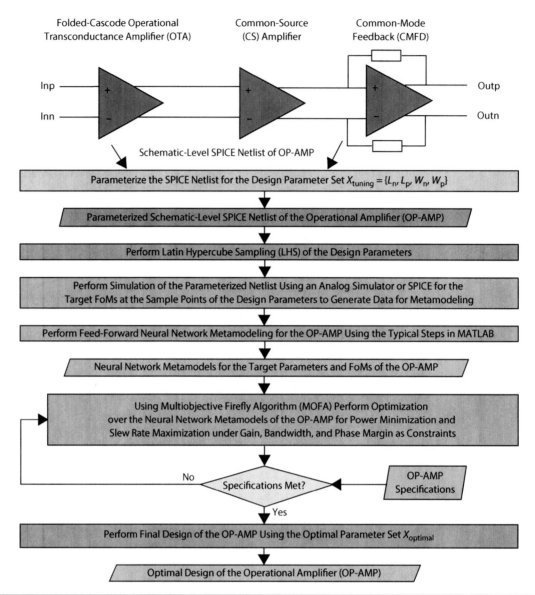

Figure 12.36 Neural network metamodel-based ultrafast optimization of an OP-AMP [84, 86].

MOFA is a metaheurstic algorithm that mimics the behavior of the tropic firefly swarms which are attracted toward flies with higher flash intensity [79, 84, 86]. The primary purpose of a flash of the fireflies is to act as a signal system to attract other fireflies. Any firefly can be attracted to any other firefly. The attractiveness is proportional to their brightness. For any two fireflies, the less bright firefly will move to the brighter firefly. The brightness can change as their distance changes. If there are no fireflies brighter than a given firefly, it will move randomly. For any given two fireflies x_i and x_j, the movement of firefly i attracted to another brighter firefly j is determined by the following expression [79]:

$$x_{i,t+1} = x_{i,t} + \beta_0 \exp(-\gamma r_{ij}^2)(x_{j,t} - x_{i,t}) + \alpha_t \varepsilon_{i,t} \quad (12.60)$$

where $x_{i,t}$ is the current location and $x_{i,t+1}$ is the next location of firefly x_i. Similarly, $x_{j,t}$ is the current location of the firefly x_j. β_0 is the attractiveness at Cartesian distance $r_{ij} = 0$. r_{ij} is the Cartesian distance between x_i and x_j, i.e., $r_{ij} = \|x_i - x_j\|$; however, other distance measures can also be used. γ characterizes the variation of the attractiveness whose value determines the speed of the convergence. α_t is a randomization parameter. $\varepsilon_{i,t}$ is a vector of random numbers following either a Gaussian distribution or a uniform distribution. The location of fireflies can be updated sequentially or each pair of them can be updated in every iteration of the algorithm. During the optimization, the brightness is associated with the objective function. Several variants of the firefly algorithm is available as follows: (1) discrete firefly algorithm, (2) Lagrangian firefly algorithm, (3) chaotic firefly algorithm, (4) hybrid firefly algorithm, and (5) MOFA. MOFA has been used in this specific NN metamodel-based optimization [84, 86]. The detailed steps of the OP-AMP optimization are shown in Algorithm 12.9 [84, 86].

ALGORITHM 12.9 iVAMS-Assisted Firefly-Based OP-AMP Multiobjective Optimization [84, 86]

1: **Input:** Baseline tuning parameters or design variables X_{tuning} with their ranges for the OP-AMP, Neural Network metamodels of the OP-AMP, Maximum Number of Iterations N_{cnt}.
2: **Output:** Optimized tunning parameters $X_{optimal}$ of the OP-AMP.
3: Initialize the iteration counter as $Counter \leftarrow 0$;
4: Initialize K samples of the multidimensional design parameter $X_{tuning,i}$ ($i = 1, 2, \ldots N_{sample}$) and call it $X_{tuning,K}$.
5: Determine the minimum bandwidth from the various samples using neural network metamodels the OP-AMP, $BW_{OP\text{-}AMP,min} \leftarrow \text{Minimum}(BW_{OP\text{-}AMP}(X_{tuning,K}))$.
6: Determine the minimum phase margin from the various samples using neural network metamodels of the OP-AMP, $PM_{OP\text{-}AMP,min} \leftarrow \text{Minimum}(PM_{OP\text{-}AMP}(X_{tuning,K}))$.
7: Determine the minimum open-loop DC gain from the various samples using neural network metamodels of the OP-AMP, $A_{DC,min} \leftarrow \text{Minimum}(A_{DC}(X_{tuning,K}))$.
8: **while** ($Counter < N_{cnt}$) **do**
9:     Evaluate performance of the OP-AMP at all sample points of $X_{tuning,K}$ using neural network metamodels.
10:    Determine the nondominated sample points as, $X_{tuning,ND} \subset X_{tuning,K}$.
11:    **for** ($i = 1 \rightarrow K$) **do**
12:       **if** (No nondominated designs are present) **then**
13:          Generate random weight $\omega \in [0, 1]$.
14:          Find the best sample point $X_{tuning,best} \in X_{tuning,K}$ that maximizes a combined FoM of OP-AMP for slew rate and power dissipation as: $\psi(X_{tuning,K}) = (1 - \omega)SR_{OP\text{-}AMP}(X_{tuning,K}) - \omega PD_{OP\text{-}AMP}(X_{tuning,K})$.
15:          Compute a move vector $\Delta X_{tuning,i}$ for $X_{tuning,i}$ toward $X_{tuning,best}$.
16:       **else**
17:          Compute a move vector $\Delta X_{tuning,i}$ for $X_{tuning,i}$ toward $X_{tuning,ND}$.
18:       **end if**
19:       $Constraint_{BW} \leftarrow BW_{OP\text{-}AMP}(X_{tuning,i} + \Delta X_{tuning,i}) > BW_{OP\text{-}AMP,min}$.
20:       $Constaint_{PM} \leftarrow PM_{OP\text{-}AMP}(X_{tuning,i} + \Delta X_{tuning,i}) > PM_{OP\text{-}AMP,min}$.
21:       $Constraint_{DC} \leftarrow A_{DC}(X_{tuning,i} + \Delta X_{tuning,i}) > A_{DC,min}$.
22:       **if** (Not all constraints are satisfied) **then**
23:          Generate a new move vector.
24:       **end if**
25:    **end for**
26:    Assume the move vector for the tuning variable set as: $\Delta X_{tuning} = \{\Delta X_{tuning,1}, \Delta X_{tuning,2}, \ldots, \Delta X_{tuning,K}\}$.
27:    Update the design variable set, $X_{tuning} \leftarrow X_{tuning} + X_{tuning}$.
28:    $Counter \leftarrow Counter + 1$.
29: **end while**
30: **return** The final tuning variable set X_{tuning} as $X_{optimal}$.

The goal of the MOFA shown in Algorithm 12.9 is to determine K Pareto points that constitute the PF through a predetermined number of iterations [84, 86]. The algorithm starts with K randomly generated parameter sample. During each iteration, the performance of the K OP-AMP designs is estimated using the NN metamodels expressed as iVAMS. If nondominated designs are determined, the move vectors from each current sample point toward the nondominated sample points will be computed based on the attractiveness and the characteristic distance. If no nondominated designs are found, a move vector toward a current best sample point is determined by the combined weighted sum of the objectives as follows [84, 86]:

$$\psi(X_{tuning,K}) = (1-\omega)SR_{OP\text{-}AMP^*}(X_{tuning,K}) - \omega PD_{OP\text{-}AMP^*}(X_{tuning,K}) \quad (12.61)$$

where ω is weight. $SR_{OP\text{-}AMP^*}$ and $PD_{OP\text{-}AMP^*}$ are normalized values of the slew rate and power dissipation, respectively. $\psi(X_{tuning,K})$ represents FoM for the OP-AMP for each sample point.

The above MOFA-based optimization can be implemented and tested for various parameters; specific values are: $K = 50$ and $N_{cnt} = 500$. The results are presented in Table 12.19 [84, 86]. For comparison purposes, the MOFA is run over SPICE netlist with an analog or SPICE engine in the loop. As evident, the NN metamodel and SPICE netlist results match and hence metamodel-based approach is accurate. A further comparative data has been presented in Table 12.20 [84, 86]. As evident from the results, NN metamodel-based flow runs $5580\times$ faster for a comparable accuracy.

12.9.2 Neural Network Metamodel-Based Variability-Aware Optimization of a PLL

This section discussed optimization of a 180nm CMOS based PLL integrated circuit over its neural network metamodels using PSO. PSO algorithm has been introduced in Chapter 11 on variability-aware AMS-SoC design methodologies. Thus, these will not be repeated for brevity. A high-level overview of the variability-aware ultra-fast design optimization flow for a PLL is presented in Fig. 12.37 [22].

In this flow, a total of 35 tuning variables consisting of W_n and W_p of transistors of the four components of the PLL (i.e., PFD, CP-LPF, LC-VCO, and FBDIV) are considered, i.e., $X_{tuning} = \{W_{nLC}, W_{pLC}, W_{nCP}, ...\}$. The target FoMs of this PLL that are considered for metamodeling are center frequency, power dissipation, locking time, horizontal jitter, and vertical jitter. Following the typical

OP-AMP Characteristics	Objectives and Constraints	From SPICE	For NN Metamodels
Power Dissipation (μW)	Minimized	85.8	85.1
Slew Rate (mV/ns)	Maximized	05.5	05.5
Bandwidth (kHz)	>50	56.7	56.8
Phase Margin (Degree)	>70	88.5	81.0
Open Loop Gaing (dB)	>43	55.7	56.4

TABLE 12.19 Results of Multiobjective Firefly Algorithm (MOFA)-Based Optimization of a 90-nm CMOS-Based OP-AMP over Its Neural Network Metamodels [84, 86]

Optimization Characteristics	From SPICE	For NN Metamodels	For NN Metamodels
Number of Iterations N_{cnt}	500	500	5000
Number of Pareto Points K	20	20	50
Algorithm Runtime	131.2 h	84.6 sec	0.57 h
Normalized Speed	1	5580×	–

TABLE 12.20 Comparison of Multiobjective Firefly Algorithm (MOFA)-Based Optimization of a 90-nm CMOS-Based OP-AMP [84, 86]

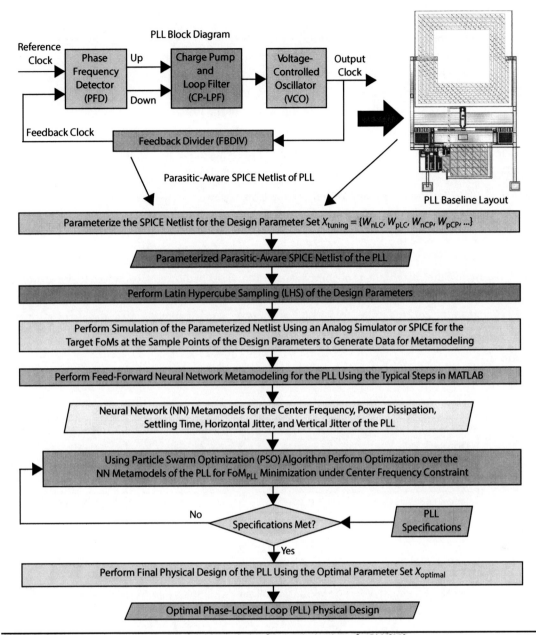

FIGURE 12.37 Neural network metamodel-based ultrafast optimization of a PLL [22].

approaches, multilayer feed-forward NN metamodels are generated for these characteristics of the 180-nm CMOS-based PLL. A PSO algorithm is used to perform minimization of using statistical distribution of the FOM of the PLL under center frequency constraints. In this PSO algorithm, each particle location contains a 35-dimensional entity in which each dimension corresponds to a parameter in X_{tuning}. The PSO algorithm calculates the statistical cost function through the steps of Algorithm 12.10 using the NN metamodels or iVAMS [22]. The statistical cost parameter (α_{cost}) is used to quantify the PDF as ($\mu + \alpha_{cost}\sigma$). For example, $\alpha_{cost} = 1$ makes it ($\mu + \sigma$), which accounts for 66.7% cases in the Gaussian distribution and is an optimistic approach for optimization. Similarly, $\alpha_{cost} = 3$ makes it ($\mu + 3\sigma$), which accounts for 99.7% cases in the Gaussian distribution and is a worst case approach. The results for this optimization flow are presented in Table 12.21 [22]. With 200 SPICE simulations used to create the NN metamodels, the overall speedup due to this design flow is 5×.

752 Chapter Twelve

ALGORITHM 12.10 Statistical Cost Function Calculation During PSO of the PLL [22]

1: **Input:** Baseline tuning parameters X_{tuning} with their ranges and probability density functions (PDFs) for the PLL, Neural network metamodels of the PLL, Statistical cost parameter (α_{cost}), Number of Monte Carlo runs N_{MC}.
2: **Output:** Value of the statistical cost function $Cost_{PLL}$ of the PLL.
3: Perform Monte Carlo simulations of N_{MC} runs over the neural network metamodels of the PLL.
4: Calculate the cost of probability density function of frequency $Cost - PDF_{frequency} \leftarrow \mu_{frequency} + \alpha_{cost}\sigma_{frequency}$.
5: **if** ($Cost - PDF_{frequency}$ < Frequency-Constraint) **then**
6:     Calculate the cost of probability density function of power dissipation $Cost - PDF_{power} \leftarrow \mu_{power} + \alpha_{cost}\sigma_{power}$.
7:     Calculate the cost of probability density function of locking time $Cost - PDF_{lockingtime} \leftarrow \mu_{lockingtime} + \alpha_{cost}\sigma_{lockingtime}$.
8:     Calculate the cost of probability density function of horizontal jitter $Cost - PDF_{hjitter} \leftarrow \mu_{hjitter} + \alpha_{cost}\sigma_{hjitter}$.
9:     Normalize the costs of the probability density functions of power dissipation, locking time, and horizontal jitter.
10:    Calculate the cost function of the PLL $Cost_{PLL}$ as the sum of above 3 normalized costs.
11:    **return** The cost function of the PLL $Cost_{PLL}$.
12: **end if**

Different PLL Characteristics	For ($\mu + \sigma$) Optimization		For ($\mu + 3\sigma$) Optimization	
	Mean (μ)	Standard Deviation (σ)	Mean (μ)	Standard Deviation (σ)
Center Frequency	2.75 GHz	28.6 MHz	2.74 GHz	29.1 MHz
Power Dissipation	0.99 mW	0.28 mW	0.98 mW	0.27 mW
Locking Time	4.7 μs	1.1 μs	4.6 μs	1.1 μs
Horizontal Jitter	5.8 ps	3.4 ps	5.9 ps	3.3 ps
Vertical Jitter	0.49 mV	0.1 mV	0.49 mV	0.1 mV

TABLE 12.21 Results of Particle Swarm Optimization over the Neural Network Metamodels for the PLL [22]

12.10 Kriging Metamodel-Based Ultrafast Design Optimization

This section demonstrates the use of Kriging metamodels for ultrafast design optimization. Specifically, simple Kriging and ordinary Kriging metamodels are demonstrated. A thermal sensor and a sense amplifier have been used as case study nano-CMOS-based circuits.

12.10.1 Simple Kriging Metamodel-Based Optimization of a Thermal Sensor

Simple Kriging metamodeling steps were discussed in a previous section for a 45-nm CMOS-based clamped bitline sense amplifier design. The same steps can be followed to generate simple Kriging metamodels for the 45-nm CMOS-based thermal sensor of this section and hence is not repeated for brevity. The detailed design of a 45-nm CMOS-based thermal sensor was discussed in Chapter 6 on sensor circuits and systems. The overall ultrafast design flow is depicted in Fig. 12.38 [49, 52, 54].

A total of six tuning parameters are considered for tuning during optimization and hence these six are used as input parameters for metamodel generation. These are based on the three components of the thermal sensor as follows: ring oscillator, binary counter, and 10-bit register. The widths of

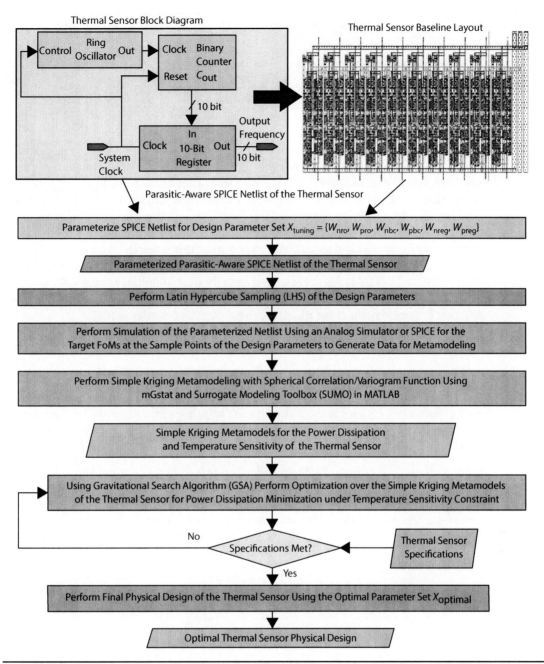

FIGURE 12.38 Simple Kriging metamodel-based ultrafast optimization of a thermal sensor [49, 52, 54].

NMOS and PMOS devices of the ring oscillator, binary counter, and 10-bit register make the parameter set: $X_{tuning} = \{W_{nro}, W_{pro}, W_{nbc}, W_{pbc}, W_{nreg}, W_{preg}\}$. As a specific case, 100 sample points are generated using LHS for these parameters. The parasitic-aware parameterized netlist of the thermal sensor is then characterized for two FoMs, such as power dissipation and sensitivity. Spherical semivariogram model has been used and typical tools have been used to generate the simple Kriging metamodels of the thermal sensor. The accuracy results of this metamodeling are presented in Table 12.22 [49, 52, 54]. A very low RMSE and the correlation coefficient R^2 very close to 1 prove that the metamodels are accurate.

The above generated simple Kriging metamodels are now used in the ultrafast design optimization flow for automatic iterations using a gravitational search algorithm (GSA) [49, 52, 54]. A high-level perspective of the GSA is shown in Fig. 12.39 [49, 52, 54, 60]. GSA is metaheuristic optimization

Metamodels Accuracy Metrics	Calculated Value
Root Mean Square Error (RMSE)	2.0×10^{-09}
Coefficient of Determination (R^2)	0.99

TABLE 12.22 Accuracy of the Simple Kriging Metamodels of the 45-nm CMOS-Based Thermal Sensor [49, 52, 54]

approach based on the Newtonian laws of gravity. In the GSA, the search agents are modeled as mass objects in which the location of the masses is the value of design parameters, and the FoMs are modeled as the mass of the objects. The heavier masses correspond to better performing agents, in other words, sample points with superior performance objectives. As the agent masses become heavier, they attract other agents that are lighter toward them by gravity force, i.e., pulling search agents toward an area with a likely optimal solution. In the process, the agents that attract other masses become heavier and move slower, concentrating in a search area with a likely optimal solution; on the other hand, the lighter masses are able to move faster exploring other search locations.

As depicted in Fig. 12.39, the search agenets for a set of design variables X_{tuning} are represented by their masses (M_1, M_2, M_3, M_4, M_5) and locations [49, 52, 54, 60]. For N_{sample} number of masses, the location or sample point of the ith mass can be expressed as follows [60]:

$$X_{tuning,i} = \left(x_{i,1}, x_{i,2}, \ldots, x_{i,d}, \ldots, x_{i,N_{tuning}}\right) \text{ for } i = 1, 2, \ldots, N_{sample} \quad (12.62)$$

where $x_{i,d}$ represents the position of the ith agent in the dth dimension, and N_{tuning} is the number of dimensions. The quality of solution is indicated by the mass size of the agent. In the figure, M_5, has the best quality while M_2 has the worst quality. The principle of the algorithm is now presented using the forces acting on search agent M_1 as an example. The attractive force on a mass object "i" from a mass object "j" is given by the following expression [49, 52, 54, 60]:

$$F_{ij,d}(t) = G(t) \left(\frac{M_{pi}(t) M_{aj}(t)}{R_{ij}(t) + \varepsilon} \right) \left(x_{j,d}(t) - x_{i,d}(t)\right) \quad (12.63)$$

where M_{aj} and M_{pi} are the active and passive gravitational mass of objects "j" and "i," respectively. $G(t)$ is a gravitational constant at time t, and R_{ij} is the Euclidean distance between both objects. The mass of each agent is updated with the following expression [49, 52, 54, 60]:

$$M_i(t) = \left(\frac{m_i(t)}{\sum_{j=1}^{N} m_j(t)} \right) \quad (12.64)$$

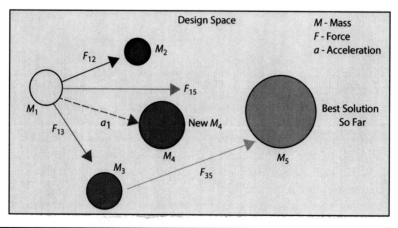

FIGURE 12.39 Illustration of gravitational search algorithm [49, 52, 54, 60].

where any m_i is calculated as follows [49, 52, 54, 60]:

$$m_i(t) = \left(\frac{\text{fit}_i(t) - \text{worst}(t)}{\text{best}(t) - \text{worst}(t)} \right) \quad (12.65)$$

where $\text{fit}_i(t)$ represents the best solution found in each iteration. Thus, the total force acting on an object is given by the following expression [49, 52, 54, 60]:

$$F_{i,d}(t) = \sum_{j=1, j \neq i}^{N} \text{rand}_j(0,1) F_{ij,d}(t) \quad (12.66)$$

where rand_j (0, 1) is a random number between 0 and 1. The use of the random number adds a stochastic flavor to the algorithm, which reduces the likelihood of optimization being stuck in local minima. The velocity is updated using the following expression [49, 52, 54, 60]:

$$v_{i,d}(t+1) = \text{rand}_i(0,1) \times v_{i,d}(t) + a_{i,d}(t) \quad (12.67)$$

where rand_i (0, 1) is a random number between 0 and 1 and acceleration $a_{i,d}(t)$ is the total force $F_{ij,d}(t)$ divided by the mass of the object $M_i(t)$. The agent location is updated using the following expression [49, 52, 54, 60]:

$$x_{i,d}(t+1) = x_{i,d}(t) + v_{i,d}(t+1) \quad (12.68)$$

In general, the mass locations that find optimal solutions gradually attract masses with poor performances, which effectively increases the chances of exploration of the design space. A convergence of the masses by the heaviest mass presents an optimal solution of the search space. One unique feature of the GSA is that it is memoryless; in other words, it does not need to remember previous best solutions but still guarantees near-optimal solution by mass acquisition. The pseudocode of the GSA is presented in Algorithm 12.11 [49, 52, 54].

As a specific example, this algorithm is executed for an initial number of $N_{\text{sample}} = 50$ search agents and a maximum iteration of $N_{\text{cnt}} = 1000$ runs over the simple Kriging metamodels. The design objective of FoM is the minimization of power dissipation of the thermal sensor. It is observed that the algorithm is able to reach an optimized solution of 184.7 μW in about 900 iterations. The optimization results of the 45-nm CMOS-based thermal sensor are provided in Table 12.23 [49, 52, 54]. As compared to the schematic baseline design of the thermal sensor, a 36.9% reduction in power dissipation is obtained with an area penalty of about 45%.

12.10.2 Ordinary Kriging Metamodel-Based Optimization of a Sense Amplifier

The overall view of an ordinary Kriging metamodel-based ultrafast design optimization flow is presented in Fig. 12.40 [49, 50]. A 45-nm CMOS-based conventional sense amplifier is considered as a case study IC. This is the same sense amplifier whose ordinary Kriging metamodels are presented in a previous section. Based on parametric analysis, width and length of NMOS devices are considered for tuning during optimization, i.e., $X_{\text{tuning}} = \{W_n, L_n\}$. As presented in the figure, the typical steps of the ordinary Kriging metamodeling are followed to generate metamodels for precharge time, sense delay, power dissipation, and sense margin of the 45-nm CMOS-based sense amplifier. As a specific example of the design optimization, the precharge time is considered for minimization, as this characteristic affects the overall speed of the memory, under power dissipation constraints.

Ant colony optimization (ACO) is a metaheuristic algorithm that is inspired by the foraging behavior of ant species [7, 12, 28, 50, 65]. In an ant colony, some ants walk to and from a food source to deposit a substance called "pheromone." Other ants sense the pheromone and follow paths where pheromone concentration is higher. Using this mechanism, the ants are able to transport food to their nest in a remarkably effective way. There are several variations of the ACO including the ant system (AS), the max-min ant system (MMAS), the ant colony system (ACS). The basic characteristics of an ACO algorithm

ALGORITHM 12.11 Gravitational Search Algorithm Optimization over the Simple Kriging Metamodels for the Thermal Sensor Circuit [49, 52, 54]

1: **Input:** Baseline tuning parameters or design variables X_{tuning} with their ranges for the thermal sensor, Simple Kriging metamodels of the thermal sensor, Maximum number of iterations N_{cnt}.
2: **Output:** Optimized tunning parameters $X_{optimal}$ of the thermal sensor.
3: Initialize iteration counter, $Counter \leftarrow 0$.
4: Initialize number of search agents N_{sample}, gravity constant $G(t)$, and velocity v.
5: Generate N_{tuning} random search nodes or design parameter set.
6: Consider the figure-of-merit (FoM) of interest, the average power of the thermal sensor P_{TS}.
7: **while**($Counter < N_{cnt}$) **do**
8:     Evaluate objective of interest for each search node.
9:     Update best and worst solutions for the objective function.
10:    Update the gravity constant $G(t)$.
11:    Calculate mass of each agent $M(t)$.
12:    Calculate acceleration a for each search node.
13:    Update velocity v for each search node.
14:    Update location $x_{i,d}$ of search nodes by applying velocity on agent $M(t)$.
15:    Increment the counter, $Counter \leftarrow Counter + 1$.
16: **end while**
17: **return** The best locations i.e. optimal $size_{optimal}$.

Thermal Sensor Designs	Average Power (P_{TS})	Temperature Sensitivity (T_{TS})
Baseline Layout	379.4 μW	9.4 MHz/°C
Optimal Layout	184.7 μW	9.4 MHz/°C
% Change	37%	44%

TABLE 12.23 Results of Gravitational Search Algorithm Optimization over the Simple Kriging Metamodels for the 45-nm CMOS-Based Thermal Sensor Circuit [49, 52, 54]

involve the incremental construction of solutions and the use of pheromone updates to guide point explorations. The basic steps for the ACO metaheuristic are the following [12, 49, 70, 76]: (1) initialization of variables and set conditions, (2) construction of ant nodes, (3) performance of optional local search, and (4) updating of the pheromones. The ACO algorithms for the continuous optimization problems differ from the discrete optimization in the selection of ant nodes. In one approach the design space is sampled for search nodes and each decision variable is a given node and not a set of nodes. In such a scenario, a solution by a node is assumed as a path traversed by an ant, which is in line with the original ACO framework. The discretization of the continuous problem space makes the ACO algorithm easy to generate optimal solutions. In the ACO, the pheromones are updated using the following expression [12, 49, 70, 76]:

$$\tau_{ij} = \rho\tau_0 + (1-\rho)\tau_{ij} \qquad (12.69)$$

where ρ is the rate of evaporation or pheromone decay coefficient. τ_0 is the initial value of the pheromone. The pheromone is updated only for the best solutions during each iteration.

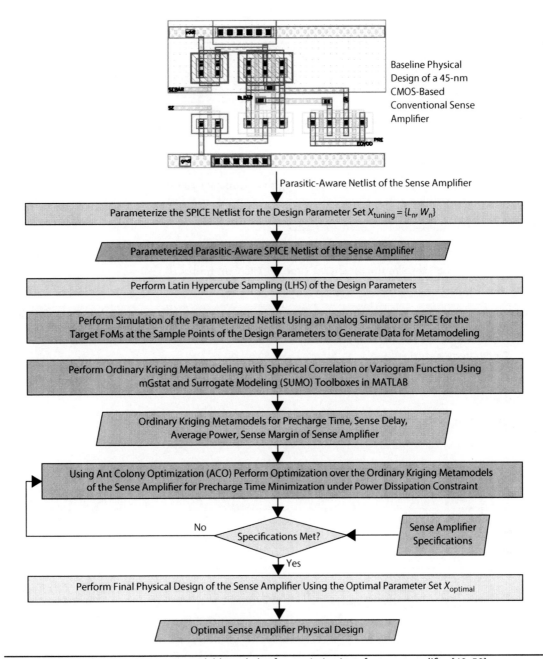

FIGURE 12.40 Ordinary Kriging metamodel-based ultrafast optimization of a sense amplifier [49, 50].

A pheromone value is associated with each possible assignment of a value to a variable. The detailed steps of the ACO for the sense amplifier over the ordinary Kriging metamodels are given in Algorithm 12.12 [49, 50].

The experiments show that for this 45-nm CMOS-based sense amplifier circuit, an average of 100 iterations are needed to converge to an optimal solution. The results of the optimization are presented in Table 12.24 [49, 50]. The final design also increases the area cost for the physical layout by 23.1%. The final design parameters L_n and W_n are 65 nm and 300 nm, respectively, for this 45-nm CMOS-based design. The average time taken for design optimization for this sense amplifier circuit is 3.9 minutes. The bulk of the time is consumed in the metamodel generation as the ACO algorithm converges in an average time of 1.36 s. The process from design space exploration to optimization is reduced by a factor of approximately $1000\times$.

ALGORITHM 12.12 Ant Colony Optimization (ACO) of a Sense Amplifier over Ordinary Kriging Metamodels [49, 50]

1: **Input:** Baseline tuning parameters or design variables X_{tuning} with their ranges for the sense amplifier, Ordinary Kriging metamodels of the sense amplifier, Maximum number of iterations N_{cnt}.
2: **Output:** Optimized tunning parameters $X_{optimal}$ of the sense amplifier.
3: Initialize the number of ants which is the solution set.
4: Initialize the iteration counter, $Counter \leftarrow 0$.
5: Start with initial baseline solution $X_{tuning,1}$.
6: Calculate the initial precharge time T_{PC_1} using the ordinary Kriging metamodels.
7: Generate random ant nodes $X_{tuning,i}$, where $i = 1, 2, \ldots, N_{ant}$.
8: Assign initial pheromone for each ant nodes τ_i.
9: **while** ($Counter < N_{cnt}$) **do**
10:    Calculate the values of the precharge time for each ant nodes, T_{PC_i}.
11:    Rank the T_{PC_i} values from best to worst.
12:    Update pheromone, increase pheromone for best solution and evaporate pheromone for all others.
13:    $X_{optimal} \leftarrow$ The sizes corresponding to the ranked T_{PC_i}.
14:    Generate new ant nodes $X_{tuning,i}$, where $i = 1, 2, \ldots, N_{ant}$.
15:    Increment the counter, $Counter = Counter + 1$.
16: **end while**
17: **return** The optimal values of design parameters $X_{optimal}$.

Design Alternative	Precharge Time T_{PC} (ns)	Sense Delay T_{SD} (ns)	Power Dissipation P_{SA} (µW)	Sense Margin V_{SM} (mV)
Baseline Layout	18.2	7.5	1.17	29.2
Optimal Layout	6.2	2.6	1.18	35.6
Improvement	65.8%	65.4%	−0.85%	21.6%

TABLE 12.24 Results of the Ant Colony Optimization (ACO) of a 45-nm CMOS-Based Sense Amplifier over Ordinary Kriging Metamodels [49, 50]

12.11 Questions

12.1 Briefly discuss the concept of metamodeling and metamodels.

12.2 Discuss three different types of metamodels of your choice.

12.3 Briefly discuss three different desired characteristics of the metamodels.

12.4 Discuss a generic flow for metamodel generation for ICs.

12.5 Briefly discuss the metrics for metamodel accuracy characterization.

12.6 Compare the two approaches for high-level circuit or system models: metamodel and macromodel.

12.7 Discuss the generic concept of metamodel-assisted ultrafast design optimization flow for ICs.

12.8 Briefly discuss the polynomial metamodeling of ICs.

12.9 Consider a schematic design of an IC of your choice and generate polynomial metamodels for three of its characteristics.

12.10 Consider a physical design of an IC of your choice and generate polynomial metamodels for three of its characteristics.

12.11 Discuss the differences between the polynomial metamodels generated from the schematic and physical designs of the same circuit.

12.12 Discuss the differences between circuit parameter metamodel (CPM) and performance metric metamodel (PMM) of a circuit.

12.13 Briefly discuss the Kriging metamodeling of ICs.

12.14 Consider a schematic design of an IC of your choice and generate Kriging metamodels for three of its characteristics. Use any Kriging method and semivariogram function of your choice.

12.15 Consider a physical design of an IC of your choice and generate Kriging metamodels for three of its characteristics. Use any Kriging method and semivariogram function of your choice.

12.16 Discuss the differences between the Kriging metamodels generated from the schematic and physical designs of the same circuit.

12.17 Briefly discuss the neural network metamodeling of ICs.

12.18 Consider a schematic design of an IC of your choice and generate neural network metamodels for three of its characteristics. Use any neural network architecture and activation function of your choice.

12.19 Consider a physical design of an IC of your choice and generate neural network metamodels for three of its characteristics. Use any neural network architecture and activation function of your choice.

12.20 Discuss the differences between the neural network metamodels generated from the schematic and physical designs of the same circuit.

12.21 Consider an IC of your choice. Perform statistical process variation analysis for three target characteristics using 1000 Monte Carlo runs using SPICE netlist. Derive the polynomial metamodels for these three characteristics. Perform statistical process variation analysis for three target characteristics using 1000 Monte Carlo runs over the polynomial metamodels. Compare the results of probability density functions (PDFs) with that of the SPICE and polynomial metamodels.

12.22 Consider an IC of your choice. Perform statistical process variation analysis for three target characteristics using 1000 Monte Carlo runs using SPICE netlist. Derive the Kriging metamodels for these three characteristics. Use any Kriging method of your choice. Perform statistical process variation analysis for three target characteristics using 1000 Monte Carlo runs over the Kriging metamodels. Compare the results of probability density functions (PDFs) with that of the SPICE and Kriging metamodels.

12.23 Consider an IC of your choice. Perform statistical process variation analysis for three target characteristics using 1000 Monte Carlo runs using SPICE netlist. Derive the neural network metamodels for these three characteristics. Use any neural network architecture of your choice. Perform statistical process variation analysis for three target characteristics using 1000 Monte Carlo runs over the neural network metamodels. Compare the results of probability density functions (PDFs) with that of the SPICE and neural network metamodels.

12.24 Consider an IC of your choice. Perform its optimization for a selected characteristic with a specific constraint using SPICE netlist. Use an algorithm of your choice. Derive the polynomial metamodels for these characteristics. Perform its optimization for a selected characteristic with a specific constraint now using the polynomial metamodels. Compare the optimization results obtained from the SPICE and polynomial metamodels.

12.25 Consider an IC of your choice. Perform its optimization for a selected characteristic with a specific constraint using SPICE netlist. Use an algorithm of your choice. Derive the Kriging metamodels for these characteristics. Use any Kriging method of your choice. Perform its optimization for a selected characteristic with a specific constraint now using the Kriging metamodels. Compare the optimization results obtained from the SPICE and Kriging metamodels.

12.26 Consider an IC of your choice. Perform its optimization for a selected characteristic with a specific constraint using SPICE netlist. Use an algorithm of your choice. Derive the neural network metamodels for these characteristics. Use any neural network architecture of your choice. Perform its optimization for a selected characteristic with a specific constraint now using the neural network metamodels. Compare the optimization results obtained from the SPICE and neural network metamodels.

12.12 References

1. mGstat: A Geostatistical Matlab Toolbox. Last Accessed on 20th July 2014.
2. Akay, B., Karaboga, D.: A Modified Artificial Bee Colony Algorithm for Real-Parameter Optimization. *Information Sciences* **192**, 120–142 (2012). DOI http://dx.doi.org/10.1016/j.ins.2010.07.015.
3. Bennour, S., Sallem, A., Kotti, M., Gaddour, E., Fakhfakh, M., Loulou, M.: Application of the PSO Technique to the Optimization of CMOS Operational Transconductance Amplifiers. In: *Proceedings of the 5th International Conference on Design and Technology of Integrated Systems in Nanoscale Era (DTIS)*, pp. 1–5 (2010). DOI 10.1109/DTIS.2010.5487582.
4. Boolchandani, D., Ahmed, A., Sahula, V.: Efficient Kernel Functions for Support Vector Machine Regression Model for Analog Circuits Performance Evaluation. *Analog Integrated Circuits and Signal Processing* **66**(1), 117–128 (2011).
5. Brunak, S., Lautrup, B.: *Neural Networks: Computers with Intuition*. World Scientific (1990). URL http://books.google.com/books?id=aPXnuEAuBuwC.
6. Casella, G.: Bayesians and Frequentists: Models, Assumptions, and Inference. http://www.stat.ufl.edu/archived/casella/Talks/BayesRefresher.pdf (2008). Accessed on 16th July 2014.
7. Chen, L., Sun, H., Wang, S.: Solving Continuous Optimization Using Ant Colony Algorithm. In: *Proceedings of the Second International Conference on Future Information Technology and Management Engineering*, pp. 92–95 (2009).
8. Clark, I.: Practical Geostatistics. Tech. rep., Geostokos (Ecosse) Limited, Alloa Business Centre, Whins Road, Alloa FK10 3SA, Scotland (2001).
9. Cressie, N.A.C.: *Statistics for Spatial Data*. Wiley, New York (1993).
10. De Bernardinis, F., Jordan, M.I., SangiovanniVincentelli, A.: Support Vector Machines for Analog Circuit Performance Representation. In: *Proceedings of the Design Automation Conference*, pp. 964–969 (2003). DOI 10.1109/DAC.2003.1219160.
11. Deng, H., Ma, Y., Shao, W., Tu, Y.: A Bayesian Meta-Modeling Approach for Gaussian Stochastic Process Models Using a Non Informative Prior. *Communications in Statistics – Theory and Methods* **41**(5), 829–850 (2012). DOI 10.1080/03610926.2010.533230.
12. Dorigo, M., Birattari, M., Stutzle, T.: Ant Colony Optimization – Artificial Ants as a Computational Intelligence Technique. *IEEE Computational Intelligence Magazine* **1**, 28–39 (2006).
13. Ebden, M.: Gaussian Processes for Regression: A Quick Introduction. Tech. rep., Department of Engineering Science, University of Oxford, Oxford, OX13PJ, United Kingdom (2008). Accessed on 16th July 2014.
14. Fang, K.T., Li, R., Sudjianto, A.: *Design and Modeling for Computer Experiments*. Chapman and Hall/CRC, 23–25 Blades Court, London SW15 2NU, UK (2006).
15. Feng, Z., Li, P.: Performance-oriented Statistical Parameter Reduction of Parameterized Systems via Reduced Rank Regression. In: *Proceedings of the IEEE/ACM International Conference Computer-Aided Design*, pp. 868–875 (2006). DOI 10.1109/ICCAD.2006.320091.
16. Garitselov, O.: *Metamodeling-based Fast Optimization of Nanoscale AMS-SoCs*. Ph.D. thesis, Computer Science and Engineering, University of North Texas, Denton, 76203, TX, USA (2012).
17. Garitselov, O., Mohanty, S.P., Kougianos, E.: Fast Optimization of Nano-CMOS Mixed-Signal Circuits Through Accurate Metamodeling. In: *Proceedings of the 12th IEEE International Symposium on Quality Electronic Design (ISQED)*, pp. 405–410 (2011).
18. Garitselov, O., Mohanty, S.P., Kougianos, E.: A Comparative Study of Metamodels for Fast and Accurate Simulation of Nano-CMOS Circuits. *IEEE Transactions on Semiconductor Manufacturing* **25**(1), 26–36 (2012). DOI 10.1109/TSM.2011.2173957.
19. Garitselov, O., Mohanty, S.P., Kougianos, E.: Accurate Polynomial Metamodeling-based Ultra-Fast Bee Colony Optimization of a Nano-CMOS Phase-Locked Loop. *Journal of Low Power Electronics* **8**(3), 317–328 (2012).
20. Garitselov, O., Mohanty, S.P., Kougianos, E.: Fast-accurate Non-polynomial Metamodeling for Nano-CMOS PLL Design Optimization. In: *Proceedings of the 25th International Conference on VLSI Design (VLSID)*, pp. 316–321 (2012). DOI 10.1109/VLSID.2012.90.
21. Garitselov, O., Mohanty, S.P., Kougianos, E., Okobiah, O.: Metamodel-Assisted Ultra-fast Memetic Optimization of a PLL for WiMax and MMDS Applications. In: *Proceedings of the 13th International Symposium on Quality Electronic Design (ISQED)*, pp. 580–585 (2012). DOI 10.1109/ISQED.2012.6187552.

22. Garitselov, O., Mohanty, S.P., Kougianos, E., Zheng, G.: Particle Swarm Optimization over Non-polynomial Metamodels for Fast Process Variation Resilient Design of Nano-CMOS PLL. In: *Proceedings of the ACM Great Lakes Symposium on VLSI*, pp. 255–258 (2012).
23. Ghai, D.: *Variability Aware Low-power Techniques for Nanoscale Mixed-Signal Circuits*. Ph.D. thesis, Department of Computer Science and Engineering, University of North Texas, Denton, TX 76207 (2009).
24. Gielen, G., Wambacq, P., Sansen, W.M.: Symbolic Analysis Methods and Applications for Analog Circuits: A Tutorial Overview. *Proceedings of the IEEE* **82**(2), 287–304 (1994). DOI 10.1109/5.265355.
25. Gopalakrishnan, C., Katkoori, S.: Tabu Search Based Behavioural Synthesis of Low Leakage Datapaths. In: *Proceedings of the IEEE Computer society Annual Symposium on VLSI*, pp. 260–261 (2004). DOI 10.1109/ISVLSI.2004.1339548.
26. Gorissen, D., Couckuyt, I., Demeester, P., Dhaene, T., Crombecq, K.: A Surrogate Modeling and Adaptive Sampling Toolbox for Computer Based Design. *Journal of Machine Learning Research* **11**, 2051–2055 (2010).
27. Gorissen, D., De Tommasi, L., Hendrickx, W., Croon, J., Dhaene, T.: RF Circuit Block Modeling via Kriging Surrogates. In: *Proceedings of the 17th International Conference on Microwaves, Radar and Wireless Communications*, pp. 1–4 (2008).
28. Hu, X.M., Zhang, J., Chung, H.S.H., Li, Y., Liu, O.: SamACO: Variable Sampling Ant Colony Optimization Algorithm for Continuous Optimization. *IEEE Transactions on Systems, Man, and Cybernetics, Part B: Cybernetics* **40**(6), 1555–1566 (2010).
29. Huelsman, L.P.: Symbolic Analysis – A Tool for Teaching Undergraduate Circuit Theory. *IEEE Transactions on Education* **39**(2), 243–250 (1996). DOI 10.1109/13.502071.
30. Isaksson, M., Wisell, D., Ronnow, D.: Wide-band Dynamic Modeling of Power Amplifiers Using Radial-Basis Function Neural Networks. *IEEE Transactions on Microwave Theory and Techniques* **53**(11), 3422–3428 (2005). DOI 10.1109/TMTT.2005.855742.
31. Jain, A.K., Mao, J., Mohiuddin, K.M.: Artificial Neural Networks: A Tutorial. *Computer* **29**(3), 31–44 (1996). DOI 10.1109/2.485891. URL http://dx.doi.org/10.1109/2.485891.
32. Jin, R., Chen, W., Simpson, T.W.: Comparative Studies of Metamodelling Techniques under Multiple Modelling Criteria. *Structural and Multidisciplinary Optimization* **23**, 1–13 (2001). URL http://dx.doi.org/10.1007/s00158-001-0160-4. 10.1007/s00158-001-0160-4.
33. Kabir, H., Wang, Y., Yu, M., Zhang, Q.J.: High-dimensional Neural-network Technique and Applications to Microwave Filter Modeling. *IEEE Transactions on Microwave Theory and Techniques* **58**(1), 145–156 (2010). DOI 10.1109/TMTT.2009.2036412.
34. Karaboga, D., Akay, B.: A Comparative Study of Artificial Bee Colony Algorithm. *Applied Mathematics and Computation* **214**(1), 108–132 (2009).
35. Kiely, T., Gielen, G.: Performance Modeling of Analog Integrated Circuits Using Least-Squares Support Vector Machines. In: *Proceedings of the Design, Automation and Test in Europe Conference and Exhibition*, pp. 448–453 (2004). DOI 10.1109/DATE.2004.1268887.
36. Kleijnen, J.P.: Kriging Metamodeling in Simulation: A Review. *European Journal of Operational Research* **192**(3), 707–716 (2009). DOI http://dx.doi.org/10.1016/j.ejor.2007.10.013.
37. Lamecki, A., Balewski, L., Mrozowski, M.: Towards Automated Full-Wave Design of Microwave Circuits. In: *Proceedings of the 17th International Conference on Microwaves, Radar and Wireless Communications*, pp. 1–2 (2008).
38. Larumbe, F., Sanso, B.: A Tabu Search Algorithm for the Location of Data Centers and Software Components in Green Cloud Computing Networks. *IEEE Transactions on Cloud Computing* **1**(1), 22–35 (2013). DOI 10.1109/TCC.2013.2.
39. Lesh, F.H.: Multi-Dimensional Least-Squares Polynomial Curve Fitting. *Communication of ACM* **2**, 29–30 (1959).
40. Li, X., Gopalakrishnan, P., Xu, Y., Pileggi, L.T.: Robust Analog/RF Circuit Design With Projection-Based Performance Modeling. *IEEE Transction on Computer-Aided Design Integrated Circuits Systems* **26**(1), 2–15 (2007).
41. Mathaikutty, D.A., Shukla, S.: *Metamodeling Driven IP Reuse for SoC Integration and Microprocessor Design*. Artech House (2009).
42. McCray, A.T., McNames, J., Abercrombie, D.: Stepwise Regression for Identifying Sources of Variation in a Semiconductor Manufacturing Process. In: *Proceedings of the IEEE Conference and Workshop Advanced Semiconductor Manufacturing*, pp. 448–452 (2004). DOI 10.1109/ASMC.2004.1309613.
43. Mohanty, S.P.: Ultra-Fast Design Exploration of Nanoscale Circuits through Metamodeling). http://www.cse.unt.edu/smohanty/Presentations/2012/Mohanty_SRC-TxACE_Talk_2012-04-27.pdf (2012). Accessed on 13th February 2014.
44. Mohanty, S.P., Kougianos, E.: *Models, Methods, and Tools for Complex Chip Design*: Selected Contributions from FDL 2012, chap. Polynomial Metamodel-Based Fast Optimization of Nanoscale PLL Components. Springer (2014).
45. Mohanty, S.P., Kougianos, E.: Polynomial Metamodel Based Fast Optimization of Nano-CMOS Oscillator Circuits. *Analog Integrated Circuits and Signal Processing* **79**(3), 437–453 (2014). DOI 10.1007/s10470-014-0284-2. URL http://dx.doi.org/10.1007/s10470-014-0284-2.
46. Mohanty, S.P., Kougianos, E., Garitselov, O., Molina, J.M.: Polynomial-Metamodel Assisted Fast Power Optimization of Nano-CMOS PLL Components. In: *Proceeding of the 2012 Forum on Specification and Design Languages*, pp. 233–238 (2012).

47. Mohanty, S.P., Ranganathan, N., Kougianos, E., Patra, P.: *Low-Power High-Level Synthesis for Nanoscale CMOS Circuits*. Springer (2008). 0387764739 and 978-0387764733.
48. Montgomery, D.C.: *Design and Analysis of Experiments*, 6th edn. John Wiley & Sons, Inc. (2005).
49. Okobiah, O.: *Geostatistical Inspired Metamodeling and Optimization of Nanoscale Analog Circuits*. Ph.D. thesis, Computer Science and Engineering, University of North Texas, Denton, 76203, TX, USA (2014).
50. Okobiah, O., Mohanty, S., Kougianos, E.: Ordinary Kriging Metamodel-Assisted Ant Colony Algorithm for Fast Analog Design Optimization. In: *Proceedings of the 13th IEEE International Symposium on Quality Electronic Design (ISQED)*, pp. 458–463 (2012).
51. Okobiah, O., Mohanty, S., Kougianos, E.: Fast Design Optimization Through Simple Kriging Metamodeling: A Sense Amplifier Case Study. *IEEE Transactions on Very Large Scale Integration (VLSI) Systems* **22**(4), 932–937 (2014). DOI 10.1109/TVLSI.2013.2256436.
52. Okobiah, O., Mohanty, S.P., Kougianos, E.: Geostatistical-Inspired Metamodeling and Optimization of Nano-CMOS Circuits. In: *Proceedings of the IEEE Computer Society Annual Symposium on VLSI (ISVLSI)*, pp. 326–331 (2012). DOI 10.1109/ISVLSI.2012.12.
53. Okobiah, O., Mohanty, S.P., Kougianos, E.: Fast Statistical Process Variation Analysis Using Universal Kriging Metamodeling: A PLL Example. In: *Proceedings of the IEEE 56th International Midwest Symposium on Circuits and Systems (MWSCAS)*, pp. 277–280 (2013). DOI 10.1109/MWSCAS.2013.6674639.
54. Okobiah, O., Mohanty, S.P., Kougianos, E.: Geostatistical-Inspired Fast Layout Optimization of a Nano-CMOS Thermal Sensor. *IET Circuits, Devices Systems* **7**(5), 253–262 (2013). DOI 10.1049/iet-cds.2012.0358.
55. Okobiah, O., Mohanty, S.P., Kougianos, E.: Exploring Kriging for Fast and Accurate Design Optimization of Nanoscale Analog Circuits. In: *Proceedings of the 13th IEEE Computer Society Annual Symposium on VLSI (ISVLSI)* (2014).
56. Okobiah, O., Mohanty, S.P., Kougianos, E.: Kriging Bootstrapped Neural Network Training for Fast and Accurate Process Variation Analysis. In: *Proceedings of the 15th International Symposium on Quality Electronic Design (ISQED)*, pp. 365–372 (2014). DOI 10.1109/ISQED.2014.6783349.
57. Okobiah, O., Mohanty, S.P., Kougianos, E., Garitselov, O.: Kriging-Assisted Ultra-Fast Simulated-Annealing Optimization of a Clamped Bitline Sense Amplifier. In: *Proceedings of the 25th IEEE International Conference on VLSI Design (VLSID)*, pp. 310–315 (2012).
58. Pandit, S., Mandal, C., Patra, A.: Systematic Methodology for High-Level Performance Modeling of Analog Systems. In: *Proceedings of the International Conference on VLSI Design*, pp. 361–366 (2009). DOI 10.1109/VLSI.Design.2009.26.
59. Park, J., Choi, K., Allstot, D.J.: Parasitic-aware Design and Optimization of a Fully Integrated CMOS Wideband Amplifier. In: *Proceedings of the Asia South Pacific Design Automation Conference*, pp. 904–907 (2003).
60. Rashedi, E., Nezamabadi-pour, H., Saryazdi, S.: GSA: A Gravitational Search Algorithm. *Information Sciences* **179**(13), 2232–2248 (2009). DOI http://dx.doi.org/10.1016/j.ins.2009.03.004.
61. Rasmussen, C.E., Williams, C.K.I.: *Gaussian Processes for Machine Learning*. Adaptative Computation and Machine Learning Series. University Press Group Limited (2006).
62. Sacks, J., Welch, W.J., Mitchell, T.J., Wynn., H.P.: Design and Analysis of Computer Experiments. *Statistical Science* **4**(4), 409–423 (1989).
63. Sallaberry, C.J., Helton, J.C., Hora, C.C.: Extension of Latin Hypercube Samples with Correlated Variables. Tech. Rep. SAND2006-6135, Department of Engineering Science, University of Oxford, Oxford, OX13PJ, United Kingdom (2006). Accessed on 16th July 2014.
64. Saranya, P.K., Sumangala, K.: ABC Optimization: A Co-Operative Learning Approach to Complex Routing Problems. *Progress in Nonlinear Dynamics and Chaos* **1**, 39–46 (2013).
65. Sarkar, M., Ghosal, P., Mohanty, S.P.: Reversible Circuit Synthesis Using ACO and SA based Quinne-McCluskey Method. In: *Proceedings of the 56th IEEE International Midwest Symposium on Circuits & Systems (MWSCAS)*, pp. 416–419 (2013). DOI 10.1109/MWSCAS.2013.6674674.
66. Shi, C.J.R., Tan, X.D.: Canonical Symbolic Analysis of Large Analog Circuits with Determinant Decision Diagrams. *IEEE Transactions on Computer-Aided Design of Integrated Circuits and Systems* **19**(1), 1–18 (2000). DOI 10.1109/43.822616.
67. Silveira, F., Flandre, D., Jespers, P.G.A.: A g_m/I_D Based Methodology for the Design of CMOS Analog Circuits and its Application to the Synthesis of a Silicon-on-Insulator Micropower OTA. *IEEE Journal of Solid-State Circuits* **31**(9), 1314–1319 (1996). DOI 10.1109/4.535416.
68. Simpson, T.W., Peplinski, J.D., Koch, P.N., Allen, J.K.: On the Use of Statistics in Design and the Implications for Deterministic Computer Experiments. In: *Proceedings of ASME Design Engineering Technical Conferences*, pp. 1–14 (1997).
69. Simpson, T.W., Poplinski, J.D., Koch, P.N., Allen, J.K.: Metamodels for Computer-based Engineering Design: Survey and Recommendations. *Engineering with Computers* **17**(2), 129–150 (2001).
70. Socha, K., Dorigo, M.: Ant Colony Optimization for Continuous Domains. *European Journal of Operational Research* **185**(3), 1155–1173 (2008).

71. Tang, B.: Orthogonal Array-Based Latin Hypercubes. *Journal of the American Statistical Association* **88**(424), 1392–1397 (1993).
72. Turner, S., Balll, T., Marshall, D.D.: Gaussian Process Metamodeling Applied to a Circulation Control Wing. In: *Proceedings of 38th Fluid Dynamics Conference and Exhibit*, pp. 1–20 (2008).
73. Van Beers, W.C.M.: Kriging Metamodeling in Discrete-Event Simulation: An Overview. In: *Proceedings of the Winter Simulation Conference*, pp. 202–208 (2005).
74. Vemuri, R., Wolfe, G.: Adaptive Sampling and Modeling of Analog Circuit Performance Parameters with Pseudo-cubic Splines. In: *Proceedings of the IEEE/ACM International Computer Aided Design Conference*, pp. 931–938 (2004).
75. Wolfe, G., Vemuri, R.: Extraction and Use of Neural Network Models in Automated Synthesis of Operational Amplifiers. *IEEE Transactions on Computer-Aided Design Integrated Circuits and Systems* **22**(2), 198–212 (2003).
76. Xue, X., Sun, W., Peng, C.: Improved Ant Colony Algorithm for Continuous Function Optimization. In: *Proceedings of the Control and Decision Conference (CCDC)*, pp. 20–24 (2010). DOI 10.1109/CCDC.2010.5499143.
77. Yang, X., Deb, S.: Engineering Optimisation by Cuckoo Search. *International Journal of Mathematical Modelling and Numerical Optimisation* **1**(4), 330–343 (2010). DOI 10.1504/IJMMNO.2010.03543.
78. Yang, X.S.: *Nature-inspired Metaheuristic Algorithms*. Luniver Press (2010).
79. Yang, X.S.: Multiobjective Firefly Algorithm for Continuous Optimization. *Engineering with Computers* **29**(3), 175–184 (2013).
80. Yelten, M., Zhu, T., Koziel, S., Franzon, P., Steer, M.: Demystifying Surrogate Modeling for Circuits and Systems. *IEEE Circuits and Systems Magazine* **12**(1), 45–63 (2012). DOI 10.1109/MCAS.2011.2181095.
81. You, H., Yang, M., Wang, D., Jia, X.: Kriging Model Combined with Latin Hypercube Sampling for Surrogate Modeling of Analog Integrated Circuit Performance. In: *Proceedings of the International Symposium on Quality Electronic Design*, pp. 554–558 (2009). DOI 10.1109/ISQED.2009.4810354.
82. Yu, G., Li, P.: Yield-aware Analog Integrated Circuit Optimization Using Geostatistics Motivated Performance Modeling. In: *Proceedings of the IEEE/ACM International Conferenceon Computer-Aided Design*, pp. 464–469 (2007).
83. Zheng, G.: *Layout-accurate Ultra-fast System-Level Design Exploration through Verilog-AMS*. Ph.D. thesis, Computer Science and Engineering, University of North Texas, Denton, 76203, TX, USA, Denton, TX 76207 (2013).
84. Zheng, G., Mohanty, S., Kougianos, E., Okobiah, O.: iVAMS: Intelligent Metamodel-integrated Verilog-AMS for Circuit-accurate System-level Mixed-signal Design Exploration. In: *Proceedings of the IEEE 24th International Conference on Application-Specific Systems, Architectures and Processors (ASAP)*, pp. 75–78 (2013). DOI 10.1109/ASAP.2013.6567553.
85. Zheng, G., Mohanty, S.P., Kougianos, E.: Metamodel-assisted Fast and Accurate Optimization of an OP-AMP for Biomedical Applications. In: *Proceedings of the IEEE Computer Society Annual Symposium on VLSI (ISVLSI)*, pp. 273–278 (2012). DOI 10.1109/ISVLSI.2012.11.
86. Zheng, G., Mohanty, S.P., Kougianos, E.: iVAMS: Intelligent Metamodel-Integrated Verilog-AMS for Fast Analog Block Optimization. In: *Work-in-Progress Session Poster, Design Automation Conference* (2013).
87. Zheng, G., Mohanty, S.P., Kougianos, E.: Verilog-AMS-POM: Verilog-AMS Integrated Polynomial Metamodelling of a Memristor-based Oscillator. In: *Work-in-Progress Session Poster, Design Automation Conference* (2013).
88. Zheng, G., Mohanty, S.P., Kougianos, E., Garitselov, O.: Verilog-AMS-PAM: Verilog-AMS Integrated with Parasitic-aware Metamodels for Ultra-fast and Layout-accurate Mixed-signal Design Exploration. In: *Proceedings of the ACM Great Lakes Symposium on VLSI*, pp. 351–356 (2012).
89. Zheng, G., Mohanty, S.P., Kougianos, E., Okobiah, O.: Polynomial Metamodel Integrated Verilog-AMS for Memristor-based Mixed-signal System Design. In: *Proceedings of the IEEE 56th International Midwest Symposium on Circuits and Systems (MWSCAS)*, pp. 916–919 (2013). DOI 10.1109/MWSCAS.2013.6674799.
90. Zhu, T., Steer, M.B., Franzon, P.D.: Surrogate Model-based Self-calibrated Design for Process and Temperature Compensation in Analog/RF Circuits. *IEEE Design Test of Computers* **29**(6), 74–83 (2012). DOI 10.1109/MDT.2012.2220332.

Index

1-bit DAC, 221, 255, 256f
1 of N code generator, 240, 245–246
1T-1C-DRAM, 340–341
1T-1C nonvolatile FeDRAM, 345, 345f
1T-DRAM, 341–342
3D transistor, 78
3D-DRAM, 340
3T-DRAM, 342, 342f
4-tap finite-impulse response (FIR) filter, 670–672
4T-DRAM, 343, 343f
4T-SRAM, 321, 321f
5-tap finite-impulse response (FIR) filter, 672–673
5Spice, 447
5T-SRAM, 321–322, 322f
6T-SRAM, 318–320, 331–334
7T-SRAM, 322–324, 334–335, 662–667
8-bit synchronous up counter, 479, 480f
8-bit up counter, 474f
8T-SRAM, 324–326
9T-SRAM, 326, 326f
10-bit binary counter, 269f
10-bit register, 269f
10T-SRAM, 326–327, 327f, 335–338
32-nm CMOS APS design, 289–291, 292t
45-nm CMOS-based flash ADC, 245–248
45-nm CMOS clamped bitline sense amplifier, 304–306
45-nm CMOS five-stage CS-VCO, 179, 182f, 182t
45-nm CMOS ring oscillator, 171–173
45-nm CMOS ring oscillator-based thermal sensor, 268–271
45-nm double-gate FinFET-based 21-stage CS-VCO, 183, 184f, 184t
50-nm CMOS CS-VCO, 179, 181f, 181t
90-nm CMOS-based flash ADC, 238–245
 1 of N code generator, 240
 characteristics, 241, 242t
 comparator bank, 238–239
 DNL plot, 244f
 dynamic characterization, 244
 FFT plot, 244f
 INL plot, 243f
 logic-level diagram, 242f
 NOR ROM, 240–241
 physical design, 241, 242f
 post-layout simulation and characterization, 241–245
 power analysis, 244, 245f, 245t
 static characterization, 243
90-nm CMOS-based fully differential OP-AMP, 704f
90-nm CMOS CS-VCO, 179, 180f, 181f
180-nm CMOS LC-VCO, 186–187, 187f, 187t
180-nm CMOS phase-locked loop, 203–204

A

A/d-based mixed-signal system, 4–5
a/D-based mixed-signal system, 5
Abacus, 52
ABB. See Adaptive body bias (ABB)
ABB method for dynamic process variation compensation, 678–679
ABC optimization. See Artificial bee colony (ABC) optimization
Absolute jitter, 167
Abstraction levels, 370, 370f
.AC control line, 451
AC small-signal analysis, 450–451
ACan, 459f
Acceleration coefficient, 658
Acceptance-rejection Monte Carlo method, 622–623
Access transistor, 318
AccuCore, 397
Accurate modeling, 499–500, 499f
ACO. See Ant colony optimization (ACO)
Acquisition time, 162, 233
ACS. See Ant colony system (ACS)
Active-matrix TFT, 85
Active pixel sensor (APS), 281, 282–284, 648–653
Active RFID tag, 46
Adaptive body bias (ABB), 522, 523, 538
Adaptive body bias method for dynamic process variation compensation, 678–679
Adaptive voltage scaling (AVS), 540
ADC. See Analog-to-digital converter (ADC)
ADDLL. See All-digital delay-locked loop (ADDLL)
ADiT, 450
ADLL. See Analog delay-locked loop (ADLL)
ADMS, 18, 392. See Automatic device model synthesizer (ADMS)
ADPLL. See All-digital phase-locked loop (ADPLL)
ADRAM. See Audio DRAM (ADRAM)
Advance MS, 18
Advanced power controller (APC), 540, 540f
AFM. See Atomic force microscope (AFM)
AGE, 136
Aging of an IC, 135
ALAP schedule. See As late as possible (ALAP) schedule
Aldec Riviera-PRO, 398, 399
Algebraic decomposition, 531
Algorithm-level design, 384
Algorithmic DAC, 229–230, 230f
Algorithmic level (digital circuit), 370, 370f
Algorithmic synthesis, 384
All-digital delay-locked loop (ADDLL), 207
All-digital phase-locked loop (ADPLL), 159, 204–206
Alliance VLSI system, 399–400
Alternative bit latch sense amplifier, 306
ALU. See Arithmetic-logic unit (ALU)
AMGIE, 392
Ambipolar electric field, 81
Ampere law, 569
Amperometric electrochemical biosensor, 28, 29f
AMS circuit synthesis, 391–393
AMS Designer, 18, 389
AMS EDA tools, 398–399, 398f, 399f, 402–403
AMS-SoC. See Analog/mixed-signal system-on-chip (AMS-SoC)
AMSHDLs. See Analog/mixed-signal hardware description languages (AMSHDLs)
ANACONDA, 392
Analog circuit, 3–4, 4f, 103–104
Analog circuit design flow, 371–381
 behavioral simulation, 372–374
 design rule check (DRC), 375–376
 electrical rule check (ERC), 377–378
 flowcharts, 371f, 372f
 layout design, 374–375
 LVS verification, 377, 377f
 parasitics extraction, 376–377
 performance optimization, 380–381
 physical design, 374–375
 physical design characterization, 378, 379f
 schematic capture, 374
 transistor-level design, 374
 transistor-level simulation and characterization, 374
 variability analysis, 379–380

Analog circuit power optimization, 523–525
Analog circuits and systems, 3, 103
Analog delay-locked loop (ADLL), 207–208, 208f
Analog design verification, 424
Analog EDA tools, 394–395, 395f, 396f, 401–402
Analog-fast SPICE, 447f, 450
Analog FastSPICE, 450
Analog library, 408, 408f
Analog/mixed-signal circuit synthesis, 391–393
Analog/mixed-signal hardware description languages (AMSHDLs), 15, 16–17, 480, 480t
Analog/mixed-signal system-on-chip (AMS-SoC), 1, 15
　big-A/small-d AMS-SoC, 496
　comparable-A/comparable-D AMS-SoC, 496
　complexity of hardware component designs, 365
　component manufacturing and development, 367–368
　component/software design, 366–367
　design. See Mixed-signal circuit and system design flow
　design (digital, logic, analog synthesis), 16
　digital circuitry, 365
　languages, 16–17
　memory. See Memory
　metamodels. See Metamodel-based fast design methodologies
　representative AMS-SoC, 2f
　small-a/big-D AMS-SoC, 496
　subsystem design, 15f
　tools, 17–18
　transistor models, 18
　unified optimization, 15–16
　variant tolerant design. See Variability-aware design methodologies
Analog output filter, 229
Analog physical design flow, 375
Analog PLL (APLL), 158, 204
Analog/RF integrated circuit characteristics, 379f
Analog routing techniques, 375
Analog signal, 215, 215f
Analog SPICE modeling, 373
Analog STB, 51
Analog synthesis, 16
Analog-to-digital converter (ADC), 217
　45-nm CMOS-based flash ADC, 245–248
　90-nm CMOS-based flash ADC, 238–245
　acquisition time, 233
　architecture selection, 224, 224f, 225t
　DNL, 234

Analog-to-digital converter (Cont.):
　DR, 232–233
　ENOB, 231–232
　flash ADC, 219–220, 219f
　folding ADC, 222–223, 223f
　gain error, 235
　high-level representation, 217f
　INL, 233–234
　integrating ADC, 220–221, 220f
　monotonicity, 235
　offset error, 235
　organic thin-film transistor-based ADC, 250–252
　overview, 218, 218f
　pipeline ADC, 221, 221f
　quantization noise, 235
　ramp-compare ADC, 219, 219f
　resolution, 231
　sampling rate, 233
　settling time, 233
　SFDR, 235
　sigma-delta ADC, 221–222, 222f
　sigma-delta modulator-based ADC, 252–256
　SINAD, 235
　single-electron-based ADC, 249–250
　SNR, 234
　successive-approximation ADC, 220, 220f
　THD, 236
　time-interleaved ADC, 222, 223f
　tracking ADC, 223–224, 224f
Analog TV tuner, 57
ANN. See Artificial neural network (ANN)
Ant colony optimization (ACO), 755–758
Ant colony system (ACS), 755
Ant system (AS), 755
Antenna rules, 375
Anti-aliasing filter, 217
Antiparallel direction reading, 356
APC. See Advanced power controller (APC)
APLL. See Analog PLL (APLL)
Apple TV, 41
Application-specific instruction processor (ASIP), 3, 3f
Application-specific integrated circuit (ASIC), 3, 3f
Applications. See Emerging systems
APS. See Active pixel sensor (APS)
APS-based motion sensor, 300f
Aptamer-modified electrolyte-gated GFET-based biosensor, 266f
Arbitrary source (ASRC), 456
ARCHGEN, 392
Architecture level (digital circuit), 370, 370f
Architecture-level design, 384–385
Architecture level library, 408, 408f
Architecture-level MATLAB simulation, 426–430

Architecture-level Simulink simulation, 435–438
Architecture parallelism, 522, 527f, 528
Architecture selection, 224, 224f, 225t
ARF. See Autoregressive filter (ARF)
ARIADNE, 392
Arithmetic-logic unit (ALU), 476, 476f
Armchair CNT, 79
Armchair GNR, 81
Arrhenius equation, 135
Artificial bee colony (ABC) optimization, 740–744
Artificial neural network (ANN), 691–692
AS. See Ant system (AS)
As late as possible (ALAP) schedule, 534, 547f
As soon as possible (ASAP) schedule, 534, 547f
ASAP schedule. See As soon as possible (ASAP) schedule
ASIC. See Application-specific integrated circuit (ASIC)
asimut, 400, 402
ASIP. See Application-specific instruction processor (ASIP)
ASRC. See Arbitrary source (ASRC)
Assura, 375, 378, 397
Asus O!Play HD2, 41
Asynchronous SRAM, 318
ATE. See Automated test equipment (ATE)
Atomic force microscope (AFM), 25–27
Audio DRAM (ADRAM), 339–340
Automated test equipment (ATE), 389
Automatic design, 383
Automatic device model synthesizer (ADMS), 501
Autoregressive filter (ARF), 536, 536f, 669
Avalanche current, 117
Average code, 206
Average-DCO, 205
Average output voltage, 237
Average power dissipation, 517
AVS. See Adaptive voltage scaling (AVS)
AZTECA, 392

B

B2.Spice A/D, 447–448
Back-end design, 383
Back-end of line (BEOL) corners, 636, 636f
Back-gated CNTFET, 80, 80f
Backlight management, 523
Band-to-band tunneling (BTBT), 3
Bandpass filter (BPF), 536, 536f, 669
Battery-assisted RFID, 46
Battery life, 515
Battery scheduling, 523
Bayesian process regression (BPR), 691

BCEs. *See* Branch constitutive-
    equations (BCEs)
BCO. *See* Bee colony
    optimization (BCO)
BeagleBoard, 433
BEDO-DRAM. *See* Burst extended
    data out (BEDO-DRAM)
Bee colony optimization (BCO),
    740–744
Behavioral level (analog circuit),
    370, 370f
Behavioral level (digital circuit),
    370, 370f
Behavioral model calibration, 374
Behavioral modeling, 423, 424
Behavioral simulation, 372–374
Behavioral synthesis, 16, 384
BeiDou navigation system, 36
BEM. *See* Boundary element
    method (BEM)
BEOL corners. *See* Back-end of line
    (BEOL) corners
Berkeley Open-Source License, 399
Berkeley short-channel insulated-gate
    field-effect transistor (IGFET)
    model (BSIM), 18, 500
Best straight-line INL, 233
Big-A/small-d AMS-SoC, 496
Bilayer GFET, 82
Binary encoder, 223
Binary-weighted DAC, 225–226, 226f
Binary-weighted principle, 229
Bioaffinity biosensor, 29
Biodetector, 30
Biosensor systems, 27–31
    applications, 29, 29f
    basic concepts, 27, 27f, 28f
    bio-element, 27, 28f
    biodetector, 30
    components, 30
    defined, 27
    DNA detection, 30
    historical overview, 29
    microfluidic system, 30, 31f
    sensor element, 28f
    types, 28
    wearable, 30
Biosensors, 264–266
Bipolar junction transistor (BJT),
    68, 138
Bit synchronizer, 223
BJT. *See* Bipolar junction transistor
    (BJT)
Black equation, 141
Blech length, 141
Blood-glucose biosensor, 29
Blu-ray Disc Association (BDA), 31
Blu-ray player, 31–32
BlueSpec Compiler, 385
Body node, 295, 296f
Body node power management
    unit, 296
Body node state controller, 296
Body sensors, 295–296
Bonding wires, 389

boog, 400, 402
boom, 400, 402
Bootstrap charge pump, 196, 196f
Bottom-gate thin-film transistor, 86f
Bottom-up design flow, 369f, 370
Boundary element method (BEM),
    568, 568f
Boyle, W., 281
BPF. *See* Bandpass filter (BPF)
BPR. *See* Bayesian process
    regression (BPR)
Branch constitutive-equations
    (BCEs), 461, 462
Breadboarding, 421
BSIM. *See* Berkeley short-channel
    insulated-gate field-effect
    transistor (IGFET) model (BSIM)
BSIM1, 500
BSIM4, 500
BSIM6.0, 500
BSIM-CMG. *See* BSIM common
    multi-gate (BSIM-CMG) FET
    model
BSIM common multi-gate (BSIM-
    CMG) FET model, 500
BSIM-IMG. *See* BSIM independent
    multi-gate (BSIM-IMG)
    FET model
BSIM independent multi-gate
    (BSIM-IMG) FET model, 501
BSIM silicon-on-insulator (BSIM-SOI)
    FET model, 501
BSIM-SOI. *See* BSIM silicon-on-
    insulator (BSIM-SOI) FET model
BTBT. *See* Band-to-band tunneling
    (BTBT)
Bulk FinFET, 77, 77f
Burst extended data out
    (BEDO-DRAM), 343–344

C

C/C++, 474
C-to-Silicon, 385
Cache DRAM (CDRAM), 340
Cache memory, 317. *See also*
    Static random-access memory
    (SRAM)
Cache tag SRAM, 318
CAD, 393. *See also* Electronic design
    automation (EDA) tools
Cadence Design Systems, 18, 394
CADICS, 392
Calibre, 375, 378, 387, 395
Calibre PERC, 378
Calibre xACT3D, 376
CAM. *See* Content-addressable
    memory (CAM)
CANCER. *See* Computer Analysis of
    Non-Linear Circuits Excluding
    Radiation (CANCER)
Capacitance components (MOSFET
    structure), 124f
Capacitive switching power
    dissipation, 112
Capacitive touchscreen, 54

Capacitive-switching power reduction,
    541–542, 541f
Capacitor-less one-transistor DRAM
    (1T-DRAM), 341–342
Capacitor-less twin transistor RAM
    (TTRAM), 347, 347f
Capture range, 161
Capture time, 286
Carbon nanotube (CNT), 79, 102
Carbon nanotube FET (CNTFET),
    79–81
Carbon nanotube FET fabrication
    process, 95
Carbon nanotube (CNT) inductor-
    based LC-VCO, 187–188,
    189f, 189t
CATALYST, 392
Catapult C, 385
CBGA joint. *See* Ceramic ball grid
    array (CBGA) joint
CCCS. *See* Current-controlled current
    source (CCCS)
CCVS. *See* Current-controlled voltage
    source (CCVS)
CDF. *See* Cumulative distribution
    function (CDF)
CDFG. *See* Control data flow graph
    (CDFG)
CDRAM. *See* Cache DRAM
    (CDRAM)
Cell phone, 55, 518–519
Cell ratio, 330
Center frequency, 162
Central processing unit (CPU), 3, 3f
Ceramic ball grid array (CBGA)
    joint, 584
Challenges:
    leakage dissipation, 9–10
    overview, 8–9, 9f, 67f
    parasitics. *See* Parasitics
    power consumption. *See* Power
        dissipation
    process variation. *See* Process
        variation
    reliability issue. *See* Reliability issue
    temperature variation, 13–14,
        133–135. *See* Temperature or
        thermal issue
    trust issue, 141–144
    yield issue, 107–109
Channel hot-electron (CHE) injection,
    135, 136f
Channel length modulation (CLM),
    66, 118, 119f
Channel punchthrough current,
    118–119
Chaotic firefly algorithm, 749
Characterization engine, 669
Charge pump (CP), 195–198
Charge-pump (CP) PLL, 159, 484–488
Charge transfer sense amplifier, 301
CHE injection. *See* Channel
    hot-electron (CHE) injection
Chemical mechanical polishing
    (CMP), 91

Chemical vapor deposition (CVD), 71, 91
Chip finishing, 375
Chip intellectual property protection issue, 143–144, 143f
Chip operating temperature, 585
Chip yield, 107–109
Chiral angle, 79
Chiral CNT, 79
Chirality, 79
Circuit (device + parasitics) extraction process, 566–567
Circuit level (analog circuit), 370, 370f
Circuit-level/device-level analog simulations. See SPICE
Circuit-level MATLAB simulation, 430–431
Circuit-level Simulink simulations, 438–440
Circuit-level Verilog-A simulation, 465–468
Circuit parameter metamodel (CPM), 702
Circuit reliability. See Reliability issue
Circuit simulation method, 106
Circuit yield, 108–109
CircuitLogix, 448
Clamped bitline sense amplifier, 304–306, 718–720
Class testing, 389
Classical CMOS manufacturing process, 93–94
Classical SiO_2/polysilicon FET, 68–70
Classical six-transistor SRAM, 318–320
Classical synchronous DRAM, 344
Classification threshold, 682
Cleanroom, 388
Client-server platform (EDA tool installation), 409–410, 409f
CLM. See Channel length modulation (CLM)
Clock divider, 200
Clock gating, 537, 537f
Clock network, 679
Clock shifter, 679
Clock stretching, 679–681
Clock stretching logic (CSL), 680
Clock tree synthesis, 387
Clocked comparator, 300
CMFB circuit. See Common-mode feedback (CMFB) circuit
CMG MOSFET. See Common multi-gate (CMG) MOSFET
CMOS. See Complementary metal-oxide semiconductor (CMOS)
CMOS device-based humidity sensor, 298–299, 299f
CMOS inverter cross-section, 93f
CMOS inverter schematic representation, 93f
CMOS nanoelectronics, 452–453, 453f, 454f
CMOX memory. See Conductive metal oxide (CMOX) memory
CMP. See Chemical mechanical polishing (CMP)
CNT. See Carbon nanotube (CNT)
CNT inductor-based LC-VCO, 187–188, 189f, 189t
CNTFET. See Carbon nanotube FET (CNTFET)
CNTFET-based chemical sensor, 294–295, 295f
CNTFET-based gas sensor, 294, 294f
Coarse encoder, 223
Coefficient of determination (R^2), 695
Column access time (tCAC), 347
Column address strobe (select) latency (tCL), 346
Column decoder, 317
Combination TV tuner, 57
Combined DOE-ILP approach, 550, 663
Comment Card, 446
Commercial EDA tools, 393–399. See also Electronic design automation (EDA) tools
Commercial EDA vendors, 393, 394
Commercial SPICE simulators, 446–447, 447f
Common-mode feedback (CMFB) circuit, 704f, 705
Common multi-gate (CMG) MOSFET, 500
Common-source amplifier (CSA), 704f, 705
Compact model generation flow, 498–500
Compact modeling, 499f, 500
Compact SPICE models, 498f, 500–501
Compact thermal modeling, 586–587
Comparable-A/comparable-D AMS-SoC, 496
Comparator, 221
Comparator bank, 238–239, 245
Complementary metal-oxide semiconductor (CMOS), 1
Complementary OTFT-based SA-ADC, 250–252
Component manufacturing and development, 367–368
Component/software design, 366–367. See also Mixed-signal circuit and system design flow
Component-wise power, 110f
Comprehensive AMS circuit design flow, 403–406
Comprehensive device model, 500
Comprehensive digital circuit design flow, 406–407
Computer-aided design (CAD), 393. See also Electronic design automation (EDA) tools
Computer Analysis of Non-Linear Circuits Excluding Radiation (CANCER), 446
Computer memory. See Memory
Concurrent mixed-signal design flow, 390
Conduction model heat transfer, 588
Conductive metal oxide (CMOX) memory, 353, 354f
Conductometric electrochemical biosensor, 28, 29f
Conjugate-gradient-based optimization (90-nm CMOS CS-VCO), 654–656
Conjugate-gradient optimization algorithm:
  DOXCMOS-based power and delay aware nano-CMOS ULC, 580
  nano-CMOS VCO, 555–556, 557
  parasitics-optimal VCO, 576, 577
  thermal-aware VCO optimization, 600
Contact-type temperature sensors, 266
Content-addressable memory (CAM), 316, 317
Contention current, 119–120
Continuous-time sigma-delta modulator (CT-SDM):
  MATLAB-based simulation, 426–429
  Simulink/Simscape-based simulations, 434–436
  Verilog-AMS simulation, 481–484
Control Card, 446
Control data flow graph (CDFG), 385, 534
Convection model heat transfer, 588
Conventional AMS and RF integrated circuit design flow, 403–404, 404f
Conversion gain, 287
Conversion speed (DAC), 237
CoolSpice, 448
Coordinate rotation digital computer based numerically controlled oscillator (CORDIC-NCO), 205
Copyright protection, 142
CORDIC-NCO. See Coordinate rotation digital computer based numerically controlled oscillator (CORDIC-NCO)
Core engine-based mixed-mode simulator, 496, 498
Corner-based method, 379, 633–638
Corner-based modeling, 500
Cougar, 400, 402
Counter-ramp ADC, 223–224, 224f
Coupled engine-based mixed-mode simulator, 496, 497–498
CP. See Charge pump (CP)
CP PLL. See Charge-pump (CP) PLL
CPM. See Circuit parameter metamodel (CPM)
CPU. See Central processing unit (CPU)

Critical path circuit block, 679
Critical path delay, 536
Critical task set (CTS), 678
Cross-point detector, 227
Cross-talk, 562
Crude Monte Carlo method, 622
Cryptography, 142
Crystal oscillator, 165
CS oscillator, 165
CS-VCO. *See* Current-starved voltage controlled oscillator (CS-VCO)
CSA. *See* Common-source amplifier (CSA)
CSL. *See* Clock stretching logic (CSL)
CSO. *See* Cuckoo search optimization (CSO)
CT-SDM. *See* Continuous-time sigma-delta modulator (CT-SDM)
CTS. *See* Critical task set (CTS)
Cuckoo search optimization (CSO), 744–746
Cumulative distribution function (CDF), 683
Current-controlled current source (CCCS), 455, 457f
Current-controlled switch (CSW), 457
Current-controlled voltage source (CCVS), 455, 457f
Current latch voltage-mode sense amplifier, 301
Current-mode R-2R ladder DAC, 228, 228f
Current-mode R-2R ladder network-based multiplying DAC, 230f
Current-mode sense amplifier, 301, 302, 302f
Current-starved voltage controlled oscillator (CS-VCO), 177–184
   45-nm CMOS, 179, 182f, 182t
   45-nm double-gate FinFET, 183, 184f, 184t
   50-nm CMOS, 179, 181f, 181t
   90-nm CMOS, 179, 180f, 181f
   basic concept, 177, 177f
   circuit design, 177–179
   conjugate-gradient-based optimization, 654–656
   dual-threshold-based circuit-level optimization, 554–557
   particle swarm optimization (PSO) approach, 657–661
Custom elements, 457
CustomSim, 450
CV^2f loss, 126
CVD. *See* Chemical vapor deposition (CVD)
Cycle jitter, 168
Cycle-to-cycle jitter, 168
Cyclic DAC, 229–230, 230f
Cygwin application software, 401

D

D flip-flop-based phase-frequency detector, 194

DAC. *See* Digital-to-analog converter (DAC)
DAEs. *See* Differential algebraic equations (DAEs)
DAHC. *See* Drain avalanche hot carrier (DAHC)
DAHC injection. *See* Drain avalanche hot-carrier (DAHC) injection
DAISY, 18, 392
Dark current, 287
Dark signal nonuniformity (DSNU), 288
Data converters. *See* Electronic signal converter circuits
Data flow graph (DFG):
   dual-oxide-based RTL optimization, 534, 545
   input generation engine, 669
   postsilicon tuning for variability-driven RTL design, 676, 676f
   TED-based variability-aware RTL optimization, 669–673
   variability-aware RTL timing optimization, 673, 674f
Data gating, 536, 537f
Data over cable service interface specifications (DOCSIS), 35
Datapath and control generation engine, 669
Datapath library, 408
DC analysis, 450
.DC control line, 450
DC-to-DC voltage-level converter (VLC), 216, 216f
DCan, 459f
DCD. *See* Duration count detector (DCD)
dcdcmp, 459f
DCO divider, 205
DCop, 459f
DCsol, 459f
DCT. *See* Discrete cosine transformation (DCT)
DCT unit. *See* Discrete cosine transformation (DCT) unit
DCtran, 459f
DCU. *See* Dynamic clocking unit (DCU)
DDLL. *See* Digital delay-locked loop (DDLL)
DDNEMS. *See* Drug-delivery nano-electro-mechanical system (DDNEMS)
DDR-SDRAM. *See* Double-data rate SDRAM (DDR-SDRAM)
DDR2-SDRAM. *See* Double-data rate type-2 SDRAM (DDR2-SDRAM)
DDR3-SDRAM. *See* Double-data rate type-3 SDRAM (DDR3-SDRAM)
ddt operator, 481
Decca Navigator, 36
Decimator, 221
Defect, 107
Delay-locked loop (DLL), 206–209

Delay yield, 668
Delta-encoded ADC, 223–224, 224f
Delta-sigma ADC, 221–222, 221f, 225t
Delta-sigma DAC, 229, 229f
Delta-sigma modulator-based ADC, 252–256
Delta-sigma modulator-based DAC, 256–257
Demodulator PLL, 159
Depletion-type MOSFET, 68
Design. *See* Mixed-signal circuit and system design flow
Design Compiler, 385, 386, 397
"Design for" approaches, 67
Design for eXcellence (DFX), 67
Design for manufacturing (DFM), 67, 375
Design for yield (DFY) technique, 109
Design libraries, 408–409, 409f
Design matrix template (DMT), 699
Design-of-experiments-assisted Monte Carlo (DOE-MC) analysis, 380, 638–641
Design of experiments (DOE) method, 628–633
   flow, 631f
   fractional factorial experiment, 632
   full-factorial experiment, 632
   high-level idea, 631f
   nano-CMOS oscillator design optimization, 633, 634f, 635f, 635t
   nano-CMOS oscillator design space sampling, 632–633
   overview, 629–632
Design optimization. *See* Ultrafast design optimization
Design rule check (DRC), 375–376
Design signoff, 387–388
Design verification, 424
Design yield, 108–109
Device-circuit cosimulation method, 106
Device extraction, 566
Device-level MATLAB simulation, 432, 432f
Device-level Simulink simulation, 440–445
Device-level variations, 644
Device-level Verilog-A simulation, 468–473
Device parasitics, 122–125
Device parasitics capacitances, 124–125
Device parasitics inductances, 125
Device parasitics resistances, 123–124
DFD. *See* Digital frequency divider (DFD)
DFF-based frequency divider circuit, 200, 201f
DFG. *See* Data flow graph (DFG)
DFM. *See* Design for manufacturing (DFM)

DFT. *See* Discrete Fourier transform (DFT)
DFX. *See* Design for eXcellence (DFX)
DFY technique. *See* Design for yield (DFY) technique
DG TFET. *See* Double-gate TFET (DG TFET)
DGFET. *See* Double-gate FET (DGFET)
DIBL. *See* Drain-induced barrier lowering (DIBL)
Dicing process, 389
Dickson charge pump, 196, 196f
Die yield, 108
Dielectrics:
  high-κ, 71t
  low-κ, 131t
Differential algebraic equations (DAEs), 445
Differential nonlinearity (DNL), 234, 636–638
Differential transconductance amplifier, 298
Diffusion current, 117
Diffusion equation, 566
Diffusion process, 92
Digital behavioral synthesis, 16
Digital camera, 49–50, 280, 281f
Digital circuit, 4, 4f, 105–107
Digital circuit design flow, 382–389
  architecture-level design, 384–385
  circuit fabrication, 388–389
  design signoff, 387–388
  engineering change order (ECO), 388
  logic-level design, 385, 386f
  packaging, 389
  physical design, 386–387, 387f
  physical verification, 387
  standard digital circuit design flow, 382f
  system-level design, 384
  testing, 389
  transistor-level design, 386
Digital circuits and systems, 3
Digital comparator, 300
Digital delay-locked loop (DDLL), 208–209, 208f, 209f
Digital EDA tools, 395–397, 402
Digital filter, 221
Digital frequency divider (DFD), 200, 200f
Digital integrated circuit verification, 424
Digital integrator, 254
Digital interpolation filter, 229
Digital library, 408–409, 408f
Digital media receiver, 40
Digital pixel sensor (DPS), 285, 285f
Digital PLL (DPLL), 159, 204
Digital signal, 215f, 216
Digital signal processing, 216f

Digital single-lens reflex camera (DSLR), 49
Digital SoC power optimization, 525–527
Digital STB, 51
Digital-to-analog converter (DAC), 217
  binary-weighted DAC, 225–226, 226f
  characteristics, 237–239
  high-level representation, 218f
  multiplying DAC, 230, 230f
  oversampling DAC, 229, 229f
  overview, 225f
  pipeline DAC, 230–231, 231f
  pulse-width modulator (PWM) DAC, 227, 227f
  R-2R ladder DAC, 227–228, 228f
  segmented DAC, 228–229, 228f
  sigma-delta DAC, 229, 229f
  sigma-delta modulator-based DAC, 256–257
  single electron transistor-based DAC, 257
  successive-approximation DAC, 229–230, 230f
  thermometer-coded DAC, 227, 227f
Digital TV tuner, 57
Digital video disc (DVD), 31
Digital video recorder (DVR), 34–35
Digital watermarking. *See* Watermarking chip
Digital watermarking low-power chip layout, 543f
DIMM. *See* Dual in-line memory module (DIMM)
Diode-based humidity sensor, 298, 298f
Direct conversion ADC, 219–220, 219f
Direct force, 139
Direct-form digital oscillator system, 205
Direct Rambus DRAM (DRDRAM), 344
Direct tunneling, 114, 115, 115f, 138f, 139
Discovery AMS, 481
Discrete-amplitude signal, 215–216, 215f
Discrete cosine transformation (DCT), 536, 536f, 669
Discrete cosine transformation (DCT) unit, 430, 430f, 541, 542f, 543, 544
Discrete firefly algorithm, 749
Discrete Fourier transform (DFT), 235
Discrete-time sigma-delta modulator (DT-SDM), 426–429
Discrete-time signal, 215, 215f
.DISTO control line, 451
Distributed thermal modeling, 589
Distributed transmission line modeling, 569, 569f
Divide and conquer approach, 391
Divide-by-two-amplifier, 230

DLL. *See* Delay-locked loop (DLL)
DMT. *See* Design matrix template (DMT)
DNA detection biosensor system, 30
DNL. *See* Differential nonlinearity (DNL)
DNM. *See* Dynamic noise margin (DNM)
DOCSIS. *See* Data over cable service interface specifications (DOCSIS)
DOE-assisted tabu search optimization (DOE-TSO), 737, 738–739
DOE-MC analysis. *See* Design-of-experiments-assisted Monte Carlo (DOE-MC) analysis
Dolphin Integration, 18
Dosimeter reader system, 45f
Double charge pump, 196, 196f
Double-data rate SDRAM (DDR-SDRAM), 340f, 344
Double-data rate type-2 SDRAM (DDR2-SDRAM), 340f, 344
Double-data rate type-3 SDRAM (DDR3-SDRAM), 340f, 344
Double-e PROM, 350
Double-ended SRAM, 318
Double-ended 7T-SRAM, 322–323
Double-gate FET (DGFET), 6, 73–75
Double-gate FinFET, 75–77, 121–122, 173
Double-gate FinFET transistor, 8f
Double-gate GFET, 83, 83f
Double-gate GNR-FET, 83, 83f
Double-gate N-type P-I-N- structure TFET, 88f
Double-gate TFET (DG TFET), 87, 87f
Downsampler, 256
DP-DRAM. *See* Dual-port DRAM (DP-DRAM)
DPLL. *See* Digital PLL (DPLL)
DPS. *See* Digital pixel sensor (DPS)
DR. *See* Dynamic range (DR)
Drain avalanche hot carrier (DAHC), 118
Drain avalanche hot-carrier (DAHC) injection, 135, 136f
Drain-induced barrier lowering (DIBL), 66, 548
DRAM. *See* Dynamic random-access memory (DRAM)
DRC. *See* Design rule check (DRC)
DRDRAM. *See* Direct Rambus DRAM (DRDRAM)
dreal, 400
Driven right leg (DRL) circuit, 35
Driver transistor, 318
druc, 402
Drug delivery, 32
Drug-delivery nano-electro-mechanical system (DDNEMS), 32–34
Drug delivery system, 32

DSLR. *See* Digital single-lens reflex camera (DSLR)
DSNU. *See* Dark signal nonuniformity (DSNU)
DT-SDM. *See* Discrete-time sigma-delta modulator (DT-SDM)
Dual-damascene process, 127
Dual gating, 537
Dual in-line memory module (DIMM), 344
Dual-oxide 90-nm CMOS-based current-starved VCO, 554–557
Dual-oxide-based logic-level leakage optimization, 531–534
Dual-oxide-based RTL optimization, 534–536, 558–560
Dual-oxide HLS technique, 558–560
Dual-oxide thickness 90-nm CMOS-based low-power VCO, 556f
Dual-oxide thickness technique, 524
Dual-port DRAM (DP-DRAM), 344
Dual-threshold-based circuit-level optimization:
  current-starved VCO, 554–557
  nano-CMOS SRAM, 550–552
  ULC, 528–531
Dual-threshold-based RTL optimization, 560f
Dual-threshold voltage, 524
Dual-threshold voltage-based nano-CMOS SRAM topologies, 552f
Dual-threshold voltage-based technique, 528
Dual-voltage/frequency-based dynamic power reduction, 542–544
Dummy polysilicon gate, 94
Dump terminals, 411
Duration count detector (DCD), 297
DVR. *See* Digital video recorder (DVR)
DVS. *See* Dynamic voltage scaling (DVS)
DxDesigner, 395
Dynamic clock stretching, 679–680, 680f
Dynamic clocking unit (DCU), 539, 539f
Dynamic compact thermal models, 586, 587f
Dynamic frequency scaling technique, 537–538, 542–544
Dynamic noise margin (DNM), 330
Dynamic parameter variations, 619, 620f
Dynamic power dissipation, 515, 515f, 516f
Dynamic power reduction techniques, 540–548
Dynamic random-access memory (DRAM), 338–347
  1T-1C-DRAM, 340–341
  1T-DRAM, 341–342
  3T-DRAM, 342, 342f

Dynamic random-access memory (*Cont.*):
  4T-DRAM, 343, 343f
  bandwidth, 347
  BEDO-DRAM, 343–344
  characteristics, 346–347
  DRAM array, 339
  DRDRAM, 344
  EDO-DRAM, 343
  FeDRAM, 345–346
  FPM-DRAM, 343
  graphics DRAM, 344–345
  schematic representation, 339f
  SDRAM, 344
  SGDRAM, 344–345
  types, 339–340, 340f
  WDRAM, 344
Dynamic range (DR):
  ADC, 232–233
  DAC, 238
  image sensor, 286–287
Dynamic thermal analysis, 590
Dynamic threshold voltage, 538
Dynamic timing control, 679, 680f
Dynamic/variable frequency clocking, 538–540
Dynamic voltage scaling (DVS), 522–523, 522f, 537–538, 537f

E

E^2PROM, 350
ECAD, 393. *See also* Electronic design automation (EDA) tools
ECB. *See* Electron tunneling from conduction band (ECB)
ECO. *See* Engineering change order (ECO)
EDA tool installation, 409–411
Edge detection unit, 430, 542
EDI System. *See* Encounter digital implementation (EDI) System
EDO-DRAM. *See* Extended data out DRAM (EDO-DRAM)
EDP. *See* Energy-delay-product (EDP)
EDRAM. *See* Enhanced DRAM (EDRAM)
EEG. *See* Electroencephalogram (EEG)
EEPROM. *See* Electrically erasable programmable read-only memory (EEPROM)
Effective number of bits (ENOB), 231–232
Eight-transistor SRAM (8T-SRAM), 324–326
eispice, 448
EKV compact model, 501
EKV MOS transistor model, 18
Eldo, 395, 448
Eldo Classic, 448
Eldo Premier, 450
Electric field integral equation (EFIE), 569
Electric™ VLSI Design System, 400

Electrical circuit extraction, 566, 567
Electrical linear network (ELN), 494
Electrical parameters, 587t
Electrical rule check (ERC), 377–378
Electrically erasable programmable read-only memory (EEPROM), 350–351, 351f, 351t
Electrochemical biosensor, 28, 29f
Electroencephalogram (EEG), 35, 36f
Electrogram, 35
Electromigration (EM), 139–141
Electron tunneling from conduction band (ECB), 114
Electron tunneling from valence band (EVB), 114
Electron wind, 139
Electron wind force, 139
Electronic CAD (ECAD), 393. *See also* Electronic design automation (EDA) tools
Electronic design automation (EDA) tools, 17–18, 393–403
  analog design, 394–395, 395f, 396f
  commercial EDA vendors, 393, 394
  digital design, 395–397
  EDA tool installation, 409–411
  free/open-source EDA tools, 399–403
  mixed-signal system design, 398–399, 398f, 399f
  variability-aware design, 643, 644f
Electronic signal converter circuits, 215–262
  ADC. *See* Analog-to-digital converter (ADC)
  concrete applications, 216–217
  DAC. *See* Digital-to-analog converter (DAC)
  signal converter types, 217
  types of electronic signals, 215f
  types of signal converters, 216, 216f
Electronic system level (ESL), 384
ElectronStorm, 398
Electroplating, 92
Element Card, 446
Elliptic wave filter (EWF), 536, 536f
Elmore delay model, 570–574
ELN. *See* Electrical linear network (ELN)
EM. *See* Electromigration (EM)
Emerging systems, 25–64
  atomic force microscope (AFM), 25–27
  biosensor systems, 27–31
  Blu-ray player, 31–32
  digital video recorder (DVD), 34–35
  drug-delivery nano-electro-mechanical system (DDNEMS), 32–34
  electroencephalogram (EEG), 35, 36f
  GPS navigation device, 36–37
  GPU-CPU hybrid (GCH) system, 37–40

Emerging systems (Cont.):
  net-centric multimedia processor (NMP), 41–44
  networked media tank (NMT), 40–41
  radiation detection system (RDS), 44–46
  RFID chip, 46–49
  secure digital camera (SDC), 49–50
  set-top box (STB), 50–51
  slate personal computer (slate PC), 52–54
  smartphone, 54–55
  software-defined radio (SDR), 56–57
  TV tuner card, 57–58
  universal remote control, 59
Empirical semivariogram, 715
Employed bee, 741
EMU. See Energy management unit (EMU)
Enclosure rules, 375
Encoder, 217
Encounter Conformal ECO Designer, 388
Encounter digital implementation (EDI) System, 386
Encounter RTL Compiler, 385
Encryption, 142
.END card, 460
.end card, 446
End-point INL, 233
Energy-critical systems, 515
Energy-delay-product (EDP), 517, 545–546
Energy dissipation. See Power dissipation
Energy management unit (EMU), 540, 540f
Engineering change order (ECO), 388
Enhanced DRAM (EDRAM), 340
Enhanced SDRAM (ESDRAM), 340
Enhancement-mode MOSFET, 68
Enhancement type NMOS, 68
ENIAC, 52
ENOB. See Effective number of bits (ENOB)
Enz-Krummenacher-Vittoz (EKV) compact model, 501
Enz-Krummenacher-Vittoz (EKV) MOS transistor model, 18
Epileptic seizure sensors, 297, 297f
EPROM. See Erasable programmable read-only memory (EPROM)
Equation-based approaches (AMS circuit synthesis), 393
Equation-formulation phase, 462
Erasable programmable read-only memory (EPROM), 349–350, 350f, 351t
ERC. See Electrical rule check (ERC)
errchk, 459f
Error amplifier, 221, 595

ESDRAM. See Enhanced SDRAM (ESDRAM)
ESL. See Electronic system level (ESL)
Etching process, 92
EVB. See Electron tunneling from valence band (EVB)
Even corner, 636
Event counter alarm, 293
EVT. See Extreme value theory (EVT)
EWF. See Elliptic wave filter (EWF)
Extended data out DRAM (EDO-DRAM), 343
Extreme event, 681
Extreme value theory (EVT), 682
Eyring equation, 141

F

Fab process variation, 99
Fabrication process. See Nanomanufacturing
Fan-in logic gates, 385
Faraday law of induction, 569
Fast AMS and RF integrated circuit design flow, 404–405, 405f
Fast asynchronous time marching technique (FATA), 592, 593f
Fast-fast (FF) corner, 635, 636f
Fast Fourier transform (FFT), 244
Fast Monte Carlo methods, 448, 638–643. See also Monte Carlo analysis
  DOE-MC method, 638–641
  Latin hypercube sampling-based Monte Carlo method, 641–642, 642f
  middle Latin hypercube sampling-based Monte Carlo method, 642, 643f
Fast numerical solvers, 689
Fast page cycle time (tPC), 346
Fast page mode (FPM) DRAM (FPM-DRAM), 343
Fast-slow (FS) corner, 635, 636f
Fast SPICE, 447f, 449–450
FATA. See Fast asynchronous time marching technique (FATA)
Fault, 107
FBB. See Forward body bias (FBB)
FBDIV. See Feedback divider (FBDIV)
FDM. See Finite difference method (FDM); Frequency division multiplexing (FDM)
FDRAM. See Fusion DRAM (FDRAM)
FED. See Fitted Elmore delay (FED)
FeDRAM, 345–346. See Ferroelectric DRAM (FeDRAM)
Feedback divider (FBDIV), 160f, 161, 487. See also Frequency divider
FEM. See Finite element method (FEM)
FEOL corners. See Front-end of line (FEOL) corners

Ferroelectric DRAM (FeDRAM), 345–346
FF corner. See Fast-fast (FF) corner
FFT. See Fast Fourier transform (FFT)
Field-programmable gate array (FPGA), 3, 3f
Field PROM (FPROM), 349
Fill factor, 286
Fin pitch, 8
FineSim, 450
FinFET, 7–8, 73, 73f, 75
FinFET-based 4T-SRAM, 321f
FinFET-based 6T-SRAM, 320, 320f
FinFET fabrication process, 95–96
Finite difference method (FDM), 590
Finite element method (FEM), 568, 568f, 589f, 590
Finite impulse response (FIR) filter, 536, 536f, 670–673
Finite-impulse response (FIR) integrator, 254
FIR filter. See Finite impulse response (FIR) filter
Firefly algorithm, 747–750
First-in and first-out (FIFO) SRAM, 318
First-principle physics approach, 106
Fitted Elmore delay (FED), 570
Five-transistor SRAM (5T-SRAM), 321–322, 322f
Fixed layer, 355
Fixed-pattern noise (FPN), 288
Fixed RFID, 47
Flash ADC, 219–220, 219f, 225t
Flash memory, 351–352, 358t
Flashmatic, 58
Flat SPICE simulator, 449
FLL. See Frequency-locked loop (FLL)
Floating gate approach, 524
Floating gate transistor, 350
Floor planning, 375, 387
Folded-cascode OTA, 704f, 705
Folding ADC, 222–223, 223f
Forward body bias (FBB), 523
Four-transistor DRAM (4T-DRAM), 343, 343f
Four-transistor SRAM (4T-SRAM), 321, 321f
fouran, 459f
Fourth-order Runge-Kutta method, 590
Fowler-Nordheim tunneling, 114, 138–139, 138f
FPGA. See Field-programmable gate array (FPGA)
FPM-DRAM. See Fast page mode (FPM) DRAM (FPM-DRAM)
FPN. See Fixed-pattern noise (FPN)
FPROM. See Field PROM (FPROM)
Fractional factorial experiment, 632
Fractional-N frequency divider, 200, 200f
Free EDA tools, 399–403
Free layer, 355

Free SPICE simulators, 446, 447f, 448–449
Frequency divider, 200–203
Frequency division multiplexing (FDM), 264
Frequency-locked loop (FLL), 204
Frequency reduction, 542
Frequency scaling, 538
Frequency-synthesizer PLL, 159
Frequentist, 691, 692f
Front-end design, 383
Front-end of line (FEOL) corners, 635, 636f
Front-gated CNTFET manufacturing, 95f
FS. See Full scale (FS)
FS corner. See Fast-slow (FS) corner
FSR. See Full scale range (FSR)
Fuji DS-1P, 49
Full-custom design, 383
Full-factorial experiment, 632
Full-latch cross-coupled sense amplifier, 301, 302f, 304f
Full scale (FS), 237
Full scale range (FSR), 237
Fully automatic design flow, 369–370, 369f
Fully differential OP-AMP, 704f, 705, 706–707
Functional level (analog circuit), 370, 370f
Functional level (digital circuit), 370, 370f
Fundamental frequency, 162
Fusion DRAM (FDRAM), 340

G

GAGAN, 37
Gain amplifier, 255, 256f
Gain error
 ADC, 235
 DAC, 237
Gain FPN, 288
Galileo positioning system, 37
Gas sensors, 294–295
Gate all around DGFET, 74
Gate-induced drain leakage (GIDL), 117, 117f
Gate last approach, 94
Gate leakage current, 66
Gate level (digital circuit), 370, 370f
Gate oxide leakage, 6, 114–115
Gate-oxide leakage reduction techniques, 553–560
Gate sheet resistance, 66
Gauss law, 569
Gauss law for magnetism, 569
Gaussian process regression (GPR), 691
GCH system. See GPU-CPU hybrid (GCH) system
GDSII, 375
gEDA, 401
Geiger counter, 45

Geiger-Mueller counter, 45
General process design kit (GPDK), 18
General-purpose semiconductor simulator (GSS), 449
Generalized Pareto distribution (GPD), 682–683
Generation current, 117
Genius device simulator, 449
Geometric model-based methods (parasitics), 568, 568f
Geostatistics-inspired metamodeling, 715. See also Kriging metamodeling
Germanium-antimony-tellurium (GST), 357
GFET. See Graphene FET (GFET)
GFET-based biosensor, 265–266, 266f
GFET-based LC-VCO, 466, 467f
GFET manufacturing, 96f
Ghost reducer, 57
GIDL. See Gate-induced drain leakage (GIDL)
Global navigation satellite system (GNSS), 36
GLObal NAvigation Satellite System (GLONASS), 36
Global system for mobile communications (GSM) call operation, 518, 518t, 519f
Global uniform variation, 99, 99f
Glucose Analyser, 29
Glued mixed-mode simulator, 496, 497–498
GNR. See Graphene nanoribbon (GNR)
GNR-FET. See Graphene-nanoribbon FET (GNR-FET)
GNR-TFET, 87–88, 88f
GNSS. See Global navigation satellite system (GNSS)
GNU General Public License (GPL), 399, 426
GNU Octave, 425–426
Gnucap, 449
GPCAD, 392
GPD. See Generalized Pareto distribution (GPD)
GPDK. See General process design kit (GPDK)
GPR. See Gaussian process regression (GPR)
GPS navigation device, 36–37
GPU. See Graphics processing unit (GPU)
GPU-CPU hybrid (GCH) system, 37–40
Graphene, 81
Graphene FET (GFET), 81–83, 103, 442–445, 471–473
Graphene FET (GFET)–based biosensor, 265–266, 266f
Graphene FET-based LC-VCO, 466, 467f
Graphene FET fabrication process, 96–97

Graphene nanoribbon (GNR), 81
Graphene-nanoribbon FET (GNR-FET), 82, 83f
Graphic signal digitizer, 157f, 158f
Graphics DRAM, 344–345
Graphics processing unit (GPU), 38
Gravitational search algorithm (GSA), 753–755, 756, 756t
Green function method-based simulation, 592–593
Green method function, 590
Green TFET (GTFET), 87, 87f
GSA. See Gravitational search algorithm (GSA)
gschem, 374, 386
GSM call operation. See Global system for mobile communications (GSM) call operation
GSS. See General-purpose semiconductor simulator (GSS)
GST. See Germanium-antimony-tellurium (GST)
GTFET. See Green TFET (GTFET)
GTKWave, 401, 402
Guardian, 397

H

HAL differential equation solver, 536, 536f
Hall effect, 82
Hall resistance, 82
Hall voltage, 82
Halogen counter, 45
Hard-switching problem, 469
Hardware-based postsilicon power reduction techniques, 522, 522f, 536–540
Hardware-based thermal monitoring, 594
Hardware description and verification language (HDVL), 424
Hardware description language (HDL), 366, 424
Hardware power monitor (HPM), 540, 540f
Hardware Trojans, 144, 144f
Hardware verification language (HVL), 424
Harmony, 389
HCI. See Hot carrier injection (HCI)
HD-DVD, 31
HDL. See Hardware description language (HDL)
HDL Coder, 434
HDLA, 16
HDVL. See Hardware description and verification language (HDVL)
Heat dissipation, 584–586
Heat sink, 584, 585f
Heat transfer, 588
Hercules, 375, 397
Hercules LVS, 378
Hierarchical fast SPICE, 449

Hierarchical floor planning, 375
HiFaDiCC, 392
High-definition DVD, 31
High frequency (HF) tag, 46
High-κ dielectrics, 71t
High-κ/metal-gate FET (HKMGFET), 70–72, 120–121
High-κ/metal-gate process technology, 555
High-κ/metal gate transistor, 7f
High-level design, 384
High-level synthesis (HLS), 384–385, 534. *See also* RTL optimization
High-resolution ADPLL, 205–206
High-speed SPICE simulators, 447f, 450
Hillocks, 140
HiPer PX, 395, 399
HiPer Simulation AMS, 389, 398, 399
HiPer Verify, 375, 395, 399
Hipex, 397
HiRoshima University STARC IGFET model (HiSIM), 501
HiSIM. *See* HiRoshima University STARC IGFET model (HiSIM)
HiSIM-SOI, 501
HiSIM2, 501
HiSIMHV, 501
HKMG fabrication, 94–95
HKMGFET. *See* High-κ/metal-gate FET (HKMGFET)
HLS. *See* High-level synthesis (HLS)
HNM. *See* Hold noise margin (HNM)
Hold noise margin (HNM), 330
Hole tunneling from valence band (HVB), 114
Home video system, 31
Horizontal double-gate FET (FinFET), 73, 73f
Hot carrier, 117, 122
Hot carrier current, 117–118
Hot carrier effects, 136
Hot carrier injection (HCI), 135–137, 448
HPM. *See* Hardware power monitor (HPM)
HSIM, 450
HSPICE, 397, 448
Humidity sensors, 297–299
HVB. *See* Hole tunneling from valence band (HVB)
HVL. *See* Hardware verification language (HVL)
Hybrid BEM and FEM method, 568, 568f
Hybrid firefly algorithm, 749
Hybrid hardware- and software-based thermal monitoring, 594–595
Hybrid octree, 591
Hybrid sampling, 638, 638f
Hybrid STB, 51
Hybrid TV tuner, 57
Hygrometer, 297

I

IC. *See* Integrated circuit (IC)
IC Compiler, 386, 397
.ic control line, 452–453
IC Validator, 387
Icarus Verilog, 401
IDAC, 17, 392
Ideal elements, 456
Ideality factor, 277
Idle mode operation (mobile phone), 518, 518t, 519f
idt operator, 481
idtmod operator, 481, 487
IG DG FinFET. *See* Independent-gate (IG) DG FinFET
IGFET. *See* Insulated-gate FET (IGFET)
IIR filter. *See* Infinite-impulse response (IIR) filter
ILP. *See* Integer linear programming (ILP)
Image sensors, 280–294
  32-nm CMOS APS design, 289–291, 292t
  active pixel sensor (APS), 282–284
  capture time, 286
  conversion gain, 287
  dark current, 287
  digital pixel sensor (DPS), 285, 285f
  dynamic range, 286–287
  fill factor, 286
  fixed-pattern noise (FPN), 288
  output voltage swing, 286
  passive pixel sensor (PPS), 282, 282f
  photo current, 287
  pixel size, 289
  power dissipation, 289
  quantum efficiency, 285–286
  resolution, 289
  responsivity, 288
  secure digital camera (SDC), 293, 294f
  signal-to-noise ratio (SNR), 289
  smart, 291–293
  temporal noise, 288–289
  types, 281, 281f
IMG MOSFET. *See* Independent multi-gate (IMG) MOSFET
Important sampling Monte Carlo method, 623
Improved Runge-Kutta-Fehlberg method, 590
Impurity atoms, 92
Incentia DesignCraft, 399
Incisive Enterprise Simulator, 397
.include control line, 453
InCyte Chip Estimator, 398
Independent double-gate VBFET, 88, 88f
Independent-gate (IG) DG FinFET, 173
Independent multi-gate (IMG) MOSFET, 500, 501
Inductance extraction, 377, 568
Infinite-impulse response (IIR) filter, 536, 536f
Infinite-impulse response (IIR) integrator, 254
Information leakage issue, 143
Information protection issue, 142
Infrared (IR) remote, 58
Initial layout design, 525
Initial schematic design, 524
Injection-locked frequency divider, 200, 200f
INL. *See* Integral nonlinearity (INL)
Input buffer, 317
Input frequency range, 162
Input generation engine, 669
Insulated-gate FET (IGFET), 68
Integer linear programming (ILP), 546, 547, 558–560, 662–667
Integral nonlinearity (INL), 233–234, 636–638
Integrated circuit (IC), 1, 11
Integrated circuit simulation, 421f. *See also* Mixed-signal circuit and system simulation
Integrated space-and-time-adaptive chip (ISAC) thermal analysis framework, 590–592
Integrating ADC, 220–221, 220f
Integration mode (PPS), 282
Integration time, 282
Integrator, 221, 254, 255f
Intellectual property blocks, 143
Intellectual property watermarking, 143–144
Intelligent battery, 523, 537f, 538
Intelligent Verilog-AMS (iVAMS), 727–731
Inter-die process variation, 12, 100f
Inter-pixel process variation, 648, 649
Interconnect parasitics, 125–133
Interconnect parasitics capacitances, 130–132
Interconnect parasitics extractions, 568
Interconnect parasitics inductances, 133
Interconnect parasitics resistances, 128–130
Interface, 479
Interface trap, 137
Interlayer capacitance, 131
Interlevel capacitance, 131
Intermetal capacitance, 131
Interpolating DAC, 229, 229f
Intertia weight, 658
Interwire capacitance, 131
Intra-die process variation, 12, 100f
Intra-pixel mismatch, 648, 649
Inverse-VTC-1-SNM, 329
Invisible insertion unit, 543
Invisible watermarked image, 543f
Ion beam lithography, 93

Ion implantation, 92
Ion-sensitive FET (ISFET) biosensor, 28
IP STB, 51
IP-TV, 42
iPad, 52
IR drop, 126
I^2R loss, 126, 133
ISAC thermal analysis framework, 590–592
ISFET biosensor, 28
Iteration-control phase, 462
Iterative module selection method, 676
iVAMS. *See* Intelligent Verilog-AMS (iVAMS)
iVAMS-assisted firefly-based OP-AMP multiobjective algorithm, 749

J

Jasper Design Automation, 394
Jitter, 162, 167–169, 167f
JK FF-based frequency divider circuit, 202f
JK flip-flop-based frequency divider, 200, 202f
Joint biasing, 174, 175t
Joule heating, 267, 585
Junction capacitance, 66
Junction diodes (nano-CMOS inverter), 116f
Junction leakage, 66, 116–117

K

KANSYS, 392
KCL. *See* Kirchhoff's current law (KCL)
Key issues. *See* Challenges
Kinetic inductance, 133
Kirchhoff's current law (KCL), 461
Kirchhoff's voltage law (KVL), 461
Knowledge-based approaches (AMS circuit synthesis), 393
Kodak DCS SLR camera, 49
Kriging metamodeling, 714–722
  advantages, 715
  clamped bitline sense amplifier, 718–720
  disadvantage, 715, 731
  generation, 717–718, 718f
  Kriging-NN metamodeling, 731, 732f, 733t, 736–737
  ordinary Kriging method, 716–717, 720, 721f, 721t, 755–758
  phase-locked loop, 720, 722f
  sense amplifier, 718–720, 721f, 721t, 755–758
  simple Kriging method, 716, 718–720, 752–755
  spatial autocorrelation, 715
  steps, 715
  theory, 715–717
  thermal sensor, 752–755
  ultrafast design optimization, 752–758

Kriging metamodeling (*Cont.*):
  ultrafast process variation analysis, 733, 734f, 734t, 736–737
  universal Kriging method, 720, 722f
Kriging-NN metamodeling, 731, 732f, 733t, 736–737
KSpice, 448
KVL. *See* Kirchhoff's voltage law (KVL)

L

L-Edit, 375, 395
LabVIEW, 434
Lagrangian firefly algorithm, 749
LaserDisc (LD), 31
LASI. *See* LAyout System for Individuals (LASI)
Latched charge pump, 196, 196f
Latchup effect (LUE), 137–138
Latin hypercube sampling-based Monte Carlo method, 641–642, 642f
Layout compaction, 387
Layout design, 374–375
Layout level (analog circuit), 370, 370f
Layout level (digital circuit), 370, 370f
Layout-level netlist, 525
Layout-level power optimization, 525
LAyout System for Individuals (LASI), 401
Layout *versus* schematic (LVS) verification, 377, 377f
lc lossless transmission line, 569
LC-tank oscillator. *See* LC-tank voltage controlled oscillator (LC-VCO)
LC-tank voltage controlled oscillator (LC-VCO), 184–190
  180-nm CMOS, 186–187
  basics, 184–186
  CNTFET, 187–188
  inductor and capacitor tank, 185f
  memristor, 188–190, 190f
  Monte Carlo analysis, 626–627
  parasitic elements, 563f
  performance metrics, 170t
  polynomial-based metamodeling, 702
  process variation tolerant design, 661–662
  ring oscillator, compared, 170t
  schematic diagram, 185f
LC-VCO. *See* LC-tank voltage controlled oscillator (LC-VCO)
LCD. *See* Liquid crystal display (LCD)
LDP. *See* Leakage delay product (LDP)
Leakage delay product (LDP), 558, 559
Leakage dissipation, 9–10, 110f. *See also* Power dissipation
Learning-strategy-based approaches (AMS circuit synthesis), 393
Least-squares regression analysis, 699
Leeson's model, 169
LER. *See* Line edge roughness (LER)
Level-down conversion, 529
Level shifter (LS), 160f, 161

Level-up conversion, 529
Levels of abstraction, 370, 370f
Levenberg-Marquardt nonlinear optimizer, 448
LFSR. *See* Linear feedback shift register (LFSR)
LHS-MC. *See* Latin hypercube sampling-based Monte Carlo method
.lib control line, 452
Likelihoodist, 691, 692f
Line edge roughness (LER), 101
Linear feedback shift register (LFSR), 293
Linear mean square (LMS), 536, 536f
Linear-mode APS, 284, 284f
Linear PLL (LPLL), 158
Linear signal flow (LSF), 494
Linear technology tools, 401
Linear variation, 99, 99f
Liquid crystal, 85
Liquid crystal display (LCD), 85
Lithography, 91, 92f, 93
LMS. *See* Linear mean square (LMS)
Load, 459f
Load transistor, 318
Local Boolean optimization, 531
Local random variation, 99f, 100
Lock-detector PLL, 159
Lock range, 161
Lock time, 162
Logarithmic-mode APS, 284, 284f
Logic level (digital circuit), 370, 370f
Logic-level design, 385, 386f
Logic-level technology mapping, 531
Logic synthesis, 16, 385, 386f
Logical level (analog circuit), 370, 370f
Logitech Fotoman, 49
Long-term jitter, 167
Look up table (LUT), 447
Look up table based numerically controlled oscillator (LUT-NCO), 205
loon, 400
Loop bandwidth, 161
Loop filter (LPF), 198–199
Loop gain, 162
LORAN, 36
Lossy transmission line, 571–574
Lot process variation, 99
Low frequency (LF) tag, 46
Low-κ dielectrics, 131t
Low-pass digital filter, 256
Low-power design, 10
Low-power (LP) DG FinFET, 173–174
Low-power relaxation oscillator, 191
Low-temperature operations, 549
LP DG FinFET. *See* Low-power (LP) DG FinFET
LPF:
  loop filter, 198–199
  low-pass filter, 467, 468
LPLL. *See* Linear PLL (LPLL)

LS. *See* Level shifter (LS)
LSF. *See* Linear signal flow (LSF)
LTSpice, 448
LTspice IV. *See* LAyout System for Individuals (LASI)
LUE. *See* Latchup effect (LUE)
Lumped compact thermal modeling, 586–587
LUT. *See* Look up table (LUT)
LUT-NCO. *See* Look up table based numerically controlled oscillator (LUT-NCO)
LVS rule check, 378
LVS verification. *See* Layout *versus* schematic verification
Lvx, 400

M

M-bit DAC, 229
Mac TV, 57
Mach8, 38
Macro level (analog circuit), 370, 370f
Macromodel, 696–697, 697f
Macromodeling-based simulation, 424
MacSpice, 448
Magic VLSI layout tool, 401
Magnetic inductance, 133
Magnetic/magnetoresistive RAM (MRAM), 355–356, 358t
Magnetic tunneling junction (MTJ), 355
Malicious design modifications issue, 144
Manchester encoder, 293
Manufacturing process. *See* Nanomanufacturing
MapleSim, 434
MARS. *See* Multivariate adaptive regression spline (MARS)
MAST, 491
Mathematical-based MOR, 570
MATLAB, 373, 384
MATLAB-based simulation, 425–433
 analog/mixed signal (AMS) design modeling, 426–429
 circuit-level simulation, 430–431
 device-level simulation, 432, 433f
 digital design modeling (watermarking chip), 429–430, 431
 hardware modeling, 465
 resistive-circuit-based simulation, 431
 system-/architecture-level simulation, 426–430
 thin-film memristor, 432
Matrix-solution phase, 462
Maximum likelihood estimation (MLE), 683
Maximum sampling rate, 237
Maxmin ant system (MMAS), 755
Maxwell equations, 569, 570
MCAST, 18, 392
MDRAM. *See* Multibank DRAM (MDRAM)
Mean data centering, 695

Mean square error (MSE), 695
Mean time to failure (MTTF):
 EM, 140
 integrated circuit, 135
 TDDB, 139
Measured equation of invariance, 568
Media player, 41
Memcapacitor, 90
Memdevices, 90
Meminductor, 90
Memory, 315–363
 CMOX, 353, 354f
 DRAM. *See* Dynamic random-access memory (DRAM)
 flash, 351–352, 358t
 magnetic/magnetoresistive RAM (MRAM), 355–356, 358t
 memristor-based nonvolatile SRAM, 354–355, 354f
 phase-change RAM (PCRAM), 356–358, 358t
 read-only memory (ROM), 349–351
 resistive random-access memory (ReRAM), 352–355, 358t
 SRAM. *See* Static random-access memory (SRAM)
 thyristor random-access memory (TRAM), 348, 348f
 types, 316–317, 316f
Memristor, 88–91, 103, 441–442
Memristor-based LC-tank oscillator, 188–190, 190f
Memristor-based nonvolatile SRAM, 354–355, 354f
Memristor-based Schmitt trigger oscillator, 192–193, 707–710
Memristor-based Wien oscillator, 436–438
Memristor fabrication process, 98
Memristor nanoelectronics, 453–455
Memristor relaxation oscillator, 191–192
MEMS. *See* Micro-electro-mechanical system (MEMS)
MEMS-based drug delivery system, 33, 33f
Mentor Graphics, 18, 394
Metal insulator semiconductor FET (MISFET), 68
Metal-mask ECO, 388
Metal-oxide semiconductor FET (MOSFET), 5
Metal tunnel junctions QCA, 67
Metallization process, 92
Metamodel, 689–697. *See also* Metamodel-based fast design methodologies
 accuracy, 693
 conceptual simplicity, 693
 defined, 690–691
 efficiency, 693
 generation flow, 693–696
 high-level concept, 690f
 macromodel, compared, 696–697, 697f

Metamodel (*Cont.*):
 robustness, 693
 transparency, 693
 types, 691–693
Metamodel-based fast design methodologies, 380, 689–763. *See also* Metamodel
 compact model generation flow, 499f, 500
 Kriging-based metamodeling. *See* Kriging metamodeling
 metamodel-assisted ultrafast AMS and RF design flow, 406, 406f
 metamodel-based ultrafast design flow, 697–698
 NN-based metamodeling. *See* Neural network-based metamodeling
 polynomial-based metamodeling. *See* Polynomial-based metamodeling
 ultrafast design optimization, 737–758
 ultrafast process variation analysis, 733–737
Metamodel-based ultrafast design flow, 697–698
mGstat, 717
Micro-Cap, 448
Micro-electro-mechanical system (MEMS), 3
Microarchitecture parallelism, 522, 527f, 528
Microchip drug reservoir, 33f
Microwave RFID tag, 46
Microwind, 394
MIDAS, 392
Middle Latin hypercube sampling-based Monte Carlo method, 642, 643f
Middle LHS. *See* Middle Latin hypercube sampling-based Monte Carlo method
MIGFET. *See* Multiple independent gate FET (MIGFET)
Miller frequency divider, 200, 200f
Minimize capacitance design, 522, 522f
Minimum area rule, 375
MISFET. *See* Metal insulator semiconductor FET (MISFET)
Mismatch, 12
Mixed-configuration platform (EDA tool installation), 410–411, 410f
Mixed-level simulator, 495
Mixed-mode circuit-level simulations, 495–498
Mixed-signal circuit, 4, 4f, 104
Mixed-signal circuit and system design flow, 365–420
 analog circuit design flow. *See* Analog circuit design flow
 analog/mixed-signal circuit synthesis, 391–393

Mixed-signal circuit and system
  design flow (*Cont.*):
  broad perspective of design
    cycle, 366f
  design libraries, 408–409, 409f
  digital circuit design flow. *See* Digital
    circuit design flow
  EDA tool installation, 409–411
  high-level design flow, 368f
  levels of abstraction, 370, 370f
  mixed-signal design flow, 389–391
  overview (complete design
    perspective), 365–369
  process design kit (PDK), 407–408
  tools. *See* Electronic design
    automation (EDA) tools
  top-down *vs.* bottom-up design,
    369–371
Mixed-signal circuit and system
  simulation, 421–511
  circuit-level/device-level analog
    simulations. *See* SPICE
  compact SPICE models, 498f,
    500–501
  digital circuits and systems,
    474–480
  MATLAB, 425–433
  mixed-mode circuit-level
    simulations, 495–498
  mixed-signal HDL-based
    simulation, 480–495
  models for circuit simulations,
    498–501
  MyHDL-based simulation,
    477–478, 478f
  OpenMAST-based simulation,
    491–493
  simulation languages, 422f, 424–425
  simulations based on abstraction
    levels, 422–423, 422f
  simulations based on design tasks,
    422f, 423–424
  simulations based on signal types,
    422f, 423
  simulations based on system models,
    422f, 423
  Simulink/Simscape-based
    simulations, 433–445
  SystemC-AMS-based simulation,
    494–495
  SystemC-based simulation, 478–480
  SystemVerilog-based simulation,
    474–475
  Verilog-A-based analog simulation,
    464–473
  Verilog-AMS-based simulation,
    481–489
  VHDL-AMS-based simulation,
    489–491
  VHDL-based simulation,
    475–477
Mixed-signal circuits and systems, 4f
Mixed-signal design flow, 389–391
Mixed-signal HDL-based simulation,
  480–495

Mixed-signal/mixed-mode
  simulator, 495
Mixed-signal system, 4, 4f
MLE. *See* Maximum likelihood
  estimation (MLE)
MLHS technique. *See* Middle Latin
  hypercube sampling-based Monte
  Carlo method
MMAS. *See* Maxmin ant system
  (MMAS)
Mobile phone, 55, 518–519
Mobile RFID, 47
Mobility degradation, 66
Model, 498
Model-based approaches (AMS circuit
  synthesis), 393
.model Card, 446
Model-evaluation phase, 462
.model line, 452
Model order reduction (MOR),
  566, 570
Models for circuit simulations,
  498–501
ModelSim, 396
MOFA. *See* Multiobjective firefly
  algorithm (MOFA)
Monolayer GFET, 82
Monotonic code, 219
Monotonicity:
  ADC, 235
  DAC, 238
Monte Carlo analysis, 380, 622–628.
  See also Fast Monte Carlo methods
  acceptance-rejection Monte Carlo
    method, 622–623
  common features, 622–623, 622f
  crude Monte Carlo method, 622
  digital logic level, 627–628,
    629f, 630f
  drawbacks, 638
  graphene LC-VCO circuit, 626–627
  important sampling Monte Carlo
    method, 623
  nano-CMOS amplifier circuits,
    624–625
  nano-CMOS oscillator circuits,
    623–624
  overview, 622–623, 622f
  SPICE, 452
  stratified Monte Carlo method, 623
Moore's law, 5
MOR. *See* Model order reduction
  (MOR)
MOSFET, 68–70. *See* Metal-oxide
  semiconductor FET (MOSFET)
Motherboard, 584
Motion sensors, 299–300
MPSoC architecture. *See*
  Multiprocessor SoC (MPSoC)
  architecture
MRAM. *See* Magnetic/
  magnetoresistive RAM (MRAM)
MSE. *See* Mean square error (MSE)
MTJ. *See* Magnetic tunneling
  junction (MTJ)

MTTF. *See* Mean time to failure
  (MTTF)
MuGFET. *See* Multiple-gate FET
  (MuGFET)
Multi-gate MOSFET, 500
Multi-RF PLL, 159
Multibank DRAM (MDRAM), 340
Multilayer feed-forward neural
  network (NN), 724, 728
Multimedia jukebox, 41
Multimedia watermarking,
  142, 143–144
Multiobjective firefly algorithm
  (MOFA), 747–750
Multiphase PLL, 159
Multiple/dual oxide thickness
  technology, 553
Multiple-gate FET (MuGFET), 72
Multiple independent gate FET
  (MIGFET), 72–81
  advantages, 72
  carbon nanotube FET (CNTFET),
    79–81
  double-gate FET (DGFET), 73–75
  double-gate FinFET, 75–77
  tri-gate/triple-gate FinFET (TGFET),
    77–79
Multiple oxide devices, 522,
  527f, 528
Multiple oxide thickness, 549
Multiple supply voltage, 521,
  527–528, 527f
Multiple supply voltage-based RTL
  technique, 544–548
Multiple threshold devices, 521–522,
  527f, 528
Multiplication ratio, 162
Multiplying DAC, 230, 230f
Multiport SRAM, 318
Multiprocessor SoC (MPSoC)
  architecture, 605–606, 605f
Multisim, 448
Multivariate adaptive regression spline
  (MARS), 691
Multivibrator, 190
Multiwall carbon nanotube
  (MWCNT), 79, 79f
Multiwall carbon nanotube (MWCNT)
  inductor-based differential
  LC-VCO, 188, 190f, 190t
MWCNT. *See* Multiwall carbon
  nanotube (MWCNT)
MWCNT inductor-based differential
  LC-VCO, 188, 190f, 190t
MxSICO compiler, 17, 392
MyHDL, 477–478
MyHDL-based simulation, 477–478,
  478f

N

N-type GNR-TFET, 88f
N-type GTFET, 88f
N-type MOSFET (NMOS), 68
N-type P-I-N structure TFET, 86f
N-well, 93

NAND-flash memory, 352, 352f, 358t
Nano-CMOS oscillator design optimization, 633, 634f, 635f, 635t
Nano-CMOS oscillator design space sampling, 632–633
Nano-electro-mechanical system (NEMS), 3
Nanoelectronic circuit parasitics. See Parasitics
Nanoelectronic devices, 67–91
 classical SiO_2/polysilicon FET, 68–70
 graphene FET (GFET), 81–83
 high-κ/metal-gate FET (HKMGFET), 70–72
 memristor, 88–91
 multiple independent gate FET (MIGFET), 72–81
 overview, 68f
 single-electron transistor (SET), 84
 thin-film transistor (TFT), 85, 86f
 tunnel FET (TFET), 86–88
 vibrating body FET (VBFET), 88
Nanoelectronic parametric variations, 619, 620f
Nanoelectronic system, 2f
Nanoelectronics, 65
 applications. See Emerging systems
 complexity of nanoscale circuits, 66
 devices. See Nanoelectronic devices
 key issues. See Challenges
Nanoelectronics-based biosensors, 264–266
Nanoelectronics-based gas sensors, 294–295
Nanoelectronics reliability and aging causes, 135f
Nanoelectronics variations, 379, 379f
Nanomanufacturing, 91–98
 carbon nanotube FET fabrication process, 95
 classical CMOS manufacturing process, 93–94
 FinFET fabrication process, 95–96
 graphene FET fabrication process, 96–97
 HKMG fabrication, 94–95
 memristor fabrication process, 98
 phases of manufacturing, 91–93
 tunnel FET fabrication process, 97–98
Nanometer, 65
Nanoscale ADC FoM, 236
Nanoscale classical SiO_2/polysilicon FET, 68–70
Nanoscale CMOS, 68f
Nanoscale MOSFET, 67
NanoSpice, 450
Nanotechnology, 65
Nanowire FinFET, 77, 77f
Native mixed-mode simulator, 496, 498
Natures, 490

Navigation system for timing and ranging (NAVSTAR), 36
NBTI. See Negative bias temperature instability (NBTI)
nDRAM. See Next-generation DRAM (nDRAM)
NEC laptop TV, 57
Negative bias temperature instability (NBTI), 137, 137f, 448
Negative binomial model (die yield analysis), 108
Negative resistance, 90
NEMS. See Nano-electro-mechanical system (NEMS)
nero, 400, 402
Net-centric multimedia processor (NMP), 41–44
Net-die-per-wafer yield, 108
net_discipline, 481
Netlist-driven mixed-signal design flow, 390
Network interface module (NIM), 51
Network transfer function (NTF), 451
Networked media tank (NMT), 40–41
Neumann formula, 569
Neural network-based metamodeling, 722–733
 generation, 725–726
 iVAMS, 727–731
 Kriging-NN metamodeling, 731, 732f, 733t
 multilayer feed-forward NN, 724
 OP-AMP, 727–731, 747–750
 PLL, 726, 727f, 750–752
 radial NN, 724
 theory, 722–724
 ultrafast design optimization, 747–752
 ultrafast process variation analysis, 733, 735f, 735t
Neural network metamodel-based ultrafast design optimization, 747–752
Newton-Raphson (NR) iterations, 462
Next-generation DRAM (nDRAM), 340
NgSpice, 402, 448, 449
NIM. See Network interface module (NIM)
Nine-transistor SRAM (9T-SRAM), 326, 326f
NMOS. See N-type MOSFET (NMOS)
NMOS gating, 537
NMOS Widlar current mirror, 271
NMP. See Net-centric multimedia processor (NMP)
NMT. See Networked media tank (NMT)
NN metamodeling. See Neural network-based metamodeling
No-custom design, 383
Node, 481
Noise analysis, 452
.NOISE control line, 452

Noise-shaping filter, 227
Noise transfer function (NTF), 428
Non-contact-type temperature sensors, 266
Non-deterministic polynomial-time hard (NP hard) problem, 638
Non-minimum channel length devices, 549
Nonlinear negative resistance, 90
Nonvolatile memory, 316, 316f, 358t
Nonvolatile resistive RAM, 352–353
NOR-flash memory, 352, 352f
NOR ROM, 240–241, 246
NP hard problem. See Non-deterministic polynomial-time hard (NP hard) problem
NPN bipolar junction transistor (BJT), 348
NR iterations. See Newton-Raphson (NR) iterations
NTF. See Network transfer function (NTF); Noise transfer function (NTF)
Nyquist frequency, 217
Nyquist rate, 217
Nyquist-Shannon sampling theorem, 217

O

OASIS. See Open artwork system interchange standard (OASIS)
OASYS, 17, 392
OBG. See Out-of-band gain (OBG)
Objects, 490
Ocean script, 395
ocp, 400, 402
ode14x solver, 440
Off-chip thermal control techniques, 596, 596f
Offset error:
 ADC, 235
 DAC, 237
Offset FPN, 288
Olympus-SoC, 386
Omega navigation system, 36
On-chip thermal control techniques, 596, 596f
On-chip thermal sensors, 267, 267f
One-memristor-based SRAM, 354–355, 354f
One-time programmable nonvolatile memory (OTP NVM), 349
One-transistor capacitor-less DRAM (1T-DRAM), 341–342
One transistor-one capacitor DRAM (1T-1C-DRAM), 340–341
Onlooker bee, 741
OP-AMP. See Operational amplifier (OP-AMP)
.OP control line, 450
OPASYN, 392
Open artwork system interchange standard (OASIS), 375
Open circuit failure, 140

Open-source EDA tools, 399–403
Open-source SPICE simulators, 446, 447f, 448–449
OpenMAST, 491–492
OpenMAST-based simulation, 491–493
OpenVera, 474
Operating frequency, 162
Operation scheduling, 523
Operational amplifier (OP-AMP):
   NN metamodel-based ultrafast design optimization, 747–750
   NN metamodeling (iVAMS), 727–731
   polynomial-metamodel-based ultrafast design optimization, 744–747
   Verilog-AMS-PoM, 702–707
Operational transconductance amplifier (OTA), 704f, 705
Optical-detection biosensor, 28
.option Card, 446
Ordinary Kriging method, 716–717, 720, 721f, 721t, 755–758
Organic solar cell, 276, 276f
Organic thin-film transistor-based ADC, 250–252
Organic thin-film transistor VCO-based ADC, 250, 250f, 251t
Oscillator circuits, 164–171
   characteristics, 165–170
   current-starved voltage controlled oscillator (CS-VCO), 177–184
   frequency pulling, 167
   frequency pushing, 167
   jitter, 167–169, 167f
   LC-tank voltage controlled oscillator (LC-VCO), 170t, 184–190
   oscillation frequency, 165–166
   output level, 169
   performance metrics, 165–170
   power consumption, 169–170
   relaxation oscillator, 190–193
   ring oscillator (RO), 170t, 171–176
   silicon size, 170
   spectral purity, 167–169
   tuning linearity, 166–167
   tuning range, 166, 167f
   tuning sensitivity, 166
   types, 164–165, 165f
OSR. *See* Oversampling ratio (OSR)
OTA. *See* Operational transconductance amplifier (OTA)
OTFT VCO-based ADC, 250, 250f, 251t
OTP NVM. *See* One-time programmable nonvolatile memory (OTP NVM)
Out-of-band gain (OBG), 426
Output amplitude, 163
Output buffer, 317
Output divider, 205
Output frequency range, 162
Output generation engine, 669

Output spectrum purity, 163
Output voltage swing, 286
Oversampling ADC, 221–222, 221f
Oversampling DAC, 229, 229f
Oversampling filter, 227
Oversampling ratio (OSR), 426
Oxidation process, 92
Oxide thickness increase, 553

P

P-I-N-based TFET, 97
P^+/N^- asymmetric DGFET, 74
P-type MOSFET (PMOS), 68
Packing, 389
Page access time, 347
Parallel direction reading, 356
Parallel SmartSpice, 450
Parallel solvers, 689
Parallel SPICE engine, 460, 460f
Parameter sweep analysis, 452
Parametric yield, 108–109, 668
Parasitics, 11–12, 122–133, 560–582
   circuit (device + parasitics) extraction process, 566–567
   cross-talk, 562
   device, 122–125
   device parasitic capacitances, 124–125
   device parasitic inductances, 125
   device parasitic resistances, 123–124
   effects, 562–564
   Elmore delay model, 570–574
   extraction techniques, 376–377, 568–569
   interconnect, 125–133
   interconnect parasitic capacitances, 130–132
   interconnect parasitic inductances, 133
   interconnect parasitic resistances, 128–130
   issues to be addressed, 562
   modeling, 569–570
   MOR, 570
   parasitic oscillations, 564
   parasitics-aware analog design flow, 574, 575f
   parasitics-aware circuit optimization, 574–582
   parasitics capacitor components, 561f
   parasitics-optimal ULC, 578–582
   parasitics-optimal VCO, 576–578
   rapid parasitics-aware design flow, 574–575, 576f
   RC-chain network of lossy long transmission line, 571–574
   types, 122, 123f, 561f
Parasitics-aware analog design flow, 574, 575f
Parasitics-aware circuit optimization, 574–582
Parasitics-aware netlist, 380

Parasitics capacitance extraction, 376–377
Parasitics extraction, 376–377, 566, 568–569
Parasitics inductance extraction, 377, 568
Parasitics model order reduction (MOR), 570
Parasitics modeling and extraction, 565–574
Parasitics oscillations, 564
Parasitics resistance extraction, 376
Pareto chart:
   defined, 632
   DOE-ILP optimization, 664f
   DOE method-based optimization of current-starved VCO, 635f
   power dissipation of APS, 653f
   voltage swing of APS, 653f
Pareto front (PF), 747
Partial element equivalent circuit (PEEC), 568–569
Particle detector, 44
Particle swarm optimization (PSO):
   90-nm CS-VCO, 657–661
   parasitic-optimal VCO, 576
   ultrafast design optimization, 750–752
Partition-driven placement approach, 604
Passive pixel sensor (PPS), 281, 282, 282f
Pattern matching methods, 568, 568f
PC RAM. *See* Phase-change RAM (PCRAM)
PCB. *See* Printed circuit board (PCB)
PCR. *See* Polymerase chain reaction (PCR)
PD. *See* Phase detector (PD)
PDF. *See* Probability density function (PDF)
PDK. *See* Process design kit (PDK)
PDP. *See* Power-delay-product (PDP)
Peak power dissipation, 517
PEEC. *See* Partial element equivalent circuit (PEEC)
Perceptual analysis unit, 430, 542
Performance metric metamodel (PMM), 702
Performance yield, 668
Personal video recorder (PVR), 34
PF. *See* Pareto front (PF)
PFD. *See* Phase-frequency detector (PFD)
PFR. *See* Power over frequency ratio (PFR)
Phase-change RAM (PCRAM), 356–358, 358t
Phase detector (PD), 679
Phase-frequency detector (PFD), 193–195
Phase-locked loop (PLL), 157–214
   180-nm CMOS PLL, 203–204

Phase-locked loop (PLL) (*Cont.*):
  all-digital phase-locked loop
    (ADPLL), 204–206
  applications, 157f
  block-level representation,
    160–161, 160f
  characteristics, 161–163
  charge pump (CP), 195–198
  current-starved voltage controlled
    oscillator (CS-VCO), 177–184
  definition, 160, 160f
  delay-locked loop (DLL), 206–209
  frequency divider, 200–203
  Kriging metamodeling, 720, 733,
    736–737
  LC-tank voltage controlled oscillator
    (LC-VCO), 184–190
  loop filter (LPF), 198–199
  NN metamodeling, 726, 727f,
    750–752
  oscillator circuits. *See* Oscillator
    circuits
  performance metrics, 161–163
  phase-frequency detector (PFD),
    193–195
  polynomial-metamodel-based
    ultrafast design optimization,
    739–744
  relaxation oscillator, 190–193
  ring oscillator (RO), 171–176
  system types, 158–159
  theory, 163–164
  ultrafast design optimization,
    739–744
  ultrafast process variation analysis,
    733–737
Phase noise, 162
Pheromone, 755, 756
Philco mystery control, 58
Photo current, 287
Photo response nonuniformity
  (PRNU), 288
Photo sensors. *See* Image sensors
Photodetector-based motion
  sensor, 300f
Photodiode-based APS, 282–283, 283f
Photoelectric cell, 273
Photoelectric effect, 274
Photogate-type APS, 283–284, 283f
Photolithography, 92f, 93, 388–389
Photomask variation, 99, 99f, 100
Photoresist, 93
Photovoltaic cell, 273
Photovoltaic effect, 274
Physical attacks on security chips, 143f
Physical design:
  analog circuit, 374–375
  digital circuit, 386–387, 387f
Physical level (analog circuit),
  370, 370f
Physical level (digital circuit), 370, 370f
Physical vapor deposition (PVD),
  71, 92
Physical verification, 387

Physical verification system (PVS)
  programmable-ERC
  (PVS-PERC), 378
Physics-based MOR, 570
PID control principle, 596
PID controller-based thermal
  monitoring, 596
Piezoelectric effect, 278
Piezoelectric sensors, 278–280
Piezoresistive sensor, 278
Pin ordering, 555
Pinned layer, 355
Pipeline ADC, 221, 221f, 225t
Pipeline DAC, 230–231, 231f
Pixel size (image sensor), 289
Placement, 387
Planar double-gate FET, 73, 73f
PlayBook, 52
PLL. *See* Phase-locked loop (PLL)
PMM. *See* Performance metric
  metamodel (PMM)
PMOS. *See* P-type MOSFET (PMOS)
PMOS gating, 537
PMOS Widlar current mirror, 271
PNP bipolar junction transistor
  (BJT), 348
Poisson model (die yield analysis),
  108
Pole-zero analysis, 451
Polymerase chain reaction (PCR), 30
Polynomial-based metamodeling,
  698–714
  generation, 699–701
  LC-VCO, 702
  memristor oscillator, 707–710
  OP-AMP, 702–707, 744–747
  PLL, 739–744
  ring oscillator, 701, 737–739
  theory, 698–699
  ultrafast design optimization,
    737–747
  Verilog-AMS-PAM, 711–714
  Verilog-AMS-PoM, 702–710
Polynomial metamodel integrated
  Verilog-AMS (Verilog-AMS-
  PoM), 705–710
Polynomial regression model, 598
Polysilicon, 71
Polysilicon gate depletion, 66
Popcorn Hour player, 41
Post-layout simulation, 424
Post-mask ECO, 388
Post nanoscale CMOS, 68f
Postlayout parasitics extraction, 575
Postsilicon power reduction
  techniques, 522, 522f, 536–540
Postsilicon tuning for variation-aware
  RTL, 675–676, 677t
Potentiometric electrochemical
  biosensor, 28, 29f
Power-aware design, 10
Power delay estimation engine, 669
Power-delay-product (PDP), 517
Power density, 517

Power dissipation, 9–10, 109–122,
  513–560
  adaptive voltage scaling (AVS), 540
  capacitative switching power
    dissipation, 112
  channel punchthrough current,
    118–119
  contention current, 119–120
  double-gate FinFET, 121–122
  dual-oxide-based logic-level leakage
    optimization, 531–534
  dual-oxide-based RTL optimization,
    534–536, 558–560
  dual-threshold-based circuit-level
    optimization (current-starved
    VCO), 554–557
  dual-threshold-based circuit-level
    optimization (nano-CMOS
    SRAM), 550–552
  dual-threshold-based circuit-level
    optimization (ULC), 528–531
  dual-voltage/frequency-based
    dynamic power reduction,
    542–544
  dynamic power reduction
    techniques, 540–548
  dynamic/variable frequency
    clocking, 538–540
  effects of, 514–515, 514f
  forms/sources, 515–516, 515f
  gate-induced drain leakage (GIDL),
    117, 117f
  gate oxide leakage, 114–115
  gate-oxide leakage reduction
    techniques, 553–560
  hot carrier current, 117–118
  illustrative power dissipation
    profile, 517f
  limits to low power design,
    520–521
  metrics, 516–517, 517f
  mobile phone, 518–519
  multiple supply voltage-based RTL
    technique, 544–548
  nanoscale classical CMOS circuits,
    111–120
  nanoscale high-κ and metal-gate
    FET, 120–121
  overview (analog circuit power
    optimization), 523–525
  overview (digital SoC power
    optimization), 525–527
  overview (power reduction
    techniques), 521–523
  postsilicon techniques, 522, 522f,
    536–540
  presilicon techniques, 521–522, 522f,
    527–536
  reversed biased junction leakage,
    116–117
  short circuit power, 116
  subthreshold leakage, 112–114
  subthreshold leakage reduction
    techniques, 548–552

Power dissipation (*Cont.*):
  why is it an issue? 110
  wireless sensor network (WSN), 519–520
Power dissipation change, 517
Power fluctuation, 517
Power gating, 537, 537f
Power over frequency ratio (PFR), 737
Power-square-delay-product (PSDP), 517
Power-to-SNM ratio (PSR), 598
Power yield, 668
PowerWise interface (PWI), 540, 540f
PPS. *See* Passive pixel sensor (PPS)
PRAM. *See* Phase-change RAM (PCRAM)
Precharge and voltage equalization time, 302–303
Predivider, 205
Prelayout parasitics extraction, 575
Presilicon power reduction techniques, 521–522, 522f, 527–536
PrimeTime, 388
PrimeTime Suite, 388
Primitive elements, 455, 457f
Primitive sources, 455–456, 457f
Printed circuit board (PCB), 584, 585f
PRNU. *See* Photo response nonuniformity (PRNU)
Probability density function (PDF):
  important sampling Monte Carlo method, 623
  neural network metamodel-based process variation analysis, 733
  PVT variations, 380
Probe DNA, 30
Problems. *See* Challenges
Process corner, 634–635
Process design kit (PDK), 407–408
Process-logic-RTL hierarchical simulation approach, 106
Process-transistor-logic-RTL (partial-physics) approach, 106
Process variation, 12, 13f, 98–107, 379, 379f
  analog circuits, 103–104
  carbon nanotube FET, 102
  classical CMOS transistor, 101
  design phase incorporation, 103–107
  digital circuits, 105–107
  FinFET, 102
  graphene FET, 103
  impact on device parameters, 100–103
  intra-die/inter-die, 12, 100f
  memristor, 103
  mixed-signal circuits, 104
  negative impact, 619
  tunnel FET, 102
  types, 99–100
  ultrafast analysis, 733–737

Process variation (*Cont.*):
  variability tolerant design. *See* Variability-aware design methodologies
Process variation-aware 90-nm CMOS-based VCO, 654–656
Process-variation aware design (analog circuits), 104f
Process-variation aware statistical hierarchical modeling (digital circuits), 106f
Process variation engine, 669
Process variation optimal current-starved VCO, 657–661
Process variation tolerant LC-VCO design, 661–662
Process, voltage, and temperature (PVT) variations, 14, 379, 379f
Process (die) yield, 108
Programmable PLL, 159
Programmable read-only memory (PROM), 349
Programmable remote control, 58
PROM. *See* Programmable read-only memory (PROM)
PSDP. *See* Power-square-delay-product (PSDP)
PSDRAM. *See* Pseudostatic DRAM (PSDRAM)
Pseudostatic DRAM (PSDRAM), 340
Pseudostatic SRAM (PSRAM), 340
PSO. *See* Particle swarm optimization (PSO)
Pspice, 448
PSR. *See* Power-to-SNM ratio (PSR)
PSRAM. *See* Pseudostatic SRAM (PSRAM)
Pull-up ratio, 330–331
Pulse-width modulator (PWM) DAC, 227, 227f
Punchthrough current, 118–119
PVD. *See* Physical vapor deposition (PVD)
PVR. *See* Personal video recorder (PVR)
PVS, 387
PVS-PERC. *See* Physical verification system (PVS) programmable-ERC (PVS-PERC)
PVT variations. *See* Process, voltage, and temperature (PVT) variations
PWI. *See* PowerWise interface (PWI)
PWM DAC. *See* Pulse-width modulator (PWM) DAC
pycircuit, 449
Python, 477
Pyxis Layout, 375
Pyxis Schematic, 374, 386
.PZ control line, 451

Q

QCA. *See* Quantum cellular automata (QCA)
QRC, 376
QRC Extraction, 395, 397

Quantization noise, 235
Quantizer, 217, 254–255, 255f
Quantum cellular automata (QCA), 67
Quantum-dot nanostructures, 67
Quantum efficiency (image sensor), 285–286
Quantum Hall effect, 82
Quantum mechanical tunneling, 114, 114f
Quartz, 165
Quartz crystal oscillator (XO), 170
Quasi floating gate, 524
Questa ADMS, 389, 481, 490
Quite universal circuit simulator (Qucs), 448

R

R-2R ladder DAC, 227–228, 228f
R^2, 695
RAAE. *See* Relative average absolute error (RAAE)
Radial basis function (RBF), 691
Radial neural network (NN), 724
Radial variation, 99–100, 99f
Radiation, 588
Radiation detection system (RDS), 44–46
Radiation dosimeter, 45
Radio barcode, 46
Radio frequency identification (RFID) chip, 46–49
Radio frequency (RF) remote, 58
RAM. *See* Random-access memory (RAM)
Rambus DRAM (RDRAM), 344
Rambus in-line memory module (RIMM), 344
Ramp-compare ADC, 219, 219f
Ramp run-up ADC, 219, 219f
Random-access memory (RAM), 316
Random access time (tRAC), 347
Random number generator unit, 543
Random walk, 568, 568f
Rapid parasitics-aware design flow, 574–575, 576f
Rare event analysis, 681–683
RAS active time (tRAS), 346
RAS to column address delay (tRCD), 346
Raspberry Pi, 434
Raytheon Company, 56
RBB. *See* Reverse body bias (RBB)
RBF. *See* Radial basis function (RBF)
RBT. *See* Read buffer transistor (RBT)
RC-based LPF circuit, 198–199, 198f
RC-chain network of lossy long transmission line, 571–574
rc lossy transmission line, 569
RC oscillator (RCO), 165
RCLK, 126
RCO. *See* RC oscillator (RCO)
RDRAM. *See* Rambus DRAM (RDRAM)

RDS. *See* Radiation detection system (RDS)
Reactance-less oscillator, 192
Read buffer transistor (RBT), 324
Read Columnline, 342
Read-modify-write cycle time, 347
Read noise margin (RNM), 330
Read-only memory (ROM), 349–351
Read-only RFID tag, 46–47
Read Rowline, 342
Read-write RFID tag, 47
Readin, 459f
Readout mode (PPS), 282
Ready tasks set (RTS), 678
Reconstruction filter, 217
Reduced instruction set computer (RISC), 366
Reduction of manual-layout steps, 689
Reference resistor, 295
Reference voltage controller, 300
Regenerative frequency divider, 200, 200f
Register-transfer-level optimization. *See* RTL optimization
Rejection Monte Carlo method, 622
Relative accuracy (ADC), 233
Relative average absolute error (RAAE), 695
Relative maximum absolute error (RMAE), 695
Relaxation oscillator, 190–193
Relaxation phase, 190
Reliability issue, 135–141
   electromigration (EM), 139–141
   hot carrier injection (HCI), 135–137
   latchup effect (LUE), 137–138
   negative bias temperature instability (NBTI), 137, 137f
   thermal stress, 141
   time-dependent dielectric breakdown (TDDB), 138–139
Remote control, 59
Rent coefficient, 125
Rent rules, 125, 126
Replacement metal gate, 94
ReplayTV, 34
ReRAM. *See* Resistive random-access memory (ReRAM)
Reset mode (PPS), 282
Reset operation, 357
Resistance components (MOSFET structure), 123f
Resistive circuit-based simulation, 431
Resistive random-access memory (ReRAM), 352–355, 358t
Resistive thermal sensors, 266
Resistive touchscreen, 54
Resistor-network binary-weighted summing amplifier DAC, 226, 226f
Resolution:
   ADC, 231
   DAC, 237
   image sensors, 289

Resonant biosensor, 28
Resonant-gate FET (RG-FET), 88
Responsivity (image sensor), 288
Retention time, 347
Reticle variation, 99, 99f, 100
Reverse body bias (RBB), 523
Reversed biased junction leakage, 116–117
RFID chip, 46–49
RFID device, 46
RFID tag, 46
RFID transponder, 46
RG-FET. *See* Resonant-gate FET (RG-FET)
RIMM. *See* Rambus in-line memory module (RIMM)
Ring, 400
Ring oscillator (RO), 171–176
   45-nm CMOS, 171–173
   carbon nanotube, 175, 176f
   high-level schematic, 171f
   LC-tank oscillator, compared, 170t
   multigate FET, 173–175
   oscillation principle, 171f
   parasitic elements, 563f
   performance metrics, 170t
   polynomial-based metamodeling, 702
   polynomial-metamodel-based ultrafast design optimization, 737–739
RISC. *See* Reduced instruction set computer (RISC)
rlc lossy transmission line, 569
RLCK, 11
RMAE. *See* Relative maximum absolute error (RMAE)
RMS. *See* Root mean square (RMS)
RMSE. *See* Root mean square error (RMSE)
RNM. *See* Read noise margin (RNM)
RO. *See* Ring oscillator (RO)
Rohrer, Heinrich, 25
ROM. *See* Read-only memory (ROM)
Root mean square (RMS), 222
Root mean square error (RMSE), 695
Routing deals, 375
Row/column scanner, 293
Row cycle time (tRC), 346
Row precharge time (tRP), 346
RRAM. *See* Resistive random-access memory (ReRAM)
RTL design, 384
RTL level (digital circuit), 370, 370f
RTL library, 408, 526
RTL optimization:
   dual-oxide-based, 534–536, 558–560
   multiple supply voltage-based RTL technique, 544–548
   thermal-aware, 604–605
   variability-aware. *See* Variability-aware RTL optimization
RTL synthesis, 384
RTS. *See* Ready tasks set (RTS)

Runge-Kutta-Fehlberg method, 590
Runge-Kutta method for thermal simulation, 590, 591f

S

S-Edit, 395
s2r, 400
SA ADC. *See* Successive-approximation (SA) ADC
Saber, 373, 491
Sacrificial polysilicon gate, 94
SAM. *See* Serial/sequential-access memory (SAM)
Sample and hold, 217
Sampler, 217
Sampling frequency, 233
Sampling rate, 233
Sandwich tunnel barrier FET, 86
SAR. *See* Successive approximation register (SAR)
SB method. *See* Statistical blockage (SB) method
Scaling and embedding factor unit, 430, 542
Scanning electron microscope (SEM), 25
Scanning force microscope (SFM), 25
Scanning probe microscope (SPM), 25
SCE. *See* Short-channel effects (SCE)
Schematic capture, 374
Schematic-driven mixed-signal design flow, 390
Schematic editor, 374
Schematic level (analog circuit), 370, 370f
Schematic-level netlist, 524, 645
Schematic-level optimization, 525
Schmitt trigger oscillator, 192–193, 707–710
Schottky diode, 138
Scicos, 434
ScicosLab, 434
Scilab, 426
SciPy, 426
Scout bee, 741
SCP. *See* Sequential conic program (SCP)
SDC. *See* Secure digital camera (SDC)
SDL Router, 395, 399
SDR. *See* Software-defined radio (SDR)
SDR-SDRAM. *See* Single-data rate SDRAM (SDR-SDRAM)
SDRAM. *See* Synchronous DRAM (SDRAM)
Secondary generated hot-electron (SGHE) injection, 136, 136f
Secure digital camera (SDC), 49–50, 293, 294f
Secure media processor (SMP), 31, 41, 41f
Segmentation, 228
Segmented DAC, 228–229, 228f
Seizure, 297

Selective collapsing, 531
Self-heating (thermal sensor), 267
SEM. *See* Scanning electron microscope (SEM)
Semi-custom design, 383
Semi-custom design flow, 369f, 370
Semi-passive RFID, 46
Semivariogram, 715
.SENS control line, 452
Sense amplifier, 300–306
   clamped bitline, 304–306, 718–720
   deployment of, in memory, 301f
   Kriging metamodeling, 718–720, 755–758
   power dissipation, 303–304
   precharge and voltage equalization time, 302–303
   sense delay, 303
   sense margin, 303
   silicon area, 304
   types, 301, 301f
Sense delay, 303
Sense margin, 303
Sensing node, 295, 296f
Sensitivity analysis, 452
Sensor circuits and systems, 263–313
   body sensors, 295–296
   epileptic seizure sensors, 297, 297f
   humidity sensors, 297–299
   image sensors. *See* Image sensors
   motion sensors, 299–300
   nanoelectronics-based biosensors, 264–266
   nanoelectronics-based gas sensors, 294–295
   piezoelectric sensors, 278–280
   sense amplifiers, 300–306
   sensor, defined, 263
   solar cells, 272–278
   thermal sensors, 266–272
   types of sensors, 263f
Separate biasing, 174, 175t
Sequential conic program (SCP), 676
Sequin, C. H., 281
Serial/sequential-access memory (SAM), 316
Serializer, 300
Series-parallel charge pump, 196, 196f
SET. *See* Single-electron transistor (SET)
SET-based ADC, 249–250, 250f
SET-based DAC, 257
Set operation, 357
Set-top box (STB), 50–51
Set-top unit (STU), 50
Settling time, 233
Seven-transistor SRAM (7T-SRAM), 322–324, 334–335, 662–667
SF corner. *See* Slow-fast (SF) corner
SFDR. *See* Spurious-free dynamic range (SFDR)
SFM. *See* Scanning force microscope (SFM)

SG DG FinFET. *See* Short-gate (SG) DG FinFET
SGDRAM. *See* Synchronous graphics DRAM (SGDRAM)
SGHE injection. *See* Secondary generated hot-electron (SGHE) injection
SHC injection. *See* Substrate hot-carrier (SHC) injection
SHE injection. *See* Substrate hot-electron (SHE) injection
SHH injection. *See* Substrate hot-hole (SHH) injection
Short-channel effect (SCE), 6, 66
Short circuit failure, 140
Short circuit power dissipation, 116
Short-circuit power reduction, 541f, 542
Short-gate (SG) DG FinFET, 173, 174
Shunt transistor, 345
Si-Fin, 7
Side-channel attack, 143
Sigma-delta ADC, 221–222, 222f, 225t
Sigma-delta DAC, 229, 229f
Sigma-delta modulator, 229
Sigma-delta modulator-based ADC, 252–256
   1-bit DAC, 255, 256f
   architecture overview, 252–253
   block diagram, 253f
   broad perspective, 252, 252f
   downsampler, 256
   gain amplifier, 255, 256f
   integrator, 254, 255f
   low-pass digital filter, 256
   quantizer, 254–255, 255f
   summing amplifier, 254, 254f
Sigma-delta modulator-based DAC, 256–257
Sigma-delta PLL, 159
Signal, 215
Signal blocking, 529
Signal converters. *See* Electronic signal converter circuits
Signal integrity, 514–515
Signal passing, 529
Signal propagation, 565–566
Signal rise/fall time reduction and equalization, 542
Signal routing, 387
Signal-to-noise-and-distortion ratio (SINAD), 235
Signal-to-noise-distortion ratio (SNDR), 235
Signal-to-noise ratio (SNR):
   ADC, 234
   image sensor, 289
Signoff, 387–388
Silicon area, 163, 304
Silicon dioxide (SiO_2), 5, 6, 71t
Silicon ingot, 388
Silicon-on-insulator (SOI) process, 7–8, 75

Silicon-on-insulator (SOI) transistor, 341
Silvaco International, 394
SIMD, multiple data (SIMD), Single instruction
SIMetrix, 448
Simple Kriging method, 716, 718–720, 752–755
Simscape, 434
Simulated annealing, 534, 535f, 576
Simulated annealing-based statistical RTL optimization, 668–669, 670f, 671
Simulation. *See* Mixed-signal circuit and system simulation
Simulation-based approaches (AMS circuit synthesis), 393
Simulation Programs with Integrated Circuit Emphasis. *See* SPICE
Simulink, 433–434
Simulink/Simscape-based simulations, 433–445
   circuit-level simulations, 438–440, 465
   CT-SDM, 434–436
   device-level simulation, 440–445
   graphene FET (GFET) device, 442–445
   memristor-based Wien oscillator, 436–438
   memristor device, 441–442
   system-/architecture-level simulation, 435–438
   Verilog-AMS, compared, 488–489, 489t
   watermarking chip, 438
SINAD. *See* Signal-to-noise-and-distortion ratio (SINAD)
Single-data rate SDRAM (SDR-SDRAM), 340f
Single-electron-based ADC, 249–250
Single-electron circuitry-based ADC, 249, 249f
Single-electron transistor (SET), 84
Single-electron transistor (SET)–based ADC, 249–250, 250f
Single electron transistor (SET)–based DAC, 257
Single-ended SRAM, 318
Single-ended 7T-SRAM, 323–324, 323f
Single instruction, multiple data (SIMD), 39
Single iteration automatic power-aware design flow, 524f
Single layer rules, 375
Single-manual iteration, 576
Single-manual iteration fast design flow, 576–578
Single manual layout iteration automatic flow (variability-aware optimization), 646–648
Single-wall carbon nanotube (SWCNT), 79, 79f

Single-wall carbon nanotube (SWCNT) ring oscillator, 175, 176f
SINM. *See* Static current noise margin (SINM)
SiO_2. *See* Silicon dioxide (SiO_2)
Six-transistor SRAM (6T-SRAM), 318–320, 331–334
Skewed corner, 636
Skin depth, 129
Skin effect, 129
Slate personal computer (slate PC), 52–54
Slave node, 295, 296f
SLDRAM. *See* SyncLink DRAM (SLDRAM)
Slow-fast (SF) corner, 635, 636f
Slow-slow (SS) corner, 635, 636f
Small-a/big-D AMS-SoC, 496
Small-signal distortion analysis, 451
Smart battery, 538
Smart camera, 291
Smart image sensor, 291–293
Smart label, 46
Smart tag, 46
Smartphone, 54–55, 518–519
SmartSpice, 397, 448
SMASH simulator, 18, 448
SMP. *See* Secure media processor (SMP)
SNDR. *See* Signal-to-noise-distortion ratio (SNDR)
SNM. *See* Static noise margin (SNM)
SNR. *See* Signal-to-noise ratio (SNR)
Soft errors, 135
Software-based temperature monitoring, 594
Software-defined radio (SDR), 56–57
Software PLL (SPLL), 159
SOI FinFET, 77, 77f
SOI process. *See* Silicon-on-insulator (SOI) process
SOI transistor. *See* Silicon-on-insulator (SOI) transistor
Solar cell:
  basic operation, 274f
  circuit models, 276–278
  defined, 273
  efficiency, 274–275
  operation, 273–274
  organic, 276, 276f
  solar energy to electrochemical energy conversion, 274f
  thin-film solar cell (TFSC), 275–276
  types, 273f
Solar cells, 272–278
Solve Elec, 448
SOP. *See* Sum of product (SOP)
Source and drain resistance, 66
Spatial autocorrelation, 715
Spatial noise, 288
Spatial thermal control techniques, 596, 596f
Spectre, 395, 397, 448

Spectre XPS, 450
SPICE, 445–463
  AC small-signal analysis, 450–451
  analog-fast SPICE simulators, 447f, 450
  behavioral simulation, 373
  CMOS nanoelectronics, 452–453, 453f, 454f
  commercial simulators, 446, 447–448, 447f
  comprehensive AMS circuit design flow, 404–406
  custom elements, 457
  DC analysis, 450
  equation-formulation phase, 462
  Fast SPICE simulators, 447f, 449–450
  free/open-source simulators, 446, 447f, 448–449
  general concept, 445f
  high-speed SPICE simulators, 447f, 450
  historical background, 446
  ideal elements, 456
  iteration-control phase, 462
  matrix-solution phase, 462
  memristor nanoelectronics, 453–455
  model-evaluation phase, 462
  Monte Carlo analysis, 452
  noise analysis, 452
  ODEs/PDEs, 480
  parameter sweep analysis, 452
  pole-zero analysis, 451
  primitive elements/sources, 455–456
  semiconductor devices, 457
  sensitivity analysis, 452
  simulation characteristics, 480t
  small-signal distortion analysis, 451
  SPICE cards, 446
  SPICE elements, 455–459
  SPICE engine, 459–460
  SPICE simulation flow, 460–463
  switches, 457
  temperature sweep analysis, 452
  terminology, 445
  transformers, 456
  transient analysis, 451
  transistor-level design, 374, 386
  transistor-level simulation and characterization, 374
  transmission lines, 456
  Verilog-A, compared, 464f
  worst-case analysis, 452
SPICE 1, 446
SPICE 2, 446
SPICE 3, 446
SPICE cards, 445, 446
SPICE deck, 445
SPICE elements, 455–459
SPICE engine, 459–460
SPICE netlist, 104, 445
SPICE simulation DC analysis flow, 462, 463f
SPICE simulation flow, 460–463

SPICE simulation transient analysis flow, 462, 463f
Spider, 397
Spin-transfer torque (STT) MRAM (STT-MRAM), 355–356, 355f, 356f
Spintronic memristor-based biosensor, 264–265, 264f
Spintronic memristor temperature sensor, 271–272
SPLL. *See* Software PLL (SPLL)
SPM. *See* Scanning probe microscope (SPM)
Spurious-free dynamic range (SFDR), 235
SRAM. *See* Static random-access memory (SRAM)
SS corner. *See* Slow-slow (SS) corner
STA, 623. *See also* Variability-aware design methodologies
Stacked devices technique, 549
Stage adaptive charge pump, 196, 196f
Standard cell library, 408, 408f
Star-RCXT, 376
StarRC, 397
Static current noise margin (SINM), 329
Static noise margin (SNM), 328–330
Static parameter variations, 619, 620f
Static power dissipation, 515, 515f
Static random-access memory (SRAM), 317–338
  4T-SRAM, 321, 321f
  5T-SRAM, 321–322, 322f
  6T-SRAM, 318–320, 331–334
  7T-SRAM, 322–324, 334–335
  8T-SRAM, 324–326
  9T-SRAM, 326, 326f
  10T-SRAM, 326–327, 327f, 335–338
  access time, 330
  cell ratio, 330
  dynamic noise margin (DNM), 330
  hold noise margin (HNM), 330
  power and SNM characterization, 331–338
  power dissipation, 331
  propagation delay, 331
  pull-up ratio, 330–331
  read noise margin (RNM), 330
  silicon area, 331
  SRAM array, 317, 317f
  static noise margin (SNM), 328–330
  types, 318, 318f
  variability-aware design, 662–667
  write noise margin (WNM), 330
Statistical blockage (SB) method, 681
Statistical characterization, 380
Statistical design optimization, 381
Statistical DOE-ILP optimization of nano-CMOS SRAM, 662–667
Statistical power analysis, 623
Statistical process variation analysis, 644f. *See also* Variability-aware design methodologies

Index

Statistical timing analysis (STA), 620
Statistically unlikely event, 681
STB. *See* Set-top box (STB)
Steady-state errors, 163
Steady-state thermal analysis, 590
Stochastic techniques (parasitics), 568, 568f
Stratified Monte Carlo method, 623
Structural modeling, 423, 424
Structure thermal dynamics, 595
STT-MRAM. *See* Spin-transfer torque (STT) MRAM (STT-MRAM)
STU. *See* Set-top unit (STU)
.subckt Card, 446
.subckt control line, 453
Subranging ADC, 221, 221f
Substrate hot-carrier (SHC) injection, 136
Substrate hot-electron (SHE) injection, 136, 136f
Substrate hot-hole (SHH) injection, 136
Subthreshold leakage, 112–114
Subthreshold leakage reduction techniques, 548–552
Subthreshold slope, 114
Subthreshold swing, 114
Successive-approximation (SA) ADC, 220, 220f, 225t
Successive-approximation DAC, 229–230, 230f
Successive approximation register (SAR), 220
Sum of product (SOP), 222
Summing amplifier, 230, 254, 254f
SUMO. *See* Surrogate modeling (SUMO)
Super PLL, 159
Superlog, 474
Supplementary model file, 643
Supply voltage hopping, 538
Supply voltage range, 163
Supply voltage reduction, 542, 549, 555
Supply voltage scaling, 542
Support vector machine (SVM), 692, 692f
Support vector machine (SVM) classifier, 682
Surrogate, 689. *See also* Metamodel
Surrogate modeling (SUMO), 717
SVM. *See* Support vector machine (SVM)
SVM classifier. *See* Support vector machine (SVM) classifier
SWCNT. *See* Single-wall carbon nanotube (SWCNT)
SWCNT RO. *See* Single-wall carbon nanotube (SWCNT) ring oscillator
Switching capacitance, 133
Switching-type MRAM, 355
syf, 400, 402
Symbol creation, 374
Symbolic analysis, 393, 423–424
Symmetrical DGFET, 74
SymSpice Turbo, 450
Synchronous DRAM (SDRAM), 344
Synchronous graphics DRAM (SGDRAM), 344–345
Synchronous SRAM, 318
SyncLink DRAM (SLDRAM), 340
Synopsys, 394
Synphony C Compiler, 385
SysML. *See* Systems Modeling Language (SysML)
System-level design, 384
System level library, 408–409, 408f
System-level MATLAB simulation, 426–430
System-level sigma-delta modulator design flow, 427f
System-level Simulink simulation, 435–438
System Verilog, 17, 373
SystemC, 478
SystemC-AMS, 373, 384, 390, 480t, 494
<systemc-ams>, 494
SystemC-AMS-based simulation, 494–495
<systemc-ams.h>, 494
SystemC-based simulation, 478–480
Systems Modeling Language (SysML), 384
SystemVerilog, 384, 474, 478f, 479f
SystemVerilog-based simulation, 474–475

T

T-Spice, 395, 398, 448
Tablet PC, 52–54
Tabu search algorithm, 737, 738–739
Taguchi methods, 632, 633
Taguchi 8 matrix, 662
Tail threshold, 682
Tanner EDA, 394
Task graph, 677–678, 677f
Taylor-series expansion diagram (TED)–based RTL optimization, 669–673
TBFF. *See* Time-borrowing flip-flop (TBFF)
tCAC. *See* Column access time (tCAC)
TCAD. *See* Technology computer-aided design (TCAD)
TCCT. *See* Thin capacitively coupled thyristor (TCCT)
tCL. *See* Column address strobe (select) latency (tCL)
TclSpice, 448
TDDB. *See* Time-dependent dielectric breakdown (TDDB)
TDF. *See* Timed data flow (TDF)
TDFDM. *See* Time-domain finite-difference method (TDFDM)
TDM. *See* Time division multiplexing (TDM)
TDP. *See* Thermal design power (TDP)
Technology computer-aided design (TCAD), 423, 499–500
Technology mapping and optimization, 531
Technology scaling:
  advantages/disadvantages, 8, 65–66
  complexity of nanoscale circuits, 66
  device and circuit parameter change, 5, 5t
TED. *See* Transformed Elmore delay (TED)
TED-based RTL optimization. *See* Taylor-series expansion diagram (TED)–based RTL optimization
TEM. *See* Transmission electron microscopy (TEM)
.TEMP control line, 452
Temperature. *See* Temperature or thermal issue
Temperature-aware floor planning, 603–604, 605, 606
Temperature control basics, 595–596, 595f, 596f
Temperature monitoring, 594–595
Temperature or thermal issue, 13–14, 133–135, 582–606
  compact thermal modeling, 586–587
  effect of temperature, 583f
  FATA, 592, 593f
  Green function-based method, 592–593
  heat dissipation, 584–586
  heat transfer, 588
  ISAC thermal analysis framework, 590–592
  issues to consider, 582
  power dissipation, performance, temperature, and reliability interactions, 582, 583f, 584
  Runge-Kutta-based method, 590, 591f
  temperature control basics, 595–596, 595f, 596f
  temperature monitoring or sensing, 594–595
  thermal analysis basics, 589–590, 589f
  thermal-aware digital synthesis, 602–603
  thermal-aware physical design, 603–604
  thermal-aware RTL optimization, 604–605
  thermal-aware SRAM optimization, 596–599
  thermal-aware system level design, 605–606
  thermal-aware VCO optimization, 599–601
  thermal moment matching (TMM) method, 594
Temperature sensing, 594–595
Temperature sensors, 266, 595. *See also* Thermal sensors

Temperature sweep analysis, 452
Temporal noise, 288–289
Temporal thermal control techniques, 596, 596f
Temporary polysilicon gate, 94
Tempus, 388
Ten-transistor SRAM (10T-SRAM), 326–327, 327f, 335–338
Terminal-state problem, 469
.TF control line, 450
TFET. *See* Tunnel FET (TFET)
TFET-based biosensor, 265, 265f
TFET-based 6T-SRAM, 320
TFET-based 8T-SRAM, 325, 325f
TFET manufacturing, 97f
TFPV. *See* Thin-film photovoltaic cell (TFPV)
TFSC. *See* Thin-film solar cell (TFSC)
TFT. *See* Thin-film transistor (TFT)
TGFET. *See* Tri-gate/triple-gate FinFET (TGFET)
TGFET-based 1T-DRAM, 341–342, 342f
THD. *See* Total harmonic distortion (THD)
Thermal acceleration, 140f
Thermal actuator, 595
Thermal analysis basics, 589–590, 589f
Thermal-aware digital synthesis, 602–603
Thermal-aware hardware-software cosynthesis-based MPSoC design flow, 605f, 606
Thermal-aware HLS synthesis, 604–605, 604f
Thermal-aware physical design, 603–604
Thermal-aware placement techniques, 604
Thermal-aware platform-based MPSoC design flow, 605–606, 605f
Thermal-aware RTL optimization, 604–605
Thermal-aware SRAM optimization, 596–599
Thermal-aware system level design, 605–606
Thermal-aware VCO optimization, 599–601
Thermal capacitance, 587
Thermal controller, 595
Thermal design power (TDP), 586
Thermal-detection biosensor, 28
Thermal interface material (TIM), 584
Thermal issue. *See* Temperature or thermal issue
Thermal-optimal 90-nm CMOS-based current-starved VCO, 601f, 601t
Thermal parameters, 587t
Thermal resistance, 586
Thermal sensors, 266–272, 752–755
  45-nm CMOS ring oscillator-based thermal sensor, 268–271
  contact-/non-contact-type sensors, 266

Thermal sensors (*Cont.*):
  on-chip temperature sensors, 267, 267f
  performance metrics, 267–268
  spintronic memristor temperature sensor, 271–272
  types, 266f
Thermal stress, 141
Thermal variation, 379, 379f
Thermometer code, 219
Thermometer-coded DAC, 227, 227f
Thermometer-coded principle, 229
Thin capacitively coupled thyristor (TCCT), 348
Thin-film memristor:
  MATLAB-based simulation, 432
  Verilog-A-based modeling, 468–471
Thin-film memristor manufacturing, 98f
Thin-film photovoltaic cell (TFPV), 275
Thin-film solar cell (TFSC), 275–276
Thin-film transistor (TFT), 85, 86f
Three-transistor DRAM (3T-DRAM), 342, 342f
Threshold inverting quantization (TIQ) technique, 238, 245
Threshold voltage fluctuation, 66
Threshold voltage gating, 538
Threshold voltage increase, 554
Threshold voltage reduction, 542
Threshold voltage variations, 644–645
Thyristor random-access memory (TRAM), 348, 348f
Ti film, 353
TiA1N. *See* Titanium aluminum nitride (TiA1N)
TightVNC, 409
TIM. *See* Thermal interface material (TIM)
Time borrowing, 679
Time-borrowing flip-flop (TBFF), 679
Time-dependent dielectric breakdown (TDDB), 138–139
Time division multiplexing (TDM), 264
Time-domain finite-difference method (TDFDM), 569f
Time-interleaved ADC, 222, 223f
Timed data flow (TDF), 494
Timing closure, 387
Timing variation-aware RTL optimization, 673–675
Timing yield, 668, 676
TINA, 448
TINA-TI, 449
TIQ comparator, 238, 239, 240f
TIQ technique. *See* Threshold inverting quantization (TIQ) technique
Titanium aluminum nitride (TiA1N), 357
Titanium dioxide (TiO_2) thin-film memristor, 89
Titanium nitride (TiN), 353
Title Card, 446
TiVo, 34
TMR ratio, 356

Tools. *See* Electronic design automation (EDA) tools
Top-down design flow, 369–370, 369f
Top-gate thin-film transistor, 86f
Top-gated CNTFET, 80, 80f
TopSpice, 448
Total harmonic distortion (THD), 236
Touch screen remote control, 58
Touchscreen, 54
tPC. *See* Fast page cycle time (tPC)
tRAC. *See* Random access time (tRAC)
Track code, 206
Track-DCO, 205
Tracking ADC, 223–224, 224f
TRAM. *See* Thyristor random-access memory (TRAM)
.tran control line, 451, 452
tranan, 459f
Transcoder, 217
Transducers, 263, 263f
Transfer function, 161
Transformed Elmore delay (TED), 570
Transient analysis, 451
Transient gate-oxide leakage power reduction, 541f, 542
Transient power, 517
Transistor level
  analog circuit, 370, 370f
  digital circuit, 370, 370f
Transistor-level design:
  analog circuit, 374
  digital circuit, 386
Transistor models, 18
Transmission electron microscopy (TEM), 25
Trap-assisted tunneling, 138f, 139
tRAS. *See* RAS active time (tRAS)
tRC. *See* Row cycle time (tRC)
tRCD. *See* RAS to column address delay (tRCD)
trcurv, 459f
Tri-gate FET (TFET)–based 6T-SRAM, 320
Tri-gate/triple-gate FinFET (TGFET), 77–79, 175
Triple-gate transistor-based 1T-DRAM, 341–342, 342f
tRP. *See* Row precharge time (tRP)
Trust issue:
  attributes of trustworthiness, 141–142
  chip intellectual property protection issue, 143–144, 143f
  information leakage issue, 143
  information protection issue, 142
  malicious design modifications issue, 144
TT corner. *See* Typical-typical (TT) corner
Tuning range, 161
Tunnel FET (TFET), 86–88, 102
Tunnel FET (TFET)–based biosensor, 265, 265f
Tunnel FET fabrication process, 97–98

Tunnel magnetoresistance (TMR), 356
Turbo-MSIM, 450
TV tuner card, 57–58
Two layer rules, 375
Two-level hybrid sampling, 638, 638f
Two-memristor-based SRAM, 354, 354f
tWR. *See* Write recovery time (tWR)
Type-I PLL, 159
Type-II PLL, 159
Typical-typical (TT) corner, 635, 636f

U

UART. *See* Universal asynchronous receiver/transmitter (UART)
UHF RDIF tag, 46
Ulam, Stanislaw, 622
ULC. *See* Universal level converter (ULC)
Ultrafast AMS and RF integrated circuit design flow, 405–406, 406f
Ultrafast design optimization:
　Kriging metamodel-based optimization, 752–758
　neural network metamodel-based optimization, 747–752
　polynomial-metamodel-based optimization, 737–747
Ultrafast process variation analysis, 733–737
　Kriging metamodeling, 733, 734f, 734t
　Kriging NN metamodeling, 736–737
　NN metamodeling, 733, 735f, 735t
Unary code, 219
Unauthentic fabrication plants, 144
Unified optimization, 15–16
Universal asynchronous receiver/transmitter (UART), 35
Universal Kriging method, 720, 722f
Universal level converter (ULC):
　energy-efficient design, 520
　parasitics, 578–582
　power dissipation, 554–557
Universal product code (UPC), 47
Universal remote control, 59
Up/Down counter, 224
UPC. *See* Universal product code (UPC)

V

Variability analysis, 379–380, 621–643. *See also* Variability-aware design methodologies
　corner-based method, 633–638
　defined, 621
　design of experiments (DOE) method, 628–633
　fast Monte Carlo methods, 638–643
　Monte Carlo analysis. *See* Monte Carlo analysis
Variability-aware active pixel sensor (APS) optimization, 648–653
Variability-aware design methodologies, 619–688
　ABB method for dynamic variation compensation, 678–679
　active pixel sensor (APS), 648–653
　clock networks, 679–681
　conjugate-gradient-based optimization, 654–656
　CS-VCO, 654–661
　EDA-non-EDA framework, 643–644, 644f
　EDA tools, 643, 644f
　LC-VCO, 661–662
　methods for variability-aware design optimization, 644–648
　methods of variability analysis. *See* Variability analysis
　overview, 621f
　particle swarm optimization (PSO) approach, 657–661
　RTL. *See* Variability-aware RTL optimization
　schematic design optimization flow, 645–646, 647f
　single manual layout iteration automatic flow, 646–648
　SRAM, 662–667
　system-level methods, 677–678, 677f
　VCO circuits, 654–662
　yield analysis, 681–683
Variability-aware RTL optimization, 667–677
　overview, 667–668, 667f
　postsilicon/dynamic method, 675–676, 677t
　simulated annealing-based statistical approach, 668–669, 670f, 671
　Taylor-series expansion diagram (TED)–based approach, 669–673
　timing optimization, 673–675
Variability-aware RTL timing optimization, 673–675
Variability-aware schematic design optimization flow, 645–646, 647f
Variability-aware single-manual-layout iteration design optimization flow, 646–648
Variability-aware system-level design, 677–678, 677f
Variability-aware 7T-SRAM optimization, 662–667
Variable frequency, 537–538, 537f, 538t
Variable supply voltage, 522–523, 522f, 537–538, 537f
Variable threshold voltage, 523, 537f, 538
Variable voltage, 538t
Variogram, 715
VASS, 392
vasy, 400
VBFET. *See* Vibrating body FET (VBFET)
VBT. *See* Vibrating body transistor (VBT)
VCCS. *See* Voltage-controlled current source (VCCS)
VCO. *See* Voltage-controlled oscillator (VCO)
VCS, 397
VCVS. *See* Voltage-controlled voltage source (VCVS)
VDRAM. *See* Video DRAM (VDRAM)
Vera, 474
Verilog-A, 17f, 373, 386, 464
Verilog-A-based analog simulation, 464–473
　circuit-level simulation, 465–468
　device-level simulation, 468–473
　graphene FET-based LC-VCO, 466, 467f
　graphene FET (GFET) device, 468–473
　low-pass filter (LPF), 467, 468
　memristor, 468–471
　VCO, 465–466
　Verilog-A features, 464
　Verilog-A *versus* Verilog modeling, 465
Verilog-AMS, 17, 17f, 373, 386, 480t, 481
　mixed-signal design flow, 390
　mixed-signal system design, 398
Verilog-AMS-based simulation, 481–489
Verilog-AMS integrated with parasitic-aware metamodel (Verilog-AMS-PAM), 711–714
Verilog-AMS-PAM, 104, 711–714
Verilog-AMS-PoM, 104, 705–710
Verilog-D, 17f
Verilog description, 397
Verilog HDL, 464
Vertical double-gate FET, 73, 73f
Vertical pocket tunnel FET, 87, 87f
Very high-speed integrated circuit (VHSIC) hardware description language (VHDL), 475, 478f, 479f
VHDL. *See* Very high-speed integrated circuit (VHSIC) hardware description language (VHDL)
VHDL-AMS, 16, 17f, 386, 480t, 489, 489t, 490, 492t, 493f
　behavioral simulation, 373
　mixed-signal design flow, 390
VHDL-AMS-based simulation, 489–491
VHDL-based simulation, 475–477
VHS, 31
Vibrating body FET (VBFET), 88
Vibrating body transistor (VBT), 67
Video compact disc (VCD), 31
Video DRAM (VDRAM), 344
Video memory, 344
Video playback operation (mobile phone), 518, 518t, 519f
Virtual STB (V-STB), 51
Virtuoso AMS Designer, 398, 481, 490
Virtuoso Analog VoltageStorm, 398

Virtuoso Chip Assembly Router, 398
Virtuoso Layout Suite, 395
Virtuoso Schematic Editor, 374, 375, 386, 395
Virtuoso UltraSim Full-Chip Simulator, 398, 450
Visible insertion unit, 543
Visible watermarked image, 543f
VisSim, 434
Visual processing unit (VPU), 38
VLC. See DC-to-DC voltage-level converter (VLC)
VLS. See Voltage-level shifter (VLS)
Voids, 140
Volatile memory, 316, 316f
Voltage and frequency scaling, 538, 538t
Voltage-controlled current source (VCCS), 455, 457f
Voltage-controlled oscillator (VCO), 160f, 161, 165, 465–466
  CS-VCO. See Current-starved voltage controlled oscillator (CS-VCO)
  LC-VCO. See LC-tank voltage controlled oscillator (LC-VCO)
  parasitics-optimal VCO, 576–578
  thermal-aware VCO optimization, 599–601
  variability-aware design, 654–662
  Verilog-A-based analog simulation, 465–466
Voltage-controlled switch (WS), 457
Voltage-controlled voltage source (VCVS), 455, 457f
Voltage latch voltage-mode sense amplifier, 301
Voltage-level shifter (VLS), 216
Voltage-mode R-2R ladder DAC, 228, 228f
Voltage-mode sense amplifier, 301
Voltage span detector (VSD), 297
Voltage variation, 379, 379f
Voltus, 388
VPU. See Visual processing unit (VPU)
VSD. See Voltage span detector (VSD)
VTC-2-SNM, 329

W

W-Edit, 395
Wafer process variation, 99, 99f, 100
Watermark adder, 293
Watermark insertion unit, 430
Watermarking:
  copyright protection, 142
  multimedia content protection, 143–144
Watermarking chip:
  dual-voltage/frequency-based dynamic power reduction, 542–544
  MATLAB-based simulation, 429–430, 431
  Simulink/Simscape-based simulations, 438
Wave propagation equation, 566
WBSN. See Wireless body sensor network (WBSN)
WD TV HD media player, 41
WDRAM. See Window DRAM (WDRAM)
Weak inversion approach, 524
Wearable biosensor system, 30
Wien oscillator, 436–438
WiFi email communication operation, 518–519, 518t, 519f
Window DRAM (WDRAM), 344
WINM. See Write current noise margin (WINM)
Winspice, 448
Wire bonding, 389
Wire operating temperature, 585
Wireless body sensor network (WBSN), 295, 296f
Wireless sensor network (WSN), 519–520
WNM. See Write noise margin (WNM)
Worker bee, 741
Workstation-server platform EDA tool installation, 410, 410f
Worst-case analysis, 452
Worst-case temperature, 598
Wrap-around DGFET, 74
wreal, 481
Write assist methods, 322
Write Columnline, 342
Write current noise margin (WINM), 330
Write noise margin (WNM), 330
Write-once, read-many (WORM) RFID tag, 47
Write recovery time (tWR), 346
Write Rowline, 342
Write trip voltage (WTV), 329
WSN. See Wireless sensor network (WSN)
WTV. See Write trip voltage (WTV)

X

XCircuit, 401
Xming, 409
XOR gate-based phase-frequency detector, 194–195
XOR unit, 300
XSPICE, 449

Y

Yield analysis, 681–683

Z

Zener tunnel current, 117
Zero-capacitor RAM (ZRAM), 341
Zero-scale error, 235
Zigzag CNT, 79
ZRAM. See Zero-capacitor RAM (ZRAM)

CPSIA information can be obtained at www.ICGtesting.com
Printed in the USA
BVOW05*2001130316

440042BV00006B/33/P